List of the Elements with Their Symbols and Atomic Masses*

Element	Symbol	Atomic Number	Atomic Mass	Element	Symbol	Atomic Number	Atomic Mass
Actinium	Ac	89	(227)	Mendelevium	Md	101	(256)
Aluminum	Al	13	26.98	Mercury	Hg	80	200.6
Americium	Am	95	(243)	Molybdenum	Mo	42	95.94
Antimony	Sb	51	121.8	Neodymium	Nd	60	144.2
Argon	Ar	18	39.95	Neon	Ne	10	20.18
Arsenic	As	33	74.92	Neptunium	Np	93	(237)
Astatine	At	85	(210)	Nickel	Ni	28	58.69
Barium	Ba	56	137.3	Niobium	Nb	41	92.91
Berkelium	Bk	97	(247)	Nitrogen	N	7	14.01
Beryllium	Be	4	9.012	Nobelium	No	102	(253)
Bismuth	Bi	83	209.0	Osmium	Os	76	190.2
Bohrium	Bh	107	(262)	Oxygen	O	8	16.00
Boron	B	5	10.81	Palladium	Pd	46	106.4
Bromine	Br	35	79.90	Phosphorus	P	15	30.97
Cadmium	Cd	48	112.4	Platinum	Pt	78	195.1
Calcium	Ca	20	40.08	Plutonium	Pu	94	(242)
Californium	Cf	98	(249)	Polonium	Po	84	(210)
Carbon	C	6	12.01	Potassium	K	19	39.10
Cerium	Ce	58	140.1	Praseodymium	Pr	59	140.9
Cesium	Cs	55	132.9	Promethium	Pm	61	(147)
Chlorine	Cl	17	35.45	Protactinium	Pa	91	(231)
Chromium	Cr	24	52.00	Radium	Ra	88	(226)
Cobalt	Co	27	58.93	Radon	Rn	86	(222)
Copernicium	Cn	112	(227)	Rhenium	Re	75	186.2
Copper	Cu	29	63.55	Rhodium	Rh	45	102.9
Curium	Cm	96	(247)	Roentgenium	Rg	111	(272)
Darmstadtium	Ds	110	(281)	Rubidium	Rb	37	85.47
Dubnium	Db	105	(262)	Ruthenium	Ru	44	101.1
Dysprosium	Dy	66	162.5	Rutherfordium	Rf	104	(257)
Einsteinium	Es	99	(254)	Samarium	Sm	62	150.4
Erbium	Er	68	167.3	Scandium	Sc	21	44.96
Europium	Eu	63	152.0	Seaborgium	Sg	106	(263)
Fermium	Fm	100	(253)	Selenium	Se	34	78.96
Flerovium	Fl	114	(285)	Silicon	Si	14	28.09
Fluorine	F	9	19.00	Silver	Ag	47	107.9
Francium	Fr	87	(223)	Sodium	Na	11	22.99
Gadolinium	Gd	64	157.3	Strontium	Sr	38	87.62
Gallium	Ga	31	69.72	Sulfur	S	16	32.07
Germanium	Ge	32	72.59	Tantalum	Ta	73	180.9
Gold	Au	79	197.0	Technetium	Tc	43	(99)
Hafnium	Hf	72	178.5	Tellurium	Te	52	127.6
Hassium	Hs	108	(265)	Terbium	Tb	65	158.9
Helium	He	2	4.003	Thallium	Tl	81	204.4
Holmium	Ho	67	164.9	Thorium	Th	90	232.0
Hydrogen	H	1	1.008	Thulium	Tm	69	168.9
Indium	In	49	114.8	Tin	Sn	50	118.7
Iodine	I	53	126.9	Titanium	Ti	22	47.88
Iridium	Ir	77	192.2	Tungsten	W	74	183.9
Iron	Fe	26	55.85	Ununoctium	Uuo	118	(294)
Krypton	Kr	36	83.80	Ununpentium	Uup	115	(288)
Lanthanum	La	57	138.9	Ununseptium	Uus	117	(294)
Lawrencium	Lr	103	(257)	Ununtrium	Uut	113	(284)
Lead	Pb	82	207.2	Uranium	U	92	238.0
Lithium	Li	3	6.941	Vanadium	V	23	50.94
Livermorium	Lv	116	(289)	Xenon	Xe	54	131.3
Lutetium	Lu	71	175.0	Ytterbium	Yb	70	173.0
Magnesium	Mg	12	24.31	Yttrium	Y	39	88.91
Manganese	Mn	25	54.94	Zinc	Zn	30	65.39
Meitnerium	Mt	109	(266)	Zirconium	Zr	40	91.22

*All atomic masses have four significant figures except radioactive elements that have no stable isotope. For these elements, masses of the longest-lived isotope are given in parentheses. These values are recommended by the Committee on Teaching of Chemistry, International Union of Pure and Applied Chemistry.

General, Organic, and Biochemistry

NINTH EDITION

Katherine J. Denniston

Towson University

Joseph J. Topping

Towson University

Danaè R. Quirk Dorr

Minnesota State University, Mankato

Robert L. Caret

University System of Maryland

Mc
Graw
Hill
Education

GENERAL, ORGANIC, AND BIOCHEMISTRY, NINTH EDITION

Published by McGraw-Hill Education, 2 Penn Plaza, New York, NY 10121. Copyright © 2017 by McGraw-Hill Education. All rights reserved. Printed in the United States of America. Previous editions © 2014, 2011, and 2008. No part of this publication may be reproduced or distributed in any form or by any means, or stored in a database or retrieval system, without the prior written consent of McGraw-Hill Education, including, but not limited to, in any network or other electronic storage or transmission, or broadcast for distance learning.

Some ancillaries, including electronic and print components, may not be available to customers outside the United States.

This book is printed on acid-free paper.

2 3 4 5 6 7 8 9 LWI 21 20 19 18 17

ISBN 978-0-07-802154-1
MHID 0-07-802154-5

Senior Vice President, Products & Markets: *Kurt L. Strand*
Vice President, General Manager, Products & Markets: *Marty Lange*
Vice President, Content Design & Delivery: *Kimberly Meriwether David*
Managing Director: *Thomas Timp*
Director: *David Spurgeon, Ph.D.*
Brand Manager: *Andrea M. Pellerito, Ph.D.*
Director, Product Development: *Rose Koos*
Product Develper: *Mary E. Hurley*
Marketing Director, Physical Sciences: *Tamara L. Hodge*
Director of Digital Content: *Shirley Hino, Ph.D.*
Digital Product Analyst: *Patrick Diller*
Director, Content Design & Delivery: *Linda Avenarius*
Program Manager: *Lora Neyens*
Content Project Managers: *Sherry Kane/Tammy Juran*
Buyer: *Sandy Ludovissy*
Design: *Matt Backhaus*
Content Licensing Specialists: *Carrie Burger/Lorraine Buczek*
Cover Image: *©ioshertz/Getty Images*
Compositor: *SPi Global*
Printer: *LSC Communications*

All credits appearing on page or at the end of the book are considered to be an extension of the copyright page.

Library of Congress Cataloging-in-Publication Data

Denniston, K. J. (Katherine J.)
  General, organic, and biochemistry.—Ninth edition / Katherine J.
Denniston, Towson University, Joseph J. Topping, Towson University, Robert L.
Caret, University of Massachusetts, Danae R. Quirk Dorr, Minnesota State
University Mankato.
     pages cm
  Includes index.
  ISBN 978-0-07-802154-1 (alk. paper)
  1. Chemistry, Organic—Textbooks. 2. Biochemistry—Textbooks. I. Topping,
Joseph J. II. Caret, Robert L., 1947- III. Quirk Dorr, Danaè R. IV. Title.
  QD253.2.D46 2017
  547—dc23

                                                                    2015011044

The Internet addresses listed in the text were accurate at the time of publication. The inclusion of a website does not indicate an endorsement by the authors or McGraw-Hill Education, and McGraw-Hill Education does not guarantee the accuracy of the information presented at these sites.

mheducation.com/highered

Brief Contents

GENERAL CHEMISTRY

ORGANIC CHEMISTRY

BIOCHEMISTRY

Contents

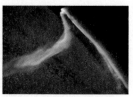

ORGANIC CHEMISTRY

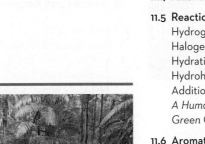

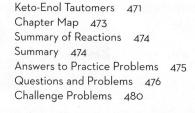

BIOCHEMISTRY

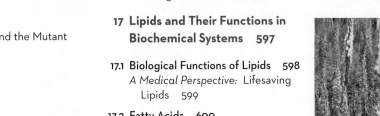

x   Contents

Perspectives

Green Chemistry

Kitchen Chemistry

Chemistry at the Crime Scene

Preface

To Our Students

Just as some researchers study chemical change, others study learning. The two are related: there are measurable changes in the brain as learning occurs. While the research on brain chemistry and learning continues, the research on learning has taught us some very successful strategies for teaching and learning chemistry. For instance, we now know that building long-term memory requires "repetitions." When you exercise to build muscle strength, you perform some number of "reps" of each exercise for each muscle that you wish to build. That is exactly what you need to do to build your long-term memory and understanding. The Center for Academic Success at the Louisiana State University has devised study tools that have allowed students to improve their performance by a full letter grade, or higher. The following is the Study Cycle with five stages that provide the "reps" needed to perform well in any course:

1. *Preview* the chapter *before* class. Either the evening before or the day of class, skim the material; pay attention to the end-of-chapter summary with boldfaced key terms, chapter map, the learning goals, and headings. Think of questions you would like the instructor to answer. Think of this 10 minutes as your "warm up."

2. *Attend* class! Be an active participant in the class, asking and answering questions and taking thoughtful, meaningful notes. Class time is much more meaningful if you have already familiarized yourself with the organization and key concepts to be discussed.

3. *Review* your notes as soon as possible after class. Fill in any gaps that exist and note any additional questions that arise. This also takes about 10 minutes; think of it as your "cool down" period.

4. *Study.* Since repetition is the key to success, The Center for Academic Success recommends 3–5 short, but intense, study sessions each day. These intense study sessions should have a very structured organization. In the first 2–5 minutes, establish your goal for the session. Spend the next 30–50 minutes studying with focus and action. Organize the material, make flash cards to help you review, draw concept maps to define the relationship among ideas, and practice problem solving. Then reward yourself with a 5–10 minute break. Call a friend, play Angry Birds, or do anything you find enjoyable. Then take 5 minutes to review the material. Finally, about once a week, perhaps on the weekend, review all of the material that you have been studying throughout the week.

5. *Assess* your progress. Are you able to solve the questions and problems at the end of the chapter? Can you explain the concepts to others? The assessment will affirm what you know well and reveal what you need to study further.

The Center for Academic Success has many other suggestions to help students learn how to learn. You can find their online tutorials and workshops at www.cas.lsu.edu.

To the Instructor

The ninth edition of *General, Organic, and Biochemistry,* like our earlier editions, has been designed to help undergraduate majors in health-related fields understand key concepts and appreciate significant connections among chemistry, health, and the treatment of disease. We have tried to strike a balance between theoretical and practical chemistry, while emphasizing material that is unique to health-related studies. We have written at a level intended for students whose professional goals do not include a mastery of chemistry, but for whom an understanding of the principles and practice of chemistry is a necessity.

Although our emphasis is the importance of chemistry to the health-related professions, we wanted this book to be appropriate for all students who need a one- or two-semester introduction to chemistry. Students learn best when they are engaged. One way to foster that engagement is to help them see clear relationships between the subject and real life. For these reasons, we have included perspectives and essays that focus on medicine and the function of the human body, as well as the environment, forensic science, and even culinary arts.

We begin that engagement with the book cover. Students may wonder why the cover has a photo of the Caucasian snowdrop (*Galanthus caucasicus*). What does this flower have to do with the study of chemistry or the practice of medicine? They will learn that Russian scientists extracted the drug galantamine from this plant in the early 1950s and others found that it was useful in treating nerve pain and poliomyelitis. More recently, it has been discovered that the drug is a reversible, competitive inhibitor of the enzyme acetylcholinesterase and that it can cross the blood-brain barrier. These characteristics have made it a useful drug for the treatment of mild to moderate Alzheimer's Disease. By inhibiting the enzyme, galantamine increases the amount of acetylcholine in the brain; this, in turn, enhances brain function, memory, and the ability to think more clearly.

The cover sets the theme for the book: chemistry is not an abstract study, but one that has an immediate impact on our lives. We try to spark student interest with an art program that uses relevant photography, clear and focused figures, and perspectives and essays that bring life to abstract ideas. We reinforce key concepts by explaining them in a clear and concise way and encouraging students to apply the concept to solve problems. We provide guidance through the inclusion of a large number of in-chapter examples that are solved in a stepwise fashion and that provide students the opportunity to test their understanding through the practice problems that follow and the suggested end-of-chapter questions and problems that apply the same concepts.

Foundations for Our Revisions

In the preparation of each edition, we have been guided by the collective wisdom of reviewers who are expert chemists and excellent teachers. They represent experience in community colleges, liberal arts colleges, comprehensive institutions, and research universities. We have followed their recommendations, while remaining true to our overriding goal of writing a readable, student-centered text. This edition has also been designed to be amenable to a variety of teaching styles. Each feature incorporated into this edition has been carefully considered with regard to how it may be used to support student learning in both the traditional classroom and the flipped learning environment.

Also for this edition, we are very pleased to have been able to incorporate real student data points and input, derived from thousands of our LearnSmart users, to help guide our revision. LearnSmart Heat Maps provided a quick visual snapshot of usage of portions of the text and the relative difficulty students experienced in mastering the content. With these data, we were able to hone not only our text content but also the LearnSmart probes.

- If the data indicated that the subject covered was more difficult than other parts of the book, as evidenced by a high proportion of students responding incorrectly, we substantively revised or reorganized the content to be as clear and illustrative as possible.

- In some sections, the data showed that a smaller percentage of the students had difficulty learning the material. In those cases, we revised the *text* to provide a clearer presentation by rewriting the section, providing additional examples to strengthen student problem-solving skills, designing new text art or figures to assist visual learners, etc.

- In other cases, one or more of the LearnSmart probes for a section was not as clear as it might be or did not appropriately reflect the content. In these cases, the *probe*, rather than the text, was edited.

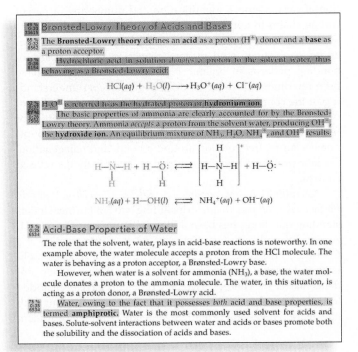

The previous image is an example of one of the heat maps from Chapter 8 that was particularly useful in guiding our revisions. The highlighted sections indicate the various levels of difficulty students experienced in learning the material. This evidence informed all of the revisions described in the "New in This Edition" section of this preface.

The following is a summary of the additions and refinements that we have included in this edition.

New in This Edition

- Chapters 4 and 8 were completely reorganized for better integration and discussion of acid-base and oxidation-reduction reactions.
- **Two new Kitchen Chemistry boxes and eight new Perspective boxes** have been added to the ninth edition to help students see the connections between chemistry and their daily lives and future careers.
- **Each of the following sections was either rewritten or significantly revised for enhanced clarity and student understanding:** 1.1, 1.2, 1.4, and 1.5; 2.1, 2.2, and 2.3; 4.4 and 4.5–4.8 (new to the chapter and revised); 5.1; 6.4; 7.4; 8.1; 9.7; 10.1, 10.2, 10.4, and 10.5; 11.5; 12.1 and 12.3–12.5; 13.1, 13.2, and 13.4; 14.1–14.4; 15.1 and 15.3; 16.2–16.4; 17.3; 18.4–18.5 and 18.7; 19.3, 19.4, and 19.6–19.8; 20.2, 20.8, and 20.10; 21.1–21.5; 23.1, 23.4, and 23.6.

Chapter 1 We have revised or added eight new learning goals to help the student identify the key concepts in the chapter. As with the last edition, each goal is used to label relevant sections and examples. Recognizing the importance of visual learning, we have revised six figures and introduced four new photos. Each of the tables, important devices for summarizing information, has also been revised. We recognize that students learn by doing and, to that end, we have paid special attention to the worked examples, with thirteen new or revised examples included. We challenge the student with in-chapter and end-of-chapter problems, forty-one of which are new or revised. The first chapter of the textbook develops fundamental skills that will be needed throughout the book, and we have revised or rewritten four of these critical sections, 1.1, 1.2, 1.4, and 1.5. Organizing and summarizing concepts is an important aspect of learning; for this reason, we have revised both the Summary and Chapter Map.

Chapter 2 We continued our focus on helping students identify key concepts by adding or revising nine learning goals focusing on the structure of the atom and the periodic table. In addition, all ten of the examples are either new or modified with reworked solutions to enhance clarity. Three of the new examples, *Determining Ion Proton and Electron Composition, Writing Shorthand Electron Configurations for Ions,* and *Determining Isoelectronic Ions and Atoms* help students understand the octet rule and ion formation. The introduction and sections featuring isotopes and electromagnetic radiation have been rewritten. Six figures and two tables are new or revised.

Chapter 3 We have introduced three new learning goals and revised four others. Figure 3.2 has been revised in order to help clarify the concept of covalent bonding. Bonding is fundamentally important to gaining a real understanding of chemistry; for that reason, we have paid special attention to Section 3.1,

Chemical Bonding, rewriting and revising where necessary, to provide a strong foundation for subsequent topics. Students must also learn to apply the concepts of bonding, structure, and the properties of ions and molecules. For that reason, we have added 30 new or revised in-chapter or end-of-chapter problems and questions. Both the Chapter Map and Summary have been revised to reflect changes in the chapter material.

Chapter 4 Chemical changes have been further developed in this chapter in conjunction with calculations and the chemical equation. Significant emphasis has been placed on problem solving beginning with the introduction of nine new and two revised learning goals, nineteen new or revised examples and forty-six new or revised questions and problems. Section 4.4, Balancing Chemical Equations, has been revised, and Sections 4.5–4.8 are new to this chapter. These sections include precipitation reactions, net-ionic equations, acid-base reactions, and oxidation-reduction reactions. Four new pictures and five figures have been added or modified, including Figure 4.10, an illustration supporting the limiting reactant concept. The Summary and Chapter Map have been revised to be consistent with the topics and learning goals of the chapter.

Chapter 5 Four new or revised learning goals have been introduced in the ninth edition to help students focus on key concepts. A comprehensive art program is critical to teaching and learning properties of gases, liquids, and solids. We have introduced five new or revised figures and nine new or revised figure captions, as well as three new photos to illustrate the effects of temperature and pressure on the behavior of the states of matter and the conversion between solids, liquids, and gases. Section 5.1, discussing the properties of gases and the ideal gas laws, has been revised to enhance clarity. Two revised examples and thirteen new or revised questions and problems were used to enhance problem-solving skills. The medical perspective, *Blood Gases and Respiration* has been moved to Chapter 6, where it accompanies the discussion of Henry's law. The Summary and Chapter Map were revised to assist students in organizing concepts as well as seeing the relationships that exist between the concepts discussed in the chapter.

Chapter 6 Several learning goals have been added or revised. Eight of the chapter examples have been modified with reworked solutions in order to enhance clarity. The discussion pertaining to osmosis, osmotic pressure, and osmolarity has been amended. Twenty new or revised questions and problems have been added to correlate to the new and revised material within the chapter. The Summary and Chapter Map have been improved for better alignment with the discussions pertaining to concentration and concentration-dependent properties.

Chapter 7 As in other chapters, we have paid special attention to the learning goals and introduction, revising where appropriate, to lead the students to understand three topics: thermodynamics, kinetics, and equilibrium. These topics are a critical part of any discussion of chemical and physical change. Opportunities for visual learning have been enhanced with three new or revised figures, six new or revised figure captions, and six new photographs. Section 7.4, dealing with equilibrium, was revised to enhance clarity. Eight new or revised questions and problems have been added to provide greater

opportunity for students to learn by doing. The Summary and Chapter Map have been revised to reflect changes in the chapter material.

Chapter 8 The emphasis of this chapter has been changed to focus primarily on acids and bases. Oxidation and reduction content has been moved to Chapter 4. Ten new learning goals have been added to correlate to the new and revised content. The Introduction and Section 8.1, Acids and Bases, have been rewritten to incorporate acids and bases commonly used in organic chemistry. Topics revised include acid and base theories, the amphiprotic nature of water, conjugate acid-base pairs, acid and base strength, self-ionization of water, and K_w. The revision includes new figures and images. Five new or revised examples, two new practice problems, and thirty-six new or revised questions and problems provide students with an opportunity to practice solving problems correlating to the learning goals emphasized. The Summary and Chapter Map have been revised in alignment with the changes to the chapter content.

Chapter 9 Two new learning goals have been added to help students identify essential concepts. The topic of nuclear chemistry can be difficult for students to conceptualize. To help overcome this problem, we have introduced three new or revised figures, twelve new or revised figure captions, and eleven new photos. Section 9.7 has been updated, including additional radiation measurement units. Thirteen new or revised questions and problems have been added, as well as four revised examples, reflecting an increased emphasis on improving the student's problem-solving skills. Both the Summary and Chapter Map have been revised to help students understand the basic concepts and their interrelationships.

Chapter 10 A new perspective, *Kitchen Chemistry: Alkanes in our Food,* including two For Further Understanding questions, has been added to the revised Chapter 10. Six new margin notes, many with associated art, have been added to help students understand line formulas, alkyl groups, the classification of carbon atoms, identification and numbering of parent carbon chains in nomenclature, and placement of substituents above and below a cycloalkane ring. A new figure has been added to facilitate student comprehension of the variety of bonding patterns in organic molecules. Several topics have been rewritten to provide students with a deeper understanding of the content. These include the discussion of families of organic compounds, functional groups, physical properties of hydrocarbons, classification of carbon atoms and alkyl groups, nomenclature, free rotation around a bond, and halogenation. Six new problems have been added to accompany the revised content.

Chapter 11 A new perspective, *Kitchen Chemistry: Pumpkin Pie Spice: An Autumn Tradition,* including two For Further Understanding questions, has been added to the revised Chapter 11. A new Example, *Writing Equations for the Hydrogenation of a Cycloalkane,* has been added, along with a set of practice problems and a set of recommended practice problems to help students master the concept. Two new problems have been added to accompany the revisions in the text. The revisions, along with new margin notes and text art, are intended to enhance student learning and understanding. Topics revised include physical characteristics, nomenclature, geometric isomers, and parts of the section on the reactions of alkenes and alkynes. A new table, and accompanying text, on saturated and unsaturated fatty acids has been added to help students recognize the practical applications of the chemistry being studied.

Chapter 12 A new Example, *Using the Common System of Nomenclature to Name Alcohols*, has been added, along with a set of practice problems and a set of recommended practice problems to help students master the concept. The *Medical Perspective: Fetal Alcohol Syndrome*, has been updated to reflect the more recently described Fetal Alcohol Spectrum Disorder. New text art has been designed to help students understand the physical properties of alcohols and the nature of intramolecular hydrogen bonding. Revision of the discussion of intramolecular hydrogen bonding, along with the new text art, provides students with a clear idea of the importance of hydrogen bonding in biological systems. The information on general anesthetics has been updated, and sections on physical properties, dehydration reactions, and oxidation of alcohols have been revised for greater clarity.

Chapter 13 A new *Human Perspective: Powerful Weak Attractions*, including two For Further Understanding questions, has been added to the revised Chapter 13. New text art has been added to the discussion of the common names of ketones and to clarify oxidation products of aldehydes under acidic or basic conditions. Other new text art clarifies the structure of hemiacetals and acetals. Three examples have been modified to include a structure of practical interest or to clarify the principle being applied. Revisions to the text included a reorganization of the discussion of structure and physical properties and additional details to clarify the IUPAC nomenclature of ketones.

Chapter 14 A new *Medical Perspective: Esters for Appetite Control*, including two For Further Understanding questions, has been added. Five new text art diagrams have been added to support the revisions of the text with regard to the structure and physical properties of carboxylic acids and esters, as well as the action of soaps and the significance of phosphoester compounds in nature. Other revisions in the text include the preparation of carboxylic acids, the properties and nomenclature of carboxylic acid salts, and the structure, physical properties, and nomenclature of esters. Unnecessary content regarding acid anhydrides has been deleted.

Chapter 15 New text art, with the associated text revisions, has been designed to assist student understanding of the physical properties and nomenclature of amines, the nomenclature of alkylammonium salts, neutralization reactions, and preparation of amides from acid chlorides. Along with revision of the nature of neutralization reactions, hydrolysis of amides, and nomenclature of amides, the synthesis and structure of primary, secondary, and tertiary amides is introduced in this edition. To complement these changes, the chapter map has been revised and three new key terms have been introduced. Four new problems have been added to allow students to test their understanding of the new materials.

Chapter 16 A new *Medical Perspective: Human Milk Oligosaccharides*, including two For Further Understanding questions, has been added. A new Example, *Identifying a Chiral Compound*, has been added, along with a set of practice problems and a set of recommended practice problems to allow students to test their mastery of the concept. A new figure (16.13) shows the action of the enzymes α-amylase, β-amylase, and maltase. The section on meso compounds has been revised completely and two new problems have been included.

Chapter 17 Section 17.2 has been reorganized so that ω-fatty acids are discussed prior to the section on prostaglandins. The reactions of fatty acids and glycerides has also been reorganized and revised for greater clarity. All text art in the section on sphingolipids has been redesigned as line formulas to enhance student understanding of the structures.

Chapter 18 Two new perspectives have been added to Chapter 18: *A Medical Perspective: Medications from Venoms* and *A Human Perspective: The New Protein*. Sections 18.4, 18.5, and 18.7 have been revised to streamline the text and clarify concepts.

Chapter 19 *A Medical Perspective: HIV Protease Inhibitors and Pharmaceutical Drug Design* has been updated to reflect the variety of new drugs available to treat the infection in adults and children. The discussion of transferases has been rewritten and new text art designed to provide students with an example that they will study later in the chapters on metabolism. Text revisions include Section 19.3 and passages in Sections 19.4, 19.6, 19.7, and 19.8. In all cases, the revisions streamline and simplify concepts to promote more effective student learning.

Chapter 20 A new *Medical Perspective: Epigenomics*, including two For Further Understanding questions, has been added. The more recently described non-invasive prenatal testing procedure has been included in *A Medical Perspective: Molecular Genetics and Detection of Human Genetic Disorders*. The sections on the chemical composition of DNA and RNA and on chromatin structure have been revised for clarity. Section 20.8, Recombinant DNA, has been rewritten to reduce some of the historical methodologies so that students will focus on the potential of more recent advances.

Chapter 21 *A Medical Perspective: High Fructose Corn Syrup* has been updated with information on the recent studies demonstrating the impact of glucose and fructose on the hypothalamus of humans. In each of Sections 21.1–21.6, the text has been revised to simplify concepts. Section 21.7 has been reorganized for greater clarity.

Chapter 22 A new *Medical Perspective, Babies with Three Parents*, including two For Further Understanding questions, has been added. Throughout the chapter, the text has been revised to streamline the writing and clarify the concepts.

Chapter 23 Section 23.5 has been revised extensively to avoid redundancy with information presented in earlier chapters. Six new problems have been added to this chapter.

Applications

Each chapter contains applications that present short stories about real-world situations involving one or more topics students will encounter within the chapter. There are over 100 applications throughout the text, so students are sure to find many topics that spark their interest. Global climate change,

DNA fingerprinting, the benefits of garlic, and gemstones are just a few examples of application topics.

- **Medical Perspectives** relate chemistry to a health concern or a diagnostic application.
- **Green Chemistry** explores environmental topics, including the impact of chemistry on the ecosystem and how these environmental changes affect human health.
- **Human Perspectives** delve into chemistry and society and include such topics as gender issues in science and historical viewpoints.
- **Chemistry at the Crime Scene** focuses on forensic chemistry, applying the principles of chemistry to help solve crimes.
- **Kitchen Chemistry** discusses the chemistry associated with everyday foods and cooking methods.

Learning Tools

In designing the original learning system we asked ourselves: "If we were students, what would help us organize and understand the material covered in this chapter?" Based on the feedback of reviewers and users of our text, we include a variety of learning tools:

- **Chapter Overview** pages begin each chapter, listing learning goals and the chapter outline. Both students and professor can see, all in one place, the plan for the chapter.
- **Learning Goal Icons** mark the sections and examples in the chapter that focus on each learning goal.
- **Chapter Cross-References** help students locate pertinent background material. These references to previous chapters, sections, and perspectives are noted in the margins of the text. Marginal cross references also alert students to upcoming topics related to the information currently being studied.
- **End-of-Chapter Questions and Problems** are arranged according to the headings in the chapter outline, with further subdivision into Foundations (basic concepts) and Applications.
- **Chapter Maps** are included just before the End-of-Chapter Summaries to provide students with an overview of the chapter—showing connections among topics, how concepts are related, and outlining the chapter hierarchy.
- **Chapter Summaries** are now a bulleted list format of chapter concepts by major sections, with the integrated bold-faced **Key Terms** appearing in context. This more succinct format helps students to quickly identify and review important chapter concepts and to make connections with the incorporated Key Terms. Each Key Term is defined and listed alphabetically in the **Glossary** at the end of the book.
- **Answers to Practice Problems** are supplied at the end of each chapter so that students can quickly check their understanding of important problem-solving skills and chapter concepts.
- **Summary of Reactions** in the organic chemistry chapters highlight each major reaction type on a tan background. Major chemical reactions are summarized by equations at the end of the chapter, facilitating review.

Problem Solving and Critical Thinking

Perhaps the best preparation for a successful and productive career is the development of problem-solving and critical thinking skills. To this end, we created a variety of problems that require recall, fundamental calculations, and complex reasoning. In this edition, we have used suggestions from our reviewers, as well as from our own experience, to enhance our 2300 problems. This edition includes new problems and hundreds of example problems with step-by-step solutions.

- **In-Chapter Examples, Solutions, and Practice Problems:** Each chapter includes examples that show the student, step-by-step, how to properly reach the correct solution to model problems. Each example contains a practice problem, as well as a referral to further practice questions. These questions allow students to test their mastery of information and to build self-confidence. The answers to the practice problems can be found at the end of each chapter so students can check their understanding.
- **Color-Coding System for In-Chapter Examples:** In this edition, we also introduced a color-coding and label system to help alleviate the confusion that students frequently have when trying to keep track of unit conversions. Introduced in Chapter 1, this color coding system has been used throughout the problem-solving chapters.

$$3.01 \ \text{mol S} \times \frac{32.06 \ \text{g S}}{1 \ \text{mol S}} = 96.5 \ \text{g S}$$

Data Given × Conversion Factor = Desired Result

- **In-Chapter and End-of-Chapter Questions and Problems:** We have created a wide variety of paired concept problems. The answers to the odd-numbered questions are found in the back of the book as reinforcement for students as they develop problem-solving skills. However, students must then be able to apply the same principles to the related even-numbered problems.
- **Challenge Problems:** Each chapter includes a set of challenge problems. These problems are intended to engage students to integrate concepts to solve more complex problems. They make a perfect complement to the classroom lecture because they provide an opportunity for in-class discussion of complex problems dealing with daily life and the health care sciences.

Over the course of the last nine editions, hundreds of reviewers have shared their knowledge and wisdom with us, as well as the reactions of their students to elements of this book. Their contributions, as well as our own continuing experience in the area of teaching and learning science, have resulted in a text that we are confident will provide a strong foundation in chemistry, while enhancing the learning experience of students.

The Art Program

Today's students are much more visually oriented than previous generations. We have built upon this observation through the use of color, figures, and three-dimensional computer-generated models. This art program enhances the readability of the text and provides alternative pathways to learning.

- **Dynamic Illustrations:** Each chapter is amply illustrated using figures, tables, and chemical formulas. All of these illustrations are carefully annotated for clarity. To help students better understand difficult concepts, there are approximately 350 illustrations and 250 photos in the ninth edition.
- **Color-Coding Scheme:** We have color-coded equations so that chemical groups being added or removed in a reaction can be quickly recognized.
  1. **Red print** is used in chemical equations or formulas to draw the reader's eye to key elements or properties in a reaction or structure.
  2. **Blue print** is used when additional features must be highlighted.
  3. **Green background** screens denote generalized chemical and mathematical equations. In the organic chemistry chapters, the Summary of Reactions at the end of the chapter is also highlighted for ease of recognition.
  4. Yellow backgrounds illustrate energy, stored either in electrons or groups of atoms, in the general and biochemistry sections of the text. In the organic chemistry section of the text, yellow background screens also reveal the parent chain of an organic compound.
  5. There are situations in which it is necessary to adopt a unique color convention tailored to the material in a particular chapter. For example, in Chapter 18, the structures of amino acids require three colors to draw attention to key features of these molecules. For consistency, blue is used to denote the acid portion of an amino acid and red is used to denote the basic portion of an amino acid. Green print is used to denote the R groups, and a yellow background screen directs the eye to the α-carbon.
- **Computer-Generated Models:** The ability of students to understand the geometry and three-dimensional structure of molecules is essential to the understanding of organic and biochemical reactions. Computer-generated models are used throughout the text because they are both accurate and easily visualized.

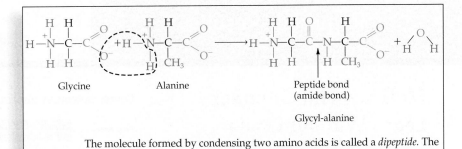

Glycine   Alanine   Peptide bond (amide bond)

Glycyl-alanine

The molecule formed by condensing two amino acids is called a *dipeptide*. The amino acid with a free α-N⁺H₃ group is known as the amino terminal, or sim-

Because amines are bases, they react with acids to form alkylammonium salts.

$$R{-}\overset{\overset{\displaystyle H}{|}}{\underset{\underset{\displaystyle H}{|}}{N}}{:}\ +\ HCl\ \longrightarrow\ R{-}\overset{\overset{\displaystyle H}{|}}{\underset{\underset{\displaystyle H}{|}}{N^+}}{-}H\ Cl^-$$

Amine   Acid   Alkylammonium salt

The reaction of methylamine with hydrochloric acid shown is typical of these reactions.

Fructose-6-phosphate   + ATP →(Phosphofructokinase)→   Fructose-1,6-bisphosphate   + ADP

α-Carbon

α-Amino group → H—N—C—C—O   α-Carboxylate group

Side-chain R group

Required=Results

McGraw-Hill Connect®
Learn Without Limits

Connect is a teaching and learning platform that is proven to deliver better results for students and instructors.

Connect empowers students by continually adapting to deliver precisely what they need, when they need it, and how they need it, so your class time is more engaging and effective.

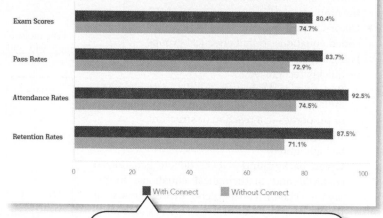

Course outcomes improve with Connect.

	With Connect	Without Connect
Exam Scores	80.4%	74.7%
Pass Rates	83.7%	72.9%
Attendance Rates	92.5%	74.5%
Retention Rates	87.5%	71.1%

Using **Connect** improves passing rates by **10.8%** and retention by **16.4%**.

88% of instructors who use **Connect** require it; instructor satisfaction **increases** by 38% when **Connect** is required.

Analytics

Connect Insight®

Connect Insight is Connect's new one-of-a-kind visual analytics dashboard—now available for both instructors and students—that provides at-a-glance information regarding student performance, which is immediately actionable. By presenting assignment, assessment, and topical performance results together with a time metric that is easily visible for aggregate or individual results, Connect Insight gives the user the ability to take a just-in-time approach to teaching and learning, which was never before available. Connect Insight presents data that empowers students and helps instructors improve class performance in a way that is efficient and effective.

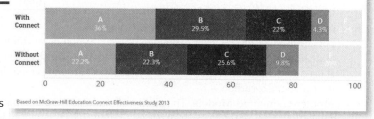

Connect helps students achieve better grades

With Connect	A 36%	B 29.5%	C 22%	D 4.3%
Without Connect	A 22.2%	B 22.3%	C 25.6%	D 9.8%

Based on McGraw-Hill Education Connect Effectiveness Study 2013

Students can view their results for any **Connect** course.

Mobile

Connect's new, intuitive mobile interface gives students and instructors flexible and convenient, anytime–anywhere access to all components of the Connect platform.

Adaptive

THE FIRST AND ONLY **ADAPTIVE READING EXPERIENCE** DESIGNED TO TRANSFORM THE WAY STUDENTS READ

> More students earn **A's** and **B's** when they use McGraw-Hill Education **Adaptive** products.

SmartBook®

Proven to help students improve grades and study more efficiently, SmartBook contains the same content within the print book, but actively tailors that content to the needs of the individual. SmartBook's adaptive technology provides precise, personalized instruction on what the student should do next, guiding the student to master and remember key concepts, targeting gaps in knowledge and offering customized feedback, driving the student toward comprehension and retention of the subject matter. Available on smartphones and tablets, SmartBook puts learning at the student's fingertips—anywhere, anytime.

> Over **4 billion questions** have been answered making McGraw-Hill Education products more intelligent, reliable precise.

STUDENTS WANT

SMARTBOOK®

95% of students reported **SmartBook** to be a more effective way of reading material

100% of students want to use the Practice Quiz feature available within **SmartBook** to help them study

100% of students reported having reliable access to off-campus wifi

90% of students say they would purchase **SmartBook** over print alone

95% reported that **SmartBook** would impact their study skills in a positive way

McGraw Hill Education

*Findings based on a 2015 focus group survey at Pellissippi State Community College administered by McGraw-Hill Education

For the Instructor

- **Instructor's Manual:** Written and developed for the ninth edition by the authors, this ancillary contains many useful suggestions for organizing flipped classrooms, lectures, instructional objectives, perspectives on readings from the text, answers to the even-numbered problems from the text, a list of each chapter's key problems and concepts, and more. The Instructor's Manual is available through the Instructor Resources in the Connect Library tab.

- **Laboratory Manual for General, Organic, and Biological Chemistry:** Authored by Applegate, Neely, and Sakuta to be the most current lab manual available for the GOB course, incorporating the most modern instrumentation and techniques. Illustrations and chemical structures were developed by the authors to conform to the most recent IUPAC conventions. A problem-solving methodology is also utilized throughout the laboratory exercises. There are two online virtual labs for Nuclear Chemistry and Gas Laws. This Laboratory Manual is also designed with flexibility in mind to meet the differing lengths of GOB courses and the variety of instrumentation available in GOB labs. Helpful instructor materials are also available on this companion website, including answers, solution recipes, best practices with common student issues and TA advice, sample syllabi, and a calculation sheet for the Density lab.

- **Presentation Tools:** Build instructional material wherever, whenever, and however you want with assets such as photos, artwork, and other media that can be used to create customized lectures, visually enhanced tests and quizzes, compelling course websites, or attractive printed support materials. The Presentation Tools can be accessed from the Instructor Resources in the Connect Library tab. Instructors can still access the animations from the OLC for use in their presentations.

- **More than 300 animations available through Connect:** They supplement the textbook material in much the same way as instructor demonstrations. However, they are only a few mouse-clicks away, any time, day or night. Because many students are visual learners and quite computer-literate, the animations add another dimension of learning; they bring a greater degree of reality to the written word.

For the Student

- **Student Study Guide/Solutions Manual:** A separate Student Study Guide/Solutions Manual, prepared by Danaè Quirk Dorr, is available. It contains the answers and complete solutions for the odd-numbered problems. It also offers students a variety of exercises and keys for testing their comprehension of basic, as well as difficult, concepts.

- **Schaum's Outline of General, Organic, and Biological Chemistry:** Written by George Odian and Ira Blei, this supplement provides students with more than 1 400 solved problems with complete solutions. It also teaches effective problem-solving techniques.

Acknowledgments

We are thankful to our families, whose patience and support made it possible for us to undertake this project. We are also grateful to our many colleagues at McGraw-Hill for their support, guidance, and assistance. In particular, we would like to thank Sherry Kane, Content Project Manager, Mary Hurley, Developmental Editor, Andrea Pellerito, Brand Manager, and Tamara Hodge, Marketing Manager.

The following individuals helped write and review learning goal-oriented content for **LearnSmart for General, Organic, & Biochemistry:**

David G. Jones, *Vistamar School*

Adam I. Keller, *Columbus State Community College*

A revision cannot move forward without the feedback of professors teaching the course. The following reviewers have our gratitude and assurance that their comments received serious consideration. The following professors provided reviews, participated in focus groups, or otherwise provided valuable advice as our textbook has evolved to its current form:

Augustine Agyeman, *Clayton State University*
Phyllis Arthasery, *Ohio University*
EJ Behrman, *The Ohio State University*
C. Bruce Bradley, *Spartanburg Community College*
Thomas Gilbert, *Northern Illinois University*
Mary Hadley, *Minnesota State University, Mankato*
Emily Halvorson, *Pima Community College*
James Hardy, *The University of Akron*
Amy Hanks, *Brigham Young University-Idaho*
Theresa Hill, *Rochester Community and Technical College*
Shirley Hino, *Santa Rosa Junior College*
Narayan Hosmane, *Northern Illinois University*
Colleen Kelley, *Pima Community College*
Myung-Hoon Kim, *Georgia Perimeter College*
Charlene Kozerow, *University of Maine*
Andrea Leonard, *University of Louisiana at Lafayette*
Lauren E. H. McMills, *Ohio University*
Jonathan McMurry, *Kennesaw State University*
Cynthia Molitor, *Lourdes College*
Matthew Morgan, *Georgia Perimeter College, Covington*
Melekeh Nasiri, *Woodland Community College*
Glenn Nomura, *Georgia Perimeter College*
Kenneth O'Connor, *Marshall University*
Dwight Patterson, *Middle Tennessee State University*
Allan Pinhas, *University of Cincinnati, Cincinnati*
Jerry Poteat, *Georgia Perimeter College*

Michael E. Rennekamp, *Columbus State Community College*

Raymond Sadeghi, *University of Texas at San Antonio*

Paul Sampson, *Kent State University*

Shirish Shah, *Towson University*

Buchang Shi, *Eastern Kentucky University*

Heather Sklenicka, *Rochester Community and Technical College*

Sara Tate, *Northeast Lakeview College*

Kimberley Taylor, *University of Arkansas at Little Rock*

Susan Tansey Thomas, *University of Texas at San Antonio*

Nathan Tice, *Eastern Kentucky University*

Steven Trail, *Elgin Community College*

David A. Tramontozzi, *Macomb Community College*

Pearl Tsang, *University of Cincinnati*

Michael Van Dyke, *Western Carolina University*

Wendy Weeks, *Pima Community College*

Gregg Wilmes, *Eastern Michigan University*

Yakov Woldman, *Valdosta State University*

METHODS AND MEASUREMENT
Chemistry

1

LEARNING GOALS

1 Explain the relationship between chemistry, matter, and energy.

2 Discuss the approach to science, the scientific method, and distinguish among the terms *hypothesis, theory,* and *scientific law.*

3 Distinguish between data and results.

4 Describe the properties of the solid, liquid, and gaseous states.

5 Classify matter according to its composition.

6 Provide specific examples of physical and chemical properties and physical and chemical changes.

7 Distinguish between intensive and extensive properties.

8 Identify the major units of measure in the English and metric systems.

9 Report data and calculate results using scientific notation and the proper number of significant figures.

10 Distinguish between *accuracy* and *precision* and their representations: *error* and *deviation.*

11 Convert between units of the English and metric systems.

12 Know the three common temperature scales, and convert values from one scale to another.

13 Use density, mass, and volume in problem solving, and calculate the specific gravity of a substance from its density.

Chemistry is the study of anything that has mass and occupies space.

OUTLINE

INTRODUCTION

Louis Pasteur, a chemist and microbiologist, said, "Chance favors the prepared mind." In the history of science and medicine, there are many examples in which individuals made important discoveries because they recognized the value of an unexpected observation.

One such example is the use of ultraviolet (UV) light to treat infant jaundice. Infant jaundice is a condition in which the skin and the whites of the eyes appear yellow because of high levels of the bile pigment bilirubin in the blood. Bilirubin is a breakdown product of the oxygen-carrying blood protein hemoglobin. If bilirubin accumulates in the body, it can cause brain damage and death. The immature liver of the baby cannot remove the bilirubin.

In 1956, an observant nurse in England noticed that when jaundiced babies were exposed to sunlight, the jaundice faded. Research based on her observation showed that the UV light changes the bilirubin into another substance, which can be excreted. To this day, jaundiced newborns undergoing phototherapy are treated with UV light. Historically, newborns were diagnosed with jaundice based only on their physical appearance. However, it has been determined that this method is not always accurate. Now, it is common to use either an instrument or a blood sample to measure the amount of bilirubin present in the serum.

In this first chapter of your study of chemistry, you will learn about the scientific method: the process of developing hypotheses to explain observations and the design of experiments to test those hypotheses.

You will also see that measurement of properties of matter, and careful observation and recording of data, are essential to scientific inquiry. So too is assessment of the precision and accuracy of measurements. Measurements (data) must be reported to allow others to determine their significance. Therefore, an understanding of significant figures, and the ability to represent data in the most meaningful units, enables other scientists to interpret data and results.

The goal of this chapter is to help you develop the skills needed to represent and communicate data and results from scientific inquiry.

1.1   The Discovery Process

Chemistry

Chemistry is the study of matter, its chemical and physical properties, the chemical and physical changes it undergoes, and the energy changes that accompany those processes.

Matter is anything that has mass and occupies space. The air we breathe, our bodies, our planet earth, our universe; all are made up of an immense variety and quantity of particles, collectively termed matter. Matter undergoes change. Sometimes this change occurs naturally or we change matter when we make new substances (creating drugs in a pharmaceutical laboratory). All of these changes involve **energy,** the ability to do work to accomplish some change. Hence, we may describe chemistry as a study of matter and energy and their interrelationship.

Chemistry is an experimental science. A traditional image of a chemist is someone wearing a white coat and safety goggles while working in solitude in a laboratory. Although much chemistry is still accomplished in a traditional laboratory setting, over the last 40 years the boundaries of the laboratory have expanded to include the power of modern technology. For example, searching the scientific literature for information no longer involves a trip to the library as it is now done very quickly via the Internet. Computers are also invaluable in the laboratory because they control sophisticated instrumentation that measures, collects, processes, and interprets information. The behavior of matter can also be modeled using sophisticated computer programs.

LEARNING GOAL

1 Explain the relationship between chemistry, matter, and energy.

Models In Chemistry, p. 4

Additionally, chemistry is a collaborative process. The solitary scientist, working in isolation, is a relic of the past. Complex problems dealing with topics such as the environment, disease, forensics, and DNA require input from other scientists and mathematicians who can bring a wide variety of expertise to problems that are chemical in nature.

The boundaries between the traditional sciences of chemistry, physics, and biology, as well as mathematics and computer science, have gradually faded. Medical practitioners, physicians, nurses, and medical technologists use therapies that contain elements of all these disciplines. The rapid expansion of the pharmaceutical industry is based on recognition of the relationship between the function of an organism and its basic chemical makeup. Function is a consequence of changes that chemical substances undergo.

For these reasons, an understanding of basic chemical principles is essential for anyone considering a medically related career; indeed, a worker in any science-related field will benefit from an understanding of the principles and applications of chemistry.

Investigating the causes of the rapid melting of glaciers is a global application of chemistry. How does this illustrate the interaction of matter and energy?

The Scientific Method

The **scientific method** is a systematic approach to the discovery of new information. How do we learn about the properties of matter, the way it behaves in nature, and how it can be modified to make useful products? Chemists do this by using the scientific method to study the way in which matter changes under carefully controlled conditions.

The scientific method is not a "cookbook recipe" that, if followed faithfully, will yield new discoveries; rather, it is an organized approach to solving scientific problems. Every scientist brings his or her own curiosity, creativity, and imagination to scientific study. Yet, scientific inquiry does involve some of the "cookbook recipe" approach. Characteristics of the scientific process include the following:

LEARNING GOAL

2 Discuss the approach to science, the scientific method, and distinguish among the terms *hypothesis, theory,* and *scientific law.*

- *Observation.* The description of, for example, the color, taste, or odor of a substance is a result of observation. The measurement of the temperature of a liquid or the size or mass of a solid results from observation.
- *Formulation of a question.* Humankind's fundamental curiosity motivates questions of why and how things work.
- *Pattern recognition.* When a cause-and-effect relationship is found, it may be the basis of a generalized explanation of substances and their behavior.
- *Theory development.* When scientists observe a phenomenon, they want to explain it. The process of explaining observed behavior begins with a hypothesis. A **hypothesis** is simply an attempt to explain an observation, or series of observations. If many experiments support a hypothesis, it may attain the status of a theory. A **theory** is a hypothesis supported by extensive testing (experimentation) that explains scientific observations and data and can accurately predict new observations and data.
- *Experimentation.* Demonstrating the correctness of hypotheses and theories is at the heart of the scientific method. This is done by carrying out carefully designed experiments that will either support or disprove the hypothesis or theory. A scientific experiment produces **data.** Each piece of data is the individual result of a single measurement or observation.

  A **result** is the outcome of an experiment. Data and results may be identical, but more often, several related pieces of data are combined, and logic is used to produce a result.
- *Information summarization.* A **scientific law** is nothing more than the summary of a large quantity of information. For example, the law of conservation of matter states that matter cannot be created or destroyed, only converted from one form to another. This statement represents a massive body of chemical information gathered from experiments.

LEARNING GOAL

3 Distinguish between data and results.

3 Distinguish between data and results.

EXAMPLE 1.1   Distinguishing Between Data and Results

In many cases, a drug is less stable in the presence of moisture, and excess moisture can hasten the breakdown of the active ingredient, leading to loss of potency. Bupropion (Wellbutrin) is an antidepressant that is moisture sensitive. Describe an experiment that will allow for the determination of the quantity of water gained by a certain quantity of bupropion when it is exposed to air.

Solution

To do this experiment, we must first weigh the buproprion sample, and then expose it to the air for a period of time and reweigh it. The change in weight,

$$[\text{weight}_{\text{final}} - \text{weight}_{\text{initial}}] = \text{weight difference}$$

indicates the weight of water taken up by the drug formulation. The initial and final weights are individual bits of *data*; by themselves they do not answer the question, but they do provide the information necessary to calculate the answer: the results. The difference in weight and the conclusions based on the observed change in weight are the *results* of the experiment.

Note: This is actually not a very good experiment because many conditions were not measured. Measurement of the temperature, humidity of the atmosphere, and the length of time that the drug was exposed to the air would make the results less ambiguous.

Practice Problem 1.1

Describe an experiment that demonstrates that the boiling point of water changes when salt (sodium chloride) is added to the water.

▶ For Further Practice: **Questions 1.35 and 1.36.**

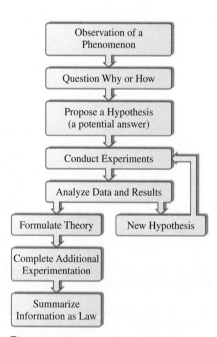

Figure 1.1 The scientific method is an organized way of doing science that incorporates a degree of trial and error. If the data analysis and results do not support the initial hypothesis, the cycle must begin again.

The scientific method involves the interactive use of hypotheses, development of theories, and thorough testing of theories using well-designed experiments. It is summarized in Figure 1.1.

Models in Chemistry

Hypotheses, theories, and laws are frequently expressed using mathematical equations. These equations may confuse all but the best of mathematicians. For this reason, a *model* of a chemical unit or system is often used to help illustrate an idea. A good model based on everyday experience, although imperfect, gives a great deal of information in a simple fashion.

Consider the fundamental unit of methane, the major component of natural gas, which is composed of one carbon (symbolized by C) atom and four hydrogen (symbolized by H) atoms.

A geometrically correct model of methane can be constructed from balls and sticks. The balls represent the individual atoms of hydrogen and carbon, and the sticks correspond to the attractive forces that hold the hydrogen and carbon together. The model consists of four balls representing hydrogen symmetrically arranged around a center ball representing carbon.

The Scientific Method

The discovery of penicillin by Alexander Fleming is an example of the scientific method at work. Fleming was studying the growth of bacteria. One day, his experiment was ruined because colonies of mold were growing on his plates. From this failed experiment, Fleming made an observation that would change the practice of medicine: Bacterial colonies could not grow in the area around the mold colonies. Fleming hypothesized that the mold was making a chemical compound that inhibited the growth of the bacteria. He performed a series of experiments designed to test this hypothesis.

The success of the scientific method is critically dependent upon carefully designed experiments that will either support or disprove the hypothesis. This is exactly what Fleming did.

In one experiment, he used two sets of tubes containing sterile nutrient broth. To one set he added mold cells. The second set (the control tubes) remained sterile. The mold was allowed to grow for several days. Then the broth from each of the tubes (experimental and control) was passed through a filter to remove any mold cells. Next, bacteria were placed in each tube. If Fleming's hypothesis was correct, the tubes in which the mold had grown would contain the chemical that inhibits growth, and the bacteria would not grow. On the other hand, the control tubes (which were never used to grow mold) would allow bacterial growth. This is exactly what Fleming observed.

Within a few years this *antibiotic*, penicillin, was being used to treat bacterial infections in patients.

A nurse administers an injection of penicillin to a young patient.

For Further Understanding

▶ What is the purpose of the control tubes used in this experiment?

▶ Match the features of this article with the flowchart items in Figure 1.1.

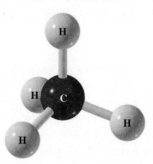

Color-coding the balls distinguishes one type of atom from another; the geometrical form of the model, all of the angles and dimensions of a tetrahedron, are the same for each methane unit found in nature. Methane is certainly not a collection of balls and sticks, but such models are valuable because they help us understand the chemical behavior of methane and other more complex substances.

The structure-properties concept has advanced so far that compounds are designed and synthesized in the laboratory with the hope that they will perform very specific functions, such as curing diseases that have been resistant to other forms of treatment. Figure 1.2 shows some of the variety of modern technology that has its roots in scientific inquiry.

Chemists and physicists have used the observed properties of matter to develop models of the individual units of matter. These models collectively make up what we now know as the atomic theory of matter, which is discussed in detail in Chapter 2.

Figure 1.2 Examples of technology originating from scientific inquiry: (a) synthesis of a new drug, (b) blood pressure app for a smartphone, (c) preparation of solid-state electronics, and (d) use of a gypsy moth sex attractant for insect control.

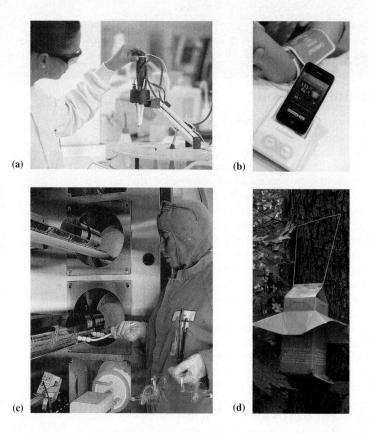

(a)

(b)

(c)

(d)

1.2   The Classification of Matter

Matter is a large and seemingly unmanageable concept because it includes everything that has mass and occupies space. Chemistry becomes manageable as we classify matter according to its **properties**—that is, the characteristics of the matter. Matter will be classified in two ways in this section, by *state* and by *composition*.

States of Matter

We will examine each of the three states of matter in detail in Chapter 5.

There are three *states of matter:* the **gaseous state,** the **liquid state,** and the **solid state.** A gas is made up of particles that are widely separated. In fact, a gas will expand to fill any container; it has no definite shape or volume. In contrast, particles of a liquid are closer together; a liquid has a definite volume but no definite shape; it takes on the shape of its container. A solid consists of particles that are close together and often have a regular and predictable pattern of particle arrangement (crystalline). The particles in a solid are much more organized than the particles in a liquid or a gas. As a result, a solid has both fixed volume and fixed shape. Attractive forces, which exist between all particles, are very pronounced in solids and much less so in gases.

Composition of Matter

We have seen that matter can be classified by its state as a solid, liquid, or gas. Another way to classify matter is by its composition. This very useful system, described in the following paragraphs and summarized in Figure 1.3, will be utilized throughout the textbook.

All matter is either a *pure substance* or a *mixture*. A **pure substance** has only one component. Pure water is a pure substance. It is made up only of particles containing two hydrogen (symbolized by H) atoms and one oxygen (symbolized by O) atom—that is, water molecules (H_2O).

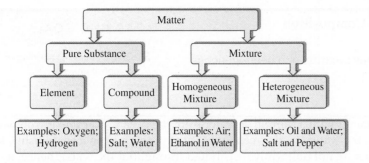

Figure 1.3 Classification of matter by composition. All matter is either a pure substance or a mixture of pure substances. Pure substances are either elements or compounds, and mixtures may be either homogeneous (uniform composition) or heterogeneous (nonuniform composition).

There are different types of pure substances. Elements and compounds are both pure substances. An **element** is a pure substance that generally cannot be changed into a simpler form of matter. Hydrogen and oxygen, for example, are elements. Alternatively, a **compound** is a substance resulting from the combination of two or more elements in a definite, reproducible way. The elements hydrogen and oxygen, as noted earlier, may combine to form the compound water, H_2O.

A **mixture** is a combination of two or more pure substances in which each substance retains its own identity. Ethanol, the alcohol found in beer, and water can be combined in a mixture. They coexist as pure substances because they do not undergo a chemical reaction. A mixture has variable composition; there are an infinite number of combinations of quantities of ethanol and water that can be mixed. For example, the mixture may contain a small amount of ethanol and a large amount of water or vice versa. Each is, however, an ethanol-water mixture.

A mixture may be either *homogeneous* or *heterogeneous* (Figure 1.4). A **homogeneous mixture** has uniform composition. Its particles are well mixed, or thoroughly intermingled. A homogeneous mixture, such as alcohol and water, is described as a *solution*. Air, a mixture of gases, is an example of a gaseous solution. A **heterogeneous mixture** has a nonuniform composition. A mixture of salt and pepper is a good example of a heterogeneous mixture. Concrete is also composed of a heterogeneous mixture of materials (various types and sizes of stone and sand with cement in a nonuniform mixture).

At present, more than 100 elements have been characterized. A complete listing of the elements and their symbols is found on the inside front cover of this textbook.

A detailed discussion of solutions (homogeneous mixtures) and their properties is presented in Chapter 6.

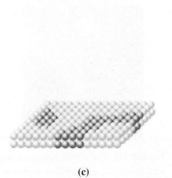

Figure 1.4 Schematic representations of some classes of matter. (a) A pure substance, water, consists of a single component. (b) A homogeneous mixture, blue dye in water, has a uniform distribution of components. The blue spheres represent the blue dye molecules. (c) The mineral orbicular jasper is an example of a heterogeneous mixture. The lack of homogeneity is apparent from its nonuniform distribution of components.

(a)   (b)   (c)

EXAMPLE 1.2	Classifying Matter by Composition

Is seawater a pure substance, a homogeneous mixture, or a heterogeneous mixture?

Solution

Imagine yourself at the beach, filling a container with a sample of water from the ocean. Examine it. You would see a variety of solid particles suspended in the water: sand, green vegetation, perhaps even a small fish! Clearly, it is a mixture, and one in which the particles are not uniformly distributed throughout the water; hence, it is a heterogeneous mixture.

Practice Problem 1.2

Is each of the following materials a pure substance, a homogeneous mixture, or a heterogeneous mixture?

  a. ethanol   c. an Alka-Seltzer tablet fizzing in water
  b. blood   d. oxygen being delivered from a hospital oxygen tank

▶ For Further Practice: **Questions 1.57 and 1.58.**

Question 1.1   Intravenous therapy may be used to introduce a saline solution into a patient's vein. Is this solution a pure substance, a homogeneous mixture, or a heterogeneous mixture?

Question 1.2   Cloudy urine can be a symptom of a bladder infection. Classify this urine as a pure substance, a homogeneous mixture, or a heterogeneous mixture.

Physical Properties and Physical Change

Water is the most common example of a substance that can exist in all three states over a reasonable temperature range (Figure 1.5). Conversion of water from one state to another constitutes a *physical change.* A **physical change** produces a recognizable difference in the appearance of a substance without causing any change in its composition or identity. For example, we can warm an ice cube and it will melt, forming liquid water. Clearly its appearance has changed; it has been transformed from

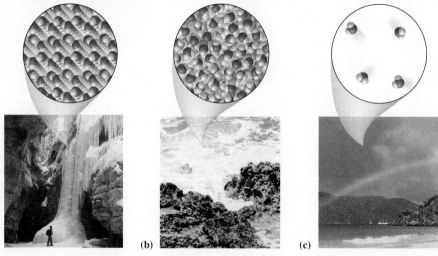

(a)   (b)   (c)

Figure 1.5   The three states of matter exhibited by water: (a) solid, as ice; (b) liquid, as ocean water; (c) gas, as humidity in the air.

the solid to the liquid state. It is, however, still water; its composition and identity remain unchanged. A physical change has occurred. We could in fact demonstrate the constancy of composition and identity by refreezing the liquid water, re-forming the ice cube. This melting and freezing cycle could be repeated over and over. This very process is a hallmark of our global weather changes. The continual interconversion of the three states of water in the environment (snow, rain, and humidity) clearly demonstrates the retention of the identity of water particles or *molecules.*

A **physical property** can be observed or measured without changing the composition or identity of a substance. As we have seen, melting ice is a physical change. We can measure the temperature when melting occurs; this is the *melting point* of water. We can also measure the *boiling point* of water, when liquid water becomes a gas. Both the melting and boiling points of water, and of any other substance, are physical properties.

A practical application of separation of materials based upon their differences in physical properties is shown in Figure 1.6.

Chemical Properties and Chemical Change

We have noted that physical properties can be exhibited, measured, or observed without any change in identity or composition. In contrast, **chemical properties** do result in a change in composition and can be observed only through chemical reactions. In a **chemical reaction,** a chemical substance is converted to one or more different substances by rearranging, removing, replacing, or adding atoms. For example, the process of photosynthesis can be shown as

$$\text{carbon dioxide} + \text{water} \xrightarrow[\text{Chlorophyll}]{\text{Light}} \text{sugar} + \text{oxygen}$$

This chemical reaction involves the conversion of carbon dioxide and water (the *reactants*) to a sugar and oxygen (the *products*). The physical properties of the reactants and products are clearly different. We know that carbon dioxide and oxygen are gases at room temperature, and water is a liquid at this temperature; the sugar is a solid white powder. A chemical property of carbon dioxide is its ability to form sugar under certain conditions. The process of formation of this sugar is the *chemical change.* The term *chemical reaction* is synonymous with **chemical change.**

Figure 1.6 An example of separation based on differences in physical properties. Magnetic iron is separated from nonmagnetic substances. A large-scale version of this process is important in the recycling industry.

Light is the energy needed to make the reaction happen. Chlorophyll is the energy absorber, converting light energy to chemical energy.

EXAMPLE 1.3 | **Classifying Change**

Can the process that takes place when an egg is fried be described as a physical or chemical change?

LEARNING GOAL

6 Provide specific examples of physical and chemical properties and physical and chemical changes.

Solution

Examine the characteristics of the egg before and after frying. Clearly, some significant change has occurred. Furthermore, the change appears irreversible. More than a simple physical change has taken place. A chemical reaction (actually, several) must be responsible; hence, there is a chemical change.

Practice Problem 1.3

Classify each of the following as either a chemical change or a physical change:

    a. water boiling to become steam   d. melting of ice in spring

    b. butter becoming rancid   e. decaying of leaves in winter

    c. burning wood

▶ For Further Practice: **Questions 1.51 and 1.52.**

Question 1.3   Classify each of the following as either a chemical property or a physical property:

    a. color   b. flammability   c. hardness

Question 1.4   Classify each of the following as either a chemical property or a physical property:

    a. odor   b. taste   c. temperature

Intensive and Extensive Properties

It is important to recognize that properties can also be classified according to whether they depend on the size of the sample. Consequently, there is a fundamental difference between properties such as color and melting point and properties such as mass and volume.

An **intensive property** is a property of matter that is *independent* of the *quantity* of the substance. Boiling and melting points are intensive properties. For example, the boiling point of one single drop of water is exactly the same as the boiling point of a liter (L) of water.

An **extensive property** *depends* on the *quantity* of a substance. Mass and volume are extensive properties. There is an obvious difference between 1 gram (g) of silver and 1 kilogram (kg) of silver; the quantities and, incidentally, the monetary values, differ substantially.

LEARNING GOAL

7 Distinguish between intensive and extensive properties.

The mass of a pediatric patient (in kg) is an extensive property that is commonly used to determine the proper dosage of medication [in milligrams (mg)] prescribed. Although the mass of the medication is also an extensive property, the dosage (in mg/kg) is an intensive property. This calculated dosage should be the same for every pediatric patient.

EXAMPLE 1.4   **Differentiating Between Intensive and Extensive Properties**

Is temperature an intensive or extensive property?

LEARNING GOAL

7 Distinguish between intensive and extensive properties.

Solution

Imagine two glasses, each containing 100 g of water, and each at 25°C. Now pour the contents of the two glasses into a larger glass. You would predict that the mass of the water in the larger glass would be 200 g (100 g + 100 g) because mass is an *extensive property*, dependent on quantity. However, we would expect the temperature of the water to remain the same (not 25°C + 25°C); hence, temperature is an *intensive property* . . . independent of quantity.

Practice Problem 1.4

Pure water freezes at 0°C. Is this an intensive or extensive property? Why?

▶ For Further Practice: **Questions 1.41 and 1.42.**

Question 1.5   Label each property as intensive or extensive:

    a. the length of my pencil   b. the color of my pencil

Question 1.6   Label each property as intensive or extensive:

    a. the shape of leaves on a tree   b. the number of leaves on a tree

LEARNING GOAL

8 Identify the major units of measure in the English and metric systems.

1.3   The Units of Measurement

The study of chemistry requires the collection of data through measurement. The quantities that are most often measured include mass, length, and volume. Measurements require the determination of an amount followed by a **unit,** which defines the basic quantity being measured. A weight of 3 *ounces* (oz) is clearly quite different than 3 *pounds* (lb). A number that is not followed by the correct unit usually conveys no useful information.

The *English system of measurement* is a collection of unrelated units used in the United States in business and industry. However, it is not used in scientific work, primarily because it is difficult to convert one unit to another. In fact, the English "system" is not really a system at all; it is simply a collection of units accumulated throughout English history. Table 1.1 shows relationships among common English units of weight, length, and volume.

The United States has begun efforts to convert to the metric system. The *metric system* is truly systematic. It is composed of a set of units that are related to each other decimally; in other words, as powers of ten. Because the metric system is a decimally based system, it is inherently simpler to use and less ambiguous. Table 1.2 shows the meaning of the prefixes used in the metric system.

The metric system was originally developed in France just before the French Revolution in 1789. The more extensive version of this system is the *Systéme International,* or *S.I. system.* Although the S.I. system has been in existence for over 50 years, it has yet to gain widespread acceptance. Because the S.I. system is truly systematic, it utilizes certain units, especially for pressure, that many find unwieldy.

In this text, we will use the metric system, not the S.I. system, and we will use the English system only to the extent of converting from it to the more systematic metric system.

Now let's look at the major metric units for mass, length, volume, and time in more detail. In each case, we will compare the unit to a familiar English unit.

The photo shows 3 oz of grapes versus a 3-lb cantaloupe. Clearly units are important.

Mass

Mass describes the quantity of matter in an object. The terms *weight* and *mass,* in common usage, are often considered synonymous. They are not, in fact. **Weight** is the force of gravity on an object:

$$\text{Weight} = \text{mass} \times \text{acceleration due to gravity}$$

LEARNING GOAL

8 Identify the major units of measure in the English and metric systems.

The mathematical process of converting between units will be covered in detail in Section 1.5.

The table of common prefixes used in the metric system relates values to the base units. For example, it defines 1 mg as being equivalent to 10^{-3} g and 1 kg as being equivalent to 10^3 g.

TABLE 1.1 Some Common Relationships Used in the English System

Weight	1 pound (lb) = 16 ounces (oz)
	1 ton (t) = 2000 pounds (lb)
Length	1 foot (ft) = 12 inches (in)
	1 yard (yd) = 3 feet (ft)
	1 mile (mi) = 5280 feet (ft)
Volume	1 quart (qt) = 32 fluid ounces (fl oz)
	1 quart (qt) = 2 pints (pt)
	1 gallon (gal) = 4 quarts (qt)

TABLE 1.2 Some Common Prefixes Used in the Metric System

Prefix	Abbreviation	Meaning	Decimal Equivalent	Equality with major metric units (g, m, or L are represented by x in each)
mega	M	10^6	1,000,000.	$1\ \text{M}x = 10^6 x$
kilo	k	10^3	1000.	$1\ \text{k}x = 10^3 x$
deka	da	10^1	10.	$1\ \text{da}x = 10^1 x$
deci	d	10^{-1}	0.1	$1\ \text{d}x = 10^{-1} x$
centi	c	10^{-2}	0.01	$1\ \text{c}x = 10^{-2} x$
milli	m	10^{-3}	0.001	$1\ \text{m}x = 10^{-3} x$
micro	μ	10^{-6}	0.000001	$1\ \mu x = 10^{-6} x$
nano	n	10^{-9}	0.000000001	$1\ \text{n}x = 10^{-9} x$

Figure 1.7 Three common balances that are useful for the measurement of mass. (a) A two-pan comparison balance for approximate mass measurement suitable for routine work requiring accuracy to 0.1 g (or perhaps 0.01 g). (b) A top-loading single-pan electronic balance that is similar in accuracy to (a) but has the advantages of speed and ease of operation. The revolution in electronics over the past 20 years has resulted in electronic balances largely supplanting the two-pan comparison balance in routine laboratory usage. (c) An analytical balance of this type is used when the highest level of precision and accuracy is required.

(a)

(b)

(c)

When gravity is constant, mass and weight are directly proportional. But gravity is not constant; it varies as a function of the distance from the center of the earth. Therefore, weight cannot be used for scientific measurement because the weight of an object may vary from one place on the earth to the next.

Mass, on the other hand, is independent of gravity; it is a result of a comparison of an unknown mass with a known mass called a *standard mass*. Balances are instruments used to measure the mass of materials.

The metric unit for mass is the gram (g). A common English unit for mass is the pound (lb).

$$1 \text{ lb} = 454 \text{ g}$$

Examples of balances commonly used for the determination of mass are shown in Figure 1.7.

Length

The standard metric unit of *length,* the distance between two points, is the meter (m). A meter is close to the English yard (yd).

$$1 \text{ yd} = 0.914 \text{ m}$$

Volume

The standard metric unit of *volume,* the space occupied by an object, is the liter (L). A liter is the volume occupied by 1000 g of water at 4 degrees Celsius (°C).

The English quart (qt) is similar to the liter.

$$1 \text{ qt} = 0.946 \text{ L} \quad \text{or} \quad 1.06 \text{ qt} = 1 \text{ L}$$

Volume can be derived using the formula

$$V = \text{length} \times \text{width} \times \text{height}$$

Therefore, volume is commonly reported with a length cubed unit. A cube with the length of each side equal to 1 m will have a volume of 1 m × 1 m × 1 m, or 1 m^3.

$$1 \text{ m}^3 = 1000 \text{ L}$$

The relationships among the units L, mL, and cm^3 are shown in Figure 1.8.

Typical laboratory devices used for volume measurement are shown in Figure 1.9. These devices are calibrated in units of milliliters (mL) or microliters (μL); 1 mL is, by definition, equal to 1 cm^3. The volumetric flask is designed to *contain* a specified volume, and the graduated cylinder, pipet, and buret *dispense* a desired volume of liquid.

Time

The standard metric unit of time is the second (s). The need for accurate measurement of time by chemists may not be as apparent as that associated with mass, length, and volume. It is necessary, however, in many applications. In fact, matter may be characterized by measuring the time required for a certain process to occur. The rate of a chemical reaction is a measure of change as a function of time.

1.4 The Numbers of Measurement

A measurement has two parts: a number and a unit. The English and metric units of mass, length, volume, and time were discussed in Section 1.3. In this section, we will learn to handle the numbers associated with the measurements.

Information-bearing figures in a number are termed *significant figures.* Data and results arising from a scientific experiment convey information about the way in which the experiment was conducted. The degree of uncertainty or doubt associated with a measurement or series of measurements is indicated by the number of figures used to represent the information.

Significant Figures

Consider the following situation: A student was asked to obtain the length of a section of wire. In the chemistry laboratory, several different types of measuring devices are usually available. Not knowing which was most appropriate, the

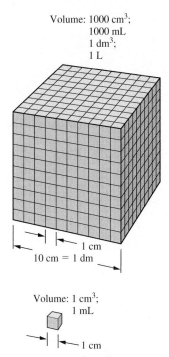

Volume: 1000 cm^3;
1000 mL
1 dm^3;
1 L

1 cm
10 cm = 1 dm

Volume: 1 cm^3;
1 mL

1 cm

Figure 1.8 The relationships among various volume units.

Figure 1.9 Common laboratory equipment used for the measurement of volume. Graduated (a) cylinders, (b) pipets, and (c) burets are used for the delivery of liquids. A graduated cylinder is usually used for measurement of approximate volume; it is less accurate and precise than either pipets or burets. (d) Volumetric flasks are used to contain a specific volume.

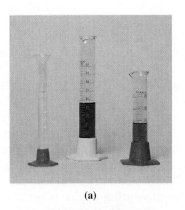

(a)

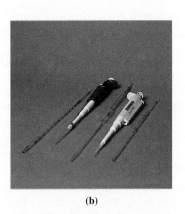

(b)

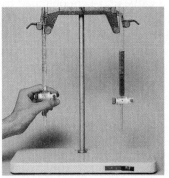

(c)

(d)

student decided to measure the object using each device that was available in the laboratory. To make each measurement, the student determined the mark nearest to the end of the wire. This is depicted in the figure below; the red bar represents the wire being measured. In each case, the student estimated one additional digit by mentally subdividing the marks into ten equal divisions. The following data were obtained:

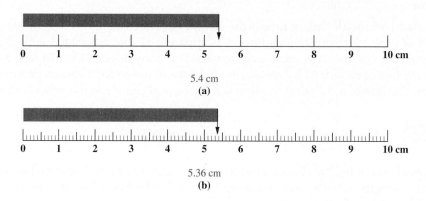

5.4 cm
(a)

5.36 cm
(b)

In case (a), we are certain that the object is at least 5 cm long and equally certain that it is *not* 6 cm long because the end of the object falls between the calibration lines 5 and 6. We can only estimate between 5 and 6, because there are no calibration indicators between 5 and 6. The end of the wire appears to be approximately four-tenths of the way between 5 and 6, hence 5.4 cm. The 5 is known with certainty, and 4 is estimated.

In case (b), the ruler is calibrated in tenths of a centimeter. The end of the wire is at least 5.3 cm and not 5.4 cm. Estimation of the second decimal place between the two closest calibration marks leads to 5.36 cm. In this case, 5.3 is certain, and the 6 is estimated (or uncertain).

Two questions should immediately come to mind:

1. Are the two answers equivalent?
2. If not, which answer is correct?

In fact, the two answers are *not* equivalent, yet *both* are correct. How do we explain this apparent discrepancy?

The data are not equivalent because each is known to a different degree of certainty. The term **significant figures** is defined to be all digits in a number representing data or results that are known with certainty *plus one uncertain digit*. The answer 5.36 cm, containing three significant figures, specifies the length of the wire more precisely than 5.4 cm, which contains only two significant figures.

Both answers are correct because each is consistent with the measuring device used to generate the data. An answer of 5.36 cm obtained from a measurement using ruler (a) would be *incorrect* because the measuring device is not capable of that precise specification. On the other hand, a value of 5.4 cm obtained from ruler (b) would be erroneous as well; in that case, the measuring device is capable of generating a higher level of certainty (more significant digits) than is actually reported.

In summary, the number of significant figures associated with a measurement is determined by the measuring device. Conversely, the number of significant figures reported is an indication of the precision of the measurement itself.

Recognition of Significant Figures

Only *significant* digits should be reported as data or results. However, are all digits, as written, significant digits? Let's look at a few examples illustrating the rules that are used to represent data and results with the proper number of significant digits.

The uncertain digit results from an estimation.

The uncertain digit represents the degree of doubt in a single measurement.

- All nonzero digits are significant.
    7.314 has *four* significant digits.
- The number of significant digits is independent of the position of the decimal point.
    73.14 has *four* significant digits, as does 7.314.
- Zeros located between nonzero digits are significant.
    60.052 has *five* significant figures.
- Zeros at the end of a number (often referred to as trailing zeros) are significant or not significant depending upon the existence of a decimal point in the number.
    ○ If there *is* a decimal point, any trailing zeros are significant.
        4.70 has *three* significant figures.
        1000. has *four* significant figures because the decimal point is included.
    ○ If the number *does not* contain a decimal point, trailing zeros are not significant.
        1000 has *one* significant figure.
- Zeros to the left of the first nonzero integer are not significant; they serve only to locate the position of the decimal point.
    0.0032 has *two* significant figures.

Question 1.7 How many significant figures are contained in each of the following numbers?
a. 7.26	c. 700.2	e. 0.0720
b. 726	d. 7.0	f. 720

Question 1.8 How many significant figures are contained in each of the following numbers?
a. 0.042	c. 24.0	e. 204
b. 4.20	d. 240	f. 2.04

Scientific Notation

It is often difficult to express very large numbers to the proper number of significant figures using conventional notation. The solution to this problem lies in the use of **scientific notation,** which involves the representation of a number that is greater than 1 and less than 10 which is multiplied by 10 raised to the power of a whole number.

The conversion is illustrated as:

$$6200 = 6.2 \times 1000 = 6.2 \times 10^3$$

If we wish to express 6200 with three significant figures, we can write it as:

$$6.20 \times 10^3$$

The trailing zero becomes significant with the existence of the decimal point in the number. Note also that the exponent of 3 has no bearing on the number of significant figures. The value of 6.20×10^{14} also contains three significant figures.

■RULE: To convert a number greater than one to scientific notation, the original decimal point is moved x places to the left, and the resulting number is multiplied by 10^x. The exponent (x) is a *positive* number equal to the number of places the original decimal point was moved.

Scientific notation is also useful in representing numbers less than one. The conversion is illustrated as:

$$0.0062 = 6.2 \times \frac{1}{1000} = 6.2 \times \frac{1}{10^3} = 6.2 \times 10^{-3}$$

LEARNING GOAL

9 Report data and calculate results using scientific notation and the proper number of significant figures.

Scientific notation is also referred to as exponential notation. When a number is not written in scientific notation, it is said to be in standard form.

By convention, in the exponential form, we represent the number with one digit to the left of the decimal point.

Scientific notation is written in the format: $y \times 10^x$, in which y represents a number between 1 and 10, and x represents a positive or negative whole number.

■RULE: To convert a number less than one to scientific notation, the original decimal point is moved x places to the right, and the resulting number is multiplied by 10^{-x}. The exponent $(-x)$ is a *negative* number equal to the number of places the original decimal point was moved.

When a number is exceedingly large or small, scientific notation must be used to enter the number into a calculator. For example, the mass of a single helium atom is a rather cumbersome number as written:

$$0.00000000000000000000000006692 \text{ g}$$

Most calculators only allow for the input of nine digits. Scientific notation would express this number as 6.692×10^{-24} g.

Question 1.9 Represent each of the following numbers in scientific notation, showing only significant digits:

a. 0.0024	c. 224	e. 72.420
b. 0.0180	d. 673,000	f. 0.83

Question 1.10 Represent each of the following numbers in scientific notation, showing only significant digits:

a. 48.20	c. 0.126	e. 0.0520
b. 480.0	d. 9,200	f. 822

Accuracy and Precision

The terms *accuracy* and *precision* are often used interchangeably in everyday conversation. However, they have very different meanings when discussing scientific measurement.

Accuracy is the degree of agreement between the true value and the measured value. The measured value may be a single number (such as the mass of an object) or the average value of a series of replicate measurements of the same quantity (reweighing the same object several times). We represent accuracy in terms of **error,** the numerical difference between the measured and true value.

Error is an unavoidable consequence of most laboratory measurements (except counted numbers, discussed on p. 17), but not for the reasons you might expect. Spills and contamination are certainly problems in a laboratory, but proper training and a great deal of practice eliminates most of these human errors. Still, errors, *systematic* and *random,* remain.

Systematic errors cause results to be generally higher than the true value or generally lower than the true value. An example would be something as simple as dust on a balance pan, causing each measurement to be higher than the true value. The causes of systematic error can often be discovered and removed. Even after correcting for systematic error we are still left with random error. Random error is an unavoidable, intrinsic consequence of measurement. Replicate measurements of the same quantity will produce some results greater than the true value and some less than the true value.

When possible, we prefer to make as many replicate measurements of the same quantity to "cancel out" the high (+) and low (−) fluctuations.

Precision is a measure of the agreement within a set of replicate measurements. Just as accuracy is measured in terms of error, precision is represented by **deviation,** the amount of variation present in a set of replicate measurements.

It is important to recognize that accuracy and precision are not the same thing. It is possible to have one without the other. However, when scientific measurements are carefully made, the two most often go hand in hand; high-quality data are characterized by high levels of precision and accuracy.

In Figure 1.10, bull's-eye (a) shows the goal of all experimentation: accuracy *and* precision. Bull's-eye (b) shows the results to be repeatable (good precision); however,

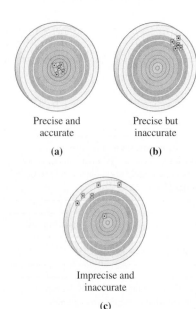

Precise and accurate

(a)

Precise but inaccurate

(b)

Imprecise and inaccurate

(c)

Figure 1.10 An illustration of precision and accuracy in replicate experiments.

some error in the experimental procedure has caused the results to center on an incorrect value. This error is systematic, occurring in each replicate measurement. Occasionally, an experiment may show "accidental" accuracy. The precision is poor, but the average of replicate measurements leads to a correct value. We don't want to rely on accidental success; the experiment should be repeated until the precision inspires faith in the accuracy of the method. Modern measuring devices in chemistry, equipped with powerful computers with immense storage capacity, are capable of making literally thousands of individual replicate measurements to enhance the quality of the result. In bull's-eye (c), we see a representation of poor precision and poor accuracy. Often, poor precision is accompanied by poor accuracy.

In summary, the presence of error and deviation in most measurements is the real basis for significant figures: all the *certain* digits plus one *uncertain* digit.

Exact (Counted) and Inexact Numbers

Inexact numbers, by definition, have uncertainty (the degree of doubt in the final significant digit). *Exact numbers,* on the other hand, have no uncertainty. Exact numbers may arise from a definition; there are *exactly* 60 min in 1 h or there are exactly 1000 mL in 1 L.

Exact numbers are a consequence of counting. Counting the number of dimes in your pocket or the number of letters in the alphabet are examples. The fact that exact numbers have no uncertainty means that they do not limit the number of significant figures in the result of a calculation.

For example, we may wish to determine the mass of three bolts purchased from the hardware store. Each bolt has a mass of 12.97 g. The total mass is determined by:

$$3 \times 12.97 \text{ g} = 38.91 \text{ g}$$

The number of significant figures in the result is governed by the data (mass of the bolt) and not by the counted (*exact*) number of bolts.

In Section 1.5, we will learn how to convert between units. A good rule of thumb to follow when performing these types of calculations is to use the measured quantity, *not* the conversion factor in order to determine the number of significant figures in the answer.

Rounding Numbers

The use of a calculator generally produces more digits for a result than are justified by the rules of significant figures on the basis of the data input. For example, your calculator may show:

$$3.84 \times 6.72 = 25.8048$$

The most correct answer would be 25.8, dropping 048. A convenient way to show this is:

$$3.84 \times 6.72 = 25.8048 \approx 25.8$$

The rule for multiplication and significant figures dictates three significant figures in the answer.

A number of acceptable conventions for rounding exist. Throughout this book, we will use the following:

■**RULE:**   When the number to be dropped is less than five, the preceding number is not changed. When the number to be dropped is five or larger, the preceding number is increased by one unit.

Question 1.11   Round each of the following numbers to three significant figures.
   a. 61.40   b. 6.171   c. 0.066494   d. 63.669   e. 8.7715

Question 1.12   Round each of the following numbers to two significant figures.
   a. 6.2262   b. 3895   c. 6.885   d. 2.2247   e. 0.0004109

9 Report data and calculate results using scientific notation and the proper number of significant figures.

Remember the distinction between the words *zero* and *nothing*. *Zero* is one of the 10 digits and conveys as much information as 1, 2, and so forth. *Nothing* implies no information; the digits in the positions indicated by *x* could be 0, 1, 2, or any other.

See the rules for rounding discussed earlier in this section.

Significant Figures in Calculation of Results

Addition and Subtraction

If we combine the following numbers:

$$37.68$$
$$108.428$$
$$6.71864$$

our calculator will show a final result of

$$152.82664$$

Clearly the answer, with eight digits, defines the total much more accurately than *any* of the individual quantities being combined. This cannot be correct; *the answer cannot have greater significance than any of the quantities that produced the answer.* We rewrite the problem:

$$37.68xxx$$
$$108.428xx$$
$$+\quad 6.71864$$
$$\overline{152.82664}\quad \text{(should be 152.83)}$$

where x = no information; x may be any integer from 0 to 9. Adding 4 to two unknown numbers (in the rightmost column) produces no information. Similar logic prevails for the previous two columns. Thus, five digits remain, all of which are significant. Conventional rules for rounding would dictate a final answer of 152.83.

> **Question 1.13**   Report the result of each of the following to the proper number of significant figures:
>   a. 4.26 + 3.831 =   b. 8.321 − 2.4 =   c. 16.262 + 4.33 − 0.40 =

> **Question 1.14**   Report the result of each of the following to the proper number of significant figures:
>   a. 7.939 + 6.26 =   b. 2.4 − 8.321 =   c. 2.333 + 1.56 − 0.29 =

Adding numbers that are in scientific notation requires a bit more consideration. The numbers must either be converted to standard form or converted to numbers that have the same exponents. Example 1.5 demonstrates this point.

EXAMPLE 1.5 | **Determining Significant Figures When Adding Numbers in Scientific Notation**

Report the result of the following addition to the proper number of significant figures and in scientific notation.

9 Report data and calculate results using scientific notation and the proper number of significant figures.

$$9.47 \times 10^{-6} + 9.3 \times 10^{-5}$$

Solution

There are two strategies that may be used in order to arrive at the correct answer.

First solution strategy.
When both numbers are converted to standard form, they can be added together. The initial answer is not the correct answer because it does not have the proper number of significant figures.

$$0.00000947$$
$$\underline{+\ 0.000093xx}$$
$$0.00010247$$

After rounding, the answer 0.000102 can then be converted to the final answer, which in scientific notation is 1.02×10^{-4}.

Second solution strategy.

When both numbers have the same power of 10 exponent, they can be added together. In this example, 9.47×10^{-6} is converted to 0.947×10^{-5}.

$$9.47 \times 10^{-6} = 0.947 \times 10^{-5}$$

$$0.947 \times 10^{-5}$$
$$\underline{+\ 9.3xx \times 10^{-5}}$$
$$10.247 \times 10^{-5}$$

As in the first solution strategy, the initial answer is rounded to the proper number of significant figures, 10.2×10^{-5} which is written in scientific notation as 1.02×10^{-4}.

Practice Problem 1.5

Report the result of the addition of $6.72 \times 10^5 + 7.4 \times 10^4$ to the proper number of significant figures and in scientific notation.

▸ For Further Practice: **Questions 1.79 b, d; 1.80 d, e.**

Question 1.15   Report the result of the following addition to the proper number of significant figures and in scientific notation.

$$8.23 \times 10^{-4} + 6.1 \times 10^{-5}$$

Question 1.16   Report the result of the following addition to the proper number of significant figures and in scientific notation.

$$4.80 \times 10^8 + 9.149 \times 10^2$$

Multiplication and Division

In the preceding discussion of addition and subtraction, the position of the decimal point in the quantities being combined had a bearing on the number of significant figures in the answer. In multiplication and division, this is not the case. The decimal point position is irrelevant when determining the number of significant figures in the answer. It is the number of significant figures in the data that is important. Consider

$$\frac{4.237 \times 1.21 \times 10^{-3} \times 0.00273}{11.125} = 1.26 \times 10^{-6}$$

The answer is limited to three significant figures; the answer can have *only* three significant figures because two numbers in the calculation, 1.21×10^{-3} and 0.00273, have three significant figures and "limit" the answer. *The answer can be no more precise than the least precise number from which the answer is derived.* The *least precise number* is the number with the fewest significant figures.

LEARNING GOAL

9 Report data and calculate results using scientific notation and the proper number of significant figures.

EXAMPLE 1.6	Determining Significant Figures When Multiplying Numbers in Scientific Notation

Report the result of the following operation to the proper number of significant figures and in scientific notation.

LEARNING GOAL

9 Report data and calculate results using scientific notation and the proper number of significant figures.

$$\frac{2.44 \times 10^4}{91}$$

Solution

Often problems that combine multiplication and addition can be broken into parts. This allows each part to be solved in a stepwise fashion.

Step 1. The numerator operation can be completed.

$$2.44 \times 10^4 = 24{,}400$$

Step 2. This value can now be divided by the value in the denominator.

$$\frac{24{,}400}{91} = 268$$

Step 3. The answer is limited to two significant figures. This is because of the two numbers in the calculation, 2.244 and 91, the number 91 has fewer significant figures and limits the number of significant figures in the answer. The answer in scientific notation is 2.7×10^2.

Practice Problem 1.6

Report the result of the following operation to the proper number of significant figures and in scientific notation.

$$\frac{837}{1.8 \times 10^{-2}}$$

▶ For Further Practice: **1.79 a, c, e and 1.80 a, b, c.**

Question 1.17 Report the result of each of the following operations using the proper number of significant figures:

a. $63.8 \times 0.80 =$

b. $\dfrac{63.8}{0.80} =$

c. $\dfrac{16.4 \times 78.11}{22.1} =$

d. $\dfrac{42.2}{21.38 \times 2.3} =$

e. $\dfrac{4.38 \times 10^8}{0.9462} =$

f. $\dfrac{6.1 \times 10^{-4}}{0.3025} =$

Question 1.18 Report the result of each of the following operations using the proper number of significant figures:

a. $\dfrac{27.2 \times 15.63}{1.84} =$

b. $\dfrac{13.6}{18.02 \times 1.6} =$

c. $\dfrac{4.79 \times 10^5}{0.7911} =$

d. $3.58 \times 4.0 =$

e. $\dfrac{3.58}{4.0} =$

f. $\dfrac{11.4 \times 10^{-4}}{0.45} =$

1.5   Unit Conversion

To convert from one unit to another, we must have a *conversion factor* or series of conversion factors that relate two units. The proper use of these conversion factors is called the *factor-label method* or *dimensional analysis*. This method is used for two kinds of conversions: to convert from one unit to another within the *same system* or to convert units from *one system to another.*

Conversion of Units within the Same System

Based on the information presented in Table 1.1, Section 1.3, we know that in the English system,

$$1 \text{ gal} = 4 \text{ qt}$$

Dividing both sides of the equation by the same term does not change its identity. These ratios are equivalent to unity (1); therefore

$$\frac{1 \text{ gal}}{1 \text{ gal}} = \frac{4 \text{ qt}}{1 \text{ gal}} = 1$$

Multiplying any other expression by either of these ratios will not change the value of the term because multiplication of any number by 1 produces the original value. However, the units will change.

Factor-Label Method

When the expressions are written as ratios, they can be used as conversion factors in the factor-label method. For example, if you were to convert 12 gal to quarts, you must decide which conversion factor to use,

$$\frac{1 \text{ gal}}{4 \text{ qt}} \quad \text{or} \quad \frac{4 \text{ qt}}{1 \text{ gal}}$$

Since you are converting 12 gal (Data Given) to qt (Desired Result), it is important to choose a conversion factor with gal in the denominator and qt in the numerator. That way, when the initial quantity (12 gal) is multiplied by the conversion factor, the original unit (gal) will cancel, leaving you with the unit qt in the answer.

$$12 \text{ g\cancel{al}} \times \frac{4 \text{ qt}}{1 \text{ g\cancel{al}}} = 48 \text{ qt}$$

Data Given × Conversion Factor = Desired Result

If the incorrect ratio was selected as a conversion factor, the answer would be incorrect.

$$12 \text{ gal} \times \frac{1 \text{ gal}}{4 \text{ qt}} = \frac{3 \text{ gal}^2}{\text{qt}} \quad \text{(incorrect units)}$$

Therefore, the factor-label method is a self-indicating system. The product will only have the correct units if the conversion factor is set up properly.

The factor-label method is also useful when more than one conversion factor is needed to convert the data given to the desired result. The use of a series of conversion factors is illustrated in Example 1.7.

The speed of an automobile is indicated in both English (mi/h) and metric (km/h) units.

Conversion factors are used to relate units through the process of the factor-label method (dimensional analysis).

EXAMPLE 1.7   **Using English System Conversion Factors**

Convert 3.28×10^4 ounces to tons.

Solution

Using the equalities provided in Table 1.1, the data given in ounces (oz) can be directly converted to a bridging data result in pounds (lb), so lb can be converted to the desired result in tons (t). The possible conversion factors are

$$\frac{1 \text{ lb}}{16 \text{ oz}} = \frac{16 \text{ oz}}{1 \text{ lb}} \quad \text{and} \quad \frac{2000 \text{ lb}}{1 \text{ t}} = \frac{1 \text{ t}}{2000 \text{ lb}}$$

Continued…

Step 1. Since the initial value is in oz, the conversion factor with oz in the denominator should be used first.

$$3.28 \times 10^4 \; \cancel{oz} \times \frac{1 \; lb}{16 \; \cancel{oz}} = 2.05 \times 10^3 \; lb$$

Data Given × Conversion Factor = Initial Data Result

If the other conversion factor relating oz and lb was used, the resulting units would have been oz^2/lb, and the answer would have been incorrect.

Step 2. Now that 3.28×10^4 oz has been converted to 2.05×10^3 lb, the conversion factor relating lb to t is used. The conversion factor with lb in the denominator and t in the numerator is the only one that leads to the correct answer.

$$2.05 \times 10^3 \; \cancel{lb} \times \frac{1 \; t}{2000 \; \cancel{lb}} = 1.03 \; t$$

Initial Data Result × Conversion Factor = Desired Result

This calculation may also be done in a single step by arranging the factors in a chain:

$$3.28 \times 10^4 \; \cancel{oz} \times \frac{1 \; \cancel{lb}}{16 \; \cancel{oz}} \times \frac{1 \; t}{2000 \; \cancel{lb}} = 1.03 \; t$$

Data Given × Conversion Factor × Conversion Factor = Desired Result

Helpful Hint: After the conversion factors have been selected and set up in the solution to the problem, it is important to also cancel the units that can be canceled. This process will allow for you to verify that you have set up the problem correctly. In addition, the unit ton represents a significantly larger quantity than the unit ounce. Therefore, one would expect a small number of tons to equal a large number of ounces.

Practice Problem 1.7

Convert 360 ft to mi.

▶ For Further Practice: **Questions 1.91 a, b and 1.92 a, b.**

Table 1.2 is located in Section 1.3.

Conversion of units within the metric system may be accomplished by using the factor-label method as well. Unit prefixes that dictate the conversion factor facilitate unit conversion (refer to Table 1.2). Example 1.8 demonstrates this process.

EXAMPLE 1.8   Using Metric System Conversion Factors

Convert 0.0047 kg to mg.

LEARNING GOAL

11 Convert between units of the English and metric systems.

Solution

Using the equalities provided in Table 1.2, the data given in kg can be directly converted to a bridging data result in g, so g can be converted to the desired result in mg. The possible conversion factors are

$$\frac{10^3 \; g}{1 \; kg} = \frac{1 \; kg}{10^3 \; g} \quad \text{and} \quad \frac{10^{-3} \; g}{1 \; mg} = \frac{1 \; mg}{10^{-3} \; g}$$

Step 1. Since the initial value is in kg, the conversion factor with kg in the denominator should be used first.

$$0.0047 \; \cancel{kg} \times \frac{10^3 \; g}{1 \; \cancel{kg}} = 4.7 \; g$$

Data Given × Conversion Factor = Initial Data Result

If the other conversion factor relating kg and g was used, the resulting units would have been kg^2/g, and the answer would have been incorrect.

Step 2. Now that 0.0047 kg has been converted to 4.7 g, the conversion factor relating g to mg is used. The conversion factor with g in the denominator and mg in the numerator is the only one that leads to the correct answer.

$$4.7 \, \cancel{g} \times \frac{1 \, mg}{10^{-3} \, \cancel{g}} = 4.7 \times 10^3 \, mg$$

Initial Data Result × Conversion Factor = Desired Result

This calculation may also be done in a single step by arranging the factors in a chain:

$$0.0047 \, \cancel{kg} \times \frac{10^3 \, \cancel{g}}{1 \, \cancel{kg}} \times \frac{1 \, mg}{10^{-3} \, \cancel{g}} = 4.7 \times 10^3 \, mg$$

Data Given × Conversion Factor × Conversion Factor = Desired Result

Helpful Hint: After the conversion factors have been selected and set up in the solution to the problem, it is important to also cancel the units that can be canceled. This process will allow for you to verify that you have set up the problem correctly. In addition, the unit mg represents a significantly smaller quantity than the unit kg. Therefore, one would expect a large number of mg to equal a small number of kg.

Practice Problem 1.8

    a. 750 cm to mm

    b. 1.5×10^8 microliters to centiliters

    c. 0.00055 Mg to kg

▶ For Further Practice: **Questions 1.95 and 1.96.**

Conversion of Units Between Systems

The conversion of a quantity expressed in a unit of one system to an equivalent quantity in the other system (English to metric or metric to English) requires the use of a relating unit, a conversion unit that relates the two systems. Examples are shown in Table 1.3.

The conversion may be represented as a three-step process:

Step 1. Conversion from the unit given in the problem to a relating unit.

Data Given × Conversion Unit = Relating Unit

Step 2. Conversion to the other system using the relating unit.

Relating Unit × Conversion Unit = Initial Data Result

Step 3. Conversion within the desired system to unit required by the problem.

Initial Data Result × Conversion Unit = Desired Result

Example 1.9 demonstrates a conversion from the English system to the metric system.

TABLE 1.3 Relationships Between Common English and Metric Units

Quantity	English		Metric
Mass	1 pound	=	454 grams
Length	1 yard	=	0.914 meter
Volume	1 quart	=	0.946 liter

EXAMPLE 1.9   **Using Both English and Metric System Conversion Factors**

LEARNING GOAL

11   Convert between units of the English and metric systems.

Convert 4.00 oz to kg.

Solution

Based on the English system relationships provided in Tables 1.1 and 1.3, the data given in oz should be converted to a relating data result in lb, so lb can be converted to the g. Then, using the prefix equalities in Table 1.2, the initial data result in g can be converted to the desired result in kg. The possible conversion factors are:

$$\frac{16\ oz}{1\ lb} = \frac{1\ lb}{16\ oz} \quad \text{and} \quad \frac{454\ g}{1\ lb} = \frac{1\ lb}{454\ g} \quad \text{and} \quad \frac{10^3\ g}{1\ kg} = \frac{1\ kg}{10^3\ g}$$

Step 1. Since the initial value is in oz, the conversion factor with oz in the denominator should be used first because it relates the data given to a relating unit.

$$4.00\ \cancel{oz} \times \frac{1\ lb}{16\ \cancel{oz}} = 0.250\ lb$$

Data Given × Conversion Factor = Relating Unit

If the other conversion factor relating oz and lb was used, the resulting units would have been oz^2/lb, and the answer would have been incorrect.

Step 2. Now that 4.00 oz has been converted to 0.250 lb, the conversion factor relating lb to g is used. The conversion factor with lb in the denominator and g in the numerator is the only one that leads to the correct answer.

$$0.250\ \cancel{lb} \times \frac{454\ g}{1\ \cancel{lb}} = 114\ g$$

Relating Unit × Conversion Factor = Initial Data Result

Step 3. In the final step of this conversion, the conversion is within the desired system of units required by the problem. The conversion factor relating g and kg with g in the denominator and kg in the numerator is the only one that leads to the correct answer.

$$114\ \cancel{g} \times \frac{1\ kg}{10^3\ \cancel{g}} = 0.114\ kg$$

Initial Data Result × Conversion Factor = Desired Result

This calculation may also be done in a single step by arranging the factors in a chain:

$$4.00\ \cancel{oz} \times \frac{1\ \cancel{lb}}{16\ \cancel{oz}} \times \frac{454\ \cancel{g}}{1\ \cancel{lb}} \times \frac{1\ kg}{10^3\ \cancel{g}} = 0.114\ kg$$

Data Given × Conversion Factor × Conversion Factor × Conversion Factor = Desired Result

Helpful Hint: After the conversion factors have been selected and set up in the solution to the problem, it is important to also cancel the units that can be canceled. This process will allow for you to verify that you have set up the problem correctly. In addition, the unit oz represents a smaller quantity than the unit kg. Therefore, one would expect a large number of oz to equal a small number of kg.

Practice Problem 1.9

Convert:

a. 0.50 in to m	d. 0.50 in to cm
b. 0.75 qt to L	e. 0.75 qt to mL
c. 56.8 g to oz	f. 56.8 mg to oz

▶ For Further Practice: **Questions 1.93 a, b and 1.94 a, b.**

A Medical Perspective

Curiosity and the Science That Leads to Discovery

Curiosity is one of the most important human traits. Small children constantly ask, "Why?" As we get older, our questions become more complex, but the curiosity remains. Curiosity is also the basis of the scientific method. A scientist observes an event, wonders why it happened, and sets out to answer the question. Dr. Eric Wieschaus's story provides an example of curiosity that led to the discovery of gene pathways that are currently the target of new medicines.

As a child, Dr. Wieschaus dreamed of being an artist, but during the summer following his junior year of high school, he took part in a science program and found his place in the laboratory. When he was a sophomore in college, he accepted a job preparing fly food in a *Drosophila* (fruit fly) lab. Then, while learning about mitosis (cell division) in his embryology course, he became excited about the process of embryonic development. He was fascinated watching how a fertilized frog egg underwent cell division with little cellular growth or differentiation until it formed an embryo. Then, when the embryo grew, the cells in the various locations within the embryo developed differently. As a direct result of his observations, he became determined to understand why certain embryonic cells developed the way they did.

Throughout graduate school, his interest in solving this mystery continued. In his search for the answer, he devised different types of experiments in order to collect data that could explain what caused certain embryonic cells to differentiate into their various shapes, sizes, and positions within the growing embryo. It is these cellular differentiations that determine which cells may become tissues, organs, muscles, or nerves. Although many of his experiments failed, some of the experiments that he completed using normal embryonic cells provided data that led to his next series of experiments in which he used mutated embryos.

After graduate school, Dr. Wieschaus and his colleague, Dr. Christiane Nüsslein-Volhard, used a trial-and-error approach to determine which of the fly's 20,000 genes were essential to embryonic development. They used a chemical to create random mutations in the flies. The mutated flies were bred, and the fly families were analyzed under a microscope. Although the fly embryos are only 0.18 mm in length, the average adult female fly is 2.5 mm long. This allowed for the physical characteristics that resulted from mutated genes to be observed. An artist at heart, Dr. Wieschaus enjoyed this visual work.

Each day was exciting because he knew that at any moment, he could find the answer that he had been seeking for so long. After many years, the team was able to find the genes that controlled the cellular development process within *Drosophila*. The hedgehog gene was one of several genes they

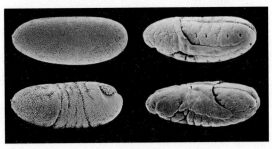

Discoveries about the Drosophila embryo have led to advances in medicine.

identified. It controls a pathway that provides cells with the information they need to develop.

Although it was initially discovered while *Drosophila* embryos were being studied, the hedgehog gene has roles in other adult animals. It has been found that if the hedgehog pathway becomes impaired in humans, basal cell carcinoma (BCC), the most common form of skin cancer, develops. The curiosity that led to the hedgehog gene also led to the discovery of an entirely new type of cancer drug, the first Food and Drug Administration (FDA) approved drug for patients with advanced BCC. This and other types of gene-controlled pathways are allowing for the creation of drugs that target specific diseases. Since these drugs can be designed to be selective, they should also have fewer side effects.

The curiosity that enabled Dr. Wieschaus to advance the field of medicine also catalyzed the development of chemistry. We will see the product of this fundamental human characteristic as we study the work of many extraordinary chemists throughout this textbook.

For Further Understanding

▶ What is the length of the fruit fly embryo in cm?
▶ What is the length of the fruit fly embryo in inches?

When a unit is raised to a power, the corresponding conversion factor must also be raised to that power. This ensures that the units cancel properly. Example 1.10 demonstrates how to convert units that are squared or cubed.

EXAMPLE 1.10   Using Conversion Factors Involving Exponents

LEARNING GOAL

Convert 1.5 m² to cm².

11   Convert between units of the English and metric systems.

Solution

This problem is similar to the conversion problems performed in the previous examples. However, in solving this problem using the factor-label method, the unit exponents must be included.

Using the metric system equalities provided in Table 1.2, the data given in m² can be directly converted to the desired result in cm². The possible conversion factors are

$$\frac{10^{-2}\ m}{1\ cm} = \frac{1\ cm}{10^{-2}\ m}$$

Since the initial value is in m², the conversion factor with cm in the denominator should be used. If the incorrect conversion factor was used, the units would not cancel and the result would be m⁴ in the numerator and cm² in the denominator.

$$1.5\ m^2 \times \frac{1\ cm}{10^{-2}\ m} \times \frac{1\ cm}{10^{-2}\ m} = 1.5 \times 10^4\ cm^2$$

Data Given × (Conversion Factor)² = Desired Result

Helpful Hint: When converting a value with a squared unit, the impact of the conversion factor is much greater than if the unit had no exponent. Without the squared unit, the two numbers would be different by a factor of 100; whereas in this example, the two numbers are different by a factor of 10,000.

Practice Problem 1.10

Convert:

    a. 1.5 cm² to m²   b. 3.6 m² to cm²

▶ For Further Practice: **Question 1.97.**

Sometimes the unit to be converted is in the denominator. Be sure to set up your conversion factor accordingly. Example 1.11 demonstrates this process.

EXAMPLE 1.11   Converting Units in the Denominator

LEARNING GOAL

The density of air is 1.29 g/L. What is the value in g/mL? (Note: Density will be discussed in more detail in Section 1.6.)

11   Convert between units of the English and metric systems.

Solution

This problem requires the use of one conversion factor. According to the metric system equalities relating mL and L provided in Table 1.2, the possible conversion factors are

$$\frac{10^{-3}\ L}{1\ mL} = \frac{1\ mL}{10^{-3}\ L}$$

Since the data given has L in the denominator, the conversion factor with L in the numerator should be used. If the incorrect conversion factor was used, the units would not cancel and the result would be g and mL in the numerator and L² in the denominator.

$$\frac{1.29 \text{ g}}{\cancel{L}} \times \frac{10^{-3}\,\cancel{L}}{1 \text{ mL}} = 1.29 \times 10^{-3}\,\frac{\text{g}}{\text{mL}}$$

Data Given × Conversion Factor = Desired Result

Helpful Hint: If the incorrect conversion factor was used, the units would not cancel and the result would be g and mL in the numerator and L^2 in the denominator.

Practice Problem 1.11

Convert 0.791 g/mL to kg/L.

▶ For Further Practice: **Question 1.98.**

It is difficult to overstate the importance of paying careful attention to units and unit conversions. Just one example of the tremendous cost that can result from a "small error," is the loss of a 125 million-dollar Mars-orbiting satellite because of failure to convert from English to metric units during one phase of its construction. As a consequence of this error, the satellite established an orbit too close to Mars and burned up in the Martian atmosphere along with 125 million dollars of the National Aeronautics and Space Administration (NASA) budget.

1.6 Additional Experimental Quantities

In Section 1.3, we introduced the experimental quantities of mass, length, volume, and time. We will now introduce other commonly measured and derived quantities.

Temperature

Temperature is the degree of "hotness" of an object. This may not sound like a very "scientific" definition, and, in a sense, it is not. Intuitively, we know the difference between a "hot" and a "cold" object, but developing a precise definition to explain this is not easy. We may think of the temperature of an object as a measure of the amount of heat in the object. However, this is not strictly true. An object increases in temperature because its heat content has increased and vice versa; however, the relationship between heat content and temperature depends on the quantity and composition of the material.

Many substances, such as mercury, expand as their temperature increases, and this expansion provides us with a way to measure temperature and temperature changes. If the mercury is contained within a sealed tube, as it is in a thermometer, the height of the mercury is proportional to the temperature. A mercury thermometer may be calibrated, or scaled, in different units, just as a ruler can be. Three common temperature scales are *Fahrenheit (°F)*, *Celsius (°C)*, and *Kelvin (K)*. Two convenient reference temperatures that are used to calibrate a thermometer are the freezing and boiling temperatures of water. Figure 1.11 shows the relationship between the scales and these reference temperatures.

Although Fahrenheit temperature is most familiar to us, Celsius and Kelvin temperatures are used exclusively in scientific measurements. It is often necessary to convert a temperature reading from one scale to another. To convert from Fahrenheit to Celsius, we use the following formula:

$$T_{°C} = \frac{T_{°F} - 32}{1.8}$$

LEARNING GOAL

12 Know the three common temperature scales, and convert values from one scale to another.

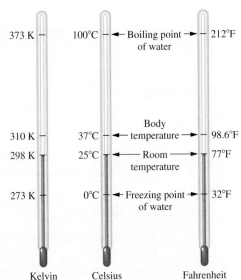

Figure 1.11 The freezing point and boiling point of water, body temperature, and room temperature expressed in the three common units of temperature.

To convert from Celsius to Fahrenheit, we solve this formula for °F, resulting in

$$T_{°F} = (1.8 \times T_{°C}) + 32$$

The Kelvin symbol does not have a degree sign. The degree sign implies a value that is *relative* to some standard. Kelvin is an *absolute* scale.

To convert from Celsius to Kelvin, we use the formula

$$T_K = T_{°C} + 273.15$$

EXAMPLE 1.12   **Converting from Fahrenheit to Celsius and Kelvin**

LEARNING GOAL

Normal body temperature is 98.6°F. Calculate the corresponding temperature in both degrees Celsius and Kelvin units and report the answer to the appropriate number of significant figures.

12 Know the three common temperature scales, and convert values from one scale to another.

Solution

Using the expression relating °C and °F,

$$T_{°C} = \frac{T_{°F} - 32}{1.8}$$

Substituting the information provided,

$$= \frac{98.6 - 32}{1.8} = \frac{66.6}{1.8}$$

results in:

$$= 37.0°C$$

To calculate the corresponding temperature in Kelvin units, use the expression relating K and °C.

$$T_K = T_{°C} + 273.15$$

Substituting the value obtained in the first part,

$$= 37.0 + 273.15$$

results in:

$$= 310.2 \text{ K}$$

According to Figure 1.11, these three temperatures are at the same place on each thermometer. Therefore, 98.6°F, 37.0°C, and 310.2 K are equivalent.

Practice Problem 1.12

  a. The freezing temperature of water is 32°F. Calculate the freezing temperature of water in Celsius units and Kelvin units.

  b. When a patient is ill, his or her temperature may increase to 104°F. Calculate the temperature of this patient in Celsius units and Kelvin units.

▶ For Further Practice: **Questions 1.115 and 1.116.**

Energy

Energy, the ability to do work, may be categorized as either **kinetic energy,** the energy of motion, or **potential energy,** the energy of position. Kinetic energy may be considered as energy in action; potential energy is stored energy. All energy is either kinetic or potential.

Another useful way of classifying energy is by form. The principal forms of energy include light, heat, electrical, mechanical, nuclear, and chemical energy. All of these forms of energy share the following set of characteristics:

- Energy cannot be created or destroyed.
- Energy may be converted from one form to another.
- Conversion of energy from one form to another always occurs with less than 100% efficiency. Energy is not lost (remember, energy cannot be destroyed) but, rather, is not useful. We use gasoline to move our cars from place to place; however, much of the energy stored in the gasoline is released as heat.
- All chemical reactions involve either a "gain" or a "loss" of energy.

Energy absorbed or liberated in chemical reactions is usually in the form of heat energy. Heat energy may be represented in units of *calories* (cal) or *joules* (J), their relationship being

$$1 \text{ cal} = 4.18 \text{ J}$$

One calorie is defined as the amount of heat energy required to increase the temperature of 1 g of water 1°C.

Heat energy measurement is a quantitative measure of heat content. It is an extensive property, dependent upon the quantity of material. Temperature, as we have mentioned, is an intensive property, independent of quantity.

Not all substances have the same capacity for holding heat; 1 g of iron and 1 g of water, even if they are at the same temperature, do *not* contain the same amount of heat energy. One gram of iron will absorb and store 0.108 cal of heat energy when the temperature is raised 1°C. In contrast, 1 g of water will absorb almost ten times as much energy, 1.00 cal, when the temperature is increased an equivalent amount.

Units for other forms of energy will be introduced in later chapters.

Question 1.19 Convert 595 cal to units of J.

Question 1.20 Convert 2.00×10^2 J to units of cal.

Concentration

Concentration is a measure of the number or mass of particles of a substance that are contained in a specified volume. Examples include:

- The concentration of oxygen in the air
- Pollen counts, given during the hay fever seasons, which are simply the number of grains of pollen contained in a measured volume of air
- The amount of an illegal drug in a certain volume of blood, indicating the extent of drug abuse
- The proper dose of an antibiotic, based on a patient's weight

We will describe many situations in which concentration is used to predict useful information about chemical reactions (Chapters 6–8, for example). In Chapter 6, we calculate a numerical value for concentration from experimental data.

Water in the environment (lakes, oceans, and streams) has a powerful effect on the climate because of its ability to store large quantities of energy. In summer, water stores heat energy and moderates temperatures of the surrounding area. In winter, some of this stored energy is released to the air as the water temperature falls; this prevents the surroundings from experiencing extreme changes in temperature.

The *kilocalorie* (kcal) is the familiar nutritional calorie. It is also known as the large Calorie (C); note that in this term the C is uppercase to distinguish it from the normal calorie. The large Calorie is 1000 normal calories. Refer to A Human Perspective: Food Calories (p. 30) for more information.

Food Calories

The body gets its energy through the processes known collectively as metabolism, which will be discussed in detail in Chapters 21–23. The primary energy sources for the body are carbohydrates, fats, and proteins, which we obtain from the foods we eat. The amount of energy available from a given foodstuff is related to the Calories (C) available in the food. Calories are a measure of energy that can be derived from food. One (food) Calorie (symbolized by C) equals 1000 (metric) calories (symbolized by cal):

$$1 \text{ C} = 1000 \text{ cal} = 1 \text{ kcal}$$

The energy available in food can be measured by completely burning the food; in other words, using the food as fuel. The energy given off in the form of heat is directly related to the amount of chemical energy, the energy stored in chemical bonds, that is available in the food. Food provides energy to the body through various metabolic pathways.

The classes of food molecules are not equally energy-rich. When oxidized via metabolic pathways, carbohydrates and proteins provide the cell with 4 C/g, whereas fats generate approximately 9 C/g.

In addition, as with all processes, not all the available energy can be efficiently extracted from the food; a certain percentage is always released to the surroundings as heat. The average person requires between 2000 and 3000 C/day to maintain normal body functions such as the regulation of body temperature and muscle movement. If a person takes in more C than the body uses, the person will gain weight. Conversely, if a person uses more C than are ingested, the individual will lose weight.

Excess C are stored in the form of fat, the form that provides the greatest amount of energy per g. Too many C leads to too much fat. Similarly, a lack of C (in the form of food) forces the body to raid its storehouse, the fat. Weight is lost in this process as the fat is consumed. Unfortunately, it always seems easier to add fat to the storehouse than to remove it.

The "rule of thumb" is that 3500 C are equivalent to approximately 1 lb of body fat. You have to take in 3500 C more than you use to gain 1 lb, and you have to expend 3500 C more than you normally use to lose 1 lb. If you eat as few as 100 C/day beyond your body's needs, you could gain about 10–11 lb per year (yr):

$$\frac{100 \ \cancel{C}}{\cancel{day}} \times \frac{365 \ \cancel{day}}{1 \text{ yr}} \times \frac{1 \text{ lb}}{3500 \ \cancel{C}} = \frac{10.4 \text{ lb}}{\text{yr}}$$

A frequently recommended procedure for increasing the rate of weight loss involves a combination of dieting (taking in fewer C) and exercise. Running, swimming, jogging, and cycling are particularly efficient forms of exercise. Running burns 0.11 C/min for every lb of body weight, and swimming burns approximately 0.05 C/min for every lb of body weight.

For Further Understanding

▸ Sarah runs 1 h each day, and Nancy swims 2 h each day. Assuming that Sarah and Nancy are the same weight, which girl burns more calories in 1 week?

▸ Would you expect a runner to burn more calories in summer or winter? Why?

Density and Specific Gravity

Both mass and volume are functions of the *amount* of material present (extensive properties). **Density,** the ratio of mass to volume,

$$\text{Density } (d) = \frac{\text{mass}}{\text{volume}} = \frac{m}{V}$$

TABLE 1.4 **Densities of Some Common Materials**

Substance	Density (g/mL)	Substance	Density (g/mL)
Air	0.00129 (at 0°C)	Mercury	13.6
Ammonia	0.000771 (at 0°C)	Methanol	0.792
Benzene	0.879	Milk	1.028–1.035
Blood	1.060	Oxygen	0.00143 (at 0°C)
Bone	1.7–2.0	Rubber	0.9–1.1
Carbon dioxide	0.001963 (at 0°C)	Turpentine	0.87
Ethanol	0.789	Urine	1.010–1.030
Gasoline	0.66–0.69	Water	1.000 (at 4°C)
Gold	19.3	Water	0.998 (at 20°C)
Hydrogen	0.000090 (at 0°C)	Wood (balsa, least dense; ebony and teak, most dense)	0.3–0.98
Kerosene	0.82		
Lead	11.3		

Figure 1.12 Density (mass/volume) is a unique property of a material. A mixture of wood, water, brass, and mercury is shown, with the cork—the least dense—floating on water. Brass, with a density greater than water but less than liquid mercury, floats on the interface between these two liquids.

is *independent* of the amount of material (intensive property). Density is a useful way to characterize or identify a substance because each substance has a unique density (Figure 1.12).

In density calculations, mass is usually represented in g, and volume is given in either mL, cm^3, or cc:

$$1 \text{ mL} = 1 \text{ cm}^3 = 1 \text{ cc}$$

The unit of density would therefore be g/mL, g/cm^3, or g/cc. It is important to recognize that because the units of density are a ratio of mass to volume, density can be used as a conversion factor when the factor-label method is used to solve for either mass or volume from density data.

A 1-mL sample of air and 1-mL sample of iron have different masses. There is much more mass in 1 mL of iron; its density is greater. Density measurements were used to distinguish between real gold and "fool's gold" during the gold rush era. Today, the measurement of the density of a substance is still a valuable analytical technique. The densities of a number of common substances are shown in Table 1.4.

EXAMPLE 1.13 | **Calculating the Density of a Solid**

A 2.00 cm^3 sample of aluminum (symbolized Al) is found to weigh 5.40 g. Calculate the density of Al in units of g/cm^3 and g/mL.

Solution

The density expression is:

$$d = \frac{m}{V}, \text{ in which mass is usually in g, and volume is in either mL or cm}^3.$$

Substituting the information given in the problem,

$$d = \frac{5.40 \text{ g}}{2.00 \text{ cm}^3} = 2.70 \text{ g/cm}^3$$

LEARNING GOAL

13 Use density, mass, and volume in problem solving, and calculate the specific gravity of a substance from its density.

Continued...

According to Table 1.2, 1 mL = 1 cm³. Therefore, we can use this identity as a conversion factor to obtain the answer in g/mL. The cm³ unit is placed in the numerator so that it cancels, and the mL unit is placed in the denominator because it will provide the correct unit for the product.

$$\frac{2.70 \text{ g}}{1 \text{ cm}^3} \times \frac{1 \text{ cm}^3}{1 \text{ mL}} = 2.70 \text{ g/mL}$$

Initial Data Result × Conversion Factor = Desired Result

The density of water, 1.0 g/mL, can be found on Table 1.4. Since aluminum is more dense than water, it should have a density greater than 1.0 g/mL.

Practice Problem 1.13

A 0.500 mL sample of a metal has a mass of 6.80 g. Calculate the density of the metal in units of g/mL and g/cm³.

▶ For Further Practice: **Questions 1.121 and 1.122.**

EXAMPLE 1.14   Using the Density to Calculate the Mass of a Liquid

LEARNING GOAL

Calculate the mass, in g, of 10.0 mL of mercury (symbolized Hg) if the density of mercury is 13.6 g/mL.

13 Use density, mass, and volume in problem solving, and calculate the specific gravity of a substance from its density.

Solution

First, it must be determined which density conversion factor is correct for this problem.

$$\frac{13.6 \text{ g Hg}}{1 \text{ mL Hg}} \quad \text{or} \quad \frac{1 \text{ mL Hg}}{13.6 \text{ g Hg}}$$

Since the data given is in mL, only the first conversion factor will result in a product with the unit g. Using the factor-label method, the answer can be calculated.

$$10.0 \text{ mL Hg} \times \frac{13.6 \text{ g Hg}}{1 \text{ mL Hg}} = 136 \text{ g Hg}$$

Data Given × Conversion Factor = Desired Result

Practice Problem 1.14

The density of ethanol (200 proof or pure alcohol) is 0.789 g/mL at 20°C. Calculate the mass of a 30.0-mL sample.

▶ For Further Practice: **Questions 1.125 and 1.126.**

EXAMPLE 1.15   Calculating the Mass of a Gas from Its Density

LEARNING GOAL

Air has a density of 0.0013 g/mL. What is the mass of a 6.0 L sample of air?

13 Use density, mass, and volume in problem solving, and calculate the specific gravity of a substance from its density.

Solution

This problem can be solved using the factor-label method. Since the data given is in L, the first conversion factor used should relate L to mL, one of the units in the density expression.

$$6.0 \, \text{L air} \times \frac{10^3 \, \text{mL air}}{1 \, \text{L air}} = 6.0 \times 10^3 \, \text{mL air}$$

Data Given × Conversion Factor = Initial Data Result

Since the units of density are in fraction form, the value of density is a ratio that can be used as a conversion factor.

$$\frac{0.0013 \, \text{g air}}{1 \, \text{mL}} \quad \text{or} \quad \frac{1 \, \text{mL}}{0.0013 \, \text{g air}}$$

The density conversion factor with mL in the denominator is the only one that will result in the product unit of g.

$$6.0 \times 10^3 \, \text{mL air} \times \frac{0.0013 \, \text{g air}}{1 \, \text{mL air}} = 7.8 \, \text{g air}$$

Initial Data Result × Conversion Factor = Desired Result

Practice Problem 1.15

What mass of air, in g, would be found in a 2.0-L party balloon?

▶ For Further Practice: **Questions 1.123 and 1.124.**

EXAMPLE 1.16 Using the Density to Calculate the Volume of a Liquid

LEARNING GOAL

13 Use density, mass, and volume in problem solving, and calculate the specific gravity of a substance from its density.

Calculate the volume, in mL, of a liquid that has a density of 1.20 g/mL and a mass of 5.00 g.

Solution

It must first be determined which density conversion factor is correct for this problem.

$$\frac{1.20 \, \text{g liquid}}{1 \, \text{mL liquid}} \quad \text{or} \quad \frac{1 \, \text{mL liquid}}{1.20 \, \text{g liquid}}$$

Since the data given for the liquid is in g, the conversion factor with g in the denominator is chosen. Using the factor-label method, the answer can be calculated.

$$5.00 \, \text{g liquid} \times \frac{1 \, \text{mL liquid}}{1.20 \, \text{g liquid}} = 4.17 \, \text{mL liquid}$$

Data Given × Conversion Factor = Desired Result

Helpful Hint: Notice in the solution that density is inverted with the volume in the numerator and the mass in the denominator. This enables the units to cancel.

Practice Problem 1.16

Calculate the volume, in mL, of 10.0 g of a saline solution that has a density of 1.05 g/mL.

▶ For Further Practice: **Questions 1.131 and 1.136.**

A Medical Perspective

Assessing Obesity: The Body-Mass Index

Density, the ratio of two extensive properties, mass and volume, is an intensive property that can provide useful information about the identity and properties of a substance. The Body-Mass Index (BMI) is also a ratio of two extensive properties, the weight and height (actually the square of the height) of an individual. As a result, the BMI is also an intensive property. It is widely used by physical trainers, medical professionals, and life insurance companies to quantify obesity, which is a predictor of a variety of potential medical problems.

In metric units, the Body-Mass Index is expressed:

$$BMI = \frac{weight\ (kg)}{height\ (m^2)}$$

This can be converted to the English system by using conversion factors. The number 703 is the commonly used conversion factor to convert from English units (in and lb) to metric units (m and kg) that are the units in the definition of BMI.

The conversion is accomplished in the following way:

Weight and height (metric) Weight and height (English)

↓ ↓

$$BMI = \frac{kg}{m^2} = \frac{lb}{in^2} \times \frac{1\ kg}{2.205\ lb} \times \left(\frac{39.37\ in}{1\ m}\right)^2$$

$$= \frac{lb}{in^2} \times 703\ \frac{kg \cdot in^2}{lb \cdot m^2}$$

The units of the conversion factor are generally not shown, and the BMI in English units is reduced to:

$$BMI = \frac{weight\ in\ lb}{(height\ in\ in)^2} \times 703$$

Online BMI calculators generally use this form of the equation.

An individual with a BMI of 25 or greater is considered overweight; if the BMI is 30 or greater, the individual is described as obese. However, for some individuals, the BMI may underestimate or overestimate body fat. For example, it is common for an athlete with a muscular build to have a high BMI that does not accurately reflect his or her body fat.

BMI values for a variety of weights and heights are shown as a function of individuals' height and weight. Once known, the BMI can be used as a guideline in the design of suitable diet and exercise programs.

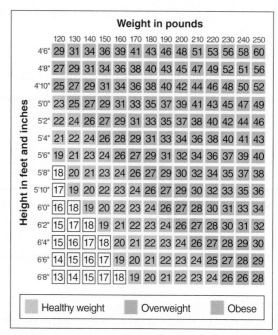

Developed by the National Center for Health Statistics in collaboration with the National Center for Chronic Disease Prevention and Health Promotion

For Further Understanding

▸ Refer to A Human Perspective: Food Calories (p. 30) and describe connections between these two perspectives.
▸ Calculate your BMI in both metric and English units. Do they agree? Explain why or why not.

Specific gravity is frequently referenced to water at 4°C, its temperature of maximum density (1.000 g/mL). Other reference temperatures may be used. However, the temperature must be specified.

For convenience, values of density are often related to a standard, well-known reference, the density of pure water at 4°C. This "referenced" density is called the **specific gravity**, the ratio of the density of the object in question to the density of pure water at 4°C.

$$Specific\ gravity = \frac{density\ of\ object\ (g/mL)}{density\ of\ water\ (g/mL)}$$

Quick and Useful Analysis

Measurement of the specific gravity of a liquid is fast, easy, and nondestructive of the sample. Changes in specific gravity over time can provide a wealth of information. Two examples follow:

Living cells carry out a wide variety of chemical reactions, which produce molecules and energy essential for the proper function of living organisms. Urine, a waste product, contains a wide variety of by-products from these chemical processes. It can be analyzed to indicate abnormalities in cell function or even unacceptable personal behavior (recall the steroid tests in Olympic competition).

Many of these tests must be performed by using sophisticated and sensitive instrumentation. However, a very simple test, the measurement of the specific gravity of urine, can be an indicator of diabetes mellitus or dehydration. The normal range for human urine specific gravity is 1.010–1.030.

A hydrometer, a weighted glass bulb inserted in a liquid, may be used to determine specific gravity. The higher it floats in the liquid, the more dense the liquid. A hydrometer that is calibrated to indicate the specific gravity of urine is called a urinometer.

Winemaking is a fermentation process (Chapter 12). The flavor, aroma, and composition of wine depend upon the extent of fermentation. As fermentation proceeds, the specific gravity of the wine gradually changes. Periodic measurement of the specific gravity during fermentation enables the wine-maker to determine when the wine has reached its optimal composition.

For Further Understanding

▶ Give reasons that may account for such a broad range of "normal" values for urine specific gravity.

▶ Could the results for a diabetes test depend on food or medicine consumed prior to the test?

Monitoring the winemaking process.

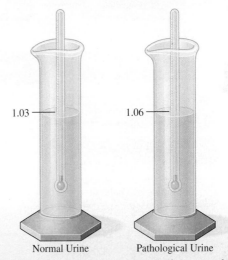
1.03   1.06

Normal Urine   Pathological Urine

A hydrometer is used in the measurement of the specific gravity of urine.

Specific gravity is a *unitless* term. Because the density of water at 4.0°C is 1.00 g/mL, the numerical values for the density and specific gravity of a substance are equal. That is, an object with a density of 2.00 g/mL has a specific gravity of 2.00 at 4°C.

Routine hospital tests involving the measurement of the specific gravity of urine and blood samples are frequently used as diagnostic tools. For example, diseases such as kidney disorders and diabetes change the composition of urine. This compositional change results in a corresponding change in the specific gravity. This change is easily measured and provides the basis for a quick preliminary diagnosis. This topic is discussed in greater detail in A Human Perspective: Quick and Useful Analysis (above).

LEARNING GOAL

13 Use density, mass, and volume in problem solving, and calculate the specific gravity of a substance from its density.

CHAPTER MAP

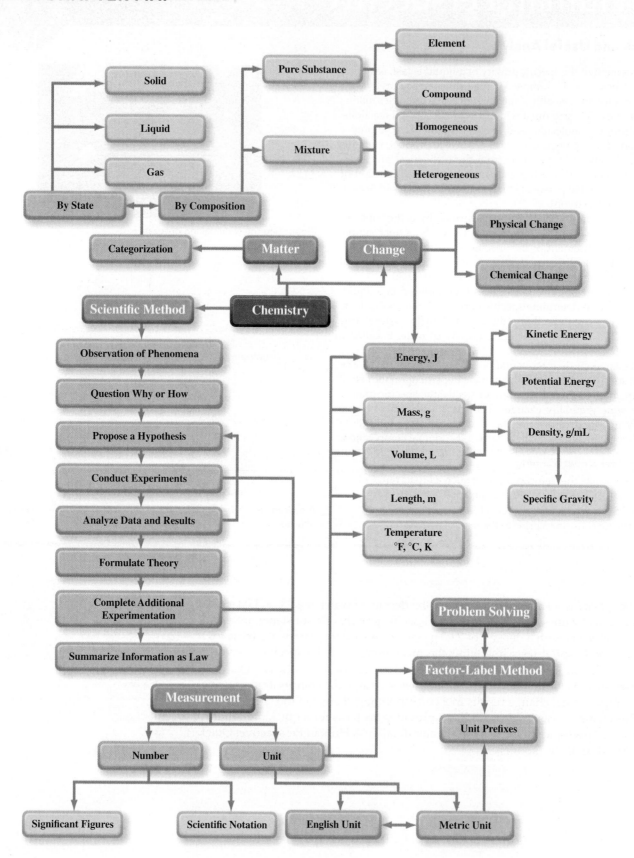

SUMMARY

1.1 The Discovery Process

▶ **Chemistry** is the study of matter, its chemical and physical properties, the chemical and physical changes it undergoes, and the energy changes that accompany those processes.

▶ **Matter** is anything that has mass and occupies space.

▶ Matter gains and loses **energy** at it undergoes change.

▶ The **scientific method** is a systematic approach to the discovery of new information. Characteristics of the scientific method include:

- Observation of a phenomenon.
- Formulation of a question concerning the observation.
- Presentation of a **hypothesis,** or answer to the question.
- Experimentation, with the collection and analysis of **data** and **results** in an attempt to support or disprove the hypothesis.
- The ultimate goal of the process is to form a **theory,** a hypothesis supported by extensive testing.
- A scientific **law** is the summary of a large quantity of information.

1.2 The Classification of Matter

▶ It is useful to classify matter according to its **properties.**

▶ Matter can be classified by state as a **solid, liquid,** or **gas.**

▶ Matter can also be classified by composition, which groups matter as either a **pure substance** or **mixture.**

▶ Pure substances can be further subdivided into **elements** or **compounds.**

▶ Mixtures can be subdivided into **homogeneous** mixtures or **heterogeneous** mixtures.

▶ Properties of matter can be classified as:

- a **physical property**—observed without changing the composition of the matter or
- a **chemical property**—observed as the matter is converted to a new substance

▶ Properties can also be classified as:

- intensive—a property that does not depend on the quantity of matter or
- extensive—a property that does depend on the quantity of matter

▶ The changes that matter can undergo can be classified as **physical change** or **chemical change.** Chemical change is synonymous with **chemical reaction.**

1.3 The Units of Measurement

▶ The **unit** of a measurement is as vital as the number.

▶ The metric system has an advantage over the English system because metric units are systematically related to each other by powers of ten.

▶ **Mass** is a measure of the quantity of matter. **Weight** is the force of gravity on an object. The metric unit for mass is the gram. A common English mass unit is the pound.

▶ **Length** is the distance between two points. The metric unit is the meter, which is similar, but not equal to the English yard.

▶ **Volume** is the space occupied by an object. The metric unit of the liter is similar to the English quart.

▶ The metric unit of time is the second.

1.4 The Numbers of Measurement

▶ **Significant figures** are all digits in a measurement known with certainty plus one uncertain digit. It is important to be able to:

- Read a measuring device to the correct number of significant figures.
- Recognize the number of significant figures in a given written measurement.
- Report answers to calculations to the correct number of significant figures. The rules for addition and subtraction are different from the rules for multiplication and division.
- Follow the rules for rounding numbers as digits are dropped from the calculation to give the correct number of significant figures.

▶ **Scientific notation** is a way to express a number as a power of ten.

▶ Measurements always have a certain degree of **deviation** and **error** associated with them. Care must be taken to obtain a measurement with minimal error and high **accuracy** and **precision.**

1.5 Unit Conversion

▶ Conversions between units can be accomplished using the factor-label method.

▶ Conversion factors are fractions in which the numerator and denominator are equivalent in magnitude, and thus, the conversion factor is equivalent to one.

▶ In the factor-label method, the undesired unit is eliminated because it cancels out with a unit in the conversion factor.

▶ When a unit is squared or cubed, the conversion factor must also be squared or cubed.

1.6 Additional Experimental Quantities

▶ **Temperature** is the degree of "hotness" of an object. Common units include Celsius, Kelvin, and Fahrenheit. Equations are used to convert between the units.

▶ **Energy,** the ability to do work, is categorized as **kinetic energy** or **potential energy.** Common units of energy are the calorie and the joule.

▶ **Concentration** is a measure of the amount of a substance contained in a specified amount of a mixture.

▶ **Density** is the ratio of mass to volume and is an intensive property.

- Density is typically reported as g/mL or g/cm^3.
- Density can be used as a conversion factor to convert between the mass and volume of an object.

▶ **Specific gravity** is the ratio of the density of an object to the density of pure water at 4°C. It is numerically equal to density but has no units.

ANSWERS TO PRACTICE PROBLEMS

1.1 Fill two beakers with identical volumes of water. Add salt (several grams) to one of the beakers of water. Insert a thermometer into each beaker, and slowly heat the beakers. Record the temperature of each liquid when boiling is observed.

1.2 **a.** Pure substance   **c.** Heterogeneous mixture
    b. Heterogeneous mixture   **d.** Pure substance

1.3 **a.** Physical change   **d.** Physical change
    b. Chemical change   **e.** Chemical change
    c. Chemical change

1.4 Intensive property; the freezing point of a glass of pure water and a gallon of pure water is the same. The freezing point is independent of the quantity of the substance.

1.5 7.46×10^5

1.6 4.7×10^4

1.7 6.82×10^{-2} mi

1.8 **a.** 7.5×10^3 mm
    b. 1.5×10^4 cL
    c. 0.55 kg

1.9 **a.** 1.3×10^{-2} m   **d.** 1.3 cm
    b. 0.71 L   **e.** 7.1×10^2 mL
    c. 2.00 oz   **f.** 2.00×10^{-3} oz

1.10 **a.** 1.5×10^{-4} m^2 **b.** 3.6×10^4 cm^2

1.11 0.791 kg/L

1.12 **a.** 0°C   **b.** 40°C
      273 K   313 K

1.13 13.6 g/mL
     13.6 g/cm^3

1.14 23.7 g ethanol

1.15 2.6 g air

1.16 9.52 mL saline

QUESTIONS AND PROBLEMS

The answers to the odd-numbered questions and problems in this section are provided at the back of the book. In addition, the Student Study Guide/Solutions Manual provides complete solutions to the odd-numbered questions and problems as well as additional problems.

The Discovery Process

Foundations

1.21 Define chemistry and explain how burning wood is related to chemistry.
1.22 Define energy and explain the importance of energy in chemistry.
1.23 Why is experimentation an important part of the scientific method?
1.24 Why is observation a critical starting point for any scientific study?
1.25 What data would be required to estimate the total cost of gasoline needed to drive from New York City to Washington, D.C.?
1.26 What data would be required to estimate the mass of planet earth?
1.27 What are the characteristics of methane emphasized by the following model?

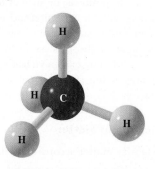

1.28 The model of methane in Question 1.27 has limitations, as do all models. What are these limitations?

Applications

1.29 Discuss the difference between *hypothesis* and *theory*.
1.30 Discuss the difference between *theory* and *scientific law*.
1.31 We use aspects of the scientific method in our everyday lives. Provide one example of formulating a hypothesis as an aid in solving a practical problem.
1.32 Describe an application of reasoning involving the scientific method that has occurred in your day-to-day life.
1.33 Experimentation has shown that stem cell research has the potential to provide replacement "parts" for the human body. Is this statement a theory of scientific law? Explain your reasoning.
1.34 Observed increases in global temperatures are caused by elevated levels of carbon dioxide. Is this statement a theory or a scientific law? Explain your reasoning.
1.35 Describe an experiment demonstrating that the freezing point of water changes when salt (sodium chloride) is added to water.
1.36 Describe an experiment that would enable you to determine the mass (g) of solids suspended in a 1-L sample of seawater.

The Classification of Matter

Foundations

1.37 List the three states of matter.
1.38 Explain the differences among the three states of matter in terms of volume and shape.

1.39 Distinguish between a pure substance and a mixture.

1.40 Give examples of pure substances and mixtures.

1.41 Describe what is meant by an intensive property and give an example.

1.42 Describe what is meant by an extensive property and give an example.

1.43 Distinguish between a homogeneous mixture and a heterogeneous mixture.

1.44 Distinguish between an intensive property and an extensive property.

1.45 Explain the difference between chemical properties and physical properties.

1.46 List the differences between chemical changes and physical changes.

1.47 Label each of the following as pertaining to either a solid, liquid, or gas.
 a. It expands to fill the container.
 b. It has a fixed shape.
 c. The particles are in a predictable pattern of arrangement.

1.48 Label each of the following as pertaining to either a solid, liquid, or gas.
 a. It has a fixed volume, but not a fixed shape.
 b. The attractive forces between particles are very pronounced.
 c. The particles are far apart.

Applications

1.49 Draw a diagram representing a heterogeneous mixture of two different substances. Use two different colored spheres to represent the two different substances.

1.50 Draw a diagram representing a homogeneous mixture of two different substances. Use two different colored spheres to represent the two different substances.

1.51 Label each of the following as either a physical change or a chemical reaction:
 a. An iron nail rusts.
 b. An ice cube melts.
 c. A limb falls from a tree.

1.52 Label each of the following as either a physical change or a chemical reaction:
 a. A puddle of water evaporates.
 b. Food is digested.
 c. Wood is burned.

1.53 Label each of the following properties of sodium as either a physical property or a chemical property:
 a. Sodium is a soft metal (can be cut with a knife).
 b. Sodium reacts violently with water to produce hydrogen gas and sodium hydroxide.

1.54 Label each of the following properties of sodium as either a physical property or a chemical property:
 a. When exposed to air, sodium forms a white oxide.
 b. The density of sodium metal at 25°C is 0.97 g/cm^3.

1.55 Label each of the following as either a pure substance or a mixture:
 a. saliva
 b. table salt (sodium chloride)
 c. wine
 d. helium inside of a balloon

1.56 Label each of the following as either a pure substance or a mixture:
 a. sucrose (table sugar) **c.** urine
 b. orange juice **d.** tears

1.57 Label each of the following as either a homogeneous mixture or a heterogeneous mixture:
 a. a carbonated soft drink **c.** gelatin
 b. a saline solution **d.** margarine

1.58 Label each of the following as either a homogeneous mixture or a heterogeneous mixture:
 a. gasoline **c.** concrete
 b. vegetable soup **d.** hot coffee

1.59 Classify the matter represented in the following diagram by state and by composition.

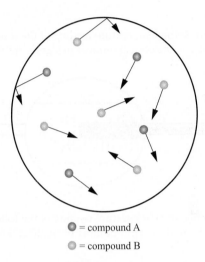

 ● = compound A
 ○ = compound B

1.60 Classify the matter represented in the following diagram by state and by composition.

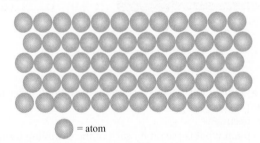

 ○ = atom

1.61 Plant varieties may be distinguished, one from another, by observing properties of their leaves.
 a. Suggest two extensive properties of leaves that would be useful in the process.
 b. Suggest two intensive properties of leaves that would be useful in the process.

1.62 Would you expect that intensive properties or extensive properties would be more useful in distinguishing among plant varieties by inspecting their leaves?

The Units of Measurement

Foundations

1.63 Mass is the measure of what property of matter?

1.64 Explain the difference between mass and weight.

1.65 Define length.

1.66 What metric unit for length is similar to the English yd?

1.67 How is the metric unit of L defined?

1.68 What English unit of volume is similar to a L?

Applications

1.69 Rank the following from shortest to longest length.

mm, km, m

1.70 Rank the following from least to greatest mass.

cg, μg, Mg

The Numbers of Measurement

Foundations

1.71 Determine the temperature reading of the following thermometer to the correct number of significant figures.

1.72 Determine the temperature reading of the following thermometer to the correct number of significant figures.

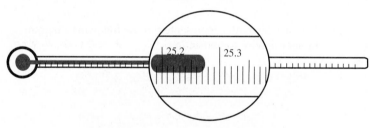

1.73 Explain what is meant by each of the following terms:
a. precision
b. accuracy

1.74 Explain what is meant by each of the following terms:
a. error
b. uncertainty

1.75 How many significant figures are represented in each of the following numbers?
a. 10.0   d. 2.062
b. 0.214   e. 10.50
c. 0.120   f. 1050

1.76 How many significant figures are represented in each of the following numbers?
a. 3.8×10^{-3} d. 24
b. 5.20×10^{2} e. 240
c. 0.00261   f. 2.40

1.77 Round the following numbers to three significant figures:
a. 3.873×10^{-3} d. 24.3387
b. 5.202×10^{-2} e. 240.1
c. 0.002616   f. 2.407

1.78 Round the following numbers to three significant figures:
a. 123700   d. 53.2995
b. 0.00285792   e. 16.96
c. 1.421×10^{-3} f. 507.5

Applications

1.79 Perform each of the following operations, reporting the answer with the proper number of significant figures:
a. (23)(657)   d. $1157.23 - 17.812$
b. $0.00521 + 0.236$ e. $\dfrac{(1.987)(298)}{0.0821}$
c. $\dfrac{18.3}{3.0576}$

1.80 Perform each of the following operations, reporting the answer with the proper number of significant figures:
a. $\dfrac{(16.0)(0.1879)}{45.3}$ d. $18 + 52.1$
b. $\dfrac{(76.32)(1.53)}{0.052}$ e. $58.17 - 57.79$
c. $(0.0063)(57.8)$

1.81 Express the following numbers in scientific notation (use the proper number of significant figures):
a. 12.3   e. 92,000,000
b. 0.0569   f. 0.005280
c. −1527   g. 1.279
d. 0.000000789   h. −531.77

1.82 Express each of the following numbers in standard notation:
a. 3.24×10^{3} e. -8.21×10^{-2}
b. 1.50×10^{4} f. 2.9979×10^{8}
c. 4.579×10^{-1} g. 1.50×10^{0}
d. -6.83×10^{5} h. 6.02×10^{23}

1.83 The following four measurements were made for an object whose true mass is 4.56 g.

4.55 g, 4.56 g, 4.56 g, 4.57 g

Describe the measurements in terms of their accuracy and their precision.

1.84 The following four measurements were made for an object whose true volume is 17.55 mL.

18.69 mL, 18.69 mL, 18.70 mL, 18.71 mL

Describe the measurements in terms of their accuracy and their precision.

Unit Conversion

Foundations

1.85 Give a reason why the metric system is a more convenient system than the English system of measurement.

1.86 Why is it important to *always* include units when recording measurements?

1.87 Give the abbreviation and meaning of the following metric prefixes:
a. kilo   c. micro
b. centi

1.88 Fill in the blank with the missing abbreviation and name the prefix.
a. 10^{6} m = 1 _____ m   c. 10^{-9} g = 1 _____ g
b. 10^{-3} L = 1 _____ L

1.89 Write the two conversion factors that can be written for the relationship between ft and in.

1.90 Write the two conversion factors that can be written for the relationship between cm and in.

1.91 Convert 2.0 lb to:
   a. oz   **d.** mg
   b. t   **e.** dag
   c. g

1.92 Convert 5.0 qt to:
   a. gal   **d.** mL
   b. pt   **e.** μL
   c. L

1.93 Convert 3.0 g to:
   a. lb   **d.** cg
   b. oz   **e.** mg
   c. kg

1.94 Convert 3.0 m to:
   a. yd   **d.** cm
   b. in   **e.** mm
   c. ft

1.95 Convert 1.50×10^4 μg to mg.

1.96 Convert 7.5×10^{-3} cm to mm.

1.97 A typical office has 144 ft² of floor space. Calculate the floor space in m² (use the proper number of significant figures).

1.98 Tire pressure is measured in units of lb/in². Convert 32 lb/in² to g/cm² (use the proper number of significant figures).

Applications

1.99 A 150 lb adult has approximately 9 pt of blood. How many L of blood does the individual have?

1.100 If a drop of blood has a volume of 0.05 mL, how many drops of blood are in the adult described in Question 1.99?

1.101 A patient's temperature is found to be 38.5°C. To what Fahrenheit temperature does this correspond?

1.102 A newborn is 21 in in length and weighs 6 lb 9 oz. Describe the baby in metric units.

1.103 Which distance is shorter: 5.0 cm or 5.0 in?

1.104 Which volume is smaller: 50.0 mL or 0.500 L?

1.105 Which mass is smaller: 5.0 mg or 5.0 μg?

1.106 Which volume is smaller: 1.0 L or 1.0 qt?

1.107 A new homeowner wished to know the perimeter of his property. He found that the front boundary and back boundary were measured in meters: 85 m and 95 m, respectively. The side boundaries were measured in feet: 435 ft and 515 ft.
   a. Describe the problem-solving strategy used to determine the perimeter in km.
   b. Calculate the perimeter in km.

1.108 Sally and Gertrude were comparing their weight-loss regimens. Sally started her diet weighing 193 lb. In 1 year she weighed 145 lb. Gertrude started her diet weighing 80 kg. At the end of the year, she weighed 65 kg. Who lost the most weight?
   a. Describe the problem-solving strategy used to determine who was more successful.
   b. Calculate the weight lost in lb and kg by Sally and Gertrude.

Additional Experimental Quantities

Foundations

1.109 List three major temperature scales.

1.110 Rank the following temperatures from coldest to hottest: zero degrees Celsius, zero degrees Fahrenheit, zero Kelvin

1.111 List and define the two subgroups of energy.

1.112 Label each of the following statements as true or false. If false, correct the statement.
   a. Energy can be created or destroyed.
   b. Energy can be converted from electrical energy to light energy.
   c. Conversion of energy from one form to another can occur with 100% efficiency.
   d. All chemical reactions involve either a gain or loss of energy.

1.113 List each of the following as an intensive or extensive property.
   a. mass   **b.** volume   **c.** density

1.114 What is the relationship between density and specific gravity?

Applications

1.115 Convert 50.0°F to:
   a. °C   **b.** K

1.116 The weather station posted that the low for the day would be −10°F. Convert −10.0°F to:
   a. °C   **b.** K

1.117 The thermostat shows that the room temperature is 20.0°C. Convert 20.0°C to:
   a. K   **b.** °F

1.118 Convert 300.0 K to:
   a. °C   **b.** °F

1.119 The combustion of a peanut releases 6 kcal of heat. Convert this energy to J.

1.120 The energy available from the world's total petroleum reserve is estimated at 2.0×10^{22} J. Convert this energy to kcal.

1.121 Calculate the density of a 3.00×10^2 g object that has a volume of 50.0 mL.

1.122 Calculate the density of 50.0 g of an isopropyl alcohol–water mixture (commercial rubbing alcohol) that has a volume of 63.6 mL.

1.123 What volume, in L, will 8.00×10^2 g of air occupy if the density of air is 1.29 g/L?

1.124 In Question 1.123, you calculated the volume of 8.00×10^2 g of air with a density of 1.29 g/L. The temperature of the air sample was lowered and the density increased to 1.50 g/L. Calculate the new volume of the air sample.

1.125 What is the mass, in g, of a piece of iron that has a volume of 1.50×10^2 mL and a density of 7.20 g/mL?

1.126 What is the mass of a femur (leg bone) having a volume of 118 cm³? The density of bone is 1.8 g/cm³.

1.127 For the treatment of cystic fibrosis, it has been found that inhaling a mist of saline solution can clear mucous from the lungs. This solution is made by dissolving 7 g of salt in 1 L of sterile water. If this is to be done twice per day for 1 year, how many g of salt will be needed?

1.128 You are given a piece of wood that is either maple, teak, or oak. The piece of wood has a volume of 1.00×10^2 cm³ and a mass of 98 g. The densities of maple, teak, and oak are as follows:

Wood	Density (g/cm³)
Maple	0.70
Teak	0.98
Oak	0.85

What is the identity of the piece of wood?

1.129 You are given three bars of metal. Each is labeled with its identity (lead, uranium, platinum). The lead bar has a mass of 5.0×10^1 g and a volume of 6.36 cm^3. The uranium bar has a mass of 75 g and a volume of 3.97 cm^3. The platinum bar has a mass of 2140 g and a volume of 1.00×10^2 cm^3. Which of these metals has the lowest density? Which has the greatest density?

1.130 Refer to Question 1.129. Suppose that each of the bars had the same mass. How could you determine which bar had the lowest density and which had the highest density?

1.131 The density of methanol at 20°C is 0.791 g/mL. What is the volume of a 10.0 g sample of methanol?

1.132 The density of methanol at 20°C is 0.791 g/mL. What is the mass of a 50.0 mL sample of methanol?

1.133 It is common for a dehydrated patient to provide a urine sample that has a specific gravity of 1.04. Using the density of water at 37°C (0.993 g/mL), calculate the density of the urine sample.

1.134 The specific gravity of a patient's urine sample was measured to be 1.008. Given that the density of water is 1.000 g/mL at 4°C, what is the density of the urine sample?

1.135 The density of grain alcohol is 0.789 g/mL. Given that the density of water at 4°C is 1.00 g/mL, what is the specific gravity of grain alcohol?

1.136 The density of mercury is 13.6 g/mL. If a sample of mercury weighs 272 g, what is the volume of the sample in mL?

1.137 The density of whole human blood in a healthy individual is 1.04 g/mL. Given that the density of water at 37°C is 0.993 g/mL, what is the specific gravity of whole human blood?

1.138 Assume the Body-Mass Index (BMI) is calculated using the expression BMI = weight (kg)/height^2 (m^2). If a patient has a height of 1.6 m and a BMI of 38 kg/m^2, what is the patient's weight in both kg and lb?

CHALLENGE PROBLEMS

1. An instrument used to detect metals in drinking water can detect as little as 1 μg of mercury in 1 L of water. Mercury is a toxic metal; it accumulates in the body and is responsible for the deterioration of brain cells. Calculate the number of mercury atoms you would consume if you drank 1 L of water that contained 1 μg of mercury. (The mass of one mercury atom is 3.3×10^{-22} g.)

2. Yesterday's temperature was 40°F. Today it is 80°F. Bill tells Sue that it is twice as hot today. Sue disagrees. Do you think Sue is correct or incorrect? Why or why not?

3. Aspirin has been recommended to minimize the chance of heart attacks in persons who have already had one or more occurrences. If a patient takes one aspirin tablet per day for 10 years, how many pounds of aspirin will the patient consume? (Assume that each tablet is approximately 325 mg.)

4. The diameter of an aluminum atom is 250 picometers (pm) (1 pm = 10^{-12} m). How many aluminum atoms must be placed end-to-end to make a "chain" of aluminum atoms 1 ft long?

TABLE 2.1 Selected Properties of the Three Basic Subatomic Particles

Name	Charge	Mass (amu)	Mass (g)
Electron (e⁻)	−1	5.486×10^{-4}	9.1094×10^{-28}
Proton (p⁺)	+1	1.007	1.6726×10^{-24}
Neutron (n)	0	1.009	1.6750×10^{-24}

We may represent an element symbolically as follows:

$$\text{Mass number} \rightarrow {}^{A}_{Z}X \leftarrow \text{Element symbol}$$
$$\text{Atomic number} \rightarrow$$

LEARNING GOAL

2 Interpret atomic symbols, and calculate the number of protons, neutrons, and electrons for atoms.

The **atomic number** (Z) is equal to the number of protons in the atom, and the **mass number** (A) is equal to the *sum* of the number of protons and neutrons (the mass of the electrons is so small that it is insignificant in comparison to the mass of the nucleus).

The atomic number is the whole number associated with the atom on the periodic table. The mass number is not found on the periodic table.

Since,

$$\text{mass number} = \text{number of protons} + \text{number of neutrons}$$

to determine the number of neutrons, we can rearrange the equation to give:

$$\text{number of neutrons} = \text{mass number} - \text{number of protons}$$

Since number of protons = atomic number

$$\text{number of neutrons} = \text{mass number} - \text{atomic number}$$

$$\text{number of neutrons} = A - Z$$

EXAMPLE 2.1   **Determining the Composition of an Atom**

LEARNING GOAL

2 Interpret atomic symbols, and calculate the number of protons, neutrons, and electrons for atoms.

Calculate the number of protons, neutrons, and electrons in an atom of fluorine. The atomic symbol for the fluorine atom is ${}^{19}_{9}F$.

Solution

The atomic symbol provided gives us both the atomic number and the mass number.

Step 1. The atomic number, Z, is 9. Therefore, there are 9 protons.

Step 2. The number of protons is equal to the number of electrons in an atom. There are 9 electrons.

Step 3. The mass number, A is equal to 19. The number of neutrons = $A - Z$. Since the number of neutrons is equal to $19 - 9 = 10$, there are 10 neutrons.

Helpful Hint: Since the mass number is the sum of the number of protons and neutrons, it is the number with the greater value provided in the atomic symbol. The atomic number, which directly provides us with the number of protons, can also be found on the periodic table.

Practice Problem 2.1

Calculate the number of protons, neutrons, and electrons in each of the following atoms:

a. ${}^{32}_{16}S$

b. ${}^{23}_{11}Na$

c. ${}^{1}_{1}H$

d. ${}^{244}_{94}Pu$

▶ For Further Practice: **Questions 2.27 and 2.28.**

3 Distinguish between the terms *atom* and *isotope*, and use isotope notations and natural abundance values to calculate atomic masses.

Isotopes

Isotopes are atoms of the same element having different masses *because they contain different numbers of neutrons.* In other words, isotopes have different mass numbers. For example, all of the following are isotopes of hydrogen:

$$^1_1H \qquad\qquad ^2_1H \qquad\qquad ^3_1H$$
$$\text{Hydrogen} \qquad \text{Deuterium} \qquad \text{Tritium}$$

Isotopes are often written with the name of the element followed by the mass number. For example, the isotopes $^{12}_6C$ and $^{14}_6C$ may be written as carbon-12 (or C-12) and carbon-14 (or C-14), respectively.

Question 2.1   How many protons, neutrons, and electrons are present in a single atom of:

a. bromine-79   b. bromine-81   c. iron-56

Question 2.2   How many protons, neutrons, and electrons are present in a single atom of:

a. phosphorus-30   b. sulfur-32   c. chlorine-35

A detailed discussion of the use of radioactive isotopes in the diagnosis and treatment of diseases is found in Chapter 9.

Isotopes are useful in many clinical situations. For example, certain isotopes (radioactive isotopes) of elements emit particles and energy that can be used to trace the behavior of biochemical systems. The chemical behavior of these isotopes is identical to the other isotopes of the same element. However, their nuclear behavior is unique. As a result, a radioactive isotope can be substituted for the "nonradioactive" isotope, and its biochemical activity can be followed by monitoring the particles or energy emitted by the isotope as it passes through the body.

As stated earlier, the atomic number is the whole number associated with every element on the periodic table. The other number associated with each element on the periodic table is its **atomic mass,** the *weighted* average of the masses of each isotope that makes up the element. Atomic mass is measured in *atomic mass units (amu).*

$$1\text{ amu} = 1.6605 \times 10^{-24}\text{ grams (g)}$$

Notice in Table 2.1 that the masses of protons and neutrons are both approximately 1.0 amu. Since the atomic mass is equal to the mass of protons and neutrons, the atomic mass of a hydrogen-2 atom containing one proton and one neutron would be 2.0 amu.

The vast majority of hydrogen on our planet is hydrogen-1. Hydrogen-2 is often termed deuterium and hydrogen-3, tritium.

If we examine the periodic table, we see that the atomic mass of hydrogen is 1.008 amu. Three natural isotopes of hydrogen exist, H-1, H-2, and H-3. The atomic mass of hydrogen, given as 1.008 amu, is not an average of 1, 2, and 3 (which would be 2), it is a weighted average. There is more H-1 in nature than there is H-2 and H-3. The percentage of each isotope, as found in nature, is called its *natural abundance.* A weighted average takes into account the natural abundance of each isotope.

Example 2.2 demonstrates the calculation of the atomic mass of chlorine.

EXAMPLE 2.2	**Determining Atomic Mass from Two Isotopes**	

Calculate the atomic mass of naturally occurring chlorine if 75.77% of the chlorine atoms are $^{35}_{17}Cl$ and 24.23% of chlorine atoms are $^{37}_{17}Cl$.

3 Distinguish between the terms *atom* and *isotope*, and use isotope notations and natural abundance values to calculate atomic masses.

Solution

The mass of each isotope is derived from the isotope notation. The atomic mass derived from the weighted average of isotope masses will be between 35 amu and 37 amu. Since there is more chlorine-35 than chlorine-37, we expect the weighted average to be closer to 35 amu.

Step 1. Convert each percentage to a decimal fraction.

$$75.77\% \text{ chlorine-35} \times \frac{1}{100\%} = 0.7577 \text{ chlorine-35}$$

$$24.23\% \text{ chlorine-37} \times \frac{1}{100\%} = 0.2423 \text{ chlorine-37}$$

Step 2. Determine the contribution of each isotope to the total atomic mass by multiplying the decimal fraction by the mass of the isotope.

For chlorine-35: $0.7577 \times 35.0 \text{ amu} = 26.5 \text{ amu}$

For chlorine-37: $0.2423 \times 37.0 \text{ amu} = 8.97 \text{ amu}$

Step 3. Add the mass contributed by each isotope.

Atomic mass $= 26.5 \text{ amu} + 8.97 \text{ amu} = 35.5 \text{ amu}$

Helpful Hint: The value obtained, 35.5 amu, is between 35.0 amu and 37.0 amu. As expected from the natural abundance values, it is closer to 35.0 amu. The periodic table provides the value of 35.45 amu for the atomic mass of chlorine.

Practice Problem 2.2

The element nitrogen has two naturally occurring isotopes. One of these has a mass of 14.0 amu and a natural abundance of 99.63%; the other isotope has a mass of 15.0 amu and a natural abundance of 0.37%. Calculate the atomic mass of nitrogen.

▶ For Further Practice: **Question 2.37.**

EXAMPLE 2.3   **Determining Atomic Mass from Three Isotopes**

LEARNING GOAL

The element neon has three naturally occurring isotopes. Neon-20 has a natural abundance of 90.48%, neon-21 has a natural abundance of 0.27%, and neon-22 has a natural abundance of 9.25%. Calculate the atomic mass of neon.

3 Distinguish between the terms *atom* and *isotope*, and use isotope notations and natural abundance values to calculate atomic masses.

Solution

Remember the mass of each isotope is derived from the isotope notation. Since the isotope of the greatest abundance is neon-20, the atomic mass value should be close to 20.0 amu. The atomic mass will be greater than 20.0 amu because the less abundant isotopes have masses greater than 20.0 amu.

Step 1. Convert each percentage to a decimal fraction.

$$90.48\% \text{ neon-20} \times \frac{1}{100\%} = 0.9048 \text{ neon-20}$$

$$0.27\% \text{ neon-21} \times \frac{1}{100\%} = 0.0027 \text{ neon-21}$$

$$9.25\% \text{ neon-22} \times \frac{1}{100\%} = 0.0925 \text{ neon-22}$$

Step 2. Determine the contribution of each isotope to the total atomic mass by multiplying the decimal fraction by the mass of the isotope.

For neon-20: $0.9048 \times 20.0 \text{ amu} = 18.1 \text{ amu}$

For neon-21: $0.0027 \times 21.0 \text{ amu} = 0.057 \text{ amu}$

For neon-22: $0.0925 \times 22.0 \text{ amu} = 2.04 \text{ amu}$

Continued…

Step 3. Add the masses contributed by each isotope.

$$\text{Atomic mass} = 18.1 \text{ amu} + 0.057 \text{ amu} + 2.04 \text{ amu} = 20.2 \text{ amu}$$

Helpful Hint: The calculated value of 20.2 amu is slightly greater than the mass of the most abundant isotope. This value is also close to the atomic mass of neon provided on the periodic table, 20.18 amu.

Practice Problem 2.3

Calculate the atomic mass of naturally occurring boron if 19.9% of the boron atoms are $^{10}_5 \text{B}$ and 80.1% are $^{11}_5 \text{B}$.

▶ For Further Practice: **Question 2.38.**

LEARNING GOAL

4 Summarize the history of the development of atomic theory, beginning with Dalton.

2.2   Development of Atomic Theory

With this overview of our current understanding of the structure of the atom, we now look at a few of the most important scientific discoveries that led to modern atomic theory.

Dalton's Theory

The first experimentally based theory of atomic structure was proposed in the early 1800s by John Dalton, an English schoolteacher. Dalton proposed the following description of atoms:

1. All matter consists of tiny particles called atoms.
2. An atom cannot be created, divided, destroyed, or converted to any other type of atom.
3. Atoms of a particular element have identical properties.
4. Atoms of different elements have different properties.
5. Atoms of different elements combine in simple whole-number ratios to produce compounds (stable combinations of atoms).
6. Chemical change involves joining, separating, or rearranging atoms.

Although Dalton's theory was founded on meager and primitive experimental information, we regard much of it as correct today. The discovery of the process of nuclear fission (fission is the process of "splitting" atoms that led to the development of the atomic bomb during World War II) disproved the postulate that atoms cannot be created or destroyed. The postulate that all the atoms of a particular element are identical was disproved by the discovery of isotopes.

Fusion, fission, radioactivity, and isotopes are discussed in some detail in Chapter 9. Figure 2.2 uses a simple model to illustrate Dalton's theory.

Evidence for Subatomic Particles: Electrons, Protons, and Neutrons

The next major discoveries occurred almost a century later (1879–1897). Although Dalton pictured atoms as indivisible, various experiments, particularly those of William Crookes and Eugene Goldstein, indicated that the atom is composed of charged (+ and −) particles.

Crookes connected two metal electrodes (metal disks connected to a source of electricity) at opposite ends of a sealed glass vacuum tube. When the electricity was turned on, rays of light were observed to travel between the two electrodes. They were called *cathode rays* because they traveled from the *cathode* (the negative electrode) to the *anode* (the positive electrode).

Atoms of element X   Atoms of element Y

(a)

Compound formed from elements X and Y

(b)

Figure 2.2 An illustration of John Dalton's atomic theory. (a) Atoms of the same element are identical but different from atoms of any other element. (b) Atoms combine in whole-number ratios to form compounds.

Chemistry at the Crime Scene

Microbial Forensics

We have learned that not all atoms of the same element are identical; usually elements are a mixture of two or more isotopes, differing in mass because they contain different numbers of neutrons. Furthermore, the atomic mass is the weighted average of the masses of these various isotopes.

It seems to follow from this that the relative amounts of these isotopes would be the same, no matter where in the world we obtain a sample of the element. In reality, this is not true. There exist small, but measurable, differences between elemental isotopic ratios; these differences correlate with their locations.

Water, of course, contains oxygen atoms, and a small fraction of these atoms are oxygen-18. We know that lakes and rivers across the United States contain oxygen-18 in different concentrations. Ocean water is richer in oxygen-18. Consequently, freshwater supplies closest to oceans are similarly enriched, perhaps because water containing oxygen-18 is a bit heavier and falls to earth more rapidly, before the clouds are carried inland, far from the coast. Scientists have been able to map the oxygen-18 distribution throughout the United States, creating contour maps, similar to weather maps, showing regions of varying concentrations of the isotope.

How can we use these facts to our advantage? The Federal Bureau of Investigation (FBI) has an interest in ascertaining the origin of microbes that may be used by terrorists. If we could carefully measure the ratio of oxygen-18 to oxygen-16 in weaponized biological materials, it might be possible to match this ratio to that found in water supplies in a specific location. We would then have an indication of the region where the material was cultured, perhaps leading to the arrest and prosecution of the perpetrators of the terrorist activity.

This is certainly not an easy task. However, the FBI has assembled an advisory board of science and forensic experts to study this and other approaches to tracking the origin of biological weapons of mass destruction. Just like the "smoking gun," these strategies, termed *microbial forensics,* could lead to more effective prosecution and a safer society for all of us.

For Further Understanding

▶ Use the Internet to determine other useful information that can be obtained by measuring the ratio of oxygen-18 to oxygen-16.

▶ How might the approach, described above, be applied to determining the origin of an oil spill at sea?

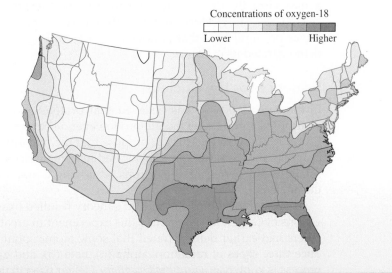

Concentrations of oxygen-18

Lower   Higher

Later experiments by J. J. Thomson, an English scientist, demonstrated the electrical and magnetic properties of cathode rays (Figure 2.3). The rays were deflected toward the positive pole of an external electric field. Because opposite charges attract, this indicates the negative character of the rays. Similar experiments with an external magnetic field showed a deflection as well; hence these cathode rays also have magnetic properties.

A change in the material used to fabricate the electrode disks brought about no change in the experimental results. This suggested that the ability to produce cathode rays is a characteristic of all materials.

Figure 2.3 Illustration of an experiment demonstrating the charge properties of cathode rays. In the absence of an electric field, cathode rays will strike the fluorescent screen at A. In the presence of an electric field, the beam will be deflected to point B. Similar deflections are observed when a magnetic field is applied.

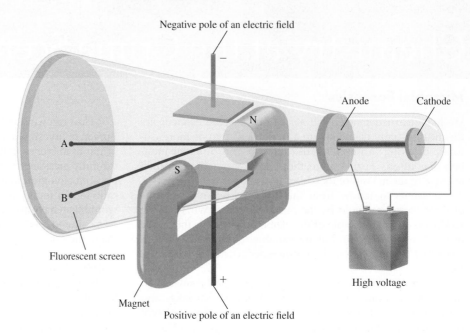

Negative pole of an electric field

Anode   Cathode

N

A

S

B

Fluorescent screen

Magnet

Positive pole of an electric field

High voltage

In 1897, Thomson announced that cathode rays are streams of negative particles of energy. These particles are *electrons*. Similar experiments, conducted by Goldstein, led to the discovery of particles that are equal in charge to the electron but opposite in sign. These particles, much heavier than electrons (actually 1837 times as heavy), are called *protons*.

As we have seen, the third fundamental atomic particle is the *neutron*. It has a mass virtually identical (it is less than 1% heavier) to that of the proton and has zero charge. The existence of the neutron was first proposed in the early 1920s, but it was not until 1932 that James Chadwick, an English physicist, experimentally demonstrated its existence.

Evidence for the Nucleus

In the early 1900s, it was believed that protons and electrons were uniformly distributed throughout the atom. However, an experiment by Hans Geiger led Ernest Rutherford (in 1911) to propose that the majority of the mass and positive charge of the atom was actually located in a small, dense region, the *nucleus*, with small, negatively charged electrons occupying a much larger volume outside of the nucleus.

To understand how Rutherford's theory resulted from the experimental observations of Geiger, let us examine this experiment in greater detail. Rutherford and others had earlier demonstrated that some atoms spontaneously "decay" to produce three types of radiation: alpha (α), beta (β), and gamma (γ) radiation. This process is known as *natural radioactivity*. Geiger used materials that were naturally radioactive, such as *radium*, as projectile sources, "firing" the alpha particles produced at a thin metal foil (gold leaf) target. He then observed the interaction of the metal and alpha particles with a detection screen (Figure 2.4) and found that:

- Most alpha particles passed through the foil without being deflected.
- A small fraction of the particles were deflected, some even *directly back to the source.*

Rutherford interpreted this to mean that most of the atom is empty space, because most alpha particles were not deflected. Further, most of the mass and positive charge must be located in a small, dense region; collision of the heavy and positively charged alpha particle with this small dense and positive region (the nucleus) caused the great deflections. Rutherford summarized his astonishment

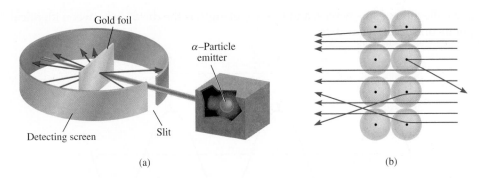

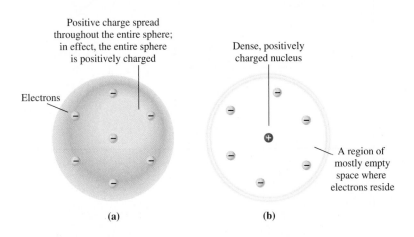

Figure 2.4 (a) The alpha-particle scattering experiment. Most alpha particles passed through the foil without being deflected; a few were deflected from their path by nuclei in the gold atoms. (b) An enlarged view of the particles passing through the atoms. The nuclei, shown as black dots, deflect some of the alpha particles.

Figure 2.5 (a) A model of the atom (credited to Thomson) prior to the work of Geiger and Rutherford. (b) A model of the atom supported by the alpha-particle scattering experiments of Geiger and Rutherford.

at observing the deflected particles: "It was almost as incredible as if you fired a 15-inch shell at a piece of tissue and it came back and hit you."

The significance of Rutherford's contribution cannot be overstated. It caused a revolutionary change in the way that scientists pictured the atom (Figure 2.5). His discovery of the nucleus is fundamental to our understanding of chemistry. Chapter 9 will provide much more information about the nucleus and its unique properties.

2.3 Light, Atomic Structure, and the Bohr Atom

The Rutherford model of the atom leaves us with a picture of a tiny, dense, positively charged nucleus containing protons and surrounded by electrons. The electron arrangement, or configuration, is not clearly detailed. More information is needed to ascertain the relationship of the electrons to each other and to the nucleus.

When dealing with dimensions on the order of 10^{-9} m (the atomic level), conventional methods for measurement of location and distance of separation become impossible. An alternative approach to determine structure involves the measurement of *energy* rather than the *position* of the atomic particles. For example, information obtained from the absorption or emission of *light* by atoms can yield valuable insight into structure. Such studies are referred to as **spectroscopy**.

Electromagnetic Radiation

Light is **electromagnetic radiation**. Electromagnetic radiation travels in *waves* from a source. The most recognizable source of this radiation is the sun. We are aware of a rainbow, in which visible white light from the sun is broken up into several characteristic bands of different colors. Similarly, visible white light, when passed through a glass prism, is separated into its various component colors (Figure 2.6). These various colors are simply light of differing *wavelengths*. Light is propagated

LEARNING GOAL

5 Describe the role of spectroscopy and the importance of electromagnetic radiation in the development of atomic theory.

Figure 2.6 The visible spectrum of light. Light passes through a prism, producing a continuous spectrum. Color results from the way in which our eyes interpret the various wavelengths.

as a collection of sine waves, and the wavelength is the distance between identical points on successive waves:

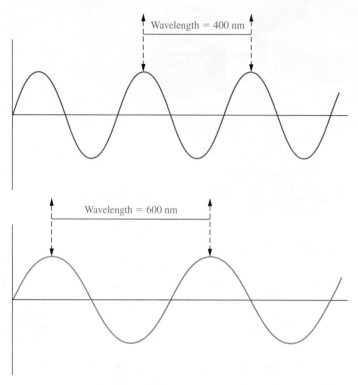

A spectrophotometer, an instrument that utilizes a prism (or similar device) and a light-sensitive detector, is capable of very accurate and precise wavelength measurement.

All electromagnetic radiation travels at a speed of 3.0×10^8 m/s, the **speed of light.** However, each wavelength of light, although traveling with identical velocity, has its own characteristic energy. A collection of all electromagnetic radiation, including each of these wavelengths, is referred to as the *electromagnetic spectrum*. For convenience in discussing this type of radiation, we subdivide electromagnetic radiation into various spectral regions, which are characterized by physical properties of the radiation, such as its *wavelength* or its *energy* (Figure 2.7). Some of these regions are quite familiar to us from our everyday experiences; the visible and microwave regions are two common examples. A variety of applications of electromagnetic radiation are discussed in Green Chemistry: Practical Applications of Electromagnetic Radiation, on p. 54.

Photons

Experimental studies by Max Planck in the early part of the twentieth century, and interpretation of his results by Albert Einstein, led us to believe that light is made up of particles we call *photons*. A **photon** is a particle of light. You may ask, "How can light that we have described as energy that moves from location to location as a wave also be described as a particle?" The best non-mathematical answer we can give is that light exhibits both wave and particle properties, depending on the type of measurement being made. We say that light has a dual nature, both wave and particle. Owing to the fact that the relationship between the wavelength and energy of a photon is known, scientists can calculate the energy of any photon when its wavelength has been measured. Photons with large wavelengths are lower in energy than photons with small wavelengths.

The Bohr Atom

If hydrogen gas is placed in an evacuated tube and an electric current is passed through the gas, light is emitted. Not all wavelengths (or energies) of light are emitted—only certain wavelengths that are characteristic of the gas in the tube. This is referred to as an *emission spectrum* (Figure 2.8). If a different gas, such as

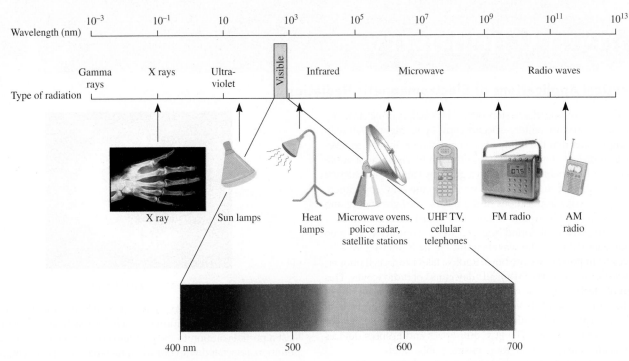

Figure 2.7 The electromagnetic spectrum. Note that the visible spectrum is only a small part of the total electromagnetic spectrum.

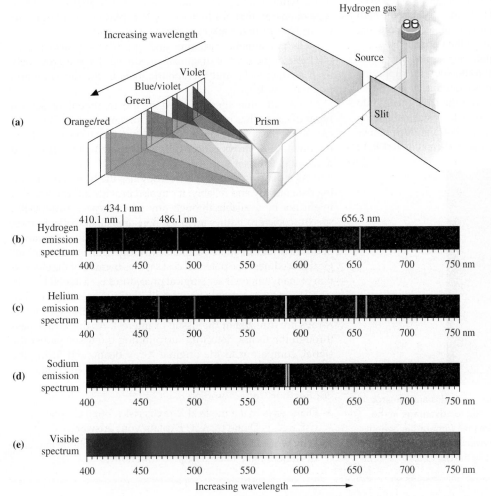

Figure 2.8 (a) The emission spectrum of hydrogen. Certain wavelengths of light, characteristic of the atom, are emitted upon electrical excitation. The line spectrum of hydrogen (b) is compared with the line spectra of helium (c), sodium (d), and the spectrum of visible light (e).

Green Chemistry

Practical Applications of Electromagnetic Radiation

From the preceding discussion of the interaction of electromagnetic radiation with matter—spectroscopy—you might be left with the impression that the utility of such radiation is limited to theoretical studies of atomic structure. Although this is a useful application that has enabled us to learn a great deal about the structure and properties of matter, it is by no means the only application. Useful, everyday applications of the theories of light energy and transmission are all around us. Let's look at just a few examples.

Transmission of sound and pictures is conducted at radio frequencies (RF) or radio wavelengths. We are immersed in radio waves from the day we are born. Radios, televisions, cell phones, and computers (via Wi-Fi) are all "detectors" of radio waves. This form of electromagnetic radiation is believed to cause no physical harm because of its very low energy. However, the Federal Communications Commission (FCC) has adopted guidelines for RF radiation exposure. The FCC authorizes and licenses devices and facilities that generate RF to ensure public safety.

X-rays are electromagnetic radiation, and they travel at the speed of light just like radio waves. However, because of their higher energy, they can pass through the human body and leave an image of the body's interior on a photographic film. X-ray photographs are invaluable for medical diagnosis. However, caution is advised in exposing oneself to X-rays, because the high energy can remove electrons from biological molecules, causing subtle and potentially harmful changes in their chemistry.

The sunlight that passes through our atmosphere provides the basis for a potentially useful technology for supplying heat and electricity: *solar energy*. Light is captured by absorbers, referred to as *solar collectors*, which convert the light energy into heat energy. This heat can be transferred to water circulating beneath the collectors to provide heat and hot water for homes or industry. Wafers of a silicon-based material can convert light

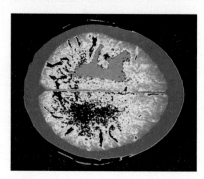

An image of a tumor detected by a CT scanner.

energy to electrical energy; many believe that if the efficiency of these processes can be improved, such approaches may provide at least a partial solution to the problems of rising energy costs and pollution associated with our fossil fuel–based energy economy.

Microwave radiation for cooking, *infrared* (IR) lamps for heating and IR photography, *ultraviolet* (UV) lamps used to kill microorganisms on environmental surfaces, *gamma radiation* from nuclear waste, the *visible* light from the lamp you are using while reading this chapter—all are forms of the same type of energy that, for better or worse, plays such a large part in our twenty-first century technological society.

Electromagnetic radiation and spectroscopy also play a vital role in the field of diagnostic medicine. They are routinely used as diagnostic and therapeutic tools in the detection and treatment of disease.

The radiation therapy used in the treatment of several types of cancer has been responsible for saving many lives and extending the span of many others. When radiation is used as a treatment, it destroys cancer cells. This topic will be discussed in detail in Chapter 9.

As a diagnostic tool, spectroscopy has the benefit of providing data quickly and reliably; it can also provide information that might not be available through any other means. Additionally, spectroscopic procedures are often nonsurgical, outpatient procedures. Such procedures involve less risk, can be more routinely performed, and are more acceptable to the general public than surgical procedures. The potential cost savings because of the elimination of many unnecessary surgical procedures is an added benefit.

The most commonly practiced technique uses the CT scanner, an acronym for *computerized tomography*. In this technique, X-rays are directed at the tissue of interest. As the X-rays pass through the tissue, detectors surrounding the tissue gather the signal, compare it to the original X-ray beam, and, using the computer, produce a three-dimensional image of the tissue.

The intensity of IR radiation from a solid or liquid is an indicator of relative temperature. This has been used to advantage in the design of IR cameras, which can obtain images without the benefit of the visible light that is necessary for conventional cameras. This IR photograph shows the coastline surrounding the city of San Francisco.

For Further Understanding

▶ Diane says that a medical X-ray is risky, but a CT scan is risk free. Is Diane correct? Explain your answer.

▶ Why would the sensors (detectors) for a cell phone and an IR camera have to be designed differently?

helium, is used, a different spectrum with different wavelengths of light is observed. The reason for this behavior was explained by Niels Bohr as he gave us our first look into the electronic structure of the atom.

Niels Bohr studied the emission spectrum of hydrogen (Figure 2.8b) and developed a model to explain the line spectrum observed in this one electron atom. (Refer to Figure 2.9 throughout this discussion.) Bohr hypothesized that surrounding each atomic nucleus were certain fixed **energy levels** that could be occupied by the electron. This is referred to as the *quantization* of energy, meaning energy can have certain values but cannot have amounts between those values. The levels were analogous to orbits of planets around the sun and were numbered according to their relative distance from the nucleus ($n = 1, 2, 3, \ldots$).

An atom is in the **ground state** when the electrons of the atom are in the lowest possible energy levels. Hydrogen is in the ground state when its electron is in the $n = 1$ level. The electron can absorb energy and be promoted to a higher energy level, farther from the nucleus; that is, elevated to an **excited state** (any $n > 1$). Once in the excited state, the electron spontaneously emits energy in the form of a photon of light of the exact energy and wavelength necessary to return to the ground state. This process is called *relaxation*. The photon of light emitted appears as a specific line (*spectral line*) on the emission spectrum. Bohr was able to use the values of the wavelengths of light emitted to calculate the energy of an electron in each of the energy levels.

The major features of the Bohr theory are summarized as follows:

- Electrons are found only in *allowed energy levels.*
- The allowed energy levels are quantized energy levels, or orbits.
- Atoms absorb energy by excitation of electrons to higher energy levels, farther from the nucleus.
- Atoms release energy by relaxation of electrons to lower energy levels, closer to the nucleus.
- Energy that is emitted upon relaxation is observed as a single wavelength of light, a collection of photons, each having the same wavelength.
- Energy differences may be calculated from the wavelengths of light emitted.
- These *spectral lines* are a result of electron transitions between allowed energy levels in the atom.

Modern Atomic Theory

The Bohr model was an immensely important contribution to the understanding of atomic structure. The idea that electrons exist in specific energy states and that the emission of energy in the form of a photon is required for an electron to move

The line spectrum of each known element is unique. Consequently, spectroscopy is a very useful tool for identifying elements.

LEARNING GOAL

6 State the basic postulates of Bohr's theory, its utility, and its limitations.

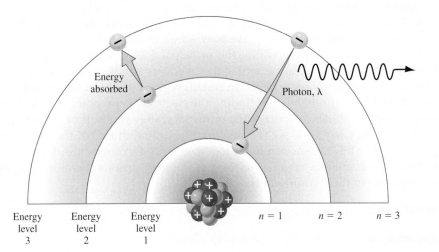

Figure 2.9 The electron, shown in yellow, absorbs energy as it is promoted from the $n = 2$ energy level to the $n = 3$ excited state. The electron then undergoes *relaxation* as it returns to the $n = 1$ orbit (the ground state). The relaxation of the electron results in the release of one photon of light with a specific wavelength associated with the $n = 3 \rightarrow n = 1$ transition.

Energy absorbed

Photon, λ

| Energy level 3 | Energy level 2 | Energy level 1 | $n = 1$ | $n = 2$ | $n = 3$ |

A Human Perspective

Atomic Spectra and the Fourth of July

At one time or another, we have all marveled at the bright, multicolored display of light and sounds characteristic of a fireworks show. These sights and sounds are produced by chemical reactions that generate the energy necessary to excite a variety of elements to their higher-energy electronic states. Light emission results from relaxation of the excited atoms to the ground state. Each atom releases light of specific wavelengths. The visible wavelengths are seen as colored light.

Fireworks use chemical reactions to produce energy. We know from experience that when oxygen reacts with a type of fuel, energy is released. The fuel in most fireworks preparations is sulfur or aluminum. Since each of these fuels reacts slowly with the oxygen in the air, a more potent solid-state source of oxygen, potassium perchlorate ($KClO_4$), is used in packaging the fireworks. The potassium perchlorate reacts with the fuel (an oxidation-reduction reaction, Chapter 4), producing a bright white flash of light. The heat produced excites the various other elements packaged with the fuel and oxidant.

Sodium salts, such as sodium chloride, furnish sodium ions, which, when excited, produce yellow light. Red colors arise from salts of strontium, which emit several shades of red corresponding to wavelengths in the 600- to 700-nm region of the visible spectrum. Copper salts produce blue radiation, because copper emits in the 400- to 500-nm spectral region.

The beauty of fireworks is a direct result of the skill of the manufacturer. Selection of the proper oxidant, fuel, and color-producing elements is critical to the production of a spectacular display. Packaging these chemicals in proper quantities so they can be stored and used safely is an equally important consideration.

For Further Understanding

▶ Explain why excited sodium emits a yellow color. (Refer to Figure 2.8.)

▶ How does this story illustrate the interconversion of potential and kinetic energy?

A fireworks display is a dramatic illustration of light emission by excited atoms.

---

from a higher to a lower energy state provided the linkage between atomic structure and atomic spectra. However, some limitations of this model quickly became apparent. Although it explained the hydrogen spectrum, it provided only a crude approximation of the spectra for atoms with more than one electron. Subsequent development of more sophisticated experimental techniques demonstrated that there are problems with the Bohr theory even in the case of hydrogen.

Although Bohr's concept of principal energy levels is still valid, restriction of electrons to fixed orbits is too rigid. All current evidence shows that electrons do *not*, in fact, orbit the nucleus. We now speak of the *probability* of finding an electron in a *region* of space within the principal energy level, referred to as an **atomic orbital.** The rapid movement of the electron spreads the charge into a *cloud* of charge. This cloud is more dense in certain regions, the **electron density** being proportional to the probability of finding the electron at any point in time. Insofar as these atomic orbitals are part of the principal energy levels, they are referred to as sublevels. In Chapter 3, we will see that the orbital model of the atom can be used to predict how atoms can bond together to form compounds. Furthermore, electron arrangement in orbitals enables us to predict various chemical and physical properties of these compounds.

Question 2.3 What is meant by the term *electron density?*

Question 2.4 How do *orbits* and *orbitals* differ?

The theory of atomic structure has progressed rapidly, from a very primitive level to its present point of sophistication. Before we proceed, let us insert a note of caution. We must not think of the present picture of the atom as final. Scientific inquiry continues, and we should view the present theory as a step in an evolutionary process. *Theories are subject to constant refinement*, as was noted in our discussion of the scientific method.

2.4 The Periodic Law and the Periodic Table

In 1869, Dmitri Mendeleev, a Russian, and Lothar Meyer, a German, working independently, found ways of arranging elements in order of increasing atomic mass such that elements with similar properties were grouped together in a *table of elements*. The **periodic law** is embodied by Mendeleev's statement, "the elements, if arranged according to their atomic weights (masses), show a distinct *periodicity* (regular variation) of their properties." The *periodic table* (Figure 2.10) is a visual representation of the periodic law.

Chemical and physical properties of elements correlate with the electronic structure of the atoms that make up the elements. In turn, the electronic structure

LEARNING GOAL

7 Recognize the important subdivisions of the periodic table: periods, groups (families), metals, and nonmetals.

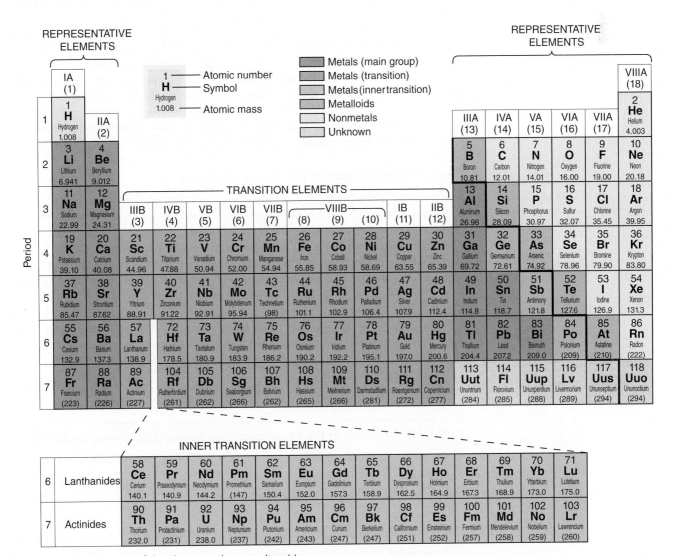

Figure 2.10 Classification of the elements: the periodic table.

correlates with the position on the periodic table. We will examine these connections in Section 2.5; however, the scientists who first arranged the elements did not know about their electronic structure.

A thorough familiarity with the arrangement of the periodic table allows us to predict electronic structure and physical and chemical properties of the various elements. It also serves as the basis for understanding chemical bonding.

The concept of "periodicity" may be illustrated by examining a portion of the modern periodic table. The elements in the second row (beginning with lithium, Li, and proceeding to the right) show a marked difference in properties. However, sodium (Na) has properties similar to those of lithium, and sodium is therefore placed below lithium; once sodium is fixed in this position, the elements Mg through Ar have properties remarkably similar (though not identical) to those of the elements just above them. The same is true throughout the complete periodic table.

Mendeleev arranged the elements in his original periodic table in order of increasing atomic mass. However, as our knowledge of atomic structure increased, atomic numbers became the basis for the organization of the table. Remarkably, his table was able to predict the existence of elements not known at the time.

The modern periodic law states that *the physical and chemical properties of the elements are periodic functions of their atomic numbers.* If we arrange the elements in order of increasing number of protons, the properties of the elements repeat at regular intervals.

We will use the periodic table as our "map," just as a traveler would use a road map. A short time spent learning how to read the map (and remembering to carry it along on your trip!) is much easier than memorizing every highway and intersection. The information learned about one element relates to an entire family of elements grouped as a recognizable unit within the table.

Not all of the elements are of equal importance to an introductory study of chemistry. Table 2.2 lists twenty of the elements that are most important to biological systems, along with their symbols and a brief description of their functions.

Numbering Groups in the Periodic Table

The periodic table created by Mendeleev has undergone numerous changes over the years. These modifications occurred as more was learned about the chemical and physical properties of the elements.

Groups or *families* are columns of elements in the periodic table. The elements of a particular group or family share many similarities, as in a human family. The similarities extend to physical and chemical properties that are related to similarities in electronic structure (that is, the way in which electrons are arranged in an atom).

The labeling of groups with Roman numerals followed by the letter *A* or *B* was standard, until 1983, in North America and Russia. However, in other parts of the world, the letters *A* and *B* were used in a different way. Consequently, two different periodic tables were in widespread use. This certainly created some confusion.

Mendeleev's original periodic table included only the elements known at the time, fewer than half of the current total.

TABLE 2.2 Summary of the Most Important Elements in Biological Systems

Roles in Biological Systems	Element Symbols
Components of major biological molecules	H, C, O, N, P, S
Responsible for fluid balance and nerve transmission	K, Na, Cl
Necessary for bones and nerve function	Ca, Mg
In very small quantities, essential for human metabolism	Zn, Sr, Fe, Cu, Co, Mn
"Heavy metals," toxic to living systems	Cd, Hg, Pb

In 1983, the International Union of Pure and Applied Chemistry (IUPAC) recommended that a third system, using numbers 1–18 to label the groups, replace both of the older systems. Unfortunately, multiple systems now exist, and this can cause confusion for both students and experienced chemists.

The periodic tables in this textbook are "double labeled." Both the old (Roman numeral) and new (1–18) systems are used to label the groups. The label that you use is simply a guide to reading the table; the real source of information is the structure of the table itself. The following sections will show you how to extract useful information from this structure.

Group A elements are called **representative elements,** and Group B elements are **transition elements.** Certain families also have common names. For example, Group IA (or 1) elements are also known as the **alkali metals**; Group IIA (or 2), the **alkaline earth metals**; Group VIIA (or 17), the **halogens**; and Group VIIIA (or 18), the **noble gases.**

Representative elements are also known as *main-group elements*. These terms are synonymous.

Periods

A **period** is a horizontal row of elements in the periodic table. The periodic table consists of seven periods containing two, eight, eight, eighteen, eighteen, thirty-two, and thirty-two elements. Note that the *lanthanide series,* a collection of fourteen elements that are chemically and physically similar to the element lanthanum, is a part of period six. It is written separately for convenience of presentation and is inserted between lanthanum (La), atomic number 57, and hafnium (Hf), atomic number 72. Similarly, the *actinide series,* consisting of fourteen elements similar to the element actinium, is inserted between actinium, atomic number 89, and rutherfordium, atomic number 104.

LEARNING GOAL

7 Recognize the important subdivisions of the periodic table: periods, groups (families), metals, and nonmetals.

Metals and Nonmetals

A **metal** is a substance whose atoms tend to lose electrons during chemical change. A **nonmetal,** on the other hand, is a substance whose atoms may gain electrons.

A close inspection of the periodic table reveals a bold zigzag line running from top to bottom, beginning to the left of boron (B) and ending between ununseptium (Uus) and ununoctium (Uuo). This line acts as the boundary between *metals,* to the left, and *nonmetals,* to the right. Elements straddling the boundary have properties intermediate between those of metals and nonmetals. These elements are referred to as **metalloids.** Commonly encountered metalloids include boron (B), silicon (Si), germanium (Ge), arsenic (As), antimony (Sb), and tellurium (Te).

Metals and nonmetals may be distinguished by differences in their physical properties in addition to their chemical tendency to lose or gain electrons. Metals have a characteristic luster and generally conduct heat and electricity well. Most (except mercury) are solids at room temperature. Nonmetals, on the other hand, are poor conductors, and several are gases at room temperature.

Note that aluminum (in spite of the fact that it borders the zigzag line) is classified as a metal, not a metalloid.

Question 2.5 Using the periodic table, write the symbol for each of the following and label as a metal, metalloid, or nonmetal.
a. sodium
b. radium
c. manganese
d. magnesium

Question 2.6 Using the periodic table, write the symbol for each of the following and label as a metal, metalloid, or nonmetal.
a. sulfur
b. oxygen
c. phosphorus
d. nitrogen

Copper is a metal that has many uses. Can you provide other uses of copper?

A Medical Perspective

Copper Deficiency and Wilson's Disease

An old adage tells us that we should consume all things in moderation. This is true for many of the trace minerals, such as copper. Too much copper in the diet causes toxicity and too little copper results in a serious deficiency disease.

Copper is extremely important for the proper functioning of the body. It aids in the absorption of iron from the intestine and facilitates iron metabolism. It is critical for the formation of hemoglobin and red blood cells in the bone marrow. Copper is also necessary for the synthesis of collagen, a protein that is a major component of the connective tissue. It is essential to the central nervous system in two important ways. First, copper is needed for the synthesis of norepinephrine and dopamine, two chemicals that are necessary for the transmission of nerve signals. Second, it is required for the formation of the myelin sheath (a layer of insulation) around nerve cells. Release of cholesterol from the liver depends on copper, as does bone development and proper function of the immune and blood clotting systems.

The estimated safe and adequate daily dietary intake (ESADDI) for adults is 1.5–3.0 milligrams (mg). Meats, cocoa, nuts, legumes, and whole grains provide significant amounts of copper.

Although getting enough copper in the diet would appear to be relatively simple, it is estimated that Americans often ingest only marginal levels of copper, and we absorb only 25–40% of that dietary copper. Despite these facts, it appears that copper deficiency is not a serious problem in the United States.

Individuals who are at risk for copper deficiency include people who are recovering from abdominal surgery, which causes decreased absorption of copper from the intestine. Others at risk are premature babies and people who are sustained solely by intravenous feedings that are deficient in copper. In addition, people who ingest high doses of antacids or take excessive supplements of zinc, iron, or vitamin C can develop copper deficiency because of reduced copper absorption. Because copper is involved in so many processes in the body, it is not surprising that the symptoms of copper deficiency are many and diverse. They include anemia; decreased red and white blood cell counts; heart disease; increased levels of serum cholesterol; loss of bone; defects in the nervous system, immune system, and connective tissue; and abnormal hair.

Just as too little copper causes serious problems, so does an excess of copper. At doses greater than about 15 mg, copper causes toxicity that results in vomiting. The effects of extended exposure to excess copper are apparent when we look at Wilson's disease. This is a genetic disorder in which excess copper cannot be removed from the body and accumulates in the cornea of the eye, liver, kidneys, and brain. The symptoms include a greenish ring around the cornea, cirrhosis of the liver, copper in the urine, dementia and paranoia, drooling, and progressive tremors. As a result of the condition, the victim generally dies in early adolescence. Wilson's disease can be treated with medication and diet modification with moderate success, if it is recognized early, before permanent damage has occurred to any tissues.

For Further Understanding

▶ Why is there an upper limit on the recommended daily amount of copper?

▶ Iron is another essential trace metal in our diet. Use the Internet to find out if upper limits exist for daily iron consumption.

LEARNING GOAL

8 Identify and use the specific information about an element that can be obtained from the periodic table.

See Section 2.1 for an explanation of atomic number and atomic mass.

Information Contained in the Periodic Table

Both the atomic number and the average atomic mass of each element are readily available from the periodic table. For example,

20	←——atomic number
Ca	←——symbol
Calcium	←——name
40.08	←——atomic mass

More detailed periodic tables may also include such information as the electron arrangement, relative sizes of atoms and ions, and most probable ion charges.

Question 2.7 Refer to the periodic table and find the following information:
a. the symbol of the element with an atomic number of 40
b. the atomic mass of the element sodium (Na)
c. the element whose atoms contain 24 protons
d. the known element that should most resemble the recently discovered element ununseptium, with an atomic number of 117

Question 2.8 Refer to the periodic table, and find the following information:
a. the symbol of the noble gas in period 3
b. the element in Group IVA (or 14) with the smallest mass
c. the only metalloid in Group IIIA (or 13)
d. the element whose atoms contain 18 protons

Question 2.9 For each of the following element symbols, give the name of the element, its atomic number, and its atomic mass.
a. He   b. F   c. Mn

Question 2.10 For each of the following element symbols, give the name of the element, its atomic number, and its atomic mass.
a. Mg   b. Ne   c. Se

2.5   Electron Arrangement and the Periodic Table

A primary objective of studying chemistry is to understand the way in which atoms join together to form chemical compounds. This *bonding process* is a direct consequence of the arrangement of the electrons in the atoms that combine. The **electron configuration** describes the arrangement of electrons in atoms. The organization of the periodic table, originally developed from careful measurement of properties of the elements, also correlates well with similarities in electron configuration. This reveals a very important fact: the properties of collections of atoms (bulk properties) are a direct consequence of their electron arrangement.

LEARNING GOAL

9 Describe the relationship between the electronic structure of an element and its position in the periodic table.

The Quantum Mechanical Atom

As we noted in Section 2.3, the success of Bohr's theory was short-lived. Emission spectra of multi-electron atoms (recall that the hydrogen atom has only one electron) could not be explained by Bohr's theory. Evidence that electrons have wave properties was problematic. Bohr stated that electrons in atoms had very specific locations, now termed *principal energy levels*. The very nature of waves, spread out in space, defies such an exact model of electrons in atoms. Furthermore, the exact model is contradictory to theory and subsequent experiments.

The basic concept of the Bohr theory, that the energy of an electron in an atom is quantized, was refined and expanded by an Austrian physicist, Erwin Schröedinger. He described electrons in atoms in probability terms, developing equations that emphasize the wavelike character of electrons. Although Schröedinger's approach was founded on complex mathematics, we can readily use models of electron probability regions to enable us to gain insight into atomic structure without needing to understand the underlying mathematics.

Schröedinger's theory, often described as quantum mechanics, incorporates Bohr's principal energy levels ($n = 1$, 2, and so forth); however, it proposes that each of these levels is made up of one or more sublevels. Each sublevel, in turn, contains one or more atomic orbitals. In the following section, we shall look at each of these regions in more detail and learn how to predict the way that electrons are arranged in stable atoms.

Principal Energy Levels, Sublevels, and Orbitals

Principal Energy Levels

Bohr concluded that electrons in atoms do not roam freely in space; rather they are confined to certain specific regions of space outside of the nucleus of an atom.

Principal energy levels (analogous to Bohr's orbits) are regions where electrons may be found, and have integral values designated $n = 1$, $n = 2$, $n = 3$, and so forth. The principal energy level is related to the average distance from the nucleus. An $n = 1$ level is closest to the nucleus; the larger the value of n, the greater the average distance of an electron in that level from the nucleus.

The maximum number of electrons that a principal energy level can hold is equal to $2(n)^2$. For example:

Principal Energy	$2(n)^2$	Maximum Number of Electrons
$n = 1$	$2(1)^2$	$2e^-$
$n = 2$	$2(2)^2$	$8e^-$
$n = 3$	$2(3)^2$	$18e^-$

Sublevels

A **sublevel** is a set of equal-energy orbitals within a principal energy level. The sublevels, or subshells, are symbolized as s, p, d, and so forth; they increase in energy in the following order:

$$s < p < d$$

We specify both the principal energy level and type of sublevel when describing the location of an electron—for example, $1s$, $2s$, $2p$. Energy level designations for the first three principal energy levels are as follows:

Figure 2.11 Representation of s orbitals.

- The first principal energy level ($n = 1$) has one possible sublevel: $1s$.
- The second principal energy level ($n = 2$) has two possible sublevels: $2s$ and $2p$.
- The third principal energy level ($n = 3$) has three possible sublevels: $3s$, $3p$, and $3d$.

Orbitals

An **atomic orbital** is a specific region of a sublevel containing a maximum of two electrons.

Figure 2.11 depicts models of three s orbitals. Each s orbital is spherically symmetrical, much like a Ping-Pong ball. Its volume represents a region where there is a high probability of finding electrons of similar energy. This probability decreases as we approach the outer region of the atom. The nucleus is at the center of the s orbital. At that point, the probability of finding the electron is zero; electrons cannot reside in the nucleus. Only one s orbital can be found in any n level. Atoms with many electrons, occupying a number of n levels, have an s orbital in each n level. Consequently $1s$, $2s$, $3s$, and so forth are possible orbitals.

Figure 2.12 illustrates the shapes of the three possible p orbitals within a given level. Each has the same shape, and that shape appears much like a dumbbell; these three orbitals differ only in the direction they extend into space. Imaginary coordinates x, y, and z are superimposed on these models to emphasize this fact. These three orbitals, termed p_x, p_y, and p_z, may coexist in a single atom.

Figure 2.13 depicts the shapes of the five possible d orbitals within a given n level. These d orbitals are important in determining the properties of the transition metals.

Higher-energy orbitals (f, g, and so forth) also exist, but they are important only in the description of the electron arrangements of the heaviest elements.

It is important to remember that the *shape* of the orbital represents regions in the atom (but outside of the nucleus) where electrons *may* be found. Not all orbitals are occupied by electrons; the shaded regions really represent the probability of finding the electrons in specific regions of the atom.

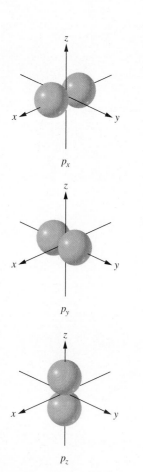

Figure 2.12 Representation of the three p orbitals, p_x, p_y, and p_z.

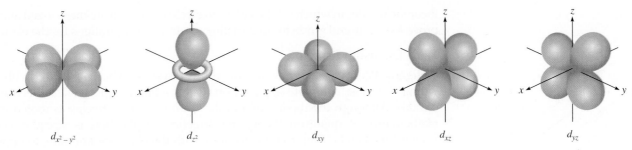

Figure 2.13 Representation of the five *d* orbitals.

Electrons in Sublevels

We can deduce the maximum electron capacity of each sublevel based on the information just given.

For the *s* sublevel:

$$1 \text{ orbital} \times \frac{2e^- \text{ capacity}}{\text{orbital}} = 2e^- \text{ capacity}$$

For the *p* sublevel:

$$3 \text{ orbitals} \times \frac{2e^- \text{ capacity}}{\text{orbital}} = 6e^- \text{ capacity}$$

For the *d* sublevel:

$$5 \text{ orbitals} \times \frac{2e^- \text{ capacity}}{\text{orbital}} = 10e^- \text{ capacity}$$

Electron Spin

As we have noted, each atomic orbital has a maximum capacity of two electrons. The electrons are perceived to *spin* on an imaginary axis, and the two electrons in the same orbital must have opposite spins: clockwise and counterclockwise. Their behavior is analogous to two ends of a magnet. Remember, electrons have magnetic properties. The electrons exhibit sufficient magnetic attraction to hold themselves together despite the natural repulsion that they "feel" for each other, owing to their similar charge (remember, like charges repel). Electrons must have opposite spins to coexist in an orbital. Two electrons in one orbital that possess opposite spins are referred to as *paired* electrons. The number and arrangement of unpaired electrons in an atom are responsible for the magnetic properties of elements (Figure 2.14).

Electron Configurations

We can write the electron arrangement (electron configuration) for an atom of any element if we know a few simple facts. We need to know the number of electrons in the atom, which is readily available from the periodic table. (Remember, the number of electrons in a neutral atom is equal to the atomic number.) We know that each principal energy level has a maximum capacity of $2(n)^2$ electrons. No more than two electrons can be placed in any orbital, and each principal energy level, n, can contain only n subshells. Additionally, the aufbau principle, the Pauli exclusion principle, and Hund's rule will complete our strategy.

Aufbau Principle

The *aufbau*, or building up, *principle* helps us to represent the electron configuration of atoms of various elements. According to this principle, electrons fill the lowest energy orbital that is available. Figure 2.15 is a useful way of depicting the

Figure 2.14 A compass is a familiar direction-finder that aligns the magnetic field of the compass needle with the earth's natural magnetic field. Unpaired electrons in iron atoms (in the needle and in the earth) are responsible for the magnetic fields.

LEARNING GOAL

10 Write electron configurations, shorthand electron configurations, and orbital diagrams for atoms and ions.

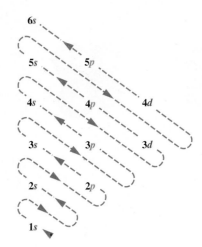

Figure 2.15 A useful way to remember the filling order for electrons in atoms. Begin adding electrons at the bottom (lowest energy) and follow the arrows. Thus, orbitals fill in the order 1s, then 2s, then 2p, then 3s, and so forth.

We can also use the periodic table to determine the order in which orbitals are filled. See Figure 2.16.

---

theoretical order in which orbitals fill with electrons. It is easily memorized and will prove to be a useful guide to constructing electron configurations for the elements.

Pauli Exclusion Principle

Wolfgang Pauli was a renowned quantum physicist who is credited with providing the most complete explanation of the relationship between electrons in orbitals. Although the theory behind the *Pauli exclusion principle* is rooted in the mathematics of quantum theory, the practical application is straightforward: Each orbital can hold up to two electrons with the electrons spinning in opposite directions (paired).

Hund's Rule

All electrons in atoms are typically in their lowest energy state, termed the *ground state,* at room temperature. *Hund's rule* is a consequence of this: When there is a set of orbitals of equal energy (for example, three *p* orbitals in the $n = 2$ level), each orbital becomes half-filled (one electron per orbital) before any become completely filled (two electrons).

Guidelines for Writing Electron Configurations of Atoms

- Obtain the total number of electrons in the atom from the atomic number found on the periodic table. The number of electrons equals the number of protons for an atom.
- Electrons in atoms occupy the lowest energy orbitals that are available, beginning with 1*s*.
- Fill subshells according to the order depicted in Figure 2.15:
  ○ 1*s*, 2*s*, 2*p*, 3*s*, 3*p*, 4*s*, 3*d*, 4*p*, 5*s*, 4*d*, 5*p*, 6*s*, . . .
- Remember:
  ○ The *s* sublevel has one orbital and can hold two electrons.
  ○ The *p* sublevel has three orbitals. The electrons will half-fill before completely filling the orbitals for a maximum of six electrons.
  ○ The *d* sublevel has five orbitals. Again, the electrons will half-fill before completely filling the orbitals for a maximum of ten electrons.

Now let us look at several examples:

Hydrogen

Hydrogen is the simplest atom; it has only one electron. That electron must be in the lowest principal energy level ($n = 1$) and the lowest orbital (*s*). We indicate the number of electrons in a region with a *superscript,* so we write $1s^1$. The electron configuration for hydrogen is $1s^1$.

An alternate way of depicting electron arrangement in atoms is the *orbital diagram.* Each orbital is represented as a box, and each electron as an arrow, indicating the relative direction of its spin. The orbital diagram for hydrogen is:

1*s*

Helium

Helium has two electrons, which will fill the lowest energy level. The ground state (lowest energy) electron configuration for helium is $1s^2$, and the corresponding orbital diagram is:

1*s*

The arrows show paired (opposite) spins, in accordance with the Pauli exclusion principle.

Lithium

Lithium has three electrons. The first two are configured the same as helium. The third must go into the orbital of the lowest energy in the second principal energy level; therefore, the configuration is $1s^2 2s^1$.
The orbital diagram is:

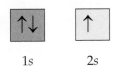

Beryllium

Beryllium has four electrons. The first three are configured the same as lithium. The fourth electron fills the 2s orbital; therefore, the configuration is $1s^2 2s^2$, and the orbital diagram is:

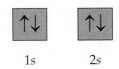

Boron

Boron has five electrons. The first four are configured the same as beryllium. The fifth electron must go into the 2p level; therefore, the configuration is $1s^2 2s^2 2p^1$.
The orbital diagram is:

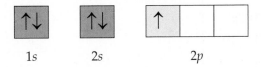

Carbon

Carbon has six electrons. The first five are configured the same as boron. The sixth electron, according to Hund's rule, enters a second 2p orbital, with the same spin as the fifth electron. The electron configuration for carbon is $1s^2 2s^2 2p^2$, and the orbital diagram is:

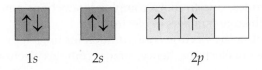

Note that Hund's rule requires the two p electrons to occupy different 2p orbitals.

Nitrogen

Nitrogen, following the same logic, is $1s^2, 2s^2, 2p^3$, and the orbital diagram is:

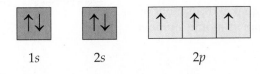

Oxygen, Fluorine, and Neon

Each of these elements adds one more electron, pairing one of the three unpaired electrons, resulting in:

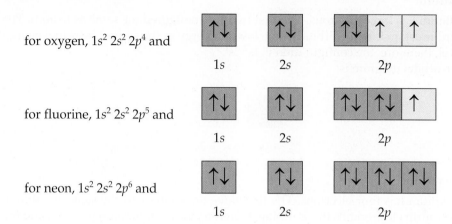

for oxygen, $1s^2\,2s^2\,2p^4$ and

       1s   2s   2p

for fluorine, $1s^2\,2s^2\,2p^5$ and

       1s   2s   2p

for neon, $1s^2\,2s^2\,2p^6$ and

       1s   2s   2p

Sodium Through Argon

Electrons in these elements retain the basic $1s^2\,2s^2\,2p^6$ arrangement of the preceding element, neon; new electrons enter the third principal energy level:

Na	$1s^2\,2s^2\,2p^6\,3s^1$
Mg	$1s^2\,2s^2\,2p^6\,3s^2$
Al	$1s^2\,2s^2\,2p^6\,3s^2\,3p^1$
Si	$1s^2\,2s^2\,2p^6\,3s^2\,3p^2$
P	$1s^2\,2s^2\,2p^6\,3s^2\,3p^3$
S	$1s^2\,2s^2\,2p^6\,3s^2\,3p^4$
Cl	$1s^2\,2s^2\,2p^6\,3s^2\,3p^5$
Ar	$1s^2\,2s^2\,2p^6\,3s^2\,3p^6$

The 3p orbital can hold a maximum of six e⁻.

We have seen that elements in the periodic table are classified as either representative or transition. Representative elements consist of all Group IA–VIIIA (or 1, 2, and 13–18) elements. All others are transition elements. The guidelines we have developed for writing electron configurations work well for representative elements. Electron configurations for transition elements include several exceptions to the rules, due to more complex interactions among electrons that occur in atoms with large numbers of electrons.

Let us now look at the electron configuration for tin, a representative element with a large number of electrons; hence, a more complex electronic structure.

EXAMPLE 2.4 **Writing the Electron Configuration**

Write the electron configuration for tin.

Solution

Since tin is a neutral atom, its number of electrons is equal to its number of protons. The number of protons can be determined from its atomic number which is found on the periodic table.

LEARNING GOAL

10 Write electron configurations, shorthand electron configurations, and orbital diagrams for atoms and ions.

Step 1. Tin (Sn) has an atomic number of 50; thus, we must place fifty electrons in atomic orbitals.

Step 2. Recall the total electron capacities of orbital types: s, two electrons; p, six electrons; and d, ten electrons. The order of filling the orbitals (Figure 2.15) is $1s$, $2s$, $2p$, $3s$, $3p$, $4s$, $3d$, $4p$, $5s$, $4d$, $5p$, $6s$.

Step 3. The electron configuration is as follows:

$$1s^2 2s^2 2p^6 3s^2 3p^6 4s^2 3d^{10} 4p^6 5s^2 4d^{10} 5p^2$$

Helpful Hint:
As a check, count electrons in the electron configuration (add all of the superscript numbers) to ensure we have accounted for all fifty electrons of the Sn atom.

Practice Problem 2.4

Write the electron configuration for an atom of:

    a. sulfur   b. calcium   c. potassium   d. phosphorus

▶ For Further Practice: **Questions 2.85 and 2.86.**

EXAMPLE 2.5 | **Writing the Electron Configuration and Orbital Diagram**

LEARNING GOAL

Write the electron configuration and orbital diagram for silicon.

10 Write electron configurations, shorthand electron configurations, and orbital diagrams for atoms and ions.

Solution

Since silicon is a neutral atom, its number of electrons is equal to its number of protons. The number of protons can be determined from its atomic number, which is found on the periodic table.

Step 1. Silicon (Si) has an atomic number of 14; thus, we must place fourteen electrons in atomic orbitals.

Step 2. Recall the total electron capacities of orbital types (s, two electrons; p, six electrons; d, ten electrons). The order of filling the orbitals (Figure 2.15) is $1s$, $2s$, $2p$, $3s$, $3p$, $4s$, $3d$, $4p$, $5s$, $4d$, $5p$, $6s$.

Step 3. The electron configuration is as follows:

$$1s^2 2s^2 2p^6 3s^2 3p^2$$

Step 4. Writing the orbital diagram follows from the electron configuration. Remember that Hund's rule tells us to enter p electrons unpaired until each orbital contains one electron.

 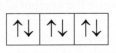   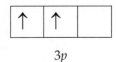

    1s   2s   2p   3s   3p

Helpful Hint:
Be sure to count the electrons (superscript numbers) in the electron configuration to ensure that all fourteen electrons are accounted for.

Practice Problem 2.5

Write the orbital diagram for an atom of:

    a. sulfur   b. calcium   c. potassium   d. phosphorus

▶ For Further Practice: **Questions 2.87 and 2.88.**

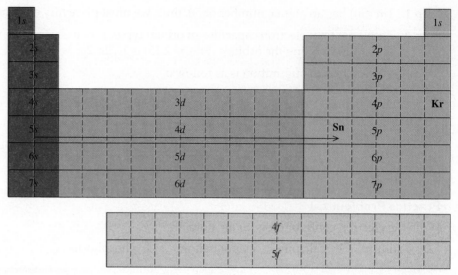

Figure 2.16 The shape of the periodic table shows four distinct regions corresponding to the type of subshell that is filling with electrons. Starting with 1s at the top left of the periodic table and proceeding from left to right and down, the same order of sublevels as shown in Figure 2.15 can be obtained. Take time to examine both figures to see the connection. The application of this figure in writing the shorthand notation for tin is detailed in Example 2.6.

Electron Configurations and the Periodic Table

Mendeleev's periodic table was based on extensive observation of physical and chemical properties. Its unusual shape is nicely explained by electron configurations resulting from atomic orbital theory. Elements in the same group (column) have the same outer shell electron configuration. Group IA(1) is ns^1, Group IIA(2) is ns^2, and so forth. The prefix n corresponds to the period (row) containing the element whose principal energy level (also n) is filling. The link between electron configuration and physical properties is, in fact, the periodic table and its unusual shape!

This relationship is shown in Figure 2.16.

Shorthand Electron Configurations

Using the periodic table and the layout of sublevels depicted in Figure 2.16, a shorthand version of the electron configuration can be obtained. To use the shorthand method, write the symbol of the nearest noble gas (with an atomic number less than the element for which you are determining the shorthand electron configuration) in square brackets. Then finish the electron configuration by writing all additional sublevels that contain electrons.

For example, the full electron configuration of a sodium atom is:

$$1s^2 2s^2 2p^6 3s^1$$

The noble gas that comes before sodium is neon. Putting [Ne] in the configuration accounts for the first ten electrons. The shorthand electron configuration is given as:

$$[Ne]3s^1$$

The advantage to this method is that the electron configurations of atoms with a large number of electrons can be written quickly by examining the periodic table. Example 2.6 demonstrates this method.

EXAMPLE 2.6 | **Writing Shorthand Electron Configurations**

LEARNING GOAL

10 Write electron configurations, shorthand electron configurations, and orbital diagrams for atoms and ions.

Use the periodic table to write the shorthand electron configuration of tin.

Solution

Step 1. Locate tin (Sn) on the periodic table. Tin has the atomic number 50.

Step 2. Krypton is the nearest noble gas that comes before Sn. The symbol for krypton (element 36) is first written in square brackets:

$$[Kr]$$

Step 3. Move to the left side of the periodic table (element 37) and put electrons in the next available sublevels, counting element blocks to determine the number of electrons in the sublevel. Finish at element Sn (Figure 2.16). This gives:

$$[Kr]\ 5s^2 4d^{10} 5p^2$$

Helpful Hint: The atomic number for the noble gas used in the shorthand electronic configuration must be less than the atomic number for tin. The difference between the two atomic numbers (for Kr and Sn) indicates the number of electrons necessary to complete the shorthand electron configuration for tin.

Practice Problem 2.6

Write the shorthand electron configuration for:

    a. sulfur   b. silicon   c. selenium   d. iron

▶ For Further Practice: **Questions 2.93 and 2.94.**

2.6 Valence Electrons and the Octet Rule

Valence Electrons

If we picture two spherical objects that we wish to join together, perhaps with glue, the glue can be applied to the surfaces and the two objects can then be brought into contact. We can extend this analogy to two atoms that are modeled as spherical objects. Although this is not a perfect analogy, it is apparent that the surface interaction is of primary importance. Although the positively charged nucleus and "interior" electrons certainly play a role in bonding, we can most easily understand the process by considering only the outermost electrons. We refer to these as *valence electrons*. **Valence electrons** are the outermost electrons in an atom, which are involved, or have the potential to become involved, in the bonding process.

Metals tend to have fewer valence electrons than nonmetals.

For representative elements, the number of valence electrons in an atom corresponds to the number of the *group* in which the atom is found. For example, elements such as hydrogen and sodium (in fact, all alkali metals, Group IA or 1) have one valence electron. From left to right in period 2, beryllium, Be (Group IIA or 2), has two valence electrons; boron, B (Group IIIA or 13), has three; carbon, C (Group IVA or 14), has four; and so forth.

The Octet Rule

LEARNING GOAL

11 Discuss the octet rule, and use it to predict the charges and the numbers of protons and electrons in cations and anions formed from neutral atoms.

Elements in the last family (Group VIIIA or 18), the noble gases, have eight valence electrons, except for helium, which has two. These elements are extremely stable and were often termed *inert gases* because they do not readily bond to other elements, although some of the larger ones (lower on the periodic table) can be made to do so under extreme experimental conditions. A full $n = 1$ energy level (as in helium) or an outer *octet* of electrons (eight valence electrons, two in the s and six in the p sublevel) is responsible for this unique stability.

EXAMPLE 2.7   Determining Electron Arrangement

Provide the total number of electrons, total number of valence electrons, and energy level in which the valence electrons are found for the silicon (Si) atom.

10 Write electron configurations, shorthand electron configurations, and orbital diagrams for atoms and ions.

Solution

Step 1. Determine the position of silicon in the periodic table. Silicon has an atomic number of 14. Silicon is found in Group IVA (or 14) and period 3 of the table.

Step 2. The atomic number provides the number of protons in an atom. Since this is a neutral atom, the number of protons is equal to the number of electrons. Silicon therefore has fourteen electrons.

Step 3. Because silicon is in Group IVA (or 14), only four of the fourteen electrons are valence electrons.

Step 4. The third period corresponds to the $n = 3$ energy level. Silicon's four valence electrons are in the third principal energy level.

Practice Problem 2.7

For each of the following elements, provide the total number of electrons and valence electrons in its atoms as well as the number of the energy level in which the valence electrons are found:

    a. Na   b. Mg   c. S   d. Cl   e. Ar

▶ For Further Practice: **Questions 2.101 and 2.102.**

Atoms of elements in other groups are more reactive than the noble gases because in the process of chemical reaction they try to achieve a more stable "noble gas" configuration by gaining or losing electrons. This is the basis of the **octet rule,** which states that elements usually react in such a way as to attain the electron configuration of the noble gas closest to them in the periodic table (a stable octet of electrons). In chemical reactions, they will gain, lose, or share the minimum number of electrons necessary to attain this more stable energy state. The octet rule, although simple in concept, is a remarkably reliable predictor of chemical change, especially for representative elements.

The shorthand electron configuration highlights the valence electrons by placing them after the noble gas core.

Ions

Ions are often formed in chemical reactions, when one or more electrons are transferred from one substance to another.

Ions are electrically charged particles that result from a gain of one or more electrons by the parent atom (forming negative ions, or **anions**) or a loss of one or more electrons from the parent atom (forming positive ions, or **cations**). Formation of an anion may occur as follows:

$$F + 1e^- \longrightarrow F^-$$

Gain of an Electron · Neutral atom · Anion formed

Alternatively, formation of a cation of sodium may proceed as follows:

$$Na \longrightarrow 1e^- + Na^+$$

Loss of an Electron · Neutral atom · Cation formed

Note that the electrons gained are written as reactants to the left of the reaction arrow, whereas the electrons lost are written as products to the right of the reaction arrow.

EXAMPLE 2.8	Determining Ion Proton and Electron Composition

LEARNING GOAL

11 Discuss the octet rule, and use it to predict the charges and the numbers of protons and electrons in cations and anions formed from neutral atoms.

Determine the number of protons and electrons in Sr^{2+}.

Solution

Step 1. Locate Sr, strontium, on the periodic table and determine its atomic number. Sr has the atomic number 38. Therefore, a neutral atom of Sr has 38 protons and 38 electrons. Since the number of protons in an ion is the same as in a neutral atom, there are 38 protons in Sr^{2+}.

Step 2. Determine whether electrons were gained or lost in order to form the ion. Sr is a metallic element, and metals lose electrons to form cations.

Step 3. Using the charge of the ion, determine how many electrons were lost. An ion with a 2^+ charge is produced by the loss of two electrons from the neutral atom.

$$Sr \longrightarrow 2e^- + Sr^{2+}$$

Since the neutral atom, Sr, has a 38 electrons, a loss of two electrons in ion formation leaves 36 electrons in the ion Sr^{2+}.

Helpful Hint: Sr is found in Group IIA (or 2). Therefore, a neutral atom of Sr has two valence electrons. Metals lose valence electrons to form cations.

Practice Problem 2.8

Determine the number of protons and electrons in As^{3-}.

▶ For Further Practice: **Questions 2.103 and 2.104.**

Question 2.11 Determine the number of protons and electrons in each of the following ions:

a. O^{2-} b. Mg^{2+} c. Fe^{3+}

Question 2.12 Determine the number of protons and electrons in each of the following ions:

a. Ni^{2+} b. Br^- c. N^{3-}

Ion Formation and the Octet Rule

LEARNING GOAL

11 Discuss the octet rule, and use it to predict the charges and the numbers of protons and electrons in cations and anions formed from neutral atoms.

Metals and nonmetals differ in the way in which they form ions. Metallic elements (located at the left of the periodic table) tend to form positively charged *cations*, by the loss of electrons to obtain a noble gas configuration.

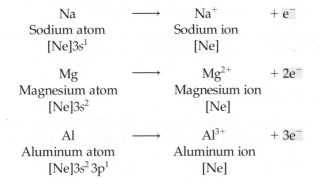

Na	$\longrightarrow$	Na$^+$	$+ e^-$
Sodium atom		Sodium ion	
[Ne]3s^1		[Ne]	

Mg	$\longrightarrow$	Mg^{2+}	$+ 2e^-$
Magnesium atom		Magnesium ion	
[Ne]3s^2		[Ne]	

Al	$\longrightarrow$	Al^{3+}	$+ 3e^-$
Aluminum atom		Aluminum ion	
[Ne]3s^{2}3p^1		[Ne]	

The prefix *iso* (Greek *isos*) means equal.

These cations are particularly stable because they are **isoelectronic** (that is, they have the same electron configuration) with their nearest noble gas neighbor, Ne, and have an octet of electrons in their outermost energy level.

Sodium is typical of each element in its group. Knowing that sodium, a Group IA (or 1) atom, forms a 1^+ ion leads to the prediction that H, Li, K, Rb, Cs, and Fr also will form 1^+ ions. Furthermore, magnesium, which forms a 2^+ ion, is typical of each element in its group (IIA or 2): Be^{2+}, Ca^{2+}, Sr^{2+}, and so forth are the resulting ions.

Nonmetallic elements, located at the right of the periodic table, tend to gain electrons to become isoelectronic with the nearest noble gas element, forming negative charged *anions*.

Consider:

$$F \quad + 1e^- \longrightarrow \quad F^-$$
Fluorine atom \qquad\qquad Fluoride ion
$[He]2s^2 2p^5$ \qquad\qquad $[He]2s^2 2p^6$ or [Ne]

$$O \quad + 2e^- \longrightarrow \quad O^{2-}$$
Oxygen atom \qquad\qquad Oxide ion
$[He]2s^2 2p^4$ \qquad\qquad $[He]2s^2 2p^6$ or [Ne]

$$N \quad + 3e^- \longrightarrow \quad N^{3-}$$
Nitrogen atom \qquad\qquad Nitride ion
$[He]2s^2 2p^3$ \qquad\qquad $[He]2s^2 2p^6$ or [Ne]

As in the case of cation formation, each of these anions has an octet of electrons in its outermost energy level.

The element fluorine, forming F^-, indicates that the other halogens, Cl, Br, and I, behave as a true group and form Cl^-, Br^-, and I^- ions. Also, oxygen and the other nonmetals in its group (VIA or 16) form 2^- ions; nitrogen and phosphorus (both of Group V or 15) form 3^- ions. It is important to recognize that ions are formed by gain or loss of *electrons*. No change occurs in the nucleus; the number of protons remains the same.

EXAMPLE 2.9   **Writing Shorthand Electron Configurations for Ions**

LEARNING GOAL

Write the electron configuration and shorthand electron configuration for Sb^{3-}.

10 Write electron configurations, shorthand electron configurations, and orbital diagrams for atoms and ions.

Solution

Step 1. Locate antimony (Sb) on the periodic table. Antimony is in Group V (or 15) and has the atomic number 51.

Step 2. Determine the number of electrons for the ion. Since Sb^{3-} indicates that the ion has three more electrons than a neutral atom of Sb, the total number of electrons in Sb^{3-} can be calculated:

$$51e^- \quad + \quad 3e^- \quad = \quad 54e^-$$

Electrons in Sb \quad Electrons gained \quad Electrons in Sb^{3-}

Step 3. Recall the total electron capacities of the orbital types: s, two electrons; p, six electrons; d, ten electrons. The order of filling the orbitals (Figure 2.15) is $1s$, $2s$, $2p$, $3s$, $3p$, $4s$, $3d$, $4p$, $5s$, $4d$, $5p$, $6s$.

Step 4: The electron configuration is as follows:

$$1s^2 2s^2 2p^6 3s^2 3p^6 4s^2 3d^{10} 4p^6 5s^2 4d^{10} 5p^6$$

Step 5: Determine if the ion is isoelectronic with its nearest noble gas neighbor. Since Sb^{3-} is isoelectronic with Xe, xenon, all fifty-four electrons of Sb^{3-} are indicated by the symbol Xe in square brackets. The shorthand electron configuration for Sb^{3-} is [Xe].

Helpful Hint: By adding together the superscripts in the electron configuration, you can confirm that fifty-four electrons were used. Since the noble gas xenon also has fifty-four electrons, it has the same electronic configuration as Sb^{3-}.

Practice Problem 2.9

Write the electron configuration and shorthand electron configuration for Cs^{1+}.

▶ For Further Practice: **Questions 2.111 and 2.112.**

Question 2.13 Write the electron configuration and shorthand electron configuration for each of the following ions:
  a. K^+
  b. Ca^{2+}
  c. Se^{2-}
  d. Br^-

Question 2.14 Write the electron configuration and shorthand electron configuration for each of the following ions:
  a. Rb^+
  b. Sr^{2+}
  c. S^{2-}
  d. I^-

EXAMPLE 2.10	Determining Isoelectronic Ions and Atoms

LEARNING GOAL

Provide the charge of the most probable ion resulting from Se. With what element is the ion isoelectronic?

11 Discuss the octet rule, and use it to predict the charges and the numbers of protons and electrons in cations and anions formed from neutral atoms.

Solution

Step 1. Locate selenium (Se) on the periodic table. Since it is in Group VIA (or 16), it is a nonmetal. Nonmetallic elements gain electrons to form anions.

Step 2. Find the noble gas that is selenium's nearest neighbor. Krypton, Kr, has an octet of electrons in its outermost level and will be isoelectronic with the anion formed by Se.

Step 3. Determine the number of electrons that Se will need to gain to become isoelectronic with Kr. A neutral atom of Se has thirty-four electrons because it has an atomic number of 34, indicating it has thirty-four protons. Kr has thirty-six electrons because it has an atomic number of 36. The difference between 36 and 34 is 2. Therefore, Se will gain two electrons to form Se^{2-}.

Helpful Hint: An element's location on the periodic table can also allow general predictions to be made regarding anion formation. Elements in Group VII (or 17) tend to form anions with a 1^- charge. Elements in Group VI (or 16) tend to form anions with a 2^- charge. Elements in Group V (or 15) tend to form anions with a 3^- charge. Since Se is in Group VI (or 16), it should form an anion with a 2^- charge.

Practice Problem 2.10

Provide the charge of the most probable ion resulting from radium. With what element is the ion isoelectronic?

▶ For Further Practice: **Questions 2.107 and 2.108.**

Question 2.15 Give the charge of the most probable ion resulting from each of the following elements. With what element is the ion isoelectronic?
  a. K
  b. Sr
  c. S
  d. Mg
  e. P
  f. Be

A Medical Perspective

Dietary Calcium

"Drink your milk!" "Eat all of your vegetables!" These imperatives are almost universal memories from our childhood. Our parents knew that calcium, present in abundance in these foods, is an essential element for the development of strong bones and healthy teeth.

Many studies, spanning the fields of biology, chemistry, and nutrition science, indicate that the benefits of calcium go far beyond bones and teeth. This element has been found to play a role in the prevention of disease throughout our bodies.

Calcium is the most abundant mineral (metal) in the body. It is ingested as the calcium ion (Ca^{2+}) either in its "free" state or "combined," as a part of a larger compound; calcium dietary supplements often contain ions in the form of calcium carbonate. The acid naturally present in the stomach produces the calcium ion:

$$CaCO_3 \;+\; 2H^+ \;\longrightarrow\; Ca^{2+} \;+\; H_2O \;+\; CO_2$$

calcium   stomach   calcium   water   carbon
carbonate   acid   ion   dioxide

Calcium is responsible for a variety of body functions including:

- transmission of nerve impulses
- release of "messenger compounds" that enable communication among nerves
- blood clotting
- hormone secretion
- growth of living cells throughout the body

The body's storehouse of calcium is bone tissue. When the supply of calcium from external sources, such as through the diet, is insufficient, the body uses a mechanism to compensate for this shortage. With vitamin D in a critical role, this mechanism removes calcium from bone to enable other functions to continue to take place. It is evident then that prolonged dietary calcium deficiency can weaken the bone structure. Unfortunately, current studies show that as much as 75% of the American population may not be consuming sufficient amounts of calcium. Developing an understanding of the role of calcium in premenstrual syndrome, cancer, and blood pressure regulation is the goal of three current research areas.

Calcium and premenstrual syndrome (PMS). Dr. Susan Thys-Jacobs, a gynecologist at St. Luke's-Roosevelt Hospital Center in New York City, and colleagues at eleven other medical centers are conducting a study of calcium's ability to relieve the discomfort of PMS. They believe that women with chronic PMS have calcium blood levels that are normal only because calcium is continually being removed from the bone to maintain an adequate supply in the blood. To complicate the situation, vitamin D levels in many young women are very low (as much as 80% of a person's vitamin D is made in the skin, upon exposure to sunlight; many of us now minimize our exposure to the sun because of concerns about ultraviolet radiation and skin cancer). Because vitamin D plays an essential role in calcium metabolism, even if sufficient calcium is consumed, it may not be used efficiently in the body.

Colon cancer. The colon is lined with a type of cell (epithelial cell) that is similar to those that form the outer layers of skin. Various studies have indicated that by-products of a high-fat diet are irritants to these epithelial cells and produce abnormal cell growth in the colon. Dr. Martin Lipkin, formerly of Rockefeller University in New York, and his colleagues have shown that calcium ions may bind with these irritants, reducing their undesirable effects. It is believed that a calcium-rich diet, low in fat, and perhaps the use of a calcium supplement can prevent or reverse this abnormal colon cell growth, delaying or preventing the onset of colon cancer.

Blood pressure regulation. Dr. David McCarron, a blood pressure specialist at the Oregon Health Sciences University, believes that dietary calcium levels may have a significant influence on hypertension (high blood pressure). Preliminary studies show that a diet rich in low-fat dairy products, fruits, and vegetables, all high in calcium, may produce a significant lowering of blood pressure in adults with mild hypertension.

The take-home lesson appears clear: a high-calcium, low-fat diet promotes good health in many ways. Once again, our parents were right!

For Further Understanding

▸ Distinguish between "free" and "combined" calcium in the diet.

▸ Why might calcium supplements be ineffective in treating all cases of calcium deficiency?

Question 2.16 Which of the following pairs of atoms and ions are isoelectronic?

a. Cl^-, Ar   c. Mg^{2+}, Na^+ e. O^{2-}, F^-

b. Na^+, Ne   d. Li^+, Ne   f. N^{3-}, Cl^-

The transition metals tend to form cations by losing electrons, just like the representative metals. Metals, whether representative or transition, share this characteristic. However, the transition elements are characterized as "variable valence" elements; depending on the type of substance with which they react, they may form more than one stable ion. For example, iron has two stable ionic forms:

$$Fe^{2+} \text{ and } Fe^{3+}$$

Copper can exist as

$$Cu^+ \text{ and } Cu^{2+}$$

and elements such as vanadium (V) and manganese (Mn) each can form four different stable ions.

Predicting the charge of an ion or the various possible ions for a given transition metal is not an easy task. Energy differences between valence electrons of transition metals are small and not easily predicted from the position of the element in the periodic table. In fact, in contrast to representative metals, the transition metals show great similarities within a *period* as well as within a *group*.

2.7 Trends in the Periodic Table

Atomic Size

Many atomic properties correlate with electronic structure; hence, with their position in the periodic table. Although correlations are not always perfect, the periodic table remains an excellent guide to the prediction of properties.

If our model of the atom is a tiny sphere whose radius is determined by the distance between the center of the nucleus and the boundary of the region where the valence electrons have a probability of being located, the size of the atom will be determined principally by two factors.

1. The energy level (*n* level) in which the outermost electrons are found increases as we go *down* a group. (Recall that the outermost *n* level correlates with period number.) Thus, the size of atoms should increase from top to bottom of the periodic table as we fill successive energy levels of the atoms with electrons (Figure 2.17).
2. Across a period, as the magnitude of the positive charge of the nucleus increases, its "pull" on the atom's valence electrons increases. This results in a contraction of the atomic radius from left to right across a period. Hence, atomic size decreases from left to right on the periodic table. Examine Figure 2.17 and notice that there are exceptions.

Ion Size

Positive ions (cations) are smaller than their parent atoms. The cation has more protons than electrons. The decrease in the number of electrons pulls the remaining electrons closer to the nucleus. Also, cation formation often results in the loss of all outer-shell electrons, resulting in a significant decrease in radius.

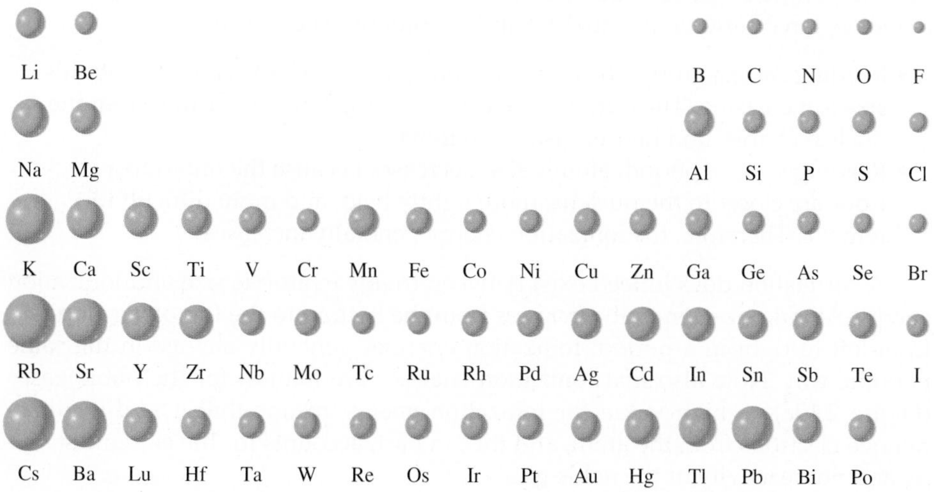

LEARNING GOAL

12 Utilize the periodic table trends to estimate the relative sizes of atoms and ions, as well as relative magnitudes of ionization energy and electron affinity.

The radius of an atom is traditionally defined as one-half of the distance between the nuclei of adjacent bonded atoms.

Figure 2.17 Variation in the size of atoms as a function of their position in the periodic table. Note the decrease in size from left to right and the increase in size as we proceed down the table, although some exceptions do exist. (Lanthanide and actinide elements are not included here.)

Figure 2.18 The relative size of ions and their parent atoms. Atomic radii are provided in units of picometers (pm).

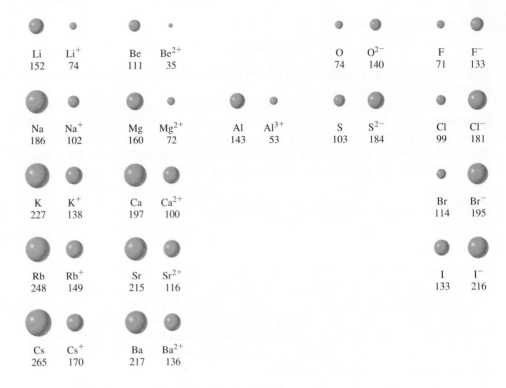

Negative ions (anions) are larger than their parent atoms. The anion has more electrons than protons. Owing to the excess negative charge, the nuclear "pull" on each individual electron is reduced. The electrons are held less tightly, resulting in a larger anion radius in contrast to the neutral atom.

Ions with multiple positive charge are even *smaller* than their corresponding monopositive ion. Thus, Cu^{2+} is smaller than Cu^+.

Figure 2.18 depicts the relative sizes of several atoms and their corresponding ions.

Ionization Energy

The energy required to remove an electron from an isolated atom is the **ionization energy.** The process for sodium is represented as follows:

$$\text{ionization energy} + \text{Na} \longrightarrow \text{Na}^+ + \text{e}^-$$

The magnitude of the ionization energy should correlate with the strength of the attractive force between the nucleus and the outermost electron.

- Reading *down* a group, the ionization energy decreases because the atom's size is increasing. The outermost electron is progressively farther from the nuclear charge, and hence, easier to remove.
- Reading *across* a period, atomic size decreases because the outermost electrons are closer to the nucleus, more tightly held, and more difficult to remove. Therefore, the ionization energy generally increases.

A correlation does indeed exist between trends in atomic size and ionization energy. Atomic size generally *decreases* from the bottom to the top of a group and from left to right in a period. Ionization energies generally *increase* in the same periodic way. Note also that ionization energies are highest for the noble gases (Figure 2.19a). A high value for ionization energy means that it is difficult to remove electrons from the atom, and this, in part, accounts for the extreme stability and nonreactivity of the noble gases.

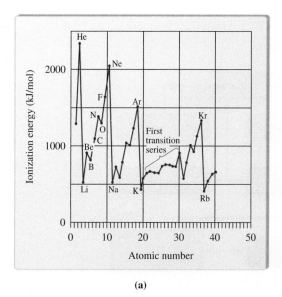

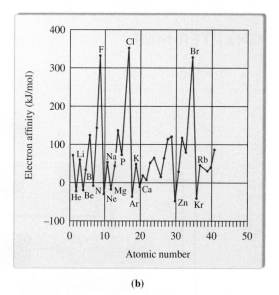

(a)   (b)

Figure 2.19 (a) The ionization energies of the first forty elements versus their atomic numbers. Note the very high values for elements located on the right in the periodic table, and low values for those on the left. Some exceptions to the trends are evident. (b) The periodic variation of electron affinity. Note the very low values for the noble gases and the elements on the far left of the periodic table. These elements do not form negative ions. In contrast, F, Cl, and Br readily form negative ions.

Electron Affinity

The energy change when a single electron is added to an isolated atom is the **electron affinity.** If we consider ionization energy in relation to positive ion formation (remember that the magnitude of the ionization energy tells us the ease of *removal* of an electron, hence the ease of forming positive ions), then electron affinity provides a measure of the ease of forming negative ions. A large electron affinity, meaning a large release of energy, indicates that the atom becomes more stable as it becomes a negative ion (through gaining an electron). Consider the gain of an electron by a bromine atom:

$$Br + e^- \longrightarrow Br^- + \text{energy}$$

Electron affinity

Periodic trends for electron affinity are as follows:

- Electron affinities generally decrease down a group.
- Electron affinities generally increase across a period.

Remember these trends are not absolute. Exceptions exist, as seen in the many irregularities in Figure 2.19b.

Question 2.17   Using periodic trends, rank Be, N, and F in order of increasing
  a. atomic size   b. ionization energy   c. electron affinity

Question 2.18   Using periodic trends, rank Br, I, and F in order of increasing
  a. atomic size   b. ionization energy   c. electron affinity

LEARNING GOAL

12   Utilize the periodic table trends to estimate the relative sizes of atoms and ions, as well as relative magnitudes of ionization energy and electron affinity.

Remember: ionization energy and electron affinity are predictable from *trends* in the periodic table. As with most trends, exceptions occur.

CHAPTER MAP

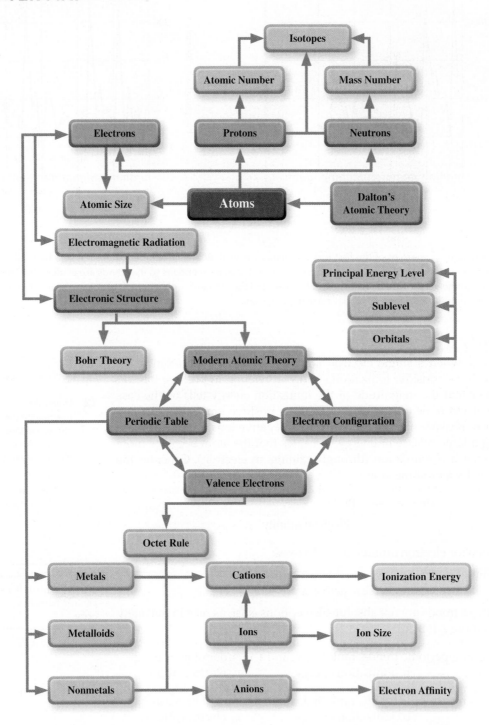

SUMMARY

2.1 Composition of the Atom

▶ The basic structural unit of an element is the **atom.**

▶ The **nucleus** is the very small and very dense core at the center of the atom containing:
  - Protons—positively charged particles. The number of protons is indicated by the **atomic number,** Z.
  - Neutrons—neutral particles. The number of neutrons is calculated from the **mass number** minus the atomic number, $A–Z$.

▶ **Electrons** are negatively charged particles that are located in a diffuse region. For a neutral atom, the number of electrons equals the number of protons.

▶ **Isotopes** are atoms of the same element that have a different number of neutrons. Isotopes of the same element have the same chemical properties.

▶ The **atomic mass** is the weighted average of the masses of the isotopes of an element in atomic mass units; 1 amu $= 1.66 \times 10^{-24}$ grams (g).

2.2 Development of Atomic Theory

▶ Dalton's atomic theory gave the first experimentally based concept of the atom.

▶ Dalton proposed that atoms were indivisible. Subsequent experiments by Thomson, Rutherford, and others led to the discovery that the atom contains subatomic particles (electrons, protons, and neutrons) and is divisible.

2.3 Light, Atomic Structure, and the Bohr Atom

▶ **Spectroscopy** is the study of the interaction of light and matter and led to the understanding of the electronic structure of the atom.

▶ Features of light (**electromagnetic radiation**) include:
  - The **speed of light** equals 3.0×10^8 m/s.
  - Light has a wave nature and the wavelength of light determines the energy of the light. The shorter the wavelength, the greater the energy.
  - Light has a particle nature. The particles are called photons.

▶ **Niels Bohr** used the emission spectrum of hydrogen to give the first model of the electronic structure of the atom. Some important features of his model include the following:
  - Electrons can exist only in certain allowed **energy levels** ($n = 1, 2, 3 \ldots$); that is, energy is *quantized*. Bohr called these levels orbits.
  - Electrons can gain energy and be promoted from the **ground state** to a higher energy level (**excited state**).
  - Electrons can lose energy in the form of a photon and return to a lower energy level.
  - The energy difference between the levels can be determined from the wavelength of the emitted light.
  - Bohr's model could not explain the emission spectra of systems with more than one electron.

▶ The modern view of the electronic structure of the atom describes **atomic orbitals,** which are regions in space with a high probability of containing electrons; that is, a high **electron density.**

2.4 The Periodic Law and the Periodic Table

▶ The **periodic law** relates the structure of elements to their chemical and physical properties. The modern periodic table groups the elements according to these properties.

▶ **Periods** are horizontal rows, numbered 1 through 7 from top to bottom. The lanthanide series is part of period 6; the actinide series is part of period 7.

▶ Vertical columns are referred to as **groups** or families. Two common numbering systems are used in this textbook:
  - Roman numeral system: IA–VIIIA for **representative elements,** IB–VIIIB for **transition elements.**
  - IUPAC system: 1–18.

▶ Four common names of groups are:
  - Alkali metals (IA, 1)
  - Alkaline earth metals (IIA, 2)
  - Halogens (VIIA, 17)
  - Noble gases (VIIIA, 18)

▶ A bold zigzag line from boron leading to the right and down separates the **metals** from the **nonmetals.**

▶ The elements bordering the line (except aluminum) are **metalloids.**

▶ The blocks on the periodic table commonly give the atomic symbol, name, atomic number, and atomic mass.

2.5 Electron Arrangement and the Periodic Table

▶ Electrons are identified by their location in the atom.

▶ **Principal energy levels,** n, give the general distance from the nucleus. Electrons that are higher in energy and further from the nucleus have larger n values.

▶ Each principal energy level, n, can have a maximum of $2(n)^2$ electrons located in it.

▶ Each principal energy level contains **sublevels,** named s, p, and d. The number of the principal energy level gives the number of sublevels.
  - For $n = 1$, there is one sublevel: $1s$.
  - For $n = 2$, there are two sublevels: $2s$, $2p$.
  - For $n = 3$, there are three sublevels: $3s$, $3p$, $3d$.

▶ Each sublevel contains atomic orbitals, named the same as the sublevel. Each atomic orbital contains a maximum of 2 electrons.
  - s sublevels contain one orbital, and a maximum of 2 electrons.
  - p sublevels contain three orbitals, and a maximum of 6 electrons.
  - d sublevels contain five orbitals, and a maximum of 10 electrons.

▶ The **electron configuration,** shorthand electron configuration, and orbital diagram are three ways to designate where all electrons are located in an atom in its ground state.

2.6 Valence Electrons and the Octet Rule

▶ **Valence electrons** are the outermost electrons in the atom. These are the electrons that are involved in bonding as elements form compounds.

▶ The **octet rule** states that atoms usually react in such a way as to obtain a noble gas configuration.

▶ The octet rule can be used to predict the charge of atoms when they become **ions.**

- Metals tend to lose electrons to become positively charged **cations.**
- Nonmetals tend to gain electrons to become negatively charged **anions.**
- Atoms gain and lose electrons to obtain a noble gas configuration; that is, to be **isoelectronic** with the nearest noble gas.

2.7 Trends in the Periodic Table

▶ **Atomic size *decreases*** from left to right and from bottom to top in the periodic table.

- Cations are smaller than their parent atoms.
- Anions are larger than their parent atoms.

▶ **Ionization energy** is the energy required to remove an electron from an isolated atom.

- Ionization energy *increases* from left to right and from bottom to top in the periodic table. The noble gases have the highest ionization energy.

▶ **Electron affinity** is the energy change when a single electron is added to an isolated atom.

- Elements with a high electron affinity will *release* a large amount of energy.
- Electron affinity *increases* from left to right and up a group. The halogens have the greatest electron affinity.

ANSWERS TO PRACTICE PROBLEMS

2.1 **a.** 16 protons, 16 electrons, 16 neutrons
 b. 11 protons, 11 electrons, 12 neutrons
 c. 1 proton, 1 electron, 0 neutrons
 d. 94 protons, 94 electrons, 150 neutrons

2.2 14.0 amu

2.3 10.8 amu

2.4 **a.** sulfur: $1s^2\,2s^2\,2p^6\,3s^2\,3p^4$
 b. calcium: $1s^2\,2s^2\,2p^6\,3s^2\,3p^6\,4s^2$
 c. potassium: $1s^2\,2s^2\,2p^6\,3s^2\,3p^6\,4s^1$
 d. phosphorus: $1s^2\,2s^2\,2p^6\,3s^2\,3p^3$

2.5 **a.**

b.

c.

d.

2.6 **a.** $[\text{Ne}]3s^2\,3p^4$ **c.** $[\text{Ar}]4s^2 3d^{10}4p^4$
 b. $[\text{Ne}]3s^2\,3p^2$ **d.** $[\text{Ar}]4s^2 3d^6$

2.7 **a.** Total electrons = 11, valence electrons = 1, energy level = 3
 b. Total electrons = 12, valence electrons = 2, energy level = 3
 c. Total electrons = 16, valence electrons = 6, energy level = 3
 d. Total electrons = 17, valence electrons = 7, energy level = 3
 e. Total electrons = 18, valence electrons = 8, energy level = 3

2.8 33 protons, 36 electrons

2.9 [Xe]

2.10 Ra^{2+} is isoelectronic with Rn.

QUESTIONS AND PROBLEMS

Composition of the Atom

Foundations

2.19 Explain the difference between the mass number and the atomic mass of an element.

2.20 Why is the number of electrons not part of the mass number of an atom?

2.21 Fill in the blanks:
 a. Isotopes of an element differ in mass because the atoms have a different number of _____.
 b. The atomic number gives the number of _____ in the nucleus.
 c. The mass number of an atom is due to the number of _____ and _____ in the nucleus.
 d. Electrons surround the _____ and have a _____ charge.

2.22 Identify which of the following isotopic symbols is incorrect.

$$^{12}_{6}\text{C} \qquad ^{13}_{7}\text{C} \qquad ^{12}\text{C}$$

2.23 Identify the major difference and the major similarity among isotopes of an element.

2.24 Label each of the following statements as true or false:
 a. An atom with an atomic number of 7 and a mass of 14 is identical to an atom with an atomic number of 6 and a mass of 14.
 b. Neutral atoms have the same number of electrons as protons.
 c. The mass number of an atom is due to the sum of the number of protons, neutrons, and electrons.

2.25 Label each of the following statements as true or false.
 a. Isotopes are atoms of the same element with different numbers of neutrons.

b. Isotopes are atoms with the same number of protons but different numbers of neutrons.

c. Isotopes are atoms with the same atomic number but different mass numbers.

2.26 The nuclei of three different atoms are depicted in the diagrams below. Which ones are isotopes, if any?

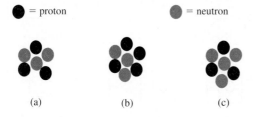

● = proton   ● = neutron

(a)   (b)   (c)

Applications

2.27 Calculate the number of protons, neutrons, and electrons in:

a. $^{136}_{56}Ba$ **b.** $^{209}_{84}Po$ **c.** $^{113}_{48}Cd$

2.28 Calculate the number of protons, neutrons, and electrons in:

a. $^{37}_{17}Cl$ **b.** $^{23}_{11}Na$ **c.** $^{84}_{36}Kr$

2.29 An atom has nine protons, ten neutrons, and nine electrons. Write the symbol of the atom.

2.30 An atom has nineteen protons, twenty neutrons, and nineteen electrons. Write the symbol of the atom.

2.31 **a.** How many protons are in the nucleus of the isotope Rn-220?

b. How many neutrons are in the nucleus of the isotope Rn-220?

2.32 **a.** How many protons are in the nucleus of the isotope In-115?

b. How many neutrons are in the nucleus of the isotope In-115?

2.33 Selenium-80 is a naturally occurring isotope used in over-the-counter supplements.

a. How many protons are found in one atom of selenium-80?

b. How many neutrons are found in one atom of selenium-80?

2.34 Iodine-131 is an isotope used in thyroid therapy.

a. How many protons are found in one atom of iodine-131?

b. How many neutrons are found in one atom of iodine-131?

2.35 Write symbols for each isotope:

a. Each atom contains one proton and zero neutrons.

b. Each atom contains six protons and eight neutrons.

2.36 Write symbols for each isotope:

a. Each atom contains one proton and two neutrons.

b. Each atom contains 92 protons and 146 neutrons.

2.37 The element copper has two naturally occurring isotopes. One of these has a mass of 62.93 amu and a natural abundance of 69.09%. A second isotope has a mass of 64.9278 amu and a natural abundance of 30.91%. Calculate the atomic mass of copper.

2.38 The element lithium has two naturally occurring isotopes. One of these has a mass of 6.0151 amu and a natural abundance of 7.49%. A second isotope has a mass of 7.0160 amu and a natural abundance of 92.51%. Calculate the atomic mass of lithium.

Development of Atomic Theory

Foundations

2.39 What are the major postulates of Dalton's atomic theory?

2.40 What points of Dalton's theory are no longer current?

2.41 Describe the experiment that provided the basis for our understanding of the nucleus.

2.42 Describe the series of experiments that characterized the electron.

2.43 Note the major accomplishment of each of the following:

a. Dalton   **c.** Chadwick

b. Crookes   **d.** Goldstein

2.44 Note the major accomplishment of each of the following:

a. Thomson   **c.** Geiger

b. Rutherford   **d.** Bohr

Applications

2.45 What is a cathode ray? Which subatomic particle is detected?

2.46 Use the concept of charges to explain why cathode rays are specifically deflected toward the positive pole by external electric fields and magnetic fields.

2.47 List at least three properties of the electron.

2.48 Use the concept of charges to explain why an alpha particle fired toward the nucleus is deflected away from the nucleus.

2.49 Prior to Rutherford's gold foil experiments and the understanding of the existence of a nucleus, how did scientists view the atom?

2.50 Explain how Rutherford concluded that the atom is principally empty space.

Light, Atomic Structure, and the Bohr Atom

Foundations

2.51 What is meant by the term *spectroscopy*?

2.52 What is meant by the term *electromagnetic spectrum*?

2.53 Describe electromagnetic radiation according to its wave nature.

2.54 Describe electromagnetic radiation according to its particle nature.

2.55 Is the following statement true or false?
Light of higher energy travels at a faster speed than light of lower energy.

2.56 What is the relationship between the energy of light and its wavelength?

2.57 Which form of radiation has greater energy, microwave or infrared? Explain your reasoning.

2.58 Which form of radiation has the longer wavelength, ultraviolet or infrared? Explain your reasoning.

2.59 Describe the process that occurs when electrical energy is applied to a sample of hydrogen gas.

2.60 When electrical energy is applied to an element in its gaseous state, light is produced. How does the light differ among elements?

Applications

2.61 Critique this statement: Electrons can exist in any position outside of the nucleus.

2.62 Critique this statement: Promotion of electrons is accompanied by a release of energy.

2.63 List three of the most important points of the Bohr theory.

2.64 Give two reasons why the Bohr theory did not stand the test of time.

2.65 What was the major contribution of Bohr's atomic model?

2.66 What was the major deficiency of Bohr's atomic model?

The Periodic Law and the Periodic Table

Foundations

2.67 Provide the atomic number, atomic mass, and name of the element represented by each of the following symbols:
  a. Na   **c.** Mg
  b. K   **d.** B

2.68 Provide the atomic number, atomic mass, and name of the element represented by each of the following symbols:
  a. Ca   **c.** Co
  b. Cu   **d.** Si

2.69 Which group of the periodic table is known as the alkali metals? List their symbols.

2.70 Which group of the periodic table is known as the alkaline earth metals? List their symbols.

2.71 Which group of the periodic table is known as the halogens? List their names.

2.72 Which group of the periodic table is known as the noble gases? List their names.

Applications

2.73 For each of the elements Na, Ni, Al, P, Cl, and Ar, provide the following information:
  a. Which are metals?
  b. Which are representative metals?
  c. Which are inert or noble gases?

2.74 For each of the elements Ca, K, Cu, Zn, Br, and Kr, provide the following information:
  a. Which are metals?
  b. Which are representative metals?
  c. Which are inert or noble gases?

2.75 Using the information below, for Group IA (1) elements:

Element	Atomic Number	Melting Point (°C)
Li	3	180.5
Na	11	97.8
K	19	63.3
Rb	37	38.9
Cs	55	28.4

Prepare a graph relating melting point and atomic number. How does this demonstrate the periodic law?

2.76 Use the graph prepared in Question 2.75 to predict the melting point of francium (Fr).

Electron Arrangement and the Periodic Table

Foundations

2.77 Distinguish between a principal energy level and a sublevel.

2.78 Distinguish between a sublevel and an orbital.

2.79 Sketch a diagram and describe our current model of an s orbital.

2.80 How is a $2s$ orbital different from a $1s$ orbital?

2.81 What is the maximum number of electrons in each of the following energy levels?
  a. $n = 1$ **b.** $n = 2$ **c.** $n = 3$

2.82 For any given principal energy level, what is the maximum number of electrons that can exist in the following subshells?
  a. s **b.** p **c.** d

2.83 State the Pauli exclusion principle. Explain how it is used to determine the number of electrons that can exist in a d subshell.

2.84 State Hund's rule. Determine whether the following orbital diagrams violate Hund's rule.

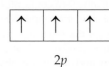

Applications

2.85 Using the periodic table, write the electron configuration of each of the following atoms:
  a. Al   **b.** Na   **c.** Sc

2.86 Using the periodic table, write the electron configuration of each of the following atoms:
  a. Ca   **b.** Fe   **c.** Cl

2.87 Using the periodic table, write the electron configuration and orbital diagram of each of the following atoms:
  a. B   **b.** S   **c.** Ar

2.88 Using the periodic table, write the electron configuration and orbital diagram of each of the following atoms:
  a. V   **b.** Cd   **c.** Te

2.89 Which of the following electron configurations are not possible? Why?
  a. $1s^2\,1p^2$ **c.** $2s^2\,2s^2\,2p^6\,2d^1$
  b. $1s^2\,2s^2\,2p^2$ **d.** $1s^2\,2s^3$

2.90 For each incorrect electron configuration in Question 2.89, assume that the number of electrons is correct, identify the element, and write the correct electron configuration.

2.91 Determine whether the following orbital diagrams are correct. If there is an error, fix the diagram.

2.92 Identify the element for each of the orbital diagrams (once corrected) in Question 2.91.

2.93 Use the periodic table and determine the shorthand electron configuration of each of the following atoms.
   a. Zr   **b.** Br   **c.** K

2.94 Use the periodic table and determine the shorthand electron configuration of each of the following atoms.
   a. I   **b.** Al   **c.** V

Valence Electrons and the Octet Rule

Foundations

2.95 How does the total number of electrons differ from the number of valence electrons in an atom?

2.96 How can the periodic table be used to determine the number of valence electrons in a representative element atom?

2.97 State the octet rule.

2.98 How can the octet rule be used to determine the number of electrons gained or lost by an atom as it becomes an ion?

2.99 Do metals tend to gain or lose electrons? Do they become cations or anions?

2.100 Do nonmetals tend to gain or lose electrons? Do they become cations or anions?

Applications

2.101 How many total electrons and valence electrons are found in an atom of each of the following elements? What is the number of the principal energy level in which the valence electrons are found?
   a. H   **d.** F
   b. Na   **e.** Ne
   c. B   **f.** He

2.102 How many total electrons and valence electrons are found in an atom of each of the following elements? What is the number of the principal energy level in which the valence electrons are found?
   a. Mg   **d.** Br
   b. K   **e.** Ar
   c. C   **f.** Xe

2.103 Determine the number of protons and electrons in each of the following ions.
   a. Cl^- **c.** Fe^{2+}
   b. Ca^{2+}

2.104 Determine the number of protons and electrons in each of the following ions.
   a. S^{2-} **c.** Cd^{2+}
   b. K^+

2.105 Predict the number of valence electrons in an atom of:
   a. calcium   **c.** cesium
   b. potassium   **d.** barium

2.106 Predict the number of valence electrons in an atom of:
   a. carbon   **c.** sulfur
   b. phosphorus   **d.** chlorine

2.107 Give the most probable ion formed from each of the following elements:
   a. Li   **c.** S
   b. Ca

2.108 Give the most probable ion formed from each of the following elements.
   a. O   **c.** Al
   b. Br

2.109 Which of the following pairs of atoms and/or ions are isoelectronic with one another?
   a. O^{2-}, Ne   **b.** S^{2-}, Cl^-

2.110 Which of the following pairs of atoms and/or ions are isoelectronic with one another?
   a. F^-, Cl^- **b.** K^+, Ar

2.111 Write the electron configuration and shorthand electron configuration of each of the following ions:
   a. I^- **c.** Se^{2-}
   b. Ba^{2+} **d.** Al^{3+}

2.112 Write the electron configuration and shorthand electron configuration of each of the following biologically important ions:
   a. Ca^{2+} **c.** K^+
   b. Mg^{2+} **d.** Cl^-

Trends in the Periodic Table

Foundations

2.113 What is the trend for atom size from left to right across a period?

2.114 What is the trend for atom size from top to bottom down a group?

2.115 Define ionization energy.

2.116 Define electron affinity.

2.117 Write an equation for the removal of an electron from a gaseous atom of sodium.

2.118 Write an equation for the addition of an electron to a gaseous atom of chlorine.

2.119 Arrange each of the following lists of elements in order of increasing atomic size:
   a. N, O, F   **c.** Cl, Br, I
   b. Li, K, Cs   **d.** Ra, Be, Mg

2.120 Arrange each of the following lists of elements in order of increasing atomic size:
   a. Al, Si, P   **c.** Sr, Ca, Ba
   b. In, Ga, Al   **d.** P, N, Sb

2.121 Arrange each of the following lists of elements in order of increasing ionization energy:
   a. N, O, F   **b.** Li, K, Cs

2.122 Arrange each of the following lists of elements in order of increasing ionization energy:
   a. Cl, Br, I   **b.** Ra, Be, Mg

2.123 Arrange each of the following lists of elements in order of decreasing electron affinity:
   a. Na, Li, K   **b.** Sr, Sn, Te

2.124 Arrange each of the following lists of elements in order of decreasing electron affinity:
   a. Mg, P, Cl   **b.** Br, I, Cl

Applications

2.125 **a.** Explain why a positive ion is always smaller than its parent atom.
   b. Explain why a fluoride ion is commonly found in nature but a fluorine atom is not.

2.126 **a.** Explain why a negative ion is always larger than its parent atom.
   b. Explain why a sodium ion is commonly found in nature but a sodium atom is not.

2.127 Cl^- and Ar are isoelectronic. Which is larger? Why?

2.128 K^+ and Ar are isoelectronic. Which is larger? Why?

CHALLENGE PROBLEMS

1. A natural sample of chromium, taken from the ground, will contain four isotopes: Cr-50, Cr-52, Cr-53, and Cr-54. Predict which isotope is in greatest abundance. Explain your reasoning.
2. Crookes's cathode ray tube experiment inadvertently supplied the basic science for a number of modern high-tech devices. List a few of these devices and describe how they involve one or more aspects of this historic experiment.
3. Name five elements that you came in contact with today. Were they in a combined form or did they exist in the form of atoms? Were they present in a pure form or in mixtures? If mixtures, were they heterogeneous or homogeneous? Locate each in the periodic table by providing the group and period designation—for example: Group IIA (or 2), period 3.
4. The periodic table is incomplete. It is possible that new elements will be discovered from experiments using high-energy particle accelerators. Predict as many properties as you can that might characterize the element that has the atomic number 118. Use the Internet to find the properties of element 118, and evaluate your predictions.
5. The element titanium is now being used as a structural material for bone and joint replacement (shoulders, knees). Predict properties that you would expect for such applications; utilize the Internet to look up the properties of titanium and evaluate your answer.

6. Imagine that you have undertaken a voyage to an alternate universe. Using your chemical skills, you find a collection of elements quite different than those found here on earth. After measuring their properties and assigning symbols for each, you wish to organize them as Mendeleev did for our elements. Design a periodic table using the information you have gathered:

Symbol	Mass (amu)	Reactivity	Electrical Conductivity
A	2.0	High	High
B	4.0	High	High
C	6.0	Moderate	Trace
D	8.0	Low	0
E	10.0	Low	0
F	12.0	High	High
G	14.0	High	High
H	16.0	Moderate	Trace
I	18.0	Low	0
J	20.0	None	0
K	22.0	High	High
L	24.0	High	High

Predict the reactivity and conductivity of an element with a mass of 30.0 amu. What element in our universe does this element most closely resemble?

Structure and Properties of Ionic and Covalent Compounds

3

LEARNING GOALS

1 Draw Lewis symbols for representative elements and their respective ions.

2 Classify compounds as having ionic, polar covalent, or nonpolar covalent bonds.

3 Write the formula of a compound when provided with the name or elemental composition of the compound.

4 Name inorganic compounds using standard naming conventions, and recall the common names of frequently used substances.

5 Predict differences in physical state, melting and boiling points, solid-state structure, and solution chemistry that result from differences in bonding.

6 Draw Lewis structures for covalent compounds and polyatomic ions.

7 Explain how the presence or absence of multiple bonding relates to bond length, bond energy, and stability.

8 Use Lewis structures to predict the geometry of molecules.

9 Describe the role that molecular geometry plays in determining the polarity of compounds.

10 Use polarity to determine solubility and predict the melting and boiling points of compounds.

Large gypsum ($CaSO_4 \cdot 2H_2O$) crystals located in a mine in Mexico. The 55-ton crystals reflect the structure seen on the atomic level.

OUTLINE

INTRODUCTION

A chemical compound is formed when two or more atoms of different elements are joined by attractive forces called chemical bonds. These bonds result from either a transfer of electrons from one atom to another (the ionic bond) or a sharing of electrons between two atoms (the covalent bond). The elements, once converted to a compound, cannot be recovered by any physical process. A chemical reaction must take place to regenerate the individual elements. The chemical and physical properties of a compound are related to the structure of the compound, and this structure is, in turn, determined by the arrangement of electrons in the atoms that produced the compound. Properties such as solubility, boiling point, and melting point correlate well with the shape and charge distribution, hence, polarity, in the individual units of the compound.

You need to learn how to properly name and write formulas for ionic and covalent compounds. You should also become familiar with some of their properties and be able to relate these properties to the structure and bonding of the compounds.

3.1   Chemical Bonding

When two or more atoms form a chemical compound, the atoms are held together in a characteristic arrangement by attractive forces. The **chemical bond** is the force of attraction between any two atoms in a compound. This force of attraction overcomes the repulsion of the positively charged nuclei of the two atoms.

Interactions involving valence electrons are responsible for the chemical bond. We shall focus our attention on these electrons and the electron arrangement of atoms both before and after bond formation.

Lewis Symbols

The **Lewis symbol,** developed by G. N. Lewis early in the twentieth century, uses the atomic symbol to represent the nucleus and core electrons and dots to represent valence electrons. Its principal advantage is that *only* valence electrons (those that may participate in bonding) are shown. Lewis symbolism is based on the octet rule that was introduced in Chapter 2.

To draw Lewis symbols, we first write the chemical symbol of the atom; this symbol represents the nucleus and all of the lower energy nonvalence electrons. The valence electrons are indicated by dots arranged around the atomic symbol.

Note that the number of dots corresponds to the number of electrons in the outermost shell of the atoms of the element (valence electrons). The four sides around the atomic symbol can each have two dots for a maximum of eight. This corresponds to the fact that a maximum of eight electrons are contained within the *s* and *p* subshells. When placing the dots around the symbol, start by placing one dot per side. When finished, each unpaired dot (representing an unpaired electron) is available to form a chemical bond with another atom.

Principal Types of Chemical Bonds: Ionic and Covalent

Two principal classes of chemical bonds exist: ionic and covalent. Both involve valence electrons.

Ionic bonding involves a transfer of one or more electrons from one atom to another. **Covalent bonding** involves a sharing of electrons.

Before discussing each type, we should recognize that the distinction between ionic and covalent bonding is not always clear-cut. Some compounds are clearly ionic, and some are clearly covalent, but many others possess both ionic and covalent characteristics.

Recall that the number of valence electrons for a representative element can be determined from the Roman numeral above the group in the periodic table (see Figure 2.10).

EXAMPLE 3.1 | Drawing Lewis Symbols for Representative Elements

LEARNING GOAL

Draw the Lewis symbol for carbon, and indicate the number of bonds that carbon will form when chemically bonded to other atoms.

1 Draw Lewis symbols for representative elements and their respective ions.

Solution

Carbon is found in Group IVA (or 14) of the periodic table. Therefore, carbon has four valence electrons.

Step 1. The chemical symbol for carbon is written as

$$C$$

Step 2. The four dots representing the four valence electrons are placed around the C, with one dot on each side.

$$\cdot \overset{\cdot}{\underset{\cdot}{C}} \cdot$$

Step 3. Carbon is able to form four bonds because each unpaired electron (dot) is available to form a bond.

Helpful Hint: Since carbon has four valence electrons, it would need four more electrons in order to satisfy the octet rule.

Practice Problem 3.1

Draw the Lewis symbol for oxygen, and indicate the number of bonds that oxygen will form when bonded to other atoms.

▶ For Further Practice: **Questions 3.13 and 3.14.**

Ionic Bonding

Representative elements form ions that obey the octet rule. Ions of opposite charge attract each other, and this attraction is the essence of the ionic bond. As a result, when metals and nonmetals combine, it is usually through the formation of an ionic bond. Consider the reaction of a sodium atom and a chlorine atom to produce sodium chloride:

$$Na + Cl \longrightarrow NaCl$$

Recall that the sodium atom, a metal, has

- a low ionization energy (it readily loses an electron) and
- a low electron affinity (it does not want more electrons).

If sodium loses its valence electron, it will become isoelectronic (same electron configuration) with neon, a very stable noble gas atom. This tells us that the sodium atom would be a good electron donor, forming the sodium cation:

$$Na \cdot \longrightarrow Na^+ + e^-$$

Recall that the chlorine atom, a nonmetal, has

- a high ionization energy (it will not easily give up an electron) and
- a high electron affinity (it readily accepts another electron).

Chlorine will gain one more electron. By doing so, it will complete an octet (eight outermost electrons) and be isoelectronic with argon, a stable noble gas. Therefore, chlorine behaves as a willing electron acceptor, forming a chloride anion:

$$:\overset{\cdot\cdot}{\underset{\cdot\cdot}{Cl}} \cdot + e^- \longrightarrow \left[:\overset{\cdot\cdot}{\underset{\cdot\cdot}{Cl}} : \right]^-$$

LEARNING GOALS

1 Draw Lewis symbols for representative elements and their respective ions.

2 Classify compounds as having ionic, polar covalent, or nonpolar covalent bonds.

Refer to Section 2.7 for a discussion of ionization energy and electron affinity.

Square brackets are placed around Lewis symbols of anions.

When a sodium atom reacts with a chlorine atom, the electron released by sodium (*electron donor*) is the electron received by chlorine (*electron acceptor*):

$$\text{Na·} \longrightarrow \text{Na}^+ + \text{e}^-$$

$$\text{e}^- + \cdot\ddot{\underset{\cdot\cdot}{\text{Cl}}}: \longrightarrow \left[:\ddot{\underset{\cdot\cdot}{\text{Cl}}}:\right]^-$$

An **ionic bond** is the *electrostatic force,* the attraction of opposite charges, between the resulting cation and anion (in this case, the Na^+ and the Cl^-). The electrostatic force is quite strong and holds the ions together as an *ion pair:* Na^+Cl^-. The essential features of ionic bonding are the following:

- Metals tend to form cations because they have low ionization energies and low electron affinities.
- Nonmetals tend to form anions because they have high ionization energies and high electron affinities.
- Ions are formed by the transfer of electrons.
- The oppositely charged ions formed are held together by an electrostatic force.
- Reactions between metals and nonmetals tend to form ionic compounds.

| EXAMPLE 3.2 | **Drawing Lewis Symbols for Ions** | LEARNING GOAL |

When there is a reaction between potassium and bromine, ions form. Using Lewis symbols, write the reactions showing how electrons are lost or gained when potassium and bromine become ions.

1   Draw Lewis symbols for representative elements and their respective ions.

Solution

The periodic table is used to determine the number of valence electrons associated with each atom.

Step 1. Potassium is in Group IA (or 1); it has one valence electron that should be represented by one dot on the atom.

$$\cdot\text{K}$$

Step 2. Potassium has low ionization energy and low electron affinity, which makes it a good electron donor. In the reaction shown below, the cation formed does not have any valence electrons.

$$\cdot\text{K} \longrightarrow \text{K}^+ + \text{e}^-$$

Step 3. Bromine is in Group VIIA (or 17); it has seven valence electrons that should be represented by seven dots on the atom.

$$\cdot\ddot{\underset{\cdot\cdot}{\text{Br}}}:$$

Step 4. Bromine has high ionization energy and high electron affinity, which make it a good electron acceptor. The anion formed will have eight valence electrons because bromine gains one electron to form a stable octet. Square brackets are often placed around Lewis symbols of anions to more easily distinguish the ion charge from the electrons.

$$\cdot\ddot{\underset{\cdot\cdot}{\text{Br}}}: + \text{e}^- \longrightarrow \left[:\ddot{\underset{\cdot\cdot}{\text{Br}}}:\right]^-$$

Helpful Hint: Lewis symbols for cations should not have any dots because the atoms lose their valence electrons in the process of ion formation. Whereas, with the exception of hydrogen, Lewis symbols for anions should contain eight dots because the atoms gained electrons in the process of attaining a stable octet.

Practice Problem 3.2

When there is a reaction between calcium and oxygen, ions form. Using Lewis symbols, write the reactions showing how electrons are lost or gained when calcium and oxygen become ions.

▶ For Further Practice: **Questions 3.23 and 3.24.**

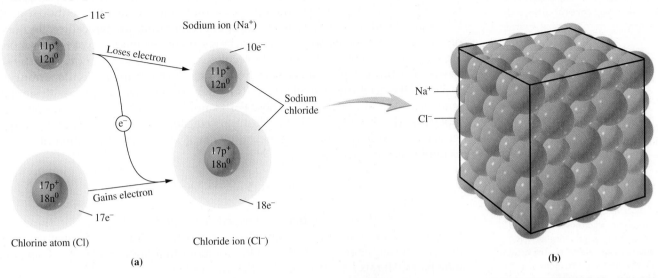

Sodium atom (Na)

Sodium ion (Na$^+$)

Sodium chloride

Na$^+$

Cl$^-$

Loses electron

Gains electron

Chlorine atom (Cl)

Chloride ion (Cl$^-$)

(a)

(b)

(c)

Figure 3.1 The arrangement of ions in a crystal of NaCl (sodium chloride, table salt). (a) A sodium atom loses one electron to become a smaller sodium ion, and a chlorine atom gains that electron, becoming a larger chloride ion. (b) Attraction of Na$^+$ and Cl$^-$ forms NaCl ion pairs that aggregate in a three-dimensional crystal lattice structure. (c) An enlarged view of NaCl crystals, magnified 400x, shows their cubic geometry. Each tiny crystal contains billions and billions of sodium and chloride ions.

Although ionic compounds are sometimes referred to as ion pairs, in the solid state these ion pairs do not actually exist as individual units. The positive ions exert attractive forces on several negative ions, and the negative ions are attracted to several positive centers. Positive and negative ions arrange themselves in a regular three-dimensional repeating array to produce a stable arrangement known as a **crystal lattice.** The lattice structure for the ionic compound sodium chloride is shown in Figure 3.1.

Covalent Bonding

The octet rule is not just for ionic compounds. In covalent bonding, atoms share electrons to complete the octet of electrons for each of the atoms participating in the bond. Consider the bond formed between two hydrogen atoms, producing the diatomic form of hydrogen: H_2.

An individual hydrogen atom is not stable; however, if it were to gain a second electron, it would be isoelectronic with the stable electron configuration of helium. Since two identical hydrogen atoms have an equal tendency to gain or lose electrons, an electron transfer from one atom to the other is unlikely to occur under normal conditions. Each atom may attain a noble gas structure only by *sharing* its electron with the other, as shown with Lewis symbols and a Lewis structure:

$$H\cdot + \cdot H \longrightarrow H : H$$

LEARNING GOAL

2 Classify compounds as having ionic, polar covalent, or nonpolar covalent bonds.

Lewis symbols represent atoms and ions. Lewis structures represent covalent compounds.

A diatomic molecule is one that is composed of two atoms joined by a covalent bond.

Fourteen valence electrons are arranged in such a way that each fluorine atom is surrounded by eight electrons. The octet rule is satisfied for each fluorine atom.

When electrons are shared rather than transferred, the *shared electron pair* is referred to as a *covalent bond* (Figure 3.2). Compounds containing only covalent bonds are called *covalent compounds*. Covalent compounds are typically collections of *molecules*. **Molecules** are neutral (uncharged) species made up of two or more atoms joined by covalent bonds. Molecules are represented by *Lewis structures*. **Lewis structures** depict valence electrons of all atoms in the molecule, arranged to satisfy the octet rule. Covalent bonds tend to form between atoms with similar tendencies to gain or lose electrons. For example, bonding within the diatomic molecules (H_2, N_2, O_2, F_2, Cl_2, Br_2, and I_2) is covalent because there can be no net tendency for electron transfer between identical atoms. The formation of F_2, for example, may be represented as

$$: \overset{..}{\underset{..}{F}} \cdot \; + \; \cdot \overset{..}{\underset{..}{F}} : \longrightarrow \; : \overset{..}{\underset{..}{F}} : \overset{..}{\underset{..}{F}} :$$

As in H_2, a single covalent bond is formed. The bonding electron pair, shown in blue, is said to be *localized*, or largely confined to the region between the two fluorine nuclei. Paired electrons around an atom that are not shared are termed **lone pairs.** The six lone pairs in F_2 are shown in red.

Two atoms do not have to be identical to form a covalent bond. Consider compounds such as the following:

$$H : \overset{..}{\underset{..}{F}} : \qquad\qquad H : \overset{..}{\underset{..}{O}} : H \qquad\qquad \begin{matrix} & H & \\ H : & \overset{..}{\underset{..}{C}} & : H \\ & H & \end{matrix} \qquad\qquad \begin{matrix} & H & \\ H : & \overset{..}{N} & : H \\ & & \end{matrix}$$

As you examine each of these structures, notice the number of electrons associated with each atom. The octet rule is satisfied for each atom. The hydrogen atoms have two electrons (shared electrons with the atom to which they are attached), and the other (non-hydrogen) atoms have eight valence electrons, which are counted as follows:

- Fluorine (in HF) has two electrons shared with hydrogen and six lone-pair electrons (3 lone pairs).
- Oxygen (in H_2O) has four electrons shared with the two hydrogen atoms and four lone-pair electrons (2 lone pairs).
- Carbon (in CH_4) has eight electrons shared with the four hydrogen atoms.
- Nitrogen (in NH_3) has six electrons shared with the three hydrogen atoms and two lone-pair electrons (1 lone pair).

In each of these cases, bond formation satisfies the octet rule. A total of eight electrons surrounds each atom other than hydrogen. Hydrogen is an exception because it has only two electrons (corresponding to the electronic structure of helium).

Polar Covalent Bonding and Electronegativity

The Polar Covalent Bond

Covalent bonding is the sharing of an electron pair by two atoms. However, just as we may observe in our day-to-day activities, sharing is not always equal. In a molecule like H_2 (or any other diatomic molecule), the electrons, on average, spend the same amount of time in the vicinity of each atom; the electrons have no preference because both atoms are identical.

Now consider a diatomic molecule composed of two different elements; HF is a common example. It has been experimentally shown that the electrons in the H—F bond between hydrogen and fluorine atoms are not equally shared; the electrons spend more time in the vicinity of the fluorine atom. This unequal sharing results in a **polar covalent bond.** One end of the bond (in this case, the F atom) is more electron rich (higher electron density), hence, more negative. The other end

Two hydrogen atoms approach at high velocity.

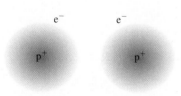

Hydrogen nuclei begin to attract each other's electrons.

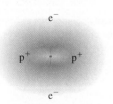

Hydrogen atoms form the hydrogen molecule; atoms are held together by sharing the bonding electron pair.

Figure 3.2 Covalent bonding in hydrogen.

of the bond (in this case, the H atom) is less electron rich (lower electron density), hence, more positive. These two ends can be described as follows (Figure 3.3):

- Somewhat positive end (partial positive), denoted with a δ^+ symbol.
- Somewhat negative end (partial negative), denoted with a δ^- symbol.

Elements whose atoms strongly attract electrons are described as electronegative elements. Linus Pauling, a chemist noted for his theories on chemical bonding, developed a scale of relative electronegativities that correlates reasonably well with the positions of the elements in the periodic table, and allows for us to use the predictive power of the periodic table to determine whether a particular bond is polar or nonpolar covalent. Electronegative elements that tend to form anions (by gaining electrons) are found to the right of the table, whereas the elements that tend to form cations (by losing electrons) are located on the left side of the table.

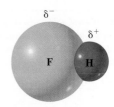

Figure 3.3 Polar covalent bonding in HF. Fluorine is electron rich (δ^-) and hydrogen is electron deficient (δ^+) due to unequal electron sharing.

Electronegativity

Electronegativity (EN) is a measure of the ability of an atom to attract electrons in a chemical bond. Elements with high EN have a greater ability to attract electrons than do elements with low EN. Pauling developed a method to assign values of EN to many of the elements in the periodic table. These values range from a low of 0.7 to a high of 4.0, with 4.0 being the most electronegative element.

Figure 3.4 shows that the most electronegative elements (excluding the non-reactive noble gas elements) are located in the upper right corner of the periodic table, whereas the least electronegative elements are found in the lower left corner of the table. In general, EN values increase as we proceed left to right and bottom to top of the table. Like other periodic trends, numerous exceptions occur.

If we picture the covalent bond as a competition for electrons between two positive centers, it is the difference in EN, ΔEN, that determines the extent of polarity. Consider the calculation of the EN difference in H_2 or H—H:

Linus Pauling is the only person to receive two Nobel Prizes in very unrelated fields; the chemistry award in 1954 and 8 years later, the Nobel Peace Prize. His career is a model of interdisciplinary science, with important contributions ranging from chemical physics to molecular biology.

$$\Delta EN = \begin{bmatrix} EN\ of \\ hydrogen \end{bmatrix} - \begin{bmatrix} EN\ of \\ hydrogen \end{bmatrix}$$

$$\Delta EN = 2.1 - 2.1 = 0 \quad \text{A nonpolar covalent bond}$$

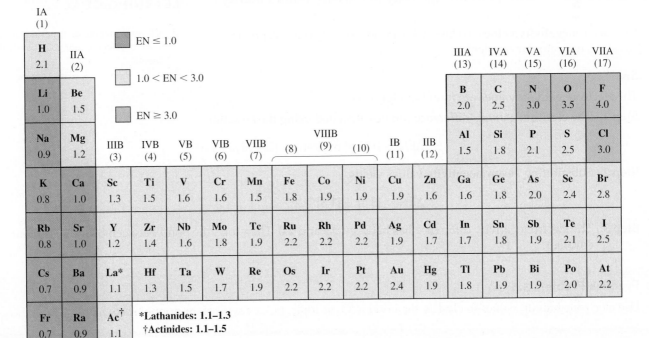

Figure 3.4 Electronegativities of the elements.

By convention, the EN difference is calculated by subtracting the less electronegative element's value from the value for the more electronegative element. In this way, negative numbers are avoided.

The bond in H_2 is nonpolar covalent. The difference in electronegativity between two identical atoms is zero. This can also be calculated for Cl_2 or Cl—Cl:

$$\Delta EN = \left[\begin{array}{c} EN\ of \\ chlorine \end{array}\right] - \left[\begin{array}{c} EN\ of \\ chlorine \end{array}\right]$$

$$\Delta EN = 3.0 - 3.0 = 0 \quad \text{A nonpolar covalent bond}$$

Bonds between *identical* atoms are *always* nonpolar covalent. Although, strictly speaking, any EN difference, no matter how small, produces a polar bond, the degree of polarity for bonds with EN differences less than 0.5 is minimal. Consequently, we shall classify these bonds as nonpolar. Now consider HCl or H—Cl:

$$\Delta EN = \left[\begin{array}{c} EN\ of \\ chlorine \end{array}\right] - \left[\begin{array}{c} EN\ of \\ hydrogen \end{array}\right]$$

$$\Delta EN = 3.0 - 2.1 = 0.9 \quad \text{A polar covalent bond}$$

The EN difference is greater than 0.5 and less than 2.0. By definition, this means that the bond in HCl is polar covalent. An EN difference of 2.0 is generally accepted as the boundary between polar covalent and ionic bonds.

When the EN difference is calculated for NaCl:

$$\Delta EN = \left[\begin{array}{c} EN\ of \\ chlorine \end{array}\right] - \left[\begin{array}{c} EN\ of \\ sodium \end{array}\right]$$

$$\Delta EN = 3.0 - 0.9 = 2.1 \quad \text{An ionic bond}$$

The EN difference of 2.1 is greater than 2.0, and the bond in NaCl is ionic. Recall that metals and nonmetals usually react to produce ionic compounds resulting from the transfer of one or more electrons from the metal to the nonmetal.

EXAMPLE 3.3   **Using Electronegativity to Classify Bond Polarity**

LEARNING GOAL

Use electronegativity values to classify the bonds in SiO_2 as ionic, polar covalent, or nonpolar covalent.

2 Classify compounds as having ionic, polar covalent, or nonpolar covalent bonds.

Solution

The electronegativity values are found in Figure 3.4.

Step 1. The electronegativity difference can be calculated using the equation

$$\Delta EN = [EN\ of\ oxygen] - [EN\ of\ silicon]$$

Step 2. Substituting electronegativity values yields,

$$\Delta EN = [3.5] - [1.8] = 1.7$$

Step 3. The electronegativity difference lies between 0.5 and 2.0. Therefore, the bond between Si and O is polar covalent.

Practice Problem 3.3

Use electronegativity values to classify the bond in O_2 as ionic, polar covalent, or nonpolar covalent.

▶ For Further Practice: **Questions 3.21 and 3.22.**

3.2   Naming Compounds and Writing Formulas of Compounds

Nomenclature is the assignment of a correct and unambiguous name to each and every chemical compound. Assigning a name to a structure or deducing the structure from a name is a necessary first step in any discussion of these compounds. The system for naming ionic compounds is different from the system for naming covalent compounds.

LEARNING GOAL

3 Write the formula of a compound when provided with the name or elemental composition of the compound.

Ionic Compounds

The **formula** is the representation of the fundamental compound using chemical symbols and numerical subscripts. It is the "shorthand" symbol for a compound such as

$$NaCl \quad \text{and} \quad MgF_2$$

The *formula* of an ionic compound is the smallest whole-number ratio of ions in the substance.

The formula identifies the number and type of the various atoms that make up the compound unit. The number of like atoms in the unit is shown by the use of a subscript. The presence of only one atom is understood when no subscript is present.

The formula NaCl indicates that each ion pair consists of one sodium cation (Na^+) and one chloride anion (Cl^-). Similarly, the formula MgF_2 indicates that one magnesium cation, Mg^{2+}, and two fluoride anions, $2\ F^-$, combine to form the compound.

Writing Formulas of Ionic Compounds from the Identities of the Component Ions

It is important to be able to write the formula of an ionic compound when provided with the identities of the ions that make up the compound. The charge of each ion can usually be determined from the group (family) on the periodic table in which the parent element is found. The cations and anions must combine in such a way that the resulting formula unit has a net charge of zero.

EXAMPLE 3.4   **Determining the Formula of an Ionic Compound**

LEARNING GOAL

3 Write the formula of a compound when provided with the name or elemental composition of the compound.

Predict the formula of the compound formed from the combination of ions of aluminum and sulfur.

Solution

Determine the charges of the ions from the periodic table. Use subscripts to create a neutral compound.

Step 1. Aluminum is in Group IIIA (or 13); it has three valence electrons. Loss of these electrons produces Al^{3+}.

$$\cdot \dot{Al} \longrightarrow Al^{3+} + 3e^-$$

Step 2. Sulfur is in Group VIA (or 16); it has six valence electrons. A gain of two electrons (to create a stable octet) produces S^{2-}.

$$\cdot \ddot{\underset{..}{S}} \cdot + 2e^- \longrightarrow \left[: \ddot{\underset{..}{S}} : \right]^{2-}$$

Step 3. In order to combine Al^{3+} and S^{2-} to yield a unit charge of zero, the least common multiple of 3 and 2, which is 6, is used. Each charge is multiplied by a different factor to achieve a charge of 6^+ and 6^-.

$$\text{For Al: } 2Al^{3+} = 6^+ \text{ charge}$$

$$\text{For S: } 3S^{2-} = 6^- \text{ charge}$$

Continued…

Step 4. The subscript 2 is used to indicate that the formula unit contains two aluminum ions. The subscript 3 is used to indicate that the formula unit contains three sulfide ions. Thus, the formula of the compound is Al_2S_3.

Helpful Hint: Electrons are transferred when this neutral ionic compound is formed. When 2 Al atoms form 2 Al^{3+} cations, there are a total of ($2 \times 3e^-$) six electrons produced. Since each S requires two electrons to form each S^{2-}, 3S atoms require a total of ($3 \times 2e^-$) six electrons.

Practice Problem 3.4

Predict the formulas of the compounds formed from the combination of ions of the following elements:

   a. calcium and nitrogen
   b. magnesium and bromine
   c. magnesium and nitrogen

▶ For Further Practice: **Questions 3.27 and 3.28.**

LEARNING GOAL

4 Name inorganic compounds using standard naming conventions, and recall the common names of frequently used substances.

Writing Names of Ionic Compounds from the Formulas of the Compounds

In naming ionic compounds, the name of the cation appears first, followed by the name of the anion. If the ionic compound consists of a metal cation and a non-metal anion, the cation has the name of the metal; the anion is named using the *stem* of the nonmetal name joined to the suffix *-ide*. Some examples follow.

Formula	Cation	and	Anion Stem	+ ide	=	Compound Name
NaCl	sodium		chlor	+ ide		sodium chloride
Na_2O	sodium		ox	+ ide		sodium oxide
Li_2S	lithium		sulf	+ ide		lithium sulfide
$AlBr_3$	aluminum		brom	+ ide		aluminum bromide
CaO	calcium		ox	+ ide		calcium oxide

The metals that exist with only one possible cation charge are the Group IA (or 1) and IIA (or 2) metals, which have 1^+ and 2^+ charges, respectively, as well as Ag^+, Cd^{2+}, Zn^{2+}, and Al^{3+}.

If the cation and anion exist in only one common charged form, there is no ambiguity between formula and name. Sodium chloride *must be* NaCl, and lithium sulfide *must be* Li_2S, so that the sum of positive and negative charges is zero. With many elements, such as the transition metals, several ions of different charge may exist. Fe^{2+}, Fe^{3+} and Cu^+, Cu^{2+} are two common examples. Clearly, an ambiguity exists if we use the name iron for both Fe^{2+} and Fe^{3+} or copper for both Cu^+ and Cu^{2+}. Two systems have been developed to avoid this problem: the *Stock system* and the *common nomenclature system.*

In the Stock system (systematic name), a Roman numeral placed immediately after the name of the ion indicates the magnitude of the cation's charge. In the older common nomenclature system, the suffix *-ous* indicates the lower ionic charge, and the suffix *-ic* indicates the higher ionic charge. Consider the examples in Table 3.1.

Systematic names are easier and less ambiguous than common names. Whenever possible, we will use this system of nomenclature. The older, common names (-ous, -ic) are less specific; furthermore, they often use the Latin names of the elements (for example, iron compounds use *ferr-*, from *ferrum*, the Latin word for iron).

EXAMPLE 3.5	Naming an Ionic Compound Using the Stock System

LEARNING GOAL

4 Name inorganic compounds using standard naming conventions, and recall the common names of frequently used substances.

Name MnO_2.

Solution

Step 1. Based on the formula, it can be concluded that this compound is ionic.

Step 2. Name the cation as the metal name followed by the anion stem with the -*ide* suffix.

Hence, manganese oxide.

Step 3. Determine if a Roman numeral is needed.

Since Mn is not a Group IA (or 1) or IIA (or 2) metal, nor is it Ag, Cd, Zn, or Al, a Roman numeral *is* needed.

Step 4. Determine the charge of the metal. This is determined by using two pieces of information: (1) the charge of the anion, and (2) the overall charge of the compound is zero.

(1) The charge of oxide is 2^-, and there are two O^{2-} ions.

$$2 \times (2^-) = 4^-$$

(2) There must be a 4^+ charge on the metal to balance the 4^- of oxide.

Step 5. Insert a Roman numeral stating this charge (IV) after the metal name. This gives manganese(IV) oxide.

Practice Problem 3.5

Name $CrBr_2$.

▶ For Further Practice: **Questions 3.39 and 3.40.**

Monatomic ions are ions consisting of a single atom. Common monatomic ions are listed in Table 3.2. The ions that are particularly important in biological systems are highlighted in red.

Polyatomic ions, such as the hydroxide ion, OH^-, are composed of two or more atoms bonded together. These ions, although bonded to other ions with ionic bonds, are themselves held together by covalent bonds.

TABLE 3.1 Systematic (Stock) and Common Names for Iron and Copper Ions

For systematic name:			
Formula	**Cation Charge**	**Cation Name**	**Systematic Name**
$FeCl_2$	2 +	Iron(II)	Iron(II) chloride
$FeCl_3$	3 +	Iron(III)	Iron(III) chloride
Cu_2O	1 +	Copper(I)	Copper(I) oxide
CuO	2 +	Copper(II)	Copper(II) oxide

For common nomenclature:			
Formula	**Cation Charge**	**Cation Name**	**Common -ous/ic Name**
$FeCl_2$	2 +	Ferr*ous*	Ferrous chloride
$FeCl_3$	3 +	Ferr*ic*	Ferric chloride
Cu_2O	1 +	Cupr*ous*	Cuprous oxide
CuO	2 +	Cupr*ic*	Cupric oxide

TABLE 3.2 Common Monatomic Cations and Anions

Cation	Name	Anion	Name
H^+	Hydrogen ion	H^-	Hydride ion
Li^+	Lithium ion	F^-	Fluoride ion
Na^+	Sodium ion	Cl^-	Chloride ion
K^+	Potassium ion	Br^-	Bromide ion
Cs^+	Cesium ion	I^-	Iodide ion
Be^{2+}	Beryllium ion	O^{2-}	Oxide ion
Mg^{2+}	Magnesium ion	S^{2-}	Sulfide ion
Ca^{2+}	Calcium ion	N^{3-}	Nitride ion
Ba^{2+}	Barium ion	P^{3-}	Phosphide ion
Al^{3+}	Aluminum ion		
Ag^+	Silver ion		

Note: The ions of principal biological importance are highlighted in red.

The polyatomic ion has an *overall* positive or negative charge. Some common polyatomic ions are listed in Table 3.3. You should memorize the formulas, charges, and names of these polyatomic ions, especially those highlighted in red.

TABLE 3.3 Common Polyatomic Cations and Anions

Ion	Name
H_3O^+	Hydronium
NH_4^+	Ammonium
NO_2^-	Nitrite
NO_3^-	Nitrate
SO_3^{2-}	Sulfite
SO_4^{2-}	Sulfate
HSO_4^-	Hydrogen sulfate
OH^-	Hydroxide
CN^-	Cyanide
PO_4^{3-}	Phosphate
HPO_4^{2-}	Hydrogen phosphate
$H_2PO_4^-$	Dihydrogen phosphate
CO_3^{2-}	Carbonate
HCO_3^-	Bicarbonate
ClO^-	Hypochlorite
ClO_2^-	Chlorite
ClO_3^-	Chlorate
ClO_4^-	Perchlorate
CH_3COO^- (or $C_2H_3O_2^-$)	Acetate
MnO_4^-	Permanganate
$Cr_2O_7^{2-}$	Dichromate
CrO_4^{2-}	Chromate
O_2^{2-}	Peroxide

Note: The most commonly encountered ions are highlighted in red.

When naming compounds containing polyatomic ions, the same rules apply. That is, name the cation and then the anion. Use Roman numerals where appropriate.

Formula	Cation	Anion	Compound Name
NH_4Cl	NH_4^+	Cl^-	ammonium chloride
$Ca(OH)_2$	Ca^{2+}	OH^-	calcium hydroxide
Na_2SO_4	Na^+	SO_4^{2-}	sodium sulfate
$NaHCO_3$	Na^+	HCO_3^-	sodium bicarbonate
$Cu(NO_3)_2$	Cu^{2+}	NO_3^-	copper(II) nitrate

Sodium bicarbonate may also be named sodium hydrogen carbonate, a preferred and less ambiguous name. Likewise, Na_2HPO_4 is named sodium hydrogen phosphate, and other ionic compounds are named similarly.

Question 3.1   Name each of the following compounds:
  a. KCN   b. MgS   c. $Mg(CH_3COO)_2$

Question 3.2   Name each of the following compounds:
  a. Li_2CO_3 b. $FeBr_2$ c. $CuSO_4$

Writing Formulas of Ionic Compounds from the Names of the Compounds

It is also important to be able to write the correct formula when given the compound name. To do this, we must be able to predict the charge of monatomic ions and remember the charge and formula of polyatomic ions. Equally important, the relative number of positive and negative ions in the unit must result in a net charge of zero for compounds that are electrically neutral. When more than one polyatomic ion needs to be represented in the formula of the neutral compound, the polyatomic ion portion of the formula is placed within parentheses before the subscript. Two examples follow.

LEARNING GOAL

3 Write the formula of a compound when provided with the name or elemental composition of the compound.

EXAMPLE 3.6 | **Writing a Formula From the Name of an Ionic Compound**

Write the formula of sodium sulfate.

LEARNING GOAL

3 Write the formula of a compound when provided with the name or elemental composition of the compound.

Solution

The charges of monatomic ions can be determined from the periodic table. The polyatomic ions should be memorized.

Step 1. Determine the charges of the ions:

sodium sulfate

Group IA (or 1) metal $\longrightarrow$ Na^+ SO_4^{2-} $\longleftarrow$ Polyatomic ion

Step 2. Balance the charges. Two positive charges are needed to balance the two negative charges of sulfate. The subscript 2 is used following Na to indicate this.
Hence, the formula is Na_2SO_4.

Helpful Hint: Overall, this compound is neutral. Therefore, when the 2 Na^+ charge component is added to the 1 SO_4^{2-} charge component, the sum should be zero.

$$(2 \times 1^+) + (1 \times 2^-) = 0$$

Continued...

Practice Problem 3.6

Write the formula for each of the following compounds:

  a. calcium carbonate   b. copper(I) sulfate

▶ For Further Practice: **Questions 3.41 and 3.42.**

EXAMPLE 3.7   **Writing a Formula From the Name of an Ionic Compound**

Write the formula of magnesium phosphate.

LEARNING GOAL

3 Write the formula of a compound when provided with the name or elemental composition of the compound.

Solution

The charges of monatomic ions can be determined from the periodic table. The polyatomic ions should be memorized.

Step 1. Determine the charges of the ions:

$$\text{magnesium phosphate}$$

$$\text{Group IIA (or 2) metal} \longrightarrow \text{Mg}^{2+}\ \ \text{PO}_4^{3-} \longleftarrow \text{Polyatomic ion}$$

Step 2. Balance the charges. The least common multiple for 2 and 3 is 6. Obtain 6^+ and 6^- charges using subscripts. Three Mg^{2+} and two PO_4^{3-} are needed. The subscript 3 is written after magnesium, and the polyatomic ion, phosphate, is placed within parentheses and followed with the subscript 2. Hence, the formula is $\text{Mg}_3(\text{PO}_4)_2$.

Helpful Hint: Overall, this compound is neutral. Therefore, when the 3 Mg^{2+} charge component is added to the 2 PO_4^{3-} charge component, the sum should be zero. $(3 \times 2^+) + (2 \times 3^-) = 0$

Practice Problem 3.7

Write the formula for each of the following compounds:

  a. iron(III) sulfide   b. calcium sulfate   c. aluminum oxide

▶ For Further Practice: **Questions 3.43 and 3.44.**

LEARNING GOAL

4 Name inorganic compounds using standard naming conventions, and recall the common names of frequently used substances.

Covalent Compounds

Naming Covalent Compounds

Most covalent compounds are formed by the reaction of nonmetals. We saw earlier that ionic compounds are not composed of single units but are a part of a massive three-dimensional crystal structure in the solid state. Covalent compounds typically exist as discrete molecules in the solid, liquid, and gas states. This is a distinctive feature of covalently bonded substances.

The conventions for naming covalent compounds follow:

1. The names of the elements are written in the order in which they appear in the formula.
2. A prefix (Table 3.4) indicating the number of each kind of atom found in the unit is placed before the name of the element.
3. If only one atom of a particular kind is the first element present in the molecule, the prefix *mono-* is usually omitted from that first element.
4. The stem of the name of the last element is used with the suffix *-ide*.
5. The final vowel in a prefix is often dropped before *oxide*.

By convention, the prefix *mono-* is often omitted from the second element as well (dinitrogen oxide, not dinitrogen monoxide). In other cases, common usage retains the prefix (carbon monoxide, not carbon oxide).

A Medical Perspective

Unwanted Crystal Formation

Conventional wisdom says that the painful symptoms associated with the presence of stones in the bladder and kidneys are just one more problem associated with aging. Unfortunately, recent observations by many clinicians and physicians across the country seem to indicate that these problems are not limited to the elderly. Children as young as 5 years old are developing kidney stones while doctors search for reasons to explain this phenomenon.

Kidney stones most often result from the combination of calcium cations (Ca^{2+}) with anions such as oxalate ($C_2O_4^{2-}$) and phosphate (PO_4^{3-}). Calcium oxalate and calcium phosphate are ionic compounds that are only sparingly soluble in water. They grow in a three-dimensional crystal lattice. When the crystals become large enough to inhibit the flow of urine

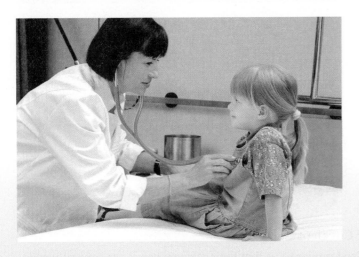

in the kidney or bladder, painful symptoms necessitate some strategy to remove the stones.

Conditions that favor stone formation are the same as those that favor the formation of the same material in a beaker: high concentrations of the ions and insufficient water to dissolve the stones.

Many physicians believe that children who tend to form stones share several behavioral characteristics. They do not drink enough water, and they consume too much salt. Salt in the diet appears to favor the transport of calcium ions to the kidney. Potato chips, French fries, and other snack foods, as well as processed foods such as frozen dinners, sandwich meats, canned soups, and sport drinks are very high in salt content. Obesity appears associated with stone formation, but this may be linked to the unhealthy diet of many children.

Although limiting calcium ion intake would appear to inhibit stone formation (remember, the stones are calcium compounds), reduction of calcium in the diet is not the answer! Dietary calcium is actually beneficial because it binds with oxalate ions before they reach the kidneys. Dietary calcium is important in preventing osteoporosis and dental caries as well. The take-home lesson? A balanced diet, in moderation, with minimal salt intake, especially in the formative years, is recommended.

For Further Understanding

Consult reliable references to support your answers to the following:

▶ Write the formulas for calcium oxalate and calcium phosphate.
▶ Describe one useful function of phosphate ions in the diet.

TABLE 3.4 Prefixes Used to Denote Numbers of Atoms in a Compound

Prefix	Number of Atoms
Mono-	1
Di-	2
Tri-	3
Tetra-	4
Penta-	5
Hexa-	6
Hepta-	7
Octa-	8
Nona-	9
Deca-	10

EXAMPLE 3.8   **Naming a Covalent Compound**

LEARNING GOAL

Name the covalent compound N_2O_4.

4 Name inorganic compounds using standard naming conventions, and recall the common names of frequently used substances.

Solution

Step 1. According to the formula provided, there are two nitrogen atoms and four oxygen atoms.

Step 2. Prefixes are used to denote the numbers of atoms in the compound. *Di-* means two, and *tetra-* means four.

Step 3. The last element listed in the formula is oxygen. The stem of the name, *ox*, is used with the suffix *-ide*.

Step 4. *Tetra-* reduces to *tetr-* by dropping the final vowel because it precedes oxide.

The name is dinitrogen tetroxide.

Practice Problem 3.8

Name each of the following compounds:

    a. B_2O_3 b. NO   c. ICl   d. PCl_3 e. PCl_5 f. P_2O_5

▶ For Further Practice: **Questions 3.51 and 3.52.**

The following are examples of other covalent compounds.

Formula	Name
N_2O	dinitrogen monoxide
NCl_3	nitrogen trichloride
SiO_2	silicon dioxide
CF_4	carbon tetrafluoride
CO	carbon monoxide
CO_2	carbon dioxide

Writing Formulas of Covalent Compounds

Many compounds are so familiar to us that their *common names* are generally used. For example, H_2O is water, NH_3 is ammonia, CH_4 is methane, C_2H_5OH (ethanol) is ethyl alcohol, and $C_6H_{12}O_6$ is glucose. It is useful to be able to correlate both systematic and common names with the corresponding molecular formula and vice versa.

When common names are used, formulas of covalent compounds can be written *only* from memory. You *must* remember that water is H_2O, ammonia is NH_3, and so forth. This is the major disadvantage of common names. Because of their widespread use, they cannot be avoided and must be memorized.

Compounds named by using numeric prefixes are easily converted to formulas. Consider the following examples.

EXAMPLE 3.9   **Writing the Formula of a Covalent Compound**

LEARNING GOAL

Write the formula of dinitrogen tetroxide.

3 Write the formula of a compound when provided with the name or elemental composition of the compound.

Solution

Use prefix meanings to determine subscripts.

Step 1. Nitrogen has the prefix *di-* which indicates two nitrogen atoms.

Step 2. Oxygen has the prefix *tetr-* which indicates four oxygen atoms.

Step 3. Hence, the formula is N_2O_4.

Practice Problem 3.9

Write the formula of each of the following compounds:

a. diphosphorus pentoxide   b. silicon dioxide

c. carbon tetrabromide   d. oxygen difluoride

▶ For Further Practice: **Questions 3.53 and 3.54.**

3.3 Properties of Ionic and Covalent Compounds

LEARNING GOAL

5 Predict differences in physical state, melting and boiling points, solid-state structure, and solution chemistry that result from differences in bonding.

The differences in ionic and covalent bonding result in markedly different properties for ionic and covalent compounds. Because covalent compounds are typically made up of molecules—distinct units—they have less of a tendency to form an extended structure in the solid state. Ionic compounds, with ions joined by electrostatic attraction, form a crystal lattice composed of enormous numbers of positive and negative ions in an extended three-dimensional network. The effects of this basic structural difference are summarized in this section.

Physical State

All ionic compounds (for example, NaCl, KCl, and $NaNO_3$) are solids at room temperature; covalent compounds may be solids (glucose), liquids (water, ethanol), or gases (carbon dioxide, methane). The three-dimensional crystal structure that is characteristic of ionic compounds holds them in a rigid, solid arrangement, whereas molecules of covalent compounds may be fixed, as in a solid, or more mobile, a characteristic of liquids and gases.

Melting and Boiling Points

The **melting point** is the temperature at which a solid is converted to a liquid, and the **boiling point** is the temperature at which a liquid is converted to a gas at a specified pressure. Considerable energy is required to break apart the uncountable numbers of ionic interactions within an ionic crystal lattice and convert the ionic substance to a liquid or a gas. As a result, the melting and boiling temperatures for ionic compounds are generally higher than those of covalent compounds, whose molecules interact less strongly in the solid state. A typical ionic compound, sodium chloride, has a melting point of 801°C; methane, a covalent compound, melts at −182°C.

Although melting and boiling temperatures for ionic compounds are generally higher than those for covalent compounds, exceptions do exist; quartz, a covalent compound of silicon dioxide with an extremely high melting point, is a familiar example.

Structure of Compounds in the Solid State

Ionic solids are *crystalline*, characterized by a regular structure, whereas covalent solids may either be crystalline or have no regular structure. In the latter case, they are said to be *amorphous*.

Rebuilding Our Teeth

Tooth decay (dental caries) is an unpleasant fact of life. Aging, diet, and improper dental hygiene lead to the breakdown of a critical building block of teeth: hydroxyapatite, $Ca_5(PO_4)_3OH$. The hydroxyapatite forms a three-dimensional crystalline structure, and this structural unit is the principal component in the exterior, the enamel part of the tooth. Hydroxyapatite is also a major component in the dentin, which is directly underneath the enamel.

Hydroxyapatite breakdown causes the super-strong enamel to become porous and weak. The pores created are ideal hiding places for bacteria that form acids from foods (especially sugars), which hasten the decay of the tooth. The traditional remedy is the familiar whine of the drill, removing the decayed material, and unfortunately, some of the unaffected tooth. This step is followed by filling with a carbon-based polymer material or metal amalgam. In extreme cases, removal of the entire tooth is the only suitable remedy.

Researchers at a variety of institutions, such as the University of California, San Francisco, are studying other alternatives, collectively termed *remineralization*. What if teeth, in the early stages of enamel decay, were exposed to solutions containing calcium, phosphate, and fluoride ions? Could the hydroxyapatite be rebuilt through the uptake of its critical components, calcium and phosphate ions? Can the hydroxide ion in the hydroxyapatite be replaced by fluoride ion, producing a more decay-resistant $Ca_5(PO_4)_3F$?

Although conventional drill and fill procedures will be with us for the foreseeable future, early remineralization research is encouraging. Already toothpastes and mouthwashes containing calcium, phosphate, and/or fluoride ions are commercially available. Regular use of these products may at least extend the life of our valuable and irreplaceable natural teeth.

For Further Understanding

▸ Examine the labels of a variety of toothpastes and mouthwashes. Do these products contain any of the ions discussed above? Describe their potential benefit.

▸ Many mouthwashes claim to kill bacteria. How can this reduce the incidence of tooth decay?

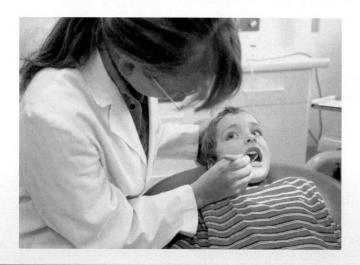

Solutions of Ionic and Covalent Compounds

In Chapter 1, we saw that mixtures are either heterogeneous or homogeneous. A homogeneous mixture is a solution. Many ionic solids dissolve in solvents, such as water. An ionic solid, if soluble, will form cations and anions in solution by **dissociation.**

The role of the solvent in the dissolution of solids is discussed in Section 6.1.

Because ions in water are capable of carrying (conducting) a current of electricity, we refer to ionic compounds dissolved in water as **electrolytes,** and the solution is termed an *electrolytic solution.* Dissolved covalent solids usually retain their neutral (molecular) character and are **nonelectrolytes.** The solution is not an electrical conductor.

3.4 Drawing Lewis Structures of Molecules and Polyatomic Ions

Lewis Structures of Molecules

In Section 3.1, we used Lewis symbols of individual atoms to help us understand the bonding process. Lewis symbols for atoms are building blocks that enable construction of *Lewis structures* for compounds and polyatomic ions that depict the role of valence electrons in bonding and structure. To begin to explain the relationship between molecular structure and molecular properties, we will first need a set of guidelines to help us write Lewis structures for molecules and polyatomic ions:

Step 1. *Use chemical symbols for the various elements to write the skeletal structure of the compound.* To accomplish this, place the bonded atoms next to one another. This is relatively easy for simple compounds; however, as the number of atoms in the compound increases, the possible number of arrangements increases dramatically. We may be told the pattern of arrangement of the atoms in advance; if not, we can make an intelligent guess and see if a reasonable Lewis structure can be constructed. Three considerations are very important here:

    1. The least electronegative atom will usually be the central atom and often the element written first in the formula.

    2. Hydrogen and fluorine must occupy terminal (non-central) positions.

    3. Carbon often forms chains of carbon-carbon covalent bonds.

Step 2. *Determine the number of valence electrons associated with each atom; combine them to determine the total number of valence electrons in the compound.* If we are representing polyatomic cations or anions, we must account for the charge on the ion. Specifically:

    1. For polyatomic cations, subtract one electron for each unit of positive charge. This accounts for the fact that the positive charge arises from electron loss.

    2. For polyatomic anions, add one electron for each unit of negative charge. This accounts for excess negative charge resulting from electron gain.

Step 3. *Connect the central atom to each of the surrounding atoms with single bonds.* Connection of the atoms uses two electrons per bond. Keep a count of how many valence electrons you placed in the Lewis structure so as not to exceed the previously determined number.

Step 4. *Place electrons as lone pairs around the terminal atoms to complete the octet for each.* Remember that hydrogen only needs two electrons. Hydrogen's electron requirement is satisfied simply by a single bond between it and the central atom. Remember to keep a count of electrons used and do not exceed the number of available electrons. Once the terminal atoms have their octet rule satisfied, do the same for the central atom (if you have enough electrons to do so).

Step 5. *If the octet rule is not satisfied for all the atoms, move one or more lone pairs on a terminal atom in to make a bond with the central atom.* Double and triple bonds may be formed to ensure that all atoms have an octet.

Step 6. *After you are satisfied with the Lewis structure that you have constructed, perform a final electron count.* Count each electron used in the structure; the total must match the number of electrons available.

Now, let us see how these guidelines are applied in the examples that follow.

LEARNING GOAL

6 Draw Lewis structures for covalent compounds and polyatomic ions.

The central atom is often the element farthest to the left and/or lowest in the periodic table.

The central atom is often the element in the compound for which there is only one atom.

Hydrogen is *never* the central atom.

Sometimes the octet rule is satisfied for all atoms but electrons are left over from the valence electron count. This problem is addressed later in this section under Lewis Structures and Exceptions to the Octet Rule.

| EXAMPLE 3.10 | **Drawing Lewis Structures of Covalent Compounds** | LEARNING GOAL |

Draw the Lewis structure of carbon dioxide, CO_2.

6 Draw Lewis structures for covalent compounds and polyatomic ions.

Solution

Step 1. Draw a skeletal structure of the molecule, arranging the atoms in their most probable order.

For CO_2, two possibilities exist:

$$C\quad O\quad O \quad \text{and} \quad O\quad C\quad O$$

Our strategy dictates that the least electronegative atom (the one written first in the formula), in this case carbon, is the central atom. Hence, the skeletal structure O   C   O may be presumed correct.

Step 2. Next, determine the number of valence electrons on each atom and add them to arrive at the total for the compound.
For CO_2,

$$1\,\text{C atom}\ \times 4\,\text{valence electrons} = \ \ 4\,e^-$$
$$2\,\text{O atoms} \times 6\,\text{valence electrons} = 12\,e^-$$
$$\overline{\phantom{2\,\text{O atoms} \times 6\,\text{valence electrons} = }16\,e^-\ \text{total}}$$

Step 3. Now use electron pairs to connect the central atom, C, to each oxygen with a single bond.

$$O:C:O$$

Step 4. Distribute the electrons around the terminal atoms in pairs, in an attempt to satisfy the octet rule.

$$:\overset{..}{\underset{..}{O}}:C:\overset{..}{\underset{..}{O}}:$$

Step 5. At this point, all sixteen electrons have been used. This structure satisfies the octet rule for each oxygen atom, but not the carbon atom (only four electrons surround the carbon). Therefore, this structure is modified by moving two electrons from each oxygen atom to a position between C and O, so that each oxygen and carbon atom is surrounded by eight electrons, and the octet rule is satisfied.

$$\overset{..}{\underset{..}{O}}::C::\overset{..}{\underset{..}{O}}$$

In this structure, four electrons (two electron pairs) are located between C and each O, and these electrons are shared in covalent bonds. Because a **single bond** is composed of two electrons (one electron pair) and four electrons "bond" the carbon atom to each oxygen atom in this structure, there must be two bonds between each oxygen atom and the carbon atom, a **double bond:**

The notation for a single bond : is equivalent to — (one pair of electrons).
The notation for a double bond : : is equivalent to = (two pairs of electrons).

The Lewis structure can be written with dashes to indicate bonds,

$$\overset{..}{\underset{..}{O}}=C=\overset{..}{\underset{..}{O}}$$

Step 6. A final electron count indicates that there are eight electron pairs (4 bonding pairs and 4 lone pairs), and they correspond to sixteen valence electrons (8 pair $\times$ 2 e^-/pair). Furthermore, there are eight electrons around each atom, and the octet rule is satisfied. Therefore,

$$\overset{..}{\underset{..}{O}}=C=\overset{..}{\underset{..}{O}}$$

is a satisfactory way to depict the structure of CO_2.

Helpful Hint: Each bond represents a bonding electron pair (2 electrons).

A Medical Perspective

Blood Pressure and the Sodium Ion/Potassium Ion Ratio

When you have a physical exam, the physician measures your blood pressure. This indicates the pressure of blood against the walls of the blood vessels each time the heart pumps. A blood pressure reading is always characterized by two numbers. With every heartbeat there is an increase in pressure; this is the systolic blood pressure. When the heart relaxes between contractions, the pressure drops; this is the diastolic pressure. Thus, the blood pressure is expressed as two values—for instance, 117/72—measured in millimeters (mm) of mercury. Hypertension is simply defined as high blood pressure. To the body it means that the heart must work too hard to pump blood, and this can lead to heart failure or heart disease.

Heart disease accounts for 50% of all deaths in the United States. Epidemiological studies correlate the following major risk factors with heart disease: heredity, gender, race, age, diabetes, cigarette smoking, high blood cholesterol, and hypertension. Obviously, we can do little about our age, gender, and genetic heritage, but we can stop smoking, limit dietary cholesterol, and maintain a normal blood pressure.

The number of Americans with hypertension is alarmingly high: sixty-seven million adults and children. Sixteen million of these individuals take medication to control blood pressure, at a cost of nearly 2.5 billion dollars each year. In many cases, blood pressure can be controlled without medication by increasing physical activity, losing weight, decreasing consumption of alcohol, and limiting intake of sodium.

It has been estimated that the average American ingests 7.5–10 grams (g) of salt (NaCl) each day. Because NaCl is about 40% (by mass) sodium ions, this amounts to 3–4 g of sodium daily. Until 1989, the Food and Nutrition Board of the National Academy of Sciences National Research Council's defined estimated safe and adequate daily dietary intake (ESADDI) of sodium ions was 1.1–3.3 g. Clearly, Americans exceed this recommendation.

Recently, studies have shown that excess sodium is not the sole consideration in the control of blood pressure. More important is the sodium ion/potassium ion (Na^+/K^+) ratio. That ratio should be about 0.6; in other words, our diet should contain about 67% more potassium than sodium. Does the typical American diet fall within this limit? Definitely not! Young American males (25–30 years old) consume a diet with a $Na^+/K^+ = 1.07$, and the diet of females of the same age range has a $Na^+/K^+ = 1.04$. It is little wonder that so many Americans suffer from hypertension.

How can we restrict sodium in the diet, while increasing potassium? A variety of foods are low in sodium and high in potassium. These include fresh fruits and vegetables and fruit juices, a variety of cereals, unsalted nuts, and cooked dried beans (legumes). Unfortunately, some high-sodium, low-potassium foods are very popular. Most of these are processed or prepared foods. This points out how difficult it can be to control sodium in the diet. The majority of the sodium we ingest comes from commercially prepared foods. The consumer must read the nutritional information printed on cans and packages to determine whether the sodium levels are within acceptable limits.

For Further Understanding

▶ Find several processed food products, and use the labels to calculate the sodium ion/potassium ion ratio.

▶ Describe each product you have chosen in terms of its suitability for inclusion in the diet of a person with moderately elevated blood pressure.

Practice Problem 3.10

Draw a Lewis structure for each of the following compounds:

    a. water   b. methane

▶ For Further Practice: **Questions 3.79 and 3.80.**

Many compounds have more than one central atom. Example 3.11 illustrates how to draw structures for these compounds.

Lewis Structures of Polyatomic Ions

The strategies for writing the Lewis structures of polyatomic ions are similar to those for neutral compounds. There is, however, one major difference: the charge on the ion must be accounted for when computing the total number of valence electrons.

LEARNING GOAL

6 Draw Lewis structures for covalent compounds and polyatomic ions.

EXAMPLE 3.11	Drawing Lewis Structures for Compounds with Multiple Central Atoms

LEARNING GOAL

Draw the Lewis structure for ethane, C_2H_4.

6 Draw Lewis structures for covalent compounds and polyatomic ions.

Solution

Step 1. Carbon often forms carbon-carbon chains, and hydrogen atoms must occupy terminal positions. This gives the skeletal structure:

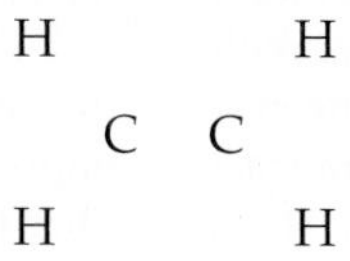

$$\begin{array}{ccc} H & & H \\ & C \quad C & \\ H & & H \end{array}$$

Step 2. Determine the number of valence electrons for each atom, and add them to obtain the total for the compound.

$$\underline{\begin{array}{l} 2\,C\ atoms \times 4\ valence\ electrons\ =\ 8\ e^- \\ 4\,H\ atoms \times 1\ valence\ electrons\ =\ 4\ e^- \end{array}}$$

$$12\ e^-\ total$$

Step 3. Connect the atoms with single bonds:

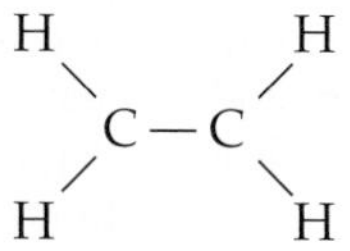

$$\begin{array}{ccc} H & & H \\ \diagdown & & \diagup \\ & C-C & \\ \diagup & & \diagdown \\ H & & H \end{array}$$

Because each bond represents a bonding electron pair, it can be determined that ten electrons (5 bonds $\times$ 2 e$^-$/bond = 10 e$^-$) are represented in this step.

Step 4. Since hydrogen's octet is satisfied by a single bond, additional electrons cannot be added to the terminal hydrogens. The two electrons remaining can be added to one of the carbons to satisfy its octet as follows:

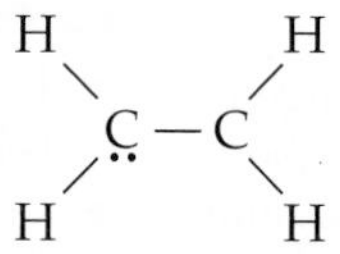

$$\begin{array}{ccc} H & & H \\ \diagdown & & \diagup \\ & \overset{..}{C}-C & \\ \diagup & & \diagdown \\ H & & H \end{array}$$

Step 5. The carbon on the right has only six valence electrons. The lone pair of electrons on the left carbon can be modified by moving them to a position where the carbon atoms form a double bond.

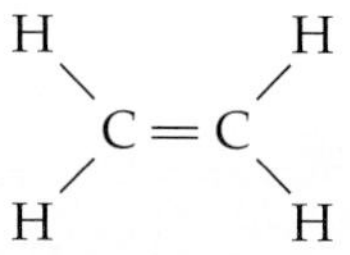

$$\begin{array}{ccc} H & & H \\ \diagdown & & \diagup \\ & C=C & \\ \diagup & & \diagdown \\ H & & H \end{array}$$

Step 6. A final electron count will verify that the number of electrons used in the structure matches the number of electrons determined in step 2. The six bonding pairs correspond to twelve valence electrons (6 bonds $\times$ 2 e$^-$/bond = 12 electrons). Therefore, the structure above is the correct Lewis structure.

Helpful Hint: Compounds that consist of only carbon and hydrogen (hydrocarbons) are often written to convey information concerning connectivity. The compound in this example can be written as CH_2CH_2 (a condensed formula). Writing it this way indicates that the carbons are connected to each other and that each carbon has two hydrogen atoms connected to it.

Practice Problem 3.11

Draw a Lewis structure for each of the following compounds:

a. N_2H_4 b. $CH_3CH_2CH_3$ c. HCN

▶ For Further Practice: **Questions 3.83 and 3.84.**

| EXAMPLE 3.12 | **Drawing Lewis Structures of Polyatomic Anions** | LEARNING GOAL |

Draw the Lewis structure of the carbonate ion, CO_3^{2-}.

6 Draw Lewis structures for covalent compounds and polyatomic ions.

Solution

Step 1. Carbon is less electronegative than oxygen. Therefore, carbon is the central atom. The carbonate ion has the following skeletal structure and charge:

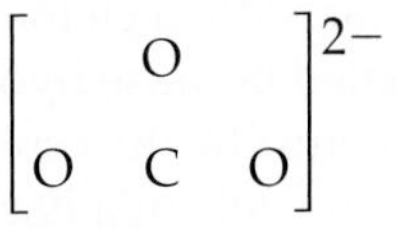

$$\begin{bmatrix} & O & \\ O & C & O \end{bmatrix}^{2-}$$

The use of square brackets with the charge outside of the brackets is standard when drawing Lewis structures for polyatomic anions.

Step 2. The total number of valence electrons is determined by adding one electron for each unit of negative charge:

$$
\begin{array}{ll}
1\text{ C atom } \times 4 \text{ valence electrons} & = \ 4\text{ e}^- \\
3\text{ O atoms} \times 6 \text{ valence electrons} & = 18\text{ e}^- \\
\underline{\text{electrons for } 2^- \text{ charge}} & = \underline{\ 2\text{ e}^-} \\
& 24\text{ e}^-\text{ total}
\end{array}
$$

Step 3. Connect the O atoms to the central atom using single bonds. This step uses six of the valence electrons.

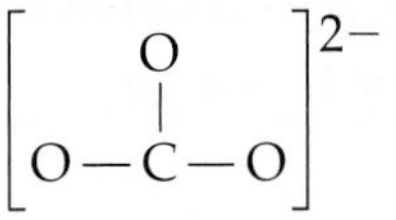

$$\begin{bmatrix} & O & \\ & | & \\ O & -C- & O \end{bmatrix}^{2-}$$

Step 4. Start placing lone-pair electrons around the terminal atoms, giving each an octet. Do not exceed the twenty-four available electrons.

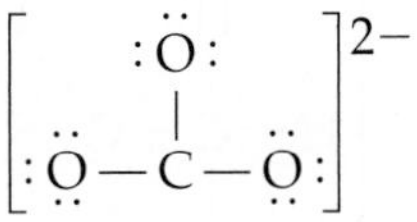

$$\begin{bmatrix} & :\ddot{O}: & \\ & | & \\ :\ddot{O} & -C- & \ddot{O}: \end{bmatrix}^{2-}$$

Step 5. The central atom only has six valence electrons surrounding it. Move a lone pair from one of the O atoms to form another bond with C. In this example, a lone pair from the top O was moved to make a second bond between it and the C.

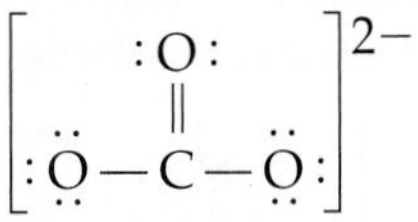

$$\begin{bmatrix} & :O: & \\ & \| & \\ :\ddot{O} & -C- & \ddot{O}: \end{bmatrix}^{2-}$$

Step 6. A final electron count indicates that each atom has eight valence electrons surrounding it. Overall, there are four bonding electron pairs (4 bonds $\times$ 2e$^-$/bond = 8e$^-$) and eight lone electron pairs (8 lone pairs $\times$ 2e$^-$/lone pair = 16e$^-$), which when combined total twenty-four electrons.

Practice Problem 3.12

Draw the Lewis structure of O_2^{2-}.

▶ For Further Practice: **Questions 3.85 and 3.86.**

EXAMPLE 3.13   **Drawing Lewis Structures of Polyatomic Anions**

LEARNING GOAL

Draw the Lewis structure of the acetate ion, CH_3COO^-.

6   Draw Lewis structures for covalent compounds and polyatomic ions.

Solution

This species has multiple central atoms and a negative charge.

Step 1. The acetate ion, a commonly encountered anion, has a skeletal structure that is more complex than any of the examples we have studied thus far. Which element should we choose as the central atom? We have three choices: H, O, and C. H is eliminated because hydrogen can never be the central atom. Oxygen is more electronegative than carbon, so carbon must be the central atom. There are two carbon atoms; often they are joined. Further clues are obtained from the formula itself; CH_3COO^- implies three hydrogen atoms attached to the first carbon atom and two oxygen atoms joined to the second carbon. A plausible skeletal structure is:

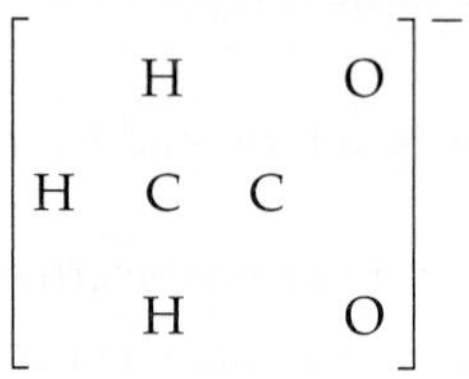

$$
\left[
\begin{array}{ccc}
 & H & O \\
H & C & C \\
 & H & O
\end{array}
\right]^-
$$

Step 2. The pool of valence electrons for anions is determined by adding one electron for each unit of negative charge:

$$
\begin{aligned}
2\ \text{C atoms} \times 4\ \text{valence electrons} &= 8\ e^- \\
3\ \text{H atoms} \times 1\ \text{valence electron} &= 3\ e^- \\
2\ \text{O atoms} \times 6\ \text{valence electrons} &= 12\ e^- \\
\underline{1\ \text{electron for } 1^-\ \text{charge}} &= \underline{1\ e^-} \\
24\ e^-\ \text{total}
\end{aligned}
$$

Step 3. Use bonds to connect the terminal atoms to the two central carbon atoms.

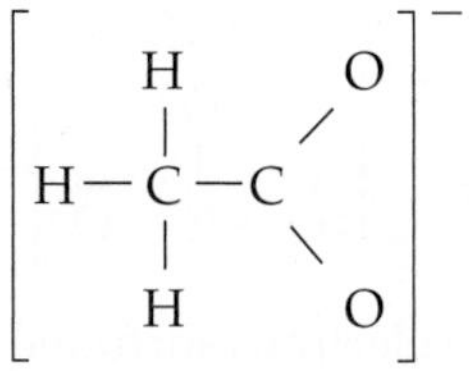

$$
\left[
\begin{array}{c}
H \\
| \\
H - C - C \underset{\diagdown}{\overset{\diagup}{}} \begin{array}{c} O \\ \\ O \end{array} \\
| \\
H
\end{array}
\right]^-
$$

This step used twelve valence electrons (6 bonds × $2e^-$/bond).

Step 4. Starting with terminal O atoms, place lone-pair electrons around the atoms to give each an octet. (Remember to keep in mind the twenty-four electrons available, and do not put lone-pair electrons around H.)

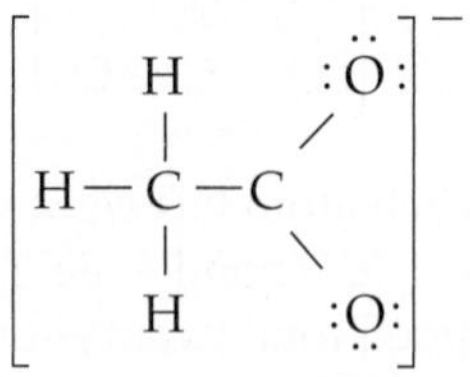

$$
\left[
\begin{array}{c}
H \\
| \\
H - C - C \underset{\diagdown}{\overset{\diagup}{}} \begin{array}{c} :\ddot{O}: \\ \\ :\ddot{O}: \end{array} \\
| \\
H
\end{array}
\right]^-
$$

This structure uses all twenty-four electrons, but one of the C atoms only has six valence electrons.

Step 5. Move one of the lone pairs from an O to create an additional bond between C and O.

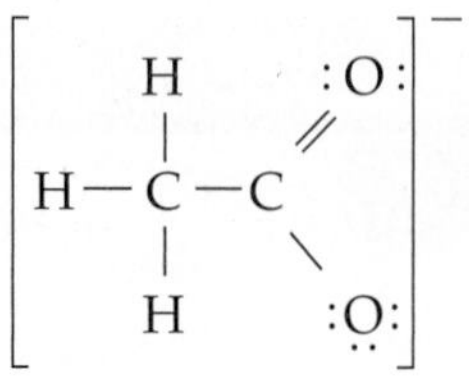

$$
\left[
\begin{array}{c}
H \\
| \\
H - C - C \underset{\diagdown}{\overset{\diagup\diagup}{}} \begin{array}{c} :O: \\ \\ :\ddot{O}: \end{array} \\
| \\
H
\end{array}
\right]^-
$$

Step 6. A final electron count indicates that there are seven bonding electron pairs (7 bonds $\times$ 2e$^-$/bond = 14e$^-$) and five lone electron pairs (5 lone pairs $\times$ 2e$^-$/lone pair = 10e$^-$), which confirms that all twenty-four electrons are used. The octet rule is satisfied for all atoms in the Lewis structure.

Practice Problem 3.13

Draw a Lewis structure illustrating the bonding in each of the following polyatomic ions:

a. the bicarbonate ion, HCO_3^-
b. the phosphate ion, PO_4^{3-}

▶ For Further Practice: **Questions 3.87 and 3.88.**

Lewis Structure, Stability, Multiple Bonds, and Bond Energies

Hydrogen, oxygen, and nitrogen are present in the atmosphere as diatomic gases, H_2, O_2, and N_2. Although they are all covalent molecules, their stability and reactivity are quite different. Hydrogen is an explosive material, sometimes used as a fuel. Oxygen, although more stable than hydrogen, reacts with fuels in combustion. The explosion of the space shuttle *Challenger* resulted from the reaction of massive amounts of hydrogen and oxygen. Nitrogen, on the other hand, is extremely nonreactive. Because nitrogen makes up about 80% of the atmosphere, it dilutes the oxygen, which accounts for only about 20% of the atmosphere.

The great difference in reactivity among these three gases can be explained, in part, in terms of their bonding characteristics. The Lewis structure for H_2 (two valence electrons, one from each atom) is

$$H - H$$

For oxygen (twelve valence electrons, six from each atom), the only Lewis structure that satisfies the octet rule is

$$\ddot{O} = \ddot{O}$$

The Lewis structure of N_2 (ten valence electrons, five from each atom) must be

$$: N \equiv N :$$

Therefore:

H_2 has a *single bond* (two bonding electrons).
O_2 has a *double bond* (four bonding electrons).
N_2 has a *triple bond* (six bonding electrons).

A **triple bond,** in which three pairs of electrons are shared by two atoms, is very stable. More energy is required to break a triple bond than a double bond, and a double bond is stronger than a single bond. Stability is related to the bond energy. The **bond energy** is the amount of energy, in units of kilocalories (kcal) or kilojoules (kJ), required to break a bond holding two atoms together. Bond energy is therefore a *measure* of stability. The values of bond energies decrease in the order *triple bond > double bond > single bond*. The H—H bond energy is 436 kJ/mol. This amount of energy is necessary to break a H—H bond. In contrast, the O=O bond energy is 499 kJ/mol, and the N≡N bond energy is 941 kJ/mol.

The bond length is related to the presence or absence of multiple bonding. The distance of separation of two nuclei is greatest for a single bond, less for a double bond, and still less for a triple bond. The *bond length* decreases in the order *single bond > double bond > triple bond.*

LEARNING GOAL

7 Explain how the presence or absence of multiple bonding relates to bond length, bond energy, and stability.

The term *bond order* is sometimes used to distinguish among single, double, and triple bonds. A bond order of one corresponds to a single bond, two corresponds to a double bond, and three corresponds to a triple bond.

The mole (mol) is simply a unit denoting quantity. Just as a *dozen* eggs represents twelve eggs, a mole of bonds is 6.022×10^{23} bonds (see Chapter 4).

pentane
(C_5H_{12})

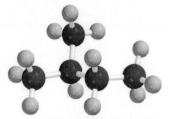

2-methylbutane
(C_5H_{12})

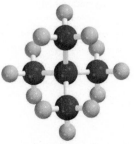

2,2-dimethylpropane
(C_5H_{12})

The compound pentane, above, is the "straight chain" isomer. Notice that the carbons are attached in a chain; however, they are not actually in a *straight* line.

Question 3.3   Contrast a single and a double bond with regard to:
  a. distance of separation of the bonded nuclei
  b. strength of the bond

How are the distance of separation and bond strength related?

Question 3.4   Two nitrogen atoms in a nitrogen molecule are held together more strongly than the two chlorine atoms in a chlorine molecule. Explain this fact by comparing their respective Lewis structures.

Isomers

Isomers are compounds that share the same molecular formula but have different structures. Hydrocarbons (compounds that contain only hydrogen and carbon atoms) frequently exhibit this property. For example, C_4H_{10} has two isomeric forms. One isomer, termed butane, has a structure characterized by the four carbon atoms being linked in a chain; the other, termed methylpropane, has a three-carbon chain, with the fourth carbon attached to the middle carbon.

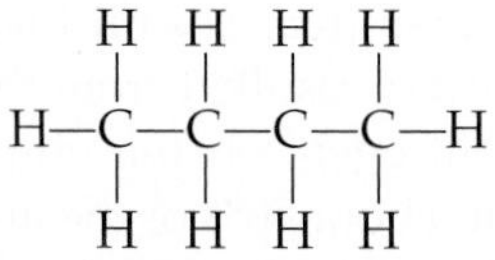

$$
\begin{array}{cccc}
 & H & H & H & H \\
 & | & | & | & | \\
H- & C- & C- & C- & C-H \\
 & | & | & | & | \\
 & H & H & H & H
\end{array}
$$

Butane (C_4H_{10})

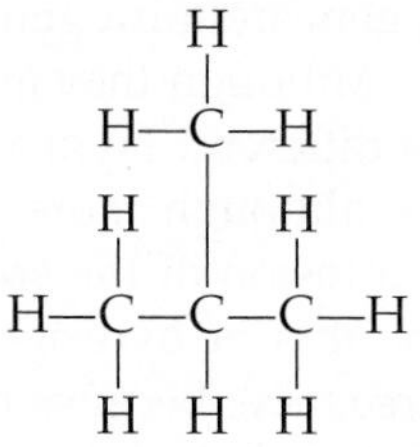

Methylpropane (C_4H_{10})

Notice that both structures, although clearly different, contain four carbon atoms and ten hydrogen atoms. Owing to their structural differences, each has a different melting point and boiling point. In fact, all of their physical properties differ. These differences in properties clearly show that, in fact, isomers are different compounds.

As the size of the hydrocarbon increases, the number of possible isomers increases dramatically. A five-carbon hydrocarbon has three isomers, but a thirty-carbon hydrocarbon has over 400 million possible isomers. Petroleum is largely a complex mixture of hydrocarbons, and the principal reason for this complexity lies in the tremendous variety of possible isomers present.

The three isomers of C_5H_{12} (a five-carbon hydrocarbon) are depicted in the figure in the margin.

Lewis Structures and Resonance

In some cases, we find that it is possible to write more than one Lewis structure that satisfies the octet rule for a particular compound. Consider sulfur dioxide, SO_2. Its skeletal structure is

O   S   O

Total valence electrons may be calculated as follows:

$$
\begin{aligned}
1 \text{ sulfur atom} \times 6 \text{ valence } e^- &= 6 \ e^- \\
+2 \text{ oxygen atoms} \times 6 \text{ valence } e^- &= 12 \ e^- \\
\hline
&18 \ e^- \text{ total}
\end{aligned}
$$

The resulting Lewis structures are

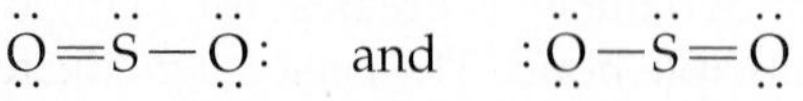

$$\ddot{O}=\ddot{S}-\ddot{O}: \quad \text{and} \quad :\ddot{O}-\ddot{S}=\ddot{O}$$

Both satisfy the octet rule. However, experimental evidence shows no double bond in SO_2. The two sulfur-oxygen bonds are equivalent. Apparently, neither structure accurately represents the structure of SO_2, and neither actually exists. The actual structure is said to be an average or *hybrid* of these two Lewis structures. When a compound has two or more Lewis structures that contribute to the real structure, we say that the compound displays the property of **resonance.** The contributing Lewis structures are *resonance forms.* The true structure, a hybrid of the resonance forms, is known as a *resonance hybrid* and may be represented as:

$$\ddot{\underset{..}{O}}=\ddot{S}-\ddot{\underset{..}{O}}: \quad \longleftrightarrow \quad :\ddot{\underset{..}{O}}-\ddot{S}=\ddot{\underset{..}{O}}$$

A common analogy might help to clarify this concept. A horse and a donkey may be crossbred to produce a hybrid, the mule. The mule doesn't look or behave exactly like either parent, yet it has attributes of both. The resonance hybrid of a molecule has properties of each resonance form but is not identical to any one form. Unlike the mule, resonance hybrids *do not actually exist.* Rather, they comprise a model that results from the failure of any one Lewis structure to agree with experimentally obtained structural information.

The presence of resonance enhances molecular stability. The more resonance forms that exist, the greater the stability of the molecule they represent. This concept is important in understanding the chemical reactions of many complex organic molecules and is used extensively in organic chemistry.

> It is a misconception to picture the real molecule as oscillating back and forth among the various resonance structures. Resonance is a modeling strategy designed to help us visualize electron arrangements too complex to be adequately explained by the simple Lewis structure.

EXAMPLE 3.14 **Drawing Resonance Hybrids of Covalently Bonded Compounds**

LEARNING GOAL

Draw the possible resonance structures of the nitrate ion, NO_3^-, and represent them as a resonance hybrid.

6 Draw Lewis structures for covalent compounds and polyatomic ions.

Solution

Step 1. Nitrogen is less electronegative than oxygen; therefore, nitrogen is the central atom and the skeletal structure is:

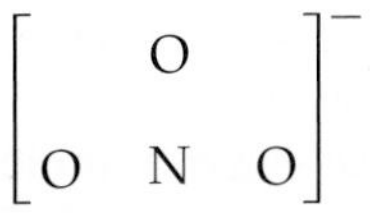

$$\left[\begin{array}{ccc} & O & \\ O & N & O \end{array}\right]^{-}$$

Step 2. The pool of valence electrons for anions is determined by adding one electron for each unit of negative charge:

$$\begin{array}{lll} 1\ \text{N atom} & \times\ 5\ \text{valence electrons} = & 5\ e^- \\ 3\ \text{O atoms} & \times\ 6\ \text{valence electrons} = & 18\ e^- \\ \underline{1\ \text{electron for 1}^-\ \text{charge}} & = & \underline{1\ e^-} \\ & & 24\ e^-\ \text{total} \end{array}$$

Step 3. First, connect O atoms to the N atom using single bonds.

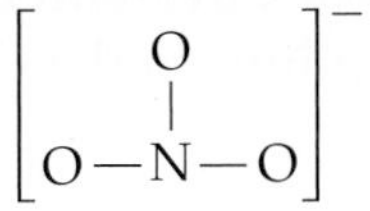

$$\left[\begin{array}{c} O \\ | \\ O-N-O \end{array}\right]^{-}$$

This step used twelve valence electrons (3 bonds $\times$ 2 e^-/bond).

Step 4. Place lone-pair electrons around the terminal O atoms until each atom has an octet. Do not exceed twenty-four electrons.

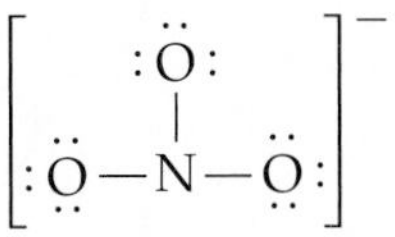

$$\left[\begin{array}{c} :\ddot{O}: \\ | \\ :\ddot{\underset{..}{O}}-N-\ddot{\underset{..}{O}}: \end{array}\right]^{-}$$

This structure uses twenty-four electrons; however, the central atom only has six valence electrons.

Continued…

Step 5. Shift a lone pair on one of the O atoms to become a bond between the O and N. To draw all possible resonance structures, this shift is drawn for each of the oxygen atoms.

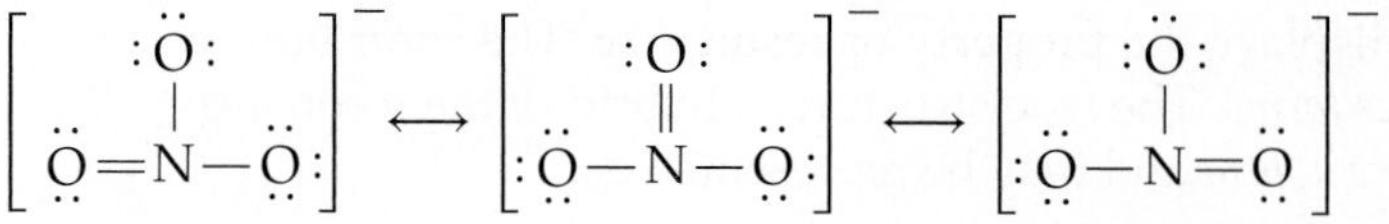

$$\left[\begin{array}{c} :\ddot{O}: \\ | \\ \ddot{O}=N-\ddot{O}: \end{array} \right]^{-} \longleftrightarrow \left[\begin{array}{c} :O: \\ \| \\ :\ddot{O}-N-\ddot{O}: \end{array} \right]^{-} \longleftrightarrow \left[\begin{array}{c} :\ddot{O}: \\ | \\ :\ddot{O}-N=\ddot{O} \end{array} \right]^{-}$$

Step 6. A final count of electrons indicates that there are four pairs of bonding electrons (4 bonds $\times$ 2 e$^-$/
bond = 8 e$^-$) and eight lone electron pairs (8 lone pairs $\times$ 2 e$^-$/lone pair = 16 e$^-$) in each of these
resonance structures. All 24 electrons are used, and the octet rule is obeyed for each atom.

Practice Problem 3.14

a. SeO$_2$, like SO$_2$, has two resonance forms. Draw their Lewis structures.

b. Explain any similarities between the structures for SeO$_2$ and SO$_2$ in light of periodic relationships.

▶ For Further Practice: **Questions 3.89 and 3.90.**

Lewis Structures and Exceptions to the Octet Rule

The octet rule is remarkable in its ability to realistically model bonding and structure in covalent compounds. But, like any model, it does not adequately describe
all systems. Beryllium, boron, and aluminum, in particular, tend to form compounds in which they are surrounded by fewer than eight electrons. This situation
is termed an *incomplete octet*.

Other molecules, such as nitric oxide:

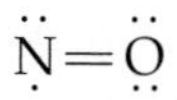

$$\ddot{N}=\ddot{O}$$

are termed *odd-electron* molecules. Note that it is impossible to pair all electrons
to achieve an octet simply because the compound contains an odd number of
valence electrons.

Elements in the third period and beyond may involve *d* orbitals and form an
expanded octet, with ten or even twelve electrons surrounding the central atom.
Example 3.15 illustrates the expanded octet.

EXAMPLE 3.15 | **Drawing Lewis Structures of Covalently Bonded
Compounds that Are Exceptions to the Octet Rule**

Draw the Lewis structure of SF$_4$.

LEARNING GOAL

6 Draw Lewis structures for covalent
compounds and polyatomic ions.

Solution

Step 1. Sulfur is the central atom with four fluorine atoms surrounding the sulfur.

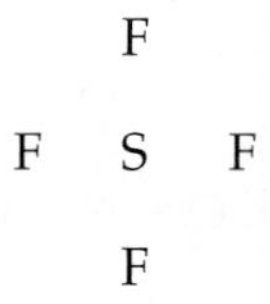

$$\begin{array}{ccc} & F & \\ F & S & F \\ & F & \end{array}$$

Step 2. The total number of valence electrons is:

$$1 \text{ sulfur atom} \quad \times 6 \text{ valence e}^- = \quad 6 \text{ e}^-$$
$$\underline{4 \text{ fluorine atoms} \times 7 \text{ valence e}^- = 28 \text{ e}^-}$$
$$= 34 \text{ e}^- \text{ total}$$

Step 3. Connect the four fluorine atoms with single bonds to the sulfur. This uses eight electrons (4 bonds $\times$ 2 e$^-$/bond).

$$\begin{array}{c} \text{F} \\ | \\ \text{F} - \text{S} - \text{F} \\ | \\ \text{F} \end{array}$$

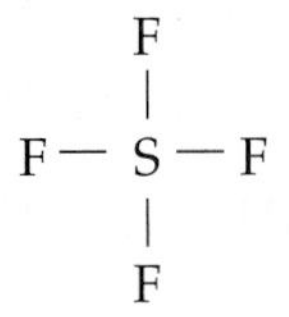

Step 4. Give each terminal atom eight valence electrons.

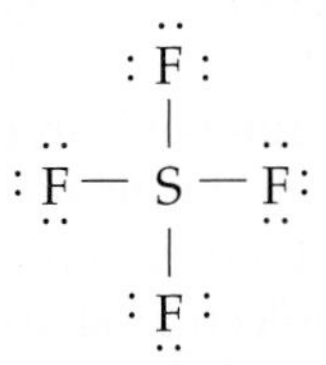

$$\begin{array}{c} :\ddot{\text{F}}: \\ | \\ :\ddot{\text{F}} - \text{S} - \ddot{\text{F}}: \\ | \\ :\ddot{\text{F}}: \end{array}$$

Step 5. The octet rule *is* satisfied for all atoms. However, we have only used thirty-two electrons and we must use all thirty-four electrons. The two extra electrons will be placed on the central atom.

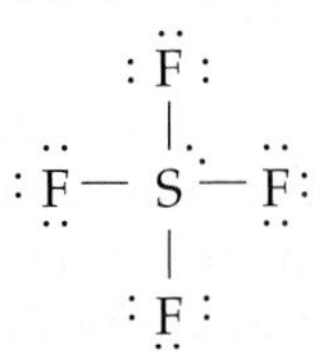

$$\begin{array}{c} :\ddot{\text{F}}: \\ |\cdot\cdot \\ :\ddot{\text{F}} - \text{S} - \ddot{\text{F}}: \\ | \\ :\ddot{\text{F}}: \end{array}$$

Step 6. A final count of electrons confirms that all thirty-four have been used. There are four pairs of bonding electrons (4 bonds $\times$ 2e$^-$/bond = 8 e$^-$) and thirteen lone electron pairs (13 lone pairs $\times$ 2e$^-$/lone pair = 26 e$^-$) in this Lewis structure. SF_4 is an example of a compound with an expanded octet. The central sulfur atom is surrounded by ten electrons.

Practice Problem 3.15

   a. Draw the Lewis structure of $SeCl_4$.
   b. Draw the Lewis structure of SF_6.
   c. Draw the Lewis structure of BCl_3.

▶ For Further Practice: **Questions 3.93–3.96.**

Lewis Structures and Molecular Geometry; VSEPR Theory

The shape of a molecule plays a large part in determining its properties and reactivity. We may predict the shapes of various molecules by inspecting their Lewis structures for the orientation of their electron pairs. The covalent bond, for instance, in which bonding electrons are localized between the nuclear centers of the atoms, is *directional*; the bond has a specific orientation in space between the bonded atoms. The specific orientation of electron pairs in covalent molecules imparts a characteristic shape to the molecules. Consider the following series of molecules whose Lewis structures are shown.

LEARNING GOAL

8   Use Lewis structures to predict the geometry of molecules.

Electrostatic forces in ionic bonds are *nondirectional*; they have no specific orientation in space.

BeH$_2$ H—Be—H   BeH$_2$ and BF$_3$ are exceptions
to the octet rule. The central
atoms are not surrounded
by eight electrons, and the

BF$_3$:Ḟ—B—Ḟ: octets are incomplete.

CH$_4$ H—C—H

NH$_3$ H—N̈—H

H$_2$O   H—Ö—H

The electron pairs around the central atom of the molecule arrange themselves to minimize electronic repulsion. This means that the electron pairs arrange themselves so that they can be as far from each other as possible. We may use this fact to predict molecular shape. This approach is termed the **valence-shell electron-pair repulsion (VSEPR) theory.**

Let's see how the VSEPR theory can be used to describe the bonding and structure of each of the preceding molecules.

BeH$_2$

As illustrated in the Lewis structure above, beryllium hydride has two bonded atoms around the beryllium atom. These bonding electron pairs have minimum repulsion if they are located as far apart as possible while still bonding the hydrogen atoms to the central atom. This results in a **linear** geometric structure or shape. The *bond angle,* the angle between the H—Be and Be—H bonds, formed by the two bonding pairs is 180° (Figure 3.5).

BeF$_2$, CS$_2$, and HCN are other examples of molecules that exhibit linear geometry. Multiple bonds are treated identically to single bonds in VSEPR theory. For example, the Lewis structure of HCN is H—C≡N:. The bonded atoms on either side of carbon are positioned 180° from each other. Instead of counting bonding electrons to predict molecular shape, it is more appropriate to count bonded atoms. The central atom has two bonded atoms.

BF$_3$

Boron trifluoride has three bonded atoms around the central atom. The Lewis structure, as illustrated above, shows boron as electron deficient. Placing the bonding electron pairs in a plane, forming a triangle, minimizes the electron-pair repulsion in this molecule, as depicted in Figure 3.6.

Such a structure is **trigonal planar,** and each F—B—F bond angle is 120°. We also find that compounds with central atoms in the same group of the periodic table have similar geometry. Aluminum, in the same group as boron, produces compounds such as AlH$_3$, which is also trigonal planar.

CH$_4$

Methane has four bonded atoms around carbon. Here, minimum electron repulsion is achieved by arranging the electrons at the corners of a tetrahedron (Figure 3.7). Each H—C—H bond angle is 109.5°. Methane has a three-dimensional **tetrahedral** structure or shape. Silicon, in the same group as carbon, forms compounds such as SiCl$_4$ and SiH$_4$ that also have tetrahedral shapes.

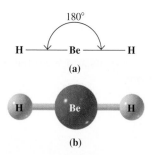

Figure 3.5 Bonding and geometry in beryllium hydride, BeH$_2$. (a) Linear geometry in BeH$_2$. (b) Computer-generated model of linear BeH$_2$.

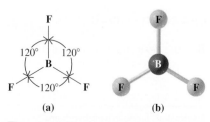

Figure 3.6 Bonding and geometry in boron trifluoride, BF$_3$. (a) Trigonal planar geometry in BF$_3$. (b) Computer-generated model of trigonal planar BF$_3$.

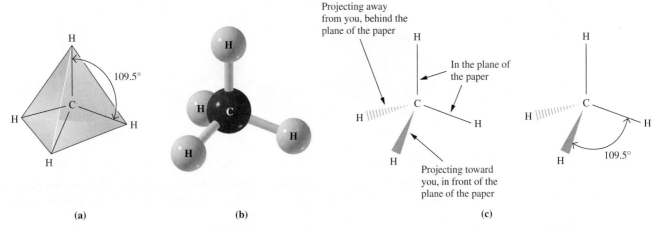

Figure 3.7 Representations of the three-dimensional structure of methane, CH_4. (a) Tetrahedral methane structure. (b) Computer-generated model of tetrahedral methane. (c) Three-dimensional representation showing the H—C—H bond angle.

NH₃

Ammonia has three bonded atoms and one lone pair about the central atom. In contrast to methane, in which all four electron pairs are bonding, ammonia has three pairs of bonding electrons and one nonbonding lone pair of electrons. We might expect CH_4 and NH_3 to have electron-pair arrangements that are similar but not identical. The lone pair in ammonia is more negative than the bonding pairs because some of the negative charge on the bonding pairs is offset by the presence of the hydrogen atoms with their positive nuclei. Thus, the arrangement of electron pairs in ammonia is distorted.

The hydrogen atoms in ammonia are pushed closer together than in methane (Figure 3.8). The bond angle is 107° because lone pair–bond pair repulsions are greater than bond pair–bond pair repulsions. The structure or shape is termed *trigonal pyramidal*, and the molecule is termed a **trigonal pyramidal** molecule.

H₂O

Water has two bonded atoms and two lone pairs about the central atom. These four electron pairs are approximately tetrahedral to each other; however, because of the difference between bonding and nonbonding electrons, noted earlier, the tetrahedral relationship is only approximate.

The **bent** structure has a bond angle of 104.5°, which is 5° smaller than the tetrahedral angle, because of the repulsive effects of the lone pairs of electrons (Figure 3.9).

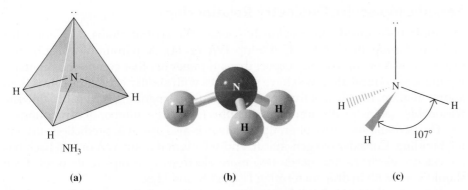

Figure 3.8 The structure of the ammonia molecule. (a) Trigonal pyramidal ammonia structure. (b) Computer-generated model of trigonal pyramidal ammonia. (c) A three-dimensional sketch showing the H—N—H bond angle.

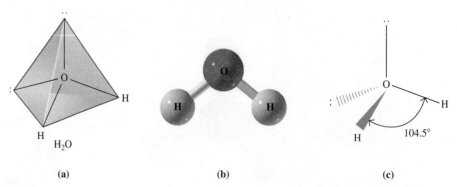

Figure 3.9 The structure of the water molecule. (a) Bent water structure. (b) Computer-generated model of bent water. (c) A three-dimensional sketch showing the H—O—H bond angle in water.

TABLE 3.5 Molecular Geometry: The Shape of a Molecule Is Affected by the Number of Bonded Atoms and the Number of Nonbonded Lone Electron Pairs Around the Central Atom

Bonded Atoms	Nonbonding Lone Electron Pairs	Bond Angle	Molecular Geometry	Example	Structure or Shape
2	0	180°	Linear	CO_2	
3	0	120°	Trigonal planar	SO_3	
2	1	<120°	Bent	SO_2	
4	0	109.5°	Tetrahedral	CH_4	
3	1	~107°	Trigonal pyramidal	NH_3	
2	2	~104.5°	Bent	H_2O	

LEARNING GOAL

8 Use Lewis structures to predict the geometry of molecules.

The characteristics of linear, trigonal planar, trigonal pyramidal, bent, and tetrahedral shapes are summarized in Table 3.5.

Periodic Molecular Geometry Relationships

The molecules considered previously contain the central atoms Be (Group IIA or 2), B (Group IIIA or 13), C (Group IVA or 14), N (Group VA or 15), and O (Group VIA or 16). We may expect that a number of other compounds containing the same central atom will have structures with similar geometries. This is an approximation, not always true, but still useful in expanding our ability to write reasonable, geometrically accurate structures for a large number of compounds.

The periodic similarity of group members is also useful in predictions involving bonding. Consider oxygen, sulfur, and selenium (Group VIA or 16). Each has six valence electrons and needs two more electrons to complete its octet. Each should react with hydrogen, forming H_2O, H_2S, and H_2Se.

If we recall that H_2O is a bent molecule with the following Lewis structure and shape,

$$H—\overset{\cdot\cdot}{\underset{\cdot\cdot}{O}}—H \qquad \overset{O}{H \quad H}$$

it follows that H_2S and H_2Se would also be bent molecules with similar Lewis structures and shapes.

$$H - \overset{..}{\underset{..}{S}} - H \qquad H\diagup\overset{S}{}\diagdown H$$

$$H - \overset{..}{\underset{..}{Se}} - H \qquad H\diagup\overset{Se}{}\diagdown H$$

This logic applies equally well to the other representative elements.

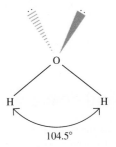

104.5°

Known bond angle, 104.5°

Question 3.5 Draw the Lewis structures and shapes of each of the following molecules. Identify the molecular geometry of each shape using VSEPR.
  a. PH_3 b. SiH_4

Question 3.6 Draw the Lewis structures and shapes of each of the following molecules. Identify the molecular geometry of each shape using VSEPR.
  a. C_2H_4 b. C_2H_2

More Complex Molecules

For compounds containing more than one central atom, geometry is determined at each central atom. Dimethyl ether, CH_3—O—CH_3, has three distinct central atoms as shown in the Lewis structure

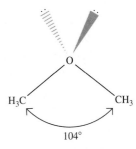

104°

Predicted bond angle is approximately 104°

$$H-\overset{\displaystyle H}{\underset{\displaystyle H}{C}} - \overset{..}{\underset{..}{O}} - \overset{\displaystyle H}{\underset{\displaystyle H}{C}}-H$$

Figure 3.10 A comparison of the bonding in water and dimethyl ether.

Each carbon atom has four bonded atoms, giving a tetrahedral geometry with bond angles of 109.5°.

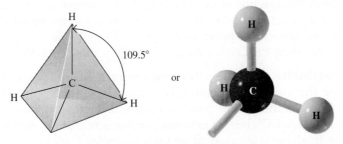

109.5°   or

The oxygen has two bonded atoms and two lone pairs, giving a bent geometry with a 104° bond angle, as shown in Figure 3.10.

Trimethylamine, $(CH_3)_3N$, is a member of the amine family. Its Lewis structure is depicted as

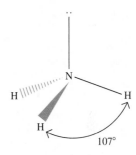

107°

Known bond angle, 107°

$$H-\overset{\displaystyle H}{\underset{\displaystyle H}{C}} - \overset{..}{N} - \overset{\displaystyle H}{\underset{\displaystyle H}{C}}-H$$
$$\underset{\displaystyle H}{\overset{|}{\underset{|}{H-C-H}}}$$

As in the case of dimethyl ether, two different central atoms are present. Carbon and nitrogen determine the geometry of amines. In this case, each carbon atom is bonded to four atoms and has a tetrahedral geometry like that shown for methane. The nitrogen atom should have the bonded carbon atoms in a pyramidal arrangement, similar to the hydrogen atoms in ammonia, as seen in Figure 3.11. This creates a pyramidal geometry around nitrogen. H—N—H bond angles in ammonia are 107°; experimental information shows a very similar C—N—C bond angle in trimethylamine.

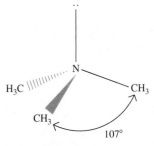

107°

Predicted bond angle is approximately 107°

Figure 3.11 A comparison of the bonding in ammonia and trimethylamine.

It is essential to represent a molecule in its correct geometric form, using the Lewis and VSEPR theories, in order to understand its physical and chemical behavior. In the rest of this section, we use these models to predict molecular behavior.

Lewis Structures and Polarity

As discussed in Section 3.1, covalent bonds can be polar or nonpolar. In this section, we will examine all the bonds that make up a molecule to determine if the *molecule* is polar or nonpolar. Polar molecules, when placed in an electric field, align themselves with the field (Figure 3.12). These molecules are said to have a *dipole* (having two "poles" or ends, one pole is more negative and the other pole is more positive). Nonpolar molecules do not align with an electric field. It is important to determine if a molecule is polar or nonpolar because this information helps predict properties of solubility as well as the melting point and boiling point. These properties will be discussed in Section 3.5.

The hydrogen molecule is the simplest nonpolar molecule. The electrons that make up the bond are shared equally between the two atoms (see Figure 3.2). Any molecule made up of only nonpolar bonds is a nonpolar molecule.

A diatomic molecule that is made up of two elements with different electronegativities will contain a polar bond. Since this bond is polar, the molecule is polar. We used the partial positive and partial negative symbols (δ^+ and δ^-) in Section 3.1 to represent the polar bond. We can also represent this polar bond using a vector (arrow) pointing in the direction of the most electronegative element in the bond.

$$\overset{\delta^+}{H}\text{—}\overset{\delta^-}{F} \qquad \overset{\longrightarrow}{H\text{—}F}$$

If a molecule contains more than one polar bond, the molecule *may* or *may not* be polar. Think of the polarity of a bond as a rope tied around the central atom pulling in the direction of the more electronegative atom. (Or in the case of lone pairs, the pull will always be in the direction of the lone pair.) Each terminal-atom to central-atom connection is a separate tug-of-war. If all the pulls are equal, the molecule is nonpolar; otherwise, the molecule is polar. To determine if a molecule with polar bonds is polar or nonpolar, draw the Lewis structure and follow these guidelines:

Molecules that have no lone pair on the central atom and where all terminal atoms are the same are *nonpolar*. For example, the molecule CO_2 has two bonded O atoms and no lone pairs on the central atom. It is linear and nonpolar. The molecule CCl_4 has no lone pairs on the central atom and four bonded Cl atoms. Its shape is tetrahedral, and it is nonpolar.

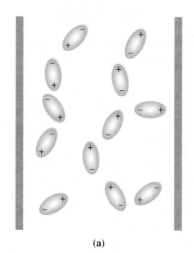

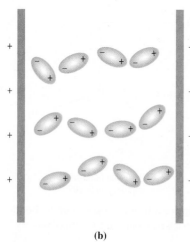

(a)

(b)

Figure 3.12 The green ovals represent polar molecules (a) in the absence of an electric field and (b) under the influence of an electric field.

Molecules with one lone pair on the central atom are *polar*. Ammonia is an example. Its nitrogen atom has three bonded H atoms and one lone pair. The shape of ammonia is trigonal pyramidal. Because the nitrogen atom is more electronegative than the H atoms, the vectors (arrows) point toward nitrogen. The pull of lone-pair electrons is always away from the H atoms, and ammonia is a polar molecule.

Molecules that have more than one lone pair on the central atom are *usually polar*, but there are exceptions. The Lewis structure of water has two lone pairs and two bonded H atoms on the central atom of oxygen. Water is a bent molecule, and it is polar.

The electron density is shifted away from the hydrogens toward oxygen in the water molecule.

Molecules that are made up of only carbon and hydrogen (hydrocarbons) are nonpolar. The C—H bond has an electronegativity difference of 0.4; hence, the C—H bond is essentially nonpolar. Furthermore, the combined effects of the numerous C—H bonds effectively cancel, rendering the molecule nonpolar.

EXAMPLE 3.16	Determining if a Molecule Is Polar Covalent or Nonpolar Covalent

Is PCl_3 a polar or nonpolar covalent compound?

LEARNING GOAL

9 Describe the role that molecular geometry plays in determining the polarity of compounds.

Solution

To determine polarity, the geometry must be determined. To determine geometry, draw the Lewis structure.

Step 1. Follow all steps to correctly draw the Lewis structure.

Skeletal Structure	Count Valence Electrons	Distribute Valence Electrons
Cl Cl P Cl	$1\,P \times 5\,e^- = 5\,e^-$ $3\,Cl \times 7\,e^- = 21\,e^-$ _____ $26\,e^-$ total	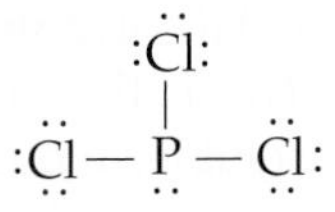 $:\!\ddot{C}l\!:$ $:\!\ddot{C}l\!-\!P\!-\!\ddot{C}l\!:$

Step 2. Examine the central atom to determine if the molecule is polar. The molecule has three chlorine atoms and one lone pair. The one lone pair makes the trigonal pyramidal molecule polar covalent.

Helpful Hint: Vectors can also be used in determining the polarity of this compound. The vectors should point in the direction of each electronegative Cl atom.

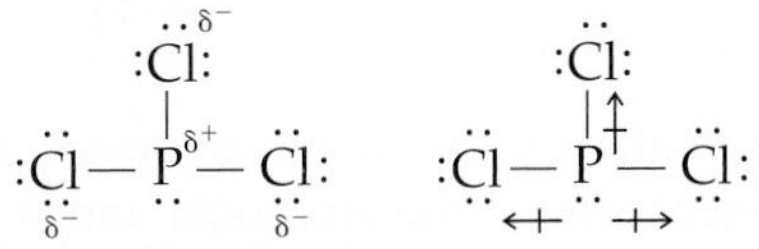

Practice Problem 3.16

Determine whether the $AsCl_3$ molecule is polar.

▶ For Further Practice: **Questions 3.99 and 3.100.**

Question 3.7 Predict which of the following bonds are polar, and, if polar, use a vector to indicate in which direction the electrons are pulled:
   a. O—S   b. C≡N   c. Cl—Cl   d. I—Cl

Question 3.8 Predict which of the following bonds are polar, and, if polar, use a vector to indicate in which direction the electrons are pulled:
   a. Si—Cl   b. S—Cl   c. H—C   d. C—C

Question 3.9 Using Lewis structures and VSEPR, predict whether each of the following molecules is polar:
   a. CS_2 b. NF_3 c. HCl   d. $SiCl_4$

Question 3.10　Using Lewis structures and VSEPR, predict whether each of the following molecules is polar:
　　a. CO_2　　b. SCl_2　　c. BrCl　　d. BCl_3

3.5　Properties Based on Molecular Geometry and Intermolecular Forces

Intramolecular forces are attractive forces *within* molecules. They are the chemical bonds that determine the shape and polarity of individual molecules. **Intermolecular forces,** on the other hand, are forces *between* molecules.

It is important to distinguish between these two kinds of forces. It is the *intermolecular* forces that determine such properties as the solubility of one substance in another and the freezing and boiling points of liquids. But, at the same time, we must realize that these forces are a direct consequence of the *intramolecular* forces within the individual units, the molecules.

In this section, we will see some of the consequences of bonding that are directly attributable to differences in intermolecular forces (solubility, boiling and melting points). In Section 5.2, we will investigate, in some detail, the nature of the intermolecular forces themselves.

Solubility

The solute is the substance that is present in a lesser quantity, and the solvent is the substance that is present in a greater amount.

Solubility is defined as the maximum amount of solute that dissolves in a given amount of solvent at a specified temperature. Polar molecules are most soluble in polar solvents, whereas nonpolar molecules are most soluble in nonpolar solvents. This is the rule of *"like dissolves like."* Substances of similar polarity are mutually soluble, and large differences in polarity lead to insolubility.

Case I: Ammonia and Water

The interaction of water and ammonia is an example of a particularly strong intermolecular force, the hydrogen bond; this phenomenon is discussed in Chapter 5.

Ammonia is soluble in water because both ammonia and water are polar molecules:

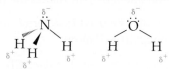

Dissolution of ammonia in water is a consequence of the intermolecular forces present among the ammonia and water molecules. The δ^- end (a nitrogen) of the ammonia molecule is attracted to the δ^+ end (a hydrogen) of the water molecule; at the same time the δ^+ end (a hydrogen) of the ammonia molecule is attracted to the δ^- end (an oxygen) of the water molecule. These attractive forces thus "pull" ammonia into water (and water into ammonia), and the ammonia molecules are randomly distributed throughout the solvent, forming a homogeneous solution (Figure 3.13).

Case II: Oil and Water

Oil and water do not mix; oil is a nonpolar substance composed primarily of molecules containing carbon and hydrogen. Water molecules, on the other hand, are quite polar. The potential solvent, water molecules, have partially charged ends, whereas the molecules of oil do not. As a result, water molecules exert their attractive forces on other water molecules, not on the molecules of oil; the oil remains insoluble, and because it is less dense than water, the oil simply floats on the surface of the water. This is illustrated in Figure 3.14.

Boiling Points of Liquids and Melting Points of Solids

Boiling a liquid requires energy. The energy is used to overcome the intermolecular attractive forces in the liquid, driving the molecules into the less-associated gas phase.

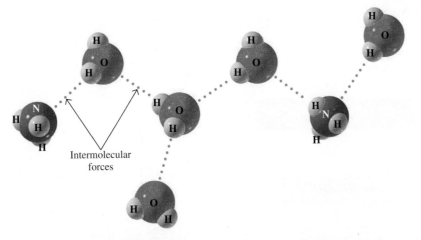

Figure 3.13 The interaction of polar covalent water molecules (the solvent) with polar covalent solute molecules, ammonia, results in the formation of a solution.

TABLE 3.6 Melting and Boiling Points of Selected Compounds in Relation to Their Bonding Type

Formula (Name)	Bonding Type	M.P. (°C)	B.P. (°C)
N_2 (nitrogen)	Nonpolar covalent	−210	−196
O_2 (oxygen)	Nonpolar covalent	−219	−183
NH_3 (ammonia)	Polar covalent	−78	−33
H_2O (water)	Polar covalent	0	100
NaCl (sodium chloride)	Ionic	801	1413
KBr (potassium bromide)	Ionic	730	1435

The amount of energy required to accomplish this is related to the boiling temperature. This, in turn, depends on the strength of the intermolecular attractive forces in the liquid, which parallels the polarity. This is not the only determinant of the boiling point. Molecular mass is also an important consideration. The larger the mass of the molecule, the more difficult it becomes to convert the collection of molecules to the gas phase.

A similar argument can be made for the melting points of solids. The ease of conversion of a solid to a liquid also depends on the magnitude of the attractive forces in the solid. The situation actually becomes very complex for ionic solids because of the complexity of the crystal lattice.

As a general rule, polar compounds have strong attractive (intermolecular) forces, and their boiling and melting points tend to be higher than those of nonpolar substances of similar molecular mass.

Melting and boiling points of a variety of substances are included in Table 3.6.

Question 3.11   Predict which compound in each of the following pairs should have the higher melting and boiling points (Hint: Write the Lewis structure and determine whether the compound is ionic, polar covalent, or nonpolar covalent.):
   a. H_2O and C_2H_4 b. CO and CH_4 c. NH_3 and N_2 d. Cl_2 and ICl

Question 3.12   Predict which compound in each of the following pairs should have the higher melting and boiling points (Hint: Write the Lewis structure and determine whether the compound is ionic, polar covalent, or nonpolar covalent.):
   a. C_2H_6 and CH_4 c. F_2 and Br_2
   b. CO and NO   d. $CHCl_3$ and Cl_2

Figure 3.14 The interaction of polar water molecules and nonpolar oil molecules. The familiar salad dressing—oil and vinegar—forms two layers. The oil does not dissolve in vinegar, an aqueous solution of acetic acid.

CHAPTER MAP

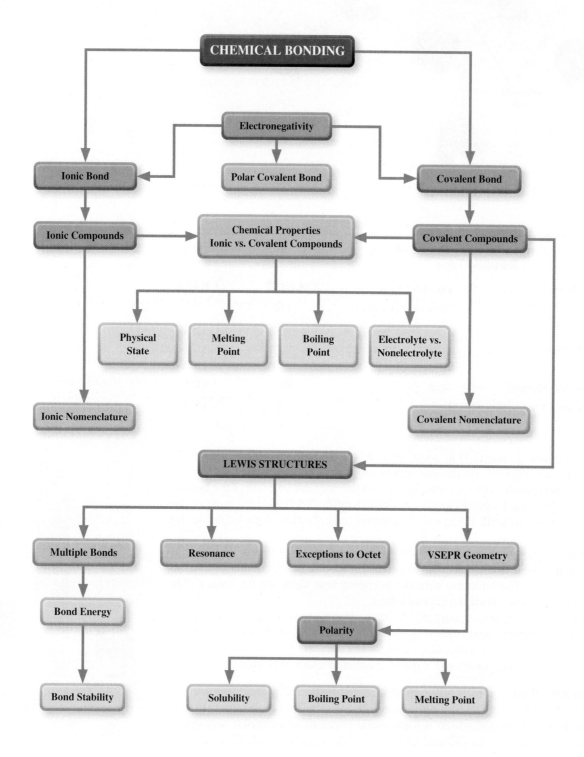

SUMMARY

3.1 Chemical Bonding

▶ **Lewis symbols** show valence electrons on atoms and ions.

▶ **Chemical bonding**—the attractive force between atoms in a compound can be classified as

- Covalent, atoms sharing electrons
  - **Polar covalent** and **nonpolar covalent**
- **Ionic,** consisting of cations and anions

▶ **Electronegativity** is a measure of the ability of an atom to attract electrons in a chemical bond, and the difference in electronegativity is used in classifying chemical bonds.

- Difference of 0.5 and less—nonpolar covalent
- Difference between 0.5 and 2.0—polar covalent
- Difference of 2.0 or larger—ionic

3.2 Naming Compounds and Writing Formulas of Compounds

▶ The system for naming compounds (**nomenclature**) is different for ionic and covalent compounds.

- Ionic compounds: name cation followed by the anion. Use Roman numeral to denote the charge on the cation, when required.
- The charge of a **monatomic ion** can be determined for main-group elements from the periodic table.
- Know the names and charges of the **polyatomic ions.**
- Covalent compounds: name first element first and then second element (with *-ide* ending). Use prefixes of di-, tri-, etc., to denote the number of atoms of each element in the compound.

▶ When writing **formulas** for neutral ionic compounds, be sure the number of positive charges equals the number of negative charges.

▶ When writing **formulas** for covalent compounds, use the prefix before the element name as the subscript for the element.

3.3 Properties of Ionic and Covalent Compounds

▶ Covalent compounds are typically made up of discrete units called **molecules.**

▶ Ionic compounds form a **crystal lattice** composed of cations and anions extended out in a three-dimensional network.

▶ Ionic compounds have higher **melting points** and **boiling points** than covalent compounds.

▶ When ionic compounds dissolve in water, the ions **dissociate** and the solution conducts electricity. These compounds are **electrolytes.**

▶ When covalent compounds dissolve in water, the compound does not dissociate. They are **nonelectrolytes.**

3.4 Drawing Lewis Structures of Molecules and Polyatomic Ions

▶ To draw a **Lewis structure,** all valence electrons are placed around the atoms of the compound as bonding electron pairs or **lone pairs** in order to satisfy the octet rule for each atom.

- The least electronegative atom is central.
- Hydrogen and fluorine occupy terminal positions.
- Carbon has a tendency to form chains.

▶ If the octet rule cannot be satisfied with **single bonds** connecting the atoms, then **double bonds** and **triple bonds** are used.

▶ **Bond energy** determines the stability of a bond. The bond energy trend is as follows:

- Single < double < triple

▶ The bond length trend is the reverse.

- Triple < double < single

▶ **Isomers** are compounds that have the same molecular formula but have different structures. Hydrocarbons have many possible isomers.

▶ When two or more Lewis structures contribute to the molecular structure, the compound displays **resonance.** The true structure is a hybrid of the drawn resonance forms. Stability of a compound increases with the number of resonance structures that can be drawn.

▶ Some molecules have exceptions to the octet rule, these include:

- Odd-electron compounds
- Incomplete octets
- Expanded octets

▶ The shapes of molecules can be predicted using **valence-shell electron-pair repulsion (VSEPR) theory.** Possible shapes include:

- Linear
- Trigonal planar
- Tetrahedral
- Bent
- Trigonal pyramidal

▶ The geometry of a molecule and the knowledge of the polarity of a bond can be used to determine if a molecule is polar (has a dipole) or is nonpolar.

3.5 Properties Based on Molecular Geometry and Intermolecular Forces

▶ **Intermolecular forces** are attractive forces between molecules. **Intramolecular forces** are chemical bonds.

▶ Polar molecules are generally attracted to each other more strongly than are nonpolar molecules.

▶ **Solubility** is the maximum amount of solute that dissolves in a given amount of solvent at a specific temperature.

- Polar molecules are soluble in polar solvents.
- Nonpolar molecules are soluble in nonpolar solvents.

▶ Polar molecules (of similar mass) have higher melting points and boiling points than nonpolar molecules.

ANSWERS TO PRACTICE PROBLEMS

3.1 $: \overset{\cdot}{\underset{\cdot}{O}} :$

Oxygen has two unpaired electrons. Therefore, it is capable of forming two bonds.

3.2 $: Ca \longrightarrow Ca^{2+} + 2e^-$

$: \overset{\cdot}{\underset{\cdot}{O}} : + 2e^- \longrightarrow [: \overset{\cdot\cdot}{\underset{\cdot\cdot}{O}} :]^{2-}$

3.3 Nonpolar covalent

3.4 **a.** Ca_3N_2 (three Ca^{2+} and two N^{3-})
 b. $MgBr_2$ (one Mg^{2+} and two Br^-)
 c. Mg_3N_2 (three Mg^{2+} and two N^{3-})

3.5 Chromium(II) bromide

3.6 **a.** $CaCO_3$ **b.** Cu_2SO_4

3.7 **a.** Fe_2S_3 **b.** $CaSO_4$ **c.** Al_2O_3

3.8 **a.** Diboron trioxide
 b. Nitrogen monoxide
 c. Iodine monochloride
 d. Phosphorus trichloride
 e. Phosphorus pentachloride
 f. Diphosphorus pentoxide

3.9 **a.** P_2O_5 **b.** SiO_2 **c.** CBr_4 **d.** OF_2

3.10 **a.**

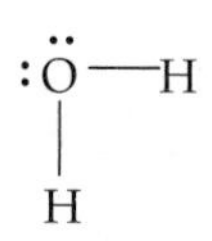

  b.

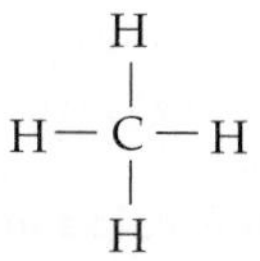

3.11 **a.**

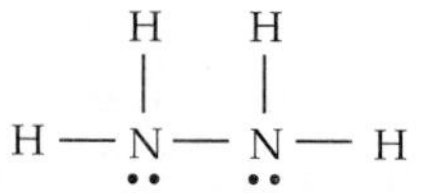

  b.

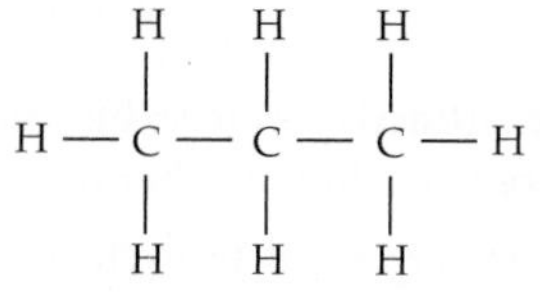

  c. $H—C{\equiv}N:$

3.12 $\left[: \overset{\cdot\cdot}{O} — \overset{\cdot\cdot}{O} :\right]^{2-}$

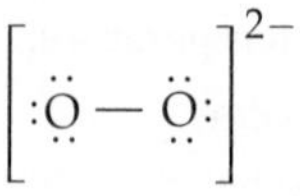

3.13 **a.**

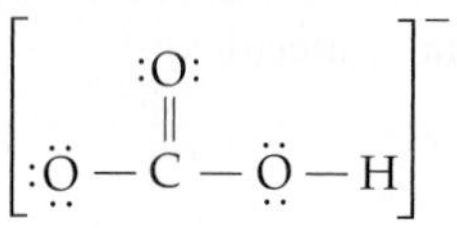

3.13 **b.**

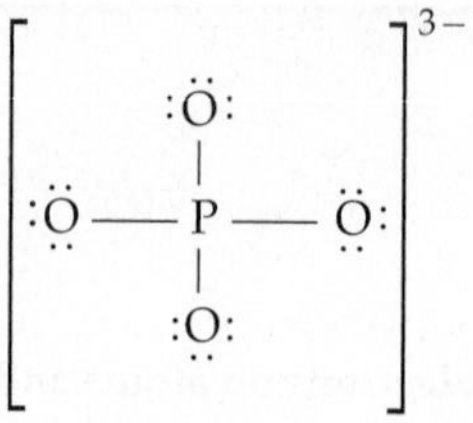

3.14 **a.** $\overset{\cdot\cdot}{O}{=}\overset{\cdot}{Se}—\overset{\cdot\cdot}{\underset{\cdot\cdot}{O}}: \longleftrightarrow :\overset{\cdot\cdot}{\underset{\cdot\cdot}{O}}—\overset{\cdot}{Se}{=}\overset{\cdot\cdot}{O}$

  b. Since both S and Se are in the same family (Group VIA or 16), they have the same number of valence electrons, and therefore, should form the same kinds of bonds.

3.15 **a.**

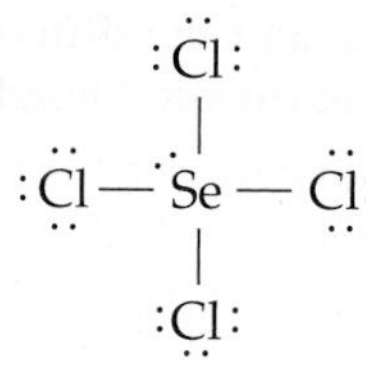

  b.

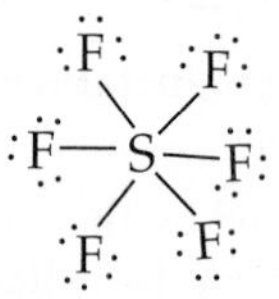

  c.

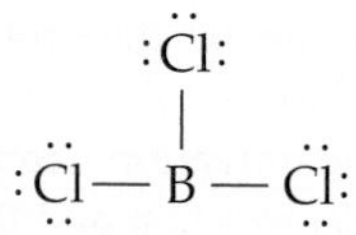

3.16 $AsCl_3$ is polar.

QUESTIONS AND PROBLEMS

Chemical Bonding

Foundations

3.13 Draw the appropriate Lewis symbol for each of the following atoms:
  a. H   **b.** He   **c.** Si   **d.** N

3.14 Draw the appropriate Lewis symbol for each of the following atoms:
  a. Be   **b.** B   **c.** F   **d.** S

3.15 Draw the appropriate Lewis symbol for each of the following ions:
  a. Li^+ **b.** Mg^{2+} **c.** Cl^- **d.** P^{3-}

3.16 Draw the appropriate Lewis symbol for each of the following ions:
  a. Be^{2+} **b.** Al^{3+} **c.** O^{2-} **d.** S^{2-}

3.17 Describe the differences between covalent bonding and ionic bonding.

3.18 Describe the difference between nonpolar covalent and polar covalent bonding.

3.19 What is the periodic trend of electronegativity?

3.20 What role does electronegativity play in determining the bonding between atoms in a compound?

3.21 Use electronegativity values to classify the bonds in each of the following compounds as ionic, polar covalent, or nonpolar covalent.
a. $MgCl_2$ **b.** CO_2 **c.** H_2S **d.** NO_2

3.22 Use electronegativity values to classify the bonds in each of the following compounds as ionic, polar covalent, or nonpolar covalent.
a. $CaCl_2$ **b.** CO **c.** ICl **d.** H_2

Applications

3.23 When there is a reaction between each of these pairs of atoms, ions form. Using Lewis symbols, write the reactions showing how electrons are lost or gained when these atoms become ions.
a. Li + Br   **b.** Mg + Cl   **c.** P + H

3.24 When there is a reaction between each of these pairs of atoms, ions form. Using Lewis symbols, write the reactions showing how electrons are lost or gained when these atoms become ions.
a. Na + O   **b.** Na + S   **c.** Si + H

3.25 Explain, using Lewis symbols and the octet rule, why helium is so nonreactive.

3.26 Explain, using Lewis symbols and the octet rule, why neon is so nonreactive.

Naming Compounds and Writing Formulas of Compounds

Foundations

3.27 Predict the formula of the compound formed from the combination of ions of magnesium and sulfur.

3.28 Predict the formula of the compound formed from the combination of ions of calcium and fluorine.

3.29 Name each of the following ions:
a. Na^+ **b.** Cu^+ **c.** Mg^{2+}

3.30 Name each of the following ions:
a. Cu^{2+} **b.** Fe^{2+} **c.** Fe^{3+}

3.31 Name each of the following ions:
a. HCO_3^- **b.** H_3O^+ **c.** CO_3^{2-}

3.32 Name each of the following ions:
a. ClO^- **b.** NH_4^+ **c.** CH_3COO^-

3.33 Write the formula for each of the following monatomic ions:
a. the potassium ion   **b.** the nickel(II) ion

3.34 Write the formula for each of the following monatomic ions:
a. the calcium ion   **b.** the chromium(VI) ion

3.35 Write the formula for each of the following polyatomic ions:
a. the sulfate ion   **b.** the nitrate ion

3.36 Write the formula for each of the following polyatomic ions:
a. the phosphate ion   **b.** the cyanide ion

Applications

3.37 Predict the formula of a compound formed from:
a. aluminum and oxygen
b. lithium and sulfur

3.38 Predict the formula of a compound formed from:
a. boron and hydrogen
b. magnesium and phosphorus

3.39 Name each of the following compounds:
a. $MgCl_2$ **b.** $AlCl_3$ **c.** $Cu(NO_3)_2$

3.40 Name each of the following compounds:
a. Na_2O **b.** $Fe(OH)_3$ **c.** $CaBr_2$

3.41 Write the correct formula for each of the following:
a. sodium chloride
b. magnesium bromide

3.42 Write the correct formula for each of the following:
a. potassium oxide   **b.** potassium nitride

3.43 Write the correct formula for each of the following:
a. silver cyanide
b. ammonium chloride

3.44 Write the correct formula for each of the following:
a. magnesium carbonate
b. magnesium bicarbonate

3.45 Write the correct formula for each of the following:
a. copper(II) oxide   **b.** iron(III) oxide

3.46 Write the correct formula for each of the following:
a. manganese(II) oxide
b. manganese(III) oxide

3.47 Write a suitable formula for:
a. sodium nitrate   **b.** magnesium nitrate

3.48 Write a suitable formula for:
a. aluminum nitrate   **b.** potassium nitrate

3.49 Write a suitable formula for:
a. ammonium iodide   **b.** ammonium sulfate

3.50 Write a suitable formula for:
a. ammonium acetate   **b.** ammonium cyanide

3.51 Name each of the following compounds:
a. NO_2 **b.** SeO_3 **c.** SO_3

3.52 Name each of the following compounds:
a. N_2O_4 **b.** CCl_4 **c.** N_2O_5

3.53 Write a suitable formula for:
a. silicon dioxide   **b.** sulfur dioxide

3.54 Write a suitable formula for:
a. diphosphorus pentoxide
b. dioxygen difluoride

Properties of Ionic and Covalent Compounds

Foundations

3.55 Contrast ionic and covalent compounds with respect to their solid-state structures.

3.56 Contrast ionic and covalent compounds with respect to their behaviors in solution.

3.57 Contrast ionic and covalent compounds with respect to their relative boiling points.

3.58 Contrast ionic and covalent compounds with respect to their relative melting points.

Applications

3.59 Would KCl be expected to be a solid at room temperature? Why?

3.60 Would CCl_4 be expected to be a solid at room temperature? Why?

3.61 Would H_2O or CCl_4 be expected to have a higher boiling point? Why?

3.62 Would H_2O or CCl_4 be expected to have a higher melting point? Why?

3.63 Would $MgCl_2$ in H_2O form an electrolytic solution? Why?

3.64 Would $C_6H_{12}O_6$ in H_2O form an electrolytic solution? Why?

Drawing Lewis Structures of Molecules and Polyatomic Ions

Foundations

3.65 When drawing a Lewis structure, how can the central atom be determined?

3.66 When drawing a Lewis structure, which elements can never be a central atom?

3.67 How is the positive charge of a polyatomic *cation* incorporated when determining the number of valence electrons to be used in a Lewis structure?

3.68 How is the negative charge of a polyatomic *anion* incorporated when determining the number of valence electrons to be used in a Lewis structure?

3.69 Rank the following in order of increasing bond energy:

single bond, double bond, triple bond

3.70 Rank the following in order of increasing bond length:

single bond, double bond, triple bond

3.71 Which will have more isomers: C_4H_{10} or C_5H_{12}? Why?

3.72 Will the number of isomers increase or decrease with the number of carbon atoms in a hydrocarbon? Explain your reasoning.

3.73 Discuss the concept of resonance, being certain to define the terms *resonance, resonance form,* and *resonance hybrid.*

3.74 Why is resonance an important concept in bonding?

3.75 What is the bond angle of a trigonal planar molecule?

3.76 What is the bond angle of a tetrahedral molecule?

3.77 True or false? Molecules that have only nonpolar bonds will always be nonpolar. Explain your reasoning.

3.78 True or false? Molecules that have only polar bonds will always be polar. Explain your reasoning.

Applications

3.79 Give the Lewis structure for each of the following compounds:
   a. NCl_3 **b.** CH_3OH **c.** CS_2 **d.** CH_2Cl_2

3.80 Give the Lewis structure for each of the following compounds:
   a. HNO_3 **b.** CCl_4 **c.** PBr_3 **d.** CH_3CH_2OH

3.81 Acetaldehyde has the molecular formula C_2H_4O. Draw the Lewis structure of acetaldehyde.

3.82 Formaldehyde, H_2CO, in water solution has been used as a preservative for biological specimens. Draw the Lewis structure of formaldehyde.

3.83 Acetone, C_3H_6O, is a common solvent. It is found in such diverse materials as nail polish remover and industrial solvents. Draw its Lewis structure if its skeletal structure is

O

C   C   C

3.84 Ethylamine is an example of an important class of organic compounds. The molecular formula of ethylamine is $CH_3CH_2NH_2$. Draw its Lewis structure.

3.85 Draw the Lewis structure of NO^+.

3.86 Draw the Lewis structure of NO_2^-.

3.87 Draw the Lewis structure of OH^-.

3.88 Draw the Lewis structure of HS^-.

3.89 The acetate ion exhibits resonance. Draw two resonance forms of the acetate ion.

3.90 Ozone, O_3, has two resonance forms. Draw each form.

3.91 All of the following Lewis structures are incorrect. Find the errors and write the correct structures.

   a. $:C=\ddot{O}:$

   b. $:\ddot{H}-\ddot{O}-\ddot{H}:$

   c. $:\ddot{O}-\ddot{C}-\ddot{O}:$

3.92 All of the following Lewis structures are incorrect. Find the errors and write the correct structures.

   a.
$$:\ddot{F}:$$
$$|$$
$$:\ddot{F}-S-\ddot{F}:$$
$$|$$
$$:\ddot{F}:$$

   b.
$$:\ddot{O}:$$
$$|$$
$$H-C-H$$

   c.
$$H$$
$$|$$
$$H-C=O=H$$
$$|$$
$$H$$

3.93 $BeCl_2$ has an incomplete octet around the beryllium atom. Draw the Lewis structure of $BeCl_2$.

3.94 $B(OH)_3$ has an incomplete octet around the boron atom. Draw the Lewis structure of $B(OH)_3$.

3.95 SeF_6 has an expanded octet around the selenium atom. Draw the Lewis structure of SeF_6.

3.96 Noble gases in the third period and beyond can undergo covalent bonding. All have an expanded octet. Draw the Lewis structure of XeF_2.

3.97 Draw the Lewis structure of each of the following compounds and predict its geometry using the VSEPR theory.
   a. SO_2 **b.** SO_3

3.98 Draw the Lewis structure of each of the following compounds and predict its geometry using the VSEPR theory.
   a. SeO_2 **b.** SeO_3

3.99 Predict the polarity of each compound in Question 3.97.

3.100 Predict the polarity of each compound in Question 3.98.

Properties Based on Molecular Geometry and Intermolecular Forces

Foundations

3.101 Which of the following compounds have polar bonds but are nonpolar covalent compounds?
   a. CO_2 **b.** NF_3 **c.** CF_4

3.102 Which of the following compounds have polar bonds but are nonpolar covalent compounds?
   a. SO_2 **b.** CF_4 **c.** NH_3

3.103 What is the relationship between the polarity of a bond and the polarity of the molecule?

3.104 What effect does polarity have on the solubility of a compound in water?

3.105 Using the VSEPR theory, predict the geometry, polarity, and water solubility of each compound in Question 3.79.

3.106 Using the VSEPR theory, predict the geometry, polarity, and water solubility of each compound in Question 3.80.

3.107 What effect does polarity have on the melting point of a pure compound?

3.108 What effect does polarity have on the boiling point of a pure compound?

Applications

3.109 Would you expect KCl to dissolve in water?

3.110 Would you expect ethylamine (Question 3.84) to dissolve in water?

3.111 In each of the following pairs of compounds, choose the compound with the higher boiling point.
  a. N_2 and NH_3 **b.** CS_2 and CF_4 **c.** NaCl and Cl_2

3.112 In each of the following pairs of compounds, choose the compound with the higher melting point.
  a. KF and F_2 **b.** CO and O_2 **c.** NBr_3 and CCl_4

CHALLENGE PROBLEMS

1. Predict differences in our global environment that may have arisen if the freezing point and boiling point of water were 20°C higher than they are.

2. Would you expect the compound $C_2S_2H_4$ to exist? Why or why not?

3. Draw the resonance forms of the carbonate ion. What conclusions, based on this exercise, can you draw about the stability of the carbonate ion?

4. Which of the following compounds would be predicted to have the higher boiling point? Explain your reasoning.

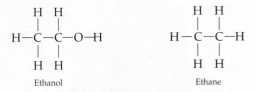

Ethanol   Ethane

5. Why does the octet rule not work well for compounds of lanthanide and actinide elements? Suggest a number other than eight that may be more suitable.

4

Calculations, Chemical Changes, and the Chemical Equation

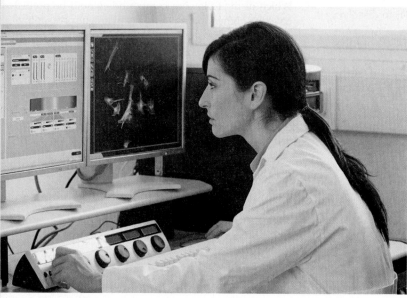

Sophisticated computer-based instrumentation is used to follow the progress of a chemical reaction.

OUTLINE

LEARNING GOALS

1 Calculate the mass of an atom using the atomic mass unit.

2 Use the relationship between Avogadro's number and the mole to perform calculations.

3 Determine molar mass, and demonstrate how it is used in mole and mass conversion calculations.

4 Use chemical formulas to calculate the formula mass and molar mass of a compound.

5 Describe the functions served by the chemical equation, the basis for chemical calculations.

6 Classify chemical reactions by type: combination, decomposition, or replacement.

7 Balance chemical equations given the identity of products and reactants.

8 Write net ionic equations, and use solubility rules to predict the formation of a precipitate.

9 Distinguish between an acid and a base.

10 Write oxidation and reduction half-reactions, and identify oxidizing agents and reducing agents.

11 Compare and contrast voltaic and electrolytic cells.

12 Describe examples of redox processes.

13 Use a chemical equation and a given number of moles or mass of a reactant or product to calculate the number of moles or mass of a reactant or product.

14 Calculate theoretical and percent yields.

INTRODUCTION

In this chapter, we define the mole (mol), the fundamental unit of measure of chemical arithmetic. We will learn to write and balance chemical equations and use these tools to perform calculations of chemical quantities. We will also see that chemical reactions are classified according to their unique patterns and characteristics.

If a pharmaceutical company wishes to manufacture 1000 kilograms (kg) of a product next year, many calculations of chemical quantities based on chemical equations would be required. The company would first have to determine the amount of each starting material that would need to be ordered. This answer would be used in determining how much the process will cost as well as the product's price per gram.

We often need to predict the quantity of a product produced from the reaction of a given amount of material. This calculation is possible. It is equally possible to calculate the amount of a material that would be necessary to produce a desired amount of product.

What is required is a recipe: a procedure to follow. The basis for our recipe is the chemical equation. A properly written chemical equation provides all of the necessary information for the chemical calculation. That critical information is the combining ratio of elements or compounds that must interact in order to produce a certain amount of product or products.

It is also necessary to understand how elements and compounds interact during a chemical reaction. When dissolved in water, some compounds react and form a product that is a solid. Other compounds react and produce salt and water, while some compounds react to produce large amounts of energy along with carbon dioxide and water. We will use chemical equations to classify reactions as precipitation, acid-base, and oxidation-reduction. We will also study some of the practical applications that are based on these reactions.

Many chemical calculations are made during the production of a pharmaceutical product.

4.1   The Mole Concept and Atoms

Atoms are exceedingly small, yet their masses have been experimentally determined for each of the elements. The unit of measurement for these determinations is the **atomic mass unit** (amu):

$$1 \text{ amu} = 1.661 \times 10^{-24} \text{ g}$$

The Mole and Avogadro's Number

The exact value of the amu is defined in relation to a standard, just as the units of the metric system represent defined quantities. The carbon-12 isotope has been chosen and is assigned a mass of exactly 12 amu. Hence, this standard reference point defines an amu as exactly one-twelfth the mass of a carbon-12 atom.

The periodic table provides atomic masses in amu. These atomic masses are average values, based on the contribution of all naturally occurring isotopes of the particular element. For example, the average mass of a carbon atom is 12.01 amu and, using the conversion factor based on the relationship between amu and grams, shown above, we can calculate the mass of a carbon atom in units of grams:

$$\frac{12.01 \text{ amu C}}{\text{C atom}} \times \frac{1.661 \times 10^{-24} \text{ g C}}{1 \text{ amu C}} = 1.995 \times 10^{-23} \frac{\text{g C}}{\text{C atom}}$$

In everyday work, chemists use much larger quantities of matter (typically g or kg). Therefore, the mole is a more practical unit than the amu for defining a "collection" of atoms. The **mole** (mol) is the amount of a substance that contains as many atoms, molecules, or ions as are found in exactly 12 g of the carbon-12

LEARNING GOAL

2 Use the relationship between Avogadro's number and the mole to perform calculations.

LEARNING GOAL

3 Determine molar mass, and demonstrate how it is used in mole and mass conversion calculations.

isotope. That number, experimentally determined, is **Avogadro's number,** named in honor of the nineteenth-century Italian scientist, Amadeo Avogadro.

Avogadro's number is expressed as:

$$1 \text{ mol of atoms} = 6.022 \times 10^{23} \text{ atoms of an element}$$

The practice of defining a unit for a quantity of small objects is common; a *dozen* eggs, a *ream* of paper, and a *gross* of pencils are well-known examples. Similarly, a mol is 6.022×10^{23} individual units of anything. We could, if we desired, speak of a mol of eggs or a mol of pencils. However, in chemistry we use the mol to represent a specific quantity of atoms, ions, or molecules.

The mol and the amu are related. The atomic mass of an element corresponds to the average mass of a single atom in amu, and the *molar mass* is the mass of a mole of atoms in grams.

The mass of 1 mol of atoms, in g, is defined as the **molar mass.** Consider this relationship for sodium in Example 4.1.

EXAMPLE 4.1   **Relating Atomic Mass Units to Molar Mass**

LEARNING GOALS

Calculate the mass, in g, of Avogadro's number of sodium atoms.

Solution

Avogadro's number represents the relationship between 6.022×10^{23} atoms and 1 mol.

1 Calculate the mass of an atom using the atomic mass unit.

2 Use the relationship between Avogadro's number and the mole to perform calculations.

Step 1. Since the data given is Avogadro's number of atoms, it should be written as a conversion factor with the number of atoms in the numerator.

$$\frac{6.022 \times 10^{23} \text{ atoms Na}}{1 \text{ mol Na}}$$

Step 2. The periodic table indicates that the average mass of one sodium atom = 22.99 amu. This can be used as a conversion factor with atoms Na in the denominator.

$$\frac{6.022 \times 10^{23} \text{ atoms Na}}{1 \text{ mol Na}} \times \frac{22.99 \text{ amu Na}}{1 \text{ atom Na}} = \frac{138.4 \times 10^{23} \text{ amu Na}}{1 \text{ mol Na}}$$

Data Given × Conversion Factor = Initial Data Result

Step 3. As previously noted, $1 \text{ amu} = 1.661 \times 10^{-24}$ g. The conversion factor written with g in the numerator and amu in the denominator is the only one that leads to the answer in g/mol, or the molar mass of sodium.

$$\frac{138.4 \times 10^{23} \text{ amu Na}}{1 \text{ mol Na}} \times \frac{1.661 \times 10^{-24} \text{ g Na}}{1 \text{ amu Na}} = \frac{22.99 \text{ g Na}}{1 \text{ mol Na}}$$

Initial Data Result × Conversion Factor = Desired Result

This calculation may also be done in a single step by arranging the factors in a chain.

$$\frac{6.022 \times 10^{23} \text{ atoms Na}}{1 \text{ mol Na}} \times \frac{22.99 \text{ amu Na}}{1 \text{ atom Na}} \times \frac{1.661 \times 10^{-24} \text{ g Na}}{1 \text{ amu Na}} = \frac{22.99 \text{ g Na}}{1 \text{ mol Na}}$$

Data Given × Conversion Factor × Conversion Factor = Desired Result

The average mass of one atom of sodium, in units of amu, is numerically identical to the mass of Avogadro's number of atoms, expressed in units of g. Hence, the molar mass of sodium is 22.99 g Na/mol.

Helpful Hint: The use of conversion factors is discussed in Section 1.5.

Practice Problem 4.1

Calculate the mass, in g, of Avogadro's number of:

  a. aluminum atoms b. mercury atoms

▶ For Further Practice: **Questions 4.21 and 4.22.**

The sodium example is not unique. The relationship holds for every element in the periodic table.

Because Avogadro's number of particles (atoms) is 1 mol, it follows that the average mass of one atom of hydrogen is 1.008 amu and the mass of 1 mol of hydrogen atoms is 1.008 g, or the average mass of one atom of carbon is 12.01 amu and the mass of 1 mol of carbon atoms is 12.01 g.

In fact, 1 mol of atoms of *any element* contains the same number, Avogadro's number, of atoms, 6.022×10^{23} atoms.

The difference in mass of a mol of two different elements can be quite striking (Figure 4.1). For example, a mol of hydrogen atoms is 1.008 g, and a mol of lead atoms is 207.19 g.

Figure 4.1 The comparison of approximately 1 mol each of silver (as Morgan and Peace dollars), gold (as Canadian Maple Leaf coins), and copper (as pennies) shows the considerable difference in mass (as well as economic value) of equivalent mol of different substances.

Calculating Atoms, Moles, and Mass

Performing calculations based on a chemical equation requires a facility for relating the number of atoms of an element to a corresponding number of mol of that element and ultimately to their mass in g. Such calculations involve the use of conversion factors. The use of conversion factors was first described in Chapter 1.

Examples 4.2–4.6 demonstrate the use of conversion factors to proceed from the information *provided* in the problem (data given) to the information *requested* by the problem (desired result).

EXAMPLE 4.2	**Converting Moles to Atoms**

LEARNING GOAL

How many iron atoms are present in 3.0 mol of iron metal?

2 Use the relationship between Avogadro's number and the mole to perform calculations.

Solution

Step 1. The calculation is based on the choice of the appropriate conversion factor. The relationship, 1 mol Fe = 6.022×10^{23} atoms Fe, can be represented with atoms Fe in the numerator and mol Fe in the denominator.

$$\frac{6.022 \times 10^{23} \text{ atoms Fe}}{1 \text{ mol Fe}}$$

Step 2. Using this conversion factor, we have

$$3.0 \ \cancel{\text{mol Fe}} \times \frac{6.022 \times 10^{23} \text{ atoms Fe}}{1 \ \cancel{\text{mol Fe}}} = 18 \times 10^{23} \text{ atoms of Fe, or}$$

Data Given × Conversion Factor = Desired Result

$$= 1.8 \times 10^{24} \text{ atoms of Fe (2 significant figures)}$$

Practice Problem 4.2

How many oxygen atoms are present in 2.50 mol of:

  a. oxygen atoms b. diatomic oxygen

▶ For Further Practice: **Questions 4.23 and 4.24.**

EXAMPLE 4.3   Converting Atoms to Moles

LEARNING GOAL

Calculate the number of mol of sulfur represented by 1.81×10^{24} atoms of sulfur.

2 Use the relationship between Avogadro's number and the mole to perform calculations.

Solution

Step 1. Just as in the previous example, the calculation is based on the choice of the appropriate conversion factor. The relationship, 1 mol S = 6.022×10^{23} atoms S, can be represented with mol S in the numerator and atoms S in the denominator.

$$\frac{1 \text{ mol S}}{6.022 \times 10^{23} \text{ atoms S}}$$

Step 2. 1.81×10^{24} $\cancel{\text{atoms S}}$ $\times \dfrac{1 \text{ mol S}}{6.022 \times 10^{23} \ \cancel{\text{atoms S}}}$ = 3.01 mol S

Data Given $\times$ Conversion Factor = Desired Result

Note that this conversion factor is the inverse of that used in Example 4.2. Remember, the conversion factor must cancel units that should not appear in the final answer.

Practice Problem 4.3

How many mol of sodium are represented by 9.03×10^{23} atoms of sodium?

▶ For Further Practice: **Questions 4.25 and 4.26.**

EXAMPLE 4.4   Converting Moles to Mass

LEARNING GOAL

What is the mass, in g, of 3.01 mol of sulfur?

3 Determine molar mass, and demonstrate how it is used in mole and mass conversion calculations.

Solution

The molar mass is the mass of a mole of atoms in grams.

Step 1. The molar mass can be calculated using information provided on the periodic table. One mol of sulfur has a mass of 32.06 g. This relationship, 1 mol S = 32.06 g S, can be represented as a conversion factor with g S in the numerator and mol S in the denominator.

$$\frac{32.06 \text{ g S}}{1 \text{ mol S}}$$

Step 2. Using this conversion factor (ensuring that the units *mol S* cancel):

$$3.01 \ \cancel{\text{mol S}} \times \frac{32.06 \text{ g S}}{1 \ \cancel{\text{mol S}}} = 96.5 \text{ g S}$$

Data Given $\times$ Conversion Factor = Desired Result

Practice Problem 4.4

What is the mass, in g, of 3.50 mol of the element helium?

▶ For Further Practice: **Questions 4.27 and 4.28.**

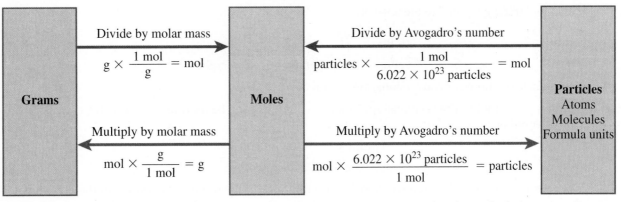

Figure 4.2 Interconversion between numbers of mol, particles, and g. The mol concept is central to chemical calculations involving measured quantities of matter.

EXAMPLE 4.5 | **Converting Mass to Moles**

Calculate the number of mol of sulfur in 1.00 kg of sulfur.

LEARNING GOAL

3 Determine molar mass, and demonstrate how it is used in mole and mass conversion calculations.

Solution

The data given has units of kg. This will need to first be converted to g.
From g, the number of mol can be calculated.

Step 1. Convert kg S to g S. Using the relationship, 1 kg S = 10^3 g S, as a conversion factor with g S in the numerator and kg S in the denominator will result in an answer in g S.

Step 2. Convert g S to the desired result in mol S using the molar mass relationship as a conversion factor. Molar mass can be determined from the periodic table, 1 mol S = 32.06 g.

Step 3. The calculation may be done in a single step by arranging the factors in a chain:

$$1.00 \ \cancel{kg \, S} \times \frac{10^3 \ \cancel{g \, S}}{1 \ \cancel{kg \, S}} \times \frac{1 \ mol \, S}{32.06 \ \cancel{g \, S}} = 31.2 \ mol \, S$$

Data Given × Conversion Factor × Conversion Factor = Desired Result

---

Practice Problem 4.5

Calculate the number of mol of silver in a silver ring that has a mass of 3.42 g.

▶ For Further Practice: **Questions 4.31 and 4.32.**

The conversion between the three principal measures of quantity of matter—the number of g (mass), the number of mol, and the number of individual particles (atoms, ions, or molecules)—is essential to the art of problem solving in chemistry. Their interrelationship is depicted in Figure 4.2.

EXAMPLE 4.6 | **Converting Grams to Number of Atoms**

Calculate the number of atoms of sulfur in 1.00 g of sulfur.

LEARNING GOALS

2 Use the relationship between Avogadro's number and the mole to perform calculations.

3 Determine molar mass, and demonstrate how it is used in mole and mass conversion calculations.

Solution

It is generally useful to map out a pattern for the required conversion. We are given the number of grams and need the number of atoms that correspond to that mass. Figure 4.2 illustrates that the mole concept is central to solving this problem.

Continued…

Begin by "tracing a path" to the answer:

$$
\boxed{\begin{matrix} \text{grams} \\ \text{sulfur} \end{matrix}} \xrightarrow[\text{1}]{\text{Step}} \boxed{\begin{matrix} \text{moles} \\ \text{sulfur} \end{matrix}} \xrightarrow[\text{2}]{\text{Step}} \boxed{\begin{matrix} \text{atoms} \\ \text{sulfur} \end{matrix}}
$$

Two transformations, or conversions, are required:

Step 1. Convert g S to mol S. The relationship, 1 mol S = 32.06 g S, can be used to complete this conversion. Consider either

$$
\frac{1 \text{ mol S}}{32.06 \text{ g S}} \quad \text{or} \quad \frac{32.06 \text{ g S}}{1 \text{ mol S}}
$$

If we want g S to cancel, the conversion factor with mol S in the numerator and g S in the denominator is the correct choice, resulting in

$$
1.00 \ \cancel{\text{g S}} \times \frac{1 \text{ mol S}}{32.06 \ \cancel{\text{g S}}} = 0.0312 \text{ mol S}
$$

Data Given × Conversion Factor = Relating Unit

Step 2. Convert mol S to atoms S.

The relationship, 1 mol S = 6.022×10^{23} atoms S, can be used to convert mol S to atoms S. The mol S must cancel; therefore

$$
0.0312 \ \cancel{\text{mol S}} \times \frac{6.022 \times 10^{23} \text{ atom S}}{1 \ \cancel{\text{mol S}}} = 1.88 \times 10^{22} \text{ atoms S}
$$

Relating Unit × Conversion Factor = Desired Result

The calculation may also be done in a single step:

$$
1.00 \ \cancel{\text{g S}} \times \frac{1 \ \cancel{\text{mol S}}}{32.06 \ \cancel{\text{g S}}} \times \frac{6.022 \times 10^{23} \text{ atoms S}}{1 \ \cancel{\text{mol S}}} = 1.88 \times 10^{22} \text{ atoms S}
$$

Data Given × Conversion Factor × Conversion Factor = Desired Result

---

Practice Problem 4.6

How many oxygen atoms are present in 40.0 g of oxygen molecules?

▶ For Further Practice: **Questions 4.35 and 4.36.**

Question 4.1   What is the mass, in g, of 1.00×10^{12} mercury (Hg) atoms?

Question 4.2   How many mol of lead (Pb) atoms are equivalent to six billion lead atoms?

4.2   The Chemical Formula, Formula Mass, and Molar Mass

The Chemical Formula

Compounds are pure substances. They are composed of two or more elements that are chemically combined. A **chemical formula** is a combination of symbols of the various elements that make up the compound. It serves as a convenient way to represent a compound. The chemical formula is based on the formula unit. The **formula unit** is the smallest collection of atoms or ions that provides two important pieces of information:

- the identity of the atoms or ions present in the compound and
- the relative numbers of each type of atom or ion.

Let's look at the following formulas:

- *Hydrogen gas*, H_2. This indicates that two atoms of hydrogen are chemically bonded forming diatomic hydrogen, hence the subscript two.
- *Water*, H_2O. Water is composed of molecules that contain two atoms of hydrogen (subscript two) and one atom of oxygen (lack of a subscript means *one* atom).
- *Sodium chloride*, NaCl. One ion of sodium and one ion of chlorine combine to make sodium chloride.
- *Calcium hydroxide*, $Ca(OH)_2$. Calcium hydroxide contains one ion of calcium and two atoms each of oxygen and hydrogen. One atom of oxygen and one atom of hydrogen are contained in one hydroxide ion (a polyatomic ion). These two hydroxide ions furnish two atoms of both oxygen and hydrogen. Recall that the subscript outside the parentheses applies to *all* atoms inside the parentheses.
- *Ammonium sulfate*, $(NH_4)_2SO_4$. Ammonium sulfate contains two ammonium ions (NH_4^+) and one sulfate ion (SO_4^{2-}). Each ammonium ion contains one nitrogen and four hydrogen atoms. The formula shows that ammonium sulfate contains two nitrogen atoms, eight hydrogen atoms, one sulfur atom, and four oxygen atoms.
- *Copper(II) sulfate pentahydrate*, $CuSO_4 \cdot 5H_2O$. This is an example of a compound that has water incorporated in its structure. Compounds containing one or more water molecules as an integral part of their structure are termed **hydrates.** Copper sulfate pentahydrate has five units of water (or ten H atoms and five O atoms) in addition to one copper atom, one sulfur atom, and four oxygen atoms for a total atomic composition of:

> 1 copper atom
>
> 1 sulfur atom
>
> 9 oxygen atoms
>
> 10 hydrogen atoms

> It is possible to determine the correct chemical formula of a compound from experimental data.

(a)   (b)

Figure 4.3 The marked difference in color of (a) hydrated and (b) anhydrous copper sulfate is clear evidence that they are, in fact, different compounds.

Note that the symbol for water is preceded by a dot, indicating that, although the water is a formula unit capable of standing alone, in this case it is a part of a larger structure. Copper sulfate also exists as a structure free of water, $CuSO_4$. This form is described as anhydrous (no water) copper sulfate. The physical and chemical properties of a hydrate often differ markedly from the anhydrous form (Figure 4.3).

Formula Mass and Molar Mass

Just as the atomic mass of an element is the average atomic mass for one atom of the naturally occurring element, expressed in amu, the **formula mass** of a compound is the sum of the atomic masses of all atoms in the compound, as represented by its formula. To calculate the formula mass of a compound we *must* know the correct chemical formula. The formula mass is expressed in amu.

When working in the laboratory, we do not deal with individual molecules; instead, we use units of mol or g. Eighteen grams of water (less than 1 ounce) contain approximately Avogadro's number of water molecules (6.022×10^{23} molecules). Defining our working units as mol and g makes good chemical sense.

We earlier concluded that the atomic mass of an element in amu from the periodic table corresponds to the mass of a mol of atoms of that element in units of grams per mole (g/mol). It follows that *molar mass*, the mass of a mol of compound, is numerically equal to the formula mass in amu.

In Example 4.7, H_2O is a covalent compound. In Example 4.8, $Ca_3(PO_4)_2$ is an ionic compound. As we have seen, it is not technically correct to describe ionic compounds as molecules; similarly, the term *molecular mass* is not appropriate for $Ca_3(PO_4)_2$. The term *formula mass* may be used to describe the formula unit of a substance, whether it is made up of ions, ion pairs, or molecules. We shall use the term *formula mass* in a general way to represent each of these species.

Figure 4.4 illustrates the difference between molecules and ion pairs.

> **LEARNING GOAL**
>
> **4** Use chemical formulas to calculate the formula mass and molar mass of a compound.

Figure 4.4 Formula units of (a) sodium chloride, an ionic compound, and (b) methane, a covalent compound.

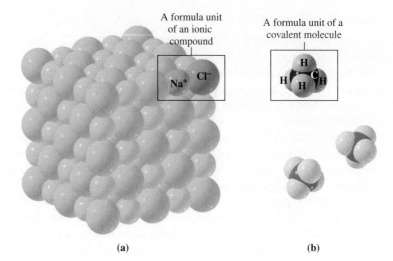

A formula unit of an ionic compound

A formula unit of a covalent molecule

(a)   (b)

EXAMPLE 4.7   **Calculating Formula Mass and Molar Mass**

LEARNING GOAL

Calculate the formula mass and molar mass of water, H_2O.

4 Use chemical formulas to calculate the formula mass and molar mass of a compound.

Solution

The formula mass is expressed in amu, and the molar mass is expressed in g/mol.

Step 1. Each water molecule contains two hydrogen atoms and one oxygen atom. The periodic table provides atomic weights in amu.

Step 2. Addition of the masses of all of the hydrogen and oxygen atoms in the water molecule results in the formula mass of water.

$$\text{2 H atoms} \times \text{1.008 amu/H atom} = 2.016 \text{ amu}$$
$$\underline{\text{1 O atom} \times \text{16.00 amu/O atom} = 16.00 \text{ amu}}$$
$$18.02 \text{ amu}$$

Step 3. The formula mass of a single molecule of H_2O is 18.02 amu. The formula mass and molar mass are related; the formula mass, in amu, is numerically equal to the molar mass, in units of g/mol. Therefore, the mass of a mol of H_2O is 18.02 g, and the molar mass is 18.02 g/mol.

Helpful Hint: Adding 2.016 and 16.00 shows a result of 18.016 on your calculator. Proper use of significant figures (Chapter 1) dictates rounding that result to 18.02.

Practice Problem 4.7

Calculate the formula mass and molar mass of NH_3 (ammonia).

▶ For Further Practice: **Questions 4.40 and 4.41.**

EXAMPLE 4.8   **Calculating Formula Mass and Molar Mass**

LEARNING GOAL

Calculate the formula mass and molar mass of calcium phosphate.

4 Use chemical formulas to calculate the formula mass and molar mass of a compound.

Solution

Step 1. The calcium ion is Ca^{2+}, and the phosphate ion is PO_4^{3-}. To form a neutral unit, three Ca^{2+} must combine with two PO_4^{3-}; $[3 \times (2^+)]$ calcium ion charges are balanced by $[2 \times (3^-)]$ phosphate ion charges.

Step 2. Thus, for calcium phosphate, $Ca_3(PO_4)_2$, the subscript two for phosphate dictates that there are two phosphorus atoms and eight oxygen atoms (2×4) in the formula unit.

Step 3. Addition of the masses of all of the calcium, phosphorus, and oxygen atoms in the compound results in the formula mass of calcium phosphate.

$$3 \text{ Ca atoms} \times 40.08 \text{ amu/Ca atom} = 120.24 \text{ amu}$$
$$2 \text{ P atoms} \times 30.97 \text{ amu/P atom} = 61.94 \text{ amu}$$
$$\underline{8 \text{ O atoms} \times 16.00 \text{ amu/O atom} = 128.00 \text{ amu}}$$
$$310.18 \text{ amu}$$

Step 4. The formula mass of calcium phosphate is 310.18 amu, and the molar mass is 310.18 g/mol.

Helpful Hint: Writing ionic formulas from compound names is explained in Section 3.2.

Practice Problem 4.8

Calculate the formula mass and molar mass of $CoCl_2 \cdot 6H_2O$ (cobalt chloride hexahydrate).

▶ For Further Practice: **Questions 4.43 and 4.44.**

Question 4.3 Caffeine occurs naturally in coffee and tea and is present in many soft drinks. The formula of caffeine is $C_8H_{10}N_4O_2$. Calculate the formula mass and molar mass of caffeine.

Question 4.4 The formula of ascorbic acid, commonly known as vitamin C, is $C_6H_8O_6$. Calculate the formula mass and molar mass of vitamin C.

4.3 The Chemical Equation and the Information It Conveys

A Recipe for Chemical Change

The **chemical equation** is the shorthand notation for a chemical reaction. It describes all the substances that react and all the products that form. **Reactants,** or starting materials, are all substances that undergo change in a chemical reaction; **products** are substances produced by a chemical reaction.

The chemical equation also describes the physical state of the reactants and products. It tells us whether the reaction occurs and identifies the solvent and experimental conditions employed, such as heat, light, or electrical energy added to the system.

Most important, the relative number of mol of reactants and products appears in the equation. According to the **law of conservation of mass,** matter cannot be either gained or lost in the process of a chemical reaction. The total mass of the products must be equal to the total mass of the reactants. In other words, the law of conservation of mass tells us that we must have a balanced chemical equation.

Features of a Chemical Equation

Consider the decomposition of calcium carbonate:

$$\underset{\text{calcium carbonate}}{CaCO_3(s)} \xrightarrow{\Delta} \underset{\text{calcium oxide}}{CaO(s)} + \underset{\text{carbon dioxide}}{CO_2(g)}$$

LEARNING GOAL

5 Describe the functions served by the chemical equation, the basis for chemical calculations.

This equation reads: One mole of solid calcium carbonate decomposes upon heating to produce one mole of solid calcium oxide and one mole of gaseous carbon dioxide.

The factors involved in writing equations of this type are described as follows:

1. *The identity of products and reactants must be specified using chemical symbols.* In some cases, it is possible to predict the products of a reaction. More often, the reactants and products must be verified by chemical analysis. (Generally, you will be given information regarding the identity of the reactants and products.)

2. *Reactants are written to the left of the reaction arrow ($\rightarrow$), and products are written to the right.* The direction in which the arrow points indicates the direction in which the reaction proceeds. In the decomposition of calcium carbonate, the reactant on the left ($CaCO_3$) is converted to products on the right ($CaO + CO_2$) during the course of the reaction.

3. *The physical states of reactants and products may be shown in parentheses.* For example:

   - $CaO(s)$ means that calcium oxide is a solid.
   - $CO_2(g)$ indicates that carbon dioxide is in the gaseous state.
   - Alternatively, (l) indicates a substance is present as a liquid, and (aq) tells us a substance is present as an aqueous solution which means it is dissolved in water.

Reactions that utilize light energy are termed *photochemical reactions.*

4. *The symbol Δ over the reaction arrow means that heat energy is necessary for the reaction to occur.* Often this and other special conditions are noted above or below the reaction arrow. For example, "light" means that a light source provides energy necessary for the reaction.

5. *The equation must be balanced.* All of the atoms of every reactant must also appear in the products, although in different compounds. We will treat this topic in detail later in this chapter.

The Experimental Basis of a Chemical Equation

The chemical equation must represent a chemical change: One or more substances are changed into new substances, with different chemical and physical properties. Evidence for the reaction may be based on observations such as

- the release of carbon dioxide gas when a carbonate is heated,
- the formation of a solid (or precipitate) when solutions of iron ions and hydroxide ions are mixed,
- the production of heat when using hot packs for treatment of injury, and
- the change in color of a solution upon addition of a second substance.

See A Medical Perspective: Hot and Cold Packs in Chapter 7.

Many reactions are not so obvious. Sophisticated instruments are available to chemists that allow the detection of subtle changes in chemical systems that would otherwise go unnoticed. Such instruments may measure

- heat or light absorbed or emitted as the result of a reaction,
- changes in the way the sample behaves in an electric or magnetic field before and after a reaction, and
- changes in electrical properties before and after a reaction.

Whether we use our senses or a million-dollar computerized instrument, the "bottom line" is the same: We are measuring a change in one or more chemical or physical properties in an effort to understand the changes taking place in a chemical system, the conversion of reactants to new products.

Disease can be described as a chemical system (actually a biochemical system) gone awry. Here, too, the underlying changes may not be obvious. Just as technology has helped chemists see subtle chemical changes in the laboratory, medical diagnosis has been revolutionized in our lifetimes using very similar technology. Some of these techniques are described in A Medical Perspective: Magnetic Resonance Imaging, in Chapter 9.

Strategies for Writing Chemical Equations

Chemical reactions generally follow one of a few simple patterns: combination, decomposition, and single- or double-replacement. Recognizing the underlying pattern will improve your ability to write chemical equations and understand chemical reactions.

LEARNING GOAL

6 Classify chemical reactions by type: combination, decomposition, or replacement.

Combination Reactions

Combination reactions involve the joining of two or more elements or compounds, producing a product of different composition. The general form of a combination reaction is

$$A + B \longrightarrow AB$$

in which A and B represent reactant elements or compounds and AB is the product.

Examples include

- combination of a metal and a nonmetal to form a salt,

$$Ca(s) + Cl_2(g) \longrightarrow CaCl_2(s)$$

- combination of hydrogen and chlorine molecules to produce hydrogen chloride,

$$H_2(g) + Cl_2(g) \longrightarrow 2HCl(g)$$

- formation of water from hydrogen and oxygen molecules,

$$2H_2(g) + O_2(g) \longrightarrow 2H_2O(g)$$

- reaction of magnesium oxide and carbon dioxide to produce magnesium carbonate,

$$MgO(s) + CO_2(g) \longrightarrow MgCO_3(s)$$

Decomposition Reactions

Decomposition reactions produce two or more products from a single reactant. The general form of these reactions is the reverse of a combination reaction:

$$AB \longrightarrow A + B$$

Some examples are

- the heating of calcium carbonate to produce calcium oxide and carbon dioxide,

$$CaCO_3(s) \xrightarrow{\Delta} CaO(s) + CO_2(g)$$

- the removal of water from a hydrated material,

$$CuSO_4 \cdot 5H_2O(s) \longrightarrow CuSO_4(s) + 5H_2O(g)$$

Hydrated compounds are described in the previous section.

Replacement Reactions

Replacement reactions include both *single-replacement* and *double-replacement*. In a **single-replacement reaction,** one atom replaces another in the compound, producing a new compound:

$$A + BC \longrightarrow AC + B$$

Examples include

- the replacement of copper by zinc in copper sulfate,

$$Zn(s) + CuSO_4\,(aq) \longrightarrow ZnSO_4\,(aq) + Cu(s)$$

In a single-replacement reaction, an element *replaces* an ion. In the first example, zinc metal is converted to Zn^{2+} and replaces Cu^{2+} in the $CuSO_4$. Cu^{2+}, in turn, is converted to copper metal.

- the replacement of aluminum by sodium in aluminum nitrate,

$$3Na(s) + Al(NO_3)_3(aq) \longrightarrow 3NaNO_3\,(aq) + Al(s)$$

A **double-replacement reaction,** on the other hand, involves *two compounds* undergoing a "change of partners." Two compounds react by exchanging atoms to produce two new compounds:

$$AB + CD \longrightarrow AD + CB$$

Examples include

Two ions *swap positions* in double-replacement reactions.

- the reaction of an acid (hydrochloric acid) and a base (sodium hydroxide) to produce water and salt, sodium chloride,

$$HCl(aq) + NaOH(aq) \longrightarrow H_2O(l) + NaCl(aq)$$

- the formation of solid barium sulfate from barium chloride and potassium sulfate,

$$BaCl_2(aq) + K_2SO_4(aq) \longrightarrow BaSO_4(s) + 2KCl(aq)$$

Question 4.5   Classify each of the following reactions as decomposition (D), combination (C), single-replacement (SR), or double-replacement (DR):

a. $HNO_3(aq) + KOH(aq) \longrightarrow KNO_3(aq) + H_2O(aq)$

b. $Al(s) + 3NiNO_3(aq) \longrightarrow Al(NO_3)_3(aq) + 3Ni(s)$

c. $KCN(aq) + HCl(aq) \longrightarrow HCN(aq) + KCl(aq)$

d. $MgCO_3(s) \longrightarrow MgO(s) + CO_2(g)$

Question 4.6   Classify each of the following reactions as decomposition (D), combination (C), single-replacement (SR), or double-replacement (DR):

a. $2Al(OH)_3(s) \xrightarrow{\Delta} Al_2O_3(s) + 3H_2O(g)$

b. $Fe_2S_3(s) \xrightarrow{\Delta} 2Fe(s) + 3S(s)$

c. $Na_2CO_3(aq) + BaCl_2(aq) \longrightarrow BaCO_3(s) + 2NaCl(aq)$

d. $C(s) + O_2(g) \xrightarrow{\Delta} CO_2(g)$

4.4 Balancing Chemical Equations

LEARNING GOAL

7 Balance chemical equations given the identity of products and reactants.

The chemical equation shows the *molar quantity* of reactants needed to produce a certain *molar quantity* of products. The relative number of mol of each product and reactant is indicated by placing a whole-number *coefficient* before the formula of each substance in the chemical equation. A coefficient of two (for example, 2Na) indicates that 2 mol of sodium are involved in the reaction. The coefficient one is understood, not written. Therefore, Cl_2 indicates 1 mol of chlorine is involved in the reaction that results in the formation of 2 mol of sodium chloride. This information is summarized in the equation:

$$2Na(s) + Cl_2(g) \longrightarrow 2NaCl(s)$$

The equation is balanced because there are two Na and two Cl on each side of the reaction arrow. Now consider the equation:

$$CaCO_3(s) \xrightarrow{\Delta} CaO(s) + CO_2(g)$$

It is balanced as written. On the reactant side we have one mol of $CaCO_3$, which is:

1 mol Ca

1 mol C

3 mol O

On the product side there are one mol of CaO and one mol of CO_2, which is:

1 mol Ca

1 mol C

3 mol O

The coefficients indicate *relative* numbers of mol: 10 mol of $CaCO_3$ produce 10 mol of CaO; 0.5 mol of $CaCO_3$ produce 0.5 mol of CaO; and so forth.

Therefore, the law of conservation of mass is obeyed, and the equation is balanced as written.

Now consider the reaction of aqueous hydrogen chloride with solid calcium metal:

$$HCl(aq) + Ca(s) \longrightarrow CaCl_2(aq) + H_2(g)$$

The equation, as written, is not balanced.

Reactants	Products
1 mol H atoms	2 mol H atoms
1 mol Cl atoms	2 mol Cl atoms
1 mol Ca atoms	1 mol Ca atoms

We need 2 mol of both H and Cl on the left, or reactant, side. We must remember that *we cannot alter any chemical substance in the process of balancing the equation*. We can *only* introduce coefficients into the equation. Changing subscripts changes the identity of the substances involved, and that is not permitted. The equation must represent the reaction accurately. The correct equation is

Coefficients placed in front of the formula indicate the relative numbers of mol of compound (represented by the formula) that are involved in the reaction. Subscripts placed to the lower right of the atomic symbol indicate the relative number of atoms in the compound.

Water (H_2O) and hydrogen peroxide (H_2O_2) illustrate the effect a subscript can have. The two compounds show marked differences in physical and chemical properties.

$$2HCl(aq) + Ca(s) \longrightarrow CaCl_2(aq) + H_2(g)$$
Balanced equation

Many equations are balanced by trial and error. After the identity of the products and reactants, the physical states, and the reaction conditions are known, the following steps provide a method for correctly balancing a chemical equation:

Do not dismantle a polyatomic ion. It will retain its identity on both reactant and product sides.

Step 1. Count the number of mol of atoms of each element on both the reactant side and the product side.

Step 2. Determine which elements are not balanced.

Generally, when using trial and error, save H and O for the very end of the process.

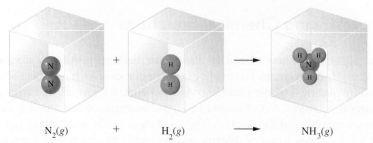

$$N_2(g) \quad + \quad H_2(g) \quad \longrightarrow \quad NH_3(g)$$

(a) Unbalanced equation. The law of conservation of mass is not obeyed.

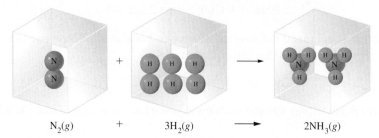

$$N_2(g) \quad + \quad 3H_2(g) \quad \longrightarrow \quad 2NH_3(g)$$

(b) Balanced equation. The species are correct, and the law of conservation of mass is obeyed.

Figure 4.5 Balancing the equation $N_2(g) + H_2(g) \longrightarrow NH_3(g)$

Step 3. Balance one element at a time using coefficients. It is often most efficient to begin by balancing the atoms in the most complicated formulas.

Step 4. When you believe that you have successfully balanced the equation, check, as in step 1, to be certain that mass conservation has been achieved.

Let us apply these steps to the reaction of nitrogen and hydrogen (Figure 4.5).

$$N_2(g) + H_2(g) \longrightarrow NH_3(g)$$

Step 1. **Reactants**   **Products**

2 mol N atoms   1 mol N atoms

2 mol H atoms   3 mol H atoms

Step 2. The numbers of mol of N and H are not balanced.

Step 3. Insertion of a 3 before $H_2(g)$ on the reactant side and a 2 before $NH_3(g)$ on the product side.

$$N_2(g) + 3H_2(g) \longrightarrow 2NH_3(g)$$

Step 4. Check to confirm that mass conservation has been achieved:

Reactants   **Products**

2 mol N atoms   2 mol N atoms

6 mol H atoms   6 mol H atoms

Hence, the equation is balanced. The balanced reaction equation used in industrial production of ammonia is illustrated in Figure 4.5.

Examples 4.9–4.10 illustrate equation-balancing strategies for a variety of commonly encountered situations.

EXAMPLE 4.9	Balancing Equations

LEARNING GOAL

7 Balance chemical equations given the identity of products and reactants.

Balance the following equation: Propane gas, C_3H_8, a fuel, reacts with oxygen gas to produce carbon dioxide and water vapor. The reaction is

$$C_3H_8(g) + O_2(g) \longrightarrow CO_2(g) + H_2O(g)$$

Solution

Step 1. Count the number of mol of atoms of each element on both the reactant side and the product side.

Reactants	Products
3 mol C atoms	1 mol C atoms
8 mol H atoms	2 mol H atoms
2 mol O atoms	3 mol O atoms

Step 2. Note that C, H, and O atoms are not balanced.

Step 3. First, balance the carbon atoms; save hydrogen and oxygen for later. There are 3 mol of carbon atoms on the left and only 1 mol of carbon atoms on the right. We need $3CO_2$ on the right side of the equation:

$$C_3H_8(g) + O_2(g) \longrightarrow 3CO_2(g) + H_2O(g)$$

Next, balance the hydrogen atoms; there are 8 mol of hydrogen atoms on the left and 2 mol of hydrogen atoms on the right. We need to place a 4 in front of H_2O on the right side of the equation:

$$C_3H_8(g) + O_2(g) \longrightarrow 3CO_2(g) + 4H_2O(g)$$

There are now 2 mol of oxygen atoms (2 mol of O from 1 mol O_2) on the left and 10 mol of oxygen atoms on the right (4 mol O from 4 mol H_2O and 6 mol O from 3 mol CO_2). To balance, we must have $5O_2$ (10 mol oxygen atoms) on the left side of the equation:

$$C_3H_8(g) + 5O_2(g) \longrightarrow 3CO_2(g) + 4H_2O(g)$$

Step 4. *Remember:* In every case, be sure to check the final equation to confirm that the law of conservation of mass is obeyed. There are 3 mol of C atoms, 8 mol of H atoms, and 10 mol of O atoms on each side of the reaction arrow.

Practice Problem 4.9

Balance the chemical equation:

$$C_2H_5OH(l) + O_2(g) \longrightarrow CO_2(g) + H_2O(g)$$

▶ For Further Practice: **Questions 4.69 and 4.70.**

EXAMPLE 4.10	Balancing Equations

LEARNING GOAL

7 Balance chemical equations given the identity of products and reactants.

Balance the following equation: Aqueous ammonium sulfate reacts with aqueous lead nitrate to produce aqueous ammonium nitrate and solid lead sulfate. The reaction is

$$(NH_4)_2SO_4(aq) + Pb(NO_3)_2(aq) \longrightarrow NH_4NO_3(aq) + PbSO_4(s)$$

Solution

In this double-replacement reaction, the polyatomic ions remain as intact units. Therefore, we can balance them as we would balance molecules rather than as atoms.

Continued…

Step 1. Count the number of mol of ions of each compound on both reactant and product side:

Reactants	Products
2 mol NH_4^+	1 mol NH_4^+
1 mol SO_4^{2-}	1 mol SO_4^{2-}
1 mol Pb^{2+}	1 mol Pb^{2+}
2 mol NO_3^-	1 mol NO_3^-

Step 2. Note that NH_4^+ and NO_3^- are not balanced.

Step 3. There are two ammonium ions on the left and only one ammonium ion on the right. To balance, the coefficient 2 is placed before NH_4NO_3 on the product side of the equation. Hence

$$(NH_4)_2SO_4(aq) + Pb(NO_3)_2(aq) \longrightarrow 2NH_4NO_3(aq) + PbSO_4(s)$$

Step 4. No further steps are necessary. The equation is now balanced. There are two ammonium ions, two nitrate ions, one lead ion, and one sulfate ion on each side of the reaction arrow.

Practice Problem 4.10

Balance the chemical equation:

$$S_2Cl_2(s) + NH_3(g) \longrightarrow N_4S_4(s) + NH_4Cl(s) + S_8(s)$$

▶ For Further Practice: **Questions 4.71 and 4.72.**

4.5   Precipitation Reactions

8   Write net ionic equations, and use solubility rules to predict the formation of a precipitate.

Precipitation reactions may be written as net ionic equations. See Section 4.6.

Formation of a precipitate, $BaSO_4$.

Precipitation reactions include any chemical change in solution that results in one or more insoluble product(s). The insoluble product is termed a **precipitate.** For aqueous solution reactions, the product is insoluble in water.

An understanding of precipitation reactions is useful in many ways. They may explain natural phenomena, such as the formation of stalagmites and stalactites in caves; these are simply precipitates in rocklike form. Kidney stones may result from the precipitation of calcium oxalate (CaC_2O_4).

How do you know whether a precipitate will form? Readily available solubility tables, such as Table 4.1, make prediction rather easy.

Example 4.11 illustrates the process.

TABLE 4.1 Solubilities of Some Common Ionic Compounds

Soluble Compounds Contain	Exceptions
Alkali metal ions (Li^+, Na^+, K^+, Rb^+, Cs^+) or the ammonium ion (NH_4^+)	
Nitrate (NO_3^-), bicarbonate (HCO_3^-), or chlorate (ClO_3^-)	
Halides (Cl^-, Br^-, I^-)	Compounds containing halides of Ag^+, Hg_2^{2+}, or Pb^{2+}
Sulfate (SO_4^{2-})	Compounds containing sulfate of Ag^+, Ca^{2+}, Sr^{2+}, Ba^{2+}, Hg_2^{2+}, or Pb^{2+}
Insoluble Compounds Contain	**Exceptions**
Carbonate (CO_3^{2-}), phosphate (PO_4^{3-}), chromate (CrO_4^{2-}), or sulfide (S^{2-})	Compounds containing alkali metal ions or the ammonium ion
Hydroxide (OH^-)	Compounds containing alkali metal ions or the Ba^{2+} ion

EXAMPLE 4.11 **Predicting Whether Precipitation Will Occur**

LEARNING GOAL

8 Write net ionic equations, and use solubility rules to predict the formation of a precipitate.

Will a precipitate form if two solutions of the soluble salts NaCl and $AgNO_3$ are mixed?

Solution

Step 1. These soluble salts form ions in aqueous solutions. If a cation from one compound reacts with an anion from another compound, it will be a double-replacement reaction, which involves a "change of partners."

$$\text{NaCl}(aq) + \text{AgNO}_3(aq) \longrightarrow \text{AgCl}(?) + \text{NaNO}_3(?)$$

Step 2. Refer to Table 4.1 to determine the solubility of AgCl and $NaNO_3$. We predict that $NaNO_3$ is soluble and AgCl is not:

$$\text{NaCl}(aq) + \text{AgNO}_3(aq) \longrightarrow \text{AgCl}(s) + \text{NaNO}_3(aq)$$

Step 3. The fact that solid AgCl is predicted to form classifies this reaction as a precipitation reaction.

Helpful Hints: (*aq*) indicates a soluble species; (*s*) indicates a solid, an insoluble species.

Practice Problem 4.11

Predict whether the following reactants, when mixed in aqueous solution, undergo a precipitation reaction. Write a balanced equation for each precipitation reaction.

a. potassium chloride and silver nitrate
b. potassium carbonate and calcium hydroxide
c. sodium hydroxide and ammonium chloride
d. sodium hydroxide and iron(II) chloride

▶ For Further Practice: **Questions 4.77 and 4.78.**

4.6 Net Ionic Equations

Writing Net Ionic Equations

The equation that we developed in Example 4.11 is written as a *molecular equation.* Although technically correct and properly balanced, it conveys no information about the way the products and reactants exist in solution, other than the precipitate. In reality, all four species are ionic compounds; both reactants and one product are actually *dissociated* in solution, existing as ions. Only the product, AgCl, is associated. Our solubility rules predicted that AgCl is a solid, crystalline precipitate. A more useful representation would show all of the reactants and products in the form in which they are actually present in solution, solid AgCl surrounded by an aqueous collection of ions. This is an *ionic equation.* Even greater clarity can be achieved by removing the ions that do not change during the course of the reaction, *spectator ions,* to produce a *net ionic equation.* The net ionic equation only shows the chemical species that actually undergo change. This is the approach that is most faithful to our original concept of a chemical equation: an accurate depiction of chemical *change.* The process we use to write a net ionic equation for a precipitation reaction is as follows:

1. Write a balanced molecular equation for the process being considered. Use the solubility rules to identify any precipitates.
2. Write the ionic equation showing all reactants and products as free ions unless they are precipitates.
3. Ions that appear on both sides of the equation are *spectator ions* (they do not undergo change) and are cancelled out. What remains is the net ionic equation.
4. Ensure that charges and numbers of atoms are balanced, just as they must be in a conventional equation.

Example 4.12 shows how this procedure is applied.

EXAMPLE 4.12   **Writing Net Ionic Equations**

LEARNING GOAL

8 Write net ionic equations, and use solubility rules to predict the formation of a precipitate.

In Example 4.11, we considered the reaction of NaCl and $AgNO_3$. We wrote the chemical equation for this process in the form of a molecular equation. Now, write this process as a net ionic equation.

Solution

Step 1. Write the balanced molecular equation. In Example 4.11, we found:

$$NaCl(aq) + AgNO_3(aq) \longrightarrow AgCl(s) + NaNO_3(aq)$$

Step 2. AgCl is a precipitate and should be written in its associated form; all others are not, and should be written in ionic form:

$$Na^+(aq) + Cl^-(aq) + Ag^+(aq) + NO_3^-(aq) \longrightarrow AgCl(s) + Na^+(aq) + NO_3^-(aq)$$

This is the form of the ionic equation.

Step 3. $Na^+(aq)$ and $NO_3^-(aq)$ appear on both sides of the equation and are spectator ions that cancel out because they are unchanged during the course of the reaction:

$$\cancel{Na^+(aq)} + Cl^-(aq) + Ag^+(aq) + \cancel{NO_3^-(aq)} \longrightarrow AgCl(s) + \cancel{Na^+(aq)} + \cancel{NO_3^-(aq)}$$

Step 4. We must now ensure that both net charge and number of atoms of each kind are the same on both sides of the reaction arrow.

Counting charges:
Reactants: One positive charge, $Ag^+(aq)$, and one negative charge, $Cl^-(aq)$, result in a net charge of zero.
Products: AgCl has a zero charge.

Counting atoms:
Reactants: One Ag and one Cl atom
Products: One Ag and one Cl atom

Charges and numbers of atoms are balanced. The net ionic equation representing the precipitation of silver chloride is:

$$Ag^+(aq) + Cl^-(aq) \longrightarrow AgCl(s)$$

Practice Problem 4.12

Write net ionic equations for the following reactions:

    a. $BaCl_2(aq) + ZnSO_4(aq) \longrightarrow$
    b. $AgNO_3(aq) + Na_2SO_4(aq) \longrightarrow$

▶ For Further Practice: **Questions 4.81 and 4.82.**

Net ionic equations are useful, not only for precipitation reactions, but for any process involving ionic compounds in aqueous solution. We will see net ionic equations representing acid-base reactions in Chapter 8.

Question 4.7   A solution of Na_2S is mixed with a solution of $CuCl_2$. A black precipitate is formed. Write the net ionic equation for the reaction, and identify the black precipitate.

Question 4.8   A solution of Na_2CO_3 is mixed with a solution of $CaCl_2$. A white precipitate is formed. Write the net ionic equation for the reaction, and identify the white precipitate.

4.7 Acid-Base Reactions

Another approach to the classification of chemical reactions is based on the gain or loss of hydrogen ions. **Acid-base reactions** involve the transfer of a *hydrogen ion, H^+,* from one reactant (the acid) to another (the base).

A common example of an acid-base reaction involves hydrochloric acid and sodium hydroxide:

$$HCl(aq) + NaOH(aq) \longrightarrow H_2O(l) + NaCl(aq)$$
$$\text{Acid} \qquad \text{Base} \qquad\qquad \text{Water} \qquad \text{Salt}$$

A hydrogen cation, H^+, is transferred from the acid to the base, producing water and a salt in solution. When equal molar quantities of acid and base are mixed, the resulting reaction is termed a **neutralization reaction.** The net ionic equation for this process is

$$H^+(aq) + OH^-(aq) \longrightarrow H_2O(l)$$

Acid and base properties and reactions are important in virtually all aspects of chemistry and biochemistry; for this reason, they are considered in much greater detail in Chapter 8.

4.8 Oxidation-Reduction Reactions

Oxidation-reduction reactions are responsible for many types of chemical change. Corrosion of metals, combustion of fossil fuels, the operation of a battery, and biochemical energy-harvesting reactions are a few examples. In this section, we explore the basic concepts underlying this class of chemical reactions.

Oxidation and Reduction

LEARNING GOAL

10 Write oxidation and reduction half-reactions, and identify oxidizing agents and reducing agents.

Oxidation is defined as a loss of electrons, loss of hydrogen atoms, or gain of oxygen atoms.

Magnesium metal, is, for example, oxidized to a *magnesium ion,* losing two electrons when it reacts with a nonmetal such as chlorine:

$$Mg \longrightarrow Mg^{2+} + 2e^-$$

Reduction is defined as a gain of electrons, gain of hydrogen atoms, or loss of oxygen atoms.

Diatomic chlorine is reduced to *chloride ions* by gaining electrons when it reacts with a metal such as magnesium:

$$Cl_2 + 2e^- \longrightarrow 2Cl^-$$

Oxidation and reduction are complementary processes and are often termed *redox reactions.* The *oxidation half-reaction* produces electrons that are the reactants for the *reduction half-reaction.* The combination of two half-reactions, one oxidation and one reduction, produces the complete reaction:

Oxidation half-reaction: $\quad Mg \qquad\quad \longrightarrow Mg^{2+} + \cancel{2e^-}$
Reduction half-reaction: $\quad Cl_2 + \cancel{2e^-} \longrightarrow 2Cl^-$
Complete redox reaction: $\quad Mg + Cl_2 \longrightarrow Mg^{2+} + 2Cl^-$

Passing an electrical current through water causes an oxidation-reduction reaction. The products are H_2 and O_2 in a 2:1 ratio. Is this ratio of products predicted by the equation for the decomposition of water, $2H_2O(l) \longrightarrow 2H_2(g) + O_2(g)$? Explain.

Half-reactions, one oxidation and one reduction, are exactly that: one-half of a complete reaction. The two half-reactions combine to produce the complete reaction. Note that the electrons cancel: in the electron transfer process, no free electrons remain.

Oxidation cannot occur without reduction, and vice versa. The reducing agent becomes oxidized and the oxidizing agent becomes reduced.

In the preceding reaction, magnesium metal is the **reducing agent.** It releases electrons for the reduction of chlorine. Chlorine is the **oxidizing agent.** It accepts electrons from the magnesium, which is oxidized.

The characteristics of oxidizing and reducing agents may be summarized as follows:

Oxidizing Agent	**Reducing Agent**
• Is reduced	• Is oxidized
• Gains electrons	• Loses electrons
• Causes oxidation	• Causes reduction

Oxidation-reduction processes are more difficult to identify using half-reactions when covalent compounds are involved. For example, consider the reaction of methane with oxygen:

$$CH_4(g) + 2O_2(g) \longrightarrow CO_2(g) + 2H_2O(g)$$

Loses H

Gains H

Our definitions of oxidation and reduction make it clear that this is an oxidation-reduction reaction. Carbon has lost hydrogen and gained oxygen, becoming CO_2 (our definition of oxidation). Oxygen has gained hydrogen, becoming water (the gain of hydrogen indicates reduction).

Consider the reaction of aluminum with atmospheric oxygen:

$$4Al(s) + 3O_2(g) \longrightarrow 2Al_2O_3(s)$$

Gains O

Aluminum is strong, and its low density makes it ideal for the wings and cabin exterior of airplanes. Aluminum oxide forms on the metal surface, and this oxide coating protects the metal from further reaction.

It is evident that aluminum has been oxidized because it gains oxygen atoms.

Question 4.9   Write the oxidation half-reaction, the reduction half-reaction, and the complete reaction for the formation of calcium sulfide from the elements Ca and S. Remember, the electron gain *must* equal the electron loss.

Question 4.10   Write the oxidation half-reaction, the reduction half-reaction, and the complete reaction for the formation of calcium iodide from calcium metal and I_2. Remember, the electron gain *must* equal the electron loss.

Question 4.11   Identify the oxidizing agent, reducing agent, substance oxidized, and substance reduced in the reaction described in Question 4.9.

Question 4.12   Identify the oxidizing agent, reducing agent, substance oxidized, and substance reduced in the reaction described in Question 4.10.

Voltaic Cells

LEARNING GOAL

11   Compare and contrast voltaic and electrolytic cells.

When zinc metal is dipped into a copper(II) sulfate solution, zinc atoms are oxidized to zinc ions and copper(II) ions are reduced to copper metal, which deposits on the surface of the zinc metal (Figure 4.6). This reaction is summarized as follows:

Oxidation/e⁻ loss

$$Zn(s) + Cu^{2+}(aq) \longrightarrow Zn^{2+}(aq) + Cu(s)$$

Reduction/e⁻ gain

In the reduction of aqueous copper(II) ions by zinc metal, electrons flow from the zinc rod directly to copper(II) ions in the solution. If electron transfer from the zinc rod to the copper ions in solution could be directed through an external

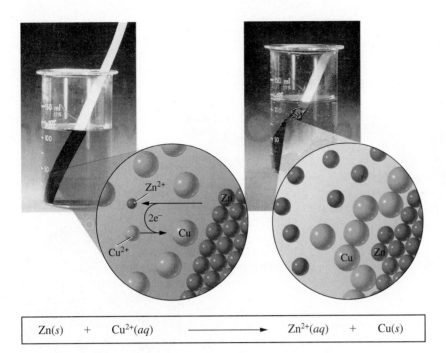

$$Zn(s) \quad + \quad Cu^{2+}(aq) \longrightarrow Zn^{2+}(aq) \quad + \quad Cu(s)$$

Figure 4.6 The spontaneous reaction of zinc metal and Cu^{2+} ions is the basis of the voltaic cell depicted in Figure 4.7.

electrical circuit, this spontaneous oxidation-reduction reaction could be used to produce an electrical current that could perform some useful function.

However, when zinc metal in one container is connected by a copper wire with a copper(II) sulfate solution in a separate container, no current flows through the wire. A complete, or continuous, circuit is necessary for current to flow. To complete the circuit, we connect the two containers with a tube filled with a solution of an electrolyte such as potassium chloride. This tube is described as a *salt bridge*.

Current now flows through the external circuit (Figure 4.7). The device shown in Figure 4.7 is an example of a *voltaic cell*. A **voltaic cell** is an *electrochemical* cell that converts stored *chemical* energy into *electrical* energy.

Solutions of ionic salts are good conductors of electricity (Chapter 6).

This cell consists of two *half-cells*. The oxidation half-reaction occurs in one half-cell, and the reduction half-reaction occurs in the other half-cell. The sum of the two half-cell reactions is the overall oxidation-reduction reaction that describes the cell. The electrode at which oxidation occurs is called the **anode,** and the electrode at which reduction occurs is the **cathode.** In the device shown in Figure 4.7, the zinc metal is the anode. At this electrode, the zinc atoms are oxidized to zinc ions:

$$\text{Anode half-reaction: } Zn(s) \longrightarrow Zn^{2+}(aq) + 2e^-$$

Electrons released at the anode travel through the external circuit to the cathode (the copper rod) where they are transferred to copper(II) ions in the solution. Copper(II) ions are reduced to copper atoms that deposit on the copper metal surface, the cathode:

$$\text{Cathode half-reaction: } Cu^{2+}(aq) + 2e^- \longrightarrow Cu(s)$$

The sum of these half-cell reactions is the voltaic cell reaction:

$$Zn(s) + Cu^{2+}(aq) \longrightarrow Zn^{2+}(aq) + Cu(s)$$

Voltaic cells are found in many aspects of our lives, as a convenient and reliable source of electrical energy, the battery. Batteries convert stored chemical energy to an electrical current to power a wide array of different commercial appliances: radios, cellphones, computers, flashlights, and a host of other useful devices.

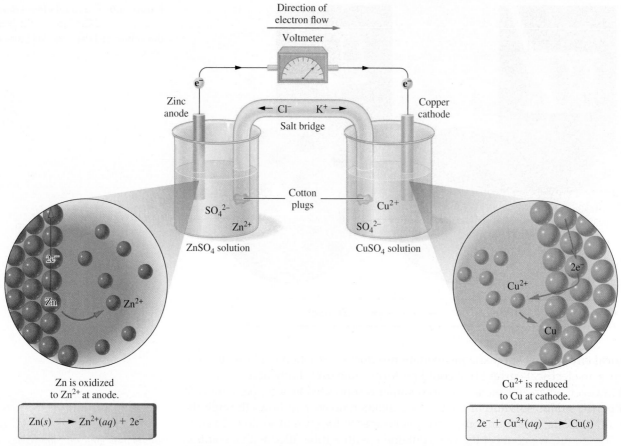

Figure 4.7 A voltaic cell generating electrical current by the reaction:

$$Zn(s) + Cu^{2+}(aq) \longrightarrow Zn^{2+}(aq) + Cu(s)$$

Each electrode consists of the pure metal, zinc or copper. Zinc is oxidized, releasing electrons that flow to the copper, reducing Cu^{2+} to Cu. The salt bridge completes the circuit, and the voltmeter displays the voltage (or chemical potential) associated with the reaction.

Technology has made modern batteries smaller, safer, and more dependable than our crudely constructed copper-zinc voltaic cell. A new generation of biocompatible batteries are now available and can be implanted in the human body as part of a pacemaker circuit used to improve heart rhythm.

Electrolysis

LEARNING GOAL

11   Compare and contrast voltaic and electrolytic cells.

Electrolytic cells are the reverse of voltaic cells.

Electrolysis reactions use electrical energy to cause nonspontaneous oxidation-reduction reactions to occur. One common application of electrolysis is the rechargeable battery. When it is being used to power a device, such as a laptop computer, it behaves as a voltaic cell. After some time, the chemical reaction approaches completion and the voltaic cell "runs down." The cell reaction is reversible, and the battery is plugged into a battery charger. The charger is really an external source of electrical energy that reverses the chemical reaction in the battery, bringing it back to its original state. The cell has been operated as an electrolytic cell. Removal of the charging device turns the cell back into a voltaic device, ready to spontaneously react to produce electrical current once again.

Another common example of electrolysis is electroplating. This process is used to cover less expensive and unappealing metal objects with a thin, chemically stable, and (usually) lustrous surface. Chromium (chrome plating) and silver (silver-plated dinnerware) are most common.

The object to be plated (for example, a spoon or fork) is the cathode (negative charge) in an electrolytic cell. The solution surrounding the cathode contains silver ions (Ag^+) that migrate to the cathode, undergoing reduction:

$$\text{Cathode half-reaction: } Ag^+(aq) + e^- \longrightarrow Ag(s)$$

The silver metal remains on the surface of the object being plated. The spoon or fork has the appearance of expensive silver at a fraction of the cost.

Question 4.13 Chrome plating involves the reduction of $Cr^{3+}(aq)$ at the surface of the electrode to be electroplated.
   a. Must the electrode that is being electroplated have a positive or negative charge? Why?
   b. Is this electrode termed the cathode or the anode?
   c. Write the reduction reaction for $Cr^{3+}(aq)$.

Question 4.14 $Cu^{2+}(aq)$ can be reduced in an electrolytic cell to prepare high-purity copper metal.
   a. Does this reduction take place at the positive or negative electrode of the electrolytic cell? Why?
   b. In this electrode termed the cathode or the anode?
   c. Write the reduction reaction for $Cu^{2+}(aq)$.

Applications of Oxidation and Reduction

Oxidation-reduction reactions are important in many areas as diverse as industrial manufacturing and biochemical processes.

Corrosion

The deterioration of metals caused by an oxidation-reduction reaction is termed **corrosion.** Metal atoms are converted to metal ions; the structure, hence the properties, changes dramatically and usually for the worse.

Millions of dollars are spent annually in an attempt to remedy damage resulting from corrosion. A current area of chemical research is concerned with the development of corrosion-inhibiting processes. In one type of corrosion, elemental iron is oxidized to iron(III) oxide (rust):

$$4Fe(s) + 3O_2(g) \longrightarrow 2Fe_2O_3(s)$$

Combustion

Burning fossil fuel (coal, oil, gasoline, and natural gas) releases energy. This energy is used to heat our homes, offices, and classrooms. The combustion of gasoline is used for transportation. The simplest fossil fuel and major component of natural gas is methane, CH_4, and its oxidation reaction is written:

$$CH_4(g) + 2O_2(g) \longrightarrow CO_2(g) + 2H_2O(g)$$

Methane is a hydrocarbon because its chemical formula contains only hydrogens and carbon. The complete oxidation of any hydrocarbon (including those in gasoline, heating oil, and liquid propane) produces carbon dioxide and water. The unseen product, energy, released by these reactions is of paramount importance. The water and carbon dioxide are viewed as waste products, and the carbon dioxide contributes to the *greenhouse effect* (Chapter 5) and global warming.

Bleaching

Bleaching agents are most often oxidizing agents. Since sodium hypochlorite (NaOCl) is an effective oxidizing agent, it is active ingredient in a variety of laundry products advertised for their stain-removing capabilities. Stains are a result of colored compounds adhering to surfaces. Oxidation of these compounds produces

LEARNING GOAL

12 Describe examples of redox processes.

At the same time that iron is oxidized, O_2 is being reduced to O^{2-} and is incorporated into the structure of iron(III) oxide. Electrons lost by iron reduce oxygen. This again shows that oxidation and reduction reactions go hand in hand.

See Chapters 21–23 for the details of these energy-harvesting cellular oxidation-reduction reactions.

Some reactions of metals with oxygen are very rapid. A dramatic example is the reaction of magnesium with oxygen.

$$2Mg(s) + O_2(g) \longrightarrow 2MgO(s)$$

products that are not colored or compounds that are subsequently easily removed from the surface, thus removing the stain.

Metabolism

When ethanol is metabolized in the liver, it is oxidized to acetaldehyde (the molecule partially responsible for hangovers). Continued oxidation of acetaldehyde produces acetic acid, which is eventually oxidized to CO_2 and H_2O. These reactions, summarized as follows, are catalyzed by liver enzymes.

$$CH_3CH_2OH \xrightarrow{\substack{liver \\ enzymes}} CH_3\overset{\overset{\displaystyle O}{\|}}{C}H \xrightarrow{\substack{liver \\ enzymes}} CH_3\overset{\overset{\displaystyle O}{\|}}{C}OH \xrightarrow{\substack{liver \\ enzymes}} CO_2 + H_2O$$

ethanol   acetaldehyde   acetic acid

It is more difficult to recognize these reactions as oxidations because neither the product nor the reactant carries a charge. In previous examples, we looked for an increase in positive charge, signifying a loss of electrons, as an indication that an oxidation had occurred. An increase in positive charge (or decrease in negative charge) would signify oxidation.

Once again our descriptions of oxidation and reduction are useful in identifying these reactions. Recall:

> Oxidation is the *gain* of oxygen or *loss* of hydrogen.
> Reduction is the *loss* of oxygen or *gain* of hydrogen.

This strategy is most useful for recognizing oxidation and reduction of *organic compounds* and organic compounds of biological interest, *biochemical compounds*.

In our example of the conversion of ethanol to acetaldehyde, ethanol has six hydrogen atoms per molecule; the product acetaldehyde has four hydrogen atoms per molecule. This represents a loss of two hydrogen atoms per molecule. Therefore, ethanol has been oxidized to acetaldehyde. In a subsequent step, the acetaldehyde is oxidized to acetic acid (the active ingredient in vinegar).

Acetaldehyde has one oxygen atom per molecule; acetic acid has two. An increase in the number of oxygen atoms per molecule indicates oxidation has occurred. In the process of fermentation of grapes to make wine, the ethanol formed may convert to acetic acid over time, and vinegar, rather than wine, is the result.

Organic compounds and their structures and reactivity are the focus of Chapters 10 through 15, and biochemical compounds are described in Chapters 16 through 23.

Chemical Control of Microbes

Many common antiseptics and disinfectants are oxidizing agents. An antiseptic is used to destroy pathogens associated with living tissue. A disinfectant is a substance that is used to kill or inhibit the growth of pathogens, disease-causing microorganisms, on environmental surfaces.

Hydrogen peroxide is an effective antiseptic that is commonly used to cleanse cuts and abrasions. We are all familiar with the furious bubbling (due to the rapid release of oxygen gas) that occurs as catalase, an enzyme from our cells, catalyzes the breakdown of H_2O_2:

$$2H_2O_2(aq) \longrightarrow 2H_2O(l) + O_2(g)$$

At higher concentrations (3–6%), H_2O_2 is used as a disinfectant. It is particularly useful for disinfection of soft contact lenses, utensils, and surgical implants because there is no residual toxicity. Concentrations of 6–25% are even used for complete sterilization of environmental surfaces. Disinfectants such as H_2O_2 solutions are critically important in the fight against outbreaks of highly infectious diseases such as Ebola and influenza.

4.9   Calculations Using the Chemical Equation

LEARNING GOAL

13   Use a chemical equation and a given number of moles or mass of a reactant or product to calculate the number of moles or mass of a reactant or product.

General Principles

The calculation of quantities of products and reactants based on a balanced chemical equation is termed *stoichiometry* and is important in many fields. The synthesis of drugs and other complex molecules on a large scale is conducted on the basis of a balanced equation. This minimizes the waste of expensive chemical compounds used in these reactions. Similarly, the ratio of fuel and air in a home furnace or automobile must be adjusted carefully, according to their combining ratio, to maximize energy conversion, minimize fuel consumption, and minimize pollution.

In carrying out chemical calculations, we apply the following guidelines:

1. The chemical formulas of all reactants and products must be known.
2. The basis for the calculations is a balanced equation because the conservation of mass must be obeyed. If the equation is not properly balanced, the calculation is meaningless.
3. The calculations are performed in terms of mol. The coefficients in the balanced equation represent the relative number of mol of products and reactants.

We have seen that the number of mol of products and reactants often differs in a balanced equation. For example,

$$C(s) + O_2(g) \longrightarrow CO_2(g)$$

is a balanced equation. Two mol of reactants combine to produce 1 mol of product:

$$1 \text{ mol C} + 1 \text{ mol O}_2 \longrightarrow 1 \text{ mol CO}_2$$

However, 1 mol of C *atoms* and 2 mol of O *atoms* produce 1 mol of C *atoms* and 2 mol of O *atoms*. In other words, the number of mol of reactants and products may differ, but the number of mol of atoms cannot. The formation of CO_2 from C and O_2 may be described as follows:

$$C(s) + O_2(g) \longrightarrow CO_2(g)$$
$$1 \text{ mol C} + 1 \text{ mol O}_2 \longrightarrow 1 \text{ mol CO}_2$$
$$12.0 \text{ g C} + 32.0 \text{ g O}_2 \longrightarrow 44.0 \text{ g CO}_2$$

Using Conversion Factors

Conversion Between Moles and Grams

The mol is the basis of our calculations. However, mol are generally measured in g. A facility for interconversion of mol and g is fundamental to chemical arithmetic (see Figure 4.2). Conversion from mol to g, and vice versa, requires only the formula mass of the compound of interest. These calculations, discussed earlier in this chapter, are reviewed in Example 4.13.

EXAMPLE 4.13	**Converting Between Moles and Mass**

a. Convert 1.00 mol of oxygen gas, O_2, to g.

Solution

Step 1.   Use the following path:

$$\text{mol O}_2 \longrightarrow \text{g O}_2$$

LEARNING GOAL

3   Determine molar mass, and demonstrate how it is used to solve mole and mass conversion calculations.

Continued...

Step 2. The molar mass of oxygen (O_2) is 32.00 g O_2, and the conversion factor becomes:

$$\frac{32.00 \text{ g } O_2}{1 \text{ mol } O_2}$$

Step 3. Using the conversion factor (ensure that mol O_2 cancel):

$$1.00 \text{ mol } O_2 \times \frac{32.00 \text{ g } O_2}{1 \text{ mol } O_2} = 32.0 \text{ g } O_2$$

Data Given × Conversion Factor = Desired Result

b. How many g of carbon dioxide are contained in 10.0 mol of carbon dioxide?

Solution

Step 1. Use the following path:

$$\boxed{\text{mol } CO_2} \longrightarrow \boxed{\text{g } CO_2}$$

Step 2. The molar mass of CO_2 is 44.01 g CO_2, and the conversion factor becomes:

$$\frac{44.01 \text{ g } CO_2}{1 \text{ mol } CO_2}$$

Step 3. Using the conversion factor (ensure that mol CO_2 cancel):

$$10.0 \text{ mol } CO_2 \times \frac{44.01 \text{ g } CO_2}{1 \text{ mol } CO_2} = 4.40 \times 10^2 \text{ g } CO_2$$

Data Given × Conversion Factor = Desired Result

Helpful Hint: Note that each conversion factor can be inverted, producing a second possible conversion factor. Only one will allow the appropriate unit cancellation.

Practice Problem 4.13

Perform each of the following conversions:

a. 5.00 mol of water to g of water   c. 1.00×10^{-5} mol of $C_6H_{12}O_6$ to micrograms (µg) of $C_6H_{12}O_6$
b. 25.0 g of LiCl to mol of LiCl   d. 35.0 g of $MgCl_2$ to mol of $MgCl_2$

▶ For Further Practice: **Questions 4.33 and 4.34.**

LEARNING GOAL

13 Use a chemical equation and a given number of moles or mass of a reactant or product to calculate the number of moles or mass of a reactant or product.

Conversion of Moles of Reactants to Moles of Products

In Example 4.9, we balanced the equation for the reaction of propane and oxygen as follows:

$$C_3H_8(g) + 5O_2(g) \longrightarrow 3CO_2(g) + 4H_2O(g)$$

In this reaction, 1 mol of C_3H_8 corresponds to, or results in,

5 mol of O_2 being consumed and
3 mol of CO_2 being formed and
4 mol of H_2O being formed.

This information may be written in the form of a conversion factor, the molar ratio:

$$\frac{1 \text{ mol C}_3\text{H}_8}{5 \text{ mol O}_2}$$ *Translated:* 1 mol of C_3H_8 reacts with 5 mol of O_2.

$$\frac{1 \text{ mol C}_3\text{H}_8}{3 \text{ mol CO}_2}$$ *Translated:* 1 mol of C_3H_8 produces 3 mol of CO_2.

$$\frac{1 \text{ mol C}_3\text{H}_8}{4 \text{ mol H}_2\text{O}}$$ *Translated:* 1 mol of C_3H_8 produces 4 mol of H_2O.

Conversion factors based on the chemical equation permit us to perform a variety of calculations.

Let us look at Examples 4.14–4.16; each is based on the combustion of propane and the equation that we balanced in Example 4.9.

EXAMPLE 4.14 | **Calculating Mass of Reactant from Moles of Reactant**

Calculate the number of g of O_2 that will react with 1.00 mol of C_3H_8. The balanced equation is: $C_3H_8(g) + 5O_2(g) \longrightarrow 3CO_2(g) + 4H_2O(g)$.

LEARNING GOAL

13 Use a chemical equation and a given number of moles or mass of a reactant or product to calculate the number of moles or mass of a reactant or product.

Solution

Step 1. Our path is:

$$\boxed{\begin{array}{c}\text{mol} \\ \text{C}_3\text{H}_8\end{array}} \longrightarrow \boxed{\begin{array}{c}\text{mol} \\ \text{O}_2\end{array}} \longrightarrow \boxed{\begin{array}{c}\text{g} \\ \text{O}_2\end{array}}$$

Step 2. Two conversion factors are necessary to solve this problem:

- conversion from mol of C_3H_8 to mol of O_2 and
- conversion of mol of O_2 to g of O_2.

The balanced equation indicates that the conversion factor relating the two reactants is based on 1 mol C_3H_8 = 5 mol O_2. The conversion factor that relates mol to g is constructed from the molar mass, 32.00 g O_2 = 1 mol O_2.

Step 3. Set up conversion factors to cancel mol C_3H_8 and mol O_2:

$$1.00 \text{ mol C}_3\text{H}_8 \times \frac{5 \text{ mol O}_2}{1 \text{ mol C}_3\text{H}_8} \times \frac{32.00 \text{ g O}_2}{1 \text{ mol O}_2} = 1.60 \times 10^2 \text{ g O}_2$$

Data Given × Conversion Factor × Conversion Factor = Desired Result

---

Practice Problem 4.14

When potassium cyanide (KCN) reacts with hydrochloric acid, hydrogen cyanide (HCN), a poisonous gas, is produced. The equation is

$$KCN(aq) + HCl(aq) \longrightarrow KCl(aq) + HCN(g)$$

Calculate the number of g of KCN that will react with 1.00 mol of HCl.

▶ For Further Practice: **Questions 4.101 and 4.102.**

EXAMPLE 4.15   **Calculating Mass of Product from Moles of Reactant**

LEARNING GOAL

13 Use a chemical equation and a given number of moles or mass of a reactant or product to calculate the number of moles or mass of a reactant or product.

Calculate the number of g of CO_2 produced from the combustion of 1.00 mol of C_3H_8 using the balanced equation from Example 4.14.

Solution

Step 1. Use the following path:

$$\text{mol } C_3H_8 \longrightarrow \text{mol } CO_2 \longrightarrow \text{g } CO_2$$

Step 2. The balanced equation indicates that the conversion factor relating the reactant C_3H_8 to the product CO_2 is based on 1 mol C_3H_8 = 3 mol CO_2. The conversion factor that relates mol to g is constructed from the molar mass, 44.01 g CO_2 = 1 mol CO_2.

Step 3. Set up conversion factors to cancel mol C_3H_8 and mol CO_2:

$$1.00 \text{ mol } C_3H_8 \times \frac{3 \text{ mol } CO_2}{1 \text{ mol } C_3H_8} \times \frac{44.01 \text{ g } CO_2}{1 \text{ mol } CO_2} = 132 \text{ g } CO_2$$

Data Given × Conversion Factor × Conversion Factor = Desired Result

Practice Problem 4.15

Fermentation is a critical step in the process of winemaking. The reaction is

$$\underset{\text{glucose}}{C_6H_{12}O_6(aq)} \longrightarrow 2\underset{\text{ethanol}}{CH_3CH_2OH(aq)} + 2CO_2(g)$$

Calculate the number of g of ethanol produced from the fermentation of 5.00 mol of glucose.

▶ For Further Practice: **Questions 4.103 and 4.104.**

EXAMPLE 4.16   **Calculating Mass of Reactant from Mass of Product**

LEARNING GOAL

13 Use a chemical equation and a given number of moles or mass of a reactant or product to calculate the number of moles or mass of a reactant or product.

Calculate the number of g of C_3H_8 required to produce 36.0 g of H_2O using the balanced equation from Example 4.14.

Solution

Step 1. Use the following path:

$$\text{g } H_2O \longrightarrow \text{mol } H_2O \longrightarrow \text{mol } C_3H_8 \longrightarrow \text{g } C_3H_8$$

Step 2. It is necessary to convert g of H_2O to mol of H_2O, mol of H_2O to mol of C_3H_8, and mol of C_3H_8 to g of C_3H_8. The molar mass of H_2O (18.0 g H_2O = 1 mol H_2O) is the conversion factor that will convert from g H_2O to mol H_2O. The balanced equation indicates that the conversion factor relating the reactant C_3H_8 to the product H_2O is based on 1 mol C_3H_8 = 4 mol H_2O. The conversion factor that relates mol to g is constructed from the molar mass, 44.09 g C_3H_8 = 1 mol C_3H_8.

A Human Perspective

The Chemistry of Automobile Air Bags

Each year, thousands of individuals are killed or seriously injured in automobile accidents. Perhaps most serious is the front-end collision. The car decelerates or stops virtually on impact; the momentum of the passengers, however, does not stop, and the driver and passengers are thrown forward toward the dashboard and the windshield. Suddenly, passive parts of the automobile, such as control knobs, the rearview mirror, the steering wheel, the dashboard, and the windshield, become lethal weapons.

Automobile engineers have been aware of these problems for a long time. They have made a series of design improvements to lessen the potential problems associated with front-end impact. Smooth switches rather than knobs, recessed hardware, and padded dashboards are examples. These changes, coupled with the use of lap and shoulder belts, which help to immobilize occupants of the car, have decreased the frequency and severity of occupant impact and lowered the death rate for this type of accident.

An almost ideal protection would be a soft, fluffy pillow, providing a cushion against impact. Such a device, an air bag inflated only on impact, is now standard equipment for the protection of the driver and front-seat passenger.

How does it work? Ideally, it inflates only when severe front-end impact occurs; it inflates very rapidly [in approximately 40 milliseconds (ms)], then deflates to provide a steady deceleration, cushioning the occupants from impact. A variety of simple chemical reactions make this a reality.

Our strategy uses sodium azide, NaN_3. When solid sodium azide is detonated by mechanical energy produced by an electrical current, it decomposes to form solid sodium and nitrogen gas:

$$2NaN_3(s) \longrightarrow 2Na(s) + 3N_2(g)$$

State trooper inspects accident with airbag deployment

The nitrogen gas inflates the air bag, cushioning the driver and front-seat passenger.

The solid sodium azide has a high density (characteristic of solids) and thus occupies a small volume. It can easily be stored in the center of a steering wheel or in the dashboard. The rate of detonation is very rapid. The reaction produces 3 mol of N_2 gas for every 2 mol of NaN_3. The N_2 gas occupies a large volume relative to the solid NaN_3 because, like all gases, its density is low.

Figuring out how much sodium azide is needed to produce enough nitrogen to properly inflate the bag is an example of a practical application of the chemical arithmetic we are learning in this chapter.

For Further Understanding

▶ Why is nitrogen gas preferred for this application?
▶ If this reaction occurs in 5 ms, how many seconds (s) does this represent?

Step 3. Set up conversion factors to cancel g H_2O, mol H_2O, and mol C_3H_8:

$$36.0 \text{ g } H_2O \times \frac{1 \text{ mol } H_2O}{18.0 \text{ g } H_2O} \times \frac{1 \text{ mol } C_3H_8}{4 \text{ mol } H_2O} \times \frac{44.09 \text{ g } C_3H_8}{1 \text{ mol } C_3H_8} = 22.0 \text{ g } C_3H_8$$

Data Given × Conversion Factor × Conversion Factor × Conversion Factor = Desired Result

Practice Problem 4.16

The balanced equation for the combustion of ethanol (ethyl alcohol) is

$$C_2H_5OH(l) + 3O_2(g) \longrightarrow 2CO_2(g) + 3H_2O(g)$$

a. How many mol of O_2 will react with 1 mol of ethanol?
b. How many g of O_2 will react with 1 mol of ethanol?
c. How many g of CO_2 will be produced by the combustion of 1 mol of ethanol?

▶ For Further Practice: **Questions 4.105 and 4.106.**

Let's consider an example that requires us to write and balance the chemical equation, use conversion factors, and calculate the amount of a reactant consumed in the chemical reaction.

EXAMPLE 4.17 Calculating Reactant Quantities

Calcium hydroxide may be used to neutralize (completely react with) aqueous hydrochloric acid. Calculate the number of g of hydrochloric acid that would be neutralized by 0.500 mol of solid calcium hydroxide.

LEARNING GOAL

13 Use a chemical equation and a given number of moles or mass of a reactant or product to calculate the number of moles or mass of a reactant or product.

Solution

Step 1. The chemical formulas for $Ca(OH)_2$ and HCl are used to write the unbalanced equation that includes the products of this acid-base reaction, calcium chloride and water:

$$Ca(OH)_2(s) + HCl(aq) \longrightarrow CaCl_2(aq) + H_2O(l)$$

Step 2. Balance the equation:

$$Ca(OH)_2(s) + 2HCl(aq) \longrightarrow CaCl_2(aq) + 2H_2O(l)$$

Step 3. Use the following path:

$$\boxed{\begin{array}{c}\text{mol}\\Ca(OH)_2\end{array}} \longrightarrow \boxed{\begin{array}{c}\text{mol}\\HCl\end{array}} \longrightarrow \boxed{\begin{array}{c}\text{g}\\HCl\end{array}}$$

Step 4. Determine the necessary conversions:

- mol of $Ca(OH)_2$ to mol of HCl and
- mol of HCl to g of HCl

The balanced equation indicates that the conversion factor relating the two reactants is based on 1 mol $Ca(OH)_2$ = 2 mol HCl. The conversion factor that relates mol HCl to g HCl is constructed from the molar mass, 36.46 g HCl = 1 mol HCl.

Step 5.

$$0.500 \text{ mol } Ca(OH)_2 \times \frac{2 \text{ mol } HCl}{1 \text{ mol } Ca(OH)_2} \times \frac{36.46 \text{ g } HCl}{1 \text{ mol } HCl} = 36.5 \text{ g } HCl$$

Data Given × Conversion Factor × Conversion Factor = Desired Result

This reaction is illustrated in Figure 4.8.

Helpful Hints:

1. Writing ionic formulas from compound names is explained in Section 3.2.

2. Recall that the reaction between an acid and a base produces a salt and water (Section 4.7 and further discussed in Chapter 8).

3. Remember to balance the chemical equation; the proper coefficients are essential parts of the subsequent calculations.

Practice Problem 4.17

Metallic iron reacts with O_2 gas to produce iron(III) oxide.

    a. Write and balance the equation.   b. Calculate the number of g of iron needed to produce 5.00 g of product.

▶ For Further Practice: **Questions 4.109 and 4.110.**

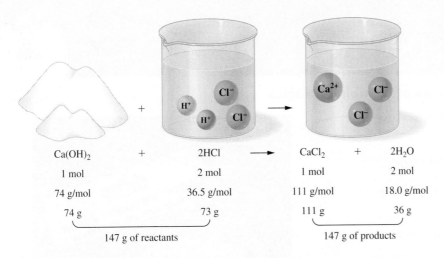

Figure 4.8 An illustration of the law of conservation of mass. In this example, 1 mol of calcium hydroxide and 2 mol of hydrogen chloride react to produce 3 mol of product (2 mol of water and 1 mol of calcium chloride). The total mass, in g, of reactant(s) consumed is equal to the total mass, in g, of product(s) formed. *Note:* In reality, HCl does not exist as discrete molecules in water. The HCl separates to form H^+ and Cl^-. Ionization in water will be discussed with the chemistry of acids and bases in Chapter 8.

$Ca(OH)_2$	+	$2HCl$	$\longrightarrow$	$CaCl_2$	+	$2H_2O$
1 mol		2 mol		1 mol		2 mol
74 g/mol		36.5 g/mol		111 g/mol		18.0 g/mol
74 g		73 g		111 g		36 g

147 g of reactants   147 g of products

EXAMPLE 4.18   **Calculating Reactant Quantities**

LEARNING GOAL

What mass of sodium hydroxide, NaOH, would be required to produce 8.00 g of the antacid milk of magnesia, $Mg(OH)_2$, by the reaction of $MgCl_2$ with NaOH?

13 Use a chemical equation and a given number of moles or mass of a reactant or product to calculate the number of moles or mass of a reactant or product.

Solution

Step 1. Write and balance the equation:

$$MgCl_2(aq) + 2NaOH(aq) \longrightarrow Mg(OH)_2(s) + 2NaCl(aq)$$

Step 2. Determine the strategy:

The equation tells us that 2 mol of NaOH form 1 mol of $Mg(OH)_2$. If we calculate the number of mol of $Mg(OH)_2$ in 8.00 g of $Mg(OH)_2$, we can determine the number of mol of NaOH necessary and then the mass of NaOH required:

$$\boxed{\text{g } Mg(OH)_2} \longrightarrow \boxed{\text{mol } Mg(OH)_2} \longrightarrow \boxed{\text{mol } NaOH} \longrightarrow \boxed{\text{g } NaOH}$$

Step 3. The conversion factor to convert g $Mg(OH)_2$ to mol $Mg(OH)_2$ is:

$$58.3 \text{ g } Mg(OH)_2 = 1 \text{ mol } Mg(OH)_2$$

Therefore, using the molar mass of $Mg(OH)_2$,

$$8.00 \text{ g } Mg(OH)_2 \times \frac{1 \text{ mol } Mg(OH)_2}{58.3 \text{ g } Mg(OH)_2} = 0.137 \text{ mol } Mg(OH)_2$$

Step 4. Two moles of NaOH react to give 1 mol of $Mg(OH)_2$. Therefore, using the molar ratio,

$$0.137 \text{ mol } Mg(OH)_2 \times \frac{2 \text{ mol } NaOH}{1 \text{ mol } Mg(OH)_2} = 0.274 \text{ mol } NaOH$$

Step 5. The conversion factor to convert mol to mass NaOH is: 40.0 g of NaOH = 1 mol of NaOH. Therefore, using the molar mass of NaOH,

$$0.274 \text{ mol } NaOH \times \frac{40.0 \text{ g } NaOH}{1 \text{ mol } NaOH} = 11.0 \text{ g } NaOH$$

This calculation may also be done in a single step:

$$8.00 \text{ g } Mg(OH)_2 \times \frac{1 \text{ mol } Mg(OH)_2}{58.3 \text{ g } Mg(OH)_2} \times \frac{2 \text{ mol } NaOH}{1 \text{ mol } Mg(OH)_2} \times \frac{40.0 \text{ g } NaOH}{1 \text{ mol } NaOH} = 11.0 \text{ g } NaOH$$

Data Given × Conversion Factor × Conversion Factor × Conversion Factor = Desired Result

Continued...

A Medical Perspective

Carbon Monoxide Poisoning: A Case of Combining Ratios

A fuel, such as methane, CH_4, burned in an excess of oxygen produces carbon dioxide and water:

$$CH_4(g) + 2O_2(g) \longrightarrow CO_2(g) + 2H_2O(g)$$

The same combustion in the presence of insufficient oxygen produces carbon monoxide and water:

$$2CH_4(g) + 3O_2(g) \longrightarrow 2CO(g) + 4H_2O(g)$$

The combustion of methane, repeated over and over in millions of gas furnaces that use natural gas as a fuel, is responsible for heating many of our homes in the winter. The furnace is designed to operate under conditions that favor the first reaction and minimize the second; excess oxygen is available from the surrounding atmosphere. Furthermore, the vast majority of exhaust gases (containing principally CO, CO_2, H_2O, and unburned fuel) are removed from the home through the chimney. However, if the chimney becomes obstructed, or the burner malfunctions, carbon monoxide levels within the home can rapidly reach hazardous levels.

Why is exposure to carbon monoxide hazardous? Hemoglobin, an iron-containing compound in our blood, binds with O_2 and transports it throughout the body. Carbon monoxide also combines with hemoglobin, thereby blocking oxygen transport. The binding affinity of hemoglobin for carbon monoxide is about 200 times greater than for O_2. Therefore, to maintain O_2 binding and transport capability, our exposure to carbon monoxide must be minimal. Proper ventilation and a suitable oxygen-to-fuel ratio are essential for any combustion process in the home, automobile, or workplace. In recent years, carbon monoxide sensors have been developed. These sensors sound an alarm when potentially toxic levels of CO are reached. Warning devices like this have helped to create a safer indoor environment.

The example we have chosen is an illustration of what is termed the *law of multiple proportions*. This law states that identical reactants may produce different products, depending on their combining ratio. The experimental conditions (in this

A carbon monoxide meter detects the concentration of the odorless and colorless gas and sounds an alarm when the concentration reaches preset levels.

case, the quantity of available oxygen) determine the preferred path of the chemical reaction.

For Further Understanding

▶ Why may new, stricter insulation standards for homes and businesses inadvertently increase the risk of carbon monoxide poisoning?

▶ Explain the link between smoking and carbon monoxide that has motivated many states and municipalities to ban smoking in restaurants, offices, and other indoor spaces.

Helpful Hint: Mass is a laboratory unit, whereas mol is a calculation unit. The laboratory balance is calibrated in units of mass (g). Although mol are essential for this type of calculation, the starting point and objective are often in mass units. As a result, our path is often g $\longrightarrow$ mol $\longrightarrow$ g.

Practice Problem 4.18

Barium carbonate decomposes upon heating to barium oxide and carbon dioxide.

  a. Write and balance the equation.
  b. Calculate the number of g of carbon dioxide produced by heating 50.0 g of barium carbonate.

▶ For Further Practice: **Questions 4.111 and 4.112.**

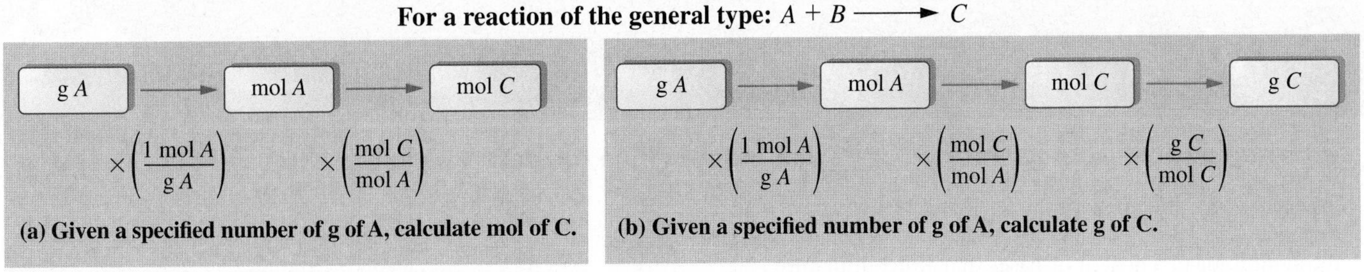
For a reaction of the general type: A + B → C
g A
mol A
mol C
× (1 mol A / g A)
× (mol C / mol A)
(a) Given a specified number of g of A, calculate mol of C.
g A
mol A
mol C
g C
× (1 mol A / g A)
× (mol C / mol A)
× (g C / mol C)
(b) Given a specified number of g of A, calculate g of C.

Figure 4.9 A general problem-solving strategy, using molar quantities.

A general problem-solving strategy is summarized in Figure 4.9. By systematically applying this strategy, you will be able to solve virtually any problem requiring calculations based on the chemical equation.

Theoretical and Percent Yield

LEARNING GOAL

14 Calculate theoretical and percent yields.

The **theoretical yield** is the *maximum* amount of product that can be produced (in an ideal world). In the "real" world, it is difficult to produce the amount calculated as the theoretical yield. This is true for a variety of reasons. Some experimental error is unavoidable. Moreover, many reactions simply do not go to completion; some amount of reactant remains at the end of the reaction, and the *actual* amount of product is less than the *theoretical* (predicted) amount. We will study these processes, termed *equilibrium reactions* in Chapter 7.

The **percent yield,** the ratio of the actual and theoretical yields multiplied by 100%, is often used to show the relationship between predicted and experimental quantities. Thus,

$$\%\ \text{yield} = \frac{\text{actual yield}}{\text{theoretical yield}} \times 100\%$$

In Example 4.15, the theoretical yield of CO_2 is 132 g. For this reaction, let's assume that a chemist actually obtained 125 g CO_2. This is the actual yield and would normally be provided as a part of the data in the problem.

Calculate the percent yield as follows:

$$\%\ \text{yield} = \frac{\text{actual yield}}{\text{theoretical yield}} \times 100\%$$

$$= \frac{125\ \cancel{\text{g}\ CO_2}}{132\ \cancel{\text{g}\ CO_2}} \times 100\% = 94.7\%$$

EXAMPLE 4.19 Calculation of Percent Yield

LEARNING GOAL

14 Calculate theoretical and percent yields.

Assume that the theoretical yield of iron in the process

$$2Al(s) + Fe_2O_3(s) \longrightarrow Al_2O_3(l) + 2Fe(l)$$

was 30.0 g. If the actual yield of iron was 25.0 g, calculate the percent yield.

Solution

$$\%\ \text{yield} = \frac{\text{actual yield}}{\text{theoretical yield}} \times 100\%$$

$$= \frac{25.0\ \cancel{\text{g}}}{30.0\ \cancel{\text{g}}} \times 100\%$$

$$= 83.3\%$$

Continued…

A Medical Perspective

Pharmaceutical Chemistry: The Practical Significance of Percent Yield

In recent years, the major pharmaceutical industries have introduced a wide variety of new drugs targeted to cure or alleviate the symptoms of a host of diseases that afflict humanity.

The vast majority of these drugs are synthetic; they are made in a laboratory or by an industrial process. These substances are complex molecules that are patiently designed and constructed from relatively simple molecules in a series of chemical reactions. A series of ten to twenty "steps," or sequential reactions, is not unusual to put together a final product that has the proper structure, geometry, and reactivity for efficacy against a particular disease.

Although a great deal of research occurs to ensure that each of the steps in the overall process is efficient (having a large percent yield), the overall process is still very inefficient (low percent yield). This inefficiency, and the research needed to minimize it, at least in part determines the cost and availability of both prescription and over-the-counter preparations.

Consider a hypothetical five-step sequential synthesis. If each step has a percent yield of 80%, our initial impression might be that this synthesis is quite efficient. However, on closer inspection we find quite the contrary to be true.

The overall yield of the five-step reaction is the product of the decimal fraction of the percent yield of each of the sequential reactions. So, if the decimal fraction corresponding to 80% is 0.80:

$$0.80 \times 0.80 \times 0.80 \times 0.80 \times 0.80 = 0.33$$

Converting the decimal fraction to percentage:

$$0.33 \times 100\% = 33\% \text{ yield}$$

Many reactions are considerably less than 80% efficient, especially those that are used to prepare large molecules with complex arrangements of atoms. Imagine a more realistic scenario in which one step is only 20% efficient (a 20% yield) and the other four steps are 50%, 60%, 70%, and 80% efficient. Repeating the calculation with these numbers (after conversion to decimal fractions):

$$0.20 \times 0.50 \times 0.60 \times 0.70 \times 0.80 = 0.0336$$

Converting the decimal fraction to a percentage:

$$0.0336 \times 100\% = 3.36\% \text{ yield}$$

A 3.36% yield refects a very inefficient process.

If we apply this logic to a fifteen- or twenty-step synthesis, we gain some appreciation of the difficulty of producing modern pharmaceutical products. Add to this the challenge of predicting the most appropriate molecular structure that will have the desired biological effect and be relatively free of side effects. All these considerations give new meaning to the term *wonder drug* that has been attached to some of the more successful synthetic products.

We will study some of the elementary steps essential to the synthesis of a wide range of pharmaceutical compounds in later chapters, beginning with Chapter 10.

For Further Understanding

▶ Explain the possible connection of this perspective to escalating costs of pharmaceutical products.

▶ Can you describe other situations, not necessarily in the field of chemistry, where multiple-step processes contribute to inefficiency?

Practice Problem 4.19

Given the reaction represented by the balanced equation

$$CH_4(g) + 3Cl_2(g) \longrightarrow 3HCl(g) + CHCl_3(g)$$

a. Calculate the number of g of $CHCl_3$ that could be produced by mixing 105 g Cl_2 with excess CH_4.

b. If 10.0 g $CHCl_3$ were actually produced, calculate the % yield.

▶ For Further Practice: **Questions 4.117 and 4.118.**

Question 4.15 It is believed that the reaction responsible for the depletion of ozone in the atmosphere is:

$$O_3(g) + NO(g) \longrightarrow O_2(g) + NO_2(g)$$

a. If 50.0 g of O_3 react with excess NO, how many g of NO_2 will be produced?
b. If the actual yield of NO_2 is 25.0 g, what is the % yield?

Question 4.16 The reaction:

$$2NO(g) + O_2(g) \longrightarrow 2NO_2(g)$$

is one step in the process of forming atmospheric smog.

a. How many g of NO_2 can be produced by the reaction of excess O_2 with 50.0 g of NO?
b. If the actual yield of NO_2 is 50.0 g, what is the % yield?

A Special Case—The Limiting Reactant

We have learned that reactants combine in molar proportions dictated by the coefficients in the balanced equation. However, we often encounter situations where the reactants are not mixed to match the theoretical proportions. In cases such as this, one reactant will be completely consumed, leaving some of the other reactant unchanged. In effect, the extent of the reaction is *limited* by the amount of the reactant that is completely consumed. This completely consumed reactant is termed the *limiting reactant*. In order to correctly calculate the amount of product formed, the *theoretical yield,* we must base the amount of product formed on the number of mol of the compound "in short supply," the limiting reactant.

Imagine that you are making cheeseburgers. You have ten buns, four slices of cheese, and five meat patties. What is the limiting ingredient? How many cheeseburgers can you make? What are the leftover ingredients? If you can answer these questions, you can do limiting reactant problems.

We know that ten buns could make ten cheeseburgers if we had ten meat patties and ten slices of cheese. But we do not, so it is obvious that we cannot make ten cheeseburgers, even though we certainly have enough buns. It can be reasoned that only four cheeseburgers are possible. We are limited by the availability of only four slices of cheese. In our example, cheese is the limiting reactant. Furthermore, five minus four, or one meat patty and ten minus four, or six buns, would be left over as shown in Figure 4.10.

The strategy outlined above is the same as that used in solving problems involving limiting reactants.

Figure 4.10 In this illustration, the number of cheese slices limits the number of cheeseburgers that can be made. The cheese slices are the limiting reactant.

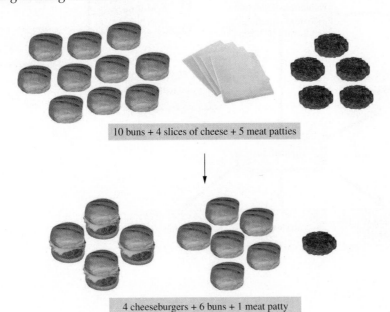

10 buns + 4 slices of cheese + 5 meat patties

4 cheeseburgers + 6 buns + 1 meat patty

CHAPTER MAP

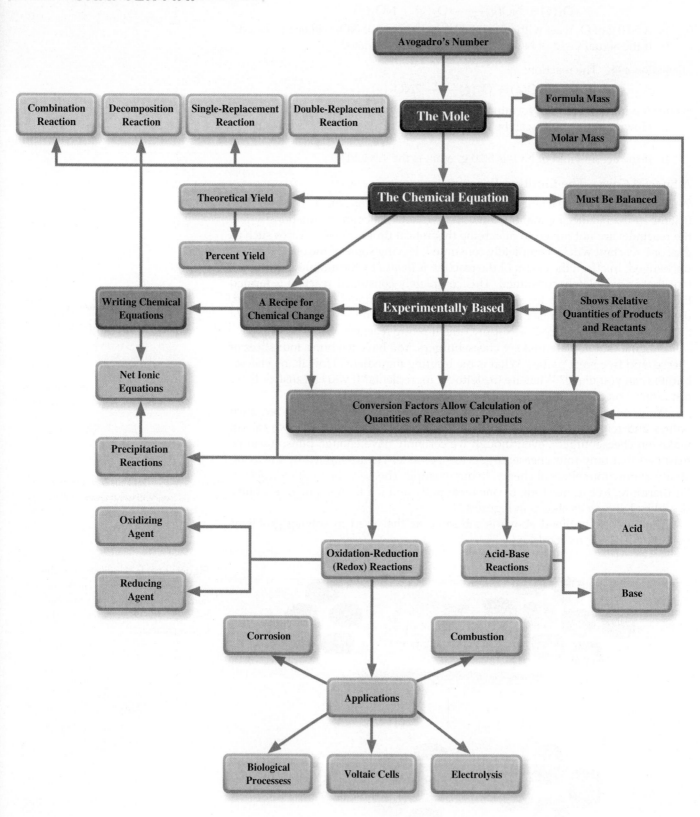

SUMMARY

4.1 The Mole Concept and Atoms

▶ Atoms are exceedingly small, yet their masses have been experimentally determined for each of the elements. The unit of measurement for these determinations is the **atomic mass unit,** (amu):

$$1 \text{ amu} = 1.661 \times 10^{-24} \text{ g}$$

▶ The periodic table provides atomic masses in amu. A more practical unit for defining a "collection" of atoms is the **mole (mol):**

1 mol of atoms = 6.022×10^{23} atoms of an element

This number is referred to as **Avogadro's number.**

▶ The **atomic mass** of a given element is the average mass of a single atom in amu. The mass of 1 mol of atoms, in grams, is termed the **molar mass** of the element. One mole of atoms of any element contains the same number, Avogadro's number, of atoms.

4.2 The Chemical Formula, Formula Mass, and Molar Mass

▶ Compounds are pure substances composed of two or more elements that are chemically combined. They are represented by their **chemical formula,** a combination of symbols of the various elements that make up the compounds. The chemical formula is based on the **formula unit.** This is the smallest collection of atoms that provides the identity of the atoms present in the compound and the relative numbers of each type of atom.

▶ Just as the mass of a mol of atoms is based on the atomic mass, the mass of a mol of a compound is based on the **formula mass.** The formula mass is calculated by adding the masses of all the atoms or ions of which the unit is composed. To calculate the formula mass, the formula unit must be known. The formula mass of 1 mol of a compound is its **molar mass** in units of g/mol.

4.3 The Chemical Equation and the Information It Conveys

▶ The **chemical equation** is the shorthand notation for a chemical reaction. It describes all of the substances that react to produce the product(s). **Reactants,** or starting materials, are all substances that undergo change in a chemical reaction; **products** are substances produced by a chemical reaction.

▶ According to the **law of conservation of mass,** matter can neither be gained nor lost in the process of a chemical reaction. The law of conservation of mass states that we must have a balanced chemical equation.

▶ Features of a chemical equation include the following:

- The identity of products and reactants must be specified.
- Reactants are written to the left of the reaction arrow ($\longrightarrow$) and products to the right.
- The physical states of reactants and products are shown in parentheses.

- If heat energy is necessary for the reaction to occur, the symbol Δ is written over the reaction arrow.
- The equation must be balanced.

▶ Chemical reactions involve the **combination** of reactants to produce products, the **decomposition** of reactant(s) into products, or the **replacement** of one or more elements in a compound to yield products. Replacement reactions are subclassified as either **single-** or **double-**replacement.

4.4 Balancing Chemical Equations

▶ The chemical equation enables us to determine the quantity of reactants needed to produce a certain molar quantity of products. The chemical equation expresses these quantities in terms of mol.

▶ The relative number of mol of each reactant and product is indicated by placing a whole-number coefficient before the formula of each substance in the chemical equation.

▶ Many equations are balanced by trial and error. If the identity of the reactants and products, the physical states, and the reaction conditions are known, the following steps provide a method for correctly balancing a chemical equation:

- Count the number of atoms of each element on both reactant and product sides.
- Determine which atoms are not balanced.
- Balance one element at a time using coefficients.
- After you believe that you have successfully balanced the equation, check to be certain that mass conservation has been achieved.

4.5 Precipitation Reactions

▶ Reactions that produce products with similar characteristics are often classified together. The formation of an insoluble solid, a **precipitate,** is very common. Such reactions are precipitation reactions and are represented using net ionic equations.

4.6 Net Ionic Equations

▶ **Net ionic equations** show the chemical species that actually undergo change. Other ions that retain their identity throughout the chemical reaction are termed *spectator ions.*

4.7 Acid-Base Reactions

▶ **Acid-base reactions** involve the transfer of a hydrogen cation, H^+, from one reactant (the acid) to another (the base).

▶ The reaction of an acid with a base to produce a salt and water is referred to as **neutralization.**

4.8 Oxidation-Reduction Reactions

▶ **Oxidation** is defined as a loss of electrons, loss of hydrogen atoms, or gain of oxygen atoms.

▶ **Reduction** is defined as a gain of electrons, gain of hydrogen atoms, or loss of oxygen atoms.

▶ Oxidation and reduction are complementary processes. The oxidation half-reaction produces one or more electrons that are the reactants for the reduction half-reaction. The combination of two half-reactions, one oxidation and one reduction, produces the complete *redox* reaction.

▶ The **reducing agent** releases electrons for the reduction of a second substance to occur. The **oxidizing agent** accepts electrons, causing the oxidation of a second substance to take place.

▶ The deterioration of metals caused by an oxidation-reduction process is termed **corrosion.**

▶ The complete oxidation of hydrocarbons produces energy, carbon dioxide, and water in a process termed **combustion.**

▶ A **voltaic cell** is an electrochemical cell that converts chemical energy into electrical energy. The best-known example of a voltaic cell is the storage battery. **Electrolysis** is the opposite of a battery. It converts electrical energy into chemical potential energy. The electrode at which oxidation occurs is called the **anode,** and the electrode at which reduction occurs is the **cathode.**

4.9 Calculations Using the Chemical Equation

▶ Calculations involving chemical quantities are based on the following requirements:

- The basis for the calculations is a balanced equation.
- The calculations are performed using mole-based conversion factors.
- The conservation of mass must be obeyed.

▶ The mol is the basis for calculations. However, masses are generally measured in g. Therefore, you must be able to interconvert mol and g to perform chemical arithmetic.

▶ The calculated amount assumes complete conversion of reactant to product. This amount is the **theoretical yield.** Most reactions are not complete; some reactant(s) remains and the actual amount is less than the theoretical amount. The **percent yield** is the ratio of the actual to theoretical yields multiplied by 100%.

- The *limiting reactant* should be completely consumed in the reaction and will limit the amount of product formed. Identifying the limiting reactant is often important in theoretical yield calculations.

ANSWERS TO PRACTICE PROBLEMS

4.1 a. $26.98 \dfrac{\text{g Al}}{\text{mol Al}}$

b. $200.59 \dfrac{\text{g Hg}}{\text{mol Hg}}$

4.2 a. 1.51×10^{24} oxygen atoms
b. 3.01×10^{24} oxygen atoms

4.3 1.50 mol Na

4.4 14.0 g He

4.5 3.17×10^{-2} mol Ag

4.6 1.51×10^{24} O atoms

4.7 17.04 amu and 17.04 g/mol

4.8 237.95 amu and 237.95 g/mol

4.9 $C_2H_5OH(l) + 3O_2(g) \longrightarrow 2CO_2(g) + 3H_2O(g)$

4.10 $6S_2Cl_2(s) + 16NH_3(g) \longrightarrow N_4S_4(s) + 12NH_4Cl(s) + S_8(s)$

4.11 a. $KCl(aq) + AgNO_3(aq) \longrightarrow KNO_3(aq) + AgCl(s)$
b. $K_2CO_3(aq) + Ca(OH)_2(aq) \longrightarrow 2KOH(aq) + CaCO_3(s)$
c. $NaOH(aq) + NH_4Cl(aq) \longrightarrow$ no reaction
d. $2NaOH(aq) + FeCl_2(aq) \longrightarrow 2NaCl(aq) + Fe(OH)_2(s)$

4.12 a. $Ba^{2+}(aq) + SO_4^{2-}(aq) \longrightarrow BaSO_4(s)$
b. $2Ag^+(aq) + SO_4^{2-}(aq) \longrightarrow Ag_2SO_4(s)$

4.13 a. 90.1 g H_2O c. 1.80×10^3 μg $C_6H_{12}O_6$
b. 0.590 mol LiCl   d. 0.368 mol $MgCl_2$

4.14 65.1 g KCN

4.15 4.61×10^2 g ethanol

4.16 a. 3 mol O_2
b. 96.00 g O_2
c. 88.0 g CO_2

4.17 a. $4Fe(s) + 3O_2(g) \longrightarrow 2Fe_2O_3(s)$
b. 3.50 g Fe

4.18 a. $BaCO_3(s) \xrightarrow{\Delta} BaO(s) + CO_2(g)$
b. 11.2 g CO_2

4.19 a. 58.9 g $CHCl_3$
b. 17.0% yield

QUESTIONS AND PROBLEMS

The Mole Concept and Atoms

Foundations

4.17 What is the average mass (in amu) of:
a. Hg   b. Kr   c. Mg
4.18 What is the average mass (in amu) of:
a. Zr   b. Cs   c. Ca
4.19 What is the average molar mass of:
a. Si   b. Ag   c. As
4.20 What is the average molar mass of:
a. S   b. Na   c. Hg

4.21 What is the mass, in g, of Avogadro's number of argon atoms?

4.22 What is the mass, in g, of Avogadro's number of iron atoms?

Applications

4.23 How many carbon atoms are present in 1.0×10^{-4} mol of carbon?

4.24 How many mercury atoms are present in 1.0×10^{-10} mol of mercury?

4.25 How many mol of arsenic correspond to 1.0×10^2 atoms of arsenic?

4.26 How many mol of sodium correspond to 1.0×10^{15} atoms of sodium?

4.27 How many g of neon are contained in 2.00 mol of neon atoms?

4.28 How many g of carbon are contained in 3.00 mol of carbon atoms?

4.29 What is the mass, in g, of 1.00 mol of helium atoms?

4.30 What is the mass, in g, of 1.00 mol of nitrogen atoms?

4.31 Calculate the number of mol corresponding to:
a. 20.0 g He   b. 0.040 kg Na   c. 3.0 g Cl_2

4.32 Calculate the number of mol corresponding to:
a. 0.10 g Ca   b. 4.00 g Fe   c. 2.00 kg N_2

4.33 What is the mass, in g, of 15.0 mol of silver?

4.34 What is the mass, in g, of 15.0 mol of carbon?

4.35 Calculate the number of silver atoms in 15.0 g of silver.

4.36 Calculate the number of carbon atoms in 15.0 g of carbon.

The Chemical Formula, Formula Mass, and Molar Mass

Foundations

4.37 Distinguish between the terms *molecule* and *ion pair*.

4.38 Distinguish between the terms *formula mass* and *molar mass*.

4.39 Calculate formula mass and the molar mass of each of the following formula units:
a. NaCl   b. Na_2SO_4 c. $Fe_3(PO_4)_2$

4.40 Calculate formula mass and the molar mass of each of the following formula units:
a. S_8 b. $(NH_4)_2SO_4$ c. CO_2

4.41 Calculate formula mass and the molar mass of oxygen gas, O_2.

4.42 Calculate formula mass and the molar mass of ozone, O_3.

4.43 Calculate formula mass and the molar mass of $CuSO_4 \cdot 5H_2O$.

4.44 Calculate formula mass and the molar mass of $CaCl_2 \cdot 2H_2O$.

Applications

4.45 Calculate the number of mol corresponding to:
a. 15.0 g NaCl   b. 15.0 g Na_2SO_4

4.46 Calculate the number of mol corresponding to:
a. 15.0 g NH_3 b. 16.0 g O_2

4.47 Calculate the mass in g corresponding to:
a. 1.000 mol H_2O c. 10.0 mol He
b. 2.000 mol NaCl   d. 1.00×10^2 mol H_2

4.48 Calculate the mass in g corresponding to:
a. 0.400 mol NH_3 c. 2.00 mol CH_4
b. 0.800 mol $BaCO_3$ d. 0.400 mol $Ca(NO_3)_2$

4.49 How many g are required to have 0.100 mol of each of the following?
a. CH_4 (methane)   c. NaOH
b. $CaCO_3$ d. H_2SO_4

4.50 How many g are required to have 0.100 mol of each of the following?
a. $C_6H_{12}O_6$ (glucose)
b. NaCl
c. C_2H_5OH (ethanol)
d. $Ca_3(PO_4)_2$

4.51 How many mol are in 50.0 g of each of the following substances?
a. KBr   c. CS_2
b. $MgSO_4$ d. $Al_2(CO_3)_3$

4.52 How many mol are in 50.0 g of each of the following substances?
a. Br_2 c. $Sr(OH)_2$
b. NH_4Cl d. $LiNO_3$

The Chemical Equation and the Information It Conveys

Foundations

4.53 What law is the ultimate basis for a balanced chemical equation?

4.54 List the general types of information that a chemical equation provides.

4.55 What is a reactant? On which side of the reaction arrow are reactants found?

4.56 What is a product? On which side of the reaction arrow are products found?

4.57 What is the meaning of Δ over the reaction arrow?

4.58 What is the meaning of (s), (l), (g), and (aq) immediately following the symbol for a chemical substance?

Applications

4.59 Classify each of the following reactions as decomposition (D), combination (C), single-replacement (SR), or double-replacement (DR):
a. $2KClO_3(s) \xrightarrow{\Delta} 2KCl(s) + 3O_2(g)$
b. $K_2CO_3(aq) + Ca(OH)_2(aq) \longrightarrow CaCO_3(s) + 2KOH(aq)$
c. $CaO(aq) + H_2O(l) \longrightarrow Ca(OH)_2(aq)$
d. $Ca(s) + Sn(NO_3)_2(aq) \longrightarrow Sn(aq) + Ca(NO_3)_2(aq)$

4.60 Classify each of the following reactions as decomposition (D), combination (C), single-replacement (SR), or double-replacement (DR):
a. $KOH(s) + CO_2(g) \longrightarrow KHCO_3(s)$
b. $K_2CO_3(aq) \xrightarrow{\Delta} K_2O(g) + CO_2(g)$
c. $H_2SO_4(aq) + 2 NaOH(aq) \longrightarrow Na_2SO_4(aq) + 2 H_2O(l)$
d. $2AgNO_3(aq) + Zn(s) \longrightarrow 2Ag(s) + Zn(NO_3)_2(aq)$

4.61 What is the meaning of the subscript in a chemical formula?

4.62 What is the meaning of the coefficient in a chemical equation?

Balancing Chemical Equations

Foundations

4.63 When you are balancing an equation, why must the subscripts in the chemical formulas remain unchanged?

4.64 Describe the process of checking to ensure that an equation is properly balanced.

Applications

4.65 Balance each of the following equations:
a. $C_2H_6(g) + O_2(g) \longrightarrow CO_2(g) + H_2O(g)$
b. $K_2O(s) + P_4O_{10}(s) \longrightarrow K_3PO_4(s)$
c. $MgBr_2(aq) + H_2SO_4(aq) \longrightarrow HBr(g) + MgSO_4(aq)$
d. $C_2H_5OH(l) + O_2(g) \longrightarrow CO_2(g) + H_2O(g)$

4.66 Balance each of the following equations:
a. $C_6H_{12}O_6(s) + O_2(g) \longrightarrow CO_2(g) + H_2O(g)$
b. $H_2O(l) + P_4O_{10}(s) \longrightarrow H_3PO_4(aq)$
c. $PCl_5(g) + H_2O(l) \longrightarrow HCl(aq) + H_3PO_4(aq)$
d. $C_6H_{12}O_6(s) \longrightarrow C_2H_6O(l) + CO_2(g)$

4.67 Complete, then balance, each of the following equations:
a. $Ca(s) + F_2(g) \longrightarrow$
b. $Mg(s) + O_2(g) \longrightarrow$
c. $H_2(g) + N_2(g) \longrightarrow$

4.68 Complete, then balance, each of the following equations:
a. $Li(s) + O_2(g) \longrightarrow$
b. $Ca(s) + N_2(g) \longrightarrow$
c. $Al(s) + S(s) \longrightarrow$

4.69 Balance each of the following equations:
a. $C_4H_{10}(g) + O_2(g) \longrightarrow H_2O(g) + CO_2(g)$
b. $Au_2S_3(s) + H_2(g) \longrightarrow Au(s) + H_2S(g)$
c. $Al(OH)_3(s) + HCl(aq) \longrightarrow AlCl_3(aq) + H_2O(l)$
d. $(NH_4)_2Cr_2O_7(s) \longrightarrow Cr_2O_3(s) + N_2(g) + H_2O(g)$

4.70 Balance each of the following equations:
a. $Fe_2O_3(s) + CO(g) \longrightarrow Fe_3O_4(s) + CO_2(g)$
b. $C_6H_6(l) + O_2(g) \longrightarrow CO_2(g) + H_2O(g)$
c. $I_4O_9(s) + I_2O_6(s) \longrightarrow I_2(s) + O_2(g)$
d. $KClO_3(s) \longrightarrow KCl(s) + O_2(g)$

4.71 Write a balanced equation for each of the following reactions:
a. Ammonia is formed by the reaction of nitrogen and hydrogen.
b. Hydrochloric acid reacts with sodium hydroxide to produce water and sodium chloride.
c. Glucose, a sugar, $C_6H_{12}O_6$, is oxidized in the body to produce water and carbon dioxide.
d. Sodium carbonate, upon heating, produces sodium oxide and carbon dioxide.

4.72 Write a balanced equation for each of the following reactions:
a. Nitric acid reacts with calcium hydroxide to produce water and calcium nitrate.
b. Butane (C_4H_{10}) reacts with oxygen to produce water and carbon dioxide.
c. Sulfur, present as an impurity in coal, is burned in oxygen to produce sulfur dioxide.
d. Hydrofluric acid (HF) reacts with glass (SiO_2) in the process of etching to produce silicon tetrafluoride and water.

Precipitation Reactions

Foundations

4.73 Which of the following ionic compounds will form a precipitate in water?
a. Na_2SO_4 c. $BaCO_3$
b. $BaSO_4$ d. K_2CO_3

4.74 Which of the following ionic compounds will form a precipitate in water?
a. $PbCO_3$ c. $Pb(NO_3)_2$
b. Na_2CO_3 d. Na_2NO_3

Applications

4.75 Will a precipitate form if solutions of the soluble salts $Pb(NO_3)_2$ and KI are mixed?

4.76 Will a precipitate form if solutions of the soluble salts $AgNO_3$ abd NaOH are mixed?

4.77 Solutions containing $(NH_4)_2CO_3(aq)$ and $CaCl_2(aq)$ are mixed. Will a precipitate form? If so, write its formula.

4.78 Solutions containing $Mg(NO_3)_2(aq)$ and $NaOH(aq)$ are mixed. Will a precipitate form? If so, write its formula.

Net Ionic Equations

Foundations

4.79 Describe the difference between the terms *ionic equation* and *net ionic equation*.

4.80 Describe the steps used in writing the net ionic equation for a reaction.

Applications

4.81 Write the net ionic equation for the reaction of $NaBr(aq)$ with $AgNO_3(aq)$.

4.82 Write the net ionic equation for the reaction of $Pb(NO_3)_2(aq)$ with $K_2S(aq)$.

Acid-Base Reactions

Foundations

4.83 Does an acid gain or lose a hydrogen cation, H^+, during an acid-base reaction?

4.84 During an acid-base reaction, what term is used to describe the reactant that gains a hydrogen cation, H^+?

Applications

4.85 Identify the acid and base in the following reaction:

$$HCN\ (aq) + KOH\ (aq) \longrightarrow KCN\ (aq) + H_2O\ (l)$$

4.86 Identify the acid and base in the following reaction:

$$HBr\ (aq) + NaOH\ (aq) \longrightarrow NaBr\ (aq) + H_2O\ (l)$$

Oxidation-Reduction Reactions

Foundations

4.87 During an oxidation process in an oxidation-reduction reaction, does the species oxidized gain or lose electrons?

4.88 During an oxidation-reduction reaction, is the oxidizing agent oxidized or reduced?

4.89 During an oxidation-reduction reaction, is the reducing agent oxidized or reduced?

4.90 Do metals tend to be good oxidizing agents or good reducing agents?

Applications

4.91 In the following reaction, identify the oxidized species, reduced species, oxidizing agent, and reducing agent:

$$Cl_2(aq) + 2KI(aq) \longrightarrow 2KCl(aq) + I_2(aq)$$

4.92 In the following reaction, identify the oxidized species, reduced species, oxidizing agent, and reducing agent:

$$Zn(s) + Cu^{2+}(aq) \longrightarrow Zn^{2+}(aq) + Cu(s)$$

4.93 Write the oxidation and reduction half-reactions for the equation in Question 4.91.

4.94 Write the oxidation and reduction half-reactions for the equation in Question 4.92.

4.95 Explain the relationship between oxidation-reduction and voltaic cells.

4.96 Compare and contrast a battery and electrolysis.

4.97 Describe one application of voltaic cells.

4.98 Describe one application of electrolytic cells.

Calculations Using the Chemical Equation

Foundations

4.99 Why is it essential to use balanced equations to solve mol problems?

4.100 Describe the steps used in the calculation of g of product resulting from the reaction of a specified number of g of reactant.

Applications

4.101 How many g of B_2H_6 will react with 3.00 mol of O_2?

$$B_2H_6(l) + 3O_2(g) \longrightarrow B_2O_3(s) + 3H_2O(l)$$

4.102 How many g of Al will react with 3.00 mol of O_2?

$$4Al(s) + 3O_2(g) \longrightarrow 2Al_2O_3(s)$$

4.103 Calculate the number of moles of $CrCl_3$ that could be produced from 50.0 g Cr_2O_3 according to the equation

$$Cr_2O_3(s) + 3CCl_4(l) \longrightarrow 2CrCl_3(s) + 3COCl_2(aq)$$

4.104 A 3.5-g sample of water reacts with PCl_3 according to the following equation:

$$3H_2O(l) + PCl_3(g) \longrightarrow H_3PO_3(aq) + 3HCl(aq)$$

How many mol of H_3PO_3 are produced?

4.105 For the reaction

$$N_2(g) + H_2(g) \longrightarrow NH_3(g)$$

a. Balance the equation.
b. How many mol of H_2 would react with 1 mol of N_2?
c. How many mol of product would form from 1 mol of N_2?
d. If 14.0 g of N_2 were initially present, calculate the number of mol of H_2 required to react with all of the N_2.
e. For conditions outlined in part (d), how many g of product would form?

4.106 Aspirin (acetylsalicylic acid) may be formed from salicylic acid and acetic acid as follows:

$$C_7H_6O_3(aq) + CH_3COOH(aq) \longrightarrow C_9H_8O_4(s) + H_2O(l)$$

  Salicylic   Acetic   Aspirin
   acid   acid

a. Is this equation balanced? If not, complete the balancing.
b. How many mol of aspirin may be produced from 1.00×10^2 mol salicylic acid?

c. How many g of aspirin may be produced from 1.00×10^2 mol salicylic acid?
d. How many g of acetic acid would be required to react completely with the 1.00×10^2 mol salicylic acid?
e. For the conditions outlined in part (d), how many g of aspirin would form?

4.107 The proteins in our bodies are composed of molecules called amino acids. One amino acid is methionine; its molecular formula is $C_5H_{11}NO_2S$. Calculate:
a. the formula mass of methionine
b. the number of oxygen atoms in a mol of this compound
c. the mass of oxygen in a mol of the compound
d. the mass of oxygen in 50.0 g of the compound

4.108 Triglycerides (Chapters 17 and 23) are used in biochemical systems to store energy; they can be formed from glycerol and fatty acids. The molecular formula of glycerol is $C_3H_8O_3$. Calculate:
a. the formula mass of glycerol
b. the number of oxygen atoms in a mol of this compound
c. the mass of oxygen in a mol of the compound
d. the mass of oxygen in 50.0 g of the compound

4.109 Joseph Priestley discovered oxygen in the eighteenth century by using heat to decompose mercury(II) oxide:

$$2HgO(s) \xrightarrow{\Delta} 2Hg(l) + O_2(g)$$

How many g of oxygen is produced from 1.00×10^2 g HgO?

4.110 Dinitrogen monoxide (also known as nitrous oxide and used as an anesthetic) can be made by heating ammonium nitrate:

$$NH_4NO_3(s) \xrightarrow{\Delta} N_2O(g) + 2H_2O(g)$$

How many g of dinitrogen monoxide can be made from 1.00×10^2 g of ammonium nitrate?

4.111 The burning of acetylene (C_2H_2) in oxygen is the reaction in the oxyacetylene torch. How many g of CO_2 is produced by burning 20.0 kg of acetylene in an excess of O_2? The unbalanced equation is

$$C_2H_2(g) + O_2(g) \xrightarrow{\Delta} CO_2(g) + H_2O(g)$$

4.112 The reaction of calcium hydride with water can be used to prepare hydrogen gas:

$$CaH_2(s) + 2H_2O(l) \longrightarrow Ca(OH)_2(aq) + 2H_2(g)$$

How many g of hydrogen gas are produced in the reaction of 1.00×10^2 g calcium hydride with water?

4.113 Various members of a class of compounds called alkenes (Chapter 11) react with hydrogen to produce a corresponding alkane (Chapter 10). Termed hydrogenation, this type of reaction is used to produce products such as margarine. A typical hydrogenation reaction is

$$C_{10}H_{20}(l) + H_2(g) \longrightarrow C_{10}H_{22}(s)$$

   Decene   Decane

How many g of decane can be produced in a reaction of excess decene with 1.00 g hydrogen?

4.114 Chemical Control of Microbes (Section 4.8) describes the breakdown of the antiseptic H_2O_2 with the balanced equation

$$2H_2O_2(aq) \longrightarrow 2H_2O(l) + O_2(g)$$

Assuming there is an unlimited amount of the enzyme, how many g of O_2 would be produced from 1.00×10^{-1} g of H_2O_2?

4.115 A rocket can be powered by the reaction between dinitrogen tetroxide and hydrazine:

$$N_2O_4(l) + 2N_2H_4(l) \longrightarrow 3N_2(g) + 4H_2O(g)$$

An engineer designed the rocket to hold 1.00 kg N_2O_4 and excess N_2H_4. How many g of N_2 would be produced according to the engineer's design?

4.116 A 4.00-g sample of Fe_3O_4 reacts with O_2 to produce Fe_2O_3:

$$4Fe_3O_4(s) + O_2(g) \longrightarrow 6Fe_2O_3(s)$$

Determine the number of g of Fe_2O_3 produced.

4.117 If the actual yield of decane in Question 4.113 is 65.4 g, what is the % yield?

4.118 If the actual yield of oxygen gas in Question 4.114 is 1.10×10^{-2} g, what is the % yield?

4.119 If the % yield of nitrogen gas in Question 4.115 is 75.0%, what is the actual yield of nitrogen?

4.120 If the % yield of Fe_2O_3 in Question 4.116 is 90.0%, what is the actual yield of Fe_2O_3?

CHALLENGE PROBLEMS

1. Which of the following has fewer mol of carbon: 100 g of $CaCO_3$ or 0.5 mol of CCl_4?
2. Which of the following has fewer mol of carbon: 6.02×10^{22} molecules of C_2H_6 or 88 g of CO_2?
3. How many molecules are found in each of the following?
   a. 1.0 lb of sucrose, $C_{12}H_{22}O_{11}$ (table sugar)
   b. 1.57 kg of N_2O (anesthetic)
4. How many molecules are found in each of the following?
   a. 4×10^5 tons (t) of SO_2 (produced by the 1980 eruption of the Mount St. Helens volcano)
   b. 25.0 lb of SiO_2 (major constituent of sand)
5. Based on the information in Questions 4.15 and 4.16, it appears that we could solve some of our atmospheric problems by reducing the amount of NO that we put into the air. Use the Internet to determine the sources of atmospheric NO. Can we control atmospheric NO? If so, how?

GASES, LIQUIDS, AND SOLIDS
States of Matter

LEARNING GOALS

1 Perform conversions between units of pressure.

2 Describe the major points of the kinetic molecular theory of gases.

3 Explain the relationship between the kinetic molecular theory and the physical properties of measurable quantities of gases.

4 Describe the behavior of gases expressed by the gas laws: Boyle's law, Charles's law, combined gas law, Avogadro's law, the ideal gas law, and Dalton's law.

5 Use gas law equations to calculate conditions and changes in conditions of gases.

6 Use molar volume and standard temperature and pressure (STP) to perform calculations.

7 Discuss the limitations to the ideal gas model as it applies to real gases.

8 Describe properties of the liquid state in terms of the properties of the individual molecules that comprise the liquid.

9 Describe the processes of melting, boiling, evaporation, condensation, and sublimation.

10 Describe the dipolar attractions known collectively as van der Waals forces.

11 Describe hydrogen bonding and its relationship to boiling and melting temperatures.

12 Relate the properties of the various classes of solids (ionic, covalent, molecular, and metallic) to the structure of these solids.

Volcanic activity is a dramatic example of interconversion among the states of matter.

OUTLINE

INTRODUCTION

Liquid nitrogen is commonly used in cryopreservation when biological samples need to be stored at temperatures below −196°C.

Section 1.2 introduces the properties of the three states of matter.

Major differences among solids, liquids, and gases are due to the relationships among particles. These relationships include:

- the average distance of separation of particles in each state,
- the kinds of interactions among the particles, and
- the degree of organization of particles.

We have already discovered that the solid state is the most organized, with particles close together, allowing significant interactions among the particles. This results in high melting and boiling points for solid substances. Large amounts of energy are needed to overcome the attractive forces and disrupt the orderly structure.

Substances that are gases, on the other hand, are disordered, with particles widely separated and weak interactions among particles. Their melting and boiling points are relatively low. Gases at room temperature must be cooled a great deal for them to liquefy or solidify. For example, the melting and boiling points of N_2 are −210°C and −196°C, respectively.

Liquids are intermediate in character. The molecules of a liquid are close together, like those of solids. However, the molecules of a liquid are disordered, like those of a gas.

Can a substance such as N_2 gas or CO_2 gas also exist as a liquid, or even as a solid? We will see that a reduction in temperature or an increase in pressure can force atoms or molecules closer together, allowing them to behave as liquids or solids. Dry ice, for example, is solid carbon dioxide.

Changes in state are described as physical changes. When a substance undergoes a change in state, many of its physical properties change. For example, when ice forms from liquid water, changes occur in density and hardness, but it is still water. Table 5.1 summarizes the important differences in physical properties among gases, liquids, and solids.

5.1   The Gaseous State

Ideal Gas Concept

An **ideal gas** is simply a model of the way that gas particles (molecules or atoms) behave at the atomic/molecular level. The behavior of the individual particles can be inferred from the measurable behavior of samples of real gases. We can easily measure temperature, volume, pressure, and quantity (number of moles) of real

TABLE 5.1 A Comparison of Physical Properties of Gases, Liquids, and Solids

	Gas	Liquid	Solid
Volume and Shape	Expands to fill the volume of its container; consequently, it takes the shape of the container	Has a fixed volume at a given mass and temperature; volume principally dependent on its mass and secondarily on temperature; it assumes the shape of its container	Has a fixed volume; volume principally dependent on its mass and secondarily on temperature; it has a definite shape
Density	Low (typically ~10^{-3} g/mL)	High (typically ~1 g/mL)	High (typically 1–10 g/mL)
Compressibility	High	Very low	Virtually incompressible
Particle Motion	Virtually unrestricted	Molecules or atoms "slide" past each other	Vibrate about a fixed position
Intermolecular Distance	Very large	Molecules or atoms are close to each other	Molecules, ions, or atoms are close to each other

gases. Similarly, when we systematically change one of these properties, we can determine the effect on each of the others. For example, putting more molecules in a balloon (the act of blowing up a balloon) causes its volume to increase in a predictable way. In fact, careful measurements show a direct proportionality between the number of molecules and the volume of the balloon, an observation made by Amadeo Avogadro more than 200 years ago.

Measurement of Properties of Gases

There are four basic gas laws:

> Boyle's law
> Charles's law
> Avogadro's law
> Dalton's law

Two laws are derived from these basic laws: the combined gas law and the ideal gas law. These laws involve the relationships among pressure (P), volume (V), temperature (T), and number of moles (n) of gas. We are already familiar with the measurements of temperature, volume, and mass (allowing the calculation of number of mol). The measurement of **pressure** is a measurement of force per unit area.

Gas pressure is a result of the force exerted by the collision of particles with the walls of the container. The pressure of a gas may be measured with a **barometer,** invented by Evangelista Torricelli in the mid-1600s.

The most common type of barometer is the mercury barometer. An early version is depicted in Figure 5.1. A tube, sealed at one end, is filled with mercury and inverted in a dish of mercury. The pressure of the atmosphere pushing down on the mercury surface in the dish supports the column of mercury. The height of the column is proportional to the atmospheric pressure. The tube can be calibrated to give a numerical reading in millimeters (mm), centimeters (cm), or inches (in) of mercury. A commonly used unit of measurement is the atmosphere (atm). One standard atmosphere (1 atm) of pressure is equivalent to a height of mercury that is equal to

> 760 mm Hg (millimeters of mercury)
> 76.0 cm Hg (centimeters of mercury)
> 1 mm of Hg is also = 1 torr, in honor of Torricelli.

The English system equivalent of the standard atmosphere is 14.7 lb/in² (pounds per square inch, abbreviated psi) or 29.9 in Hg (inches of mercury). A recommended, yet less frequently used, systematic unit is the pascal (or kilopascal), named in honor of Blaise Pascal, a seventeenth-century French mathematician and scientist:

$$1 \text{ atm} = 1.01 \times 10^5 \text{ Pa (pascal)} = 101 \text{ kPa (kilopascal)}$$

Atmospheric pressure is due to the cumulative force of the molecules of air (N_2 and O_2, for the most part) that are attracted to the earth's surface by gravity.

Question 5.1 Express each of the following in units of atm:
  a. 725 mm Hg   b. 29.0 cm Hg   c. 555 torr   d. 95 psi

Question 5.2 Express each of the following in units of atm:
  a. 10.0 torr   b. 61.0 cm Hg   c. 275 mm Hg   d. 124 psi

Kinetic Molecular Theory of Gases

The kinetic molecular theory of gases provides a reasonable explanation of the behavior of gases. The bulk properties of a gas result from the action of the individual molecules comprising the gas.

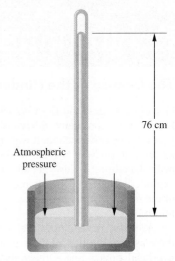

Figure 5.1 A mercury barometer of the type invented by Torricelli. The mercury in the tube is supported by atmospheric pressure, and the height of the column of mercury is a function of the magnitude of the surrounding atmospheric pressure.

76 cm

Atmospheric pressure

LEARNING GOAL

1 Perform conversions between units of pressure.

Pressure equivalencies are used to construct factors that allow conversions from one pressure unit to another. The use of conversion factors was introduced in Chapter 1.

A Human Perspective

The Demise of the Hindenburg

One of the largest and most luxurious airships of the 1930s, the Hindenburg, completed thirty-six transatlantic flights within a year after its construction. It was the flagship of a new era of air travel. But, on May 6, 1937, while making a landing approach near Lakehurst, New Jersey, the hydrogen-filled airship exploded and burst into flames. In this tragedy, thirty-seven of the ninety-six passengers were killed and many others were injured.

We may never know the exact cause. Many believe that the massive ship [it was more than 800 feet (ft) long] struck an overhead power line. Others speculate that lightning ignited the hydrogen, and some believe that sabotage may have been involved.

In retrospect, such an accident was inevitable. Hydrogen gas is very reactive, it combines with oxygen readily and rapidly, and this reaction liberates a large amount of energy. Explosions are the result of rapid, energy-releasing reactions.

Why was hydrogen chosen? Hydrogen is the lightest element. One mole of hydrogen has a mass of 2 grams (g). Hydrogen can be easily prepared in pure form, an essential requirement; more than seven million cubic feet (ft^3) of hydrogen were needed for each airship. Hydrogen has a low density; hence, it provides great lift. The lifting power of a gas is based on the difference in density of the gas and the surrounding air (air is composed of gases with much greater molar masses; N_2 is 28 g/mol and O_2 is 32 g/mol). Engineers believed that the hydrogen would be safe when enclosed by the hull of the airship.

Today, airships are filled with helium (molar mass is 4 g/mol), which is far less reactive than hydrogen, and are used

The Hindenburg

principally for advertising and television. A blimp, with its corporate logo prominently displayed, can be seen hovering over almost every significant outdoor sporting event.

For Further Understanding

▸ Would a gas such as methane (CH_4) provide lifting power for an airship? Why or why not?

▸ What property would immediately rule out methane as a replacement gas for hydrogen in an airship?

LEARNING GOAL

2 Describe the major points of the kinetic molecular theory of gases.

Kinetic energy (K.E.) is equal to $\frac{1}{2}mv^2$, in which m = mass and v = velocity. Thus, increased velocity at higher temperature correlates with an increase in kinetic energy.

The **kinetic molecular theory** can be summarized as follows:

1. Gases are made up of tiny atoms or molecules that are in constant, random motion. The particles are moving along a linear path, changing direction only as a result of collisions.
2. The distance of separation among these atoms or molecules is very large in comparison to the size of the individual atoms or molecules. In other words, a gas is mostly empty space.
3. All of the atoms and molecules behave independently. No attractive or repulsive forces exist between atoms or molecules in a gas.
4. Atoms and molecules collide with each other and with the walls of the container without *losing* energy. The energy is *transferred* from one atom or molecule to another. These collisions cause random changes in direction.
5. The average kinetic energy of the atoms or molecules increases or decreases in proportion to absolute temperature. As the temperature increases, the speed and kinetic energy of the atoms or molecules increase.

Properties of Gases and the Kinetic Molecular Theory

We know that gases are easily *compressible*. The reason is that a gas is mostly empty space, providing space for the particles to be pushed closer together.

Gases will *expand* to fill any available volume because they move freely with sufficient energy to overcome their attractive forces.

Gases readily *diffuse* through each other simply because they are in continuous motion and paths are readily available owing to the large space between adjacent atoms or molecules. Light molecules diffuse rapidly; heavier molecules diffuse more slowly (Figure 5.2).

Gases have a *low density*. Density is defined as mass per volume. Because gases are mostly empty space, they have a low mass per volume.

Gases exert *pressure* on their containers. Pressure is a force per unit area resulting from collisions of gas particles with the walls of their container.

Gases behave most *ideally at low pressures and high temperatures*. At low pressures, the average distance of separation among atoms or molecules is greatest, minimizing interactive forces. At high temperatures, the atoms and molecules are in rapid motion and are able to overcome interactive forces more easily.

LEARNING GOAL

3 Explain the relationship between the kinetic molecular theory and the physical properties of measurable quantities of gases.

Boyle's Law

The Irish scientist Robert Boyle found that the volume of a gas varies *inversely* with the pressure exerted by the gas if the number of mol and temperature of gas are held constant. This relationship is known as **Boyle's law.**

Mathematically, the *product* of pressure (P) and volume (V) is a constant, k_b:

$$PV = k_b$$

LEARNING GOAL

4 Describe the behavior of gases expressed by the gas laws: Boyle's law, Charles's law, combined gas law, Avogadro's law, the ideal gas law, and Dalton's law.

This relationship is illustrated in Figure 5.3.

Boyle's law is often used to calculate the volume resulting from a pressure change or vice versa. We consider

$$P_i V_i = k_b$$

with the subscript *i* representing the *initial* condition and

$$P_f V_f = k_b$$

with the subscript *f* representing the *final* condition. Because *PV*, initial or final, is constant and is equal to k_b,

$$P_i V_i = P_f V_f$$

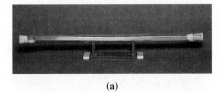

(a)

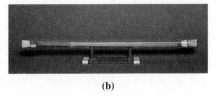

(b)

Figure 5.2 Gaseous diffusion. (a) Ammonia (17.0 g/mol) and hydrogen chloride (36.5 g/mol) are introduced into the ends of a glass tube containing indicating paper. Red indicates the presence of hydrogen chloride and green indicates ammonia. (b) Note that ammonia has diffused much farther than hydrogen chloride in the same amount of time. This is a verification of the kinetic molecular theory. Light molecules move faster than heavier molecules at a specified temperature.

Consider a gas occupying a volume of 10.0 liters (L) at 1.00 atm of pressure. The product, $P_i V_i = (1.00 \text{ atm})(10.0 \text{ L})$, is a constant, k_b, that is equal to $10.0 \text{ L} \cdot \text{atm}$. Doubling the pressure, to 2.00 atm, decreases the volume by a factor of two:

$$(2.00 \text{ atm})(V_f) = 10.0 \text{ L} \cdot \text{atm}$$
$$V_f = 5.00 \text{ L}$$

Tripling the pressure decreases the volume by a factor of three:

$$(3.00 \text{ atm})(V_f) = 10.0 \text{ L} \cdot \text{atm}$$
$$V_f = 3.33 \text{ L}$$

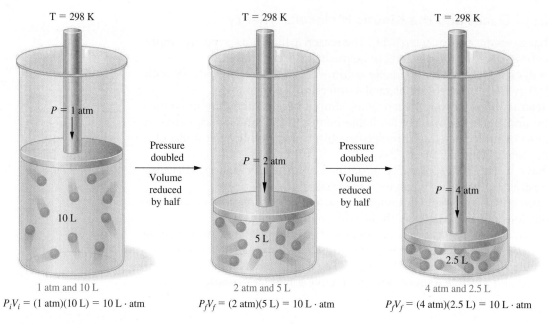

T = 298 K   T = 298 K   T = 298 K

$P = 1$ atm

10 L

Pressure
doubled

Volume
reduced
by half

$P = 2$ atm

5 L

Pressure
doubled

Volume
reduced
by half

$P = 4$ atm

2.5 L

| 1 atm and 10 L | 2 atm and 5 L | 4 atm and 2.5 L |

$P_iV_i = (1 \text{ atm})(10 \text{ L}) = 10 \text{ L} \cdot \text{atm}$ $P_fV_f = (2 \text{ atm})(5 \text{ L}) = 10 \text{ L} \cdot \text{atm}$ $P_fV_f = (4 \text{ atm})(2.5 \text{ L}) = 10 \text{ L} \cdot \text{atm}$

Figure 5.3 An illustration of Boyle's law. Note the inverse relationship of pressure and volume. Since *PV* is a constant, increases in pressure decrease the volume.

EXAMPLE 5.1   **Calculating a Final Pressure**

LEARNING GOAL

A sample of oxygen, at 25°C, occupies a volume of 5.00×10^2 milliliters (mL) at 1.50 atm pressure. What pressure must be applied to compress the gas to a volume of 1.50×10^2 mL, with no temperature change?

5 Use gas law equations to calculate conditions and changes in conditions of gases.

Solution

Step 1. Boyle's law applies directly, because there is no change in temperature or number of mol (no gas enters or leaves the container).

Step 2. Begin by identifying each term in the Boyle's law expression:

$$P_i = 1.50 \text{ atm}$$
$$P_f = ?$$
$$V_i = 5.00 \times 10^2 \text{ mL}$$
$$V_f = 1.50 \times 10^2 \text{ mL}$$

Step 3. The Boyle's law expression is:

$$P_iV_i = P_fV_f$$

Step 4. Solving for P_f:

$$P_f = \frac{P_iV_i}{V_f}$$

Step 5. Substituting:

$$P_f = \frac{(1.50 \text{ atm})(5.00 \times 10^2 \text{ mL})}{1.50 \times 10^2 \text{ mL}}$$
$$= 5.00 \text{ atm}$$

Helpful Hint: The calculation can be done with any volume units. It is important only that the initial and final volume units be the *same*.

Practice Problem 5.1

Complete the following table:

Sample Number	Initial Pressure (atm)	Final Pressure (atm)	Initial Volume (L)	Final Volume (L)
1	X	5.0	1.0	7.5
2	5.0	X	1.0	0.20
3	1.0	0.50	X	0.30
4	1.0	2.0	0.75	X

▶ For Further Practice: **Questions 5.39 and 5.40.**

Charles's Law

Jacques Charles, a French scientist, studied the relationship between gas volume and temperature. This relationship, **Charles's law,** states that the volume of a gas varies *directly* with the absolute temperature (K) if pressure and number of mol of gas are constant.

Mathematically, the *ratio* of volume (V) and temperature (T) is a constant, k_c:

$$\frac{V}{T} = k_c$$

This relationship is illustrated in Figure 5.4. In a way analogous to Boyle's law, we may establish a set of initial conditions represented with the subscript i,

$$\frac{V_i}{T_i} = k_c$$

and final conditions represented with the subscript f,

$$\frac{V_f}{T_f} = k_c$$

LEARNING GOAL

4 Describe the behavior of gases expressed by the gas laws: Boyle's law, Charles's law, combined gas law, Avogadro's law, the ideal gas law, and Dalton's law.

Temperature is a measure of the energy of molecular motion. The Kelvin scale is *absolute*; that is, directly proportional to molecular motion. Celsius and Fahrenheit are simply numerical scales based on the melting and boiling points of water. It is for this reason that Kelvin is used for energy-dependent relationships such as the gas laws.

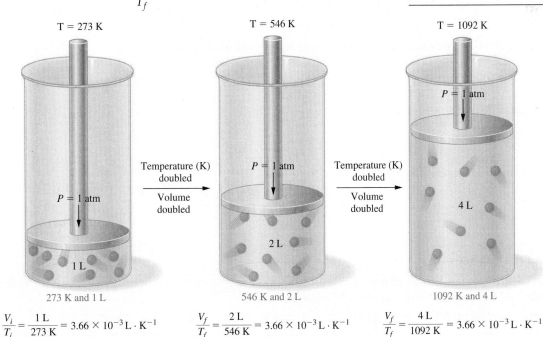

$$\frac{V_i}{T_i} = \frac{1\ L}{273\ K} = 3.66 \times 10^{-3}\ L \cdot K^{-1} \qquad \frac{V_f}{T_f} = \frac{2\ L}{546\ K} = 3.66 \times 10^{-3}\ L \cdot K^{-1} \qquad \frac{V_f}{T_f} = \frac{4\ L}{1092\ K} = 3.66 \times 10^{-3}\ L \cdot K^{-1}$$

Figure 5.4 An illustration of Charles's law. Note the direct relationship between temperature and volume. Since V/T is a constant, increases in temperature will increase the volume.

Remember that this equation is only valid at constant number of mol (n) and pressure (P).

Because k_c is a constant, we may equate them, resulting in

$$\frac{V_i}{T_i} = \frac{V_f}{T_f}$$

and we may use this expression to solve some practical problems.

Consider a gas occupying a volume of 10.0 L at 273 K. The ratio V/T is a constant, k_c. Doubling the temperature, to 546 K, increases the volume to 20.0 L, as shown here:

$$\frac{10.0 \text{ L}}{273 \text{ K}} = \frac{V_f}{546 \text{ K}}$$
$$V_f = 20.0 \text{ L}$$

Tripling the temperature, to 819 K, increases the volume by a factor of three:

$$\frac{10.0 \text{ L}}{273 \text{ K}} = \frac{V_f}{819 \text{ K}}$$
$$V_f = 30.0 \text{ L}$$

EXAMPLE 5.2   **Calculating a Final Volume**

A balloon filled with helium has a volume of 4.0×10^3 L at 25°C. What volume will the balloon occupy at 50°C if the pressure surrounding the balloon remains constant?

LEARNING GOAL

5 Use gas law equations to calculate conditions and changes in conditions of gases.

Solution

Step 1. Summarize the data, remembering that the temperature must be converted to Kelvin before Charles's law is applied:

$$T_i = 25°C + 273 = 298 \text{ K}$$
$$T_f = 50°C + 273 = 323 \text{ K}$$
$$V_i = 4.0 \times 10^3 \text{ L}$$
$$V_f = ?$$

Step 2. Using the Charles's law expression relating initial and final conditions:

$$\frac{V_i}{T_i} = \frac{V_f}{T_f}$$

Step 3. Rearrange and solve for V_f:

$$V_f = \frac{V_i T_f}{T_i}$$

Step 4. Substituting our data, we get

$$V_f = \frac{V_i T_f}{T_i} = \frac{(4.0 \times 10^3 \text{ L})(323 \text{ K})}{298 \text{ K}} = 4.3 \times 10^3 \text{ L}$$

Practice Problem 5.2

A sample of nitrogen gas has a volume of 3.00 L at 25°C. What volume will it occupy at each of the following temperatures if the pressure and number of mol are constant?

   a. 100°C   b. 150°F   c. 273 K   d. 546 K   e. 0°C   f. 373 K

▶ For Further Practice: **Questions 5.47 and 5.48.**

The behavior of a fixed-volume hot-air balloon is a commonplace consequence of Charles's law. The balloon rises because air expands when heated (Figure 5.5). The volume of the balloon is fixed because the balloon is made of an inelastic material; as a result, when the air expands as described by Charles' law, some of it must be forced out. Hence, the density of the remaining air is less (less mass contained in the same volume), and the balloon rises. Turning down the heat reverses the process, and the balloon descends.

Combined Gas Law

Boyle's law describes the inverse proportional relationship between volume and pressure; Charles's law shows the direct proportional relationship between volume and temperature. Often, a sample of gas (a fixed number of mol of gas) undergoes change involving volume, pressure, and temperature simultaneously.

The **combined gas law** is one equation that describes such processes. It can be derived from Boyle's law and Charles's law and takes the form:

$$\frac{P_i V_i}{T_i} = \frac{P_f V_f}{T_f}$$

Let's look at two examples that use this expression.

Figure 5.5 Charles's law predicts that the volume of air in the balloon will increase when heated.

EXAMPLE 5.3	**Using the Combined Gas Law**

LEARNING GOAL

If 0.100 L of N_2 at 27.0°C and 1.00 atm is compressed to a pressure of 10.0 atm at 77.0°C, calculate the new volume of N_2.

5 Use gas law equations to calculate conditions and changes in conditions of gases.

Solution

Step 1. Summarize the data, remembering that the temperature must be converted to Kelvin before the combined gas law is applied:

$$P_i = 1.00 \text{ atm} \qquad\qquad P_f = 10.0 \text{ atm}$$
$$V_i = 0.100 \text{ L} \qquad\qquad V_f = ? \text{ L}$$
$$T_i = 27.0°C + 273.15 = 300.2 \text{ K} \qquad T_f = 77.0°C + 273.15 = 350.2 \text{ K}$$

Step 2. The combined gas law expression is:

$$\frac{P_i V_i}{T_i} = \frac{P_f V_f}{T_f}$$

Step 3. Rearrange:

$$P_f V_f T_i = P_i V_i T_f$$

and solve for V_f:

$$V_f = \frac{P_i V_i T_f}{P_f T_i}$$

Step 4. Substituting gives

$$V_f = \frac{(1.00 \text{ atm})(0.100 \text{ L})(350.2 \text{ K})}{(10.0 \text{ atm})(300.2 \text{ K})}$$
$$= 0.0117 \text{ L}$$

Continued…

Practice Problem 5.3

Hydrogen sulfide, H_2S, has the characteristic odor of rotten eggs. If a sample of H_2S gas at 760.0 torr and 25.0°C in a 2.00-L container is allowed to expand into a 10.0-L container at 25.0°C, what is the pressure, in atmospheres, in the 10.0-L container?

▶ For Further Practice: **Questions 5.57 and 5.58.**

EXAMPLE 5.4   **Using the Combined Gas Law**

LEARNING GOAL

A sample of helium gas has a volume of 1.27 L at 149 K and 5.00 atm. When the gas is compressed to 0.320 L at 50.0 atm, the temperature increases markedly. What is the final temperature?

5 Use gas law equations to calculate conditions and changes in conditions of gases.

Solution

Step 1. Summarize the data:

$$P_i = 5.00 \text{ atm} \quad P_f = 50.0 \text{ atm}$$
$$V_i = 1.27 \text{ L} \quad V_f = 0.320 \text{ L}$$
$$T_i = 149 \text{ K} \quad T_f = ? \text{ K}$$

Step 2. The combined gas law expression is

$$\frac{P_i V_i}{T_i} = \frac{P_f V_f}{T_f}$$

Step 3. Rearrange:

$$P_f V_f T_i = P_i V_i T_f$$

and solve for T_f:

$$T_f = \frac{P_f V_f T_i}{P_i V_i}$$

Step 4. Substituting gives

$$T_f = \frac{(50.0 \text{ atm})(0.320 \text{ L})(149 \text{ K})}{(5.00 \text{ atm})(1.27 \text{ L})}$$
$$= 375 \text{ K}$$

Practice Problem 5.4

Cyclopropane, C_3H_6, is used as a general anesthetic. If a sample of cyclopropane stored in a 2.00-L container at 10.0 atm and 25.0°C is transferred to a 5.00-L container at 5.00 atm, what is the resulting temperature?

▶ For Further Practice: **Questions 5.59 and 5.60.**

Avogadro's Law

LEARNING GOAL

4 Describe the behavior of gases expressed by the gas laws: Boyle's law, Charles's law, combined gas law, Avogadro's law, the ideal gas law, and Dalton's law.

The relationship between the volume and number of mol of a gas at constant temperature and pressure is known as **Avogadro's law.** It states that equal volumes of any ideal gas contain the same number of mol if measured under the same conditions of temperature and pressure.

Mathematically, the *ratio* of volume (V) to number of mol (n) is a constant, k_a:

$$\frac{V}{n} = k_a$$

Consider 1 mol of gas occupying a volume of 10.0 L; using logic similar to the application of Boyle's and Charles's laws, 2 mol of the gas would occupy 20.0 L,

3 mol would occupy 30.0 L, and so forth. As we have done with the previous laws, we can formulate a useful expression relating initial and final conditions:

$$\frac{V_i}{n_i} = \frac{V_f}{n_f}$$

EXAMPLE 5.5   Using Avogadro's Law

LEARNING GOAL

If 5.50 mol of CO occupy 20.6 L, how many L will 16.5 mol of CO occupy at the same temperature and pressure?

5 Use gas law equations to calculate conditions and changes in conditions of gases.

Solution

Step 1. The quantities volume and number of moles are related through Avogadro's law. Summarizing the data:

$$V_i = 20.6 \text{ L} \qquad V_f = ? \text{ L}$$
$$n_i = 5.50 \text{ mol} \quad n_f = 16.5 \text{ mol}$$

Step 2. Using the mathematical expression for Avogadro's law:

$$\frac{V_i}{n_i} = \frac{V_f}{n_f}$$

Step 3. Rearranging to solve for V_f:

$$V_f = \frac{V_i n_f}{n_i}$$

Step 4. Substitution yields:

$$V_f = \frac{(20.6 \text{ L})(16.5 \text{ mol})}{(5.50 \text{ mol})}$$
$$= 61.8 \text{ L of CO}$$

Practice Problem 5.5

a. A 1.00-mol sample of hydrogen gas occupies 22.4 L. How many mol of hydrogen are needed to fill a 100.0-L container at the same pressure and temperature?

b. How many mol of hydrogen are needed to triple the volume occupied by 0.25 mol of hydrogen, assuming no changes in pressure or temperature?

▶ For Further Practice: **Questions 5.63 and 5.64.**

Molar Volume of a Gas

The volume occupied by *1 mol* of any gas is referred to as its **molar volume.** At **standard temperature and pressure (STP),** the molar volume of any gas is 22.4 L. STP conditions are defined as follows:

LEARNING GOAL

6 Use molar volume and standard temperature and pressure (STP) to perform calculations.

$$T = 273 \text{ K (or 0°C)}$$
$$P = 1 \text{ atm}$$

Thus, 1 mol of N_2, O_2, CO_2, H_2, or He all occupy the *same volume, 22.4 L,* at STP.

Gas Densities

It is also possible to compute the density of various gases at STP. If we recall that density is the mass/unit volume,

Gas densities are most often expressed in units of g/L. Recall that units of g/mL are generally preferred for solids and liquids.

$$d = \frac{m}{V}$$

and that 1 mol of helium weighs 4.00 g,

$$d_{He} = \frac{4.00 \text{ g}}{22.4 \text{ L}} = 0.178 \text{ g/L at STP}$$

or, because 1 mol of nitrogen weighs 28.0 g, then

$$d_{N_2} = \frac{28.0 \text{ g}}{22.4 \text{ L}} = 1.25 \text{ g/L at STP}$$

Heating a gas, such as air, will decrease its density and have a lifting effect as well.

The large difference in gas densities of helium and nitrogen (which makes up about 80% of the air) accounts for the lifting power of helium. A balloon filled with helium will rise through a predominantly nitrogen atmosphere because its gas density is less than 15% of the density of the surrounding atmosphere:

$$\frac{d_{He}}{d_{N_2}} \times 100\% = \% \text{ density}$$

$$\frac{0.178 \cancel{\text{ g/L}}}{1.25 \cancel{\text{ g/L}}} \times 100\% = 14.2\%$$

The Ideal Gas Law

Boyle's law (relating volume and pressure), Charles's law (relating volume and temperature), and Avogadro's law (relating volume to the number of mol) may be combined into a single expression relating all four terms. This expression is the **ideal gas law:**

$$PV = nRT$$

in which R is a constant based on k_b, k_c, and k_a (Boyle's, Charles's, and Avogadro's law constants) and is referred to as the *ideal gas constant:*

$$R = 0.0821 \text{ L} \cdot \text{atm} \cdot \text{K}^{-1} \cdot \text{mol}^{-1}$$

which is identical to

$$R = 0.0821 \frac{\text{L} \cdot \text{atm}}{\text{K} \cdot \text{mol}}$$

if the units

    atmospheres for P,

    liters for V,

    moles for n,

    and

    Kelvin for T

are used.

Consider some examples of the application of the ideal gas equation.

LEARNING GOAL

4 Describe the behavior of gases expressed by the gas laws: Boyle's law, Charles's law, combined gas law, Avogadro's law, the ideal gas law, and Dalton's law.

EXAMPLE 5.6	**Calculating a Molar Volume**

Show by calculation that the molar volume of oxygen gas at STP is 22.4 L.

Solution

Step 1. At STP,

$$T = 273 \text{ K}$$
$$P = 1.00 \text{ atm}$$

and the other terms are

$$n = 1.00 \text{ mol}$$
$$R = 0.0821 \text{ L} \cdot \text{atm} \cdot \text{K}^{-1} \cdot \text{mol}^{-1}$$

LEARNING GOALS

5 Use gas law equations to calculate conditions and changes in conditions of gases.

6 Use molar volume and standard temperature and pressure (STP) to perform calculations.

Step 2. The ideal gas expression is:

$$PV = nRT$$

Step 3. Rearrange and solve for V:

$$V = \frac{nRT}{P}$$

Step 4. Substitute and solve:

$$V = \frac{(1.00 \text{ mol})(0.0821 \text{ L} \cdot \text{atm} \cdot \text{K}^{-1} \cdot \text{mol}^{-1})(273 \text{ K})}{(1.00 \text{ atm})}$$

$$= 22.4 \text{ L}$$

Practice Problem 5.6

Explain why the molar volume of helium (or any other ideal gas) is 22.4 L.

▶ For Further Practice: **Questions 5.65 and 5.66.**

EXAMPLE 5.7 | **Calculating the Number of Moles of a Gas**

LEARNING GOAL

Calculate the number of mol of helium in a 1.00-L balloon at 27°C and 1.00 atm of pressure.

5 Use gas law equations to calculate conditions and changes in conditions of gases.

Solution

Step 1. The data are:

$$P = 1.00 \text{ atm}$$
$$V = 1.00 \text{ L}$$
$$T = 27°C + 273.15 = 3.00 \times 10^2 \text{ K}$$
$$R = 0.0821 \text{ L} \cdot \text{atm} \cdot \text{K}^{-1} \cdot \text{mol}^{-1}$$
$$n = ?$$

Step 2. The ideal gas expression is:

$$PV = nRT$$

Step 3. Rearrange and solve for n:

$$n = \frac{PV}{RT}$$

Step 4. Substitute and solve:

$$n = \frac{(1.00 \text{ atm})(1.00 \text{ L})}{(0.0821 \text{ L} \cdot \text{atm} \cdot \text{K}^{-1} \cdot \text{mol}^{-1})(3.00 \times 10^2 \text{ K})}$$

$$n = 0.0406 \text{ or } 4.06 \times 10^{-2} \text{ mol}$$

Practice Problem 5.7

How many mol of N_2 gas will occupy a 5.00-L container at STP?

▶ For Further Practice: **Questions 5.69 and 5.70.**

EXAMPLE 5.8 Converting Mass to Volume

Oxygen used in hospitals and laboratories is often obtained from cylinders containing liquefied oxygen. If a cylinder contains 1.00×10^2 kilograms (kg) of liquid oxygen, how many L of oxygen can be produced at 1.00 atm of pressure at room temperature (20.0°C)?

5 Use gas law equations to calculate conditions and changes in conditions of gases.

Solution

Step 1. Summarize the data:

mass of oxygen (O_2) = 1.00×10^2 kg
$T = 20.0°C$
$P = 1.00$ atm
V of O_2 = ?

Step 2. The number of moles of O_2 (n) is obtained by using two conversion factors based on 1000 g = 1 kg and the molar mass of O_2 (32.0 g/mol):

$$n = 1.00 \times 10^2 \ \text{kg } O_2 \times \frac{10^3 \ \text{g } O_2}{1 \ \text{kg } O_2} \times \frac{1 \ \text{mol } O_2}{32.0 \ \text{g } O_2} = 3.13 \times 10^3 \ \text{mol } O_2$$

Step 3. Convert °C to K:

$$T = 20.0°C + 273.15 = 293.2 \ K$$

Step 4. The ideal gas expression is:

$$PV = nRT$$

Step 5. Rearrange and solve for V:

$$V = \frac{nRT}{P}$$

Step 6. Substitute and solve:

$$V = \frac{(3.13 \times 10^3 \ \text{mol})(0.0821 \ \text{L} \cdot \text{atm} \cdot K^{-1} \cdot \text{mol}^{-1})(293.2 \ K)}{1.00 \ \text{atm}}$$

$$= 7.53 \times 10^4 \ \text{L}$$

Practice Problem 5.8

What volume is occupied by 10.0 g N_2 at 30.0°C and a pressure of 750 torr?

▶ For Further Practice: **Questions 5.75 and 5.76.**

Question 5.3   A 20.0-L gas cylinder contains 4.80 g H_2 at 25°C. What is the pressure of this gas?

Question 5.4   At what temperature will 2.00 mol of He fill a 2.00-L container at standard pressure?

Dalton's Law of Partial Pressures

LEARNING GOAL

4 Describe the behavior of gases expressed by the gas laws: Boyle's law, Charles's law, combined gas law, Avogadro's law, the ideal gas law, and Dalton's law.

Our discussion of gases so far has presumed that we are working with a single pure gas. A *mixture* of gases exerts a pressure that is the *sum* of the pressures that each gas would exert if it were present alone under the same conditions. This is known as **Dalton's law** of partial pressures. Dalton's law is based on the assumption that the behavior of each gas in a mixture of gases is independent of all of the other gases.

Green Chemistry

The Greenhouse Effect and Global Climate Change

A greenhouse is a bright, warm, and humid environment for growing plants, vegetables, and flowers even during the cold winter months. It functions as a closed system in which the concentration of water vapor is elevated and visible light streams through the windows; this creates an ideal climate for plant growth.

Some of the visible light is absorbed by plants and soil in the greenhouse and released as infrared radiation. This radiated energy is blocked by the glass or absorbed by water vapor and carbon dioxide (CO_2). This trapped energy warms the greenhouse and is a form of solar heating: light energy is converted to heat energy. Hence, water vapor and carbon dioxide are termed *greenhouse gases.*

On a global scale, the same process takes place. Although more than half of the sunlight that strikes the earth's surface is reflected back into space, the fraction of light that is absorbed produces sufficient heat to sustain life. How does this happen? Greenhouse gases, such as CO_2, trap energy radiated from the earth's surface and store it in the atmosphere. This moderates our climate. The earth's surface would be much colder and more inhospitable if the atmosphere was not able to capture some reasonable amount of solar energy.

Can we have too much of a good thing? It appears so. Since 1900, the atmospheric concentration of CO_2 has increased from 296 parts per million (ppm) to over 350 ppm (approximately 17% increase). The energy demands of technological and population growth have caused massive increases in the combustion

of organic matter and carbon-based fuels (coal, oil, and natural gas), adding over 50 billion tons of CO_2 to that already present in the atmosphere. Photosynthesis naturally removes CO_2 from the atmosphere. However, the removal of forestland to create living space and cropland has decreased the amount of vegetation available to consume atmospheric CO_2 through photosynthesis. The rapid destruction of the Amazon rain forest is just the latest of many examples.

Many gases, in addition to H_2O and CO_2, behave as greenhouse gases. Any molecule in the gas phase that is capable of absorbing infrared radiation may behave as a greenhouse gas. Many, however, are unimportant in the global climate change discussion simply because they are not present in the atmosphere in significant quantity. Methane (CH_4), an infrared absorber, exists at higher levels and is a potent greenhouse gas. It has received little attention principally because we have relatively little control over its atmospheric levels.

If our greenhouse model is a correct representation of our atmosphere, an increase in CO_2 levels should contribute to global warming, perhaps changing our climate in unforeseen and undesirable ways.

For Further Understanding

▶ What steps might be taken to decrease levels of CO_2 in the atmosphere over time?

▶ In what ways might our climate and our lives change as a consequence of significant global warming?

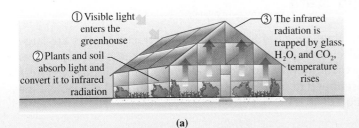

(a)

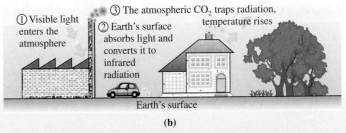

(b)

(a) A greenhouse traps solar radiation as heat. (b) Our atmosphere also acts as a solar collector. Carbon dioxide, like the windows of a greenhouse, allows the visible light to enter and traps the heat.

Stated another way, the total pressure of a mixture of gases is the sum of the **partial pressures.** That is,

$$P_t = p_1 + p_2 + p_3 + \cdots$$

in which P_t = total pressure and $p_1, p_2, p_3, \ldots$, are the partial pressures of the component gases. For example, the total pressure of our atmosphere is equal to the sum of the pressures of N_2 and O_2 (the principal components of air):

$$P_{air} = p_{N_2} + p_{O_2}$$

The ideal gas law applies to mixtures of gases as well as pure gases.

Other gases, such as argon (Ar), carbon dioxide (CO_2), carbon monoxide (CO), and methane (CH_4) are present in the atmosphere at very low partial pressures. However, their presence may result in dramatic consequences; one such gas is carbon dioxide. Classified as a "greenhouse gas," it exerts a significant effect on our climate. Its role is described in Green Chemistry: The Greenhouse Effect and Global Climate Change.

Ideal Gases Versus Real Gases

LEARNING GOAL

7 Discuss the limitations to the ideal gas model as it applies to real gases.

See Sections 3.5 and 5.2 for a discussion of interactions of polar molecules.

Gases behave less ideally as their temperature decreases; they become more like the liquid phase as they approach their condensation temperature.

To this point we have assumed, in both theory and calculations, that all gases behave as ideal gases. However, in reality there is no such thing as an ideal gas. As we noted at the beginning of this section, the ideal gas is a model (a very useful one) that describes the behavior of individual atoms and molecules; this behavior translates to the collective properties of measurable quantities of these atoms and molecules. Limitations of the model arise from the fact that interactive forces, even between the widely spaced particles of gas, are not totally absent in any sample of gas.

Gases comprised of polar molecules have stronger attractive forces than gases made up of nonpolar molecules. Nonuniform charge distribution on polar molecules creates positive and negative regions, resulting in electrostatic attraction and deviation from ideality.

Calculations involving polar gases such as HF, NO, and SO_2 based on ideal gas equations (which presume no such interactions) are approximations. However, at low pressures, such approximations certainly provide useful information. Nonpolar molecules, on the other hand, are only weakly attracted to each other and behave much more ideally in the gas phase.

Question 5.5 Radon and nitrogen dioxide are gases at 25°C. Which exhibits more ideal behavior? Explain your answer.

Question 5.6 Hydrogen sulfide (H_2S) is a gas at 0°C. When its temperature is decreased, does it behave more or less ideally? Explain your answer.

5.2   The Liquid State

LEARNING GOAL

8 Describe properties of the liquid state in terms of the properties of the individual molecules that comprise the liquid.

Molecules in the liquid state are close to one another. Attractive forces are large enough to keep the molecules together, in contrast to gases, whose cohesive forces are so low that gases expand to fill any volume. However, these attractive forces in liquids are not large enough to restrict movement, as in solids. Let's look at the various properties of liquids in more detail.

Compressibility

Liquids are practically incompressible. In fact, the molecules are so close to one another that even the application of very high pressure does not significantly decrease the volume. This makes liquids ideal for the transmission of force, as in the brake lines of an automobile. The force applied by the driver's foot on the brake pedal does not compress the brake fluid in the lines; rather, it transmits the force directly to the brake pads, and the friction between the brake pads and rotors (that are attached to the wheels) stops the car.

Viscosity

The **viscosity** of a liquid is a measure of its resistance to flow. Viscosity is a function of both the attractive forces between molecules and molecular geometry.

Molecules with complex structures, which do not "slide" smoothly past each other, and polar molecules tend to have higher viscosity than less structurally complex, less polar liquids. Glycerol, which is used in a variety of skin treatments, has the structural formula:

$$
\begin{array}{c}
H \\
| \\
H-C-O-H \\
| \\
H-C-O-H \\
| \\
H-C-O-H \\
| \\
H
\end{array}
$$

It is quite viscous, owing to its polar nature and its significant intermolecular attractive forces. This is certainly desirable in a skin treatment because its viscosity keeps it on the area being treated. Gasoline, on the other hand, is much less viscous and readily flows through the gas lines of your auto; it is composed of nonpolar molecules.

Viscosity generally decreases with increasing temperature. The increased kinetic energy at higher temperatures overcomes some of the intermolecular attractive forces. The temperature effect is an important consideration in the design of products that must remain fluid at low temperatures, such as motor oils and transmission fluids found in automobiles.

Surface Tension

The **surface tension** of a liquid is a measure of the attractive forces exerted among molecules at the surface of the liquid. It is only the surface molecules that are not totally surrounded by other liquid molecules (the top of the molecule faces the atmosphere). These surface molecules are surrounded and attracted by fewer liquid molecules than the interior molecules. Hence, the net attractive forces on surface molecules are greater (greater force per molecule) because each surface molecule shares its attractive forces with fewer molecules. The resulting stronger attractive forces pull the surface molecules downward, into the body of the liquid. As a result, the surface molecules behave as a tight "skin" that covers the interior.

This increased surface force is responsible for the spherical shape of drops of liquid. Drops of water "beading" on a polished surface, such as a waxed automobile, illustrate this effect.

Because surface tension is related to the attractive forces exerted among molecules, surface tension generally decreases with an increase in temperature or a decrease in the polarity of molecules that make up the liquid.

A **surfactant** is a substance that can be added to a liquid to decrease surface tension. Surfactants have polar and nonpolar regions at opposite ends of their molecules. The polar ends of the molecules interact with polar liquids to decrease attractive forces at the surface (hence, lowering surface tension). Common surfactants include soaps and detergents that reduce water's surface tension; this promotes the interaction of water with grease and dirt, making them easier to remove. For more information on these interesting molecules, see Kitchen Chemistry: Solubility, Surfactants and the Dishwasher, Chapter 6.

Question 5.7 What molecular properties favor high viscosity?

Question 5.8 What molecular properties favor high surface tension?

Vapor Pressure of a Liquid

Evaporation, condensation, and the meaning of the term *boiling point* are all related to the concept of liquid vapor pressure. Consider the following example. A liquid,

Skipping stones is possible due to the surface tension of water. Explain.

Compounds may be detected and identified because they have a measurable vapor pressure, as you will see in the discussion Chemistry at the Crime Scene: Explosives at the Airport.

LEARNING GOAL

9 Describe the processes of melting, boiling, evaporation, condensation, and sublimation.

such as water, is placed in a sealed container. After a time, the contents of the container are analyzed. Both liquid water and water vapor are found at room temperature, when we might expect water to be found only as a liquid. In this closed system, some of the liquid water was converted to a gas:

$$\text{energy} + H_2O(l) \longrightarrow H_2O(g)$$

How did this happen? The temperature is too low for conversion of a liquid to a gas by boiling. According to the kinetic molecular theory, liquid molecules are in continuous motion, with their *average* kinetic energy directly proportional to the Kelvin temperature. The word *average* is the key. Although the average kinetic energy is too low to allow "average" molecules to escape from the liquid phase to the gas phase, there exists a range of molecules with different energies, some low and some high, that make up the "average" (Figure 5.6). Thus, some of these high-energy molecules possess sufficient energy to escape from the bulk liquid.

At the same time, a fraction of these gaseous molecules lose energy (perhaps by collision with the walls of the container) and return to the liquid state:

$$H_2O(g) \longrightarrow H_2O(l) + \text{energy}$$

The process of conversion of liquid to gas, at a temperature too low to boil, is **evaporation.** The reverse process, conversion of the gas to the liquid state, is **condensation.** After some time, the rates of evaporation and condensation become *equal,* and this sets up a dynamic equilibrium between liquid and vapor states. The **vapor pressure of a liquid** is defined as the pressure exerted by the vapor *at equilibrium.*

$$H_2O(g) \rightleftharpoons H_2O(l)$$

The equilibrium process of evaporation and condensation of water is depicted in Figure 5.7.

Boiling Point and Vapor Pressure

The boiling point of a liquid is defined as the temperature at which the vapor pressure of the liquid becomes equal to the atmospheric pressure. The "normal" atmospheric pressure is 760 torr, or 1 atm, and the **normal boiling point** is the temperature at which the vapor pressure of the liquid is equal to 1 atm.

It follows from the definition that the boiling point of a liquid is not constant. It depends on the atmospheric pressure. At high altitudes, where the atmospheric pressure is low, the boiling point of a liquid, such as water, is lower than the normal boiling point (for water, 100°C). Alternatively, high atmospheric pressure increases the boiling point.

Apart from its dependence on the surrounding atmospheric pressure, the boiling point depends on the nature of the attractive forces between the liquid molecules. Polar liquids, such as water, with large intermolecular attractive forces have *higher* boiling points than nonpolar liquids, such as gasoline, which exhibit weak attractive forces.

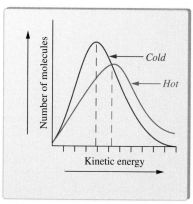

Figure 5.6 The range of molecules with different kinetic energies is illustrated. The small number of high-energy molecules possess sufficient energy to evaporate. Note that the average values are indicated by dashed lines.

The process of evaporation of perspiration from the skin produces a cooling effect, because heat is absorbed and carried away by the evaporating molecules.

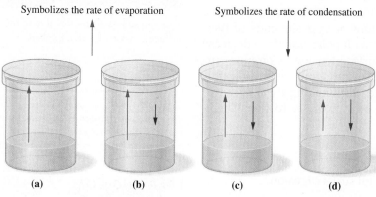

Figure 5.7 Liquid water in equilibrium with water vapor. (a) Initiation: process of evaporation exclusively. (b, c) After a time, both evaporation and condensation occur, but evaporation predominates. (d) Dynamic equilibrium established. Rates of evaporation and condensation are equal.

Question 5.9   Distinguish between the terms *evaporation* and *condensation*.

Question 5.10   Distinguish between the terms *evaporation* and *boiling*.

van der Waals Forces

Many physical properties of liquids, such as melting and boiling temperatures, can be explained in terms of their intermolecular forces. Attractive forces between polar molecules, **dipole-dipole interactions,** significantly decrease vapor pressure and increase the boiling point. However, nonpolar substances can exist as liquids as well; many are liquids and even solids at room temperature. What is the nature of the attractive forces in these nonpolar compounds?

In 1930, Fritz London demonstrated the presence of a weak attractive force between any two molecules, whether polar or nonpolar. He postulated that the electron distribution in molecules is not fixed; electrons are in continuous motion, relative to the nucleus. So, for a short time a nonpolar molecule could experience an *instantaneous dipole,* a short-lived polarity caused by a temporary dislocation of the electron cloud. These temporary dipoles could interact with other temporary dipoles, just as permanent dipoles interact in polar molecules (Figure 5.8). We now call these intermolecular forces **London dispersion forces.**

London dispersion forces and dipole-dipole interactions are collectively known as **van der Waals forces.** London dispersion forces exist among polar and nonpolar molecules because electrons are in constant motion in all molecules. Dipole-dipole attractions occur only among polar molecules. In addition to van der Waals forces, a special type of dipole-dipole force, the *hydrogen bond,* has a very significant effect on molecular properties, particularly in biological systems.

Hydrogen Bonding

Typical forces in polar liquids, discussed earlier, are only about 1–2% as strong as ionic and covalent bonds. However, certain liquids have boiling points that are much higher than we would predict from these polar interactions alone. This indicates the presence of some strong intermolecular force. This attractive force is due to **hydrogen bonding.** Molecules in which a hydrogen atom is bonded to a small, highly electronegative atom such as nitrogen, oxygen, or fluorine exhibit this effect. The presence of a highly electronegative atom bonded to a hydrogen atom creates a large dipole:

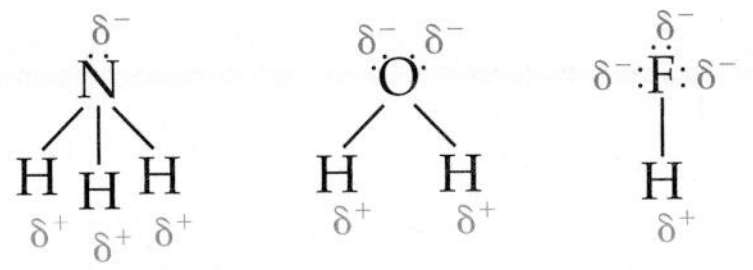

Although the hydrogen bond is weaker than bonds formed *within* molecules (covalent and polar covalent *intra*molecular forces), it is the strongest attractive force *between* molecules (*inter*molecular force).

Consider the boiling points of four small molecules:

CH_4	NH_3	H_2O	HF
molar mass	molar mass	molar mass	molar mass
16 g/mol	17 g/mol	18 g/mol	20 g/mol
$-161°C$	$-33°C$	$+100°C$	$+19.5°C$

LEARNING GOAL

10 Describe the dipolar attractions known collectively as van der Waals forces.

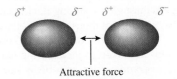

Attractive force

Figure 5.8 London dispersion forces. A temporary dipole results when the electron distribution is unsymmetrical. The nucleus has a partial positive charge represented by δ^+, and the electrons have a partial negative charge represented by δ^-.

LEARNING GOAL

11 Describe hydrogen bonding and its relationship to boiling and melting temperatures.

Recall that the most electronegative elements are in the upper right corner of the periodic table, and these elements exert strong electron attraction in molecules, as described in Chapter 3.

Chemistry at the Crime Scene

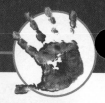

Explosives at the Airport

The images flash across our television screens: a "bomb sniffing" dog being led through an airport or train station, pausing to sniff packages or passengers, looking for anything of a suspicious nature. Or, perhaps, we see a long line of people waiting to pass through a scanning device surrounded by what appears to be hundreds of thousands of dollars worth of electronic gadgetry.

At one level, we certainly know what is happening. These steps are taken to increase the likelihood that our trip, as well as everyone else's, will be as safe and worry-free as possible. From a scientific standpoint, we may wonder how these steps actually detect explosive materials. What do the dog and some electronic devices have in common? How can a dog sniff a solid or a liquid? Surely everyone knows that the nose can only sense gases, and explosive devices are solids or liquids, or a combination of the two.

One potential strategy is based on the concept of vapor pressure. We know that liquids, such as water, have a measurable vapor pressure at room temperature. In fact, most liquids and many solids have vapor pressures large enough to allow detection of the molecules in the gas phase. The challenge is finding devices that are sufficiently sensitive and selective, enabling them to detect low concentrations of molecules characteristic of explosives, without becoming confused by thousands of other compounds routinely present in the air.

Each explosive device has its own "signature," a unique mix of chemicals used in its manufacture and assembly. If only one, or perhaps a few, of these compounds has a measurable vapor pressure, it may be detected with a sensitive measuring device.

Dogs are renowned for their keen sense of smell, and some breeds are better than others. Dogs can be trained to signal the presence of certain scents by barking or exhibiting unusual agitation. A qualified handler can recognize these cues and alert appropriate authorities.

Scientific instruments are designed to mimic the scenario described here. A device, the mass spectrometer, can detect very low concentrations of molecules in the air. Additionally,

Specially trained dogs detect the specific scents of several common explosive materials.

it can distinguish certain "target" molecules, because each different compound has its own unique molar mass. Detection of molecules of interest generates an electrical signal, and an alarm is sounded.

Compounds with high vapor pressures are most easily detected. Active areas of forensic research involve designing a new generation of instruments that are even more sensitive and selective than those currently available. Decreased cost and increased portability and reliability will enable many sites not currently being monitored to have the same level of protection as major transit facilities.

For Further Understanding

▸ Would you expect nonpolar or polar molecules of similar mass to be more easily detected? Why?

▸ Why must an explosives detection device be highly selective?

Clearly, ammonia, water, and hydrogen fluoride boil at significantly higher temperatures than methane. The N—H, O—H, and F—H bonds are far more polar than the C—H bond, owing to the high electronegativity of N, O, and F.

It is interesting to note that the boiling points increase as the electronegativity of the element bonded to hydrogen increases, with one exception: Fluorine, with the highest electronegativity, should cause HF to have the highest boiling point. This is not the case. The order of boiling points is

methane < ammonia < hydrogen fluoride < water

Why? To answer this question, we must look at the *number of potential bonding sites* in each molecule. Water has two partial positive sites (located at each hydrogen atom) and two partial negative sites (two lone pairs of electrons on the oxygen

Intramolecular hydrogen bonding between polar regions helps keep proteins folded in their proper three-dimensional structure. See Chapter 18.

atom); it can form hydrogen bonds at each site. This results in a complex network of attractive forces among water molecules in the liquid state, and the strength of the forces holding this network together accounts for water's unusually high boiling point. This network is depicted in Figure 5.9.

Ammonia and hydrogen fluoride can form only one hydrogen bond per molecule. Ammonia has three partial positive sites (three hydrogen atoms bonded to nitrogen) but only one partial negative site (the lone pair); the single partial negative site is the limiting factor. Hydrogen fluoride has only one partial positive site and three partial negative sites (three lone pairs); the single partial positive site is the limiting factor. Consequently, hydrogen fluoride, like ammonia, can form only one hydrogen bond per molecule. The network of attractive forces in ammonia and hydrogen fluoride is, therefore, much less extensive than that found in water, and their boiling points are considerably lower than that of water.

Hydrogen bonding has an extremely important influence on the behavior of many biological systems. Molecules such as proteins and DNA require extensive hydrogen bonding to maintain their structures and hence functions. DNA (deoxyribonucleic acid, Section 20.2) is a giant among molecules, with intertwined chains of atoms held together by thousands of hydrogen bonds.

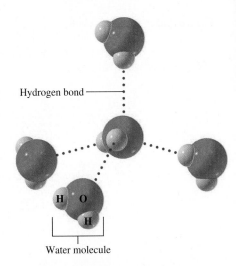

Figure 5.9 Hydrogen bonding in water. Note that the central water molecule is hydrogen bonded to four other water molecules. The attractive force between the hydrogen (δ^+) part of one water molecule and the oxygen (δ^-) part of another water molecule constitutes the hydrogen bond.

Question 5.11 Arrange the following compounds in order of increasing boiling point:

$$CO_2 \quad CH_3OH \quad CH_3Cl$$

Explain your logic.

Question 5.12 Explain the large difference in boiling point (b.p.) for the isomers butanol and diethyl ether.

butanol
b.p. = 117°C

diethyl ether
b.p. = 34.5°C

5.3 The Solid State

The close packing of the particles of a solid results from attractive forces that are strong enough to restrict motion. This occurs because the kinetic energy of the particles is insufficient to overcome the attractive forces among particles. The particles are "locked" together in a defined and highly organized fashion. This results in a fixed shape and volume, and at the atomic level, only vibrational motion is observed.

Properties of Solids

Solids are virtually incompressible, owing to the small distance between particles. Most will convert to liquids at a higher temperature, when the increased heat energy overcomes some of the attractive forces within the solid. The temperature at which a solid is converted to the liquid phase is its *melting point*. The melting point depends on the strength of the attractive forces in the solid, hence its structure. As we might expect, polar solids have higher melting points than nonpolar solids of the same molecular weight.

LEARNING GOAL

12 Relate the properties of the various classes of solids (ionic, covalent, molecular, and metallic) to the structure of these solids.

Figure 5.10 Crystalline solids.

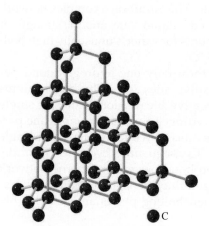

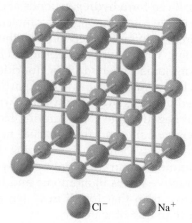

(a) The crystal structure of diamond.

(b) The crystal structure of sodium chloride.

Cl⁻   Na⁺

(c) The crystal structure of methane, a frozen molecular solid. Only one methane molecule is shown in detail.

(d) The crystal structure of a metallic solid. The gray area represents mobile electrons around fixed metal cations.

The chemical compositions and crystal-line structures of various gemstones are discussed in A Human Perspective: Gemstones.

A solid may be a **crystalline solid,** having a regular repeating structure, or an **amorphous solid,** having no organized structure. Diamond and sodium chloride (Figure 5.10) are examples of crystalline substances; glass, plastic, and concrete are examples of amorphous solids.

Types of Crystalline Solids

Crystalline solids may exist in one of four general groups:

1. *Ionic solids.* The units that comprise an **ionic solid** are positive and negative ions. Electrostatic forces hold the crystal together. Ionic solids generally have high melting points and are hard and brittle. A common example of an ionic solid is sodium chloride.
2. *Covalent solids.* The units that comprise a **covalent solid** are atoms held together by covalent bonds. Covalent solids have very high melting points (1200°C to 2000°C or more is not unusual) and are extremely hard. They are insoluble in most solvents. Diamond is a covalent solid composed of covalently bonded carbon atoms. Diamonds are used for industrial cutting because they are so hard and as gemstones because of their crystalline beauty.

Intermolecular forces are also discussed in Sections 3.5 and 5.2.

3. *Molecular solids.* The units that make up a **molecular solid,** molecules, are held together by intermolecular attractive forces (London dispersion forces, dipole-dipole interactions, and hydrogen bonding). Molecular solids are usually soft and have low melting points. They are frequently volatile and are poor electrical conductors. A common example is ice (solid water; Figure 5.11).

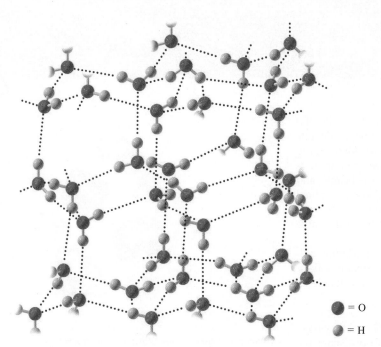

= O
= H

Figure 5.11 The structure of ice, a molecular solid. Hydrogen bonding among water molecules produces a regular open structure that is less dense than liquid water.

4. *Metallic solids.* The units that comprise a **metallic solid** are metal atoms held together by metallic bonds. *Metallic bonds* are formed by the overlap of orbitals of metal atoms, resulting in regions of high electron density surrounding the positive metal nuclei. Electrons in these regions are extremely mobile. They are able to move freely from atom to atom through pathways that are, in reality, overlapping atomic orbitals. This results in the high *conductivity* (ability to carry electrical current) exhibited by many metallic solids. Silver and copper are common examples of metallic solids. Metals are easily shaped and are used for a variety of purposes. Most of these are practical applications such as hardware, cookware, and surgical and dental tools. Others are purely for enjoyment and decoration, such as silver and gold jewelry.

Solid-state chemistry is critically important to the semiconductor industry. "Chips," crystalline solids made of silicon and traces of other elements such as arsenic and germanium, are the "brains" behind our computers, cell phones, and a host of other electronic devices.

Sublimation of Solids

Sublimation is the process in which some molecules in the solid state convert directly to the gaseous state.

As we saw in the discussion of the development of new devices for airport security (Chemistry at the Crime Scene: Explosives at the Airport, page 190), some molecular solids possess measurable vapor pressures. Recall that vapor pressure of a liquid is the key concept in explaining the conversion of a liquid to a gas by evaporation or by boiling. Certain molecular solids such as dry ice (frozen carbon dioxide) and moth balls (naphthalene) will, at room temperature, convert to a gas without passing through the liquid state because they have a sufficiently high vapor pressure. This process is termed *sublimation.*

If you live in a very cold climate, you may have observed that, over a period of several days, snow and ice "disappear," even though the temperature is continuously below the melting point of water. Snow and ice, molecular solids, have simply sublimed, converting from solid to water vapor.

LEARNING GOAL

9 Describe the processes of melting, boiling, evaporation, condensation, and sublimation.

Question 5.13
a. What properties are associated with ionic solids?
b. Provide two examples of ionic solids.

Question 5.14
a. What properties are associated with molecular solids?
b. Provide two examples of molecular solids.

Gemstones

When we think of the solid state, we may think of ice, the solid form of water; coal, a major energy source; or, perhaps, concrete and steel, bulwarks of the construction industry. All are commonplace materials with well-defined properties and applications. However, if we turn our thoughts toward beauty and value, diamonds often come to mind. Diamonds are valued for their sparkle, durability, and rarity. Diamonds are the most famous, but certainly not the only gemstones with desirable properties and high price tags.

The diamond market has a value of over twelve billion dollars per year and colored gemstones another six billion dollars. What are these gems made of, and why are they so valuable?

Diamond

Most of the carbon deposits on earth are amorphous; they have no organized structure. Consequently, they do not share the properties of solids with regular, repeating structure. Much of this carbon is coal. A tiny fraction of the world's solid carbon exists as covalent solids. Recall that covalent solids have strong covalent bonds and a regular, repeating crystal structure. A diamond is a covalent solid comprised of carbon atoms arranged in a 3-dimensional crystalline structure (Figure 5.10a).

A one-carat diamond, which is only 0.20 g, has a retail value of thousands of dollars. Just how many thousands is a function of the regularity of the repeating pattern of carbon atoms and their freedom from the inclusion of impurities. The number of diamonds that meet the highest standards is very small.

Rubies and Sapphires

These gemstones are variations of the mineral corundum, which is essentially Al_2O_3. Rubies and sapphires have similar crystalline structures; however, ruby is red and sapphire is blue. Why? Trace amounts of other elements make all the difference. Chromium replaces some aluminum ions in ruby (Cr^{3+} replaces Al^{3+}). This alters the absorption spectrum of the material, favoring the transmission of red light (610 nm); hence, the ruby red color. Sapphire appears blue because of the presence of traces of Fe^{2+} and Ti^{4+}. Light at the red end of the visible spectrum is absorbed by compounds formed from these trace ions. The blue light is transmitted and the sapphire appears blue.

Emerald

The Cr^{3+} ion is also responsible for the green color of emerald. It is a trace element incorporated into the base structure of the

A variety of gemstones. Variations in crystal structure, as well as the presence of trace metal ions, are responsible for differences in color and lustre.

mineral beryl (the origin of the name is beryllium, a key element in its structure). Beryl is $Be_3Al_2Si_6O_{18}$ and Cr^{3+} (and vanadium ions as well) replaces Al^{3+} in the crystal structure of the beryl. The resulting color is the familiar emerald green.

Tanzanite

This blue gem is a product of a mineral with the formula $Ca_2Al_2(SiO_4)(Si_2O_7)O(OH)$. In contrast to other gems we have discussed, tanzanite was not discovered and characterized until the 1960s. It was first found in Tanzania, for which it is named. Once again, trace elements make all the difference between worthless rock and valuable gemstones. Vanadium ions, substituted for aluminum ions, are responsible for its deep blue color.

We owe the origin of gemstones to the pressure and temperature events associated with earth's formation. We have learned to make synthetic gemstones in the laboratory, but the gem connoisseurs of the world continue to show a preference for the natural product. Not surprisingly, they owe their value to their rarity and intrinsic beauty.

For Further Understanding

▸ Predict the color of a gemstone that absorbs light mostly in the wavelength range 500–700 nm.

▸ Suggest a reason why a collection of sapphires may appear to be different shades of blue.

CHAPTER MAP

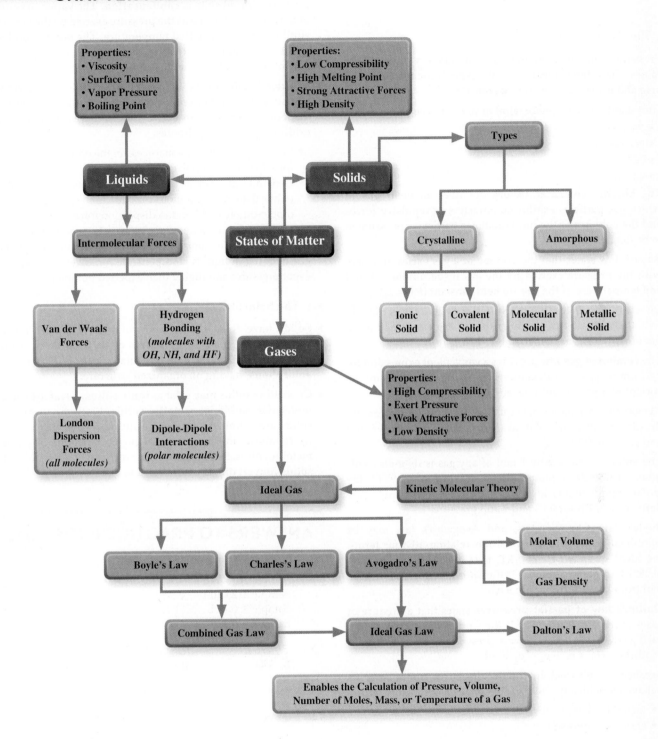

SUMMARY

5.1 The Gaseous State

▶ Pressure, volume, temperature, and quantity (number of moles) characterize ideal gases. **Pressure** is the force per unit area and is measured using a **barometer.**

▶ One standard atmosphere (atm) of pressure is equivalent to
  - 76.0 cm Hg
  - 760 mm Hg
  - 760 torr
  - 14.7 psi

▶ The **kinetic molecular theory** describes an **ideal gas** in which gas particles exhibit no attractive or repulsive forces and the volumes of the individual gas particles are assumed to be negligible.

▶ **Boyle's law** states that the volume of a gas varies inversely with the pressure exerted by the gas if the number of mol and temperature of the gas are held constant ($PV = k_b$).

▶ **Charles's law** states that the volume of a gas varies directly with the absolute temperature (Kelvin) if pressure and number of mol of gas are constant ($V/T = k_c$).

▶ The **combined gas law** provides a convenient expression for performing gas law calculations involving the most common variables: pressure, volume, and temperature.

▶ **Avogadro's law** states that equal volumes of any gas contain the same number of mol if measured at constant temperature and pressure ($V/n = k_a$).

▶ The volume occupied by 1 mol of any gas is its **molar volume.** At **standard temperature and pressure** (STP), the molar volume of any ideal gas is 22.4 L. STP conditions are defined as 273 K (or 0°C) and 1 atm pressure.

▶ Boyle's law, Charles's law, and Avogadro's law may be combined into a single expression relating all four terms, the **ideal gas law:** $PV = nRT$. R is the ideal gas constant ($0.0821 \text{ L} \cdot \text{atm} \cdot \text{K}^{-1} \cdot \text{mol}^{-1}$) if the units atm (for P), L (for V), mol (for n), and K (for T) are used.

▶ **Dalton's law of partial pressures** states that a mixture of gases exerts a pressure that is the sum of the pressures that each gas would exert if it were present alone under similar conditions ($P_t = p_1 + p_2 + p_3 + \cdots$).

▶ Because of their weak intermolecular forces, nonpolar gases behave more ideally. Polar gases behave more ideally with
  - Increasing temperature
  - Decreasing pressure

5.2 The Liquid State

▶ Liquids are practically incompressible because of the closeness of their molecules. The **viscosity** of a liquid is a measure of its resistance to flow. Viscosity generally decreases with increasing temperature. The **surface tension** of a liquid is a measure of the attractive forces at the surface of a liquid. **Surfactants** decrease surface tension.

▶ The conversion of liquid to vapor at a temperature below the boiling point of the liquid is **evaporation.** Conversion of the gas to the liquid state is **condensation.** The **vapor pressure** of the liquid is defined as the pressure exerted by the vapor at equilibrium at a specified temperature. The **normal boiling point** of a liquid is the temperature at which the vapor pressure of the liquid is equal to 1 atm.

▶ **Van der Waals forces** include **dipole-dipole interactions,** attractive forces between polar molecules, and **London dispersion forces,** attractive forces between molecules exhibiting temporary dipoles.
  - Weak London dispersion forces are the only intermolecular force between nonpolar molecules.
  - Generally, polar molecules have higher melting and boiling points than nonpolar molecules due to the presence of both dipole-dipole and London dispersion forces.

▶ Molecules in which a hydrogen atom is bonded to a nitrogen, oxygen, or fluorine atom exhibit **hydrogen bonding.** Hydrogen bonding in liquids is responsible for lower than expected vapor pressures and higher than expected boiling points.

5.3 The Solid State

▶ Solids have fixed shapes and volumes. They are incompressible, owing to the closeness of their particles. Solids may be **crystalline,** having a regular, repeating structure, or **amorphous,** having no organized structure.

▶ Crystalline solids may exist as **ionic solids, covalent solids, molecular solids,** or **metallic solids.** Electrons in metallic solids are extremely mobile, resulting in the high *conductivity* (ability to carry electrical current) exhibited by many metallic solids. **Sublimation** is a process whereby molecular solids convert directly from solid to gas.

ANSWERS TO PRACTICE PROBLEMS

5.1   Sample 1: 38 atm
       Sample 2: 25 atm
       Sample 3: 0.15 L
       Sample 4: 0.38 L

5.2   **a.** 3.76 L   **c.** 2.75 L   **e.** 2.75 L
       b. 3.41 L   **d.** 5.50 L   **f.** 3.76 L

5.3   0.200 atm

5.4   99.5°C

5.5   **a.** 4.46 mol H_2 **b.** 0.75 mol H_2

5.6   The molar volume is based on 1 mol of *any* ideal gas. For any ideal gas, all quantities substituted in the ideal gas equation are independent of the identity of the ideal gas.

5.7   0.223 mol N_2

5.8   9.00 L

QUESTIONS AND PROBLEMS

Measurement of Properties of Gases

Foundations

5.15 Explain how the pressure of O_2 gas may be measured.
5.16 Describe the molecular/atomic basis of gas pressure.

Applications

5.17 Express each of the following in units of atm:
  a. 94.4 cm Hg   **c.** 150 mm Hg
  b. 72.5 torr   **d.** 124 kPa
5.18 Express each of the following in units of atm:
  a. 128 cm Hg   **c.** 1405 mm Hg
  b. 255 torr   **d.** 303 kPa
5.19 Express each of the following in units of psi:
  a. 54.0 cm Hg   **c.** 800 mm Hg
  b. 155 torr   **d.** 1.50 atm
5.20 Express each of the following in units of psi:
  a. 12.5 cm Hg   **c.** 254 mm Hg
  b. 46.0 torr   **d.** 0.48 atm

Kinetic Molecular Theory of Gases

Foundations

5.21 Compare and contrast the gas, liquid, and solid states with regard to the average distance of particle separation.
5.22 Compare and contrast the gas, liquid, and solid states with regard to the nature of the interactions among the particles.

Applications

5.23 Why are gases easily compressible?
5.24 Why are gas densities much lower than those of liquids or solids?
5.25 Why do gases expand to fill any available volume?
5.26 Why do gases with lower molar masses diffuse more rapidly than gases with higher molar masses?
5.27 Do gases exhibit more ideal behavior at low or high pressures? Why?
5.28 Do gases exhibit more ideal behavior at low or high temperatures? Why?
5.29 Use the kinetic molecular theory to explain why dissimilar gases mix more rapidly at high temperatures than at low temperatures.
5.30 Use the kinetic molecular theory to explain why aerosol cans carry instructions warning against heating or disposing of the container in a fire.

Boyle's Law

Foundations

5.31 State Boyle's law in words.
5.32 State Boyle's law in equation form.
5.33 The pressure on a fixed mass of a gas is tripled at constant temperature. Will the volume increase, decrease, or remain the same?
5.34 By what factor will the volume of the gas in Question 5.33 change?

Applications

A sample of helium gas was placed in a cylinder, and the volume of the gas was measured as the pressure was slowly increased. The results of this experiment are shown graphically.

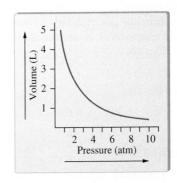

Questions 5.35–5.38 are based on this experiment.

5.35 At what pressure does the gas occupy a volume of 5 L?
5.36 What is the volume of the gas at a pressure of 5 atm?
5.37 Calculate the Boyle's law constant at a volume of 2 L.
5.38 Calculate the Boyle's law constant at a pressure of 2 atm.
5.39 Calculate the pressure, in atm, required to compress a sample of helium gas from 20.9 L (at 1.00 atm) to 4.00 L.
5.40 A balloon filled with helium gas at 1.00 atm occupies 15.6 L. What volume would the balloon occupy in the upper atmosphere at a pressure of 0.150 atm?

Charles's Law

Foundations

5.41 State Charles's law in words.
5.42 State Charles's law in equation form.
5.43 Explain why the Kelvin scale is used for gas law calculations.
5.44 The temperature on a summer day may be 90°F. Convert this value to Kelvin units.

Applications

5.45 The temperature of a gas is raised from 25°C to 50°C. Will the volume double if mass and pressure do not change? Why or why not?
5.46 Verify your answer to Question 5.45 by calculating the temperature needed to double the volume of the gas.
5.47 Determine the change in volume that takes place when a 2.00-L sample of $N_2(g)$ is heated from 250°C to 500°C.
5.48 Determine the change in volume that takes place when a 2.00-L sample of $N_2(g)$ is heated from 250 K to 500 K.
5.49 A balloon containing a sample of helium gas is warmed in an oven. If the balloon measures 1.25 L at room temperature (20°C), what is its volume at 80°C?
5.50 The balloon described in Question 5.49 was then placed in a refrigerator at 39°F. Calculate its new volume.
5.51 A balloon, filled with N_2, has a volume of 2.00 L at an indoor temperature of 68°F. When placed outdoors, the volume was observed to increase to 2.20 L. What is the outdoor temperature in °F?
5.52 A balloon, filled with an ideal gas, has a volume of 5.00 L at 50°F. At what temperature (°F) would the balloon's volume double?

Combined Gas Law

Foundations

5.53 Will the volume of gas increase, decrease, or remain the same if the temperature is increased and the pressure is decreased? Explain.

5.54 Will the volume of gas increase, decrease, or remain the same if the temperature is decreased and the pressure is increased? Explain.

Applications

Use the combined gas law,

$$\frac{P_i V_i}{T_i} = \frac{P_f V_f}{T_f}$$

to answer Questions 5.55 and 5.56.

5.55 Solve the combined gas law expression for the final volume.

5.56 Solve the combined gas law expression for the final temperature.

5.57 If 2.25 L of a gas at 16°C and 1.00 atm is compressed at a pressure of 125 atm at 20°C, calculate the new volume of the gas.

5.58 A sealed balloon filled with helium gas occupies 2.50 L at 25°C and 1.00 atm. When released, it rises to an altitude where the temperature is 20°C and the pressure is only 0.800 atm. Calculate the new volume of the balloon.

5.59 A 5.00-L balloon exerts a pressure of 2.00 atm at 30.0°C. What is the pressure that the sealed balloon exerts if the volume has increased to 7.0 L at 40°C?

5.60 If we double the pressure and temperature of the balloon in Question 5.59, what will its new volume be?

Avogadro's Law

Foundations

5.61 State Avogadro's law in words.

5.62 State Avogadro's law in equation form.

Applications

5.63 If 5.00 g helium gas is added to a 1.00-L balloon containing 1.00 g of helium gas, what is the new volume of the balloon? Assume no change in temperature or pressure.

5.64 How many g of helium must be added to a balloon containing 8.00 g helium gas to double its volume? Assume no change in temperature or pressure.

Molar Volume and the Ideal Gas Law

Foundations

5.65 Will 1.00 mol of a gas always occupy 22.4 L?

5.66 Calculate the molar volume of O_2 gas at STP.

5.67 What are the units and numerical value of standard temperature?

5.68 What are the units and numerical value of standard pressure?

Applications

5.69 A sample of nitrogen gas, stored in a 4.0-L container at 32°C, exerts a pressure of 5.0 atm. Calculate the number of mol of nitrogen gas in the container.

5.70 Calculate the pressure, in atmosphere, of 7.0 mol of carbon monoxide stored in a 30.0-L container at 65°C.

5.71 Calculate the volume of 44.0 g of carbon monoxide at STP.

5.72 Calculate the volume of 44.0 g of carbon dioxide at STP.

5.73 Calculate the density of carbon monoxide at STP.

5.74 Calculate the density of carbon dioxide at STP.

5.75 Calculate the number of mol of a gas that is present in a 7.55-L container at 45°C, if the gas exerts a pressure of 725 mm Hg.

5.76 Calculate the pressure (atm) exerted by 1.00 mol of gas contained in a 7.55-L cylinder at 45°C.

5.77 A sample of argon (Ar) gas occupies 65.0 mL at 22°C and 750 torr. What is the volume of this Ar gas sample at STP?

5.78 A sample of O_2 gas occupies 257 mL at 20°C and 1.20 atm. What is the volume of this O_2 gas sample at STP?

5.79 What is the temperature (°C) of 1.75 g of O_2 gas occupying 2.00 L at 1.00 atm?

5.80 How many g of O_2 gas occupy 10.0 L at STP?

5.81 Calculate the volume of 4.00 mol Ar gas at 8.25 torr and 27°C.

5.82 Calculate the volume of 6.00 mol O_2 gas at 30 cm Hg and 72°F.

Dalton's Law of Partial Pressures

Foundations

5.83 State Dalton's law in words.

5.84 State Dalton's law in equation form.

Applications

5.85 A gas mixture has three components: N_2, F_2, and He. Their partial pressures are 0.40 atm, 0.16 atm, and 0.18 atm, respectively. What is the pressure of the gas mixture?

5.86 A gas mixture has three components, N_2, F_2, and He. The partial pressure of N_2 is 0.35 atm and F_2 is 0.45 atm. If the total pressure is 1.20 atm, what is the partial pressure of helium?

5.87 A gas mixture has a total pressure of 0.56 atm and consists of He and Ne. If the partial pressure of the He in the mixture is 0.27 atm, what is the partial pressure of the Ne in the mixture?

5.88 If we were to remove all of the helium from the mixture described in Question 5.86, what would the partial pressures of N_2 and F_2 be? Why? What is the new total pressure?

Ideal Gases Versus Real Gases

Foundations

5.89 Explain when limitations to the ideal gas model are observed.

5.90 H_2O and CH_4 are gases at 150°C. Which exhibits more ideal behavior? Why?

Applications

5.91 Would CO behave more like an ideal gas at 5 K or 50 K? Explain your reasoning.

5.92 Would CO behave more like an ideal gas at 2 atm or 20 atm? Explain your reasoning.

The Liquid State

Foundations

5.93 Compare the strength of intermolecular forces in liquids with those in gases.

5.94 Compare the strength of intermolecular forces in liquids with those in solids.

5.95 What is the relationship between the temperature of a liquid and the vapor pressure of that liquid?

5.96 What is the relationship between the strength of the attractive forces in a liquid and its vapor pressure?

5.97 Describe the process occurring at the molecular level that accounts for the property of viscosity.

5.98 Describe the process occurring at the molecular level that accounts for the property of surface tension.

Applications

Questions 5.99–5.102 are based on the following:

methane   chloromethane   methanol

5.99 Which of these molecules exhibit London dispersion forces? Why?

5.100 Which of these molecules exhibit dipole-dipole forces? Why?

5.101 Which of these molecules exhibit hydrogen bonding? Why?

5.102 Which of these molecules would you expect to have the highest boiling point? Why?

Questions 5.103 and 5.104 are based on the following:

propane   isopropyl alcohol   propylene glycol

5.103 Predict the compound expected to have the greatest viscosity in the liquid state.

5.104 Predict the compound expected to have the greatest surface tension in the liquid state.

The Solid State

Foundations

5.105 Explain why solids are essentially incompressible.

5.106 Distinguish between amorphous and crystalline solids.

5.107 Describe one property that is characteristic of:
   a. ionic solids
   b. covalent solids

5.108 Describe one property that is characteristic of:
   a. molecular solids
   b. metallic solids

Applications

5.109 Predict whether beryllium or carbon would be a better conductor of electricity in the solid state. Why?

5.110 Why is diamond used as an industrial cutting tool?

5.111 Mercury and chromium are toxic substances. Which element is more likely to be an air pollutant? Why?

5.112 Why is the melting point of silicon much higher than that of argon, even though argon has a greater molar mass?

CHALLENGE PROBLEMS

1. An elodea plant, commonly found in tropical fish aquaria, was found to produce 5.0×10^{22} molecules of oxygen per hour (h). What volume of oxygen (at STP) would be produced in an 8-h period?

2. A chemist measures the volume of 1.00 mol of helium gas at STP and obtains a value of 22.4 L. After changing the temperature to 137 K, the experimental value was found to be 11.05 L. Verify the chemist's results using the ideal gas law and explain any apparent discrepancies.

3. A chemist measures the volumes of 1.00 mol of H_2 and 1.00 mol of CO and finds that they differ by 0.10 L. Which gas produced the larger volume? Do the results contradict the ideal gas law? Why or why not?

4. A 100.0-g sample of water was decomposed using an electric current (electrolysis), producing hydrogen gas and oxygen gas. Write the balanced equation for the process and calculate the volume of each gas produced (at STP). Explain any relationship you may observe between the volumes obtained and the balanced equation for the process.

5. An autoclave is used to sterilize surgical equipment. It is far more effective than steam produced from boiling water in the open atmosphere because it generates steam at a pressure of 2 atm. Explain why an autoclave is such an efficient sterilization device.

6. Imagine you have been asked to design a new solid material to be used in replacement bones and joints. What physical and chemical properties would you deem important?

6

Solutions

Carbonated beverages are a commonplace example of a solution of a gas (solute) dissolved in a liquid (solvent). Based on your everyday experience, can you predict whether the solubility of carbon dioxide in water (or cola) would increase or decrease as the temperature of the solution increases?

LEARNING GOALS

1 Distinguish among the terms *solution, solute,* and *solvent.*

2 Describe the properties and composition of various kinds of solutions.

3 Explain which factors influence the degree of solubility, and use trends to make predictions.

4 Describe the relationship between solubility and equilibrium.

5 Use Henry's law to calculate equilibrium solubility values for gases.

6 Calculate solution concentration in mass/volume percent, mass/mass percent, parts per thousand, and parts per million.

7 Determine the quantity of solute or solution from the concentration of solution.

8 Calculate the molarity of solution from mass or moles of solute.

9 Perform dilution calculations.

10 Describe and explain concentration-dependent solution properties.

11 Perform calculations involving colligative properties.

12 Describe why the chemical and physical properties of water make it a truly unique solvent.

13 Interconvert molar concentration of ions and milliequivalents/liter.

14 Explain the role of electrolytes in blood and their relationship to the process of dialysis.

OUTLINE

INTRODUCTION

A significant deterrent to achieving optimal athletic performance in strenuous competitive sports, especially in warm weather, is dehydration—the loss of body fluids through perspiration. Perspiration is an aqueous solution, and the loss of ions dissolved in water (the solvent) has a more negative impact on bodily function than the loss of the water itself.

This fact was not lost on researchers at the University of Florida in the 1960s. The University of Florida football team (the Florida Gators) plays much of its schedule in a very warm climate. This, coupled with the intense physical effort required and the heavy padding that must be worn to prevent injury, makes dehydration of the athletes a real concern.

The remedy, at that time, was to drink large volumes of water while consuming salt tablets, to replace the lost ions, or to consume high-sugar foods (often oranges) to provide energy. How much of each was pure guesswork. Often, the cure was worse than the problem, and cramping resulted.

The team doctor, along with medical researchers at the university, had an idea. Would it be possible to mix all three components in some proportion to produce a solution that achieved the desired effect without the unwanted side effect? The result of this research was a solution of ionic compounds and sugar dissolved in water, which was similar to the composition of the perspiration being lost, with flavoring added to make it palatable. The identity of the ions: sodium, potassium, and chloride, was certainly important, as were the concentrations of the ions in producing a solution that achieved the desired effect without inducing cramps.

The football team began using the "Gator-aid" solution and was victorious in the 1966 Orange Bowl, offering some proof that the solution had a positive effect on the team's performance. Now named Gatorade, it was the first "sports drink." Currently, several competing brands are available in the marketplace, differing slightly in composition, concentration, and flavor. They are widely used by college and professional teams, as well as the "weekend athlete." We should not forget that the most important ingredient in all sports drinks is the solvent: water. In fact, some argue that pure water is just as efficient a cell hydrator as sports drinks.

In this chapter, we will learn more about solutions, their composition and their concentration. We will see why some substances are soluble in water and others are not. We will see why the concentration of ions, such as sodium and potassium, is critical to the function and integrity of the cells in our bodies.

Many sports drink labels claim the product will replenish electrolytes. What does that mean?

6.1 Properties of Solutions

A **solution** is a homogeneous mixture of two or more substances. A solution is composed of one or more *solutes,* dissolved in a *solvent.* The **solute** is a component of a solution that is present in lesser quantity than the solvent. The **solvent** is the solution component present in the largest quantity. For example, when sugar (the solute) is added to water (the solvent), the sugar dissolves in the water to produce a solution. In those instances in which the solvent is water, we refer to the homogeneous mixture as an **aqueous solution,** from the Latin *aqua,* meaning "water."

The dissolution of a solid in a liquid is perhaps the most common example of solution formation. However, it is also possible to form solutions in gases and solids as well as in liquids. For example:

- Air is a gaseous mixture, but it is also a solution; oxygen and a number of trace gases are dissolved in the gaseous solvent, nitrogen.
- Metallic items, such as rings and bracelets, are homogeneous mixtures of two or more kinds of metal atoms in the solid state. These homogeneous mixtures are termed *alloys.*

LEARNING GOAL

1 Distinguish among the terms *solution, solute,* and *solvent.*

Although solid and gaseous solutions are important in many applications, our emphasis will be on *liquid solutions* because so many important chemical reactions take place in liquid solutions.

General Properties of Liquid Solutions

Liquid solutions are clear and transparent with no visible particles of solute. They may be colored or colorless, depending on the properties of the solute and solvent. Note that the terms *clear* and *colorless* do not mean the same thing; a clear solution has only one state of matter that can be detected; *colorless* simply means the absence of color.

Recall that solutions of **electrolytes** are formed from solutes that are soluble *ionic* compounds. These compounds dissociate in solution to produce ions that behave as charge carriers. Solutions of electrolytes are good conductors of electricity. For example, sodium chloride dissolving in water:

$$NaCl(s) \xrightarrow{\text{H}_2\text{O}} Na^+(aq) + Cl^-(aq)$$

Solid sodium   Dissolved sodium
    chloride   chloride

In contrast, solutions of **nonelectrolytes** are formed from nondissociating *molecular* solutes (nonelectrolytes), and these solutions are nonconducting. For example, dissolving sugar in water:

$$C_6H_{12}O_6(s) \xrightarrow{\text{H}_2\text{O}} C_6H_{12}O_6(aq)$$

Solid glucose   Dissolved glucose

A **true solution** is a homogeneous mixture with uniform properties throughout. In a true solution, the solute cannot be isolated from the solution by filtration. The particle size of the solute is about the same as that of the solvent, and solvent and solute pass directly through the filter paper. Furthermore, solute particles will not "settle out" after a time. All of the molecules of solute and solvent are intimately mixed. The continuous particle motion in solution maintains the homogeneous, random distribution of solute and solvent particles.

Volumes of solute and solvent are not additive; 1 liter (L) of alcohol mixed with 1 L of water does not result in exactly 2 L of solution. The volume of pure liquid is determined not only by the size of the individual molecules but also by the way in which the individual molecules "fit together." When two or more kinds of molecules are mixed, the interactions become more complex. Solvent interacts with solvent, solute interacts with solvent, and solute may interact with other solute. For example, mixing 1 L of water with 1 L of alcohol results in a solution volume measurably smaller than the anticipated 2 L.

True Solutions, Colloidal Dispersions, and Suspensions

How can you recognize a solution? A clear liquid in a beaker may be a pure substance, a true solution, or a colloidal dispersion. Only chemical analysis, determining the identity of all substances in the liquid, can distinguish between a pure substance and a solution. A pure substance has *one* component, pure water being an example. A true solution will contain *more than one substance*, with the tiny particles homogeneously intermingled.

A **colloidal dispersion** also consists of solute particles distributed throughout a solvent. However, the distribution is not completely homogeneous, owing to the size of the colloidal particles. Particles with diameters of 1×10^{-9} meter (m) to 2×10^{-7} m are colloids. [Recall that 1×10^{-9} m = 1 nanometer (nm), therefore 2×10^{-7} m = 200 nm.] Particles smaller than 1 nm are solution particles; those larger than 200 nm are particles that are large enough to eventually settle to the

---

LEARNING GOAL

2 Describe the properties and composition of various kinds of solutions.

Particles in electrolyte solutions are ions, making the solution an electrical conductor.

Particles in nonelectrolyte solutions are individual molecules. No ions are formed in the dissolution process.

Section 3.5 relates properties and molecular geometry.

bottom of the container. The settled particles are large enough to be observed by the naked eye; the collection of particles is termed a **precipitate,** a solid in contact with solvent.

To the naked eye, a colloidal dispersion and a true solution appear identical; neither solute nor colloid can be seen. However, a simple experiment, using only a bright light source, can readily make the distinction based upon differences in their interaction with light. Colloid particles are large enough to scatter light; solute particles are not. When a beam of light passes through a colloidal dispersion, the large particles scatter light, and the liquid appears hazy. We see this effect in sunlight passing through fog. Fog is a colloidal dispersion of tiny particles of liquid water dispersed throughout a gas, air. The haze is light scattered by droplets of water. You may have noticed that your automobile headlights are not very helpful in foggy weather. Visibility becomes worse rather than better because light scattering increases.

The light-scattering ability of colloidal dispersions is termed the *Tyndall effect.* True solutions, with very tiny particles, do not scatter light—no haze is observed—and true solutions are easily distinguished from colloidal dispersions by observing their light-scattering properties (Figure 6.1).

A **suspension** is a heterogeneous mixture that contains particles much larger than a colloidal dispersion; over time, these particles may settle, forming a second phase. A suspension is not a true solution, nor is it a precipitate.

Question 6.1 Describe how you would distinguish experimentally between a pure substance and a true solution.

Question 6.2 Describe how you would distinguish experimentally between a true solution and a colloidal dispersion.

Degree of Solubility

In our discussion of the relationship of polarity and solubility, the rule *"like dissolves like"* was described as the fundamental condition for solubility. Polar solutes are soluble in polar solvents, and nonpolar solutes are soluble in nonpolar solvents. Thus, knowing a little bit about the structure of the molecule enables us to predict qualitatively the solubility of the compound.

The *degree* of **solubility,** *how much* solute can dissolve in a given volume of solvent, is a quantitative measure of solubility. It is difficult to predict the solubility of each and every compound. However, general solubility trends are based on the following considerations:

- *The magnitude of difference between polarity of solute and solvent.* The greater the difference, the less soluble is the solute.
- *Temperature.* An increase in temperature usually, but not always, increases solubility (Figure 6.2). Often, the effect is dramatic. For example, an increase in temperature from 0°C to 100°C increases the water solubility of KNO_3 from 10 grams per 100 grams H_2O (10 g/100 g H_2O) to 240 g/100 g H_2O.
- *Pressure.* Pressure has little effect on the solubility of solids and liquids in liquids. However, the solubility of a gas in liquid is directly proportional to the applied pressure. Carbonated beverages, for example, are made by dissolving carbon dioxide in the beverage under high pressure (hence the term *carbonated*).

When a solution contains all the solute that can be dissolved at a particular temperature, it is a **saturated solution.** When solubility values are given—for example, 13.3 g of potassium nitrate in 100 g of water at 1°C—they refer to the concentration of a saturated solution.

See Section 4.5 for more information on precipitates.

Figure 6.1 The Tyndall effect. The sample on the right is a colloidal dispersion, which scatters the light. This scattered light is visible as a haze. The sample on the left is a true solution; no scattered light is observed.

LEARNING GOAL

3 Explain which factors influence the degree of solubility, and use trends to make predictions.

Section 3.5 describes solute-solvent interactions in detail.

The term *qualitative* implies identity, and the term *quantitative* relates to quantity.

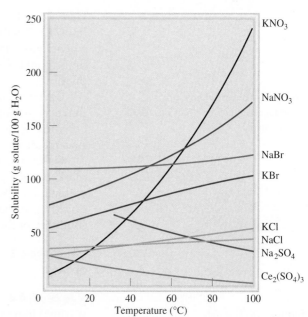

Figure 6.2 The solubility of a variety of ionic compounds in water as a function of temperature. Note that Na_2SO_4 and $Ce_2(SO_4)_3$ become less soluble at higher temperatures.

As we have already noted, *increasing* the temperature generally increases the amount of solute a given solution may hold. Conversely, *cooling* a saturated solution often results in a decrease in the amount of solute in solution. The excess solute falls to the bottom of the container as a *precipitate* (a solid in contact with the solution). Occasionally, on cooling, the excess solute may remain in solution for a time. Such a solution is described as a **supersaturated solution.** This type of solution is inherently unstable. With time, excess solute will precipitate, and the solution will revert to a saturated solution, which is stable.

Solubility and Equilibrium

When an excess of solute (beyond the solubility limit) is added to a solvent, it begins to dissolve and continues until it establishes a *dynamic equilibrium* between dissolved and undissolved solute.

Initially, the rate of dissolution is large. After a time, the rate of the reverse process, precipitation, increases. The rates of dissolution and precipitation eventually become equal, and there is no further change in the composition of the solution. There is, however, a continual exchange of solute particles between solid and liquid phases because particles are in constant motion. The solution is saturated. The most precise definition of a saturated solution is a solution that is in equilibrium with undissolved solute.

Solubility of Gases: Henry's Law

When a liquid and a gas are allowed to come to equilibrium, the amount of gas dissolved in the liquid reaches some maximum level. This quantity can be predicted from a very simple relationship. **Henry's law** states that the number of moles (mol) of a gas dissolved in a liquid at a given temperature is proportional to the pressure of the gas. In other words, the gas solubility is directly proportional to the pressure of that gas in the atmosphere that is in contact with the liquid.

Henry's law is expressed mathematically as

$$M = kP$$

Here, M is the molar concentration of the gas in the liquid in units of moles/liter (mol/L). P is the pressure (in atm) of the gas over the solution at equilibrium. For a given gas, k is a constant that depends only on temperature. The constant, k, has units of mol/L · atm. In the event that more than one gas is present, P is the partial pressure.

Carbonated beverages are bottled at high pressures of carbon dioxide. When the cap is removed, the fizzing results from the fact that the partial pressure of carbon dioxide in the atmosphere is much less than that used in the bottling process. As a result, the equilibrium quickly shifts to one of lower gas solubility.

Gases are most soluble at low temperatures, and the gas solubility decreases markedly at higher temperatures (Figure 6.3). This explains many common observations. For example, a chilled container of carbonated beverage that is opened quickly goes flat as it warms to room temperature. As the beverage warms up, the solubility of the carbon dioxide decreases.

Question 6.3   Explain why, over time, a bottle of soft drink goes "flat" after it is opened.

Question 6.4   Would the soft drink in Question 6.3 go "flat" faster if the bottle warmed to room temperature? Why?

The Henry's law constant, k, for CO_2 in aqueous solution is 3.1×10^{-2} mol/(L · atm) at 25°C. Use this information to answer Questions 6.5 and 6.6.

The concept of equilibrium was introduced in Section 5.2 and will be discussed in detail in Section 7.4.

LEARNING GOAL

4   Describe the relationship between solubility and equilibrium.

LEARNING GOAL

5   Use Henry's law to calculate equilibrium solubility values for gases.

The concept of partial pressure is a consequence of Dalton's law, discussed in Section 5.1.

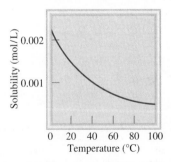

Figure 6.3 The water solubility of O_2 gas decreases markedly as the temperature of the water increases; this may have significant environmental implications.

A Human Perspective

Scuba Diving: Nitrogen and the Bends

A deep-water diver's worst fear is the interruption of the oxygen supply through equipment malfunction, forcing his or her rapid rise to the surface in search of air. If a diver must ascend too rapidly, he or she may suffer a condition known as "the bends."

Key to understanding this problem is recognition of the tremendous increase in pressure that divers withstand as they descend, because of the weight of the water above them. At the surface, the pressure is approximately 1 atmosphere (atm). At a depth of 200 feet (ft), the pressure is approximately six times as great; hence, the blood concentration of N_2 increases dramatically. Oxygen solubility increases as well, although its effect is less serious (O_2 is 20% of air, N_2 is 80%). Recall that Henry's law tells us that the number of mol of gas dissolved in blood is directly proportional to the pressure of the gas.

As the diver quickly rises, the pressure decreases rapidly, and the nitrogen "boils" out of the blood, stopping blood flow and impairing nerve transmission. The joints of the body lock in a bent position, hence the name of the condition: the bends.

To minimize the problem, scuba tanks may be filled with mixtures of helium and oxygen rather than nitrogen and oxygen. Helium has a much lower solubility in blood and, like nitrogen, is inert. To avoid the bends, divers are advised to make several decompression stops while ascending to the water's surface. Such "time-outs" allow a more gradual equilibration of nitrogen concentration in the blood; the ideal is to

Scuba diving

achieve normal (or close to normal) levels upon reaching the surface. Tables and charts relating pressures and decompression times have been developed to aid divers.

For Further Understanding

▶ Why are divers who slowly rise to the surface less likely to be adversely affected?

▶ What design features would be essential in deep-water manned exploration vessels?

Question 6.5 An unopened bottle of soda contains CO_2 gas at 6.0 atm. Calculate the equilibrium solubility of CO_2 in the unopened soda at 25°C in units of mol/L.

Question 6.6 After the soda in Question 6.5 is opened, the "fizz" shows a loss of CO_2. If the partial pressure of CO_2 in the atmosphere is 5.0×10^{-4} atm, calculate the equilibrium CO_2 concentration in mol/L in the open bottle of soda.

Henry's Law and Respiration

Henry's law helps to explain the process of respiration. Respiration depends on a rapid and efficient exchange of oxygen and carbon dioxide between the atmosphere and the blood. This transfer occurs through the lungs. The process, oxygen entering the blood and carbon dioxide released to the atmosphere, is accomplished in air sacs called *alveoli,* which are surrounded by an extensive capillary system. Equilibrium is quickly established between alveolar air and the capillary blood. The temperature of the blood is effectively constant. Therefore, the equilibrium concentrations of both oxygen and carbon dioxide are determined by the partial pressures of the gases (Henry's law). The oxygen is transported to cells, a variety of reactions take place, and the waste product of respiration, carbon dioxide, is brought back to the lungs to be expelled into the atmosphere.

See A Medical Perspective: Blood Gases and Respiration.

A Medical Perspective

Blood Gases and Respiration

Respiration must deliver oxygen to cells and carbon dioxide, the waste product, to the lungs to be exhaled. Henry's law helps to explain the way in which this process occurs.

Gases (such as O_2 and CO_2) move from a region of higher partial pressure to one of lower partial pressure in an effort to establish an equilibrium. At the interface of the lung, the membrane barrier between the blood and the surrounding atmosphere, the following situation exists: Atmospheric O_2 partial pressure is high, and atmospheric CO_2 partial pressure is low. The reverse is true on the other side of the membrane (blood). Thus, CO_2 is efficiently removed from the blood, and O_2 is efficiently moved into the bloodstream.

At the other end of the line, capillaries are distributed in close proximity to the cells that need to expel CO_2 and gain O_2. The partial pressure of CO_2 is high in these cells, and the partial pressure of O_2 is low, having been used up by the energy-harvesting reaction, the oxidation of glucose:

$$C_6H_{12}O_6 + 6O_2 \longrightarrow 6CO_2 + 6H_2O + \text{energy}$$

The O_2 diffuses into the cells (from a region of high to low partial pressure), and the CO_2 diffuses from the cells to the blood (again from a region of high to low partial pressure).

With each breath we take, oxygen is distributed to the cells and used to generate energy, and the waste product, CO_2, is expelled by the lungs.

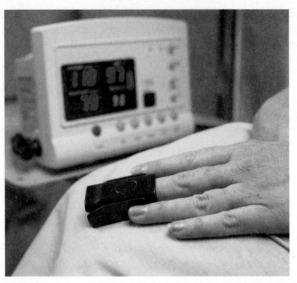

A pulse oximeter measures oxygen saturation which can be used to determine how well oxygen is being distributed to tissues.

For Further Understanding

Use the Internet to investigate pulse oximetry and:

- Determine what the range of oxygen saturation should be for a healthy individual.
- Explain what happens as the partial pressure of carbon dioxide in the blood increases.

6.2 Concentration Based on Mass

Solution **concentration** is defined as the amount of solute dissolved in a given amount of solution. The concentration of a solution has a profound effect on the properties of a solution, both *physical* (melting and boiling points) and *chemical* (solution reactivity). Solution concentration may be expressed in many different units. Here, we consider concentration units based on mass.

Mass/Volume Percent

The concentration of a solution is defined as the amount of solute dissolved in a specified amount of solution,

$$\text{concentration} = \frac{\text{amount of solute}}{\text{amount of solution}}$$

If we define the amount of solute as the *mass* of solute (in g) and the amount of solution in *volume* units (mL), concentration is expressed as the ratio

$$\text{concentration} = \frac{\text{g of solute}}{\text{mL of solution}}$$

LEARNING GOAL

6 Calculate solution concentration in mass/volume percent, mass/mass percent, parts per thousand, and parts per million.

This concentration can then be expressed as a percentage by multiplying the ratio by the factor 100%. This results in

$$\% \text{ concentration} = \frac{\text{g of solute}}{\text{mL of solution}} \times 100\%$$

The percent concentration expressed in this way is called **mass/volume percent,** or **% (m/V).** Thus

$$\% \left(\frac{m}{V}\right) = \frac{\text{g of solute}}{\text{mL of solution}} \times 100\%$$

If the units of mass are other than g, or if the solution volume is in units other than mL, the proper conversion factor must be used to arrive at the units used in the equation.

Units and unit conversions were discussed in Chapter 1.

Consider the following examples.

EXAMPLE 6.1 | **Calculating Mass/Volume Percent**

LEARNING GOAL

Calculate the mass/volume percent composition, or % (m/V), of 0.300 L of solution containing 15.0 g of glucose.

6 Calculate solution concentration in mass/volume percent, mass/mass percent, parts per thousand, and parts per million.

Solution

Step 1. The expression for mass/volume percent is:

$$\% \left(\frac{m}{V}\right) = \frac{\text{g of solute}}{\text{mL of solution}} \times 100\%$$

Step 2. We must convert L to mL using the conversion factor based on 1000 mL = 1 L:

$$0.300 \, \cancel{L} \times \frac{10^3 \, \text{mL}}{1 \, \cancel{L}} = 3.00 \times 10^2 \, \text{mL}$$

Step 3. There are 15.0 g of glucose, the solute, and 3.00×10^2 mL of total solution. Therefore, substituting in our expression for mass/volume percent:

$$\% \left(\frac{m}{V}\right) = \frac{15.0 \, \text{g glucose}}{3.00 \times 10^2 \, \text{mL solution}} \times 100\%$$

$$= 5.00\% \left(\frac{m}{V}\right) \text{ glucose}$$

Practice Problem 6.1

   a. Calculate the % (m/V) of 0.0600 L of solution containing 10.0 g NaCl.
   b. Calculate the % (m/V) of 0.200 L of solution containing 15.0 g KCl.
   c. 20.0 g of oxygen gas are diluted with 80.0 g of nitrogen gas in a 78.0-L container at standard temperature and pressure. Calculate the % (m/V) of oxygen gas.
   d. 50.0 g of argon gas are diluted with 80.0 g of helium gas in a 476-L container at standard temperature and pressure. Calculate the % (m/V) of argon gas.

▶ For Further Practice: **Questions 6.31 and 6.32.**

EXAMPLE 6.2	**Calculating the Mass of Solute from a Mass/Volume Percent**

Calculate the number of g of NaCl in 5.00×10^2 mL of a 10.0% $\left(\dfrac{m}{V}\right)$ solution.

7 Determine the quantity of solute or solution from the concentration of solution.

Solution

Step 1. The expression for mass/volume percent is:

$$\% \left(\frac{m}{V}\right) = \frac{\text{g of solute}}{\text{mL of solution}} \times 100\%$$

Step 2. Substitute the data from the problem:

$$10.0\% \left(\frac{m}{V}\right) = \frac{X \text{ g NaCl}}{5.00 \times 10^2 \text{ mL solution}} \times 100\%$$

Step 3. Multiply both sides by 5.00×10^2 mL solution to simplify:

$$X \text{ g NaCl} \times 100\% = \left[10.0\% \left(\frac{m}{V}\right)\right] (5.00 \times 10^2 \text{ mL solution})$$

Remember % (m/V) represents units % (g/mL). Therefore, after this calculation is complete, the only unit remaining is the mass unit, g.

Step 4. Divide both sides by 100% to isolate g NaCl on the left side of the equation:

$$X \text{ g NaCl} = 50.0 \text{ g NaCl}$$

Helpful Hint: The concentration % (m/V) may be used as a conversion factor to arrive at the same result. Two possibilities exist:

$$\frac{10.0\% \text{ g NaCl}}{1 \text{ mL solution}} \quad \text{and} \quad \frac{1 \text{ mL solution}}{10.0\% \text{ g NaCl}}$$

Only the former, 10.0% g NaCl/mL solution will result in the product unit, g NaCl.

$$(5.00 \times 10^2 \text{ mL solution}) \times \left(\frac{10.0\% \text{ g NaCl}}{1 \text{ mL solution}}\right) \times \left(\frac{1}{100\%}\right) = 50.0 \text{ g NaCl}$$

Data Given × Conversion Factor = Desired Result

Practice Problem 6.2

a. Calculate the mass (in g) of sodium hydroxide required to make 2.00 L of a 1.00% (m/V) solution.
b. Calculate the volume (in mL) of a 25.0% (m/V) solution containing 10.0 g NaCl.

▶ For Further Practice: **Questions 6.39 and 6.40.**

Mass/Mass Percent

The **mass/mass percent**, or **% (m/m)**, is most useful for mixtures of solids, whose masses are easily obtained. The expression used to calculate mass/mass percentage is analogous in form to % (m/V):

$$\% \left(\frac{m}{m}\right) = \frac{\text{g solute}}{\text{g solution}} \times 100\%$$

| EXAMPLE 6.3 | Calculating Mass/Mass Percent |

Calculate the % (m/m) of platinum in a gold ring that contains 14.00 g gold and 4.500 g platinum.

Solution

Step 1. Using our definition of mass/mass percent

$$\% \left(\frac{m}{m}\right) = \frac{g \text{ solute}}{g \text{ solution}} \times 100\%$$

Step 2. Substituting,

$$\% \left(\frac{m}{m}\right) = \frac{4.500 \text{ g platinum}}{4.500 \text{ g platinum} + 14.00 \text{ g gold}} \times 100\%$$

$$= \frac{4.500 \text{ g platinum}}{18.50 \text{ g ring}} \times 100\%$$

$$= 24.32\% \text{ platinum}$$

Practice Problem 6.3

    a. Calculate the % (m/m) of oxygen gas in a mixture containing 20.0 g of oxygen gas and 80.0 g of nitrogen gas.

    b. Calculate the % (m/m) of argon gas in a mixture containing 50.0 g of argon gas and 80.0 g of helium gas.

▶ For Further Practice: **Questions 6.35 and 6.36.**

Parts per Thousand (ppt) and Parts per Million (ppm)

The calculation of concentration in *parts per thousand* (ppt) or *parts per million* (ppm) is based on the same logic as mass/mass percent. Percentage is actually the number of parts of solute in 100 parts of solution. For example, a 5.00% (m/m) solution is made up of 5.00 g solute in 100 g solution.

$$5.00\% \left(\frac{m}{m}\right) = \frac{5.00 \text{ g solute}}{1.00 \times 10^2 \text{ g solution}} \times 100\%$$

It follows that a 5.00 ppt solution is made up of 5.00 g solute in 1000 g solution.

$$5.00 \text{ ppt} = \frac{5.00 \text{ g solute}}{1.00 \times 10^3 \text{ g solution}} \times 10^3 \text{ ppt}$$

Using similar logic, a 5.00 ppm solution is made up of 5.00 g solute in 1,000,000 g solution.

$$5.00 \text{ ppm} = \frac{5.00 \text{ g solute}}{1.00 \times 10^6 \text{ g solution}} \times 10^6 \text{ ppm}$$

The general expressions are:

$$\text{ppt} = \frac{g \text{ solute}}{g \text{ solution}} \times 10^3 \text{ ppt}$$

and

$$\text{ppm} = \frac{g \text{ solute}}{g \text{ solution}} \times 10^6 \text{ ppm}$$

Units of ppt and ppm are most often used for expressing the concentrations of very dilute solutions.

EXAMPLE 6.4	Calculating ppt and ppm

A 1.00-g sample of stream water was found to contain 1.0×10^{-6} g lead. Calculate the concentration of lead in the stream water in units of % (m/m), ppt, and ppm. Which is the most suitable unit?

LEARNING GOAL

6 Calculate solution concentration in mass/volume percent, mass/mass percent, parts per thousand, and parts per million.

Solution

The expressions for mass/mass percent, parts per thousand, and parts per million are used.

Step 1. mass/mass percent:

$$\% \left(\frac{m}{m} \right) = \frac{g \text{ solute}}{g \text{ solution}} \times 100\%$$

$$= \frac{1.0 \times 10^{-6} \text{ g Pb}}{1.00 \text{ g solution stream water}} \times 100\%$$

$$= 1.0 \times 10^{-4}\% \ (m/m)$$

Step 2. parts per thousand:

$$ppt = \frac{g \text{ solute}}{g \text{ solution}} \times 10^3 \text{ ppt}$$

$$= \frac{1.0 \times 10^{-6} \text{ g Pb}}{1.00 \text{ g solution stream water}} \times 10^3 \text{ ppt}$$

$$= 1.0 \times 10^{-3} \text{ ppt}$$

Step 3. parts per million:

$$ppm = \frac{g \text{ solute}}{g \text{ solution}} \times 10^6 \text{ ppm}$$

$$= \frac{1.0 \times 10^{-6} \text{ g Pb}}{1.00 \text{ g solution stream water}} \times 10^6 \text{ ppm}$$

$$= 1.0 \text{ ppm}$$

Parts per million is the most reasonable unit because exponents are not required to express the numerical value.

Practice Problem 6.4

a. Calculate the ppt and ppm of oxygen gas in a mixture containing 20.0 g of oxygen gas and 80.0 g of nitrogen gas.
b. Calculate the ppt and ppm of argon gas in a mixture containing 50.0 g of argon gas and 80.0 g of helium gas.

▶ For Further Practice: **Questions 6.43 and 6.44.**

6.3   Concentration Based on Moles

LEARNING GOAL

8 Calculate the molarity of solution from mass or moles of solute.

In our discussion of the chemical arithmetic of reactions in Chapter 4, we saw that the chemical equation represents the relative number of *moles* of reactants producing products. When chemical reactions occur in solution, it is most useful to represent their concentrations on a *molar* basis.

Molarity

The most common mole-based concentration unit is molarity. **Molarity,** symbolized *M*, is defined as the number of mol of solute per L of solution, or

$$M = \frac{\text{mol solute}}{\text{L solution}}$$

EXAMPLE 6.5 **Calculating Molarity from Moles**

Calculate the molarity of 2.0 L of solution containing 5.0 mol NaOH.

Solution

Using our expression for molarity

$$M = \frac{\text{mol solute}}{\text{L solution}}$$

Substituting,

$$M_{\text{NaOH}} = \frac{5.0 \text{ mol NaOH}}{2.0 \text{ L solution}}$$
$$= 2.5\ M$$

Practice Problem 6.5

Calculate the molarity of 2.5 L of solution containing 0.75 mol $MgCl_2$.

▶ For Further Practice: **Questions 6.47 and 6.48.**

Remember the need for conversion factors to convert from mass to number of mol. Consider the following example:

Sections 1.3 and 1.5 discuss units and unit conversion.

EXAMPLE 6.6 **Calculating Molarity from Mass**

If 5.00 g glucose are dissolved in 1.00×10^2 mL of solution, calculate the molarity, M, of the glucose solution.

Solution

Step 1. To use our expression for molarity, it is necessary to convert from units of g of glucose to mol of glucose. The molar mass of glucose is 1.80×10^2 g/mol. Therefore

$$5.00 \text{ g glucose} \times \frac{1 \text{ mol glucose}}{1.80 \times 10^2 \text{ g glucose}} = 2.78 \times 10^{-2} \text{ mol glucose}$$

Step 2. We must convert mL to L:

$$1.00 \times 10^2 \text{ mL solution} \times \frac{1 \text{ L solution}}{10^3 \text{ mL solution}} = 1.00 \times 10^{-1} \text{ L solution}$$

Step 3. Substituting these quantities:

$$M_{\text{glucose}} = \frac{2.78 \times 10^{-2} \text{ mol glucose}}{1.00 \times 10^{-1} \text{ L solution}}$$
$$= 2.78 \times 10^{-1}\ M$$

Practice Problem 6.6

Calculate the molarity, M, of KCl when 2.33 g KCl are dissolved in 2.50×10^3 mL of solution.

▶ For Further Practice: **Questions 6.53 and 6.54.**

EXAMPLE 6.7	Calculating Volume from Molarity

LEARNING GOAL

7 Determine the quantity of solute or solution from the concentration of solution.

Calculate the volume of a 0.750 M sulfuric acid (H_2SO_4) solution containing 0.120 mol of solute.

Solution

Step 1. Substituting in our basic expression for molarity, we obtain

$$0.750 \ M \ H_2SO_4 = \frac{0.120 \ \text{mol} \ H_2SO_4}{X \ \text{L solution}}$$

Step 2. Rearranging to solve for volume (L):

$$X \ \text{L solution} = \frac{0.120 \ \text{mol} \ H_2SO_4}{0.750 \ \text{M} \ H_2SO_4}$$

Remember that M represents mol/L.

$$X \ \text{L} = 0.160 \ \text{L}$$

Helpful Hint: Since the units of molarity are in fraction form, the value of molarity is a ratio that produces two possible conversion factors: $\dfrac{0.750 \ \text{mol} \ H_2SO_4}{1 \ \text{L solution}}$ and $\dfrac{1 \ \text{L solution}}{0.750 \ \text{mol} \ H_2SO_4}$

Only one will result in the product unit of L.

$$0.120 \ \cancel{\text{mol} \ H_2SO_4} \times \left(\frac{1 \ \text{L solution}}{0.750 \ \cancel{\text{mol} \ H_2SO_4}} \right) = 0.160 \ \text{L solution}$$

Data Given × Conversion Factor = Desired Result

Practice Problem 6.7

Calculate the volume of a 0.200 M KCl solution containing 5.00×10^{-2} mol of solute.

▶ For Further Practice: **Questions 6.57 and 6.58.**

Question 6.7   Calculate the number of mol of solute in 5.00×10^2 mL of 0.250 M HCl.

Question 6.8   Calculate the number of g of silver nitrate required to prepare 2.00 L of 0.500 M $AgNO_3$.

Dilution

LEARNING GOAL

9 Perform dilution calculations.

Laboratory reagents are often purchased as concentrated solutions (for example, 12 M HCl or 6 M NaOH) for reasons of safety, economy, and space limitations. We must often *dilute* such a solution to a larger volume to prepare a less concentrated solution for the experiment at hand. The approach to such a calculation is as follows.
    We define

$$M_1 = \text{molarity of solution } \textit{before} \text{ dilution}$$

$$M_2 = \text{molarity of solution } \textit{after} \text{ dilution}$$

$$V_1 = \text{volume of solution } \textit{before} \text{ dilution}$$

$$V_2 = \text{volume of solution } \textit{after} \text{ dilution}$$

and

$$M = \frac{\text{mol solute}}{\text{L solution}}$$

This equation can be rearranged as:

$$\text{mol solute} = (M)(\text{L solution})$$

The number of mol of solute *before* and *after* dilution is unchanged, because dilution involves only the addition of extra solvent:

$$\text{mol}_1 \text{ solute} = \text{mol}_2 \text{ solute}$$
Initial   Final
condition   condition

or

$$(M_1)(\text{L}_1 \text{ solution}) = (M_2)(\text{L}_2 \text{ solution})$$

$$(M_1)(V_1) = (M_2)(V_2)$$

Knowing any three of these terms enables us to calculate the fourth.

EXAMPLE 6.8   **Calculating Molarity after Dilution**

LEARNING GOAL

Calculate the molarity of a solution made by diluting 0.050 L of 0.10 M HCl solution to a volume of 1.0 L.

9 Perform dilution calculations.

Solution

Step 1. Summarize the information provided in the problem:

$$M_1 = 0.10 \, M$$
$$M_2 = \text{Desired Result}$$
$$V_1 = 0.050 \, \text{L}$$
$$V_2 = 1.0 \, \text{L}$$

Step 2. Use the dilution expression:

$$(M_1)(V_1) = (M_2)(V_2)$$

Step 3. Solve for M_2, the final solution concentration:

$$M_2 = \frac{(M_1)(V_1)}{V_2}$$

Step 4. Substituting,

$$M_2 = \frac{(0.10 \, M)(0.050 \, \cancel{L})}{(1.0 \, \cancel{L})}$$
$$= 0.0050 \, M \quad \text{or} \quad 5.0 \times 10^{-3} \, M \text{ HCl}$$

Practice Problem 6.8

What volume of 0.200 M sugar solution can be prepared from 50.0 mL of 0.400 M solution?

▶ For Further Practice: **Questions 6.60 and 6.61.**

EXAMPLE 6.9   **Calculating a Dilution Volume**

LEARNING GOAL

Calculate the volume, in L, of water that must be added to dilute 20.0 mL of 12.0 M HCl to 0.100 M HCl.

9 Perform dilution calculations.

Solution

Step 1. Summarize the information provided in the problem:

$$M_1 = 12.0\ M$$
$$M_2 = 0.100\ M$$
$$V_1 = 20.0\ mL = 0.0200\ L$$
$$V_2 = \text{Desired Result}$$

Step 2. Then, using the dilution expression:

$$(M_1)(V_1) = (M_2)(V_2)$$

Step 3. Solve for V_2, the final volume:

$$V_2 = \frac{(M_1)(V_1)}{(M_2)}$$

Step 4. Substituting,

$$V_2 = \frac{(12.0\ M)(0.0200\ L)}{0.100\ M}$$
$$= 2.40\ \text{L solution}$$

Note that this is the *total final volume*. The amount of water added equals this volume *minus* the original solution volume, or

$$2.40\ L - 0.0200\ L = 2.38\ \text{L water}$$

Practice Problem 6.9

How would you prepare 1.0×10^2 mL of 2.0 M HCl, starting with concentrated (12.0 M) HCl?

▶ For Further Practice: **Questions 6.59 and 6.62.**

The dilution equation is valid with any concentration units, such as % (m/V) *as well as* molarity, which are used in Examples 6.8 and 6.9. However, you must use the same units for both initial *and* final concentration values. Only in this way can you cancel units properly.

6.4   Concentration-Dependent Solution Properties

LEARNING GOAL

10 Describe and explain concentration-dependent solution properties.

Colligative properties are solution properties that depend on the *concentration of the solute particles*, rather than the *identity of the solute*.

There are four colligative properties of solutions:

• vapor pressure lowering
• freezing point depression
• boiling point elevation
• osmotic pressure

Each of these properties has widespread practical applications. We look at each in some detail in this section.

Vapor Pressure Lowering

Raoult's law states that, when a nonvolatile solute is added to a solvent, the vapor pressure of the solvent decreases in proportion to the concentration of the solute.

Perhaps the most important consequence of Raoult's law is the effect of the solute on the boiling point of a solution.

When a nonvolatile solute is added to a solvent, the boiling point of the solution is found to increase because it requires a higher temperature to form the gaseous state.

Raoult's law may be explained in molecular terms by using the following logic: Vapor pressure of a solution results from the escape of solvent molecules from the liquid to the gas phase, thus increasing the partial pressure of the gas-phase solvent molecules until the equilibrium vapor pressure is reached. Presence of solute molecules hinders the escape of solvent molecules, thus lowering the equilibrium vapor pressure (Figure 6.4).

Figure 6.4 An illustration of Raoult's law: lowering of vapor pressure by addition of solute molecules. White units represent solvent molecules, and red units are solute molecules. Solute molecules present a barrier to escape of solvent molecules, thus decreasing the vapor pressure.

Freezing Point Depression and Boiling Point Elevation

Freezing point depression may be explained by examining the equilibrium between liquid and solid states. At the freezing point, ice is in equilibrium with liquid water:

$$H_2O(l) \underset{(r)}{\overset{(f)}{\rightleftharpoons}} H_2O(s)$$

Solute molecules interfere with the rate at which liquid water molecules associate to form the solid state, decreasing the rate of the forward reaction. For a true equilibrium, the rate of the forward (f) and reverse (r) processes must be equal. Lowering the temperature eventually slows the rate of the reverse (r) process sufficiently to match the rate of the forward reaction. At the lower temperature, equilibrium is established, and the solution freezes.

Recall that the concept of liquid vapor pressure was discussed in Section 5.2.

Boiling point elevation can be explained by considering the definition of the boiling point; that is, the temperature at which the vapor pressure of the liquid equals the atmospheric pressure. Raoult's law states that the vapor pressure of a solution is decreased by the presence of a solute. Therefore, a higher temperature is necessary to raise the vapor pressure to the atmospheric pressure, hence the boiling point elevation.

Section 7.4 discusses equilibrium.

The extent of the freezing point depression (ΔT_f) is proportional to the solute concentration over a limited range of concentration:

$$\Delta T_f = k_f \times (\text{solute particle concentration})$$

The boiling point elevation (ΔT_b) is also proportional to the solute concentration:

$$\Delta T_b = k_b \times (\text{solute particle concentration})$$

If the value of the proportionality factor $(k_f$ or $k_b)$ is known for the solvent of interest, the magnitude of the freezing point depression or boiling point elevation can be calculated for a solution of known concentration.

Solute concentration must be in *mole*-based units. The number of *particles* (molecules or ions) is critical here, not the *mass* of solute. One high mass molecule will have exactly the same effect on the freezing or boiling point as one low mass molecule. A mole-based unit, because it is related directly to Avogadro's number, will correctly represent the number of particles in solution.

We have already worked with one mole-based unit, *molarity,* and this concentration unit can be used to calculate either the freezing point depression or the boiling point elevation.

A second mole-based concentration unit, molality, is more commonly used in these types of situations. **Molality** (symbolized m) is defined as the number of mol of solute per kilogram (kg) of solvent in a solution:

$$m = \frac{\text{mol solute}}{\text{kg solvent}}$$

Molarity is temperature dependent simply because it is expressed as mol/L. Volume (L) is temperature dependent—most liquids expand measurably when heated and contract when cooled. Molality is mol/kg; both mol and mass (kg) are temperature independent.

Molality does not vary with temperature, whereas molarity is temperature dependent. For this reason, molality is the preferred concentration unit for studies such as freezing point depression and boiling point elevation, in which measurement of *change* in temperature is critical.

Practical applications that take advantage of freezing point depression of solutions by solutes include the following:

- Salt is spread on roads to melt ice in winter. The salt lowers the freezing point of the water, so it exists in the liquid phase below its normal freezing point, 0°C or 32°F.
- Solutes such as ethylene glycol, "antifreeze," are added to car radiators to prevent freezing by lowering the freezing point of the coolant.

We refer to the concentration of *particles* in our discussion of colligative properties. Why do we stress this term? The reason is that there is a very important difference between electrolytes and nonelectrolytes. That difference is the way that they behave when they dissolve. For example, if we dissolve 1 mol of glucose ($C_6H_{12}O_6$) in 1 kg of water,

$$1 \; C_6H_{12}O_6(s) \xrightarrow{\;H_2O\;} 1 \; C_6H_{12}O_6(aq)$$

1 mol (Avogadro's number, 6.022×10^{23} particles) of glucose is present in solution. *Glucose is a covalently bonded nonelectrolyte.* Dissolving 1 mol of sodium chloride in 1 kg of water,

$$1 \; NaCl(s) \xrightarrow{\;H_2O\;} 1 \; Na^+(aq) + 1 \; Cl^-(aq)$$

produces 2 mol of particles (1 mol of sodium ions and 1 mol of chloride ions). *Sodium chloride is an ionic electrolyte.*

$$1 \text{ mol glucose} \longrightarrow 1 \text{ mol of particles in solution}$$
$$1 \text{ mol sodium chloride} \longrightarrow 2 \text{ mol of particles in solution}$$

It follows that 1 mol of sodium chloride will decrease the vapor pressure, increase the boiling point, or depress the freezing point of 1 kg of water *twice as much* as 1 mol of glucose in the same quantity of water.

Question 6.9 Comparing pure water and a 0.10 m glucose solution, which has the higher freezing point?

Question 6.10 Comparing pure water and a 0.10 m glucose solution, which has the higher boiling point?

Calculating Freezing Points and Boiling Points of Aqueous Solutions

LEARNING GOAL

11 Perform calculations involving colligative properties.

The freezing point depression constant for aqueous solutions is:

$$k_f = \frac{1.86°C}{m}$$

The expression for calculating the change (decrease) in freezing point of an aqueous solution is equal to the product of the freezing point depression constant and the molality of the particles in the solution:

$$\Delta T_f = (k_f)\,(m \text{ particles})$$

Correspondingly, for boiling point elevation:

$$k_b = \frac{0.52°C}{m}$$

and

$$\Delta T_b = (k_b)(m \text{ particles})$$

If we know the solution molality (mol solute/kg solvent), we can determine the particle molality,

$$m \text{ particles} = \frac{\text{mol solute}}{\text{kg solvent}} \times \frac{\text{mol particles}}{\text{mol solute}}$$

and calculate the decrease (freezing) or increase (boiling) in freezing or boiling temperature. Recalling that pure water freezes at 0°C and boils at 100°C, we can subtract ΔT_f from 0°C to obtain the solution freezing point and add ΔT_b to 100°C, resulting in the solution boiling point.

Let's review calculations of this type in Examples 6.10 and 6.11.

EXAMPLE 6.10 | **Calculating Freezing and Boiling Points of Aqueous Solutions of Covalent, Nondissociating Solutes**

LEARNING GOAL

11 Perform calculations involving colligative properties.

Ethylene glycol, H—O—C—C—O—H (with H H above and H H below), is a widely used automobile antifreeze.

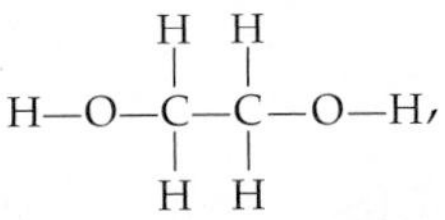

a. Calculate the freezing point of an 8.38 m aqueous solution of ethylene glycol.
b. Calculate the boiling point of an 8.38 m aqueous solution of ethylene glycol.

Solution

We begin by recognizing that ethylene glycol is a nondissociating covalent compound. Consequently, 8.38 m ethylene glycol solute is equivalent to 8.38 m particles.

For part a,

Step 1. Using our expression for freezing point depression:

$$\Delta T_f = \left(\frac{1.86°C}{m}\right)(m \text{ particles})$$

and substituting molality of particles,

$$\Delta T_f = \left(\frac{1.86°C}{m}\right)(8.38\ m) = 15.6°C$$

Step 2. The solution freezing point is 15.6°C *below* 0.0°C, the freezing point of pure water. Therefore

$$\text{freezing point} = 0.0°C - 15.6°C = -15.6°C$$

For part b,

Step 1. Substituting our value for molality of particles into the boiling point elevation equation:

$$\Delta T_b = \left(\frac{0.52°C}{m}\right)(m \text{ particles})$$

$$= \left(\frac{0.52°C}{m}\right)(8.38\ m) = 4.4°C$$

Step 2. This means that the solution boiling point is 4.4°C *above* the boiling point of pure water. Therefore

$$\text{boiling point} = 100.0°C + 4.4°C = 104.4°C$$

Continued…

Practice Problem 6.10

Calculate the boiling temperature and freezing temperature of a 1.5 m solution of glucose ($C_6H_{12}O_6$). Remember that glucose is a covalent, nondissociating solute.

▶ For Further Practice: **Questions 6.73a and 6.74a.**

EXAMPLE 6.11   **Calculating Freezing and Boiling Points of Aqueous Solutions of Ionic, Dissociating Solutes**

LEARNING GOAL

11   Perform calculations involving colligative properties.

Calcium chloride is used on roads and sidewalks to prevent ice formation in wintertime.

    a. Calculate the freezing point of a 3.00 m aqueous solution of $CaCl_2$.
    b. Calculate the boiling point of a 3.00 m aqueous solution of $CaCl_2$.

Solution

Calcium chloride dissociates in water to form three particles for each $CaCl_2$:

$$\underbrace{CaCl_2(s)}_{\text{1 particle}} \xrightarrow{\text{H}_2\text{O}} \underbrace{Ca^{2+}(aq) + 2\,Cl^-(aq)}_{\text{3 particles}}$$

Consequently, a 3.00 m $CaCl_2$ aqueous solution is

$$\frac{3\ m\ \text{particles}}{1\ m\ \cancel{CaCl_2}} \times 3.00\ \cancel{m\ CaCl_2} = 9.00\ m\ \text{particles}$$

For part a,

Step 1. Using our expression for freezing point depression:

$$\Delta T_f = \left(\frac{1.86°C}{m}\right)(m\ \text{particles})$$

and substituting molality of particles

$$= \left(\frac{1.86°C}{\cancel{m}}\right)(9.00\ \cancel{m}) = 16.7°C$$

Step 2. The solution freezing point is 16.7°C *below* 0.0°C, the freezing point of pure water. Therefore

$$\text{Freezing point} = 0.0°C - 16.7°C = -16.7°C$$

For part b,

Step 1. Substituting our value for molality of particles into the boiling point elevation equation:

$$\Delta T_b = \left(\frac{0.52°C}{m}\right)(m\ \text{particles})$$

$$= \left(\frac{0.52°C}{\cancel{m}}\right)(9.00\ \cancel{m}) = 4.7°C$$

Step 2. This means that the solution boiling point is 4.7°C *above* the boiling point of pure water. Therefore

$$\text{Boiling point} = 100.0°C + 4.7°C = 104.7°C$$

Practice Problem 6.11

Calculate the boiling temperature and freezing temperature of a 1.5 m solution of KCl. (Remember that 1 mol of KCl produces 2 mol of particles.)

▶ For Further Practice: **Questions 6.73b and 6.74b.**

Osmosis, Osmotic Pressure, and Osmolarity

The biological cell membrane mediates the interaction of the cell with its environment and is responsible for the controlled passage of material into and out of the cell. One of the principal means of transport is termed *diffusion*. **Diffusion** is the net movement of solute or solvent molecules from an area of high concentration to an area of low concentration. This region where the concentration decreases over a distance is termed the **concentration gradient.** Because of the structure of the cell membrane, only small molecules are able to diffuse freely across this barrier. Large molecules and highly charged ions are restricted by the barrier. In other words, the cell membrane is behaving in a selective fashion. Such membranes are termed **selectively permeable membranes.**

Because a cell membrane is selectively permeable, it is not always possible for solutes to pass through it in response to a concentration gradient. In such cases, the solvent diffuses through the membrane. Such membranes, permeable to solvent but not to solute, are specifically called **semipermeable membranes.**

Osmosis

Osmosis is the diffusion of a solvent (water in biological systems) through a semipermeable membrane in response to a (water) concentration gradient.

Suppose that we place a 0.5 *M* glucose solution in a dialysis bag that is composed of a membrane with pores that allow the passage of water molecules but not glucose molecules. Consider what will happen when we place this bag into a beaker of pure water. We have created a gradient in which there is a higher concentration of glucose inside the bag than outside, but the glucose cannot diffuse through the bag to achieve equal concentration on both sides of the membrane.

Now let's think about this situation in another way. We have a higher concentration of water molecules outside the bag (where there is only pure water) than inside the bag (where some of the water molecules are occupied in dipole-dipole interactions with solute particles and are consequently unable to move freely in the system). Because water can diffuse through the membrane, a net diffusion of water will occur through the membrane into the bag. This is the process of osmosis (Figure 6.5).

As you have probably already guessed, this system can never reach equilibrium (equal concentrations inside and outside the bag). Regardless of how much water diffuses into the bag, diluting the glucose solution, the concentration of glucose will always be higher inside the bag (and the accompanying free water concentration will always be lower).

Osmotic Pressure and Osmolarity

What happens when the bag has taken in as much water as it can, when it has expanded as much as possible? Now the walls of the bag exert a force that will stop the *net* flow of water into the bag. **Osmotic pressure** is the pressure that must be exerted to stop the flow of water across a selectively permeable membrane by osmosis. Stated more precisely, the osmotic pressure of a solution is the net pressure with which water enters it by osmosis from a pure water compartment when the two compartments are separated by a semipermeable membrane.

The osmotic pressure can be calculated from the solution concentration at any temperature. How do we determine "solution concentration"? Recall that osmosis is a colligative property, dependent on the concentration of solute particles. Again, it becomes necessary to distinguish between solutions of electrolytes and nonelectrolytes. For example, a 1 *M*

LEARNING GOAL

10 Describe and explain concentration-dependent solution properties.

The terms *selectively permeable* or *differentially permeable* are used to describe biological membranes because such membranes restrict the passage of particles based both on size and charge. Even small ions, such as H^+, cannot pass freely across a cell membrane.

Cellophane is a familiar example of a semipermeable membrane.

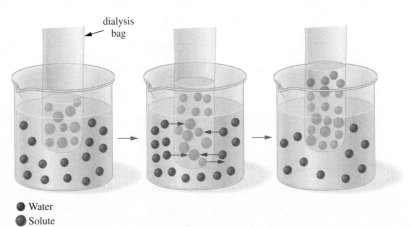

● Water
● Solute

Figure 6.5 Osmosis across a membrane. The solvent, water, diffuses from outside the bag (the membrane) to the region of high solute concentration (inside the bag).

glucose solution consists of 1 mol of particles per L; glucose is a nonelectrolyte. A solution of 1 M NaCl produces 2 mol of particles per L (1 mol of Na^+ and 1 mol of Cl^-). A 1 M $CaCl_2$ solution is 3 M in particles (1 mol of Ca^{2+} and 2 mol of Cl^- per L).

Osmolarity, the molarity of particles in solution, is used for osmotic pressure calculations. The equation relating the solution molarity to osmolarity is:

$$\text{osmolarity} = i \times M$$

where i = number of mol of particles/mol of solute

M = molar concentration of the solute, in units of mol/L

We can use the value for the solution concentration of particles in conjunction with the ideal gas constant and the temperature of the solution to calculate the osmotic pressure of a solution, symbolized by π, using the following equation:

$$\pi = \text{osmolarity} \times R \times T$$

We know, $\text{osmolarity} = i \times M$

Substituting, $\pi = iMRT$

where,

π = osmotic pressure of solution
i = number of mol of particles/mol solute
R = ideal gas constant (units of L · atm/K · mol)
T = solution temperature (units of K)

EXAMPLE 6.12   Calculating Solution Osmolarity

Determine the osmolarity of 5.0×10^{-3} M Na_3PO_4.

LEARNING GOAL

11 Perform calculations involving colligative properties.

Solution

Step 1. Na_3PO_4 is an ionic compound and produces an electrolytic solution:

$$Na_3PO_4 \xrightarrow{H_2O} 3Na^+ + PO_4^{3-}$$

1 mol of Na_3PO_4 yields 4 mol of product ions; consequently, i = 4 mol particles/1 mol Na_3PO_4.

Step 2. Using our equation relating osmolarity and molarity (mol solute/L), and substituting,

$$\text{osmolarity} = i \times M$$

$$= \frac{4 \text{ mol particles}}{1 \text{ mol } Na_3PO_4} \times 5.0 \times 10^{-3} \frac{\text{mol } Na_3PO_4}{L}$$

$$= 2.0 \times 10^{-2} \text{ mol particles/L}$$

Consequently, the osmolarity of the solution is 2.0×10^{-2} mol particles/L.

Practice Problem 6.12

Determine the osmolarity of the following solutions:

   a. 5.0×10^{-3} M NH_4NO_3 (electrolyte)   b. 5.0×10^{-3} M $C_6H_{12}O_6$ (nonelectrolyte)

▶ For Further Practice: **Questions 6.83 and 6.84.**

EXAMPLE 6.13 | Calculating Osmotic Pressure

LEARNING GOAL

11 Perform calculations involving colligative properties.

Calculate the osmotic pressure of a 5.0×10^{-2} M solution of NaCl at 25°C (298 K).

Solution

Step 1. Using our definition of osmotic pressure, π:

$$\pi = i\,MRT$$

Step 2. One mol of NaCl produces two mol of particles (Na^+ and Cl^-). Therefore, $i = 2$ mol particles/1 mol NaCl.

Step 3. Substituting in our osmotic pressure expression:

$$= \frac{2 \text{ mol particles}}{1 \text{ mol NaCl}} \times 5.0 \times 10^{-2} \frac{\text{mol NaCl}}{L} \times 0.0821 \frac{L \cdot atm}{K \cdot mol} \times 298\,K$$

$$= 2.4 \text{ atm}$$

Practice Problem 6.13

Calculate the osmotic pressure of each solution described in Practice Problem 6.12. Assume that the solutions are at 298 K.

▶ For Further Practice: **Questions 6.85 and 6.86.**

Osmotic Pressure and Osmolarity in Red Blood Cells

Blood plasma has an osmolarity equivalent to a 0.30 M glucose solution or a 0.15 M NaCl solution. The latter is true because NaCl in solution dissociates into Na^+ and Cl^- and thus contributes twice the number of solute particles as a glucose molecule that does not ionize. If red blood cells, which have an osmolarity equal to blood plasma, are placed in a 0.30 M glucose solution, no net osmosis will occur because the osmolarity and water concentration inside the red blood cells are equal to those of the 0.30 M glucose solution. The solutions inside and outside the red blood cells are said to be **isotonic** (*iso* means "same," and *tonic* means "strength") **solutions.** Because the osmolarity is the same inside and outside, the red blood cells will remain the same size (Figure 6.6b).

What happens if we now place the red blood cells into a **hypotonic solution,** in other words, a solution having a lower osmolarity than the cytoplasm of the cells? In this situation, there will be a net movement of water into the cells as water diffuses down its concentration gradient. The membranes of the red blood cells do not have the strength to exert a sufficient pressure to stop this flow of water, and the cells will swell and burst (Figure 6.6c). Alternatively, if we place the red blood cells into a **hypertonic solution** (one with a greater osmolarity than the cells), water will pass out of the cells, and they will shrink dramatically (Figure 6.6a).

These principles have important applications in the delivery of intravenous (IV) solutions into an individual (Figure 6.7). Normally, any fluids infused intravenously must have the correct osmolarity; they must be isotonic with the blood cells and the blood plasma. Such infusions are frequently either 5.5% dextrose (glucose) or "normal saline." The first solution is composed of 5.5 g of glucose per 100 mL of solution (0.30 M), and the latter of 9.0 g of NaCl per 100 mL of solution (0.15 M). In either case, they have the same osmotic pressure and osmolarity as the plasma and blood cells and can therefore be safely administered without upsetting the osmotic balance between the blood and the blood cells.

Practical examples of osmosis abound, including the following:

• A sailor, lost at sea in a lifeboat, dies of dehydration while surrounded by water. Seawater, because of its high salt concentration, dehydrates the cells of the body as a result of the large osmotic pressure difference between it and intracellular fluids.

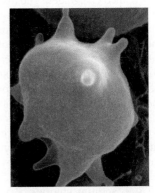
(a)

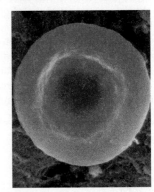
(b)

(c)

Figure 6.6 Scanning electron micrographs of red blood cells exposed to (a) hypertonic, (b) isotonic, and (c) hypotonic solutions.

Oral Rehydration Therapy

Diarrhea kills millions of children before they reach the age of 5 years. This is particularly true in third-world countries where sanitation, water supplies, and medical care are poor. In the case of diarrhea, death results from fluid loss, electrolyte imbalance, and hypovolemic shock (multiple organ failure due to insufficient perfusion). Cholera is one of the best-understood bacterial diarrheas. The organism *Vibrio cholera* survives passage through the stomach and reproduces in the intestine, where it produces a toxin called choleragen. The toxin causes the excessive excretion of Na^+, Cl^-, and HCO_3^- from epithelial cells lining the intestine. The increased ion concentration (hypertonic solution) outside the cell results in the movement of massive quantities of water into the intestinal lumen. This causes the severe, abundant, clear vomit and diarrhea that can result in the loss of 10–15 L of fluid per day. Over the 4- to 6-day progress of the disease, a patient may lose from one to two times his or her body mass!

The need for fluid replacement is obvious. Oral rehydration is preferred over intravenous administration of fluids and electrolytes since it is noninvasive. In many third-world countries, it is the only therapy available in remote areas. The rehydration formula includes 50–80 g/L rice (or other starch), 3.5 g/L sodium chloride, 2.5 g/L sodium bicarbonate, and 1.5 g/L potassium chloride. Oral rehydration takes advantage of the cotransport of Na^+ and glucose across the cells lining the intestine. Thus, the glucose enters the cells, and Na^+ is carried along. Movement of these materials into the cells helps alleviate the osmotic imbalance, reduces the diarrhea, and corrects the fluid and electrolyte imbalance.

The disease runs its course in less than a week. In fact, antibiotics are not used to combat cholera. The only effective

A woman is shown filtering water through sari cloth.

therapy is oral rehydration, which reduces mortality to less than 1%. A much better option is prevention. In the photo above, a woman is shown filtering water through sari cloth. This simple practice has been shown to reduce the incidence of cholera significantly.

For Further Understanding

▶ Explain dehydration in terms of osmosis.
▶ Explain why even severely dehydrated individuals continue to experience further fluid loss.

- A cucumber, soaked in brine, shrivels into a pickle. The water in the cucumber is drawn into the brine (salt) solution because of a difference in osmotic pressure (Figure 6.8).
- A Medical Perspective: Oral Rehydration Therapy describes one of the most lethal and pervasive examples of cellular fluid imbalance.

Question 6.11 Refer to the intravenous solution in Figure 6.7 and calculate the osmolarity of its contents.

Question 6.12 Is the solution osmolarity that you calculated in Question 6.11 isotonic with blood plasma? Why or why not?

6.5 Aqueous Solutions

Water as a Solvent

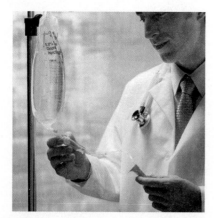

Figure 6.7 Composition and concentration are critically important in medical intervention. This solution is 0.15% (m/V) KCl and 5% (m/V) glucose.

Water is by far the most abundant substance on earth. It is an excellent solvent for most inorganic substances and is the principal biological solvent. Approximately 60% of the adult human body is water, and maintenance of this level is essential for survival. These characteristics are a direct consequence of the molecular structure of water.

As we saw in Chapter 3, water is a bent molecule with a 104.5° bond angle. This angular structure, resulting from the effect of the two lone pairs of electrons around the oxygen atom, is responsible for the polar nature of water. The polarity, in turn, gives water its unique properties.

Because water molecules are polar, water is an excellent solvent for other polar substances ("like dissolves like"). Because much of the matter on earth is polar, hence at least somewhat water soluble, water has been described as the universal solvent. The term *universal solvent* is historical; for centuries, water was known to dissolve many commonly used substances. The solvent in blood is water, and it dissolves and transports compounds as diverse as potassium chloride, glucose, and protein throughout our bodies.

Water is readily accessible and easily purified. It is nontoxic and quite nonreactive. The high boiling point of water, 100°C, compared with molecules of similar size such as N_2 (b.p. $= -196$°C), is also explained by water's polar character. Strong dipole-dipole interactions between a δ^+ hydrogen of one molecule and δ^- oxygen of a second, referred to as hydrogen bonding, create an interactive molecular network in the liquid phase (see Figure 5.9). The strength of these interactions requires more energy (higher temperature) for water to boil. The higher-than-expected boiling point enhances water's value as a solvent; often, reactions are carried out at higher temperatures to increase their rates. Other solvents, with lower boiling points, would simply boil away, and the reaction would stop.

This idea is easily extended to our own chemistry—because 60% of our bodies is water, we should appreciate the polarity of water on a hot day. As a biological solvent in the human body, water is involved in the transport of ions, nutrients, and waste into and out of cells. Water is also the solvent for biochemical reactions in cells and the digestive tract. Water is a reactant or product in some biochemical processes.

LEARNING GOAL

12 Describe why the chemical and physical properties of water make it a truly unique solvent.

See A Human Perspective: An Extraordinary Molecule, Chapter 7.

Refer to Sections 3.4 and 3.5 for a more complete description of the bonding, structure, and polarity of water.

Recall the discussion of intermolecular forces in Chapters 3 and 5.

EXAMPLE 6.14 Predicting Structure from Observable Properties

Sucrose is a common sugar, and we know that it is used as a sweetener when dissolved in many beverages. What does this allow us to predict about the structure of sucrose?

LEARNING GOAL

12 Describe why the chemical and physical properties of water make it a truly unique solvent.

Solution

Sucrose is used as a sweetener in teas, coffee, and a host of soft drinks. The solvent in all of these beverages is water, a polar molecule. The rule "like dissolves like" implies that sucrose must also be a polar molecule. Without even knowing the formula or structure of sucrose, we can infer this important information from a simple experiment—dissolving sugar in our morning cup of coffee.

Practice Problem 6.14

Predict whether ammonia or methane would be more soluble in water. Explain your answer. (Hint: Refer to Section 5.2, the discussion of interactions in the liquid state.)

▶ For Further Practice: **Questions 6.101 and 6.102.**

(a)

(b)

Figure 6.8 A cucumber (a) in an acidic salt solution undergoes considerable shrinkage on its way to becoming a pickle (b) because of osmosis.

Kitchen Chemistry

Solubility, Surfactants, and the Dishwasher

Each day we create a new supply of dirty dishes, glasses, pots and pans, and silverware. We may hand wash them, using one of the many commercially available detergents. Or, we may load them in an automatic dishwasher along with a solid or liquid detergent. Close the door, push a few buttons, and, in less than 1 hour, the dishes are clean and reasonably bacteria-free. How does this happen? It is a case of chemistry in action.

The fats, grease, and oils on the dishes are the potential solutes in the solvent water. The dissolution process is unlikely to happen if the solvent is pure water. Remember our rule "like dissolves like." The food residue is composed principally of nonpolar or very slightly polar molecules. Water, on the other hand, is extremely polar. We need some other substance to facilitate the interaction of the polar water and nonpolar food. That substance is a surfactant. Dishwashing detergents are a mixture of many components. Some have nothing to do with the cleaning process itself. Perfumes and dyes may be added to make the product more attractive to the consumer. Some components improve shelf life. The principal *active* ingredient is a surfactant. The role of the surfactant is to form a "bridge" between polar and nonpolar substances. Let us see how this happens.

Many surfactants exist, with a variety of chemical structures. However, all share one structural feature. They are molecules that have both a polar and nonpolar end. The polar end interacts with the solvent water, and the nonpolar end dissolves in the less polar (or nonpolar) food residue. Consequently, the surface tension of the water, which serves as a barrier between the water and food particles, is decreased (see Section 5.2). Agitation of the water removes the water/surfactant/residue aggregate from the surface that is being cleaned. This aggregate remains dissolved in the water until the rinse cycle sends it down the drain.

The structure of the typical surfactant, cocamido DEA,

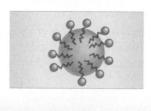

(nonpolar region)   (polar region)

is shown above. Note the combination of polar and nonpolar regions in the same molecule:

The −OH groups interact with water; the carbon–hydrogen region (termed *hydrocarbon*) dissolves in the food residue.

The existence of polar and nonpolar sites within the same molecule is certainly the most important feature of a molecule used in a washing formulation. However, it is not the only desirable feature. The molecule should be nontoxic, inexpensive, readily soluble in water, and have a minimal adverse effect on the environment. It should be biodegradable, enabling bacteria, naturally occurring in water, to attack the bonds, destroying the molecules before they can damage aquatic ecosystems. Organic chemists (who study compounds containing mainly carbon and hydrogen atoms) are very adept at synthesizing molecules that optimize desirable properties while minimizing undesirable properties.

Thus, we see that carefully designed surfactants enable us to use clean, attractive, and sanitary utensils to prepare and serve our meals while minimizing the negative impact of our wastewater on the aquatic environment.

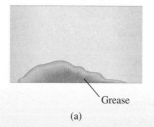
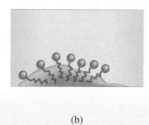
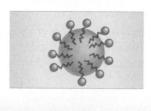

Grease

(a)   (b)   (c)

A surfactant at work. (a) Food residue (nonpolar) is not soluble in water. (b) Nonpolar ends of surfactant molecules dissolve in the food residue. (c) The food residue is removed from the surface and is solubilized in water.

Concentration of Electrolytes in Solution

Recall (Sections 3.3 and 6.1) that an electrolyte is a material that dissolves in water to produce an *electrolytic solution* that conducts an electrical current. The behavior of many biological systems is critically dependent on the concentration of electrolytic solutions that are a part of these systems.

Molarity (mol/L) is the most commonly used concentration unit; however, another useful concentration unit is based on the *equivalent*. The **equivalent** (eq) is the number of moles of an ion corresponding to Avogadro's number of electrical charges. The resulting concentration unit is eq/L.

When discussing solutions of ionic compounds, molarity emphasizes the number of individual ions. A 1 molar solution of Na^+ contains Avogadro's number, 6.022×10^{23}, of Na^+ per L. In contrast, eq/L emphasize charge; 1 eq of Na^+ contains Avogadro's number of positive charges.

The concentration of any ion in aqueous solution, in units of eq/L, is equal to the product of the equivalents/mol of the ion and the molar concentration of the ion:

$$eq/L = \left(\frac{eq}{mol\ ion}\right)\left(\frac{mol\ ion}{L}\right)$$

Since

$$\frac{mol\ ion}{L} = M$$

$$eq/L = \left(\frac{eq}{mol\ ion}\right)(M)$$

The eq/mol of the ion is simply the number of charges on the ion, regardless of whether that charge is positive or negative. For example:

For Na^+, eq/mol is equal to 1
  Cl^-, eq/mol is equal to 1
  Mg^{2+}, eq/mol is equal to 2
  CO_3^{2-}, eq/mol is equal to 2

and so forth.

The units milliequivalents/liter (meq/L) are often used when describing small amounts or a low concentration of ions. These units are routinely used when describing ions in blood, urine, and blood plasma.

We can convert between eq/L and meq/L using conversion factors based on the identity:

$$1\ eq = 10^3\ meq$$

LEARNING GOAL

13 Interconvert molar concentrations of ions and milliequivalents/liter.

EXAMPLE 6.15 | Calculating Ion Concentration

For a solution that is $5.0 \times 10^{-3}\ M$ in phosphate ion (PO_4^{3-})
  a. Calculate the concentration of phosphate ion in units of eq/L.
  b. Calculate the concentration of phosphate ion in units of meq/L.

LEARNING GOAL

13 Interconvert molar concentration of ions and milliequivalents/liter.

Solution

For part a,

For PO_4^{3-}, the eq/mol is equal to 3. When the molarity concentration of the phosphate ion is expressed in mol/L, unit cancellation leads to the answer in eq/L. Substituting in the equation:

$$eq/L = \left(\frac{eq}{mol\ ion}\right)(M)$$

$$= \left(\frac{3\ eq}{mol\ PO_4^{3-}}\right)\left(\frac{5.0 \times 10^{-3}\ mol\ PO_4^{3-}}{L}\right)$$

$$= 1.5 \times 10^{-2}\ eq/L$$

Continued…

Helpful Hint: The eq of charge in this equation is always positive even if the ion is negative. It represents the number of eq of charge, not the sign of the charge.

For part b,

The conversion factor is based on the relationship $1 \text{ eq} = 10^3 \text{ meq}$:

$$\text{meq/L} = 1.5 \times 10^{-2} \frac{\cancel{\text{eq}}}{\text{L}} \times \frac{10^3 \text{ meq}}{\cancel{\text{eq}}} = 15 \text{ meq/L}$$

Practice Problem 6.15

Calculate the number of eq/L of carbonate ion, CO_3^{2-}, in a solution that is $6.4 \times 10^{-4} \, M$ carbonate ion.

▶ For Further Practice: **Questions 6.109 and 6.110.**

EXAMPLE 6.16	**Calculating Electrolyte Concentrations**

LEARNING GOAL

A typical concentration of calcium ion in blood plasma is 4 meq/L. Represent this concentration in mol/L.

13 Interconvert molar concentration of ions and milliequivalents/liter.

Solution

Step 1. The data are given in meq/L, and the desired result is in mol/L (M). Therefore, meq/L must first be converted to eq/L using the conversion factor based on $1 \text{ eq} = 10^3 \text{ meq}$.

$$\frac{4 \text{ meq } \cancel{Ca^{2+}}}{1 \text{ L}} \times \frac{1 \text{ eq } Ca^{2+}}{10^3 \text{ meq } \cancel{Ca^{2+}}} = 4 \times 10^{-3} \frac{\text{eq } Ca^{2+}}{\text{L}}$$

Step 2. Our relationship between eq/L and mol/L (M) is:

$$\frac{\text{eq}}{\text{L}} = \left(\frac{\text{eq}}{\text{mol ion}} \right) (M)$$

Step 3. Rearranging and solving for M,

$$M = \left(\frac{\text{eq}}{\text{L}} \right) \left(\frac{\text{mol ion}}{\text{eq}} \right)$$

Step 4. The calcium ion has a 2^+ charge (recall that calcium is in Group IIA of the periodic table; hence, there is a 2^+ charge on the calcium ion). Substituting,

$$M = \left(\frac{4 \times 10^{-3} \text{ eq } \cancel{Ca^{2+}}}{\text{L}} \right) \left(\frac{1 \text{ mol } Ca^{2+}}{2 \text{ eq } \cancel{Ca^{2+}}} \right)$$

$$= \frac{2 \times 10^{-3} \text{ mol } Ca^{2+}}{\text{L}}$$

Practice Problem 6.16

Sodium chloride [0.9% (m/V)] is a solution administered intravenously to replace fluid loss. It is frequently used to avoid dehydration. The sodium ion concentration is 15.4 meq/L. Calculate the sodium ion concentration in mol/L.

▶ For Further Practice: **Questions 6.113 and 6.114.**

Question 6.13 A typical concentration of chloride ion in blood plasma is 110 meq/L. Represent this concentration in mol/L.

Question 6.14 A typical concentration of magnesium ion in certain types of intravenous solutions is 3 meq/L. Represent this concentration in mol/L.

Biological Effects of Electrolytes in Solution

The concentrations of cations, anions, and other substances in biological fluids are critical to health. Consequently, the osmolarity of body fluids is carefully regulated by the kidneys.

The two most important cations in body fluids are Na^+ and K^+. The sodium ion is the most abundant cation in the blood and intercellular fluids, whereas the potassium ion is the most abundant intracellular cation. In blood and intercellular fluid, the Na^+ concentration is 135 meq/L and the K^+ concentration is 3.5–5.0 meq/L. Inside the cell, the situation is reversed. The K^+ concentration is 125 meq/L and the Na^+ concentration is 10 meq/L.

If osmosis and simple diffusion were the only mechanisms for transporting water and ions across cell membranes, these concentration differences would not occur. One positive ion would be just as good as any other. However, the situation is more complex than this. Large protein molecules embedded in cell membranes actively pump sodium ions to the outside of the cell and potassium ions into the cell. This is termed *active transport* because cellular energy must be expended to transport those ions. Proper cell function in the regulation of muscles and the nervous system depends on the sodium ion/potassium ion ratio inside and outside of the cell.

If the Na^+ concentration in the blood becomes too low, urine output decreases, the mouth feels dry, the skin becomes flushed, and a fever may develop. The blood level of Na^+ may be elevated when large amounts of water are lost. Diabetes, certain high-protein diets, and diarrhea may cause elevated blood Na^+ level. In extreme cases, elevated Na^+ levels may cause confusion, stupor, or coma.

Concentrations of K^+ in the blood may rise to dangerously high levels following any injury that causes large numbers of cells to rupture, releasing their intracellular K^+. This may lead to death by heart failure. Similarly, very low levels of K^+ in the blood may also cause death from heart failure. This may occur following prolonged exercise that results in excessive sweating. When this happens, both body fluids and electrolytes must be replaced. Salt tablets containing both NaCl and KCl taken with water, and drinks such as Gatorade, effectively provide water and electrolytes and prevent serious symptoms.

The cationic charge in blood is neutralized by two major anions, Cl^- and HCO_3^-. The chloride ion plays a role in acid-base balance, maintenance of osmotic pressure within an acceptable range, and oxygen transport by hemoglobin. The bicarbonate anion is the form in which most waste CO_2 is carried in the blood.

A variety of proteins are also found in the blood. Because of their larger size, they exist in colloidal dispersion. These proteins include blood-clotting factors, immunoglobulins (antibodies) that help us fight infection, and albumins that act as carriers of nonpolar, hydrophobic substances (fatty acids and steroid hormones) that cannot dissolve in water.

Additionally, blood is the medium for exchange of nutrients and waste products. Nutrients, such as the polar sugar glucose, enter the blood from the intestine or the liver. Because glucose molecules are polar, they dissolve in body fluids and are circulated to tissues throughout the body. As noted above, nonpolar nutrients are transported with the help of carrier proteins. Similarly, nitrogen-containing waste products, such as urea, are passed from cells to the blood. They are continuously and efficiently removed from the blood by the kidneys.

In cases of loss of kidney function, mechanical devices—dialysis machines—mimic the action of the kidney. The process of blood dialysis—hemodialysis—is discussed in A Medical Perspective: Hemodialysis.

LEARNING GOAL

14 Explain the role of electrolytes in blood and their relationship to the process of dialysis.

A Medical Perspective

Hemodialysis

As we have seen in Section 6.5, blood is the medium for exchange of both nutrients and waste products. The membranes of the kidneys remove waste materials such as urea and uric acid (Chapter 22), excess salts, and large quantities of water. This process of waste removal is termed **dialysis,** a process similar in function to osmosis (Section 6.4). Semipermeable membranes in the kidneys, dialyzing membranes, allow small molecules (principally water and urea) and ions in solution to pass through and ultimately collect in the bladder. From there they can be eliminated from the body.

Unfortunately, a variety of diseases can cause partial or complete kidney failure. Should the kidneys fail to perform their primary function, dialysis of waste products, urea and other waste products rapidly increase in concentration in the blood. This can become a life-threatening situation in a very short time.

The most effective treatment of kidney failure is the use of a machine, an artificial kidney, that mimics the function of the kidney. The artificial kidney removes waste from the blood using the process of hemodialysis (blood dialysis). The blood is pumped through a long semipermeable membrane, the dialysis membrane. The dialysis process is similar to osmosis. However, in addition to water molecules, larger molecules (including the waste products in the blood) and ions can pass across the membrane from the blood into a dialyzing fluid. The dialyzing fluid is isotonic with normal blood; it also is similar in its concentration of all other essential blood components. The waste materials move across the dialysis membrane (from a higher to a lower concentration, as in diffusion). A successful dialysis procedure selectively removes the waste from the body without upsetting the critical electrolyte balance in the blood.

Hemodialysis, although lifesaving, is not by any means a pleasant experience. The patient's water intake must be severely limited to minimize the number of times each week that treatment must be used. Many dialysis patients require two or three treatments per week, and each session may

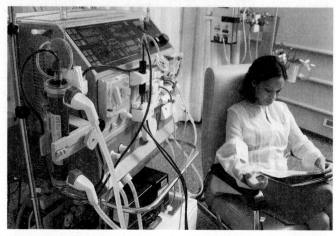

Dialysis patient

require one-half (or more) day of hospitalization, especially when the patient suffers from complicating conditions such as diabetes.

Improvements in technology, as well as the growth and sophistication of our health care delivery systems over the past several years, have made dialysis treatment much more patient friendly. Dialysis centers, specializing in the treatment of kidney patients, are now found in most major population centers. Smaller, more automated dialysis units are available for home use, under the supervision of a nurse. With the remarkable progress in kidney transplant success, dialysis is becoming, more and more, a temporary solution, sustaining life until a suitable kidney donor match can be found.

For Further Understanding

▶ In what way is dialysis similar to osmosis?
▶ How does dialysis differ from osmosis?

CHAPTER MAP

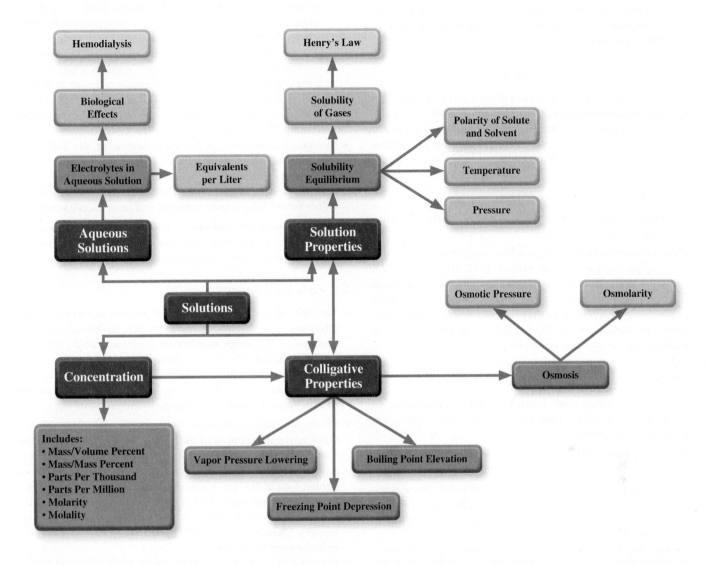

SUMMARY

6.1 Properties of Solutions

▶ A **true solution** is a homogeneous mixture of two or more substances. A solution is composed of one or more **solutes,** dissolved in a **solvent.** When the solvent is water, the solution is called an **aqueous solution.**

▶ Liquid solutions are clear and transparent with no visible particles of solute. They may be colored or colorless, depending on the properties of the solute and solvent.

▶ In solutions of **electrolytes,** the solutes are ionic compounds that dissociate in solution to produce ions. They are good conductors of electricity. Solutions of nonelectrolytes are formed from nondissociating molecular solutes (nonelectrolytes), and their solutions do not conduct electricity.

▶ The rule "like dissolves like" is the fundamental condition for **solubility.** Polar solutes are soluble in polar solvents, and nonpolar solutes are soluble in nonpolar solvents.

▶ The degree of solubility depends on the difference between the polarity of solute and solvent, the temperature, and the pressure. Pressure considerations are significant only for solutions of gases.

▶ When a solution contains all the solute that can be dissolved at a particular temperature, it is **saturated.** Excess solute falls to the bottom of the container as a **precipitate.** Occasionally, on cooling, the excess solute may remain in solution for a time before precipitation. Such a solution is a **supersaturated solution.**

▶ When excess solute, the precipitate, contacts solvent, the dissolution process reaches a state of dynamic equilibrium.

▶ **Colloidal dispersions** have particle sizes between those of true solutions and precipitates. A **suspension** is a heterogeneous mixture that contains particles much larger than a colloidal dispersion. Over time, these particles may settle, forming a second phase.

▶ **Henry's law** describes the solubility of gases in liquids. At a given temperature, the solubility of a gas is proportional to the partial pressure of the gas.

6.2 Concentration Based on Mass

▶ The amount of solute dissolved in a given amount of solution is the solution **concentration.** The more widely used percentage-based concentration units are **mass/volume percent** and **mass/mass percent. Parts per thousand** (ppt) and **parts per million** (ppm) are used with very dilute solutions.

6.3 Concentration Based on Moles

▶ **Molarity,** symbolized M, is defined as the number of moles of solute per liter of solution.

▶ Dilution is often used to prepare less concentrated solutions. The expression for this calculation is $(M_1)(V_1) = (M_2)(V_2)$. Knowing any three of these terms enables us to calculate the fourth. The concentration of solute may be represented as mol/L (molarity) or any other suitable concentration units. However, both concentrations must be in the same units when using the dilution equation.

6.4 Concentration-Dependent Solution Properties

▶ Solution properties that depend on the concentration of solute particles, rather than the identity of the solute, are **colligative properties.**

▶ There are four colligative properties of solutions, all of which depend on the concentration of *particles* in solution:

- *Vapor pressure lowering. Raoult's law* states that when a nonvolatile solute is added to a solvent, the vapor pressure of the solvent decreases in proportion to the concentration of the solute.

- *Freezing point depression* and *boiling point elevation.* When a nonvolatile solid is added to a solvent, the freezing point of the resulting solution decreases, and the boiling point increases. The magnitudes of both the freezing point depression (ΔT_f) and the boiling point elevation (ΔT_b) are proportional to the solute particle concentration. Molality is commonly used in calculations involving colligative properties, because molality is temperature independent. **Molality** (m) of a solution is the number of mol of solute per kg of solvent.

- **Osmosis and osmotic pressure.** The cell membrane mediates the interaction of the cell with its environment and is responsible for the controlled passage of material into and out of the cell. One of the principal means of transport is termed *diffusion.* **Diffusion** is the net movement of solute or solvent molecules from an area of high concentration to an area of low concentration. This region where the concentration decreases over a distance is termed the **concentration gradient.** Because of the structure of the cell membrane, only small molecules are able to diffuse freely across this barrier. Large molecules and highly

charged ions are restricted by the barrier. In other words, the cell membrane is behaving in a selective fashion. Such membranes are termed *selectively permeable membranes.* Osmosis is the movement of solvent (water) from a dilute solution to a more concentrated solution through a **semipermeable membrane.** The pressure that must be applied to the more concentrated solution to stop this flow is the **osmotic pressure.** The osmotic pressure, like the pressure exerted by a gas, may be treated quantitatively by using an equation similar in form to the ideal gas equation: $\pi = iMRT$. By convention, the molarity of particles used for osmotic pressure calculations is termed *osmolarity (mol particles/L).*

▶ In biological systems, if the concentration of the fluid surrounding red blood cells is higher than that inside the cell (a **hypertonic solution**), water flows from the cell, causing it to collapse. Too low a concentration of this fluid relative to the solution within the cell (a **hypotonic solution**) will cause the cell to rupture.

▶ Two solutions are **isotonic** if they have identical osmotic pressures. In that way, the osmotic pressure differential across the cell is zero, and no cell disruption occurs.

6.5 Aqueous Solutions

▶ The role of water in the solution process deserves special attention. It is often referred to as the "universal solvent" because of the large number of ionic and polar covalent compounds that are at least partially soluble in water. It is the principal biological solvent. These characteristics are a direct consequence of the molecular geometry and structure of water and its ability to undergo hydrogen bonding.

▶ When discussing solutions of ionic compounds, molarity emphasizes the number of individual ions. A $1\,M$ solution of Na^+ contains Avogadro's number of sodium ions. In contrast, equivalents per liter (eq/L) emphasizes charge; a solution containing 1 eq of Na^+ per L contains Avogadro's number of positive charge. One equivalent of an ion is the number of moles of the ion corresponding to Avogadro's number of electrical charges. Changing from mol/L to eq/L (or the reverse) is done using conversion factors.

▶ Concentration of charge in these electrolytic solutions is even more important than the molar concentration of the ions that are responsible for the charge. Consequently, we often use concentration units (eq/L and meq/L) that emphasize charge rather than number of ions in solution.

▶ The concentrations of cations, anions, and other substances in biological fluids are critical to health. As a result, the osmolarity of body fluids is carefully regulated by the kidney using the process of dialysis.

ANSWERS TO PRACTICE PROBLEMS

6.1 **a.** 16.7% (m/V) NaCl
   b. 7.50% (m/V) KCl
   c. 2.56×10^{-2}% (m/V) oxygen
   d. 1.05×10^{-2}% (m/V) argon

6.2 **a.** 20.0 g NaOH
b. 40.0 mL

6.3 **a.** 20.0% (m/m) oxygen
b. 38.5% (m/m) argon

6.4 **a.** 2.00×10^2 ppt oxygen gas
and
2.00×10^5 ppm oxygen gas
b. 3.85×10^2 ppt argon gas
and
3.85×10^5 ppm argon gas

6.5 0.30 M

6.6 $1.25 \times 10^{-2} M$

6.7 0.250 L

6.8 1.00×10^{-1} L (or 1.00×10^2 mL) of 0.200 M sugar solution

6.9 To prepare the solution, dilute 1.7×10^{-2} L of 12 M HCl with sufficient water to produce 1.0×10^2 mL of total solution.

6.10 Boiling point = 100.8°C
Freezing point = −2.8°C

6.11 Boiling point = 101.6°C
Freezing point = −5.6°C

6.12 **a.** 1.0×10^{-2} mol particles/L
b. 5.0×10^{-3} mol particles/L

6.13 **a.** 0.24 atm
b. 0.12 atm

6.14 Ammonia is a polar substance, as is water. The rule "like dissolves like" predicts that ammonia would be water soluble. Methane, a nonpolar substance, would not be soluble in water.

6.15

$$\frac{1.3 \times 10^{-3} \text{ eq CO}_3{}^{2-}}{\text{L}}$$

6.16

$$1.54 \times 10^{-2} \frac{\text{mol Na}^+}{\text{L}}$$

6.17 Which of the following solute(s) would form an electrolytic solution in water? Explain your reasoning.
a. $NaNO_3$
b. $C_6H_{12}O_6$
c. $FeCl_3$

6.18 Which of the following solute(s) would form an electrolytic solution in water? Explain your reasoning.
a. HCl
b. Na_2SO_4
c. Ethanol (CH_3CH_2OH)

6.19 Distinguish among the terms true solution, colloidal dispersion, and suspension.

6.20 Describe how you would distinguish experimentally between a colloidal dispersion and a suspension.

6.21 What is the difference between saturated and supersaturated solutions?

6.22 What happens when additional solute is added to a saturated solution that is being heated?

6.23 Intravenous (IV) therapy often involves colloidal dispersions. Albumin is an example. Describe some of the different properties expected for a colloidal dispersion of albumin versus a saline solution of NaCl.

6.24 An isotope of technetium, mixed with sulfur and colloidally dispersed in water, is frequently used in diagnosing various medical conditions because it is readily taken up by various tissues prior to excretion. Explain why this important mixture is not a true solution.

6.25 Is CCl_4 more likely to form a solution in water or benzene (C_6H_6)? Explain your reasoning.

6.26 Is CH_3OH more likely to form a solution in water or benzene (C_6H_6)? Explain your reasoning.

Applications

6.27 Fly fishermen in the Northeast have known for a long time that their chances of catching trout are much greater in early spring than in mid-August. Suggest an explanation based on solubility trends.

6.28 Fish kills (the sudden death of thousands of fish) often occur during periods of prolonged elevated temperatures. Pollution is often, but not always, the cause. Suggest another reason, based on solubility trends.

6.29 The Henry's law constant, k, for O_2 in aqueous solution is 1.3×10^{-3} mol/(L · atm) at 25°C. When a diver is at a depth of 240 m, the pressure is approximately 25 atm. Calculate the equilibrium solubility of O_2 at this depth (25°C) in units of mol/L.

6.30 The Henry's law constant, k, for N_2 in aqueous solution is 6.1×10^{-4} mol/(L · atm) at 25°C. When a diver is at a depth of 240 m, the pressure is approximately 25 atm. Calculate the equilibrium solubility of N_2 at this depth (25°C) in units of mol/L.

QUESTIONS AND PROBLEMS

Properties of Solutions

Foundations

6.15 Can a solution be both clear and red? Explain.

6.16 Two liters of liquid A are mixed with two liters of liquid B. The resulting volume is only 3.95 L. Explain what happened on the molecular level.

Concentration Based on Mass

Foundations

6.31 Calculate the composition of each of the following solutions in mass/volume %:
a. 33.0 g sugar, $C_6H_{12}O_6$, in 5.00×10^2 mL solution
b. 20.0 g NaCl in 1.00 L solution

6.32 Calculate the composition of each of the following solutions in mass/volume %:
   a. 0.700 g KCl in 1.00 mL solution
   b. 95.2 g MgCl$_2$ in 0.250 L solution

6.33 Calculate the composition of each of the following solutions in mass/volume %:
   a. 50.0 g ethanol dissolved in 5.00 × 10^2 mL solution
   b. 50.0 g ethanol dissolved in 1.00 L solution

6.34 Calculate the composition of each of the following solutions in mass/volume %:
   a. 20.0 g benzene dissolved in 1.00 × 10^2 mL solution
   b. 20.0 g acetic acid dissolved in 2.50 L solution

6.35 Calculate the composition of each of the following solutions in mass/mass %:
   a. 21.0 g NaCl in 1.00 × 10^2 g solution
   b. 21.0 g NaCl in 5.00 × 10^2 mL solution (d = 1.12 g/mL)

6.36 Calculate the composition of each of the following solutions in mass/mass %:
   a. 1.00 g KCl in 1.00 × 10^2 g solution
   b. 50.0 g KCl in 5.00 × 10^2 mL solution (d = 1.14 g/mL)

Applications

6.37 A solution was prepared by dissolving 14.6 g of KNO$_3$ in sufficient water to produce 75.0 mL of solution. What is the mass/volume % of this solution?

6.38 A solution was prepared by dissolving 12.4 g of NaNO$_3$ in sufficient water to produce 95.0 mL of solution. What is the mass/volume % of this solution?

6.39 How many g of sugar would you use to prepare 100 mL of a 1.00 mass/volume % solution?

6.40 How many mL of 4.0 mass/volume % Mg(NO$_3$)$_2$ solution would contain 1.2 g of magnesium nitrate?

6.41 How many g of solute are needed to prepare each of the following solutions?
   a. 2.50 × 10^2 g of 0.900% (m/m) NaCl
   b. 2.50 × 10^2 g of 1.25% (m/m) NaC$_2$H$_3$O$_2$ (sodium acetate)

6.42 How many g of solute are needed to prepare each of the following solutions?
   a. 2.50 × 10^2 g of 5.00% (m/m) NH$_4$Cl (ammonium chloride)
   b. 2.50 × 10^2 g of 3.50% (m/m) Na$_2$CO$_3$

6.43 A solution contains 1.0 mg of Cu^{2+} per 0.50 kg solution. Calculate the concentration in ppt.

6.44 A solution contains 1.0 mg of Cu^{2+} per 0.50 kg solution. Calculate the concentration in ppm.

6.45 Which solution is more concentrated: a 0.04% (m/m) solution or a 50 ppm solution?

6.46 Which solution is more concentrated: a 20 ppt solution or a 200 ppm solution?

Concentration Based on Moles

Foundations

6.47 Calculate the molarity of 5.0 L of solution containing 2.5 mol HNO$_3$.

6.48 Calculate the molarity of 2.75 L of solution containing 1.35 × 10^{-2} mol HCl.

6.49 Calculate the molarity of a solution that contains 2.25 mol of NaNO$_3$ dissolved in 2.50 L.

6.50 Calculate the molarity of a solution that contains 1.75 mol of KNO$_3$ dissolved in 3.00 L.

6.51 Why is it often necessary to dilute solutions in the laboratory?

6.52 Write the dilution expression and define each term.

Applications

6.53 Calculate the number of g of solute that would be needed to make each of the following solutions:
   a. 2.50 × 10^2 mL of 0.100 M NaCl
   b. 2.50 × 10^2 mL of 0.200 M C$_6$H$_{12}$O$_6$ (glucose)

6.54 Calculate the number of g of solute that would be needed to make each of the following solutions:
   a. 2.50 × 10^2 mL of 0.100 M NaBr
   b. 2.50 × 10^2 mL of 0.200 M KOH

6.55 How many g of glucose (C$_6$H$_{12}$O$_6$) are present in 1.75 L of a 0.500 M solution?

6.56 How many g of sodium hydroxide are present in 675 mL of a 0.500 M solution?

6.57 Calculate the volume of a 0.500 M sucrose solution (table sugar, C$_{12}$H$_{22}$O$_{11}$) containing 0.133 mol of solute.

6.58 Calculate the volume of a 1.00 × 10^{-2} M KOH solution containing 3.00 × 10^{-1} mol of solute.

6.59 It is desired to prepare 0.500 L of a 0.100 M solution of NaCl from a 1.00 M stock solution. How many mL of the stock solution must be taken for the dilution?

6.60 A 50.0-mL sample of a 0.250 M sucrose solution was diluted to 5.00 × 10^2 mL. What is the molar concentration of the resulting solution?

6.61 A 50.0-mL portion of a stock solution was diluted to 500.0 mL. If the resulting solution was 2.00 M, what was the molarity of the original stock solution?

6.62 A 6.00-mL portion of an 8.00 M stock solution is to be diluted to 0.400 M. What will be the final volume after dilution?

6.63 A 50.0-mL sample of 0.500 M NaOH was diluted to 500.0 mL. What is the new molarity?

6.64 A 300.0-mL portion of H$_2$O is added to 300.0 mL of 0.250 M H$_2$SO$_4$. What is the new molarity?

Concentration-Dependent Solution Properties

Foundations

6.65 What is meant by the term *colligative property?*

6.66 Name and describe four colligative solution properties.

6.67 Explain, in terms of solution properties, why salt is used to melt ice in the winter.

6.68 Explain, in terms of solution properties, why a wilted plant regains its "health" when watered.

6.69 State Raoult's law.

6.70 What is the major importance of Raoult's law?

6.71 Why does 1 mol of CaCl$_2$ lower the freezing point of water more than 1 mol of NaCl?

6.72 Using salt to try to melt ice on a day when the temperature is −20°C will be unsuccessful. Why?

6.73 **a.** Calculate the freezing temperature of 1.50 *m* urea, N$_2$H$_4$CO. Urea is a covalent compound.
   b. Calculate the freezing temperature of 1.50 *m* LiBr, an ionic compound.

6.74 **a.** Calculate the boiling temperature of 1.50 *m* urea, N$_2$H$_4$CO. Urea is a covalent compound.
   b. Calculate the boiling temperature of 1.50 *m* LiBr, an ionic compound.

Applications

Answer Questions 6.75–6.78 by comparing two solutions: 0.50 *M* sodium chloride (an ionic compound) and 0.50 *M* sucrose (a covalent compound).

6.75 Calculate the freezing temperature of each solution. Assume that the molality of each solution is 0.50 m. (The molar and molal concentrations of dilute aqueous solutions are often identical, to two significant figures.)

6.76 Calculate the boiling temperature of each solution. Assume that the molality of each solution is 0.50 m. (The molar and molal concentrations of dilute aqueous solutions are often identical, to two significant figures.)

6.77 Which solution has the higher vapor pressure?

6.78 Each solution is separated from water by a semipermeable membrane. Which solution has the higher osmotic pressure?

Answer Questions 6.79–6.82 based on the following scenario: Two solutions, A and B, are separated by a semipermeable membrane. For each case, predict whether there will be a net flow of water in one direction and, if so, which direction.

6.79 A is pure water and B is 5% glucose.

6.80 A is 0.10 M glucose and B is 0.10 M KCl.

6.81 A is 0.10 M NaCl and B is 0.10 M KCl.

6.82 A is 0.10 M NaCl and B is 0.20 M glucose.

6.83 Determine the osmolarity of 5.0×10^{-4} M KNO_3 (electrolyte).

6.84 Determine the osmolarity of 2.5×10^{-4} M $C_6H_{12}O_6$ (nonelectrolyte).

6.85 Calculate the osmotic pressure of 0.50 M KNO_3 (electrolyte).

6.86 Calculate the osmotic pressure of 0.50 M $C_6H_{12}O_6$ (nonelectrolyte).

In Questions 6.87–6.90, label each solution as isotonic, hypotonic, or hypertonic in comparison to 0.9% (m/V) NaCl (0.15 M NaCl).

6.87 0.15 M $CaCl_2$

6.88 0.35 M glucose

6.89 0.15 M glucose

6.90 3% (m/V) NaCl

Aqueous Solutions

Foundations

6.91 What properties make water such a useful solvent?

6.92 Sketch the "interactive network" of water molecules in the liquid state.

6.93 Sketch the interaction of water with an ammonia molecule.

6.94 Sketch the interaction of water with an ethanol, CH_3CH_2OH, molecule.

6.95 Why is it important to distinguish between electrolytes and nonelectrolytes when discussing colligative properties?

6.96 Name the two most important cations in biological fluids.

6.97 Explain why a dialysis solution must have a low sodium ion concentration if it is designed to remove excess sodium ions from the blood.

6.98 Explain why a dialysis solution must have an elevated potassium ion concentration when loss of potassium ions from the blood is a concern.

Applications

6.99 Solutions of ammonia in water are sold as window cleaner. Why do these solutions have a long "shelf life"?

6.100 Why does water's abnormally high boiling point help to make it a desirable solvent?

6.101 Sketch the interaction of a water molecule with a sodium ion.

6.102 Sketch the interaction of a water molecule with a chloride ion.

6.103 What type of solute dissolves readily in water?

6.104 What type of solute dissolves readily in benzene (C_6H_6)?

6.105 Describe the clinical effects of elevated concentrations of sodium ions in the blood.

6.106 Describe the clinical effects of depressed concentrations of potassium ions in the blood.

6.107 Describe conditions that can lead to elevated concentrations of sodium in the blood.

6.108 Describe conditions that can lead to dangerously low concentrations of potassium in the blood.

6.109 Calculate the number of eq/L of Ca^{2+} in a solution that is 5.0×10^{-2} M in Ca^{2+}.

6.110 Calculate the number of eq/L of SO_4^{2-} in a solution that is 2.5×10^{-3} M in SO_4^{2-}.

6.111 A physiological saline solution is labeled in part "154 meq/L of Na^+ and 154 meq/L of Cl^-."
  a. Calculate the number of mol of Na^+ in 1.00 L of solution.
  b. Calculate the number of mol of Cl^- in 1.00 L of solution.

6.112 A physiological solution designed to replace a patient's lost K^+ is 40 meq/L in K^+ and 40 meq/L in Cl^-.
  a. Calculate the number of mol of K^+ in 1.00L of solution.
  b. Calculate the number of mol of Cl^- in 1.00L of solution.

6.113 A potassium chloride solution that also contains 5% (m/V) dextrose is administered intravenously to treat some forms of malnutrition. The potassium ion concentration in this solution is 40 meq/L. Calculate the potassium ion concentration in mol/L.

6.114 If the potassium ion concentration in the solution described in Question 6.113 was only 35 meq/L, calculate the potassium ion concentration in units of mol/L.

CHALLENGE PROBLEMS

1. Which of the following compounds would cause the greater freezing point depression, per mol, in H_2O: $C_6H_{12}O_6$ (glucose) or NaCl?

2. Which of the following compounds would cause the greater boiling point elevation, per mol, in H_2O: $MgCl_2$ or $HOCH_2CH_2OH$ (ethylene glycol, antifreeze)?
   (Hint: $HOCH_2CH_2OH$ is covalent.)

3. Analytical chemists often take advantage of differences in solubility to separate ions. For example, adding Cl^- to a solution of Cu^{2+} and Ag^+ causes AgCl to precipitate; Cu^{2+} remains in solution. Filtering the solution results in a separation. Design a scheme to separate the cations Ca^{2+} and Pb^{2+}.

4. Using the strategy outlined in the above problem, design a scheme to separate the anions S^{2-} and CO_3^{2-}.

5. Design an experiment that would enable you to measure the degree of solubility of a salt such as KI in water.

6. How could you experimentally distinguish between a saturated solution and a supersaturated solution?

7. Blood is essentially an aqueous solution, but it must transport a variety of nonpolar substances (hormones, for example). Colloidal proteins, termed *albumins*, facilitate this transport. Must these albumins be polar or nonpolar? Why?

7

Energy, Rate, and Equilibrium

GENERAL CHEMISTRY

Describe the relationship between windmills and the topics discussed in this chapter.

OUTLINE

LEARNING GOALS

1. Correlate the terms *endothermic* and *exothermic* with heat flow between a *system* and its *surroundings*.
2. Explain what is meant by *enthalpy*, *entropy*, and *free energy* and demonstrate their implications.
3. Describe experiments that yield thermochemical information, and use experimental data to calculate the quantity of energy involved in reactions.
4. Describe the concept of reaction rate and the role of kinetics in chemical and physical change.
5. Describe the importance of *activation energy* and the *activated complex* in determining reaction rate.
6. Predict the way reactant structure, concentration, temperature, and catalysis affect the rate of a chemical reaction.
7. Write rate laws, and use these equations to calculate the effect of concentration on rate.
8. Recognize and describe equilibrium situations.
9. Write equilibrium constant expressions, and use these expressions to calculate equilibrium constants or equilibrium concentrations.
10. Use LeChatelier's principle to predict changes in equilibrium position.

INTRODUCTION

In Chapter 4, we calculated quantities of matter involved in chemical change assuming that the reactions went to *completion*; that is, at least one reactant was *completely* used up and only product(s) remained at the end of the reaction. Often, this is not the case. There are many reactions, including processes occurring in living systems, that form product(s) but also have significant quantities of reactants still remaining. Additionally, we should be aware that all chemical reactions do not occur instantaneously. Reactions may occur rapidly (an explosion is one example) or over a period of minutes, hours, or even days. Some reactions are so slow that we may not observe their completion in our lifetime (corrosion is a common example).

Two concepts play important roles in determining the extent and speed of a chemical reaction: (1) thermodynamics, which deals with energy changes in chemical reactions, and (2) kinetics, which describes the rate or speed of a chemical reaction.

Although both thermodynamics and kinetics involve energy, they are two separate considerations. The laws of thermodynamics may predict that a reaction will occur, but the process may not be observed because the reaction is so slow; conversely, a reaction may be very fast because it is kinetically favorable yet produce very little product because it is thermodynamically unfavorable.

In this chapter, we investigate the fundamentals of thermodynamics and kinetics, with an emphasis on the critical role that energy changes play in chemical reactions. We consider physical change and chemical change, including the conversions that take place among the states of matter (solid, liquid, and gas). We use these concepts to explain the behavior of reactions that do not go to completion. These are termed *equilibrium reactions*. We develop the equilibrium constant expression and use LeChatelier's principle to demonstrate how equilibrium composition can be altered.

The Deepwater Horizon explosion was the result of a reaction between methane gas and oxygen gas.

7.1 Thermodynamics

Thermodynamics is the study of energy, work, and heat. It may be applied to chemical change, such as the calculation of the quantity of heat obtainable from the combustion of 1 gallon (gal) of fuel oil. Similarly, energy released or consumed in physical change, such as the boiling or freezing of water, may be determined.

There are three basic laws of thermodynamics:

1. Energy cannot be created or destroyed, only converted from one form to another.
2. The universe spontaneously tends toward increasing disorder or randomness.
3. The disorder of a pure, perfect crystal at absolute zero (0 Kelvin) is zero.

Only the first two laws of thermodynamics will be of concern here. They help us to understand why some chemical reactions may occur spontaneously and others do not. Whether we are synthesizing compounds in the laboratory, manufacturing industrial chemicals, or trying to determine causes of cancer, the ability to predict spontaneity is essential.

The Chemical Reaction and Energy

John Dalton believed that chemical change involved joining, separating, or rearranging atoms. Two centuries later, this statement stands as an accurate description of chemical reactions. However, we now know much more about the energy changes that are an essential part of every reaction.

The kinetic molecular theory of gases was introduced in Section 5.1. The basic ideas of this theory will be useful to remember throughout the discussion

Energy, its various forms, and commonly used energy units were introduced in Chapter 1.

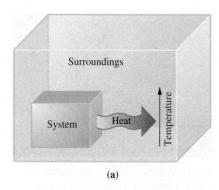

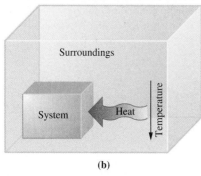

Figure 7.1 Illustration of heat flow in (a) exothermic and (b) endothermic reactions.

1 Correlate the terms *endothermic* and *exothermic* with heat flow between a *system* and its *surroundings*.

of thermodynamics and kinetics. Keep in mind that molecules and atoms are in constant, random motion and frequently collide with each other. Remember that the average kinetic energy of the atoms or molecules increases with increasing temperature. In addition to these statements of the kinetic molecular theory, we will add the following concepts as they pertain to chemical reactions:

- some collisions, those with sufficient energy, will break bonds in molecules, and
- when reactant bonds are broken, new bonds may be formed and products result.

It is worth noting that we cannot measure an absolute value for energy stored in a chemical system. We can only measure the *change* in energy (energy absorbed or released) as a chemical reaction occurs. Also, it is often both convenient and necessary to establish a boundary between the *system* and its *surroundings*.

The **system** contains the process under study. The **surroundings** encompass the rest of the universe. Our universe is composed of the system and its surroundings. Energy is lost from the system to the surroundings or energy is gained by the system at the expense of the surroundings. This energy change, most often in the form of heat, may be experimentally determined because the temperature of the system or surroundings will change, and this change can be measured. This process is illustrated in Figure 7.1. *Heat flow* is a term that describes the transfer of heat when the direction of transfer is specified. For example, we often use the term *heat flow* to describe the transfer of thermal energy from a hot object to a cold one.

Consider the combustion of methane; that is, the reaction of methane and oxygen to form carbon dioxide and water. If we define the reaction as the system, the temperature of the air surrounding the reaction increases, indicating that some of the potential energy (energy stored in the bonds) is converted to kinetic energy, causing the molecules surrounding the reaction to speed up. This type of kinetic energy is called *thermal energy*. The transfer of thermal energy to the surroundings is known as **heat** and is sometimes called *heat flow*.

Now, an exact temperature measurement of the air before and after the reaction is difficult. However, if we could insulate a portion of the surroundings, to isolate and trap the heat, we could calculate a useful quantity, the heat of the reaction. Experimental strategies for measuring temperature change and calculating heats of reactions, termed *calorimetry*, are discussed in Section 7.2.

The First Law of Thermodynamics

Exothermic and Endothermic Reactions

The first law of thermodynamics states that the energy of the universe is constant. It is the law of conservation of energy. The study of energy changes that occur in chemical reactions is a very practical application of the first law. Consider, for example, the generalized reaction:

$$A{-}B + C{-}D \longrightarrow A{-}D + C{-}B$$

An **exothermic reaction** releases energy to the surroundings. The surroundings become warmer.

Each chemical bond is stored chemical energy (potential energy). For the reaction to take place, bond $A{-}B$ and bond $C{-}D$ must break; this process *always* requires energy. At the same time, bonds $A{-}D$ and $C{-}B$ must form; this process always releases energy.

If the energy required to break the $A{-}B$ and $C{-}D$ bonds is *less* than the energy given off when the $A{-}D$ and $C{-}B$ bonds form, the reaction will release the excess energy. The energy is a *product,* and the reaction is called an exothermic (Greek *exo,* "out," and Greek *therm,* "heat") reaction.

An example of an exothermic reaction is the combustion of methane, represented by a *thermochemical equation*, a chemical equation that also shows energy as a product or reactant.

$$CH_4(g) + 2O_2(g) \longrightarrow CO_2(g) + 2H_2O(g) + 211 \text{ kilocalories (kcal)}$$
Exothermic reaction

> In an exothermic reaction, heat is released *from the system* to the surroundings. In an endothermic reaction, heat is absorbed *by the system* from the surroundings.

This thermochemical equation reads: the combustion of 1 mole (mol) of methane releases 211 kcal of heat.

An **endothermic reaction** absorbs energy from the surroundings. The surroundings become colder.

If the energy required to break the A—B and C—D bonds is *greater* than the energy released when the A—D and C—B bonds form, the reaction will need an external supply of energy. Insufficient energy is available in the system to initiate the bond-breaking process. Such a reaction is called an endothermic (Greek *endo*, "within," and Greek *therm*, "heat") reaction, and energy is a *reactant*.

> Bond breaking is an endothermic process. Bond formation is an exothermic process.

The decomposition of ammonia into nitrogen and hydrogen is one example of an endothermic reaction:

$$22 \text{ kcal} + 2NH_3(g) \longrightarrow N_2(g) + 3H_2(g)$$
Endothermic reaction

This thermochemical equation reads: the decomposition of 2 mol of ammonia requires 22 kcal of heat.

The examples used here show the energy absorbed or released as heat energy. Depending on the reaction and the conditions under which the reaction is run, the energy may also take the form of light energy or electrical energy. A firefly releases energy as a soft glow of light on a summer evening. An electrical current results from a chemical reaction in a battery, enabling your car to start.

EXAMPLE 7.1 | **Determining Whether a Process Is Exothermic or Endothermic**

An ice cube is dropped into a glass of water at room temperature. The ice cube melts. Is the melting of the ice exothermic or endothermic?

LEARNING GOAL

1 Correlate the terms *endothermic* and *exothermic* with heat flow between a *system* and its *surroundings*.

Solution

Step 1. Consider the ice cube to be the system and the water, the surroundings.
Step 2. Recognize that the water (the surroundings) will decrease in temperature. The surroundings are transferring heat to the system, the ice cube. The ice cube uses the heat to overcome hydrogen bonding among molecules, and the ice melts.
Step 3. The heat flow is from surroundings to system.
Step 4. The system gains energy; hence, the melting process (physical change) is endothermic.

Practice Problem 7.1

Are the following processes exothermic or endothermic?

   a. Fuel oil is burned in a furnace.
   b. When solid NaOH is dissolved in water, the solution temperature increases.

▶ For Further Practice: **Questions 7.23 and 7.24.**

LEARNING GOAL

2 Explain what is meant by *enthalpy*, *entropy*, and *free energy* and demonstrate their implications.

Enthalpy of Reactions

A chemical reaction may involve the breaking and forming of many bonds. Our interest is often focused on the total amount of heat released or absorbed by the overall reaction.

Green Chemistry

Twenty-First Century Energy

When we purchase gasoline for our automobiles or oil for the furnace, we are certainly buying matter. But that matter is only a storage device; we are really purchasing the energy stored in the chemical bonds. Combustion, burning in oxygen, releases the stored energy (potential energy) in a form suited to its function: mechanical energy to power a vehicle or heat energy to warm a home.

Energy release is a consequence of change. In fuel combustion, this change results in the production of waste products that may be detrimental to our environment. This necessitates the expenditure of time, money, and *more* energy to clean up our surroundings.

If we are paying a considerable price for our energy supply, it would be nice to believe that we are at least getting full value for our expenditure. Even that is not the case. Removal of energy from molecules also extracts a price. For example, a properly tuned automobile engine is perhaps 30% efficient. That means that less than one-third of the available energy actually moves the car. The other two-thirds is released into the atmosphere as wasted energy, mostly heat energy. The law of conservation of energy tells us that the energy is not destroyed, but it is certainly no longer available to us in a useful form.

Can we build a 100%-efficient energy transfer system? Is there such a thing as cost-free energy? The answer to these questions is no, on both counts. It is theoretically impossible, and the laws of thermodynamics tell us why this is so.

Although 100% energy conversion is not possible, there is still room for improvement in our current energy usage. Consuming vast quantities of nonrenewable resources (fossil fuels: coal, oil, and natural gas) and expelling the waste products (carbon dioxide and sulfur oxides) into the atmosphere may not be the best energy generation system that human ingenuity can devise.

The process of generating electricity is simply finding ways to move electrons. Converting water to steam, then using the steam pressure to drive a turbine (a mechanical device to "push" electrons) requires massive quantities of heat to boil the water and produce steam. Nuclear reactions produce heat, which is the basis of a nuclear power plant. Combustion of coal, oil, and natural gas (exothermic reactions) also produces heat. So we see that the essential difference among the various types of power plants lies in the identity and technological complexity of the heat source.

Newer approaches to moving electrons involve wind energy and solar energy. It has been estimated that twenty-first century,

Harvesting energy directly from the sun and converting it to electricity is a safe, nonpolluting strategy to, at least partially, satisfy our growing energy needs.

high-tech windmills erected in unpopulated areas (the shallow waters of the Atlantic Ocean and the plains of the western portion of the United States), coupled with a modern electrical transmission grid, could satisfy the electrical demands of virtually the entire country. Solar energy, collected by photovoltaic cells arranged in solar panels, is converted to electrical energy.

Despite the tremendous progress in the design and development of new technologies, many problems remain. It will take decades to move away from nonrenewable energy sources. Nuclear energy has serious safety questions, raised by the disasters at Three Mile Island, Chernobyl, and Fukushima. Solar energy systems suffer from low efficiencies and cloudy days; they require storage devices that increase the cost of building and maintaining these systems. Wind turbines are perceived by some as unattractive and are most effective in areas with strong prevailing winds. We must remember that electric automobiles replace gasoline with electricity and the electricity (or at least a portion of it) comes from the combustion of fossil fuels.

However, improvement in the design and efficiency of these alternate energy sources holds the promise of reducing dependence on polluting, nonrenewable resources.

For Further Understanding

▶ Solar and wind power have been mentioned as energy alternatives. Suggest a third alternative.

▶ What do you see as potential advantages (and potential shortcomings) of the alternative you have suggested?

Enthalpy is the term used to represent heat and is symbolized as H. The *change in enthalpy* is the energy difference between the products and reactants of a chemical reaction and is symbolized as ΔH. By convention, energy released is represented with a negative sign (indicating an exothermic reaction), and energy absorbed is shown with a positive sign (indicating an endothermic reaction). The change in enthalpy is represented by the relationship:

$$\Delta H_{reaction} = H_{products} - H_{reactants}$$

- If $H_{reactants} > H_{products}$, ΔH must be negative and the reaction is exothermic.
- If $H_{reactants} < H_{products}$, ΔH must be positive and the reaction is endothermic.

For the combustion of methane, an exothermic process, energy is a *product* in the thermochemical equation:

$$CH_4(g) + 2O_2(g) \longrightarrow CO_2(g) + 2H_2O(g) + \boxed{211 \text{ kcal}}$$

and

$$\Delta H = -211 \text{ kcal}$$

For the decomposition of ammonia, an endothermic process, energy is a *reactant* in the thermochemical equation:

$$\boxed{22 \text{ kcal}} + 2NH_3(g) \longrightarrow N_2(g) + 3H_2(g)$$

and

$$\Delta H = +22 \text{ kcal}$$

Diagrams representing changes in enthalpy for exothermic (a) and endothermic (b) reactions are shown in Figure 7.2.

Question 7.1   Are the following processes exothermic or endothermic?
a. $C_6H_{12}O_6(s) \longrightarrow 2C_2H_5OH(l) + 2CO_2(g)$, $\Delta H = -16$ kcal
b. $N_2O_5(g) + H_2O(l) \longrightarrow 2HNO_3(l) + 18.3$ kcal

Question 7.2   Are the following processes exothermic or endothermic?
a. $S(s) + O_2(g) \longrightarrow SO_2(g)$, $\Delta H = -71$ kcal
b. $N_2(g) + 2O_2(g) + 16.2$ kcal $\longrightarrow 2NO_2(g)$

Spontaneous and Nonspontaneous Reactions

It seems that all exothermic reactions should be spontaneous. After all, an external supply of energy does not appear to be necessary; in fact, energy is a product of the reaction. It also seems that all endothermic reactions should be nonspontaneous; energy is a reactant that we must provide. However, these hypotheses are not supported by experimentation.

Experimental measurement has shown that most *but not all* exothermic reactions are spontaneous. Likewise, most *but not all*, endothermic reactions are not spontaneous. There must be some factor in addition to enthalpy that will help us to explain the less obvious cases of nonspontaneous exothermic reactions and spontaneous endothermic reactions. This other factor is entropy.

The Second Law of Thermodynamics

The first law of thermodynamics considers the enthalpy of chemical reactions. The second law states that the universe spontaneously tends toward increasing disorder or randomness.

Entropy

A measure of the randomness of a chemical system is its **entropy.** The entropy of a substance is represented by the symbol S. A random, or disordered, system is characterized by *high entropy*; a well-organized system has *low entropy*.

What do we mean by disorder in chemical systems? Disorder is simply the absence of a regular repeating pattern. Disorder or randomness increases as we convert from the solid to the liquid to the gaseous state. As we have seen, solids often have an ordered crystalline structure, liquids have, at best, a loose arrangement, and gas particles are virtually random in their distribution. Therefore, gases have high entropy, and crystalline solids have very low entropy. Figures 7.3 and 7.4 illustrate properties of entropy in systems.

In these discussions, we consider the enthalpy change and energy change to be identical. This is true for most common reactions carried out in a lab, with minimal volume change.

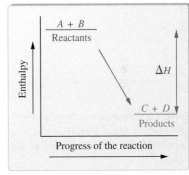

(a)

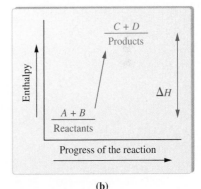

(b)

Figure 7.2 (a) An exothermic reaction. ΔH represents the energy released during the progress of the exothermic reaction: $A + B \longrightarrow C + D + \Delta H$. (b) An endothermic reaction. ΔH represents the energy absorbed during the progress of the endothermic reaction: $\Delta H + A + B \longrightarrow C + D$.

LEARNING GOAL

2 Explain what is meant by *enthalpy*, *entropy*, and *free energy* and demonstrate their implications.

Figure 7.3 (a) Gas particles, trapped in the left chamber, spontaneously diffuse into the right chamber, initially under vacuum, when the valve is opened. (b) It is unimaginable that the gas particles will rearrange themselves and reverse the process to create a vacuum. This can only be accomplished using a pump; that is, by doing work on the system.

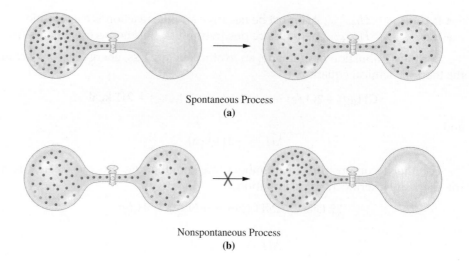

Spontaneous Process
(a)

Nonspontaneous Process
(b)

The second law describes the entire universe or any isolated system within the universe. On a more personal level, we all fall victim to the law of increasing disorder. Chaos in our room or workplace is certainly not our intent! It happens almost effortlessly. However, reversal of this process requires work and energy. The same is true at the molecular level. The gradual deterioration of our cities'

Figure 7.4 Processes such as (a) melting, (b) vaporization, and (c) dissolution increase entropy, or randomness, of the particles.

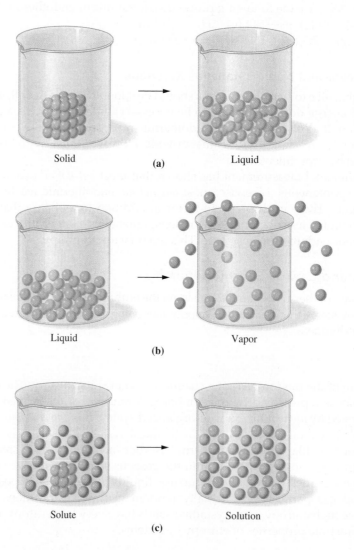

Solid   **(a)**   Liquid

Liquid   Vapor
(b)

Solute   Solution
(c)

infrastructure (roads, bridges, water mains, and so forth) is an all-too-familiar example. Millions of dollars (translated into energy and work) are needed annually just to try to maintain the status quo.

The entropy of a reaction is measured as a difference, ΔS, between the entropies, S, of products and reactants.

The drive toward increased entropy, along with a tendency to achieve a lower potential energy, is responsible for spontaneous chemical reactions. Reactions that are exothermic and whose products are more disordered (higher in entropy) will occur spontaneously, whereas endothermic reactions producing products of lower entropy will not be spontaneous. If they are to take place at all, they will need some energy input.

Question 7.3 Which substance has the greatest entropy, $He(g)$ or $Na(s)$? Explain your reasoning.

Question 7.4 Which substance has the greatest entropy, $H_2O(l)$ or $H_2O(g)$? Explain your reasoning.

Free Energy

The two situations just described are clear-cut and unambiguous. In any other situation, the reaction may or may not be spontaneous. It depends on the relative size of the enthalpy and entropy values.

Free energy, symbolized by ΔG, represents the combined contribution of the enthalpy *and* entropy values for a chemical reaction. Thus, free energy is the ultimate predictor of reaction spontaneity and is expressed as

$$\Delta G = \Delta H - T\Delta S$$

ΔH represents the change in enthalpy between products and reactants, ΔS represents the change in entropy between products and reactants, and T is the Kelvin temperature of the reaction.

- A reaction with a negative value of ΔG will *always* be spontaneous.
- A reaction with a positive ΔG will *always* be nonspontaneous.

We need to know both ΔH and ΔS in order to predict the sign of ΔG and make a definitive statement regarding the spontaneity of the reaction. Additionally, the temperature may determine the direction of spontaneity. Consider the four possible situations:

- ΔH positive and ΔS negative: ΔG is always positive, regardless of the temperature. The reaction is always nonspontaneous.
- ΔH negative and ΔS positive: ΔG is always negative, regardless of the temperature. The reaction is always spontaneous.
- Both ΔH and ΔS positive: The sign of ΔG depends on the temperature.
- Both ΔH and ΔS negative: The sign of ΔG depends on the temperature.

LEARNING GOAL

2 Explain what is meant by *enthalpy*, *entropy*, and *free energy* and demonstrate their implications.

EXAMPLE 7.2 | **Determining the Sign of ΔG**

Determine the sign of ΔG for the following reaction. Is the reaction spontaneous, nonspontaneous, or temperature dependent?

$$42.5 \text{ kcal} + CaCO_3(s) \longrightarrow CaO(s) + CO_2(g)$$

LEARNING GOAL

2 Explain what is meant by *enthalpy*, *entropy*, and *free energy* and demonstrate their implications.

Solution

Step 1. Determine the sign of ΔH. Because heat is added as a reactant, the reaction is endothermic. This gives a positive sign for ΔH.

Continued...

A Medical Perspective

Hot and Cold Packs

Hot packs provide "instant warmth" for hikers and skiers and are used in treatment of injuries such as pulled muscles. Cold packs are in common use today for the treatment of injuries and the reduction of swelling. These useful items are an excellent example of basic science producing a technologically useful product.

Both hot and cold packs depend on large energy changes taking place during a chemical reaction. Cold packs rely on an endothermic reaction, and hot packs generate heat energy from an exothermic reaction.

A cold pack is fabricated as two separate compartments within a single package. One compartment contains NH_4NO_3, and the other contains water. When the package is squeezed, the inner boundary between the two compartments ruptures, allowing the components to mix, and the following reaction occurs:

$$6.7 \text{ kcal} + NH_4NO_3(s) \longrightarrow NH_4^+(aq) + NO_3^-(aq)$$

This reaction is endothermic; heat taken from the surroundings produces the cooling effect.

The design of a hot pack is similar. Here, finely divided iron powder is mixed with oxygen. Production of iron oxide results in the creation of heat:

$$4Fe(s) + 3O_2(g) \longrightarrow 2Fe_2O_3(s) + 198 \text{ kcal}$$

This reaction occurs via an oxidation-reduction mechanism (see Chapter 4). The iron atoms are oxidized, and O_2 is reduced. Electrons are transferred from the iron atoms to O_2, and Fe_2O_3

A hot pack

forms exothermically. The rate of the reaction is slow; therefore, the heat is liberated gradually over a period of several hours.

For Further Understanding

▶ What is the sign of ΔH for each equation in this story?
▶ Predict whether entropy increases or decreases for each equation in this story.

Step 2. Determine the sign of ΔS. This reaction starts with only solid and produces some gas. Therefore, the reaction begins with a small amount of disorder (the solid) and produces a product, $CO_2(g)$, that has a random distribution of molecules. Thus, there is an increase in the disorder and ΔS is positive.

Step 3. Consider the equation $\Delta G = \Delta H - T\Delta S$. Since ΔH and ΔS are both positive, ΔG is temperature dependent. The reaction is only spontaneous if the $T\Delta S$ term becomes so large that when it is subtracted from ΔH, the value for ΔG will be negative. Therefore, this reaction is temperature dependent and is spontaneous at high temperature.

Practice Problem 7.2

Determine the sign of ΔG for the following reaction. Is the reaction spontaneous, nonspontaneous, or temperature dependent?

$$2H_2(g) + O_2(g) \longrightarrow 2H_2O(l) + 137 \text{ kcal}$$

▶ For Further Practice: **Questions 7.33 and 7.34.**

The introduction to this chapter stated that thermodynamics is used to predict whether or not a reaction will occur. The value of ΔG determines if a reaction is thermodynamically favorable or unfavorable.

Question 7.5 Predict whether a reaction with positive ΔH and negative ΔS will be spontaneous, nonspontaneous, or temperature dependent. Explain your reasoning.

Question 7.6 Predict whether a reaction with positive ΔH and positive ΔS will be spontaneous, nonspontaneous, or temperature dependent. Explain your reasoning.

7.2 Experimental Determination of Energy Change in Reactions

The measurement of heat energy changes in a chemical reaction is **calorimetry.** This technique involves the measurement of the change in the temperature of a quantity of water or solution that is in contact with the reaction of interest and isolated from the surroundings. A device used for these measurements is a *calorimeter,* which measures heat changes in calories (cal) or joules (J).

Think of a calorimeter as a self-contained "universe" where heat can exchange between the system and its surroundings but cannot escape the calorimeter. At the same time, external heat is prevented from entering the calorimeter.

A Styrofoam coffee cup is a simple design for a calorimeter, and it produces surprisingly accurate results. It is a good insulator, and, when filled with solution, it can be used to measure temperature changes taking place as the result of a chemical reaction occurring in that solution (Figure 7.5). The change in the temperature of the solution, caused by the reaction, can be used to calculate the gain or loss of heat energy for the reaction.

For an exothermic reaction, heat released by the reaction is absorbed by the surrounding solution. For an endothermic reaction, the reactants absorb heat from the solution.

The **specific heat** of a substance is defined as the number of calories of heat needed to raise the temperature of 1 gram of the substance 1 degree Celsius (°C). Knowing the specific heat of the water or the aqueous solution along with the total number of g of solution and the temperature increase (measured as the difference between the final and initial temperatures of the solution) enables the experimenter to calculate the heat released during the reaction.

The solution behaves as a "trap" or "sink" for energy released in an exothermic process. The temperature increase indicates a gain in heat energy. Endothermic reactions, on the other hand, take heat energy away from the solution, lowering its temperature.

The quantity of heat absorbed or released by the reaction (Q) is the product of the mass of solution in the calorimeter (m_s), the specific heat of the solution (SH_s), and the change in temperature (ΔT_s) of the solution as the reaction proceeds from the initial to final state.

The heat is calculated by using the following equation:

$$Q = m_s \times \Delta T_s \times SH_s$$

with units

$$\mathrm{cal} = \cancel{g} \times \cancel{°C} \times \frac{\mathrm{cal}}{\cancel{g} \cdot \cancel{°C}}$$

The details of the experimental approach are illustrated in Examples 7.3 and 7.4.

LEARNING GOAL

3 Describe experiments that yield thermochemical information, and use experimental data to calculate the quantity of energy involved in reactions.

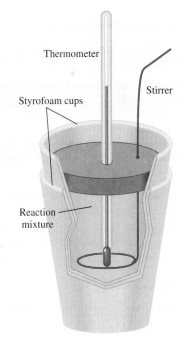

Figure 7.5 A "coffee cup" calorimeter used for the measurement of heat change in chemical reactions. The concentric Styrofoam cups insulate the solution. Heat released by an exothermic chemical reaction (the system) enters the solution (the surroundings), raising its temperature, which is measured by using a thermometer.

EXAMPLE 7.3 | **Calculating Energy Involved in Calorimeter Reactions**

LEARNING GOAL

3 Describe experiments that yield thermochemical information, and use experimental data to calculate the quantity of energy involved in reactions.

When 0.050 mol of hydrochloric acid is mixed with 0.050 mol of sodium hydroxide in a "coffee cup" calorimeter, the temperature of 1.00×10^2 g of the resulting solution increases from 25.0°C to 31.5°C. If the specific heat of the solution is 1.00 cal/g · °C, calculate the quantity of energy involved in the reaction. Is the reaction endothermic or exothermic?

Solution

Step 1. The calorimetry expression is:

$$Q = m_s \times \Delta T_s \times SH_s$$

The reaction is the system; the solution is the surroundings.

Step 2. The change in temperature is

$$\Delta T_s = T_{s\,final} - T_{s\,initial}$$

$$= 31.5°C - 25.0°C = 6.5°C$$

The temperature of the solution has increased by 6.5°C; therefore, the reaction is exothermic.

Step 3. Substituting these values into the calorimetry expression:

$$Q = 1.00 \times 10^2 \; \cancel{\text{g solution}} \times 6.5 \; \cancel{°C} \times \frac{1.00 \; cal}{\cancel{\text{g solution}} \cdot \cancel{°C}}$$

$$= 6.5 \times 10^2 \, cal$$

6.5×10^2 cal (or 0.65 kcal) of heat energy were released by this acid-base reaction to the surroundings, the solution; the reaction is exothermic.

Practice Problem 7.3

Calculate the temperature change that would have been observed if 50.0 g of solution were in the calorimeter instead of 1.00×10^2 g of solution.

▶ For Further Practice: **Question 7.43.**

EXAMPLE 7.4 | **Calculating Energy Involved in Calorimeter Reactions**

LEARNING GOAL

3 Describe experiments that yield thermochemical information, and use experimental data to calculate the quantity of energy involved in reactions.

When 0.10 mol of ammonium chloride (NH_4Cl) is dissolved in water producing 1.00×10^2 g solution, the water temperature decreases from 25.0°C to 18.0°C. The specific heat of the resulting solution is 1.00 cal/g · °C. Calculate the quantity of energy involved in the process. Is the dissolution of ammonium chloride endothermic or exothermic?

Solution

Step 1. The calorimetry expression is:

$$Q = m_s \times \Delta T_s \times SH_s$$

The reaction is the system; the solution is the surroundings.

Step 2. The change in temperature is

$$\Delta T = T_{s\,final} - T_{s\,initial}$$

$$= 18.0°C - 25.0°C = -7.0°C$$

The temperature of the solution has decreased by 7.0°C; therefore, the dissolution is endothermic.

Step 3. Substituting these values into the calorimetry expression:

$$Q = 1.00 \times 10^2 \text{ g solution} \times (-7.0°C) \times \frac{1.00 \text{ cal}}{\text{g solution} \cdot °C}$$

$$= -7.0 \times 10^2 \text{ cal}$$

7.0×10^2 cal (or 0.70 kcal) of heat energy were absorbed by the dissolution process because the solution lost 7.0×10^2 cal of heat energy to the system. The dissolution of ammonium chloride is endothermic.

Practice Problem 7.4

Calculate the temperature change that would have been observed if 1.00×10^2 g of another liquid, producing a solution with a specific heat of 0.800 cal/g · °C, were substituted for the water in the calorimeter.

▶ For Further Practice: **Question 7.44.**

Question 7.7 Using the conversion factor in Chapter 1, convert the energy released in Example 7.3 to joules (J).

Question 7.8 Using the conversion factor in Chapter 1, convert the energy absorbed in Example 7.4 to joules (J).

Note: Refer to A Human Perspective: Food Calories, Section 1.6.

Many chemical reactions that produce heat are combustion reactions. In our bodies, many food substances (carbohydrates, proteins, and fats; Chapters 21–23) are oxidized to release energy. **Fuel value** is the amount of energy per g of food.

The fuel value of food is an important concept in nutrition science. The fuel value is generally reported in units of *nutritional Calories*. One **nutritional Calorie (Cal)** is equivalent to one kilocalorie (1000 cal). It is also known as the *large Calorie* (uppercase C).

Energy necessary for our daily activity and bodily function comes largely from the reaction of oxygen with carbohydrates. Chemical energy from foods that is not used to maintain normal body temperature or in muscular activity is stored in the bonds of chemical compounds known collectively as fat. Thus, consumption of "high-calorie" foods is implicated in the problem of obesity.

A special type of calorimeter, a *bomb calorimeter*, is useful for the measurement of the fuel value (Cal) of foods. Such a device is illustrated in Figure 7.6. Its design

LEARNING GOAL

3 Describe experiments that yield thermochemical information, and use experimental data to calculate the quantity of energy involved in reactions.

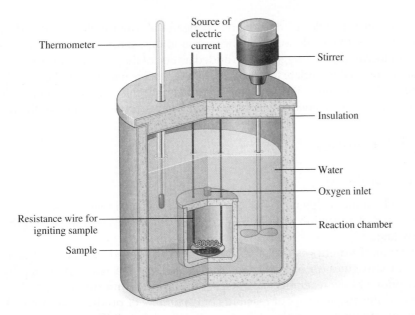

Thermometer
Source of electric current
Stirrer
Insulation
Water
Oxygen inlet
Resistance wire for igniting sample
Reaction chamber
Sample

Figure 7.6 A bomb calorimeter that may be used to measure heat released upon combustion of a sample. This device is commonly used to determine the fuel value of foods. The bomb calorimeter is similar to the "coffee cup" calorimeter. However, note the electrical component necessary to initiate the combustion reaction.

EXAMPLE 7.5   **Calculating the Fuel Value of Foods**

A 1-g sample of glucose (a common sugar or carbohydrate) was burned in a bomb calorimeter. The temperature of 1.00×10^3 g H_2O was raised from 25.0°C to 28.8°C ($\Delta T_s = 3.8$°C). Calculate the fuel value of glucose.

3 Describe experiments that yield thermochemical information, and use experimental data to calculate the quantity of energy involved in reactions.

Solution

Step 1. Recall that the fuel value is the number of nutritional Cal liberated by the combustion of 1 g of material, and 1 g of material was burned in the calorimeter.

Step 2. Then, the fuel value may be equated with the calorimetry expression:

$$\text{Fuel value} = Q = m_s \times \Delta T_s \times SH_s$$

Step 3. Water is the surroundings in the calorimeter; it has a specific heat capacity equal to 1.00 cal/g $H_2O \cdot$ °C. Substituting the values provided in the problem into our expression for fuel value:

$$\text{Fuel value} = \text{g } H_2O \times °C \times \frac{1.00 \text{ cal}}{\text{g } H_2O \cdot °C}$$

$$= 1.00 \times 10^3 \text{ g } H_2O \times 3.8 °C \times \frac{1.00 \text{ cal}}{\text{g } H_2O \cdot °C}$$

$$= 3.8 \times 10^3 \text{ cal}$$

Step 4. Converting from cal to nutritional Cal:

$$3.8 \times 10^3 \text{ cal} \times \frac{1 \text{ nutritional Cal}}{10^3 \text{ cal}} = 3.8 \text{ Cal (nutritional Calories, or kcal)}$$

Since the quantity of glucose burned in the calorimeter is 1 g, the fuel value of glucose is 3.8 kcal/g.

Practice Problem 7.5

A 1.0-g sample of a candy bar (which contains lots of sugar and fats!) was burned in a bomb calorimeter. A 3.0°C temperature increase was observed for 1.00×10^3 g of water. The entire candy bar weighed 2.5 ounces (oz). Calculate the fuel value (in Cal) of the sample and the total caloric content of the candy bar.

▶ For Further Practice: **Questions 7.45 and 7.46.**

is similar, in principle, to that of the "coffee cup" calorimeter discussed earlier. It incorporates the insulation from the surroundings, solution pool, reaction chamber, and thermometer. Oxygen gas is added as one of the reactants, and an electrical igniter is inserted to initiate the reaction. However, it is not open to the atmosphere. In the sealed container, the reaction continues until the sample is completely oxidized. All of the heat energy released during the reaction is captured in the water.

4 Describe the concept of reaction rate and the role of kinetics in chemical and physical change.

7.3   Kinetics

Chemical Kinetics

Thermodynamics helps us to decide whether a chemical reaction is spontaneous. Knowing that a reaction can occur spontaneously tells us nothing about the time it may take.

Chemical **kinetics** is the study of the **rate** (or speed) of chemical reactions. Kinetics also gives an indication of the *mechanism* of a reaction, a step-by-step description of how reactants become products. Kinetic information may be represented as the *disappearance* of reactants or *appearance* of product over time. A typical graph of number of molecules versus time is shown in Figure 7.7.

Information about the rate at which various chemical processes occur is useful. For example, what is the "shelf life" of processed foods? When will slow changes in composition make food unappealing or even unsafe? Many drugs lose their potency with time because the active ingredient decomposes into other substances. The rate of hardening of dental filling material (via a chemical reaction) influences the dentist's technique. Our very lives depend on the efficient transport of oxygen to each of our cells and the rapid use of the oxygen for energy-harvesting reactions.

The diagram in Figure 7.8 is a useful way of depicting the kinetics of a reaction at the molecular level.

Often a color change, over time, can be measured. Such changes are useful in assessing the rate of a chemical reaction (Figure 7.9).

Let's see what actually happens when two chemical compounds react and what experimental conditions affect the rate of a reaction.

Activation Energy and the Activated Complex

Consider the exothermic reaction we discussed in Section 7.1:

$$CH_4(g) + 2O_2(g) \longrightarrow CO_2(g) + 2H_2O(g) + 211 \text{ kcal}$$

For the reaction to proceed, four C—H and two O=O bonds must be broken, and two C=O and four H—O bonds must be formed. Sufficient energy must be available to cause the bonds to break if the reaction is to take place. This energy is provided by the collision of molecules. If sufficient energy is available at the temperature of the reaction, one or more bonds will break, and the atoms will recombine in a lower energy arrangement, in this case as carbon dioxide and water. A collision producing product molecules is termed an *effective collision*. An effective collision requires sufficient energy and, in the case of complex molecules, the proper orientation of reacting molecules. Only effective collisions lead to chemical reaction.

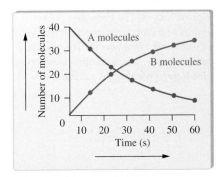

Figure 7.7 For a hypothetical reaction A $\longrightarrow$ B the number of A molecules (reactant molecules) decreases over time and B molecules (product molecules) increase in number over time.

LEARNING GOAL

5 Describe the importance of *activation energy* and the *activated complex* in determining reaction rate.

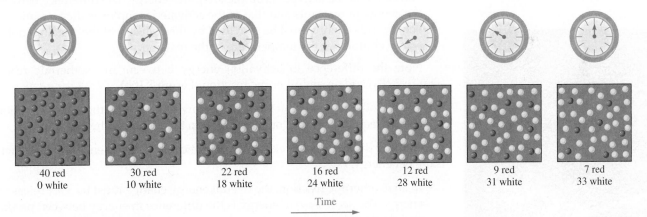

40 red
0 white

30 red
10 white

22 red
18 white

16 red
24 white

12 red
28 white

9 red
31 white

7 red
33 white

Time

Figure 7.8 An alternate way of representing the information contained in Figure 7.7.

Time

Figure 7.9 The conversion of reddish brown Br_2 in solution to colorless Br^- over time. Figure 7.8 represents this reaction, A $\longrightarrow$ B, on a molecular level.

Figure 7.10 (a) The change in potential energy as a function of reaction time for an exothermic reaction. (b) The change in potential energy as a function of reaction time for an endothermic reaction.

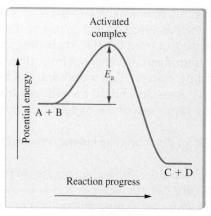

(a) Exothermic reaction

(b) Endothermic reaction

The minimum amount of energy required to initiate a chemical reaction is called the **activation energy,** E_a, for the reaction.

We can picture the chemical reaction in terms of the changes in potential energy that occur during the reaction. Figure 7.10a graphically shows these changes for an exothermic reaction, and Figure 7.10b shows these changes for an endothermic reaction. Important characteristics of these graphs include the following:

- The reaction proceeds from reactants to products through an extremely unstable state that we call the **activated complex.** The activated complex cannot be isolated from the reaction mixture but may be thought of as a short-lived group of atoms structured in such a way that it quickly and easily breaks apart into the products of the reaction.
- Formation of the activated complex requires energy. The difference between the energy of reactants and that of the activated complex is the activation energy. This energy must be provided by the collision of the reacting molecules or atoms at the temperature of the reaction.

Note the difference in activation energy between an exothermic reaction (Figure 7.10a) and an endothermic reaction (Figure 7.10b).

- The endothermic reaction takes place slowly because of the large activation energy required for the conversion of reactants into products.

Another difference between an exothermic reaction and endothermic reaction is the net energy.

- In an exothermic reaction, the overall energy change must be a *net* release of energy. The *net* release of energy is the difference in energy between products and reactants.

Question 7.9   The act of striking a match illustrates the role of activation energy in a chemical reaction. Explain.

Question 7.10   Distinguish between the terms *net energy* and *activation energy*.

Factors that Affect Reaction Rate

The following six factors influence reaction rate:

• Structure of the Reacting Species

Reactions among ions in solution are usually very rapid. Ionic compounds in solution are dissociated; consequently, their bonds are already broken, and the activation energy for their reaction should be very low. On the other hand,

How does a match illustrate the concept of activation energy?

reactions involving covalently bonded compounds may proceed more slowly. Covalent bonds must be broken and new bonds formed. The activation energy for this process would be significantly higher than that for the reaction of free ions. Bond strengths certainly play a role in determining reaction rates because the magnitude of the activation energy, or energy barrier, is related to bond strength.

• Molecular Shape and Orientation

The size and shape of reactant molecules influence the rate of the reaction. Large molecules, containing bulky groups of atoms, may block the reactive part of the molecule from interacting with another reactive substance, causing the reaction to proceed slowly. Only molecular collisions that have the correct collision orientation, as well as sufficient energy, lead to product formation. These collisions are termed *effective collisions*.

• The Concentration of Reactants

The rate of a chemical reaction is often a complex function of the concentration of one or more of the reacting substances. The rate will generally *increase* as concentration *increases* simply because a higher concentration means more reactant molecules in a given volume and therefore a greater number of collisions per unit time. If we assume that other variables are held constant, a larger number of collisions leads to a larger number of effective collisions. For example, the rate at which a fire burns depends on the concentration of oxygen in the atmosphere surrounding the fire, as well as the concentration of the fuel (perhaps methane or propane). A common fire-fighting strategy is the use of fire extinguishers filled with carbon dioxide. The carbon dioxide dilutes the oxygen to a level where the combustion process can no longer be sustained.

• The Temperature of Reactants

The rate of a reaction increases as the temperature increases, because the average kinetic energy of the reacting particles is directly proportional to the Kelvin temperature. Increasing the speed of particles increases the likelihood of collision, and the higher kinetic energy means that a higher percentage of these collisions will result in product formation (effective collisions). A 10°C rise in temperature has often been found to double the reaction rate.

• The Physical State of Reactants

The rate of a reaction depends on the physical state of the reactants: solid, liquid, or gas. For a reaction to occur, the reactants must collide frequently and have sufficient energy to react. In the solid state, the atoms, ions, or molecules are restricted in their motion. In the gaseous and liquid states, the particles have both free motion and proximity to each other. Hence, reactions tend to be fastest in the gaseous and liquid states and slowest in the solid state.

• The Presence of a Catalyst

A **catalyst** is a substance that *increases* the reaction rate. If added to a reaction mixture, the catalytic substance undergoes no net change, nor does it alter the outcome of the reaction. However, the catalyst interacts with the reactants to create an alternative pathway for production of products. This alternative path has a lower activation energy. This makes it easier for the reaction to take place and thus increases the rate. This effect is illustrated in Figure 7.11.

Catalysis is important industrially; it may often make the difference between profit and loss in the sale of a product. For example, catalysis is useful in converting double bonds to single bonds. An important application of this principle involves the process of hydrogenation. Hydrogenation converts one or more of

Concentration is introduced in Section 1.6, and units and calculations are discussed in Sections 6.2 and 6.3.

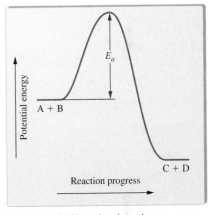

(a) Uncatalyzed reaction

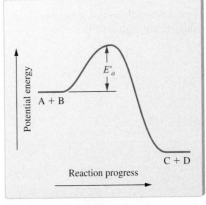

(b) Catalyzed reaction

Figure 7.11 The effect of a catalyst on the magnitude of the activation energy of a chemical reaction: (a) uncatalyzed reaction, (b) catalyzed reaction. Note that the presence of a catalyst decreases the activation energy ($E'_a < E_a$), thus increasing the rate of the reaction.

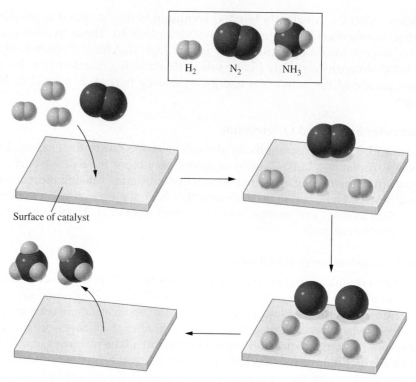

Figure 7.12 The synthesis of ammonia, an important industrial product, is facilitated by a solid-phase catalyst (the Haber process). H_2 and N_2 bind to the surface, their bonds are weakened, dissociation and reformation as ammonia occur, and the newly formed ammonia molecules leave the surface. This process is repeated over and over, with no change in the catalyst.

the carbon-carbon double bonds of unsaturated fats (e.g., corn oil, olive oil) to single bonds characteristic of saturated fats (such as margarine). The use of a metal catalyst, such as nickel, in contact with the reaction mixture dramatically increases the rate of the reaction.

Thousands of essential biochemical reactions in our bodies are controlled and speeded up by biological catalysts called *enzymes*.

A molecular-level view of the action of a solid catalyst widely used in industrial synthesis of ammonia is presented in Figure 7.12.

For a detailed discussion of enzyme catalysis, see Chapter 19.

Question 7.11 Would you imagine that a substance might act as a poison if it interfered with the function of an enzyme? Explain your reasoning.

Question 7.12 Bacterial growth decreases markedly in a refrigerator. Why?

Mathematical Representation of Reaction Rate

LEARNING GOAL

7 Write rate laws, and use these equations to calculate the effect of concentration on rate.

Consider the decomposition reaction of N_2O_5 (dinitrogen pentoxide) in the gas phase. When heated, N_2O_5 decomposes and forms two products: NO_2 (nitrogen dioxide) and O_2 (diatomic oxygen). The balanced chemical equation for the reaction is

$$2N_2O_5(g) \xrightarrow{\Delta} 4NO_2(g) + O_2(g)$$

When all of the factors that affect the rate of the reaction (except concentration) are held constant (that is, the nature of the reactant, temperature and physical

A Human Perspective

Too Fast or Too Slow?

Kinetics plays a vital role in the development of a wide range of commercial products. We will consider a few of these applications that contribute to our health, our enjoyment, and the preservation of our environment.

Refrigeration

Throughout history, few technologies have had a more profound benefit to human health than the development of inexpensive and efficient refrigeration, allowing food to be stored for long periods without spoiling. At room temperature, unwanted chemical reactions, with the help of bacteria, rapidly decompose many foods, especially meats and fish, making them unsuitable for human consumption. Storing food at low temperatures slows the rate of decomposition by slowing the rate of bacterial reproduction and the rate of the destructive chemical reactions. Refrigeration is a straightforward application of kinetics: Rates are halved for each 10°C decrease in temperature.

Dentistry

The most widely used strategy for whitening teeth is bleaching with hydrogen peroxide (H_2O_2). Because the reaction is slow, manufacturers of home-use products use devices to trap the H_2O_2 and hold it in contact with the teeth. A better strategy involves higher concentrations of H_2O_2. (Remember, the rate of a reaction is proportional to the reactant concentration.) However, these treatments must be administered by a dentist; used incorrectly, high concentrations of H_2O_2 can damage sensitive tissues, such as the gums and the interior of the mouth.

Artwork and Photography

Old paintings and photographs are very susceptible to the effects of ultraviolet (UV) light. Chemical reactions that produce discoloration are energized by UV light. More molecules achieve the necessary activation energy, and the quality and value of the artwork degrade more rapidly. Glass that absorbs UV radiation (conservation glass) robs these reactions of needed energy, and the lifetime of these valuable materials is extended.

The Environment

An essential requirement for herbicides and pesticides is biodegradability. Ideally, these materials enter soil or vegetation, quickly complete their task of killing weeds or bugs, and rapidly decompose, often with the help of bacteria. Most people are aware of the problems caused by earlier, nonbiodegradable pesticides, most famously DDT (dichlorodiphenyltrichloroethane). DDT degrades very slowly in the environment. It is fat-soluble, bioaccumulates in animal tissue, and is responsible for many detrimental environmental effects, most notably its interference with bird reproduction. Even though it has been banned in the United States for decades, DDT can still be found in the environment. It has been replaced with compounds that exhibit more favorable decay kinetics.

Automobiles

The catalytic converter in modern automobiles helps to speed up the rate of conversion of harmful emissions (CO and NO,

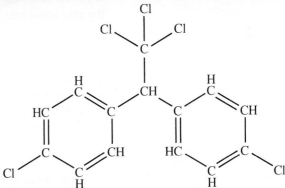

Egg shell thinning is one of the chronic impacts that DDT has had on the environment.

for example) to less harmful products (CO_2 and N_2 gas). The converter contains the metals platinum and palladium, which serve as solid catalysts, enabling the almost complete conversion of the auto emissions before they leave the tailpipe.

Nuclear Waste

Conversion of nuclear waste to harmless products is *not* a chemical process. We wish that chemical kinetics applied to radioactive waste. Perhaps we could heat the waste and it would convert to harmless products. Not so! Radioactive decay (Chapter 9) is a nuclear process and is immune to all of the strategies routinely used to accelerate the rate of chemical reactions. That leaves us with a serious radioactive waste disposal problem. For more details, refer to Green Chemistry: Nuclear Waste Disposal in Chapter 9.

For Further Understanding

▶ Provide another example of a process that has a less-than-optimal rate. What changes in conditions would improve the rate of this process?

▶ Use the Internet to find the current market values of platinum and palladium. How does this help to explain the high cost of a replacement catalytic converter?

state of the reactant, and the presence or absence of a catalyst), the rate of the reaction is proportional to the concentration of N_2O_5, the reactant.

$$\text{Rate} \propto \text{concentration } N_2O_5$$

We will represent the concentration of N_2O_5 in units of molarity and represent molar concentration using brackets.

$$\text{Concentration } N_2O_5 = [N_2O_5]$$

Then,

$$\text{Rate} \propto [N_2O_5]$$

Laboratory measurement shows that the rate of the reaction depends on the molar concentration raised to an experimentally determined exponent, which we will symbolize as n, the *reaction order*. The **reaction order** represents the number of molecules that are involved in the formation of product.

$$\text{Rate} \propto [N_2O_5]^n$$

In expressions such as the one shown, the proportionality symbol, α, may be replaced by an equality sign and a proportionality constant that we represent as k, the **rate constant.**

$$\text{Rate} = k[N_2O_5]^n$$

Equations that follow this format, the rate being equal to the rate constant multiplied by the reactant concentration raised to an exponent that is the reaction order, are termed **rate laws.**

For the reaction described here, n has been experimentally determined and is equal to 1; hence, the reaction is first order in N_2O_5 and the rate law for the reaction is:

$$\text{Rate} = k[N_2O_5]$$

In general, the rate of reaction for an equation of the general form:

$$A \longrightarrow product$$

is represented by the rate law

$$\text{Rate} = k[A]^n$$

in which

$$n = \text{the reaction order}$$
$$k = \text{the rate constant of the reaction}$$
$$[A] = \text{the molar concentration of the reactant}$$

An equation involving two reactants

$$A + B \longrightarrow products$$

has a rate law of the form

$$\text{Rate} = k[A]^n[B]^{n'}$$

Both the value of the rate constant and the reaction order are deduced from a series of experiments. We cannot predict them by simply looking at the chemical equation.

Knowledge of the form of the rate law, coupled with the experimental determination of the value of the rate constant, k, and the order, n, are valuable in a number of ways. Industrial chemists use this information to establish optimum conditions for preparing a product in the shortest practical time. The design of an entire manufacturing facility may, in part, depend on the rates of the critical reactions.

Note that the exponent, n, in the rate law is not the same as the coefficient of N_2O_5 in the balanced equation. However, in some reactions the coefficient in the balanced equation and the exponent n (the order of the reaction) are numerically the same.

In Section 7.4, we will see how the rate law forms the basis for describing equilibrium reactions.

EXAMPLE 7.6	Writing Rate Laws	LEARNING GOAL

Write the rate law for the oxidation of ethanol (C_2H_5OH). The reaction has been experimentally determined to be first order in ethanol and third order in oxygen (O_2). The chemical reaction for this process is:

7 Write rate laws, and use these equations to calculate the effect of concentration on rate.

$$C_2H_5OH(l) + 3O_2(g) \longrightarrow 2CO_2(g) + 3H_2O(l)$$

Solution

Step 1. The rate law involves only the reactants, C_2H_5OH and O_2. Depict their concentrations as

$$[C_2H_5OH][O_2]$$

Step 2. Now raise each to an exponent corresponding to its experimentally determined order

$$[C_2H_5OH][O_2]^3$$

Step 3. This is proportional to the rate:

$$\text{Rate} \propto [C_2H_5OH][O_2]^3$$

Step 4. Proportionality (α) is incorporated into an equation using a proportionality constant, k.

$$\text{Rate} = k[C_2H_5OH][O_2]^3$$

Helpful Hint: Remember that 1 is understood as an exponent; $[C_2H_5OH]$ is correct and $[C_2H_5OH]^1$ is not.

Practice Problem 7.6

Write the general form of the rate law for each of the following processes. Represent the order as n, n', and so forth.

a. $N_2(g) + O_2(g) \longrightarrow 2NO(g)$ c. $CH_4(g) + 2O_2(g) \longrightarrow CO_2(g) + 2H_2O(g)$

b. $2C_4H_6(g) \longrightarrow C_8H_{12}(g)$ d. $2NO_2(g) \longrightarrow 2NO(g) + O_2(g)$

▶ For Further Practice: **Questions 7.63 and 7.64.**

7.4 Equilibrium

Physical Equilibrium

Recall from Chapter 1 that change can be described as physical or chemical. While a chemical change is a consequence of chemical reactions, a physical change produces a recognizable difference in the appearance of a substance without causing any change in the composition or identity of the substance. Consequently, a **physical equilibrium** is one that occurs between two phases of the same substance. Examples of physical equilibria include:

LEARNING GOAL

8 Recognize and describe equilibrium situations.

- Liquid water in equilibrium with either ice or water vapor
- A saturated solution (Section 6.1) consisting of an ionic solid, such as silver chloride, in equilibrium with a solution of silver ions and chloride ions
- A saturated solution consisting of a covalent solid, such as glucose, in equilibrium with a solution of glucose molecules.

Dissolution of sugar in water, producing a saturated solution, is a convenient illustration of a state of *dynamic equilibrium*.

A **dynamic equilibrium** is a situation in which the rate of the forward process in a reversible reaction is exactly balanced by the rate of the reverse process.

A physical equilibrium, such as sugar dissolving in water, is a reversible reaction. A **reversible reaction** is a process that can occur in both directions. It is indicated by using the equilibrium arrows ($\rightleftharpoons$) symbol.

Sugar in Water

Imagine that you mix a small amount of sugar (2 or 3 g) in 100 milliliters (mL) of water. After you stir it for a short time, all of the sugar dissolves; there is no residual solid sugar because the sugar has dissolved *completely*. The reaction clearly has converted all solid sugar to its dissolved state, an aqueous solution of sugar, or

$$\text{sugar}(s) \longrightarrow \text{sugar}(aq)$$

Now, suppose that you add a very large amount of sugar (100 g), more than can possibly dissolve, to the same volume of water. As you stir the mixture, you observe more and more sugar dissolving. After some time, the amount of solid sugar remaining in contact with the solution appears constant. Over time, you observe no further decrease in the amount of undissolved sugar. Although nothing further appears to be happening, in reality a great deal of activity is taking place!

An equilibrium situation has been established. Over time, the amount of sugar dissolved in the measured volume of water (the concentration of sugar in water) does not change. Hence, the amount of undissolved sugar remains the same. However, if you could look at the individual sugar molecules, you would see something quite amazing. Rather than sugar molecules in the solid simply staying in place, you would see them continuing to leave the solid state and go into solution. At the same time, a like number of dissolved sugar molecules would leave the water and form more solid. This active process is described as a *dynamic equilibrium*. The reaction is proceeding in a forward (left to right) and a reverse (right to left) direction at the same time, making it a reversible reaction:

$$\text{sugar}(s) \rightleftharpoons \text{sugar}(aq)$$

The rates of the forward and reverse reactions are identical, and the amount of sugar present as solid and in solution is constant. The system is at equilibrium. The equilibrium arrows serve as

- an indicator of a reversible process,
- an indicator of an equilibrium process, and
- a reminder of the dynamic nature of the process.

Examples of physical equilibria abound in nature. Many environmental systems depend on fragile equilibria. The amount of oxygen dissolved in a certain volume of lake water (the oxygen concentration) is governed by the principles of equilibrium because solubility equilibria are temperature dependent (Henry's law, Section 6.1). The lives of plants and animals within this system are critically related to the levels of dissolved oxygen in the water.

Dynamic equilibrium may be dangerous for living cells in certain situations because it represents a process in which nothing is getting done. There is no gain. Let's consider an exothermic reaction designed to produce a net gain of energy for the cell. In a dynamic equilibrium, the rate of the forward (energy-releasing) reaction is equal to the rate of the backward (energy-requiring) reaction. Thus, there is no net gain of energy to fuel cellular activity, and the cell will die.

Question 7.13   Correlate a busy restaurant at lunchtime to dynamic equilibrium.

Question 7.14   A certain change in reaction conditions for an equilibrium process was found to increase the rate of the forward reaction much more than that of the reverse reaction. Did the amount of product increase, decrease, or remain the same? Explain your reasoning.

Chemical Equilibrium

We have assumed that most chemical reactions considered thus far proceed to completion. A complete reaction is one in which all reactants have been converted to products. However, many important chemical reactions do not go to completion. As a result, after no further obvious change is taking place, measurable quantities of reactants and products remain. Reactions of this type (incomplete reactions) are called *equilibrium reactions*. **Chemical equilibrium** is the state of a reaction in which the rates of the forward and reverse reactions are equal.

The Reaction of N_2 and H_2

When we mix nitrogen gas (N_2) and hydrogen gas (H_2) at an elevated temperature (perhaps 500°C), some of the molecules will collide with sufficient energy and proper orientation to break $N \equiv N$ and $H—H$ bonds. Rearrangement of the atoms will produce the product (NH_3):

$$N_2(g) + 3H_2(g) \rightleftharpoons 2NH_3(g)$$

Beginning with a mixture of hydrogen and nitrogen, the rate of the reaction is initially rapid, because the reactant concentration is high; as the reaction proceeds, the concentration of reactants decreases. At the same time, the concentration of the product, ammonia, is increasing. At equilibrium, the *rate of depletion* of hydrogen and nitrogen *is equal* to the *rate of depletion* of ammonia. In other words, *the rates of the forward and reverse reactions are equal.*

The concentration of the various species is fixed at equilibrium because product is being *consumed and formed at the same rate.* In other words, the reaction continues indefinitely (dynamic), but the concentration of products and reactants is fixed (equilibrium). This is a *dynamic equilibrium.* The rate of this reaction as a function of time is depicted in Figure 7.13.

For systems at equilibrium, an **equilibrium constant** expression can be written; it summarizes the relationship between the concentration of reactants and products in an equilibrium reaction.

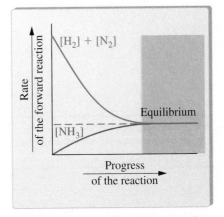

Figure 7.13 The change of the rate of reaction of H_2 and N_2 as a function of time. The rate of reaction, initially rapid, decreases as the concentration of reactants decreases and product increases. The rate approaches a limiting value at equilibrium.

The Generalized Equilibrium Constant Expression for a Chemical Reaction

We write the general form of an equilibrium chemical reaction as

$$aA + bB \rightleftharpoons cC + dD$$

in which A and B represent reactants, C and D represent products, and a, b, c, and d are the coefficients of the balanced equation. The equilibrium constant expression for this general case is

$$K_{eq} = \frac{[C]^c[D]^d}{[A]^a[B]^b}$$

For the ammonia system, it follows that the appropriate equilibrium expression is:

$$K_{eq} = \frac{[NH_3]^2}{[N_2][H_2]^3}$$

It does not matter what initial amounts (concentrations) of reactants or products we choose. When the system reaches equilibrium, the calculated value of K_{eq} will not change. The magnitude of K_{eq} can be altered only by changing the temperature. Thus, K_{eq} is temperature dependent. The chemical industry uses this fact to its advantage by choosing a reaction temperature that will maximize the yield of a desired product.

Question 7.15   How could we determine when a reaction has reached equilibrium?

Question 7.16   Does the attainment of equilibrium imply that no further change is taking place in the system?

Writing Equilibrium Constant Expressions

An equilibrium constant expression can be written only after a correct, balanced chemical equation that describes the equilibrium system has been developed.

LEARNING GOAL

9 Write equilibrium constant expressions, and use these expressions to calculate equilibrium constants or equilibrium concentrations.

A balanced equation is essential because the *coefficients* in the equation become the *exponents* in the equilibrium constant expression.

Each chemical reaction has a unique equilibrium constant value at a specified temperature. Equilibrium constants listed in the chemical literature are often reported at 25°C, to allow comparison of one system with any other. For any equilibrium reaction, the value of the equilibrium constant changes with temperature.

The brackets represent molar concentration or molarity; recall that molarity has units of moles per liter (mol/L). In our discussion of equilibrium, all equilibrium constants are shown as *unitless*.

A properly written equilibrium constant expression may not include all of the terms in the chemical equation upon which it is based. Only the concentrations of gases and substances in solution are shown, because their concentrations can change. Concentration terms for liquids and solids are *not* shown. The concentration of a liquid is constant. Most often, the liquid is the solvent for the reaction under consideration. A solid also has a fixed concentration and, for solution reactions, is not really a part of the solution. When a solid is formed, it exists as a solid phase in contact with a liquid phase (the solution).

Products of the overall equilibrium reaction are in the numerator, and reactants are in the denominator.

EXAMPLE 7.7   Writing an Equilibrium Constant Expression

LEARNING GOAL

Write an equilibrium constant expression for the reversible reaction:

$$H_2(g) + F_2(g) \rightleftharpoons 2HF(g)$$

9 Write equilibrium constant expressions, and use these expressions to calculate equilibrium constants or equilibrium concentrations.

Solution

Step 1. Inspection of the chemical equation reveals that no solids or liquids are present. Hence, all reactants and products appear in the equilibrium constant expression.

Step 2. The numerator term is the product term $[HF]^2$. The exponent for [HF] is identical to the coefficient of HF in the balanced equation.

Step 3. The denominator terms are the reactants $[H_2]$ and $[F_2]$. Note that each term contains an exponent identical to the corresponding coefficient in the balanced equation.

Step 4. Arranging the numerator and denominator terms as a fraction and setting the fraction equal to K_{eq} yields

$$K_{eq} = \frac{[HF]^2}{[H_2][F_2]}$$

Practice Problem 7.7

Write an equilibrium constant expression for each of the following reversible reactions.

  a. $2NO_2(g) \rightleftharpoons N_2(g) + 2O_2(g)$ b. $2H_2O(l) \rightleftharpoons 2H_2(g) + O_2(g)$

▶ For Further Practice: **Questions 7.83 and 7.84.**

Interpreting Equilibrium Constants

What utility does the equilibrium constant have? The reversible arrow in the chemical equation alerts us to the fact that an equilibrium exists. Some measurable quantity of the product and reactant remain. However, there is no indication whether products predominate, reactants predominate, or significant concentrations of both products and reactants are present at equilibrium.

EXAMPLE 7.8 Writing an Equilibrium Constant Expression

LEARNING GOAL

9 Write equilibrium constant expressions, and use these expressions to calculate equilibrium constants or equilibrium concentrations.

Write an equilibrium constant expression for the reversible reaction:

$$MnO_2(s) + 4H^+(aq) + 2Cl^-(aq) \rightleftharpoons Mn^{2+}(aq) + Cl_2(g) + 2H_2O(l)$$

Solution

Step 1. MnO_2 is a solid and H_2O is a liquid. Thus, they are not written in the equilibrium constant expression.

$$MnO_2(s) + 4H^+(aq) + 2Cl^-(aq) \rightleftharpoons Mn^{2+}(aq) + Cl_2(g) + 2H_2O(l)$$

Not a part of the K_{eq} expression

Step 2. The numerator term includes the remaining products:

$$[Mn^{2+}] \quad \text{and} \quad [Cl_2]$$

Step 3. The denominator term includes the remaining reactants:

$$[H^+]^4 \quad \text{and} \quad [Cl^-]^2$$

Note that each exponent is identical to the corresponding coefficient in the chemical equation.

Step 4. Arranging the numerator and denominator terms as a ratio and setting the ratio equal to K_{eq} yields

$$K_{eq} = \frac{[Mn^{2+}][Cl_2]}{[H^+]^4[Cl^-]^2}$$

Practice Problem 7.8

Write an equilibrium constant expression for each of the following reversible reactions.

a. $AgCl(s) \rightleftharpoons Ag^+(aq) + Cl^-(aq)$ b. $PCl_5(s) \rightleftharpoons PCl_3(g) + Cl_2(g)$

▶ For Further Practice: **Questions 7.85 and 7.87.**

The numerical value of the equilibrium constant provides this additional information. It tells us the extent to which reactants have converted to products. This is important information for anyone who wants to manufacture and sell the product. It also is important to anyone who studies the effect of equilibrium reactions on environmental systems and living organisms.

Although an absolute interpretation of the numerical value of the equilibrium constant depends on the form of the equilibrium constant expression, the following generalizations are useful:

- K_{eq} greater than 1×10^3. A large numerical value of K_{eq} indicates that the numerator (product term) is much larger than the denominator (reactant term) and that at equilibrium mostly product is present.
- K_{eq} less than 1×10^{-3}. A small numerical value of K_{eq} indicates that the numerator (product term) is much smaller than the denominator (reactant term) and that at equilibrium mostly reactant is present.
- K_{eq} between 1×10^{-3} and 1×10^3. In this case, the equilibrium mixture contains significant concentrations of both reactants and products.

Question 7.17 At a given temperature, the equilibrium constant for a certain reaction is 1×10^{20}. Does this equilibrium favor products or reactants? Why?

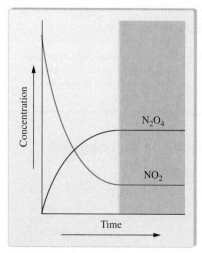

Figure 7.14 For the reaction $2NO_2(g) \rightleftharpoons N_2O_4(g)$, the concentration of reactant (NO_2) diminishes rapidly while the N_2O_4 concentration builds. Eventually, the concentrations of both reactant and product become constant over time (blue area). The equilibrium condition has been attained.

LEARNING GOAL

9 Write equilibrium constant expressions, and use these expressions to calculate equilibrium constants or equilibrium concentrations.

Question 7.18   At a given temperature, the equilibrium constant for a certain reaction is 1×10^{-18}. Does this equilibrium favor products or reactants? Why?

Calculating Equilibrium Constants

The magnitude of the equilibrium constant for a chemical reaction is determined experimentally. The reaction under study is allowed to proceed until the concentration of products and reactants no longer changes (Figure 7.14). This may be a matter of seconds, minutes, hours, or even months or years, depending on the rate of the reaction. The reaction mixture is then analyzed to determine the molar concentration of each of the products and reactants. These concentrations are substituted in the equilibrium constant expression, and the equilibrium constant is calculated. The following example illustrates this process.

EXAMPLE 7.9   **Calculating an Equilibrium Constant**

Hydrogen iodide is placed in a sealed container and allowed to come to equilibrium. The equilibrium reaction is:

$$2HI(g) \rightleftharpoons H_2(g) + I_2(g)$$

and the equilibrium concentrations are:

$$[HI] = 0.54 \ M$$
$$[H_2] = 1.72 \ M$$
$$[I_2] = 1.72 \ M$$

Calculate the equilibrium constant.

Solution

Step 1. Write the equilibrium constant expression:

$$K_{eq} = \frac{[H_2][I_2]}{[HI]^2}$$

Step 2. Substitute the equilibrium concentrations of products and reactants to obtain

$$K_{eq} = \frac{[1.72][1.72]}{[0.54]^2} = \frac{2.96}{0.29}$$

$$= 10.1 \text{ or } 1.0 \times 10^1 \text{ (two significant figures)}$$

Practice Problem 7.9

A container holds the following mixture at equilibrium:

$$[NH_3] = 0.25 \ M$$
$$[N_2] = 0.11 \ M$$
$$[H_2] = 1.91 \ M$$

If the reaction is:

$$N_2(g) + 3H_2(g) \rightleftharpoons 2NH_3(g)$$

calculate the equilibrium constant.

▶ For Further Practice: **Questions 7.88.**

Using Equilibrium Constants

We have seen that the equilibrium constant for a reaction can be calculated if we know the equilibrium concentrations of all of the reactants and products. Once known, the equilibrium constant can be used to obtain equilibrium concentrations of one or more reactants or products for a variety of different situations. These calculations can be quite complex. Let's look at one straightforward but useful case, one where the equilibrium concentration of reactants is known and we wish to calculate the product concentration.

EXAMPLE 7.10 **Using an Equilibrium Constant**

LEARNING GOAL

Given the equilibrium reaction studied in Example 7.9:

$$2HI(g) \rightleftharpoons H_2(g) + I_2(g)$$

9 Write equilibrium constant expressions, and use these expressions to calculate equilibrium constants or equilibrium concentrations.

A sample mixture of HI, H_2, and I_2, at equilibrium, was found to have $[H_2] = 1.0 \times 10^{-2}\ M$ and $[HI] = 4.0 \times 10^{-2}\ M$. Calculate the molar concentration of I_2 in the equilibrium mixture.

Solution

Step 1. From Example 7.9, the equilibrium expression and equilibrium constant are:

$$K_{eq} = \frac{[H_2][I_2]}{[HI]^2};\ K_{eq} = 1.0 \times 10^1$$

Step 2. To solve the equilibrium expression for $[I_2]$, first multiply both sides of the equation by $[HI]^2$:

$$[H_2][I_2] = K_{eq}[HI]^2$$

Then, divide both sides by $[H_2]$:

$$[I_2] = \frac{K_{eq}[HI]^2}{[H_2]}$$

Step 3. Substitute the values:

$$K_{eq} = 1.0 \times 10^1$$
$$[H_2] = 1.0 \times 10^{-2}\ M$$
$$[HI] = 4.0 \times 10^{-2}\ M$$

Step 4. Solve:

$$[I_2] = \frac{[1.0 \times 10^1][4.0 \times 10^{-2}]^2}{[1.0 \times 10^{-2}]}$$
$$= 1.6\ M$$

Practice Problem 7.10

Using the reaction given above, calculate the $[I_2]$ if both $[H_2]$ and $[HI]$ were $1.0 \times 10^{-4}\ M$.

▶ For Further Practice: **Questions 7.89 and 7.90.**

LEARNING GOAL

10   Use LeChatelier's principle to predict changes in equilibrium position.

Addition of products or reactants may have a profound effect on the composition of a reaction mixture but *does not* affect the value of the equilibrium constant.

(a)   (b)   (c)

Figure 7.15 The effect of concentration on the equilibrium composition of the reaction:

$$FeSCN^{2+}(aq) \rightleftharpoons Fe^{3+}(aq) + SCN^-(aq)$$
(red)   (yellow)   (colorless)

Solution (a) represents this reaction at equilibrium; (b) addition of SCN^- shifts the position of the equilibrium to the left, intensifying the red color. (c) Removal of SCN^- shifts the position of the equilibrium to the right, shown by the disappearance of the red color.

LeChatelier's Principle

In the nineteenth century, the French chemist H. L. LeChatelier discovered that changes in equilibrium depend on the amount of "stress" applied to the system. The stress may take the form of an increase or decrease of the temperature of the system at equilibrium, or perhaps a change in the amount of reactant or product present in a fixed volume (the concentration of reactant or product).

LeChatelier's principle states that if a stress is placed on a system at equilibrium, the system will respond by altering the equilibrium composition in such a way as to minimize the stress.

Consider the equilibrium situation discussed earlier:

$$N_2(g) + 3H_2(g) \rightleftharpoons 2NH_3(g)$$

If the reactants and products are present in a fixed volume, such as 1 liter (L), and more NH_3 (the *product*) is introduced into the container, the system will be stressed—the equilibrium will be disturbed. The system will try to alleviate the stress (as we all do) by *removing* as much of the added material as possible. How can it accomplish this? By converting some NH_3 to H_2 and N_2. The position of the equilibrium shifts to the left, and a new dynamic equilibrium is soon established.

Adding extra H_2 or N_2 would apply the stress to the reactant side of the equilibrium. To minimize the stress, the system would "use up" some of the excess H_2 or N_2 to make product, NH_3. The position of the equilibrium would shift to the right.

In summary,

$$N_2(g) + 3H_2(g) \rightleftharpoons 2NH_3(g)$$

Product introduced: position of the equilibrium shifted left

Reactant introduced: position of the equilibrium shifted right

What would happen if some of the ammonia molecules were *removed* from the system? The loss of ammonia represents a stress on the system; to relieve that stress, the ammonia would be replenished by the reaction of hydrogen and nitrogen. The position of the equilibrium would shift to the right.

Product removed: position of the equilibrium shifted right

Reactant removed: position of the equilibrium shifted left

Effect of Concentration

Addition of extra product or reactant to a fixed reaction volume is just another way of saying that we have increased the concentration of product or reactant. Removal of material from a fixed volume decreases the concentration. Therefore, changing the concentration of one or more components of a reaction mixture is a way to alter the equilibrium composition of an equilibrium mixture (Figure 7.15). Let's look at some additional experimental variables that may change the equilibrium composition.

Effect of Heat

The change in equilibrium composition caused by the addition or removal of heat from an equilibrium mixture can be explained by treating heat as a product or reactant. The reaction of nitrogen and hydrogen is an exothermic reaction:

$$N_2(g) + 3H_2(g) \rightleftharpoons 2NH_3(g) + 22 \text{ kcal}$$

Adding heat by raising the temperature is similar to increasing the amount of product. The position of the equilibrium will shift to the left, increasing the amounts of N_2 and H_2 and decreasing the amount of NH_3. If the reaction takes place in a fixed volume, the concentrations of N_2 and H_2 increase and the NH_3 concentration decreases.

Removal of heat by lowering the temperature produces the reverse effect. More ammonia is produced from N_2 and H_2, and the concentrations of these reactants must decrease.

In the case of an endothermic reaction, such as

$$39 \text{ kcal} + 2N_2(g) + O_2(g) \rightleftharpoons 2N_2O(g)$$

addition of heat is analogous to the addition of reactant, and the position of the equilibrium shifts to the right. Removal of heat would shift the reaction to the left, favoring the formation of reactants.

The dramatic effect of heat on the position of equilibrium is shown in Figure 7.16.

Effect of Pressure

Only gases are affected significantly by changes in pressure because gases are free to expand and compress in accordance with Boyle's law. However, liquids and solids are not compressible, so their volumes are unaffected by pressure.

Therefore, pressure changes will alter equilibrium composition only in reactions that involve a gas or variety of gases as products and/or reactants. Again, consider the ammonia example,

$$N_2(g) + 3H_2(g) \rightleftharpoons 2NH_3(g)$$

One mole of N_2 and three moles of H_2 (total of 4 mol of reactants) react to form two moles of NH_3 (2 mol of product). The equilibrium mixture (perhaps at 25°C) contains N_2, H_2, and NH_3. An increase in pressure will cause a stress to the system because the pressure is greater than the equilibrium pressure. To relieve the stress, the system will shift toward the side of the reaction that has fewer mol of gas. As the number of mol decreases, the pressure will decrease. In the example of ammonia, an increase in pressure will shift the position of the equilibrium to the right, producing more NH_3 at the expense of N_2 and H_2.

A decrease in pressure will be countered by the reaction shifting toward the side that contains more mol of gas because the pressure is less than the equilibrium pressure. As the number of mol of gas increases, the pressure will increase. Therefore, in the equilibrium reaction of ammonia, a decrease in pressure will cause the position of the equilibrium to shift to the left, and ammonia decomposes to form more nitrogen and hydrogen.

In contrast, the decomposition of hydrogen iodide,

$$2HI(g) \rightleftharpoons H_2(g) + I_2(g)$$

is unaffected by pressure. The number of mol of gaseous product and reactant are identical. No volume advantage is gained by a shift in equilibrium composition. In summary:

- Pressure affects the equilibrium composition only of reactions that involve at least one gaseous substance.
- Additionally, the relative number of mol of gaseous products and reactants must differ.
- The equilibrium composition will shift to increase the number of mol of gas when the pressure decreases; it will shift to decrease the number of mol of gas when the pressure increases.

Figure 7.16 The effect of heat on the equilibrium position. For the reaction:

$$CoCl_4{}^{2-}(aq) + 6H_2O(l) \rightleftharpoons$$
(blue)
$$Co(H_2O)_6{}^{2+}(aq) + 4Cl^-(aq)$$
(pink)

heating the solution favors the blue $CoCl_4{}^{2-}$ species; cooling favors the pink $Co(H_2O)_6{}^{2+}$ species.

Expansion and compression of gases and Boyle's law are discussed in Section 5.1.

The industrial process for preparing ammonia, the Haber process, uses pressures of several hundred atmospheres (atm) to increase the yield.

Effect of a Catalyst

A catalyst has no effect on the equilibrium composition. A catalyst increases the rates of both forward and reverse reactions to the same extent. The equilibrium composition *and* equilibrium concentration do not change when a catalyst is used, but the equilibrium composition is achieved in a shorter time. The role of a solid-phase catalyst in the synthesis of ammonia is shown in Figure 7.12.

EXAMPLE 7.11   **Predicting Changes in Equilibrium Composition**

LEARNING GOAL

A geologically important reaction, shown below, is critical for the formation of the stalactites and stalagmites in caves.

10   Use LeChatelier's principle to predict changes in equilibrium position.

$$4.67 \text{ kcal} + Ca^{2+}(aq) + 2HCO_3^-(aq) \rightleftharpoons CaCO_3(s) + CO_2(aq) + H_2O(l)$$

Predict the effect on the equilibrium composition for each of the following changes.
- a. The concentration of Ca^{2+} is increased by addition of some $CaCl_2$.
- b. Some solid $CaCO_3$ is removed from the mixture.
- c. The concentration of HCO_3^- is decreased.
- d. The temperature of the system is increased.
- e. A catalyst is added.

Solution
- a. The addition of Ca^{2+} ions will cause the position of the equilibrium to shift to the right, and more products will form.
- b. Because $CaCO_3$ is a solid, addition or removal of the substance will have no effect on the equilibrium composition.
- c. If $HCO_3^-(aq)$ is removed from the system, the position of the equilibrium will shift to the left to produce more HCO_3^- and other reactants.
- d. The reaction is endothermic. The addition of heat by the raising of the temperature will shift the position of the equilibrium to the right, and more products will form.
- e. The addition of a catalyst has no effect on the equilibrium composition.

Practice Problem 7.11

The reaction

$$382 \text{ cal} + Pb(s) + 2H^+(aq) \rightleftharpoons Pb^{2+}(aq) + H_2(g)$$

is carried out at constant pressure. Predict whether the volume of hydrogen gas would increase, decrease, or remain the same for each of the following changes.

- a. Addition of $Pb(s)$.
- b. Addition of Pb^{2+} ions by adding some $Pb(NO_3)_2$ to the solution.
- c. Removal of H^+ ions.
- d. Lowering of the temperature.
- e. Addition of a catalyst.

▶ For Further Practice: **Questions 7.91 and 7.92.**

An Extraordinary Molecule

Think for a moment: What is the only common molecule that exists in all three physical states of matter (solid, liquid, and gas) under natural conditions on earth? This molecule is absolutely essential for life; in fact, life probably arose in this substance. It is the most abundant molecule in the cells of living organisms (70–95%) and covers 75% of the earth's surface. Without it, cells quickly die, and without it the earth would not be a fit environment in which to live. By now, you have guessed that we are talking about the water molecule. It is so abundant on earth that we take this deceptively simple molecule for granted.

As we finish our discussion of thermodynamics, kinetics, and equilibrium, it is fitting that we take a further look at water and its unique properties. Enormous quantities of water continuously convert from solid $\rightleftharpoons$ liquid $\rightleftharpoons$ gas in our environment; equally large energy changes (storage and release) are essential to these processes. This is thermodynamics in action! One of our best examples of physical equilibrium is change of state (ice $\rightleftharpoons$ liquid $\rightleftharpoons$ vapor). This equilibrium process can have a profound influence on our weather.

Life can exist only within a fairly narrow range of temperatures. Above or below that range, the chemical reactions necessary for life, and thus life itself, will cease. Water can moderate temperature fluctuation and maintain the range necessary for life, and one property that allows it to do so is its unusually high specific heat, 1 cal/g · °C. This means that water can absorb or lose more heat energy than many other substances without a significant temperature change. This is because in the liquid state, every water molecule is hydrogen bonded to other water molecules. Because a temperature increase is really just a measure of increased (more rapid) molecular movement, we must get the water molecules moving more rapidly, independent of one another, to register a temperature increase. Before we can achieve this independent, increased activity, the hydrogen bonds between molecules must be broken. Much of the heat energy that water absorbs is involved in breaking hydrogen bonds and is *not* used to increase molecular movement. Thus, a great deal of heat is needed to raise the temperature of water even a little bit.

Water also has a very high heat of vaporization. It takes 540 calories (cal) to change 1 g of liquid water at 100°C to a gas, and even more, 603 cal/g, when the water is at 37°C, human body temperature. That is about twice the heat of vaporization of alcohol. As water molecules evaporate, the surface of the liquid cools because only the highest-energy (or "hottest") molecules leave as a gas. Only the "hottest" molecules have enough energy to break the hydrogen bonds that bind them to other water molecules. Indeed, evaporation of water molecules from the surfaces of lakes and oceans helps to maintain stable temperatures in those bodies of water. Similarly, evaporation of

Beauty is also a property of water.

perspiration from body surfaces helps to prevent overheating on a hot day or during strenuous exercise.

Even the process of freezing helps stabilize and moderate temperatures. This is especially true in the fall. Water releases heat when hydrogen bonds are formed. This is an example of an exothermic process. Thus, when water freezes, solidifying into ice, additional hydrogen bonds are formed, and heat is released into the environment. As a result, the temperature change between summer and winter is more gradual, allowing organisms to adjust to the change.

One last feature that we take for granted is the fact that when we put ice in our iced tea on a hot summer day, the ice floats. This means that the solid state of water is actually *less* dense than the liquid state! In fact, it is about 10% less dense, having an open lattice structure with each molecule hydrogen bonded to the maximum of four other water molecules. What would happen if ice did sink? All bodies of water, including the mighty oceans would eventually freeze solid, killing all aquatic and marine plant and animal life. Even in the heat of summer, only a few inches of ice at the surface would thaw. Instead, the ice forms at the surface and provides a layer of insulation that prevents the water below from freezing.

As we continue our study of chemistry, we will refer again and again to this amazing molecule. In other Human Perspective features, we will examine properties of water that make it essential to life.

For Further Understanding

▶ Why is the high heat of vaporization of water important to our bodies?

▶ Why is it cooler at the ocean shore than in the desert during summer?

CHAPTER MAP

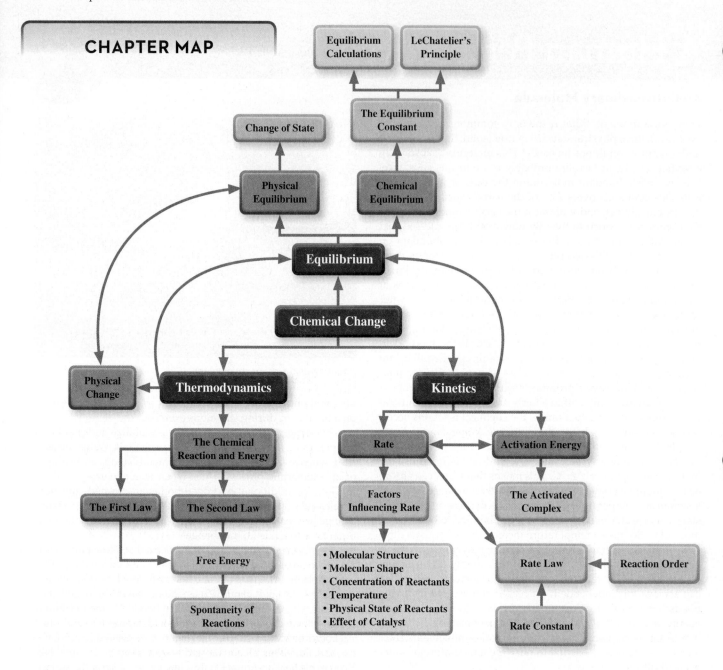

SUMMARY

7.1 Thermodynamics

▶ **Thermodynamics** is the study of energy, work, and **heat,** and is useful in predicting the spontaneity of chemical change.

▶ Thermodynamics can be applied to the study of chemical reactions because we can determine the quantity of heat flow (by measuring the temperature change) between the **system** and the **surroundings.**

▶ **Exothermic reactions** release energy and produce products that are lower in energy than the reactants.

▶ **Endothermic reactions** require energy input. Heat energy is represented as **enthalpy,** *H.*

▶ The energy gain or loss is the change in enthalpy, Δ*H,* and is one factor that is useful in predicting whether a reaction is spontaneous or nonspontaneous.

▶ **Entropy,** *S,* is a measure of the randomness of a system. A random or disordered system has high entropy; a well-ordered system has low entropy. The change in entropy in a chemical reaction, Δ*S,* is also a factor in predicting reaction spontaneity.

▶ **Free energy,** Δ*G,* incorporates both factors, enthalpy and entropy; as such, it is an absolute predictor of the spontaneity of a chemical reaction.

7.2 Experimental Determination of Energy Change in Reactions

▶ A **calorimeter** measures heat changes (in cal or J) that occur in chemical reactions.

▶ The **specific heat** of a substance is the number of cal of heat needed to raise the temperature of 1 g of the substance 1°C.

▶ The amount of energy per g of food is referred to as its **fuel value.** Fuel values are commonly reported in units of **nutritional Calories** (1 nutritional Cal = 1 kcal). A bomb calorimeter is useful for measurement of the fuel value of foods.

7.3 Kinetics

▶ Chemical **kinetics** is the study of the **rate** or speed of a chemical reaction. Energy for reactions is provided by molecular collisions. If this energy is sufficient, bonds may break, and atoms may recombine in a different arrangement to produce product. A collision producing one or more product molecules is termed an *effective collision.*

▶ The minimum amount of energy needed for a reaction is the **activation energy.** The reaction proceeds from reactants to products through an intermediate state, the **activated complex.**

▶ Experimental conditions influencing the reaction rate include the structure and shape of the reacting species, the concentration of reactants, the temperature of reactants, the physical state of reactants, and the presence or absence of a catalyst.

▶ A **catalyst** increases the rate of a reaction. The catalytic substance undergoes no net change in the reaction, nor does it alter the outcome of the reaction.

▶ The **rate law** relates the rate of a reaction to the **rate constant** multiplied by the concentration of the reactants raised to a power that is the **reaction order.**

7.4 Equilibrium

▶ Incomplete reactions are *equilibrium reactions.*

▶ Physical changes are often **reversible reactions. A physical equilibrium** occurs between two phases of the same substance.

▶ Many chemical reactions do not completely convert reactants to products. A mixture of products and reactants exists, and its composition will remain constant until the experimental conditions are changed. This mixture is in a state of **chemical equilibrium.**

▶ The position of the equilibrium is indicated by the **equilibrium constant.**

▶ An equilibrium reaction continues indefinitely (dynamic), but the concentrations of products and reactants are fixed (equilibrium) because the rates of the forward and reverse reactions are equal. This is a **dynamic equilibrium.**

▶ **LeChatelier's principle** states that if a stress is placed on an equilibrium system, the system will respond by altering the equilibrium in such a way as to minimize the stress.

ANSWERS TO PRACTICE PROBLEMS

7.1   **a.** Exothermic
      b. Exothermic

7.2 ΔG is temperature dependent.

7.3   13°C

7.4   −8.8°C

7.5   3.0 nutritional Cal in 1.0 g of candy and

$$\frac{2.1 \times 10^2 \text{ nutritional Cal}}{\text{candy bar}}$$

7.6   **a.** rate $= k[N_2]^n[O_2]^{n'}$
      b. rate $= k[C_4H_6]^n$
      c. rate $= k[CH_4]^n[O_2]^{n'}$
      d. rate $= k[NO_2]^n$

7.7   **a.** $K_{eq} = \dfrac{[N_2][O_2]^2}{[NO_2]^2}$

      b. $K_{eq} = [H_2]^2[O_2]$

7.8   **a.** $K_{eq} = [Ag^+][Cl^-]$
      b. $K_{eq} = [PCl_3][Cl_2]$

7.9 $K_{eq} = 8.2 \times 10^{-2}$

7.10 $1 \times 10^{-3} M$

7.11   **a.** The addition of Pb(s) would have no effect on the volume of hydrogen gas. This is because the substance is a solid, and addition of a solid has no effect on the equilibrium composition.
      b. The addition of Pb^{2+} (aq) would result in a decrease in the volume of hydrogen gas.
      c. The removal of H^+ would result in a decrease in the volume of hydrogen gas.
      d. Lowering the temperature would result in a decrease in the volume of hydrogen gas.
      e. Addition of a catalyst would have no effect on the volume of hydrogen gas.

QUESTIONS AND PROBLEMS

Thermodynamics

Foundations

7.19   State the first law of thermodynamics.
7.20   State the second law of thermodynamics.
7.21   Describe what is meant by an exothermic reaction.
7.22   Describe what is meant by an endothermic reaction.

7.23 The combustion of fuels (coal, oil, gasoline) are exothermic reactions. Why?

7.24 Provide an explanation for the fact that most decomposition reactions are endothermic but most combination reactions are exothermic.

7.25 Explain what is meant by the term *free energy*. When free energy is a positive value, what does it indicate about the spontaneity of the reaction?

7.26 Write the expression for free energy. When ΔG is a negative value, what does it indicate about the spontaneity of the reaction?

7.27 Explain what is meant by the term *enthalpy*.

7.28 Explain what is meant by the term *entropy*.

Applications

7.29 Predict whether each of the following processes increases or decreases entropy, and explain your reasoning.
 a. melting of a solid metal
 b. boiling of water

7.30 Predict whether each of the following processes increases or decreases entropy, and explain your reasoning.
 a. burning a log in a fireplace
 b. condensing of water vapor on a cold surface

7.31 Isopropyl alcohol, commonly known as rubbing alcohol, feels cool when applied to the skin. Explain why.

7.32 Energy is required to break chemical bonds during the course of a reaction. When is energy released?

7.33 Predict whether a reaction with a negative ΔH and a positive ΔS will be spontaneous, nonspontaneous, or temperature dependent. Explain your reasoning.

7.34 Predict whether a reaction with a negative ΔH and a negative ΔS will be spontaneous, nonspontaneous, or temperature dependent. Explain your reasoning.

Experimental Determination of Energy Change in Reactions

Foundations

7.35 Explain what is meant by fuel value.

7.36 Explain what is meant by the term *specific heat*.

7.37 Describe how a calorimeter is used to distinguish between exothermic and endothermic reactions.

7.38 Construct a diagram of a coffee-cup calorimeter.

7.39 What are the energy units most commonly employed in chemistry?

7.40 What energy unit is commonly employed in nutrition science?

7.41 Why does a calorimeter have a "double-walled" container?

7.42 Explain why the fuel value of foods is an important factor in nutrition science.

Applications

7.43 A 5.00-g sample of octane is burned in a bomb calorimeter containing 2.00×10^2 g H_2O. How much energy, in cal, is released if the water temperature increases 6.00°C?

7.44 A 0.325-mol sample of ammonium nitrate was dissolved in water producing a 4.00×10^2 g solution. The temperature decreased from 25.0°C to 15.7°C. If the specific heat of the resulting solution is 1.00 cal/g · °C, calculate the quantity of energy absorbed in the process. Is the dissolution of ammonium nitrate endothermic or exothermic?

7.45 A 30-g sample of chips is burned in a bomb calorimeter containing 2.50×10^2 g H_2O. What is the fuel value (in nutritional Cal) if the temperature of the water increased from 25.0°C to 34.6°C?

7.46 A 0.0500-mol sample of a nutrient substance is burned in a bomb calorimeter containing 2.00×10^2 g H_2O. If the formula weight of this nutrient substance is 114 g/mol, what is the fuel value (in nutritional Cal) if the temperature of the water increased 5.70°C?

Kinetics

Foundations

7.47 Provide an example of a reaction that is extremely slow, taking days, weeks, or years to complete.

7.48 Provide an example of a reaction that is extremely fast, perhaps quicker than the eye can perceive.

7.49 Define the term *activated complex* and explain its significance in a chemical reaction.

7.50 Define and explain the term *activation energy* as it applies to chemical reactions.

7.51 Distinguish among the terms *rate, rate constant,* and *reaction order*.

7.52 Distinguish between the terms *kinetics* and *thermodynamics*.

7.53 Describe the general characteristics of a catalyst.

7.54 Select one enzyme from a later chapter in this book and describe its biochemical importance.

7.55 Describe how an increase in the concentration of reactants increases the rate of a reaction.

7.56 Describe how an increase in the temperature of reactants increases the rate of a reaction.

7.57 Describe how a catalyst speeds up a chemical reaction.

7.58 Explain how a catalyst can be involved in a chemical reaction without being consumed in the process.

7.59 Sketch a potential energy diagram for a reaction that shows the effect of a catalyst on an exothermic reaction.

7.60 Sketch a potential energy diagram for a reaction that shows the effect of a catalyst on an endothermic reaction.

Applications

7.61 Write the rate law for:

$$CH_4(g) + 2O_2(g) \longrightarrow 2H_2O(l) + CO_2(g)$$

if the order of all reactants is one.
Will the rate of the reaction increase, decrease, or remain the same if the rate constant doubles?

7.62 Will the rate of the reaction in Question 7.61 increase, decrease, or remain the same if the concentration of methane increases?

7.63 Write the rate law for the reaction:

$$N_2O_4(g) \rightleftharpoons 2NO_2(g)$$

Represent the order as n, n', and so forth.

7.64 Write the rate law for the reaction:

$$H_2S(aq) + Cl_2(aq) \rightleftharpoons S(s) + 2HCl(aq)$$

Represent the order as n, n', and so forth.

7.65 For the equilibrium

$$2I(g) \rightleftharpoons I_2(g)$$

the rate law is:

$$\text{rate} = k[I]^2$$

at 23°C, $k = 7.0 \times 10^9$ M^{-1} s^{-1}. What effect will doubling the [I] have on the rate?

7.66 For the reaction

$$2H_2O_2(aq) \longrightarrow 2H_2O(l) + O_2(g)$$

the rate law is:

$$\text{rate} = k[H_2O_2]$$

at 25°C, $k = 3.1 \times 10^{-3}\,s^{-1}$. What effect would doubling the $[H_2O_2]$ have on the rate?

Equilibrium

Foundations

7.67 Explain LeChatelier's principle.

7.68 How can LeChatelier's principle help us to increase yields of chemical reactions?

7.69 Distinguish between a physical equilibrium and a chemical equilibrium.

7.70 Distinguish between the rate constant and the equilibrium constant for a reaction.

7.71 Does a large equilibrium constant mean that products or reactants are favored?

7.72 Does a large equilibrium constant mean that the reaction must be rapid?

7.73 Label each of the following statements as true or false and explain why.
   a. A slow reaction is an incomplete reaction.
   b. The rates of forward and reverse reactions are never the same.

7.74 Label each of the following statements as true or false and explain why.
   a. A reaction is at equilibrium when no reactants remain.
   b. A reaction at equilibrium is undergoing continual change.

7.75 The following diagram represents the reversible reaction $A(g) \rightleftharpoons 2B(g)$ at equilibrium with a total pressure of P.

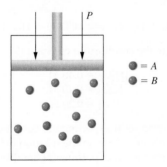

The pressure is increased to P'. Which of the following diagrams would represent the system once equilibrium is reestablished?

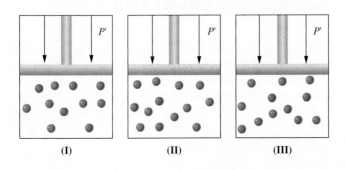

(I)   (II)   (III)

7.76 The following diagram represents the endothermic reaction at equilibrium at 25°C: heat $+ A(g) \rightleftharpoons 2B(g)$.

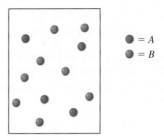

   a. The temperature is increased to 50°C. Which of the following diagrams would represent the system once equilibrium is reestablished?
   b. The temperature is decreased to 15°C. Which of the following diagrams would represent the system once equilibrium is reestablished?

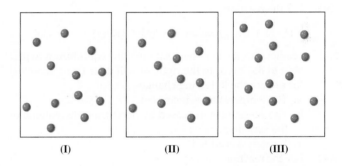

(I)   (II)   (III)

7.77 Describe the meaning of the term *dynamic equilibrium*.

7.78 What is the relationship between the forward and reverse rates for a reaction at equilibrium?

7.79 Describe the meaning of the term *position of the equilibrium*.

7.80 Can a catalyst alter the position of the equilibrium?

7.81 Name three factors that can shift the position of an equilibrium.

7.82 A change in pressure could have the greatest effect on which type of equilibria: gaseous, liquid, or solid?

Applications

7.83 Write a valid equilibrium constant expression for the reaction shown in Question 7.63.

7.84 Write a valid equilibrium constant expression for the reaction shown in Question 7.64.

7.85 Write the equilibrium constant expression for the reaction:

$$N_2(g) + 3H_2(g) \rightleftharpoons 2NH_3(g)$$

7.86 Using the equilibrium constant expression in Question 7.85, calculate the equilibrium constant if:

$$[N_2] = 0.071\,M$$
$$[H_2] = 9.2 \times 10^{-3}\,M$$
$$[NH_3] = 1.8 \times 10^{-4}\,M$$

7.87 Write the equilibrium constant expression for the reaction:

$$2H_2(g) + S_2(g) \rightleftharpoons 2H_2S(g)$$

7.88 Using the equilibrium constant expression in Question 7.87, calculate the equilibrium constant if:

$$[H_2] = 2.1 \times 10^{-1} M$$
$$[S_2] = 1.1 \times 10^{-6} M$$
$$[H_2S] = 7.3 \times 10^{-1} M$$

7.89 Use the equilibrium constant expression you wrote in Question 7.85 and the equilibrium constant you calculated in Question 7.86 to determine the equilibrium concentration of NH_3 if:

$$[N_2] = 8.0 \times 10^{-2} M$$
$$[H_2] = 5.0 \times 10^{-3} M$$

7.90 Use the equilibrium constant expression you wrote in Question 7.87 and the equilibrium constant you calculated in Question 7.88 to determine the equilibrium concentration of H_2S if:

$$[H_2] = 1.0 \times 10^{-1} M$$
$$[S_2] = 1.0 \times 10^{-5} M$$

7.91 For the reaction

$$CH_4(g) + Cl_2(g) \rightleftharpoons CH_3Cl(g) + HCl(g) + 26.4 \text{ kcal}$$

predict the effect on the position of the equilibrium (will it shift to the left or to the right, or will there be no change?) for each of the following changes.
  a. The temperature is increased.
  b. The pressure is increased by decreasing the volume of the container.
  c. A catalyst is added.

7.92 For the reaction

$$47 \text{ kcal} + 2SO_3(g) \rightleftharpoons 2SO_2(g) + O_2(g)$$

predict the effect on the position of the equilibrium (will it shift to the left or to the right, or will there be no change?) for each of the following changes.
  a. The temperature is increased.
  b. The pressure is increased by decreasing the volume of the container.
  c. A catalyst is added.

7.93 Use LeChatelier's principle to predict whether the amount of PCl_3 in a 1.00-L container is increased, is decreased, or remains the same for the equilibrium

$$PCl_3(g) + Cl_2(g) \rightleftharpoons PCl_5(g) + \text{heat}$$

when each of the following changes is made.
  a. PCl_5 is added.
  b. Cl_2 is added.
  c. PCl_5 is removed.
  d. The temperature is decreased.
  e. A catalyst is added.

7.94 Use LeChatelier's principle to predict the effects, if any, of each of the following changes on the equilibrium system, described below, in a closed container.

$$C(s) + 2H_2(g) \rightleftharpoons CH_4(g) + 18 \text{ kcal}$$

  a. C is added.
  b. H_2 is added.
  c. CH_4 is removed.

  d. The temperature is increased.
  e. A catalyst is added.

7.95 Will an increase in pressure increase, decrease, or have no effect on the concentration of $H_2(g)$ in the reaction:

$$C(s) + H_2O(g) \rightleftharpoons CO(g) + H_2(g)$$

7.96 Will an increase in pressure increase, decrease, or have no effect on the concentration of $NO(g)$ in the reaction:

$$N_2(g) + O_2(g) \rightleftharpoons 2NO(g)$$

7.97 Write the equilibrium constant expression for the reaction described in Question 7.95.

7.98 Write the equilibrium constant expression for the reaction described in Question 7.96.

7.99 True or false: The position of the equilibrium will shift to the right when a catalyst is added to the mixture described in Question 7.95. Explain your reasoning.

7.100 True or false: The position of the equilibrium for an endothermic reaction will shift to the right when the reaction mixture is heated. Explain your reasoning.

7.101 Why is it dangerous to heat an unopened bottle of cola?

7.102 Carbonated beverages quickly go flat (lose CO_2) when heated. Explain, using LeChatelier's principle.

7.103 **a.** Write the equilibrium constant expression for the reaction:

$$2SO_2(g) + O_2(g) \rightleftharpoons 2SO_3(g)$$

  b. Calculate the equilibrium constant if

$$[SO_2] = 0.10 \ M$$
$$[O_2] = 0.12 \ M$$
$$[SO_3] = 0.60 \ M$$

7.104 **a.** Write the equilibrium constant expression for the reaction:

$$C(s) + H_2O(g) \rightleftharpoons CO(g) + H_2(g)$$

  b. Calculate the equilibrium constant if

$$[H_2O] = 0.40 \ M$$
$$[CO] = 0.40 \ M$$
$$[H_2] = 0.20 \ M$$

7.105 Suggest a change in experimental conditions that would increase the yield of SO_3 in the reaction in Question 7.103.

7.106 Suggest a change in experimental conditions that would increase the yield of H_2 in the reaction in Question 7.104.

CHALLENGE PROBLEMS

1. Predict the sign of ΔG for perspiration evaporating. Would you expect the ΔH term or the ΔS term to be more dominant? Explain your reasoning.

2. Can the following statement ever be true? "Heating a reaction mixture increases the rate of a certain reaction but decreases the yield of product from the reaction." Explain why or why not.

3. Molecules must collide for a reaction to take place. Sketch a model of the orientation and interaction of HI and Cl that is most favorable for the reaction:

$$HI(g) + Cl(g) \longrightarrow HCl(g) + I(g)$$

4. Refer to Question 7.92. Choose the statements that are correct.
   a. K_{eq} (at 25°C) > K_{eq} (at 50°C)
   b. K_{eq} (at 25°C) < K_{eq} (at 50°C)
   c. K_{eq} (at 25°C) = K_{eq} (at 50°C)
   d. K_{eq} (at 25°C) > K_{eq} (at 15°C)
   e. K_{eq} (at 25°C) < K_{eq} (at 15°C)
   f. K_{eq} (at 25°C) = K_{eq} (at 15°C)

5. Silver ion reacts with chloride ion to form the precipitate silver chloride:

$$Ag^+(aq) + Cl^-(aq) \rightleftharpoons AgCl(s)$$

After the reaction reached equilibrium, the chemist filtered 99% of the solid silver chloride from the solution, hoping to shift the equilibrium to the right, to form more product. Critique the chemist's experiment.

6. Human behavior often follows LeChatelier's principle. Provide one example and explain in terms of LeChatelier's principle.

7. A clever device found in some homes is a figurine that is blue on dry, sunny days and pink on damp, rainy days. These figurines are coated with substances containing chemical species that undergo the following equilibrium reaction:

$$Co(H_2O)_6{}^{2+}(aq) + 4Cl^-(aq) \rightleftharpoons CoCl_4{}^{2-}(aq) + 6H_2O(l)$$

   a. Which substance is blue?
   b. Which substance is pink?
   c. How is LeChatelier's principle applied here?

8. You have spent the entire morning in a 20°C classroom. As you ride the elevator to the cafeteria, six persons enter the elevator after being outside on a subfreezing day. You suddenly feel chilled. Explain the heat flow situation in the elevator in thermodynamic terms.

8

Acids and Bases

Solution properties, including clarity and bacteria levels, are often pH dependent.

OUTLINE

LEARNING GOALS

1 Classify compounds with acid-base properties as acids, bases, or amphiprotic.
2 Write equations illustrating the role of water in acid-base reactions.
3 Identify conjugate acid-base pairs.
4 Describe the relationship between acid and base strength and dissociation.
5 Use the ion product constant for water to solve for hydronium and hydroxide ion concentrations.
6 Calculate pH from solution concentration data.
7 Calculate hydronium and/or hydroxide ion concentration from pH data.
8 Describe the meaning and utility of neutralization reactions.
9 Use titration data to determine the molar concentration of an unknown solution.
10 Demonstrate the reactions and dissociation of polyprotic substances.
11 Describe the effects of adding acid or base to a buffer system.
12 Calculate the pH of buffer solutions.
13 Explain the role of buffers in the control of blood pH under various conditions.

INTRODUCTION

Acids and bases include some of the most important compounds in nature. Historically, it was recognized that certain compounds, acids, had a sour taste, were able to dissolve some metals, and caused vegetable dyes to change color. Bases have long been recognized by their bitter taste, slippery feel, and corrosive nature. Bases react strongly with acids and cause many metal ions in solution to form a solid precipitate.

Digestion of proteins is aided by stomach acid (hydrochloric acid), and many biochemical processes such as enzyme catalysis depend on the proper level of acidity. Indeed, a wide variety of chemical reactions critically depend on the acid-base composition of the solution. This is especially true of the biochemical reactions occurring in the cells of our bodies. For this reason, the level of acidity must be very carefully regulated. This is done with substances called buffers.

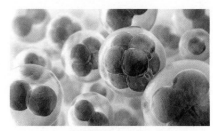

Various processes that occur within our cells are pH dependent.

8.1 Acids and Bases

Properties specific to acids and bases are due to unique characteristics of their chemical structures. Two important theories help us to understand the behavior of these classes of compounds in aqueous solution.

Acid and Base Theories

Arrhenius Theory

The Arrhenius theory was proposed in the late 1800s to describe general characteristics of acids and bases. According to this early theory, when an **Arrhenius acid** dissolves in water, it dissociates to form *hydrogen ions* or *protons* **(H^+),** and when an **Arrhenius base** dissolves in water, it dissociates to form *hydroxide ions* **(OH^-).** For example, hydrochloric acid dissociates in aqueous solution according to the reaction:

$$HCl(aq) \longrightarrow H^+(aq) + Cl^-(aq)$$

Sodium hydroxide, a base, produces hydroxide ions in aqueous solution:

$$NaOH(aq) \longrightarrow Na^+(aq) + OH^-(aq)$$

The Arrhenius theory also explains neutralization of acids and bases. When an acid and base react, a *salt* (an ionic compound) and water form. The water is formed from the proton generated by the acid and the hydroxide ion generated by the base.

Although the Arrhenius theory explains the behavior of many acids and bases, it does not explain substances with basic properties that do not contain OH^-. For example, a substance such as ammonia (NH_3) has basic properties that cannot be explained as an Arrhenius base because it does not contain OH^-. Another observation not explained by the Arrhenius theory is that protons do not exist as H^+ in aqueous solutions. Rather, they interact with water and form **hydronium ions, H_3O^+.**

Brønsted-Lowry Theory

The Brønsted-Lowry theory was developed in the 1920s from the Arrhenius theory, and it explains the acid-base chemistry that could not be explained by the Arrhenius theory. The expanded nature of the Brønsted-Lowry theory considers the central role of the solvent water in the dissociation process.

According to this inclusive theory, a **Brønsted-Lowry acid** is defined as a proton donor, and a **Brønsted-Lowry base** is defined as a proton acceptor.

As a Brønsted-Lowry acid, hydrochloric acid in aqueous solution donates a proton to the solvent, water, which results in the formation of the hydronium ion, H_3O^+. A single arrow in the following equation is used because this acid-base reaction is essentially an irreversible process.

$$HCl(aq) + H_2O(l) \longrightarrow H_3O^+(aq) + Cl^-(aq)$$

LEARNING GOAL

1 Classify compounds with acid-base properties as acids, bases, or amphiprotic.

LEARNING GOAL

2 Write equations illustrating the role of water in acid-base reactions.

$$HCl \quad + \quad H_2O \quad \longrightarrow \quad H_3O^+ \quad + \quad Cl^-$$

Only this hydrogen can be donated as a proton, H^+.

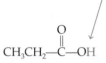

Figure 8.1 This organic acid contributes to the flavor of Swiss cheese.

Organic acids (those principally comprised of carbon, hydrogen, and oxygen) are classified as Brønsted-Lowry acids. Although most organic acids contain several hydrogen atoms, only a specific hydrogen can be donated as a proton. When evaluating the structures of these organic acids, it is essential to examine the bonds. There are often hydrogen atoms directly attached to carbon atoms along with a carbon atom that shares a double bond with one oxygen atom and is also bonded to an OH. Only the hydrogen that is directly bonded to this electronegative oxygen can be donated as a proton (Figure 8.1). Below, you can see how such an organic acid donates its proton to water.

$$CH_3CH_2COOH(aq) + H_2O(l) \rightleftharpoons H_3O^+(aq) + CH_3CH_2COO^-(aq)$$

The double arrows in the above equation are used because this acid-base reaction is seen as a reversible equilibrium process. In this dynamic equilibrium (see Section 7.4), a mixture of CH_3CH_2COOH, H_2O, H_3O^+, and $CH_3CH_2COO^-$ results.

The basic properties of ammonia are also accounted for by this expanded theory. As a Brønsted-Lowry base, ammonia accepts a proton from the solvent water and produces the hydroxide ion, OH^-.

$$NH_3(aq) + H_2O(l) \rightleftharpoons NH_4^+(aq) + OH^-(aq)$$

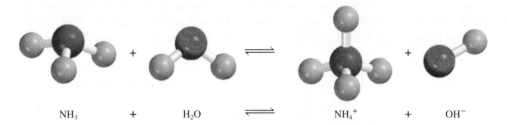

$$NH_3 \quad + \quad H_2O \quad \rightleftharpoons \quad NH_4^+ \quad + \quad OH^-$$

Many organic bases are derivatives of ammonia. When the nitrogen atom in these molecules has only three bonds and a lone pair of electrons, it can function as a Brønsted-Lowry base and form a fourth bond upon accepting a proton.

EXAMPLE 8.1   **Classifying a Compound as an Acid or Base**

LEARNING GOALS

Classify NH_4^+ as a Brønsted-Lowry acid or base, and write the reaction of NH_4^+ with water.

1 Classify compounds with acid-base properties as acids, bases, or amphiprotic.

Solution

2 Write equations illustrating the role of water in acid-base reactions.

Step 1. The central atom, nitrogen, contains four bonds. Nitrogen cannot accept an additional proton because it cannot form a fifth bond. Since the nitrogen in this molecule is directly attached to four hydrogens and carries a positive charge, it can donate a proton. Therefore, NH_4^+ is a Brønsted-Lowry acid.

Step 2. Since NH_4^+ is a Brønsted-Lowry acid, it donates a proton when it reacts with water.

$$NH_4^+ + H_2O \rightleftharpoons NH_3 + H_3O^+$$

Helpful Hint: When the ammonium ion, NH_4^+, donates a proton, the H^+ is transferred to water to form a hydronium ion. In this reaction, water serves as a Brønsted-Lowry base because it accepts the proton.

Practice Problem 8.1

Classify CH_3COO^- as a Brønsted-Lowry acid or base, and write the reaction of CH_3COO^- with water.

▶ For Further Practice: **Questions 8.23 and 8.24.**

Question 8.1 Classify each of the following compounds as a Brønsted-Lowry acid or base.
a. $HClO_4$ b. $HCOOH$ c. ClO_4^- d. $C_6H_5COO^-$

Question 8.2 Classify each of the following compounds as a Brønsted-Lowry acid or base.
a. PO_4^{3-} b. $CH_3NH_3^+$ c. HI d. H_3PO_4

Amphiprotic Nature of Water

The role that the solvent, water, plays in acid-base reactions is noteworthy. As shown in a previous example, the water molecule accepts a proton from the HCl molecule. The water is behaving as a proton acceptor, a Brønsted-Lowry base.

However, when water is a solvent for ammonia (NH_3), a Brønsted-Lowry base, the water molecule donates a proton to the ammonia molecule. The water, in this situation, is acting as a proton donor, a Brønsted-Lowry acid.

Water, owing to the fact that it possesses *both* acidic and basic properties, is termed **amphiprotic.** Water is the most commonly used solvent for acids and bases. In addition to its amphiprotic properties, the solute-solvent interactions between water with acids (or bases) promote both the solubility and the dissociation of acids and bases.

Question 8.3 Write an equation for the reversible reactions of each of the following with water.
a. HF b. $C_6H_5COO^-$

Question 8.4 Write an equation for the reversible reactions of each of the following with water.
a. H_3PO_4 b. CH_3NH_2

LEARNING GOAL

2 Write equations illustrating the role of water in acid-base reactions.

The bicarbonate ion, HCO_3^-, is an example of another amphiprotic compound. It can serve as a proton donor and a proton acceptor.

Conjugate Acid-Base Pairs

LEARNING GOAL

3 Identify conjugate acid-base pairs.

When a Brønsted-Lowry acid donates a proton to a Brønsted-Lowry base, the base that accepts the proton becomes a **conjugate acid.** Alternatively, the acid that donated the proton becomes a **conjugate base.** The product acids and bases are termed conjugate acids and bases.

Consider the model equation below in which HX is an acid and Y is a base. In the forward reaction, the acid (HX) donates a proton to the base (Y) leading to the formation of a conjugate acid (HY^+) and a conjugate base (X^-).

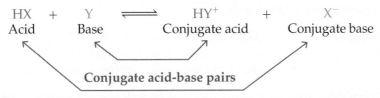

In the reverse reaction, it is the conjugate acid (HY^+) that behaves as an acid; it donates its proton to X^-. Therefore, X^- is a base in its own right because it accepts the proton. An acid and base on the opposite sides of the equation are collectively

Conjugate acid-base pairs are usually notated by writing the acid first (even if it is a conjugate acid from the product side of the reaction equation). The acid is then separated from the base with a "/."

termed a **conjugate acid-base pair.** The two conjugate acid-base pairs in the above equation are HX/X^- and HY^+/Y.

When hydrogen sulfide, H_2S, reacts with water, H_2S donates a proton to water and forms the conjugate base HS^-. Water accepts the proton and forms the conjugate acid, H_3O^+.

$$H_2S(aq) \quad + \quad H_2O(l) \quad \rightleftharpoons \quad H_3O^+(aq) \quad + \quad HS^-(aq)$$

Acid   Base   Conjugate acid   Conjugate base

Conjugate acid-base pairs

The two conjugate acid-base pairs in this reaction are H_2S/HS^- and H_3O^+/H_2O.

Question 8.5   Identify the conjugate acid-base pairs for the reversible reactions in Question 8.3.

Question 8.6   Identify the conjugate acid-base pairs for the reversible reactions in Question 8.4.

Acid and Base Strength

4 Describe the relationship between acid and base strength and dissociation.

Acids and bases are electrolytes. When dissolved in water, the dissociation of the acid (or base) produces ions that can conduct an electrical current. The strength of an acid (or base) is measured by the degree of dissociation of the acid (or base) in solution. As a result of the differences in the degree of dissociation, *strong acids and bases are strong electrolytes while weak acids and bases are weak electrolytes.* It is important to note that acid (or base) strength is independent of acid (or base) concentration. Recall that concentration refers to the amount of solute (in this case, the quantity of acid or base) per quantity of solution.

The strength of an acid (or base) in an aqueous solution depends on the extent to which it reacts with the solvent, water. Although in several reactions, we show the forward and reverse arrows to indicate the reversibility of the reaction, seldom are the two processes "equal but opposite." One reaction, either forward or reverse, is usually favored. Consider the reaction of hydrochloric acid in water:

Significant
$$HCl(aq) \quad + \quad H_2O(l) \quad \rightleftharpoons \quad H_3O^+(aq) \quad + \quad Cl^-(aq)$$
Insignificant

In this reaction, the forward reaction predominates, the reverse reaction is inconsequential, and hydrochloric acid is termed a strong acid.

Acids and bases are classified as *strong* when the reaction with water is virtually 100% complete and as *weak* when the reaction with water is much less than 100% complete. Since strong acids and bases are 100% dissociated in water, we use only a single forward arrow to represent its behavior.

Important strong acids include:

HCl is present in our stomach acid. HNO_3 is utilized in the manufacturing of ammonium nitrate fertilizers. H_2SO_4 is found in lead-acid automobile batteries.

Hydrochloric acid $HCl(aq) + H_2O(l) \longrightarrow H_3O^+(aq) + Cl^-(aq)$
Nitric acid $HNO_3(aq) + H_2O(l) \longrightarrow H_3O^+(aq) + NO_3^-(aq)$
Sulfuric acid $H_2SO_4(aq) + H_2O(l) \longrightarrow H_3O^+(aq) + HSO_4^-(aq)$

Note that the equation for the dissociation of each of these acids is written with a single forward arrow to represent their behavior. This indicates that the reaction has little or no tendency to proceed in the reverse direction to establish equilibrium. Each of these acids is virtually completely dissociated in water, forming ions.

All common strong bases are *metal hydroxides*. Strong bases completely dissociate in aqueous solution to produce hydroxide ions and metal cations. Of the common metal hydroxides, only NaOH and KOH are soluble in water.

Sodium hydroxide	$NaOH(aq) \longrightarrow Na^+(aq) + OH^-(aq)$
Potassium hydroxide	$KOH(aq) \longrightarrow K^+(aq) + OH^-(aq)$

Both NaOH and KOH are used in the production of soap from animal fats and vegetable oils.

Weak acids and weak bases dissolve in water principally in the molecular form. Only a small percentage of the molecules dissociate to form the hydronium or hydroxide ion. All organic acids are weak acids.

Two important weak acids are:

Acetic acid	$CH_3COOH(aq) + H_2O(l) \rightleftharpoons H_3O^+(aq) + CH_3COO^-(aq)$
Carbonic acid	$HOCOOH(aq) + H_2O(l) \rightleftharpoons H_3O^+(aq) + HOCOO^-(aq)$

Diluted acetic acid is in vinegar, and carbonic acid is used in make bubbly beverages like soda.

We have already mentioned the most common weak base, ammonia. Many organic derivatives of ammonia function as weak bases. Two examples of weak bases are:

Aniline	$C_6H_5NH_2(aq) + H_2O(l) \rightleftharpoons C_6H_5NH_3^+(aq) + OH^-(aq)$
Methylamine	$CH_3NH_2(aq) + H_2O(l) \rightleftharpoons CH_3NH_3^+(aq) + OH^-(aq)$

The chemistry of organic acids will be discussed in Chapter 14, and the chemistry of organic bases will be discussed in Chapter 15.

The fundamental chemical difference between strong and weak acids (or bases) is their equilibrium ion concentration. A strong acid in aqueous solution, such as HCl, does not exist to any measurable degree in equilibrium with its ions, H_3O^+ and Cl^-. On the other hand, a weak acid in aqueous solution, such as acetic acid, establishes a dynamic equilibrium with its ions, H_3O^+ and CH_3COO^-, as illustrated in Figure 8.2.

This explains the inverse correlation that exists between conjugate acid-base pairs. The strongest acids and bases have the weakest conjugate bases and acids. Conjugate acid-base pairs are listed in the middle columns of Figure 8.3.

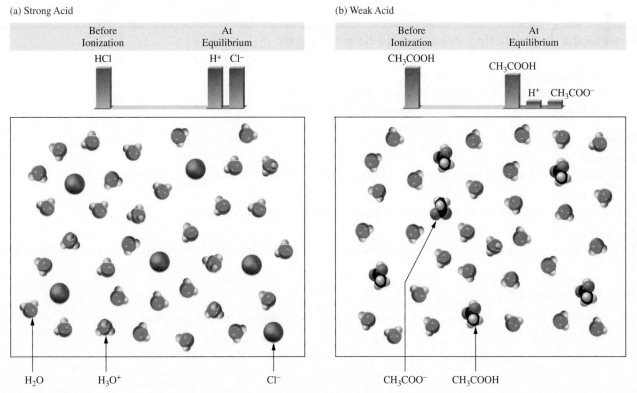

Figure 8.2 Initially, there were six HCl and six CH_3COOH molecules present. (a) The strong acid is completely dissociated in aqueous solution, whereas (b) the weak acid exists in equilibrium with its ions.

Figure 8.3 Conjugate acid-base pairs. Strong acids have weak conjugate bases; strong bases have weak conjugate acids. Note the complementary nature of the conjugate acid-base pairs. In every case, the conjugate base has one fewer H^+ than the corresponding conjugate acid.

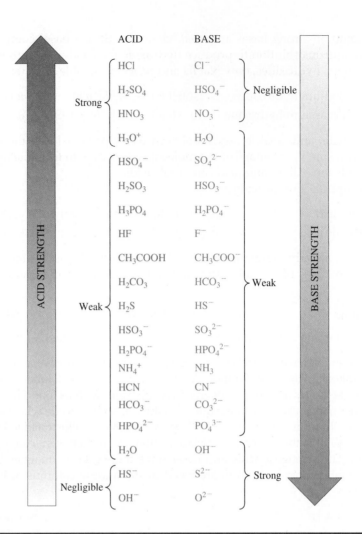

EXAMPLE 8.2   Predicting Relative Acid-Base Strengths

a. Write the conjugate acid of HS^-.
b. Using Figure 8.3, identify the stronger base, HS^- or F^-.
c. Using Figure 8.3, identify the stronger acid, H_2S or HF.

Solution

For part a,
The conjugate acid may be constructed by adding a proton (H^+) to the base structure, consequently, H_2S.

For part b,
HS^- is the stronger base because it is located farther down the right-hand column.

For part c,
HF is the stronger acid because its conjugate base is weaker *and* because it is located farther up the left-hand column.

Practice Problem 8.2

a. In each pair, write the conjugate base of each acid and identify the stronger acid.
   H_2O or NH_4^+ H_2SO_4 or H_2SO_3
b. In each pair, write the conjugate acid of each base and identify the stronger base.
   CO_3^{2-} or PO_4^{3-} HCO_3^- or HPO_4^{2-}

▶ For Further Practice: **Questions 8.37–8.40.**

As illustrated, the strongest acids (top left) are directly adjacent to the weakest bases (top right), and the weakest acids (bottom left) are directly adjacent to the strongest bases (bottom right).

Self-Ionization of Water and K_w

Although pure water is virtually 100% molecular, a small number of water molecules do ionize. This process, called **self-ionization,** occurs by the transfer of a proton from one water molecule to another, producing a hydronium ion and a hydroxide ion:

$$H_2O(l) + H_2O(l) \rightleftharpoons H_3O^+(aq) + OH^-(aq)$$

Based on the self-ionization reaction, the generalized equilibrium constant expression is:

$$K_{eq} = [H_3O^+][OH^-]$$

The water term is not included in the equilibrium concentration expression because water is a liquid and its concentration does not change (see Equilibrium, Section 7.4). As a result, the equilibrium constant for water is the product of the hydronium and hydroxide ion concentrations and is called the **ion product constant for water,** symbolized by K_w (the subscript w refers to water).

$$K_w = [H_3O^+][OH^-]$$

It has been determined that at 25°C, pure water has a hydronium ion concentration of 1.0×10^{-7} M. When two water molecules react, one hydroxide ion is produced for each hydronium ion. Therefore, the hydroxide ion concentration is also 1.0×10^{-7} M. As we saw in Chapter 7, the molar equilibrium concentration is conveniently indicated by brackets around the species whose concentration is represented:

$$[H_3O^+] = [OH^-] = 1.0 \times 10^{-7}\,M \quad \text{(at 25°C)}$$

These molar equilibrium concentrations can be used to solve for K_w.

$$K_w = [H_3O^+][OH^-] = (1.0 \times 10^{-7}\,M)(1.0 \times 10^{-7}\,M) = 1.0 \times 10^{-14}\,M$$

The ion product constant for water is a temperature-dependent quantity but it does not depend on the identity or concentration of the solute. When solutes are added to water, they alter the relative concentrations of the hydronium and hydroxide ions, but not the product $[H_3O^+][OH^-]$. *The product,* $[H_3O^+][OH^-]$, *always equals* 1.0×10^{-14} *M (at 25°C)*. This product relationship is the basis for the pH scale which is useful in the measurement of the level of acidity (or basicity) of solutions.

LEARNING GOAL

5 Use the ion product constant for water to solve for hydronium and hydroxide ion concentrations.

Since the ion product, $[H_3O^+][OH^-]$, is constant, increasing the concentration of one results in a decreased concentration of the other.

EXAMPLE 8.3 | **Using the Ion Product Constant for Water to Calculate an Unknown Hydroxide Ion Concentration**

A solution of green tea has a hydronium ion concentration of 4.0×10^{-4} M (at 25°C). What is the concentration of hydroxide ions in this solution?

Solution

Step 1. Write the expression for the ion product constant for water.

$$K_w = [H_3O^+][OH^-] = 1.0 \times 10^{-14}\,M$$

LEARNING GOAL

5 Use the ion product constant for water to solve for hydronium and hydroxide ion concentrations.

Continued...

Step 2. To solve the expression for [OH⁻], divide both sides of the equation by [H₃O⁺].

$$\frac{K_w}{[H_3O^+]} = [OH^-]$$

Step 3. Substitute the values:

$$\frac{1.0 \times 10^{-14}\,M}{4.0 \times 10^{-4}\,M} = [OH^-]$$

Step 4. Solve:

$$[OH^-] = 2.5 \times 10^{-11}\,M$$

Practice Problem 8.3

Analysis of a patient's blood sample indicated that the hydroxide ion concentration was $3.0 \times 10^{-7}\,M$ (at 25°C). What was the hydronium ion concentration in the blood sample?

▶ For Further Practice: **Questions 8.51 and 8.52.**

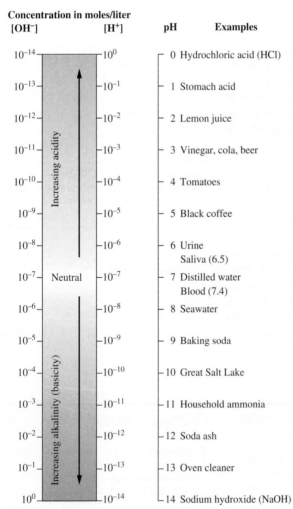

Figure 8.4 The pH scale. A pH of 7 is neutral ([H₃O⁺] = [OH⁻]). Values less than 7 are acidic (H₃O⁺ predominates), and values greater than 7 are basic (OH⁻ predominates).

Question 8.7 Calculate the hydroxide ion concentration in an aqueous solution of lemon juice that has a hydronium ion concentration of 0.025 M.

Question 8.8 The hydroxide ion concentration in a sample of urine was determined to be of $1.0 \times 10^{-8}\,M$ (at 25°C). Calculate the hydronium ion concentration of this aqueous solution.

8.2 pH: A Measurement Scale for Acids and Bases

A Definition of pH

The **pH scale** gauges the hydronium ion concentration and reflects the degree of acidity or basicity of a solution. The pH scale is somewhat analogous to the temperature scale used for assignment of relative levels of hot or cold. The temperature scale was developed to allow us to indicate how cold or how hot an object is. The pH scale specifies how acidic or how basic a solution is. The pH scale has values that range from 0 (very acidic) to 14 (very basic). A pH of 7, the middle of the scale, is neutral, neither acidic nor basic. Figure 8.4 provides a convenient overview of solution pH.

To help us develop a concept of pH, let's consider the following:

- Addition of an acid (proton donor) to water *increases* the [H₃O⁺] and decreases the [OH⁻].
- Addition of a base (proton acceptor) to water *decreases* the [H₃O⁺] by increasing the [OH⁻].
- [H₃O⁺] = [OH⁻] when *equal* amounts of acid and base are present.

In all three cases, [H₃O⁺][OH⁻] = 1.0×10^{-14} = the ion product for water at 25°C.

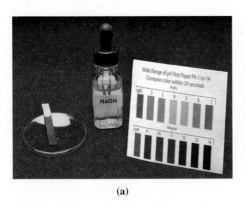
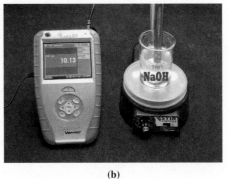

(a)   (b)

Figure 8.5 The measurement of pH. (a) A strip of test paper impregnated with indicator (a material that changes color as the acidity of the surroundings changes) is put in contact with the solution of interest. The resulting color is matched with a standard color chart (colors shown as a function of pH) to obtain the approximate pH. (b) A pH meter uses a sensor (a pH electrode) that develops an electrical potential that is proportional to the pH of the solution.

Measuring pH

The pH of a solution can be calculated if the concentration of either H_3O^+ or OH^- is known. Alternatively, measurement of pH allows the calculation of the H_3O^+ or OH^- concentration. The pH of aqueous solutions may be approximated by using indicating paper (pH paper) that develops a color related to the solution pH. Alternatively, a pH meter can give us a much more exact pH measurement. A sensor measures an electrical property of a solution that is proportional to pH (Figure 8.5).

pH values greater than 14 and less than zero are possible, but largely meaningless, due to ion association characteristics of very concentrated solutions.

Calculating pH

One of our objectives in this chapter is to calculate the pH of a solution when the hydronium or hydroxide ion concentration is known, and to calculate $[H_3O^+]$ or $[OH^-]$ from the measured pH.

The pH of a solution is defined as the negative logarithm of the molar concentration of the hydronium ion:

$$pH = -\log [H_3O^+]$$

The proper use of the logarithm function often appears to contradict our understanding of significant figures. This is because the logarithm is an exponent that contains two kinds of information: the information in the measurement and the position of the decimal point.

The operative rule is that the number of decimal places in the logarithm is equal to the number of significant figures in the original number. In Examples 8.4 and 8.5, two significant figures in concentration correspond to two decimal places in pH (the logarithm).

LEARNING GOAL

6 Calculate pH from solution concentration data.

EXAMPLE 8.4   **Calculating pH from Acid Molarity**

LEARNING GOAL

6 Calculate pH from solution concentration data.

Calculate the pH of a $1.0 \times 10^{-3} M$ solution of HCl.

Solution

Step 1. Recognize that HCl is a strong acid; consequently, it is 100% dissociated.

Step 2. One H_3O^+ is produced for each HCl. Therefore, a $1.0 \times 10^{-3} M$ HCl solution has $[H_3O^+] = 1.0 \times 10^{-3} M$.

Step 3. Using the expression for pH:

$$pH = -\log [H_3O^+]$$

Continued...

Step 4. Substituting for $[H_3O^+]$:

$$pH = -\log(1.0 \times 10^{-3})$$
$$= -(-3.00) = 3.00$$

Practice Problem 8.4

Calculate the pH of a $1.0 \times 10^{-4}\,M$ solution of HNO_3.

▶ For Further Practice: **Questions 8.55 and 8.56.**

EXAMPLE 8.5   Calculating $[H_3O^+]$ from pH

Calculate the $[H_3O^+]$ of a solution of hydrochloric acid with pH = 4.00.

Solution

Step 1. Based on the pH expression:

$$pH = -\log[H_3O^+]$$

an alternative mathematical form of this equation is:

$$[H_3O^+] = 10^{-pH}$$

Step 2. Substituting for pH:

$$[H_3O^+] = 10^{-4.00}$$
$$= 1.0 \times 10^{-4}\,M$$

LEARNING GOAL

7 Calculate hydronium and/or hydroxide ion concentration from pH data.

Practice Problem 8.5

Calculate the $[H_3O^+]$ of a solution of HNO_3 that has a pH = 5.00.

▶ For Further Practice: **Questions 8.57 and 8.58.**

It is important to remember that in the case of a base you must convert the $[OH^-]$ to $[H_3O^+]$, using the expression for the ion product for the solvent, water. It is useful to be aware that, because pH is a base 10 logarithmic function, each tenfold change in concentration changes the pH by one unit. A tenfold change in concentration is equivalent to moving the decimal point one place.

EXAMPLE 8.6   Calculating the pH of a Base

Calculate the pH of a $1.0 \times 10^{-5}\,M$ solution of NaOH.

Solution

Step 1. Recognize that NaOH is a strong base; consequently, it is 100% dissociated.

Step 2. One $[OH^-]$ is produced for each NaOH. Therefore, a $1.0 \times 10^{-5}\,M$ NaOH solution has $[OH^-] = 1.0 \times 10^{-5}\,M$.

Step 3. To calculate pH, we need $[H_3O^+]$. Recall that

$$[H_3O^+][OH^-] = 1.0 \times 10^{-14}$$

LEARNING GOAL

6 Calculate pH from solution concentration data.

Step 4. Solving this equation for $[H_3O^+]$,

$$[H_3O^+] = \frac{1.0 \times 10^{-14}}{[OH^-]}$$

Step 5. Substituting the information provided in the problem,

$$[H_3O^+] = \frac{1.0 \times 10^{-14}}{1.0 \times 10^{-5}}$$
$$= 1.0 \times 10^{-9}\,M$$

Step 6. The solution is now similar to that in Example 8.4:

$$pH = -\log[H_3O^+]$$
$$= -\log(1.0 \times 10^{-9})$$
$$= 9.00$$

Practice Problem 8.6

a. Calculate the pH corresponding to a $1.0 \times 10^{-2}\,M$ solution of sodium hydroxide.
b. Calculate the pH corresponding to a $1.0 \times 10^{-6}\,M$ solution of sodium hydroxide.

▶ For Further Practice: **Questions 8.59 and 8.60.**

EXAMPLE 8.7   **Calculating Both Hydronium and Hydroxide Ion Concentrations from pH**

LEARNING GOAL

7 Calculate hydronium and/or hydroxide ion concentration from pH data.

Calculate the $[H_3O^+]$ and $[OH^-]$ of a sodium hydroxide solution with a pH = 10.00.

Solution

Step 1. First, calculate $[H_3O^+]$ using our expression for pH:

$$pH = -\log[H_3O^+]$$
$$[H_3O^+] = 10^{-pH}$$
$$= 10^{-10}\,M$$

Step 2. To calculate the $[OH^-]$, we need to solve for $[OH^-]$ by using the following expression:

$$K_w = [H_3O^+][OH^-] = 1.0 \times 10^{-14}$$
$$[OH^-] = \frac{1.0 \times 10^{-14}}{[H_3O^+]}$$

Step 3. Substituting the $[H_3O^+]$ from the first step, we have

$$[OH^-] = \frac{1.0 \times 10^{-14}}{1.0 \times 10^{-10}}$$
$$= 1.0 \times 10^{-4}\,M$$

Practice Problem 8.7

Calculate the $[H_3O^+]$ and $[OH^-]$ of a potassium hydroxide solution with a pH = 8.00.

▶ For Further Practice: **Questions 8.61 and 8.62.**

Often, the pH or $[H_3O^+]$ will not be a whole number (pH = 1.5, pH = 5.3, $[H_3O^+]$ = 1.5 × 10^{-3} and so forth) as shown in Examples 8.4–8.7. Consider Examples 8.8 and 8.9.

EXAMPLE 8.8   **Calculating pH with Noninteger Numbers**

LEARNING GOAL

Calculate the pH of a sample of lake water that has a $[H_3O^+]$ = 6.5 × 10^{-5} M.

6 Calculate pH from solution concentration data.

Solution

Step 1. Use the expression for pH:

$$pH = -\log[H_3O^+]$$

Step 2. Substituting the hydronium ion concentration provided into the problem,

$$pH = -\log(6.5 \times 10^{-5})$$
$$= 4.19$$

Note: The pH, 4.19, is low enough to suspect acid rain. (See Green Chemistry: Acid Rain in this chapter.)

Practice Problem 8.8

Calculate the pH of a sample of blood that has a $[H_3O^+]$ = 3.3 × 10^{-8} M.

▶ For Further Practice: **Questions 8.73 and 8.74.**

EXAMPLE 8.9   **Calculating $[H_3O^+]$ from pH**

LEARNING GOAL

The measured pH of a sample of stream water is 6.40. Calculate $[H_3O^+]$.

7 Calculate hydronium and/or hydroxide ion concentration from pH data.

Solution

Step 1. An alternative mathematical form of

$$pH = -\log[H_3O^+]$$

is the expression

$$[H_3O^+] = 10^{-pH}$$

Step 2. Substituting the pH provided into this expression to solve for $[H_3O^+]$:

$$[H_3O^+] = 10^{-6.40}$$
$$= 3.98 \times 10^{-7} \quad \text{or} \quad 4.0 \times 10^{-7} M$$

Practice Problem 8.9

   a. Calculate the $[H_3O^+]$ corresponding to pH = 8.50.
   b. Calculate the $[H_3O^+]$ corresponding to pH = 4.50.

▶ For Further Practice: **Questions 8.69 and 8.70.**

A Medical Perspective

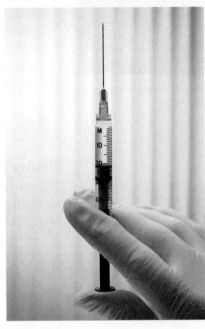

Drug Delivery

When a doctor prescribes medicine to treat a disease or relieve its symptoms, the medication may be administered in a variety of ways. Drugs may be taken orally, injected into a muscle or a vein, or absorbed through the skin. Specific instructions are often provided to regulate the particular combination of drugs that can or cannot be taken. The diet, both before and during the drug therapy, may be of special importance.

To appreciate why drugs are administered in a specific way, it is necessary to understand a few basic facts about medications and how they interact with the body.

Drugs function by undergoing one or more chemical reactions in the body. Few compounds react in only one way, to produce a limited set of products, even in the simple environment of a beaker or flask. Imagine the number of possible reactions that a drug can undergo in a complex chemical factory like the human body. In many cases, a drug can react in a variety of ways other than its intended path. These alternative paths are side reactions, sometimes producing *side effects* such as nausea, vomiting, insomnia, or drowsiness. Side effects may be unpleasant and may actually interfere with the primary function of the drug.

The development of safe, effective medication, with minimal side effects, is a slow and painstaking process, and determining the best drug delivery system is a critical step. For example, a drug that undergoes an unwanted side reaction in an acidic solution would not be very effective if administered orally. The acidic digestive fluids in the stomach could prevent the drug from even reaching the intended organ, let alone retaining its potency. The drug could be administered through a vein into the blood; blood is not acidic, in contrast to digestive fluids. In this way, the drug may be delivered intact to the intended site in the body, where it is free to undergo its primary reaction.

Drug delivery has become a science in its own right. Pharmacology, the study of drugs and their uses in the treatment of disease, has a goal of creating drugs that are highly selective. In other words, they will undergo only one reaction, the intended reaction. Encapsulation of drugs, enclosing them within larger molecules or collections of molecules, may protect them from unwanted reactions as they are transported to their intended site.

For Further Understanding

▶ Certain drugs lose potency when consumed with grapefruit juice. Can you propose a possible reason?

▶ Would you expect that the drugs referred to (above) are basic, acidic, or neutral? Explain your reasoning.

Question 8.9 Calculate the $[OH^-]$ of a $1.0 \times 10^{-3}\,M$ solution of HCl.

Question 8.10 Calculate the $[OH^-]$ of a solution of hydrochloric acid with pH = 4.00.

The Importance of pH and pH Control

Solution pH and pH control play a major role in many facets of our lives. Consider a few examples:

- *Agriculture:* Crops grow best in a soil of proper pH. Proper fertilization involves the maintenance of a suitable pH.
- *Physiology:* If the pH of our blood were to shift by one unit, we would die. Many biochemical reactions in living organisms are extremely pH dependent.

Healthy lakes and streams generally have a pH between 6.5 and 7.5.

See Green Chemistry: Acid Rain in this chapter.

Equation balancing and net ionic equations are discussed in Chapter 4.

- *Industry:* From the manufacture of processed foods to the manufacture of automobiles, industrial processes often require rigorous pH control.
- *Municipal services:* Purification of drinking water and treatment of sewage must be carried out at their optimum pH.
- *Acid rain:* Nitric acid and sulfuric acid, resulting largely from reactions of components of vehicle emissions and electric power generation (nitrogen and sulfur oxides) with water, are carried down by precipitation and enter aquatic systems (lakes and streams), lowering the pH of the water. A less than optimum pH poses serious problems for native fish populations.

In summary, any change that takes place in aqueous solution generally has at least some pH dependence.

8.3   Reactions between Acids and Bases

Neutralization

The reaction of an acid with a base to produce a salt and water is referred to as **neutralization.** In the strictest sense, neutralization requires equal numbers of moles (mol) of H_3O^+ and OH^- to produce a neutral solution (no excess acid or base).

Consider the reaction of a solution of hydrochloric acid and sodium hydroxide:

$$\underset{\text{Acid}}{HCl(aq)} + \underset{\text{Base}}{NaOH(aq)} \longrightarrow \underset{\text{Salt}}{NaCl(aq)} + \underset{\text{Water}}{H_2O(l)}$$

Our objective is to make the balanced equation represent the process actually occurring. We recognize that HCl, NaOH, and NaCl are dissociated in solution:

$$H^+(aq) + Cl^-(aq) + Na^+(aq) + OH^-(aq) \longrightarrow Na^+(aq) + Cl^-(aq) + H_2O(l)$$

We also know that Na^+ and Cl^- are unchanged in the reaction; they are termed *spectator ions.* If we write only those components that actually change, ignoring the spectator ions, we produce a *net ionic equation:*

$$H^+(aq) + OH^-(aq) \longrightarrow H_2O(l)$$

If we realize that the H^+ occurs in aqueous solution as the hydronium ion, H_3O^+, the most correct form of the net ionic equation is

$$H_3O^+(aq) + OH^-(aq) \longrightarrow 2H_2O(l)$$

The equation for any strong acid/strong base neutralization reaction is the same as this equation.

A neutralization reaction may be used to determine the concentration of an unknown acid or base solution. The technique of **titration** involves the addition of measured amounts of a **standard solution** (one whose concentration is known with certainty) to neutralize the second, unknown solution. From the volumes of the two solutions and the concentration of the standard solution, the concentration of the unknown solution may be determined.

A strategy for carrying out an acid-base titration is summarized in Table 8.1. The calculations involved in an acid-base titration are illustrated in Example 8.10.

Polyprotic Substances

Not all acid-base reactions occur in a 1:1 combining ratio (as hydrochloric acid and sodium hydroxide in Example 8.10). Acid-base reactions with other than 1:1

TABLE 8.1 Conducting an Acid-Base Titration

1. A known volume, perhaps 25.00 milliliters (mL), of the unknown acid of unknown concentration is measured into a flask using a pipet.

2. An **indicator,** a substance that changes color as the solution reaches a certain pH, is added to the unknown solution. We must know, from prior experience, the expected pH at the equivalence point (see step 4). For this titration, two indicators, phenol red or phenolphthalein, would be suitable choices.

3. A solution of sodium hydroxide, perhaps 0.1000 M, is carefully added to the unknown solution using a **buret** (Figure 8.6), which is a long glass tube calibrated in mL. A stopcock at the bottom of the buret regulates the amount of liquid dispensed. The standard solution is added until the indicator changes color.

4. At this point, the **equivalence point,** the number of mol of hydroxide ion added is equal to the number of mol of hydronium ion present in the unknown acid. The solution is neutral, with a pH equal to 7.

5. The volume dispensed by the buret, perhaps 35.00 mL, is measured.

6. Using the data from the experiment (volume of the unknown, volume of the titrant, and molarity of the titrant), the molar concentration of the unknown substance is calculated.

Phenol red, a commonly used indicator, is yellow in acid solution and turns red after all acid is neutralized. Phenolphthalein is colorless in acid solution and turns pink after all acid is neutralized. Phenolphthalein is often the indicator of choice because it is easier to discern a change of colorless to a color, rather than one color to another.

(a)

(b)

Figure 8.6 An acid-base titration. (a) An exact volume of a standard solution (in this example, a base) is added to a solution of unknown concentration (in this example, an acid). (b) From the volume (read from the buret) and concentration of the standard solution, coupled with the mass or volume of the unknown, the molar concentration of the unknown may be calculated.

EXAMPLE 8.10 | Determining the Concentration of a Solution of Hydrochloric Acid

LEARNING GOAL

9 Use titration data to determine the molar concentration of an unknown solution.

A 25.00-mL sample of a hydrochloric acid solution of unknown concentration was transferred to a flask. A few drops of the indicator phenolphthalein were added, and the resulting solution was titrated with 0.1000 M sodium hydroxide solution. After 35.00 mL of sodium hydroxide solution were added, the indicator turned pink, signaling that the unknown and titrant had reached their equivalence point. Calculate the M of the HCl solution.

Continued…

Solution

Step 1. Pertinent information for this titration includes:

Volume of the unknown HCl solution, 25.00 mL

Volume of the NaOH solution added, 35.00 mL

Concentration of the NaOH solution, 0.1000 M

Step 2. From the balanced equation, we know that 1 mol of HCl will react with 1 mol of NaOH:

$$HCl(aq) + NaOH(aq) \longrightarrow NaCl(aq) + H_2O(l)$$

Note: The net ionic equation for this reaction provides the same information; 1 mol of H_3O^+ reacts with 1 mol of OH^-.

$$H_3O^+(aq) + OH^-(aq) \longrightarrow 2H_2O(l)$$

Step 3. The number of mol NaOH can be calculated from the volume of NaOH using conversion factors (1000 mL = 1 L and M = mol/L):

$$35.00 \text{ mL NaOH} \times \frac{1 \text{ L NaOH}}{10^3 \text{ mL NaOH}} \times \frac{0.1000 \text{ mol NaOH}}{1 \text{ L NaOH}} = 3.500 \times 10^{-3} \text{ mol NaOH}$$

$\qquad$ Data Given $\qquad$ Conversion Factor $\quad$ Conversion Factor $\qquad$ Relating Unit

Step 4. Knowing that HCl and NaOH undergo a 1:1 reaction, the number of mol HCl in the unknown solution can be calculated from the number of mol NaOH added.

$$3.500 \times 10^{-3} \text{ mol NaOH} \times \frac{1 \text{ mol HCl}}{1 \text{ mol NaOH}} = 3.500 \times 10^{-3} \text{ mol HCl}$$

$\qquad\quad$ Relating Unit $\qquad\qquad$ Conversion Factor $\quad$ Initial Data Result

Step 5. The concentration of HCl can be calculated from the number of mol HCl in 25.00 mL.

$$\frac{3.500 \times 10^{-3} \text{ mol HCl}}{25.00 \text{ mL HCl soln}} \times \frac{10^3 \text{ mL HCl soln}}{1 \text{ L HCl soln}} = 1.400 \times 10^{-1} \text{ mol HCl/L HCl soln} = 0.1400 \ M$$

$\quad$ Initial Data Result $\qquad$ Conversion Factor $\qquad\qquad$ Desired Result

Practice Problem 8.10

a. Calculate the molar concentration of a sodium hydroxide solution if 40.00 mL of this solution were required to neutralize 20.00 mL of a 0.2000 M solution of hydrochloric acid.

b. Calculate the molar concentration of a sodium hydroxide solution if 36.00 mL of this solution were required to neutralize 25.00 mL of a 0.2000 M solution of hydrochloric acid.

▶ For Further Practice: **Questions 8.87 and 8.88.**

combining ratios occur between what are termed *polyprotic substances*. **Polyprotic substances** donate (as acids) or accept (as bases) more than one proton per formula unit.

Reactions of Polyprotic Substances

HCl dissociates to produce one H^+ ion for each HCl. For this reason, it is termed a *monoprotic acid*. Its reaction with sodium hydroxide is:

$$HCl(aq) + NaOH(aq) \longrightarrow H_2O(l) + Na^+(aq) + Cl^-(aq)$$

Sulfuric acid, in contrast, is a *diprotic acid*. Each unit of H_2SO_4 produces two H^+ ions (the prefix *di-* indicating two). Its reaction with sodium hydroxide is:

$$H_2SO_4(aq) + 2NaOH(aq) \longrightarrow 2H_2O(l) + 2Na^+(aq) + SO_4^{2-}(aq)$$

Green Chemistry

Hydrangea, pH, and Soil Chemistry

It is difficult to pass a group of hydrangea in a park or garden without stopping, at least briefly, to take note of their beauty. Most are blue or pink; occasionally, one flower shows a mixture of two colors. They are usually the variety *Hydrangea macrophylla*, a plant native to Japan.

It was originally believed that the pink and blue-flowered plants were different species. It was subsequently determined that they were, in fact, the same species, and the color that developed was a function of the type of soil used. Shortly thereafter, soil pH was implicated; low pH favored blue coloration, and high pH favored pink. The mechanism was probably protonation and deprotonation of one or more molecules whose color depended on the extent of protonation.

We learned in Chapter 1 that theories are subject to further investigation and this investigation may lead to revised theories. This is certainly true with the hydrangea. Recent studies have shown that a compound containing Al^{3+} must form in the petals; this compound is responsible for the blue color. Absence of Al^{3+} leads to pink petals.

Where must the Al^{3+} originate? Certainly, it must be found in the soil. Aluminum in soil is present as compounds, such as Al_2O_3 or $Al(OH)_3$. Aluminum compounds are notoriously insoluble in neutral or slightly basic soils, but their solubility increases markedly in acidic soils.

So, low pH indicates high acid concentration, increased solubility of aluminum-containing soil compounds, higher concentration of $Al^{3+}(aq)$ transported from the soil to the petals, and, as a final consequence, beautiful blue hydrangea.

For Further Understanding

▶ Explain how a hydrangea can function as an indicator of soil pH.

▶ The petals of some hydrangea appear purple. Construct a possible explanation for this phenomenon.

Phosphoric acid is a *triprotic acid.* Each unit of H_3PO_4 produces three H^+ ions. Its reaction with sodium hydroxide is:

$$H_3PO_4(aq) + 3NaOH(aq) \longrightarrow 3H_2O(l) + 3Na^+(aq) + PO_4^{3-}(aq)$$

Dissociation of Polyprotic Substances

Sulfuric acid, and other diprotic acids, dissociate in two steps:

Step 1. $H_2SO_4(aq) + H_2O(l) \longrightarrow H_3O^+(aq) + HSO_4^-(aq)$

Step 2. $HSO_4^-(aq) + H_2O(l) \rightleftharpoons H_3O^+(aq) + SO_4^{2-}(aq)$

Notice that H_2SO_4 behaves as a strong acid (step 1) and HSO_4^- behaves as a weak acid, indicated by a double arrow (step 2).

Phosphoric acid dissociates in three steps, all forms behaving as weak acids.

Step 1. $H_3PO_4(aq) + H_2O(l) \rightleftharpoons H_3O^+(aq) + H_2PO_4^-(aq)$

Step 2. $H_2PO_4^+(aq) + H_2O(l) \rightleftharpoons H_3O^+(aq) + HPO_4^{2-}(aq)$

Step 3. $HPO_4^{2-}(aq) + H_2O(l) \rightleftharpoons H_3O^+(aq) + PO_4^{3-}(aq)$

Bases exhibit this property as well.

287

NaOH produces one OH^- ion per formula unit:

$$NaOH(aq) \longrightarrow Na^+(aq) + OH^-(aq)$$

$Ba(OH)_2$, barium hydroxide, produces two OH^- ions per formula unit:

$$Ba(OH)_2(aq) \longrightarrow Ba(OH)^+(aq) + OH^-(aq)$$

and

$$Ba(OH)^+(aq) \longrightarrow Ba^{2+}(aq) + OH^-(aq)$$

8.4   Acid-Base Buffers

A **buffer solution** contains components that enable the solution to resist large changes in pH when either acids or bases are added. Buffer solutions may be prepared in the laboratory to maintain optimum conditions for a chemical reaction. Buffers are routinely used in commercial products to maintain optimum conditions for product behavior.

Buffer solutions also occur naturally. Blood, for example, is a complex natural buffer solution maintaining a pH of approximately 7.4, optimum for oxygen transport. The major buffering agent in blood is the mixture of carbonic acid (H_2CO_3) and bicarbonate ions (HCO_3^-).

The Buffer Process

The basis of buffer action is the establishment of an equilibrium between either a weak acid and its conjugate base or a weak base and its conjugate acid. Let's consider the case of a weak acid and its conjugate base.

A common buffer solution may be prepared from acetic acid (CH_3COOH) and sodium acetate (CH_3COONa). Sodium acetate is a salt that is the source of the conjugate base CH_3COO^-. An *equilibrium* is established in solution between the weak acid and the conjugate base.

$$\underset{\substack{\text{Acetic acid} \\ \text{(weak acid)}}}{CH_3COOH(aq)} + \underset{\text{Water}}{H_2O(l)} \rightleftharpoons \underset{\text{Hydronium ion}}{H_3O^+(aq)} + \underset{\substack{\text{Acetate ion} \\ \text{(conjugate base)}}}{CH_3COO^-(aq)}$$

> We ignore Na^+ in the description of the buffer. Na^+ does not actively participate in the reaction and is termed a *spectator ion*.

> The acetate ion is the conjugate base of acetic acid.

> *See Section 7.4 for a discussion of LeChatelier's principle.*

A buffer solution functions in accordance with LeChatelier's principle, which states that an equilibrium system, when stressed, will shift its equilibrium position to relieve that stress. This principle is illustrated by the following examples.

Addition of Base or Acid to a Buffer Solution

Addition of a basic substance to our buffer solution causes the following changes.

- OH^- from the base reacts with H_3O^+, producing water.
- Molecular acetic acid *dissociates* to replace the H_3O^+ consumed by the base, maintaining the pH close to the initial level.

This is an example of LeChatelier's principle, because the loss of H_3O^+ (the *stress*) is compensated for by the dissociation of acetic acid to produce more H_3O^+.

Addition of an acidic solution to the buffer results in the following changes.

- H_3O^+ from the acid increases the overall $[H_3O^+]$.
- The system reacts to this stress, in accordance with LeChatelier's principle, to form more molecular acetic acid; the acetate ion combines with H_3O^+. Thus, the H_3O^+ concentration and the pH remain close to the initial level.

This follows LeChatelier's principle because the added H_3O^+ (the *stress*) is consumed by the acetate ion (conjugate base) to produce more undissociated acetic acid.

> ### LEARNING GOAL
>
> **11** Describe the effects of adding acid or base to a buffer system.

These effects may be summarized as follows:

$$CH_3COOH(aq) + H_2O(l) \rightleftharpoons H_3O^+(aq) + CH_3COO^-(aq)$$

OH⁻ added, equilibrium position shifts to the right →

← H₃O⁺ added, equilibrium position shifts to the left

The weak acid is the key component in buffering against excess OH^-, and the salt of the weak acid (the conjugate base) is key to buffering against excess H_3O^+.

Buffer Capacity

Buffer capacity is a measure of the ability of a solution to resist large changes in pH when a strong acid or strong base is added. More specifically, buffer capacity is described as the amount of strong acid or strong base that can be added to a buffer solution without significantly changing its pH. Buffering capacity against base is a function of the concentration of the weak acid (in this case, CH_3COOH). Buffering capacity against acid is dependent on the concentration of the anion of the salt, the conjugate base (CH_3COO^- in this example). Buffer solutions are often designed to have an identical buffer capacity for both acids and bases. This is achieved when, in the above example, $[CH_3COO^-]/[CH_3COOH] = 1$. As an added bonus, making the $[CH_3COO^-]$ and $[CH_3COOH]$ as large as is practical ensures a high buffer capacity for both added acid and added base.

Determining Buffer Solution pH

It is useful to understand how to determine the pH of a buffer solution. Many chemical reactions produce the largest amount of product only when they are run at an optimal, constant pH. The study of biologically important processes in the laboratory often requires conditions that approximate the composition of biological fluids. A constant pH would certainly be essential.

The buffer process is an equilibrium reaction and is described by an equilibrium constant expression. For acids, the equilibrium constant is represented as K_a (the subscript a implying an acid equilibrium). For example, the acetic acid/acetate ion system is described by

$$CH_3COOH(aq) + H_2O(l) \rightleftharpoons H_3O^+(aq) + CH_3COO^-(aq)$$

and

$$K_a = \frac{[H_3O^+][CH_3COO^-]}{[CH_3COOH]}$$

Using a few mathematical maneuvers, we can turn this equilibrium constant expression into one that will allow us to calculate the pH of the buffer if we know how much acid (acetic acid) and conjugate base (acetate ion) are present in a known volume of the solution.

First, multiply both sides of the equation by the concentration of acetic acid, $[CH_3COOH]$. This will eliminate the denominator on the right side of the equation.

$$[CH_3COOH]K_a = \frac{[H_3O^+][CH_3COO^-]\cancel{[CH_3COOH]}}{\cancel{[CH_3COOH]}}$$

or

$$[CH_3COOH]K_a = [H_3O^+][CH_3COO^-]$$

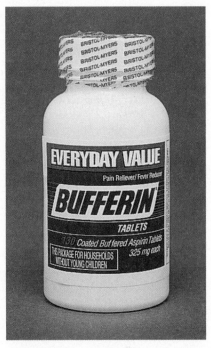

Some commercial products claim improved function owing to their ability to control pH. This product contains antacids (calcium carbonate and magnesium oxide) in addition to aspirin, an acid. Can you name other products whose performance is pH dependent?

Now, dividing both sides of the equation by the acetate ion concentration $[CH_3COO^-]$ will give us an expression for the hydronium ion concentration $[H_3O^+]$:

$$\frac{[CH_3COOH]K_a}{[CH_3COO^-]} = [H_3O^+]$$

The calculation of pH from $[H_3O^+]$ is discussed in Section 8.2.

Once we know the value for $[H_3O^+]$, we can easily find the pH. To use this equation:

• assume that $[CH_3COOH]$ represents the concentration of the acid component of the buffer.
• assume that $[CH_3COO^-]$ represents the concentration of the conjugate base (principally from the dissociation of the salt, sodium acetate) component of the buffer.

$$\frac{[CH_3COOH]K_a}{[CH_3COO^-]} = [H_3O^+]$$

$$\frac{[acid]K_a}{[conjugate\ base]} = [H_3O^+]$$

Let's look at examples of practical applications of this equation.

EXAMPLE 8.11   **Calculating the pH of a Buffer Solution**

LEARNING GOAL

Calculate the pH of a buffer solution in which both the acetic acid and sodium acetate concentrations are 1.0×10^{-1} M. The equilibrium constant, K_a, for acetic acid is 1.8×10^{-5}.

12   Calculate the pH of buffer solutions.

Solution

Step 1.   Acetic acid is the acid; $[acid] = 1.0 \times 10^{-1}$ M.
Sodium acetate is the salt, furnishing the conjugate base; $[conjugate\ base] = 1.0 \times 10^{-1}$ M.

Step 2.   The equilibrium equation is

$$CH_3COOH(aq) + H_2O(l) \rightleftharpoons H_3O^+(aq) + CH_3COO^-(aq)$$

$\qquad\qquad$ Acid $\qquad\qquad\qquad\qquad\qquad\qquad$ Conjugate base

Step 3.   The hydronium ion concentration is expressed as

$$[H_3O^+] = \frac{[acid]K_a}{[conjugate\ base]}$$

Step 4.   Substituting the values given in the problem:

$$[H_3O^+] = \frac{(1.0 \times 10^{-1})(1.8 \times 10^{-5})}{1.0 \times 10^{-1}}$$

$$= 1.8 \times 10^{-5}$$

Step 5.   This hydronium ion concentration is substituted into the expression for pH:

$$pH = -\log[H_3O^+]$$

$$= -\log(1.8 \times 10^{-5})$$

$$= 4.74$$

The pH of the buffer solution is 4.74.

Practice Problem 8.11

A buffer solution is prepared in such a way that the concentrations of propanoic acid and sodium propanoate are each 2.00×10^{-1} M. If the buffer equilibrium is described by

$$C_2H_5COOH(aq) + H_2O(l) \rightleftharpoons H_3O^+(aq) + C_2H_5COO^-(aq)$$
Propanoic acid \qquad\qquad\qquad\qquad Propanoate anion

with $K_a = 1.34 \times 10^{-5}$, calculate the pH of the solution.

▶ For Further Practice: **Questions 8.101 and 8.102.**

| **EXAMPLE 8.12** | **Calculating the pH of a Buffer Solution** | LEARNING GOAL |

Calculate the pH of a buffer solution similar to that described in Example 8.11, except that the acid concentration is doubled, while the salt concentration remains the same.

12 Calculate the pH of buffer solutions.

Solution

Step 1. Acetic acid is the acid; $[\text{acid}] = 2.0 \times 10^{-1}$ M (remember, the acid concentration is twice that of Example 8.11; $2 \times [1.0 \times 10^{-1}] = 2.0 \times 10^{-1}$ M).

Sodium acetate is the salt, furnishing the conjugate base; $[\text{conjugate base}] = 1.0 \times 10^{-1}$ M.

K_a for acetic acid is 1.8×10^{-5}.

Step 2. The equilibrium equation is

$$CH_3COOH(aq) + H_2O(l) \rightleftharpoons H_3O^+(aq) + CH_3COO^-(aq)$$
Acid \qquad\qquad\qquad\qquad Conjugate base

Step 3. The hydronium ion concentration is expressed as,

$$[H_3O^+] = \frac{[\text{acid}]K_a}{[\text{conjugate base}]}$$

Step 4. Substituting the values from step 1:

$$[H_3O^+] = \frac{(2.0 \times 10^{-1})(1.8 \times 10^{-5})}{1.00 \times 10^{-1}}$$

$$= 3.60 \times 10^{-5}$$

Step 5. This hydronium ion concentration is substituted into the expression for pH:

$$pH = -\log [H_3O^+]$$
$$= -\log (3.60 \times 10^{-5})$$
$$= 4.44$$

The pH of the buffer solution is 4.44.

Practice Problem 8.12

A buffer solution is prepared in such a way that the concentration of propanoic acid is 2.00×10^{-1} M and the concentration of sodium propanoate is 4.00×10^{-1} M. If the buffer equilibrium is described by

$$C_2H_5COOH(aq) + H_2O(l) \rightleftharpoons H_3O^+(aq) + C_2H_5COO^-(aq)$$
Propanoic acid \qquad\qquad\qquad\qquad Propanoate anion

with $K_a = 1.34 \times 10^{-5}$, calculate the pH of the solution.

▶ For Further Practice: **Questions 8.105 and 8.106.**

A comparison of the two solutions described in Examples 8.11 and 8.12 demonstrates a buffer solution's most significant attribute: the ability to stabilize pH. Although the acid concentration of these solutions differs by a factor of two, the difference in their pH is only 0.30 units.

The Henderson-Hasselbalch Equation

The solution of the equilibrium constant expression and the pH are sometimes combined into one operation. The combined expression is termed the **Henderson-Hasselbalch equation.**

For the acetic acid/sodium acetate buffer system,

$$CH_3COOH(aq) + H_2O(l) \rightleftharpoons H_3O^+(aq) + CH_3COO^-(aq)$$

$$K_a = \frac{[H_3O^+][CH_3COO^-]}{[CH_3COOH]}$$

Taking the $-\log$ of both sides of the equation:

$$-\log K_a = -\log [H_3O^+] - \log \frac{[CH_3COO^-]}{[CH_3COOH]}$$

$$pK_a = pH - \log \frac{[CH_3COO^-]}{[CH_3COOH]}$$

$pK_a = -\log K_a$, which is analogous to $pH = -\log[H_3O^+]$.

the Henderson-Hasselbalch expression is:

$$pH = pK_a + \log \frac{[CH_3COO^-]}{[CH_3COOH]}$$

The form of this equation is especially amenable to buffer problem calculations. In this expression, $[CH_3COOH]$ represents the molar concentration of the weak acid and $[CH_3COO^-]$ is the molar concentration of the conjugate base of the weak acid. The generalized expression is:

$$pH = pK_a + \log \frac{[\text{conjugate base}]}{[\text{acid}]}$$

Substituting concentrations along with the value for the pK_a of the acid allows the calculation of the pH of the buffer solution in problems such as those shown in Examples 8.11 and 8.12.

Question 8.11   Use the Henderson-Hasselbalch equation to calculate the pH of a buffer solution in which both the acetic acid and the sodium acetate concentrations are 1.0×10^{-1} M. The equilibrium constant, K_a, for acetic acid is 1.8×10^{-5}.

Question 8.12   Use the Henderson-Hasselbalch equation to calculate the pH of a buffer solution in which the acetic acid concentration is 2.0×10^{-1} M and the sodium acetate concentration is 1.0×10^{-1} M. The equilibrium constant, K_a, for acetic acid is 1.8×10^{-5}.

Question 8.13   A buffer solution is prepared in such a way that the concentrations of propanoic acid and sodium propanoate are each 2.00×10^{-1} M. If the buffer equilibrium is described by

$$C_2H_5COOH(aq) + H_2O(l) \rightleftharpoons H_3O^+(aq) + C_2H_5COO^-(aq)$$

Propanoic acid   Propanoate anion

$$K_a = 1.34 \times 10^{-5}$$

use the Henderson-Hasselbalch equation to calculate the pH of the solution.

Question 8.14 A buffer solution is prepared in such a way that the concentration of propanoic acid is $2.00 \times 10^{-1}\ M$ and the concentration of sodium propanoate is $4.00 \times 10^{-1}\ M$. If the buffer equilibrium is described by

$$C_2H_5COOH(aq) + H_2O(l) \rightleftharpoons H_3O^+(aq) + C_2H_5COO^-(aq)$$
$$\text{Propanoic acid} \qquad\qquad\qquad \text{Propanoate anion}$$

$$K_a = 1.34 \times 10^{-5}$$

use the Henderson-Hasselbalch equation to calculate the pH of the solution.

Control of Blood pH

A pH of 7.4 is maintained in blood partly by a carbonic acid–bicarbonate buffer system based on the following equilibrium:

$$H_2CO_3(aq) + H_2O(l) \rightleftharpoons H_3O^+(aq) + HCO_3^-(aq)$$
$$\text{Carbonic acid} \qquad\qquad\qquad \text{Bicarbonate ion}$$
$$\text{(weak acid)} \qquad\qquad\qquad \text{(conjugate base)}$$

LEARNING GOAL

13 Explain the role of buffers in the control of blood pH under various conditions.

The regulation process based on LeChatelier's principle is similar to the acetic acid–sodium acetate buffer equilibrium, which we have already discussed.

Red blood cells transport O_2, bound to hemoglobin, to the cells of body tissue. The metabolic waste product, CO_2, is picked up by the blood and delivered to the lungs.

The CO_2 in the blood also participates in the carbonic acid–bicarbonate buffer equilibrium. Carbon dioxide reacts with water in the blood to form carbonic acid:

$$CO_2(aq) + H_2O(l) \rightleftharpoons H_2CO_3(aq)$$

As a result, the buffer equilibrium becomes more complex:

$$CO_2(aq) + 2H_2O(l) \rightleftharpoons H_2CO_3(aq) + H_2O(l) \rightleftharpoons H_3O^+(aq) + HCO_3^-(aq)$$

Through this sequence of relationships, the concentration of CO_2 in the blood affects the blood pH.

Higher than normal CO_2 concentrations shift the above equilibrium to the right (LeChatelier's principle), increasing $[H_3O^+]$ and lowering the pH. The blood becomes too acidic, leading to numerous medical problems. A situation of high blood CO_2 levels and low pH is termed *acidosis*. Respiratory acidosis results from various diseases (emphysema, pneumonia) that restrict the breathing process, causing the buildup of waste CO_2 in the blood.

Lower than normal CO_2 levels, on the other hand, shift the equilibrium to the left, decreasing $[H_3O^+]$ and making the pH more basic. This condition is termed *alkalosis* (from "alkali," implying basic). Hyperventilation, or rapid breathing, is a common cause of respiratory alkalosis.

Question 8.15 Explain how the molar concentration of H_2CO_3 in the blood would change if the partial pressure of CO_2 in the lungs were to increase.

Question 8.16 Explain how the molar concentration of H_2CO_3 in the blood would change if the partial pressure of CO_2 in the lungs were to decrease.

Question 8.17 Explain how the molar concentration of hydronium ions in the blood would change under each of the conditions described in Questions 8.15 and 8.16.

Question 8.18 Explain how the pH of blood would change under each of the conditions described in Questions 8.15 and 8.16.

Question 8.19 Write the Henderson-Hasselbalch expression for the equilibrium between carbonic acid and the bicarbonate ion.

Question 8.20 Calculate the $[HCO_3^-]/[H_2CO_3]$ that corresponds to a pH of 7.4. The K_a for carbonic acid is 4.2×10^{-7}.

Green Chemistry

Acid Rain

Acid rain is a global environmental problem, caused by our industrial society, that has raised public awareness of the chemicals polluting the air. Normal rain has a pH of about 5.6 as a result of the chemical reaction between carbon dioxide gas and water in the atmosphere. The following equation shows this reaction:

$$CO_2(g) + H_2O(l) \rightleftharpoons H_2CO_3(aq)$$
Carbon dioxide   Water   Carbonic acid

Acid rain refers to conditions that are much more acidic than this. In upstate New York, the rain has as much as twenty-five times the acidity of normal rainfall. One rainstorm, recorded in West Virginia, produced rainfall that measured 1.5 on the pH scale. This is approximately the pH of stomach acid or about ten thousand times more acidic than "normal rain" (remember that the pH scale is logarithmic; a 1 pH unit decrease represents a tenfold increase in hydronium ion concentration).

Acid rain is destroying life in streams and lakes. More than half the highland lakes in the western Adirondack Mountains now have no native game fish. In addition to these 300 lakes, 140 lakes in Ontario have suffered a similar fate. It is estimated that 48,000 other lakes in Ontario and countless others in the northeastern and central United States are threatened. Our forests are endangered as well. The acid rain decreases soil pH, which in turn alters the solubility of minerals needed by plants. Studies have shown that about 40% of the red spruce and maple trees in New England have died. Increased acidity of rainfall appears to be the major culprit.

What is the cause of this acid rain? The combustion of fossil fuels (gas, oil, and coal) by power plants produces oxides of sulfur and nitrogen. Nitrogen oxides, in excess of normal levels, arise mainly from conversion of atmospheric nitrogen to nitrogen oxides in the engines of gasoline and diesel-powered vehicles. Sulfur oxides result from the oxidation of sulfur in fossil fuels. The sulfur atoms were originally a part of the amino acids and proteins of plants and animals that became, over the millennia, our fuel. These react with water, as does the CO_2 in normal rain, but the products are strong acids: sulfuric and nitric acids. Let's look at the equations for these processes.

In the atmosphere, nitric oxide (NO) can react with oxygen to produce nitrogen dioxide as shown:

$$2NO(g) + O_2(g) \longrightarrow 2NO_2(g)$$
Nitric oxide   Oxygen   Nitrogen dioxide

Nitrogen dioxide (which causes the brown color of smog) then reacts with water to form nitric acid:

$$3NO_2(g) + H_2O(l) \longrightarrow 2HNO_3(aq) + NO(g)$$

A similar chemistry is seen with the sulfur oxides. Coal may contain as much as 3% sulfur. When the coal is burned, the sulfur also burns. This produces choking, acrid sulfur dioxide gas:

$$S(s) + O_2(g) \longrightarrow SO_2(g)$$

By itself, sulfur dioxide can cause serious respiratory problems for people with asthma or other lung diseases, but matters are worsened by the reaction of SO_2 with atmospheric oxygen:

$$2SO_2(g) + O_2(g) \longrightarrow 2SO_3(g)$$

Sulfur trioxide will react with water in the atmosphere:

$$SO_3(g) + H_2O(l) \longrightarrow H_2SO_4(aq)$$

The product, sulfuric acid, is even more irritating to the respiratory tract. When the acid rain created by the reactions shown above falls to earth, the biological impact is significant, as we have already noted.

It is easy to balance these chemical equations, but decades could be required to balance the ecological systems that we have disrupted by our massive consumption of fossil fuels. A sudden decrease of even 25% in the use of fossil fuels would lead to worldwide financial chaos. Development of alternative fuel sources, such as solar energy and safe nuclear power, are helping to reduce our dependence on fossil fuels and helping us to balance the global equation.

For Further Understanding

▶ Criticize this statement: "Passing and enforcing strong legislation against sulfur and nitrogen oxide emission will solve the problem of acid rain in the United States."

▶ Use the Internet to determine the percentage of electricity that is produced from coal in your state of residence.

pH Values for a Variety of Substances Compared with the pH of Acid Rain

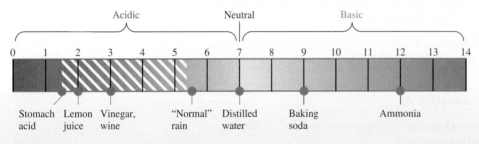

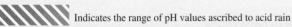 Indicates the range of pH values ascribed to acid rain

Damage caused by acid rain.

CHAPTER MAP

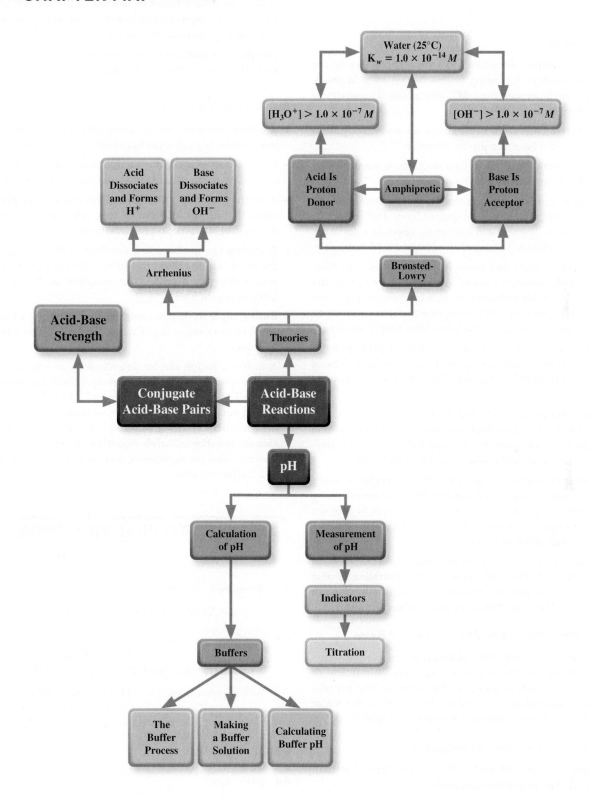

SUMMARY

8.1 Acids and Bases

▶ One of the earliest definitions of acids and bases is the **Arrhenius theory.** According to this theory, an acid dissociates to form hydrogen ions, H^+, and a base dissociates to form hydroxide ions, OH^-.

▶ The **Brønsted-Lowry theory** defines an acid as a proton (H^+) donor and a base as a proton acceptor.

▶ Water, the solvent in many acid-base reactions, is **amphiprotic.** It has both acid and base properties.

▶ A **conjugate acid** is the species formed when a base accepts a proton. A **conjugate base** is the species formed when an acid donates a proton. The acid and base on the opposite sides of the equation are collectively termed a **conjugate acid-base pair.**

▶ The strength of acids and bases in water depends on their degree of dissociation, the extent to which they react with the solvent, water. Acids and bases are strong when their reaction with water is virtually 100% complete and are weak when their reaction with water is much less than 100% complete.

▶ Weak acids and weak bases dissolve in water principally in the molecular form. Only a small percentage of the molecules dissociate to form the **hydronium ion (H_3O^+)** or hydroxide ion (OH^-).

▶ Aqueous solutions of acids and bases are electrolytes. The dissociation of the acid or base produces ions, which conduct an electrical current. Strong acids and bases are strong electrolytes. Weak acids and bases are weak electrolytes.

▶ Although pure water is virtually 100% molecular, a small number of water molecules do ionize. This process occurs by the transfer of a proton from one water molecule to another, producing a *hydronium ion* and a *hydroxide ion*. This process is the **self-ionization** of water.

▶ Pure water at 25°C has a hydronium ion concentration of 1.0×10^{-7} M. The hydroxide ion concentration is also 1.0×10^{-7} M. The product of hydronium and hydroxide ion concentration (1.0×10^{-14} M) is the **ion product for water,** K_w.

8.2 pH: A Measurement Scale for Acids and Bases

▶ The **pH scale** correlates the hydronium ion concentration with a number, the pH, that serves as a useful indicator of the degree of acidity or basicity of a solution. The pH of a solution is defined as the negative logarithm of the molar concentration of the hydronium ion ($pH = -\log[H_3O^+]$).

8.3 Reactions between Acids and Bases

▶ The reaction of an acid with a base to produce a salt and water is referred to as **neutralization.**

▶ Neutralization requires equal numbers of mol of H_3O^+ and OH^- to produce a neutral solution (no excess acid or base).

▶ A neutralization reaction may be used to determine the concentration of an unknown acid or base solution. The technique of **titration** involves the addition of measured amounts of a **standard solution** (one whose concentration is known) from a **buret** to neutralize the second, unknown solution. The **equivalence point** is signaled by an **indicator.** This signal is termed the *end point*.

▶ **Polyprotic substances** donate (as acids) or accept (as bases) more than one proton per formula unit.

8.4 Acid-Base Buffers

▶ A **buffer solution** contains components that enable the solution to resist large changes in pH when acids or bases are added.

▶ Buffer action relies on the equilibrium between either a weak acid and its conjugate base or a weak base and its conjugate acid.

▶ A buffer solution follows LeChatelier's principle, which states that an equilibrium system, when stressed, will shift its equilibrium position to alleviate that stress.

▶ Buffering against base is a function of the concentration of the weak acid for an acidic buffer, whereas buffering against acid is dependent on the concentration of the anion of the salt (conjugate base).

▶ **Buffer capacity** is a measure of the ability of a solution to resist large changes in pH when a strong acid or strong base is added.

▶ A buffer solution can be described by an equilibrium constant expression. The equilibrium constant expression for an acidic system can be rearranged and solved for $[H_3O^+]$. In that way, the pH of a buffer solution can be obtained, if the composition of the solution is known. Alternatively, the **Henderson-Hasselbalch equation,** derived from the equilibrium constant expression, may be used to calculate the pH of a buffer solution.

ANSWERS TO PRACTICE PROBLEMS

8.1 Brønsted-Lowry base with water.
$$CH_3COO^- + H_2O \rightleftharpoons CH_3COOH + OH^-$$

8.2 **a.** Conjugate base: OH^- and NH_3
                               HSO_4^- and HSO_3^-
  Stronger acid of each pair: NH_4^+ and H_2SO_4

  b. Conjugate acid: HCO_3^- and HPO_4^{2-}
                               H_2CO_3 and $H_2PO_4^-$
  Stronger base of each pair: PO_4^{3-} and HPO_4^{2-}

8.3 3.3×10^{-8} M

8.4 pH = 4.00

8.5 $[H_3O^+] = 1.0 \times 10^{-5}$ M

8.6 **a.** pH = 12.00
  b. pH = 8.00

8.7 $[H_3O^+] = 1.0 \times 10^{-8} M$
$[OH^-] = 1.0 \times 10^{-6} M$

8.8 pH = 7.48

8.9 **a.** $[H_3O^+] = 3.2 \times 10^{-9} M$
b. $[H_3O^+] = 3.2 \times 10^{-5} M$

8.10 **a.** 0.1000 M
b. 0.1389 M

8.11 pH = 4.87

8.12 pH = 5.17

QUESTIONS AND PROBLEMS

Acids and Bases

Foundations

8.21 **a.** Define an acid according to the Arrhenius theory.
b. Define an acid according to the Brønsted-Lowry theory.

8.22 **a.** Define a base according to the Arrhenius theory.
b. Define a base according to the Brønsted-Lowry theory.

8.23 What are the essential differences between the Arrhenius and Brønsted-Lowry theories?

8.24 Why is ammonia described as a Brønsted-Lowry base and not an Arrhenius base?

8.25 Classify each of the following as either a Brønsted-Lowry acid, base, or as amphiprotic.
a. H_3O^+ **b.** OH^- **c.** H_2O

8.26 Classify each of the following as either a Brønsted-Lowry acid, base, or as amphiprotic.
a. NH_4^+ **b.** NH_3 **c.** $CH_3CH_2CH_2NH_3^+$

8.27 Classify each of the following as either a Brønsted-Lowry acid, base, or as amphiprotic.
a. HOCOOH   **b.** HCO_3^- **c.** CO_3^{2-}

8.28 Classify each of the following as either a Brønsted-Lowry acid, base, or as amphiprotic.
a. H_2SO_4 **b.** HSO_4^- **c.** SO_4^{2-}

Applications

8.29 Write an equation for the reaction of each of the following with water:
a. HNO_2 **b.** HCN   **c.** $CH_3CH_2CH_2COO^-$

8.30 Write an equation for the reaction of each of the following with water:
a. HNO_3 **b.** HCOOH   **c.** $CH_3CH_2CH_2NH_2$

8.31 Write the formula of the conjugate acid of CN^-.

8.32 Write the formula of the conjugate acid of Br^-.

8.33 Write the formula of the conjugate base of HI.

8.34 Write the formula of the conjugate base of HCOOH.

8.35 Write the formula of the conjugate acid of NO_3^-.

8.36 Write the formula of the conjugate acid of F^-.

If necessary, consult Figure 8.3 when solving Questions 8.37–8.40.

8.37 Which is the stronger base, NO_3^- or CN^-?

8.38 Which is the stronger acid, HNO_3 or HCN?

8.39 Which is the stronger acid, HF or CH_3COOH?

8.40 Which is the stronger base, F^- or CH_3COO^-?

8.41 Identify the conjugate acid-base pairs in each of the following chemical equations:
a. $NH_4^+(aq) + CN^-(aq) \rightleftharpoons NH_3(aq) + HCN(aq)$
b. $CO_3^{2-}(aq) + HCl(aq) \rightleftharpoons HCO_3^-(aq) + Cl^-(aq)$

8.42 Identify the conjugate acid-base pairs in each of the following chemical equations:
a. $HCOOH(aq) + NH_3(aq) \rightleftharpoons HCOO^-(aq) + NH_4^+(aq)$
b. $HCl(aq) + OH^-(aq) \rightleftharpoons H_2O(l) + Cl^-(aq)$

8.43 Distinguish between the terms acid-base *strength* and acid-base *concentration*.

8.44 Of the following diagrams, which one represents:
a. a concentrated strong acid
b. a dilute strong acid
c. a concentrated weak acid
d. a dilute weak acid

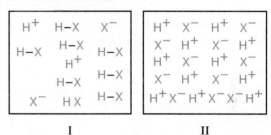

I   II

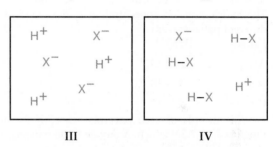

III   IV

8.45 Label each of the following as a strong or weak acid:
a. H_2SO_3 **b.** H_2CO_3 **c.** H_3PO_4

8.46 Label each of the following as a strong or weak base:
a. KOH   **b.** CN^- **c.** SO_4^{2-}

8.47 Calculate the $[H_3O^+]$ of an aqueous solution that is:
a. $1.0 \times 10^{-7} M$ in OH^- **b.** $1.0 \times 10^{-3} M$ in OH^-

8.48 Calculate the $[H_3O^+]$ of an aqueous solution that is:
a. $1.0 \times 10^{-9} M$ in OH^- **b.** $1.0 \times 10^{-5} M$ in OH^-

8.49 Calculate the $[OH^-]$ of an aqueous solution that is:
a. $1.0 \times 10^{-4} M$ in H_3O^+ **b.** $1.0 \times 10^{-2} M$ in H_3O^+

8.50 Calculate the $[OH^-]$ of an aqueous solution that is:
a. $1.0 \times 10^{-6} M$ in H_3O^+ **b.** $1.0 \times 10^{-8} M$ in H_3O^+

8.51 The concentration of hydronium ions in a sample of pomegranate juice is $6.0 \times 10^{-4} M$. What is the concentration of hydroxide ions in the juice sample?

8.52 What is the concentration of hydronium ions in an aqueous solution of acetaminophen if the concentration of hydroxide ions is $2.5 \times 10^{-9} M$?

pH: A Measurement Scale for Acids and Bases

Foundations

8.53 Consider two beakers, one containing 0.10 M HCl and the other, 0.10 M CH_3COOH. Which solution has the greater pH? Why?

8.54 Consider two beakers, one containing 0.10 M NaOH and the other, 0.10 M NH_3. Which solution has the greater pH? Why?

8.55 Calculate the pH of a solution that is:
   a. 1.0×10^{-2} M in HCl   **b.** 1.0×10^{-4} M in HNO_3

8.56 Calculate the pH of a solution that is:
   a. 1.0×10^{-1} M in HCl   **b.** 1.0×10^{-5} M in HNO_3

8.57 Calculate $[H_3O^+]$ for a solution of nitric acid for which:
   a. pH = 1.00   **b.** pH = 5.00

8.58 Calculate $[H_3O^+]$ for a solution of hydrochloric acid for which:
   a. pH = 2.00   **b.** pH = 3.00

8.59 Calculate the pH of a 1.0×10^{-3} M solution of KOH.

8.60 Calculate the pH of a 1.0×10^{-5} M solution of NaOH.

8.61 Calculate both $[H_3O^+]$ and $[OH^-]$ for a solution for which:
   a. pH = 1.30   **b.** pH = 9.70

8.62 Calculate both $[H_3O^+]$ and $[OH^-]$ for a solution for which:
   a. pH = 5.50   **b.** pH = 7.00

8.63 What is a neutralization reaction?

8.64 Describe the purpose of a titration.

Applications

8.65 The pH of urine may vary between 4.5 and 8.2. Determine the H_3O^+ concentration and OH^- concentration if the measured pH is:
   a. 6.00   **b.** 5.20   **c.** 7.80

8.66 The hydronium ion concentration in the blood of three different patients was:

Patient	$[H_3O^+]$
A	5.0×10^{-8}
B	3.1×10^{-8}
C	3.2×10^{-8}

What is the pH of each patient's blood? If the normal range is 7.30–7.50, which, if any, of these patients has an abnormal blood pH?

8.67 Criticize the following statement: A lakewater sample of pH = 3.0 is twice as acidic as a lakewater sample of pH = 6.0.

8.68 Can a dilute solution of a strong acid ever have a higher pH than a more concentrated solution of a weak acid? Why or why not?

8.69 What is the H_3O^+ concentration of a solution with a pH of:
   a. 5.00   **b.** 12.00   **c.** 5.50

8.70 What is the H_3O^+ concentration of a solution with a pH of:
   a. 6.80   **b.** 4.60   **c.** 2.70

8.71 Calculate the pH of a solution with a H_3O^+ concentration of:
   a. 1.0×10^{-6} M **b.** 1.0×10^{-8} M **c.** 5.6×10^{-4} M

8.72 What is the OH^- concentration of each solution in Question 8.71?

8.73 Calculate the pH of a solution that has $[H_3O^+] = 7.5 \times 10^{-4}$ M.

8.74 Calculate the pH of a solution that has $[H_3O^+] = 6.6 \times 10^{-5}$ M.

8.75 Calculate the pH of a solution that has $[OH^-] = 5.5 \times 10^{-4}$ M.

8.76 Calculate the pH of a solution that has $[OH^-] = 6.7 \times 10^{-9}$ M.

Reactions between Acids and Bases

Foundations

8.77 In a neutralization reaction, how many mol of HCl are needed to react with 4 mol of NaOH?

8.78 In a neutralization reaction, how many mol of NaOH are needed to react with 3 mol of HCl?

8.79 What function does an indicator perform?

8.80 What are the products of a neutralization reaction?

8.81 Write an equation to represent the neutralization of an aqueous solution of HNO_3 with an aqueous solution of NaOH.

8.82 Write an equation to represent the neutralization of an aqueous solution of HCl with an aqueous solution of KOH.

8.83 Rewrite the equation in Question 8.81 as a net, balanced ionic equation.

8.84 Rewrite the equation in Question 8.82 as a net, balanced ionic equation.

8.85 Carbonic acid, H_2CO_3, is a polyprotic acid. How many protons can it donate?

8.86 Chromic acid, H_2CrO_4, is a polyprotic acid. How many protons can it donate?

Applications

8.87 Titration of 15.00 mL of HCl solution requires 22.50 mL of 0.1200 M NaOH solution. What is the molarity of the HCl solution?

8.88 Titration of 17.85 mL of HNO_3 solution requires 16.00 mL of 0.1600 M KOH solution. What is the molarity of the HNO_3 solution?

8.89 What volume of 0.1500 M NaOH is required to titrate 20.00 mL of 0.1000 M HCl?

8.90 What volume of 0.2000 M KOH is required to titrate 25.00 mL of 0.1500 M HNO_3?

8.91 Write out each step of the dissociation of carbonic acid, H_2CO_3.

8.92 Write out each step of the dissociation of chromic acid, H_2CrO_4.

Acid-Base Buffers

Foundations

8.93 Which of the following are capable of forming a buffer solution?
   a. NH_3 and NH_4Cl **b.** HNO_3 and KNO_3

8.94 Which of the following are capable of forming a buffer solution?
   a. HBr and $MgCl_2$ **b.** H_2CO_3 and $NaHCO_3$

8.95 Consider two beakers, one containing a mixture of HCl and NaCl (each solute is 0.10 M) and the other containing a mixture of CH_3COOH and CH_3COONa (each solute is also 0.10 M). If we add 0.10 M NaOH (10.0 mL) to each beaker, which solution will have the greater change in pH? Why?

8.96 Consider two beakers, one containing a mixture of NaOH and NaCl (each solute is 0.10 M) and the other containing a mixture of NH_3 and NH_4Cl (each solute is also 0.10 M). If we add 0.10 M HNO_3 (10.0 mL) to each beaker, which solution will have the greater change in pH? Why?

8.97 Explain how the molar concentration of carbonic acid in the blood would be changed during a situation of acidosis.

8.98 Explain how the molar concentration of carbonic acid in the blood would be changed during a situation of alkalosis.

Applications

8.99 For the equilibrium situation involving acetic acid,
$CH_3COOH(aq) + H_2O(l) \rightleftharpoons CH_3COO^-(aq) + H_3O^+(aq)$,
explain the equilibrium shift occurring for the following changes:
a. A strong acid is added to the solution.
b. The solution is diluted with water.

8.100 For the equilibrium situation involving acetic acid,
$CH_3COOH(aq) + H_2O(l) \rightleftharpoons CH_3COO^-(aq) + H_3O^+(aq)$,
explain the equilibrium shift occurring for the following changes:
a. A strong base is added to the solution.
b. More acetic acid is added to the solution.

8.101 What is $[H_3O^+]$ for a buffer solution that is 0.200 M in acid and 0.500 M in the corresponding salt if the weak acid $K_a = 5.80 \times 10^{-7}$?

8.102 What is the pH of the solution described in Question 8.101?

8.103 For the buffer system described in Question 8.99, which substance is responsible for buffering capacity against added hydrochloric acid? Explain.

8.104 For the buffer system described in Question 8.99, which substance is responsible for buffering capacity against added sodium hydroxide? Explain.

8.105 Calculate the pH of a buffer system containing 1.0 M CH_3COOH and 1.0 M CH_3COONa. (K_a of acetic acid, CH_3COOH, is 1.8×10^{-5}.)

8.106 Calculate the pH of a buffer system containing 1.0 M NH_3 and 1.0 M NH_4Cl. (K_a of NH_4^+, the acid in this system, is 5.6×10^{-10}.)

8.107 The pH of blood plasma is 7.40. The principal buffer system is HCO_3^-/H_2CO_3. Calculate the ratio $[HCO_3^-]/[H_2CO_3]$ in blood plasma. (K_a of H_2CO_3, carbonic acid, is 4.5×10^{-7}.)

8.108 The pH of blood plasma from a patient was found to be 7.6, a life-threatening situation. Calculate the ratio $[HCO_3^-]/[H_2CO_3]$ in this sample of blood plasma. (K_a of H_2CO_3, carbonic acid, is 4.5×10^{-7}.)

CHALLENGE PROBLEMS

1. Acid rain is a threat to our environment because it can increase the concentration of toxic metal ions, such as Cd^{2+} and Cr^{3+}, in rivers and streams. If cadmium and chromium are present in sediment as $Cd(OH)_2$ and $Cr(OH)_3$, write reactions that demonstrate the effect of acid rain. Use the Internet to find the properties of cadmium and chromium responsible for their environmental impact.

2. Aluminum carbonate is soluble in acidic solution, forming aluminum cations. Write a reaction (or series of reactions) that explains this observation.

3. Carbon dioxide reacts with the hydroxide ion to produce the bicarbonate anion. Write the Lewis structure for each reactant and product. Label each as a Brønsted acid or base. Explain the reaction using the Brønsted theory. Why would the Arrhenius theory provide an inadequate description of this reaction?

4. Maalox is an antacid composed of $Mg(OH)_2$ and $Al(OH)_3$. Explain the origin of the trade name Maalox. Write chemical reactions that demonstrate the antacid activity of Maalox.

5. Acid rain has been described as a regional problem, whereas the greenhouse effect is a global problem. Do you agree with this statement? Why or why not?

9

GENERAL CHEMISTRY

The Nucleus, Radioactivity, and Nuclear Medicine

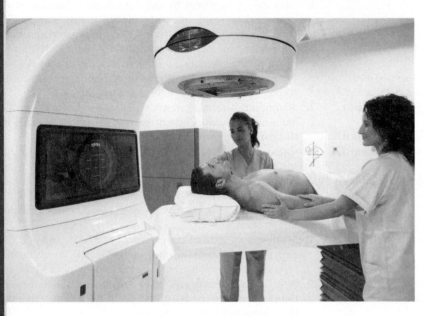

Nuclear technology has revolutionized the practice of medicine.

OUTLINE

LEARNING GOALS

1   Use nuclear symbols to represent isotopes and nuclides.
2   Enumerate characteristics of alpha, beta, positron, and gamma radiation.
3   Write balanced equations for nuclear processes.
4   Calculate the amount of radioactive substance remaining after a specified time has elapsed.
5   Explain the process of radiocarbon dating.
6   Describe how nuclear energy can generate electricity: fission, fusion, and the breeder reactor.
7   Cite examples of the use of radioactive isotopes in medicine.
8   Describe the use of ionizing radiation in cancer therapy.
9   Discuss the preparation and use of radioisotopes in diagnostic imaging studies.
10   Explain the difference between natural and artificial radioactivity.
11   Describe characteristics of radioactive materials that relate to radiation exposure and safety.
12   Be familiar with common techniques for the detection of radioactivity.
13   Interpret common units of radiation intensity and discuss their biological implications.

INTRODUCTION

Our discussion of the atom and atomic structure in Chapter 2 revealed a nucleus containing protons and neutrons surrounded by electrons. Until now, we have treated the nucleus as simply a region of positive charge in the center of the atom. The focus of our interest has been the electrons and their arrangement around the nucleus. Electron arrangement is an essential part of a discussion of bonding or chemical change.

In this chapter, we consider the nucleus and nuclear properties. The behavior of nuclei may have as great an effect on our everyday lives as any of the thousands of synthetic compounds developed over the past several decades. Examples of nuclear technology range from everyday items (smoke detectors) to sophisticated instruments for medical diagnosis and treatment and electric power generation (nuclear power plants).

Beginning in 1896 with Becquerel's discovery of radiation emitted from uranium ore, the technology arising from this and related findings has produced both risks and benefits. Although early discoveries of radioactivity and its properties expanded our fundamental knowledge and brought fame to the investigators, it was not accomplished without a price. Several early investigators died prematurely of cancer and other diseases caused by the radiation they studied.

Even today, the existence of nuclear energy and its associated technology is a mixed blessing. On one side, the horrors of Nagasaki and Hiroshima, the fear of nuclear war, and potential contamination of populated areas resulting from the peaceful application of nuclear energy are critical problems facing society. Conversely, hundreds of thousands of lives have been saved because of the early detection of disease such as cancer by diagnosis based on the interaction of radiation and the body. The use of techniques such as cobalt-60 therapy is considered a routine part of comprehensive treatment for a variety of tumors. Furthermore, nuclear energy is an alternative energy source, providing an opportunity for us to compensate for the depletion of nonrenewable oil reserves.

High energy radiation is used to target cancer cells.

9.1 Natural Radioactivity

Radioactivity is the process by which some atoms emit energy and particles. The energy and particles are termed *radiation*. Nuclear radiation occurs as a result of an alteration in nuclear composition or structure. This process occurs in a nucleus that is unstable and hence radioactive. Radioactivity is a nuclear event: *matter and energy released during this process come from the nucleus.*

We shall designate the nucleus using *nuclear symbols*, analogous to the *atomic symbols* that were discussed in Chapter 2. The nuclear symbols consist of the *symbol* of the element, the *atomic number* (the number of protons in the nucleus), and the *mass number*, which is defined as the sum of neutrons and protons in the nucleus.

With the use of nuclear symbols, the fluorine nucleus is represented as

$$\text{Mass number} \longrightarrow {}^{19}_{9}\text{F} \longleftarrow \text{Element symbol}$$
$$\text{Atomic number} \longrightarrow$$
$$\text{(or nuclear charge)}$$

This symbol is equivalent to writing fluorine-19. This alternative representation is frequently used to denote specific isotopes of elements.

Not all nuclei are unstable. Only unstable nuclei undergo change and produce radioactivity, the process of *radioactive decay*. Recall that *isotopes* can exist when atoms of the same element have different masses. One isotope of an element may be radioactive, whereas others of the same element may be quite stable. It is important to distinguish between the terms *isotope* and *nuclide*. The term *isotope* refers to any atoms that have the same atomic number but different mass

Be careful not to confuse the mass number (a simple count of the neutrons and protons) with the atomic mass, which includes the contribution of electrons and is a true *mass* figure.

LEARNING GOAL

1 Use nuclear symbols to represent isotopes and nuclides.

Isotopes are introduced in Section 2.1.

Figure 9.1 Three isotopes of carbon. Each nucleus contains the same number of protons (depicted in blue). Only the number of neutrons (depicted in red) is different; hence, each isotope has a different mass.

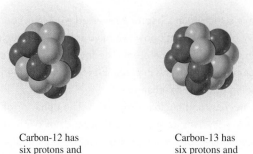

Carbon-12 has
six protons and
six neutrons

Carbon-13 has
six protons and
seven neutrons

Carbon-14 has
six protons and
eight neutrons

Carbon has three isotopes, three distinct nuclides, C-12, C-13, and C-14.

Hydrogen has three isotopes, three distinct nuclides, H-1, H-2, and H-3.

Other radiation particles, such as neutrinos and deuterons, will not be discussed here.

LEARNING GOAL

2 Enumerate characteristics of alpha, beta, positron, and gamma radiation.

numbers. The term **nuclide** refers to any atom characterized by an atomic number and a mass number.

Many elements in the periodic table occur in nature as mixtures of isotopes. Two common examples include carbon (Figure 9.1),

$$^{12}_{6}C \qquad ^{13}_{6}C \qquad ^{14}_{6}C$$
Carbon-12   Carbon-13   Carbon-14

and hydrogen,

$$^{1}_{1}H \qquad ^{2}_{1}H \qquad ^{3}_{1}H$$
Hydrogen-1   Hydrogen-2   Hydrogen-3
Protium   Deuterium   Tritium

Protium is a stable isotope and makes up more than 99.9% of naturally occurring hydrogen. Deuterium (D) can be isolated from hydrogen; it can form compounds such as "heavy water," D_2O. Tritium (T) is rare and unstable, and hence radioactive.

In writing the symbols for a nuclear process, it is essential to indicate the particular isotope involved. This is why the mass number and atomic number are used. These values tell us the number of neutrons in the species, and hence the isotope's identity.

Natural radiation emitted by unstable nuclei includes *alpha particles, beta particles, positrons,* and *gamma rays.*

Alpha Particles

An **alpha particle** (α) contains two protons and two neutrons. An alpha particle is identical to the nucleus of the helium atom (He) or a *helium ion* (He^{2+}), which also contains two protons (atomic number = 2) and two neutrons (mass number − atomic number = 2). Having no electrons to counterbalance the nuclear charge, the alpha particle may be symbolized as

$$^{4}_{2}He^{2+} \qquad or \qquad ^{4}_{2}He \qquad or \qquad \alpha$$

Alpha particles have a relatively large mass compared to other nuclear particles. Consequently, alpha particles emitted by radioisotopes are relatively slow-moving particles (approximately 10% of the speed of light), and they are stopped by barriers as thin as a few pages of this book.

Beta Particles and Positrons

The **beta particle** (β), in contrast, is a fast-moving electron traveling at approximately 90% of the speed of light as it leaves the nucleus. It is formed in the nucleus by the conversion of a neutron into a proton. The beta particle is represented as

$$^{0}_{-1}e \qquad or \qquad ^{0}_{-1}\beta \qquad or \qquad \beta$$

The subscript −1 is written in the same position as the atomic number and, like the atomic number (number of protons), indicates the charge of the particle.

Beta particles are smaller and faster than alpha particles. They are more penetrating and are stopped only by denser materials such as wood, metal, or several layers of clothing.

A **positron** has the same mass as a beta particle but carries a positive charge and is symbolized as

$$_{+1}^{0}e \quad \text{or} \quad _{+1}^{0}\beta$$

Positrons are produced by the conversion of a proton to a neutron in the nucleus of the isotope. The proton, in effect, loses its positive charge, as well as a tiny bit of mass. This positively charged mass that is released is the positron.

Gamma Rays

Gamma rays (γ) are the most energetic part of the electromagnetic spectrum (see Section 2.3) and result from nuclear processes; in contrast, alpha radiation and beta radiation are matter. Because electromagnetic radiation has no protons, neutrons, or electrons, the symbol for a gamma ray is simply

$$\gamma$$

Gamma radiation is highly energetic and is the most penetrating form of nuclear radiation. Barriers of lead, concrete, or, more often, a combination of the two are required for protection from this type of radiation.

Properties of Alpha, Beta, Positron, and Gamma Radiation

Important properties of alpha, beta, positron, and gamma radiation are summarized in Table 9.1.

Alpha, beta, positron, and gamma radiation are collectively termed *ionizing radiation*. **Ionizing radiation** produces a trail of ions throughout the material that it penetrates. The ionization process changes the chemical composition of the material. When the material is living tissue, radiation-induced illness may result (Section 9.6).

The penetrating power of alpha radiation is very low. Damage to internal organs from this form of radiation is negligible except when an alpha particle emitter is actually ingested. Beta particles and positrons have much higher velocities than alpha particles; still, they have limited penetrating power. They cause skin and eye damage and, to a lesser extent, damage to internal organs. The great penetrating power and high energy of gamma radiation can damage internal organs.

Anyone working with any type of radiation must take precautions. Radiation safety is required, monitored, and enforced in the United States by the Occupational Safety and Health Administration (OSHA).

LEARNING GOAL

2 Enumerate characteristics of alpha, beta, positron, and gamma radiation.

Question 9.1 Gamma radiation is a form of *electromagnetic radiation*. Provide examples of other forms of electromagnetic radiation.

Question 9.2 How does the energy of gamma radiation compare with that of other regions of the electromagnetic spectrum?

TABLE 9.1 A Summary of the Major Properties of Alpha, Beta, Positron, and Gamma Radiation

Name and Symbol	Identity	Charge	Mass (amu)	Velocity	Penetration
Alpha (α)	Helium nucleus	+2	4.0026	5–10% of the speed of light	Low
Beta ($_{-1}^{0}\beta$)	Electron	−1	0.000549	Up to 90% of the speed of light	Medium
Positron ($_{+1}^{0}\beta$)	Electron	+1	0.000549	Up to 90% of the speed of light	Medium
Gamma (γ)	Radiant energy	0	0	Speed of light	High

Origin of the Elements

The current, most widely held theory of the origin of the universe is the "big bang" theory. An explosion of very dense matter was followed by expansion into space of the fragments resulting from this explosion. This is one of the scenarios that have been created by scientists fascinated by the origins of matter, the stars and planets, and life as we know it today.

The first fragments, or particles, were protons and neutrons moving with tremendous velocity and possessing large amounts of energy. Collisions involving these high-energy protons and neutrons formed deuterium atoms (^2_1H), which are isotopes of hydrogen. As the universe expanded and cooled, tritium (^3_1H), another hydrogen isotope, formed as a result of collisions of neutrons with deuterium atoms. Subsequent capture of a proton produced helium (He). Scientists theorize that a universe that was principally composed of hydrogen atoms and helium atoms persisted for perhaps 100,000 years until the temperature decreased sufficiently to allow the formation of a simple unit, diatomic hydrogen, two atoms of hydrogen bonded together (H_2).

Many millions of years later, the effect of gravity caused these small units to coalesce, first into clouds and eventually into stars, with temperatures of millions of degrees. In this setting, these small collections of protons and neutrons combined to form larger atoms such as carbon (C) and oxygen (O), then sodium (Na), neon (Ne), magnesium (Mg), silicon (Si), and so forth. Subsequent explosions of stars provided the conditions that formed many larger atoms. These fragments, gathered together by the force of gravity, are the most probable origin of the planets in our own solar system.

The reactions that formed the elements as we know them today were a result of a series of *fusion reactions,* the joining of nuclei to produce larger atoms at very high temperatures (millions of degrees Celsius). These fusion reactions are similar to processes that are currently being studied as a possible alternative source of nuclear power.

Nuclear reactions of this type do not naturally occur on the earth today. The temperature is simply too low. As a result we have, for the most part, a collection of stable elements existing as chemical compounds, atoms joined together by chemical bonds while retaining their identity even in the combined state.

New stars are forming by fusion reactions within the clouds of dust and gas of the Orion nebula.

Silicon exists all around us as sand and soil in a combined form, silicon dioxide; most metals exist as a part of a chemical compound, such as iron ore.

The wonder and complexity of our universe is a legacy of events that happened a long time ago. As our understanding of bonding, molecular structure, and structure-property relationships has developed, our ability to design and synthesize useful new compounds has dramatically increased.

For Further Understanding

▶ How does tritium differ from "normal" hydrogen?

▶ Would you expect to find similar atoms on other planets? Explain your reasoning.

9.2 Writing a Balanced Nuclear Equation

LEARNING GOAL

3 Write balanced equations for nuclear processes.

Nuclear equations represent nuclear change in much the same way as chemical equations represent chemical change.

A **nuclear equation** can be used to represent the process of radioactive decay. In radioactive decay, a *nuclide* breaks down, producing a *new nuclide, smaller particles, and/or energy.* The concept of mass balance, required when writing chemical equations, is also essential for nuclear equations. When writing a balanced equation, remember that:

- the total mass on each side of the reaction arrow must be identical, and
- the sum of the atomic numbers on each side of the reaction arrow must be identical.

Alpha Decay

Consider the decay of one isotope of uranium, $^{238}_{92}U$, into thorium and an alpha particle. Because an alpha particle is lost in this process, this decay is called *alpha decay*.

Examine the balanced equation for this nuclear reaction:

$$^{238}_{92}U \longrightarrow\ ^{234}_{90}Th\ +\ ^{4}_{2}He$$

Uranium-238   Thorium-234   Helium-4

The sum of the mass numbers on the right ($234 + 4 = 238$) is equal to the mass number on the left. The sum of atomic numbers on the right ($90 + 2 = 92$) is equal to the atomic number on the left.

Beta Decay

Beta decay is illustrated by the decay of one of the less-abundant nitrogen isotopes, $^{16}_{7}N$. Upon decomposition, nitrogen-16 produces oxygen-16 and a beta particle. Conceptually, a neutron = proton + electron. In beta decay, one neutron in nitrogen-16 is converted to a proton, and the electron, the beta particle, is released. The reaction is represented as

$$^{16}_{7}N \longrightarrow\ ^{16}_{8}O\ +\ ^{0}_{-1}e$$

or

$$^{16}_{7}N \longrightarrow\ ^{16}_{8}O\ +\ \beta$$

Note that the mass number of the beta particle is zero, because the electron has no protons or neutrons. Sixteen nuclear particles are accounted for on both sides of the reaction arrow. Note also that the product nuclide has the same mass number as the parent nuclide but the atomic number has *increased* by one unit.

The atomic number on the left (7) is counterbalanced by [8 + (−1)] or (7) on the right. Therefore, the equation is correctly balanced.

Beta decay *increases the atomic number of the nuclide by one unit.*

Positron Emission

The decay of carbon-11 to a stable isotope, boron-11, is one example of *positron emission*.

$$^{11}_{6}C \longrightarrow\ ^{11}_{5}B\ +\ ^{0}_{+1}e$$

or

$$^{11}_{6}C \longrightarrow\ ^{11}_{5}B\ +\ ^{0}_{+1}\beta$$

A positron has the same mass as an electron, or beta particle, but opposite (+1) charge. In contrast to beta emission, the product nuclide has the same mass number as the parent nuclide, but the atomic number has *decreased* by one unit.

The atomic number on the left (6) is counterbalanced by [5 + (+1)] or (6) on the right. Therefore, the equation is correctly balanced.

Positron emission *decreases the atomic number of the nuclide by one unit.*

Gamma Production

If *gamma radiation* were the only product of nuclear decay, there would be no measurable change in the mass or identity of the radioactive nuclei. This is so because the gamma emitter has simply gone to a lower energy state. An example of an isotope that decays in this way is technetium-99m. It is described as a *metastable isotope*, meaning that it is unstable and increases its stability through gamma decay without change in the mass or charge of the isotope. The letter *m* is used to denote a metastable isotope. The decay equation for $^{99m}_{43}\text{Tc}$ is

Gamma production produces *no change* in the atomic number of the nuclide.

$$^{99m}_{43}\text{Tc} \longrightarrow {}^{99}_{43}\text{Tc} + \gamma$$

More often, gamma radiation is produced along with other products. For example, iodine-131 decays as follows:

$$^{131}_{53}\text{I} \longrightarrow {}^{131}_{54}\text{Xe} + {}^{0}_{-1}\text{e} + \gamma$$
Iodine-131   Xenon-131   Beta   Gamma
                                       particle   ray

This reaction may also be represented as

$$^{131}_{53}\text{I} \longrightarrow {}^{131}_{54}\text{Xe} + {}^{0}_{-1}\beta + \gamma$$

An isotope of xenon, a beta particle, and gamma radiation are produced.

Predicting Products of Nuclear Decay

It is possible to use a nuclear equation to predict one of the products of a nuclear reaction if the others are known. Consider the following example, in which we represent the unknown product as ?:

$$^{40}_{19}\text{K} \longrightarrow ? + {}^{0}_{-1}\text{e}$$

Step 1. The mass number of this isotope of potassium is 40. Therefore, the sum of the mass number of the products must also be 40. Since the mass number of the beta particle is 0, the unknown mass number must be 40, because [40 + 0 = 40].

Step 2. Likewise, the atomic number on the left is 19, and the sum of the unknown atomic number plus the charge of the beta particle (−1) must equal 19.

Therefore, the unknown atomic number must be 20, because [20 + (−1) = 19].

Step 3. If we consult the periodic table, the element that has atomic number 20 is calcium; therefore, the unknown product, ?, is $^{40}_{20}\text{Ca}$.

Step 4. Verify the result. Use the complete equation

$$^{40}_{19}\text{K} \longrightarrow {}^{40}_{20}\text{Ca} + {}^{0}_{-1}\text{e}$$

to confirm that the mass number of the reactant is equal to the mass number of the products (40). In addition, the atomic number of the reactant must be equal to the atomic number of the products (19).

Mass number of reactants = 40
Mass number of products = 40 + 0 = 40

and

Atomic number of reactants = 19
Atomic number of products = 20 + (−1) = 19

EXAMPLE 9.1	Predicting the Product of Alpha Decay	LEARNING GOAL

Determine the identity of the unknown product of the alpha decay of curium-245:

$$^{245}_{96}Cm \longrightarrow {}^{4}_{2}He + ?$$

3 Write balanced equations for nuclear processes.

Solution

Step 1. The mass number of the curium isotope is 245. The sum of the mass numbers of the products must also be 245. Since the mass number of the alpha particle is 4, the unknown mass number must be 241, because [241 + 4 = 245].

Step 2. Likewise, the atomic number on the left is 96, and the sum of the unknown atomic number plus the atomic number of the alpha particle (2) must equal 96. Therefore, the unknown atomic number must be 94, because [94 + 2 = 96].

Step 3. Referring to the periodic table, we find that the element that has atomic number 94 is plutonium; therefore, the unknown product, ?, is $^{241}_{94}Pu$.

Step 4. Verify the result. The complete equation is:

$$^{245}_{96}Cm \longrightarrow {}^{4}_{2}He + {}^{241}_{94}Pu$$

Mass number of reactants = 245
Mass number of products = 4 + 241 = 245

and

Atomic number of reactant = 96
Atomic number of products = 2 + 94 = 96

Practice Problem 9.1

Complete each of the following nuclear equations:

a. $? \longrightarrow {}^{4}_{2}He + {}^{222}_{86}Rn$

b. $^{11}_{5}B \longrightarrow {}^{7}_{3}Li + ?$

▶ For Further Practice: **Questions 9.44 and 9.46.**

Question 9.3 Neodymium-144 is a rare earth isotope that undergoes alpha decay. Write a balanced nuclear equation for this process.

Question 9.4 Samarium-147 is one of many rare earth isotopes that undergo alpha decay. Write a balanced nuclear equation for this process.

EXAMPLE 9.2	Predicting the Product of Beta Decay	LEARNING GOAL

Determine the identity of the unknown product of the beta decay of chromium-55.

$$^{55}_{24}Cr \longrightarrow {}^{0}_{-1}e + ?$$

3 Write balanced equations for nuclear processes.

Solution

Step 1. The mass number of the chromium isotope is 55. Therefore, the sum of the mass numbers of the products must also be 55. Since the mass number of the beta particle is 0, the unknown mass number must be 55, because [55 + 0 = 55].

Step 2. Likewise, the atomic number on the left is 24, and the sum of the unknown atomic number plus (−1), representing the beta particle, equals the atomic number of the chromium isotope.

Unknown atomic number + (−1) = 24
Unknown atomic number = 24 + 1 = 25

Continued…

Step 3. Referring to the periodic table, we find that the element having atomic number 25 is manganese; therefore, the unknown product, ?, is $^{55}_{25}$Mn.

Step 4. Verify the result. The complete equation is:

$$^{55}_{24}\text{Cr} \longrightarrow \ ^{0}_{-1}\text{e} + \ ^{55}_{25}\text{Mn}$$

Mass number of reactants = 55
Mass number of products = 0 + 55 = 55

and

Atomic number of reactants = 24
Atomic number of products = (−1) + 25 = 24

Practice Problem 9.2

Complete each of the following nuclear equations:

a. $^{85}_{36}\text{Kr} \longrightarrow ? + \ ^{0}_{-1}\text{e}$
b. $^{239}_{92}\text{U} \longrightarrow ? + \ ^{0}_{-1}\text{e}$

▶ For Further Practice: **Questions 9.43 and 9.45.**

Question 9.5 Iodine-131, useful in the treatment of certain kinds of thyroid disease, decays by beta emission. Write a balanced nuclear equation for this process.

Question 9.6 Phosphorus-31, known to bioaccumulate in the liver, decays by beta emission. Write a balanced nuclear equation for this process.

EXAMPLE 9.3	Predicting the Product of Positron Emission

LEARNING GOAL

Determine the identity of the unknown product of positron emission from xenon-118.

3 Write balanced equations for nuclear processes.

$$^{118}_{54}\text{Xe} \longrightarrow \ ^{0}_{+1}\text{e} + ?$$

Solution

Step 1. The mass number of the xenon isotope is 118. Therefore, the sum of the mass numbers of the products must also be 118. Since the mass number of the positron is 0, the unknown mass number must be 118, because [118 + 0 = 118].

Step 2. Likewise, the atomic number on the left is 54, and the sum of the unknown atomic number plus (+1), representing the positron, equals the atomic number of the xenon isotope.

Unknown atomic number + (+1) = 54

Unknown atomic number = 54 − 1 = 53

Step 3. Referring to the periodic table, we find that the element having atomic number 53 is iodine; therefore, the unknown product, ?, is $^{118}_{53}$I.

Step 4. Verify the result. The complete equation is:

$$^{118}_{54}\text{Xe} \longrightarrow \ ^{0}_{+1}\text{e} + \ ^{118}_{53}\text{I}$$

Mass number of reactants = 118
Mass number of products = 118 + 0 = 118

and

Atomic number of reactant = 54
Atomic number of products = (+1) + 53 = 54

Practice Problem 9.3

Complete each of the following nuclear equations:

a. $^{79}_{37}\text{Rb} \longrightarrow ? + ^{0}_{+1}\text{e}$

b. $^{38}_{20}\text{Ca} \longrightarrow ? + ^{0}_{+1}\text{e}$

▶ For Further Practice: **Questions 9.51 and 9.52.**

Question 9.7 In what way do beta particles and positrons differ?

Question 9.8 In what way are beta particles and positrons similar?

9.3 Properties of Radioisotopes

Why are some isotopes radioactive but others are not? Do all radioactive isotopes decay at the same rate? Are all radioactive materials equally hazardous? We address these and other questions in this section.

Nuclear Structure and Stability

A measure of nuclear stability is the **binding energy** of the nucleus. The binding energy of the nucleus is the energy required to break up a nucleus into its component protons and neutrons. This must be very large, because identically charged protons in the nucleus exert extreme repulsive forces on one another. These forces must be overcome if the nucleus is to be stable. When a nuclide decays, some energy is released because the products are more stable than the parent nuclide. The energy released serves as the basis for much of our nuclear technology.

Why are some isotopes more stable than others? The answer to this question is not completely clear. Evidence obtained so far points to several important factors that describe stable nuclei:

- Nuclear stability correlates with the ratio of neutrons to protons in the isotope. For example, for light atoms a neutron:proton ratio of 1:1 characterizes a stable atom.
- All isotopes (except hydrogen-1) with more protons than neutrons are unstable. However, the reverse is not true.
- Nuclei with large numbers of protons (84 or more) tend to be unstable.
- Naturally occurring isotopes containing 2, 8, 20, 50, 82, or 126 protons or neutrons are stable. These *magic numbers* seem to indicate the presence of energy levels in the nucleus, analogous to electronic energy levels in the atom.
- Isotopes with even numbers of protons or neutrons are generally more stable than those with odd numbers of protons or neutrons.

Half-Life

The **half-life ($t_{1/2}$)** of an isotope is the time required for one-half of a given quantity of the radioactive isotope to undergo change. Half-life is symbolized as $t_{1/2}$. Not all radioactive isotopes decay at the same rate. The rate of nuclear decay is generally represented in terms of the half-life of the isotope. Each isotope has its own characteristic half-life that may be as short as a few millionths of a second or as long as billions of years. Half-lives of some naturally occurring and synthetic isotopes are given in Table 9.2.

The stability of an isotope is indicated by the isotope's half-life. Isotopes with short half-lives decay rapidly; they are very unstable. This is not meant to imply that substances with long half-lives are less hazardous. Often, just the reverse is true.

LEARNING GOAL

4 Calculate the amount of radioactive substance remaining after a specified time has elapsed.

Refer to the discussion of radiation exposure and safety in Sections 9.6 and 9.7.

TABLE 9.2 Half-Lives of Selected Radioisotopes

Name	Symbol	Half-Life
Carbon-14	$^{14}_{6}C$	5730 years
Cobalt-60	$^{60}_{27}Co$	5.3 years
Hydrogen-3	$^{3}_{1}H$	12.3 years
Iodine-131	$^{131}_{53}I$	8.1 days
Iron-59	$^{59}_{26}Fe$	45 days
Molybdenum-99	$^{99}_{42}Mo$	67 hours
Sodium-24	$^{24}_{11}Na$	15 hours
Strontium-90	$^{90}_{38}Sr$	28 years
Technetium-99m	$^{99m}_{43}Tc$	6 hours
Uranium-235	$^{235}_{92}U$	710 million years

Figure 9.2 The decay curve for the medically useful radioisotope technetium-99m. Note that the number of radioactive atoms remaining—hence the radioactivity—approaches zero.

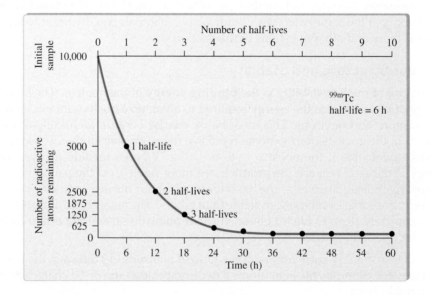

Imagine that we begin with 100 milligrams (mg) of a radioactive isotope that has a half-life of 24 hours (h). After one half-life, or 24 h, 1/2 of 100 mg will have decayed to other products, and 50 mg remain. After two half-lives (48 h), 1/2 of the remaining material has decayed, leaving 25 mg, and so forth:

$$100 \text{ mg} \xrightarrow[\substack{\text{Half-life} \\ (24 \text{ h})}]{\text{First}} 50 \text{ mg} \xrightarrow[\substack{\text{Half-life} \\ (48 \text{ h total})}]{\text{Second}} 25 \text{ mg} \longrightarrow \text{etc.}$$

Decay of a radioisotope that has a reasonably short $t_{1/2}$ is experimentally determined by following its activity as a function of time. Graphing the results produces a radioactive decay curve, as shown in Figure 9.2.

The mass of any radioactive substance remaining after a period may be calculated with a knowledge of the initial mass and the half-life of the isotope, following the scheme just outlined. The general equation for this process is:

$$m_f = m_i (0.5)^n$$

where m_f = final or remaining mass
m_i = initial mass
n = number of half-lives

EXAMPLE 9.4	Predicting the Extent of Radioactive Decay

LEARNING GOAL

4 Calculate the amount of radioactive substance remaining after a specified time has elapsed.

A 50.0-mg supply of iodine-131, used in hospitals in the treatment of hyperthyroidism, was stored for 32.4 days (d). If the half-life of iodine-131 is 8.1 d, how many mg remain?

Solution

Step 1. The general equation for calculating remaining mass is:

$$m_f = m_i (0.5)^n$$

Step 2. The number of half-lives elapsed, n, is calculated using the half-life as a conversion factor:

$$n = 32.4 \, \cancel{d} \times \frac{1 \text{ half-life}}{8.1 \, \cancel{d}} = 4.0 \text{ half-lives}$$

Step 3. Substituting the initial mass, m_i, provided in the problem and the number of half-lives, n, calculated into the equation:

$$m_f = (50.0 \text{ mg})(0.5)^4$$
$$m_f = 3.13 \text{ mg of iodine-131 remain after 32.4 d}$$

Helpful Hint: Alternatively, the calculated number of half-lives can be used to determine the final mass stepwise by using an illustration.

$$\begin{array}{ccccccccc}
& \text{First} & & \text{Second} & & \text{Third} & & \text{Fourth} & \\
50.0 \text{ mg} & \longrightarrow & 25.0 \text{ mg} & \longrightarrow & 12.5 \text{ mg} & \longrightarrow & 6.25 \text{ mg} & \longrightarrow & 3.13 \text{ mg} \\
& \text{Half-life} & & \text{Half-life} & & \text{Half-life} & & \text{Half-life} &
\end{array}$$

Practice Problem 9.4

a. A 100.0 nanogram (ng) sample of sodium-24 was stored in a lead-lined cabinet for 2.5 d. How much sodium-24 remained? See Table 9.2 for the half-life of sodium-24.

b. If a patient is administered 10 ng of technetium-99m, how much will remain 1 day later, assuming that no technetium has been eliminated by any other process? See Table 9.2 for the half-life of technetium-99m.

▶ For Further Practice: **Questions 9.63 and 9.64.**

Question 9.9 The half-life of molybdenum-99 is 67 h. A 200 microgram (μg) quantity decays, over time, to 25 μg. How much time has elapsed?

Question 9.10 The half-life of strontium-87 is 2.8 h. What percentage of this isotope will remain after 8 h and 24 minutes (min)?

Radiocarbon Dating

LEARNING GOAL

5 Explain the process of radiocarbon dating.

Natural radioactivity is useful in establishing the approximate age of objects of archaeological, anthropological, or historical interest. **Radiocarbon dating** is the estimation of the age of objects through measurement of isotopic ratios of carbon.

Radiocarbon dating is based on the measurement of the relative amounts (or ratio) of $^{14}_{6}C$ and $^{12}_{6}C$ present in an object. The $^{14}_{6}C$ is formed in the upper atmosphere by the bombardment of $^{14}_{7}N$ by high-speed neutrons (cosmic radiation):

$$^{14}_{7}N + ^{1}_{0}n \longrightarrow ^{14}_{6}C + ^{1}_{1}H$$

The carbon-14, along with the more abundant carbon-12, is converted into living plant material by the process of photosynthesis. Carbon proceeds up the

A Human Perspective

An Extraordinary Woman in Science

The path to a successful career in science, or any other field for that matter, is seldom smooth or straight. That was certainly true for Madame Marie Sklodowska Curie. Her lifelong ambition was to raise a family and do something interesting for a career. This was a lofty goal for a nineteenth-century woman.

The political climate in Poland, coupled with the prevailing attitudes toward women and careers, especially careers in science, certainly did not make it any easier for Mme. Curie. To support herself and her sister, she toiled at menial jobs until moving to Paris to resume her studies.

It was in Paris that she met her future husband and fellow researcher, Pierre Curie. Working with crude equipment in a laboratory that was primitive, even by the standards of the time, she and Pierre made a most revolutionary discovery only 2 years after Henri Becquerel discovered radioactivity. Radioactivity, the emission of energy from certain substances, was released from *inside* the atom and was independent of the molecular form of the substance. The absolute proof of this assertion came only after the Curies processed over 1 *ton* (t) of a material (pitchblende) to isolate less than a gram (g) of pure radium. The difficult conditions under which this feat was accomplished are perhaps best stated by Sharon Bertsch McGrayne in her book *Nobel Prize Women in Science* (Birch Lane Press, New York, 2001):

> The only space large enough at the school was an abandoned dissection shed. The shack was stifling hot in summer and freezing cold in winter. It had no ventilation system for removing poisonous fumes, and its roof leaked. A chemist accustomed to Germany's modern laboratories called it "a cross between a stable and a potato cellar and, if I had not seen the work table with the chemical apparatus, I would have thought it a practical joke." This ramshackle shed became the symbol of the Marie Curie legend.

The pale green glow emanating from the radium was beautiful to behold. Mme. Curie would go to the shed in the middle of the night to bask in the light of her accomplishment. She did not realize that this wonderful accomplishment would, in time, be responsible for her death.

Mme. Curie received not one, but two Nobel Prizes, one in physics and one in chemistry. She was the first woman in France to earn the rank of professor.

Mme. Curie (left) standing beside her daughter, Irene.

For Further Understanding

▸ Paradoxically, radiation can both cause and cure cancer. Use the Internet to develop an explanation for this paradox.

▸ What is the message that you, personally, can take away from this story?

---

food chain as the plants are consumed by animals, including humans. When a plant or animal dies, the uptake of both carbon-14 and carbon-12 ceases. However, the amount of carbon-14 slowly decreases because carbon-14 is radioactive ($t_{1/2} = 5730$ years). Consequently, the ratio of $^{14}_{6}C/^{12}_{6}C$ decreases over time in a predictable way.

Carbon-14 decay produces nitrogen and a beta particle:

$$^{14}_{6}C \longrightarrow \ ^{14}_{7}N + \ ^{0}_{-1}e$$

TABLE 9.3 Isotopes Useful in Radioactive Dating

Isotope	Half-Life (years)	Upper Limit (years)	Dating Applications
Carbon-14	5730	5×10^4	Charcoal, organic material, artwork
Hydrogen-3	12.3	1×10^2	Aged wines, artwork
Potassium-40	1.3×10^9	Age of earth (4×10^9)	Rocks, planetary material
Rhenium-187	4.3×10^{10}	Age of earth (4×10^9)	Meteorites
Uranium-238	4.5×10^9	Age of earth (4×10^9)	Rocks, earth's crust

When an artifact is found and studied, the relative amounts of carbon-14 and carbon-12 are determined. By using suitable equations involving the $t_{1/2}$ of carbon-14, it is possible to approximate the age of the artifact.

This technique has been widely used to increase our knowledge about the history of the earth, to establish the age of objects (Figure 9.3), and even to detect art forgeries. Early paintings were made with inks fabricated from vegetable dyes (plant material that, while alive, metabolized carbon).

The carbon-14 dating technique is limited to objects that are less than fifty thousand years old, or approximately nine half-lives, which is a practical upper limit. Older objects that have geological or archaeological significance may be dated using naturally occurring isotopes having much longer half-lives.

Examples of isotopes that are useful in radioactive dating are listed in Table 9.3.

Figure 9.3 Radiocarbon dating was used in the authentication study of the Shroud of Turin. It is a minimally destructive technique and is valuable in estimating the age of historical artifacts. It was found that the Shroud is approximately 700 years old, negating the claim that it was used to wrap the body of Jesus.

9.4 Nuclear Power

Energy Production

Einstein predicted that a small amount of nuclear mass corresponds to a very large amount of energy that is released when the nucleus breaks apart. Einstein's equation is

$$E = mc^2$$

in which

$$E = \text{energy}$$
$$m = \text{mass}$$
$$c = \text{speed of light}$$

This energy, when rapidly released, is the basis for the greatest instruments of destruction developed by humankind, nuclear bombs. However, when nuclear energy is released in a controlled fashion, as in a nuclear power plant, the heat from the nuclear reaction converts liquid water into steam. The steam, in turn, drives an electrical generator, which produces electricity. The entire process takes place within a **nuclear reactor.**

Nuclear Fission

Fission (splitting) occurs when a heavy nuclear particle is split into lighter nuclei by collision with a smaller nuclear particle (such as a neutron). This splitting process is accompanied by the release of large amounts of energy.

A nuclear power plant uses a fissionable material (capable of undergoing fission), such as uranium-235, as fuel. The energy released by the fission process in the nuclear core heats water in an adjoining chamber, producing steam. The high pressure of the steam drives a turbine, which converts this heat energy into

The Sequoyah Nuclear Power Plant in Tennessee uses water to transform nuclear energy into electrical energy.

electricity using an electric power generator. The energy transformation may be summarized as follows:

$$\text{Nuclear energy} \longrightarrow \text{heat energy} \longrightarrow \text{mechanical energy} \longrightarrow \text{electrical energy}$$

Reactor core   Steam   Turbine   Electricity

The fission reaction, once initiated, is self-perpetuating. For example, neutrons are used to initiate the reaction:

$$\underset{\text{Fuel}}{{}_{0}^{1}n + {}_{92}^{235}U} \longrightarrow \underset{\text{Unstable}}{{}_{92}^{236}U} \longrightarrow \underset{\text{Products of reaction}}{{}_{36}^{92}Kr + {}_{56}^{141}Ba + 3{}_{0}^{1}n + \text{energy}}$$

Note that three neutrons are released as product for each single reacting neutron. Each of the three neutrons produced is available to initiate another fission process. Nine neutrons are released from this process. These, in turn, react with other nuclei. The fission process continues and intensifies, producing very large amounts of energy (Figure 9.4). This process of intensification is referred to as a **chain reaction.**

To maintain control over the process and to prevent dangerous overheating, rods fabricated from cadmium or boron are inserted into the core. These rods, which are controlled by the reactor's main operating system, absorb free neutrons as needed, thereby moderating the reaction.

A nuclear fission reactor may be represented as a series of energy transfer zones, as depicted in Figure 9.5. The head of a fission reactor is shown in Figure 9.6.

Nuclear Fusion

Fusion (meaning *to join together*) results from the combination of two small nuclei to form a larger nucleus with the concurrent release of large amounts of energy. The best example of a fusion reactor is the sun. Continuous fusion processes furnish our solar system with light and heat.

An example of a fusion reaction is the combination of two isotopes of hydrogen, deuterium (${}_{1}^{2}H$) and tritium (${}_{1}^{3}H$), to produce helium, a neutron, and energy:

$$ {}_{1}^{2}H + {}_{1}^{3}H \longrightarrow {}_{2}^{4}He + {}_{0}^{1}n + \text{energy} $$

Although fusion is capable of producing tremendous amounts of energy, no commercially successful fusion plant exists in the United States. Safety concerns relating to problems of containment of the reaction, resulting directly from the technological problems associated with containing high temperatures (millions of degrees) and pressures required to sustain a fusion process, have slowed the development of fusion reactors.

Breeder Reactors

A **breeder reactor** is a variation of a fission reactor that literally manufactures its own fuel. A perceived shortage of fissionable isotopes makes the breeder an attractive alternative to conventional fission reactors. A breeder reactor uses ${}_{92}^{238}U$, which is abundant but nonfissionable. In a series of steps, the uranium-238 is converted to plutonium-239, which *is* fissionable and undergoes a fission chain reaction, producing energy. The attractiveness of a reactor that makes its own fuel from abundant starting materials is offset by the high cost of the system, potential environmental damage, and fear of plutonium proliferation. Plutonium can be readily used to manufacture nuclear bombs. Currently, only a few countries operate breeder reactors for electrical power generation.

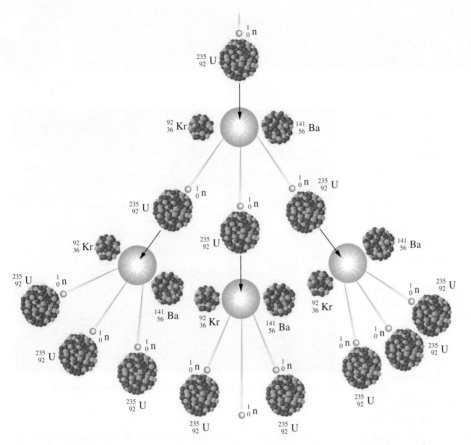

Figure 9.4 The fission of uranium-235 producing a chain reaction. Note that the number of available neutrons, which "trigger" the decomposition of the fissionable nuclei to release energy, increases at each step in the "chain." In this way, the reaction builds in intensity. Control rods stabilize (or limit) the extent of the chain reaction to a safe level.

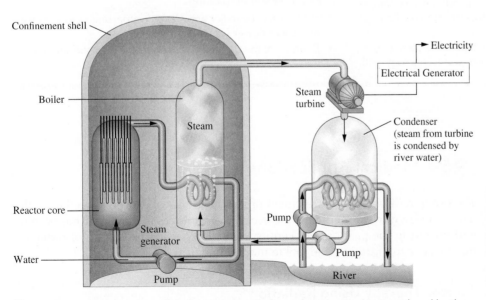

Figure 9.5 A representation of the "energy zones" of a nuclear reactor. Heat produced by the reactor core (red zone) is carried by water to a boiler (blue zone). Water in the boiler is converted to steam, which drives a turbine to convert heat energy to electrical energy. The isolation of these zones from each other allows heat energy transfer without actual physical mixing. This minimizes the transport of radioactive material into the environment (green zone).

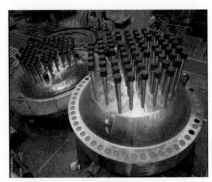

Figure 9.6 The nuclear reactor vessel contains coolant and the reactor core.

Green Chemistry

Nuclear Waste Disposal

Nuclear waste arises from a variety of sources. A major source is the spent fuel from nuclear power plants. Medical laboratories generate significant amounts of low-level waste from tracers and therapy. Even household items with limited lifetimes, such as certain types of smoke detectors, use a tiny amount of radioactive material, americium-241.

Virtually everyone is aware, through news reports and the Internet, of the problems of solid waste (nonnuclear) disposal that our society faces. For the most part, this material will degrade in some reasonable amount of time. Still, we are disposing of trash and garbage at a rate that far exceeds nature's ability to recycle it.

Now imagine the problem with nuclear waste. We cannot alter the rate at which it decays. This is defined by the half-life. We can't heat it, stir it, or add a catalyst to speed up the process as we can with chemical reactions. Furthermore, the half-lives of many nuclear waste products are very long: plutonium, for example, has a half-life in excess of 24,000 years. Ten half-lives represents the approximate time required for the radioactivity of a substance to reach background levels. So we are talking about a *very* long storage time, on the order of 250,000 years!

Where on earth can something so very hazardous be contained and stored with reasonable assurance that it will lie undisturbed for a quarter of a million years? Perhaps this is a rhetorical question. Scientists, engineers, and politicians have debated this question for almost 50 years. As yet, no permanent disposal site has been agreed upon. Most agree that the best solution is burial in a stable rock formation, but there is no firm agreement on the location. Fear of earthquakes, which may release large quantities of radioactive materials into the underground water system, is the most serious consideration. Such a disaster could render large sections of the country unfit for habitation.

Many argue for the continuation of temporary storage sites with the hope that the progress of science and technology will,

Spent nuclear fuel is temporarily stored in a storage pond.

in the years ahead, provide a safer and more satisfactory long-term solution.

The nuclear waste problem, important for its own sake, also affects the development of future societal uses of nuclear chemistry. Before we can fully enjoy its benefits, we must learn to use and dispose of its waste products safely.

For Further Understanding

▶ Summarize the major arguments supporting expanded use of nuclear power for electrical energy.

▶ Enumerate the characteristics of an "ideal" solution to the nuclear waste problem.

LEARNING GOAL

7 Cite examples of the use of radioactive isotopes in medicine.

9.5 Medical Applications of Radioactivity

The use of radiation in the treatment of various forms of cancer, as well as in the newer area of **nuclear medicine,** the use of radioisotopes in diagnosis, has become widespread in the past quarter century. Let's look at the properties of radiation that make it an indispensable tool in modern medical care.

Cancer Therapy Using Radiation

LEARNING GOAL

8 Describe the use of ionizing radiation in cancer therapy.

When high-energy radiation, such as gamma radiation, passes through a cell, it may collide with one of the molecules in the cell and cause it to lose one or more electrons, causing a series of events that result in the production of ion pairs. For this reason, such radiation is termed *ionizing radiation* (Section 9.1).

Ions produced in this way are highly energetic. Consequently, they may damage biological molecules and cause changes in cellular biochemical processes. Interaction of ionizing radiation with intracellular water produces free electrons and other particles that can damage DNA. This may result in diminished or altered cell function or, in extreme cases, the death of the cell.

An organ that is cancerous is composed of both healthy cells and malignant cells. Tumor cells are more susceptible to the effects of gamma radiation than normal cells because they are undergoing cell division more frequently. Consequently, exposure of the tumor area to carefully targeted and controlled dosages of high-energy gamma radiation from cobalt-60 (a high-energy gamma ray source) will kill a higher percentage of abnormal cells than normal cells. If the dosage is administered correctly, a sufficient number of malignant cells will die, destroying the tumor, and enough normal cells will survive to maintain the function of the affected organ.

Gamma radiation can cure cancer. Paradoxically, the exposure of healthy cells to gamma radiation can actually cause cancer. For this reason, radiation therapy for cancer is a treatment that requires unusual care and sophistication.

Nuclear Medicine

The diagnosis of a host of biochemical irregularities or diseases of the human body has been made routine through the use of radioactive tracers. Medical **tracers** are small amounts of radioactive substances used as probes to study internal organs. Medical techniques involving tracers are **nuclear imaging** procedures.

A small amount of the tracer, an isotope of an element that is known to be attracted to the organ of interest, is administered to the patient. For a variety of reasons, such as ease of administration of the isotope to the patient and targeting the organ of interest, the isotope is often a part of a larger molecule or ion. Because the isotope is radioactive, its path may be followed by using suitable detection devices. A radioimage, or "picture," of the organ is obtained, often far more detailed than is possible with conventional X-rays. Such techniques are noninvasive; that is, surgery is not required to investigate the condition of the internal organ, eliminating the risk associated with an operation.

The radioactive isotope of an element chosen for tracer studies has a chemical behavior similar to any other isotope of the same element. For example, iodine-127, the most abundant nonradioactive isotope of iodine, is used by the body in the synthesis of thyroid hormones and tends to concentrate in the thyroid gland. Both iodine-127 and radioactive iodine-131 behave in the same way, making it possible to use iodine-131 to study the thyroid. The rate of uptake of the radioactive isotope gives valuable information regarding underactivity or overactivity (hypoactive or hyperactive thyroid).

Isotopes with short half-lives are preferred for tracer studies. These isotopes emit their radiation in a more concentrated burst (short half-life materials have greater activity), facilitating their detection. If the radioactive decay is easily detected, the method is more sensitive and thus capable of providing more information. Furthermore, an isotope with a short half-life decays to background more rapidly. This is a mechanism for removal of the radioactivity from the body. If the radioactive element is also rapidly metabolized and excreted, this is beneficial as well.

The following examples illustrate the use of imaging procedures for diagnosis of disease.

- *Bone disease and injury.* The most widely used isotope for bone studies is technetium-99m, which is incorporated into a variety of ions and molecules that direct the isotope to the tissue being investigated. Technetium compounds containing phosphate are preferentially adsorbed on the surface of bone. New bone formation (common to virtually all bone injuries) increases the incorporation of the technetium compound. As a result, an enhanced image appears at the site of the injury. Bone tumors behave in a similar fashion.

LEARNING GOAL

9 Discuss the preparation and use of radioisotopes in diagnostic imaging studies.

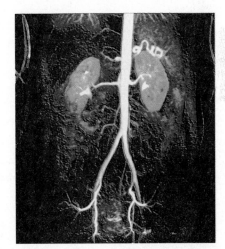

A radioimage of human kidneys (in light blue)

- *Cardiovascular disease.* Thallium-201 is used in the diagnosis of coronary artery disease. The isotope is administered intravenously and delivered to the heart muscle in proportion to the blood flow. Areas of restricted flow are observed as having lower levels of radioactivity, indicating some type of blockage.
- *Pulmonary disease.* Xenon is one of the noble gases. Radioactive xenon-133 may be inhaled by the patient. The radioactive isotope will be transported from the lungs and distributed through the circulatory system. Monitoring the distribution, as well as the reverse process, the removal of the isotope from the body (exhalation), can provide evidence of obstructive pulmonary disease, such as cancer or emphysema.

Examples of useful isotopes and the area(s) of the body in which they tend to concentrate are summarized in Table 9.4.

For many years, imaging with radioactive tracers was used exclusively for diagnosis. Recent applications have expanded to other areas of medicine. Imaging is now used extensively to guide surgery, assist in planning radiation therapy, and support the technique of angioplasty.

Question 9.11   Technetium-99m is used in diagnostic imaging studies involving the brain. What fraction of the radioisotope remains after 12 h have elapsed? See Table 9.2 for the half-life of technetium-99m.

Question 9.12   Barium-131 is a radioisotope used to study bone formation. A patient ingested barium-131. How much time will elapse until only one-fourth of the barium-131 remains, assuming that none of the isotope is eliminated from the body through normal processes? The half-life of barium-131 is 11.6 min.

Making Isotopes for Medical Applications

In early experiments with radioactivity, the radioactive isotopes were naturally occurring. For this reason, the radioactivity produced by these unstable isotopes is described as **natural radioactivity.** If, on the other hand, a normally stable, nonradioactive nucleus is made radioactive, the resulting radioactivity is termed **artificial radioactivity.** The stable nucleus is made unstable by the introduction of "extra" protons, neutrons, or both.

The process of forming radioactive substances is often accomplished in the core of a *nuclear reactor,* in which an abundance of small nuclear particles,

LEARNING GOAL

10   Explain the difference between natural and artificial radioactivity.

TABLE 9.4 Isotopes Commonly Used in Nuclear Medicine

Area of Body	Isotope	Use
Blood	Chromium-51	Determine blood volume in body
Bone	*Technetium-99m, barium-131, strontium-87	Allow early detection of the extent of bone tumors and active sites of rheumatoid arthritis
Brain	*Technetium-99m	Detect and locate brain tumors and stroke
Coronary artery	Thallium-201	Determine the presence and location of obstructions in coronary arteries
Heart	*Technetium-99m	Determine cardiac output, size, and shape
Kidney	*Technetium-99m	Determine renal function and location of cysts; a common follow-up procedure for kidney transplant patients
Liver-spleen	*Technetium-99m	Determine size and shape of liver and spleen; location of tumors
Lung	Xenon-133	Determine whether lung fills properly; locate region of reduced ventilation and tumors
Thyroid	Iodine-131	Determine rate of iodine uptake by thyroid

*The destination of this isotope is determined by the identity of the compound in which it is incorporated.

particularly neutrons, is available. Alternatively, extremely high velocity charged particles (such as alpha and beta particles) may be produced in *particle accelerators,* such as a cyclotron. Accelerators are extremely large and use magnetic and electric fields to "push and pull" charged particles toward their target at very high speeds. A portion of the accelerator at the Brookhaven National Laboratory is shown in Figure 9.7.

Many isotopes that are useful in medicine are produced by particle bombardment. A few examples include the following:

- Gold-198, used as a tracer in the liver, is prepared by neutron bombardment.

$$^{197}_{79}\text{Au} + {}^{1}_{0}\text{n} \longrightarrow {}^{198}_{79}\text{Au}$$

- Gallium-67, used in the diagnosis of Hodgkin's disease, is prepared by proton bombardment.

$$^{66}_{30}\text{Zn} + {}^{1}_{1}\text{p} \longrightarrow {}^{67}_{31}\text{Ga}$$

Some medically useful isotopes with short half-lives must be prepared near the site of the clinical test. Preparation and shipment from a reactor site would waste time and result in an isotopic solution that had already undergone significant decay, resulting in diminished activity.

A common example is technetium-99m. It has a half-life of only 6 h. It is prepared in a small generator, often housed in a hospital's radiology laboratory (Figure 9.8). The generator contains radioactive molybdate ion (MoO_4^{2-}). Molybdenum-99 is more stable than technetium-99m; it has a half-life of 67 h.

The molybdenum in molybdate ion decays according to the following nuclear equation:

$$^{99}_{42}\text{Mo} \longrightarrow {}^{99\text{m}}_{43}\text{Tc} + {}^{0}_{-1}\text{e}$$

The *m* in technetium-99m means that the isotope is *metastable*, indicating that, over time, the isotope decays to a more stable form of the same isotope. In this case:

$$^{99\text{m}}_{43}\text{Tc} \longrightarrow {}^{99}_{43}\text{Tc} + \gamma$$

Figure 9.7 A portion of a linear accelerator located at Brookhaven National Laboratory in New York. Particles can be accelerated at velocities close to the speed of light and accurately strike small "target" nuclei. At such facilities, rare isotopes can be synthesized and their properties studied.

A Medical Perspective

Magnetic Resonance Imaging

The Nobel Prize in physics was awarded to Otto Stern in 1943 and to Isidor Rabi in 1944. They discovered that certain atomic nuclei have a property known as spin, analogous to the spin associated with electrons, which we discussed in Chapter 2. The spin of electrons is responsible for the magnetic properties of atoms. Spinning nuclei behave as tiny magnets, producing magnetic fields as well.

One very important aspect of this phenomenon is the fact that the atoms in close proximity to the spinning nucleus (its chemical environment) exert an effect on the nuclear spin. In effect, measurable differences in spin are indicators of their surroundings. This relationship has been exhaustively studied for one atom in particular, hydrogen, and magnetic resonance techniques have become useful tools for the study of molecules containing hydrogen.

Human organs and tissues are made up of compounds containing hydrogen atoms. In the 1970s and 1980s, the experimental technique was extended beyond tiny laboratory samples of pure compounds to the most complex sample possible—the human body. The result of these experiments is termed *magnetic resonance imaging (MRI)*.

MRI is noninvasive to the body, requires no use of radioactive substances, and is quick, safe, and painless. A person is placed in a cavity surrounded by a magnetic field, and an image (based on the extent of radio frequency energy absorption) is generated, stored, and sorted in a computer. Differences between normal and malignant tissue, atherosclerotic thickening of an aortal wall, and a host of other problems may be seen clearly in the final image.

Advances in MRI technology have provided medical practitioners with a powerful tool in diagnostic medicine. This is

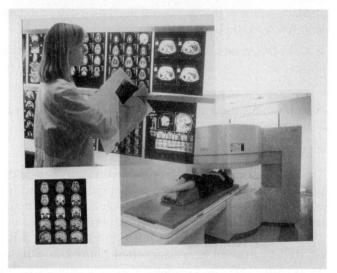

A patient is in an MRI scanner while a doctor studies brain scan images.

but one more example of basic science leading to technological advancement.

For Further Understanding

▶ Why is hydrogen a useful atom to study in biological systems?
▶ Why would MRI provide minimal information about bone tissue?

Figure 9.8 Preparation of technetium-99m. (a) A diagram depicting the conversion of $^{99}MoO_4^{2-}$ to $^{99m}TcO_4^-$ through radioactive decay. The radioactive pertechnetate ion is periodically removed from the generator in saline solution and used in tracer studies. (b) A photograph of a commercially available technetium-99m generator suitable for use in a hospital laboratory.

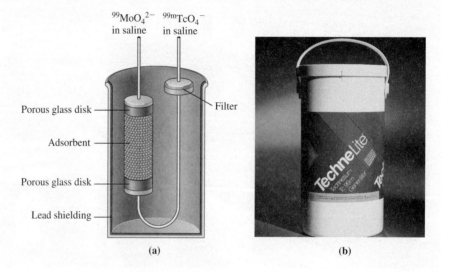

$^{99}MoO_4^{2-}$ in saline $^{99m}TcO_4^-$ in saline

Porous glass disk

Filter

Adsorbent

Porous glass disk

Lead shielding

(a)   (b)

Chemically, radioactive molybdate MoO_4^{2-} converts to radioactive pertechnetate ion (TcO_4^-). The radioactive TcO_4^- is removed from the generator when needed. It is administered to the patient as an aqueous salt solution that has an osmotic pressure identical to that of human blood.

9.6   Biological Effects of Radiation

It is necessary to use suitable precautions in working with radioactive substances. The chosen protocol is based on an understanding of the effects of radiation, dosage levels and "tolerable levels," the way in which radiation is detected and measured, and the basic precepts of radiation safety.

Radiation Exposure and Safety

In working with radioactive materials, the following factors must be considered.

LEARNING GOAL

11   Describe characteristics of radioactive materials that relate to radiation exposure and safety.

The Magnitude of the Half-Life

Short-half-life radioisotopes produce a larger amount of radioactivity per unit time than a long-half-life substance. For example, consider equal amounts of hypothetical isotopes that produce alpha particles. One has a half-life of 10 days; the other has a half-life of 100 days. In one half-life, each substance will produce exactly the same number of alpha particles. However, the first substance generates the alpha particles in only one-tenth of the time, and hence emits ten times as much radiation per unit time. Equal exposure times will result in higher levels of radiation exposure for substances with short half-lives, and lower levels for substances with long half-lives.

On the other hand, materials with short half-lives (weeks, days, or less) may be safer to work with, especially if an accident occurs. Over time (depending on the magnitude of the half-life), radioactive isotopes will decay to **background radiation** levels. This is the level of radiation attributable to our surroundings on a day-to-day basis.

Virtually all matter is composed of both radioactive and nonradioactive isotopes. Small amounts of radioactive material in the air, water, soil, and so forth make up a part of the background levels. Cosmic rays from outer space continually bombard us with radiation, contributing to the total background. Owing to the inevitability of background radiation, there can be no situation on earth where we observe zero radiation levels.

An isotope with a short half-life, for example 5.0 min, may decay to background in less than 1 h:

$$10 \; \text{half-lives} \times \frac{5.0 \; \text{min}}{1 \; \text{half-life}} = 50 \; \text{min}$$

A spill of such material could be treated by evacuating the area and waiting ten half-lives, perhaps by going to lunch. When you return to the laboratory, the material that was spilled will be no more radioactive than the floor itself. An accident with plutonium-239, which has a half-life of 24,000 years, would be quite a different matter! After 50 min, virtually all of the plutonium-239 would still remain. Long-half-life isotopes, by-products of nuclear technology, pose the greatest problems for safe disposal. Finding a site that will remain undisturbed "forever" is quite a formidable task.

See Green Chemistry: Nuclear Waste Disposal.

Question 9.13   Describe the advantage of using isotopes with short half-lives for tracer applications in a medical laboratory.

Question 9.14   Can you think of any disadvantage associated with the use of isotopes described in Question 9.13? Explain.

Shielding

Alpha and beta particles, being relatively low in penetrating power, require low-level *shielding*. **Shielding** is protection from harmful radiation. A lab coat and gloves are generally sufficient protection from this low-penetration radiation. On the other hand, shielding made of lead, concrete, or both is required for gamma

Figure 9.9 Photograph of the construction of one-million-gallon (gal) capacity storage tanks for radioactive waste. Located in Hanford, Washington, they are now covered with 6–8 feet (ft) of earth.

rays (and X-rays, which are also high-energy radiation). Extensive manipulation of gamma emitters is often accomplished in laboratory and industrial settings by using robotic control: computer-controlled mechanical devices that can be programmed to perform virtually all manipulations normally carried out by humans.

Distance from the Radioactive Source

Radiation intensity varies *inversely* with the *square* of the distance from the source. Doubling the distance from the source *decreases* the intensity by a factor of four (2^2). Again, the use of robot manipulators is advantageous, allowing a greater distance between the operator and the radioactive source.

Time of Exposure

The effects of radiation are cumulative. Generally, potential damage is directly proportional to the time of exposure. Workers exposed to moderately high levels of radiation on the job may be limited in the time that they can perform that task. For example, workers involved in the cleanup of the Fukushima Daiichi nuclear plant, incapacitated in 2011, observed strict limits on the amount of time that they could be involved in the cleanup activities.

Types of Radiation Emitted

Alpha and beta emitters are generally less hazardous than gamma emitters, owing to differences in energy and penetrating power that require less shielding. However, ingestion or inhalation of an alpha emitter or beta emitter can, over time, cause serious tissue damage; the radioactive substance is in direct contact with sensitive tissue. Green Chemistry: Radon and Indoor Air Pollution expands on this problem.

Waste Disposal

Virtually all applications of nuclear chemistry create radioactive waste and, along with it, the problems of safe handling and disposal. Most disposal sites, at present, are considered temporary, until a long-term safe solution can be found. Figure 9.9 conveys a sense of the enormity of the problem. Green Chemistry: Nuclear Waste Disposal examines this problem in more detail.

LEARNING GOAL

12 Be familiar with common techniques for the detection of radioactivity.

9.7 Measurement of Radiation

The changes that take place when radiation interacts with matter (such as photographic film) provide the basis of operation for various radiation detection devices.

The principal detection methods involve the use of either photographic film to create an image of the location of the radioactive substance or a counter that allows the measurement of intensity of radiation emitted from some source by converting the radiation energy to an electrical signal.

Photographic Imaging

This approach is often used in nuclear medicine. An isotope, perhaps iodine-131, is administered to a patient to study the thyroid gland, and the isotope begins to concentrate in the organ of interest. Nuclear images (photographs) of that region of the body are taken at periodic intervals using a special type of film. The emission of radiation from the radioactive substance creates the image, in much the same way as light causes the formation of images on conventional film in a camera. Upon development of the series of photographs, a record of the organ's uptake of the isotope over time enables the radiologist to assess the condition of the organ.

Computer Imaging

The coupling of rapid developments in the technology of television and computers, resulting in the marriage of these two devices, has brought about a versatile alternative to photographic imaging.

A specialized television camera, sensitive to emitted radiation from a radioactive substance administered to a patient, develops a continuous and instantaneous record of the voyage of the isotope throughout the body. The signal, transmitted to the computer, is stored, sorted, and portrayed on a monitor. Advantages include increased sensitivity, allowing a lower dose of the isotope, speed through elimination of the developing step, and versatility of application, limited perhaps only by the creativity of the medical practitioners.

A particular type of computer imaging, useful in diagnostic medicine, is the CT scanner. The CT scanner measures the interaction of X-rays with biological tissue, gathering huge amounts of data and processing the data to produce detailed information, all in a relatively short time. Such a device may be less hazardous than conventional X-ray techniques because it generates more useful information per unit of radiation. It often produces a superior image. Images of the human brain taken by a CT scanner are shown in Figure 9.10.

CT represents *computerized tomography:* the computer reconstructs a series of measured images of tissue density (tomography). Small differences in tissue density may indicate the presence of a tumor.

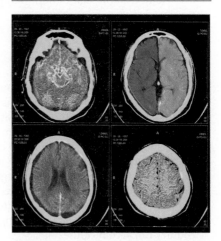

Figure 9.10 CT scanned images of the human brain can allow for detection of disease.

The Geiger Counter

A Geiger counter is an instrument that detects ionizing radiation (Figure 9.11). Ions, produced by radiation passing through a tube filled with an ionizable gas, can conduct an electrical current between two electrodes. This current flow can be measured and is proportional to the level of radiation (Figure 9.12). Such devices, which were routinely used in laboratory and industrial monitoring, have been largely replaced by more sophisticated devices, often used in conjunction with a computer.

Film Badges

A common sight in any hospital or medical laboratory or any laboratory that routinely uses radioisotopes is the film badge worn by all staff members exposed in any way to low-level radioactivity.

A film badge is merely a piece of photographic film that is sensitive to energies corresponding to radioactive emissions. It is shielded from light, which would interfere, and mounted in a clip-on plastic holder that can be worn throughout the workday. The badges are periodically collected and developed. The degree of darkening is proportional to the amount of radiation to which the worker has been exposed, just as a conventional camera produces images on film in proportion to the amount of light that it "sees."

Proper record keeping thus allows the laboratory using radioactive substances to maintain an ongoing history of each individual's exposure and, at the same time, promptly pinpoint any hazards that might otherwise go unnoticed.

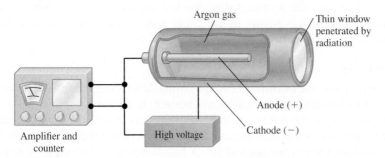

Figure 9.12 The design of a Geiger counter used for the measurement of radioactivity.

Argon gas

Thin window penetrated by radiation

Anode (+)

Cathode (−)

High voltage

Amplifier and counter

Figure 9.11 A worker in a protective suit is measuring the ground using a Geiger counter.

Green Chemistry

Radon and Indoor Air Pollution

Marie and Pierre Curie first discovered that air in contact with radium compounds became radioactive. Later experiments by Ernest Rutherford and others isolated the radioactive substance from the air. This substance was an isotope of the noble gas radon (Rn).

We now know that radium (Ra) produces radon by spontaneous decay:

$$^{226}_{88}Ra \longrightarrow {}^{4}_{2}He + {}^{222}_{86}Rn$$

Radium in trace quantities is found in soil and rock and is unequally distributed in the soil. The decay product, radon, is emitted from the soil to the surrounding atmosphere. Radon is also found in higher concentrations where uranium is found in the soil. This is not surprising, because radium is formed as a part of the stepwise decay of uranium.

If someone constructs a building over soil or rock that has a high radium content (or uses stone with a high radium content to build the foundation!), the radon gas can percolate through the basement and accumulate in the house. Couple this with the need to build more energy-efficient, well-insulated dwellings, and the radon levels in buildings in some regions of the country can become quite high.

Radon itself is radioactive; however, its radiation is not the major problem. Because it is a gas and chemically inert, it is rapidly exhaled after breathing. However, radon decays to polonium:

$$^{222}_{86}Rn \longrightarrow {}^{4}_{2}He + {}^{218}_{84}Po$$

This polonium isotope is radioactive and is a nonvolatile heavy metal that can attach itself to bronchial or lung tissue, emitting hazardous radiation and producing other isotopes that are also radioactive.

In the United States, homes are now being tested and monitored for radon. In many states, proof of acceptable levels of radon is a condition of sale of the property. Studies continue to

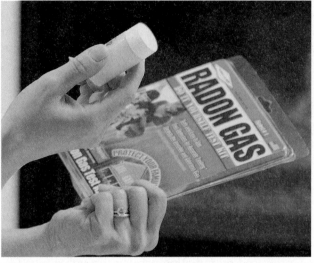

A radon gas detector can be used by a homeowner to quickly determine the level of radon present in a home. However, the collection kit must be sent to a laboratory for analysis.

attempt to find reasonable solutions to the radon problem. Current recommendations include sealing cracks and openings in basements, increasing ventilation, and evaluating sites before construction of buildings. Debate continues within the scientific community regarding a safe and attainable indoor air quality standard for radon.

For Further Understanding

▶ Why is indoor radon more hazardous than outdoor radon?
▶ Polonium-218 has a very long half-life. Explain why this constitutes a potential health problem.

LEARNING GOAL

13 Interpret common units of radiation intensity and discuss their biological implications.

Units of Radiation Measurement

The amount of radiation emitted by a source or received by an individual is reported in a variety of ways, using units that describe different aspects of radiation. The *curie* and the *becquerel* describe the intensity of the emitted radiation, the *roentgen* describes exposure to radiation, and the *rad, gray, rem,* and the *sievert* describe the biological effects of radiation.

Radioactivity

The **curie** (Ci) and the **becquerel** (Bq) are units describing the measurement of the amount of radioactivity in a radioactive source. Both the curie and the becquerel are independent of the nature of the radiation and its effect on biological tissue. A curie is defined as the amount of radioactive material that produces 3.7×10^{10} atomic disintegrations per second (s). A becquerel is the amount of radioactive material that produces 1 atomic disintegration per second. Therefore, one Ci is equivalent to 3.7×10^{10} Bq.

Exposure

The **roentgen** is a measure of very high energy ionizing radiation (X-ray and gamma ray) only. The roentgen is defined as the amount of radiation needed to produce 2×10^9 ion pairs when passing through 1 cubic centimeter (cm^3) of air at 0°C. The roentgen is a measure of radiation's interaction with air and gives no information about the effect on biological tissue. Since the measurement is based on air ionization, it can be used to determine exposure. Generally, it would be lethal if a human were exposed to 500 roentgens in 5 hours.

Absorbed Dosage

The **rad,** or *radiation absorbed dosage,* provides more meaningful information than either of the previous units of measure. It takes into account the nature of the absorbing material. It is defined as the dosage of radiation able to transfer 2.4×10^{-3} calories (cal) of energy to 1 kilogram (kg) of matter. The **gray** (Gy) is also used to measure an absorbed dosage. The Gy is commonly used in radiation therapy. It is defined as the absorption of 1 joule (J) of energy by 1 kg of matter.

Dose Equivalent

The **rem,** or *roentgen equivalent for man,* describes the biological damage caused by the absorption of different kinds of radiation by the human body. The *rem* is obtained by multiplication of the *rad* by a factor called the *relative biological effect (RBE).* The RBE is a function of the type of radiation. Although a beta particle is more penetrating than an alpha particle, an alpha particle is approximately ten times more damaging to biological tissue. As a result, the RBE is ten for alpha particles and one for beta particles. The **sievert,** Sv, describes the biological effect that results when one Gy of radiation energy is absorbed by human tissue. One Sv is equivalent to 100 rem.

The **lethal dose (LD$_{50}$)** of radiation is defined as the acute dosage of radiation that would be fatal for 50% of the exposed population within 30 days. An estimated lethal dose is 500 rems, or 5 Sv. Some biological effects, however, may be detectable at a level as low as 25 rems, or 0.25 Sv. Relative yearly background radiation dosages received by Americans are shown in Figure 9.13.

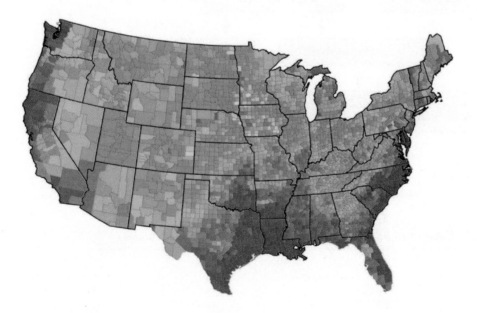

Figure 9.13 Relative yearly radiation dosages for individuals in the continental United States. Red, yellow, and green shading indicates higher levels of background radiation. Blue shading indicates regions of lower background exposure.

Question 9.15   From a clinical standpoint, what advantages does expressing radiation in rems have over the use of other radiation units?

Question 9.16   Is the roentgen unit used in the measurement of alpha particle radiation? Why or why not?

CHAPTER MAP

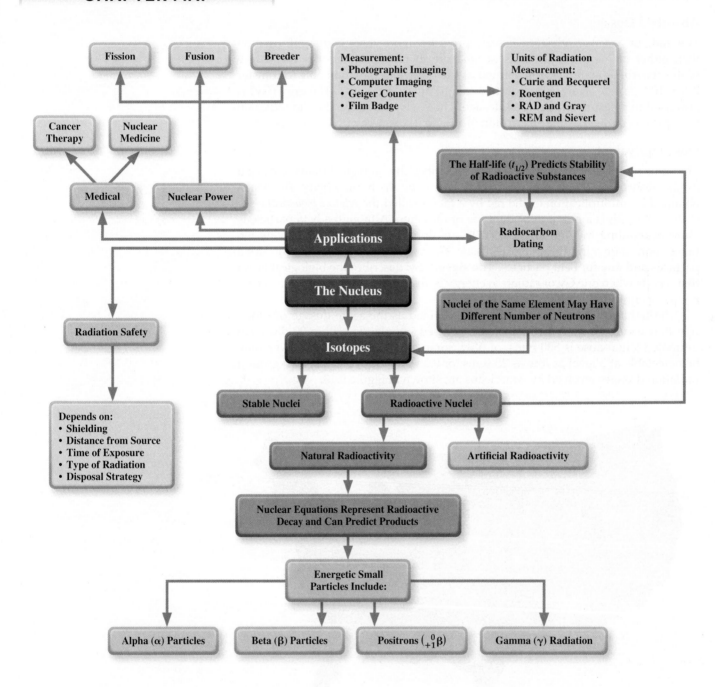

SUMMARY

9.1 Natural Radioactivity

▶ **Radioactivity** is the process by which atoms emit energetic, ionizing particles or rays. These particles or rays are termed radiation.

▶ Nuclear radiation occurs because the nucleus is unstable, hence radioactive.

▶ Nuclear symbols consist of the elemental symbol, the atomic number, and the mass number.

▶ Not all **nuclides** are unstable. Only unstable nuclides undergo change and produce radioactivity in the process of radioactive decay.

▶ Natural radiation emitted by unstable nuclei includes **alpha particles (α), beta particles (β), positrons ($_{+1}^{0}\beta$),** and **gamma rays (γ).**

▶ This radiation is collectively termed **ionizing radiation.**

9.2 Writing a Balanced Nuclear Equation

▶ A **nuclear equation** represents a nuclear process such as radioactive decay.

▶ In a nuclear equation, the total of the mass numbers on each side of the reaction arrow must be identical, and the sum of the atomic numbers of the reactants must equal the sum of the atomic numbers of the products.

▶ Nuclear equations can be used to predict products of nuclear reactions.

9.3 Properties of Radioisotopes

▶ The **binding energy** of the nucleus is a measure of nuclear stability. When an isotope decays, energy is released.

▶ Nuclear stability correlates with the ratio of neutrons to protons in the isotope. Isotopes (except hydrogen-1) with more protons than neutrons are unstable. Nuclei with large numbers of protons tend to be unstable, and isotopes containing 2, 8, 20, 50, 82, or 126 protons or neutrons (magic numbers) are stable. Also, isotopes with even numbers of protons or neutrons are generally more stable than those with odd numbers of protons or neutrons.

▶ The **half-life**, $t_{1/2}$, is the time required for one-half of a given quantity of a substance to undergo change. Each isotope has its own characteristic half-life. The degree of stability of an isotope is indicated by the isotope's half-life. Isotopes with short half-lives decay rapidly; they are very unstable.

▶ **Radiocarbon dating** is based on the measurement of the relative amounts of carbon-12 and carbon-14 present in an object. The ratio of the masses of these isotopes changes slowly over time, making it useful in determining the age of objects containing carbon.

9.4 Nuclear Power

▶ Einstein predicted that a small amount of nuclear mass would convert to a very large amount of energy when the nucleus breaks apart. His prediction formed the basis for the modern **nuclear reactor.**

▶ **Fission** reactors, relying on a **chain reaction,** are used to generate electrical power. Technological problems with **fusion** and **breeder reactors** have prevented their commercialization in the United States.

9.5 Medical Applications of Radioactivity

▶ The use of radiation in the treatment of various forms of cancer, and in the newer area of **nuclear medicine,** has become widespread in the past quarter century.

▶ **Ionizing radiation** causes changes in cellular biochemical processes that may damage or kill the cell.

▶ A cancerous organ is composed of both healthy and malignant cells. Exposure of the tumor area to controlled dosages of high-energy gamma radiation from cobalt-60 will kill a higher percentage of abnormal cells than normal cells and is a valuable cancer therapy.

▶ The diagnosis of a host of biochemical irregularities or diseases of the human body has been made routine through the use of radioactive tracers.

▶ **Tracers** are small amounts of radioactive substances used as probes to study internal organs. Because the isotope is radioactive, its path may be followed by using suitable detection devices. An image of the organ is obtained, far more detailed than is possible with conventional X-rays. Medical techniques involving tracers are **nuclear imaging** procedures.

▶ The radioactivity produced by unstable isotopes is described as **natural radioactivity.** A normally stable, nonradioactive nucleus can be made radioactive, and this is termed **artificial radioactivity** (the process produces synthetic isotopes).

▶ Synthetic isotopes are often used in clinical situations. Isotopic synthesis may be carried out in the core of a nuclear reactor or in a *particle accelerator*. Short-lived isotopes, such as technetium-99m, are often produced directly at the site of the clinical testing.

9.6 Biological Effects of Radiation

▶ Safety considerations are based on the magnitude of the **half-life, shielding,** distance from the radioactive source, time of exposure, and type of radiation emitted.

▶ We are never entirely free of the effects of radioactivity. **Background radiation** is normal radiation attributable to our surroundings.

▶ Virtually all applications of nuclear chemistry create radioactive waste and, along with it, the problems of safe handling and disposal. Most disposal sites are considered temporary, until a long-term safe solution can be found.

9.7 Measurement of Radiation

▶ The changes that take place when radiation interacts with matter provide the basis for various radiation detection devices. Photographic imaging, computer imaging, the

Geiger counter, and film badges represent the most frequently used devices for detecting and measuring radiation.

▶ Commonly used radiation units include the **curie** and **becquerel,** measurements of the amount of radioactivity in a radioactive source; the **roentgen,** a measurement of high-energy radiation (X-ray and gamma ray); the **rad** (radiation absorbed dosage) and **gray,** which take into account the nature of the absorbing material; and the **rem** (roentgen equivalent for man) and **sievert,** which describe the biological damage caused by the absorption of different kinds of radiation by the human body.

▶ The **lethal dose** of radiation, LD_{50}, is defined as the dose that would be fatal for 50% of the exposed population within 30 days.

ANSWERS TO PRACTICE PROBLEMS

9.1 a. $^{226}_{88}Ra \longrightarrow \, ^{4}_{2}He + \, ^{222}_{86}Rn$ **b.** $^{11}_{5}B \longrightarrow \, ^{7}_{3}Li + \, ^{4}_{2}He$

9.2 a. $^{85}_{36}Kr \longrightarrow \, ^{85}_{37}Rb + \, ^{0}_{-1}e$ **b.** $^{239}_{92}U \longrightarrow \, ^{239}_{93}Np + \, ^{0}_{-1}e$

9.3 a. $^{79}_{37}Rb \longrightarrow \, ^{79}_{36}Kr + \, ^{0}_{+1}e$ **b.** $^{38}_{20}Ca \longrightarrow \, ^{38}_{19}K + \, ^{0}_{+1}e$

9.4 a. 6.3 ng of sodium-24 remains after 2.5 days
 b. 0.6 ng of technetium-99m remains after 1 day

QUESTIONS AND PROBLEMS

Natural Radioactivity

Foundations

9.17 Describe the meaning of the term *natural radioactivity.*
9.18 Describe what is meant by the term *nuclide.*
9.19 What is the composition of an alpha particle?
9.20 What is alpha decay?
9.21 What is the composition of a beta particle?
9.22 What is the composition of a positron?
9.23 What are the major differences between alpha and beta particles?
9.24 What are the major differences between alpha particles and gamma radiation?
9.25 How do nuclear reactions and chemical reactions differ?
9.26 We can control the rate of chemical reactions. Can we control the rate of natural radiation?
9.27 Write the nuclear symbol for an alpha particle.
9.28 Write the nuclear symbol for a beta particle.
9.29 How does an alpha particle differ from a helium atom?
9.30 What is the major difference between beta and gamma radiation?

9.31 Compare and contrast the four major types of radiation produced by nuclear decay.
9.32 Rank the four major types of radiation in order of size, speed, and penetrating power.

Applications

9.33 Write the nuclear symbol for nitrogen-15.
9.34 Write the nuclear symbol for carbon-14.
9.35 Write the nuclear symbol for uranium-235.
9.36 How many protons and neutrons are contained in the nucleus of uranium-235?
9.37 How many protons and neutrons are contained in each of the three isotopes of hydrogen?
9.38 How many protons and neutrons are contained in each of the three isotopes of carbon?

Writing a Balanced Nuclear Equation

Foundations

9.39 Write a nuclear equation to represent cobalt-60 decaying to nickel-60 plus a beta particle plus a gamma ray.
9.40 Write a nuclear equation to represent radium-226 decaying to radon-222 plus an alpha particle.
9.41 Complete the following nuclear equation:

$$^{23}_{11}Na + \, ^{2}_{1}H \longrightarrow ? + \, ^{1}_{1}H$$

9.42 Complete the following nuclear equation:

$$^{238}_{92}U + \, ^{14}_{7}N \longrightarrow ? + 6 \, ^{1}_{0}n$$

9.43 Complete the following nuclear equation:

$$^{24}_{10}Ne \longrightarrow \beta + ?$$

9.44 Complete the following nuclear equation:

$$^{190}_{78}Pt \longrightarrow \alpha + ?$$

9.45 Complete the following nuclear equation:

$$? \longrightarrow \, ^{140}_{56}Ba + \, ^{0}_{-1}e$$

9.46 Complete the following nuclear equation:

$$? \longrightarrow \, ^{214}_{90}Th + \, ^{4}_{2}He$$

Applications

9.47 Element 107 was synthesized by bombarding bismuth-209 with chromium-54. Write the equation for this process if one product is a neutron.
9.48 Element 109 was synthesized by bombarding bismuth-209 with iron-58. Write the equation for this process if one product is a neutron.
9.49 Write a balanced nuclear equation for beta emission by magnesium-27.
9.50 Write a balanced nuclear equation for alpha decay of bismuth-212.
9.51 Write a balanced nuclear equation for positron emission by nitrogen-12.

9.52 Write a balanced nuclear equation for positron emission by manganese-52.

9.53 Americium-241, found in many home smoke detectors, decays by alpha emission. The alpha particle ionizes surrounding air molecules, and ions produced conduct an electric current. Smoke particles block this process and the change in current triggers an alarm. Write the balanced nuclear equation for the decay of americium-241.

9.54 Element 106 was named seaborgium (Sg) in honor of Glenn T. Seaborg, a pioneer in the discovery of lanthanide and actinide elements. Seaborgium-263 decays by alpha emission. Write a balanced nuclear equation for this process.

Properties of Radioisotopes

Foundations

9.55 Summarize the major characteristics of nuclei for which we predict a high degree of stability.

9.56 Explain why the binding energy of a nucleus is expected to be large.

9.57 Sodium-24 has a half-life of 15 h. How many half-lives occur after 225 h?

9.58 Cobalt-60 has a half-life of 5.3 years. How many half-lives occur after 21.2 years?

Applications

9.59 Would you predict oxygen-20 to be stable? Explain your reasoning.

9.60 Would you predict cobalt-59 to be stable? Explain your reasoning.

9.61 Would you predict chromium-48 to be stable? Explain your reasoning.

9.62 Would you predict lithium-9 to be stable? Explain your reasoning.

9.63 If 3.2 mg of the radioisotope iodine-131 is administered to a patient, how much will remain in the body after 24 days, assuming that no iodine has been eliminated from the body by any other process? (See Table 9.2 for the half-life of iodine-131.)

9.64 A patient receives 9.0 ng of a radioisotope with a half-life of 12 h. How much will remain in the body after 2.0 days, assuming that radioactive decay is the only path for removal of the isotope from the body?

9.65 A sample containing 1.00×10^2 mg of iron-59 is stored for 135 days. What mass of iron-59 will remain at the end of the storage period? (See Table 9.2 for the half-life of iron-59.)

9.66 An instrument for cancer treatment containing a cobalt-60 source was manufactured in 1988. In 1995, it was removed from service and, in error, was buried in a landfill with the source still in place. What percentage of its initial radioactivity will remain in the year 2020? (See Table 9.2 for the half-life of cobalt-60.)

9.67 Describe the process used to determine the age of the wooden coffin of King Tut.

9.68 What property of carbon enables us to assess the age of a painting?

Nuclear Power

Foundations

9.69 Which type of nuclear process splits nuclei to release energy?

9.70 Which type of nuclear process combines small nuclei to release energy?

9.71 a. Describe the process of fission.
b. How is this reaction useful as the basis for the production of electrical energy?

9.72 a. Describe the process of fusion.
b. How could this process be used for the production of electrical energy?

Applications

9.73 Write a balanced nuclear equation for a fusion reaction.

9.74 What are the major disadvantages of a fission reactor for electrical energy production?

9.75 What is meant by the term *breeder reactor?*

9.76 What are the potential advantages and disadvantages of breeder reactors?

9.77 Describe what is meant by the term *chain reaction.*

9.78 Why are cadmium rods used in a fission reactor?

9.79 What is the greatest barrier to development of fusion reactors?

9.80 What type of nuclear reaction fuels our solar system?

Medical Applications of Radioactivity

Foundations

9.81 Why is radiation therapy an effective treatment for certain types of cancer?

9.82 Describe how radioactive tracers are used in the diagnosis of disease.

9.83 What is the difference between natural radioactivity and artificial radioactivity?

9.84 Describe how medically useful isotopes may be prepared.

Applications

9.85 Describe an application of each of the following isotopes:
a. technetium-99m
b. xenon-133

9.86 Describe an application of each of the following isotopes:
a. iodine-131
b. thallium-201

9.87 The isotope indium-111 is used in medical laboratories as a label for blood platelets. To prepare indium-111, silver-108 is bombarded with an alpha particle, forming an intermediate isotope of indium. Write a nuclear equation for the process, and identify the intermediate isotope of indium.

9.88 Radioactive molybdenum-99 is used to produce the tracer isotope, technetium-99m. Write a nuclear equation for the formation of molybdenum-99 from stable molybdenum-98 bombarded with neutrons.

Biological Effects of Radiation

Foundations

9.89 What is the source of background radiation?

9.90 Why do high-altitude jet flights increase a person's exposure to background radiation?

Applications

Answer Questions 9.91 through 9.98 based on the assumption that you are employed by a clinical laboratory that prepares radioactive isotopes for medical diagnostic tests. Consider alpha, beta, positron, and gamma emission.

9.91 What would be the effect on your level of radiation exposure if you increase your average distance from a radioactive source?

9.92 Would wearing gloves have any significant effect? Why?

9.93 Would limiting your time of exposure have a positive effect? Why?

9.94 Would wearing a lab apron lined with thin sheets of lead have a positive effect? Why?

9.95 Would the use of robotic manipulation of samples have an effect? Why?

9.96 Would the use of concrete rather than wood paneling help to protect workers in other parts of the clinic? Why?

9.97 Would the thickness of the concrete in Question 9.96 be an important consideration? Why?

9.98 Suggest a protocol for radioactive waste disposal.

Measurement of Radiation

Foundations

9.99 What is meant by the term *relative biological effect?*

9.100 What is meant by the term *lethal dose* of radiation?

9.101 Define each of the following units:
  a. curie
  b. roentgen
  c. becquerel

9.102 Define each of the following radiation units:
  a. rad
  b. rem
  c. gray

Applications

9.103 X-ray technicians often wear badges containing photographic film. How is this film used to indicate exposure to X-rays?

9.104 Why would a Geiger counter be preferred to film for assessing the immediate danger resulting from a spill of some solution containing a radioisotope?

CHALLENGE PROBLEMS

1. Isotopes used as radioactive tracers have chemical properties that are similar to those of a nonradioactive isotope of the same element. Explain why this is a critical consideration in their use.

2. A chemist proposes a research project to discover a catalyst that will speed up the decay of radioactive isotopes that are waste products of a medical laboratory. Such a discovery would be a potential solution to the problem of nuclear waste disposal. Critique this proposal.

3. A controversial solution to the disposal of nuclear waste involves burial in sealed chambers far below the earth's surface. Describe the potential pros and cons of this approach.

4. What type of radioactive decay is favored if the number of protons in the nucleus is much greater than the number of neutrons? Explain your reasoning.

5. If the proton-to-neutron ratio in Problem 4 (above) were reversed, what radioactive decay process would be favored? Explain.

6. Radioactive isotopes are often used as "tracers" to follow an atom through a chemical reaction, and the following is an example. Acetic acid reacts with methanol by eliminating a molecule of water to form methyl acetate. Explain how you would use the radioactive isotope oxygen-18 to show whether the oxygen atom in the water product comes from the —OH of acetic acid or the —OH of methanol.

$$\underset{\text{Acetic acid}}{H_3C\overset{\displaystyle O}{\overset{\|}{C}}-OH} + \underset{\text{Methanol}}{HO-CH_3} \longrightarrow \underset{\text{Methyl acetate}}{H_3C\overset{\displaystyle O}{\overset{\|}{C}}-O-CH_3} + H_2O$$

7. Chromium-51, used as a tracer to study red blood cells, decays by electron capture. In this process, a proton in the chromium nucleus is, in effect, converted to a neutron by combining with (capturing) an electron. Write a balanced equation for this process. What is the identity of the product?

An Introduction to Organic Chemistry

LEARNING GOALS

1 Compare and contrast organic and inorganic compounds.

2 Recognize structures that represent each of the families of organic compounds.

3 Write the names and draw the structures of the common functional groups that characterize the families of organic compounds.

4 Write condensed, structural, and line formulas for saturated hydrocarbons.

5 Describe the relationship between the structure and physical properties of saturated hydrocarbons.

6 Use the basic rules of the IUPAC nomenclature system to name alkanes and substituted alkanes.

7 From the IUPAC name of an alkane or substituted alkane, be able to draw the structure.

8 Draw constitutional (structural) isomers of simple organic compounds.

9 Write the names and draw the structures of simple cycloalkanes.

10 Draw cis- and trans-isomers of cycloalkanes.

11 Describe conformations of alkanes.

12 Draw the chair and boat conformations of cyclohexane.

13 Write balanced equations for combustion reactions of alkanes.

14 Write balanced equations for halogenation reactions of alkanes.

The origins of fossil fuels

OUTLINE

INTRODUCTION

Organic chemistry is the study of carbon-containing compounds. The term *organic* was coined in 1807 by the Swedish chemist Jöns Jakob Berzelius. At that time, it was thought that all organic compounds, such as fats, sugars, coal, and petroleum, were formed by living or once living organisms. All early attempts to synthesize these compounds in the laboratory failed, and it was thought that a vital force, available only in living cells, was needed for their formation.

This idea began to change in 1828 when a 27-year-old German physician, whose first love was chemistry, synthesized the organic molecule urea from inorganic starting materials. This man was Friedrich Wöhler, the "father of organic chemistry."

As a child, Wöhler didn't do particularly well in school because he spent so much time doing chemistry experiments at home. Eventually, he did earn his medical degree, but he decided to study chemistry in the laboratory of Berzelius rather than practice medicine.

After a year he returned to Germany to teach and, as it turned out, to do the experiment that made him famous. The goal of the experiment was to prepare ammonium cyanate from a mixture of potassium cyanate and ammonium sulfate. Wöhler heated a solution of the two salts and crystallized the product. But the product didn't look like ammonium cyanate. It was a white crystalline material that looked exactly like urea! Urea is a waste product of protein breakdown in the body and is excreted in the urine. Wöhler recognized urea crystals because he had previously purified them from the urine of dogs and humans. Excited about his accidental discovery, he wrote to his teacher and friend Berzelius, "I can make urea without the necessity of a kidney, or even an animal, whether man or dog."

$$NH_4^+ \; [N{=}C{=}O]^-$$

$$\begin{matrix} & O \\ & \| \\ & C \\ H_2N & \diagup \quad \diagdown \; NH_2 \end{matrix}$$

Ammonium cyanate (inorganic salt)	Urea (organic compound)

Ironically, Wöhler, the first person to synthesize an organic compound from inorganic substances, devoted the rest of his career to inorganic chemistry. In his words, "Organic chemistry nowadays almost drives me mad. To me it appears like a primeval tropical forest full of the most remarkable things, a dreadful endless jungle into which one does not dare enter, for there seems to be no way out." However, other chemists continued this work, and as a result, the "vital force theory" was laid to rest, and modern organic chemistry was born.

In this and later chapters, we will travel through Wohler's "primeval tropical forest" as we study the amazing array of organic molecules (molecules made up of carbon, hydrogen, and a few other elements). Many of these molecules are essential to life. As we will see, all of the structural and functional molecules of the cell, including the phospholipids that make up the cell membrane and the enzymes that speed up biological reactions, are organic molecules. Smaller organic molecules, such as the sugars glucose and fructose, are used as fuel by our cells. Others, such as penicillin, aspirin, and the Alzheimer's drug galantamine featured on the cover of this book are useful in the treatment of disease. Although the number of organic molecules may seem intimidatingly large, we will see that their structure and behavior can be understood by mastering a few basic principles.

10.1   The Chemistry of Carbon

The number of possible carbon-containing compounds is almost limitless. The importance of these organic compounds is reflected in the fact that over half of this book is devoted to the study of molecules made with this single element.

Why are there so many organic compounds? There are several reasons. First, carbon can form *stable, covalent* bonds with other carbon atoms. In fact, carbon can form up to four covalent bonds with other carbon atoms. The resulting molecules may be linear, branched, or cyclic (Figure 10.1).

A second reason for the vast number of organic compounds is that carbon atoms can form stable bonds with other elements. Several families of organic compounds (alcohols, aldehydes, ketones, esters, and ethers) contain oxygen atoms bonded to carbon. Others contain nitrogen, sulfur, or halogens [Group VIIA(17) elements]. The presence of these elements confers a wide variety of new chemical and physical properties on an organic compound.

Third, carbon can form double or triple covalent bonds with other carbon atoms to produce a variety of organic molecules with very different properties. Finally, the number of ways in which carbon and other atoms can be arranged is nearly limitless. As shown in Figure 10.1, linear chains of carbon atoms, ring structures, and branched chains are common. Two organic compounds may even have the same number and kinds of atoms but completely different structures and thus different properties. Such organic molecules are called *isomers.* Although the number of organic molecules may seem intimidatingly large, we will see that their structures and behavior can be understood by mastering a few basic principles.

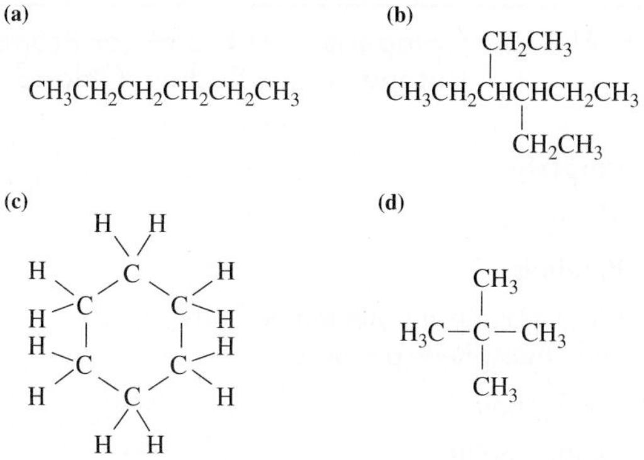

Figure 10.1 Carbon forms stable, covalent bonds with other carbon atoms. The resultant molecules may take many structural forms, including (a) linear molecules, (b) branched molecules, and (c) cyclic molecules. (d) Carbon can form bonds with up to four other carbon atoms.

Important Differences between Organic and Inorganic Compounds

The bonds between carbon and another atom are almost always *covalent bonds,* whereas the bonds in many inorganic compounds are *ionic bonds.* Ionic bonds form when one or more electrons transfer from one atom to another. The atom donating the electron becomes positively charged (a cation) and the atom receiving the electron becomes negatively charged (an anion). The ionic bond is the electrostatic attraction between these positive and negative ions. Covalent bonds are formed by sharing one or more pairs of electrons.

Ionic compounds often form three-dimensional crystals made up of many positive and negative ions. Covalent compounds exist as individual units called molecules. Water-soluble ionic compounds often dissociate in water to form ions and are called electrolytes. Most covalent compounds are nonelectrolytes, remaining intact in solution.

As a result of these differences, ionic substances usually have much higher melting and boiling points than covalent compounds. They are more likely to dissolve in water than in a less-polar solvent, whereas organic compounds, which are typically nonpolar or only moderately polar, are less soluble, or insoluble in water. In Table 10.1, the physical properties of the organic compound butane are compared with those of sodium chloride, an inorganic compound of similar molar mass.

LEARNING GOAL

1 Compare and contrast organic and inorganic compounds.

Polar covalent compounds, such as HCl, dissociate in water and, thus, are electrolytes. Carboxylic acids, the family of organic compounds we will study in Chapter 14, are weak electrolytes when dissolved in water.

Question 10.1 A student is presented with a sample of an unknown substance and asked to classify the substance as organic or inorganic. What tests should the student carry out?

Question 10.2 What results would the student expect if the sample in Question 10.1 were an inorganic compound? What results would the student expect if it were an organic compound?

Families of Organic Compounds

The most general classification of organic compounds divides them into hydrocarbons and substituted hydrocarbons. A **hydrocarbon** molecule contains only

LEARNING GOAL

2 Recognize structures that represent each of the families of organic compounds.

TABLE 10.1 Comparison of the Major Properties of a Typical Organic and an Inorganic Compound: Butane versus Sodium Chloride

Property	Organic Compound (e.g., Butane)	Inorganic Compound (e.g., Sodium Chloride)
Molar mass	58	58.5
Bonding	Covalent (C_4H_{10})	Ionic (Na^+ and Cl^- ions)
Physical state at room temperature and atmospheric pressure	Gas	Solid
Boiling point	Low ($-0.5°C$)	High ($1413°C$)
Melting point	Low ($-139°C$)	High ($801°C$)
Solubility in water	Insoluble	High ($36\ g/100\ mL$)
Solubility in organic solvents (e.g., hexane)	High	Insoluble
Flammability	Flammable	Nonflammable
Electrical conductivity	Nonconductor	Conducts electricity in solution and in molten liquid

carbon and hydrogen. A **substituted hydrocarbon** is one in which one or more hydrogen atoms is replaced by another atom or group of atoms.

The hydrocarbons can be further subdivided into aliphatic and aromatic hydrocarbons (Figure 10.2). These terms were coined in Wohler's era when organic chemists primarily studied molecules from nature. The term *aliphatic* comes from a Greek word meaning fat (*aleiphar*). Many *aromatic* compounds were so called because they were obtained from pleasant-smelling plant extracts.

There are four major classes of aliphatic compounds: alkanes, cycloalkanes, alkenes, and alkynes. Alkanes and cycloalkanes are **saturated hydrocarbons.** They are composed of only carbon and hydrogen and have only carbon-hydrogen and carbon-carbon single bonds. Alkenes and alkynes are **unsaturated hydrocarbons.** Alkenes have at least one carbon-carbon double bond and alkynes have at least one carbon-carbon triple bond.

Recall that each of the lines in these structures represents a shared pair of electrons. See Chapter 3.

Saturated hydrocarbon
(Alkane)

Unsaturated hydrocarbon
(Alkene)

Unsaturated hydrocarbon
(Alkyne)

Figure 10.2 The family of hydrocarbons is divided into two major classes: aliphatic and aromatic. The aliphatic hydrocarbons are further subdivided into four major subclasses: alkanes, cycloalkanes, alkenes, and alkynes.

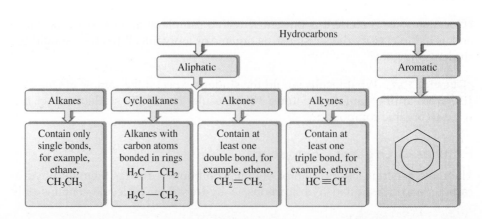

Frozen Methane: Treasure or Threat?

Methane is the simplest hydrocarbon, but it has some unusual behaviors. One of these is the ability to form a clathrate, which is an unusual type of matter in which molecules of one substance form a cage around molecules of another substance. For instance, water molecules can form a latticework around methane molecules to form frozen methane hydrate, possibly one of the biggest reservoirs of fossil fuel on earth.

Typically, we wouldn't expect a nonpolar molecule, such as methane, to interact with a polar molecule, such as water. So, then, how is this structure formed? As we have studied earlier, water molecules interact with one another by strong hydrogen bonding. In the frozen state, these hydrogen-bonded water molecules form an open latticework. The nonpolar methane molecule is simply trapped inside one of the spaces within the lattice.

Vast regions of the ocean floor are covered by ice fields of frozen methane. Scientists would like to "mine" this ice to use the methane as a fuel. In fact, the U.S. Geological Survey estimates that the amount of methane hydrate in the United States is worth over 200 times the conventional natural gas resources in this country!

But is it safe to harvest the methane from this ice? Caution will certainly be required. Methane is flammable, and, like carbon dioxide, it is a greenhouse gas. In fact, it is about twenty times more efficient at trapping heat than carbon dioxide. (Greenhouse gases are discussed in greater detail in Green Chemistry: The Greenhouse Effect and Global Climate Change in Chapter 5.) So the U.S. Department of Energy, which is working with industry to develop ways to harvest the methane, must figure out how to do that without releasing much into the atmosphere where it could intensify global warming.

It may be that a huge release of methane from these frozen reserves was responsible for a major global warming that occurred fifty-five million years ago and lasted for one hundred thousand years. NASA scientists using computer simulations hypothesize that a shift of the continental plates may have released vast amounts of methane gas from the ocean floor. This methane raised the temperature of earth by about 13°F. In fact, the persistence of the methane in the atmosphere warmed earth enough to melt the ice in the oceans and at polar caps and completely change the global climate. This theory, if it turns out

Three-dimensional structure of methane hydrate

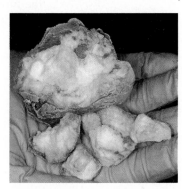

A sample of frozen methane hydrate

to be true, highlights the importance of controlling the amount of methane, as well as carbon dioxide, that we release into the air. Certainly, harvesting the frozen methane of the oceans, if we choose to do it, must be done with great care.

For Further Understanding

► News stories alternatively describe frozen methane as a "New Frontier" and "Armageddon." Explain these opposing views.

► What are the ethical considerations involved in mining frozen methane?

Some hydrocarbons are cyclic. Cycloalkanes consist of carbon atoms bonded to one another to produce a ring. **Aromatic hydrocarbons** contain a benzene ring or a derivative of the benzene ring.

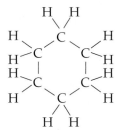

A cycloalkane
(Cyclohexane)

Benzene

LEARNING GOAL

3 Write the names and draw the structures of the common functional groups that characterize the families of organic compounds.

A substituted hydrocarbon is produced by replacing one or more hydrogen atoms with a functional group. A **functional group** is an atom or group of atoms arranged in a particular way that is primarily responsible for the chemical and physical properties of the molecule in which it is found. A functional group may be a single atom, such as chlorine, or more complex, such as the amino group ($-NH_2$). They are important because hydrocarbons have little biological activity. However, the addition of a functional group confers unique and interesting properties that give the molecule important biological or medical properties.

All compounds that have a particular functional group are members of the same family. We have just seen that the alkenes are characterized by the presence of carbon-to-carbon double bonds. Similarly, all alcohols contain a hydroxyl group (—OH). Since this group is polar and can form hydrogen bonds, an alcohol such as ethanol (CH_3CH_2OH) is liquid at room temperature and highly soluble in water, while the alkane of similar molar mass, propane ($CH_3CH_2CH_3$), is nonpolar and is a gas at room temperature and completely insoluble in water. Other common functional groups are shown in Table 10.2, along with an example of a molecule from each family.

The chemistry of organic and biological molecules is usually controlled by the functional group found in the molecule. Just as members of the same family of the periodic table exhibit similar chemistry, organic molecules with the same

TABLE 10.2 Common Functional Groups

Type of Compound	Functional Group	Structural Formula	Condensed Formula	Example Structural Formula	Example IUPAC Name	Common Name
Alcohol	Hydroxyl	*R—O—H	ROH	CH_3CH_2—O—H	Ethanol	Ethyl alcohol
Aldehyde	Carbonyl	R—C(=O)—H	RCHO	CH_3C(=O)—H	Ethanal	Acetaldehyde
Amide	Carboxamide	R—C(=O)—N(H)—H	$RCONH_2$	CH_3C(=O)—N(H)—H	Ethanamide	Acetamide
Amine	Amino	R—N(H)—H	RNH_2	CH_3CH_2N(H)—H	Ethanamine	Ethylamine
Carboxylic acid	Carboxyl	R—C(=O)—O—H	RCOOH	CH_3C(=O)—O—H	Ethanoic acid	Acetic acid
Ester	Ester	R—C(=O)—O—R′	RCOOR′	CH_3C(=O)—OCH_3	Methyl ethanoate	Methyl acetate
Ether	Ether	†R—O—R′	ROR′	CH_3OCH_3	Methoxymethane	Dimethyl ether
Halide	Halogen atom	R—Cl (or —Br, —F, —I)	RCl	CH_3CH_2Cl	Chloroethane	Ethyl chloride
Ketone	Carbonyl	†R—C(=O)—R′	RCOR′	CH_3CCH_3 (=O)	Propanone	Acetone
Alkene	Double bond	R₂C=CR₂	R_2CCR_2	$CH_3CH=CH_2$	Propene	Propylene
Alkyne	Triple bond	R—C≡C—R	RCCR	$CH_3C≡CH$	Propyne	Methyl acetylene

*R and R′ represent an alkyl group, an aryl group, or H.
†In a ketone or an ether, R and R′ must be alkyl or aryl groups.

functional group exhibit similar chemistry. Although it would be impossible to learn the chemistry of each organic molecule, it is relatively easy to learn the chemistry of each functional group. In this way, you can learn the chemistry of all members of a family of organic compounds, or biological molecules, just by learning the chemistry of its characteristic functional group or groups.

10.2 Alkanes

Alkanes are saturated hydrocarbons; that is, alkanes contain only carbon and hydrogen bonded together through carbon-hydrogen and carbon-carbon single bonds. C_nH_{2n+2} is the general formula for alkanes. In this formula, n is the number of carbon atoms in the molecule.

Structure

Four types of formulas, each providing different information, are used in organic chemistry: the molecular formula, the structural formula, the condensed formula, and the line formula.

The **molecular formula** tells the kind and number of each type of atom in a molecule but does not show the bonding pattern. Consider the molecular formulas for simple alkanes:

4 Write condensed, structural, and line formulas for saturated hydrocarbons.

CH_4	C_2H_6	C_3H_8	C_4H_{10}
Methane	Ethane	Propane	Butane

For the first three compounds, there is only one possible arrangement of the atoms. However, for C_4H_{10} there are two possible arrangements. How do we know which is correct? The problem is solved by using the **structural formula,** which shows each atom and bond in a molecule. The following are the structural formulas for methane, ethane, propane, and the two isomers of butane:

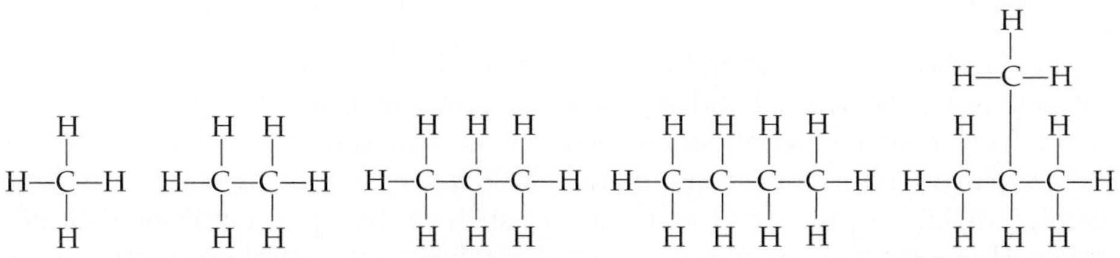

| Methane | Ethane | Propane | Butane | Methylpropane (isobutane) |

Recall that a covalent bond, representing a pair of shared electrons, can be drawn as a line between two atoms. For the structure to be correct, each carbon atom must show four pairs of shared electrons.

Although a structural formula shows the complete structure of a molecule, for large molecules they are time-consuming to draw and take too much space. This problem is solved by using the **condensed formula.** It shows all the atoms in a molecule and places them in a sequential order that indicates which atoms are bonded to which. The following are the condensed formulas for the preceding five compounds.

CH_4	CH_3CH_3	$CH_3CH_2CH_3$	$CH_3(CH_2)_2CH_3$	$(CH_3)_3CH$
Methane	Ethane	Propane	Butane	Methylpropane (isobutane)

The names and formulas of the first ten straight-chain alkanes are shown in Table 10.3.

The simplest representation of a molecule is the **line formula.** In the line formula, we assume that there is a carbon atom at any location where two or more lines intersect. We also assume that there is a methyl group at the end of any line and that each carbon in the structure is bonded to the correct number of hydrogen

TABLE 10.3 Names and Formulas of the First Ten Straight-Chain Alkanes

Name	Molecular Formula	Condensed Formula	Melting Point, °C*	Boiling Point, °C*
Alkanes	C_nH_{2n+2}			
Methane	CH_4	CH_4	−182.5	−162.2
Ethane	C_2H_6	CH_3CH_3	−183.9	−88.6
Propane	C_3H_8	$CH_3CH_2CH_3$	−187.6	−42.1
Butane	C_4H_{10}	$CH_3CH_2CH_2CH_3$ or $CH_3(CH_2)_2CH_3$	−137.2	−0.5
Pentane	C_5H_{12}	$CH_3CH_2CH_2CH_2CH_3$ or $CH_3(CH_2)_3CH_3$	−129.8	36.1
Hexane	C_6H_{14}	$CH_3CH_2CH_2CH_2CH_2CH_3$ or $CH_3(CH_2)_4CH_3$	−95.2	68.8
Heptane	C_7H_{16}	$CH_3CH_2CH_2CH_2CH_2CH_2CH_3$ or $CH_3(CH_2)_5CH_3$	−90.6	98.4
Octane	C_8H_{18}	$CH_3CH_2CH_2CH_2CH_2CH_2CH_2CH_3$ or $CH_3(CH_2)_6CH_3$	−56.9	125.6
Nonane	C_9H_{20}	$CH_3CH_2CH_2CH_2CH_2CH_2CH_2CH_2CH_3$ or $CH_3(CH_2)_7CH_3$	−53.6	150.7
Decane	$C_{10}H_{22}$	$CH_3CH_2CH_2CH_2CH_2CH_2CH_2CH_2CH_2CH_3$ or $CH_3(CH_2)_8CH_3$	−29.8	174.0

*Melting and boiling points as reported in the National Institute of Standards and Technology Chemistry Webbook, which can be found at http://webbook.nist.gov/.

atoms. Compare the structural and line formulas for butane and methylpropane, shown here:

Interpreting Line Formulas

Methyl groups

Carbon atoms

Butane
(Straight chain alkane)

Methylpropane
(Branched alkane)

Each carbon atom can form four single covalent bonds, and each hydrogen atom can form a single covalent bond. Figure 10.3a shows the Lewis structure of the simplest alkane, methane, with four covalent bonds (shared pairs of electrons). When carbon is involved in four single bonds, the *bond angle,* the angle between three atoms, is 109.5°, as predicted by the valence-shell electron-pair repulsion (VSEPR) theory. Thus, alkanes contain carbon atoms that have tetrahedral geometry.

A tetrahedron is a geometric solid having the structure shown in Figure 10.3b. There are many different ways to draw the tetrahedral carbon (Figure 10.3c–e). In Figure 10.3c, solid lines, dashes, and wedges are used to represent the structure of methane. Dashed wedges go back into the page away from you; solid wedges come out of the page toward you; and solid lines are in the plane of the page. The structure in Figure 10.3d is the same as that in Figure 10.3c; it just leaves a lot more to the imagination. Figure 10.3e is a ball-and-stick model of the methane molecule. Three-dimensional drawings of two other simple alkanes are shown in Figure 10.4.

Molecular geometry is described in Section 3.4.

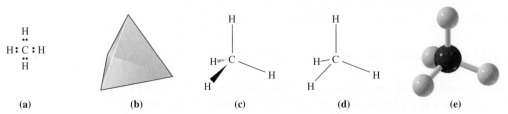

(a)   (b)   (c)   (d)   (e)

Figure 10.3 The tetrahedral carbon atom: (a) Lewis structure; (b) a tetrahedron; (c) the tetrahedral carbon drawn with dashed and solid wedges; (d) the stick drawing of the tetrahedral carbon atom; (e) ball-and-stick model of methane.

EXAMPLE 10.1 Using Different Types of Formulas to Represent Organic Compounds

LEARNING GOAL

4 Write condensed, structural, and line formulas for saturated hydrocarbons.

The following line structure represents 2,2,4-trimethylpentane (also called isooctane), which is the standard of excellence used in determining the octane rating of gasoline. See also Green Chemistry: The Petroleum Industry and Gasoline Production later in this chapter.

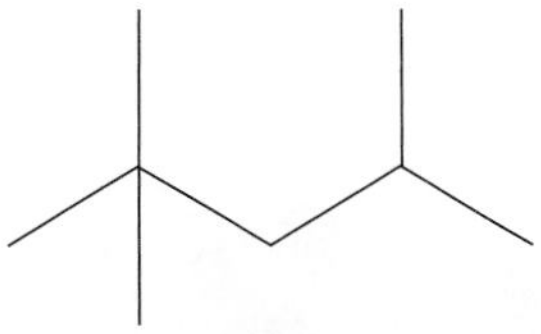

2,2,4-Trimethylpentane
(isooctane)

Draw the structural and condensed formulas of this molecule.

Solution

Remember that each intersection of lines represents a carbon atom and that each line ends in a methyl group. This gives us the following carbon skeleton:

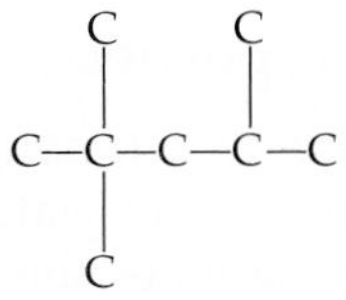

By adding the correct number of hydrogen atoms to the carbon skeleton, we are able to complete the structural formula of this compound.

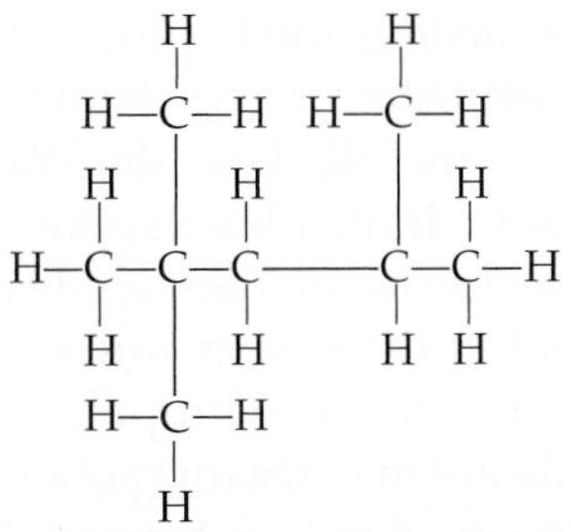

From the structural formula, we can write the condensed formula as follows:

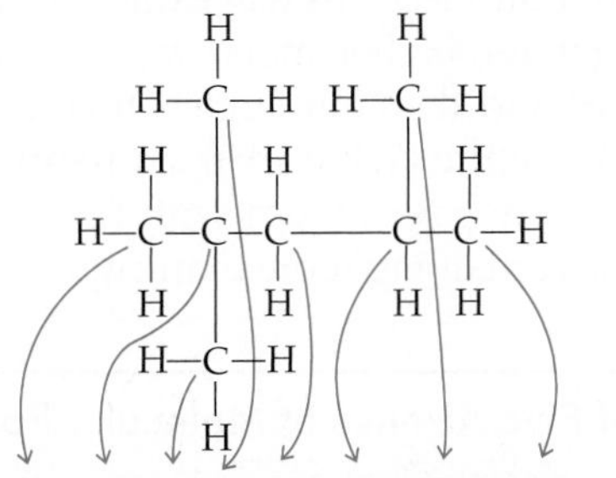

$$CH_3C(CH_3)_2CH_2CH(CH_3)CH_3$$

Practice Problem 10.1

For each of the molecules in Table 10.4, draw the structural and the line formulas.

▶ For Further Practice: **Questions 10.20 and 10.21.**

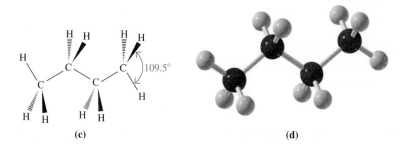

(a)

(b)

Figure 10.4 (a) A drawing and (b) ball-and-stick model of ethane. All the carbon atoms have a tetrahedral arrangement, and all bond angles are approximately 109.5°. (c) A drawing and (d) ball-and-stick model of a more complex alkane, butane.

(c)

(d)

LEARNING GOAL

5 Describe the relationship between the structure and physical properties of saturated hydrocarbons.

See Section 5.2 for a discussion of the forces responsible for the physical properties of a substance.

Physical Properties

All hydrocarbons are nonpolar molecules. Because water is a polar molecule, hydrocarbons are not water soluble. They are soluble only in nonpolar organic solvents. Furthermore, they have relatively low melting points and boiling points and are generally less dense than water. In general, the longer the hydrocarbon chain (greater the molar mass), the higher the melting and boiling points and the greater the density (see Table 10.3).

The melting and boiling points of a substance are determined by the intermolecular attractive forces between molecules. Attractive forces between neutral molecules are called van der Waals forces, which include dipole-dipole attractions (such as hydrogen bonds) and London dispersion forces. In alkanes, the major attractive forces are London dispersion forces.

London dispersion forces result from the attraction of two molecules that experience short-lived dipoles. These dipoles can exist because the electrons in a molecule are in constant motion. If the electron cloud undergoes a transient shift, a temporary dipole is formed. When this dipole interacts with other temporary dipoles, the result is an attractive force. As you can imagine, a larger molecule with more electrons will exhibit a stronger attraction. We would predict, then, that the longer, higher molar mass alkanes would have stronger London dispersion forces and therefore higher boiling and melting points. This is exactly what we see in nature (Table 10.3). At room temperature, alkanes with one to four carbon atoms are gases, those with five to seventeen carbons are colorless liquids, and those containing more than eighteen carbon atoms are white, waxy solids.

TABLE 10.4 Melting and Boiling Points of Five Alkanes of Molecular Formula C_6H_{14}

Name	Condensed Formula	Boiling Point* °C	Melting Point* °C
Hexane	$CH_3(CH_2)_4CH_3$	68.8	−95.2
2-Methylpentane	$CH_3CH(CH_3)(CH_2)_2CH_3$	60.9	−153.2
3-Methylpentane	$CH_3CH_2CH(CH_3)CH_2CH_3$	63.3	−118
2,3-Dimethylbutane	$CH_3CH(CH_3)CH(CH_3)CH_3$	58.1	−130.2
2,2-Dimethylbutane	$CH_3C(CH_3)_2CH_2CH_3$	49.8	−100.2

*Melting and boiling points as reported in the National Institute of Standards and Technology Chemistry Webbook, which can be found at http://webbook.nist.gov/.

Some alkanes have one or more carbon atoms branching from the main carbon chain. Compare the following two molecules with molecular formula C_9H_{20}.

Nonane 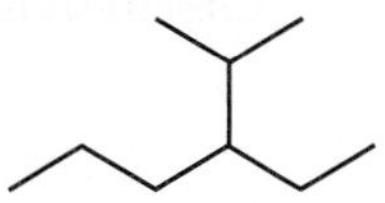   3-Isopropylhexane

The branched-chain form, 3-isopropylhexane, has a much smaller surface area than the straight chain. As a result, the London dispersion forces attracting the molecules to one another are less strong and these molecules have lower melting and boiling points than the straight-chain isomers. The melting and boiling points of the five structural isomers of C_6H_{14} are found in Table 10.4.

Alkyl Groups

Branched chains are formed when alkyl groups are bonded to one of the carbons in the hydrocarbon chain. **Alkyl groups** are alkanes with one fewer hydrogen atom. The name of the alkyl group is derived from the name of the alkane containing the same number of carbon atoms. The -*ane* ending of the alkane name is replaced by the -*yl* ending. Thus, $—CH_3$ is a methyl group and $—CH_2CH_3$ is an ethyl group. The dash at the end of these two structures represents the point at which the alkyl group can bond to another atom. The first five continuous-chain alkyl groups are presented in Table 10.5.

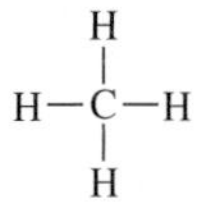

Carbon atoms are classified according to the number of other carbon atoms to which they are attached. A **primary (1°) carbon** is directly bonded to one other carbon. A **secondary (2°) carbon** is bonded to two other carbon atoms; a **tertiary (3°) carbon** is bonded to three other carbon atoms, and a **quaternary (4°) carbon** to four.

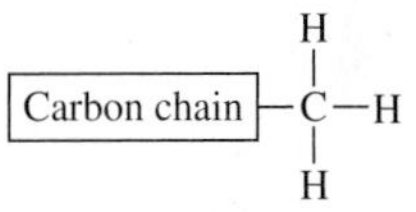

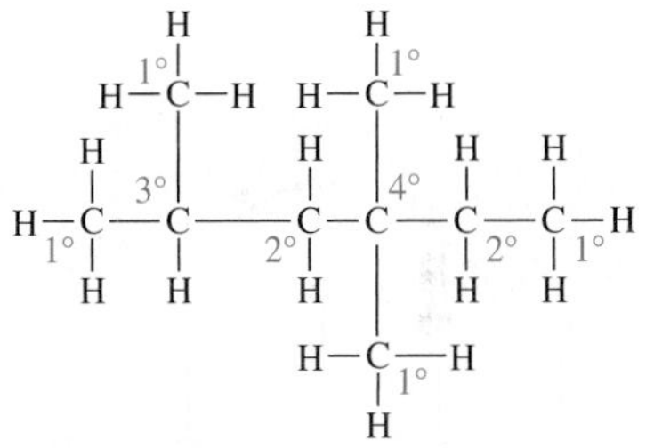

Using this classification scheme, alkyl groups are also designated as primary, secondary, or tertiary based on the number of carbons attached to the carbon atom that joins the alkyl group to a molecule.

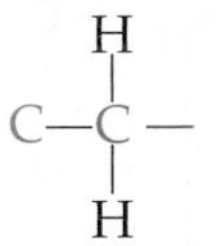

Primary   Secondary   Tertiary
alkyl group   alkyl group   alkyl group

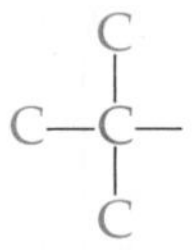

All of the linear alkyl groups are primary alkyl groups (see Table 10.5). Several branched-chain alkyl groups are shown in Table 10.6. The carbon atom that attaches the alkyl group to the parent compound is highlighted in red in the structures above and in Table 10.6. Notice that the isopropyl and *sec*-butyl groups are secondary alkyl groups; the isobutyl group is a primary alkyl group; and the *t*-butyl (*tert*-butyl) is a tertiary alkyl group.

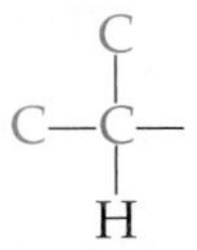

Methane

Methyl group

A branched-chain alkane with the carbon atoms identified as primary (1°), secondary (2°), tertiary (3°), or quaternary (4°)

TABLE 10.5 Names and Formulas of the First Five Continuous-Chain Alkyl Groups

Alkyl Group Structure	Name
$—CH_3$	Methyl
$—CH_2CH_3$	Ethyl
$—CH_2CH_2CH_3$	Propyl
$—CH_2CH_2CH_2CH_3$	Butyl
$—CH_2CH_2CH_2CH_2CH_3$	Pentyl

TABLE 10.6 Structures and Names of Some Branched-Chain Alkyl Groups

Structure	Classification	Common Name	IUPAC Name

CH₃CH—
     |
     CH₃
	2°	Isopropyl*	1-Methylethyl

CH₃ | CH₃CHCH₂—	1°	Isobutyl*	2-Methylpropyl
CH₃ | CH₃CH₂CH—	2°	sec-Butyl†	1-Methylpropyl

CH₃
     |
 CH₃C—
     |
     CH₃
	3°	t-Butyl or tert-Butyl‡	1,1-Dimethylethyl

*The prefix iso- (isomeric) is used when there are two methyl groups at the end of the alkyl group.
†The prefix sec- (secondary) indicates that there are two carbons bonded to the carbon that attaches the alkyl group to the parent molecule.
‡The prefix t- or tert- (tertiary) means that three carbons are attached to the carbon that attaches the alkyl group to the parent molecule.

Question 10.3 Classify each of the carbon atoms in the following structures as primary, secondary, tertiary, or quaternary.

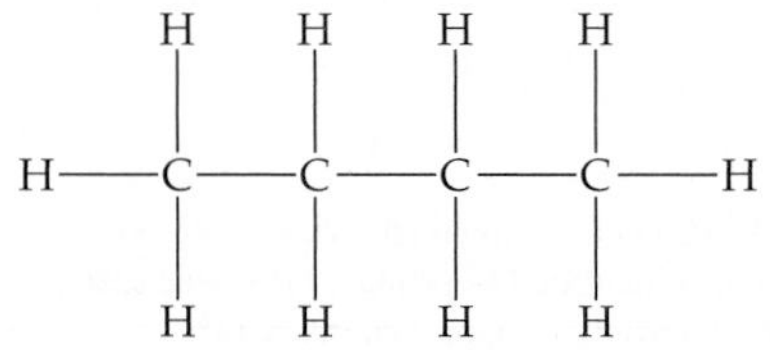

a.

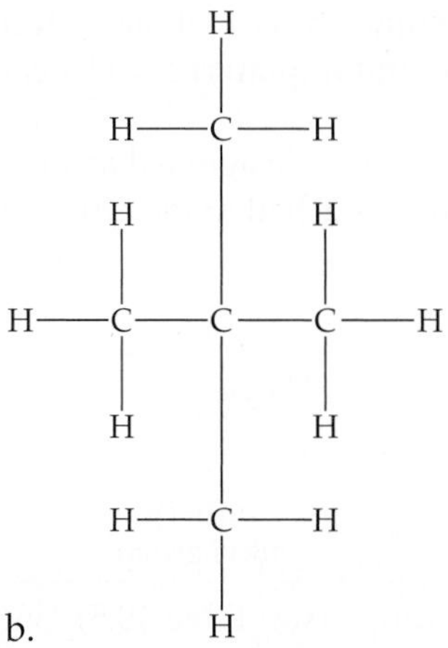

b.

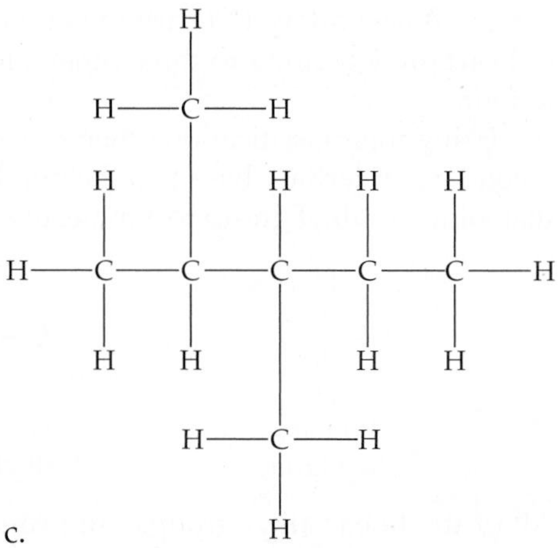

c.

Question 10.4 Classify each of the carbon atoms in the following structures as primary, secondary, tertiary, or quaternary.
    a. CH₃CH₂C(CH₃)₂CH₂CH₃
    b. CH₃CH₂CH₂CH₂CH(CH₃)CH(CH₃)CH₃

c.

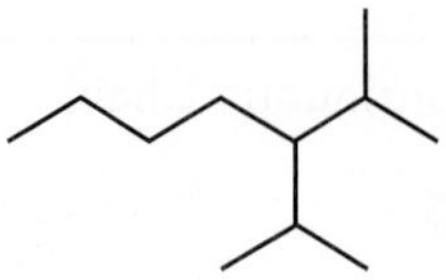

d.

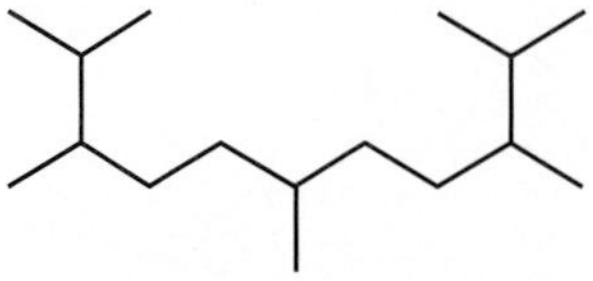

Kitchen Chemistry

Alkanes in Our Food

When discussing vegetable oils, we generally describe the types of lipids from which they are composed. These lipids, termed triglycerides, are composed of three fatty acids bonded to a glycerol molecule, as shown here.

We characterize lipids by the levels of saturated, monounsaturated, and polyunsaturated fatty acids that they contain. Generally, solid fats contain a higher proportion of saturated fatty acids than oils do. For instance, butter is composed of about 61% saturated fatty acids and 33% mono- and polyunsaturated fatty acids, while canola oil has only 6% saturated fatty acids and 92% mono- and polyunsaturated fatty acids.

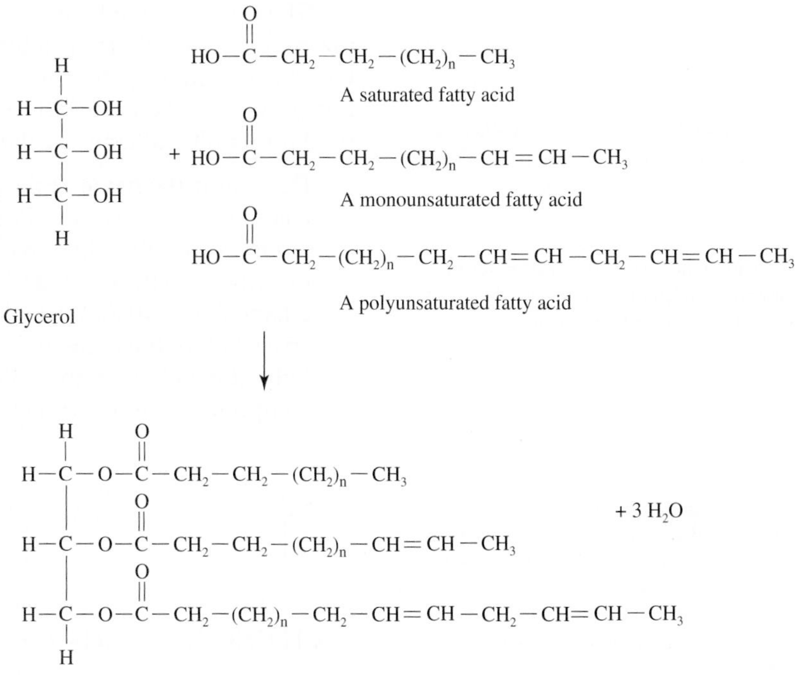

When we discuss alkanes, we generally think of molecules derived from crude oil, not something we want to find in our food! As it turns out, plants make straight-chain alkanes by removing the carboxyl group of saturated fatty acids. One recent study found that all of the edible oils they studied contained straight chain alkanes. They found that walnut oil had the lowest concentration of alkanes and sunflower oil had the highest. They also found that different types of oil had different alkane profiles and that the majority of these alkanes had parent chains composed of an odd number of carbon atoms. All of the oils studied, with the exception of olive oil, included alkanes with chain lengths of 27, 29, and 31 carbon atoms. Olive oil primarily included alkanes of 23, 25, and 27 carbon atoms. The results of the study suggested that the food industry could characterize a specific oil based on its alkane profile.

A second study found similar results with a wide range of oils; but the focus of their study was the contamination of natural vegetable oils with mineral oils (derived from the distillation of crude oil) through transport, processing, and packaging of the oils. Since there was a small concern about the toxicity of the mineral oils, the research group wanted to know whether they could detect the presence of such contamination by analyzing the straight-chain alkane profiles. That is exactly what they were able to do. Their research has provided another tool to ensure that our food supply is safe.

For Further Understanding

▶ What is the functional group that characterizes all of the fatty acids? To what family of compounds do they belong? What other functional group do you find in the monounsaturated and polyunsaturated fatty acids?

▶ What functional group characterizes glycerol? To what family of compounds does it belong?

▶ Write the molecular formula for each of the alkanes mentioned in this perspective.

Nomenclature

Historically, organic compounds were named by the chemist who discovered them. Often the names reflected the source of the compound. For instance, the antibiotic penicillin is named for the mold *Penicillium notatum,* which produces it. The pain reliever aspirin was made by adding an acetate group to a compound first purified from the bark of a willow tree and later from the meadowsweet plant (*Spirea ulmaria*). Thus, the name aspirin comes from *a* (acetate) and *spir* (genus of meadowsweet).

These names are easy for us to remember because we come into contact with these compounds often. However, as the number of compounds increased, organic chemists realized that historical names were not adequate because they revealed nothing about the structure of a compound. Thousands of such compounds and their common names had to be memorized! What was needed was a set of nomenclature (naming) rules that would produce a unique name for every organic compound. Furthermore, the name should be so descriptive that, by knowing the name, a student or scientist could write the structure.

The International Union of Pure and Applied Chemistry (IUPAC) is the organization responsible for establishing and maintaining a standard, universal system for naming organic compounds. The system of nomenclature developed by this group is called the **IUPAC Nomenclature System.** The following rules are used for naming alkanes by the IUPAC system.

1. Determine the name of the **parent compound,** the longest continuous carbon chain in the compound. Refer to Tables 10.3 and 10.7 to determine the parent name. Notice that these names are made up of a prefix related to the number of carbons in the chain and the suffix -*ane*, indicating that the molecule is an alkane (Table 10.7). Write down the name of the parent compound, leaving space before the name to identify the substituents. Parent chains are highlighted in yellow in the following examples, and the names of the parent compounds are shown below each structure.

$$\underset{\text{1 \quad 2 \quad 3}}{CH_3CHCH_3} \atop \underset{}{|} \atop CH_3$$

$$\underset{\text{5 \quad 4 \quad 3 \quad 2 \quad 1}}{CH_3CH_2CHCH_2CH_3} \atop \underset{}{|} \atop CH_3$$

$$\underset{\text{9 \quad 8 \quad 7 \quad 6 \quad 5 \quad 4 \quad 3}}{CH_3CH_2CH_2CH_2CH_2CH_2CHCH_3} \atop \underset{}{|} \atop \underset{\text{2 \quad 1}}{CH_2CH_3}$$

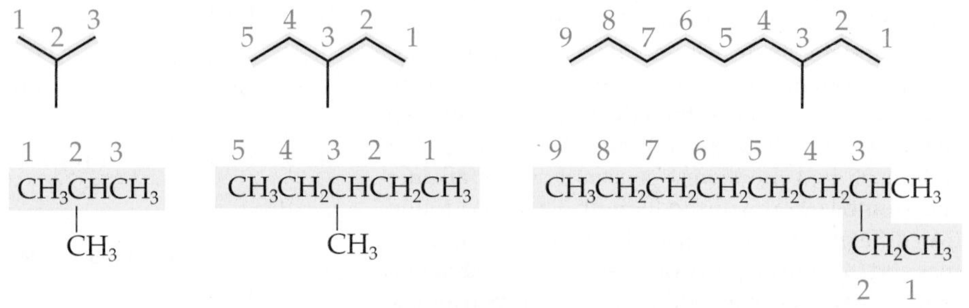

Propane   Pentane   Nonane

2. Number the parent chain to give the lowest number to the carbon bonded to the first group encountered on the parent chain, regardless of the numbers that result for the other substituents.

3. Name and number each atom or group attached to the parent compound. The number tells you the position of the group on the main chain, and the name tells you what type of substituent is present at that position. For example, it may be one of the halogens [F—fluoro), Cl—(chloro), Br—(bromo), and

LEARNING GOAL

6 Use the basic rules of the IUPAC Nomenclature System to name alkanes and substituted alkanes.

It is important to learn the prefixes for the carbon chain lengths. We will use them in the nomenclature for all organic molecules.

In the examples to the left, it doesn't matter whether the propane or pentane chains are numbered from left or from right, since the methyl group is symmetrically located. Nonane is numbered from right to left to give the methyl group the lowest possible number, carbon-3. If we had numbered from the opposite end of the chain, it would have been located on carbon-7, which is incorrect.

TABLE 10.7 Carbon Chain Length and Prefixes Used in the IUPAC Nomenclature System

Carbon Chain Length	Prefix	Alkane Name
1	Meth-	Methane
2	Eth-	Ethane
3	Prop-	Propane
4	But-	Butane
5	Pent-	Pentane
6	Hex-	Hexane
7	Hept-	Heptane
8	Oct-	Octane
9	Non-	Nonane
10	Dec-	Decane

Green Chemistry

Biofuels: A Renewable Resource

What do a high school science project and a train in the Midwest have in common: biofuels. For their senior project, two young Florida women re-engineered one of the school's maintenance trucks to run on the used oil from a local Chinese restaurant. Aside from the aroma of egg rolls frying, the truck performs very well. There are complications, however. The oil is too viscous to use until the truck engine has warmed up. So the driver needs to start the truck with diesel fuel, warm up the engine, then switch to the cooking oil fuel. Similarly, when preparing to stop the truck, the driver must switch back to diesel fuel so that the biofuel is no longer in the engine when it cools.

In 2010, Amtrak unveiled its first biodiesel-fueled train. The biodiesel fuel is a B20 blend, which means that it is **20%** biodiesel and **80%** regular diesel. The biodiesel is a complex mixture of fatty acids that are a by-product of the beef industry. So it is fitting that this train runs the Oklahoma City to Fort Worth route. This source works well in Texas and Oklahoma, which both produce large numbers of cattle, but what of other locations? Some have suggested corn or soybeans for biodiesel production. The question there is whether it is wise to rely on food crops that, in the event of widespread famine, may be needed for food.

The options are as creative as the minds that have envisioned them. As it turns out, 15 million pounds of alligator fat are dumped into landfills each year as a by-product of processing the meat. Over 60% of that fat can be extracted by microwaving, and the resultant fatty acid profile makes it an excellent biodiesel fuel. NASA scientists are developing biodiesel from halophytic (salt-loving) plants. These plants can grow in the desert and can be irrigated with seawater. Thus they take advantage of environments and conditions that traditionally have not been usefully employed. The scientists even predict that this would create a cooler, wetter land surface that would promote rainfall in areas that are extremely arid.

Boeing Corporation scientists think that biodiesel could cut aircraft emissions by 60–80%, which is a considerable improvement when you consider that 2% of all human emissions are from the aviation industry. Several test flights have shown that aircraft fly very well on fuels that are a mixture of 50% jet fuel and 50% biodiesel produced from a variety of sources including algae, the jatropha plant, and *Camelina* seed.

Biodiesel is much more common in Europe than elsewhere in the world and is produced from palm oil, rapeseed, flax, sunflower, and jatropha. These oils are treated by a transesterification process that produces fatty acid methyl or ethyl esters (FAMES). The resultant fuels are considered safer than traditional diesel because they are biodegradable, are ten times less toxic than table salt, and have high flash points.

We have much to learn about biofuels and their production before we can hope to replace the use of fossil fuels. However, knowing that fossil fuels are a limited resource, it is important to study renewable energy sources so that we can make wise decisions about those processes that provide the greatest benefit with the least negative impact on our environment.

For Further Understanding

▶ What are some of the environmental issues that must be studied when considering biodiesel from a variety of sources?

▶ What are some of the economic issues that need to be analyzed when considering biodiesel as a replacement for fossil fuels?

I—(iodo)] or an alkyl group (Tables 10.5 and 10.6). In the following examples, the parent chain is highlighted in yellow:

Substituent:	2-Bromo	3-Methyl	4-Ethyl
IUPAC name:	2-Bromopropane	3-Methylpentane	4-Ethyloctane

4. If the same substituent occurs more than once in the compound, a separate position number is given for each, and the prefixes *di-, tri-, tetra-, penta-*, and so forth are used, as shown in the following examples:

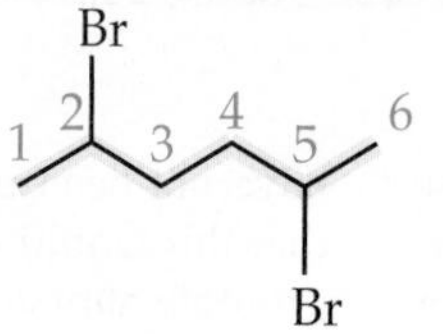

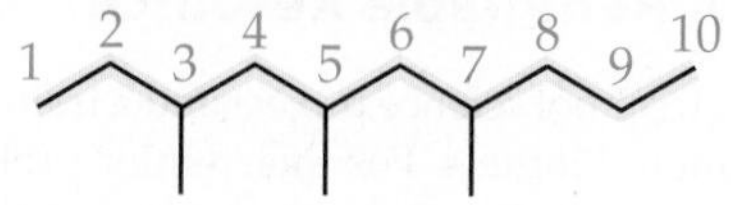

Br
   2   4
1 / 2 \ 3 / 4 \ 5 / 6
              |
              Br

2   4   6   8   10
1 / 2 \ 3 / \ 5 / \ 7 \ 9 /
         | |
         3

Br   Br
| |
CH₃CHCH₂CH₂CHCH₃
1   2   3   4   5   6

2,5-Dibromo
2,5-Dibromohexane

         CH₃   CH₃   CH₃
          | | |
CH₃CH₂CHCH₂CHCH₂CHCH₂CH₂CH₃
1   2   3   4   5   6   7   8   9   10
10   9   8   7   6   5   4   3   2   1

3,5,7-Trimethyldecane
Not 4,6,8-Trimethyldecane

5. Place the names of the substituents in alphabetical order before the name of the parent compound, which you wrote down in step 1. Numbers are separated by commas, and numbers are separated from names by hyphens. By convention, halogen substituents are placed before alkyl substituents.

          CH₃
           |
1   2   3| 4   5
CH₃CHCCH₂CH₃
    | |
    Br CH₃

2-Bromo-3,3-dimethylpentane
Not 3,3-Dimethyl-2-bromopentane

             CH₂CH₃
              |
1   2   3   4| 5   6   7   8
CH₃CH₂CHCHCH₂CH₂CH₂CH₃
        |
        CH₃

4-Ethyl-3-methyloctane
Not 3-Methyl-4-ethyloctane

EXAMPLE 10.2   **Naming Substituted Alkanes Using the IUPAC System**

a. What is the IUPAC name of the molecule below, which is commonly called Freon-12? This compound is a chlorofluorocarbon (CFC) once used as a refrigerant and aerosol propellant. It has not been manufactured in the United States since 1995 because of the CFC damage to the ozone layer.

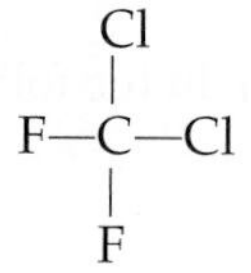

LEARNING GOAL

6 Use the basic rules of the IUPAC Nomenclature System to name alkanes and substituted alkanes.

        Cl
        |
F—C—Cl
        |
        F

Solution

Helpful Hint: No numbers are necessary if there is only one carbon or if the numbering is clear cut.

Parent chain: methane

Substituents: dichlorodifluoro (no numbers are necessary)

Name: Dichlorodifluoromethane

b. What is the IUPAC name of the following molecule, which is a component of the tsetse fly pheromone? Molecules such as this are used as attractants in tsetse fly control measures.

             CH₃
              |
CH₃(CH₂)₁₄CHCH₃

Solution

Helpful Hint: For alkanes of between eleven and nineteen carbons, a prefix is used before the word decane (see Table 10.8).

> Parent chain: heptadecane
> Substituent: 2-methyl
> Name: 2-Methylheptadecane

c.   What is the IUPAC name of the following molecule, which is the standard of excellence used in determining the octane rating of gasoline?

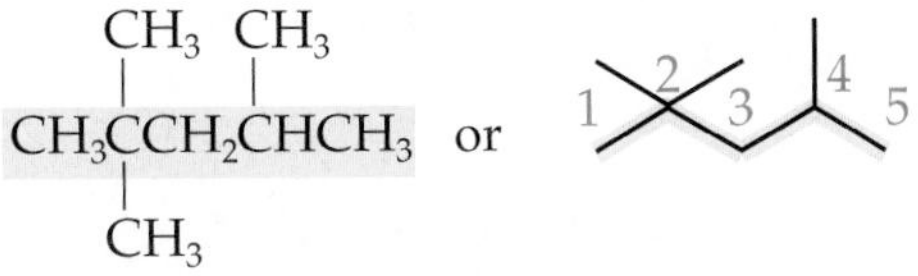

$$\underset{\underset{CH_3}{|}}{\overset{\overset{CH_3\ \ CH_3}{|\ \ \ \ |}}{CH_3CCH_2CHCH_3}} \quad \text{or} \quad$$

Solution

> Parent chain: pentane
> Substituents: 2,2,4-trimethyl
> Name: 2,2,4-Trimethylpentane

Practice Problem 10.2

Determine the IUPAC name for each of the following molecules.

a. 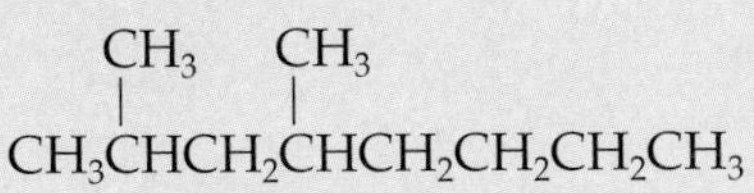 $\underset{}{\overset{\overset{CH_3\ \ \ \ CH_3}{|\ \ \ \ \ \ \ |}}{CH_3CHCH_2CHCH_2CH_2CH_2CH_3}}$

b. 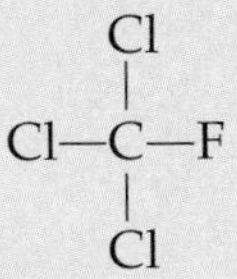 $\underset{\underset{Cl}{|}}{\overset{\overset{Cl}{|}}{Cl-C-F}}$

c. $\overset{\overset{F\ \ Cl\ \ CH_3}{|\ \ \ |\ \ \ \ |}}{CH_3CH_2CH_2CHCHCHCH_3}$

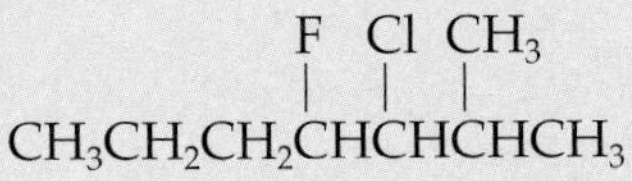

▶ For Further Practice: **Questions 10.59–10.62.**

Having learned to name a compound using the IUPAC system, we can easily write the structural formula of a compound, given its name. First, draw and number the parent carbon chain. Add the substituent groups to the correct carbon, and finish the structure by adding the correct number of hydrogen atoms.

TABLE 10.8 IUPAC Nomenclature for Alkane Parent Chains Longer than Ten Carbons

Number of Carbons	IUPAC Name
11	*Undecane*
12	*Dodecane*
13	*Tridecane*
14	*Tetradecane*
15	*Pentadecane*
16	*Hexadecane*
17	*Heptadecane*
18	*Octadecane*
19	*Nonadecane*

EXAMPLE 10.3 | **Drawing the Structure of a Compound Using the IUPAC Name**

LEARNING GOAL

7 From the IUPAC name of an alkane or substituted alkane, be able to draw the structure.

Draw the structural formula for 1-bromo-4-methylhexane.

Solution

Hexane has six carbons. Begin to draw the structure by drawing the six-carbon parent chain and indicating the four bonds for each carbon atom.

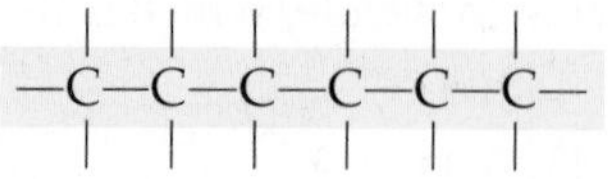

$$-C-C-C-C-C-C-$$

Next, number each carbon atom:

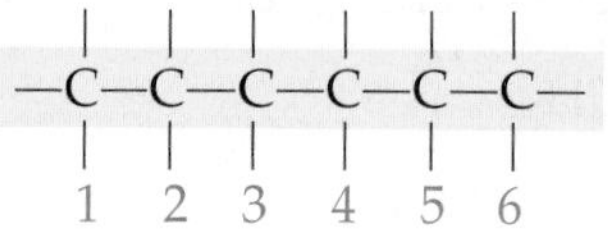

$$-\underset{1}{C}-\underset{2}{C}-\underset{3}{C}-\underset{4}{C}-\underset{5}{C}-\underset{6}{C}-$$

Now add the substituents. In this example, a bromine atom is bonded to carbon-1, and a methyl group is bonded to carbon-4:

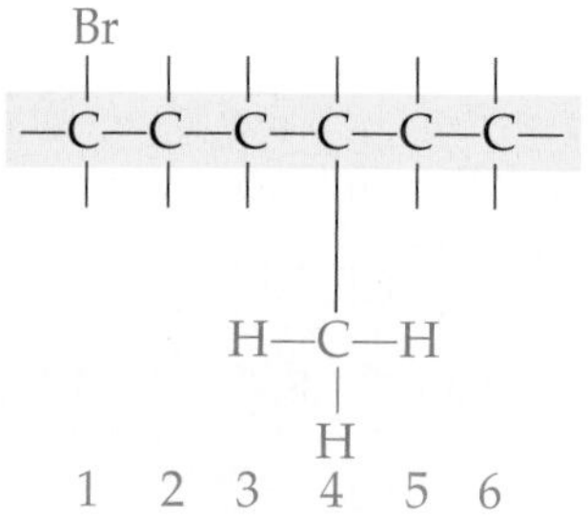

Finally, add the correct number of hydrogen atoms so that each carbon has four covalent bonds:

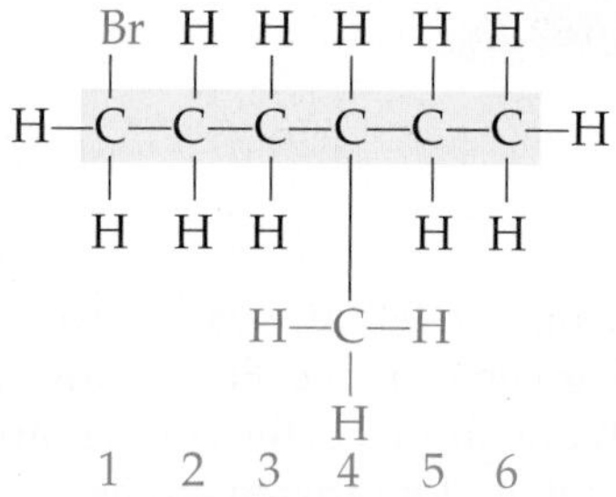

As a final check of your accuracy, use the IUPAC system to name the compound you have just drawn, and compare the name with that in the original problem.

The molecular, condensed, and line formulas can be written from the structural formula shown. The molecular formula is $C_7H_{15}Br$, the condensed formula is $BrCH_2(CH_2)_2CH(CH_3)CH_2CH_3$, and the line formula for 1-bromo-4-methylhexane is

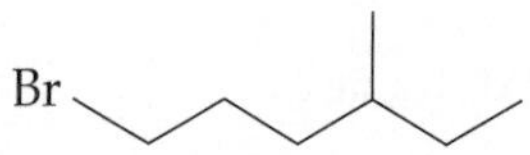

Practice Problem 10.3

Draw the condensed formula of each of the following compounds:

a. 1-Bromo-2-chlorohexane
b. 2,3-Dimethylpentane
c. 1,3,5-Trichloroheptane
d. 3-Chloro-5-iodo-4-methyloctane
e. 1,2-Dibromo-3-chlorobutane
f. Trifluorochloromethane

▶ For Further Practice: **Questions 10.55–10.58.**

Constitutional or Structural Isomers

As we saw earlier, there are two arrangements of the atoms represented by the molecular formula C_4H_{10}: butane and methylpropane. Molecules having the same molecular formula but a different arrangement of atoms are called **constitutional,** or **structural, isomers.** These isomers are unique compounds because of their structural differences, and they have different physical and chemical properties. Branched-chain isomers have a smaller surface area than the linear molecules. With less surface area for intermolecular attractions (London dispersion forces), the branched-chain molecules are not as strongly attracted to one another. As a result, they have lower boiling and melting points than the straight-chain isomer. These differences reflect the different shapes of the molecules, as shown in Table 10.4 for the isomers of hexane.

LEARNING GOAL

8 Draw constitutional (structural) isomers of simple organic compounds.

EXAMPLE 10.4	**Drawing Constitutional or Structural Isomers of Alkanes**

Write all the constitutional isomers having the molecular formula C_6H_{14}.

LEARNING GOAL

8 Draw constitutional (structural) isomers of simple organic compounds.

Solution

1. Begin with the continuous six-carbon chain structure:

$$\overset{1}{C}H_3\overset{2}{C}H_2\overset{3}{C}H_2\overset{4}{C}H_2\overset{5}{C}H_2\overset{6}{C}H_3$$

Isomer A

2. Now try five-carbon chain structures with a methyl group attached to one of the internal carbon atoms of the chain:

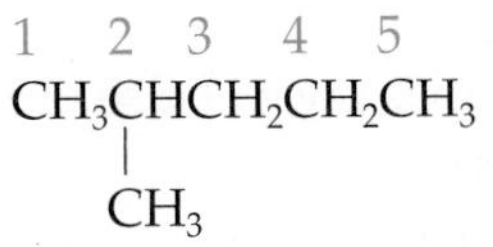

$$\overset{1}{C}H_3\overset{2}{C}H\overset{3}{C}H_2\overset{4}{C}H_2\overset{5}{C}H_3 \quad \text{and} \quad \overset{1}{C}H_3\overset{2}{C}H_2\overset{3}{C}H\overset{4}{C}H_2\overset{5}{C}H_3$$
$$\quad\quad | \quad\quad\quad\quad\quad\quad\quad\quad\quad\quad\quad | $$
$$\quad\quad CH_3 \quad\quad\quad\quad\quad\quad\quad\quad\quad\quad CH_3$$

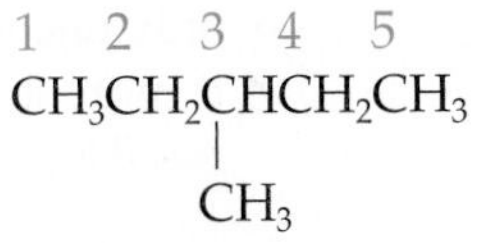

Isomer B   Isomer C

3. Next consider the possibilities for a four-carbon structure to which two methyl groups (—CH_3) may be attached:

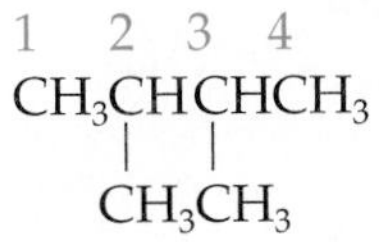

$$\quad\quad\quad\quad\quad\quad\quad\quad\quad\quad\quad\quad\quad CH_3$$
$$\overset{1}{C}H_3\overset{2}{C}H\overset{3}{C}H\overset{4}{C}H_3 \quad \text{and} \quad \overset{1}{C}H_3\overset{2}{C}\overset{3}{C}H_2\overset{4}{C}H_3$$
$$\quad\quad | \quad | \quad\quad\quad\quad\quad\quad\quad\quad | $$
$$\quad\quad CH_3 CH_3 \quad\quad\quad\quad\quad\quad CH_3$$

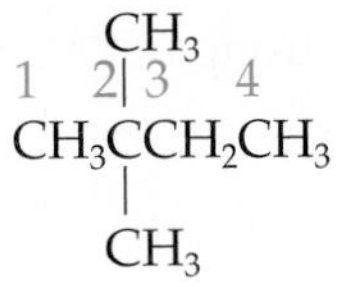

Isomer D   Isomer E

These are the five possible constitutional isomers of C_6H_{14}. At first, it may seem that other isomers are also possible. But careful comparison will show that they are duplicates of those already constructed. For example, rather than add two methyl groups, a single ethyl group (—CH_2CH_3) could be added to the four-carbon chain:

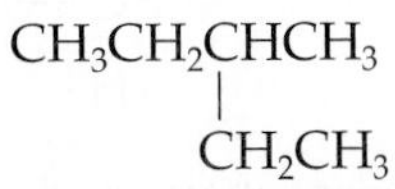

$$CH_3CH_2CHCH_3$$
$$\quad\quad\quad | $$
$$\quad\quad CH_2CH_3$$

But close examination will show that this is identical to isomer C. Perhaps we could add one ethyl group and one methyl group to a three-carbon parent chain, with the following result:

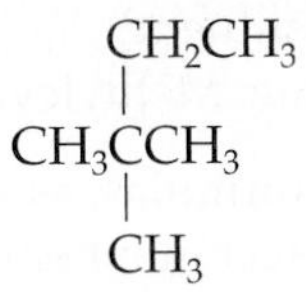

$$\quad\quad\quad CH_2CH_3$$
$$\quad\quad\quad | $$
$$CH_3CCH_3$$
$$\quad\quad | $$
$$\quad CH_3$$

Again we find that this structure is the same as one of the isomers we have already identified, isomer E.

Continued...

To check whether you have accidentally made duplicate isomers, name them using the IUPAC system. All isomers must have different IUPAC names. So if two names are identical, the structures are also identical. Use the IUPAC system to name the isomers in this example, and prove to yourself that the last two structures are simply duplicates of two of the original five isomers.

Practice Problem 10.4

Heptane is a very poor fuel and is given a zero on the octane rating scale. Gasoline octane ratings are determined in test engines by comparison with 2,2,4-trimethylpentane (the standard of high quality) and heptane (standard of poor quality). Draw the line formula and give the IUPAC name for each of the nine isomers of heptane.

▶ For Further Practice: **Questions 10.66 and 10.67.**

10.3 Cycloalkanes

The **cycloalkanes** are a family having C—C single bonds in a ring structure. They have the general molecular formula C_nH_{2n} and thus have two fewer hydrogen atoms than the corresponding alkane (C_nH_{2n+2}). The structures and names of some simple cycloalkanes are shown in Figure 10.5.

In the IUPAC system, the cycloalkanes are named by applying the following simple rules.

- Count the number of carbon atoms in the ring. Determine the name of the alkane with the same number of carbon atoms and add the prefix *cyclo-*. For example, cyclopentane is the cycloalkane that has five carbon atoms.
- If the cycloalkane is substituted, place the names of the groups in alphabetical order before the name of the cycloalkane. No number is needed if there is only one substituent.
- If more than one group is present, use numbers that result in the *lowest possible position numbers.*

EXAMPLE 10.5 **Naming a Substituted Cycloalkane Using the IUPAC Nomenclature System**

Name the following cycloalkanes using IUPAC nomenclature.

LEARNING GOAL

9 Write the names and draw the structures of simple cycloalkanes.

Solution

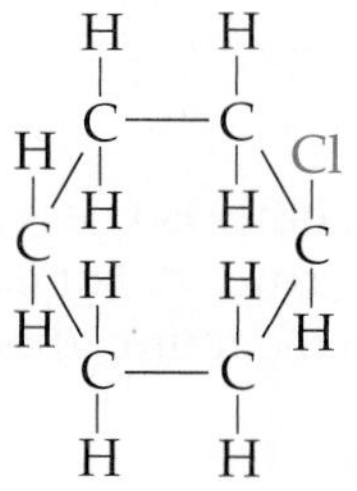

Parent chain: cyclohexane
Substituent: chloro (no number is required because there is only one substituent)
Name: Chlorocyclohexane

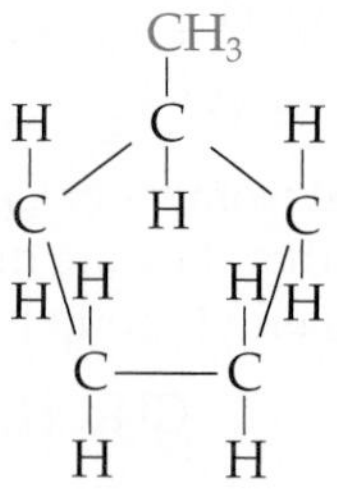

Parent chain: cyclopentane
Substituent: methyl (no number is required because there is only one substituent)
Name: Methylcyclopentane

These cycloalkanes could also be shown as line formulas, as shown on the next page. Each line represents a carbon-carbon bond. A carbon atom and the correct number of hydrogen atoms are assumed to be at the point where the lines meet and at the end of a line.

Chlorocyclohexane   Methylcyclopentane

Practice Problem 10.5

Name each of the following cycloalkanes using IUPAC nomenclature:

a.   b.   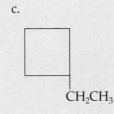   c.   d.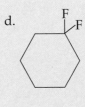

For Further Practice: **Questions 10.81 and 10.82.**

(a)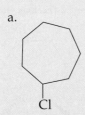

(b)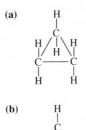

(c)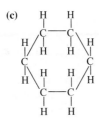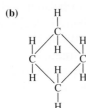

Figure 10.5 Cycloalkanes: (a) cyclopropane; (b) cyclobutane; (c) cyclohexane. All of the cycloalkanes are shown using structural formulas (left column), and line formulas (right column).

cis-trans Isomerism in Cycloalkanes

Atoms of an alkane can rotate freely around the carbon-carbon single bond, resulting in an unlimited number of arrangements. (See Figure 10.6 for examples of two possible arrangements of the molecules ethane and butane.) However, rotation around the bonds in a cyclic structure is limited by the fact that the carbons of the ring are all bonded to another carbon within the ring. The formation of ***cis-trans* isomers,** or **geometric isomers,** is a consequence of the absence of free rotation.

Geometric isomers are a type of *stereoisomer.* **Stereoisomers** are molecules that have the same structural formulas and bonding patterns but different arrangements of atoms in space. The *cis-trans* isomers of cycloalkanes are stereoisomers

LEARNING GOAL

10   Draw *cis-* and *trans-*isomers of cycloalkanes.

Stereoisomers are discussed in detail in Section 16.3.

that differ from one another in the arrangement of substituents in space. Consider the following two views of the *cis*- and *trans*-isomers of 1,2-dichlorocyclohexane:

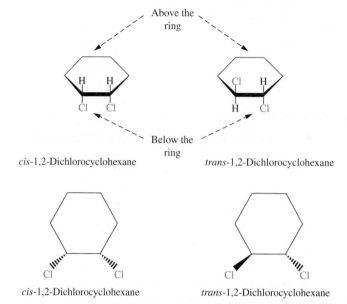

Because there is no free rotation around the carbon atoms of a cycloalkane, the groups shown above or below the ring will always remain in that position.

cis-1,2-Dichlorocyclohexane *trans*-1,2-Dichlorocyclohexane

cis-1,2-Dichlorocyclohexane *trans*-1,2-Dichlorocyclohexane

In the wedge diagram at the top, it is easy to imagine that you are viewing the ring structures as if an edge were projecting toward you. This will help you understand the more common structural formulas, shown beneath them. In the structure on the left, both Cl atoms are beneath the ring. They are termed *cis* (*Latin,* "on the same side"). The complete name for this compound is *cis*-1,2-dichlorocyclohexane. In the structure on the right, one Cl is above the ring and the other is below it. They are said to be *trans* (*Latin,* "across from") to one another, and the complete name of this compound is *trans*-1,2-dichlorocyclohexane. In the line structures, the wedges tell us the same information. The dashed wedges indicate that the bond projects away from you. The solid wedges indicate that the bond is projecting toward you.

Geometric isomers do not readily interconvert. The cyclic structure prevents unrestricted free rotation and, thus, prevents interconversion. Only by breaking carbon-carbon bonds of the ring could interconversion occur. As a result, geometric isomers may be separated from one another in the laboratory.

EXAMPLE 10.6 Naming *cis-trans* Isomers of Substituted Cycloalkanes

LEARNING GOAL

10 Draw *cis*- and *trans*-isomers of cycloalkanes.

Determine whether the following substituted cycloalkanes are *cis*- or *trans*-isomers, and write the complete name for each.

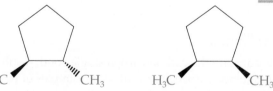

Solution

Both molecules are cyclopentanes having two methyl group substituents. Thus, both would be named 1,2-dimethylcyclopentane. In the structure on the left, one methyl group is above the ring and the other is below the ring; they are in the *trans* configuration, and the structure is named *trans*-1,2-dimethylcyclopentane. In the structure on the right, both methyl groups are on the same side of the ring (above it, in this case); they are *cis* to one another, and the complete name of this compound is *cis*-1,2-dimethylcyclopentane.

Practice Problem 10.6

Determine whether each of the following is a *cis*- or a *trans*-isomer.

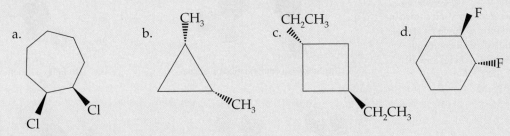

▶ For Further Practice: **Questions 10.89 and 10.90.**

EXAMPLE 10.7 | **Naming a Cycloalkane Having Two Substituents Using the IUPAC Nomenclature System**

Name the following cycloalkanes using IUPAC nomenclature.

LEARNING GOAL

9 Write the names and draw the structures of simple cycloalkanes.

Solution

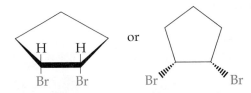 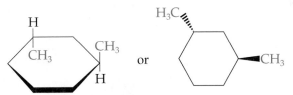

Parent chain: cyclopentane
Substituent: 1,2-dibromo
Isomer: *cis*
Since both bromine atoms are below the ring, the name is:
cis-1,2-Dibromocyclopentane

Parent chain: cyclohexane
Substituent: 1,3-dimethyl
Isomer: *trans*
Since one of the methyl groups is above the ring and the other is below, the name is:
trans-1,3-Dimethylcyclohexane

Practice Problem 10.7

Write the complete IUPAC name for each of the following cycloalkanes.

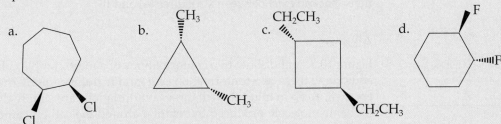

▶ For Further Practice: **Questions 10.87 and 10.88.**

10.4   Conformations of Alkanes and Cycloalkanes

Because there is *free rotation* around a carbon-carbon single bond, even a very simple alkane, like ethane, can exist in an unlimited number of forms (Figure 10.6a and b). These different arrangements are called **conformations,** or **conformers.**

LEARNING GOAL

11   Describe conformations of alkanes.

Figure 10.6 Conformational isomers of ethane and butane. The hydrogen atoms are much more crowded in the eclipsed conformation, depicted in (b) compared with the staggered conformation shown in (a). The staggered form is the most stable. The staggered and eclipsed conformations of butane are shown in (c) and (d).

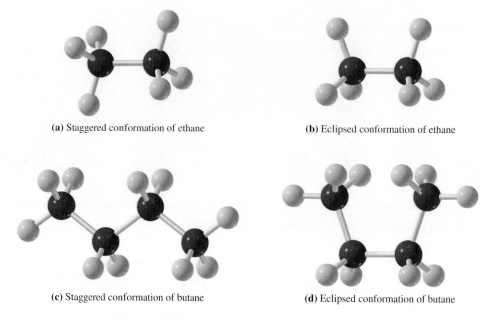

(a) Staggered conformation of ethane   **(b)** Eclipsed conformation of ethane

(c) Staggered conformation of butane   **(d)** Eclipsed conformation of butane

To get a sense of what is meant by free rotation, it is useful to visualize a simpler molecule, hydrogen peroxide (H_2O_2).

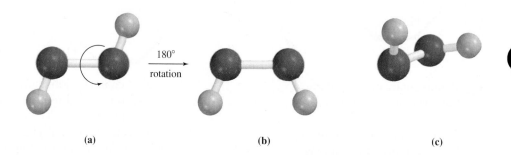

180°
rotation

(a)   **(b)**   **(c)**

Structure (a) is one possible conformation. The arrow indicates that there is free rotation around the oxygen-oxygen bond. Structure (b) is the same molecule, but with a 180° rotation. Structure (c) is one of the infinite number of conformations that can exist between structures (a) and (b).

Alkanes

Figure 10.6c and d show two conformations of a more complex alkane, butane. In addition to these two conformations, an infinite number of intermediate conformers exist. Keep in mind that all these conformations are simply different forms of the same molecule produced by rotation around the carbon-carbon single bonds. Even at room temperature these conformers interconvert rapidly. As a result, they cannot be separated from one another.

Although all conformations can be found in a population of molecules, the *staggered* conformation (see Figure 10.6a and c) is the most common. One reason for this is that the bonding electrons are farthest from one another in this conformation. Because this minimizes the repulsion between these bonding electrons, the staggered conformation is the most stable (most energetically favorable) of the possible conformations. In the *eclipsed* conformation (Figure 10.6b and d), the hydrogen atoms are closest to one another. This maximizes the repulsion between electrons. As a result, the eclipsed conformation is the least stable of the conformations.

Make a ball-and-stick model of butane and demonstrate these rotational changes for yourself.

Green Chemistry

The Petroleum Industry and Gasoline Production

Petroleum consists primarily of alkanes and small amounts of alkenes and aromatic hydrocarbons. Substituted hydrocarbons, such as phenol, are also present in very small quantities. Although the composition of petroleum varies with the source (United States, Persian Gulf, etc.), the mixture of hydrocarbons can be separated into its component parts on the basis of differences in the boiling points of various hydrocarbons (distillation). Often, several successive distillations of various fractions of the original mixture are required to completely purify the desired component. In the first distillation, the petroleum is separated into several fractions, each of which consists of a mix of hydrocarbons. Each fraction can be further purified by successive distillations. On an industrial scale, these distillations are carried out in columns that may be hundreds of feet (ft) in height.

The gasoline fraction of petroleum, called straight-run gasoline, consists primarily of alkanes and cycloalkanes with six to twelve carbon atoms in the skeleton. This fraction has very poor fuel performance. In fact, branched-chain alkanes are superior to straight-chain alkanes as fuels because they are more volatile, burn less rapidly in the cylinder, and thus reduce "knocking." Alkenes and aromatic hydrocarbons are also good fuels. Methods have been developed to convert hydrocarbons of higher and lower molar masses than gasoline to the appropriate molar mass range and to convert straight-chain hydrocarbons into branched ones. *Catalytic cracking* fragments a large hydrocarbon into smaller ones. *Catalytic re-forming* results in the rearrangement of a hydrocarbon into a more useful form.

The antiknock quality of a fuel is measured as its octane rating. Heptane is a very poor fuel and is given an octane rating of zero. 2,2,4-Trimethylpentane (commonly called isooctane) is an excellent fuel and is given an octane rating of one hundred.

Mining the sea for hydrocarbons

Gasoline octane ratings are experimentally determined by comparison with these two compounds in test engines.

For Further Understanding

▶ Explain why the mixture of hydrocarbons in crude oil can be separated by distillation.

▶ Draw the structures of heptane and 2,2,4-trimethylpentane (isooctane).

Cycloalkanes

Cycloalkanes also exist in different conformations. The only exception to this is cyclopropane. Because it has only three carbon atoms, it is always planar.

The conformations of six-member rings have been the most thoroughly studied. One reason is that many important and abundant biological molecules have six-member ring structures. Among these is the simple sugar glucose, also called *blood sugar*. Glucose is the most important sugar in the human body. It is absorbed by the cells of the body and broken down to provide energy for cellular activities.

The most stable (energetically favorable) conformation for a six-member ring is the **chair conformation.** In this conformation, the hydrogen atoms are perfectly staggered; that is, they are as far from one another as possible. In addition, the bond angle between carbons is 109.5°, exactly the angle expected for tetrahedral carbon atoms. Because the hydrogen atoms are as far from one another as possible, the repulsion between the bonding electrons is minimized. As a result, the chair conformation is the most stable conformation of cyclohexane.

LEARNING GOAL

12 Draw the chair and boat conformations of cyclohexane.

The structure of glucose is found in Section 16.4. The physiological roles of glucose are discussed in Chapters 21 and 22.

Compare the structure of this deck chair to the conformation of cyclohexane shown to the right. Explain why this conformation is called the chair conformation.

Compare the structure of these boats to the conformation of cyclohexane shown to the left. Explain why this conformation is called the boat conformation.

Chair conformation

Six-member rings can also exist in a **boat conformation,** so-called because it resembles a rowboat. This form is much less stable than the chair conformation because the hydrogen atoms are much more crowded, creating much more repulsion between the electrons.

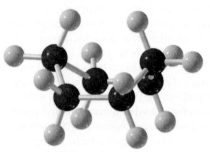

Boat conformation

10.5 Reactions of Alkanes and Cycloalkanes

Combustion

Alkanes, cycloalkanes, and other hydrocarbons can be oxidized (by burning) in the presence of excess molecular oxygen. In this reaction, called **combustion,** they burn at high temperatures, producing carbon dioxide and water and releasing large amounts of energy as heat.

$$C_nH_{2n+2} \quad + \quad O_2 \quad \longrightarrow \quad CO_2 \quad + \quad H_2O \quad + \quad \text{heat energy}$$

Alkane   Oxygen $\longrightarrow$ Carbon dioxide   Water

The following examples show a combustion reaction for a simple alkane and a simple cycloalkane:

$$CH_4 + 2O_2 \longrightarrow CO_2 + 2H_2O + \text{heat energy}$$
Methane

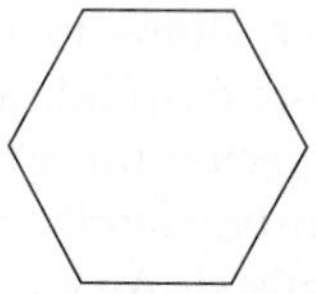

(or C_6H_{12}) + $9O_2 \longrightarrow 6CO_2 + 6H_2O$ + heat energy

Cyclohexane

LEARNING GOAL

13   Write balanced equations for combustion reactions of alkanes.

Combustion reactions are discussed in Section 4.8.

| EXAMPLE 10.8 | Balancing Equations for the Combustion of Alkanes |

Balance the following equation for the combustion of hexane:

$$C_6H_{14} + O_2 \longrightarrow CO_2 + H_2O$$

Solution

First, balance the carbon atoms; there are 6 moles (mol) of carbon atoms on the left and only 1 mol of carbon atoms on the right:

$$C_6H_{14} + O_2 \longrightarrow 6CO_2 + H_2O$$

Next, balance hydrogen atoms; there are 14 mol of hydrogen atoms on the left and only 2 mol of hydrogen atoms on the right:

$$C_6H_{14} + O_2 \longrightarrow 6CO_2 + 7H_2O$$

Now there are 19 mol of oxygen atoms on the right and only 2 mol of oxygen atoms on the left. Therefore, a coefficient of 9.5 is needed for O_2.

$$C_6H_{14} + 9.5O_2 \longrightarrow 6CO_2 + 7H_2O$$

Although decimal coefficients are sometimes used, it is preferable to have all integer coefficients. Multiplying each term in the equation by 2 will satisfy this requirement, giving us the following balanced equation:

$$2C_6H_{14} + 19O_2 \longrightarrow 12CO_2 + 14H_2O$$

The equation is now balanced with 12 mol of carbon atoms, 28 mol of hydrogen atoms, and 38 mol of oxygen atoms on each side of the equation.

Practice Problem 10.8

Write a balanced equation for the complete combustion of each of the following hydrocarbons:

    a. cyclobutane   b. ethane   c. decane

▶ For Further Practice: **Questions 10.97 and 10.98.**

LEARNING GOAL

13 Write balanced equations for combustion reactions of alkanes.

Explain why the butane found in this lighter is a liquid only when maintained under pressure. Write an equation representing the complete combustion of butane.

The energy released, along with their availability and relatively low cost, makes hydrocarbons very useful as fuels. In fact, combustion is essential to our very existence. It is the process by which we heat our homes, run our cars, and generate electricity. Although combustion of fossil fuels is vital to industry and society, it also represents a threat to the environment. The buildup of CO_2 may contribute to global warming and change the face of the earth in future generations.

See Green Chemistry: The Greenhouse Effect and Global Climate Change in Chapter 5.

Halogenation

Alkanes and cycloalkanes can also react with a halogen (usually chlorine or bromine) in a reaction called **halogenation.** Halogenation is a **substitution reaction,** that is, a reaction that results in the replacement of one group for another. In this reaction, a halogen atom is substituted for one of the hydrogen atoms in the alkane. The products of this reaction are an **alkyl halide** or *haloalkane* and a hydrogen halide.

LEARNING GOAL

14 Write balanced equations for halogenation reactions of alkanes.

Halogenation can occur only in the presence of heat and/or light, as indicated by the reaction conditions noted over the reaction arrows. The general equation for the halogenation of an alkane follows. The R in the general structure for the alkane may be either a hydrogen atom or an alkyl group.

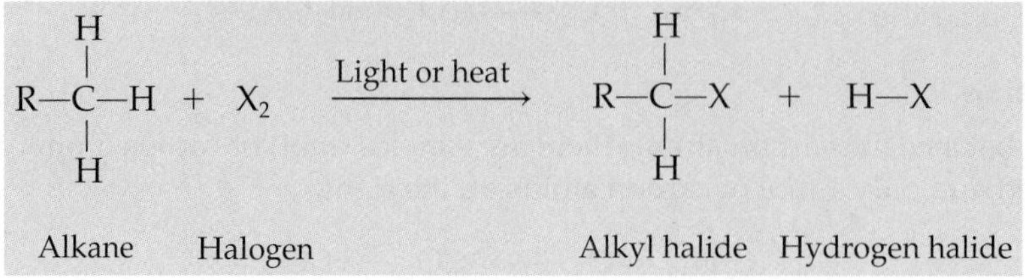

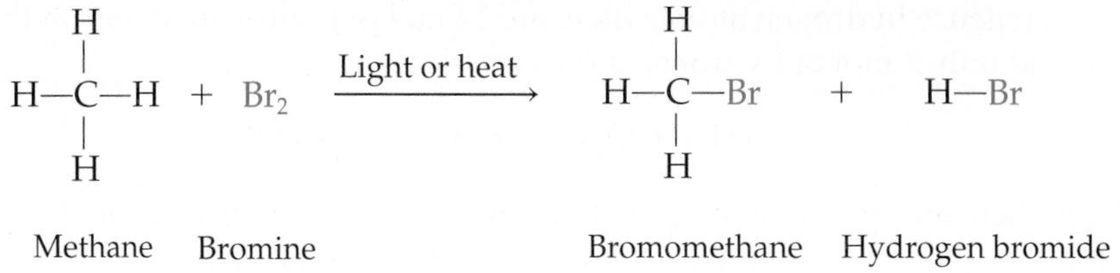

$$
\underset{\text{Alkane}}{R-\overset{\displaystyle H}{\underset{\displaystyle H}{C}}-H} + \underset{\text{Halogen}}{X_2} \xrightarrow{\text{Light or heat}} \underset{\text{Alkyl halide}}{R-\overset{\displaystyle H}{\underset{\displaystyle H}{C}}-X} + \underset{\text{Hydrogen halide}}{H-X}
$$

$$
\underset{\text{Methane}}{H-\overset{\displaystyle H}{\underset{\displaystyle H}{C}}-H} + \underset{\text{Bromine}}{Br_2} \xrightarrow{\text{Light or heat}} \underset{\text{Bromomethane}}{H-\overset{\displaystyle H}{\underset{\displaystyle H}{C}}-Br} + \underset{\text{Hydrogen bromide}}{H-Br}
$$

$$
\underset{\text{Ethane}}{CH_3CH_3} + \underset{\text{Chlorine}}{Cl_2} \xrightarrow{\text{Light}} \underset{\text{Chloroethane}}{CH_3CH_2-Cl} + \underset{\text{Hydrogen chloride}}{H-Cl}
$$

$$
\underset{\text{Cyclohexane}}{\bigcirc\!\!\!C\!\!\overset{H}{\underset{H}{\diagup}}} + \underset{\text{Chlorine}}{Cl_2} \xrightarrow{\text{Heat}} \underset{\text{Chlorocyclohexane}}{\bigcirc\!\!\!C\!\!\overset{H}{\underset{Cl}{\diagup}}} + \underset{\text{Hydrogen chloride}}{HCl}
$$

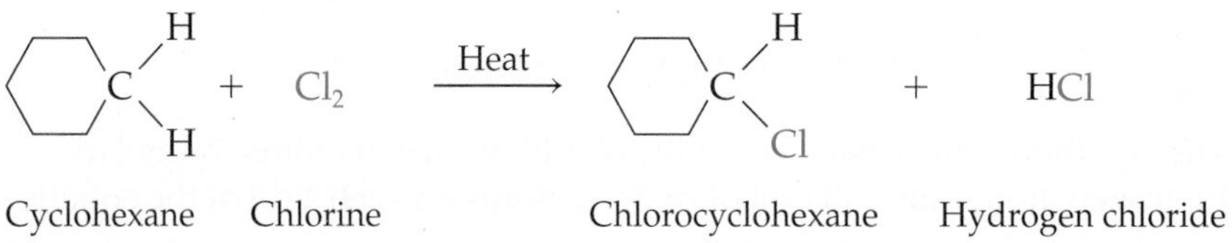

The alkyl halide may continue to react, forming a mixture of products substituted at multiple sites or substituted multiple times at the same site.

If the halogenation reaction is allowed to continue, the alkyl halide formed may react with other halogen atoms. When this happens, a mixture of products may be formed. For instance, bromination of methane will produce bromomethane (CH_3Br), dibromomethane (CH_2Br_2), tribromomethane ($CHBr_3$), and tetrabromomethane (CBr_4).

In more complex alkanes, including branched alkanes, halogenation can occur to some extent at all positions to give a mixture of monosubstituted products. For example, bromination of propane produces a mixture of 1-bromopropane and 2-bromopropane. Halogenation of the branched alkane 2-methylpropane results in two alkyl halide products. If the hydrogen atom on carbon-2 is replaced (by chlorine, for example), the product will be 2-chloro-2-methylpropane. If any other hydrogen atom is replaced, the product will be 1-chloro-2-methylpropane.

$$
\underset{\underset{\displaystyle CH_3}{|}}{CH_3CHCH_3} + Cl_2 \xrightarrow{\text{light or heat}} \underset{\substack{\displaystyle CH_3 \\ \text{2-Chloro-} \\ \text{2-methylpropane}}}{\overset{\displaystyle Cl}{\underset{|}{\overset{|}{CH_3CCH_3}}}} + HCl \ \text{ or } \ \underset{\substack{\displaystyle CH_3 \\ \text{1-Chloro-} \\ \text{2-methylpropane}}}{\underset{|}{CH_3CHCH_2Cl}} + HCl
$$

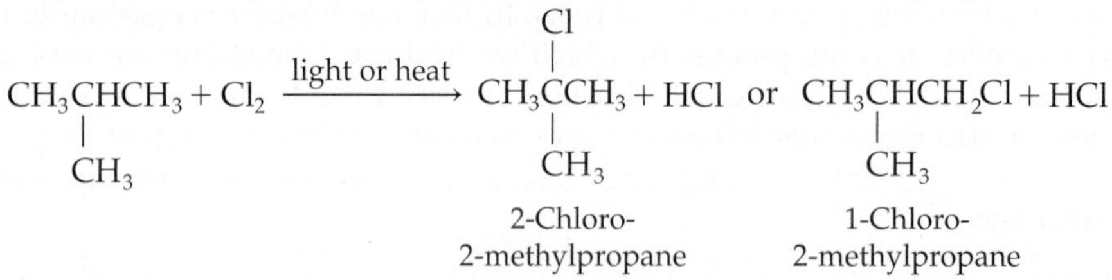

Alkanes are not very reactive molecules. However, alkyl halides are very useful reactants for the synthesis of other organic compounds. Thus, the halogenation reaction is of great value because it converts unreactive alkanes into versatile starting materials for the synthesis of desired compounds. This is important in the pharmaceutical industry for the synthesis of some drugs. In addition, alkyl halides having two or more halogen atoms are useful solvents, insecticides, and herbicides.

A Medical Perspective

Polyhalogenated Hydrocarbons Used as Anesthetics

Polyhalogenated hydrocarbons are hydrocarbons containing two or more halogen atoms. Some polyhalogenated compounds are notorious for the problems they have caused. For instance, some insecticides such as DDT, chlordane, kepone, and lindane do not break down rapidly in the environment. As a result, these toxic compounds accumulate in biological tissue of a variety of animals, including humans, and may cause neurological damage, birth defects, or even death.

Other halogenated hydrocarbons are very useful in medicine. They were among the first anesthetics (pain relievers) used routinely in medical practice. These chemicals played a central role as the studies of medicine and dentistry advanced into modern times.

CH_3CH_2Cl CH_3Cl

Chloroethane   Chloromethane

(ethyl chloride)   (methyl chloride)

Chloroethane and chloromethane are local anesthetics. A local anesthetic deadens the feeling in a portion of the body. Applied topically (on the skin), chloroethane and chloromethane numb the area. Rapid evaporation of these anesthetics lowers the skin temperature, deadening the local nerve endings. They act rapidly, but the effect is brief, and feeling is restored quickly.

$CHCl_3$

Trichloromethane

(chloroform)

In the past, chloroform was used as both a general and a local anesthetic. When administered through inhalation, it rapidly causes loss of consciousness. However, the effects of this powerful anesthetic are of short duration. Chloroform is no longer used because it was shown to be carcinogenic.

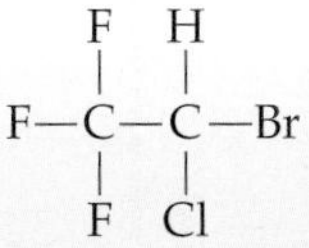

F—C—C—Br with F, F below first C and H, Cl on second C

2-Bromo-2-chloro-1,1,1-trifluoroethane

(Halothane)

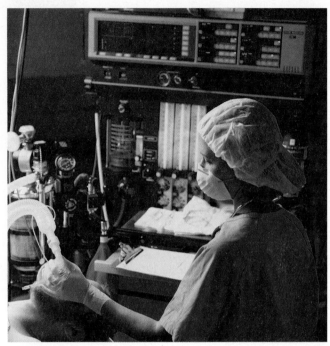

Halothane is administered to a patient.

Halothane is a general anesthetic that is administered by inhalation. It is considered to be a very safe anesthetic and is widely used.

For Further Understanding

▶ In the first 24 hours (h) following administration, 70% of the halothane is eliminated from the body in exhaled gases. Explain why halothane is so readily eliminated in exhaled gases.

▶ Rapid evaporation of chloroethane from the skin surface results in cooling that causes local deadening of nerve endings. Explain why the skin surface cools dramatically as a result of evaporation of chloroethane.

Question 10.5 Write a balanced equation for each of the following reactions. Show all possible products.

a. the monobromination of propane
b. the monochlorination of butane
c. the monochlorination of cyclobutane
d. the monobromination of pentane

Question 10.6 Provide the IUPAC names for the products of the reactions in Question 10.5.

CHAPTER MAP

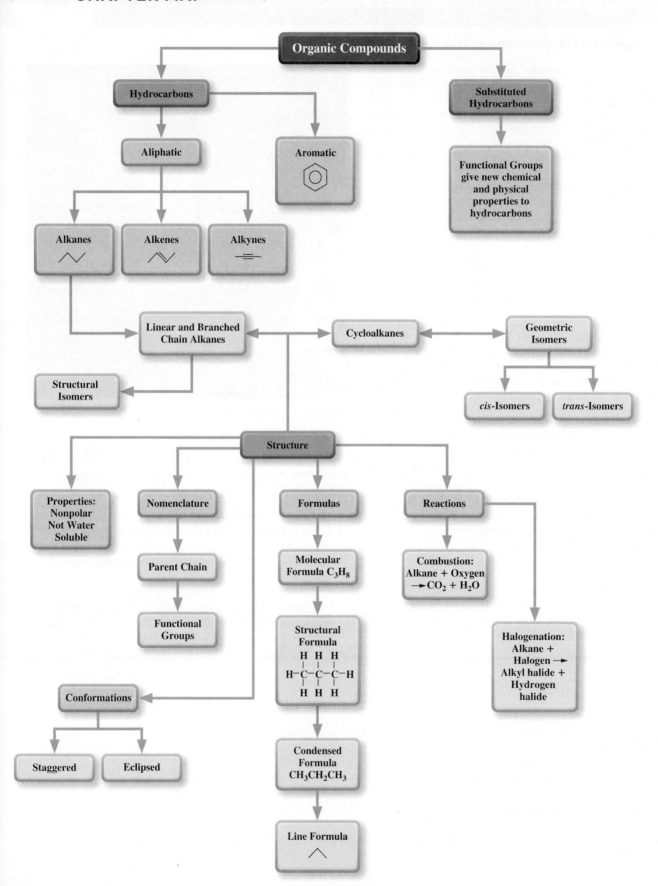

SUMMARY OF REACTIONS

Reactions of Alkanes
Combustion:

$$C_nH_{2n+2} + O_2 \longrightarrow CO_2 + H_2O + \text{heat energy}$$

Alkane Oxygen   Carbon Water
                  dioxide

Halogenation:

$$\begin{array}{c}
\quad\text{H} \qquad\qquad\qquad\qquad\qquad \text{H} \\
\quad| \qquad\qquad\qquad\qquad\qquad\quad | \\
\text{R}\!-\!\text{C}\!-\!\text{H} + \text{X}_2 \xrightarrow{\text{light or heat}} \text{R}\!-\!\text{C}\!-\!\text{X} + \text{H}\!-\!\text{X} \\
\quad| \qquad\qquad\qquad\qquad\qquad\quad | \\
\quad\text{H} \qquad\qquad\qquad\qquad\qquad \text{H}
\end{array}$$

Alkane   Halogen   Alkyl   Hydrogen
                         halide   halide

SUMMARY

10.1 The Chemistry of Carbon

▶ Organic chemistry is the study of carbon-containing compounds.

▶ All organic compounds are classified as **hydrocarbons** or **substituted hydrocarbons.**

▶ Hydrocarbons contain only carbon and hydrogen atoms and may be **aliphatic** (alkanes, alkenes, and alkynes) or **aromatic** (containing a benzene ring).

▶ Aliphatic hydrocarbons may be **saturated** (only C—C and C—H single bonds) or **unsaturated** (at least one C—C double or triple bond).

▶ Substituted hydrocarbons include a **functional group** that gives the molecule particular chemical and physical properties.

10.2 Alkanes

▶ **Alkanes** are **saturated hydrocarbons** with the general formula C_nH_{2n+2}.

▶ Four types of formulas are used to represent organic molecules:
  - **Molecular formulas**
  - **Structural formulas**
  - **Condensed formulas**
  - **Line formulas**

▶ Two organic molecules with the same molecular formula but different bonding patterns have different physical and chemical properties. Such molecules are **structural** or **constitutional isomers.**

▶ Alkanes are nonpolar, water-insoluble, and have low melting and boiling points.

▶ The **IUPAC Nomenclature System** is the universal system for naming organic compounds.
  - The **parent compound,** the longest carbon chain with the characteristic functional group, gives the molecule its primary name.
  - Attached to the parent chain, **alkyl groups** (alkanes with one fewer hydrogen atom), halogens, or other functional groups may replace one or more of the hydrogen atoms in the alkane.

▶ The carbon atoms within a molecule are classified as **primary (1°), secondary (2°), tertiary (3°), or quaternary (4°),** depending on the number of carbon atoms to which they are attached.

10.3 Cycloalkanes

▶ **Cycloalkanes** are organic molecules having C—C single bonds in a ring structure.

▶ There is very limited rotation around the C—C bonds in a cycloalkane. As a result, **geometric isomers,** also called *cis-trans* isomers, occur.

▶ Geometric isomers are a type of stereoisomers, which are molecules with the same molecular formulas and bonding patterns, but different arrangements of atoms in space.

10.4 Conformations of Alkanes and Cycloalkanes

▶ **Conformations** or **conformers** result from the free rotation of the C—C bonds in alkanes and the limited rotation of C—C bonds in **cycloalkanes.**

▶ In cyclohexane, two common conformers are the **chair** and **boat conformations.** The chair conformation is the most stable and therefore is the most common conformation.

10.5 Reactions of Alkanes and Cycloalkanes

▶ Common reactions of alkanes and cycloalkanes are **combustion** and **halogenation.**

▶ Complete combustion produces CO_2, H_2O, and heat.

▶ Halogenation is a **substitution reaction** in which a halogen atom replaces one of the hydrogen atoms in the molecule, producing an **alkyl halide.**

ANSWERS TO PRACTICE PROBLEMS

10.1   Structural Formulas:

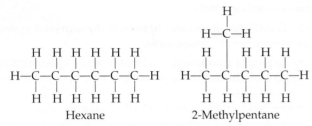

Hexane   2-Methylpentane

3-Methylpentane

2,3-Dimethylbutane   2,2-Dimethylbutane

Line Formulas:

Hexane   2-Methylpentane   3-Methylpentane

2,3-Dimethylbutane   2,2-Dimethylbutane

10.2   **a.** 2,4-Dimethyloctane
          b. Trichlorofluoromethane
          c. 3-Chloro-4-fluoro-2-methylheptane

10.3   **a.** $CH_2BrCHCl(CH_2)_3CH_3$
          b. $CH_3CH(CH_3)CH(CH_3)CH_2CH_3$
          c. $CH_2ClCH_2CHClCH_2CHClCH_2CH_3$
          d. $CH_3CH_2CHClCH(CH_3)CHI(CH_2)_2CH_3$
          e. $CH_2BrCHBrCHClCH_3$
          f. CF_3Cl

10.4   The following are the line structures and IUPAC names of the nine isomers of heptane:

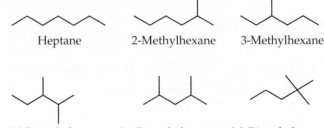

Heptane   2-Methylhexane   3-Methylhexane

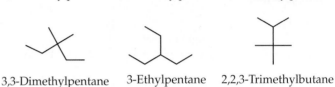

2,3-Dimethylpentane 2,4-Dimethylpentane 2,2-Dimethylpentane

3,3-Dimethylpentane   3-Ethylpentane   2,2,3-Trimethylbutane

10.5   **a.** Chlorocycloheptane   **c.** Ethylcyclobutane
          b. Methylcyclopropane   **d.** 1,1-Difluorocyclohexane

10.6   **a.** *cis*-isomer   **c.** *trans*-isomer
          b. *cis*-isomer   **d.** *trans*-isomer

10.7   **a.** *cis*-1,2-Dichlorocycloheptane
          b. *cis*-1,2-Dimethylcyclopropane
          c. *trans*-1,3-Diethylcyclobutane
          d. *trans*-1,2-Difluorocyclohexane

10.8   **a.** The complete combustion of cyclobutane:

$$\square + 6O_2 \longrightarrow 4CO_2 + 4H_2O + \text{heat energy}$$

          b. The complete combustion of ethane:

$$2CH_3CH_3 + 7O_2 \rightarrow 4CO_2 + 6H_2O + \text{heat energy}$$

          c. The complete combustion of decane:

$$2C_{10}H_{22} + 31O_2 \rightarrow 20CO_2 + 22H_2O + \text{heat energy}$$

QUESTIONS AND PROBLEMS

The Chemistry of Carbon

Foundations

10.7   Why is the number of organic compounds nearly limitless?
10.8   How do linear, continuous-chain alkanes differ from branched-chain alkanes?
10.9   Why do ionic substances generally have higher melting and boiling points than covalent substances?
10.10   Why are ionic substances more likely to be water-soluble?

Applications

10.11   Based on intermolecular forces or ionic interactions, rank the following compounds from highest to lowest boiling points:
          a. H_2O CH_4 LiCl
          b. C_2H_6 C_3H_8 NaCl

10.12 Based on intermolecular forces or ionic interactions, rank the following compounds from highest to lowest melting points:
  a. H_2O_2 CH_4 KCl
  b. C_6H_{14} $C_{18}H_{38}$ NaCl

10.13 What would the physical state of each of the compounds in Question 10.11 be at room temperature?

10.14 Which of the compounds in Question 10.12 would be soluble in water?

10.15 Consider the differences between organic and inorganic compounds as you answer each of the following questions.
  a. Which compounds make good electrolytes?
  b. Which compounds exhibit ionic bonding?
  c. Which compounds have lower melting points?
  d. Which compounds are more likely to be soluble in water?
  e. Which compounds are flammable?

10.16 Describe the major differences between ionic and covalent bonds.

10.17 For centuries, fishermen have used shark liver oil to treat a variety of ailments, including general weakness, wounds, and inflammation of the respiratory and gastrointestinal tracts. In fact, it is an ingredient in the hemorrhoid cream Preparation H. In addition to chemicals such as vitamins A and D, omega-3 fatty acids, and triglycerides, shark liver oil contains approximately 14% 2,6,10,14-tetramethyl-pentadecane, also known as pristane.
  a. Write the molecular formula for pristane.
  b. Draw the line and condensed formulas for pristane.
  c. Calculate the molar mass of pristane.

10.18 The tsetse fly *Glossina morsitans* is a large biting fly found in regions of Africa. They carry sleeping sickness, a deadly disease caused by a parasitic protozoan. The pheromone secreted by the tsetse fly contains four straight-chain alkanes: 2-methylheptadecane ($C_{18}H_{38}$), 17,21-dimethylheptatriacontane ($C_{39}H_{80}$), 15,19-dimethylheptatriacontane ($C_{39}H_{80}$), and 15,19,23-trimethylheptatriacontane ($C_{40}H_{82}$). Because this pheromone works by smell over long distances, it has proved useful as an agent to control tsetse fly populations.
  a. Draw the line formula for each of these alkanes.
  b. Calculate the molar mass of each of these alkanes.

10.19 Condense each of the following structural formulas:

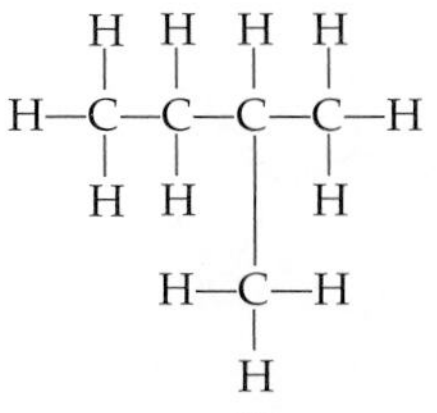

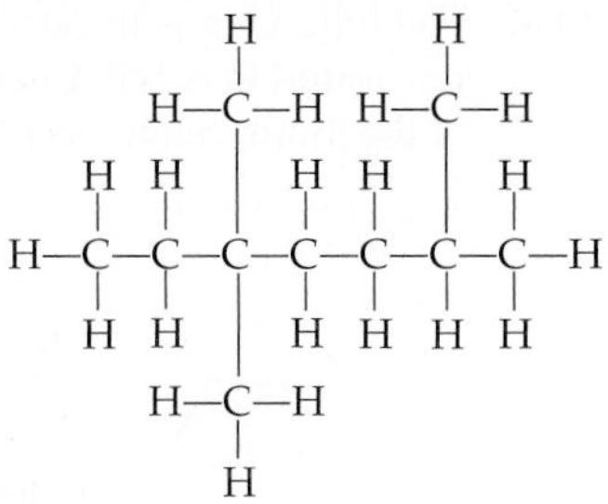

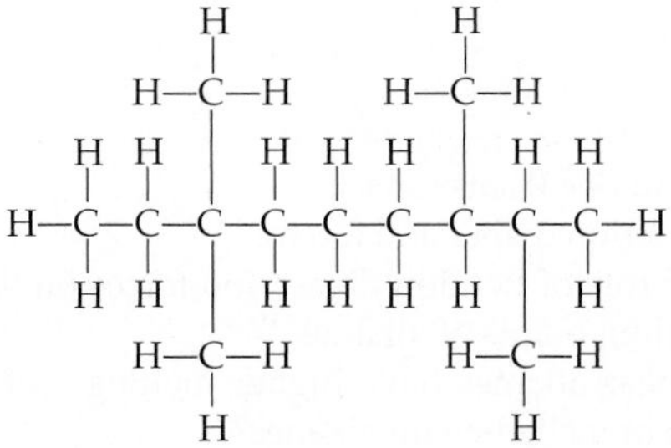

10.20 Condense each of the following structural formulas:

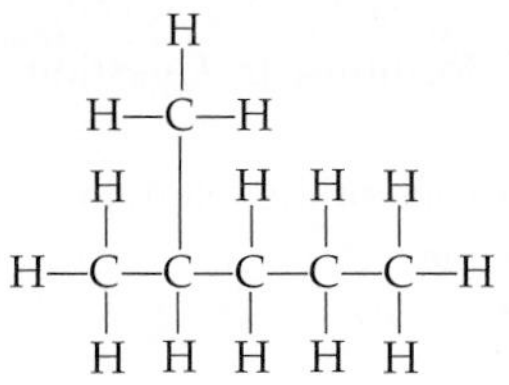

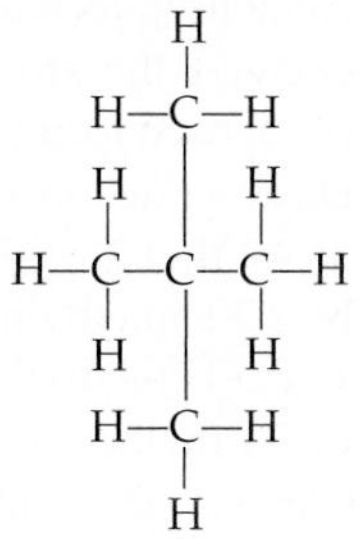

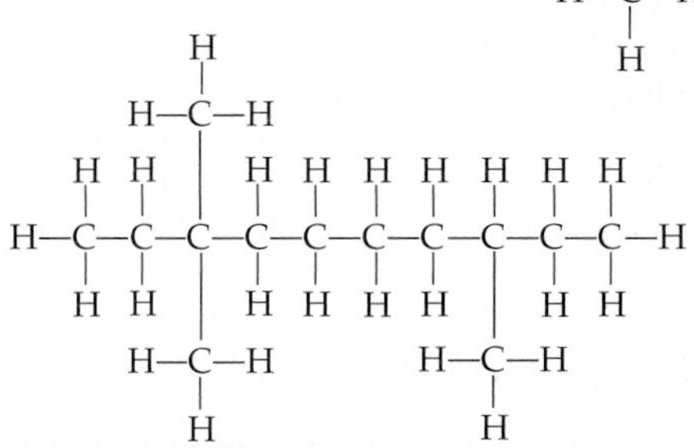

10.21 Convert the following structural formulas into line formulas:

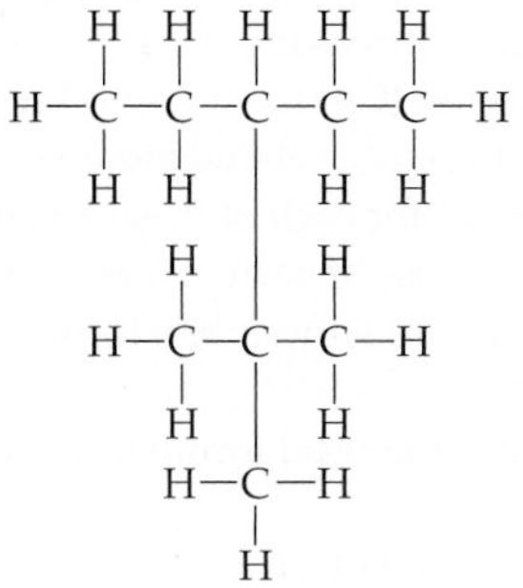

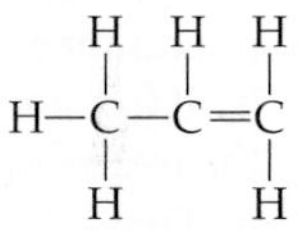

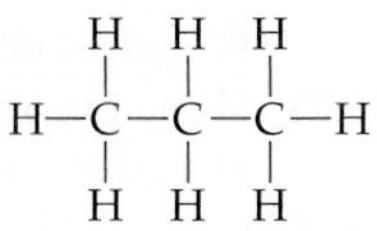

10.22 Convert the following structural formulas into line formulas:

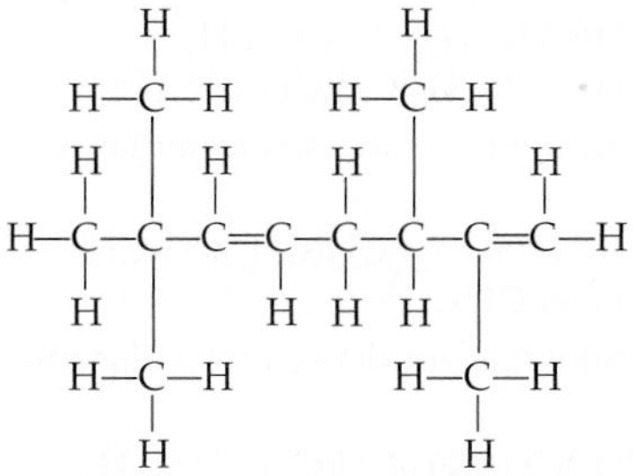

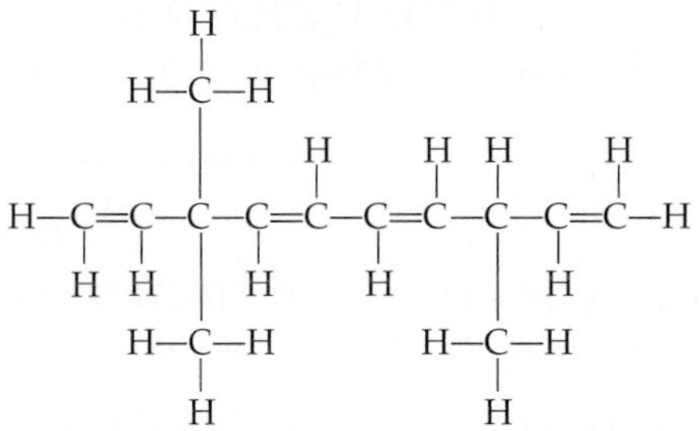

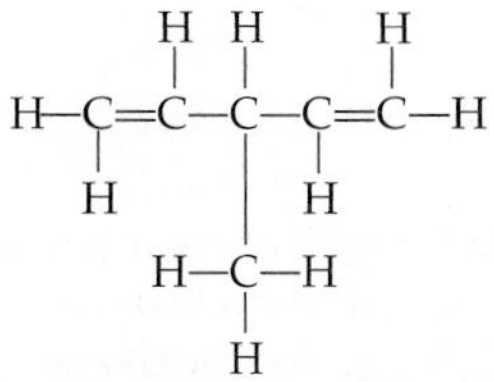

10.23 Convert the structural formulas in Question 10.21 into condensed formulas.

10.24 Convert the structural formulas in Question 10.22 into condensed formulas.

10.25 Draw a line formula for each of the following alkanes:
  a. 4-Ethyl-2-methylhexane
  b. 2,3-Dimethylhexane
  c. 3,3-Dimethylhexane
  d. 3-Ethylpentane

10.26 Convert the following line formulas into condensed formulas:

a. 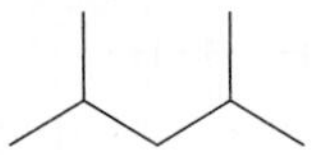   b.

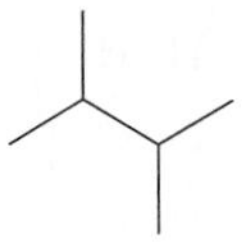

c.    d.

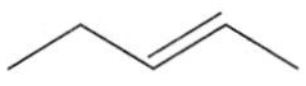

10.27 Certain spider orchids can only be pollinated by a particular sand bee. To ensure pollination, the spider orchid emits a mixture of straight-chain alkanes that is identical to the sex pheromone produced by the female sand bee. The pheromone, a mixture of tricosane ($C_{23}H_{48}$), pentacosane ($C_{25}H_{52}$), and heptacosane ($C_{27}H_{56}$) in the ratio 3:3:1 lures the male sand bee to the spider orchid blooms.
  a. Draw the line formula for each of these alkanes.
  b. Calculate the molar mass of each of these alkanes.

10.28 Using the octet rule, explain why carbon forms four bonds in a stable compound.

10.29 Convert the following condensed formulas into structural formulas:
  a. $CH_3CH(CH_3)CH(CH_3)CH_2CH_3$
  b. $CH_3CH_2CH_2CH_2CH_3$

10.30 Convert the following condensed formulas into structural formulas:
  a. $CH_3CH_2CH(CH_2CH_3)CH_2CH_2CH_3$
  b. $CH_3CH(CH_3)CH(CH_3)CH_2CH_2CH_2CH_3$

10.31 Convert the following condensed formulas into structural formulas:
  a. $CH_3CH_2CH(CH_3)(CH_2)_3CH(CH_3)CH_2CH_3$
  b. $CH_3C(CH_3)_2CH_2CH_3$

10.32 Convert the following condensed formulas into structural formulas:
  a. $CH_3CH(CH_3)CH(CH_3)CH(CH_3)CH_2CH_3$
  b. $CH_3C(CH_3)_2CH(CH_2CH_3)CH_2CH_2CH_3$

10.33 Name the functional group in each of the following molecules:
  a. $CH_3CH_2CH_2OH$
  b. $CH_3CH_2CH_2NH_2$
  c. $CH_3CH_2CH_2C=O$ with H below
  d. $CH_3CH_2CH_2C=O$ with OH below
  e. $CH_3CH_2CH_2C=O$ with OCH_2CH_3 below
  f. $CH_3CH_2OCH_2CH_3$
  g. $CH_3CH_2CH_2I$

10.34 Convert the condensed structures in Question 10.33 into line formulas.

10.35 Give the general formula for each of the following:
  a. An alkane
  b. An alkyne
  c. An alkene
  d. A cycloalkane
  e. A cycloalkene

10.36 Of the classes of compounds listed in Question 10.35, which are saturated? Which are unsaturated?

10.37 What major structural feature distinguishes the alkanes, alkenes, and alkynes? Give examples.

10.38 What is the major structural feature that distinguishes between saturated and unsaturated hydrocarbons?

10.39 The following formula represents the structure of Lisinopril, one of the angiotensin-converting enzyme (ACE) inhibitor drugs used in the treatment of high blood pressure and congestive heart failure. Circle and identify the functional groups in this molecule.

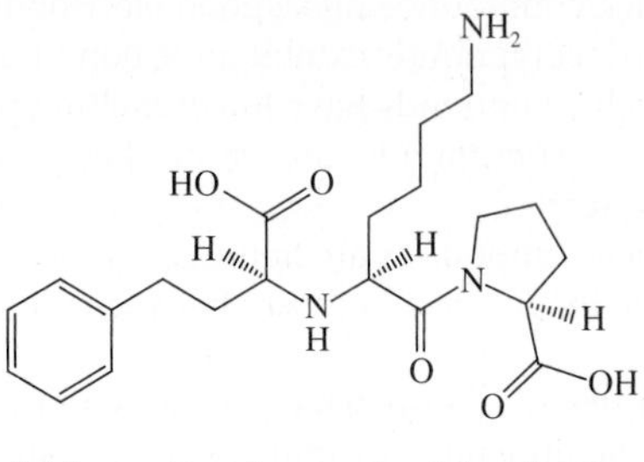

10.40 Folic acid is a vitamin required by the body for nucleic acid synthesis. The structure of folic acid is given below. Circle and identify as many functional groups as possible.

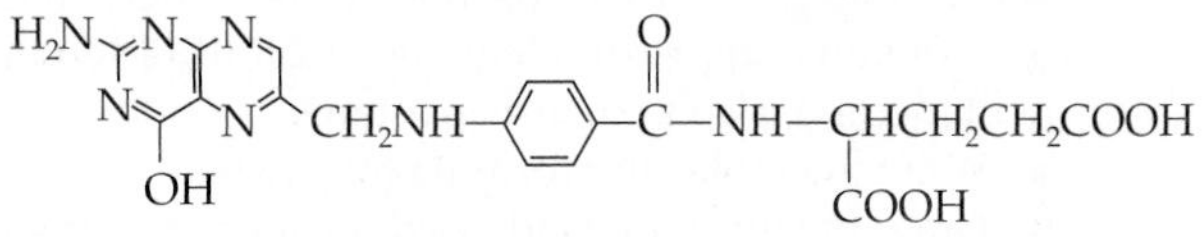

Folic acid

10.41 The following structure is the artificial sweetener aspartame, found in Equal. Circle and name the functional groups found in this molecule.

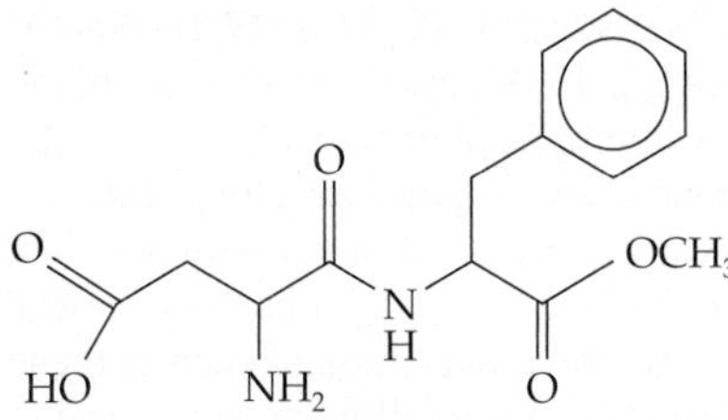

10.42 The following is the structure of the pain reliever ibuprofen, found in Advil. Circle and label the functional groups of the ibuprofen molecule.

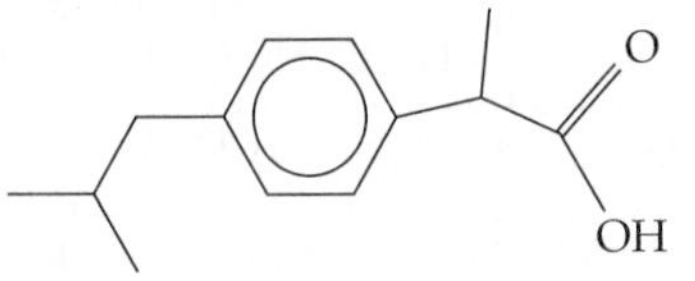

Ibuprofen

Alkanes

Foundations

10.43 What are van der Waals forces?

10.44 Describe London dispersion forces.

10.45 Explain the role of London dispersion forces on the melting and boiling points of alkanes.

10.46 Why do linear alkanes have higher melting and boiling points than branched-chain alkanes?

10.47 Why are hydrocarbons not water-soluble?

10.48 Describe the relationship between the length of hydrocarbon chains and the melting points of the compounds.

Applications

10.49 Based on intermolecular forces, rank the following compounds from highest to lowest boiling points:
  a. heptane butane hexane ethane
  b. $CH_3CH_2CH_2CH_2CH_3$ $CH_3CH_2CH_3$
  $CH_3CH_2CH_2CH_2CH_2CH_2CH_2CH_3$

10.50 Based on intermolecular forces, rank the following compounds from highest to lowest melting points:
  a. decane propane methane ethane
  b. $CH_3CH_2CH_2CH_2CH_3$ $CH_3(CH_2)_8CH_3$ $CH_3(CH_2)_6CH_3$

10.51 What would the physical state of each of the compounds in Question 10.49 be at room temperature?

10.52 What would the physical state of each of the compounds in Question 10.50 be at room temperature?

10.53 Name each of the compounds in Question 10.49b.

10.54 Name each of the compounds in Question 10.50b.

10.55 Draw each of the following using line formulas:
  a. 2-Bromobutane
  b. 2-Chloro-2-methylpropane
  c. 2,2-Dimethylhexane

10.56 Draw each of the following using line formulas:
  a. Dichlorodiiodomethane
  b. 1,4-Diethylcyclohexane
  c. 2-Iodo-2,4,4-trimethylpentane

10.57 Draw each of the following compounds using structural formulas:
  a. 2,2-Dibromobutane   **c.** 1,2-Dichloropentane
  b. 2-Iododecane   **d.** 1-Bromo-2-methylpentane

10.58 Draw each of the following compounds using structural formulas:
  a. 1,1,1-Trichlorodecane
  b. 1,2-Dibromo-1,1,2-trifluoroethane
  c. 3,3,5-Trimethylheptane
  d. 1,3,5-Trifluoropentane

10.59 Name each of the following using the IUPAC Nomenclature System:

  a. 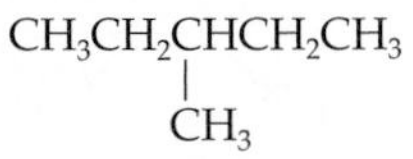 CH₃CH₂CHCH₂CH₃
          |
          CH₃

  c. 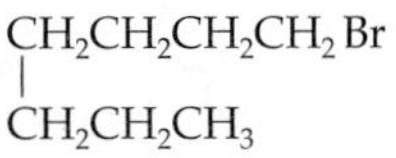 CH₂CH₂CH₂CH₂ Br
          |
          CH₂CH₂CH₃

  b. 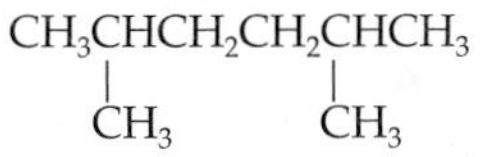 CH₃CHCH₂CH₂CHCH₃
          | |
          CH₃   CH₃

  d. 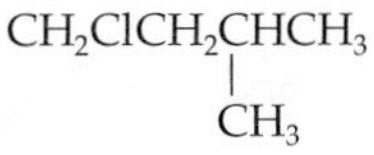 CH₂ClCH₂CHCH₃
                    |
                    CH₃

10.60 Provide the IUPAC name for each of the following compounds:

  a. 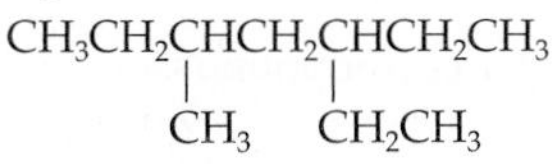 CH₃CH₂CHCH₂CHCH₂CH₃
          | |
          CH₃   CH₂CH₃

  c. 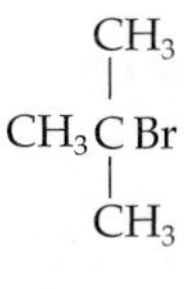   CH₃
               |
          CH₃C Br
               |
               CH₃

  b. CH₃CHClCH₂CH₂CH₂Cl

10.61 Give the IUPAC name for each of the following:

  a. 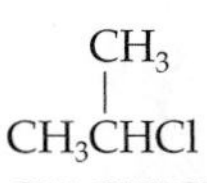   CH₃
           |
        CH₃CHCl

  d. 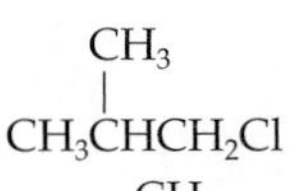   CH₃
            |
        CH₃CHCH₂Cl

  b. CH₃CHICH₂CH₃

  e. 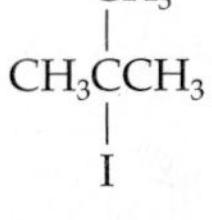   CH₃
              |
         CH₃CCH₃
              |
              I

  c. CH₃CBr₂CH₃

10.62 Name the following using the IUPAC Nomenclature System:

  a. 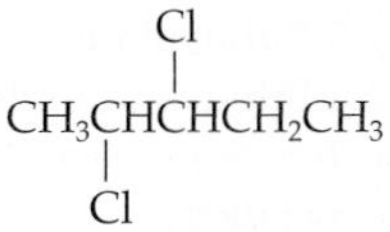   Cl
               |
        CH₃CHCHCH₂CH₃
               |
               Cl

  c. 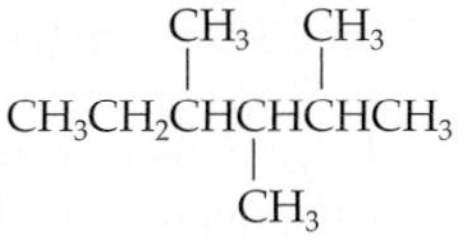   CH₃   CH₃
               | |
        CH₃CH₂CHCHCHCH₃
                    |
                    CH₃

  b. 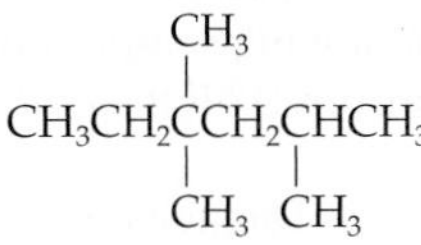   CH₃
               |
        CH₃CH₂CCH₂CHCH₃
               | |
               CH₃   CH₃

  d. 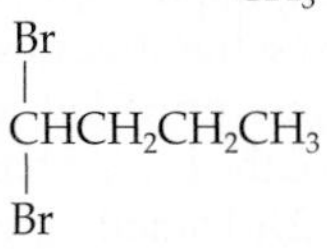   Br
               |
        CHCH₂CH₂CH₃
               |
               Br

10.63 Draw a complete structural formula for each of the straight-chain isomers of the following alkanes:
  a. C_4H_9Br **b.** $C_4H_8Br_2$

10.64 Name all of the isomers that you obtained in Question 10.63.

10.65 Name the following using the IUPAC Nomenclature System:
  a. $CH_3(CH_2)_3CHClCH_3$ **c.** $CH_3CH_2CHClCH_2CH_3$
  b. $CH_2Br(CH_2)_2CH_2Br$ **d.** $CH_3CH(CH_3)(CH_2)_4CH_3$

10.66 Which of the following pairs of compounds are identical? Which are constitutional isomers? Which are completely unrelated?

  a. 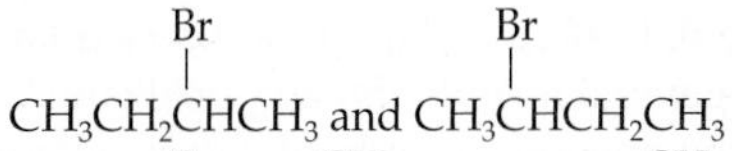   Br   Br
               | |
        CH₃CH₂CHCH₃ and CH₃CHCH₂CH₃

  b. 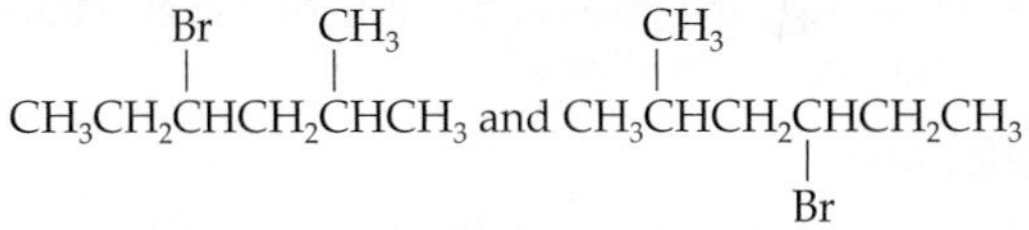   Br   CH₃   CH₃
               | | |
        CH₃CH₂CHCH₂CHCH₃ and CH₃CHCH₂CHCH₂CH₃
                                         |
                                         Br

  c. 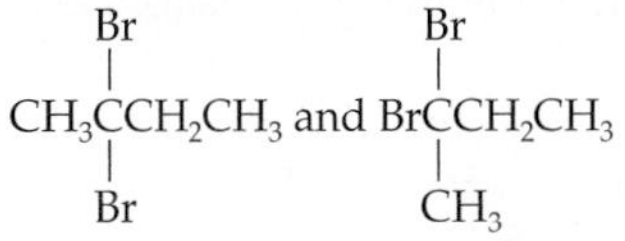   Br   Br
          | |
        CH₃CCH₂CH₃ and BrCCH₂CH₃
          | |
          Br   CH₃

  d. 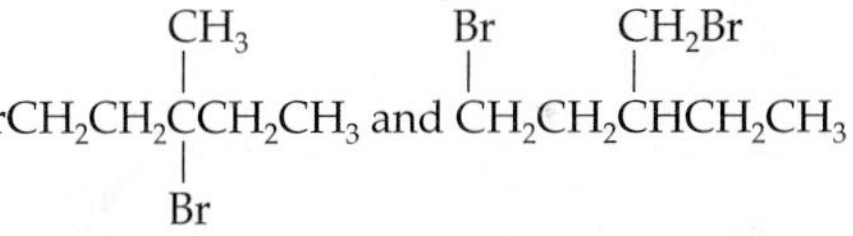   CH₃   Br   CH₂Br
               | | |
        BrCH₂CH₂CCH₂CH₃ and CH₂CH₂CHCH₂CH₃
               |
               Br

10.67 Which of the following pairs of molecules are identical compounds? Which are constitutional isomers?
  a. 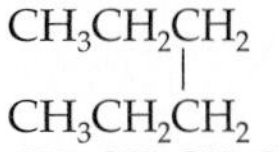 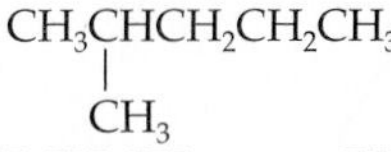 CH₃CH₂CH₂   CH₃CHCH₂CH₂CH₃
          | |
        CH₃CH₂CH₂   CH₃
  b. CH₃CH₂CH₂CH₂CH₂CH₂CH₃   CH₃CH₂CH₂CH₂CH₂
                                                  |
                                                  CH₃CH₂

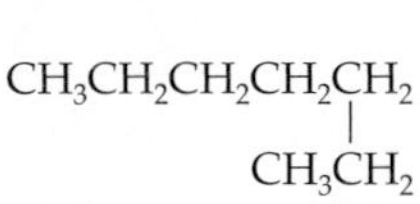

10.68 Which of the following pairs of molecules are identical compounds? Which are constitutional isomers?
  a. CH₃CH₂CH(CH₃)CH₂CH₃   CH₃C(CH₃)₂CH₂CH₂CH₃
  b. CH₃CH₂CH₂CH₂C(CH₃)₂CH₂CH(CH₃)CH₃
        CH₃CH(CH₃)CH₂CH₂C(CH₃)₂CH₂CH₂CH₃

10.69 Which of the following structures are incorrect?

  a. 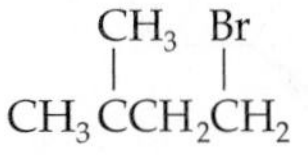   CH₃ Br
          | |
        CH₃CCH₂CH₂

  c. 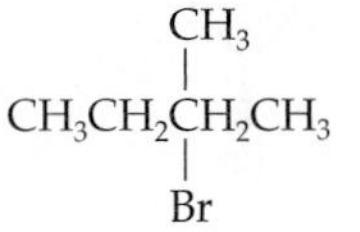   CH₃
               |
        CH₃CH₂CH₂CH₃
               |
               Br

  b. 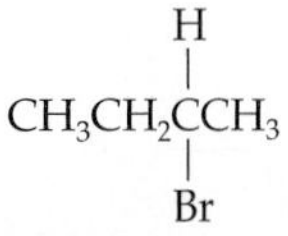   H
               |
        CH₃CH₂CCH₃
               |
               Br

  d. 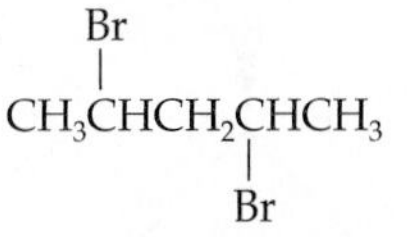   Br
               |
        CH₃CHCH₂CHCH₃
                    |
                    Br

10.70 Which of the following structures are incorrect? Describe the problem in those that are incorrect.
  a. CH₃CH(CH₃)₂CH₂CH₃   **c.** CH₂CH₂C(CH₃)CH₂CH₂CH₃
  b. CH₃(CH₂)₅CH₃   **d.** CH₃CH₂CH(CH₃)₃CH₃

10.71 Are the following names correct or incorrect? If they are incorrect, give the correct name.
  a. 1,3-Dimethylpentane   **c.** 3-Butylbutane
  b. 2-Ethylpropane   **d.** 3-Ethyl-4-methyloctane
10.72 In your own words, describe the steps used to name a compound, using IUPAC nomenclature.
10.73 Draw the structures of the following compounds. Are the names provided correct or incorrect? If they are incorrect, give the correct name.
  a. 2,4-Dimethylpentane   **d.** 1,4-Diethylheptane
  b. 1,3-Dimethylhexane   **e.** 1,6-Dibromo-6-methyloctane
  c. 1,5-Diiodopentane
10.74 Draw the structures of the following compounds. Are the names provided correct or incorrect? If they are incorrect, give the correct name.
  a. 1,4-Dimethylbutane   **c.** 2,3-Dimethylbutane
  b. 1,2-Dichlorohexane   **d.** 1,2-Diethylethane

Cycloalkanes

Foundations

10.75 Describe the structure of a cycloalkane.
10.76 Describe the IUPAC rules for naming cycloalkanes.
10.77 What is the general formula for a cycloalkane?
10.78 How does the general formula of a cycloalkane compare with that of an alkane?

Applications

10.79 Name each of the following cycloalkanes, using the IUPAC system:

a.

c.

b.

d.

10.80 Name each of the following cycloalkanes, using the IUPAC system:

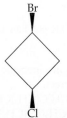
a.

c.

b.

d.

10.81 Draw the structure of each of the following cycloalkanes:
  a. 1-Bromo-2-methylcyclobutane
  b. Iodocyclopropane
  c. 1-Bromo-3-chlorocyclopentane
  d. 1,2-Dibromo-3-methylcyclohexane
10.82 Draw each of the following cycloalkanes:
  a. 1,2,3-Trichlorocyclopropane
  b. 1,1-Dibromo-3-chlorocyclobutane
  c. 1,2,4-Trimethylcycloheptane
  d. 1,2-Dichloro-3,3-dimethylcyclohexane
10.83 How many geometric and structural isomers of dichlorocyclopropane can you construct? Use a set of molecular models to construct the isomers and to contrast their differences. Draw all these isomers.
10.84 How many isomers of dibromocyclobutane can you construct? As in Question 10.83, use a set of molecular models to construct the isomers and then draw them.
10.85 Which of the following names are correct and which are incorrect? If incorrect, write the correct name.
  a. 2,3-Dibromocyclobutane **c.** 1,2-Dimethylcyclopropane
  b. 1,4-Diethylcyclobutane   **d.** 4,5,6-Trichlorocyclohexane
10.86 Which of the following names are correct and which are incorrect? If incorrect, write the correct name.
  a. 1,4,5-Tetrabromocyclohexane
  b. 1,3-Dimethylcyclobutane
  c. 1,2-Dichlorocyclopentane
  d. 3-Bromocyclopentane
10.87 Draw the structures of each of the following compounds:
  a. *cis*-1,3-Dibromocyclopentane
  b. *trans*-1,2-Dimethylcyclobutane
  c. *cis*-1,2-Dichlorocyclopropane
  d. *trans*-1,4-Diethylcyclohexane
10.88 Draw the structures of each of the following compounds:
  a. *trans*-1,4-Dimethylcyclooctane
  b. *cis*-1,3-Dichlorocyclohexane
  c. *cis*-1,3-Dibromocyclobutane
10.89 Name each of the following compounds:

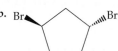
a.

c.

b.

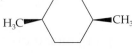
d.

10.90 Name each of the following compounds:

a.

c.

b.

d.

Conformations of Alkanes and Cycloalkanes

Foundations

10.91 What are conformational isomers?

10.92 Why is the staggered conformation of ethane more stable than the eclipsed conformation?

Applications

10.93 Make a model of cyclohexane and compare the boat and chair conformations. Use your model to explain why the chair conformation is more stable.

10.94 What is meant by free rotation around a carbon-carbon single bond? Why can't conformations be separated from one another?

Reactions of Alkanes and Cycloalkanes

Foundations

10.95 Define the term *combustion*.

10.96 Explain why halogenation of an alkane is a substitution reaction.

Applications

10.97 Write a balanced equation for the complete combustion of each of the following:
  a. propane
  b. heptane
  c. nonane
  d. decane

10.98 Write a balanced equation for the complete combustion of each of the following:
  a. pentane
  b. hexane
  c. octane
  d. ethane

10.99 Just as an octane rating is applied to gasoline, a cetane number is used as the measure of the combustion quality of diesel fuel. Cetane, or hexadecane, is an unbranched alkane that ignites easily in a combustion engine. Write a balanced equation for the complete combustion of cetane.

10.100 For gasoline, isooctane, 2,2,4-trimethylpentane, is the standard of excellence. Write a balanced equation for the complete combustion of isooctane.

10.101 Complete each of the following equations by supplying the missing reactant or product as indicated by a question mark:

  a. $2CH_3CH_2CH_2CH_3 + 13O_2 \xrightarrow{\text{Heat}}$? (Complete combustion)

  b. $CH_3CH(CH_3)_2 + Br_2 \xrightarrow{\text{Light}}$? (Give all possible monobrominated products)

  c. 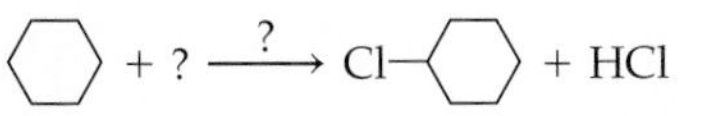 ⬡ + ? $\xrightarrow{?}$ Cl–⬡ + HCl

10.102 Give all the possible monochlorinated products for the following reaction:

$$CH_3CH(CH_3)CH_2CH_3 + Cl_2 \xrightarrow{\text{Light}} ?$$

Name the products, using IUPAC nomenclature.

10.103 Draw the constitutional isomers of molecular formula C_6H_{14} and name each using the IUPAC system.
  a. Which one gives two and only two monobromo derivatives when it reacts with Br_2 and light? Name the products, using the IUPAC system.
  b. Which give three and only three monobromo products? Name the products, using the IUPAC system.
  c. Which give four and only four monobromo products? Name the products, using the IUPAC system.

10.104 **a.** Draw and name all of the isomeric products obtained from the monobromination of propane with Br_2/light. If halogenation were a completely random reaction and had an equal probability of occurring at any of the C—H bonds in a molecule, what percentage of each of these monobromo products would be expected?
  b. Answer part (a) using 2-methylpropane as the starting material.

10.105 A mol of hydrocarbon formed 8 mol of CO_2 and 8 mol of H_2O upon combustion. Determine the molecular formula of the hydrocarbon and write the balanced equation for the combustion reaction. Use line formulas.

10.106 Highly substituted alkyl fluorides, called perfluoroalkanes, are often used as artificial blood substitutes. These perfluoroalkanes have the ability to transport O_2 through the bloodstream as blood does. Some even have twice the O_2 transport capability and are used to treat gangrenous tissue. The structure of perfluorodecalin is shown below. How many mol of fluorine must be reacted with 1 mol of decalin to produce perfluorodecalin?

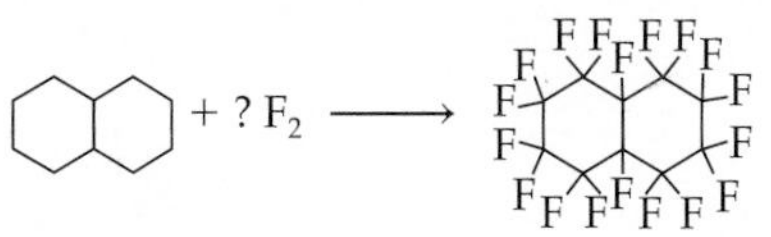

Decalin   Perfluorodecalin

CHALLENGE PROBLEMS

1. You are given two unlabeled bottles, each of which contains a colorless liquid. One contains hexane and the other contains water. What physical properties could you use to identify the two liquids? What chemical property could you use to identify them?

2. You are given two beakers, each of which contains a white crystalline solid. Both are soluble in water. How would you determine which of the two solids is an ionic compound and which is a covalent compound?

3. Chlorofluorocarbons (CFCs) are man-made compounds made up of carbon and the halogens fluorine and chlorine. One of the most widely used is Freon-12 (CCl_2F_2). It was introduced as a refrigerant in the 1930s. This was an important advance because Freon-12 replaced ammonia and sulfur dioxide, two toxic chemicals that were previously used in refrigeration systems. Freon-12 was hailed as a perfect replacement because it has a boiling point of −30°C

and is almost completely inert. To what family of organic molecules do CFCs belong? Design a strategy for the synthesis of Freon-12.

4. Over time, CFC production increased dramatically as their uses increased. They were used as propellants in spray cans, as gases to expand plastic foam, and in many other applications. By 1985, production of CFCs reached 850,000 tons (t). Much of this leaked into the atmosphere and in that year the concentration of CFCs reached 0.6 parts per billion (ppb). Another observation was made by groups of concerned scientists: as the level of CFCs rose, the ozone level in the upper atmosphere declined. Does this correlation between CFC levels and ozone levels prove a relationship between these two phenomena? Explain your reasoning.

5. Although manufacture of CFCs was banned on December 31, 1995, the C—F and C—Cl bonds of CFCs are so strong that the molecules may remain in the atmosphere for 120 years. Within 5 years, they diffuse into the upper stratosphere where ultraviolet photons can break the C—Cl

bonds. This process releases chlorine atoms, as shown here for Freon-12:

$$CCl_2F_2 + photon \longrightarrow CClF_2 + Cl$$

The chlorine atoms are extremely reactive because of their strong tendency to acquire a stable octet of electrons. The following reactions occur when a chlorine atom reacts with an ozone molecule (O_3). First, chlorine pulls an oxygen atom away from ozone:

$$Cl + O_3 \longrightarrow ClO + O_2$$

Then ClO, a highly reactive molecule, reacts with an oxygen atom:

$$ClO + O \longrightarrow Cl + O_2$$

Write an equation representing the overall reaction (sum of the two reactions). How would you describe the role of Cl in these reactions?

ALKENES, ALKYNES, AND AROMATICS

The Unsaturated Hydrocarbons

LEARNING GOALS

1 Describe the physical properties of alkenes and alkynes.

2 Draw the structures and write the IUPAC names for simple alkenes and alkynes.

3 Write the names and draw the structures of simple geometric isomers of alkenes.

4 Write equations predicting the products of addition reactions of alkenes and alkynes: hydrogenation, halogenation, hydration, and hydrohalogenation.

5 Apply Markovnikov's rule to predict the major and minor products of the hydration and hydrohalogenation reactions of unsymmetrical alkenes.

6 Write equations representing the formation of addition polymers of alkenes.

7 Draw the structures and write the names of common aromatic hydrocarbons.

8 Write equations for substitution reactions involving benzene.

9 Describe heterocyclic aromatic compounds and list several biological molecules in which they are found.

Ethene gas causes fruit ripening.

OUTLINE

Oleic acid, the major fatty acid in olive oil, is a monounsaturated fatty acid. What does the term "monounsaturated" mean?

INTRODUCTION

For many years, it was suspected that there existed a gas that stimulated fruit ripening and had other effects on plants. The ancient Chinese observed that their fruit ripened more quickly if incense was burned in the same room. Early in the last century, shippers realized that they could not store oranges and bananas on the same ships because some "emanation" given off by the oranges caused the bananas to ripen too early.

Puerto Rican pineapple growers and Philippine mango growers independently developed a traditional practice of building bonfires near their crops. They believed that the smoke caused the plants to bloom synchronously.

In the mid-nineteenth century, streetlights were fueled with natural gas. Occasionally the pipes leaked, releasing gas into the atmosphere. On some of these occasions, the leaves fell from all the shade trees in the region surrounding the gas leak.

What is the gas responsible for these diverse effects on plants? In 1934, R. Gane demonstrated that the simple alkene ethene (ethylene) was the "emanation" responsible for fruit ripening. More recently, it has been shown that ethene induces and synchronizes flowering in pineapples and mangoes, induces senescence (aging) and loss of leaves in trees, and effects a wide variety of other responses in various plants.

We can be grateful to ethene for the fresh, unbruised fruits that we purchase at the grocery store. These fruits are picked when they are not yet ripe, while they are still firm. They then can be shipped great distances and gassed with ethene when they reach their destination. Under the influence of ethene, the fruit ripens for display in the store.

In this chapter, we will study the **unsaturated hydrocarbons.** This group of organic compounds includes the alkenes, such as ethene, which all contain at least one carbon-carbon double bond; the alkynes, which all contain at least one carbon-carbon triple bond; and aromatic compounds, particularly stable compounds that contain a benzene ring. The benzene ring is often depicted as having alternating double and single bonds. This arrangement is called a *conjugated system* of double bonds.

11.1   Alkenes and Alkynes: Structure and Physical Properties

LEARNING GOAL

1 Describe the physical properties of alkenes and alkynes.

Fatty acids are long hydrocarbon chains having a carboxyl group at the end. Thus by definition they are carboxylic acids. See Chapters 14 and 17.

Many important biological molecules are characterized by the presence of double bonds or a linear or cyclic conjugated system of double bonds (Figure 11.1). For instance, we classify fatty acids as either monounsaturated (having one double bond), polyunsaturated (having two or more double bonds), or saturated (having single bonds only). Vitamin A (retinol), a vitamin required for vision, contains a nine-carbon conjugated hydrocarbon chain. Vitamin K, a vitamin required for blood clotting, contains an aromatic ring.

Alkenes and **alkynes** are unsaturated hydrocarbons. The characteristic functional group of an alkene is the carbon-carbon double bond. The functional group that characterizes the alkynes is the carbon-carbon triple bond. The general formulas shown below compare the structures of alkanes, alkenes, and alkynes.

	Alkane	Alkene	Alkyne
General formulas:	C_nH_{2n+2}	C_nH_{2n}	C_nH_{2n-2}
Structural formulas:	$H-\overset{\displaystyle H}{\underset{\displaystyle H}{C}}-\overset{\displaystyle H}{\underset{\displaystyle H}{C}}-H$	$\overset{H}{_{H}}C=C\overset{H}{_{H}}$	$H-C\equiv C-H$
	Ethane (ethane)	Ethene (ethylene)	Ethyne (acetylene)
Molecular formulas:	C_2H_6	C_2H_4	C_2H_2
Condensed formulas:	CH_3CH_3	$H_2C=CH_2$	$HC\equiv CH$

Figure 11.1 (a) Line formula of the eighteen-carbon monounsaturated fatty acid oleic acid. (b) Line formula of vitamin A, which is required for vision. Notice that the carbon chain of vitamin A is a conjugated system of double bonds. (c) Line formula of vitamin K, a lipid-soluble vitamin required for blood clotting. The six-member ring with the circle represents a benzene ring. See Figure 11.5 for other representations of the benzene ring.

Alkanes, alkenes, and alkynes of the same carbon chain length differ in the number of hydrogen atoms. Alkenes contain two fewer hydrogens than alkanes of the same carbon chain length because electrons of two carbon atoms form the second bond between them, rather than forming bonds with two hydrogen atoms. Similarly, alkynes contain two fewer hydrogens than alkenes of the same carbon chain length.

In alkanes, the central carbon has four single bonds, giving it a tetrahedral molecular geometry. When carbon is bonded by one double bond and two single bonds, as in ethene (an alkene), the molecule is *trigonal planar,* because all atoms lie in a single plane. Each bond angle is approximately 120°. When two carbon atoms are bonded by a triple bond, as in ethyne (an alkyne), each bond angle is 180°. Thus, the molecule is linear, and all atoms are positioned in a straight line. For comparison, examples of a five-carbon alkane, alkene, and alkyne are shown in Figure 11.2.

As we saw for alkanes, the melting points and boiling points of the alkenes and alkynes are the result of London dispersion forces. These attractive forces increase as the molar mass of the molecule and the number of electrons increase. Thus, the London dispersion forces are stronger and the attraction is greater for molecules having a larger surface area. We can predict that the longer, higher molar mass alkenes and alkynes exhibit stronger London dispersion forces and therefore higher boiling and melting points. The melting points and boiling points of several alkenes and alkynes presented in Table 11.1 reveal that our prediction is true. Ethene, with

Bananas are picked and shipped while still green. They are treated with the fruit-ripening agent, ethene, once they reach the grocery store. Describe the geometry and bonding of ethene. What is the common name of ethene?

TABLE 11.1 Physical Properties of Selected Alkenes and Alkynes

Name	Molecular Formula	Structural Formula	Melting Point (°C)	Boiling Point (°C)
Ethene	C_2H_4	$CH_2{=}CH_2$	−169.1	−103.7
Propene	C_3H_6	$CH_2{=}CHCH_3$	−185.0	−47.6
1-Butene	C_4H_8	$CH_2{=}CHCH_2CH_3$	−185.0	−6.1
Methylpropene	C_4H_8	$CH_2{=}C(CH_3)_2$	−140.0	−6.6
Ethyne	C_2H_2	$HC{\equiv}CH$	−81.8	−84.0
Propyne	C_3H_4	$HC{\equiv}CCH_3$	−101.5	−23.2
1-Butyne	C_4H_6	$HC{\equiv}CCH_2CH_3$	−125.9	8.1
2-Butyne	C_4H_6	$CH_3C{\equiv}CCH_3$	−32.3	27.0

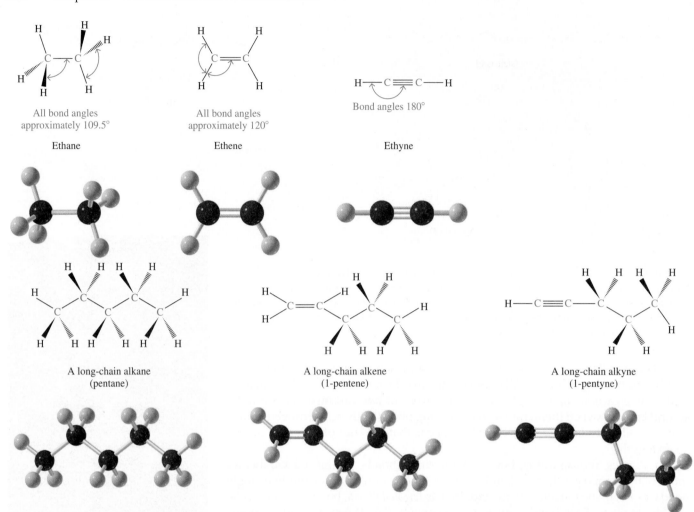

Figure 11.2 Three-dimensional drawings and ball-and-stick models of typical alkanes, alkenes, and alkynes.

a boiling point of −103.7°C, is a gas at room temperature. The boiling point of the ten-carbon alkene, 1-decene, is 172°C, and it is a liquid at room temperature.

Alkenes, alkynes, and aromatic compounds are less dense than water and are nonpolar. Since they are nonpolar, the "like dissolves like" rule tells us that they are not soluble in water. Like the alkanes, they are very soluble in nonpolar solvents such as other hydrocarbons.

11.2   Alkenes and Alkynes: Nomenclature

Alkene and alkyne names are derived from the names of the alkanes of the same carbon chain length. The same prefixes are used (*meth-*, *eth-*, etc.), but the suffixes are different (*-ene* for alkenes and *-yne* for alkynes). To determine the name of an alkene or alkyne using the IUPAC Nomenclature System, use the following simple rules:

- Count the number of carbon atoms in the longest continuous carbon chain containing the double bond (alkenes) or triple bond (alkynes). Name the alkane with the same number of carbon atoms. This is the parent compound.
- Replace the *-ane* ending of the alkane with the *-ene* ending for an alkene or the *-yne* ending for an alkyne. For example:

2 Draw the structures and write the IUPAC names for simple alkenes and alkynes.

$$CH_3-CH_3 \qquad CH_2=CH_2 \qquad CH\equiv CH$$
Ethane   Ethene   Ethyne

$$CH_3-CH_2-CH_3 \qquad CH_2=CH-CH_3 \qquad CH\equiv C-CH_3$$
Propane   Propene   Propyne

- Number the parent chain to give the double or triple bond the lowest number. For example:

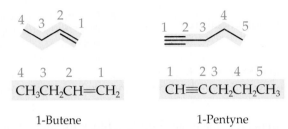

$$CH_3CH_2CH{=}CH_2$$

1-Butene

$$CH{\equiv}CCH_2CH_2CH_3$$

1-Pentyne

- Determine the name and carbon number of each group bonded to the parent alkene or alkyne, and place the name and number in front of the name of the parent compound. Remember that with alkenes and alkynes the double or triple bond takes precedence over a halogen or alkyl group, as shown in the following examples:

Remember, it is the position of the double bond, not the substituent, that determines the numbering of the carbon chain.

$$CH_3CH{=}CClCH_3$$

2-Chloro-2-butene

$$CH_3CHBr\,C{\equiv}CCH_2CH_3$$

2-Bromo-3-hexyne

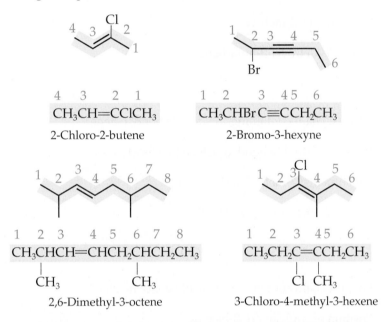

$$CH_3CHCH{=}CHCH_2CHCH_2CH_3$$
$$\quad |\qquad\qquad\qquad |$$
$$\quad CH_3\qquad\qquad\quad CH_3$$

2,6-Dimethyl-3-octene

$$CH_3CH_2C{=}CCH_2CH_3$$
$$\qquad\qquad | \ |$$
$$\qquad\qquad Cl\ CH_3$$

3-Chloro-4-methyl-3-hexene

- Alkenes having more than one double bond are called alkadienes (two double bonds) or alkatrienes (three double bonds). Determine which double bond is nearest one end of the chain. Number the carbon chain by using the number of the first carbon of each double bond, using the lowest numbers. Use the appropriate suffix (-diene or -triene), as seen in these examples:

Alkenes with many double bonds are often referred to as polyenes (poly—many, enes— double bonds).

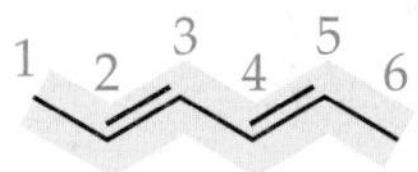

$$CH_3CH{=}CHCH{=}CHCH_3$$

2,4-Hexadiene

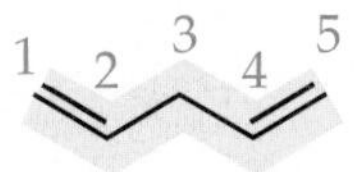

$$CH_2{=}CHCH_2CH{=}CH_2$$

1,4-Pentadiene

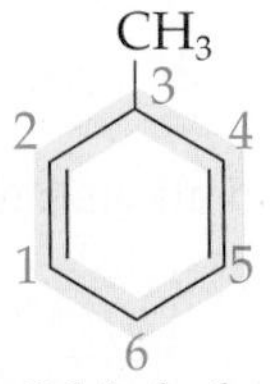

3-Methyl-
1,4-cyclohexadiene

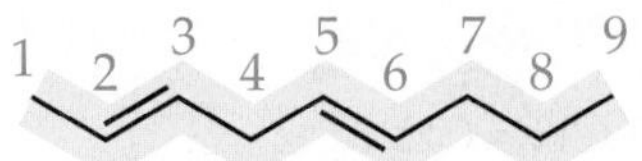

$$CH_3CH{=}CHCH_2CH{=}CHCH_2CH_2CH_3$$

2,5-Nonadiene

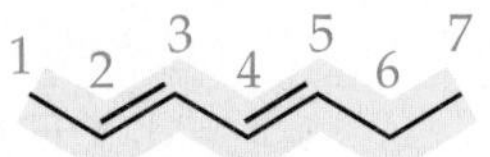

$$CH_3CH{=}CHCH{=}CHCH_2CH_3$$

2,4-Heptadiene

EXAMPLE 11.1 | Naming Alkenes and Alkynes Using IUPAC Nomenclature

LEARNING GOAL

Name the following alkenes and alkynes using IUPAC nomenclature.

2 Draw the structures and write the IUPAC names for simple alkenes and alkynes.

Solution

To determine the IUPAC name, identify the parent chain, number it to give the lowest possible positions for the carbon or carbons with the double bonds, and finally, identify and number each functional group.

The common name of the first molecule below is isoprene. It is the major building block for many important biological molecules, including cholesterol and other steroids, β-carotene, and vitamins A, D, E, and K.

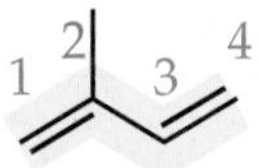

$CH_2=CCH=CH_2$
$\quad 1\quad 23\quad 4$
with CH_3 on carbon 2

Isoprene

Longest chain containing the double bond: butene

Positions of the double bonds: 1,3-butadiene

Substituents: 2-methyl

Name: 2-Methyl-1,3-butadiene

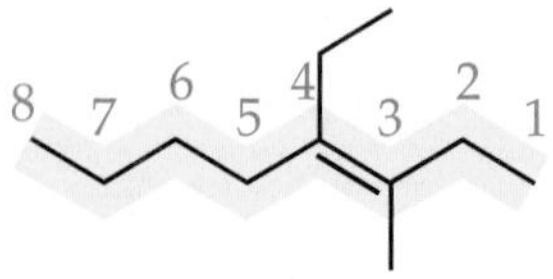

$CH_3CH_2CH_2CH_2C=CCH_2CH_3$
$\quad 8\quad 7\quad 6\quad 5\quad 4|\quad 32\quad 1$
with CH_2CH_3 on carbon 4 and CH_3 on carbon 3

Longest chain containing the double bond: octene

Position of double bond: 3-octene (*not* 5-octene)

Substituents: 3-methyl and 4-ethyl

Name: 4-Ethyl-3-methyl-3-octene

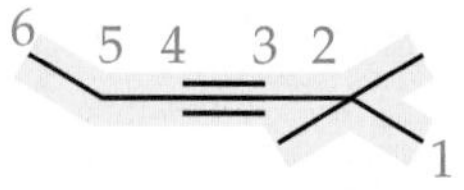

$CH_3CH_2C\equiv CCCH_3$
$\quad 6\quad 5\quad 4\quad 32|1$
with CH_3 groups (2,2-dimethyl)

Longest chain containing the triple bond: hexyne

Position of triple bond: 3-hexyne

Substituents: 2,2-dimethyl

Name: 2,2-Dimethyl-3-hexyne

Practice Problem 11.1

Name each of the following alkenes and alkynes using IUPAC nomenclature.

a. $CH_3CH=CHCHCH_2CHCH_3$ (with CH_3 and Cl substituents)

b. $CH\equiv CCH_2C\equiv CH$

c. $CH_3CH_2C\equiv CCHCH_2CH_3$ (with Br substituent)

d. $CH_3CH=CHCH_2CH_2CH=CH_2$

▶ For Further Practice: **Questions 11.43a–c and 11.44a–c.**

| EXAMPLE 11.2 | Naming Cycloalkenes Using IUPAC Nomenclature |

LEARNING GOAL

2 Draw the structures and write the IUPAC names for simple alkenes and alkynes.

Name the following cycloalkenes using IUPAC nomenclature.

Solution

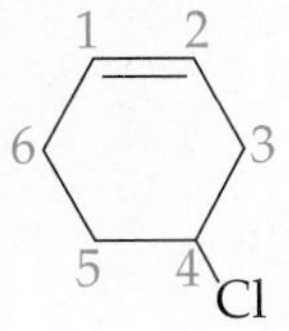

Parent chain: cyclohexene

Position of double bond: carbon-1 (carbons of the double bond are numbered 1 and 2)

Substituents: 4-chloro

Name: 4-Chlorocyclohexene

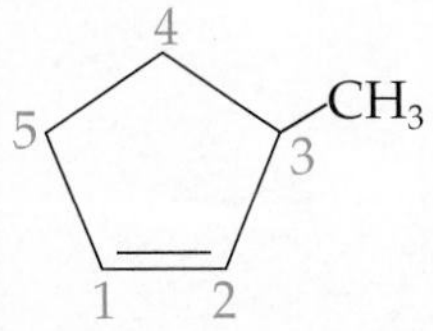

Parent chain: cyclopentene

Position of double bond: carbon-1

Substituent: 3-methyl

Name: 3-Methylcyclopentene

Practice Problem 11.2

Name each of the following cycloalkenes using IUPAC nomenclature.

a. 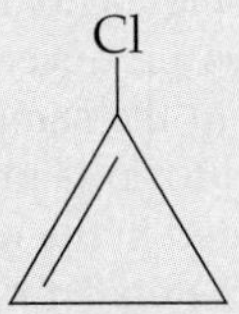   b. 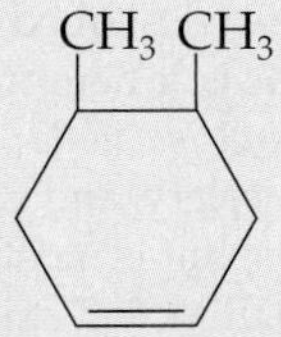 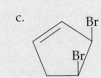   c.   d.

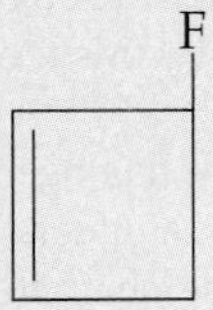

▶ For Further Practice: **Questions 11.43d and 11.44d.**

Question 11.1 Draw a condensed formula and line formula for each of the following compounds:

a. 1-Bromo-3-hexyne
b. 2-Butyne
c. Dichloroethyne
d. 9-Iodo-1-nonyne

Question 11.2 Name the following compounds using the IUPAC Nomenclature System:

a. $CH_3C{\equiv}CCH_2CH_3$
b. $CH_3CH_2CHBrCHBrCH_2C{\equiv}CH$
c. $CH_3CH(CH_3)CCl{=}C(CH_3)CH(CH_3)_2$
d. $CH_3CH(CH_2CH_3)C{\equiv}CCHClCH_3$

11.3 Geometric Isomers: A Consequence of Unsaturation

LEARNING GOAL

3 Write the names and draw the structures of simple geometric isomers of alkenes.

The carbon-carbon double bond is rigid because of the shapes of the orbitals involved in its formation. As a result, rotation around the carbon-carbon double bond is restricted. In Section 10.3, we saw that the rotation around the carbon-carbon bonds of cycloalkanes is also restricted. As a consequence, these molecules form geometric or *cis-trans* isomers. The *cis*-isomers of cycloalkanes have substituent groups on the same side of the ring (*Latin, cis,* "on the same side").

A Medical Perspective

Killer Alkynes in Nature

There are many examples of alkynes that are beneficial to humans. Among these are *parsalmide*, a pain reliever, *pargyline*, an antihypertensive, and *17-ethynylestradiol*, a synthetic estrogen that is used as an oral contraceptive.

But in addition to these medically useful alkynes, there are in nature a number that are toxic. Some are extremely toxic to mammals, including humans; others are toxic to fungi, fish, or insects. All of these compounds are plant products that may help protect the plant from destruction by predators.

Capillin is produced by the oriental wormwood plant. Research has shown that a dilute solution of capillin inhibits the growth of certain fungi. Since fungal growth can damage or destroy a plant, the ability to make capillin may provide a survival advantage to the plants. Perhaps it may one day be developed to combat fungal infections in humans.

Ichthyothereol is a fast-acting poison commonly found in plants referred to as fish-poison plants. Ichthyothereol is a very toxic polyacetylenic alcohol that inhibits energy production in the mitochondria. Latin American native tribes use these plants to coat the tips of the arrows used to catch fish. Although ichthyothereol is poisonous to the fish, fish caught by this method pose no risk to the people who eat them!

An extract of the leaves of English ivy has been reported to have antibacterial, analgesic, and sedative effects. The compound thought to be responsible for these characteristics, as well as antifungal activity, is *falcarinol*. Falcarinol, isolated from a tree in Panama, also has been reported by the Molecular Targets Drug Discovery Program to have antitumor activity. Perhaps one day this compound, or a derivative of it, will be useful in treating cancer in humans.

Falcarinol is extracted from English ivy, like that covering this stone house.

Cicutoxin has been described as the most lethal toxin native to North America. It is a neurotoxin that is produced by the water hemlock (*Cicuta maculata*), which is in the same family of plants as parsley, celery, and carrots. Cicutoxin is present in all parts of the plants, but is most concentrated in the root. Eating a portion as small as 2–3 square centimeters (cm^2) can be fatal to adults. Cicutoxin acts directly on the nervous system. Signs and symptoms of cicutoxin poisoning include dilation of pupils, muscle twitching, rapid pulse and breathing, violent convulsions, coma, and death. Onset of symptoms is rapid and death may occur within 2 to 3 hours (h). No antidote exists for

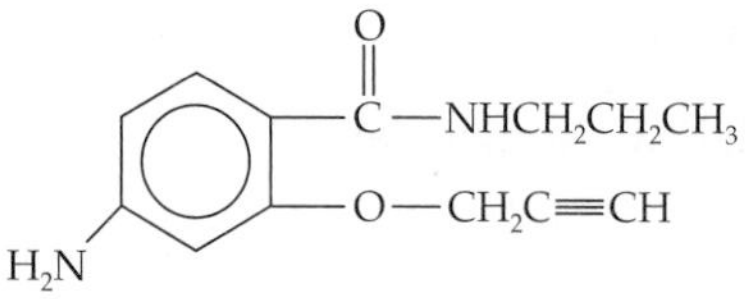

Parsalmide

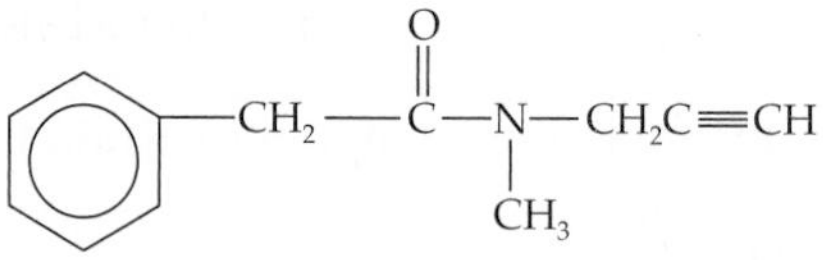

Pargyline

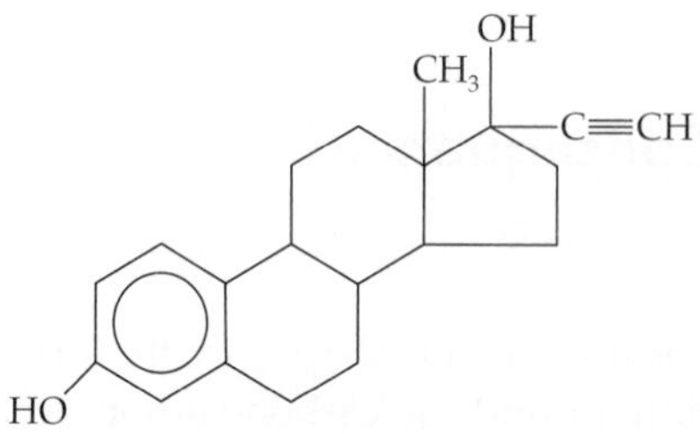

17-Ethynylestradiol

Alkynes used for medicinal purposes.

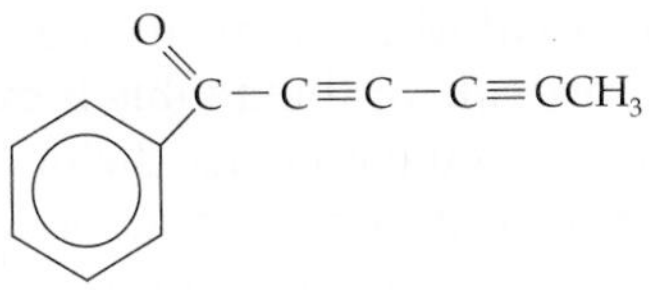

Capillin

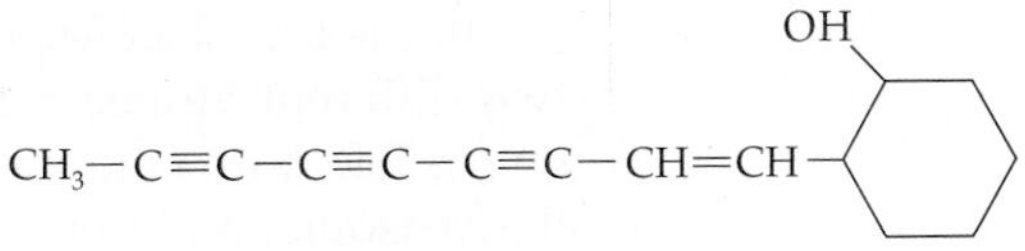

Ichthyothereol

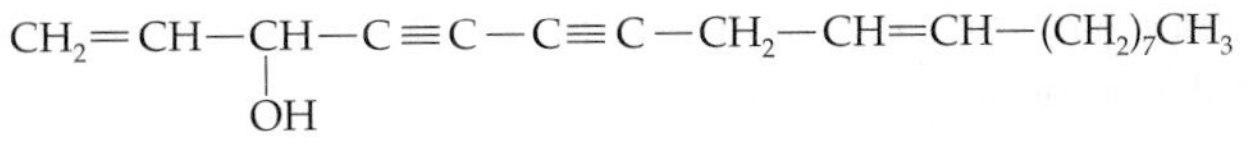

$$CH_2=CH-CH-C\equiv C-C\equiv C-CH_2-CH=CH-(CH_2)_7CH_3$$
$$\overset{|}{OH}$$

Falcarinol

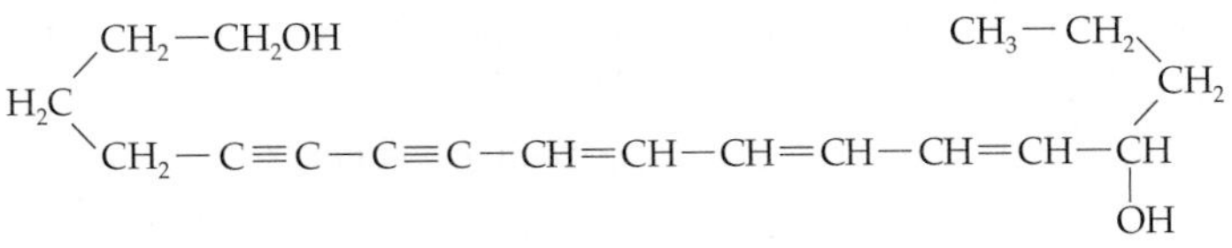

Cicutoxin

Alkynes that exhibit toxic activity.

Cicuta maculata, or water hemlock, produces the most deadly toxin indigenous to North America.

cicutoxin poisoning. The only treatment involves controlling convulsions and seizures in order to preserve normal heart and lung function. Fortunately, cicutoxin poisoning is a very rare occurrence. Occasionally animals may graze on the plants in the spring, resulting in death within 15 minutes (min). Humans seldom come into contact with the water hemlock. The most recent cases have involved individuals foraging for wild ginseng, or other wild roots, and mistaking the water hemlock root for an edible plant.

For Further Understanding

▸ Circle and name the functional groups in parsalmide and pargyline.
▸ The fungus *Tinea pedis* causes athlete's foot. Describe an experiment you might carry out to determine whether capillin might be effective against athlete's foot.

The *trans*-isomers of cycloalkanes have substituent groups located on opposite sides of the ring (*Latin, trans,* "across from").

In alkenes, **geometric isomers** occur when there are two different groups or atoms on each of the carbon atoms attached by the double bond. If both groups or atoms are on the same side of the double bond, the molecule is a *cis*-isomer. If the groups or atoms are on opposite sides of the double bond, the molecule is a *trans*-isomer.

Consider the two isomers of 1,2-dichloroethene:

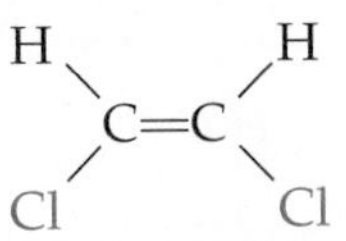

> Restricted rotation around double bonds is partially responsible for the conformation and hence the activity of many biological molecules that we will study later.

cis-1,2-Dichloroethene 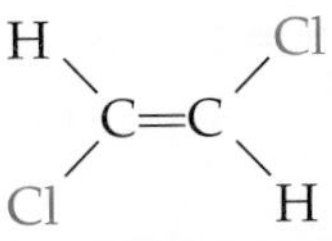   *trans*-1,2-Dichloroethene

In these molecules, each carbon atom of the double bond is also bonded to two different atoms: a hydrogen atom and a chlorine atom. In the molecule on the left, both chlorine atoms are on the same side of the double bond; this is the *cis*-isomer and the complete name for this molecule is *cis*-1,2-dichloroethene. In the molecule on the right, the chlorine atoms are on opposite sides of the double bond; this is the *trans*-isomer, and the complete name of this molecule is *trans*-1,2-dichloroethene.

If one of the two carbon atoms of the double bond has two identical substituents, there are no *cis-trans* isomers for that molecule. Consider the example of 1,1-dichloroethene:

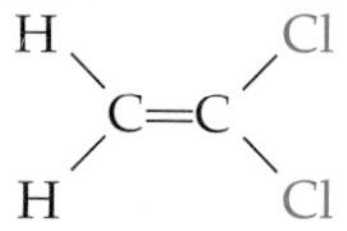

$$
\begin{array}{ccc}
H & & Cl \\
 & C=C & \\
H & & Cl \\
\end{array}
$$

1,1-Dichloroethene

EXAMPLE 11.3   Identifying *cis*- and *trans*-Isomers of Alkenes

LEARNING GOAL

Two isomers of 2-butene are shown below. Which is the *cis*-isomer and which is the *trans*-isomer?

3 Write the names and draw the structures of simple geometric isomers of alkenes.

$$
\begin{array}{ccc}
H & & H \\
 & C=C & \\
H_3C & & CH_3 \\
\end{array}
$$

$$
\begin{array}{ccc}
H & & CH_3 \\
 & C=C & \\
H_3C & & H \\
\end{array}
$$

Solution

As we saw with cycloalkanes, the prefixes *cis* and *trans* refer to the placement of the substituents attached to a bond that cannot undergo free rotation. In the case of alkenes, it is the groups attached to the carbon-carbon double bond (in this example, the H and CH$_3$ groups). When the groups are on the same side of the double bond, as in the structure on the left, the prefix *cis* is used. When the groups are on the opposite sides of the double bond, as in the structure on the right, *trans* is the appropriate prefix.

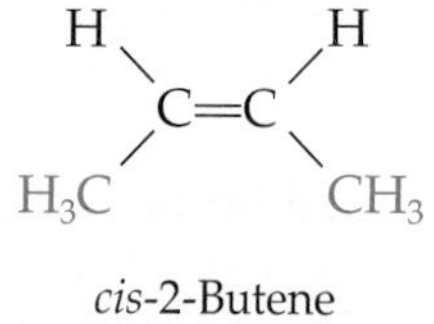

$$
\begin{array}{ccc}
H & & H \\
 & C=C & \\
H_3C & & CH_3 \\
\end{array}
$$
cis-2-Butene

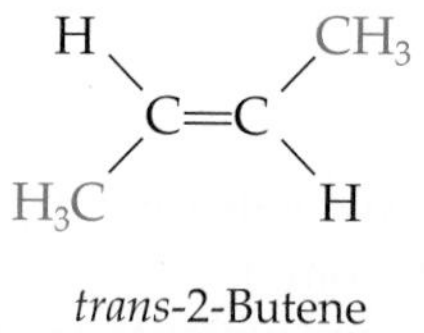

$$
\begin{array}{ccc}
H & & CH_3 \\
 & C=C & \\
H_3C & & H \\
\end{array}
$$
trans-2-Butene

Practice Problem 11.3

Which of the following alkenes are *cis*-isomers and which are *trans*-isomers?

a. 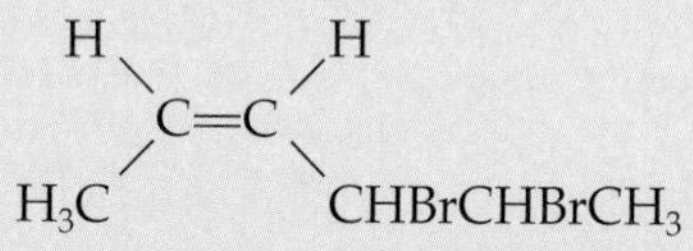
$$
\begin{array}{ccc}
H & & H \\
 & C=C & \\
H_3C & & CHBrCHBrCH_3 \\
\end{array}
$$

b.
$$
\begin{array}{ccc}
CH_3CH_2 & & Br \\
 & C=C & \\
Br & & CH_2CH_3 \\
\end{array}
$$

c.
$$
\begin{array}{ccc}
Cl & & Cl \\
 & C=C & \\
H_3C & & CH_2CH_2CH_3 \\
\end{array}
$$

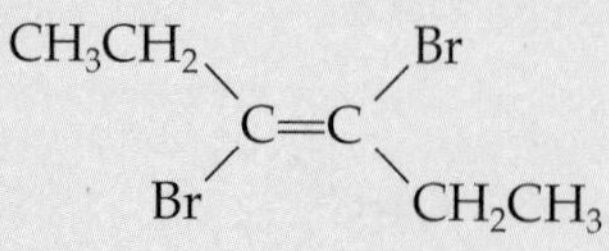

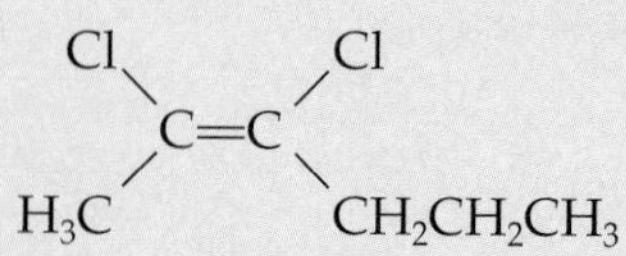

▶ For Further Practice: **Questions 11.47 and 11.49.**

In alkenes, the orientation of the carbon atoms of the parent chain relative to the double bond determines the *cis-* or *trans-* configuration of the alkene. If the carbon atoms of the parent chain are on opposite sides of the double bond, the alkene is in the *trans-* configuration.

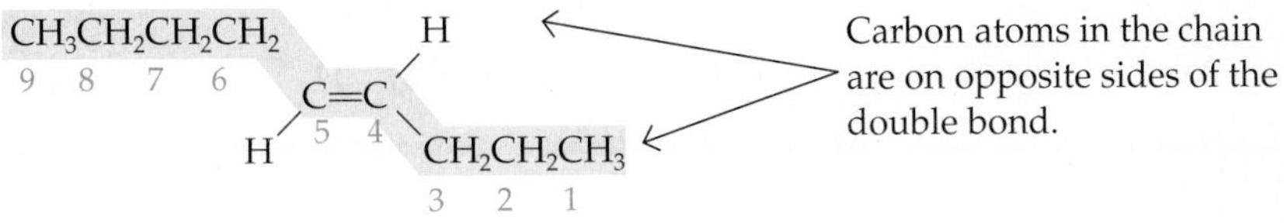

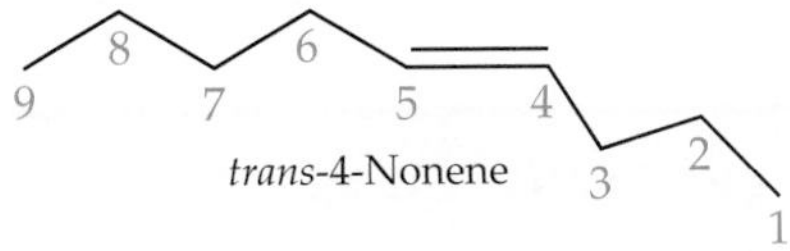

trans-4-Nonene

If the carbon atoms of the parent chain are on the same side of the double bond, the alkene is in the *cis-* configuration. This can be seen in the *cis-* and *trans-* configurations of 4-nonene shown here:

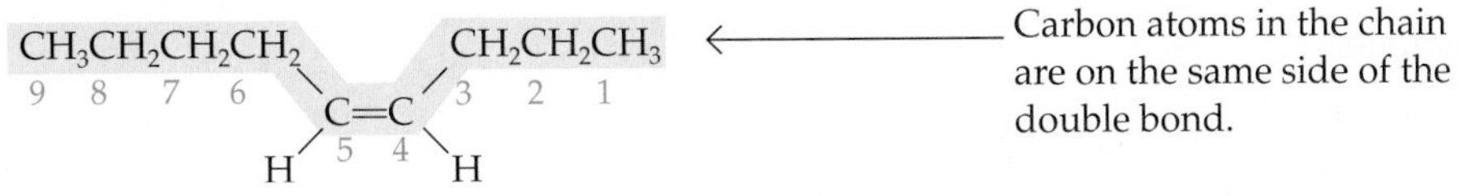

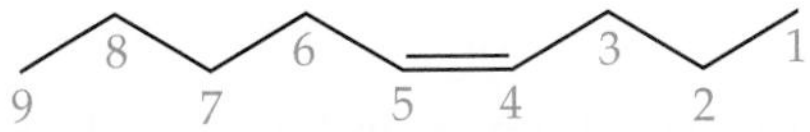

cis-4-Nonene

EXAMPLE 11.4   **Naming *cis-* and *trans*-Isomers of Alkenes**

LEARNING GOAL

Name the following geometric isomers.

3 Write the names and draw the structures of simple geometric isomers of alkenes.

Solution

The longest chain of carbon atoms in each of the following molecules is highlighted in yellow. *The chain must also contain the carbon-carbon double bond.* The orientation of the carbon atoms of the parent chain relative to the double bond is used in determining the appropriate prefix, *cis* or *trans*, to be used in naming each of the molecules.

Parent chain: heptene

Position of double bond: 3-

Substituents: 3,4-dichloro

Configuration: *trans* (the carbon atoms of the parent chain are on opposite sides of the double bond)

Name: *trans*-3,4-Dichloro-3-heptene

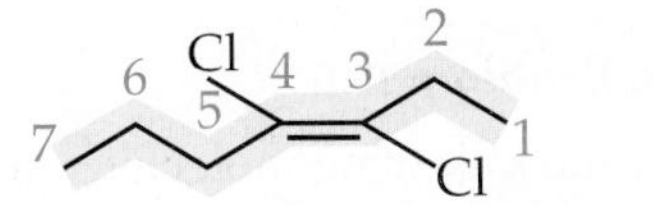

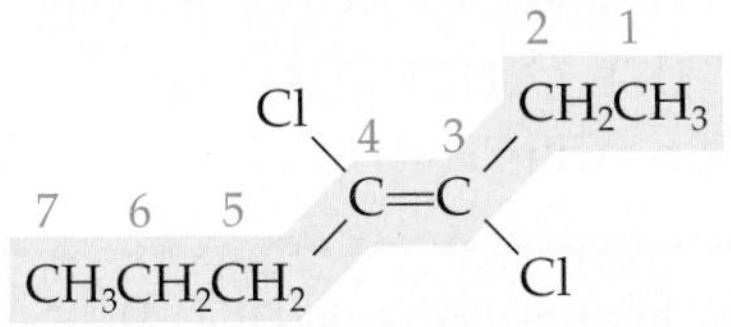

Parent chain: octene

Position of double bond: 3-

Substituents: 3-methyl

Configuration: *cis* (the carbon atoms of the parent chain are on the same side of the double bond)

Name: *cis*-3-Methyl-3-octene

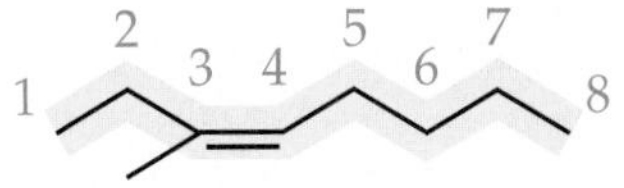

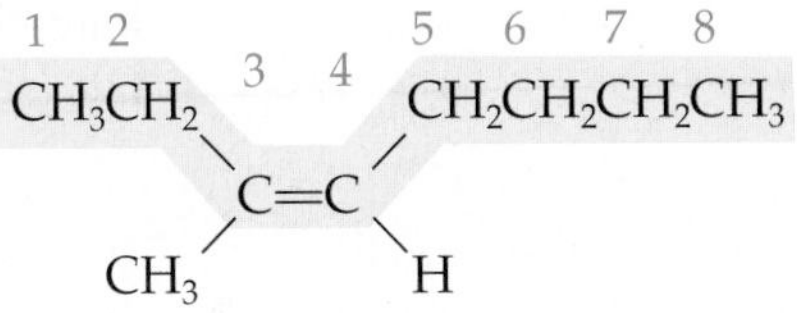

Continued...

Practice Problem 11.4

Name each of the following geometric isomers:

a. 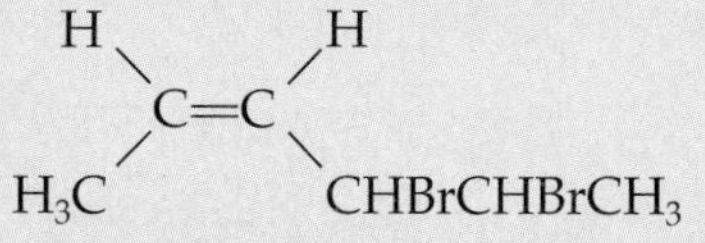
$$\underset{H_3C}{\overset{H}{\diagdown}}C{=}C\underset{CHBrCHBrCH_3}{\overset{H}{\diagup}}$$

b. 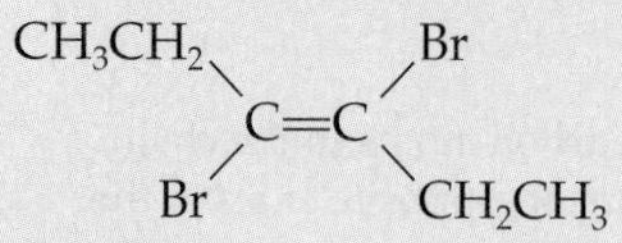
$$\underset{Br}{\overset{CH_3CH_2}{\diagdown}}C{=}C\underset{CH_2CH_3}{\overset{Br}{\diagup}}$$

c.
$$\underset{H_3C}{\overset{Cl}{\diagdown}}C{=}C\underset{CH_2CH_2CH_3}{\overset{Cl_3}{\diagup}}$$

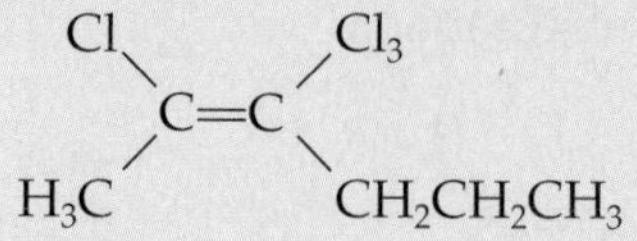

▶ For Further Practice: **Questions 11.51 and 11.52.**

EXAMPLE 11.5   **Identifying Geometric Isomers**

LEARNING GOAL

Determine whether each of the following molecules can exist as *cis-trans* isomers: (a) 1-pentene and (b) 3-methyl-2-pentene.

3 Write the names and draw the structures of simple geometric isomers of alkenes.

Solution

a. Examine the structure of 1-pentene,

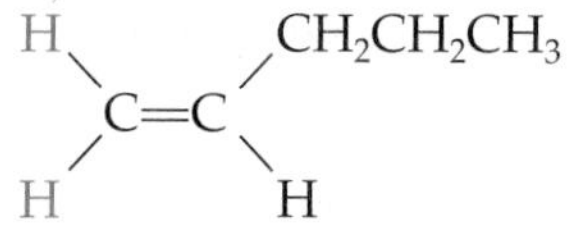

$$\underset{H}{\overset{H}{\diagdown}}C{=}C\underset{H}{\overset{CH_2CH_2CH_3}{\diagup}}$$

We see that carbon-1 is bonded to two hydrogen atoms, rather than to two different substituents. In this case, there can be no *cis-trans* isomers.

b. Examination of the structure of 3-methyl-2-pentene reveals that both a *cis-* and *trans*-isomer can be drawn.

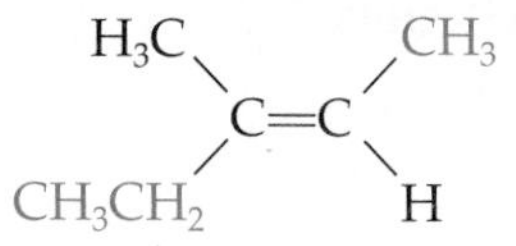

$$\underset{CH_3CH_2}{\overset{H_3C}{\diagdown}}C{=}C\underset{H}{\overset{CH_3}{\diagup}}\qquad\underset{CH_3CH_2}{\overset{H_3C}{\diagdown}}C{=}C\underset{CH_3}{\overset{H}{\diagup}}$$

trans-3-Methyl-2-pentene 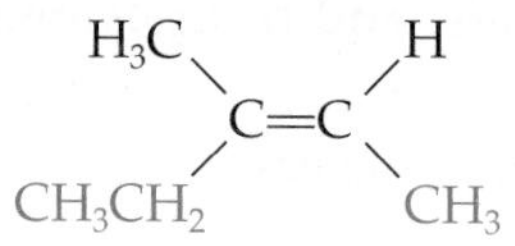   *cis*-3-Methyl-2-pentene

Each of the carbon atoms involved in the double bond is attached to two different groups. In *trans*-3-methyl-2-pentene, the carbons of the parent chain are on opposite sides of the double bond. In the *cis*-configuration, the carbons of the parent chain are on the same side of the double bond.

Practice Problem 11.5

Which of the following molecules can exist as *cis*- and *trans*-isomers? Draw the *cis*- and *trans*-isomers, where possible. For those molecules that cannot exist as *cis*- and *trans*-isomers, explain why.

a. $CH_3CH{=}CHCH_2CH_3$

b. $CBr_2{=}CBrCH_2CH_2CH_3$

c. $CH_2{=}CHCH_3$

d. $CH_3CBr{=}CBrCH_2CH_2CH_2CH_3$

▶ For Further Practice: **Questions 11.49 and 11.50.**

The recent debate over the presence of *cis*- and *trans*-isomers of fatty acids in our diet points out the relevance of geometric isomers to our lives. Fatty acids are long-chain carboxylic acids found in vegetable oils (unsaturated fats) and animal fats (saturated fats). Oleic acid,

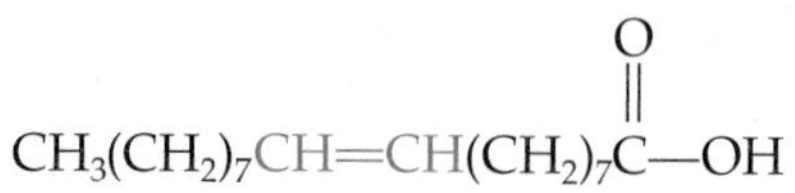

$$CH_3(CH_2)_7CH{=}CH(CH_2)_7\overset{\overset{O}{\|}}{C}{-}OH$$

is the naturally occurring fatty acid in olive oil. Its IUPAC name, *cis*-9-octadecenoic acid, reveals that this is a *cis*-fatty acid.

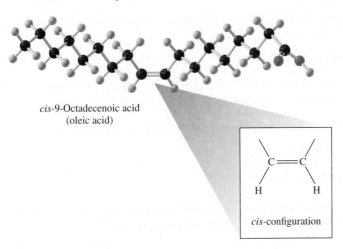

cis-9-Octadecenoic acid
(oleic acid)

cis-configuration

The *cis*-isomer of oleic acid is V-shaped as a result of the configuration of the double bond, and the molecule is more flexible. Its geometric isomer, *trans*-9-octadecenoic acid is a rigid linear molecule.

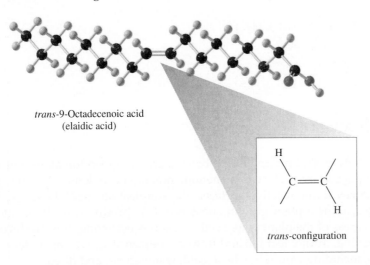

trans-9-Octadecenoic acid
(elaidic acid)

trans-configuration

The majority of *trans*-fatty acids found in the diet result from hydrogenation, which is a reaction used to convert oils into solid fats, such as margarine. It has recently been reported that *trans*-fatty acids in the diet elevate levels of "bad" or LDL cholesterol and lower the levels of "good" or HDL cholesterol, thereby increasing the risk of heart disease. Other studies suggest that *trans*-fatty acids may also increase the risk of type 2 diabetes.

cis- and trans-*Fatty acids will be discussed in greater detail in Chapter 17. Hydrogenation is described in Section 11.5.*

Question 11.3 In each of the following pairs of molecules, identify the *cis*-isomer and the *trans*-isomer.

a.

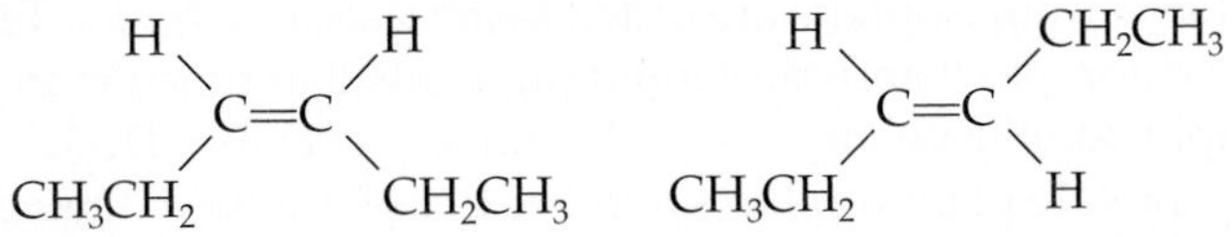

$$\underset{\text{CH}_3\text{CH}_2}{\overset{\text{H}}{\diagdown}}\text{C}=\text{C}\underset{\text{CH}_2\text{CH}_3}{\overset{\text{H}}{\diagup}} \qquad \underset{\text{CH}_3\text{CH}_2}{\overset{\text{H}}{\diagdown}}\text{C}=\text{C}\underset{\text{H}}{\overset{\text{CH}_2\text{CH}_3}{\diagup}}$$

b.

$$\underset{\text{H}_3\text{C}}{\overset{\text{Br}}{\diagdown}}\text{C}=\text{C}\underset{\text{Br}}{\overset{\text{CH}_3}{\diagup}} \qquad \underset{\text{H}_3\text{C}}{\overset{\text{Br}}{\diagdown}}\text{C}=\text{C}\underset{\text{CH}_3}{\overset{\text{Br}}{\diagup}}$$

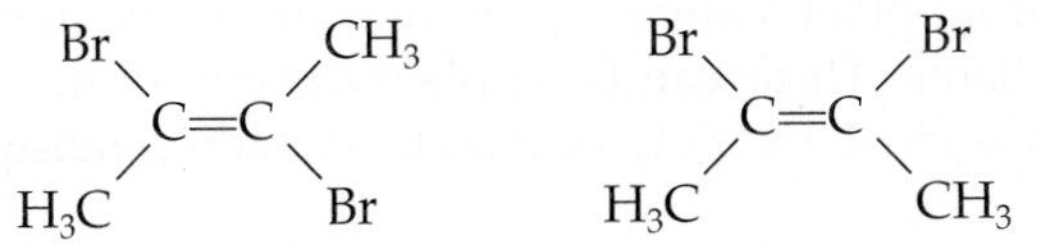

Question 11.4   Provide the complete IUPAC name for each of the compounds in Question 11.3.

Question 11.5   Which of the following molecules can exist as both *cis-* and *trans-* isomers? Explain your reasoning.

   a. $CH_3CH_2CCl= C(CH_2CH_3)_2$ b. $CH_3CH_2CBr= CBr_2$ c. $CH_3CCl= CClCH_3$

Question 11.6   Draw each of the *cis-trans* isomers in Question 11.5, and provide the complete names using the IUPAC Nomenclature System.

Question 11.7   Draw condensed formulas for each of the following compounds:

   a. *cis*-3-Octene
   b. *trans*-5-Chloro-2-hexene
   c. *trans*-2,3-Dichloro-2-butene

Question 11.8   Name each of the following compounds, using the IUPAC system. Be sure to indicate *cis* or *trans* where applicable.

a. 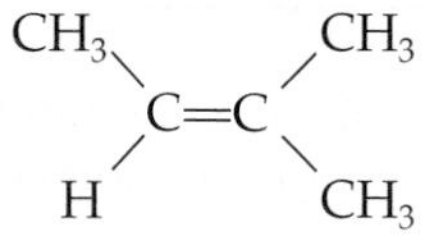
$$CH_3 \quad\quad CH_3$$
$$\diagdown C=C \diagup$$
$$H \quad\quad CH_3$$

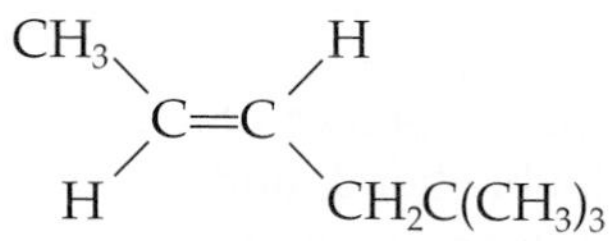

b. 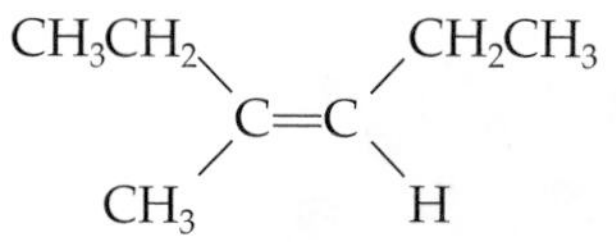
$$CH_3CH_2 \quad\quad CH_2CH_3$$
$$\diagdown C=C \diagup$$
$$CH_3 \quad\quad H$$

c.
$$CH_3 \quad\quad H$$
$$\diagdown C=C \diagup$$
$$H \quad\quad CH_2C(CH_3)_3$$

11.4   Alkenes in Nature

Folklore tells us that placing a ripe banana among green tomatoes will speed up the ripening process. In fact, this phenomenon has been demonstrated experimentally. The key to the reaction is *ethene,* the simplest alkene. Ethene, produced by ripening fruit, is a plant growth substance. It is produced in the greatest abundance in areas of the plant where cell division is occurring. It is produced during fruit ripening, during leaf fall and flower senescence, as well as under conditions of stress, including wounding, heat, cold, water stress, and disease.

   Polyenes are alkenes with several double bonds, and there are a surprising number of these molecules found in nature. Although they have wildly different properties and functions, they are built from one or more five-carbon units called *isoprene.*

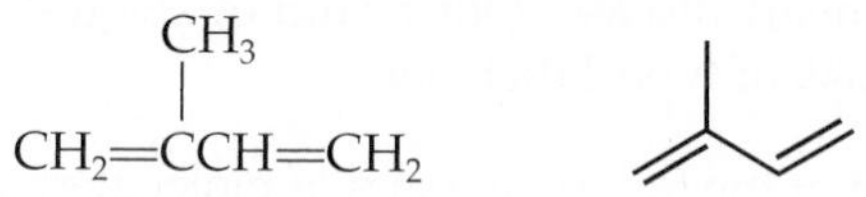

$$CH_3$$
$$|$$
$$CH_2=CCH=CH_2$$

Isoprene

The molecules produced from isoprene are called *isoprenoids,* or *terpenes.* Terpenes include the steroids; chlorophyll and carotenoid pigments that function in photosynthesis; and the lipid-soluble vitamins A, D, E, and K (see Figure 11.1).

   Many other terpenes are plant products familiar to us because of their distinctive aromas. *Geraniol,* the familiar scent of geraniums, is a molecule made up of two isoprene units. Purified from plant sources, geraniol is the active ingredient in several natural insect repellants. These can be applied directly to the skin to provide 4 hours of protection against a variety of insects, including mosquitoes, ticks, and fire ants.

Geraniol
(Roses and geraniums)

D-*Limonene* is the most abundant component of the oil extracted from the rind of citrus fruits. Because of its pleasing orange aroma, D-limonene is used as a flavor and fragrance additive in foods. However, the most rapidly expanding use of the compound is as a solvent. In this role, D-limonene can be used in place of more toxic solvents, such as mineral spirits, methyl ethyl ketone, acetone, toluene, and fluorinated and chlorinated organic solvents. It can also be formulated as a water-based cleaning product, such as Orange Glo, that can be used in place of more caustic cleaning solutions. There is a form of limonene that is a molecular mirror image of D-limonene. It is called L-limonene and has a pine or turpentine aroma.

D-Limonene
(Oil of lemon and orange)

The terpene *myrcene* is found in bayberry. It is used in perfumes and scented candles because it adds a refreshing, spicy aroma to them. Trace amounts of myrcene may be used as a flavor component in root beer.

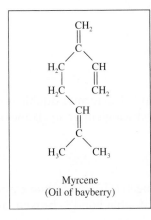

Myrcene
(Oil of bayberry)

Farnesol is a terpene found in roses, orange blossom, wild cyclamen, and lily of the valley. Cosmetics companies began to use farnesol in skin care products in the early 1990s. It is claimed that farnesol smoothes wrinkles and increases skin elasticity. It is also thought to reduce skin aging by promoting regeneration of cells and activation of the synthesis of molecules, such as collagen, that are required for healthy skin.

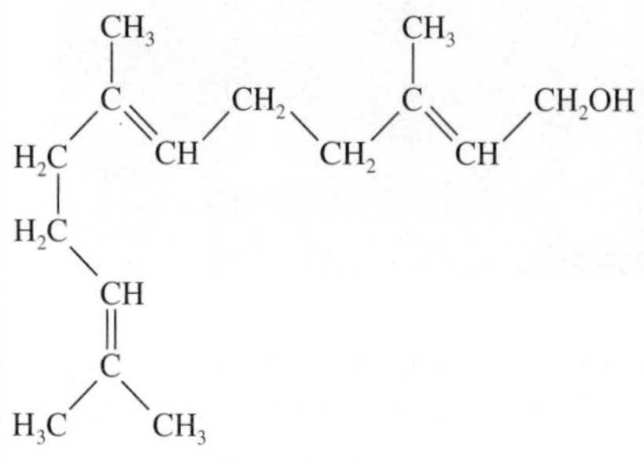

Farnesol
(Lily of the valley)

Another terpene, *retinol*, is a form of vitamin A (see Figure 11.1). It is able to penetrate the outer layers of skin and stimulate the formation of collagen and elastin. This reduces wrinkles by creating skin that is firmer and smoother.

11.5   Reactions Involving Alkenes and Alkynes

LEARNING GOAL

4 Write equations predicting the products of addition reactions of alkenes and alkynes: hydrogenation, halogenation, hydration, and hydrohalogenation.

Reactions of alkenes involve the carbon-carbon double bond. The key reaction of the double bond is the **addition reaction.** This involves the addition of two atoms or groups of atoms to a double bond.

The major alkene addition reactions are:

1. *Hydrogenation*—the addition of a *hydrogen* atom (H_2)
2. *Halogenation*—the addition of a *halogen* atom (Br_2 or Cl_2)
3. *Hydration*—the addition of a *water* molecule (H_2O)
4. *Hydrohalogenation*—the addition of a *hydrogen halide* molecule (HBr or HCl)

A generalized addition reaction is shown here. The R in these structures represents a hydrogen atom or any alkyl or aryl group.

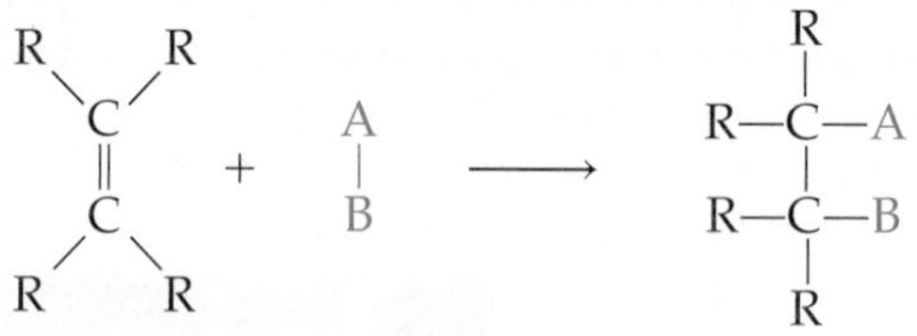

R = an alkyl or
aryl group

Note that the double bond is replaced by a single bond. The former double bond carbons receive a new single bond to a new atom, producing either an alkane or a substituted alkane.

Hydrogenation: Addition of H_2

Hydrogenation is the addition of a molecule of hydrogen (H_2) to a carbon-carbon double bond to give an alkane. In this reaction, the double bond is broken, and two new C—H single bonds result. One hydrogen molecule is required for each double bond in the alkene. Platinum, palladium, or nickel is needed as a catalyst to speed up the reaction. Heat and/or pressure may also be required. These reaction conditions are noted above and below the arrow in the general equation shown here.

Recall that a catalyst itself undergoes no net change in the course of a chemical reaction (see Section 7.3).

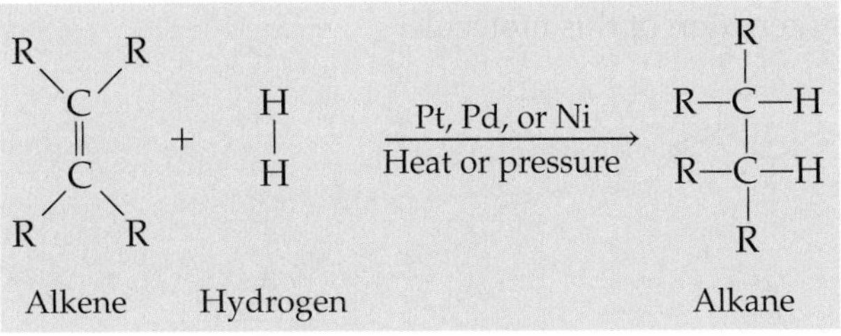

Note that the alkene is gaining two hydrogens. Thus, hydrogenation is a reduction reaction (see Sections 4.8 and 12.5).

EXAMPLE 11.6 | **Writing Equations for the Hydrogenation of Alkenes**

LEARNING GOAL

Linoleic acid (*cis, cis*-9,12-octadecadienoic acid) is an essential fatty acid. This means that we must obtain it in the diet, because we cannot make it. Fortunately, it is found in abundance in sunflower, safflower, and corn oil. These oils are often hydrogenated to produce margarine. Write a balanced equation showing the hydrogenation of linoleic acid.

4 Write equations predicting the products of addition reactions of alkenes and alkynes: hydrogenation, halogenation, hydration, and hydrohalogenation.

Solution

First, notice that there are two double bonds. This means we will need two molecules of H_2 for every molecule of linoleic acid that is hydrogenated. Remember to note the presence of a catalyst, Ni in this example, and heat.

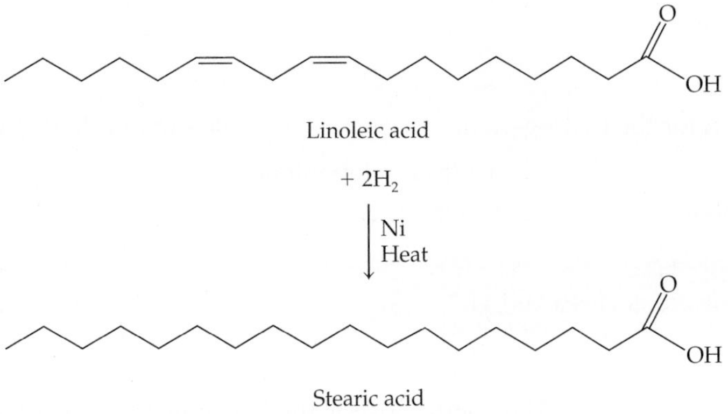

Linoleic acid

$+ 2H_2$

Ni
Heat

Stearic acid

The liquid linoleic acid is converted to stearic acid, which is a rather hard solid. In margarine production today, partial hydrogenation is generally used because it produces a fat that can be more easily spread and has a better "mouth feel." Producers may then add butter flavoring, milk solids, salt, an emulsifying agent, preservatives, vitamin A for nutritional value, and a bit of β-carotene for color.

Practice Problem 11.6

Write a balanced equation for the hydrogenation of each of the following alkenes.

   a. *cis*-2-Heptene   c. 3-Hexene
   b. *trans*-2-Pentene   d. Propene

▶ For Further Practice: **Questions 11.55, 11.67a, and 11.85.**

EXAMPLE 11.7 | **Writing Equations for the Hydrogenation of a Cycloalkene**

LEARNING GOAL

4 Write equations predicting the products of addition reactions of alkenes and alkynes: hydrogenation, halogenation, hydration, and hydrohalogenation.

Bisabolines are a group of closely related compounds called sesquiterpenes. They are found in the essential oils of a number of plants, including lemon, oregano, and the bergamot orange. The structure of one bisaboline is found here. Write a balanced equation showing the hydrogenation of this molecule.

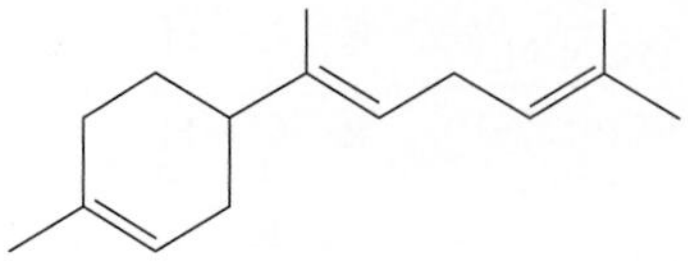

Solution

There are three double bonds in this molecule. As a result, we will need three molecules of hydrogen for every bisaboline molecule that is hydrogenated. Remember to note the requirement for a catalyst (Pt is shown here) and heat.

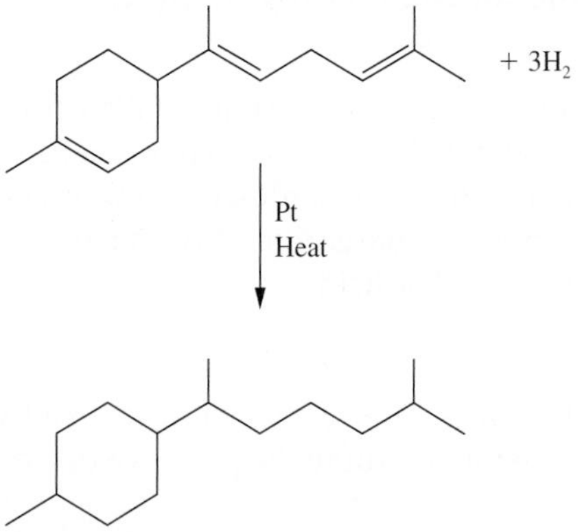

$+ 3H_2$

Pt
Heat

Practice Problem 11.7

Write a balanced equation for the hydrogenation of each of the following cycloalkenes:

  a. Cycloheptene
  b. 3-Methylcycloheptene
  c. 1-Ethylcyclopentene
  d. Cyclobutene

▶ For Further Practice: **Questions 11.70a and 11.71a.**

The conditions for the hydrogenation of alkynes are very similar to those for the hydrogenation of alkenes. Two molecules of hydrogen add to the triple bond of the alkyne to produce an alkane, as seen in the following general reaction:

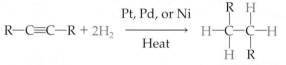

$$R-C\equiv C-R + 2H_2 \xrightarrow[\text{Heat}]{\text{Pt, Pd, or Ni}} \underset{\begin{subarray}{c} H & R \end{subarray}}{\overset{\begin{subarray}{c} R & H \end{subarray}}{H-C-C-H}}$$

Alkyne   Hydrogen   Alkane

Margarine and solid shortening are made by partial hydrogenation of vegetable oils. While natural oils contain only *cis*-fatty acids, the hydrogenation reaction produces *trans*-fatty acids. What are the health risks associated with *trans*-fats?

Question 11.9   The *trans*-isomer of 2-pentene was used in Practice Problem 11.6b. Would the result be any different if the *cis*-isomer had been used?

Question 11.10   Write balanced equations for the hydrogenation of 1-butene and *cis*-2-butene.

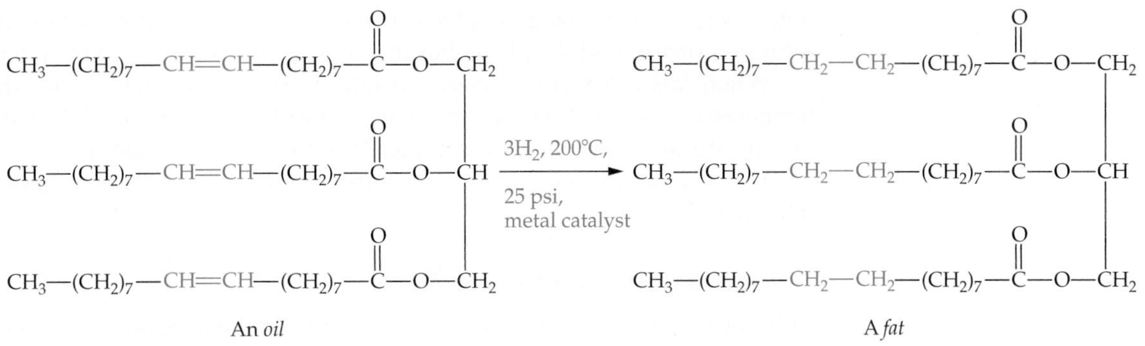

$$CH_3-(CH_2)_7-CH=CH-(CH_2)_7-\overset{\overset{\displaystyle O}{\|}}{C}-O-CH_2$$

$$CH_3-(CH_2)_7-CH=CH-(CH_2)_7-\overset{\overset{\displaystyle O}{\|}}{C}-O-CH \xrightarrow[\substack{25 \text{ psi,} \\ \text{metal catalyst}}]{3H_2, 200°C,} CH_3-(CH_2)_7-CH_2-CH_2-(CH_2)_7-\overset{\overset{\displaystyle O}{\|}}{C}-O-CH$$

$$CH_3-(CH_2)_7-CH=CH-(CH_2)_7-\overset{\overset{\displaystyle O}{\|}}{C}-O-CH_2$$

$$CH_3-(CH_2)_7-CH_2-CH_2-(CH_2)_7-\overset{\overset{\displaystyle O}{\|}}{C}-O-CH_2$$

$$CH_3-(CH_2)_7-CH_2-CH_2-(CH_2)_7-\overset{\overset{\displaystyle O}{\|}}{C}-O-CH_2$$

An *oil*   A *fat*

Figure 11.3 Conversion of a typical oil to a fat involves hydrogenation. In this example, triolein (an oil) is converted to tristearin (a fat). Since there are three unsaturated fatty acids in triolein, three molecules of hydrogen are required for complete hydrogenation of the molecule.

Question 11.11   Write balanced equations for the complete hydrogenation of each of the following alkynes:
  a. $H_3CC\equiv CCH_3$ b. $H_3CC\equiv CCH_2CH_3$

Question 11.12   Using the IUPAC Nomenclature System, name each of the products and reactants in the reactions described in Question 11.11.

Hydrogenation is used in the food industry to produce items such as margarine and Crisco, which are mixtures of hydrogenated vegetable oils (Figure 11.3). Vegetable oils are polyunsaturated; that is, they contain many double bonds and as a result have low melting points and are liquid at room temperature. The hydrogenation of these double bonds to single bonds increases the melting point of these oils and results in a fat, such as Crisco, that remains solid at room temperature. A similar process with highly refined corn oil and added milk solids produces corn oil margarine. As we saw in Section 11.3, such margarine may contain *trans*-fatty acids as a result of hydrogenation.

Because saturated fatty acids are extended, rigid chains, they tend to stack. This increases the intermolecular attractions along the surface of the molecule (London dispersion forces) and results in higher melting points, producing a fatty acid that is solid at room temperature. The naturally occurring *cis*-fatty acids are more flexible because the double bond introduces a "kink" or bend in the molecule. This reduces the tendency of the molecules to stack and decreases the attractive forces between the molecules. As a result, they have lower melting temperatures and are typically liquids at room temperature. The data in Table 11.2 demonstrate these differences.

Saturated and unsaturated dietary fats are discussed in Section 17.2.

TABLE 11.2 Comparison of the Melting Points of Several Saturated and Unsaturated Fatty Acids

Fatty Acids	Carbon Chain Length	Number of Double Bonds	Melting Point (°C)
Saturated			
Palmitic acid	16	0	63
Stearic acid	18	0	70
Arachidic acid	20	0	77
Unsaturated			
Palmitoleic acid	16	1	0
Oleic acid	18	1	16
Linoleic acid	20	2	5
Linolenic acid	20	3	−11
Arachidonic acid	20	4	−50

Both stearic acid and oleic acid have 18-carbon chains, but the melting point of the saturated stearic acid is 54°C higher than that of the unsaturated oleic acid.

When fats are formed from saturated fatty acids, they are solids at room temperature. Beef fat and bacon fat are examples of saturated fats. Fats containing unsaturated fatty acids are typically liquid at room temperature. Olive oil, peanut oil, and most other plant oils are examples of fats containing unsaturated fatty acids.

Halogenation: Addition of X_2

Chlorine (Cl_2) or bromine (Br_2) can be added to a double bond. This reaction, called **halogenation,** proceeds readily and does not require a catalyst:

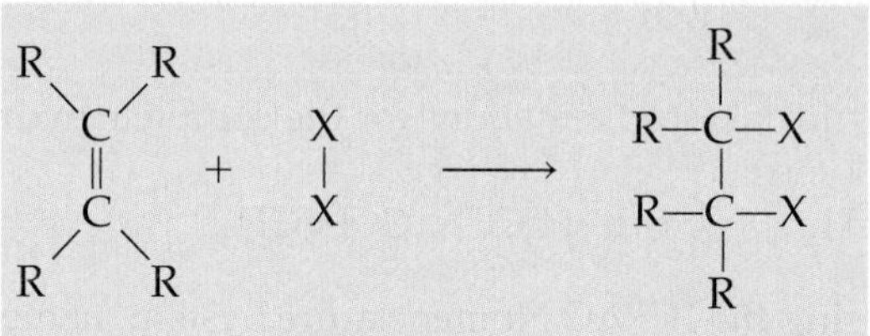

Alkene   Halogen   Alkyl dihalide

R = an alkyl or
aryl group or a hydrogen atom

EXAMPLE 11.8   **Writing Equations for the Halogenation of Alkenes**

LEARNING GOAL

Write a balanced equation showing (a) the chlorination of 1-pentene and (b) the bromination of *trans*-2-butene.

4 Write equations predicting the products of addition reactions of alkenes and alkynes: hydrogenation, halogenation, hydration, and hydrohalogenation.

Solution

a. Begin by drawing the structure of 1-pentene and of diatomic chlorine (Cl_2).

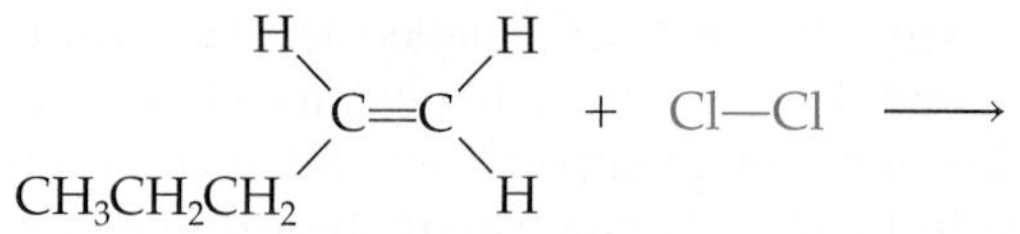

1-Pentene   Chlorine

Knowing that one chlorine atom will form a covalent bond with each of the carbon atoms of the carbon-carbon double bond, we can write the product and complete the equation.

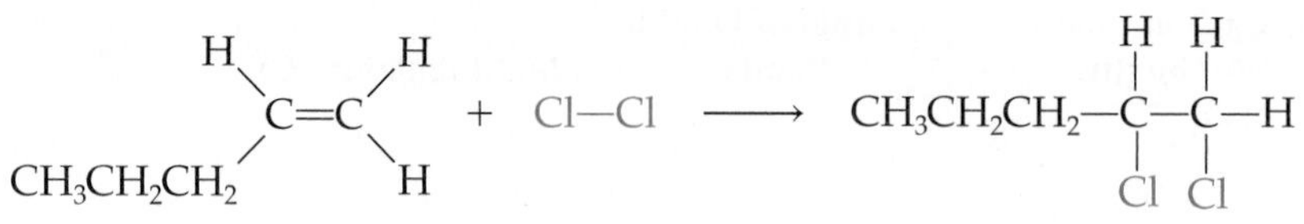

1-Pentene   Chlorine   1,2-Dichloropentane

b. Begin by drawing the structure of *trans*-2-butene and of diatomic bromine (Br_2).

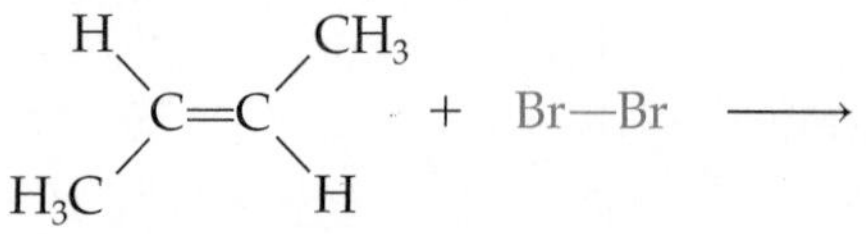

trans-2-Butene   Bromine

Knowing that one bromine atom will form a covalent bond with each of the carbon atoms of the carbon-carbon double bond, we can write the product and complete the equation.

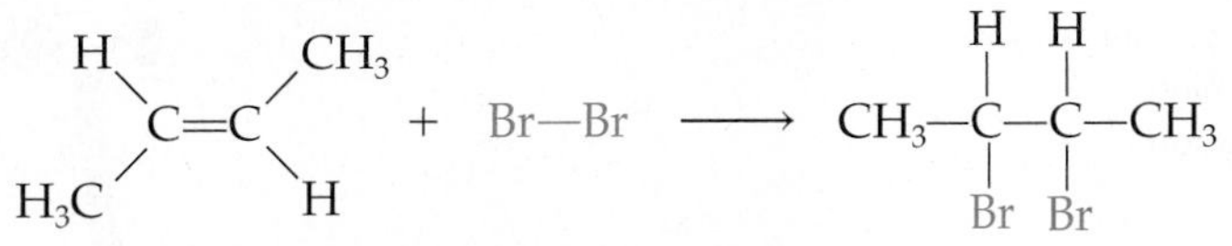

$$\underset{\substack{\text{trans-2-Butene}}}{\underset{H_3C}{\overset{H}{>}}C=C\underset{H}{\overset{CH_3}{<}}} \; + \; \underset{\text{Bromine}}{Br-Br} \; \longrightarrow \; \underset{\substack{\text{2,3-Dibromobutane}}}{CH_3-\underset{\underset{Br}{|}}{\overset{\overset{H}{|}}{C}}-\underset{\underset{Br}{|}}{\overset{\overset{H}{|}}{C}}-CH_3}$$

Practice Problem 11.8

Write a balanced equation for the chlorination of each of the following alkenes.

a. 2,4-Hexadiene
b. 1,5-Dibromo-2-pentene

c. 2,3-Dimethyl-1-butene
d. 4,6-Dimethyl-2-heptene

▶ For Further Practice: **Questions 11.57, 11.67b, and 11.86.**

Alkynes also react with the halogens bromine or chlorine. Two molecules of halogen add to the triple bond to produce a tetrahaloalkane:

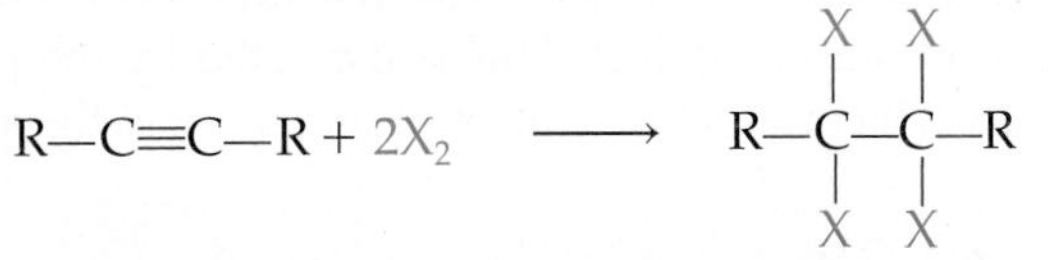

$$\underset{\text{Alkyne}}{R-C\equiv C-R} + \underset{\text{Halogen}}{2X_2} \; \longrightarrow \; \underset{\text{Tetrahaloalkane}}{R-\underset{\underset{X}{|}}{\overset{\overset{X}{|}}{C}}-\underset{\underset{X}{|}}{\overset{\overset{X}{|}}{C}}-R}$$

Question 11.13 Write a balanced equation for the addition of bromine to each of the following alkenes. Draw the products and reactants for each reaction.

a. $CH_3CH=CH_2$
b. $CH_3CH=CHCH_3$

Question 11.14 Using the IUPAC Nomenclature System, name each of the products and reactants in the reactions described in Question 11.13.

Question 11.15 Write balanced equations for the complete chlorination of each of the following alkynes:

a. $H_3CC\equiv CCH_3$
b. $H_3CC\equiv CCH_2CH_3$

Question 11.16 Using the IUPAC Nomenclature System, name each of the products and reactants in the reactions described in Question 11.15.

The following equation represents the bromination of 1-pentene:

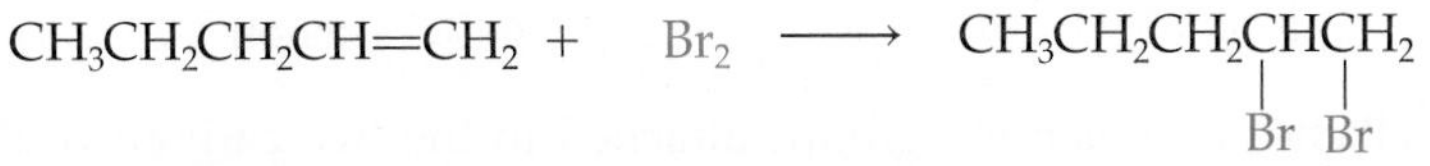

$$\underset{\substack{\text{1-Pentene} \\ \text{(colorless)}}}{CH_3CH_2CH_2CH=CH_2} + \underset{\substack{\text{Bromine} \\ \text{(red)}}}{Br_2} \; \longrightarrow \; \underset{\substack{\text{1,2-Dibromopentane} \\ \text{(colorless)}}}{CH_3CH_2CH_2\underset{\underset{Br}{|}}{C}H\underset{\underset{Br}{|}}{C}H_2}$$

This reaction is represented in Figure 11.4. Notice that the solution of reactants is red because of the presence of bromine. However, the product is colorless because the bromine has been used up in the reaction.

This bromination reaction can be used to show the presence of double or triple bonds in an organic compound. A reaction mixture of an alkane and Br_2 and a reaction mixture of an alkene and Br_2 would both be red. However, only the reaction mixture with the alkene would lose the red color as the Br_2 is used up in the halogenation reaction. Not only does this test distinguish saturated from

Figure 11.4 Bromination of an alkene. The solution on the left is red because of the presence of bromine and absence of an alkene or alkyne. In the presence of an unsaturated hydrocarbon, the bromine is used in the reaction and the solution becomes colorless.

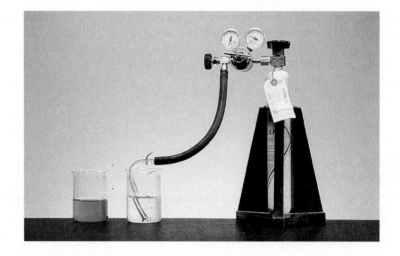

unsaturated hydrocarbons, it can provide a measure of the degree of saturation. For instance, a diene (having two double bonds) or an alkyne (having a triple bond) would consume twice as much bromine as an alkene with a single double bond.

Hydration: Addition of H₂O

A water molecule can be added to an alkene. This reaction, termed **hydration,** requires a trace of strong acid (H^+) as a catalyst. This is indicated by the presence of the H^+ over the reaction arrow. The product is an alcohol, as shown in the following equation:

$$
\underset{\text{Alkene}}{\overset{R}{\underset{R}{\overset{C}{\parallel}}}\overset{R}{\underset{R}{C}}} + \underset{\text{Water}}{\overset{H}{\underset{OH}{|}}} \xrightarrow{H^+} \underset{\text{Alcohol}}{\overset{R}{\underset{R}{\overset{R-C-H}{|}}}\overset{|}{R-C-OH}}
$$

R = an alkyl or
aryl group or hydrogen atom

The following equation shows the hydration of ethene to produce ethanol.

$$
\underset{\text{Ethene}}{\overset{H\quad H}{\underset{H\quad H}{\overset{C}{\parallel}C}}} + \underset{\text{Water}}{\overset{H}{\underset{OH}{|}}} \xrightarrow{H^+} \underset{\substack{\text{Ethanol}\\ \text{(ethyl alcohol)}}}{\overset{H}{\underset{H}{\overset{H-C-H}{|}}\overset{|}{H-C-OH}}}
$$

With alkenes in which the groups attached to the two carbons of the double bond are different (unsymmetrical alkenes), two products are possible. For example:

LEARNING GOAL

5 Apply Markovnikov's rule to predict the major and minor products of the hydration and hydrohalogenation reactions of unsymmetrical alkenes.

$$
\underset{\substack{\text{Propene}\\ \text{(propylene)}}}{\overset{3\quad 2\quad 1}{\underset{H\quad H}{\overset{H\quad H}{H-C-C=C-H}}}} + H-OH \xrightarrow{H^+} \underset{\substack{\text{Major product}\\ \text{2-Propanol}\\ \text{(isopropyl alcohol)}}}{\overset{3\ \ 2\ \ 1}{\underset{H\ OH\ H}{\overset{H\ H\ H}{H-C-C-C-H}}}} + \underset{\substack{\text{Minor product}\\ \text{1-Propanol}\\ \text{(propyl alcohol)}}}{\overset{3\ \ 2\ \ 1}{\underset{H\ H\ OH}{\overset{H\ H\ H}{H-C-C-C-H}}}}
$$

When hydration of an unsymmetrical alkene, such as propene, is carried out in the laboratory, one product is favored over the other. In this example, 2-propanol is the major product. The Russian chemist Vladimir Markovnikov studied many such reactions and came up with a rule that can be used to predict the major product of such a reaction. **Markovnikov's rule** tells us that the carbon of the carbon-carbon double bond that originally has more hydrogen atoms receives the hydrogen atom being added to the double bond. The remaining carbon forms a bond with the —OH. Simply stated, "the rich get richer"—the carbon with the greater number of hydrogens gets the new one as well. In the preceding example, carbon-1 has two C—H bonds originally, and carbon-2 has only one. The major product, 2-propanol, results from the new C—H bond forming on carbon-1 and the new C—OH bond on carbon-2.

Addition of water to a double bond is a reaction that we find in several biochemical pathways. For instance, the citric acid cycle is a key metabolic pathway in the complete oxidation of the sugar glucose and the release of the majority of the energy used by the body. The citric acid cycle is also the source of starting materials for the synthesis of the biological molecules needed for life. The next-to-last reaction in the citric acid cycle is the hydration of a molecule of fumarate to produce a molecule called malate.

> The carbon with the greatest number of hydrogen atoms is referred to as the least substituted carbon. The other carbon, bonded to a hydrogen and other alkyl or aryl groups, is the most substituted.

> We have seen that hydration of a double bond requires a trace of acid as a catalyst. In the cell, this reaction is catalyzed by an enzyme, or biological catalyst, called fumarase.

$$
\begin{array}{ccc}
\text{COO}^- & & \text{COO}^- \\
| & & | \\
\text{C—H} & & \text{HO—C—H} \\
|| & \xrightarrow{\text{Fumarase}} & | \\
\text{H—C} \quad + \text{H}_2\text{O} & & \text{H—C—H} \\
| & & | \\
\text{COO}^- & & \text{COO}^-
\end{array}
$$

Fumarate   Malate

EXAMPLE 11.9 **Writing Equations for the Hydration of Symmetrical and Unsymmetrical Alkenes**

LEARNING GOAL

4 Write equations predicting the products of addition reactions of alkenes and alkynes: hydrogenation, halogenation, hydration, and hydrohalogenation.

a. More and more frequently we see ethanol used as an additive in the gasoline we buy. A mixture of 10% ethanol and 90% gasoline significantly raises the octane rating of the fuel. In fact, some cities where auto emissions may reach harmful levels require the use of 10% ethanol gasoline. Bioethanol is produced by fermentation of crops such as corn. However, it can also be manufactured by the hydration of ethene. Write a balanced equation showing the hydration of ethene.

Solution

$$
\underset{\text{Ethene}}{\overset{\displaystyle H}{\underset{\displaystyle H}{>}}C=C\overset{\displaystyle H}{\underset{\displaystyle H}{<}}} \quad + \quad \text{HOH} \quad \xrightarrow{\text{H}^+} \quad \underset{\text{Ethanol}}{H-\overset{\displaystyle H}{\underset{\displaystyle H}{C}}-\overset{\displaystyle H}{\underset{\displaystyle OH}{C}}-H}
$$

b. Write an equation showing all the products of the hydration of the unsymmetrical alkene 1-pentene.

LEARNING GOAL

5 Apply Markovnikov's rule to predict the major and minor products of the hydration and hydrohalogenation reactions of unsymmetrical alkenes.

Solution

Begin by drawing the structure of 1-pentene and of water and indicating the catalyst.

Continued…

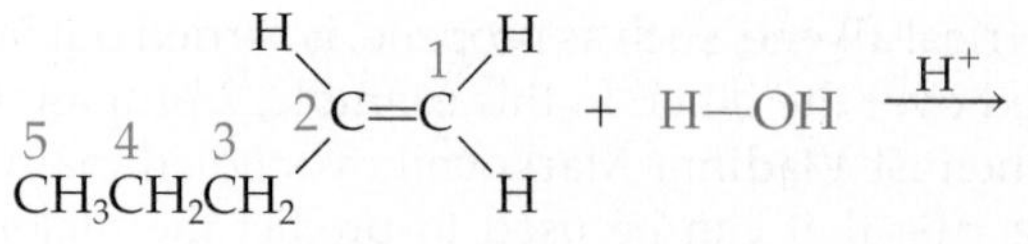

$$\underset{\underset{\text{1-Pentene}}{\underset{5\quad4\quad3}{CH_3CH_2CH_2}}}{\overset{\overset{H}{|}}{\underset{|}{C}}}\!\!\!=\!\!\!\overset{\overset{H}{|}}{\underset{|}{\underset{1}{C}}}\!\!\!-\!H \quad + \quad \underset{\text{Water}}{H-OH} \quad \xrightarrow{H^+}$$

Markovnikov's rule tells us that the carbon atom that is already bonded to the greater number of hydrogen atoms is more likely to receive the hydrogen atom from the water molecule. The other carbon atom is more likely to become bonded to the hydroxyl group. Thus, we can predict that the major product of this reaction will be 2-pentanol and that the minor product will be 1-pentanol. Now we can complete the equation by showing the products:

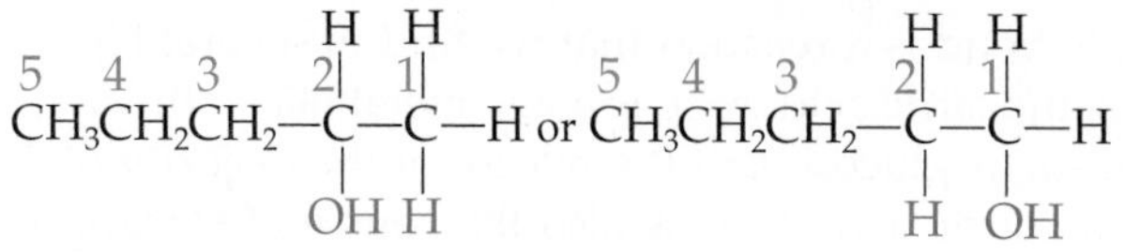

$$\underset{\underset{\text{(major product)}}{\text{2-Pentanol}}}{\underset{5\quad4\quad3}{CH_3CH_2CH_2}\!-\!\overset{\overset{H}{|}}{\underset{\underset{OH}{|}}{\underset{2}{C}}}\!-\!\overset{\overset{H}{|}}{\underset{\underset{H}{|}}{\underset{1}{C}}}\!-\!H} \text{ or } \underset{\underset{\text{(minor product)}}{\text{1-Pentanol}}}{\underset{5\quad4\quad3}{CH_3CH_2CH_2}\!-\!\overset{\overset{H}{|}}{\underset{\underset{H}{|}}{\underset{2}{C}}}\!-\!\overset{\overset{H}{|}}{\underset{\underset{OH}{|}}{\underset{1}{C}}}\!-\!H}$$

Practice Problem 11.9

Write a balanced equation for the hydration of each of the following alkenes.

   a. 1-Butene

   b. 2-Methyl-3-hexene

     c. Propene

     d. 1,4-Dichloro-2-butene

▶ For Further Practice: **Questions 11.59, 11.68, and 11.81.**

Hydration of an alkyne is a more complex process because the initial product is not stable and is rapidly isomerized. As you would expect, the product is an alcohol but, in this case, one in which the hydroxyl group is bonded to one of the carbons of a carbon-carbon double bond. This type of molecule is called an *enol* because it is both an alkene (*ene*) and an alcohol (*ol*). The enol is immediately isomerized into either an aldehyde or ketone, as shown in the following general reaction:

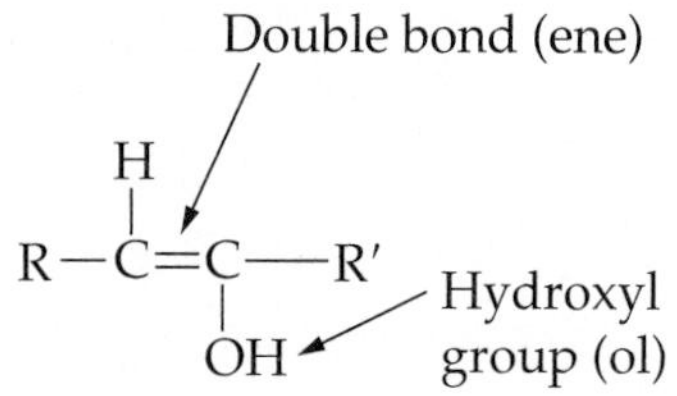

Double bond (ene)

$$R\!-\!\overset{\overset{H}{|}}{C}\!\!=\!\!\overset{}{\underset{\underset{OH}{|}}{C}}\!-\!R'$$ Hydroxyl group (ol)

An enol is an alkene and an alcohol

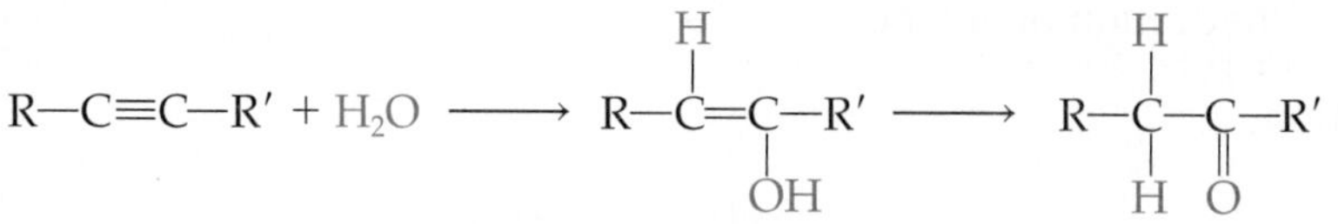

$$R\!-\!C\!\equiv\!C\!-\!R' + H_2O \longrightarrow \underset{\text{Enol}}{R\!-\!\overset{\overset{H}{|}}{C}\!\!=\!\!\overset{}{\underset{\underset{OH}{|}}{C}}\!-\!R'} \longrightarrow R\!-\!\overset{\overset{H}{|}}{\underset{\underset{H}{|}}{C}}\!-\!\overset{}{\underset{\underset{O}{\|}}{C}}\!-\!R'$$

Alkyne   Water   Enol   Aldehyde if R′ = H
Ketone if R′ = alkyl group

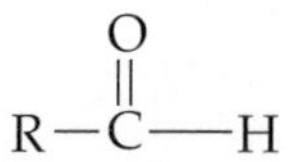

$$R\!-\!\overset{\overset{O}{\|}}{C}\!-\!H$$

An aldehyde

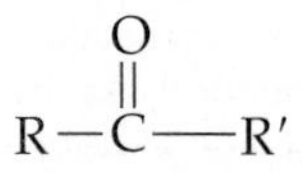

Question 11.17 Write an equation for the hydration of each of the following alkenes. Predict the major product of each of the reactions.

   a. $CH_3CH\!=\!CHCH_3$

   b. $CH_2\!=\!CHCH_2CH_2CH(CH_3)_2$

   c. $CH_3CH_2CH_2CH\!=\!CHCH_2CH_3$

   d. $CH_3CHClCH\!=\!CHCHClCH_3$

$$R\!-\!\overset{\overset{O}{\|}}{C}\!-\!R'$$

A ketone

Question 11.18 Write an equation for the hydration of each of the following alkenes. Predict the major product of each of the reactions.

   a. $CH_2\!=\!CHCH_2CH_2CH_3$

   b. $CH_3CH_2CH_2CH\!=\!CHCH_3$

   c. $CH_3CHBrCH_2CH\!=\!CHCH_2Cl$

   d. $CH_3CH_2CH_2CH_2CH_2CH\!=\!CHCH_3$

Question 11.19 Write equations for the complete hydration of each of the following alkynes:

a. $H_3CC{\equiv}CH$

b. $H_3CC{\equiv}CCH_2CH_3$

Question 11.20 Is the final product in each of the reactions in Question 11.19 an aldehyde or a ketone?

Hydrohalogenation: Addition of HX

A hydrogen halide (HBr, HCl, or HI) also can be added to an alkene. The product of this reaction, called **hydrohalogenation,** is an alkyl halide:

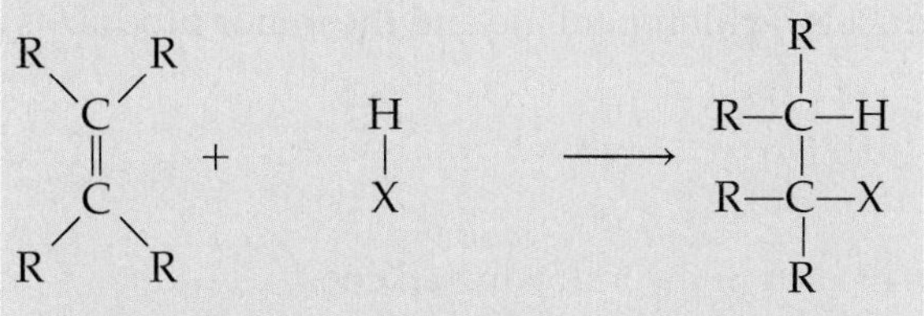

Alkene   Hydrogen halide   Alkyl halide

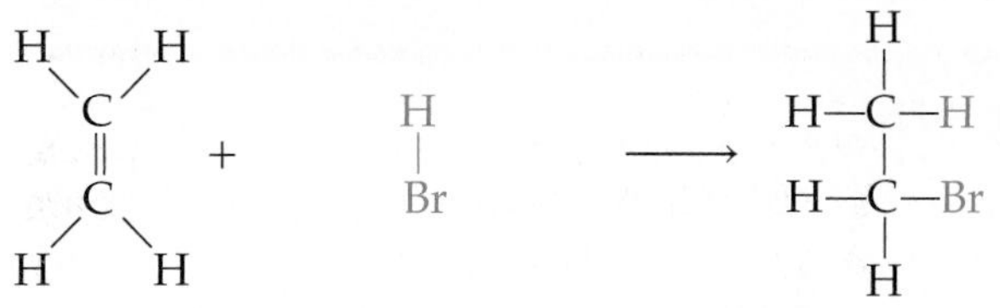

Ethene   Hydrogen bromide   Bromoethane

This reaction also follows Markovnikov's rule. That is, if HX is added to an unsymmetrical alkene, the hydrogen atom will be added preferentially to the carbon atom that originally had the most hydrogen atoms. Consider the following example:

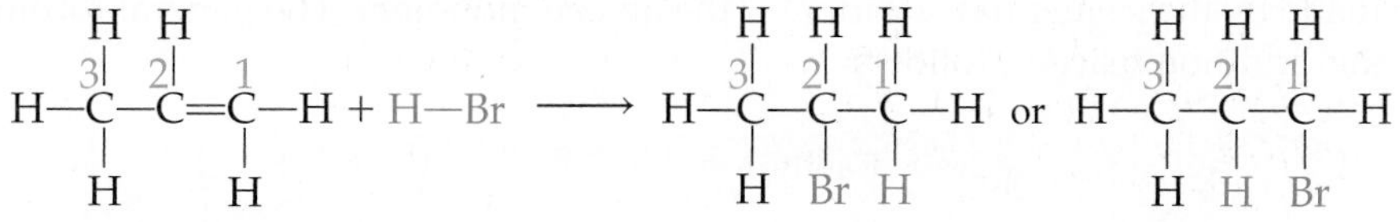

Propene   Major product   Minor product
                 2-Bromopropane   1-Bromopropane

> Recall that the carbon with the most hydrogens is referred to as the least substituted carbon. The other carbon, bonded to both hydrogen and other groups, is designated as the most substituted.

EXAMPLE 11.10   **Writing Equations for the Hydrohalogenation of Alkenes**

Write an equation showing all the products of the hydrohalogenation of 1-pentene with HCl.

Solution

Begin by drawing the structure of 1-pentene and of hydrochloric acid.

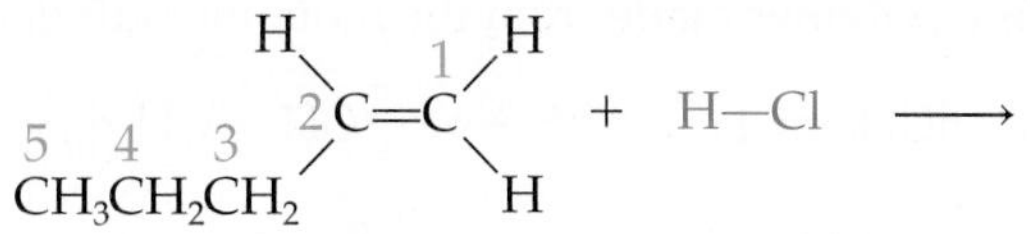

1-Pentene   Hydrochloric
                 acid

LEARNING GOAL

4 Write equations predicting the products of addition reactions of alkenes and alkynes: hydrogenation, halogenation, hydration, and hydrohalogenation.

LEARNING GOAL

5 Apply Markovnikov's rule to predict the major and minor products of the hydration and hydrohalogenation reactions of unsymmetrical alkenes.

Continued...

Markovnikov's rule tells us that the carbon atom that is already bonded to the greater number of hydrogen atoms is more likely to receive the hydrogen atom of the hydrochloric acid molecule. The other carbon atom is more likely to become bonded to the chlorine atom. Thus, we can complete the equation by writing the major and minor products.

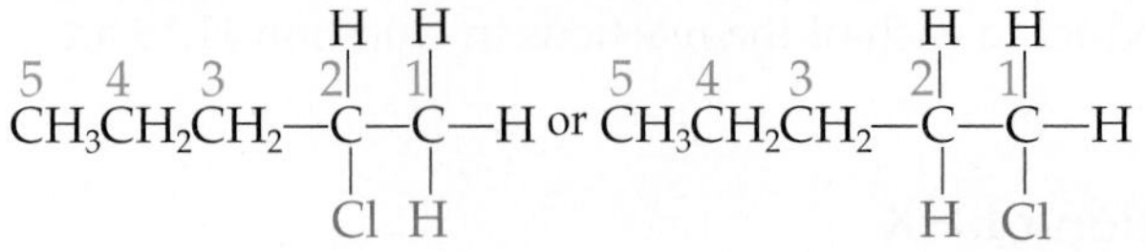

$$
\underset{5}{CH_3}\underset{4}{CH_2}\underset{3}{CH_2}-\underset{\underset{Cl}{|}}{\overset{\overset{H}{|}}{\underset{2}{C}}}-\underset{\underset{H}{|}}{\overset{\overset{H}{|}}{\underset{1}{C}}}-H \quad \text{or} \quad \underset{5}{CH_3}\underset{4}{CH_2}\underset{3}{CH_2}-\underset{\underset{H}{|}}{\overset{\overset{H}{|}}{\underset{2}{C}}}-\underset{\underset{Cl}{|}}{\overset{\overset{H}{|}}{\underset{1}{C}}}-H
$$

2-Chloropentane   1-Chloropentane
(major product)   (minor product)

As predicted by Markovnikov's rule, the major product is 2-chloropentane and the minor product is 1-chloropentane.

Practice Problem 11.10

Write a balanced equation for the hydrobromination of each of the following alkenes.

a. 2-Pentene   c. 3-Heptene

b. Propene   d. Ethene

▶ For Further Practice: **Questions 11.69d, 11.70d, and 11.75.**

LEARNING GOAL

6 Write equations representing the formation of addition polymers of alkenes.

Addition Polymers of Alkenes

Polymers are macromolecules composed of repeating structural units called **monomers.** A polymer may be made up of several thousand monomers. Many commercially important plastics and fibers are addition polymers made from alkenes or substituted alkenes. They are called **addition polymers** because they are made by the sequential addition of the alkene monomer. The general formula for this addition reaction follows:

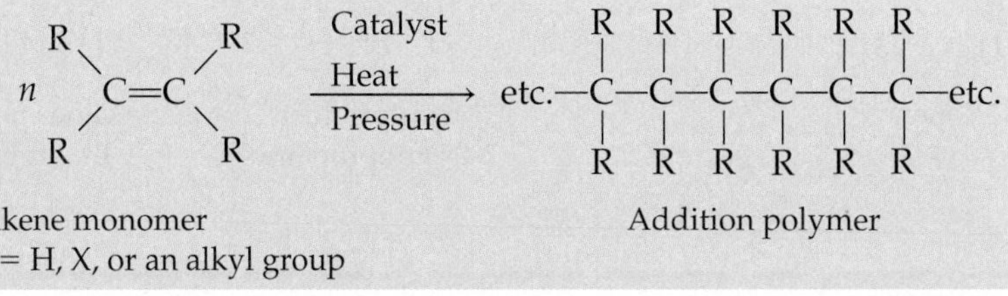

$$
n \; \underset{R}{\overset{R}{\diagdown}}C=C\underset{R}{\overset{R}{\diagup}} \quad \xrightarrow[\text{Pressure}]{\text{Catalyst} \atop \text{Heat}} \quad \text{etc.} -\overset{\overset{R}{|}}{\underset{\underset{R}{|}}{C}}-\overset{\overset{R}{|}}{\underset{\underset{R}{|}}{C}}-\overset{\overset{R}{|}}{\underset{\underset{R}{|}}{C}}-\overset{\overset{R}{|}}{\underset{\underset{R}{|}}{C}}-\overset{\overset{R}{|}}{\underset{\underset{R}{|}}{C}}-\overset{\overset{R}{|}}{\underset{\underset{R}{|}}{C}}-\text{etc.}
$$

Alkene monomer   Addition polymer
R = H, X, or an alkyl group

The product of the reaction is generally represented in a simplified manner:

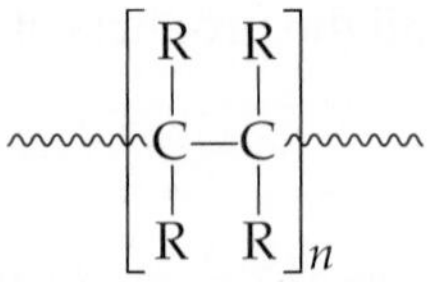

$$
\sim\sim\sim\!\!\left[\overset{\overset{R\quad R}{|\quad\;\;|}}{\underset{\underset{R\quad R}{|\quad\;\;|}}{C-C}}\right]_n\!\!\sim\sim\sim
$$

Polyethylene is a polymer made from the monomer ethylene (ethene):

$$
n\,CH_2{=}CH_2 \longrightarrow \sim\sim\sim\!\!\left[CH_2{-}CH_2\right]_n\!\!\sim\sim\sim
$$

Ethene   Polyethylene
(ethylene)

It is used to make bottles, injection-molded toys and housewares, and wire coverings.

Life without Polymers?

What do Nike Air-Sole shoes, disposable diapers, tires, shampoo, and artificial joints and skin share in common? These products and a great many other items we use every day are composed of synthetic or natural polymers. Indeed, the field of polymer chemistry has come a long way since the 1920s and 1930s when DuPont chemists invented nylon and Teflon.

Consider the disposable diaper. The outer, waterproof layer is composed of polyethylene. The polymerization reaction that produces polyethylene is shown in Section 11.5. The diapers have elastic to prevent leaking. The elastic is made of a natural polymer, rubber. The monomer from which natural rubber is formed is 2-methyl-1,3-butadiene. The common name of this monomer is *isoprene*. As we will see in coming chapters, isoprene is an important monomer in the synthesis of many natural polymers.

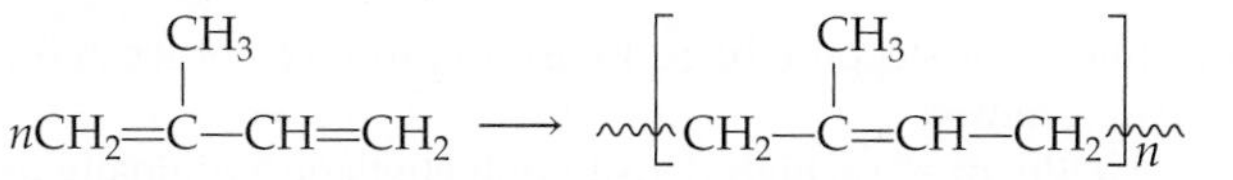

$$n CH_2=\overset{\overset{\displaystyle CH_3}{|}}{C}-CH=CH_2 \longrightarrow \left[CH_2-\overset{\overset{\displaystyle CH_3}{|}}{C}=CH-CH_2\right]_n$$

2-Methyl-1,3-butadiene   Rubber polymer
(isoprene)

The diaper is filled with a synthetic polymer called poly(acrylic acid). This polymer has the remarkable ability to absorb many times its own weight in liquid. Polymers that have this ability are called superabsorbers, but polymer chemists have no idea why they have this property! The acrylic acid monomer and resulting poly(acrylic acid) polymer are shown here:

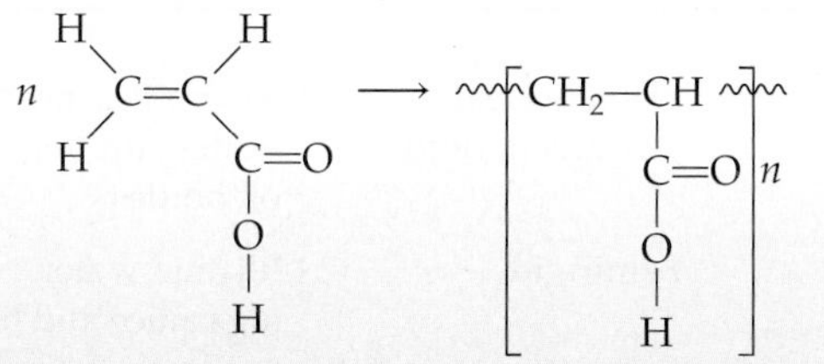

$$n \underset{H}{\overset{H}{C}}=\underset{\underset{\underset{\underset{H}{|}}{O}}{\overset{|}{C=O}}}{\overset{H}{C}} \longrightarrow \left[CH_2-\underset{\underset{\underset{\underset{H}{|}}{O}}{\overset{|}{C=O}}}{CH}\right]_n$$

Acrylic acid monomer   Poly(acrylic acid)

Another example of a useful polymer is GORE-TEX. This amazing polymer is made by stretching Teflon. Teflon is produced from the monomer tetrafluoroethene, as seen in the following equation:

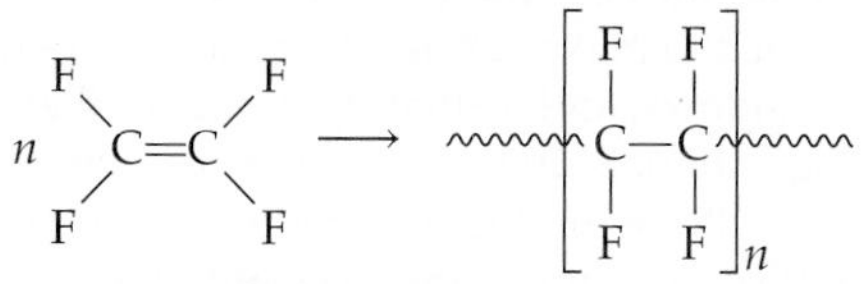

$$n \underset{F}{\overset{F}{C}}=\underset{F}{\overset{F}{C}} \longrightarrow \left[\underset{\underset{F}{|}}{\overset{\overset{F}{|}}{C}}-\underset{\underset{F}{|}}{\overset{\overset{F}{|}}{C}}\right]_n$$

Tetrafluoroethene   Teflon

Clothing made from this fabric is used to protect firefighters because of its fire resistance. Because it also insulates, GORE-TEX clothing is used by military forces and by many amateur athletes, for protection during strenuous activity in the cold. In addition to its use in protective clothing, GORE-TEX has been used in millions of medical procedures for sutures, synthetic blood vessels, and tissue reconstruction.

For Further Understanding

▶ Visit The Macrogalleria, www.pslc.ws/macrog/index.htm, an Internet site maintained by the Department of Polymer Science of the University of Southern Mississippi, to help you answer the following questions:
  ♦ Why does shrink wrap shrink?
  ♦ What are optical fibers made of and how do they transmit light?

Polypropylene is a plastic made from propylene (propene). It is used to make indoor-outdoor carpeting, packaging materials, toys, and housewares. When propylene polymerizes, a methyl group is located on every other carbon of the main chain:

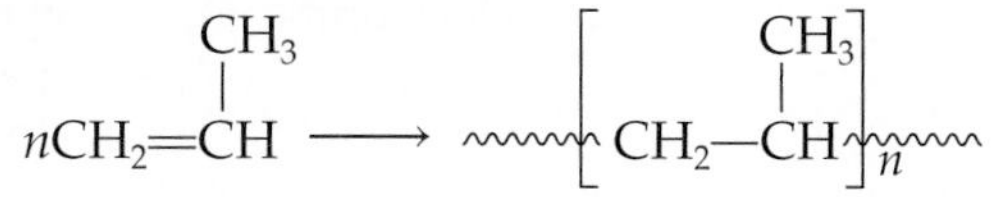

$$n CH_2=\overset{\overset{\displaystyle CH_3}{|}}{CH} \longrightarrow \left[CH_2-\overset{\overset{\displaystyle CH_3}{|}}{CH}\right]_n$$

or

$$CH_2-\overset{\overset{\displaystyle CH_3}{|}}{CH}-CH_2-\overset{\overset{\displaystyle CH_3}{|}}{CH}-CH_2-\overset{\overset{\displaystyle CH_3}{|}}{CH}$$

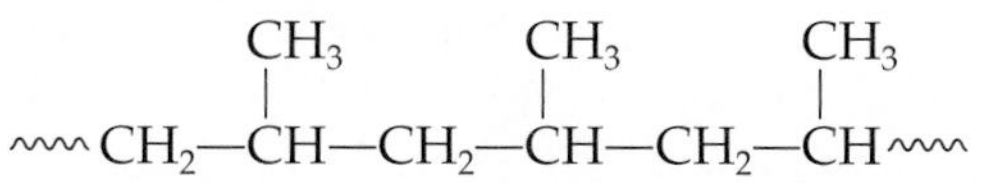

Green Chemistry

Plastic Recycling

Plastics, first developed by British inventor Alexander Parkes in 1862, are amazing substances. Some serve as containers for many of our foods and drinks, keeping them fresh for long periods. Other plastics serve as containers for detergents and cleansers or are formed into pipes for our plumbing systems. We have learned to make strong, clear sheets of plastic that can be used as windows, and feather-light plastics that can be used as packaging materials. In the United States alone, seventy-five billion pounds (lb) of plastics are produced each year.

But plastics, amazing in their versatility, are a mixed blessing. One characteristic that makes them so useful, their stability, has created an environmental problem. It may take 40 to 50 years for plastics discarded into landfill sites to degrade. Concern that we could soon be knee-deep in plastic worldwide inspired the development of the plastic recycling industry.

Since there are so many types of plastics, it is necessary to identify, sort, and recycle them separately. To help with this sorting process, manufacturers place recycling symbols on their plastic wares. As you can see in the accompanying table, each symbol corresponds to a different type of plastic.

Polyethylene terephthalate, also known as PETE or simply #1, is a form of polyester often used to make bottles and jars to contain food. When collected, it is ground up into flakes and formed into pellets. The most common use for recycled PETE is the manufacture of polyester carpets. But it may also be spun into a cotton candy–like form that can be used as a fiber filling

for pillows or sleeping bags. Reuse to produce bottles and jars is also common.

HDPE, or #2, is high-density polyethylene. Originally used for milk and detergent bottles, recycled HDPE is used to produce pipes, plastic lumber, trash cans, or bottles for storage of materials other than food. Low-density polyethylene (LDPE), or #4, is very similar to HDPE chemically. Because it is a more highly branched polymer, it is less dense and more flexible. Originally used to produce plastic bags, recycled LDPE is also used to make trash bags, grocery bags, and plastic tubing and lumber.

Code	Type	Name	Formula	Description	Examples
PETE	1	Polyethylene terephthalate	$-CH_2-CH_2-O-\overset{O}{\underset{O}{C}}-\bigcirc-C-O-$	Usually clear or green, rigid	Soda bottles, peanut butter jars, vegetable oil bottles
HDPE	2	High-density polyethylene	$-CH_2-CH_2-$	Semirigid	Milk and water jugs, juice and bleach bottles
PVC	3	Polyvinyl chloride	$-CH-CH_2-$ with Cl	Semirigid	Detergent and cleanser bottles, pipes
LDPE	4	Low-density polyethylene	$-CH_2-CH_2-$	Flexible, not crinkly	Six-pack rings, bread bags, sandwich bags
PP	5	Polypropylene	$-CH-CH_2-$ with CH_3	Semirigid	Margarine tubs, straws, screw-on lids
PS	6	Polystyrene	$-CH-CH_2-$ with phenyl ring	Often brittle	Styrofoam, packing peanuts, egg cartons, foam cups
Other	7	Multilayer plastics	N/A	Squeezable	Ketchup and syrup bottles

PVC, or #3, is one of the less commonly recycled plastics in the United States, although it is actively recycled in Europe. The recycled material is used to make non-food-bearing containers, shoe soles, flooring, sweaters, and pipes. Polypropylene, PP or #5, is found in margarine tubs, fabrics, and carpets. Recycled polypropylene has many uses, including fabrication of gardening implements.

You probably come into contact with polystyrene, PS or #6, almost every day. It is used to make foam egg cartons and meat trays, serving containers for fast food chains, CD "jewel boxes," and "peanuts" used as packing material. At the current time,

polystyrene food containers are not recycled. PS from nonfood products can be melted down and converted into pellets that are used to manufacture office desktop accessories, hangers, and plastic trays used to hold plants.

For Further Understanding

▸ Use the Internet or other resources to investigate recycling efforts in your area.

▸ In some areas, efforts to recycle plastics and paper have been abandoned. What factors contributed to this?

TABLE 11.3 Some Important Addition Polymers of Alkenes

Monomer Name	Formula	Polymer	Uses
Styrene	$CH_2=CH-C_6H_5$ 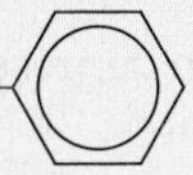	Polystyrene	Styrofoam containers
Acrylonitrile	$CH_2=CHCN$	Polyacrylonitrile (Orlon)	Clothing
Methyl methacrylate	$CH_2=C(CH_3)-\overset{\overset{O}{\|\|}}{C}OCH_3$	Polymethyl methacrylate (Plexiglas, Lucite)	Basketball backboards
Vinyl chloride	$CH_2=CHCl$	Polyvinyl chloride (PVC)	Plastic pipe, credit cards
Tetrafluoroethene	$CF_2=CF_2$	Polytetrafluoroethylene (Teflon)	Nonstick surfaces

Polymers made from alkenes or substituted alkenes are simply very large alkanes or substituted alkanes. Like the alkanes, they are typically inert. This chemical inertness makes these polymers ideal for making containers to hold juices, chemicals, and fluids used medically. They are also used to make sutures, catheters, and other indwelling devices. A variety of polymers made from substituted alkenes are listed in Table 11.3.

11.6 Aromatic Hydrocarbons

In the early part of the nineteenth century, chemists began to discover organic compounds with chemical properties quite distinct from the alkanes, alkenes, and alkynes. They called these substances *aromatic compounds* because many of the first examples were isolated from the pleasant-smelling resins of tropical trees. The carbon/hydrogen ratio of these compounds suggested a very high degree of unsaturation, similar to the alkenes and alkynes. Imagine, then, how puzzled these early organic chemists must have been when they discovered that these compounds do not undergo the kinds of addition reactions common to the alkenes and alkynes.

$$CH_2=CH_2 + Br_2 \longrightarrow \underset{\underset{Br}{\|}}{CH_2}-\underset{\underset{Br}{\|}}{CH_2}$$

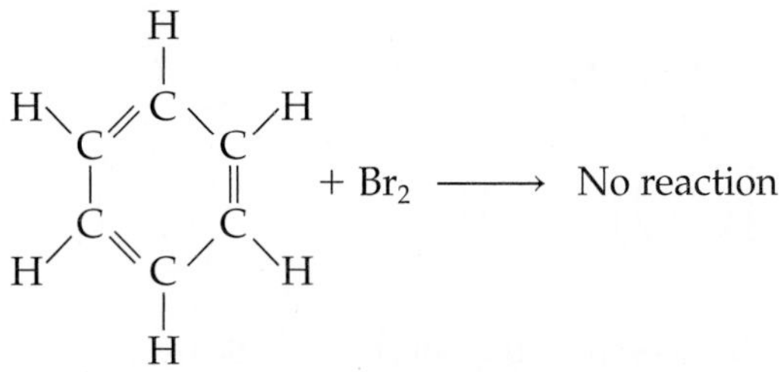

$$+ \ Br_2 \longrightarrow \text{No reaction}$$

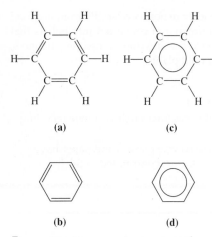

(a)   **(c)**

(b)   **(d)**

Figure 11.5 Four ways to represent the benzene molecule. Structure (b) is a simplified diagram of structure (a). Structure (d), a simplified diagram of structure (c), is the most commonly used representation.

Resonance models are described in Section 3.4.

We no longer define aromatic compounds as those having a pleasant aroma; in fact, many do not. We now recognize **aromatic hydrocarbons** as those that exhibit a much higher degree of chemical stability than their chemical composition would predict. The most common group of aromatic compounds is based on the six-member aromatic ring, the benzene ring. The structure of the benzene ring is represented in various ways in Figure 11.5.

Structure and Properties

The benzene ring consists of six carbon atoms joined in a planar hexagonal arrangement. Each carbon atom is bonded to one hydrogen atom. Friedrich Kekulé offered up a model for the structure of benzene in 1865. He proposed that single and double bonds alternated around the ring (a conjugated system of double bonds). To explain why benzene did not decolorize bromine—in other words, didn't react like an unsaturated compound—he suggested that the double and single bonds shift positions rapidly.

Actually, the most accurate way to represent the benzene molecules is as a resonance hybrid of two Kekulé structures:

Benzene as a resonance hybrid

The current model of the structure of benzene is based on the idea of overlapping orbitals. Each carbon is bonded to two others by sharing a pair of electrons. Each carbon atom also shares a pair of electrons with a hydrogen atom. The remaining six electrons are located in p orbitals that are perpendicular to the plane of the ring. These p orbitals overlap laterally to form a cloud of electrons above and below the ring (Figure 11.6). Because of the cloud of electrons above and below the ring, there is an even charge distribution over the entire molecule. This allows benzene to exist as a flat, planar molecule.

Two symbols are commonly used to represent the benzene ring. The representation in Figure 11.5b is the structure proposed by Kekulé. The structure in Figure 11.5d uses a circle to represent the electron cloud.

Nomenclature

Most simple aromatic compounds are named as derivatives of benzene. Thus, benzene is the parent ring, and the name of any atom or group bonded to benzene is used as a prefix, as in these examples:

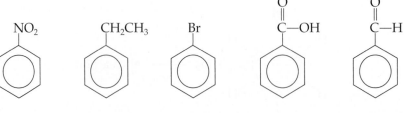

Nitrobenzene   Ethylbenzene   Bromobenzene   Benzoic acid   Benzaldehyde

Other members of this family have unique names based on history rather than logic:

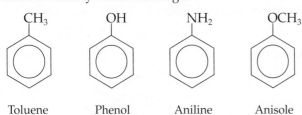

Toluene   Phenol   Aniline   Anisole

Other common names include xylene, a benzene ring with two methyl substituents, and cresol, a benzene ring bonded to a methyl group and a hydroxyl group.

When two groups are present on the ring, three possible orientations exist, and they may be named by either the IUPAC Nomenclature System or the common system of nomenclature. If the groups or atoms are located on two adjacent carbons, they are referred to as *ortho* (*o*) in the common system or with the prefix 1,2- in the IUPAC system. If they are on carbons separated by one carbon atom, they are termed *meta* (*m*) in the common system or 1,3- in the IUPAC system. Finally, if the substituents are on carbons separated by two carbon atoms, they are said to be *para* (*p*) in the common system or 1,4- in the IUPAC system. The three orientations of xylene and cresol are shown here:

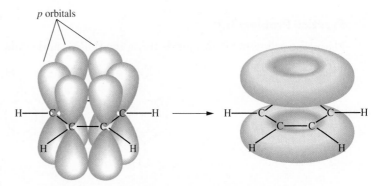

p orbitals

Figure 11.6 The current model of the bonding in benzene. The overlapping *p* orbitals form a cloud of electrons above and below the ring.

ortho-Xylene   *meta*-Xylene   *para*-Xylene   *ortho*-Cresol   *meta*-Cresol   *para*-Cresol

---

EXAMPLE 11.11   **Naming Derivatives of Benzene**

Name the following compounds using the IUPAC and common systems of nomenclature.

LEARNING GOAL

7 Draw the structures and write the names of common aromatic hydrocarbons.

a.   CH₃
      Cl

b.   NO₂
      OH

c.   CH₂CH₃
      NH₂

Solution

IUPAC names:

Parent compound:	toluene	phenol	aniline
Substituents:	2-chloro	4-nitro	3-ethyl
Name:	2-Chlorotoluene	4-Nitrophenol	3-Ethylaniline

Common names:

Parent compound:	toluene	phenol	aniline
Substituents:	*ortho*-chloro	*para*-nitro	*meta*-ethyl
Name:	*ortho*-Chlorotoluene	*para*-Nitrophenol	*meta*-Ethylaniline
Abbreviated name:	*o*-Chlorotoluene	*p*-Nitrophenol	*m*-Ethylaniline

Continued…

Practice Problem 11.11

Name the following compounds using the IUPAC and common nomenclature systems.

a. 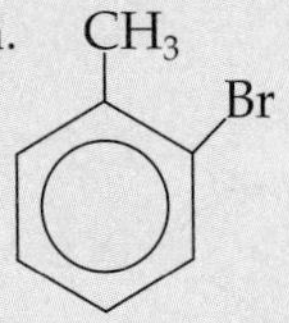 CH₃
Br

b. 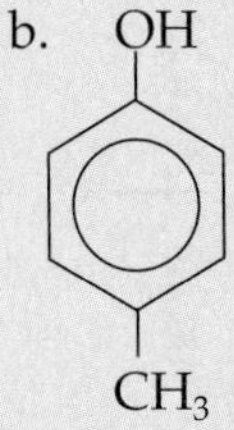 OH

CH₃

c. 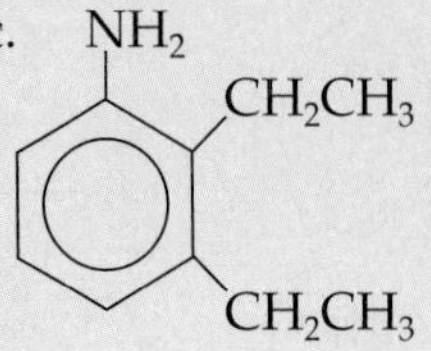 NH₂
CH₂CH₃

CH₂CH₃

d. 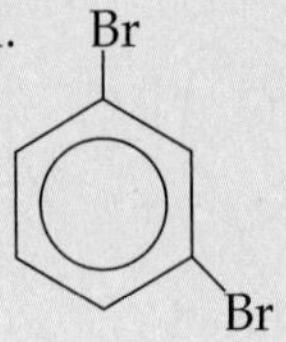 Br

Br

▶ For Further Practice: **Questions 11.92–11.94.**

If three or more groups are attached to the benzene ring, numbers must be used to describe their location. The names of the substituents are given in alphabetical order.

In IUPAC nomenclature, the group derived by removing one hydrogen from benzene (—C₆H₅), is called the **phenyl group.** An aromatic hydrocarbon attached to a long aliphatic group is named as a phenyl substituted hydrocarbon. For example:

1   2   3   4
CH₃CHCH₂CH₃

2-Phenylbutane

4   3   2   1
CH₃CHCH=CH₂

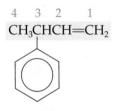

3-Phenyl-1-butene

One final special name that occurs frequently in aromatic compounds is the benzyl group:

C₆H₅CH₂—   or   —CH₂—

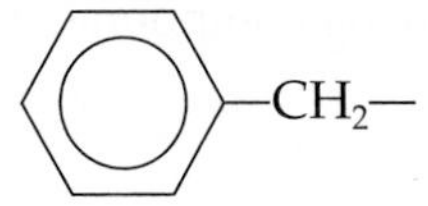

The use of this group name is illustrated by:

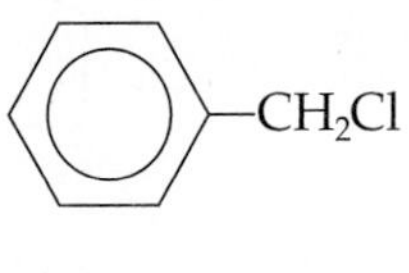

 —CH₂Cl

Benzyl chloride

—CH₂OH

Benzyl alcohol

Question 11.21   Draw each of the following compounds:

a. 1,3,5-Trichlorobenzene
b. *ortho*-Cresol
c. 2,5-Dibromophenol

d. *para*-Dinitrobenzene
e. 2-Nitroaniline
f. *meta*-Nitrotoluene

Question 11.22   Draw each of the following compounds:

a. 2,3-Dichlorotoluene
b. 3-Bromoaniline
c. 1-Bromo-3-ethylbenzene

d. *o*-Nitrotoluene
e. *p*-Xylene
f. *o*-Dibromobenzene

Kitchen Chemistry

Pumpkin Pie Spice: An Autumn Tradition

You know autumn is on the way when you see the "Pumpkin Spice Latte Is Back" sign in the Starbucks shop. Then, you visit the grocery store and notice pumpkin spice products everywhere: pumpkin spice Oreo cookies, English muffins, marshmallows, and coffee creamer! Since their debut in 2003, 200 million pumpkin spice lattes have been sold by Starbucks and sales of other pumpkin spice products reached nearly $350 million in 2013.

The flavor of pumpkin pie is the result of a combination of cooked pumpkin (actually a sweet squash that is less watery and fibrous than pumpkin) and a blend of spices, including nutmeg, cinnamon, ginger, and clove or allspice. It is estimated that there are about 340 flavor compounds that make up the characteristic flavor of a pumpkin pie. So how does Starbucks mimic that flavor in a latte? Actually, the recipe is a secret, but scientists have a good idea of some of the components. Many of them are alkenes, isoprenoids, or aromatic compounds (benzene derivatives).

Cinnamaldehydes provide the cinnamon flavor. Eugenol, a benzene derivative, provides the flavors of clove or allspice. Sabinene, an isoprenoid, adds the taste of nutmeg, and another isoprenoid, zingiberene, adds the zip of the ginger. But just these molecules alone cannot simulate the rich flavor of a pumpkin pie. What brings the depth and richness to the flavor? Food chemists think that it is cyclotene, a compound that is found in almost all sugar-containing products that are roasted. It is frequently added to maple syrup products and brings a deep caramel-like flavor and aroma.

The next time you sip a pumpkin spice latte or munch on a pumpkin spice Oreo cookie, think about some of the amazing compounds that bring lovely flavors and aromas to our lives. Think also of the clever chemists who work so hard in their test kitchens and labs to mimic the bounty of nature.

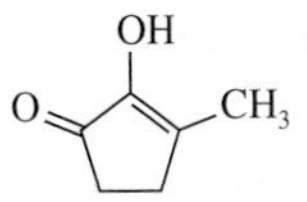

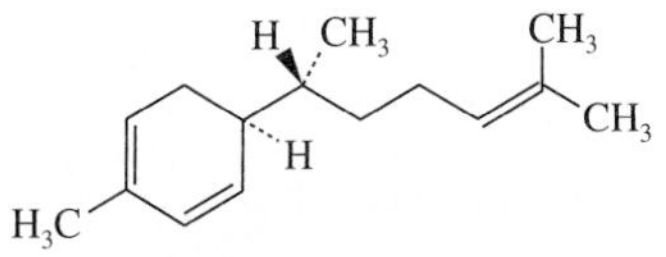

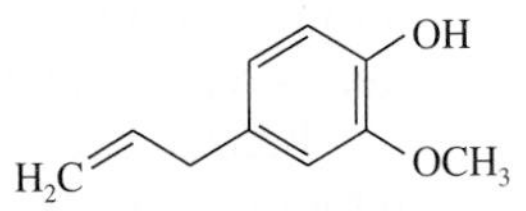

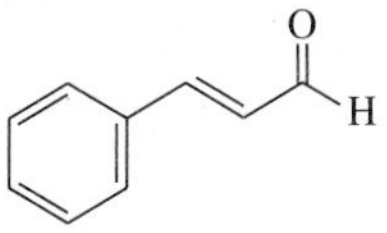

For Further Understanding

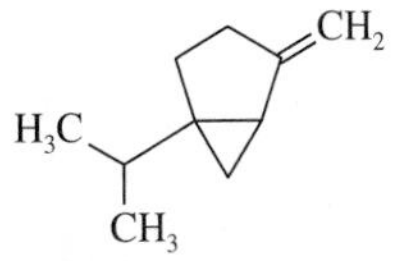

▸ List the functional groups for each of the flavor molecules shown above.

▸ Go online to find the sources of each of these flavor molecules and other uses that they have in the food industry.

Polynuclear Aromatic Hydrocarbons

The polynuclear aromatic hydrocarbons (PAH) are composed of two or more aromatic rings joined together. Many of them have been shown to be carcinogenic; that is, they cause cancer.

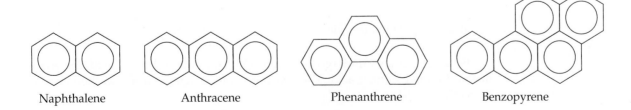

Naphthalene   Anthracene   Phenanthrene   Benzopyrene

Naphthalene has a distinctive aroma. It has been frequently used as mothballs and may cause hemolytic anemia (a condition causing breakdown of red blood cells) in humans, but has not been associated with human or animal cancers. Anthracene, derived from coal tar, is the parent compound of many dyes and pigments. Phenanthrene is common in the environment, although there is no industrial use of the compound. It is a product of incomplete combustion of fossil fuels and wood. Although anthracene is a suspected carcinogen, phenanthrene has not been shown to be one. Benzopyrene is found in tobacco smoke, smokestack effluents, charcoal-grilled meat, and automobile exhaust. It is one of the most potent carcinogens known.

Reactions Involving Benzene

LEARNING GOAL

8 Write equations for substitution reactions involving benzene.

As we have noted, benzene does not readily undergo addition reactions as the alkenes and alkynes do. The typical reactions of benzene are **substitution reactions,** in which a hydrogen atom is replaced by another atom or group of atoms. Unlike addition reactions of alkenes, no carbon-carbon bonds are broken. In substitution reactions of benzene, the carbon-hydrogen bond is broken and the hydrogen is replaced (substituted) with another atom or group of atoms.

Benzene can react (by substitution) with Cl_2 or Br_2. These reactions require either iron or an iron halide as a catalyst. For example:

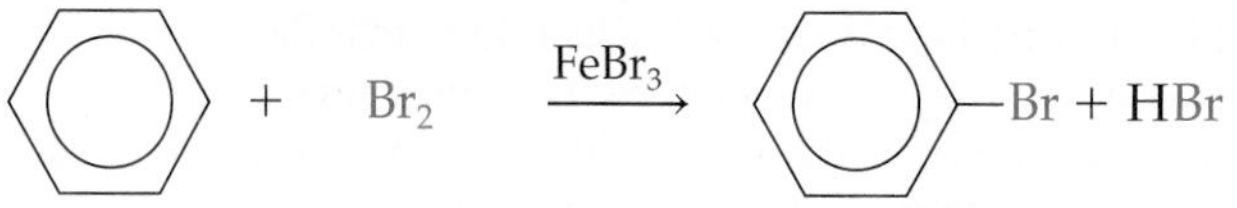

| Benzene | Bromine | Bromobenzene |

Notice that only one hydrogen atom is replaced in the reaction. The other product is a hydrogen halide. In the reaction above, the two products are bromobenzene and hydrogen bromide. When a second equivalent of the halogen is added, three isomers—*para, ortho,* and *meta*—are formed.

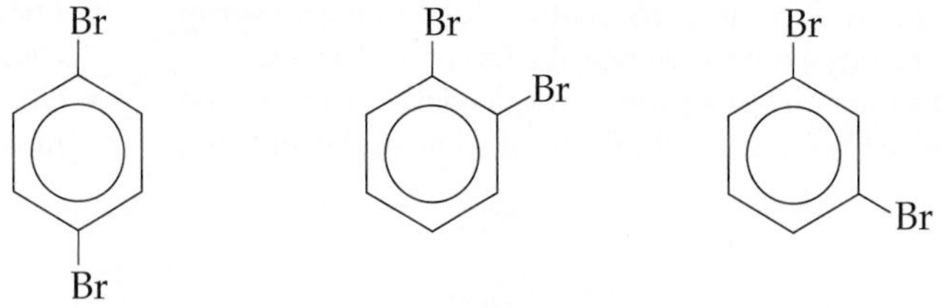

| *para*-Dibromobenzene | *ortho*-Dibromobenzene | *meta*-Dibromobenzene |
| 1,4-Dibromobenzene | 1,2-Dibromobenzene | 1,3-Dibromobenzene |

Benzene also reacts with sulfur trioxide by substitution. Concentrated sulfuric acid is required as the catalyst. Benzenesulfonic acid, a strong acid, is the product:

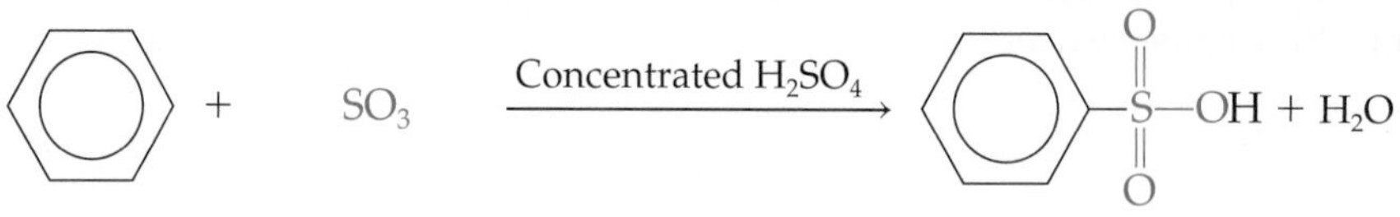

| Benzene | Sulfur trioxide | Benzenesulfonic acid |

Benzene can also undergo nitration with concentrated nitric acid dissolved in concentrated sulfuric acid. This reaction requires temperatures in the range of 50–55°C.

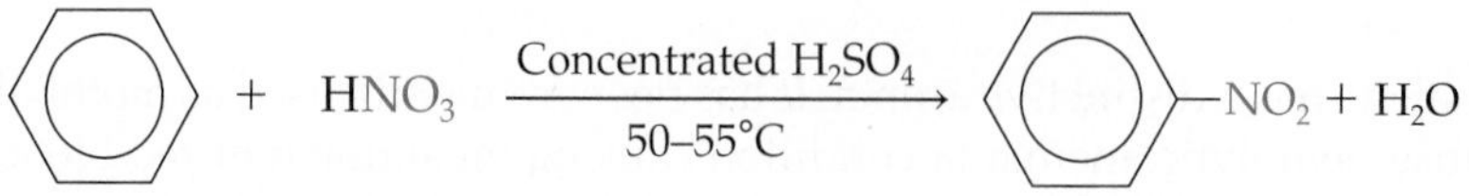

| Benzene | Nitric acid | Nitrobenzene |

11.7 Heterocyclic Aromatic Compounds

LEARNING GOAL

9 Describe heterocyclic aromatic compounds and list several biological molecules in which they are found.

Heterocyclic aromatic compounds are those having at least one atom other than carbon as part of the structure of the aromatic ring. The structures and common names of several heterocyclic aromatic compounds are shown:

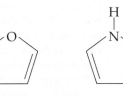

Pyridine Pyrimidine Purine

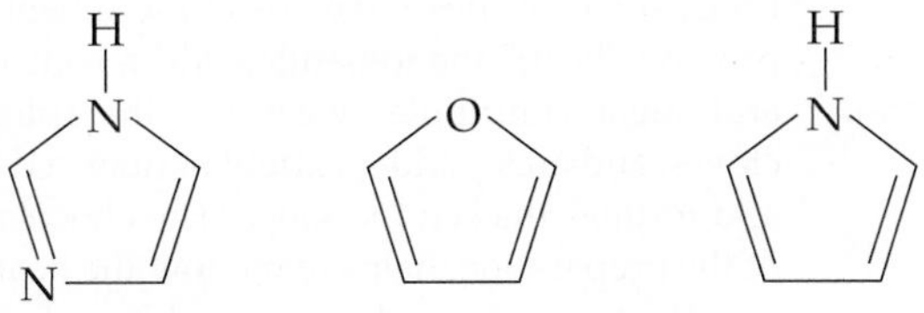

Imidazole Furan Pyrrole

All these compounds are more similar to benzene in stability and chemical behavior than they are to the alkenes. Many of these compounds are components of molecules that have significant effects on biological systems. For instance, the purines and pyrimidines are components of DNA (deoxyribonucleic acid) and RNA (ribonucleic acid). DNA and RNA are the molecules responsible for storing and expressing the genetic information of an organism. The pyridine ring is found in nicotine, the addictive compound in tobacco. The pyrrole ring is a component of the porphyrin ring found in hemoglobin and chlorophyll.

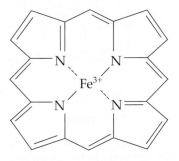

Porphyrin

The imidazole ring is a component of cimetidine, a drug used in the treatment of stomach ulcers. The structure of cimetidine is shown below:

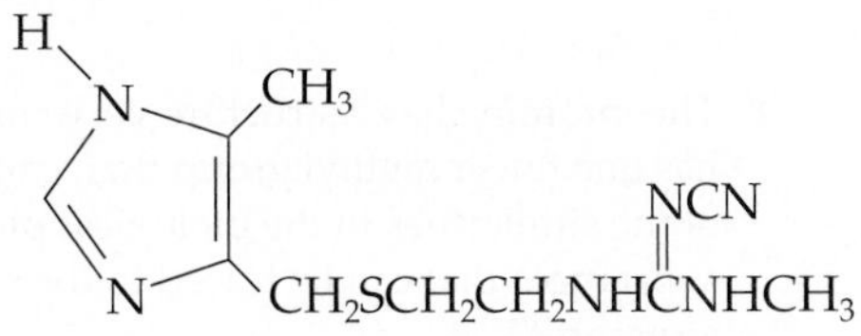

Cimetidine

We will discuss a subset of the heterocyclic aromatic compounds, the heterocyclic amines, in Chapter 15.

Amazing Chocolate

Chocolate has been consumed for over 3000 years. The Mayans and Aztecs enjoyed a frothy drink made from ground-up cacao beans. The Mayans preferred it hot and the Aztecs drank it cold. Although the Mayans could grow the cacao tree, the Aztecs couldn't. As a result, the beans became very valuable and were even used as money. The drink that the Aztecs and Mayans enjoyed was very bitter and was sometimes enhanced by adding ground chili peppers or vanilla. They thought that the drink would fight fatigue. Both the bitter taste and the mood enhancement that lifted fatigue are caused by theobromine.

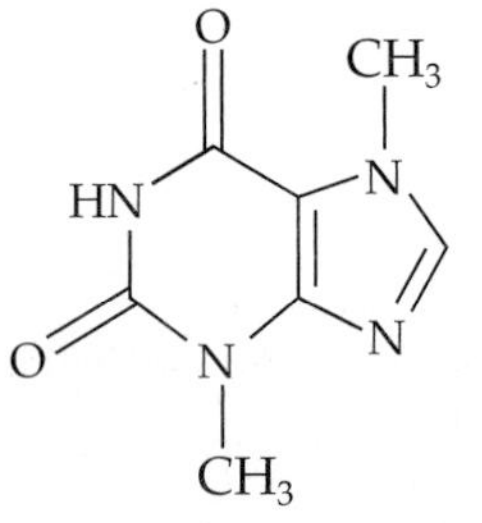

Theobromine

Theobromine is an alkaloid that, in pure form, is a crystalline, bitter powder. Medically, it can be used as a vasodilator, a diuretic, and a heart stimulant. One study concluded that theobromine was a better cough suppressant than codeine, and it has been found helpful in the treatment of asthma. As with any pharmacologically active substance, an excess of theobromine can cause theobromine poisoning, which has been observed in some, particularly the elderly, who eat chocolate to excess!

The chocolate we enjoy is produced by first allowing the cacao seeds to ferment to reduce the bitter taste. The shells are then removed and the inner nib is dried and ground into a powder that is pure chocolate. The powder is generally liquefied, placed in molds to harden, and then processed into cocoa solids and cocoa butter. We can purchase chocolate in many forms. Baking chocolate is generally bitter. No sugar has been added and it primarily consists of the cocoa solids and cocoa butter. Milk chocolate, dark chocolate, and white chocolate are produced by combining the cocoa butter and solids in varying amounts and adding other ingredients, particularly sugar and milk or condensed milk. Dark chocolate has the greatest amount of cocoa solids and, at the extreme, white chocolate has no cocoa solids and is enjoyed by those who have allergies to cocoa solids.

A Mexican sauce made with chili peppers and chocolate brings to mind the drink of the Aztecs. It is a mole sauce (pronounced *mole lay*). One legend tells us that this sauce was invented by the nuns in Santa Rosa Convent in Puebla, Mexico. The nuns were expecting a visit from the Archbishop. Being poor, they were concerned about the meal they would prepare for him.

After praying about it, they simply took what they had, some chili peppers, nuts, fruit, bread, tomatillos, spices, and chocolate. They mixed these ingredients into a sauce and served it over turkey. The Archbishop was impressed with the meal and thus mole sauce over poultry has been a favorite in Mexican cuisine for centuries. It has been said that the secret to the mole sauce is that it combines so many different flavors and sensations; some recipes have 20–30 ingredients! The chili peppers provide "heat;" the tomatillos add a sour note; the dried fruits and sugar contribute sweetness; the spices including cumin, cloves, and anise, add a variety of nuanced flavors; and the nuts and tortillas thicken the sauce. The chocolate, added at the end of the preparation, helps to mellow the heat of the peppers.

Recently scientists have published data to suggest that dark chocolate can reduce blood pressure. Other articles claim that dark chocolate can prevent cancer and heart disease. The molecules responsible for these observed effects are a variety of catechins and derivatives of phenol (see the structures below). Found also in tea, these compounds are antioxidants. It is thought they prevent the oxidation of lipids in the blood, which, in turn, is thought to prevent them from sticking to the surfaces of blood vessels. This reduces the risk of atherosclerosis or hardening of the arteries.

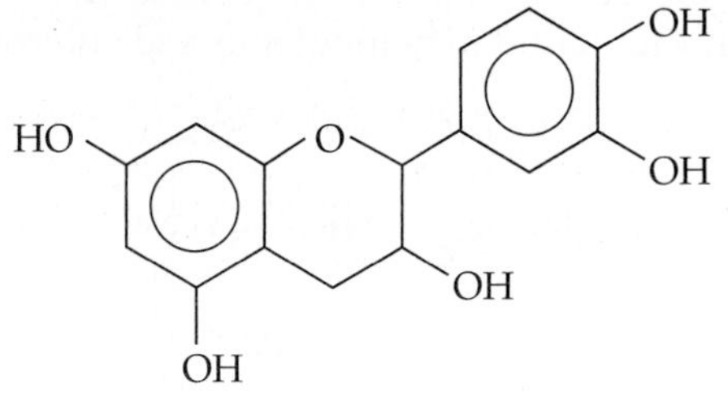

Catechin 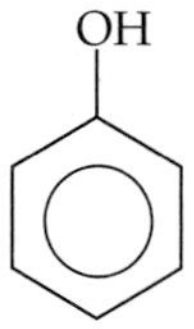   Phenol

We enjoy chocolate in dark, bittersweet form, as creamy milk chocolate, in hot cocoa, and in mole sauces. Whether it is the wide variety of tastes, the melt-in-your mouth sensation, the mild enhancement of mood and energy, or the reported health benefits, chocolate has been enjoyed over the ages and will continue to be a treat in years to come.

For Further Understanding

▸ Theobromine has a structure very similar to caffeine. It has one fewer methyl group than caffeine. What are some of the similarities in the biological properties of these two substances that can be traced to the similarity in their structures?

▸ Why is dark chocolate a richer source of antioxidants than milk chocolate or white chocolate?

CHAPTER MAP

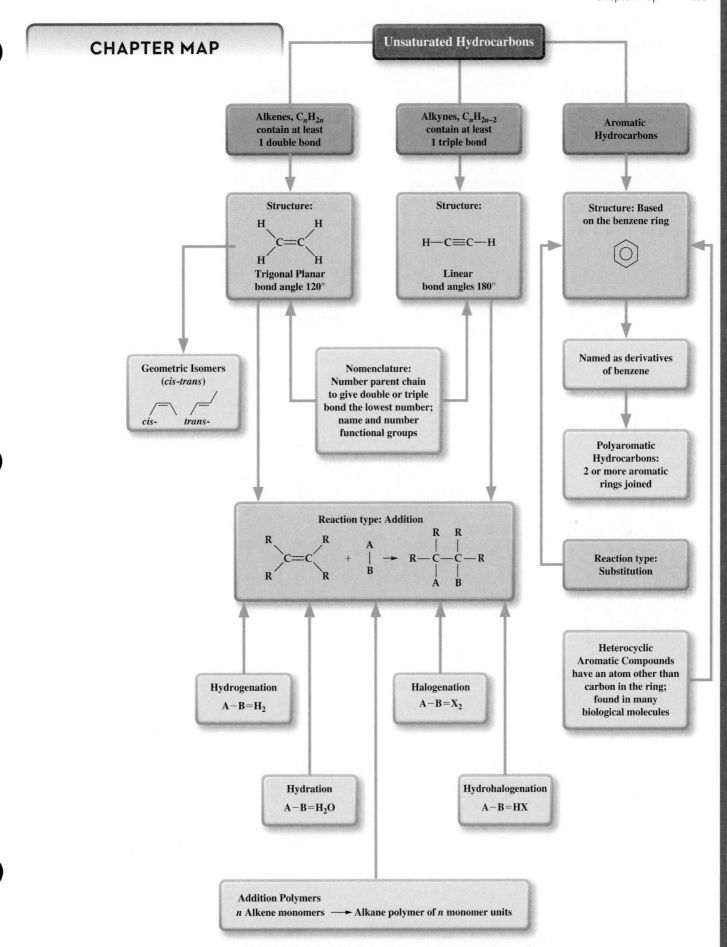

Unsaturated Hydrocarbons

Alkenes, C_nH_{2n} contain at least 1 double bond

Alkynes, C_nH_{2n-2} contain at least 1 triple bond

Aromatic Hydrocarbons

Structure:

H   H
 \   /
  C=C
 /   \
H   H

Trigonal Planar bond angle 120°

Structure:

H—C≡C—H

Linear bond angles 180°

Structure: Based on the benzene ring

Geometric Isomers (*cis-trans*)

cis-   *trans-*

Nomenclature: Number parent chain to give double or triple bond the lowest number; name and number functional groups

Named as derivatives of benzene

Polyaromatic Hydrocarbons: 2 or more aromatic rings joined

Reaction type: Addition

R   R
 \   /
  C=C   A
 /   \   + | →   R—C—C—R
R   R   B | |
                        A   B

with groups R, R on top carbons

Reaction type: Substitution

Hydrogenation A−B=H₂

Halogenation A−B=X₂

Heterocyclic Aromatic Compounds have an atom other than carbon in the ring; found in many biological molecules

Hydration A−B=H₂O

Hydrohalogenation A−B=HX

Addition Polymers
n Alkene monomers ⟶ Alkane polymer of *n* monomer units

SUMMARY OF REACTIONS

Addition Reactions of Alkenes

Hydrogenation:

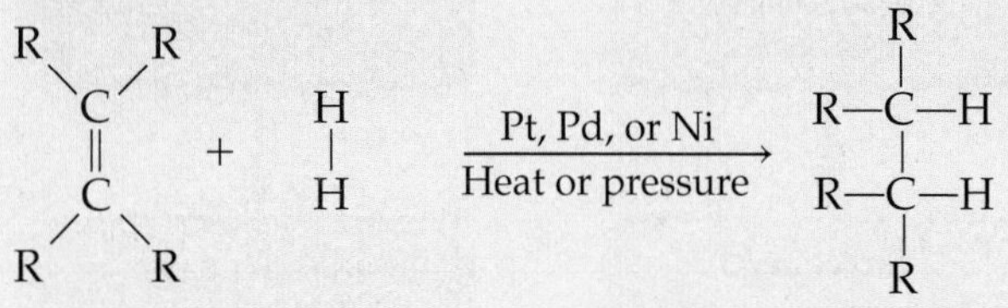

Alkene   Hydrogen   Alkane

Hydration:

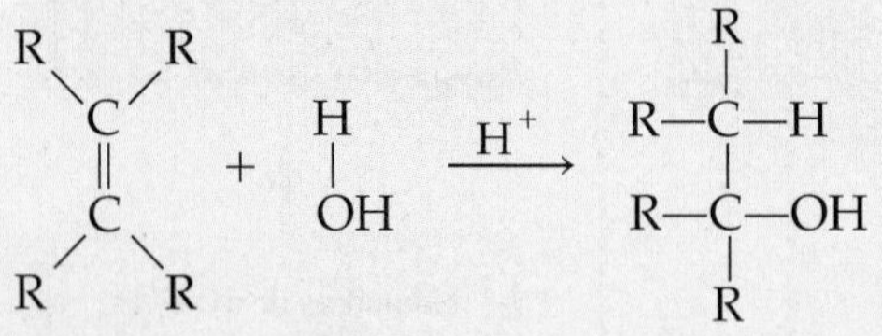

Alkene   Water   Alcohol

Halogenation:

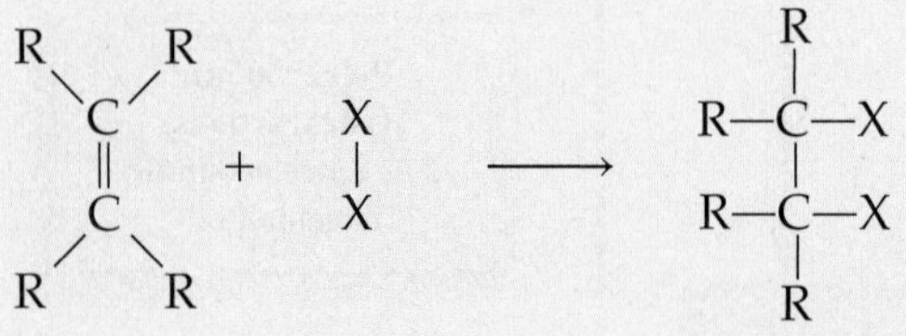

Alkene   Halogen   Alkyl dihalide

Hydrohalogenation:

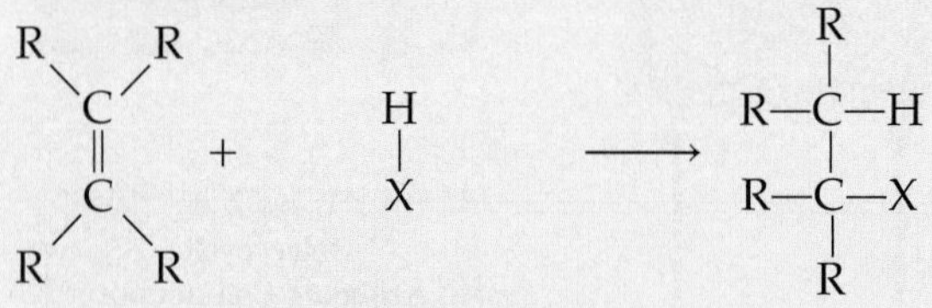

Alkene   Hydrogen halide   Alkyl halide

Reactions of Benzene

Halogenation:

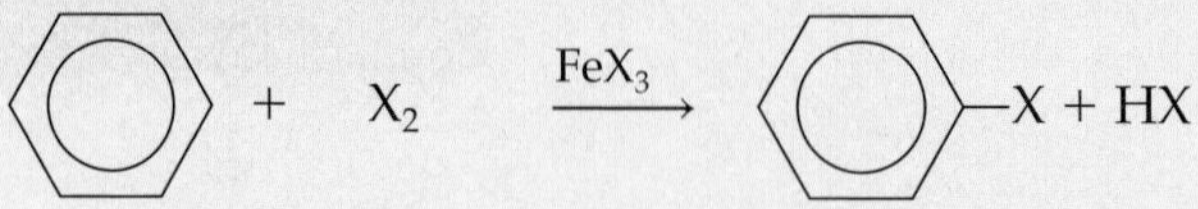

Benzene   Halogen   Halobenzene

Sulfonation:

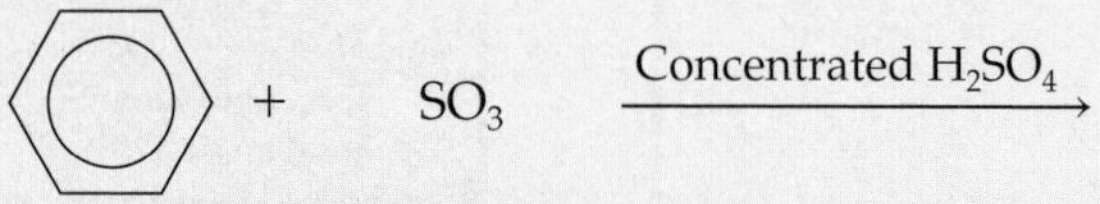

Benzene   Sulfur trioxide

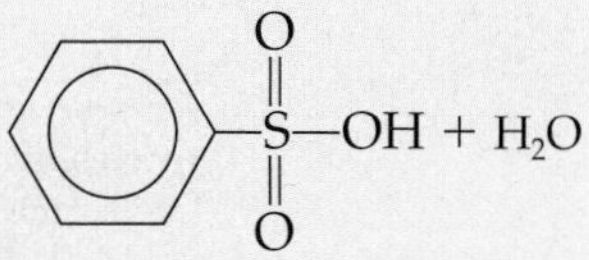

Benzenesulfonic acid

Nitration:

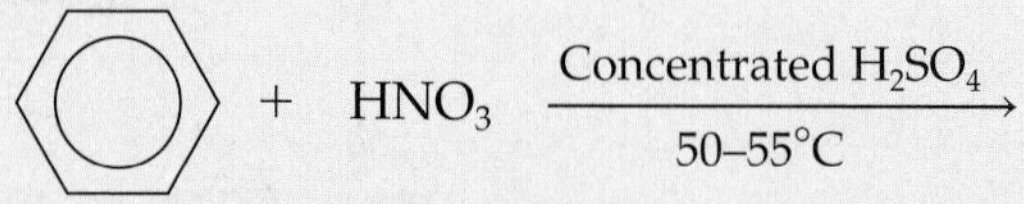

Benzene   Nitric acid

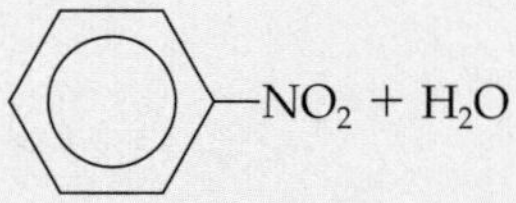

Nitrobenzene

Addition Polymers of Alkenes

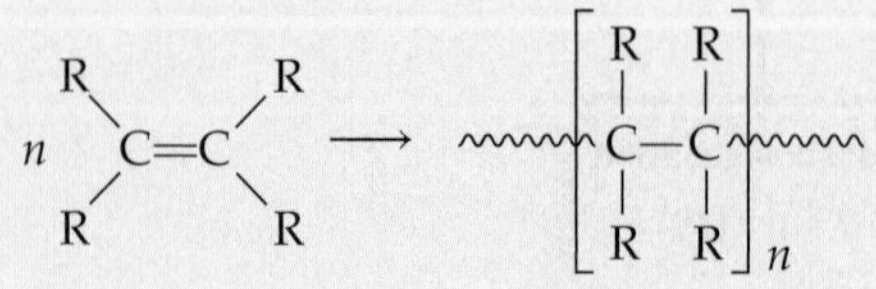

Alkene monomer   Addition polymer

SUMMARY

11.1 Alkenes and Alkynes: Structure and Physical Properties

▶ **Alkenes** and **alkynes** are **unsaturated** hydrocarbons because they have at least one C—C double bond (alkenes) or C—C triple bond (alkynes).

▶ Alkenes have the general formula C_nH_{2n} and alkynes have the general formula C_nH_{2n-2}.

▶ Alkenes and alkynes have physical properties like alkanes, but very different chemical properties.

11.2 Alkenes and Alkynes: Nomenclature

▶ Identify the parent compound and replace the *-ane* ending with *-ene* for alkenes and *-yne* for alkynes.

▶ Number the parent chain to give the lowest number to the double or triple bond.

▶ Name and number the other groups and place them alphabetically in front of the parent alkene or alkyne name.

11.3 Geometric isomers: A Consequence of Unsaturation

▶ Because the carbon-carbon double bond is rigid, **geometric isomers** occur when two different groups are bonded to each of the carbons of the double bond.

▶ When identical groups are on the same side of the double bond, the prefix *cis-* is used. When identical groups are on opposite sides of the double bond, the prefix *trans-* is used.

11.4 Alkenes in Nature

▶ Alkenes and polyenes (alkenes with several carbon-carbon double bonds) are common in nature.

▶ Isoprenoids, or terpenes, include steroids, chlorophyll and other photosynthetic pigments, and vitamins A, D, E, and K.

11.5 Reactions Involving Alkenes and Alkynes

▶ Alkenes and alkynes undergo **addition reactions** in which two atoms or groups of atoms add to the C—C double or triple bond.

▶ Addition reactions include: **halogenation, hydrohalogenation, hydration,** and **hydrogenation.**

▶ Markovnikov's rule predicts the most abundant products in addition reactions involving unsymmetrical alkenes.

▶ **Polymers** are formed by the sequential addition of alkene **monomers.** These are called **addition polymers.**

11.6 Aromatic Hydrocarbons

▶ **Aromatic hydrocarbons** contain benzene rings.

▶ Simple aromatic compounds are named as derivatives of benzene. Others have historical common names.

▶ A **phenyl group** is a benzene ring with one hydrogen atom removed. The term may be used in the nomenclature of molecules that include the benzene ring as a substituent group.

▶ Benzene participates in substitution reactions in which a hydrogen atom is replaced by another atom or group.

▶ Polynuclear aromatic hydrocarbons consist of two or more benzene rings joined.

11.7 Heterocyclic Aromatic Compounds

▶ **Heterocyclic aromatic compounds** have at least one atom other than carbon in the structure of the aromatic ring.

▶ Many of these compounds are important components of biological molecules, including DNA, RNA, and hemoglobin.

ANSWERS TO PRACTICE PROBLEMS

11.1   a. 6-Chloro-4-methyl-2-heptene   c. 5-Bromo-3-heptyne
       b. 1,4-Pentadiyne   d. 1,5-Heptadiene

11.2   a. l-Chlorocyclopropene   c. 3,4-Dibromocyclopentene
       b. 4,5-Dimethylcyclohexene   d. 3-Fluorocyclobutene

11.3   a. *cis*-isomer   b. *trans*-isomer   c. *cis*-isomer

11.4   a. *cis*-4,5-Dibromo-2-hexene
       b. *trans*-3,4-Dibromo-3-hexene
       c. *cis*-2,3-Dichloro-2-hexene

11.5   a. 2-Pentene can exist as both *cis*- and *trans*-isomers.

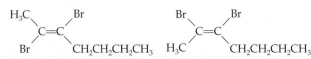

              cis-2-Pentene   *trans*-2-Pentene

   b. 1,1,2-Tribromo-1-pentene cannot exist as *cis*- and *trans*-isomers because one of the carbons involved in the double bond is also bonded to two identical atoms (Br).

   c. Propene cannot exist as *cis*- and *trans*-isomers because one of the carbons involved in the double bond is also bonded to two identical atoms (H).

   d. 2,3-Dibromo-2-heptene can exist as both *cis*- and *trans*-isomers.

trans-2,3-Dibromo-2-heptene   *cis*-2,3-Dibromo-2-heptene

11.6 **a.**

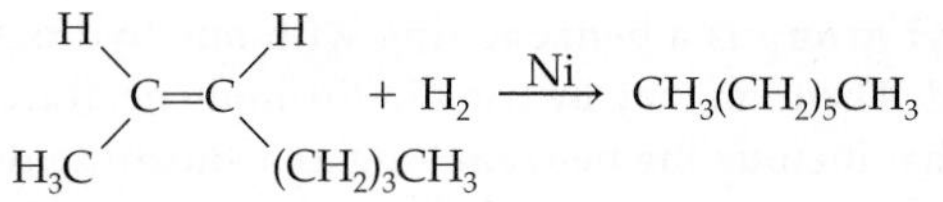

cis-2-Heptene   Heptane

$+ H_2 \xrightarrow{Ni} CH_3(CH_2)_5CH_3$

b.

$+ H_2 \xrightarrow{Ni} CH_3CH_2CH_2CH_2CH_3$

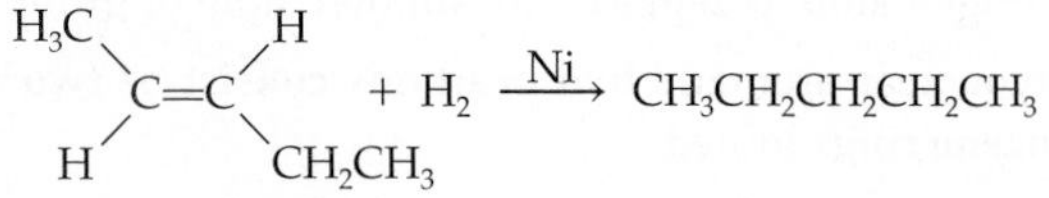

trans-2-Pentene   Pentane

c. $CH_3CH_2CH=CHCH_2CH_3 + H_2 \xrightarrow{Pt} CH_3CH_2CH_2CH_2CH_2CH_3$

3-Hexene   Hexane

The product, hexane, would be the same regardless of whether this were cis- or trans-3-hexene.

d. $CH_2=CHCH_3 + H_2 \xrightarrow{Pd} CH_3CH_2CH_3$

Propene   Propane

11.7 **a.**

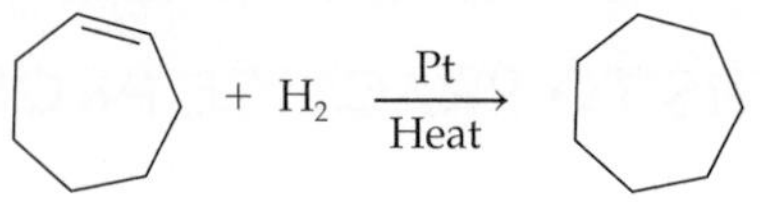

$+ H_2 \xrightarrow[Heat]{Pt}$

Cycloheptene   Cycloheptane

b.

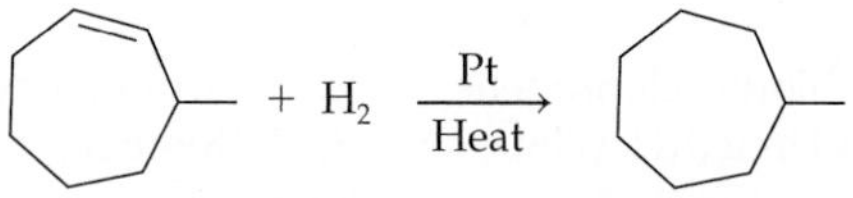

$+ H_2 \xrightarrow[Heat]{Pt}$

3-Methylcycloheptene   Methylcycloheptane

c.

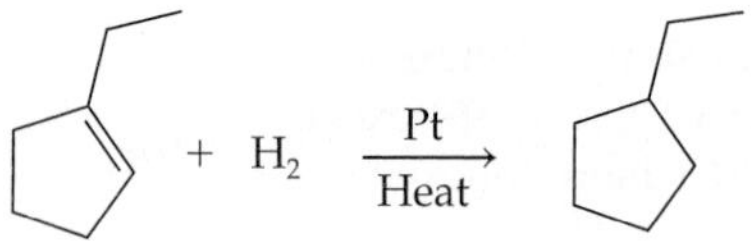

$+ H_2 \xrightarrow[Heat]{Pt}$

1-Ethylcyclopentene   Ethylcyclopentane

d.

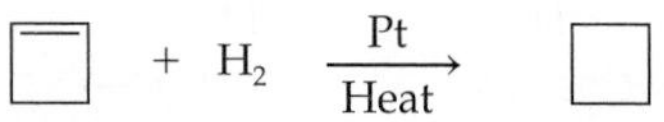

$+ H_2 \xrightarrow[Heat]{Pt}$

Cyclobutene   Cyclobutane

11.8 **a.**

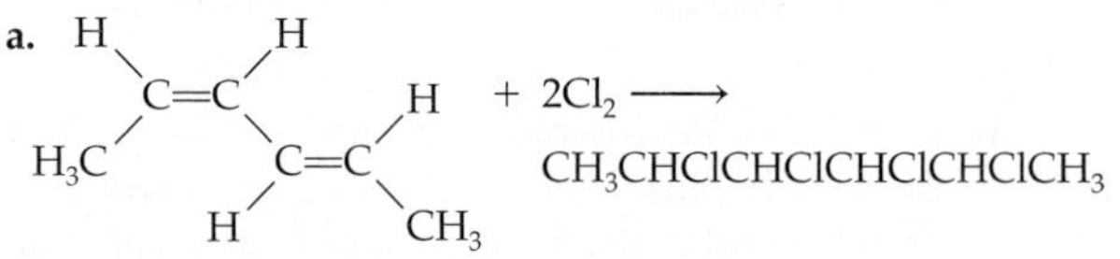

$+ 2Cl_2 \longrightarrow$

$CH_3CHClCHClCHClCHClCH_3$

2,4-Hexadiene   2,3,4,5-Tetrachlorohexane

b. Both cis- and trans-1,5-dibromo-2-pentene would produce 1,5-dibromo-2,3-dichloropentane when chlorinated. The trans-isomer is shown here:

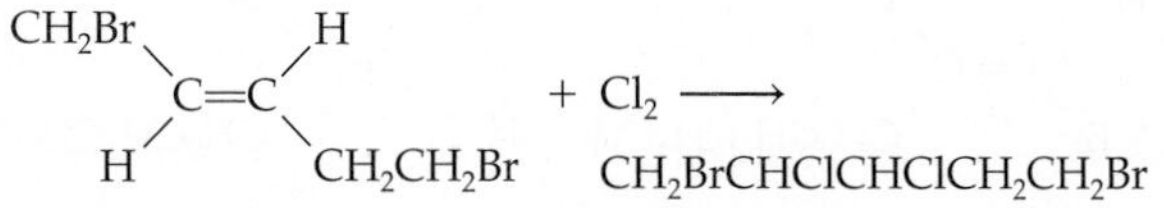

$+ Cl_2 \longrightarrow$

$CH_2BrCHClCHClCH_2CH_2Br$

trans-1,5-Dibromo-2-pentene 1,5-Dibromo-2,3-dichloropentane

c.

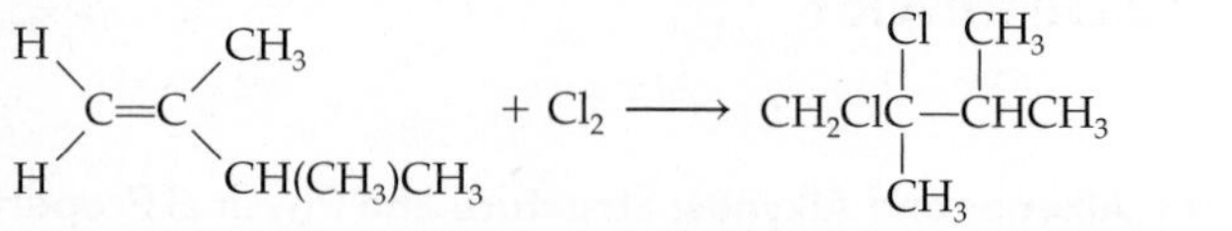

$+ Cl_2 \longrightarrow$

$\underset{\underset{CH_3}{|}}{CH_2ClC}Cl-\underset{\underset{CH_3}{|}}{CHCH_3}$

2,3-Dimethyl-1-butene   1,2-Dichloro-2,3 dimethylbutane

d. Both cis- and trans-4,6-dimethyl-2-heptene would produce 2,3-dichloro-4,6-dimethylheptane when chlorinated. The cis-isomer is shown here:

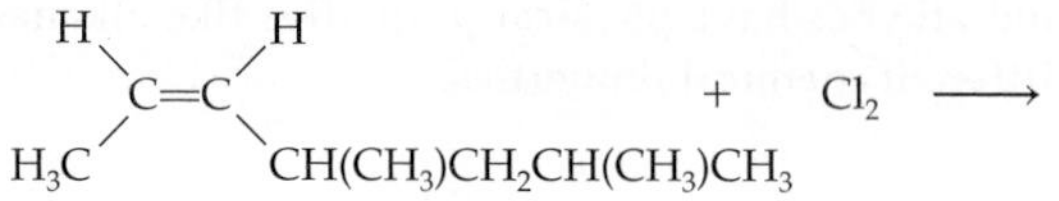

$+ Cl_2 \longrightarrow$

$CH_3CHClCHClCH(CH_3)CH_2CH(CH_3)CH_3$

cis-4,6-Dimethyl-2-heptene   2,3-Dichloro-4,6-dimethylheptane

11.9 **a.**

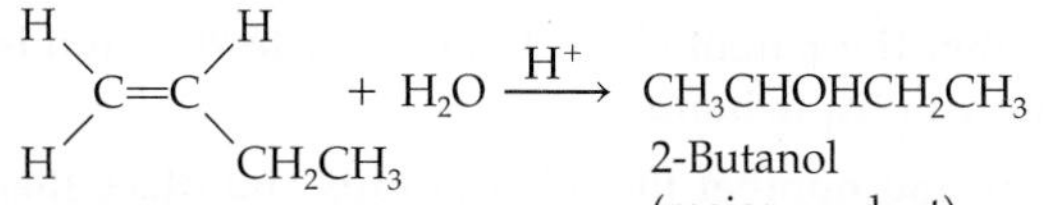

$+ H_2O \xrightarrow{H^+} CH_3CHOHCH_2CH_3$

2-Butanol
(major product)

1-Butene

$+ \quad CH_2OHCH_2CH_2CH_3$

1-Butanol
(minor product)

b. 2-Methyl-3-hexanol and 5-methyl-3-hexanol would be produced in equal amounts.

$CH_3CH(CH_3)CH=CHCH_2CH_3 + H_2O \xrightarrow{H^+}$

$CH_3CH(CH_3)CHOHCH_2CH_2CH_3$

2-Methyl-3-hexene   2-Methyl-3-hexanol

$CH_3CH(CH_3)CH=CHCH_2CH_3 + H_2O \xrightarrow{H^+}$

$CH_3CH(CH_3)CH_2CHOHCH_2CH_3$

2-Methyl-3-hexene   5-Methyl-3-hexanol

c. $CH_2=CHCH_3 + H_2O \xrightarrow{H^+} CH_3CHOHCH_3$

Propene   2-Propanol
(major product)

$+ \quad CH_2OHCH_2CH_3$

1-Propanol
(minor product)

d. $CH_2ClCH=CHCH_2Cl + H_2O \xrightarrow{H^+} CH_2ClCHOHCH_2CH_2Cl$

1,4-Dichloro-2-butene   1,4-Dichloro-2-butanol
(only product)

11.10 a. 2-Bromopentane and 3-bromopentane would be produced in equal amounts.

$$CH_2CH=CHCH_2CH_3 + HBr \longrightarrow CH_2CHBrCH_2CH_2CH_3$$

2-Pentene   2-Bromopentane

$$+ CH_3CH_2CHBrCH_2CH_3$$

3-Bromopentane

b.

$$CH_2=CHCH_3 + HBr \longrightarrow CH_3CHBrCH_3 + CH_2BrCH_2CH_3$$

Propene   2-Bromopropane   1-Bromopropane
                                      (major product)   (minor product)

c. 3-Bromoheptane and 4-bromoheptane would be produced in equal amounts.

$$CH_3CH_2CH=CHCH_2CH_2CH_3 + HBr \longrightarrow$$

3-Heptene

$$CH_3CH_2CHBr(CH_2)_3CH_3$$

3-Bromoheptane

$$+ CH_3(CH_2)_2CHBr(CH_2)_2CH_3$$

4-Bromoheptane

11.11 a. IUPAC Name: 2-Bromotoluene
    Common name: *ortho*-Bromotoluene or *o*-bromotoluene
b. IUPAC Name: 4-Methylphenol
    Common name: *para*-Cresol or *p*-cresol
c. IUPAC Name: 2,3-Diethylaniline
    Common name: 2,3-Diethylaniline (Numbers must be used since there are three substituents.)
d. IUPAC Name: 1,3-Dibromobenzene
    Common name: *meta*-Dibromobenzene or *m*-dibromobenzene

QUESTIONS AND PROBLEMS

Alkenes and Alkynes: Structure and Physical Properties

Foundations

11.23 Explain why the boiling points of alkynes increase as the length of the hydrocarbon chains increases.

11.24 Saturated fatty acids have higher melting points than monounsaturated fatty acids. Polyunsaturated fatty acids have even lower melting points. Develop a hypothesis to explain this observation.

11.25 Write the general formulas for alkanes, alkenes, and alkynes.

11.26 What are the characteristic functional groups of alkenes and alkynes?

Applications

11.27 Describe the geometry of ethene.
11.28 What are the bond angles in ethene?
11.29 Compare the bond angles in ethene with those in ethane.
11.30 Explain the bond angles of ethene in terms of the valence-shell electron-pair repulsion (VSEPR) theory.

11.31 Describe the geometry of ethyne.
11.32 What are the bond angles in ethyne?
11.33 Compare the bond angles in ethane, ethene, and ethyne.
11.34 Explain the bond angles of ethyne in terms of the valence-shell electron-pair repulsion (VSEPR) theory.
11.35 Arrange the following groups of molecules from the highest to lowest boiling points:
    a. ethyne   propyne   2-pentyne
    b. 2-butene   3-decene   ethene
11.36 Arrange the following groups of molecules from the highest to the lowest melting points.

a.

b.

Alkenes and Alkynes: Nomenclature

Foundations

11.37 Briefly describe the rules for naming alkenes and alkynes.
11.38 What is meant by a geometric isomer?
11.39 Describe what is meant by a *cis*-isomer of an alkene.
11.40 Describe what is meant by a *trans*-isomer of an alkene.

Applications

11.41 Draw a condensed formula for each of the following compounds:
    a. 2-Methyl-2-hexene
    b. *trans*-3-Heptene
    c. *cis*-1-Chloro-2-pentene
    d. *cis*-2-Chloro-2-methyl-3-heptene
    e. *trans*-5-Bromo-2,6-dimethyl-3-octene
11.42 Draw a condensed formula for each of the following compounds:
    a. 2-Hexyne
    b. 4-Methyl-1-pentyne
    c. 1-Chloro-4,4,5-trimethyl-2-heptyne
    d. 2-Bromo-3-chloro-7,8-dimethyl-4-decyne
11.43 Name each of the following using the IUPAC Nomenclature System:

a. $CH_3CH_2CH(CH_3)CH=CH_2$

b. $CH_2CH_2CH_2CH_2Br$
        |
        $CH_2CH=CH_2$

c. $CH_3CH_2CH=CHCHBrCH_2CH_3$

d.

11.44 Name each of the following using the IUPAC Nomenclature System:

    a. $CH_3CH(CH_3)CH_2CH=C(CH_3)_2$

    b. $CH_2ClCH(CH_3)C≡CH$

    c. $CH_3CHClCH_2CH_2CH_2C≡CH$

    d.

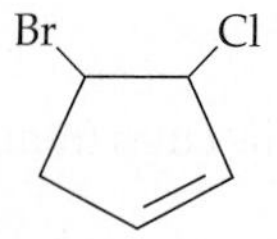

11.45 Draw each of the following compounds using condensed formulas:
    a. 1,3,5-Trifluoropentane
    b. *cis*-2-Octene
    c. Dipropylacetylene

11.46 Draw each of the following compounds using condensed formulas:
    a. 3,3,5-Trimethyl-1-hexene
    b. 1-Bromo-3-chloro-1-heptyne
    c. 3-Heptyne

Geometric Isomers: A Consequence of Unsaturation

11.47 Which of the following alkenes can exist as *cis-trans* isomers? Explain your reasoning.
    a. 1-Heptene   **d.** 2-Methyl-2-hexene
    b. 2-Heptene   **e.** 3-Methyl-2-hexene
    c. 3-Heptene

11.48 Draw the line formula for each of the alkenes in Question 11.47.

11.49 Which of the following alkenes would not exhibit *cis-trans* geometric isomerism?

    a.

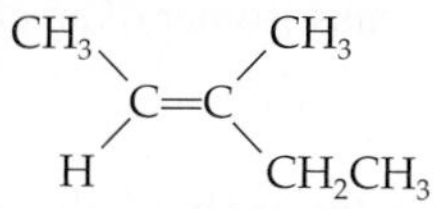

    b.

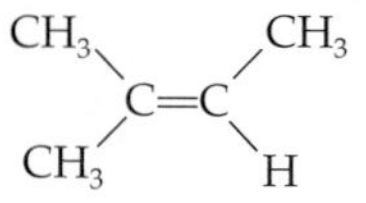

    c.

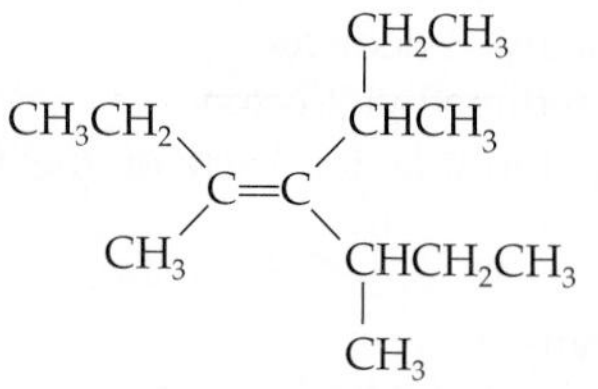

    d.

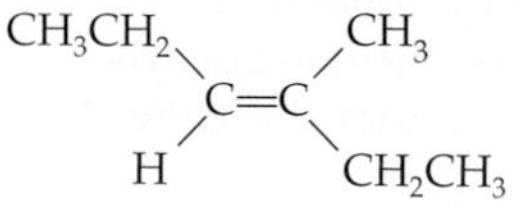

11.50 Which of the following structures have incorrect IUPAC names? If incorrect, give the correct IUPAC name.

    a. $CH_3C≡CCH_2CH(CH_3)_2$

    2-Methyl-4-hexyne

    b.

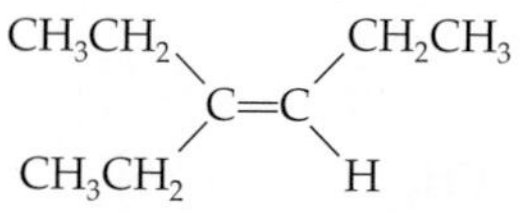

    3-Ethyl-3-hexene

    c. $CH_3CH(CH_3)CH_2C≡CCH_2CH(CH_3)CH_2CH_3$

    2-Ethyl-7-methyl-4-octyne

    d.

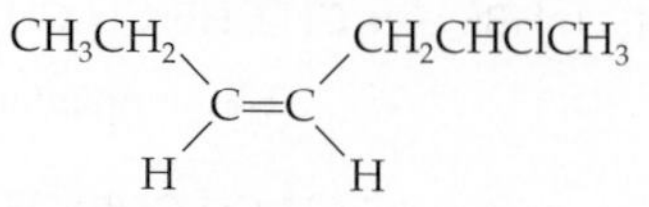

    trans-6-Chloro-3-heptene

    e.

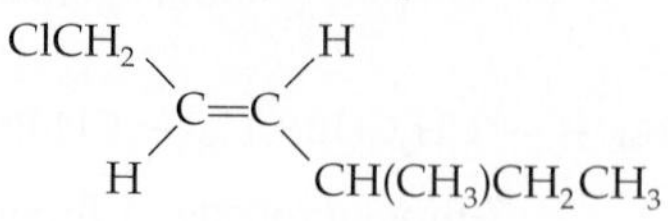

    1-Chloro-5-methyl-2-hexene

11.51 Which of the following can exist as *cis-* and *trans*-isomers?
    a. $H_2C=CH_2$
    b. $CH_3CH=CHCH_3$
    c. $Cl_2C=CBr_2$
    d. $ClBrC=CClBr$
    e. $(CH_3)_2C=C(CH_3)_2$

11.52 Draw and name all the *cis-* and *trans*-isomers in Question 11.51.

11.53 Provide the IUPAC name for each of the following molecules:

    a. $CH_2=CHCH_2CH_2CH=CHCH_2CH_2CH_3$
    b. $CH_2=CHCH_2CH=CHCH_2CH=CHCH_3$
    c. $CH_3CH=CHCH_2CH=CHCH_2CH_3$
    d. $CH_3CH=CHCH(CH_3)CH=CHCH_3$

11.54 Provide the IUPAC name for each of the following molecules:

    a. $CH_3CBr=CHCH(CH_3)CH=CBrCH_3$

    b. $CH_2=CHCH(CH_3)CH=CHCH(CH_2CH_3)CH_2CH_3$

    c. $CH_2=CHC(CH_3)_2CH=CHCH_2CH=CHCH(CH_3)_2$

    d. $CH_3CH(CH_3)CH(CH_2CH_3)CH=CHCH_2CH=CHCH(CH_3)CH_2CH_3$

Reactions Involving Alkenes and Alkynes

Foundations

11.55 Write a general equation representing the hydrogenation of an alkene.

11.56 Write a general equation representing the hydrogenation of an alkyne.

11.57 Write a general equation representing the halogenation of an alkene.

11.58 Write a general equation representing the halogenation of an alkyne.

11.59 Write a general equation representing the hydration of an alkene.

11.60 Write a general equation representing the hydration of an alkyne.

11.61 What is the principal difference between the hydrogenation of an alkene and the hydrogenation of an alkyne?

11.62 What is the major difference between the hydration of an alkene and the hydration of an alkyne?

Applications

11.63 Write an equation representing each of the following reactions:
    a. 1-Heptene + H_2O (H^+)
    b. 2-Heptene + HBr
    c. 3-Heptene + H_2
    d. 2-Methyl-2-hexene + HCl

11.64 Write a balanced equation for each of the following reactions:
   a. 1,4-Pentadiene + H_2
   b. 3-Methyl-1,4-cyclohexadiene + Cl_2
   c. 2,4-Heptadiene + Br_2
   d. 3-Methylcyclopentene + H_2O (H^+)

11.65 Complete each of the following reactions by supplying the missing reactant or product(s) as indicated by question marks:

   a. $CH_3CH_2CH{=}CHCH_2CH_3 + ? \longrightarrow CH_3(CH_2)_4CH_3$

   b. 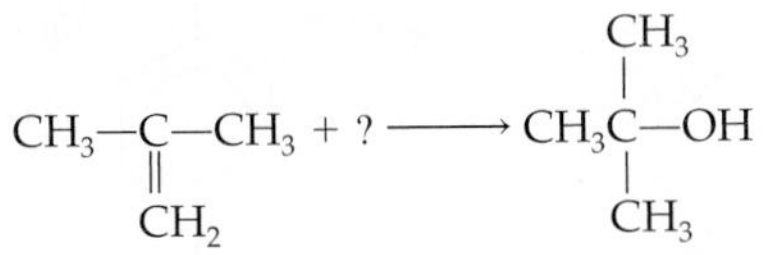
$$CH_3-\underset{\underset{CH_2}{\|}}{C}-CH_3 + ? \longrightarrow CH_3\underset{\underset{CH_3}{|}}{\overset{\overset{CH_3}{|}}{C}}-OH$$

   c. 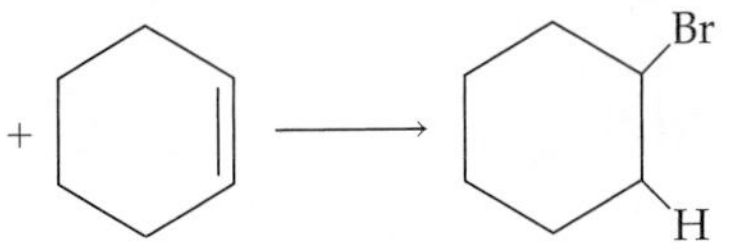 ? +

   d. $2CH_3CH_2CH_2CH_2CH_2CH_3 + ?O_2 \xrightarrow{\text{Heat}} ? + ?$
        (complete combustion)

   e. 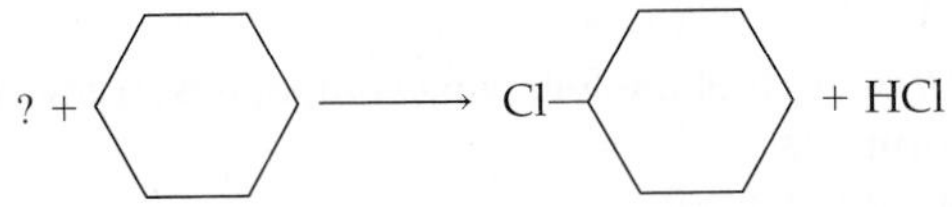 ? +
+ HCl

   f. 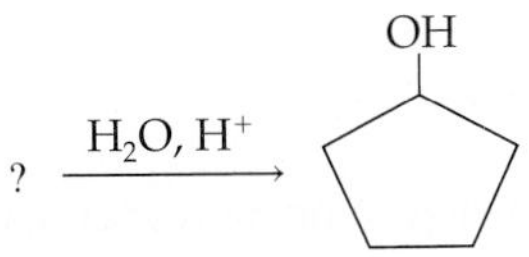 ? $\xrightarrow{H_2O, H^+}$

11.66 Draw and name the product in each of the following reactions:
   a. Cyclopentene + H_2O (H^+)
   b. Cyclopentene + HCl
   c. Cyclopentene + H_2
   d. Cyclopentene + HI

11.67 Write a balanced equation for each of the following reactions:
   a. Hydrogenation of 2-butyne
   b. Halogenation of 2-pentyne

11.68 Write a balanced equation for each of the following reactions:
   a. Hydration of 1-butyne
   b. Hydration of 2-butyne

11.69 Predict the major product in each of the following reactions. Name the alkene reactant and the product, using IUPAC nomenclature.

   a. 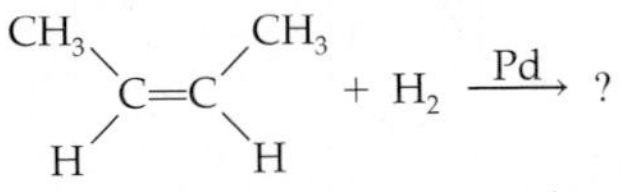
$+ H_2 \xrightarrow{Pd} ?$

   b. $CH_3CH_2CH{=}CH_2 + H_2O \xrightarrow{H^+} ?$
   c. $CH_3CH{=}CHCH_3 + Cl_2 \longrightarrow ?$
   d. $CH_3CH_2CH_2CH{=}CH_2 + HBr \longrightarrow ?$

11.70 Predict the major product in each of the following reactions. Name the alkene reactant and the product, using IUPAC nomenclature.

   a. 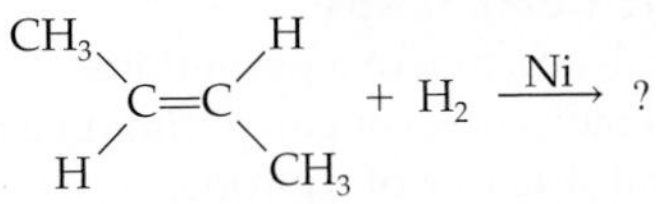
$+ H_2 \xrightarrow{Ni} ?$

   b. $(CH_3)_2C{=}CHCH_2CH_2CH_3 + H_2O \xrightarrow{H^+} ?$

   c. $(CH_3)_2C{=}CHCH_3 + Br_2 \longrightarrow ?$
   d. $CH_3C(CH_3)_2CH{=}CH_2 + HCl \longrightarrow ?$

11.71 Write a balanced equation for each of the following reactions:
   a. Hydrogenation of 4-chlorocyclooctene
   b. Halogenation of 1,3-cyclooctadiene
   c. Hydration of 3-methylcyclobutene
   d. Hydrohalogenation of cyclopentene

11.72 Write a balanced equation for each of the following reactions:
   a. Hydrogenation of 3,4-dichlorocycloheptene
   b. Halogenation of 3-methylcyclopentene
   c. Hydration of cyclobutene
   d. Hydrohalogenation of cyclohexene

11.73 A hydrocarbon with the formula C_5H_{10} decolorized Br_2 and consumed 1 mol of hydrogen upon hydrogenation. Draw all the isomers of C_5H_{10} that are possible based on the above information.

11.74 Triple bonds react in a manner analogous to that of double bonds. The extra pair of electrons in the triple bond, however, generally allows 2 mol of a given reactant to *add* to the triple bond in contrast to 1 mol with the double bond. The "rich get richer" rule holds. Predict the major product in each of the following reactions:
   a. Acetylene with 2 mol HCl
   b. Propyne with 2 mol HBr
   c. 2-Butyne with 2 mol HI

11.75 Complete each of the following by supplying the missing product indicated by the question mark:

   a. 2-Butene $\xrightarrow{HBr} ?$
   b. 3-Methyl-2-hexene $\xrightarrow{HI} ?$

   c. 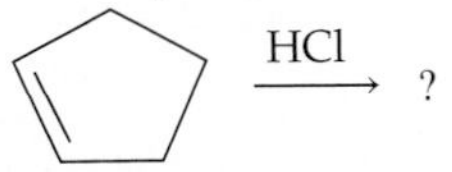
$\xrightarrow{HCl} ?$

11.76 Bromine is often used as a laboratory spot test for unsaturation in an aliphatic hydrocarbon. Bromine in CCl_4 is red. When bromine reacts with an alkene or alkyne, the alkyl halide formed is colorless; hence, a disappearance of the red color is a positive test for unsaturation. A student tested the contents of two vials, A and B, both containing compounds with a molecular formula, C_6H_{12}. Vial A decolorized bromine, but vial B did not. How may the results for vial B be explained? What class of compound would account for this?

11.77 What is meant by the term *polymer*?

11.78 What is meant by the term *monomer*?

11.79 Write an equation representing the synthesis of polyvinyl chloride from vinyl chloride. (*Hint:* Refer to Table 11.3.) What are some of the uses for polyvinyl chloride?

11.80 Write an equation representing the synthesis of polypropylene from propene. What are some uses for polypropylene?

11.81 Provide the IUPAC name for each of the following molecules. Write a balanced equation for the hydration of each.
   a. $CH_3CH{=}CHCH_2CH_3$
   b. $CH_2BrCH{=}CH_2$

   c.

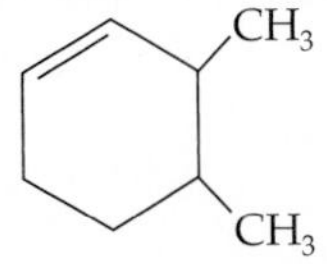

11.82 Provide the IUPAC name for each of the following molecules. Write equations for the hydration of each.

a.

—CH₂CH₃

b. CH₃CH=CHCH₂CH=CHCH₂CH=CHCH₃
c. CH₃CH=CHC(CH₃)₂CH₂CH₃

11.83 Write an equation for the addition reaction that produced each of the following molecules:

a. CH₂OHCH₂CH₂CH(CH₃)₂

b. CH₃CH₂CHBr(CH₂)₂CH₃

c.

Br

CH₃

d.

—CH₂CH₃

OH

11.84 Write an equation for the addition reaction that produced each of the following molecules:

a. CH₃CH₂CHOHCH(CH₃)CH₂CH₃

b. CH₃CHOHCH₂CH₃

c. HO   CH₃

CH₃

11.85 Draw the structure of each of the following compounds and write a balanced equation for the complete hydrogenation of each:

a. 1,4-Hexadiene   c. 1,3-Cyclohexadiene
b. 2,4,6-Octatriene   d. 1,3,5-Cyclooctatriene

11.86 Draw the structure of each of the following compounds and write a balanced equation for the bromination of each:
a. 3-Methyl-1,4-hexadiene
b. 4-Bromo-1,3-pentadiene
c. 3-Chloro-2,4-hexadiene
d. 3-Bromo-1,3-cyclohexadiene

Aromatic Hydrocarbons

Foundations

11.87 Where did the term *aromatic hydrocarbon* first originate?
11.88 What chemical characteristic of the aromatic hydrocarbons is most distinctive?
11.89 What is meant by the term *resonance hybrid*?
11.90 Draw a pair of structures to represent the benzene resonance hybrid.

Applications

11.91 Draw the structure for each of the following compounds:
a. 2,4-Dibromotoluene
b. 1,2,4-Triethylbenzene
c. Isopropylbenzene
d. 2-Bromo-5-chlorotoluene

11.92 Name each of the following compounds, using the IUPAC system.

a.   CH₃
        CH₃

d.   Br

CH₃   Cl

b.   NO₂
        NO₂

e. O₂N   CH₃   NO₂

Br

c.   CH₂CH₃

CH₃

11.93 Draw each of the following compounds, using condensed formulas:
a. *meta*-Cresol
b. Propylbenzene
c. 1,3,5-Trinitrobenzene
d. *m*-Chlorotoluene
11.94 Draw each of the following compounds, using condensed formulas:
a. *p*-Xylene
b. Isopropylbenzene
c. *m*-Nitroanisole
d. *p*-Methylbenzaldehyde
11.95 Describe the Kekulé model for the structure of benzene.
11.96 Describe the current model for the structure of benzene.
11.97 How does a substitution reaction differ from an addition reaction?
11.98 Give an example of a substitution reaction and of an addition reaction.
11.99 Write equations for the reactions that would produce each of the following products:

Br   Cl   NO₂

a.   b.   c.

11.100 Draw all of the products that could be formed in a reaction of benzene with each of the following:
a. 2Cl₂ in the presence of FeCl₃
b. 2SO₃ in the presence of concentrated sulfuric acid
c. 2HNO₃ in the presence of concentrated sulfuric acid

Heterocyclic Aromatic Compounds
11.101 Draw the general structure of a pyrimidine.
11.102 What biological molecules contain pyrimidine rings?
11.103 Draw the general structure of a purine.
11.104 What biological molecules contain purine rings?

CHALLENGE PROBLEMS

1. There is a plastic polymer called polyvinylidene difluoride (PVDF) that can be used to sense a baby's breath and thus be used to prevent sudden infant death syndrome (SIDS). The secret is that this polymer can be specially processed so that it becomes piezoelectric (produces an electrical current when it is physically deformed) and pyroelectric (develops an electrical potential when its temperature changes). When a PVDF film is placed beside a sleeping baby, it will set off an alarm if the baby stops breathing. The structure of this polymer is shown here:

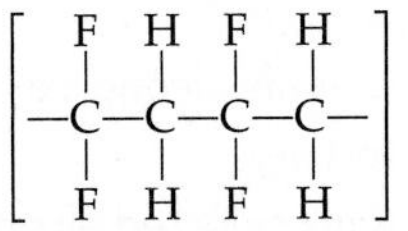

$$\begin{bmatrix} & F & H & F & H \\ -C & -C & -C & -C- \\ & F & H & F & H \end{bmatrix}$$

Go to the library and investigate some of the other amazing uses of PVDF. Draw the structure of the alkene from which this compound is produced.

2. Isoprene is the repeating unit of the natural polymer rubber. It is also the starting material for the synthesis of cholesterol and several of the lipid-soluble vitamins, including vitamin A and vitamin K. The structure of isoprene is shown below.

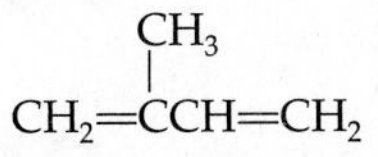

$$CH_2=CCH=CH_2$$
with CH_3 attached

What is the IUPAC name for isoprene?

3. When polyacrylonitrile is burned, toxic gases are released. In fact, in airplane fires, more passengers die from inhalation of toxic fumes than from burns. Refer to Table 11.3 for the structure of acrylonitrile. What toxic gas would you predict to be the product of the combustion of these polymers?

4. If a molecule of polystyrene consists of 25,000 monomers, what is the molar mass of the molecule?

5. A factory produces one million tons (t) of polypropylene. How many moles (mol) of propene would be required to produce this amount? What is the volume of this amount of propene at 25°C and 1 atmosphere (atm)?

12

Alcohols, Phenols, Thiols, and Ethers

Sugar-free chocolates are not calorie free!

OUTLINE

LEARNING GOALS

1 Classify alcohols as primary, secondary, or tertiary.

2 Rank selected alcohols by relative water solubility, boiling points, or melting points.

3 Write the names and draw the structures for common alcohols.

4 Discuss the biological, medical, or environmental significance of several alcohols.

5 Write equations representing the preparation of alcohols by the hydration of an alkene.

6 Write equations representing the preparation of alcohols by hydrogenation (reduction) of aldehydes or ketones.

7 Write equations showing the dehydration of an alcohol.

8 Write equations representing the oxidation of alcohols.

9 Discuss the role of oxidation and reduction reactions in the chemistry of living systems.

10 Discuss the use of phenols as germicides.

11 Write names and draw structures for common ethers and discuss their use in medicine.

12 Write equations representing the condensation reaction between two alcohol molecules to form an ether.

13 Write names and draw structures for simple thiols and discuss their biological significance.

INTRODUCTION

Research tells us that even as babies we prefer sweet tastes over all others, and the sugar molecules in our diet that satisfy this sweet tooth are all alcohols and are characterized by the presence of a hydroxyl group (−OH). Glucose (blood sugar) and fructose (fruit sugar), sweet individually, can be chemically combined to produce sucrose, or table sugar. The U.S. Department of Agriculture reports that the average American consumes about 152 pounds (lb) of sucrose each year! It is little wonder that physicians, nutritionists, and dentists are concerned about the obesity and tooth decay caused by so much sugar in our diets.

As a result of these concerns, the food chemistry industry has invested billions of dollars in the synthesis of non-nutritive sugar substitutes, such as aspartame (Equal) and Splenda. You can also find a variety of candies, soft drinks, and gums that are labeled "sugar-free." A quick check of the nutritional label reveals that, although they are sugar-free, they are not calorie-free. These sweets contain sugar alcohols, such as sorbitol or mannitol, instead of sucrose. Sorbitol and mannitol molecules look very much like glucose and fructose, except that they are missing the carbonyl (C=O) group.

List the types of molecules containing the hydroxyl group that are important elements in the meal shown in this photograph.

D-Glucose   D-Fructose   D-Sorbitol   D-Mannitol

Compared to sucrose, they range in sweetness from about half to nearly the same. They also have fewer calories than table sugar (about one-third to one-half the calories). Sugar alcohols actually absorb heat from the surroundings when they dissolve. As a result, they cause a cooling sensation in the mouth. You may have noticed this when eating certain breath-freshening mints and gums.

In biological systems, the hydroxyl group (−OH) is important to the structures of sugars, fats, and proteins. It allows these biological molecules to undergo a variety of reactions such as oxidation, reduction, hydration, and dehydration, that are essential for life.

In this chapter, we will study alcohols and phenols, both of which are characterized by the presence of the hydroxyl group. The difference is that alcohols contain an alkyl group bonded to the hydroxyl group, while phenols have the hydroxyl group bonded to a benzene molecule (aryl group).

An aryl group is an aromatic ring with one hydrogen removed.

General formulas: Example:

Alcohol   Phenol   Methanol
(methyl alcohol)

We will also be studying ethers, which have two alkyl or aryl groups bonded to oxygen, and thiols, which are similar to alcohols except that the oxygen atom has been replaced by a sulfur atom.

R and R' = alkyl or aryl groups

Ether

Methoxymethane
(dimethyl ether)

R—SH

Thiol

CH_3—SH

Methanethiol

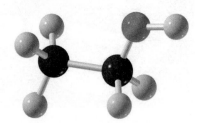

Figure 12.1 Ball-and-stick model of the simple alcohol ethanol.

LEARNING GOAL

1   Classify alcohols as primary, secondary, or tertiary.

12.1   Alcohols: Structure and Physical Properties

An **alcohol** is an organic compound that contains a **hydroxyl group** (—OH) attached to an alkyl group (Figure 12.1). The R—O—H portion of an alcohol is similar to the structure of water. The oxygen and the two atoms bonded to it lie in the same plane, and the R—O—H bond angle is approximately 104°, which is very similar to the H—O—H bond angle of water.

Alcohols are classified as **primary (1°)**, **secondary (2°)**, or **tertiary (3°)**, depending on the number of alkyl groups attached to the **carbinol carbon**, the carbon bearing the hydroxyl (—OH) group. If no alkyl groups are attached, the alcohol is methyl alcohol; if there is a single alkyl group, the alcohol is a primary alcohol; an alcohol with two alkyl groups bonded to the carbon bearing the hydroxyl group is a secondary alcohol, and if three alkyl groups are attached, the alcohol is a tertiary alcohol.

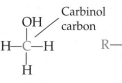

Methyl alcohol   1° Alcohol   2° Alcohol   3° Alcohol

Methanol
(methyl alcohol)

Ethanol
(1° alcohol)

2-Propanol
(2° alcohol)

2-Methyl-2-propanol
(3° alcohol)

EXAMPLE 12.1   Classifying Alcohols

LEARNING GOAL

1   Classify alcohols as primary, secondary, or tertiary.

Classify each of the following alcohols as primary, secondary, or tertiary.

Solution

a.   Two alcohols contribute to the distinctive flavor of mushrooms. These are 1-octanol and 3-octanol. In each of the structures shown below, the carbinol carbon is shown in red:

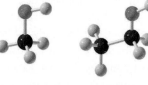

OH

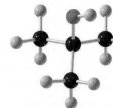

$CH_3(CH_2)_6CH_2OH$

This alcohol, 1-octanol, is a primary alcohol because there is one alkyl group attached to the carbinol carbon.

$CH_3CH_2CH(CH_2)_4CH_3$
                      |
                      OH

This alcohol, 3-octanol, is a secondary alcohol because there are two alkyl groups attached to the carbinol carbon.

b.

$$CH_3CCH_3$$ with CH_3 on top and OH on bottom

This alcohol, 2-methyl-2-propanol, is a tertiary alcohol because there are three alkyl groups attached to the carbinol carbon.

Practice Problem 12.1

Classify each of the following alcohols as 1°, 2°, 3°, or aromatic (phenol).

a. $CH_3CH_2CH_2CH_2OH$

b. $CH_3CH_2CHCH_2CH_3$ with OH below

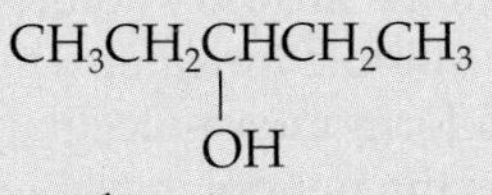

c.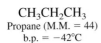

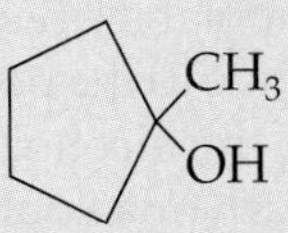

d.

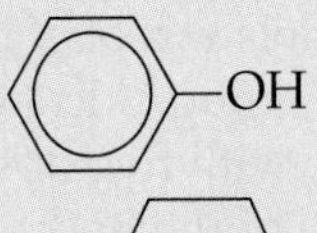

e.

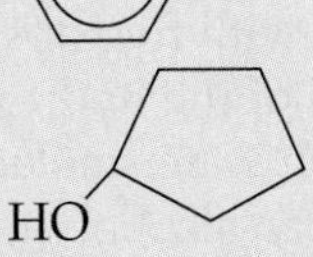

▶ For Further Practice: **Questions 12.17 and 12.19.**

When we considered the physical properties of the alkanes and alkenes, we noted that the major attractive forces between molecules were London dispersion forces. We noted that these attractive forces increase as the molar mass of the molecule and the number of electrons increase. Thus, we could predict that the longer, higher-molar-mass hydrocarbons would have higher boiling and melting points but that alkanes and alkenes of the same or similar molar mass would have very similar boiling and melting points.

However, these principles don't seem to work when we compare the boiling points of propane and ethanol. As seen here, they have nearly the same molar masses, but there is a 120°C difference in their boiling points!

Electronegativity is discussed in Section 3.1. Hydrogen bonding is described in detail in Section 5.2.

LEARNING GOAL

2 Rank selected alcohols by relative water solubility, boiling points, or melting points.

$CH_3CH_2CH_3$
Propane (M.M. = 44)
b.p. = −42°C

CH_3CH_2OH
Ethanol (M.M. = 46)
b.p. = 78°C

How can we explain this observation? By looking at the structures we see that the major difference between propane and ethanol is the presence of the hydroxyl group. By understanding the properties of the hydroxyl group, we can understand the large difference between their boiling points.

The hydroxyl group of alcohols is very polar because the oxygen atom has a significantly higher electronegativity than the hydrogen atom. In addition, the oxygen atom has two unshared pairs of electrons. This results in a very polar bond in which the oxygen atom carries a partially negative charge (δ^-) and the hydrogen carries a partially positive charge (δ^+). This structure allows hydrogen bonds,

the attractive force between a hydrogen atom covalently bonded to a highly electronegative atom with unshared pairs of electrons, to form.

It is the intermolecular (between molecules) hydrogen bonds that cause alcohols to boil at much higher temperatures than hydrocarbons of similar molar
mass. These higher boiling points are caused by the large amount of heat needed
to break the hydrogen bonds that attract molecules to one another (Figure 12.2a).

Alcohols are also more water soluble than alkanes as a result of the polar
hydroxyl group. Alcohols of one to four carbon atoms are very soluble in water,
and those with five or six carbons are moderately soluble in water. This is due to
the ability of the hydroxyl group of the alcohol to form intermolecular hydrogen
bonds with water molecules (see Figure 12.2b). As the nonpolar, or hydrophobic,
portion of an alcohol (the carbon chain) becomes larger relative to the polar, hydrophilic region (the hydroxyl group), the water solubility of the alcohol decreases.
As a result, alcohols of seven carbon atoms or more are nearly insoluble in water.
The term *hydrophobic*, which literally means "water fearing," is used to describe
a molecule or a region of a molecule that is nonpolar and, thus, more soluble in
nonpolar solvents than in water. Similarly, the term *hydrophilic*, meaning "water
loving," is used to describe a polar molecule or region of a molecule that is more
soluble in the polar solvent water than in a nonpolar solvent.

An increase in the number of hydroxyl groups along a carbon chain will
increase the influence of the polar hydroxyl group. It follows, then, that diols and
triols are more water-soluble than alcohols with only a single hydroxyl group.

Large biological molecules, such as proteins and nucleic acids, have large
numbers of hydroxyl groups in their structure. Hydrogen bonds can form between
the hydroxyl groups *within* one of these molecules. These are called *intramolecular*
hydrogen bonds because they are within a single molecule. Imagine that the blue
ribbon shown on the next page is part of a large protein strand that is composed
of many amino acids. Some of the amino acids have hydroxyl groups. The hydrogen bonds between these hydroxyl groups are holding the chain in a folded loop

Intermolecular hydrogen bonds are attractive
forces between two separate molecules.
Intramolecular hydrogen bonds are attractive
forces between polar groups within the same
molecule.

Figure 12.2 (a) Hydrogen bonding
between alcohol molecules. (b) Hydrogen
bonding between alcohol molecules and
water molecules.

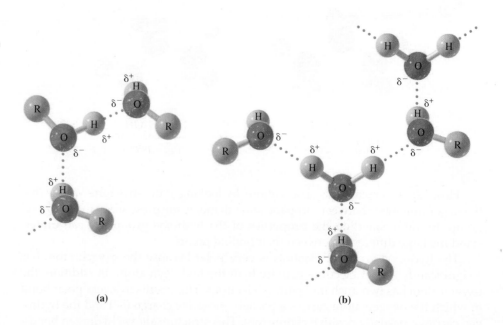

(a)   (b)

structure, rather than allowing it to remain as an extended long strand. Hydrogen bonds are one of the several weak interactions that hold proteins into the three-dimensional folded shape that is needed for them to function properly. Hydrogen bonds also hold the two strands of DNA together.

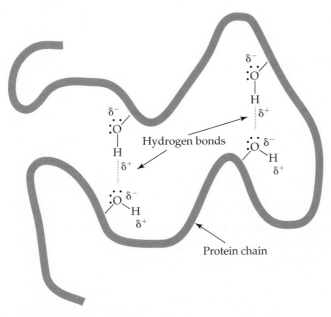

12.2 Alcohols: Nomenclature

IUPAC Names

In the IUPAC Nomenclature System, alcohols are named according to the following steps:

LEARNING GOAL

3 Write the names and draw the structures for common alcohols.

- Determine the name of the *parent compound*, the longest continuous carbon chain or the ring containing the —OH group.
- Replace the *-e* ending of the alkane chain with the *-ol* ending of the alcohol. Following this pattern, an alkane becomes an alkanol. For instance, etha*ne* becomes etha*nol*, and propa*ne* becomes propa*nol*.
- Number the parent chain to give the carbon bearing the hydroxyl group the lowest possible number.
- Name and number all substituents, and add them as prefixes to the "alkanol" name.
- Alcohols containing two hydroxyl groups are named *-diols*. Those bearing three hydroxyl groups are called *-triols*. A number giving the position of each of the hydroxyl groups is needed in these cases and the final *-e* of the parent compound is retained.

The way to determine the parent compound was described in Section 10.2.

EXAMPLE 12.2 | **Using IUPAC Nomenclature to Name Alcohols**

LEARNING GOAL

3 Write the names and draw the structures for common alcohols.

a. The molecule shown here is an alarm-defense pheromone in certain populations of leaf-cutter ants. Provide the IUPAC name for this molecule.

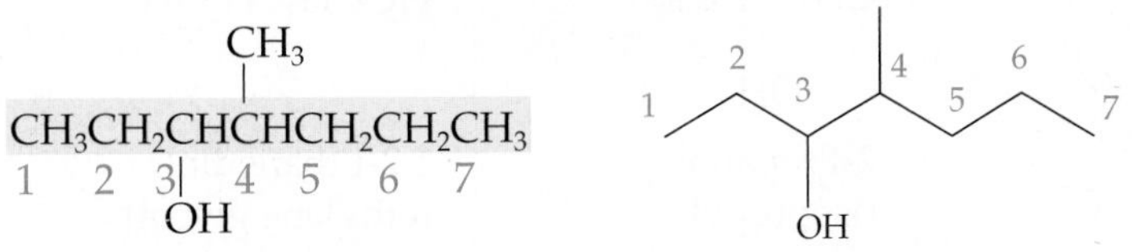

$$\underset{\underset{1 \quad 2 \quad 3 \quad 4 \quad 5 \quad 6 \quad 7}{|}{\underset{OH}{CH_3CH_2CHCHCH_2CH_2CH_3}}}{\overset{CH_3}{|}}$$

Continued…

Solution

Parent compound: heptane (becomes heptanol)

Position of —OH: carbon-3 (*Not* carbon 5 because the parent chain must be numbered to give the lowest possible number for the carbinol carbon.)

Substituents: 4-methyl

Name: 4-Methyl-3-heptanol

b. Name the following cyclic alcohol using IUPAC nomenclature.

Solution

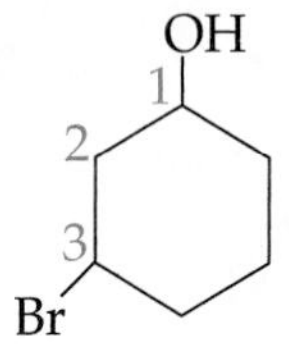

Parent compound: cyclohexane (becomes cyclohexanol)

Position of —OH: carbon-1 (*Not* carbon-3 because the carbinol carbon is defined as carbon-1.)

Substituents: 3-bromo (*Not* 5-bromo because the lowest possible numbering should be used.)

Name: 3-Bromocyclohexanol (It is assumed that the —OH is on carbon-1 in cyclic structures.)

Practice Problem 12.2

Use the IUPAC Nomenclature System to name each of the following compounds.

a. CH₃CHCH₂CH₂CH₂OH
        |
        CH₃

c. CH₂CHCH₂
        | |
        OH   OH
    (Common name: Glycerol)

b. CH₃CHCH₂CHCH₃
        | |
        OH   CH₂CH₃

d. CH₃CH₂CHCHCH₂CH₂OH
            | |
            Cl CH₃

▶ For Further Practice: **Questions 12.29 and 12.30.**

See Section 10.2 for the names of the common alkyl groups.

Common Names

The common names for alcohols are derived from the alkyl group corresponding to the parent compound. The name of the alkyl group is followed by the word *alcohol*. For some alcohols, such as ethylene glycol and glycerol, historical names are used. The following examples provide the IUPAC and common names of several alcohols:

CH₃CHCH₃
    |
    OH

2-Propanol
(isopropyl alcohol)

HOCH₂CH₂OH

1,2-Ethanediol
(ethylene glycol)

CH₃CH₂OH

Ethanol
(ethyl alcohol)

EXAMPLE 12.3 | **Using the Common System of Nomenclature to Name Alcohols**

LEARNING GOAL

3 Write the names and draw the structures for common alcohols.

Provide the common names of the following four alcohols.

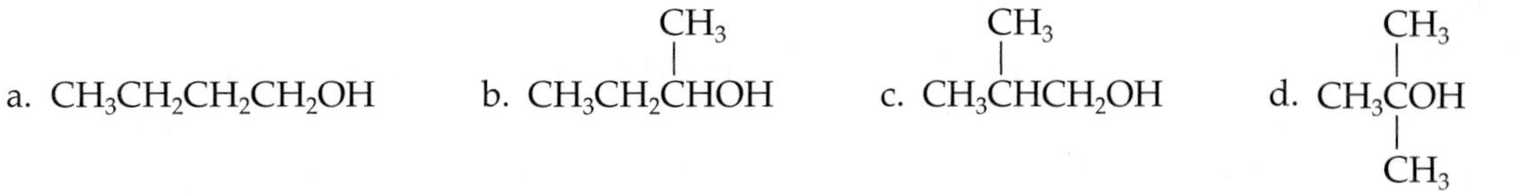

a. $CH_3CH_2CH_2CH_2OH$

b. $CH_3CH_2\overset{\underset{|}{CH_3}}{C}HOH$

c. $CH_3\overset{\underset{|}{CH_3}}{C}HCH_2OH$

d. $CH_3\overset{\underset{|}{CH_3}}{\underset{|}{C}}OH$ with CH_3

Common names are derived from the alkyl group that corresponds to the parent alkane. The name of the alkyl group is followed by the word *alcohol.* First, identify the carbinol carbon and the alkyl groups of which they are a part.

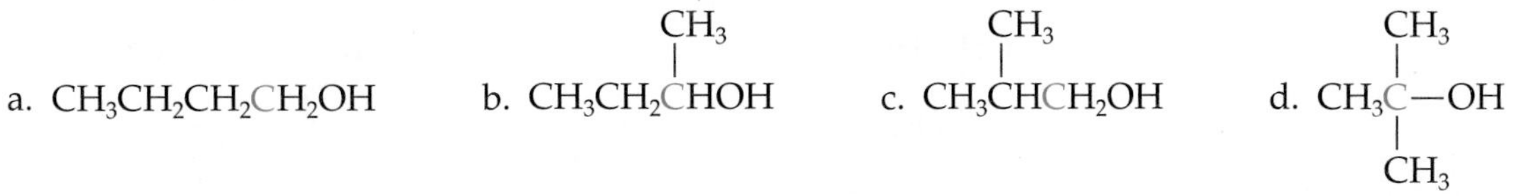

a. $CH_3CH_2CH_2CH_2OH$

b. $CH_3CH_2\overset{\underset{|}{CH_3}}{C}HOH$

c. $CH_3\overset{\underset{|}{CH_3}}{C}HCH_2OH$

d. $CH_3\overset{\underset{|}{CH_3}}{\underset{|}{C}}{-}OH$ with CH_3

In the structures above, the carbinol carbon is highlighted in red. You may wish to review the names of the continuous- and branched-chain alkyl groups in Tables 10.5 and 10.6.

a. In structure a, the carbinol carbon is part of a four-carbon continuous chain alkyl group. Referring to Table 10.5, we see that this is the butyl group. The common name for this alcohol is butyl alcohol.

$$CH_3CH_2CH_2CH_2OH$$
Butyl alcohol

b. In structure b, the carbinol carbon is bonded to *two carbon atoms.* From Table 10.6, we see that this is the *sec*-butyl group. The common name of this structure is *sec*-butyl alcohol.

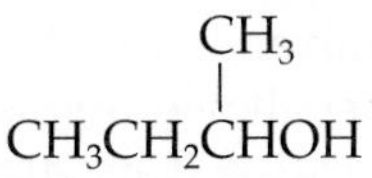

$$CH_3CH_2\overset{\underset{|}{CH_3}}{C}HOH$$
sec-Butyl alcohol

c. In structure c, the alkyl group has *two methyl groups.* From Table 10.6, we see that this is an *iso*-alkyl group. In this case, it is the isobutyl group and the alcohol is isobutyl alcohol.

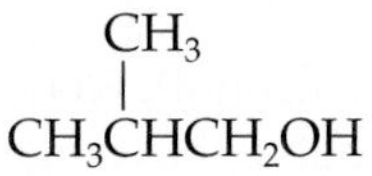

$$CH_3\overset{\underset{|}{CH_3}}{C}HCH_2OH$$
Isobutyl alcohol

d. In structure d, the carbinol carbon is attached to three carbon atoms. From Table 10.6, we see that this is the *tert*- or *t*-butyl group. This alcohol is *tert*-butyl alcohol.

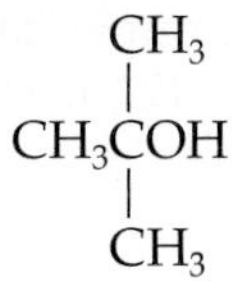

$$CH_3\overset{\underset{|}{CH_3}}{\underset{|}{C}}OH \text{ with } CH_3$$
tert-Butyl alcohol or
t-butyl alcohol

Practice Problem 12.3

Draw line formulas for each of the following alcohols:

a. *sec*-Hexyl alcohol   b. Isopentyl alcohol   c. Cyclobutyl alcohol   d. Hexyl alcohol

▶ For Further Practice: **Questions 12.35 and 12.36.**

Question 12.1 Draw structures for each of the following alcohols.
a. 2-Methyl-1-propanol   c. 2,4-Dimethylcyclohexanol
b. 2-Chlorocyclopentanol   d. 2,3-Dichloro-3-hexanol

Question 12.2 Give the common name and the IUPAC name for each of the following compounds.

a. $CH_3CH_2CH_2CH_2CH_2CH_2CH_2OH$

b. CH_3CHCH_3
        |
        OH

c.

d.

12.3 Medically Important Alcohols

LEARNING GOAL

4 Discuss the biological, medical, or environmental significance of several alcohols.

Methanol

Methanol (methyl alcohol), CH_3OH, is a colorless and odorless liquid that is used as a solvent and as the starting material for the synthesis of methanal (formaldehyde). Methanol is often called *wood alcohol* because it can be made by heating wood in the absence of air. In fact, ancient Egyptians produced methanol by this process and, mixed with other substances, used it for embalming. It was not until 1661 that Robert Boyle first isolated pure methanol, which he called *spirit of box*, because he purified it by distillation from boxwood. Methanol is toxic and can cause blindness and perhaps death if ingested. Methanol may also be used as fuel, especially for "formula" racing cars.

Figure 12.3 Champagne, a sparkling wine, results when fermentation is carried out in a sealed bottle. Under these conditions, the CO_2 produced during fermentation is trapped in the wine.

Fermentation reactions are described in detail in Section 21.4 and in A Human Perspective: Fermentations: The Good, the Bad, and the Ugly.

Ethanol

Ethanol (ethyl alcohol), CH_3CH_2OH, is a colorless and odorless liquid and is the alcohol in alcoholic beverages. It is also widely used as a solvent and as a raw material for the preparation of other organic chemicals.

The ethanol used in alcoholic beverages comes from the **fermentation** of carbohydrates (sugars and starches). The beverage produced depends on the starting material and the fermentation process: scotch (grain), bourbon (corn), burgundy wine (grapes and grape skins), and chablis wine (grapes without red skins) (Figure 12.3). The following equation summarizes the fermentation process:

$$C_6H_{12}O_6 \xrightarrow[\text{enzyme action}]{\substack{\text{Several steps} \\ \text{involving}}} 2CH_3CH_2OH + 2CO_2$$

Sugar   Ethanol
(glucose)   (ethyl alcohol)

Distillation is the separation of compounds in a mixture based on differences in boiling points.

The alcoholic beverages listed have quite different alcohol concentrations. Wines are generally 12–13% alcohol because the yeasts that produce the ethanol are killed by ethanol concentrations of 12–13%. To produce bourbon or scotch with an alcohol concentration of 40–45% ethanol (80 or 90 proof), the original fermentation products must be distilled.

A Medical Perspective

Fetal Alcohol Syndrome

The first months of pregnancy are a time of great joy and anticipation but are not without moments of anxiety. On her first visit to the obstetrician, the mother-to-be is tested for previous exposure to a number of infectious diseases that could damage the fetus. She is provided with information about diet, weight gain, and drugs that could harm the baby. Among the drugs that should be avoided are alcoholic beverages.

The use of alcoholic beverages by a pregnant woman can cause *fetal alcohol syndrome* (*FAS*). A *syndrome* is a set of symptoms that occur together and are characteristic of a particular disease. In this case, physicians have observed that infants born to women with chronic alcoholism showed a reproducible set of abnormalities including mental retardation, poor growth before and after birth, and facial malformations. In the United States and Europe, the incidence of FAS is estimated to be between 0.2 and 2 in every 1000 live births.

FAS is just one disorder that is included in the broader description of Fetal Alcohol Spectrum Disorders (FASD), which affects approximately 1% of live births in the United States. FASD includes a continuum of permanent birth defects, ranging from mild to severe, that result from maternal consumption of alcohol.

How does alcohol consumption cause these varied symptoms? No one is exactly sure, but it is well known that the alcohol consumed by the mother crosses the placenta and enters the bloodstream of the fetus. Within about 15 minutes (min), the concentration of alcohol in the blood of the fetus is as high as that of the mother! However, the mother has enzymes to detoxify the alcohol in her blood; the fetus does not. Now consider that alcohol can cause cell division to stop or be radically altered. In fact, studies suggest that ethanol toxicity damages a number of aspects of fetal nervous system development. Among these are cell division, differentiation, and migration of axons.

This raises the question, "How much alcohol can a pregnant woman safely drink?" As we have seen, the severity of the symptoms seems to increase with the amount of alcohol consumed by the mother. However, it is virtually impossible to do the scientific studies that would conclusively determine the risk to the fetus caused by different amounts of alcohol. There

The American Medical Association recommends abstaining from alcohol during pregnancy.

is some evidence that suggests there is a risk associated with drinking even 1 ounce (oz) of absolute (100%) alcohol each day. Because of these facts and uncertainties, the American Medical Association and the U.S. Surgeon General recommend that pregnant women completely abstain from alcohol.

For Further Understanding

▸ In October 2005, the U.S. Centers for Disease Control issued a document entitled *Guidelines for Identifying and Referring Persons with Fetal Alcohol Syndrome*, which can be found at the following Web address: www.cdc.gov/ncbddd/fasd/index.html. Refer to this document when answering the following questions.

♣ What is the estimate of the number of babies born each year with fetal alcohol syndrome, and why is a more accurate number so difficult to determine?

♣ What are some of the issues, practical and ethical, involved in intervention efforts to prevent fetal alcohol syndrome?

The sale and use of pure ethanol (100% ethanol) are regulated by the federal government. To prevent illegal use of pure ethanol, it is *denatured* by the addition of a denaturing agent, which makes it unfit to drink but suitable for many laboratory applications.

2-Propanol

2-Propanol (isopropyl alcohol),

$$CH_3CHCH_3$$
$$|$$
$$OH$$

was commonly called *rubbing alcohol* because patients with high fevers were often given alcohol baths to reduce body temperature. Rapid evaporation of the alcohol results in skin cooling. This practice is no longer commonly used.

Figure 12.4 Isopropyl alcohol, or rubbing alcohol, is used as an antiseptic before and after an injection or blood test.

It is also used as an antiseptic (Figure 12.4), an astringent (skin-drying agent), an industrial solvent, and a raw material in the synthesis of organic chemicals. It is colorless, has a very slight odor, and is toxic when ingested.

1,2-Ethanediol

1,2-Ethanediol (ethylene glycol),

$$\underset{\underset{OH}{|}}{CH_2}-\underset{\underset{OH}{|}}{CH_2}$$

is used as automobile antifreeze. When added to water in the radiator, the ethylene glycol solute lowers the freezing point and raises the boiling point of the water. Ethylene glycol has a sweet taste but is extremely poisonous. For this reason, color additives are used in antifreeze to ensure that it is properly identified.

1,2,3-Propanetriol

1,2,3-Propanetriol (glycerol),

$$\underset{\underset{OH}{|}}{CH_2}-\underset{\underset{OH}{|}}{CH}-\underset{\underset{OH}{|}}{CH_2}$$

is a viscous, sweet-tasting, nontoxic liquid. It is very soluble in water and is used in cosmetics, pharmaceuticals, and lubricants. Glycerol is obtained as a by-product of the hydrolysis of fats.

12.4   Reactions Involving Alcohols

Preparation of Alcohols

LEARNING GOAL

5 Write equations representing the preparation of alcohols by the hydration of an alkene.

Hydration of alkenes is described in Section 11.5.

As we saw in the last chapter, the most important reactions of alkenes are *addition reactions*. Addition of a water molecule to the carbon-carbon double bond of an alkene produces an alcohol. This reaction, called **hydration**, requires a trace of acid (H^+) as a catalyst, as shown in the following equation:

Alkene + Water $\xrightarrow{H^+}$ Alcohol

Hydrogenation of aldehydes and ketones will also be discussed in Section 13.4.

Alcohols may also be prepared via the hydrogenation (reduction) of aldehydes and ketones. In organic and biochemical reactions, reduction is recognized as the loss of oxygen or the gain of hydrogen. In the hydrogenation of aldehydes and ketones, it is the gain of hydrogen atoms that allows us to recognize that reduction has occurred. The general equations for these reactions can be summarized as follows:

Aldehyde + Hydrogen $\xrightarrow{Catalyst}$ Primary Alcohol   Ketone + Hydrogen $\xrightarrow{Catalyst}$ Secondary Alcohol

EXAMPLE 12.4

Writing an Equation Representing the Preparation of an Alcohol by the Hydrogenation (Reduction) of an Aldehyde

LEARNING GOAL

6 Write equations representing the preparation of alcohols by hydrogenation (reduction) of aldehydes or ketones.

Write an equation representing the preparation of 1-propanol from propanal.

Solution

Begin by writing the structure of propanal. Propanal is a three-carbon aldehyde. Aldehydes are characterized by the presence of a carbonyl group (—C=O) attached to the end of the carbon chain of the molecule (Table 10.2). After you have drawn the structure of propanal, add diatomic hydrogen to the equation.

$$
\underset{\text{Propanal}}{\text{H}-\overset{\overset{\text{H}}{|}}{\underset{\underset{\text{H}}{|}}{\text{C}}}-\overset{\overset{\text{H}}{|}}{\underset{\underset{\text{H}}{|}}{\text{C}}}-\overset{\displaystyle\diagup\!\!\!\!\diagup O}{\underset{\diagdown\text{H}}{\text{C}}} \quad + \quad \underset{\text{Hydrogen}}{\text{H}-\text{H}} \quad \xrightarrow{\text{catalyst}}
$$

Notice that the general equation reveals this reaction to be an example of a hydrogenation reaction. As the hydrogens are added to the carbon-oxygen double bond, it is converted to a carbon-oxygen single bond, and the carbonyl oxygen becomes a hydroxyl group.

$$
\underset{\text{Propanal}}{\text{H}-\overset{\overset{\text{H}}{|}}{\underset{\underset{\text{H}}{|}}{\text{C}}}-\overset{\overset{\text{H}}{|}}{\underset{\underset{\text{H}}{|}}{\text{C}}}-\overset{\displaystyle\diagup\!\!\!\!\diagup O}{\underset{\diagdown\text{H}}{\text{C}}} \quad + \quad \underset{\text{Hydrogen}}{\text{H}-\text{H}} \quad \xrightarrow{\text{catalyst}} \quad \underset{\text{1-Propanol}}{\text{H}-\overset{\overset{\text{H}}{|}}{\underset{\underset{\text{H}}{|}}{\text{C}}}-\overset{\overset{\text{H}}{|}}{\underset{\underset{\text{H}}{|}}{\text{C}}}-\overset{\overset{\text{H}}{|}}{\underset{\underset{\text{H}}{|}}{\text{C}}}-\text{OH}}
$$

Practice Problem 12.4

Write an equation representing the reduction of butanal. Provide the structures and names for the reactants and products. *Hint:* Butanal is a four-carbon aldehyde with the structure $CH_3CH_2CH_2C\overset{\displaystyle\diagup\!\!\!\!\diagup O}{\diagdown H}$.

▶ For Further Practice: **Questions 12.65a and c and 12.66a and c.**

EXAMPLE 12.5

Writing an Equation Representing the Preparation of an Alcohol by the Hydrogenation (Reduction) of a Ketone

LEARNING GOAL

6 Write equations representing the preparation of alcohols by hydrogenation (reduction) of aldehydes or ketones.

Write an equation representing the preparation of 2-propanol from propanone.

Solution

Begin by writing the structure of propanone. Propanone is a three-carbon ketone. Ketones are characterized by the presence of a carbonyl group (—C=O) located anywhere within the carbon chain of the molecule (Table 10.2). In the structure of propanone, the carbonyl group must be associated with the center carbon. After you have drawn the structure of propanone, add diatomic hydrogen to the equation.

$$
\underset{\text{Propanone}}{\text{H}-\overset{\overset{\text{H}}{|}}{\underset{\underset{\text{H}}{|}}{\text{C}}}-\overset{\overset{\displaystyle O}{\|}}{\text{C}}-\overset{\overset{\text{H}}{|}}{\underset{\underset{\text{H}}{|}}{\text{C}}}-\text{H}} \quad + \quad \underset{\text{Hydrogen}}{\text{H}-\text{H}} \quad \xrightarrow{\text{catalyst}}
$$

Continued…

Notice that this reaction is an example of a hydrogenation reaction. As the hydrogens are added to the carbon-oxygen double bond, it is converted to a carbon-oxygen single bond, and the carbonyl oxygen becomes a hydroxyl group.

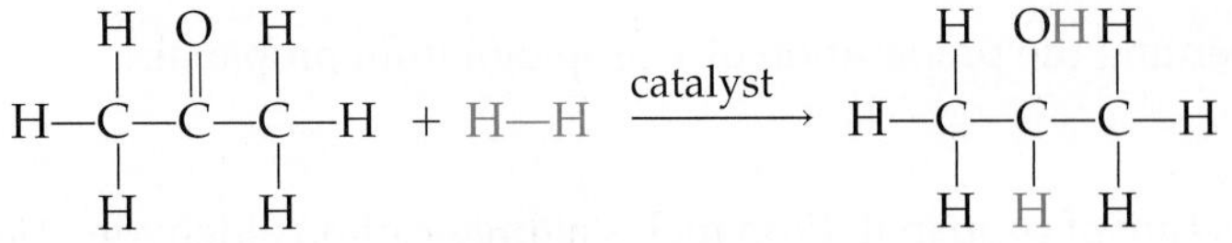

$$\underset{\text{Propanone}}{\begin{array}{c} H \;\; O \;\; H \\ | \quad \; || \quad | \\ H{-}C{-}C{-}C{-}H \\ | \qquad\quad | \\ H \qquad\quad H \end{array}} + \underset{\text{Hydrogen}}{H{-}H} \;\xrightarrow{\text{catalyst}}\; \underset{\text{2-Propanol}}{\begin{array}{c} H \;\; OH\,H \\ | \quad \; | \quad | \\ H{-}C{-}C{-}C{-}H \\ | \quad \; | \quad | \\ H \;\; H \;\; H \end{array}}$$

Practice Problem 12.5

Write an equation representing the reduction of butanone. *Hint:* Butanone is a four-carbon ketone with the structure

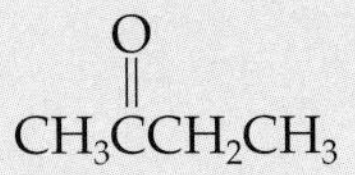

$$\underset{}{\overset{\displaystyle O}{\overset{\displaystyle ||}{CH_3CCH_2CH_3}}}$$

▶ For Further Practice: **Questions 12.65b and d and 12.66b and d.**

Dehydration of Alcohols

Alcohols can undergo dehydration reactions. A **dehydration reaction** is one in which a water molecule is lost. The dehydration of an alcohol requires heat and concentrated sulfuric acid (H_2SO_4) or phosphoric acid (H_3PO_4). This is seen in the following general reaction and the examples that follow:

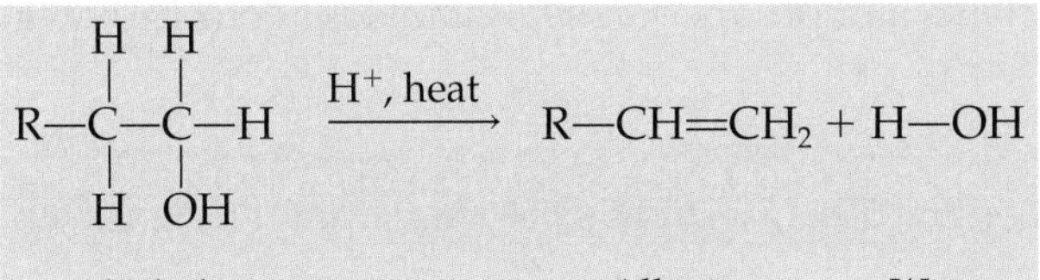

$$\underset{\text{Alcohol}}{\begin{array}{c} H \;\; H \\ | \quad\; | \\ R{-}C{-}C{-}H \\ | \quad\; | \\ H \;\; OH \end{array}} \;\xrightarrow{H^+,\,\text{heat}}\; \underset{\text{Alkene}}{R{-}CH{=}CH_2} + \underset{\text{Water}}{H{-}OH}$$

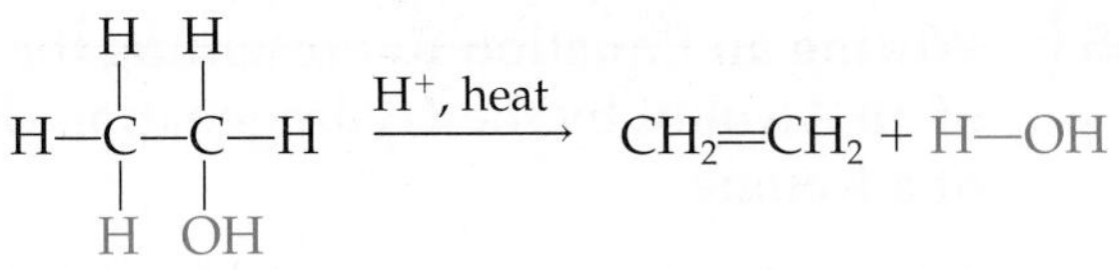

$$\underset{\substack{\text{Ethanol}\\ \text{(ethyl alcohol)}}}{\begin{array}{c} H \;\; H \\ | \quad\; | \\ H{-}C{-}C{-}H \\ | \quad\; | \\ H \;\; OH \end{array}} \;\xrightarrow{H^+,\,\text{heat}}\; \underset{\substack{\text{Ethene}\\ \text{(ethylene)}}}{CH_2{=}CH_2} + H{-}OH$$

$$\underset{\substack{\text{1-Propanol}\\ \text{(propyl alcohol)}}}{CH_3CH_2CH_2OH} \;\xrightarrow{H^+,\,\text{heat}}\; \underset{\substack{\text{Propene}\\ \text{(propylene)}}}{CH_3CH{=}CH_2} + H_2O$$

Dehydration is an example of an **elimination reaction**; that is, a reaction in which a molecule loses atoms or ions from its structure. In this case, the —OH and —H are "eliminated" from adjacent carbons in the alcohol to produce an alkene and water. We have just seen that alkenes can be hydrated to give alcohols. Dehydration is simply the reverse process: the conversion of an alcohol back to an alkene.

In some cases, dehydration of alcohols produces a mixture of products, as seen in the following example:

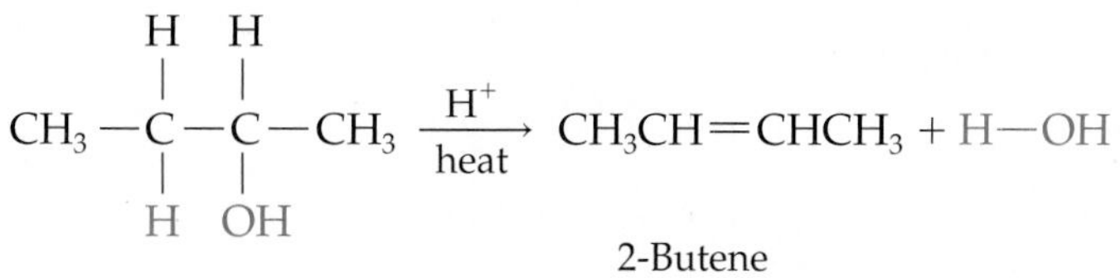

$$CH_3-\underset{\underset{H}{|}}{\overset{\overset{H}{|}}{C}}-\underset{\underset{OH}{|}}{\overset{\overset{H}{|}}{C}}-CH_3 \xrightarrow[\text{heat}]{H^+} CH_3CH=CHCH_3 + H-OH$$

2-Butanol $\qquad$ 2-Butene
(major product)

$$CH_3-\underset{\underset{H}{|}}{\overset{\overset{H}{|}}{C}}-\underset{\underset{OH}{|}}{\overset{\overset{H}{|}}{C}}-CH_3 \xrightarrow[\text{heat}]{H^+} CH_3CH_2CH=CH_2 + H-OH$$

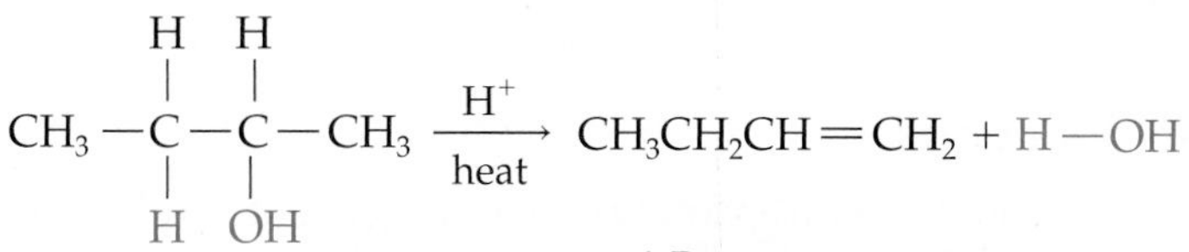

2-Butanol $\qquad$ 1-Butene
(minor product)

Notice in these equations and in Example 12.6, the major product is the more highly substituted alkene. In 1875, the Russian chemist Alexander Zaitsev developed a rule to describe such reactions. **Zaitsev's rule** states that in an elimination reaction, the alkene with the greatest number of alkyl groups on the double bonded carbons (the more highly substituted alkene) is the major product of the reaction.

EXAMPLE 12.6 | **Predicting the Products of Alcohol Dehydration**

LEARNING GOAL

Predict the products of the dehydration of 3-methyl-2-butanol.

7 Write equations showing the dehydration of an alcohol.

Solution

The product(s) of dehydration of an alcohol will contain a double bond in which one of the carbons was the original carbinol carbon—the carbon to which the hydroxyl group is attached. Consider the following reaction:

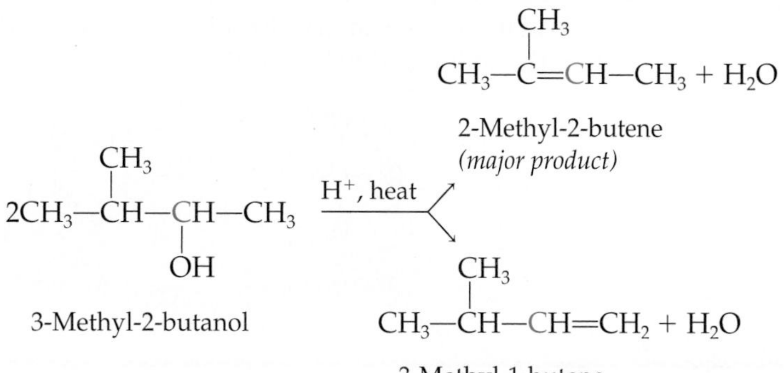

$$CH_3-\overset{\overset{CH_3}{|}}{C}=CH-CH_3 + H_2O$$

2-Methyl-2-butene
(major product)

$$2CH_3-\overset{\overset{CH_3}{|}}{CH}-\underset{\underset{OH}{|}}{CH}-CH_3 \xrightarrow{H^+,\ heat}$$

3-Methyl-2-butanol

$$CH_3-\overset{\overset{CH_3}{|}}{CH}-CH=CH_2 + H_2O$$

3-Methyl-1-butene
(minor product)

It is clear that both the major and minor products have a double bond to carbon number 2, the carbinol carbon in the original alcohol (shown in red). Zaitsev's rule tells us that in dehydration reactions with more than one product possible, the more highly branched alkene predominates. In the reaction shown, 2-methyl-2-butene has three alkyl groups at the double bond, whereas 3-methyl-1-butene has only one alkyl group at the double bond. The more highly branched alkene is more stable and thus is the major product.

Practice Problem 12.6

Write an equation showing the dehydration of each of the following alcohols. If there are two possible alkene products, indicate which is the major product and which is the minor product.

Continued…

a. CH_3CH_2OH

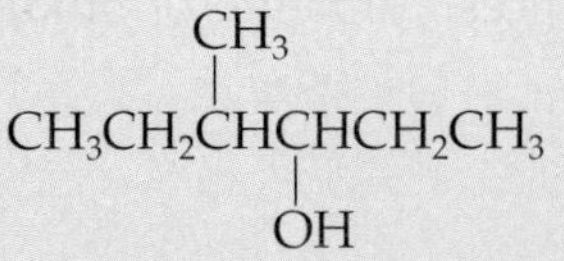

b. 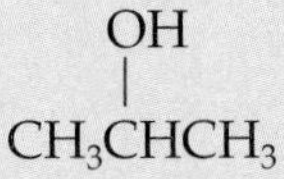 CH_3CHCH_3 with OH group

c. $CH_3CH_2CHCHCH_2CH_3$ with CH_3 and OH groups

d. 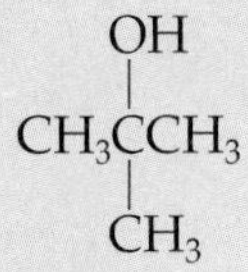 CH_3CCH_3 with OH and CH_3 groups

▶ For Further Practice: **Questions 12.52 and 12.54.**

LEARNING GOAL

8 Write equations representing the oxidation of alcohols.

Note that the symbol [O] is used throughout this book to designate any oxidizing agent.

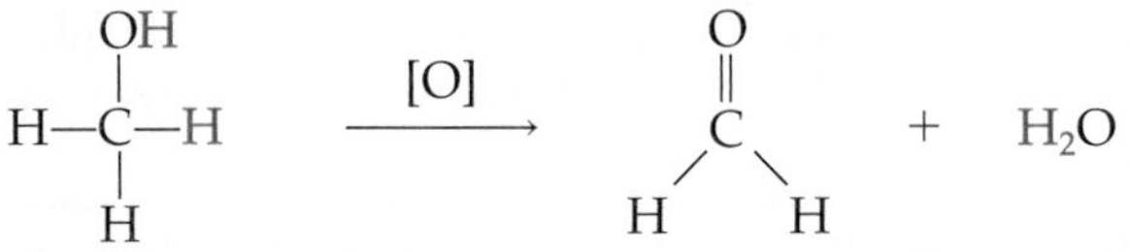

As we will see in Section 13.4, aldehydes can undergo further oxidation to produce carboxylic acids.

Oxidation Reactions

Alcohols may be oxidized with a variety of oxidizing agents to aldehydes, ketones, and carboxylic acids. The most commonly used oxidizing agents are solutions of basic potassium permanganate ($KMnO_4/OH^-$) and chromic acid (H_2CrO_4). The symbol [O] over the reaction arrow is used throughout this book to designate any general oxidizing agent.

Oxidation of methanol produces the aldehyde methanal:

Methanol
(methyl alcohol)
An alcohol

Methanal
(formaldehyde)
An aldehyde

In organic and biochemical systems, oxidation is recognized by the loss of hydrogen or the gain of oxygen. In the oxidation of alcohols, two hydrogens are removed from the alcohol. One is removed from the hydroxyl group, and a second is removed to form the carbinol carbon. In the process, a carbonyl group (—C=O) is formed.

Oxidation of a primary alcohol produces an aldehyde, as seen in the following general equation and Example 12.7:

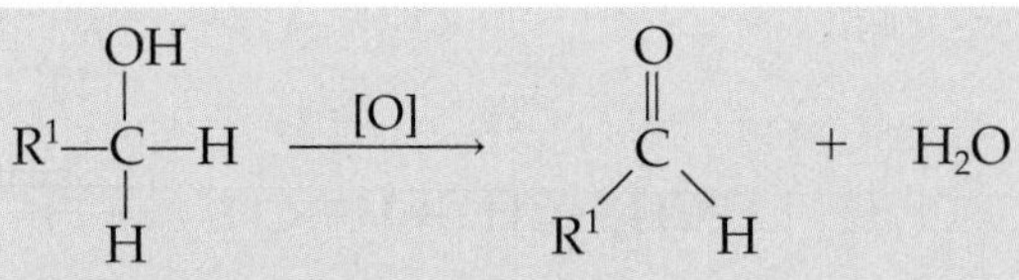

1° Alcohol   An aldehyde

| EXAMPLE 12.7 | **Writing an Equation Representing the Oxidation of a Primary Alcohol** |

LEARNING GOAL

8 Write equations representing the oxidation of alcohols.

Write an equation showing the oxidation of 2,2-dimethylpropanol to produce 2,2-dimethylpropanal.

Solution

Begin by writing the structure of the reactant, 2,2-dimethylpropanol, and indicate the need for an oxidizing agent by placing the designation [O] over the reaction arrow:

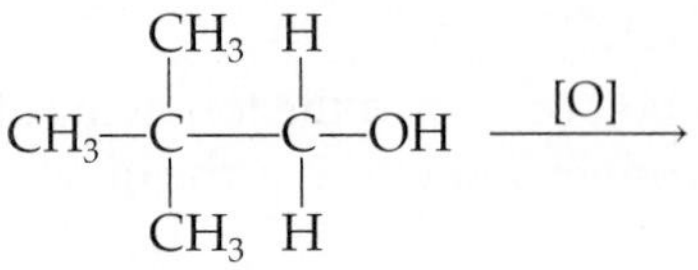

2,2-Dimethylpropanol

Now show the oxidation of the hydroxyl group to the aldehyde carbonyl group, and the loss of a hydrogen from the carbinol carbon. The products are an aldehyde and water.

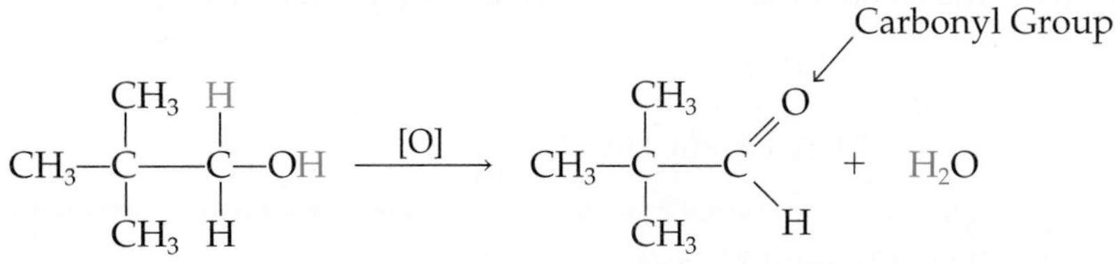

2,2-Dimethylpropanol   2,2-Dimethylpropanal

Practice Problem 12.7

Write an equation showing the oxidation of the following primary alcohols:

   a. $CH_3C(CH_3)_2CH_2CH_2OH$ b. CH_3CH_2OH

▶ For Further Practice: **Questions 12.48 and 12.56.**

Oxidation of a secondary alcohol produces a ketone:

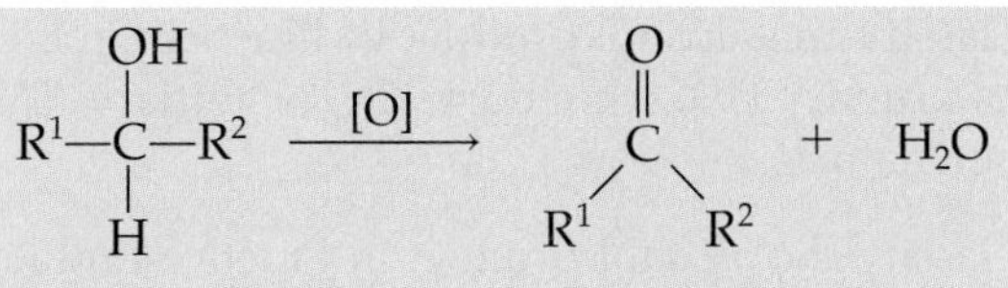

2° Alcohol   A ketone

EXAMPLE 12.8	**Writing an Equation Representing the Oxidation of a Secondary Alcohol**

LEARNING GOAL

Write an equation showing the oxidation of 2-propanol to produce propanone.

8 Write equations representing the oxidation of alcohols.

Solution

Begin by writing the structure of the reactant, 2-propanol, and indicate the need for an oxidizing agent by placing the designation [O] over the reaction arrow:

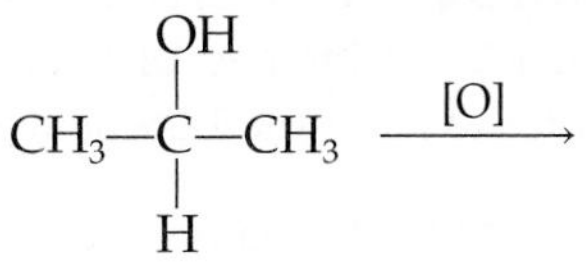

2-Propanol

Now show the oxidation of the hydroxyl group to the ketone carbonyl group, and the loss of a hydrogen from the carbinol carbon. The products are a ketone and water.

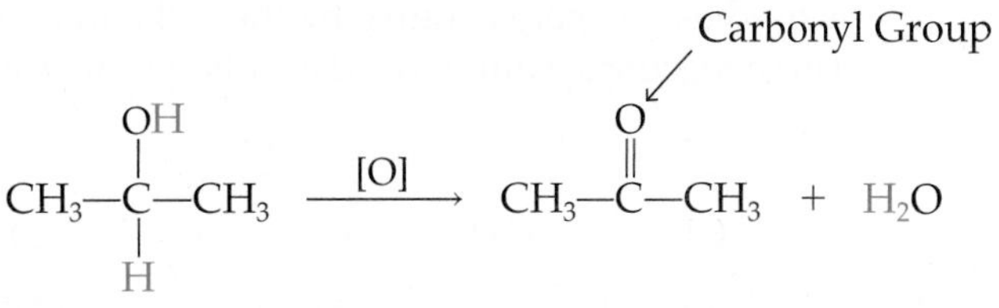

2-Propanol   Propanone

Continued...

Practice Problem 12.8

Write an equation showing the oxidation of the following secondary alcohols:

a. 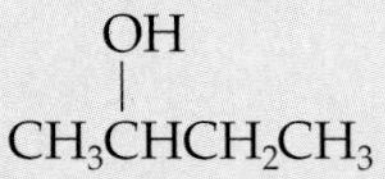
$$
\begin{array}{c}
OH \\
| \\
CH_3CHCH_2CH_3
\end{array}
$$

b. 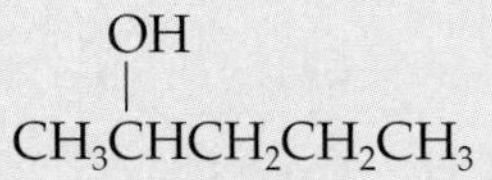
$$
\begin{array}{c}
OH \\
| \\
CH_3CHCH_2CH_2CH_3
\end{array}
$$

▶ For Further Practice: **Questions 12.49 and 12.55a–c.**

Tertiary alcohols cannot be oxidized:

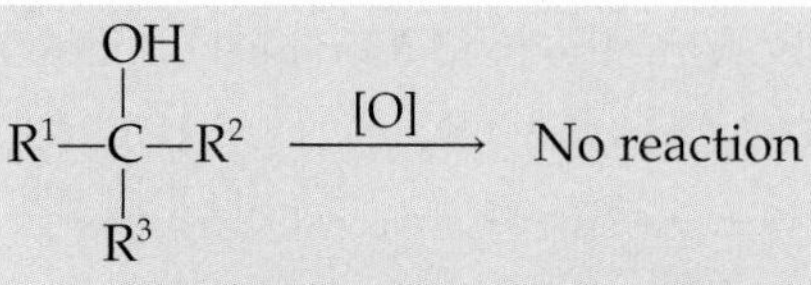

$$
\begin{array}{c}
OH \\
| \\
R^1-C-R^2 \\
| \\
R^3
\end{array}
\xrightarrow{[O]} \text{No reaction}
$$

$3°$ Alcohol

For the oxidation reaction to occur, the carbinol carbon must contain at least one C—H bond. Because tertiary alcohols contain three C—C bonds to the carbinol carbon, they cannot undergo oxidation.

Question 12.3   Classify the alcohol product in Practice Problem 12.4 at the end of Example 12.4 as primary ($1°$), secondary ($2°$), or tertiary ($3°$), and provide the IUPAC and common names.

Question 12.4   Classify the alcohol product in Practice Problem 12.5, at the end of Example 12.5, as a primary ($1°$), secondary ($2°$), or tertiary ($3°$) alcohol, and provide the IUPAC name.

Question 12.5   Name the alcohol reactants and alkene products in each of the reactions in Practice Problem 12.6, at the end of Example 12.6, using the IUPAC Nomenclature System, and classify each of these alcohols as primary ($1°$), secondary ($2°$), or tertiary ($3°$).

Question 12.6   Name each of the reactant alcohols and product aldehydes in Practice Problem 12.7, at the end of Example 12.7, using the IUPAC Nomenclature System. *Hint:* Refer to Example 12.7, as well as to Section 13.2, to name the aldehyde products.

Question 12.7   Name each of the reactant alcohols and product ketones in Practice Problem 12.8, at the end of Example 12.8, using the IUPAC Nomenclature System. *Hint:* Refer to Example 12.8, as well as to Section 13.2, to name the ketone products.

Question 12.8   Explain why a tertiary alcohol cannot undergo oxidation.

When ethanol is metabolized in the liver, it is oxidized to ethanal (acetaldehyde). If too much ethanol is present in the body, an overabundance of ethanal is formed, which causes many of the adverse effects of the "morning-after hangover." Continued oxidation of ethanal produces ethanoic acid (acetic acid), which is used as an energy source by the cell and eventually oxidized to CO_2 and H_2O. These reactions, summarized as follows, are catalyzed by liver enzymes.

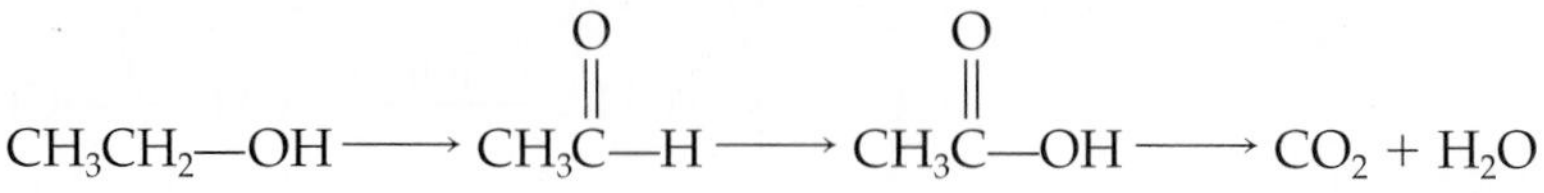

$$
CH_3CH_2-OH \longrightarrow
\begin{array}{c}
O \\
\| \\
CH_3C-H
\end{array}
\longrightarrow
\begin{array}{c}
O \\
\| \\
CH_3C-OH
\end{array}
\longrightarrow CO_2 + H_2O
$$

Ethanol   Ethanal   Ethanoic acid
(ethyl alcohol) (acetaldehyde)   (acetic acid)

12.5 Oxidation and Reduction in Living Systems

Before beginning a discussion of oxidation and reduction in living systems, we must understand how to recognize **oxidation** (loss of electrons) and **reduction** (gain of electrons) in organic compounds. It is easy to determine when an oxidation or a reduction occurs in inorganic compounds because the process is accompanied by a change in charge. For example,

$$Ag^0 \longrightarrow Ag^+ + 1e^-$$

With the loss of an electron, the neutral atom is converted to a positive ion, which is oxidation. In contrast,

$$:\overset{..}{\underset{..}{Br}}\cdot + e^- \longrightarrow :\overset{..}{\underset{..}{Br}}:^-$$

With the gain of one electron, the bromine atom is converted to a negative ion, which is reduction.

When organic compounds are involved, however, there may be no change in charge, and it is often difficult to determine whether oxidation or reduction has occurred. The following simplified view may help.

In organic systems, *oxidation* may be recognized as a gain of oxygen or a loss of hydrogen. A *reduction* reaction may involve a loss of oxygen or gain of hydrogen.

Consider the following compounds. A primary or secondary alcohol may be oxidized to an aldehyde or ketone, respectively, by the loss of hydrogen. An aldehyde may be oxidized to a carboxylic acid by gaining an oxygen.

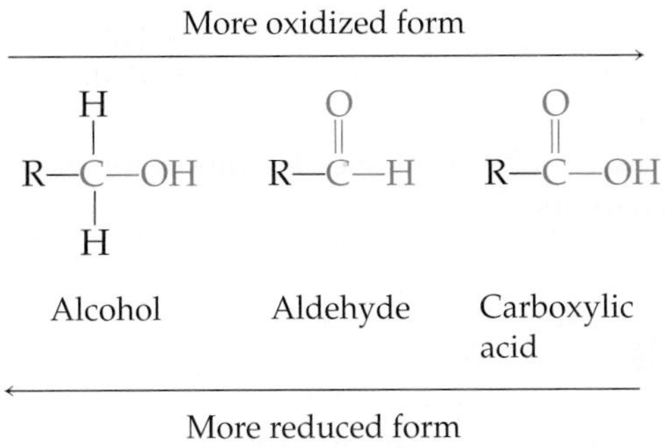

More oxidized form →

$$\underset{\text{Alcohol}}{R-\overset{\displaystyle H}{\underset{\displaystyle H}{C}}-OH} \qquad \underset{\text{Aldehyde}}{R-\overset{\displaystyle O}{C}-H} \qquad \underset{\substack{\text{Carboxylic}\\\text{acid}}}{R-\overset{\displaystyle O}{C}-OH}$$

← More reduced form

Thus, the conversion of an alcohol to a carbonyl compound and of a carbonyl compound (aldehyde) to a carboxylic acid are both examples of oxidations. Conversions in the opposite direction are reductions.

Oxidation and reduction reactions also play an important role in the chemistry of living systems. In living systems, these reactions are catalyzed by the action of various enzymes called *oxidoreductases*. These enzymes require compounds called *coenzymes* to accept or donate hydrogen in the reactions they catalyze.

Nicotinamide adenine dinucleotide, NAD^+, is a coenzyme commonly involved in biological oxidation-reduction reactions (Figure 12.5). We see NAD^+ in action in the final reaction of the citric acid cycle, an energy-harvesting pathway essential to life. In this reaction, catalyzed by the enzyme malate dehydrogenase, malate is oxidized to produce oxaloacetate:

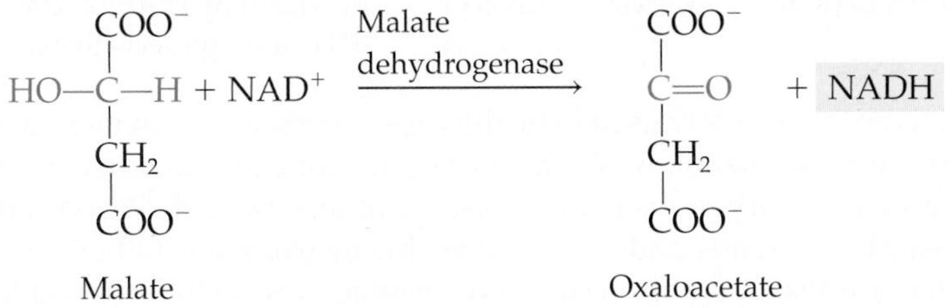

LEARNING GOAL

9 Discuss the role of oxidation and reduction reactions in the chemistry of living systems.

$$\underset{\text{Malate}}{\overset{\displaystyle COO^-}{\underset{\displaystyle \overset{\displaystyle |}{\underset{\displaystyle COO^-}{CH_2}}}{\underset{\displaystyle |}{HO-C-H}}}} + NAD^+ \xrightarrow{\substack{\text{Malate}\\\text{dehydrogenase}}} \underset{\text{Oxaloacetate}}{\overset{\displaystyle COO^-}{\underset{\displaystyle \overset{\displaystyle |}{\underset{\displaystyle COO^-}{CH_2}}}{\underset{\displaystyle |}{C=O}}}} + \boxed{NADH}$$

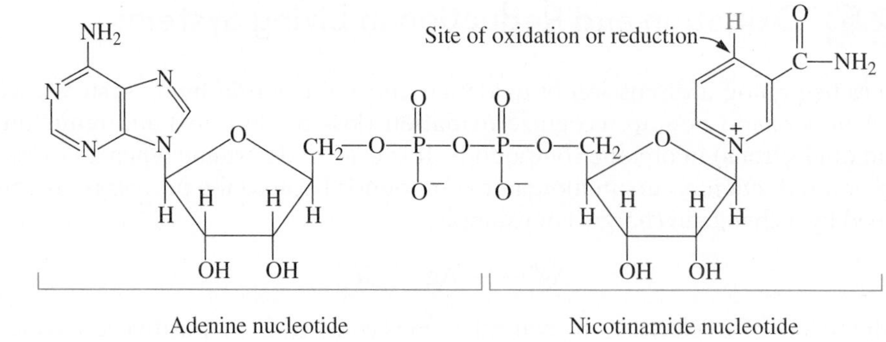

Figure 12.5 Nicotinamide adenine dinucleotide.

Adenine nucleotide   Nicotinamide nucleotide

Nicotinamide adenine dinucleotide (NAD$^+$)

NAD^+ actually accepts a hydride anion, H$^-$, hydrogen with two electrons.

NAD^+ participates by accepting hydrogen from the malate. As malate is oxidized, NAD^+ is reduced to NADH.

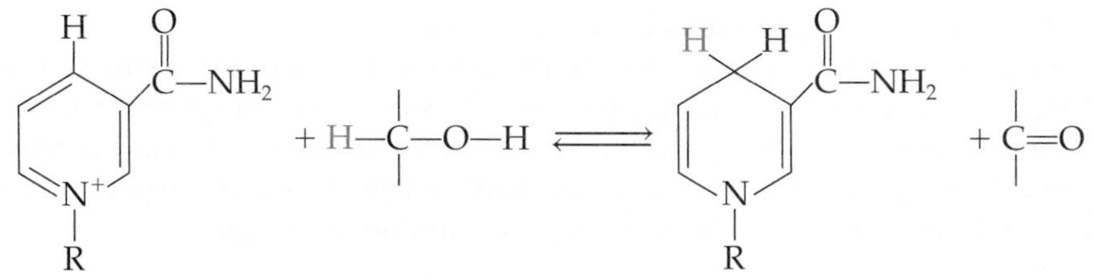

NAD$^+$
Oxidized form

NADH
Reduced form

We will study many other biologically important oxidation-reduction reactions in upcoming chapters.

LEARNING GOAL

10 Discuss the use of phenols as germicides.

Figure 12.6 Ball-and-stick model of phenol. Keep in mind that this model is not completely accurate because it cannot show the cloud of shared electrons above and below the benzene ring. Review Section 11.6 for a more accurate description of the benzene ring.

A dilute solution of phenol must be used because concentrated phenol causes severe burns and because phenol is not highly soluble in water.

12.6 Phenols

Phenols are compounds in which the hydroxyl group is attached to a benzene ring (Figure 12.6). Like alcohols, they are polar compounds because of the polar hydroxyl group. Thus, the simpler phenols are somewhat soluble in water. They are found in flavorings and fragrances (mint and savory) and are used as preservatives (butylated hydroxytoluene, BHT). Examples include:

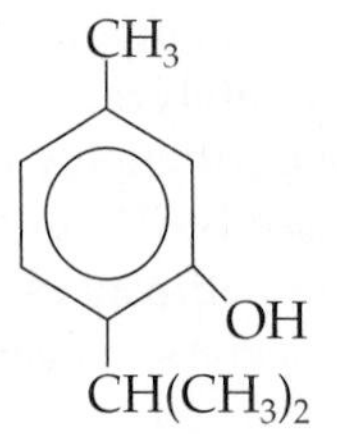

Thymol (mint)   Carvacrol (savory)   Butylated hydroxytoluene, BHT (food preservative)

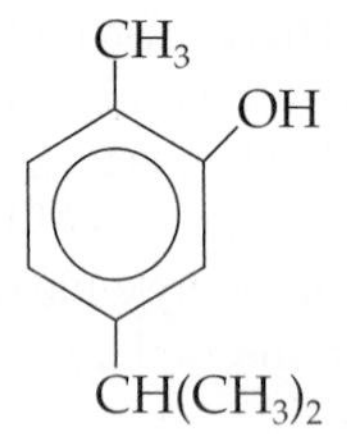

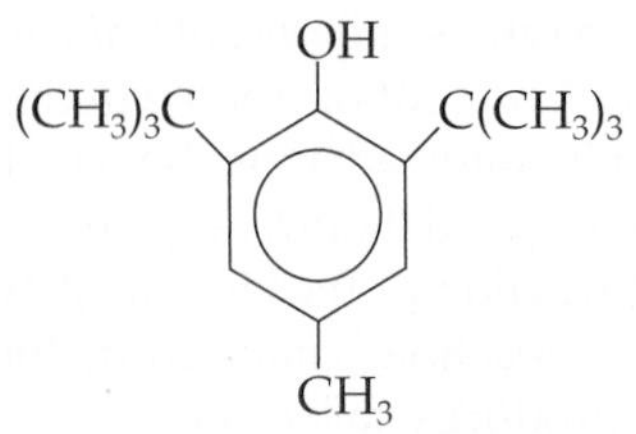

Phenols are also widely used in health care as germicides. In fact, carbolic acid, a dilute solution of phenol, was used as an antiseptic and disinfectant by Joseph Lister in his early work to decrease postsurgical infections. He used carbolic acid to bathe surgical wounds and to "sterilize" his instruments. Other derivatives of phenol that are used as antiseptics and disinfectants include hexachlorophene,

Kitchen Chemistry

Spicy Phenols

Did you ever wonder why chili peppers are so "hot" or why drinking milk soothes the burning sensation? Why does fresh ginger have such a bite to it? The answer is in the structure of some of the phenols that are found in these foods. In the case of chili peppers, the culprit is capsaicin, which irritates mucous membranes and causes the burning sensation. In the pepper plant, the purpose of capsaicin, and the related molecules found in the peppers, is to protect the plant from herbivores.

The burning sensation when eating foods with capsaicin is caused by binding of the molecule to a receptor on the surface of our sensory nerve cells or neurons. Binding to the receptor causes the same reaction that occurs when the nerve cells are stimulated by excessive heat or abrasion. The reason that milk helps alleviate these symptoms is that capsaicin is hydrophobic and soluble in the milk fat. This removes it from the area of the neurons and eases the burning sensation.

The interaction of capsaicin with the sensory nerve cells is the key to the use of capsaicin to treat a variety of peripheral pain such as that associated with diabetic neuropathy and neuropathy that may follow a shingles infection, as well as pain associated with osteoarthritis and rheumatoid arthritis. Initial use of a capsaicin cream will cause a burning or itching sensation as the sensory neurons are stimulated. This stimulation depletes a neurotransmitter that the neurons need to signal the burning sensation to the brain. When the neurotransmitter is depleted, the pain signals stop. Regular use can significantly decrease chronic pain.

Capsaicins are also used in riot control and personal defense sprays. When the spray comes into contact with the skin and especially with the mucous membranes and eyes, the resulting pain incapacitates the would-be assailant, allowing the victim to run for help.

Ginger is the ingredient that imparts a bite to ginger ale, gingerbread, and ginger cookies. The flavor of ginger is largely due to gingerol, a phenolic compound with a structure very similar to capsaicin. When ginger is cooked, the gingerol is converted into zingerone, which has a less piquant flavor and a spicy-sweet aroma. Zingerone is related structurally to vanillin and is added to some perfumes to give them a spicy aroma. When ginger is dried, the gingerol is converted into shagaol, which is twice as piquant as gingerol. For this reason, dried, powdered ginger is spicier than fresh. Ginger is used in cuisine around the world. It has also been used as a folk medicine, prescribed for gastrointestinal discomfort, including seasickness and morning sickness. In other cultures, it is used to prevent the flu or combat a cold.

Chili peppers and ginger are just two examples of the amazing larder nature provides for us. They bring spice into our lives and, sometimes, relief from our pain.

Capsaicin

Gingerol

Shagaol

Zingerone

For Further Understanding

► Chili oil is a popular, spicy condiment used in Chinese cooking and sometimes as a dipping sauce. It is prepared by mixing dried chili peppers with oil. Explain why the oil that has been infused with the dried peppers becomes a hot sauce.

► What functional groups do capsaicin, gingerol, shagaol, and zingerone share in common?

A Medical Perspective

Resveratrol: Fountain of Youth?

Will a daily snack of dark chocolate and red wine allow us to live longer, healthier lives? Recently, such claims have been made based on a compound that is found in both. That compound is resveratrol, which is a type of natural phenol that is produced by plants to protect them from bacterial or fungal infections.

Resveratrol

For quite some time, people have thought that red wine in moderation was associated with cardiovascular health. Resveratrol is one of the components in red wine that might help reduce the levels of LDL cholesterol, sometimes referred to as "bad" cholesterol, prevent blood clots and prevent damage to the lining of blood vessels. But how good is the evidence?

Consider that most of the research has been carried out in animals, not in people. There are studies that indicate that resveratrol protected mice from diabetes and obesity, both of which are factors associated with heart disease. These types of experiments have not been reproduced in humans. In fact, to reproduce this experiment with humans would require that each test subject drink 60 liters (L) of wine each day!

Other studies have shown that resveratrol extended the lifespan of yeast, some worms, fruit flies, and a short-lived fish. However, nothing is known about the effect of resveratrol on the human physiology or lifespan. Clinical studies with humans are needed to ensure that there are no negative side effects from consumption of large doses and to determine whether the potential health benefits seen experimentally occur in humans, as well.

As with all nutritional claims, it is important to look at the evidence on which the claims are made before taking a nutritional supplement or changing your lifestyle. Until the data are in, however, it couldn't hurt to have an occasional glass of red wine or piece of dark chocolate.

For Further Understanding

▶ Resveratrol is described as an antioxidant. Go online to look up further information on what is meant by an antioxidant and how they are thought to act in the body.

▶ What other types of compounds are described as antioxidants?

---

hexylresorcinol, and *o*-phenylphenol. The structures of these compounds are shown below:

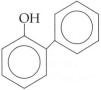

Phenol
(carbolic acid;
phenol dissolved
in water;
antiseptic)

Hexachlorophene
(antiseptic)

Hexylresorcinol
(antiseptic)

o-Phenylphenol
(antiseptic)

To learn more about nutritional aspects of polyphenols, see A Medical Perspective: Resveratrol: Fountain of Youth? in this chapter and Kitchen Chemistry: Amazing Chocolate in Chapter 11.

In recent years, there has been a great deal of discussion about the health benefits of phenols and polyphenols (compounds composed of a large number of phenol rings) from natural sources such as green tea, chocolate, and red wine.

Green tea extracts are thought to have antioxidant, anti-inflammatory, and anti-carcinogenic properties. Two of the polyphenols in green tea are epicatechin gallate (ECG) and epigallocatechin gallate (EGCG) shown here:

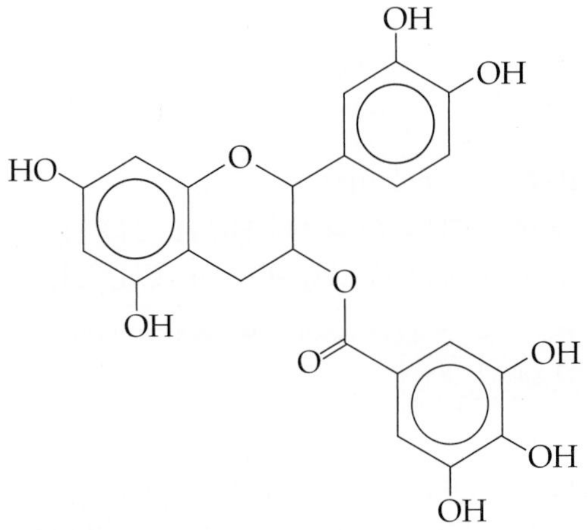

Epicatechin gallate 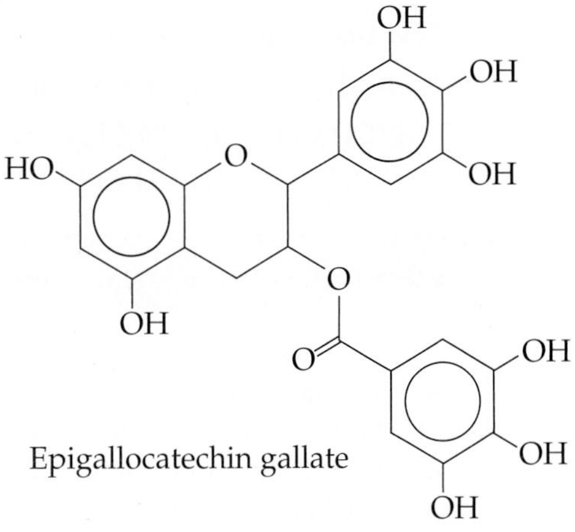   Epigallocatechin gallate

Question 12.9 Why are simple phenols somewhat soluble in water?

Question 12.10 What is carbolic acid? How did Joseph Lister use carbolic acid?

12.7 Ethers

Ethers have the general formula R—O—R, and thus they are structurally related to alcohols (R—O—H). The C—O bonds of ethers are polar, so ether molecules are polar (Figure 12.7). However, ethers do not form hydrogen bonds with one another because there is no —OH group. Therefore, they have much lower boiling points than alcohols of similar molar mass but higher boiling points than alkanes of similar molar mass. Compare the following examples:

LEARNING GOAL

11 Write names and draw structures for common ethers and discuss their use in medicine.

CH₃CH₂CH₂CH₃	CH₃—O—CH₂CH₃	CH₃CH₂CH₂OH

$CH_3CH_2CH_2CH_3$

Butane
(butane)
M.M. = 58
b.p. = −0.5°C

CH_3—O—CH_2CH_3

Methoxyethane
(ethyl methyl ether)
M.M. = 60
b.p. = 7.9°C

$CH_3CH_2CH_2OH$

1-Propanol
(propyl alcohol)
M.M. = 60
b.p. = 97.2°C

In the IUPAC system of naming ethers, one alkyl group of the molecule is named as a hydrocarbon chain and the simpler alkyl group is named as an alkoxy group bonded to that chain. The alkoxy group has the structure —OR. This is analogous to the name *hydroxy* for the —OH group. Thus, CH_3—O— is methoxy, CH_3CH_2—O— is ethoxy, and so on.

An alkoxy group is an alkyl group bonded to an oxygen atom (—OR).

EXAMPLE 12.9 **Using IUPAC Nomenclature to Name an Ether**

Name the following ether using IUPAC nomenclature.

Solution

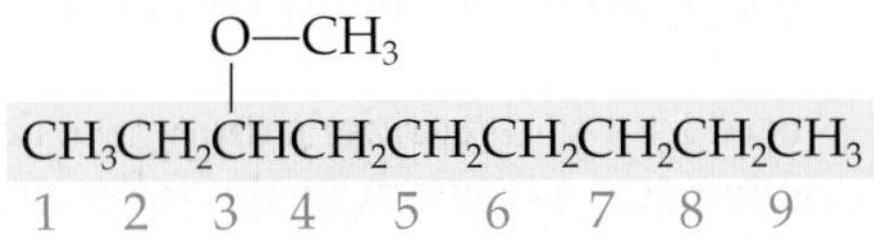

$$O\text{—}CH_3$$
$$CH_3CH_2CHCH_2CH_2CH_2CH_2CH_2CH_3$$
$$1 \quad 2 \quad 3 \quad 4 \quad 5 \quad 6 \quad 7 \quad 8 \quad 9$$

LEARNING GOAL

11 Write names and draw structures for common ethers and discuss their use in medicine.

Continued…

Figure 12.7 Ball-and-stick model of the ether methoxymethane (dimethyl ether).

Parent compound: nonane

Position of alkoxy group: carbon-3 (*Not* carbon-7 because the chain must be numbered to give the lowest number to the carbon bonded to the alkoxy group.)

Substituents: 3-methoxy

Name: 3-Methoxynonane

Practice Problem 12.9

Name the following ethers using IUPAC nomenclature:

a. CH_3CH_2—O—$CH_2CH_2CH_3$ c. CH_3CH_2—O—$CH_2CH_2CH_2CH_2CH_3$

b. CH_3—O—$CH_2CH_2CH_3$ d. $CH_3CH_2CH_2CH_2$—O—$CH_2CH_2CH_3$

▸ For Further Practice: **Questions 12.83 and 12.85.**

In the common system of nomenclature, ethers are named by placing the names of the two alkyl groups attached to the ether oxygen as prefixes in front of the word *ether*. The names of the two groups can be placed either alphabetically or by size (smaller to larger), as seen in the following examples:

CH_3—O—CH_3 CH_3—O—CH_2CH_3 CH_3CH_2—O—$CH(CH_3)_2$

Dimethyl ether   Ethyl methyl ether   Ethyl isopropyl ether

or methyl ether   or methyl ethyl ether

EXAMPLE 12.10   **Naming Ethers Using the Common Nomenclature System**

Write the common name for each of the following ethers.

Solution

	CH_3CH_2—O—CH_2CH_3	CH_3—O—$CH_2CH_2CH_3$
Alkyl groups:	two ethyl groups	methyl and propyl
Name:	Diethyl ether	Methyl propyl ether

LEARNING GOAL

11 Write names and draw structures for common ethers and discuss their use in medicine.

Notice that there is only one correct name for methyl propyl ether because the methyl group is smaller than the propyl group and it would be first in an alphabetical listing also.

Practice Problem 12.10

Write the common name for each of the following ethers:

a. CH_3CH_2—O—$CH_2CH_2CH_3$ c. CH_3CH_2—O—$CH_2CH_2CH_2CH_2CH_3$

b. CH_3—O—$CH_2CH_2CH_3$ d. $CH_3CH_2CH_2CH_2$—O—$CH_2CH_2CH_3$

▸ For Further Practice: **Questions 12.84 and 12.86.**

Chemically, ethers are moderately inert. They do not react with reducing agents or bases under normal conditions. However, they are extremely volatile and highly flammable (easily oxidized in air) and hence must always be treated with great care.

Ethers may be prepared by a condensation reaction (removal of water) between two alcohol molecules, as shown in the following general reaction. The reaction requires heat and acid.

$$R^1-OH + R^2-OH \xrightarrow[\text{heat}]{H^+} R^1-O-R^2 + H_2O$$

Alcohol   Alcohol   Ether   Water

EXAMPLE 12.11 | **Writing an Equation Representing the Synthesis of an Ether via a Condensation Reaction**

LEARNING GOAL

12 Write equations representing the condensation reaction between two alcohol molecules to form an ether.

Write an equation showing the synthesis of dimethyl ether.

Solution

The alkyl substituents of this ether are two methyl groups. Thus, the alcohol that must undergo condensation to produce dimethyl ether is methanol.

$$CH_3OH + CH_3OH \xrightarrow[\text{heat}]{H^+} CH_3-O-CH_3 + H_2O$$

Methanol   Methanol   Dimethyl ether   Water

Practice Problem 12.11

a. Write an equation showing the condensation reaction that would produce diethyl ether. Provide structures and names for all reactants and products.

b. Write an equation showing the condensation reaction between two molecules of 2-propanol. Provide structures and names for all reactants and products.

▶ For Further Practice: **Questions 12.81 and 12.82.**

Diethyl ether was the first general anesthetic used. The dentist Dr. William Morton is credited with its introduction in the 1800s. Diethyl ether functions as an anesthetic by interacting with the central nervous system. It appears that diethyl ether (and many other general anesthetics) functions by accumulating in the lipid material of the nerve cells, thereby interfering with nerve impulse transmission. This results in analgesia, a lessened perception of pain.

Halogenated ethers are also routinely used as general anesthetics (Figure 12.8). They are less flammable than diethyl ether and are therefore safer to store and use. Desflurane, sevoflurane, and isoflurane are three of the commonly used members of this family:

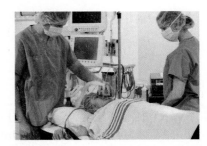

Figure 12.8 An anesthesiologist administers isoflurane to a surgical patient.

Desflurane   Sevoflurane   Isoflurane

Question 12.11 Why do ethers have much lower boiling points than alcohols?

Question 12.12 Describe the structure of ether molecules.

12.8 Thiols

LEARNING GOAL

13 Write names and draw structures for simple thiols and discuss their biological significance.

Compounds that contain the sulfhydryl group (—SH) are called **thiols**. They are similar to alcohols in structure, but the sulfur atom replaces the oxygen atom.

Figure 12.9 This skunk on a bed of roses is surrounded by scent molecules. The two most common compounds in the defense spray of the striped skunk are the thiols *trans*-2-butene-1-thiol and 3-methyl-1-butanethiol. Two alcohols, 2-phenylethanol and geraniol, are the major components of the scent of roses.

Thiols and many other sulfur compounds have nauseating aromas. They are found in substances as different as the defense spray of the North American striped skunk, onions, and garlic. The structures of the two most common compounds in the defense spray of the striped skunk, *trans*-2-butene-1-thiol and 3-methyl-1-butanethiol, are shown in Figure 12.9. These structures are contrasted with the structures of the two molecules that make up the far more pleasant scent of roses: geraniol, an unsaturated alcohol, and 2-phenylethanol, an aromatic alcohol.

The IUPAC rules for naming thiols are similar to those for naming alcohols, except that the full name of the alkane is retained. The suffix *-thiol* follows the name of the parent compound.

EXAMPLE 12.12   **Naming Thiols Using the IUPAC Nomenclature System**

Write the IUPAC names for the thiols shown below.

LEARNING GOAL

13 Write names and draw structures for simple thiols and discuss their biological significance.

Solution

Retain the full name of the parent compound and add the suffix *-thiol*.

	$CH_3CH_2—SH$	$HS—CH_2CH_2—SH$
Parent compound:	ethane	ethane
Position of —SH:	carbon-1	carbon-1 and carbon-2
Name:	Ethanethiol	1,2-Ethanedithiol

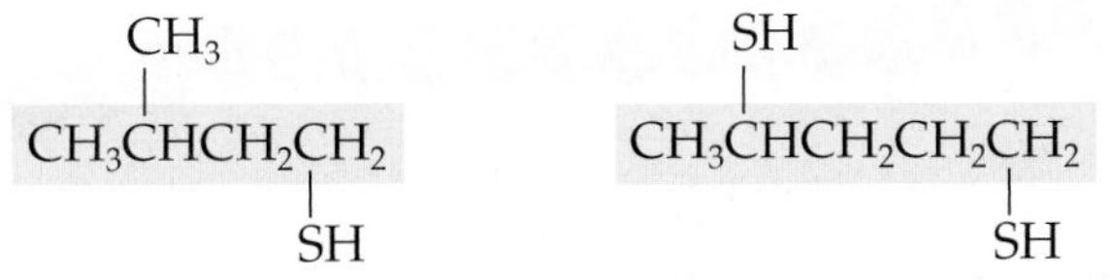
CH3
CH3CHCH2CH2
SH
SH
CH3CHCH2CH2CH2
SH

Parent compound:	butane	pentane
Position of —SH:	carbon-1	carbon-1 and carbon-4
Substituent:	3-methyl	
Name:	3-Methyl-1-butanethiol	1,4-Pentanedithiol

Practice Problem 12.12

Draw the structures of each of the following thiols.

a. 1,3-Butanedithiol
b. 2-Methyl-2-pentanethiol
c. 2-Chloro-2-propanethiol
d. Cyclopentanethiol

▶ For Further Practice: **Questions 12.91 and 12.92.**

The amino acid cysteine is a thiol that plays an important role in the structure and shape of many proteins. Two cysteine molecules can undergo oxidation to form cystine. The new bond formed is called a **disulfide** (—S—S—) bond.

Amino acids are the subunits from which proteins are made. A protein is a long polymer, or chain, of many amino acids bonded to one another.

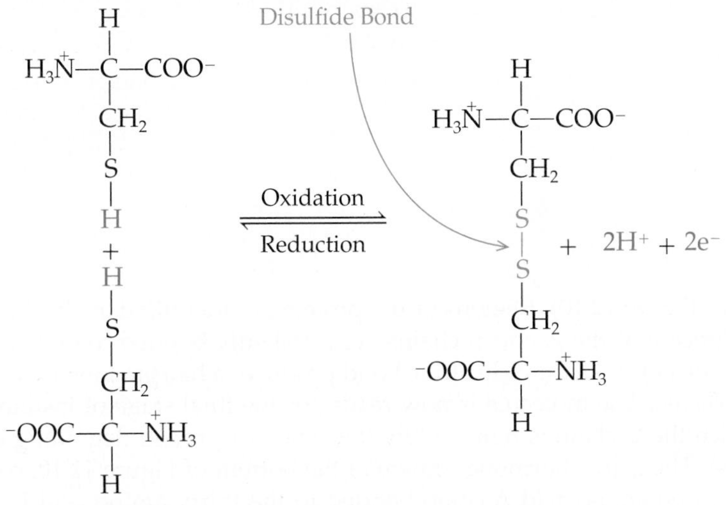
Disulfide Bond
Oxidation
Reduction
+ 2H+ + 2e−

2 Cysteine   Cystine

If the two cysteines are in different protein chains, the disulfide bond between them forms a bridge joining them together (Figure 12.10). If the two cysteines are in the same protein chain, a loop is formed. An example of the importance of disulfide bonds is seen in the production and structure of the protein hormone insulin, which controls blood sugar levels in the body. Insulin is initially produced as a protein called preproinsulin. Enzymatic removal of twenty-four amino acids and formation of disulfide bonds between cysteine amino acids produce

Figure 12.10 Two steps in the synthesis of insulin. Disulfide bonds hold the A and B chains together and form a hairpin loop in the A chain.

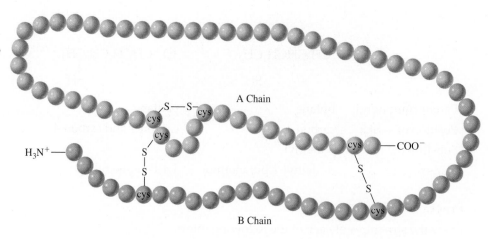

Preproinsulin

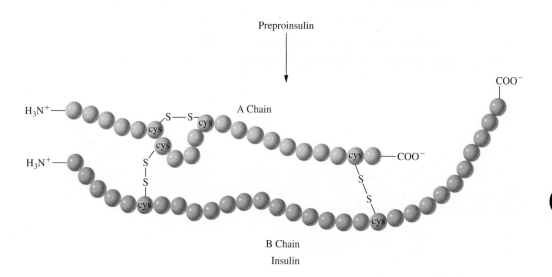

Insulin

The reactions involving coenzyme A are discussed in detail in Chapters 21, 22, and 23.

A high-energy bond is one that releases a great deal of energy when it is broken.

proinsulin (Figure 12.10). Regions of the protein are identified as the A, B, and C chains. Notice that the A and B chains are covalently bonded to one another by two disulfide bonds. A third disulfide bond produces a hairpin loop in the A chain of the molecule. The molecule is now ready for the final stage of insulin synthesis in which the C chain is removed by the action of protein degrading enzymes (proteases). The active hormone, shown at the bottom of Figure 12.10, consists of the twenty-one amino acid A chain bonded to the thirty amino acid B chain by two disulfide bonds. Without these disulfide bonds, functional insulin molecules could not exist because there would be no way to keep the two chains together in the proper shape.

Coenzyme A is a thiol that serves as a "carrier" of acetyl groups (CH_3CO—) in biochemical reactions. It plays a central role in metabolism by shuttling acetyl groups from one reaction to another. When the two-carbon acetate group is attached to coenzyme A, the product is acetyl coenzyme A (acetyl CoA). The bond between coenzyme A and the acetyl group is a high-energy *thioester bond*. In effect, formation of the high-energy thioester bond "energizes" the acetyl group so that it can participate in other biochemical reactions.

Kitchen Chemistry

The Magic of Garlic

Chefs have known for centuries that whole roasted garlic cloves impart a sweet flavor to foods and can even be used to make ice cream! But if you chop or crush the clove of garlic before cooking, the flavors are entirely different, much stronger. The chemistry behind this difference was not understood until the 1940s, when two researchers from the Sandoz Company investigated the chemistry of garlic.

They found that the difference between whole cooked garlic cloves and crushed garlic lies in the presence of the compound alliin in certain cells of the garlic clove and the presence of the enzyme allinase in other cells. When the cells are crushed, the enzyme and its substrate are brought together, initiating a chain of chemical reactions that begins with the production of allicin and increases in complexity with time and further cooking. The chemistry becomes even more complex depending on the other ingredients used in the recipe. One set of compounds will form if the garlic is heated in water. Using butter or olive oil or milk will produce quite different products and, thus, quite different flavors in the food.

Some of the compounds produced in these reactions have strong biological activity. For instance, allicin is an antibiotic, having one-fiftieth the power of penicillin and one-tenth the activity of tetracycline. However, allicin is quickly broken down during cooking to produce a complex mixture of compounds that includes propenyl disulfide, which gives onions their aroma, and propenyl sulfenic acid, which causes tears when you cut into an onion. Also among these breakdown products is diallyl disulfide, which has been suggested to have anticancer properties and to be of benefit in reducing cholesterol levels.

For Further Understanding

▸ Design an experiment to test the suspected anticancer activity of diallyl disulfide.

▸ Explain why using a variety of ingredients, such as butter or olive oil, will result in different flavors and aromas in the food that is prepared.

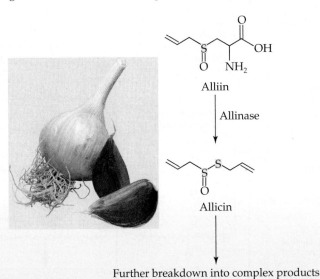

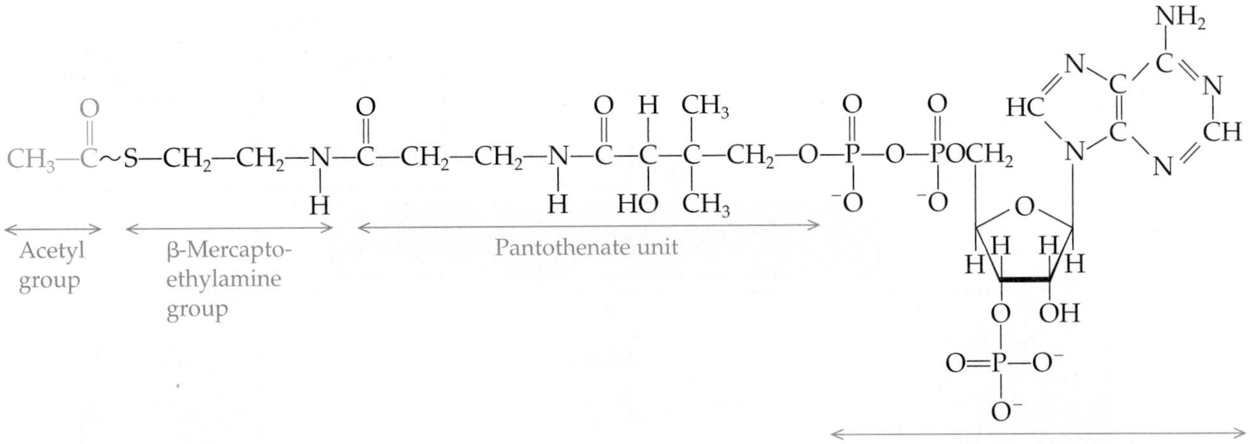

Alliin

Allinase

Allicin

Further breakdown into complex products

Acetyl coenzyme A
(acetyl CoA)

Acetyl CoA is made and used in the energy-producing reactions that provide most of the energy for life processes. It is also required for the biosynthesis of many biological molecules.

441

CHAPTER MAP

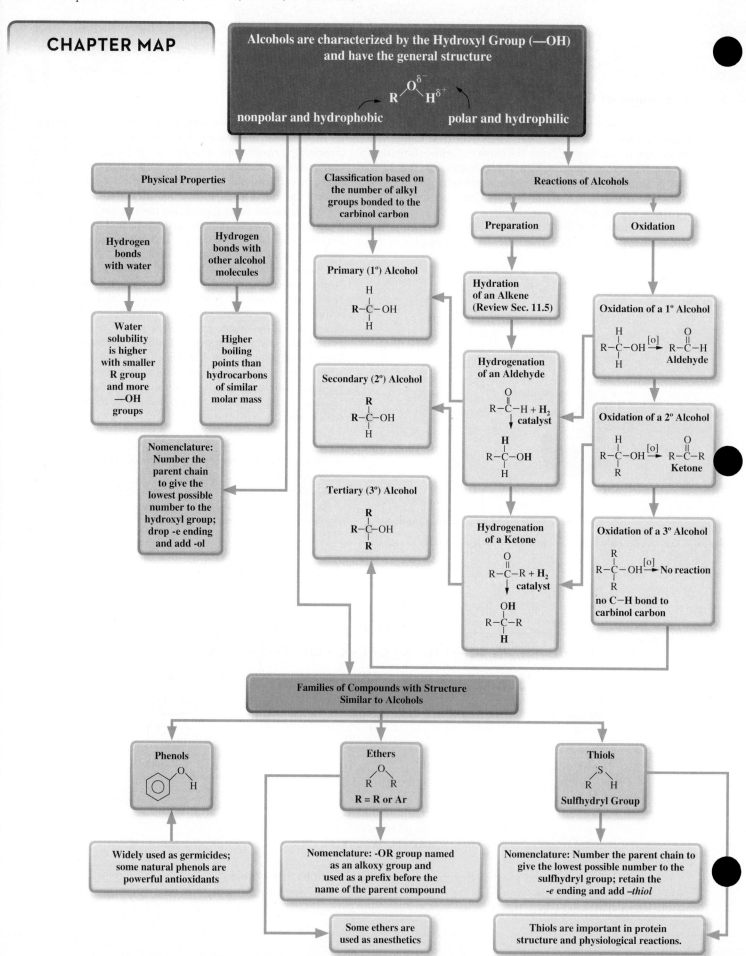

Alcohols are characterized by the Hydroxyl Group (—OH) and have the general structure

$$R\!-\!\overset{\delta-}{O}\!-\!H^{\delta+}$$

nonpolar and hydrophobic   polar and hydrophilic

Physical Properties

- Hydrogen bonds with water
- Hydrogen bonds with other alcohol molecules

Water solubility is higher with smaller R group and more —OH groups

Higher boiling points than hydrocarbons of similar molar mass

Nomenclature: Number the parent chain to give the lowest possible number to the hydroxyl group; drop -e ending and add -ol

Classification based on the number of alkyl groups bonded to the carbinol carbon

Primary (1°) Alcohol

$$R\!-\!\overset{\overset{\displaystyle H}{|}}{\underset{\underset{\displaystyle H}{|}}{C}}\!-\!OH$$

Secondary (2°) Alcohol

$$R\!-\!\overset{\overset{\displaystyle R}{|}}{\underset{\underset{\displaystyle H}{|}}{C}}\!-\!OH$$

Tertiary (3°) Alcohol

$$R\!-\!\overset{\overset{\displaystyle R}{|}}{\underset{\underset{\displaystyle R}{|}}{C}}\!-\!OH$$

Reactions of Alcohols

Preparation

Hydration of an Alkene (Review Sec. 11.5)

Hydrogenation of an Aldehyde

$$R\!-\!\overset{\overset{\displaystyle O}{\|}}{C}\!-\!H + H_2 \xrightarrow{\text{catalyst}}$$

$$R\!-\!\overset{\overset{\displaystyle H}{|}}{\underset{\underset{\displaystyle H}{|}}{C}}\!-\!OH$$

Hydrogenation of a Ketone

$$R\!-\!\overset{\overset{\displaystyle O}{\|}}{C}\!-\!R + H_2 \xrightarrow{\text{catalyst}}$$

$$R\!-\!\overset{\overset{\displaystyle OH}{|}}{\underset{\underset{\displaystyle H}{|}}{C}}\!-\!R$$

Oxidation

Oxidation of a 1° Alcohol

$$R\!-\!\overset{\overset{\displaystyle H}{|}}{\underset{\underset{\displaystyle H}{|}}{C}}\!-\!OH \xrightarrow{[o]} R\!-\!\overset{\overset{\displaystyle O}{\|}}{C}\!-\!H \quad \text{Aldehyde}$$

Oxidation of a 2° Alcohol

$$R\!-\!\overset{\overset{\displaystyle H}{|}}{\underset{\underset{\displaystyle R}{|}}{C}}\!-\!OH \xrightarrow{[o]} R\!-\!\overset{\overset{\displaystyle O}{\|}}{C}\!-\!R \quad \text{Ketone}$$

Oxidation of a 3° Alcohol

$$R\!-\!\overset{\overset{\displaystyle R}{|}}{\underset{\underset{\displaystyle R}{|}}{C}}\!-\!OH \xrightarrow{[o]} \text{No reaction}$$

no C—H bond to carbinol carbon

Families of Compounds with Structure Similar to Alcohols

Phenols

Widely used as germicides; some natural phenols are powerful antioxidants

Ethers

$$\overset{\displaystyle O}{\underset{R \quad R}{\diagup \diagdown}}$$

R = R or Ar

Nomenclature: -OR group named as an alkoxy group and used as a prefix before the name of the parent compound

Some ethers are used as anesthetics

Thiols

$$\overset{\displaystyle S}{\underset{R \quad H}{\diagup \diagdown}}$$

Sulfhydryl Group

Nomenclature: Number the parent chain to give the lowest possible number to the sulfhydryl group; retain the -e ending and add –thiol

Thiols are important in protein structure and physiological reactions.

SUMMARY OF REACTIONS

Preparation of Alcohols

Hydration of alkenes:

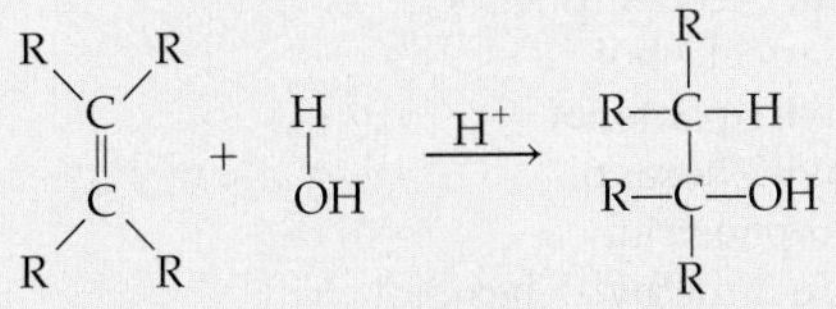

Alkene   Water   Alcohol

Reduction of an aldehyde or ketone:

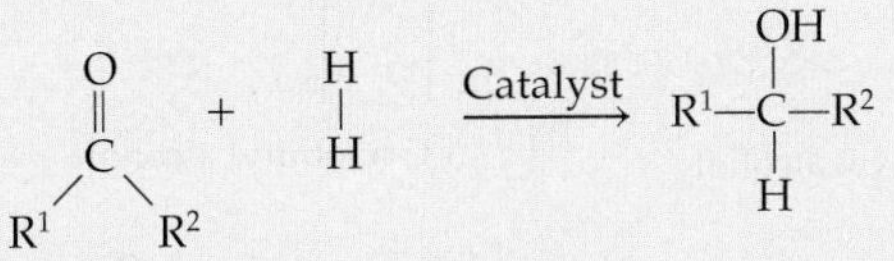

Aldehyde   Hydrogen   Alcohol
or
ketone

Dehydration of Alcohols

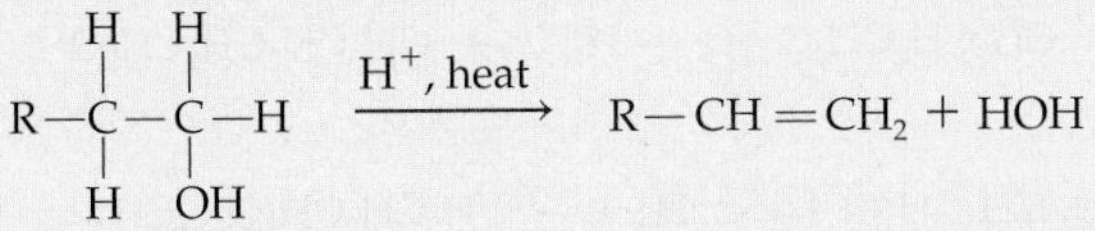

Alcohol   Alkene   Water

Oxidation Reactions

Oxidation of a primary alcohol:

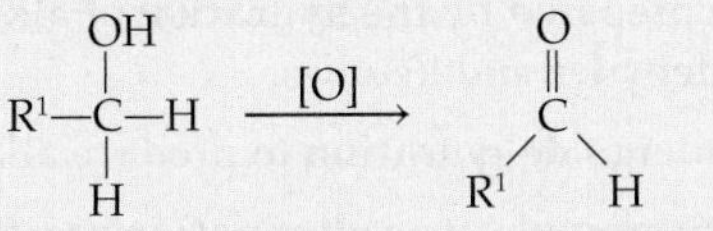

1° Alcohol   An aldehyde

Oxidation of a secondary alcohol:

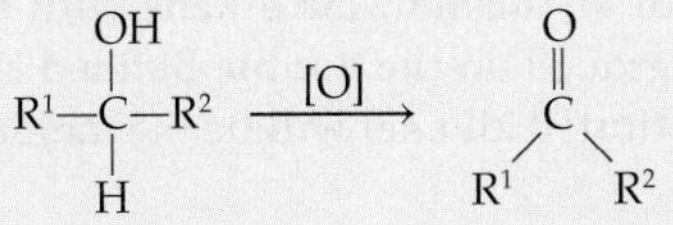

2° Alcohol   A ketone

Oxidation of a tertiary alcohol:

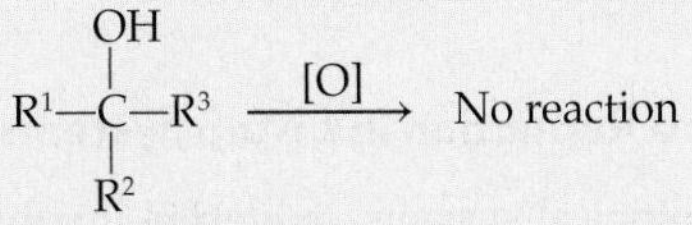

3° Alcohol

Condensation Synthesis of an Ether

$$R^1-OH + R^2-OH \xrightarrow[\text{heat}]{H^+} R^1-O-R^2 + H_2O$$

Alcohol   Alcohol   Ether   Water

SUMMARY

12.1 Alcohols: Structure and Physical Properties

▶ **Alcohols** are characterized by the **hydroxyl group (—OH).**

▶ Alcohols have the general structure R—OH.

▶ Alcohols are classified as primary, secondary, or tertiary depending on the number of alkyl groups attached to the **carbinol carbon.**

▶ A **primary alcohol** has one alkyl group attached to the carbinol carbon.

▶ A **secondary alcohol** has two alkyl groups attached to the carbinol carbon.

▶ A **tertiary alcohol** has three alkyl groups attached to the carbinol carbon.

▶ They are very polar because the hydroxyl group is polar.

▶ Alcohols form intermolecular hydrogen bonds and as a result have higher boiling points than hydrocarbons of comparable molar mass.

▶ Smaller alcohols are very water-soluble.

12.2 Alcohols: Nomenclature

▶ In the IUPAC system, alcohols are named by determining the **parent compound** and replacing the -e ending with -ol.

▶ The parent chain is numbered to give the hydroxyl group the lowest possible number.

▶ Common names are derived from the alkyl group corresponding to the parent compound.

12.3 Medically Important Alcohols

▶ Methanol is a toxic alcohol that is used as a solvent.

▶ Ethanol is the alcohol consumed in beer, wine, and distilled liquors. It is produced by the alcohol **fermentation** of sugars.

▶ Isopropanol is used as an antiseptic.

▶ Ethylene glycol (1,2-ethanediol) is used as antifreeze.

▶ Glycerol (1,2,3-propanetriol) is used in cosmetics and pharmaceuticals.

12.4 Reactions Involving Alcohols

▶ Alcohols can be prepared by the **hydration** of alkenes or the **reduction** of aldehydes and ketones.

▶ Alcohols can undergo **dehydration** to produce alkenes.

▶ Dehydration is an example of an **elimination reaction;** that is, one in which a molecule loses atoms or ions from its structure.

▶ **Zaitsev's rule** states that in an elimination reaction, such as the dehydration of an alcohol, the alkene with the greatest number of alkyl groups on the double-bonded carbons (the more highly substituted alkene) will be the major product in the reaction.

▶ Primary and secondary alcohols can undergo **oxidation** reactions to yield aldehydes and ketones, respectively.

▶ Tertiary alcohols do not undergo oxidation.

12.5 Oxidation and Reduction in Living Systems

▶ In organic and biological systems, **oxidation** involves the gain of oxygen or the loss of hydrogen, while **reduction** involves the loss of oxygen or the gain of hydrogen.

▶ Nicotinamide adenine dinucleotide, NAD^+, is a coenzyme involved in many biological oxidation and reduction reactions.

12.6 Phenols

▶ **Phenols** are compounds in which the hydroxyl group is attached to a benzene ring.

▶ Phenols have the general structure Ar—OH.

▶ Many phenols are important as antiseptics and disinfectants.

12.7 Ethers

▶ **Ethers** are characterized by the R—O—R functional group.

▶ Ethers are generally nonreactive but are extremely flammable.

▶ Diethyl ether was the first general anesthetic used in medical practice.

▶ Desflurane, sevoflurane, and isoflurane, which are much less flammable, have replaced the use of diethyl ether.

12.8 Thiols

▶ **Thiols** are characterized by the sulfhydryl group (—SH).

▶ The amino acid cysteine is a thiol that is extremely important for maintaining the correct shapes of proteins by forming **disulfide bonds** with other cysteine molecules within the same or another protein.

▶ Coenzyme A is a thiol that serves as a "carrier" of acetyl groups in important biological reactions.

ANSWERS TO PRACTICE PROBLEMS

12.1 a. Primary alcohol
b. Secondary alcohol
c. Tertiary alcohol
d. Aromatic alcohol (phenol)
e. Secondary alcohol

12.2 a. 4-Methyl-1-pentanol
b. 4-Methyl-2-hexanol
c. 1,2,3-Propanetriol
d. 4-Chloro-3-methyl-1-hexanol

12.3 a.

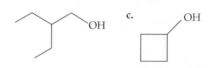

sec-Hexyl alcohol

c.

Cyclobutyl alcohol

b.

Isopentyl alcohol (also known as isoamyl alcohol)

d.

Hexyl alcohol

12.4 $$CH_3CH_2CH_2\overset{\displaystyle O}{\overset{\|}{C}}\!-\!H + H_2 \longrightarrow CH_3CH_2CH_2CH_2OH$$

12.5 $$CH_3CH_2\overset{\displaystyle O}{\overset{\|}{C}}CH_3 + H_2 \longrightarrow CH_3CH_2CH(OH)CH_3$$

12.6 a. $CH_3CH_2OH \xrightarrow{H^+, \text{ heat}} CH_2{=}CH_2 + H_2O$

b. $CH_3CHOHCH_3 \xrightarrow{H^+, \text{ heat}} CH_3CH{=}CH_2 + H_2O$

c. There are two possible products.

$$CH_3CH_2\overset{\displaystyle CH_3}{\overset{|}{C}}HCHCH_2CH_3 \xrightarrow{H^+, \text{ heat}} CH_3CH_2\overset{\displaystyle CH_3}{\overset{|}{C}}{=}CHCH_2CH_3 + H_2O$$
(OH below)

$$CH_3CH_2\overset{\displaystyle CH_3}{\overset{|}{C}}HCHCH_2CH_3 \xrightarrow{H^+, \text{ heat}} CH_3CH_2\overset{\displaystyle CH_3}{\overset{|}{C}}HCH{=}CHCH_3$$
(OH below)

d. $$CH_3\overset{\displaystyle OH}{\overset{|}{C}}CH_3 \xrightarrow{H^+\text{heat}} CH_2{=}\overset{\displaystyle CH_3}{\overset{|}{C}}CH_3 + H_2O$$
(CH$_3$ below)

12.7 a.

$$CH_3\overset{\displaystyle CH_3}{\overset{|}{C}}CH_2CH_2OH \xrightarrow{[O]} CH_3\overset{\displaystyle CH_3}{\overset{|}{C}}CH_2\overset{\displaystyle O}{\overset{\|}{C}}\!-\!H$$

b.

$$CH_3CH_2OH \xrightarrow{[O]} CH_3\overset{\displaystyle O}{\overset{\|}{C}}\!-\!H$$

12.8 a.

$$CH_3\overset{\underset{|}{OH}}{CH}CH_2CH_3 \longrightarrow CH_3\overset{\underset{\|}{O}}{C}CH_2CH_3$$

b.

$$CH_3\overset{\underset{|}{OH}}{CH}CH_2CH_2CH_3 \longrightarrow CH_3\overset{\underset{\|}{O}}{C}CH_2CH_2CH_3$$

12.9 a. 1-Ethoxypropane   **c.** 1-Ethoxypentane
     b. 1-Methoxypropane   **d.** 1-Propoxybutane

12.10 a. Ethyl propyl ether   **c.** Ethyl pentyl ether
      b. Methyl propyl ether   **d.** Propyl butyl ether

12.11

a.

$$CH_3CH_2OH + CH_3CH_2OH \xrightarrow[\text{heat}]{H^+} CH_3CH_2-O-CH_2CH_3 + H_2O$$

Ethanol   Diethyl ether   Water

b.

$$2CH_3\overset{\underset{|}{OH}}{CH}CH_3 \xrightarrow[\text{heat}]{H^+} CH_3\underset{\underset{CH_3}{|}}{CH}-O-\underset{\underset{CH_3}{|}}{CH}CH_3 + H_2O$$

2-Propanol   Diisopropyl ether

12.12

  a. 1,3-Butanedithiol

$$\underset{\underset{SH}{|}}{CH_2}CH_2\underset{\underset{SH}{|}}{CH}CH_3$$

  b. 2-Methyl-2-pentanethiol

$$CH_3\underset{\underset{CH_3}{|}}{\overset{\overset{SH}{|}}{C}}CH_2CH_2CH_3$$

  c. 2-Chloro-2-propanethiol

$$CH_3\underset{\underset{Cl}{|}}{\overset{\overset{SH}{|}}{C}}CH_3$$

  d. Cyclopentanethiol

QUESTIONS AND PROBLEMS

Alcohols: Structure and Physical Properties

Foundations

12.13 Describe the relationship between the water solubility of alcohols and their hydrocarbon chain length.

12.14 Explain the relationship between the water solubility of alcohols and the number of hydroxyl groups in the molecule.

12.15 Define the term *carbinol carbon.*

12.16 Define the terms *primary, secondary,* and *tertiary alcohol,* and draw a general structure for each.

Applications

12.17 Classify each of the following as a 1°, 2°, or 3° alcohol:
  a. 3-Methyl-1-butanol
  b. 2-Methylcyclopentanol
  c. *t*-Butyl alcohol
  d. 1-Methylcyclopentanol
  e. 2-Methyl-2-pentanol

12.18 1-Heptanol has a pleasant aroma and is sometimes used in cosmetics to enhance the fragrance. Draw the condensed formula for 1-heptanol. Is this a primary, secondary, or tertiary alcohol?

12.19 Classify each of the following as a 1°, 2°, or 3° alcohol:
  a. $CH_3CH_2CH_2CH_2CH_2CH_2CH_2OH$

  b. $CH_3\underset{\underset{OH}{|}}{CH}CH_3$
          CH_3

  c. $CH_3\overset{\overset{CH_3}{|}}{\underset{\underset{CH_2OH}{|}}{C}}CH_3$

  d. $CH_3CH_2\underset{\underset{CH_3}{|}}{\overset{\overset{Br}{|}}{CH}}CH_2CH_2CH_2OH$

  e. $CH_3\underset{\underset{OH}{|}}{CH}-\underset{\underset{CH_3}{|}}{C}CH_2CH_2CH_3$

12.20 Classify each of the following as a primary, secondary, or tertiary alcohol:

  a. $CH_3CH_2\underset{\underset{CH_3}{|}}{\overset{\overset{CH_2CH_3}{|}}{C}}OH$

  b. $CH_3\underset{\underset{OH}{|}}{CH}CH_2\underset{\underset{Br}{|}}{CH}CH_3$

  c. $CH_3CH_2CH_2OH$

  d. $CH_3\underset{\underset{OH}{|}}{\overset{\overset{CH_3}{|}}{C}}CH_2CH_2CH_3$

12.21 Stingless bees use complex systems to communicate. One aspect of this communication is chemical: the bees produce 2-nonanol, 2-heptanol, and 2-undecanol in their mouthparts (mandibles) to direct other bees to pollen sources. Draw the condensed structure of each of these alcohols. Classify each as a primary, secondary, or tertiary alcohol.

12.22 Classify each of the following as a primary, secondary, or tertiary alcohol:
  a. 2-Methyl-2-butanol
  b. 1,2-Dimethylcyclohexanol
  c. 2,3,4-Trimethylcyclopentanol
  d. 3,3-Dimethyl-2-pentanol

12.23 Arrange the following compounds in order of increasing boiling point, beginning with the lowest:

    a. $CH_3CH_2CH_2CH_2CH_3$ **c.** $CH_3CHCH_2CH_2CH_3$
$$\overset{|}{OH}$$

    b. $CH_3CHCH_2CHCH_3$ **d.** $CH_3CH_2CH_2{-}O{-}CH_2CH_3$
$$\overset{|}{OH}\quad\overset{|}{OH}$$

12.24 Why do alcohols have higher boiling points than alkanes?

12.25 Which member of each of the following pairs is more soluble in water?

    a. CH_3CH_2OH or $CH_3CH_2CH_2CH_2OH$

    b. $CH_3CH_2CH_2CH_2CH_3$ or $CH_3CH_2CH_2CH_2OH$

    c.   OH   or CH_3CHCH_3
$$\overset{|}{OH}$$

12.26 Arrange the three alcohols in each of the following sets in order of increasing solubility in water:

    a. $CH_3CH_2CH_2CH_2CH_2OH$ $CH_3CHCH_2CHCH_2CH_3$
$$\overset{|}{OH}\quad\overset{|}{OH}$$

    $CH_3CHCH_2CHCH_2CH_2OH$
$$\overset{|}{OH}\quad\overset{|}{OH}$$

    b. Pentyl alcohol 1-Hexanol Ethylene glycol

Alcohols: Nomenclature

Foundations

12.27 Briefly describe the IUPAC rules for naming alcohols.

12.28 Briefly describe the rules for determining the common names for alcohols.

Applications

12.29 Give the IUPAC name for each of the following compounds:

    a. $CH_3CH_2CH(OH)CH_2CH_2CH_2OH$

    b. $CH_3CH(OH)CH(OH)CH_2CH_3$

    c. $CH_3CH(CH_3)CH(OH)CH_2CH_3$

12.30 Give the IUPAC name for each of the following compounds:

    a. $CH_3C(CH_2CH_3)_2CH(CH_3)CH_2CH_2CH_2OH$

    b. $CH_3CH_2CH(OH)CH_2CH_2CH_3$

    c. $CH_3CH_2CHClCHBrCH_2CH_2OH$

12.31 Draw each of the following, using complete structural formulas and line formulas:

    a. 2-Pentanol

    b. 1,2,4-Heptanetriol

    c. 2-Methylcyclopentanol

12.32 Draw each of the following, using condensed formulas and line formulas:

    a. 3-Methyl-4-ethyl-3-hexanol

    b. 1-Bromo-2-methyl-3-pentanol

    c. 2,4-Dimethylcyclohexanol

12.33 Give the IUPAC name for each of the following compounds:

    a.   **b.** OH   **c.** OH

12.34 Draw each of the following alcohols:

    a. 1-Iodo-2-butanol

    b. 1,2-Butanediol

    c. Cyclobutanol

12.35 Give the common name for each of the following compounds:

    a. CH_3OH **c.** $CH_2{-}CH_2$
$$\overset{|}{OH}\quad\overset{|}{OH}$$

    b. CH_3CH_2OH **d.** $CH_3CH_2CH_2OH$

12.36 Draw the structure of each of the following compounds:

    a. Pentyl alcohol

    b. Isopropyl alcohol

    c. Octyl alcohol

    d. Propyl alcohol

12.37 Draw a condensed formula for each of the following compounds:

    a. 4-Methyl-2-hexanol

    b. Isobutyl alcohol

    c. 1,5-Pentanediol

    d. 2-Nonanol

    e. 1,3,5-Cyclohexanetriol

12.38 Name each of the following alcohols using the IUPAC Nomenclature System:

    a. OH   **c.** $CH_3CHCH_2CHCH_2CHCH_3$
$$\qquad\qquad\overset{|}{OH}\quad\overset{|}{OH}\quad\overset{|}{OH}$$

                CH_3

                              OH

    b.   OH   **d.** $CH_3CH_2CHCHCHCH_3$
$$\qquad\qquad\qquad\qquad\overset{|}{OH}\quad\overset{|}{OH}$$

        Br

Medically Important Alcohols

12.39 What is denatured alcohol? Why is alcohol denatured?

12.40 What are the principal uses of methanol, ethanol, and isopropyl alcohol?

12.41 What is fermentation?

12.42 Why do wines typically have an alcohol concentration of 12–13%?

12.43 Why must fermentation products be distilled to produce liquors such as scotch?

12.44 If a bottle of distilled alcoholic spirits—for example, scotch whiskey—is labeled as 80 proof, what is the percentage of alcohol in the scotch?

Reactions Involving Alcohols

Foundations

12.45 Write a general equation representing the preparation of an alcohol by hydration of an alkene.

12.46 Write a general equation representing the preparation of an alcohol by hydrogenation of an aldehyde or a ketone.

12.47 Write a general equation representing the dehydration of an alcohol.

12.48 Write a general equation representing the oxidation of a 1° alcohol.

12.49 Write a general equation representing the oxidation of a 2° alcohol.

12.50 Write a general equation representing the oxidation of a 3° alcohol.

Applications

12.51 Predict the products formed by the hydration of the following alkenes:
  a. 1-Hexene
  b. 2-Hexene
  c. 2-Methyl-3-hexene
  d. 2,2-Dimethyl-3-heptene

12.52 Draw the alkene products of the dehydration of the following alcohols:
  a. 2-Butanol
  b. 1-Butanol
  c. 2-Propanol
  d. 4-Bromo-2-hexanol

12.53 Write an equation showing the hydration of each of the following alkenes. Name each of the products using the IUPAC Nomenclature System.
  a. 2-Hexene
  b. Cyclopentene
  c. 1-Octene
  d. 1-Methylcyclohexene

12.54 Write an equation showing the dehydration of each of the following alcohols. Name each of the reactants and products using the IUPAC Nomenclature System.

  a. $CH_3CHCH_2CH_3$
         $\quad\ \ |$
         $\quad\ \ OH$
  b.

12.55 What product(s) would result from the oxidation of each of the following alcohols with, for example, potassium permanganate? If no reaction occurs, write N.R.
  a. 2-Butanol
  b. 2-Methyl-2-hexanol
  c. Cyclohexanol
  d. 1-Methyl-1-cyclopentanol

12.56 We have seen that ethanol is metabolized to ethanal (acetaldehyde) in the liver. What would be the product formed, under the same conditions, from each of the following alcohols?
  a. CH_3OH
  b. $CH_3CH_2CH_2OH$
  c. $CH_3CH_2CH_2CH_2OH$

12.57 Using the IUPAC system, name each of the following alcohols and the product formed when it is oxidized. If no reaction occurs, write N.R.

  $\qquad\quad OH$
  $\qquad\quad |$
  a. $CH_3CH_2CHCH_2CH_3$
  b. $CH_3CH_2CH_2OH$
  $\qquad\quad OH\quad CH_3$
  $\qquad\quad |\qquad\ |$
  c. $CH_3CHCH_2CHCH_3$

  $\qquad\qquad OH$
  $\qquad\qquad |$
  d. $CH_3CCH_2CH_3$
  $\qquad\qquad |$
  $\qquad\qquad CH_3$

  e.

12.58 Write an equation demonstrating each of the following chemical transformations:
  a. Oxidation of an alcohol to an aldehyde
  b. Oxidation of an alcohol to a ketone
  c. Dehydration of a cyclic alcohol to a cycloalkene
  d. Hydrogenation of an alkene to an alkane

12.59 Write the reaction, occurring in the liver, that causes the oxidation of ethanol. What is the product of this reaction and what symptoms are caused by the product?

12.60 Write the reaction, occurring in the liver, that causes the oxidation of methanol. What is the product of this reaction, and what is the possible result of the accumulation of the product in the body?

12.61 Write an equation for the preparation of 2-butanol from 1-butene. What type of reaction is involved?

12.62 Write a general equation for the preparation of an alcohol from an aldehyde or ketone. What type of reaction is involved?

12.63 Show how acetone can be prepared from propene.

$$\begin{array}{c} O \\ || \\ CH_3CCH_3 \\ \text{Acetone} \end{array}$$

12.64 Give the oxidation product for cholesterol.

Cholesterol

12.65 Write a balanced equation for the hydrogenation of each of the following:
  a. Hexanal (a six-carbon aldehyde)
  b. 2-Hexanone (a six-carbon ketone)
  c. 2-Methylbutanal (an aldehyde with a four-carbon parent chain)
  d. 6-Ethyl-2-octanone (a ketone with an eight-carbon parent chain)

12.66 Write a balanced equation for the hydrogenation of each of the following:
  a. Propanal (a three-carbon aldehyde)
  b. Propanone (a three-carbon ketone)
  c. 2,3-Dimethylheptanal (an aldehyde with a seven-carbon parent chain)
  d. 3-Methyl-4-heptanone (a ketone with a seven-carbon parent chain)

Oxidation and Reduction in Living Systems

Foundations

12.67 Define the terms *oxidation* and *reduction*.

12.68 How do we recognize oxidation and reduction in organic compounds?

Applications

12.69 Arrange the following compounds from the most reduced to the most oxidized:

$$CH_3CH_2\overset{\overset{\displaystyle O}{\|}}{C}-OH \qquad CH_3CH_2CH_3$$

$$CH_3CH_2\overset{\overset{\displaystyle O}{\|}}{C}-H \qquad CH_3CH_2CH_2OH$$

12.70 What is the role of the coenzyme nicotinamide adenine dinucleotide (NAD^+) in enzyme-catalyzed oxidation-reduction reactions?

Phenols

Foundations

12.71 What are phenols?
12.72 Describe the water solubility of phenols.

Applications

12.73 2,4,6-Trinitrophenol is known by the common name *picric acid*. Picric acid is a solid but is readily soluble in water. In solution, it is used as a biological tissue stain. As a solid, it is also known to be unstable and may explode. In this way, it is similar to 2,4,6-trinitrotoluene (TNT). Draw the structures of picric acid and TNT. Why is picric acid readily soluble in water, whereas TNT is not?
12.74 Name the following aromatic compounds using the IUPAC system:

a.

OH
NO$_2$

c. HO

Br
Cl

b. CH$_3$ CH$_3$
CH

OH

d.

OH
Br   CH$_3$

12.75 List some phenol compounds that are commonly used as antiseptics or disinfectants.
12.76 Why must a dilute solution of phenol be used for disinfecting environmental surfaces?

Ethers

Foundations

12.77 Describe the physical properties of ethers.
12.78 Compare the water solubility of ethers and alcohols.

Applications

12.79 Draw all of the alcohols and ethers of molecular formula $C_4H_{10}O$.
12.80 Name each of the isomers drawn for Question 12.79.

12.81 Ethers may be prepared by the removal of water (condensation) between two alcohols, as shown. Give the structure(s) of the ethers formed by the reaction of the following alcohol(s) under acidic conditions with heat. Provide the IUPAC and common names of each reactant and product.

Example: $CH_3OH + HOCH_3 \xrightarrow[\text{Heat}]{H^+} CH_3OCH_3 + H_2O$

a. $2CH_3CH_2OH \longrightarrow ?$
b. $CH_3OH + CH_3CH_2OH \longrightarrow ?$
c. $(CH_3)_2CHOH + CH_3OH \longrightarrow ?$
d.

2 [pentane ring]—CH$_2$OH $\longrightarrow$?

12.82 Write an equation showing the condensation reaction that would produce each of the following ethers:
a. Diethyl ether   c. Dibutyl ether
b. Ethyl propyl ether   d. Heptyl hexyl ether

12.83 Name each of the following ethers using the IUPAC Nomenclature System:

a. CH$_3$CHCH$_2$CH$_2$CH$_3$
OCH$_2$CH$_3$

b. CH$_3$CH$_2$CHCH$_3$
OCH$_3$

c. CH$_3$CH$_2$CH$_2$CH$_2$
OCH$_2$CH$_3$

d. [pentane ring]—OCH$_3$

12.84 Provide the common names for each of the following ethers:
a. $CH_3CH_2-O-CH_2CH_2CH_3$
b. $CH_3-O-CH_2CH_2CH_2CH_2CH_2CH_2CH_3$
c. $CH_3CH_2CH_2CH_2-O-CH_2CH_2CH_2CH_3$
d. $CH_3CH_2-O-CH_2CH_2CH_2CH_2CH_2CH_3$

12.85 Draw the structural formula and line formula for each of the following ethers:
a. Dibutyl ether
b. Ethyl heptyl ether
c. Propyl pentyl ether
d. *t*-Butyl hexyl ether

12.86 Write the IUPAC and common name for each of the following ethers:
a. $CH_3CH_2-O-CH_2CH_2CH_2CH_3$
b. $CH_3-O-CH_2CH_2CH_2CH_2CH_3$
c. $CH_3CH_2CH_2-O-CH_2CH_2CH_2CH_3$
d. $CH_3CH_2-O-CH_2CH_2CH_2CH_2CH_3$

Thiols

Foundations

12.87 Compare the structure of thiols and alcohols.
12.88 Describe the IUPAC rules for naming thiols.

Applications

12.89 Cystine is an amino acid formed from the oxidation of two cysteine molecules to form a disulfide bond. The molecular formula of cystine is $C_6H_{12}O_4N_2S_2$. Draw the structural formula of cystine. (*Hint:* For the structure of cysteine, see page 439.)

12.90 Describe the role of disulfide bonds in the synthesis and structure of insulin.

12.91 Give the IUPAC name for each of the following thiols.

a. $CH_3CH_2CH_2$—SH

c. $CH_3\overset{\overset{\displaystyle CH_2CH_3}{|}}{\underset{\underset{\displaystyle SH}{|}}{C}}CH_3$

b. $CH_3\overset{}{\underset{\underset{\displaystyle SH}{|}}{C}}HCH_2CH_3$

d. HS—⟨ ⟩—SH

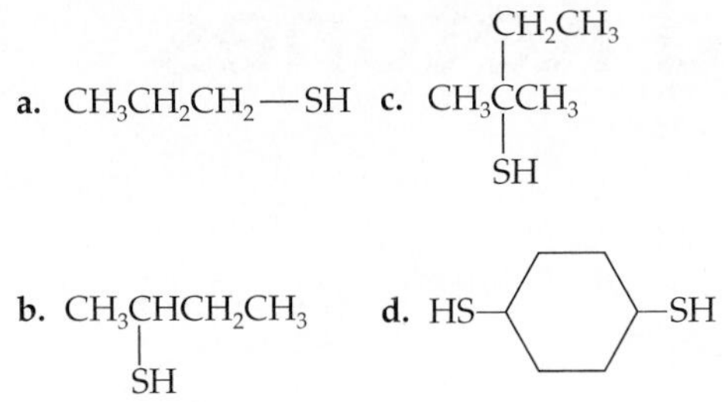

12.92 Give the IUPAC name for each of the following thiols.

a. $CH_2\overset{}{\underset{\underset{\displaystyle SH}{|}}{C}}HCH_3$ with SH on first carbon

$\overset{}{\underset{\underset{\displaystyle SH\ SH}{|\ \ |}}{CH_2CHCH_3}}$

c. $CH_3\overset{}{\underset{\underset{\displaystyle SH}{|}}{C}}HCH_2CH_2CH_3$

b. ⬡—SH

d. $CH_3CH_2CH_2CH_2CH_2CH_2CH_2SH$

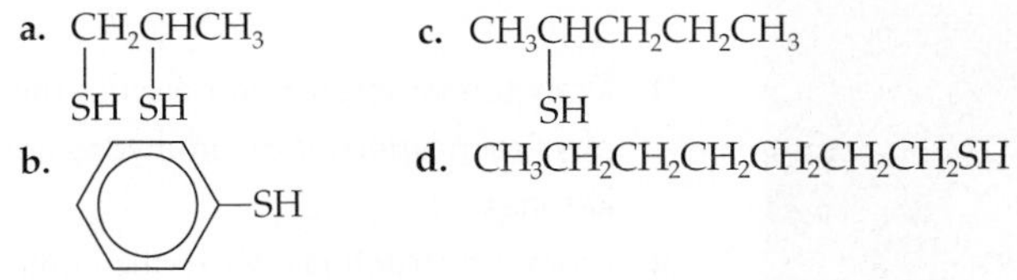

CHALLENGE PROBLEMS

1. You are provided with two solvents: water (H_2O) and hexane ($CH_3CH_2CH_2CH_2CH_2CH_3$). You are also provided with two biological molecules whose structures are shown here:

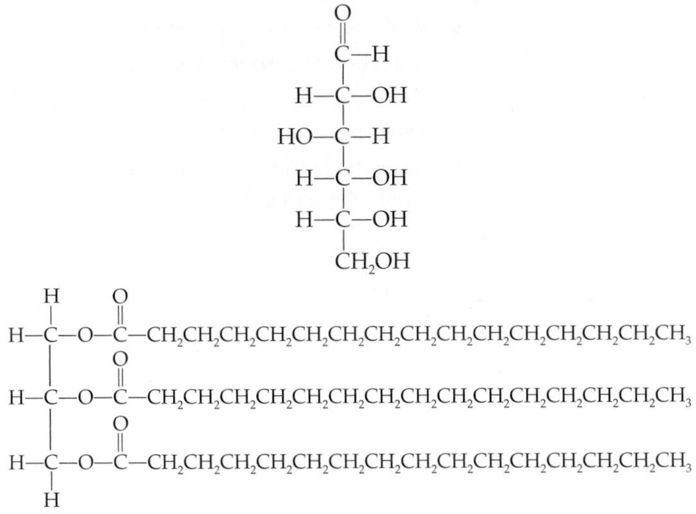

$$\begin{array}{c} O \\ \| \\ C-H \\ | \\ H-C-OH \\ | \\ HO-C-H \\ | \\ H-C-OH \\ | \\ H-C-OH \\ | \\ CH_2OH \end{array}$$

$$\begin{array}{c} H\ \ \ \ O \\ | \ \ \ \| \\ H-C-O-C-CH_2CH_2CH_2CH_2CH_2CH_2CH_2CH_2CH_2CH_2CH_2CH_2CH_2CH_3 \\ | \ \ \ \ O \\ | \ \ \ \| \\ H-C-O-C-CH_2CH_2CH_2CH_2CH_2CH_2CH_2CH_2CH_2CH_2CH_2CH_2CH_2CH_3 \\ | \ \ \ \ O \\ | \ \ \ \| \\ H-C-O-C-CH_2CH_2CH_2CH_2CH_2CH_2CH_2CH_2CH_2CH_2CH_2CH_2CH_2CH_3 \\ | \\ H \end{array}$$

Predict which biological molecule would be more soluble in water and which would be more soluble in hexane. Defend your prediction. Design a careful experiment to test your hypothesis.

Consider the digestion of dietary molecules in the digestive tract. Which of the two biological molecules shown in this problem would be more easily digested under the conditions present in the digestive tract?

2. Cholesterol is an alcohol and a steroid (Chapter 17). Diets that contain large amounts of cholesterol have been linked to heart disease and atherosclerosis, hardening of the arteries. The narrowing of the artery, caused by plaque buildup, is very apparent. Cholesterol is directly involved in this buildup. Describe the various functional groups and principal structural features of the cholesterol molecule. Would you use a polar or nonpolar solvent to dissolve cholesterol? Explain your reasoning.

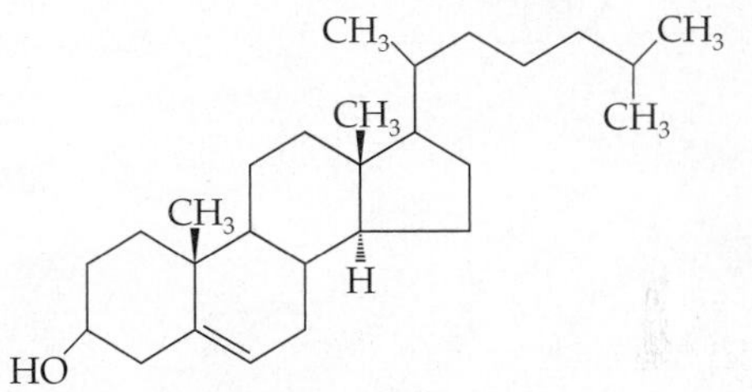

Cholesterol

3. An unknown compound A is known to be an alcohol with the molecular formula $C_4H_{10}O$. When dehydrated, compound A gave only one alkene product, C_4H_8, compound B. Compound A could not be oxidized. What are the identities of compound A and compound B?

4. Sulfides are the sulfur analogs of ethers; that is, ethers in which oxygen has been substituted by a sulfur atom. They are named in an analogous manner to the ethers with the term *sulfide* replacing *ether*. For example, CH_3—S—CH_3 is dimethyl sulfide. Draw the sulfides that correspond to the following ethers and name them:
   a. diethyl ether
   b. methyl propyl ether
   c. dibutyl ether
   d. ethyl phenyl ether

5. Dimethyl sulfoxide (DMSO) has been used by many sports enthusiasts as a linament for sore joints; it acts as an anti-inflammatory agent and a mild analgesic (painkiller). However, it is no longer recommended for this purpose because it carries toxic impurities into the blood. DMSO is a sulfoxide—it contains the S=O functional group. DMSO is prepared from dimethyl sulfide by mild oxidation, and it has the molecular formula C_2H_6SO. Draw the structure of DMSO.

13

Aldehydes and Ketones

Cinnamon sticks

LEARNING GOALS

1 Draw the structures and discuss the physical properties of aldehydes and ketones.
2 From the structures, write the common and IUPAC names of aldehydes and ketones.
3 List several aldehydes and ketones that are of natural, commercial, health, and environmental interest and describe their significance.
4 Write equations for the preparation of aldehydes and ketones by the oxidation of alcohols.
5 Write equations representing the oxidation of carbonyl compounds.
6 Write equations representing the reduction of carbonyl compounds.
7 Write equations for the preparation of hemiacetals and acetals.
8 Draw the keto and enol forms of aldehydes and ketones.

OUTLINE

INTRODUCTION

For centuries, we have used bloodhounds to locate missing persons or criminals. This works because of the amazing ability of bloodhounds to detect scent molecules, and because individuals have characteristic odor prints that are as unique as their fingerprints or DNA. Forensic scientists are now making the next step in developing the technology for more accurate detection of people associated with a crime scene. The first step is to identify the components of the odor print and an appropriate source of the sample. Some scientists suggest that odor molecules should be collected from the hands, since this is the part of the body that would handle a gun, bomb, or other materials at a crime scene. When samples are collected from hands, complex mixtures of compounds are collected. Among the molecules identified as prominent in these mixtures are some large and complex members of the two groups of compounds we will study in this chapter, the aldehydes and ketones. Among these are the aldehydes nonanal, decanal, and undecanal, and the ketones 6-methyl-5-hepten-2-one and 6,10-dimethyl-5,9-undecadien-2-one. The next step in using this information to identify individuals will be the development of an instrument to detect, identify, and quantify these and other components of the human odor print samples found at the scene of a crime so that pattern can be compared with suspects in the case.

Nonanal

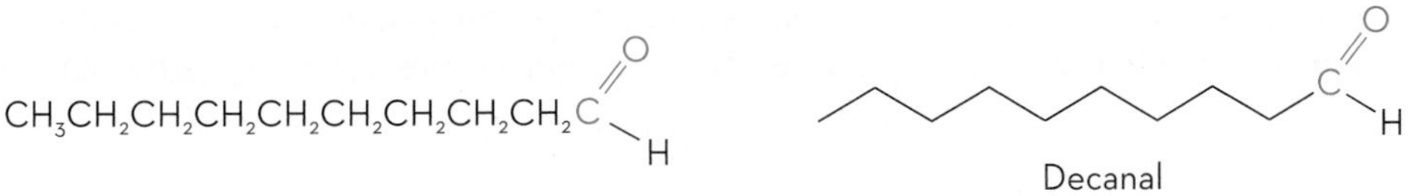

Decanal

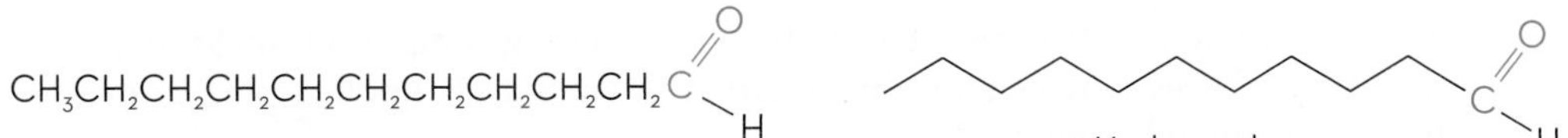

Undecanal

6-Methyl-5-hepten-2-one

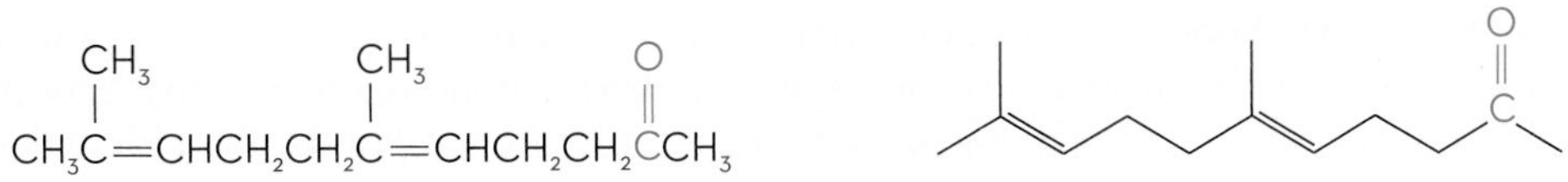

6,10-Dimethyl-5,9-undecadien-2-one

The aldehydes and ketones are characterized by the presence of the **carbonyl group**, highlighted in red in the structures above, which is made up of a carbon atom bonded to an oxygen atom by a double bond.

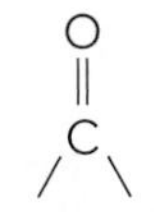

Carbonyl group

Compounds containing a carbonyl group are called carbonyl compounds. These include the aldehydes and ketones covered in this chapter, as well as the carboxylic acids and amides discussed in Chapters 14 and 15.

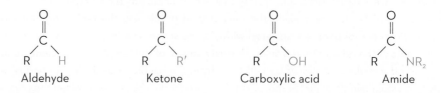

Aldehyde Ketone Carboxylic acid Amide

13.1 Structure and Physical Properties

LEARNING GOAL

1 Draw the structures and discuss the physical properties of aldehydes and ketones.

Aldehydes and **ketones** are carbonyl group–containing compounds distinguished by the location of the carbonyl group within the carbon chain. In aldehydes, the carbonyl group is always located at the end of the carbon chain (carbon-1). In ketones, the carbonyl group is located within the carbon chain of the molecule. Thus, in ketones the carbonyl carbon is attached to two other carbon atoms. However, in aldehydes the carbonyl carbon is attached to at least one hydrogen atom; the second atom attached to the carbonyl carbon of an aldehyde may be another hydrogen or a carbon atom (Figure 13.1).

The carbonyl group of an aldehyde or a ketone is polar because oxygen is more electronegative than carbon (3.5 versus 2.5). This produces a dipole in which the oxygen carries a partial negative charge and the carbon carries a partial positive charge.

$$\overset{\delta^-}{O} = \overset{\delta^+}{C}$$

Thus, the attractive forces between carbonyl-containing compounds include London dispersion forces between the hydrocarbon chains and dipole-dipole attractions between carbonyl groups.

$$\overset{\delta^+}{C} = \overset{\delta^-}{O} \cdots \overset{\delta^+}{C} = \overset{\delta^-}{O} \cdots \overset{\delta^+}{C} = \overset{\delta^-}{O}$$

Dipole-dipole attraction

Because the dipole-dipole attractions between molecules are stronger than London Dispersion forces, aldehydes and ketones boil at higher temperatures than hydrocarbons of equivalent molar mass. Aldehydes and ketones cannot hydrogen bond because there is no oxygen-hydrogen bond in the carbonyl group. Since dipole-dipole attractions are weaker than hydrogen bonds, the aldehydes

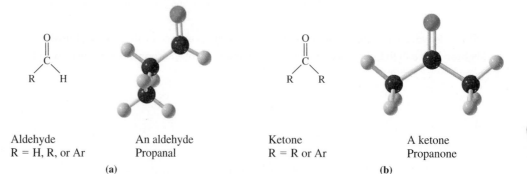

Figure 13.1 The structures of aldehydes and ketones. (a) The general structure of an aldehyde and a ball-and-stick model of the aldehyde propanal. (b) The general structure of a ketone and a ball-and-stick model of the ketone propanone.

Aldehyde
R = H, R, or Ar

An aldehyde
Propanal

Ketone
R = R or Ar

A ketone
Propanone

(a) (b)

Powerful Weak Attractions

Have you ever wondered how a gecko can walk up a polished surface, like the glass in this photograph? It is tempting to think that they have suction cups on the bottoms of their toes or that their feet are sticky. But neither can be true, because a small gecko can run up a wall at a rate of one meter/second. That means geckos have to be able to detach their feet more than twenty times each second!

In fact, every square millimeter of gecko footpads are covered with 14,000 hairlike setae, microscopic bristles that are much finer than a human hair. The tip of each of these bristles has a spatula-shaped structure made primarily of β-keratin, a protein very similar to that found in hair or fingernails. These structures allow maximal surface area contact between the gecko footpad and the environmental surface.

Research biologist Kellar Autumn and his colleagues have discovered that it is the van der Waals forces between the β-keratin protein and the surface that allow the gecko to adhere to even polished surfaces like Teflon or glass. Because they are weak attractions, they can be readily broken. This explains the ability for rapid attachment and detachment needed for the gecko to scurry away from predators. van der Waals forces include dipole-dipole interactions (such as those we see between the polar carbonyl groups of aldehydes or ketones) and London dispersion forces (the induced, short-lived dipole attractions that we see between hydrophobic hydrocarbon

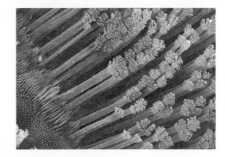

chains). It seems quite remarkable that these weak attractions are the basis of the amazing talents of geckos.

Others have studied gecko adhesion from the point of view of development of novel adhesives. Drs. Al Crosby and Duncan Irschick, along with their research group, at the University of Massachusetts, Amherst, developed an adhesive that is so powerful that a piece the size of an index card can hold a 700-pound weight on a glass surface. Called Geckskin, the adhesive consists of a soft pad (mimicking the gecko setae) backed with a stiff fabric (mimicking the gecko skin and tendons). While it holds very firmly when attached, if you want to remove it, you simply have to gently twist the stiff fabric and Geckskin releases. Geckskin can be used over and over, because the van der Waals forces that allow the adhesion will form again with a new surface. Because of the novel properties and potential uses for Geckskin, it was named one of the 2013–2014 Top 10 Textile Innovations.

It is tempting for the public to question the use of federal research funds to support interesting studies that don't seem to have an immediate technological application. The study of gecko adhesion and the resulting revolutionary adhesive that it has inspired demonstrate the importance of such basic research to provide the knowledge and tools to develop new products and technologies.

For Further Understanding

▶ Which of the following molecules would exhibit London dispersion forces and which would exhibit dipole-dipole attractions?

$$CH_3CH_2CH_2CH_3 \quad CH_3CH_2\overset{\displaystyle O}{\overset{\displaystyle \|}{C}}CH_3 \quad CH_3CH_2CH_2\overset{\displaystyle O}{\overset{\displaystyle \|}{C}}H$$

▶ The novel features of Geckskin (https://geckskin.umass.edu/) are that it is a powerful adhesive that can be detached easily and it leaves no residue. Why does Geckskin leave no residue?

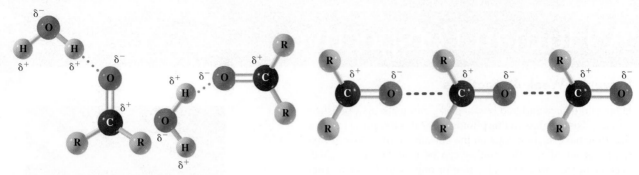

Figure 13.2 (a) Hydrogen bonding between the carbonyl group of an aldehyde or ketone and water. (b) Polar interactions between carbonyl groups of aldehydes or ketones.

(a)   (b)

and ketones boil at lower temperatures than alcohols of similar molar mass. These trends are clearly demonstrated in the following examples:

$CH_3CH_2CH_2CH_3$ $CH_3CH_2CH_2$—OH CH_3CH_2—$\overset{\displaystyle O}{\overset{\|}{C}}$—H CH_3—$\overset{\displaystyle O}{\overset{\|}{C}}$—$CH_3$

Butane   1-Propanol   Propanal   Propanone
(butane)   (propyl alcohol)   (propionaldehyde)   (acetone)
M.M. = 58   M.M. = 60   M.M. = 58   M.M. = 58
b.p. −0.5°C   b.p. 97.2°C   b.p. 49°C   b.p. 56°C

Although aldehyde and ketone molecules cannot hydrogen bond to one another, they can form intermolecular hydrogen bonds with water (Figure 13.2). As a result, the smaller members of the two families (five or fewer carbon atoms) are reasonably soluble in water. However, as the carbon chain length increases, the compounds become less polar and more hydrocarbonlike. These larger compounds are soluble in nonpolar organic solvents.

Question 13.1   Which member in each of the following pairs will be more water-soluble?

a. $CH_3(CH_2)_2CH_3$ or $CH_3\overset{\displaystyle O}{\overset{\|}{C}}CH_3$

b. $CH_3\overset{\displaystyle O}{\overset{\|}{C}}CH_2CH_2CH_3$ or $CH_3CHOHCH_2CH_2CH_3$

Question 13.2   Which member in each of the following pairs will be more water-soluble?

a.   CH₃   or   H—C=O

[cyclopentane structures]

b. $HOCH_2CH_2OH$ or $H\overset{\displaystyle O\ O}{\overset{\|\ \|}{C}}CH$

Question 13.3   Which member in each of the following pairs would have a higher boiling point?

a. $CH_3CH_2\overset{\displaystyle O}{\overset{\|}{C}}OH$ or $CH_3CH_2\overset{\displaystyle O}{\overset{\|}{C}}H$

b. $CH_3\overset{\displaystyle O}{\overset{\|}{C}}OH$ or $CH_3\overset{\displaystyle O}{\overset{\|}{C}}CH_3$

Question 13.4   Which member in each of the following pairs would have a higher boiling point?

a. CH_3CH_2OH or $CH_3\overset{\displaystyle O}{\overset{\|}{C}}H$
b. $CH_3(CH_2)_6CH_3$ or $CH_3(CH_2)_5\overset{\displaystyle O}{\overset{\|}{C}}H$

13.2   IUPAC Nomenclature and Common Names

Naming Aldehydes

In the IUPAC system, aldehydes are named according to the following set of rules:

- Determine the parent compound; that is, the longest continuous carbon chain containing the carbonyl group.
- Replace the final -e of the parent alkane with -al.
- Number the chain beginning with the carbonyl carbon (or aldehyde group) as carbon-1.
- Number and name all substituents as usual. No number is used for the position of the carbonyl group because it is always at the end of the parent chain. Therefore, it must be carbon-1.

Several examples are provided here with common names given in parentheses:

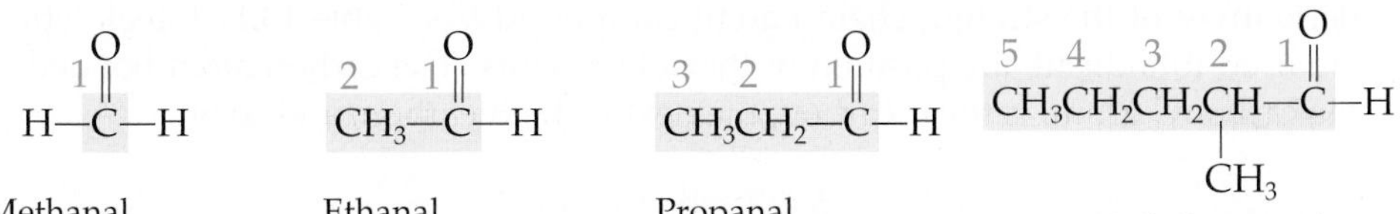

$$\underset{\substack{\text{Methanal}\\(\text{formaldehyde})}}{\overset{\overset{\displaystyle O}{\overset{\|}{\underset{1}{}}}}{\text{H—C—H}}} \qquad \underset{\substack{\text{Ethanal}\\(\text{acetaldehyde})}}{\overset{\overset{\displaystyle O}{\overset{\|}{\underset{1}{}}}}{\underset{2}{\text{CH}_3}\text{—C—H}}} \qquad \underset{\substack{\text{Propanal}\\(\text{propionaldehyde})}}{\overset{\overset{\displaystyle O}{\overset{\|}{\underset{1}{}}}}{\underset{3\ \ 2}{\text{CH}_3\text{CH}_2}\text{—C—H}}}$$

$$\underset{\substack{\text{2-Methylpentanal}\\(\alpha\text{-methylvaleraldehyde})}}{\overset{5\ \ \ 4\ \ \ 3\ \ \ 2\ \ \ 1}{\text{CH}_3\text{CH}_2\text{CH}_2\underset{\underset{\displaystyle \text{CH}_3}{|}}{\text{CH}}\overset{\overset{\displaystyle O}{\|}}{\text{C}}\text{—H}}}$$

EXAMPLE 13.1 **Using the IUPAC Nomenclature System to Name Aldehydes**

For many years, scientists were puzzled about what compounds give the nutty flavor to expensive, aged cheddar cheeses. In 2004, Dr. Mary Anne Drake of North Carolina State University solved this puzzle. She identified a group of aldehydes that impart this desirable flavor. Use the IUPAC Nomenclature System to name two of these aldehydes shown below.

Solution

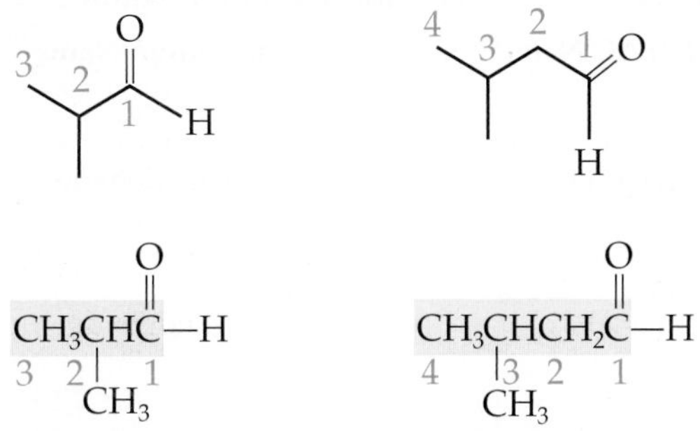

$$\underset{\text{CH}_3\underset{3\ \ 2}{\text{CH}}\overset{\overset{\displaystyle O}{\|}}{\underset{1}{\text{C}}}\text{—H}}{} \qquad \qquad$$

$$\underset{\underset{4\ \ \ \ 3\ \ \ \ 2\ \ \ 1}{\text{CH}_3\text{CHC—H}}}{\overset{\overset{\displaystyle O}{\|}}{}} \qquad \underset{\underset{4\ \ \ 3\ \ 2\ \ 1}{\text{CH}_3\text{CHCH}_2\text{C—H}}}{\overset{\overset{\displaystyle O}{\|}}{}}$$

Parent compound:	propane (becomes propanal)	butane (becomes butanal)
Position of carbonyl group:	carbon-1	carbon-1
Substituents:	2-methyl	3-methyl
Name:	2-Methylpropanal	3-Methylbutanal

Continued...

Another aldehyde associated with the nutty flavor of cheddar cheese is 2-methylbutanal. Draw the condensed formula of this molecule.

By definition, the carbonyl group is located at the end of the carbon chain of an aldehyde. Thus, the carbonyl carbon is always carbon-1, and it is not necessary to include the number in the name of the compound.

Practice Problem 13.1

Many molecules contribute to the complex flavors of olive oils. Among these are hexanal and *trans*-2-hexenal, which have flavors described as "green, grassy" and "green, bitter," respectively. Draw the structures of these two compounds.

▶ For Further Practice: **Questions 13.5, 13.29, and 13.36a, b, and c.**

Carboxylic acid nomenclature is described in Section 14.1.

The common names of the aldehydes are derived from the same Latin roots as the corresponding carboxylic acids. The common names of the first five aldehydes are presented in Table 13.1.

In the common system of nomenclature, substituted aldehydes are named as derivatives of the straight-chain parent compound (see Table 13.1). Greek letters are used to indicate the position of the substituents. The carbon atom bonded to the carbonyl group is the α-carbon, the next is the β-carbon, and so on.

$$\overset{\delta}{-C}-\overset{\gamma}{C}-\overset{\beta}{C}-\overset{\alpha}{C}-\overset{O}{\overset{\|}{C}}-H$$

Consider the following examples:

2-Methylpentanal
(α-methylvaleraldehyde)

3-Methylpentanal
(β-methylvaleraldehyde)

TABLE 13.1 IUPAC and Common Names and Formulas for Several Aldehydes

IUPAC Name	Common Name	Formula
Methanal	Formaldehyde	$H-\overset{O}{\overset{\|}{C}}-H$
Ethanal	Acetaldehyde	$CH_3\overset{O}{\overset{\|}{C}}-H$
Propanal	Propionaldehyde	$CH_3CH_2\overset{O}{\overset{\|}{C}}-H$
Butanal	Butyraldehyde	$CH_3CH_2CH_2\overset{O}{\overset{\|}{C}}-H$
Pentanal	Valeraldehyde	$CH_3CH_2CH_2CH_2\overset{O}{\overset{\|}{C}}-H$

EXAMPLE 13.2 **Comparing the Common and IUPAC Nomenclature Systems to Name Aldehydes**

Name the two aldehydes represented by the following condensed formulas.

LEARNING GOAL

2 From the structures, write the common and IUPAC names of aldehydes and ketones.

Solution

$$\overset{\delta}{C}H_3\overset{\gamma}{C}H\overset{\beta}{C}H_2\overset{\alpha}{C}H_2\overset{O}{\overset{\|}{C}}-H \qquad \overset{CH_3\ O}{\overset{\beta}{C}H_3\overset{\alpha}{C}H\overset{\alpha}{C}H_2\overset{\|}{C}-H}$$
$$\quad\ \overset{|}{Br} \qquad\qquad\qquad\qquad\qquad \overset{|}{CH_3}$$

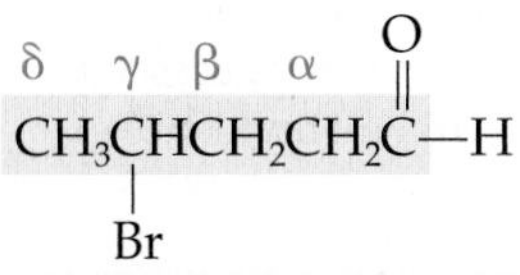

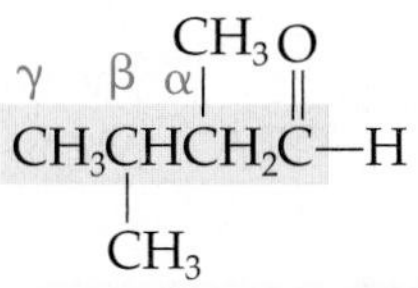

Parent compound:	pentane (becomes valeraldehyde)	butane (becomes butyraldehyde)
Position of carbonyl group:	carbon-1	carbon-1
Substituents:	γ-bromo	α, β-dimethyl
Common Name:	γ-Bromovaleraldehyde	α, β-Dimethylbutyraldehyde

In the common system, the carbon atom bonded to the carbonyl group is called the α-carbon, the next is the β-carbon, etc. Greek letters are used to indicate the position of the substituents.

Parent compound:	pentane (becomes pentanal)	butane (becomes butanal)
Position of carbonyl group:	carbon-1	carbon-1
Substituents:	4-bromo	2,3-dimethyl
IUPAC name:	4-Bromopentanal	2,3 Dimethylbutanal

In the IUPAC system, the carbonyl carbon is defined as carbon-1. Any substituents are numbered from that point in the chain to identify their locations.

Also remember that, by definition, the carbonyl group is located at the beginning of the carbon chain of an aldehyde. Thus, it is not necessary to include the position of the carbonyl group in the name of the compound.

Practice Problem 13.2

Use the common nomenclature system to name each of the following compounds.

a. 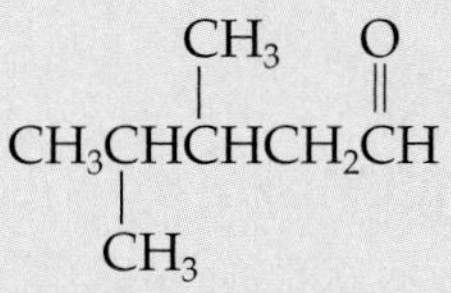 $\overset{CH_3}{\overset{|}{CH_3CHCHCH_2CH}}\overset{O}{\overset{\|}{}}$ b. 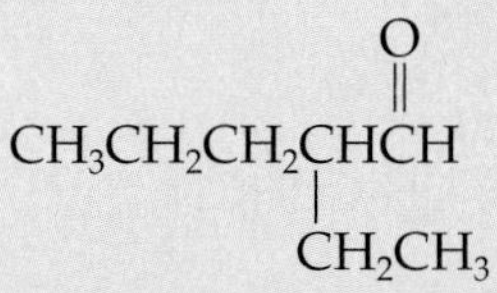 $CH_3CH_2CH_2\overset{O}{\overset{\|}{CHCH}}$ c. 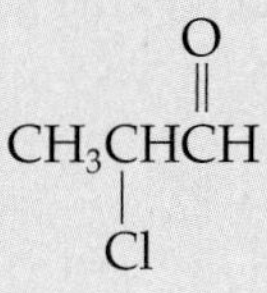 $CH_3\overset{O}{\overset{\|}{CHCH}}$ d. 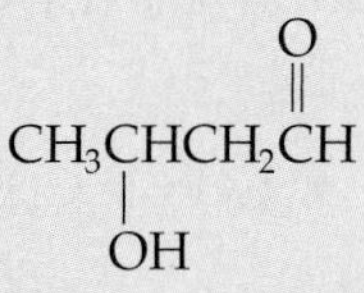 $CH_3\overset{O}{\overset{\|}{CHCH_2CH}}$
     $\overset{|}{CH_3}$ $\overset{|}{CH_2CH_3}$ $\overset{|}{Cl}$ $\overset{|}{OH}$

▶ For Further Practice: **Questions 13.6b and d, 13.41c and d, 13.42b.**

Naming Ketones

The rules for naming ketones in the IUPAC Nomenclature System are directly analogous to those for naming aldehydes.

- Determine the parent compound; that is, the longest carbon chain containing the carbonyl group.
- Replace the *–e* ending of the parent alkane with the *–one* suffix of the ketone family.
- Number the carbon chain to give the carbonyl carbon the lowest possible number.

LEARNING GOAL

2 From the structures, write the common and IUPAC names of aldehydes and ketones.

Several examples are provided here, along with the common names of these ketones in parentheses:

$$\underset{\substack{1\quad\;2\quad\;3}}{CH_3-\overset{\displaystyle O}{\overset{\|}{C}}-CH_3}\qquad\underset{\substack{4\quad\;\;3\quad\;2\quad1}}{CH_3CH_2-\overset{\displaystyle O}{\overset{\|}{C}}-CH_3}\qquad\underset{\substack{8\;\;7\;\;6\;\;5\quad\;4\;\;3\;\;2\;\;1}}{CH_3CH_2CH_2CH_2-\overset{\displaystyle O}{\overset{\|}{C}}-CH_2CH_2CH_3}$$

Propanone   Butanone   4-Octanone
(no number necessary)   (no number necessary)   (*not* 5-octanone)
(acetone)   (ethyl methyl ketone)   (butyl propyl ketone)

EXAMPLE 13.3   **Using the IUPAC Nomenclature System to Name Ketones**

LEARNING GOAL

Name the following ketones using the IUPAC Nomenclature System.

Solution

2 From the structures, write the common and IUPAC names of aldehydes and ketones.

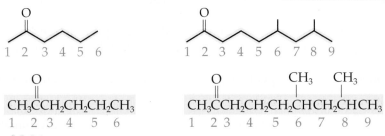

$$\underset{\substack{1\;\;2\;\;3\;\;4\;\;\;5\;\;\;6}}{CH_3\overset{\displaystyle O}{\overset{\|}{C}}CH_2CH_2CH_2CH_3}\qquad\underset{\substack{1\;\;2\;3\;\;\;4\;\;\;5\;\;6\;\;7\;\;8\;\;9}}{CH_3\overset{\displaystyle O}{\overset{\|}{C}}CH_2CH_2CH_2\overset{\displaystyle CH_3}{\overset{|}{C}H}CH_2\overset{\displaystyle CH_3}{\overset{|}{C}H}CH_3}$$

Parent compound:	hexane (becomes hexanone)	nonane (becomes nonanone)
Position of carbonyl group:	carbon-2	carbon-2
Substituents:	None	6, 8-dimethyl
IUPAC name:	2-Hexanone	6, 8-Dimethyl-2-nonanone

2-Hexanone has recently been discovered to be one of several compounds that attract bedbugs and signal them to colonize.

Practice Problem 13.3

Provide the IUPAC name for each of the following ketones.

a. $CH_3CH_2\underset{\underset{\displaystyle CH_2CH_3}{|}}{C}HCH_2CH_2CH_2\overset{\displaystyle O}{\overset{\|}{C}}CH_3$

b. $CH_3CH_2CH_2\overset{\displaystyle O}{\overset{\|}{C}}\underset{\underset{\displaystyle CH_3}{|}}{C}HCH_3$

c. $CH_3\underset{\underset{\displaystyle CH_3}{|}}{C}H\overset{\displaystyle O}{\overset{\|}{C}}\underset{\underset{\displaystyle CH_3}{|}}{C}HCH_3$

▶ For Further Practice: **Questions 13.7, 13.33a, 13.34a, 13.35b, 13.38, and 13.39.**

The common names of ketones are derived by naming the alkyl groups that are bonded to the carbonyl carbon. These are used as prefixes followed by the word *ketone*. The alkyl groups may be arranged alphabetically or by size (smaller to larger), as demonstrated in the following examples:

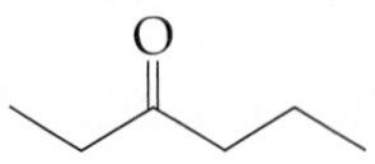

Ethyl propyl ketone 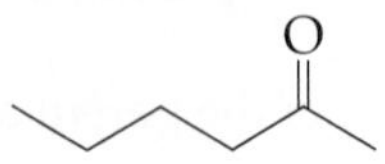   Methyl butyl ketone (by size)
                                            or
                            Butyl methyl ketone (alphabetically)

EXAMPLE 13.4 | **Using the Common Nomenclature System to Name Ketones**

Name the ketones represented by the following condensed and line formulas.

Solution

Identify the alkyl groups that are bonded to the carbonyl carbon.

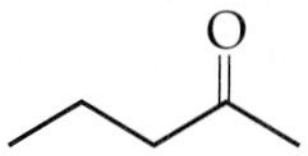

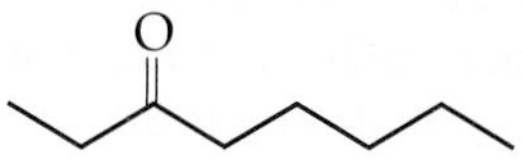

$$CH_3CH_2CH_2-\overset{\overset{\displaystyle O}{\|}}{C}-CH_3 \qquad CH_3CH_2-\overset{\overset{\displaystyle O}{\|}}{C}-CH_2CH_2CH_2CH_2CH_3$$

| Alkyl groups: | propyl and methyl | ethyl and pentyl |
| Common name: | Methyl propyl ketone | Ethyl pentyl ketone |

LEARNING GOAL

2 From the structures, write the common and IUPAC names of aldehydes and ketones.

Two molecules that contribute to the aroma of blue cheese are 2-heptanone and 2-nonanone. Draw the condensed formula for each of these molecules.

Practice Problem 13.4

Provide the common names for each of the following ketones:

a. $CH_3CH_2\overset{\overset{\displaystyle O}{\|}}{C}CH_2CH_3$

b. $CH_3CH_2CH_2CH_2\overset{\overset{\displaystyle O}{\|}}{C}CH_2CH_3$

c. $CH_3CH_2CH_2CH_2CH_2CH_2\overset{\overset{\displaystyle O}{\|}}{C}CH_3$

d. $CH_3\overset{}{\underset{\underset{\displaystyle CH_3}{|}}{C}H}\overset{\overset{\displaystyle O}{\|}}{C}\overset{}{\underset{\underset{\displaystyle CH_3}{|}}{C}H}CH_3$

▶ For Further Practice: **Questions 13.41a, b, and e and 13.42a and c.**

Question 13.5   From the IUPAC names, draw the condensed formula for each of the following aldehydes.
   a. 2,3-Dichloropentanal
   b. 2-Bromobutanal
   c. 4-Methylhexanal
   d. Butanal
   e. 2,4-Dimethylpentanal

Question 13.6   Write the condensed formula for each of the following compounds.
   a. 3-Methylnonanal
   b. β-Bromovaleraldehyde
   c. 4-Fluorohexanal
   d. α,β-Dimethylbutyraldehyde

Question 13.7   Use the IUPAC Nomenclature System to name each of the following compounds.

a. $CH_3\overset{}{\underset{\underset{\displaystyle I}{|}}{C}H}\overset{\overset{\displaystyle O}{\|}}{C}CH_3$

b. $CH_3\overset{}{\underset{\underset{\displaystyle CH_2CH_2CH_2CH_3}{|}}{C}H}CH_2\overset{\overset{\displaystyle O}{\|}}{C}CH_3$

c. $CH_3\overset{}{\underset{\underset{\displaystyle CH_3}{|}}{C}H}\overset{\overset{\displaystyle O}{\|}}{C}CH_3$

d. $CH_3\overset{}{\underset{\underset{\displaystyle CH_3}{|}}{C}H}\overset{\overset{\displaystyle O}{\|}}{C}CH_2CH_3$

e. $CH_3\overset{}{\underset{\underset{\displaystyle F}{|}}{C}H}\overset{\overset{\displaystyle O}{\|}}{C}CH_2CH_3$

Question 13.8   Write the condensed formula for each of the following compounds.
a. Methyl isopropyl ketone (What is the IUPAC name for this compound?)
b. 4-Heptanone
c. 2-Fluorocyclohexanone
d. Hexachloroacetone (What is the IUPAC name of this compound?)

LEARNING GOAL

3   List several aldehydes and ketones that are of natural, commercial, health, and environmental interest and describe their significance.

13.3   Important Aldehydes and Ketones

Methanal (formaldehyde) is a gas (b.p. −21°C). It is available commercially as an aqueous solution called *formalin*. Formalin has been used as a preservative for tissues and as an embalming fluid.

Ethanal (acetaldehyde) is produced from ethanol in the liver. Ethanol is oxidized in this reaction, which is catalyzed by the liver enzyme alcohol dehydrogenase. The ethanal that is produced in this reaction is responsible for the symptoms of a hangover.

Propanone (acetone), the simplest ketone, is an important and versatile solvent for organic compounds. It has the ability to dissolve organic compounds and is also soluble in water. As a result, it has a number of industrial applications and is used as a solvent in adhesives, paints, cleaning solvents, nail polish, and nail polish remover. Propanone is flammable and should therefore be treated with appropriate care. *Butanone*, a four-carbon ketone, is also an important industrial solvent in the manufacture of plastics, paint remover, varnish, and glue.

Many aldehydes and ketones are produced industrially as food and fragrance chemicals, medicinals, and agricultural chemicals. They are particularly important to the food industry, in which they are used as artificial and/or natural additives to food. Vanillin, a principal component of natural vanilla, is shown in Figure 13.3.

O
‖
C
H   H
Methanal

O
‖
C
CH₃   H
Ethanal

O
‖
C
CH₃   CH₃
Propanone

O
‖
C
CH₃   CH₂CH₃
Butanone

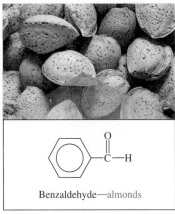
Benzaldehyde—almonds

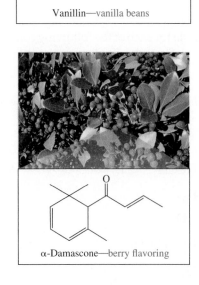
Vanillin—vanilla beans

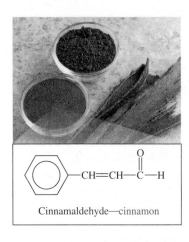
Cinnamaldehyde—cinnamon

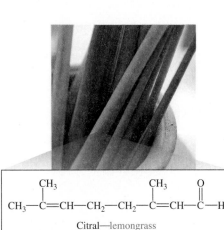
Citral—lemongrass

α-Damascone—berry flavoring

2-Octanone—mushroom flavoring

Figure 13.3   Important aldehydes and ketones.

Artificial vanilla flavoring is a dilute solution of synthetic vanillin dissolved in ethanol. Figure 13.3 also shows other examples of important aldehydes and ketones.

Question 13.9 Draw the structure of the aldehyde synthesized from ethanol in the liver.

Question 13.10 Draw the structure of a ketone that is an important, versatile solvent for organic compounds.

13.4 Reactions Involving Aldehydes and Ketones

Preparation of Aldehydes and Ketones

Aldehydes and ketones are prepared primarily by the **oxidation** of the corresponding alcohol. In organic systems, oxidation is recognized as a loss of hydrogen atoms or the gain of oxygen atoms. As we saw in Chapter 12, the oxidation of methyl alcohol gives methanal (formaldehyde). The oxidation of a primary alcohol produces an aldehyde, and the oxidation of a secondary alcohol yields a ketone. Tertiary alcohols do not undergo oxidation under the conditions normally used.

LEARNING GOAL

4 Write equations for the preparation of aldehydes and ketones by the oxidation of alcohols.

EXAMPLE 13.5 **Differentiating the Oxidation of Primary, Secondary, and Tertiary Alcohols**

LEARNING GOAL

4 Write equations for the preparation of aldehydes and ketones by the oxidation of alcohols.

Use specific examples to show the oxidation of a primary, a secondary, and a tertiary alcohol.

Solution

The oxidation of a primary alcohol, 1-butanol, to an aldehyde, butanal:

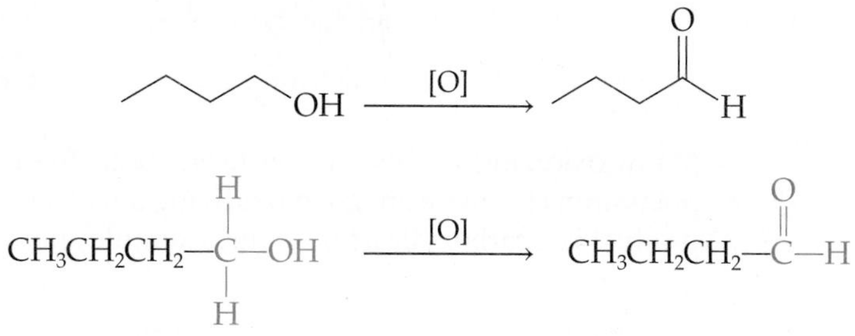

$$CH_3CH_2CH_2-\underset{\underset{H}{|}}{\overset{\overset{H}{|}}{C}}-OH \xrightarrow{[O]} CH_3CH_2CH_2-\overset{\overset{O}{\|}}{C}-H$$

1-Butanol   Butanal

The oxidation of a secondary alcohol, 2-hexanol, to a ketone, 2-hexanone:

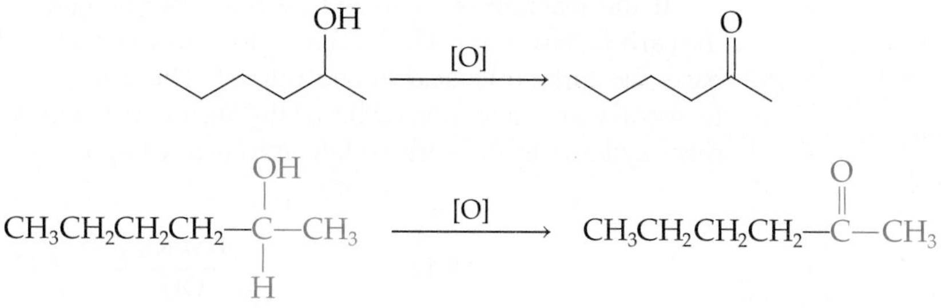

$$CH_3CH_2CH_2CH_2-\underset{\underset{H}{|}}{\overset{\overset{OH}{|}}{C}}-CH_3 \xrightarrow{[O]} CH_3CH_2CH_2CH_2-\overset{\overset{O}{\|}}{C}-CH_3$$

2-Hexanol   2-Hexanone

Continued…

Tertiary alcohols, such as 2-methyl-2-pentanol, cannot undergo oxidation:

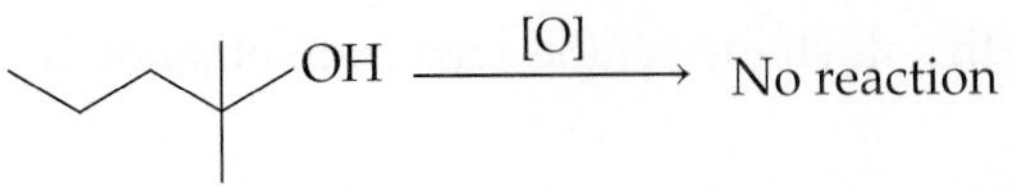

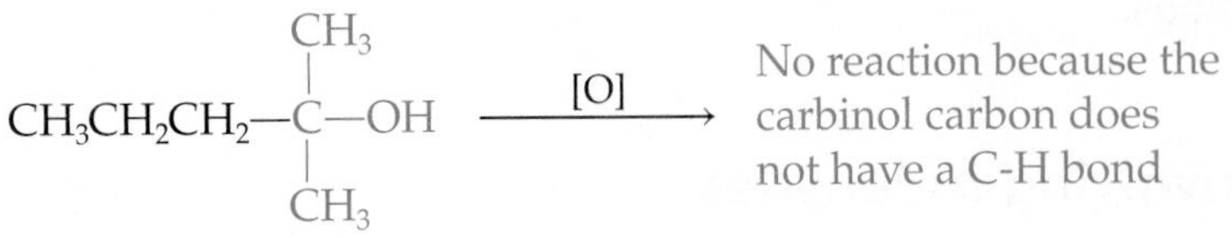

$$\text{(structure)} \xrightarrow{[O]} \text{No reaction}$$

$$CH_3CH_2CH_2-\overset{\overset{\displaystyle CH_3}{|}}{\underset{\underset{\displaystyle CH_3}{|}}{C}}-OH \xrightarrow{[O]} \text{No reaction because the carbinol carbon does not have a C-H bond}$$

2-Methyl-2-pentanol

In all these examples, the symbol [O] is used to represent the oxidizing agent.

Practice Problem 13.5

Write equations showing the oxidation of (a) 1-propanol and (b) 2-butanol.

▶ For Further Practice: **Questions 13.57 and 13.58.**

Oxidation Reactions

Aldehydes are oxidized to carboxylic acids, whereas ketones do not generally undergo further oxidation. The reason is that a carbon-hydrogen bond, present in the aldehyde but not in the ketone, is needed for the reaction to occur. The following example shows a general equation for the oxidation of an aldehyde to a carboxylic acid:

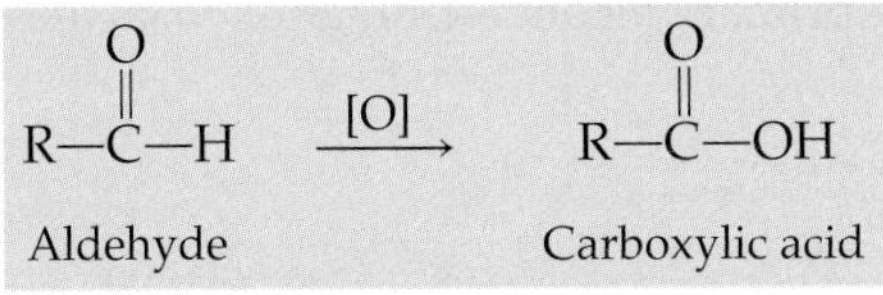

$$R-\overset{\overset{\displaystyle O}{\|}}{C}-H \xrightarrow{[O]} R-\overset{\overset{\displaystyle O}{\|}}{C}-OH$$

Aldehyde   Carboxylic acid

Many oxidizing agents can be used. Both basic potassium permanganate and potassium chromate are good oxidizing agents. If potassium chromate is used, the product is a carboxylic acid, as this example shows:

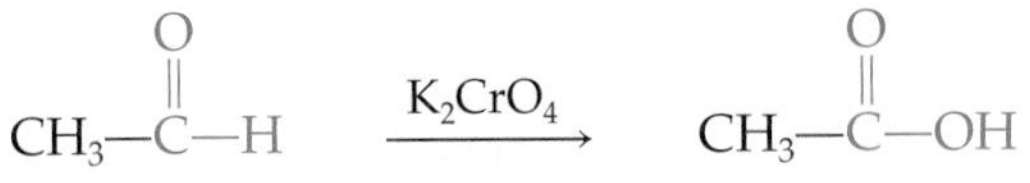

$$CH_3-\overset{\overset{\displaystyle O}{\|}}{C}-H \xrightarrow{K_2CrO_4} CH_3-\overset{\overset{\displaystyle O}{\|}}{C}-OH$$

Ethanal   Ethanoic acid
(acetaldehyde)   (acetic acid)

If the reaction is catalyzed by basic permanganate, however, the product is the carboxylate anion. Carboxylic acids are weak acids. In the presence of a strong base, the carboxylic acid is neutralized. The acid proton is removed by the OH^- to form water and the carboxylate anion. We will learn the rules for naming carboxylic acids and carboxylate anions in Chapter 14.

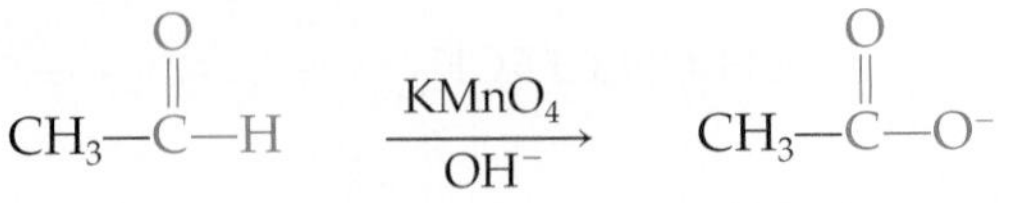

$$CH_3-\overset{\overset{\displaystyle O}{\|}}{C}-H \xrightarrow[OH^-]{KMnO_4} CH_3-\overset{\overset{\displaystyle O}{\|}}{C}-O^-$$

The rules for naming carboxylic acid anions are described in Section 14.1.

Ethanal   Ethanoate anion
(acetaldehyde)   (acetate anion)

The oxidation of benzaldehyde to benzoic acid is an example of the conversion of an aromatic aldehyde to the corresponding aromatic carboxylic acid:

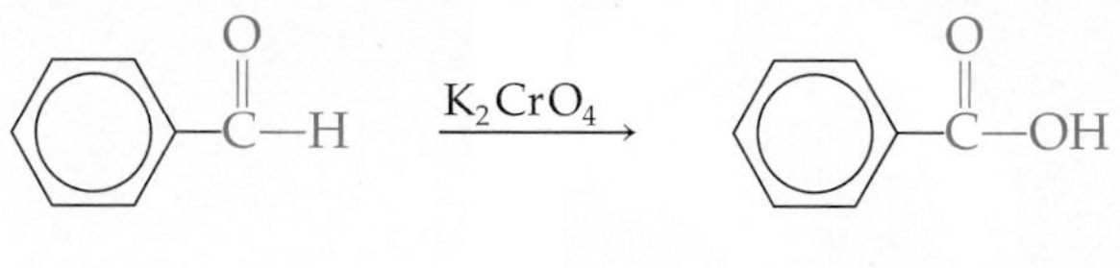

Benzaldehyde   Benzoic acid

Aldehydes and ketones can be distinguished on the basis of the fact that aldehydes are readily oxidized and ketones are not. The most common laboratory test for aldehydes is the **Tollens' test.** When exposed to the Tollens' reagent, a basic solution of $Ag(NH_3)_2^+$, an aldehyde undergoes oxidation. The silver ion (Ag^+) is reduced to silver metal (Ag^0) as the aldehyde is oxidized to a carboxylate anion.

Silver ions are very mild oxidizing agents. They will oxidize aldehydes but not alcohols.

$$R{-}\overset{\overset{\Large O}{\|}}{C}{-}H + Ag(NH_3)_2^+ \longrightarrow R{-}\overset{\overset{\Large O}{\|}}{C}{-}O^- + Ag^0$$

Aldehyde; Silver ammonia complex— Tollens' reagent; Carboxylate anion; Silver metal mirror

Silver metal precipitates from solution and coats the flask, producing a smooth silver mirror, as seen in Figure 13.4. The test is therefore often called the Tollens' silver mirror test. The commercial manufacture of silver mirrors uses a similar process. Ketones cannot be oxidized to carboxylic acids and do not react with the Tollens' reagent.

EXAMPLE 13.6 Writing Equations for the Reaction of an Aldehyde and of a Ketone with Tollens' Reagent

LEARNING GOAL

5 Write equations representing the oxidation of carbonyl compounds.

Write equations for the reaction of propanal and 2-pentanone with Tollens' reagent.

Solution

$$CH_3CH_2\overset{\overset{\Large O}{\|}}{C}{-}H + Ag(NH_3)_2^+ \longrightarrow CH_3CH_2\overset{\overset{\Large O}{\|}}{C}{-}O^- + Ag^0$$

Propanal   Propanoate anion

$$CH_3CH_2CH_2\overset{\overset{\Large O}{\|}}{C}CH_3 + Ag(NH_3)_2^+ \longrightarrow \text{No reaction}$$

2-Pentanone

Practice Problem 13.6

Write equations for the reaction of (a) ethanal and (b) propanone with Tollens' reagent.

▸ For Further Practice: **Questions 13.62 and 13.67.**

Another test that is used to distinguish between aldehydes and ketones is **Benedict's test.** Here, a buffered aqueous solution of copper(II) hydroxide and sodium citrate reacts to oxidize aldehydes but does not generally react with ketones. Cu^{2+} is reduced to Cu^+ in the process. Cu^{2+} is soluble and gives a blue solution, whereas the Cu^+ precipitates as the red solid copper(I) oxide, Cu_2O.

Cu(II) is an even milder oxidizing agent than silver ions.

All simple sugars (monosaccharides) are either aldehydes or ketones. Glucose is an aldehyde sugar that is commonly called *blood sugar* because it is the sugar

(a)   (b)   (c)

Figure 13.4 The silver precipitate produced by the Tollens' reaction is deposited on glass. The progress of the reaction is visualized in panels (a) through (c). Silver mirrors are made in a similar process.

Figure 13.5 The amount of precipitate formed and thus the color change observed in the Benedict's test are directly proportional to the amount of reducing sugar in the sample.

found transported in the blood and used for energy by many cells. In uncontrolled diabetes, glucose may be found in the urine. One early method used to determine the amount of glucose in the urine was to observe the color change of the Benedict's test. The amount of precipitate formed is directly proportional to the amount of glucose in the urine (Figure 13.5). The reaction of glucose with the Benedict's reagent is represented in the following equation:

$$
\begin{array}{c}
\underset{\displaystyle C}{\overset{\displaystyle O\diagdown\,\,H}{\|}} \\
\mathrm{H-C-OH} \\
\mathrm{HO-C-H} \\
\mathrm{H-C-OH} \\
\mathrm{H-C-OH} \\
\mathrm{CH_2OH}
\end{array}
\;+\;2Cu^{2+}\;\xrightarrow{\;OH^-\;}\;
\begin{array}{c}
\underset{\displaystyle C}{\overset{\displaystyle O\diagdown\,\,O^-}{\|}} \\
\mathrm{H-C-OH} \\
\mathrm{HO-C-H} \\
\mathrm{H-C-OH} \\
\mathrm{H-C-OH} \\
\mathrm{CH_2OH}
\end{array}
\;+\;Cu_2O
$$

Glucose

We should also note that when the carbonyl group of a ketone is bonded to a —CH_2OH group, the molecule will give a positive Benedict's test. This occurs because such ketones are converted to aldehydes under basic conditions. In Chapter 16, we will see that this applies to the ketone sugars, as well. They are converted to aldehyde sugars and react with Benedict's reagent.

Reduction Reactions

Aldehydes and ketones are both readily reduced to the corresponding alcohol by a variety of reducing agents. Throughout the text, the symbol [H] over the reaction arrow represents a reducing agent.

The classical method of aldehyde or ketone reduction is **hydrogenation.** The carbonyl compound is reacted with hydrogen gas and a catalyst (nickel, platinum, or palladium metal) in a pressurized reaction vessel. Heating may also be necessary. The carbon-oxygen double bond (the carbonyl group) is reduced to a

LEARNING GOAL

6 Write equations representing the reduction of carbonyl compounds.

One way to recognize reduction, particularly in organic chemistry, is the gain of hydrogen. Oxidation and reduction are discussed in Section 12.5.

A Human Perspective

Alcohol Abuse and Antabuse

According to a study carried out by the Centers for Disease Control and Prevention,[1] more than 75,000 Americans die each year as a result of alcohol abuse. Of these, nearly 35,000 people died of cirrhosis of the liver, cancer, or other drinking-related diseases. The remaining nearly 41,000 died in alcohol-related automobile accidents. Of those who died, 72% were men and 6% were under the age of twenty-one. In fact, a separate study has estimated that 1400 college-age students die each year of alcohol-related causes.

These numbers are striking. Alcohol abuse is now the third leading cause of preventable death in the United States, outranked only by tobacco use and poor diet and exercise habits. As the study concluded, "These results emphasize the importance of adopting effective strategies to reduce excessive drinking, including increasing alcohol excise taxes and screening for alcohol misuse in clinical settings."

Tetraethylthiuram disulfide
(disulfiram)
Antabuse

One approach to treatment of alcohol abuse, the drug tetraethylthiuram disulfide or disulfiram, has been used since 1951. The activity of this drug, generally known by the trade name Antabuse, was discovered accidentally by a group of Danish researchers who were testing it for antiparasitic properties. They made the observation that those who had taken disulfiram became violently ill after consuming any alcoholic beverage.

Further research revealed that this compound inhibits one of the liver enzymes in the pathway for the oxidation of alcohols.

In Chapter 12, we saw that ethanol is oxidized to ethanal (acetaldehyde) in the liver. This reaction is catalyzed by the enzyme alcohol dehydrogenase. Acetaldehyde, which is more toxic than ethanol, is responsible for many of the symptoms of a hangover. The enzyme acetaldehyde dehydrogenase oxidizes acetaldehyde into ethanoic acid (acetic acid), which then is used in biochemical pathways that harvest energy for cellular work or that synthesize fats.

Antabuse inhibits acetaldehyde dehydrogenase. This inhibition occurs within 1 to 2 hours (h) of taking the drug and continues up to 14 days. When a person who has taken Antabuse drinks an alcoholic beverage, the level of acetaldehyde quickly reaches levels that are five to ten times higher than would normally occur after a drink. Within just a few minutes, the symptoms of a severe hangover are experienced and may continue for several hours.

Experts in drug and alcohol abuse have learned that drugs such as Antabuse are generally not effective on their own. However, when used in combination with support groups and/or psychotherapy to solve underlying behavioral or psychological problems, Antabuse is an effective deterrent to alcohol abuse.

1. Alcohol-Attributable Deaths and Years of Potential Life Lost—United States, 2001, Morbidity and Mortality Weekly Report, 53 (37): 866–870, September 24, 2004, also available at www.cdc.gov/mmwr/preview/mmwrhtml/mm5337a2.htm.

For Further Understanding

▸ Antabuse alone is not a cure for alcoholism. Consider some of the reasons why this is so.

▸ Write equations showing the oxidation of ethanal to ethanoic acid as a pathway with the product of the first reaction serving as the reactant for the second. Explain the physiological effects of Antabuse in terms of these chemical reactions.

---

carbon-oxygen single bond. The addition of hydrogen to a carbon-oxygen double bond is shown in the following general equation:

Aldehyde Hydrogen   Alcohol
or ketone

The hydrogenation (reduction) of a ketone produces a secondary alcohol, as seen in the following equation showing the reduction of the ketone, 3-octanone:

Hydrogenation was first discussed in Section 11.5 for the hydrogenation of alkenes.

3-Octanone   Hydrogen   3-Octanol
(A ketone)   (A secondary alcohol)

EXAMPLE 13.7	Writing an Equation Representing the Hydrogenation of a Ketone

LEARNING GOAL

6 Write equations representing the reduction of carbonyl compounds.

Write an equation showing the hydrogenation of 3-pentanone.

Solution

The product of the reduction of a ketone is a secondary alcohol, in this case, 3-pentanol.

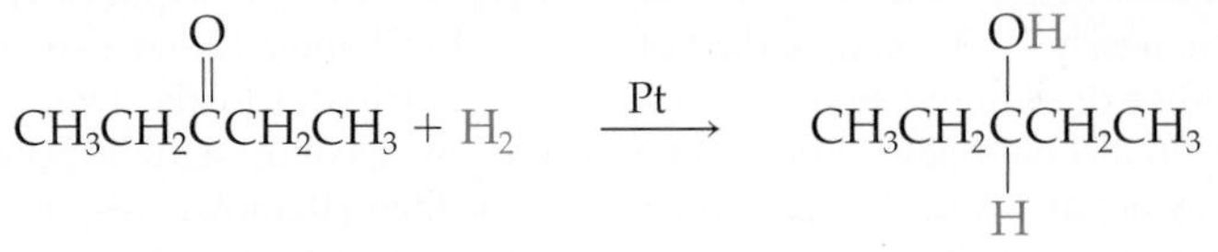

$$\underset{\text{3-Pentanone}}{CH_3CH_2\overset{\overset{\displaystyle O}{\|}}{C}CH_2CH_3} + H_2 \xrightarrow{Pt} \underset{\text{3-Pentanol}}{CH_3CH_2\underset{\underset{\displaystyle H}{|}}{\overset{\overset{\displaystyle OH}{|}}{C}}CH_2CH_3}$$

Practice Problem 13.7

Write an equation for the hydrogenation of (a) propanone and (b) butanone.

▶ For Further Practice: **Question 13.64.**

The hydrogenation of an aldehyde results in the production of a primary alcohol, as seen in the following equation showing the reduction of the aldehyde, butanal:

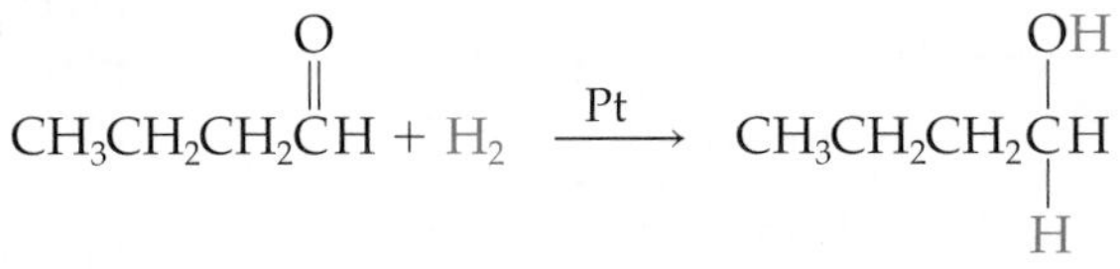

$$\underset{\substack{\text{Butanal} \\ \text{(An aldehyde)}}}{CH_3CH_2CH_2\overset{\overset{\displaystyle O}{\|}}{C}H} + \underset{\text{Hydrogen}}{H_2} \xrightarrow{Pt} \underset{\substack{\text{1-Butanol} \\ \text{(A primary alcohol)}}}{CH_3CH_2CH_2\underset{\underset{\displaystyle H}{|}}{\overset{\overset{\displaystyle OH}{|}}{C}}H}$$

EXAMPLE 13.8	Writing an Equation Representing the Hydrogenation of an Aldehyde

LEARNING GOAL

6 Write equations representing the reduction of carbonyl compounds.

Write an equation showing the hydrogenation of 3-methylbutanal.

Solution

Recall that the reduction of an aldehyde results in the production of a primary alcohol, in this case, 3-methyl-1-butanol.

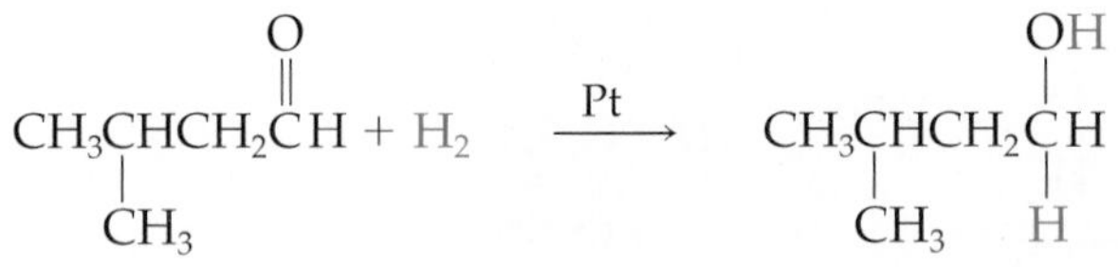

$$\underset{\text{3-Methylbutanal}}{CH_3\underset{\underset{\displaystyle CH_3}{|}}{C}HCH_2\overset{\overset{\displaystyle O}{\|}}{C}H} + H_2 \xrightarrow{Pt} \underset{\text{3-Methyl-1-butanol}}{CH_3\underset{\underset{\displaystyle CH_3}{|}}{C}HCH_2\underset{\underset{\displaystyle H}{|}}{\overset{\overset{\displaystyle OH}{|}}{C}}H}$$

Practice Problem 13.8

Write equations for the hydrogenation of 3,4-dimethylhexanal and 2-chloropentanal.

▶ For Further Practice: **Questions 13.65 and 13.66.**

In organic and biological systems, oxidation is recognized as the loss of hydrogen or the gain of oxygen, and reduction is recognized as the gain of hydrogen or the loss of oxygen. The following scheme summarizes the relationship among alkanes, alcohols, aldehydes (or ketones), and carboxylic acids with regard to their relative oxidation or reduction.

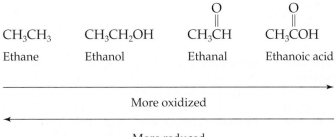

CH₃CH₃ — Ethane

CH₃CH₂OH — Ethanol

$$CH_3\overset{\displaystyle O}{\overset{\|}{C}}H$$ — Ethanal

$$CH_3\overset{\displaystyle O}{\overset{\|}{C}}OH$$ — Ethanoic acid

More oxidized →

← More reduced

A biological example of the reduction of a ketone occurs in the body, particularly during strenuous exercise when the lungs and circulatory system may not be able to provide enough oxygen to the muscles. Under these circumstances, the lactate fermentation begins. In this reaction, the enzyme *lactate dehydrogenase* reduces pyruvate, the product of glycolysis (a pathway for the breakdown of glucose) into lactate. The reducing agent for this reaction is nicotinamide adenine dinucleotide (NADH, which is oxidized in the reaction to form NAD⁺).

The role of the lactate fermentation in exercise is discussed in greater detail in Section 21.4.

$$CH_3-\overset{O}{\overset{\|}{C}}-\overset{O}{\overset{\|}{C}}-O^- \xrightarrow[\text{NADH} \quad \text{NAD}^+]{\text{Lactate dehydrogenase}} CH_3-\overset{OH}{\underset{H}{\overset{|}{C}}}-\overset{O}{\overset{\|}{C}}-O^-$$

Pyruvate   Lactate

Question 13.11 Label each of the following as an oxidation or a reduction reaction.
a. Ethanal to ethanol
b. Benzoic acid to benzaldehyde
c. Cyclohexanone to cyclohexanol
d. 2-Propanol to propanone
e. 2,3-Butanedione (found in butter) to 2,3-butanediol

Question 13.12 Write an equation for each of the reactions in Question 13.11.

Addition Reactions

The principal reaction of the carbonyl group is the **addition reaction** across the polar carbon-oxygen double bond. This reaction is very similar to some that we have already studied, addition across the carbon-carbon double bond of alkenes. Such reactions require that a catalytic amount of acid be present in solution, as shown by the H⁺ over the arrow for the reactions shown in the following examples.

An example of an addition reaction is the reaction of aldehydes and ketones with alcohols in the presence of catalytic amounts of acid. In this reaction, the hydrogen of the alcohol adds to the carbonyl oxygen. The alkoxy group of the alcohol (—OR) adds to the carbonyl carbon. The predicted product is a **hemiacetal**. A hemiacetal is characterized by a carbon that is bonded to one hydroxyl group (—OH) and one alkoxy group (—OR).

LEARNING GOAL

7 Write equations for the preparation of hemiacetals and acetals.

Addition reactions of alkenes are described in detail in Section 11.5.

$$R-\overset{OH}{\underset{R}{\overset{|}{\underset{|}{C}}}}-OR$$

↗ H and alkoxy group from the alcohol

↙ Remainder of the structure is from the aldehyde or ketone

General structure of a hemiacetal

Kitchen Chemistry

The Allure of Truffles

Truffles are the fruiting body of fungi of the genus *Tuber*. They are remarkably ugly, roughly round and warty in appearance, and have an aroma variously described as "nutty," "musky," and "stinky." Nonetheless, truffles are one of the most highly prized ingredients in most of the finest kitchens in the world. In fact, the white truffle has been described as the "diamond of the kitchen."

Truffles grow under a limited variety of trees, including oak, beech, and poplar. They may be as deep as 1 meter (m) underground and only grow in the fall and winter of the year. Traditionally, they were harvested using truffle-sniffing pigs. These pigs were able to detect the smell of the chemicals the truffle produces to attract animals who will eat them and spread their spores to other sites through their feces. Today, trained dogs are used to find the truffles because of a tendency of the pigs to eat the truffles before they could be harvested! With prices as high as $500 per pound (lb) for black truffles and $2000 per lb for white truffles, harvesters couldn't afford to have the pigs feasting on them!

The flavor of the truffles is a direct result of their aroma. The compounds that produce that aroma are only produced as the spores mature, which is why it is useful to use animals to locate the scent of the truffles when they are at their peak.

Many have studied the compounds in truffles that produce the aroma, and hence the flavor. They have found as many as thirty-six components of that flavor profile, including alcohols, aldehydes, ketones, carboxylic acids, esters, amines, aromatic compounds, and hydrocarbons—virtually an entire organic chemistry laboratory! Two alcohols and two aldehydes are found in the highest concentrations. These are 2-methylbutanol, 3-methylbutanol, 2-methylbutanal, and 3-methylbutanal. The flavor profile is rounded out by several sulfur compounds, including dimethylsulfide, 2,4-pentanedithiol, and 2,4,6-heptanetrithiol. In fact, it is the dimethylsulfide that attracts animals to the fruiting bodies.

As we have learned in this chapter, alcohols and aldehydes such as these in the truffle flavor profile are easily oxidized. When that happens, the truffle loses the molecules that are characteristic of its aroma and flavor. Fresh truffles must be used quickly and stored carefully to retain the flavor. Often, truffles are sold in a preserved form that has a shelf-life as long as 2 years.

The truffle is becoming much more rare. While it was reported in 1890 that 2200 tons of truffles were harvested, now only about 150 tons are harvested. The truffle can only grow in association with certain trees and those trees only support truffles for about 15–30 years. To renew the harvest, tree saplings that have been inoculated with spores are being planted in areas with the right type of soil. However, it requires about 7 years before these trees begin to produce the precious fruiting bodies.

Truffles will continue to be a much sought-after delicacy, and science is helping us understand the chemical complexity of this unlovely fungus, as well as ways to ensure that we will be able to enjoy its remarkable flavors in the centuries to come.

For Further Understanding

▶ Draw the structures of the molecules that are responsible for the flavor profile of truffles.

▶ Write equations for the oxidation reactions that destroy the complex aroma and flavor of truffles.

In the following reactions, notice that the alkoxy group (—OR) of the alcohol adds to the carbonyl carbon and that the hydrogen of the hydroxyl group adds to the carbonyl oxygen. The following equations show the addition of ethanol to cyclohexylmethanal and propanone:

$$\text{Cyclohexyl}-\overset{\overset{\displaystyle O}{\|}}{\underset{\underset{\displaystyle H}{|}}{C}}-H \;+\; \overset{\overset{\displaystyle H}{|}}{O}-CH_2CH_3 \;\underset{}{\overset{H^+}{\rightleftharpoons}}\; \text{Cyclohexyl}-\overset{\overset{\displaystyle O-H}{|}}{\underset{\underset{\displaystyle H}{|}}{C}}-OCH_2CH_3$$

Cyclohexylmethanal   Ethanol   A hemiacetal

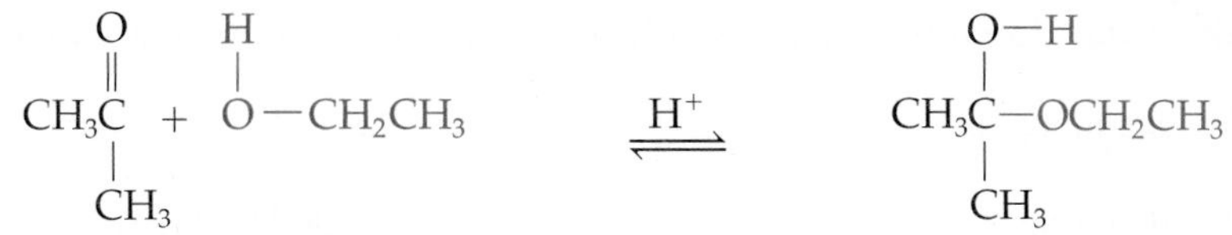

$$
\underset{\substack{\text{Propanone}}}{\underset{\underset{CH_3}{|}}{\overset{\overset{O}{\|}}{CH_3C}}} + \underset{\text{Ethanol}}{\overset{H}{\underset{}{O-CH_2CH_3}}} \quad \underset{}{\overset{H^+}{\rightleftharpoons}} \quad \underset{\substack{\text{A hemiacetal}}}{\underset{\underset{CH_3}{|}}{\overset{\overset{O-H}{|}}{CH_3C-OCH_2CH_3}}}
$$

Hemiacetals are generally unstable. In the presence of acid and excess alcohol, they undergo a substitution reaction in which the —OH group of the hemiacetal is exchanged for another —OR group from the alcohol. The product of this reaction is an **acetal.** Acetal formation is a reversible reaction, as the general equation shows:

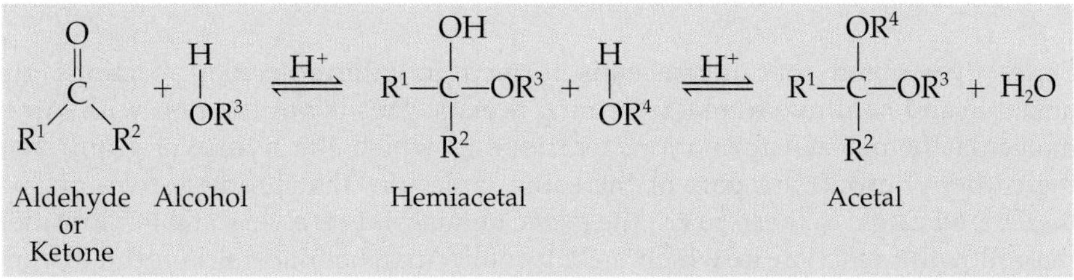

$$
\underset{\substack{\text{Aldehyde} \\ \text{or} \\ \text{Ketone}}}{\underset{R^1}{\overset{O}{\underset{}{\overset{\|}{C}}}}\underset{R^2}{}} + \underset{\substack{\text{Alcohol}}}{\overset{H}{\underset{OR^3}{|}}} \overset{H^+}{\rightleftharpoons} \underset{\substack{\text{Hemiacetal}}}{R^1-\underset{R^2}{\overset{OH}{\underset{|}{\overset{|}{C}}}}-OR^3} + \overset{H}{\underset{OR^4}{|}} \overset{H^+}{\rightleftharpoons} \underset{\substack{\text{Acetal}}}{R^1-\underset{R^2}{\overset{OR^4}{\underset{|}{\overset{|}{C}}}}-OR^3} + H_2O
$$

Notice that the acetal is characterized by a carbon bonded to two alkoxy groups (—OR).

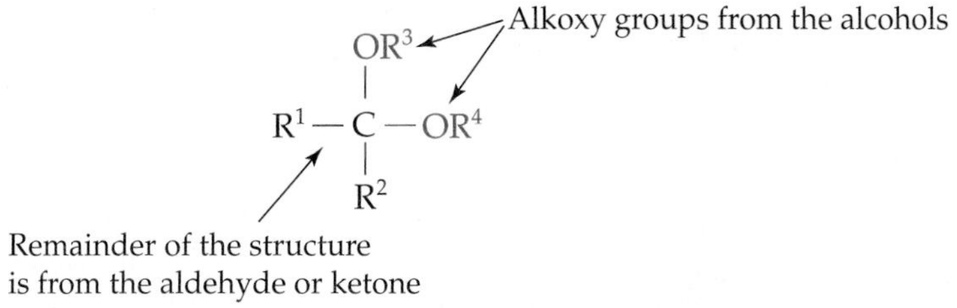

$$
\overset{\text{Alkoxy groups from the alcohols}}{\underset{\underset{\substack{\text{Remainder of the structure} \\ \text{is from the aldehyde or ketone}}}{R^2}}{R^1-\underset{|}{C}-OR^4}}\overset{OR^3}{|}
$$

General structure of an acetal

Acetal formation is seen in the following equations that represent the acid-catalyzed reactions between propanal and methanol and between propanone and ethanol.

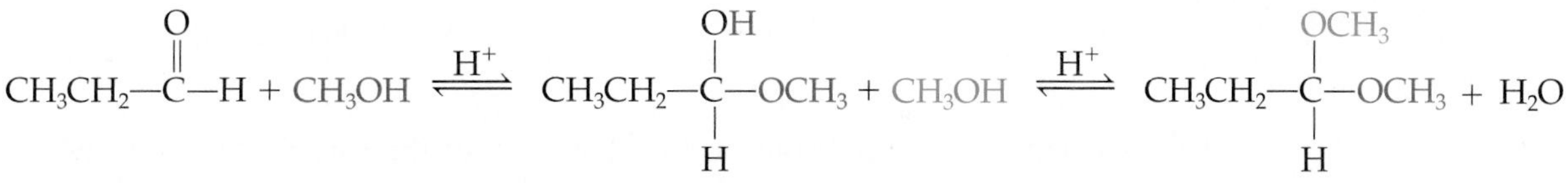

$$
\underset{\substack{\text{Propanal}}}{\overset{O}{\underset{}{CH_3CH_2-\overset{\|}{C}-H}}} + \underset{\substack{\text{Methanol}}}{CH_3OH} \overset{H^+}{\rightleftharpoons} \underset{\substack{\text{Hemiacetal}}}{CH_3CH_2-\underset{H}{\overset{OH}{\underset{|}{\overset{|}{C}}}}-OCH_3} + CH_3OH \overset{H^+}{\rightleftharpoons} \underset{\substack{\text{Acetal}}}{CH_3CH_2-\underset{H}{\overset{OCH_3}{\underset{|}{\overset{|}{C}}}}-OCH_3} + H_2O
$$

$$
\underset{\substack{\text{Propanone}}}{\overset{O}{\underset{}{CH_3-\overset{\|}{C}-CH_3}}} + \underset{\substack{\text{Ethanol}}}{CH_3CH_2OH} \overset{H^+}{\rightleftharpoons} \underset{\substack{\text{Hemiacetal}}}{CH_3-\underset{CH_3}{\overset{OH}{\underset{|}{\overset{|}{C}}}}-OCH_2CH_3} + CH_3CH_2OH \overset{H^+}{\rightleftharpoons} \underset{\substack{\text{Acetal}}}{CH_3-\underset{CH_3}{\overset{OCH_2CH_3}{\underset{|}{\overset{|}{C}}}}-OCH_2CH_3} + H_2O
$$

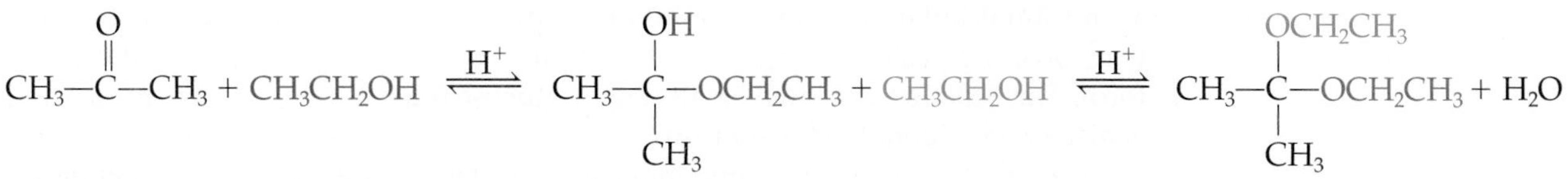

Question 13.13 Identify each of the following structures as a hemiacetal or acetal.

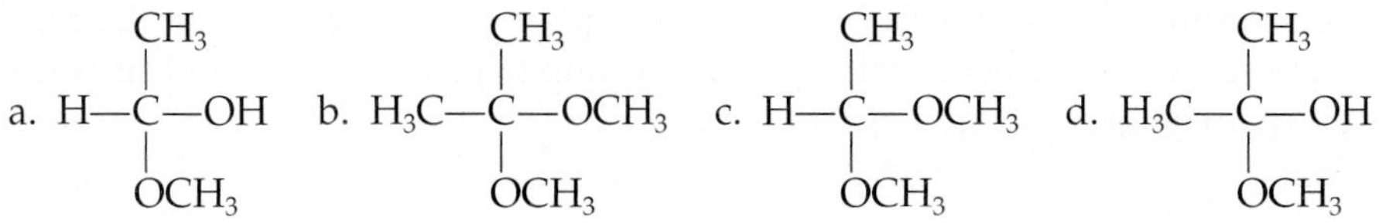

a. $H-\underset{OCH_3}{\overset{CH_3}{\underset{|}{\overset{|}{C}}}}-OH$ b. $H_3C-\underset{OCH_3}{\overset{CH_3}{\underset{|}{\overset{|}{C}}}}-OCH_3$ c. $H-\underset{OCH_3}{\overset{CH_3}{\underset{|}{\overset{|}{C}}}}-OCH_3$ d. $H_3C-\underset{OCH_3}{\overset{CH_3}{\underset{|}{\overset{|}{C}}}}-OH$

Question 13.14 Identify each of the following structures as a hemiacetal or acetal.

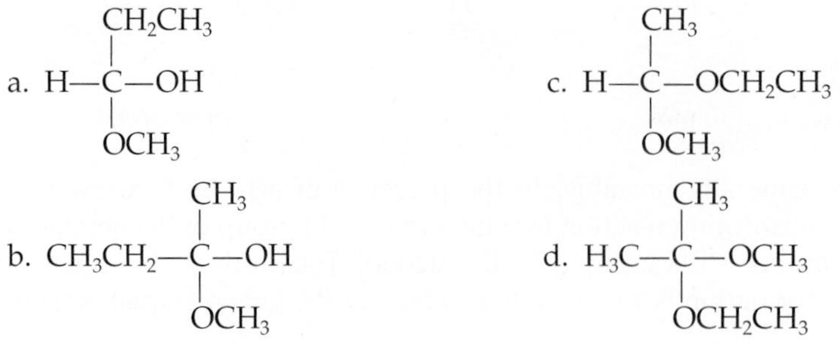

a.
$$
\begin{array}{c}
CH_2CH_3 \\
| \\
H-C-OH \\
| \\
OCH_3
\end{array}
$$

c.
$$
\begin{array}{c}
CH_3 \\
| \\
H-C-OCH_2CH_3 \\
| \\
OCH_3
\end{array}
$$

b.
$$
\begin{array}{c}
CH_3 \\
| \\
CH_3CH_2-C-OH \\
| \\
OCH_3
\end{array}
$$

d.
$$
\begin{array}{c}
CH_3 \\
| \\
H_3C-C-OCH_3 \\
| \\
OCH_2CH_3
\end{array}
$$

Earlier, we noted that hemiacetals formed in intermolecular reactions are unstable and continue to react, forming acetals. This is not the case with intramolecular hemiacetal formation, reactions in which the hydroxyl group and the carbonyl group are part of the same molecule, that produce five- or six-membered rings. In these cases, the cyclic hemiacetals are very stable. Consider the following reaction in which an intramolecular hemiacetal is formed from 4-hydroxyheptanal:

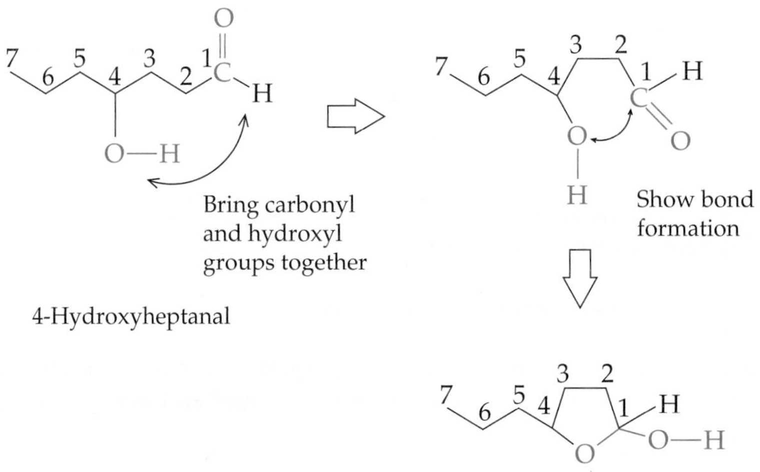

Bring carbonyl and hydroxyl groups together

4-Hydroxyheptanal

Show bond formation

A cyclic hemiacetal

This reaction is very important in the chemistry of the carbohydrates, in which hemiacetals are readily formed. Monosaccharides contain several hydroxyl groups and one carbonyl group. The linear form of a monosaccharide quickly undergoes an intramolecular reaction in solution to produce a cyclic hemiacetal. In these reactions, the cyclic or ring form of the molecule is more stable than the linear form. This reaction is shown for the sugar glucose (blood sugar) in Figure 13.6 and is discussed in detail in Section 16.2.

When the hemiacetal of one monosaccharide reacts with the hydroxyl group of another monosaccharide, the product is an acetal. A sugar molecule made up of two monosaccharides is called a *disaccharide*. The C—O—C or acetal bond between the two monosaccharides is called a *glycosidic bond*. An example of this is found in Figure 13.7, which depicts the formation of the disaccharide lactose (milk sugar) from the monosaccharides glucose and galactose. Notice that the acetal is formed between the hemiacetal hydroxyl group of galactose (in blue) and an alcohol hydroxyl group of glucose (in red).

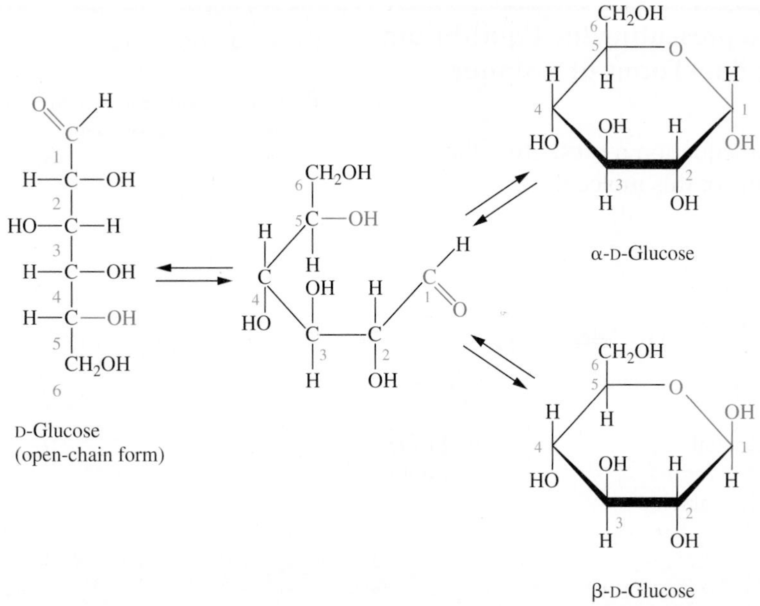

Figure 13.6 Hemiacetal formation in sugars, shown for the intramolecular reaction of D-glucose.

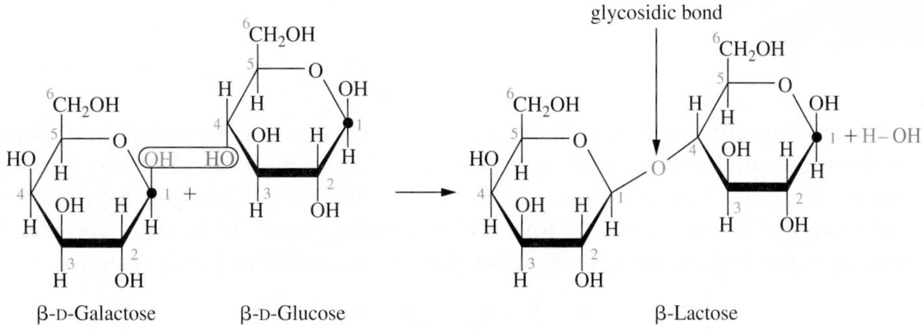

Figure 13.7 Acetal formation, demonstrated in the formation of the disaccharide lactose, milk sugar. The reaction between the hemiacetal hydroxyl group of the monosaccharide galactose (blue) and an alcohol hydroxyl group of the monosaccharide glucose (red) produces the acetal lactose. The bond between the two sugars is a glycosidic bond.

Keto-Enol Tautomers

Many aldehydes and ketones may exist in an equilibrium mixture of two constitutional or structural isomers called tautomers. **Tautomers** differ from one another in the placement of a hydrogen atom and a double bond. One tautomer is the *keto form* (on the left in the following equation). The keto form has the structure typical of an aldehyde or ketone. The other form is called the *enol form* (on the right in the following equation). The **enol** form has a structure containing a carbon-carbon double bond (*en*) and a hydroxyl group, the functional group characteristic of alcohols (*ol*).

LEARNING GOAL

8 Draw the keto and enol forms of aldehydes and ketones.

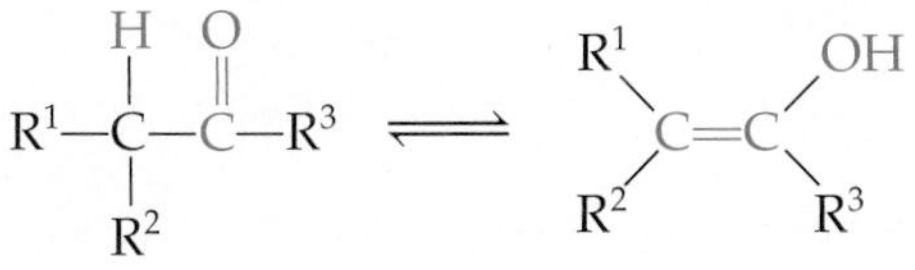

Keto form   Enol form

Because the keto form of most simple aldehydes and ketones is more stable, they exist mainly in that form.

EXAMPLE 13.9   **Writing an Equation Representing the Equilibrium between the Keto and Enol Forms of a Simple Aldehyde**

LEARNING GOAL

8 Draw the keto and enol forms of aldehydes and ketones.

Draw the keto form of ethanal, and write an equation representing the equilibrium between the keto and enol forms of this molecule.

Solution

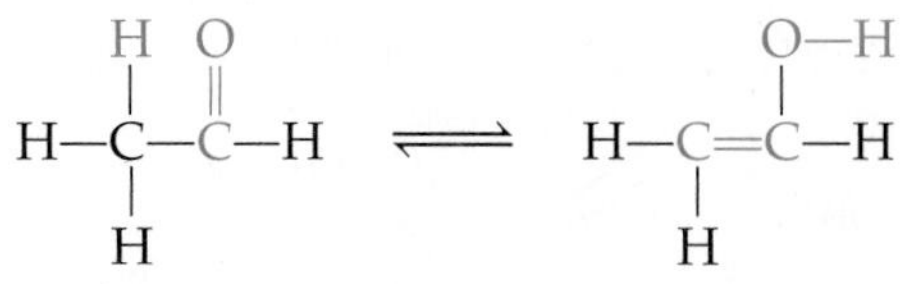

$$
\begin{array}{ccc}
\underset{\underset{\displaystyle H}{|}}{\overset{\displaystyle H}{\overset{|}{H-C}}} - \overset{\displaystyle O}{\overset{\|}{C}} - H
&\rightleftharpoons&
H - \overset{}{\underset{\underset{\displaystyle H}{|}}{C}} = \overset{\displaystyle O-H}{\overset{|}{C}} - H
\end{array}
$$

Ethanal   Enol form
Keto form   Less stable
More stable

Practice Problem 13.9

Draw the keto and enol forms of (a) propanal and (b) 3-pentanone.

▶ For Further Practice: **Questions 13.79 and 13.80.**

Phosphoenolpyruvate is a biologically important phosphorylated enol. Note that the molecule has a phosphoryl group ($-PO_4^{-2}$) in place of a hydroxyl group (—OH). The bond between the carbon and the phosphoryl group is represented by a "squiggle" (~) to show that this is a high-energy bond. In fact, phosphoenolpyruvate is the highest energy phosphorylated compound in living systems.

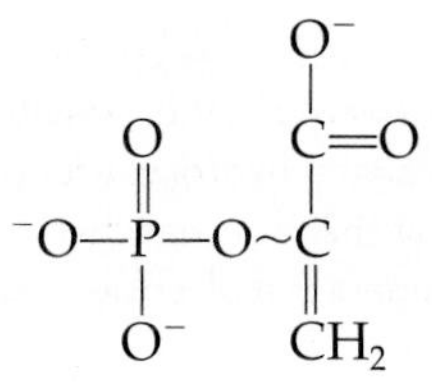

$$
\begin{array}{ccc}
&& \overset{\displaystyle O^-}{\underset{}{|}} \\
\overset{\displaystyle O}{\overset{\|}{}} && \overset{\displaystyle C=O}{\underset{}{|}} \\
{}^-O-\overset{}{\underset{\underset{\displaystyle O^-}{|}}{P}}-O\sim\overset{}{\underset{\underset{\displaystyle CH_2}{\|}}{C}}
\end{array}
$$

Phosphoenolpyruvate

The glycolysis pathway is discussed in detail in Chapter 21.

Phosphoenolpyruvate is produced in the next-to-last step in the metabolic pathway called *glycolysis*, which is the first stage of carbohydrate breakdown. In the final reaction of glycolysis, the phosphoryl group from phosphoenolpyruvate, along with energy from the high-energy bond, are transferred to adenosine diphosphate (ADP). The reaction produces ATP, the major energy currency of the cell.

CHAPTER MAP

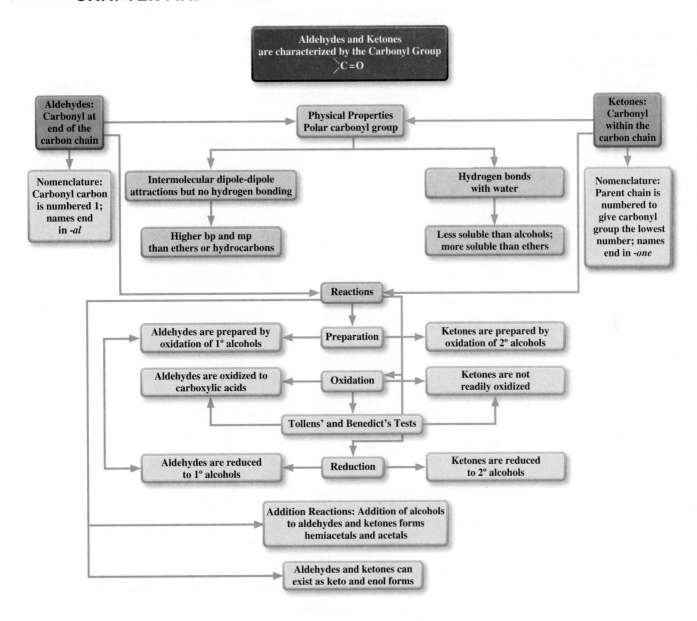

SUMMARY OF REACTIONS

Aldehydes and Ketones

Oxidation of an Aldehyde

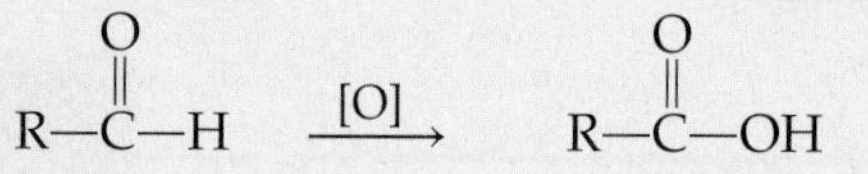

Aldehyde   Carboxylic acid

Reduction of Aldehydes and Ketones

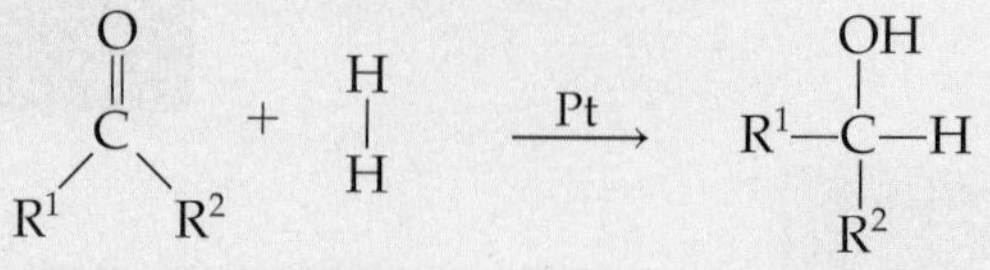

Aldehyde   Hydrogen   Alcohol
or Ketone

Addition Reactions

Addition of an alcohol to an aldehyde or a ketone—acetal formation:

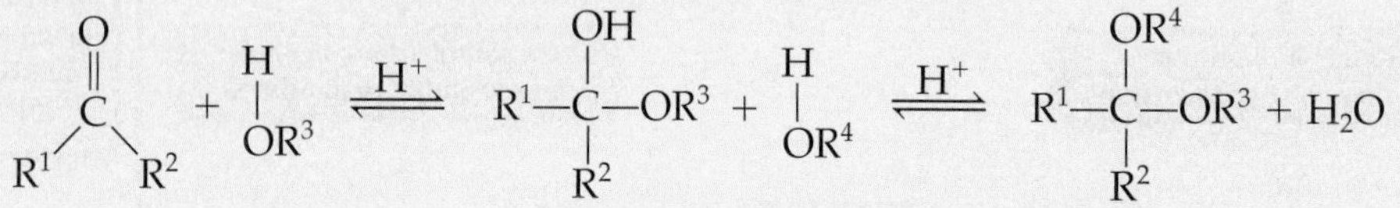

Aldehyde   Alcohol   Hemiacetal   Acetal
or
Ketone

Keto-enol Tautomerization

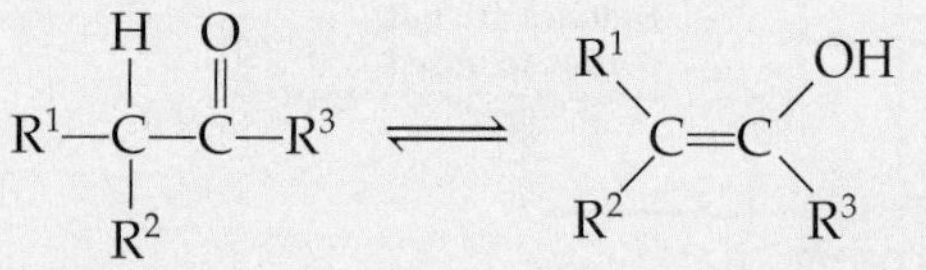

Keto form   Enol form

SUMMARY

13.1 Structure and Physical Properties

▶ The **carbonyl group** is characteristic of the **aldehydes** and **ketones.**

▶ The carbonyl group and the two groups attached to it are coplanar.

▶ In ketones, the carbonyl carbon is attached to two carbon-containing groups.

▶ In aldehydes, the carbonyl carbon is attached to at least one hydrogen; the second group may be another hydrogen or a carbon-containing group.

▶ Because of the polar carbonyl group, aldehydes and ketones are polar compounds. However, they cannot form hydrogen bonds with one another. As a result, they have higher boiling points than comparable hydrocarbons, but lower boiling points than comparable alcohols.

▶ Aldehydes and ketones can form hydrogen bonds with water and as a result are reasonably soluble in water.

▶ Larger carbonyl-containing compounds are less polar and thus are more soluble in nonpolar organic solvents.

13.2 IUPAC Nomenclature and Common Names

▶ In the IUPAC Nomenclature System, aldehydes are named by determining the parent compound and replacing the final -*e* of the parent alkane with -*al.*

   • The chain is numbered beginning with the carbonyl carbon as carbon-1.

▶ In the IUPAC Nomenclature System, ketones are named by determining the parent compound and replacing the -*e* of the parent alkane with the -*one* suffix of the ketone family.

   • The longest carbon chain is numbered to give the carbonyl carbon the lowest possible number.

▶ In the common system of nomenclature, substituted aldehydes are named as derivatives of the parent compound. Greek letters indicate the position of substituents.

▶ Common names of ketones are derived by naming the alkyl groups bonded to the carbonyl carbon. These names are followed by the word *ketone.*

13.3 Important Aldehydes and Ketones

▶ Members of the aldehyde and ketone families are important as food and fragrance chemicals, and as medicinal and agricultural chemicals.

▶ Methanal (formaldehyde) is used to preserve tissue.

▶ Ethanal causes the symptoms of a hangover and is oxidized to produce acetic acid commercially.

▶ Propanone (acetone) is a useful and versatile solvent for organic compounds.

13.4 Reactions Involving Aldehydes and Ketones

▶ In the laboratory, aldehydes and ketones are prepared by the oxidation of alcohols.

- Oxidation of a primary alcohol produces an aldehyde.
- Oxidation of a secondary alcohol produces a ketone.

▶ Aldehydes and ketones can be distinguished from one another on the basis of their ability to undergo **oxidation** reactions. The **Tollens' test** and **Benedict's test** are the most common tests to distinguish aldehydes and ketones.

- Aldehydes are relatively easily oxidized to carboxylic acids.
- Ketones do not undergo further oxidation reactions.

▶ Aldehydes and ketones are readily reduced to alcohols by **hydrogenation.**

▶ The most common reaction of the carbonyl group is **addition** across the highly polar carbon-oxygen double bond.

- The addition of an alcohol to an aldehyde or ketone produces a **hemiacetal.**
- The hemiacetal may react with a second alcohol to form an **acetal.**

▶ Aldehydes and ketones may exist in an equilibrium mixture of keto and enol tautomers.

- An **enol** is a molecule having both a carbon-carbon double bond and a hydroxyl group (—OH).
- **Tautomers** are isomers that differ from one another in the placement of a hydrogen and a double bond.

ANSWERS TO PRACTICE PROBLEMS

13.1

Hexanal: $CH_3CH_2CH_2CH_2CH_2\overset{\displaystyle O}{\overset{\|}{C}}\!-\!H$

trans-2-Hexanal:

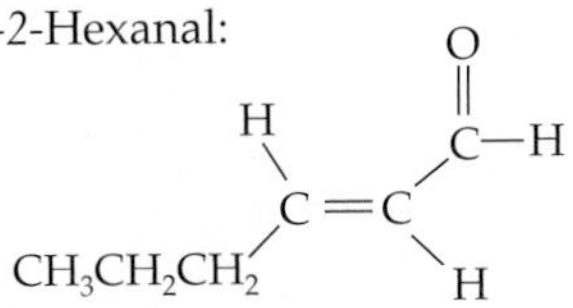

13.2 **a.** β,γ-Dimethylvaleraldehyde
b. α-Ethylvaleraldehyde
c. α-Chloropropionaldehyde
d. β-Hydroxybutyraldehyde

13.3 **a.** 6-Ethyl-2-octanone
b. 2-Methyl-3-hexanone
c. 2,4-Dimethyl-3-pentanone

13.4 **a.** Diethyl ketone   **c.** Methyl hexyl ketone
b. Ethyl butyl ketone   **d.** Diisopropyl ketone

13.5 **a.** The following equation represents the oxidation of 1-propanol to form propanal:

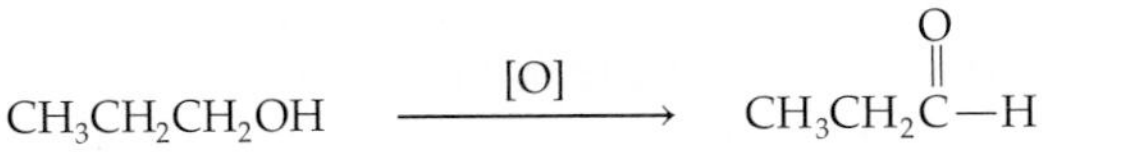

$$CH_3CH_2CH_2OH \xrightarrow{[O]} CH_3CH_2\overset{\displaystyle O}{\overset{\|}{C}}\!-\!H$$

Note that propanal may be further oxidized to form propanoic acid (a carboxylic acid).

b. The following equation represents the oxidation of 2-butanol to form butanone:

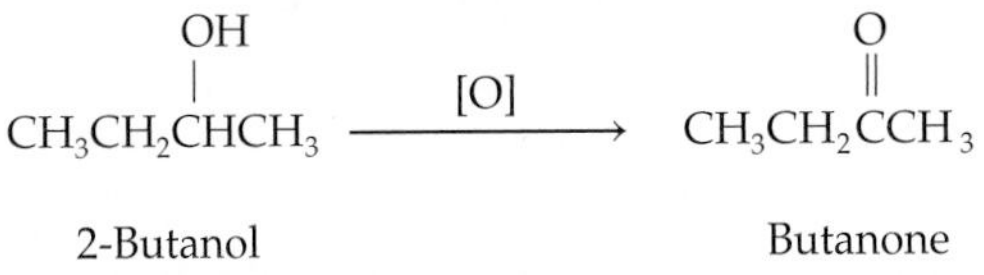

$$CH_3CH_2\overset{\displaystyle OH}{\underset{|}{C}}HCH_3 \xrightarrow{[O]} CH_3CH_2\overset{\displaystyle O}{\overset{\|}{C}}CH_3$$

2-Butanol   Butanone

13.6 **a.** The following equation represents the reaction between ethanal and Tollens' reagent:

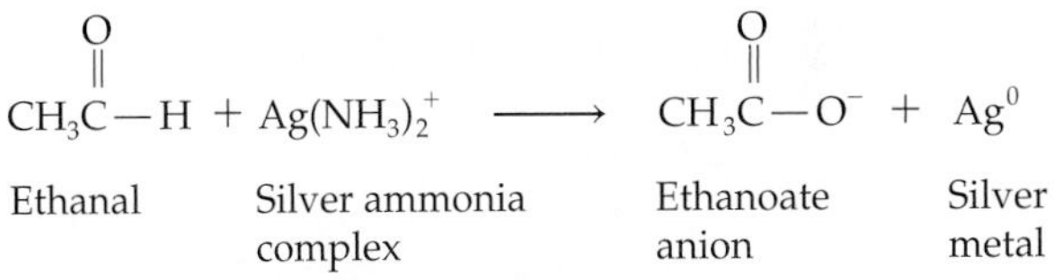

$$CH_3\overset{\displaystyle O}{\overset{\|}{C}}\!-\!H + Ag(NH_3)_2^+ \longrightarrow CH_3\overset{\displaystyle O}{\overset{\|}{C}}\!-\!O^- + Ag^0$$

Ethanal   Silver ammonia   Ethanoate   Silver
             complex   anion   metal

b. Tollens' reagent reacts with aldehydes and not ketones. Therefore, there would be no reaction between propanone and Tollens' reagent.

13.7 **a.** The following equation represents the hydrogenation of propanone:

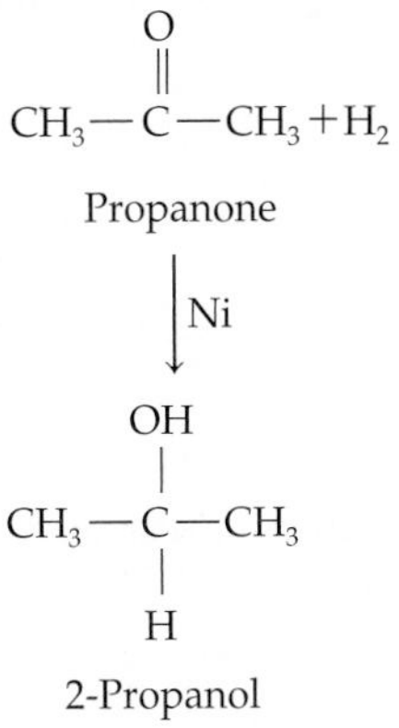

$$CH_3\!-\!\overset{\displaystyle O}{\overset{\|}{C}}\!-\!CH_3 + H_2$$

Propanone

$$\Big\downarrow Ni$$

$$CH_3\!-\!\overset{\displaystyle OH}{\underset{\displaystyle H}{\overset{|}{\underset{|}{C}}}}\!-\!CH_3$$

2-Propanol

b. The following equation represents the hydrogenation of butanone:

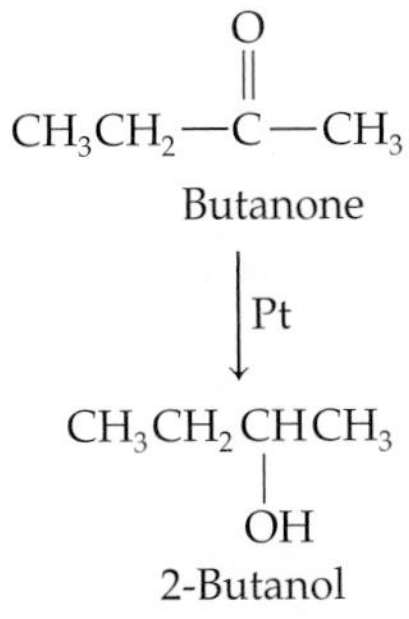

$$CH_3CH_2\!-\!\overset{\displaystyle O}{\overset{\|}{C}}\!-\!CH_3 + H_2$$

Butanone

$$\Big\downarrow Pt$$

$$CH_3CH_2\underset{\displaystyle OH}{\overset{|}{\underset{|}{C}}}HCH_3$$

2-Butanol

13.8 a. The following equation shows the hydrogenation of 3,4-dimethylhexanal, which produces 3,4-dimethyl-l-hexanol.

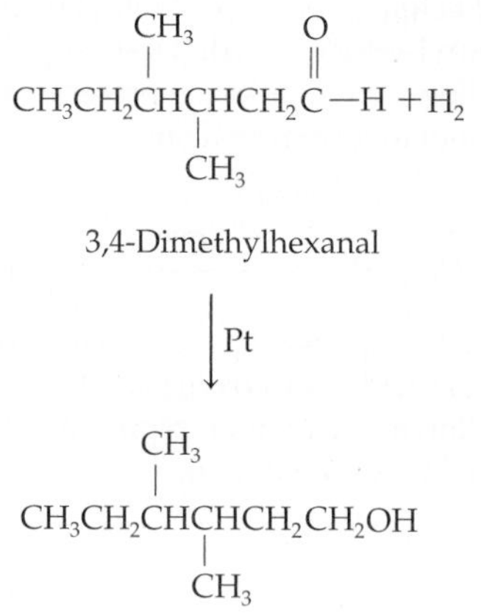

$$CH_3CH_2\underset{\underset{CH_3}{|}}{\overset{\overset{CH_3}{|}}{CH}}CHCH_2\overset{\overset{O}{\|}}{C}-H + H_2$$

3,4-Dimethylhexanal

$$\downarrow Pt$$

$$CH_3CH_2\underset{\underset{CH_3}{|}}{\overset{\overset{CH_3}{|}}{CH}}CHCH_2CH_2OH$$

3,4-Dimethyl-1-hexanol

b. The following equation shows the hydrogenation of 2-chloropentanal, which produces 2-chloro-l-pentanol.

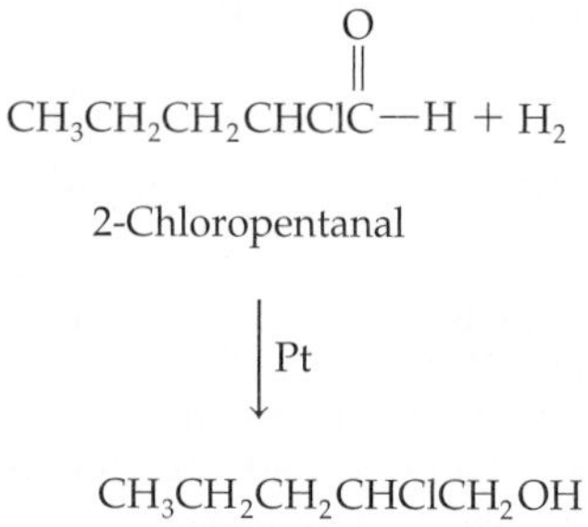

$$CH_3CH_2CH_2\overset{\overset{O}{\|}}{CHClC}-H + H_2$$

2-Chloropentanal

$$\downarrow Pt$$

$$CH_3CH_2CH_2CHClCH_2OH$$

2-Chloro-1-pentanol

13.9 a. The following structures show the keto and enol forms of propanal:

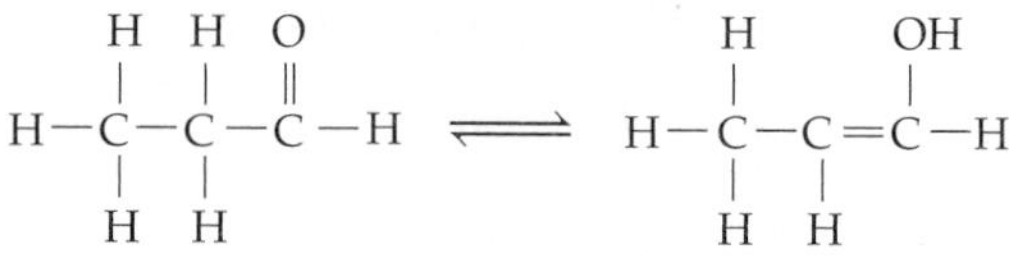

$$H-\underset{\underset{H}{|}}{\overset{\overset{H}{|}}{C}}-\underset{\underset{H}{|}}{\overset{\overset{H}{|}}{C}}-\overset{\overset{O}{\|}}{C}-H \rightleftharpoons H-\underset{\underset{H}{|}}{\overset{\overset{H}{|}}{C}}-\underset{\underset{H}{|}}{\overset{}{C}}=\overset{\overset{OH}{|}}{C}-H$$

Propanal   Propanal
Keto form   Enol form

b. The following structures show the keto and enol forms of 3-pentanone:

$$H-\underset{\underset{H}{|}}{\overset{\overset{H}{|}}{C}}-\underset{\underset{H}{|}}{\overset{\overset{H}{|}}{C}}-\overset{\overset{O}{\|}}{C}-\underset{\underset{H}{|}}{\overset{\overset{H}{|}}{C}}-\underset{\underset{H}{|}}{\overset{\overset{H}{|}}{C}}-H \rightleftharpoons H-\underset{\underset{H}{|}}{\overset{\overset{H}{|}}{C}}-\underset{\underset{H}{|}}{\overset{}{C}}=\overset{\overset{OH}{|}}{C}-\underset{\underset{H}{|}}{\overset{\overset{H}{|}}{C}}-\underset{\underset{H}{|}}{\overset{\overset{H}{|}}{C}}-H$$

3-Pentanone   3-Pentanone
(keto form)   (enol form)

QUESTIONS AND PROBLEMS

Structure and Physical Properties

Foundations

13.15 Explain the relationship between carbon chain length and water solubility of aldehydes or ketones.

13.16 Explain the dipole-dipole interactions that occur between molecules containing carbonyl groups.

Applications

13.17 Simple ketones (for example, acetone) are often used as industrial solvents for many organically based products such as adhesives and paints. They are often considered "universal solvents," because they dissolve so many diverse materials. Why are these chemicals such good solvents?

13.18 Explain briefly why simple (containing fewer than five carbon atoms) aldehydes and ketones exhibit appreciable solubility in water.

13.19 Draw intermolecular hydrogen bonding between ethanal and water.

13.20 Draw the polar interactions that occur between acetone molecules.

13.21 Why do alcohols have higher boiling points than aldehydes or ketones of comparable molar mass?

13.22 Why do hydrocarbons have lower boiling points than aldehydes or ketones of comparable molar mass?

13.23 Rank the following from highest to lowest boiling points:

a.

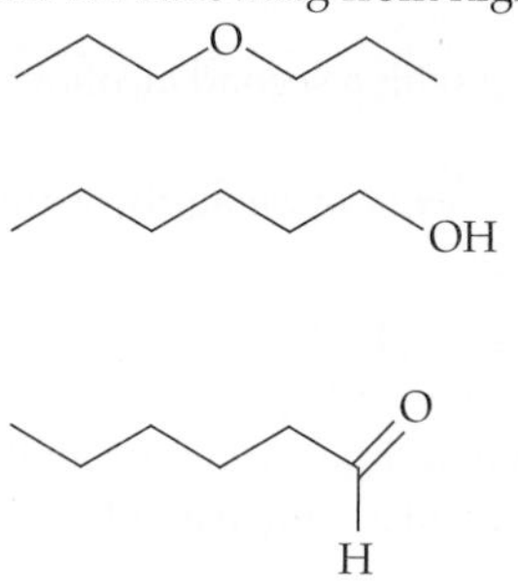

b.

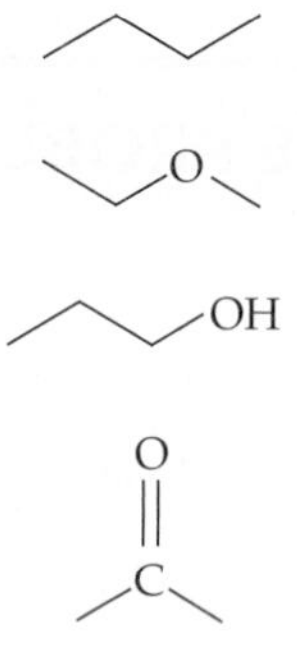

13.24 Rank the following from highest to lowest water solubility:

a.

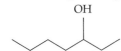

OH

OH   OH

O
‖
C
H

O

b.

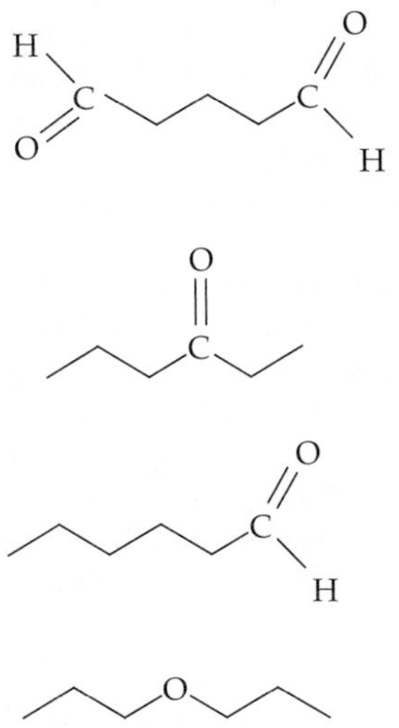

H
C
O

O
C
H

O
C

O
C
H

O

IUPAC Nomenclature and Common Names

Foundations

13.25 Briefly describe the rules of the IUPAC Nomenclature System for naming aldehydes.

13.26 Briefly describe the rules of the IUPAC Nomenclature System for naming ketones.

13.27 Briefly describe how to determine the common name of an aldehyde.

13.28 Briefly describe how to determine the common name of a ketone.

Applications

13.29 Draw each of the following using condensed formulas and line formulas:
  a. Ethanal
  b. 3,4-Dimethylpentanal
  c. 2-Ethylheptanal
  d. 5,7-Dichloroheptanal

13.30 Draw each of the following using condensed formulas and line formulas:
  a. 2-Nonanone
  b. 4-Methyl-2-heptanone
  c. 4,6-Diethyl-3-octanone
  d. 5-Bromo-4-octanone

13.31 Draw each of the following using condensed formulas and line formulas:
  a. Ethyl isopropyl ketone
  b. Ethyl propyl ketone
  c. Dibutyl ketone
  d. Heptyl hexyl ketone

13.32 Draw each of the following using condensed formulas and line formulas:
  a. β-Methylbutyraldehyde
  b. α-Hydroxypropionaldehyde
  c. α,β-Dimethylvaleraldehyde
  d. γ-Chlorovaleraldehyde

13.33 Use the IUPAC Nomenclature System to name each of the following compounds:

$$\text{O}$$
a. 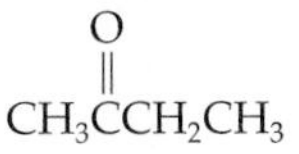 $CH_3CCH_2CH_3$

$$\text{O}$$
b. 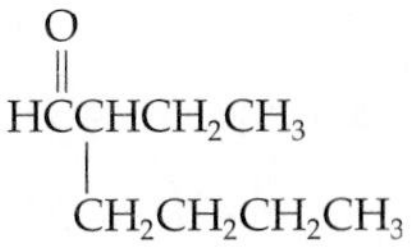 $HCCHCH_2CH_3$
        $CH_2CH_2CH_2CH_3$

13.34 Name each of the following using the IUPAC Nomenclature System:

a. 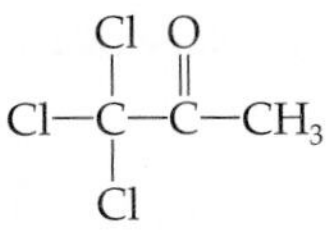
Cl O
Cl—C—C—CH₃
    Cl

b.

Cl

13.35 Name each of the following using the IUPAC Nomenclature System:

a. 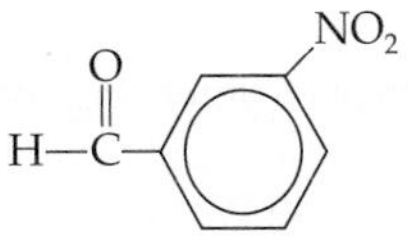
O
‖
H—C
        NO₂

b.

O

HO   OH

13.36 Name each of the following using the IUPAC Nomenclature System:

O
‖
a.  $CH_3CH_2CH_2CH$

O
‖
c. 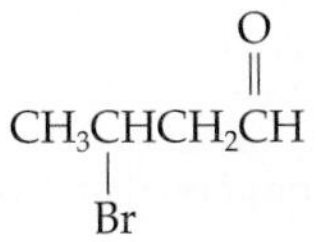 CH_3CHCH_2CH
       Br

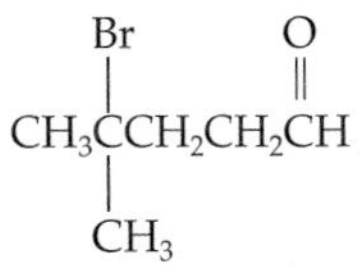

Br   O
b. $CH_3CCH_2CH_2CH$
    CH₃

CH₃ O
d. 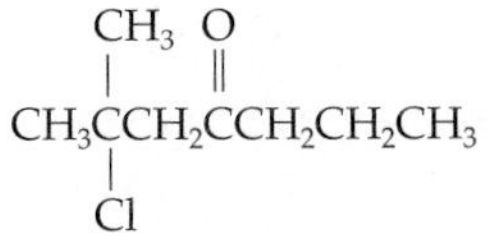 $CH_3CCH_2CCH_2CH_2CH_3$
    Cl

13.37 The molecule shown below has a lovely aroma of lily-of-the-valley. Discovered in 1908, it has been used in hundreds of perfumes. What is the IUPAC name of this molecule?

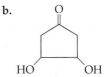

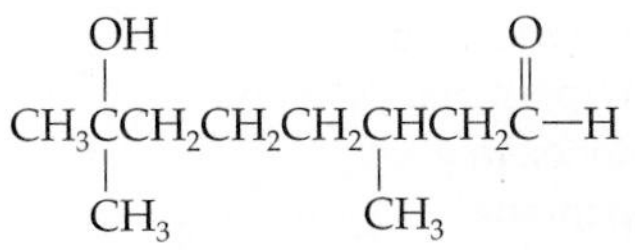

OH   O
‖
CH₃CCH₂CH₂CH₂CHCH₂C—H
    CH₃   CH₃

13.38 Give the IUPAC name for each of the following compounds:

a. 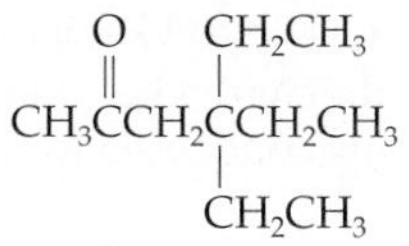 $CH_3CCH_2CCH_2CH_3$
with O double bond on first C, and CH_2CH_3 groups on the second C (one above, one below)

b. 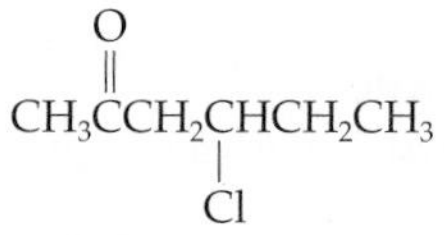 $CH_3CCH_2CHCH_2CH_3$
with O double bond on first C, and Cl on the CH below

13.39 Give the IUPAC name for each of the following compounds:

a. 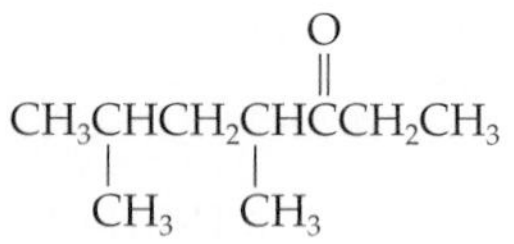 $CH_3CHCH_2CHCCH_2CH_3$
with O double bond on the C, CH_3 groups below two carbons

b. 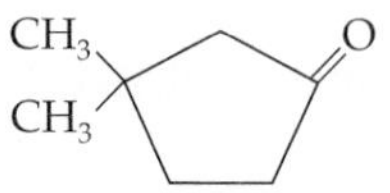 CH_3, CH_3 on a cyclopentane ring with O double bond

13.40 Give the IUPAC name for each of the following compounds:

a. 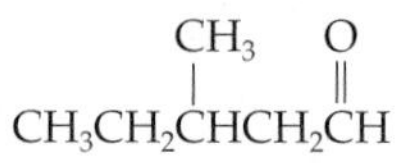 $CH_3CH_2CHCH_2CH$
with CH_3 above and O double bond on last C

b. 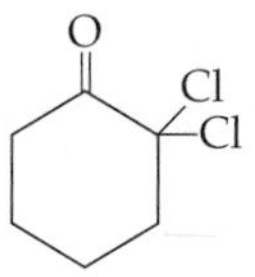 cyclohexanone ring with Cl and Cl on the carbon adjacent to the carbonyl

13.41 Give the common name for each of the following compounds:

a. 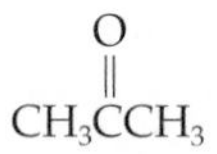 CH_3CCH_3 with O double bond

b. 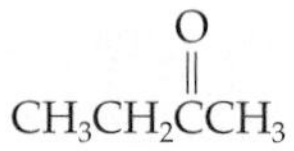 $CH_3CH_2CCH_3$ with O double bond

c. 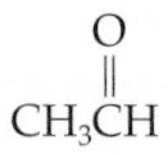 CH_3CH with O double bond

d. 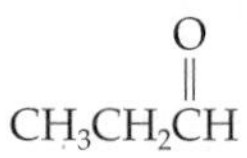 CH_3CH_2CH with O double bond

e. 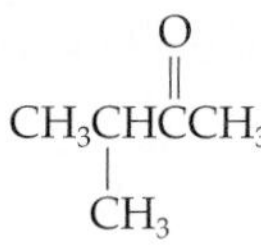 CH_3CHCCH_3 with O double bond and CH_3 below

13.42 Give the common name for each of the following compounds:

a. 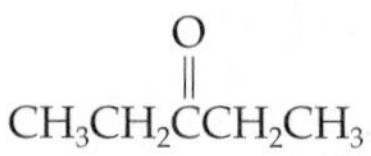 $CH_3CH_2CCH_2CH_3$ with O double bond

b. 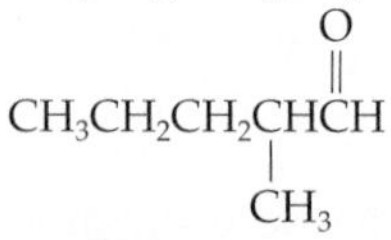 $CH_3CH_2CH_2CHCH$ with O double bond and CH_3 below

c.  $CH_3CCH_2CH_2CH_3$ with O double bond

13.43 Draw the structure of each of the following compounds:
  a. 3-Hydroxybutanal
  b. 2-Methylpentanal
  c. 4-Bromohexanal
  d. 3-Iodopentanal
  e. 2-Hydroxy-3-methylheptanal

13.44 Draw the structure of each of the following compounds:
  a. Propanone
  b. 2-Pentanone
  c. 3-Heptanone
  d. 2,4-Dimethyl-3-pentanone

Important Aldehydes and Ketones

13.45 Why is acetone a good solvent for many organic compounds?
13.46 List several uses for formaldehyde.
13.47 Ethanal is produced by the oxidation of ethanol. Where does this reaction occur in the body?
13.48 List several aldehydes and ketones used as food or fragrance chemicals.

Reactions Involving Aldehydes and Ketones

Foundations

13.49 Explain what is meant by oxidation in organic molecules and provide an example of an oxidation reaction involving an aldehyde.
13.50 Explain what is meant by reduction in organic reactions and provide an example of a reduction reaction involving an aldehyde or ketone.
13.51 Define the term *addition reaction*.
13.52 Provide an example of an addition reaction involving an aldehyde or ketone.
13.53 Write a general equation representing the oxidation of an aldehyde. What is the product of this reaction?
13.54 Write a general equation representing the reduction of an aldehyde. What is the product of this reaction?
13.55 Write a general equation representing the addition of one alcohol molecule to an aldehyde or a ketone.
13.56 Write a general equation representing the addition of two alcohol molecules to an aldehyde or a ketone.

Applications

13.57 Draw the structure of each of the following alcohols. Then draw and name the product you would expect to produce by the oxidation of each.
  a. 4-Methyl-2-heptanol
  b. 3,4-Dimethyl-1-pentanol
  c. 4-Ethyl-2-heptanol
  d. 5,7-Dichloro-3-heptanol
13.58 Draw the structure of each of the following alcohols. Then draw and name the product you would expect to produce by the oxidation of each.
  a. 1-Nonanol
  b. 4-Methyl-1-heptanol
  c. 4,6-Diethyl-3-methyl-3-octanol
  d. 5-Bromo-4-octanol
13.59 Draw the generalized equation for the oxidation of a primary alcohol.
13.60 Draw the generalized equation for the oxidation of a secondary alcohol.
13.61 Draw the structures of the reactants and products for each of the following reactions. Label each as an oxidation or a reduction reaction:
  a. Ethanal to ethanol
  b. Cyclohexanone to cyclohexanol
  c. 2-Propanol to propanone

13.62 An unknown has been determined to be one of the following three compounds:

$$\underset{\text{3-Pentanone}}{CH_3CH_2\overset{\overset{\displaystyle O}{\|}}{C}CH_2CH_3} \qquad \underset{\text{Pentanal}}{CH_3CH_2CH_2CH_2\overset{\overset{\displaystyle O}{\|}}{C}H}$$

$$\underset{\text{Pentane}}{CH_3CH_2CH_2CH_2CH_3}$$

The unknown is fairly soluble in water and produces a silver mirror when treated with the silver ammonia complex. A red precipitate appears when it is treated with the Benedict's reagent. Which of the compounds is the correct structure for the unknown? Explain your reasoning.

13.63 Write a balanced equation for the hydrogenation of each of the following aldehydes:
   a. 3-Methylpentanal   c. 2,3-Dimethylpentanal
   b. 2-Hydroxypropanal   d. 4-Chloropentanal

13.64 Write a balanced equation for the hydrogenation of each of the following ketones:
   a. 2-Methyl-3-pentanone c. 5-Nonanone
   b. 3-Hexanone   d. 7-Tetradecanone

13.65 Write a balanced equation for the hydrogenation of each of the following aldehydes:
   a. Butanal   c. 2-Methylpropanal
   b. 3-Methylpentanal

13.66 Write a balanced equation for the hydrogenation of each of the following aldehydes:
   a. 3-Methylbutanal   c. Propanal
   b. 4-Bromopentanal

13.67 Which of the following compounds would be expected to give a positive Tollens' test?
   a. 3-Pentanone   d. Cyclopentanol
   b. Cyclohexanone   e. 2,2-Dimethyl-1-pentanol
   c. 3-Methylbutanal   f. Acetaldehyde

13.68 Write an equation representing the reaction of glucose with the Benedict's reagent. How was this test used in medicine?

13.69 Write an equation for the addition of one ethanol molecule to each of the following aldehydes:

$$\textbf{a. } CH_3CH_2\overset{\overset{\displaystyle O}{\|}}{C}H \qquad \textbf{b. } CH_3\overset{\overset{\displaystyle O}{\|}}{C}H$$

13.70 Write an equation for the addition of one ethanol molecule to each of the following ketones:

$$\textbf{a. } CH_3\overset{\overset{\displaystyle O}{\|}}{C}CH_3 \qquad \textbf{b. } CH_3\overset{\overset{\displaystyle O}{\|}}{C}CH_2CH_2CH_3$$

13.71 Write an equation for the addition of two methanol molecules to each of the following aldehydes:

$$\textbf{a. } CH_3CH_2\overset{\overset{\displaystyle O}{\|}}{C}H \qquad \textbf{b. } CH_3\overset{\overset{\displaystyle O}{\|}}{C}H$$

13.72 Write an equation for the addition of two methanol molecules to each of the following ketones:

$$\textbf{a. } CH_3\overset{\overset{\displaystyle O}{\|}}{C}CH_3 \qquad \textbf{b. } CH_3\overset{\overset{\displaystyle O}{\|}}{C}CH_2CH_2CH_3$$

13.73 An aldehyde can be oxidized to produce a carboxylic acid. Draw the carboxylic acid that would be produced by the oxidation of each of the following aldehydes:
   a. Pentanal   c. Heptanal
   b. Hexanal   d. Octanal

13.74 An aldehyde can be oxidized to produce a carboxylic acid. Draw the carboxylic acid that would be produced by the oxidation of each of the following aldehydes:
   a. 3-Methylpentanal   c. 2,4-Diethylhexanal
   b. 2,3-Dichlorobutanal   d. 2-Methylpropanal

13.75 An alcohol can be oxidized to produce an aldehyde or a ketone. What aldehyde or ketone is produced by the oxidation of each of the following alcohols?
   a. Methanol   b. 1-Propanol

13.76 An alcohol can be oxidized to produce an aldehyde or a ketone. What aldehyde or ketone is produced by the oxidation of each of the following alcohols?
   a. 3-Pentanol   b. 2-Methyl-2-butanol

13.77 Indicate whether each of the following statements is true or false.
   a. Aldehydes and ketones can be oxidized to produce carboxylic acids.
   b. Oxidation of a primary alcohol produces an aldehyde.
   c. Oxidation of a tertiary alcohol produces a ketone.
   d. Alcohols can be produced by the oxidation of an aldehyde or ketone.

13.78 Indicate whether each of the following statements is true or false.
   a. Ketones, but not aldehydes, react in the Tollens' silver mirror test.
   b. Addition of one alcohol molecule to an aldehyde results in formation of a hemiacetal.
   c. The cyclic forms of monosaccharides are intramolecular hemiacetals.
   d. Disaccharides (sugars composed of two covalently joined monosaccharides) are acetals.

13.79 Draw the keto and enol forms of propanone.

13.80 Draw the keto and enol forms of butanone.

13.81 Draw the hemiacetal that results from the reaction of each of the following aldehydes or ketones with ethanol:

$$\textbf{a. } CH_3CH_2CH_2\overset{\overset{\displaystyle O}{\|}}{C}CH_3$$

b. 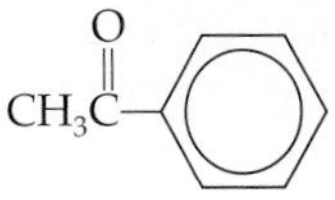 $CH_3\overset{\overset{\displaystyle O}{\|}}{C}-$

c.

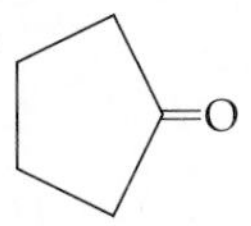

13.82 Identify each of the following compounds as a hemiacetal or acetal:

a.

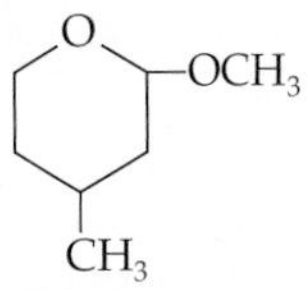

b.

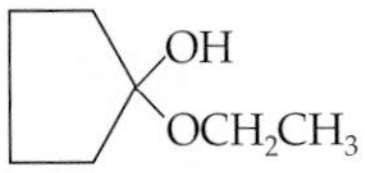

c. $CH_3\overset{\overset{\displaystyle OH}{|}}{\underset{\underset{\displaystyle OCH_2CH_3}{|}}{C}}CH_3$

d.

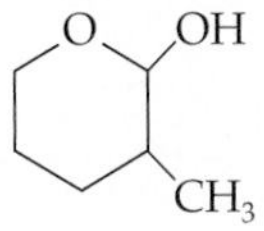

e. $CH_3\overset{\overset{\displaystyle OCH_3}{|}}{\underset{\underset{\displaystyle OCH_2CH_3}{|}}{C}}CH_3$

f. $CH_3CH=CH\overset{\overset{\displaystyle OCH_3}{|}}{\underset{\underset{\displaystyle OH}{|}}{C}}CH_3$

13.83 Complete the following synthesis by supplying the missing reactant(s), reagent(s), or product(s) indicated by the question marks:

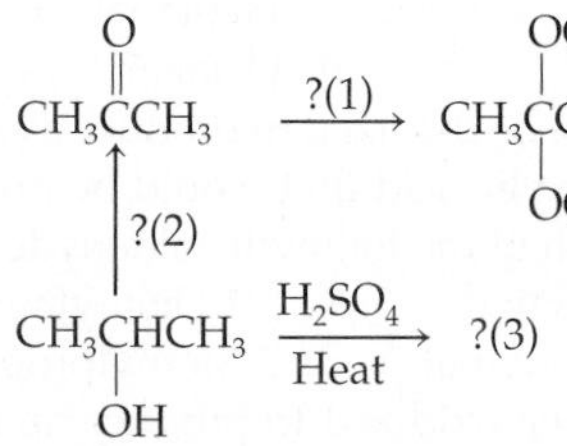

13.84 Which alcohol would you oxidize to produce each of the following compounds?

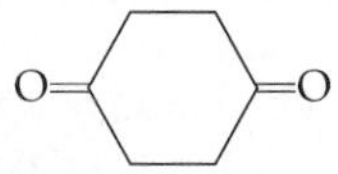

CHALLENGE PROBLEMS

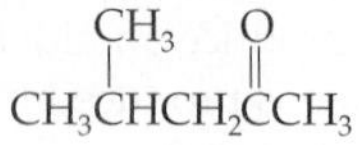

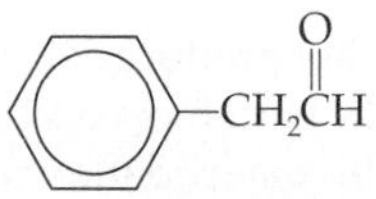

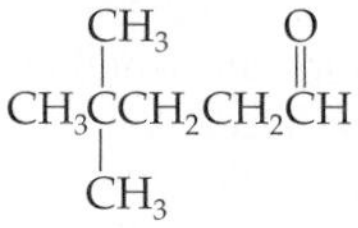

1. Review the material on the chemistry of vision found on the Web at www.mhhe.com/denniston, and, with respect to the isomers of retinal, discuss the changes in structure that occur as the nerve impulses that result in vision are produced. Provide condensed formulas of the retinal isomers that you discuss.

2. Classify the structure of β-D-fructose as a hemiacetal or acetal. Explain your choice.

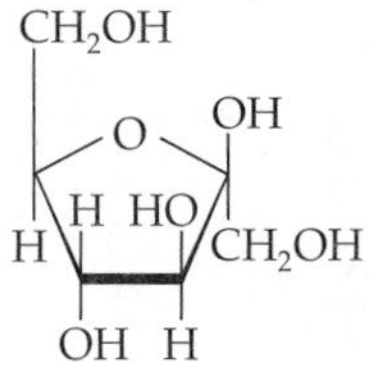

3. Design a synthesis for each of the following compounds, using any inorganic reagent of your choice and any hydrocarbon or alkyl halide of your choice:
   a. Octanal
   b. Cyclohexanone
   c. 2-Phenylethanoic acid

4. When alkenes react with ozone, O_3, the double bond is cleaved, and an aldehyde and/or a ketone is produced. The reaction, called *ozonolysis*, is shown in general as:

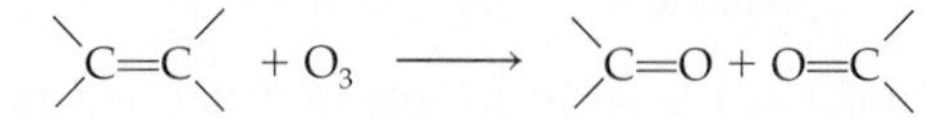

   Predict the ozonolysis products formed when each of the following alkenes is reacted with ozone:
   a. 1-Butene
   b. 2-Hexene
   c. *cis*-3,6-Dimethyl-3-heptene

5. Lactose is the major sugar found in mammalian milk. It is a disaccharide composed of the monosaccharides glucose and galactose:

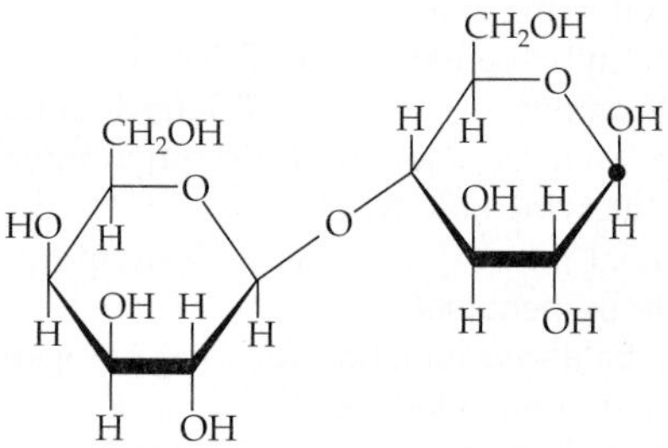

   Is lactose a hemiacetal or an acetal? Explain your choice.

6. The following are the keto and enol tautomers of phenol:

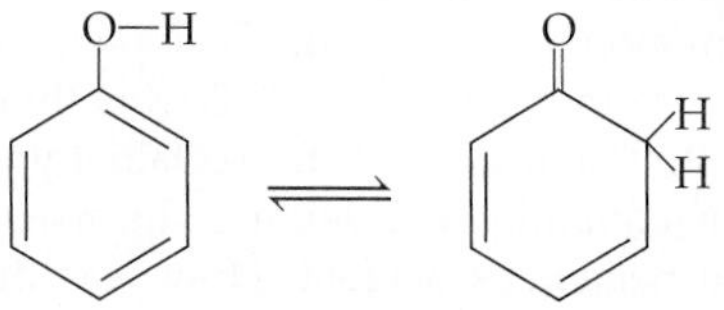

   Enol form   Keto form
   of phenol   of phenol

   We have seen that most simple aldehydes and ketones exist mainly in the keto form because it is more stable. Phenol is an exception, existing primarily in the enol form. Propose a hypothesis to explain this.

Carboxylic Acids and Carboxylic Acid Derivatives

LEARNING GOALS

1 Write structures and describe the physical properties of carboxylic acids.

2 Determine the common and IUPAC names of carboxylic acids.

3 Describe the biological, medical, or environmental significance of several carboxylic acids.

4 Write equations that show the synthesis of a carboxylic acid.

5 Write equations representing acid-base reactions of carboxylic acids.

6 Write equations representing the preparation of an ester.

7 Write structures and describe the physical properties of esters.

8 Determine the common and IUPAC names of esters.

9 Write equations representing the hydrolysis of an ester.

10 Define the term *saponification,* and describe how soap works in the emulsification of grease and oil.

11 Determine the common and IUPAC names of acid chlorides.

12 Determine the common and IUPAC names of acid anhydrides.

13 Write equations representing the synthesis of acid anhydrides.

14 Discuss the significance of thioesters and phosphoesters in biological systems.

The salad in this figure is part of a healthy diet. Using the Internet, make a list of the organic molecules found in these foods and their role in diet and health.

OUTLINE

INTRODUCTION

Carboxylic acids and their derivatives, the esters, are a common part of our daily lives. You may encounter many of them in the chef's salad you have for lunch. The two-carbon carboxylic acid acetic acid in aqueous solution, or vinegar, adds that tartness to the Italian dressing on your salad. Propionic acid, with three carbons, gives that tangy flavor to the Swiss cheese on your salad.

Long-chain carboxylic acids are called fatty acids, and they form esters when they react with an alcohol, such as glycerol. The olive oil in your salad dressing and the solid fat in the meat and cheese on your salad are examples of such triglycerides.

Perhaps you had a fruit salad for breakfast. If so, you may have enjoyed a number of sweet, fruity-tasting esters, such as 2-methylbutyl ethanoate (bananas) or methyl thiobutanoate (strawberries).

Some carboxylic acids are key elements in exercise physiology, and others are used in the treatment of diseases. For instance, lactic acid is a product of our metabolism that builds up in muscles and blood when we are exercising so strenuously that we cannot provide sufficient oxygen to working muscle. We will learn more about the lactate fermentation and its role in exercising muscle in Chapter 21. A four-carbon carboxylic acid, butanoic acid (butyric acid), is being used to treat sickle cell anemia. Although we don't yet understand how, this compound "turns on" the gene for fetal hemoglobin, protecting some patients from the harmful effects of this genetic disorder.

Carboxylic acids (Figure 14.1a) have the following general structure:

$$Ar = aromatic \qquad Ar-\underset{\underset{O}{\parallel}}{C}-OH \qquad R-\underset{\underset{O}{\parallel}}{C}-OH \qquad R = aliphatic$$

Aromatic carboxylic acid   Aliphatic carboxylic acid

They are characterized by the carboxyl group, shown in red, which may also be written in condensed form as —COOH or —CO$_2$H. The name *carboxylic acid* describes this family of compounds quite well. The term *carboxylic* is taken from the terms *carbonyl* and *hydroxyl*, the two structural units that make up the carboxyl group. The word *acid* in the name tells us one of the more important properties of these molecules: they dissociate in water to release protons. Thus, they are acids.

In fact, carboxylic acids are weak acids because they partially dissociate in water.

In this chapter, we also will study the esters (Figure 14.1b), which have the following general structures:

$$R-\overset{O}{\underset{\parallel}{C}}-O-R \qquad Ar-\overset{O}{\underset{\parallel}{C}}-O-Ar \qquad Ar-\overset{O}{\underset{\parallel}{C}}-O-R$$

Examples of aliphatic and aromatic esters

The group shown in red is called the acyl group. The acyl group is part of the functional group of the carboxylic acid derivatives, including the esters, acid chlorides, acid anhydrides, and amides.

Amides are discussed in Chapter 15.

Figure 14.1 Ball-and-stick models of (a) a carboxylic acid, propanoic acid, and (b) an ester, methyl ethanoate.

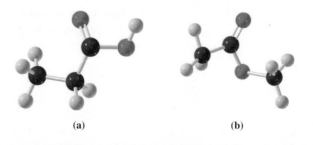

(a)   (b)

14.1 Carboxylic Acids

LEARNING GOAL

1 Write structures and describe the physical properties of carboxylic acids.

Structure and Physical Properties

Carboxylic acids are very polar compounds. This is due to the **carboxyl group,** which consists of two very polar functional groups, the *carbonyl group* and the *hydroxyl group.*

$$
\underset{\text{Carboxyl group}}{\overset{\displaystyle \overset{O}{\underset{\|}{}}}{-C-OH}}
\quad
\begin{array}{l}
\text{Carbonyl group} \\
\text{Hydroxyl group}
\end{array}
$$

Recall from Chapter 13 that the carbonyl group is polar because oxygen is more electronegative than carbon. This produces a dipole in which the oxygen carries a partial negative charge and the carbon carries a partial positive charge. This allows intermolecular dipole-dipole attractions:

$$
\overset{\delta^+}{>}C=\overset{\delta^-}{O}\cdots\overset{\delta^+}{>}C=\overset{\delta^-}{O}\cdots\overset{\delta^+}{>}C=\overset{\delta^-}{O}
$$

In Chapter 12, we saw that the hydroxyl group is very polar because the oxygen atom has a significantly higher electronegativity than the hydrogen atom. In addition, the oxygen atom has two unshared pairs of electrons. This results in a very polar bond in which the oxygen atom carries a partially negative charge and the hydrogen carries a partially positive charge. This structure allows hydrogen bonds, the attractive force between a hydrogen atom covalently bonded to a highly electronegative atom with unshared pairs of electrons, to form:

$$
\overset{\delta^-}{\underset{\text{:O}}{}}-\overset{\delta^+}{H}\cdots\cdots\overset{\overset{\delta^+}{H}}{\underset{\text{:O}}{}}\,\delta^-
$$

Thus, carboxylic acids can hydrogen bond to one another and to molecules of a polar solvent, such as water (Figure 14.2).

Because of the intermolecular hydrogen bonding and the strong dipole-dipole attractions, carboxylic acids boil at higher temperatures than aldehydes, ketones, or alcohols of comparable molecular mass. This can be seen in the comparison of the boiling points of an alkane, alcohol, ether, aldehyde, ketone, and carboxylic acid of comparable molar mass shown below:

$CH_3CH_2CH_2CH_3$	$CH_3-O-CH_2CH_3$	$CH_3CH_2CH_2-OH$
Butane (butane)	Methoxyethane (ethyl methyl ether)	1-Propanol (propyl alcohol)
M.M. = 58	M.M. = 60	M.M. = 60
b.p. −0.5°C	b.p. 7.0°C	b.p. 97.2°C

$\overset{\displaystyle O}{\underset{\|}{CH_3CH_2C}}-H$	$\overset{\displaystyle O}{\underset{\|}{CH_3C}}-CH_3$	$\overset{\displaystyle O}{\underset{\|}{CH_3C}}-OH$
Propanal (propionaldehyde)	Propanone (acetone)	Ethanoic acid (acetic acid)
M.M. = 58	M.M. = 58	M.M. = 60
b.p. 49°C	b.p. 56°C	b.p. 118°C

As with alcohols, the smaller carboxylic acids are soluble in water (Figure 14.2b). However, solubility falls off dramatically as the carbon content of the carboxylic

Formic acid causes the burning sensation at the site of an ant bite. What is the IUPAC name for formic acid? Why does treating the bite with baking soda reduce the burning sensation?

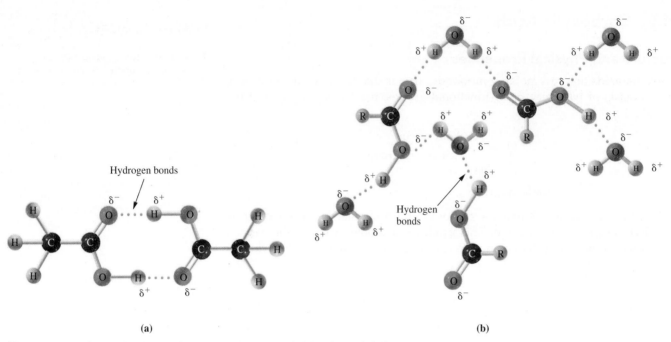

Figure 14.2 Hydrogen bonding (a) between carboxylic acid molecules and (b) between carboxylic acid molecules and water molecules.

acid increases because the molecules become more hydrocarbonlike and less polar. For example, acetic acid (the two-carbon carboxylic acid found in vinegar) is completely soluble in water, but hexadecanoic acid (a sixteen-carbon carboxylic acid found in palm oil) is insoluble in water.

The lower-molar-mass carboxylic acids have sharp, sour tastes and unpleasant aromas. Formic acid, HCOOH, is used as a chemical defense by ants and causes the burning sensation of the ant bite. Acetic acid, CH_3COOH, is found in vinegar; propionic acid, CH_3CH_2COOH, is responsible for the tangy flavor of Swiss cheese; and butyric acid, $CH_3CH_2CH_2COOH$, causes the stench associated with rancid butter and gas gangrene.

Fatty acid structure and properties are covered in Section 17.2.

The longer-chain carboxylic acids are generally called **fatty acids** and are important components of biological membranes and triglycerides, the major lipid storage form in the body.

Question 14.1   Assuming that each of the following pairs of molecules has the same carbon chain length, which member of each of the following pairs has the lower boiling point?
   a. a carboxylic acid or a ketone
   b. a ketone or an alcohol
   c. an alcohol or an alkane

Question 14.2   Assuming that each of the following pairs of molecules has the same carbon chain length, which member of each of the following pairs has the lower boiling point?
   a. an ether or an aldehyde
   b. an aldehyde or a carboxylic acid
   c. an ether or an alcohol

Question 14.3   Why would you predict that a carboxylic acid would be more polar and have a higher boiling point than an aldehyde of comparable molar mass?

Question 14.4   Why would you predict that a carboxylic acid would be more polar and have a higher boiling point than an alcohol of comparable molar mass?

Nomenclature

In the IUPAC Nomenclature System, carboxylic acids are named according to the following set of rules:

- Determine the parent compound, the longest continuous carbon chain bearing the carboxyl group.
- Number the chain so that the carboxyl carbon is carbon-1.
- Replace the *-e* ending of the parent alkane with the suffix *-oic acid.* If there are two carboxyl groups, the suffix *-dioic acid* is used and the *-e* ending of the parent alkane is *not* dropped.
- Name and number substituents in the usual way. When a hydroxyl group (—OH) is one of the substituents, it is numbered and the term *hydroxy* is used in the prefix.

The following examples illustrate the naming of carboxylic acids with one carboxyl group:

$$\underset{2}{\text{CH}_3}\underset{1}{\overset{\text{O}}{\overset{\|}{\text{C}}}}\text{—OH}$$

$$\underset{3}{\text{CH}_3}\underset{2}{\underset{\underset{\text{OH}}{|}}{\text{CH}}}\underset{1}{\overset{\text{O}}{\overset{\|}{\text{C}}}}\text{—OH}$$

$$\underset{4}{\text{CH}_3}\underset{3}{\underset{\underset{\text{CH}_3}{|}}{\text{CH}}}\underset{2}{\text{CH}_2}\underset{1}{\overset{\text{O}}{\overset{\|}{\text{C}}}}\text{—OH}$$

Ethanoic acid   2-Hydroxypropanoic acid   3-Methylbutanoic acid
(acetic acid)   (α-Hydroxypropionic acid)   (β-methylbutyric acid)

The following examples illustrate naming carboxylic acids with two carboxyl groups:

$$\overset{\text{O}}{\overset{\|}{\text{HO—C}}}\underset{1}{}\underset{2}{\text{—CH}_2}\underset{3}{\text{CH}_2}\underset{4}{\text{CH}_2}\underset{5}{\text{CH}_2}\underset{6}{\text{—}}\overset{\text{O}}{\overset{\|}{\text{C}}}\text{—OH}$$

$$\overset{\text{O}}{\overset{\|}{\text{HO—C}}}\underset{1}{}\underset{2}{\text{—CH}_2}\underset{3}{\text{—}}\overset{\text{O}}{\overset{\|}{\text{C}}}\text{—OH}$$

Hexanedioic acid   Propanedioic acid
(adipic acid)   (malonic acid)

EXAMPLE 14.1   **Use the IUPAC Nomenclature System to Name a Carboxylic Acid**

a. The following compound is one of the monomers from which a biodegradable plastic called Biopol is made (see Green Chemistry: Garbage Bags from Potato Peels? later in this chapter). Name this carboxylic acid using the IUPAC nomenclature system.

$$\underset{5}{\text{CH}_3}\underset{4}{\text{CH}_2}\underset{3}{\underset{\underset{\text{OH}}{|}}{\text{CH}}}\underset{2}{\text{CH}_2}\underset{1}{\overset{\text{O}}{\overset{\|}{\text{C}}}}\text{—OH}$$

Solution

Parent compound: pentane (becomes pentanoic acid)
Position of —COOH: carbon-1 (must be!)
Substituent: 3-hydroxy
Name: 3-Hydroxypentanoic acid

b. Name the following carboxylic acid:

$$\underset{8}{}\underset{7}{}\underset{6}{}\underset{5}{}\underset{4}{}\underset{3}{}\underset{2}{}\underset{1}{}\text{COOH}$$
$$\text{Br}\qquad\text{Br}\qquad\text{Br}$$

$$\underset{8}{\text{CH}_3}\underset{7}{\underset{\underset{\text{Br}}{|}}{\text{CH}}}\underset{6}{\text{CH}_2}\underset{5}{\underset{\underset{\text{Br}}{|}}{\text{CH}}}\underset{4}{\text{CH}_2}\underset{3}{\underset{\underset{\text{Br}}{|}}{\text{CH}}}\underset{2}{\text{CH}_2}\underset{1}{\text{—}}\overset{\text{O}}{\overset{\|}{\text{C}}}\text{—OH}$$

Continued…

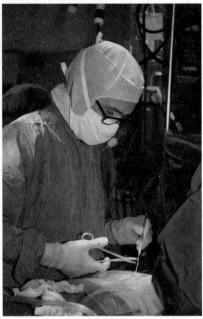

Polymers of lactic acid are used as bio-degradable sutures. What is the IUPAC name of lactic acid, shown below?

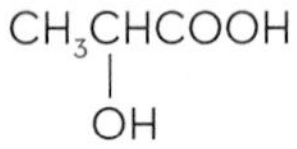

$$CH_3CHCOOH$$
$$|$$
$$OH$$

Solution

Parent compound: octane (becomes octanoic acid)
Position of —COOH: carbon-1 (must be!)
Substituents: 3,5,7-bromo
Name: 3,5,7-Tribromooctanoic acid

Practice Problem 14.1

Determine the IUPAC name for each of the following structures. Remember that —COOH is an alternative way to represent the carboxyl group.

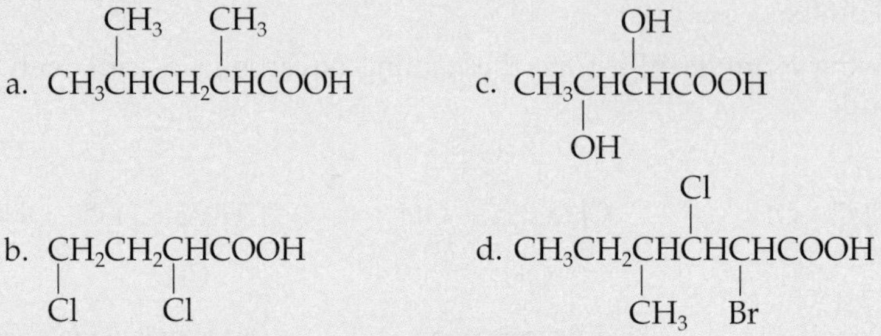

a. $CH_3CHCH_2CHCOOH$ (with CH_3 substituents)

b. $CH_2CH_2CHCOOH$ (with Cl, Cl substituents)

c. $CH_3CHCHCOOH$ (with OH, OH substituents)

d. $CH_3CH_2CHCHCHCOOH$ (with Cl, CH_3, Br substituents)

▶ For Further Practice: **Questions 14.32 and 14.35.**

The carboxylic acid derivatives of cycloalkanes are named by adding the suffix *carboxylic acid* to the name of the cycloalkane or substituted cycloalkane. The carboxyl group is always on carbon-1, and other substituents are named and numbered as usual.

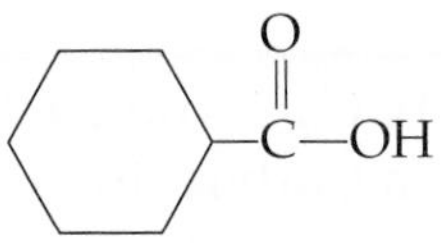

Cyclohexanecarboxylic acid

Question 14.5 Determine the IUPAC name for each of the following structures.

a. 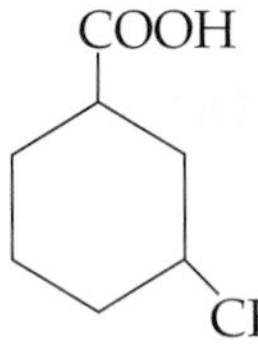 COOH (cyclohexane with CH_3)

b. 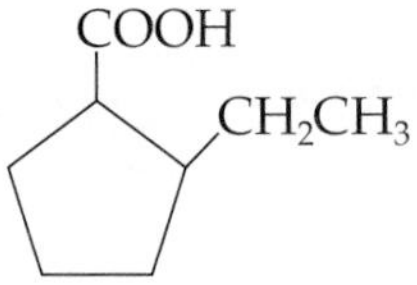 COOH (cyclopentane with CH_2CH_3)

Question 14.6 Write the structure for each of the following carboxylic acids.

a. 1,4-Cyclohexanedicarboxylic acid
b. 4-Hydroxycyclohexanecarboxylic acid

As we have seen so often, the use of common names, rather than systematic names, persists. Often, these names have evolved from the source of a given compound. This is certainly true of the carboxylic acids. Table 14.1 shows the IUPAC and common names of several carboxylic acids, as well as their sources and the Latin or Greek words that gave rise to the common names. Not only are the

LEARNING GOAL

2 Determine the common and IUPAC names of carboxylic acids.

TABLE 14.1 Names and Sources of Some Common Carboxylic Acids

Name	Structure	Source	Root
Formic acid (methanoic acid)	HCOOH	Ants	L: *formica*, ant
Acetic acid (ethanoic acid)	CH₃COOH	Vinegar	L: *acetum*, vinegar
Propionic acid (propanoic acid)	CH₃CH₂COOH	Swiss cheese	Gk: *protos*, first; *pion*, fat
Butyric acid (butanoic acid)	CH₃(CH₂)₂COOH	Rancid butter	L: *butyrum*, butter
Valeric acid (pentanoic acid)	CH₃(CH₂)₃COOH	Valerian root	
Caproic acid (hexanoic acid)	CH₃(CH₂)₄COOH	Goat fat	L: *caper*, goat
Caprylic acid (octanoic acid)	CH₃(CH₂)₆COOH	Goat fat	L: *caper*, goat
Capric acid (decanoic acid)	CH₃(CH₂)₈COOH	Goat fat	L: *caper*, goat
Palmitic acid (hexadecanoic acid)	CH₃(CH₂)₁₄COOH	Palm oil	
Stearic acid (octadecanoic acid)	CH₃(CH₂)₁₆COOH	Tallow (beef fat)	Gk: *stear*, tallow

Note: IUPAC names are shown in parentheses.

prefixes different than those used in the IUPAC system, the suffix is different as well. Common names end in *-ic acid* rather than *-oic acid*.

In the common system of nomenclature, substituted carboxylic acids are named as derivatives of the parent compound (see Table 14.1). Greek letters are used to indicate the position of the substituent. The carbon atom bonded to the carboxyl group is the α-carbon, the next is the β-carbon, and so on.

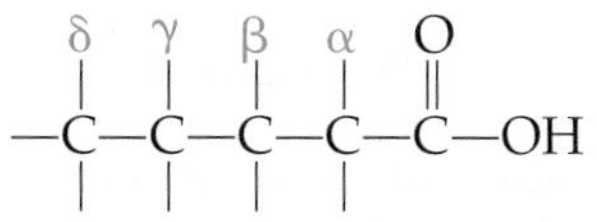

Some examples of common names are

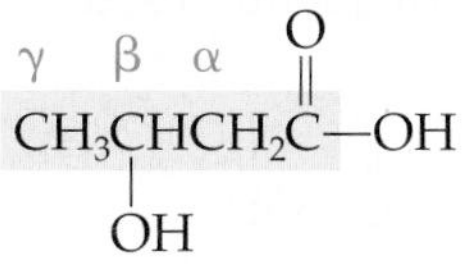

β-Hydroxybutyric acid 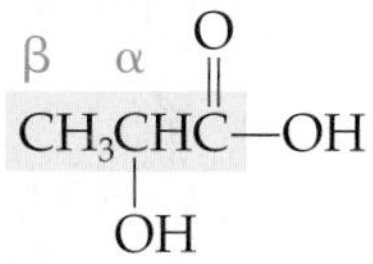   α-Hydroxypropionic acid

β-Hydroxybutyric acid is the other monomer used to make the biodegradable plastic Biopol (see Example 14.1).

EXAMPLE 14.2 | **Naming Carboxylic Acids Using the Common System of Nomenclature**

Write the common name for each of the following carboxylic acids.

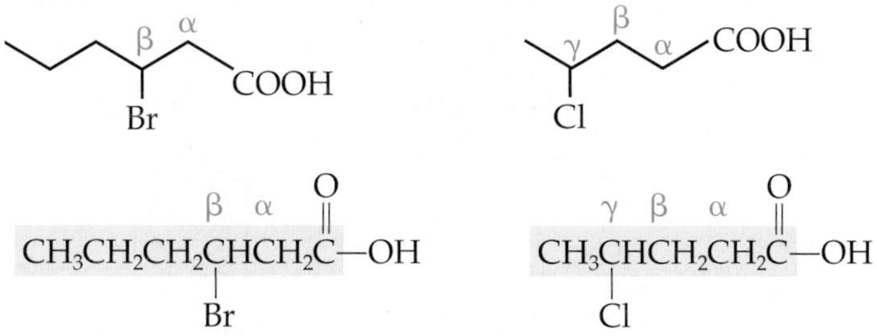

LEARNING GOAL

2 Determine the common and IUPAC names of carboxylic acids.

Continued…

Solution

Parent compound:	caproic acid	valeric acid
Substituents:	β-bromo	γ-chloro
Name:	β-Bromocaproic acid	γ-Chlorovaleric acid

Practice Problem 14.2

Provide the common name for each of the following molecules. Keep in mind that the carboxyl group can be represented as —COOH.

a. 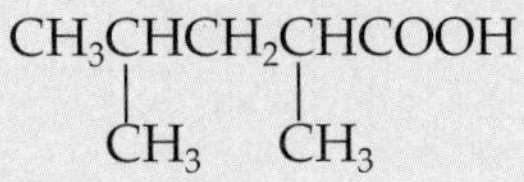 CH₃CHCH₂CHCOOH
        | |
       CH₃   CH₃

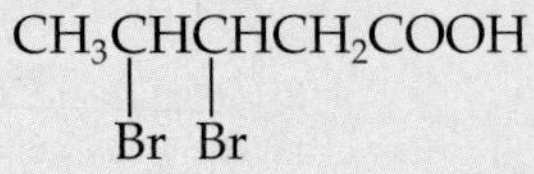

b. 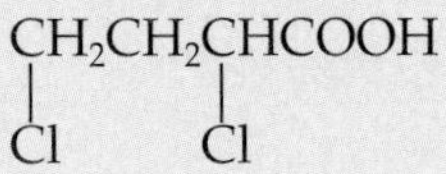 CH₂CH₂CHCOOH
        | |
        Cl   Cl

c. CH₃CHCHCH₂COOH
          | |
          Br Br

d. 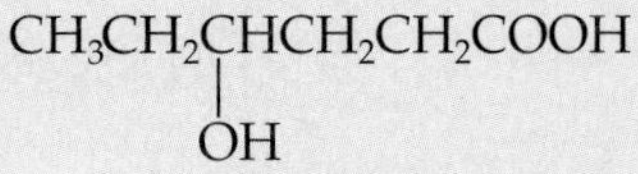 CH₃CH₂CHCH₂CH₂COOH
                |
                OH

▶ For Further Practice: **Questions 14.31, 14.37, and 14.38.**

Benzoic acid is the simplest aromatic carboxylic acid.

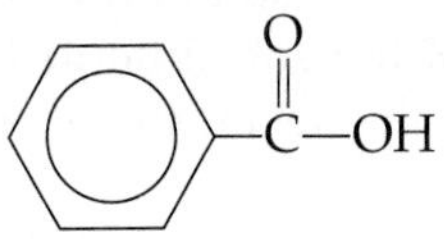

$$\underset{\text{Benzoic acid}}{\bigcirc\!\!-\!\!\overset{\displaystyle O}{\underset{\displaystyle \|}{C}}\!\!-\!\!OH}$$

Nomenclature of aromatic compounds is described in Section 11.6.

In many cases, the aromatic carboxylic acids are named, in either system, as derivatives of benzoic acid. Generally, the *-oic acid* suffix is used in the IUPAC system and the *-ic acid* suffix is used in the common system and is attached to the appropriate prefix. However, "common names" of substituted benzoic acids (for example, toluic acid and phthalic acid) are frequently used.

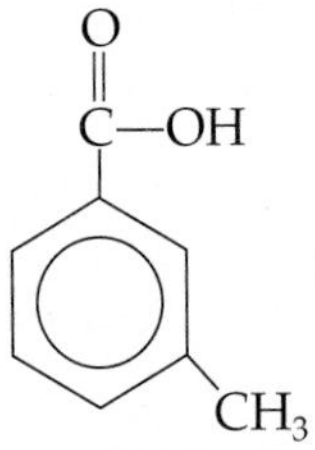

The phenyl group is benzene with one hydrogen removed.

Phenyl group

m-Toluic acid 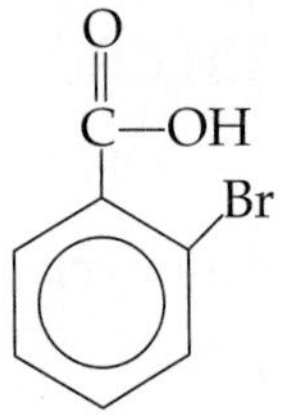   *o*-Bromobenzoic acid   *m*-Iodobenzoic acid   Phthalic acid

The benzyl group is toluene with one hydrogen removed from the methyl group:

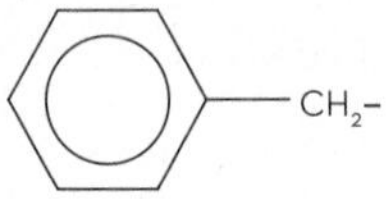

— CH₂–

Benzyl group

Often, the phenyl group is treated as a substituent, and the name is derived from the appropriate alkanoic acid parent chain. For example:

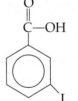

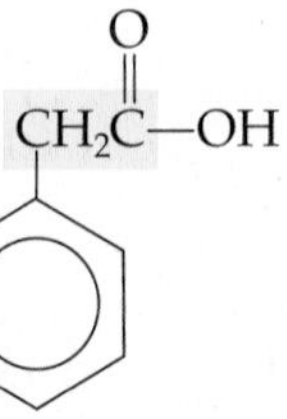

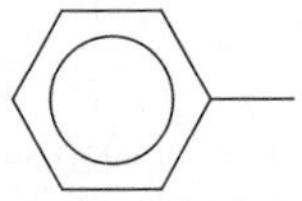

2-Phenylethanoic acid
(α-phenylacetic acid)

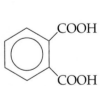

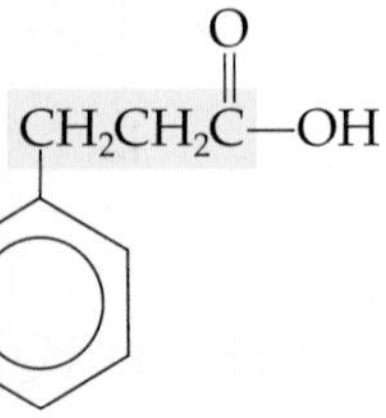

3-Phenylpropanoic acid
(β-phenylpropionic acid)

Chemistry at the Crime Scene

Carboxylic Acids and the Body Farm

Dr. Arpad Vass of Oak Ridge National Laboratory explains his unusual vocation in the following way, "Each year there are about 90,000 homicides in this country and for every one, you need to know when the person died. Our research can help answer the where and when, and that should help law enforcement officials solve crimes." Time and location of death are key pieces of information in a criminal investigation. Whether a suspect is convicted or goes free may rest on an accurate estimate of the time since death (TSD). Dr. Vass and graduate student Jennifer Love study decomposing bodies on a research plot in Tennessee. Although its official name is University of Tennessee Anthropological Research Facility, it has been called the "Body Farm" since Patricia Cornwell published a novel of that name in 1994.

Vass and Love are trying to develop an instrument that can sample the air around a corpse to determine the concentrations of certain carboxylic acids, also referred to as volatile fatty acids. What they envision is a "tricorder"-like device that will sample the air for valeric acid, propionic acid, and the straight and branched types of butyric acid. Because the ratio of these carboxylic acids changes in a predictable way following death, measurement of their relative concentrations, from air or from the soil under the corpse, could provide an accurate TSD.

The research is done by taking daily measurements of the air around decomposing bodies at the "Body Farm." At death, the proteins and lipids of the body begin to break down. This decomposition produces, among other substances, the carboxylic acids being studied. The bodies may be placed in various environments so that the researchers can study the effects of different temperature and moisture levels on the process of decomposition. The more that is learned about the amounts and types of organic compounds produced during decomposition under different conditions, the more precise the TSD determinations will be. In fact, Vass considers that these processes are a chemical "clock" and is working to refine the accuracy of that clock so that, eventually, TSD may be measured accurately in hours, rather than days.

For Further Understanding

▶ Draw the structures of the carboxylic acids described in this article and write their IUPAC names.

▶ The bones of a murdered 9-year-old boy were found lying on his father's property. Immediately the father was suspected of having killed the child. A sample of the soil under the boy's bones was sent to Dr. Vass for analysis. No volatile fatty acids were found in the soil sample. What can you conclude from these data?

EXAMPLE 14.3 | Naming Aromatic Carboxylic Acids

a. Name the following aromatic carboxylic acid.

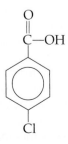

LEARNING GOAL

2 Determine the common and IUPAC names of carboxylic acids.

Continued…

Solution

It is simplest to name the compound as a derivative of benzoic acid. The substituent, Cl, is attached to carbon-4 of the benzene ring. This compound is 4-chlorobenzoic acid or *p*-chlorobenzoic acid.

b. Name the following aromatic carboxylic acid.

Solution

This compound is most easily named by treating the phenyl group as a substituent. The phenyl group is bonded to carbon-4 (or the γ-carbon, in the common system of nomenclature). The parent compound is pentanoic acid (valeric acid in the common system). Hence, the name of this compound is 4-phenylpentanoic acid or γ-phenylvaleric acid.

Practice Problem 14.3

Draw structures for each of the following compounds.

a. *o*-Toluic acid
b. 2,4,6-Tribromobenzoic acid
c. 2,2,2-Triphenylethanoic acid
d. *p*-Toluic acid
e. 3-Phenylhexanoic acid
f. 3-Phenylcyclohexanecarboxylic acid

▶ For Further Practice: **Questions 14.39 and 14.40.**

Some Important Carboxylic Acids

LEARNING GOAL

3 Describe the biological, medical, or environmental significance of several carboxylic acids.

As Table 14.1 shows, many carboxylic acids occur in nature. The stinging sensation of an ant bite is caused by methanoic (formic) acid, and ethanoic (acetic) acid provides the acidic zip to vinegars. Propanoic (propionic) acid is the product of bacterial fermentation of milk products and is most notable as a tangy component in the characteristic flavor of Swiss cheese.

Several of the larger carboxylic acids have foul odors. For instance, butanoic (butyric) acid is the odor associated with rancid butter and is produced by the bacteria that cause gas gangrene, contributing to the characteristic smell of the necrotic tissue. Pentanoic (valeric) acid is associated with the valerian plant, which has long been known to have an aroma alternately described as over-ripe cheese or a wet dog. Nonetheless, extracts of valerian have been used for thousands of years as a natural sedative. Hexanoic (caproic) acid was first isolated from goats and, fittingly, is described as smelling like a goat. Heptanoic (enanthic) acid is also foul smelling and is associated with the odor of rancid oil.

To learn more about the fragrances and flavors of esters, see A Human Perspective: The Chemistry of Flavor and Fragrance later in this chapter.

These foul-smelling carboxylic acids have a far more pleasant potential, however. When carboxylic acids react with alcohols, the products are esters, which contribute to the lovely fragrance and flavor of many fruits.

Octanoic (caprylic) acid has an interesting function in the chemistry of human appetite. The hormone ghrelin, produced in the stomach, is sometimes called the "hunger hormone" because it stimulates the hypothalamus of the brain to signal that the body is hungry. However, the hormone alone does not have this effect. Ghrelin must be covalently bonded to a molecule of octanoic acid in order to have the hunger-stimulating effect on the hypothalamus.

A triglyceride is a molecule of glycerol (Section 12.3) bonded to three fatty acid molecules by esterification (Sections 14.2 and 17.3).

Fatty acids are long-chain monocarboxylic acids and can be isolated from a variety of sources including palm oil, coconut oil, butter, milk, lard, and animal

Green Chemistry

Garbage Bags from Potato Peels?

One problem facing society is its enormous accumulation of trash. To try to control the mountains of garbage that we produce, many institutions and towns practice recycling of aluminum, paper, and plastics. One problem that remains, however, is the plastic trash bag. We stuff the trash bag full of biodegradable garbage and bury it in a landfill; but soil bacteria can't break down the plastic to get to the biodegradable materials inside. Imagine a twenty-fourth century archeologist excavating one of these monuments to our society!

The good news is that laboratory research and products of bacterial metabolism are providing new materials that have the properties of plastics, but are readily biodegradable. For instance, sheets of plastic can be made by making polymers of lactic acid, which is a natural carboxylic acid produced by fermentation of sugars, particularly in milk and working muscle. Because many common soil bacteria can break down polylactic acid (PLA), trash bags made from this polymer would be quickly broken down in landfill soil.

Making plastic from lactic acid requires a huge supply of this carboxylic acid. As it turns out, we can produce an enormous quantity of lactic acid from garbage. When French fries are produced, nearly half of the potato is wasted. That amounts to about ten billion pounds (lb) of potato waste each year. When cheese is made, the curds are separated from the whey, and several billion gallons (gal) of whey are poured down the drain each year. Potato waste and whey can easily be broken down to produce glucose, which, in turn, can be converted into lactic acid used to make biodegradable plastics. PLA plastics have been available since the early 1990s and have been used successfully for sutures, medical implants, and drug delivery systems.

Other researchers have experimented with heteropolymers, polymers composed of two or more different monomers. One such polymer with useful properties is a heteropolymer of β-hydroxybutyric acid and β-hydroxyvaleric acid, which has been given the name Biopol.

β-Hydroxybutyric acid   β-Hydroxyvaleric acid

Biopol—a heteropolymer

Biopol has properties that make it commercially useful, and it is completely broken down into carbon dioxide and water by microorganisms in the soil. Thus, it is completely biodegradable. Because bacteria produced a low yield of Biopol, scientists produced transgenic plants in hopes that they would produce high yields of the polymer. This approach also encountered problems.

For the time being, biodegradable plastics cannot outcompete their nonbiodegradable counterparts. Future research and development will be required to reduce the cost of commercial production and fulfill the promise of an "environmentally friendly" garbage bag.

For Further Understanding

▶ Why are biodegradable plastics useful as sutures?
▶ Two hydrogen atoms are lost in each reaction that adds a carboxylic acid to the polymers described here. What type of chemical reaction is this?

Lactic acid   Polylactic acid (PLA)

---

fat. These fatty acids, in the form of triglycerides, are the major energy storage form in mammals and many plants. When the fatty acids in the triglyceride are saturated, the result is solid fat, as in animal fats. When the fatty acids are unsaturated, the result is a liquid, as in olive oil and canola oil.

Several common dicarboxylic acids are shown in Table 14.2. Oxalic acid is a dicarboxylic acid found in spinach and rhubarb. Human kidney stones are often formed from the calcium salt of oxalic acid. In fact, it is toxic in high concentrations, and foods with high levels of oxalic acid must be boiled before being eaten. While oxalic acid is used in industry as a bleaching agent and spot remover, the potassium salt is used in clinical laboratories to prevent blood samples from coagulating.

TABLE 14.2 Common Dicarboxylic Acids

Common Name	IUPAC Name	Condensed Formula	Line Formula
Oxalic acid	Ethanedioic acid	HOOCCOOH	
Malonic acid	Propanedioic acid	HOOCCH$_2$COOH	
Succinic acid	Butanedioic acid	HOOC(CH$_2$)$_2$COOH	
Glutaric acid	Pentanedioic acid	HOOC(CH$_2$)$_3$COOH	
Adipic acid	Hexanedioic acid	HOOC(CH$_2$)$_4$COOH	

Malonic acid is used in the synthesis of barbiturates. Succinic acid is one of the intermediates in the citric acid cycle, a metabolic pathway involved in the breakdown of carbohydrates, lipids, and proteins to harvest energy for cellular functions. The name comes from the Latin word for amber (L. *succinum*) because it was first isolated from crushed amber. Glutaric acid is used in the production of several types of condensation polymers, including polyester, polyols, and polyamides. It is useful because it has an odd number of carbons in the chain, which reduces the elasticity of the polymer. Adipic acid (hexanedioic acid) gives tartness to soft drinks and helps retard food spoilage. However, the greatest demand for adipic acid (2.5 billion kilograms (kg) per year) is for the synthesis of nylon and polyurethane, and of plasticizers that are particularly useful in the manufacture of polyvinyl chloride (PVC).

More complex carboxylic acids are found in a variety of foods. For example, citric acid is found in citrus fruits and is often used to give the sharp taste to sour candies. It is also added to foods as a preservative and antioxidant.

Bacteria in milk produce lactic acid as a product of fermentation of sugars. Lactic acid contributes a tangy flavor to yogurt and buttermilk. It is also used as a food preservative to lower the pH to a level that retards microbial growth that causes food spoilage. Lactic acid is produced in muscle cells when an individual is exercising strenuously. If the level of lactic acid in the muscle and bloodstream becomes high enough, the muscle can't continue to work.

Tartaric acid is used in baking powder because it will undergo a reaction with carbonates in the dough, producing CO_2 that will cause the bread or cake to rise. It has also been used as a laxative. Malic acid gives the sour taste to green apples. Since the amount of malic acid decreases as a fruit ripens, the fruit becomes sweeter and less tart as it ripens.

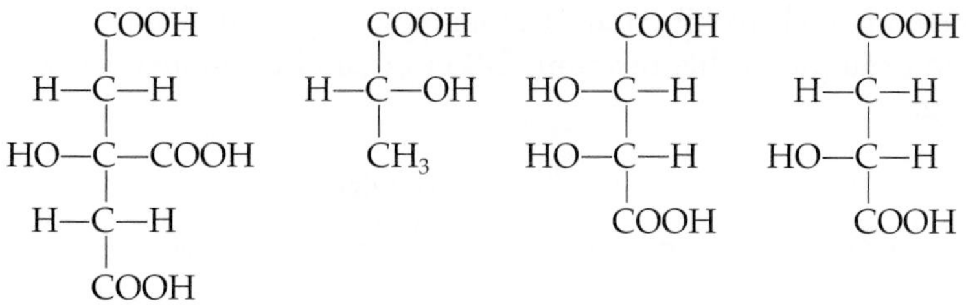

$$
\begin{array}{cccc}
\text{COOH} & \text{COOH} & \text{COOH} & \text{COOH} \\
| & | & | & | \\
\text{H—C—H} & \text{H—C—OH} & \text{HO—C—H} & \text{H—C—H} \\
| & | & | & | \\
\text{HO—C—COOH} & \text{CH}_3 & \text{HO—C—H} & \text{HO—C—H} \\
| & & | & | \\
\text{H—C—H} & & \text{COOH} & \text{COOH} \\
| & & & \\
\text{COOH} & & & \\
\end{array}
$$

Citric acid   Lactic acid   Tartaric acid   Malic acid

Several aromatic carboxylic acids are also of medical interest. The sodium salt of benzoic acid (sodium benzoate) is used as a preservative in soft drinks, pickles, jellies, and many other foods and some cosmetics. It is of value as a preservative because it is colorless, odorless, and tasteless, and will kill bacteria at a concentration of only 0.1%.

Salicylic acid is used as a disinfectant and, in fact, is superior to phenol. It is also used in ointments to remove corns or warts because it causes the top layer of the skin to flake off, leaving the underlying living skin undamaged.

Acetylsalicylic acid is aspirin. As early as the fifth century BC, the revered physician Hippocrates described a bitter extract from willow bark that could reduce fevers and relieve pain. Ancient texts from the Middle East reveal that Egyptian and Sumerian physicians appreciated the medicinal value of willow bark, and Native Americans used it to treat headache, fever, and chills, as well as sore muscles. In 1828, Henri Leroux isolated crystals of the compound that came to be called salicin. Nearly 70 years later, in 1897, chemists at Bayer and Company chemically added an acetyl group to salicin, synthesizing acetylsalicylic acid, a derivative that did not produce the severe gastrointestinal side effects caused by salicin. Thus, aspirin became the first synthetic drug, launching the pharmaceutical industry. Today, aspirin is recommended in low daily doses by the American Heart Association as a preventive measure against heart attacks and strokes caused by blood clots.

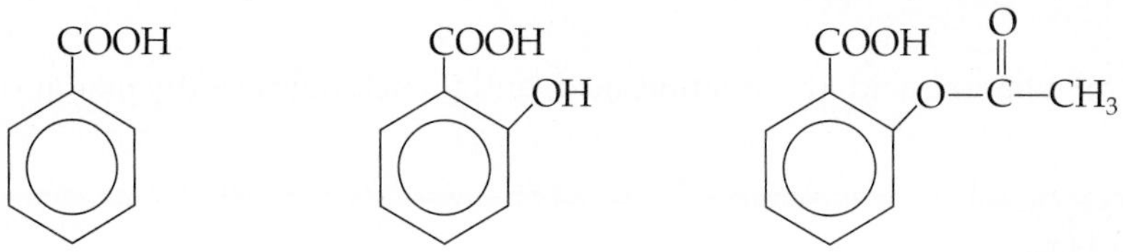

Benzoic acid   Salicylic acid   Acetylsalicylic acid

As we will see in the next section, terephthalic acid is primarily used to synthesize polyethylene terephthalate (PETE). Because this polymer is so useful, in excess of 30 million tons (t) of terephthalic acid is needed each year.

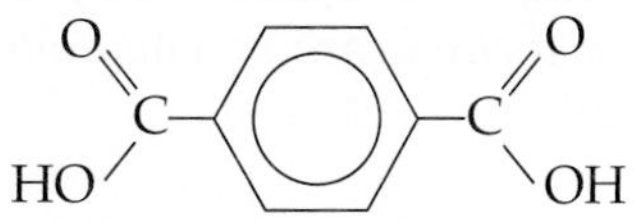

Terephthalic acid

Reactions Involving Carboxylic Acids

Preparation of Carboxylic Acids

Simple carboxylic acids can be made by **oxidation** of the appropriate primary alcohol or aldehyde. A variety of oxidizing agents, including oxygen, can be used. The general reaction is represented in the following equation:

$$
\underset{\text{Primary alcohol}}{R-CH_2OH} \xrightarrow{[O]} \underset{\text{Aldehyde}}{R-\overset{\displaystyle O}{\overset{\|}{C}}-H} \xrightarrow{[O]} \underset{\text{Carboxylic acid}}{R-\overset{\displaystyle O}{\overset{\|}{C}}-OH}
$$

LEARNING GOAL

4 Write equations that show the synthesis of a carboxylic acid.

Oxidation reactions involving aldehydes and primary alcohols were discussed in Sections 12.4 and 13.4.

The commercial production of ethanoic acid (acetic acid), found in vinegar, is an example of this reaction. Either ethanol or ethanal can be used, as show here:

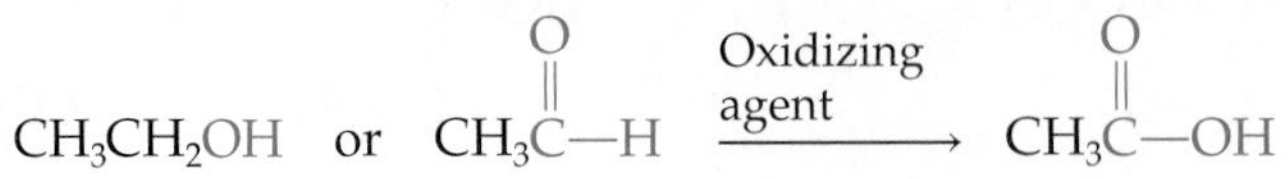

$$CH_3CH_2OH \quad or \quad CH_3\overset{\overset{O}{\|}}{C}-H \xrightarrow{\text{Oxidizing agent}} CH_3\overset{\overset{O}{\|}}{C}-OH$$

Ethanol   Ethanal   Ethanoic acid

EXAMPLE 14.4   **Writing Equations for the Oxidation of a Primary Alcohol to a Carboxylic Acid**

Write an equation showing the oxidation of 1-propanol to propanoic acid.

LEARNING GOAL

4 Write equations that show the synthesis of a carboxylic acid.

Solution

Recall that the alcohol will first be oxidized to the aldehyde, which will then be oxidized to the carboxylic acid. Show these reactions in two steps, and indicate the need for an oxidizing agent by adding [O] above the arrow.

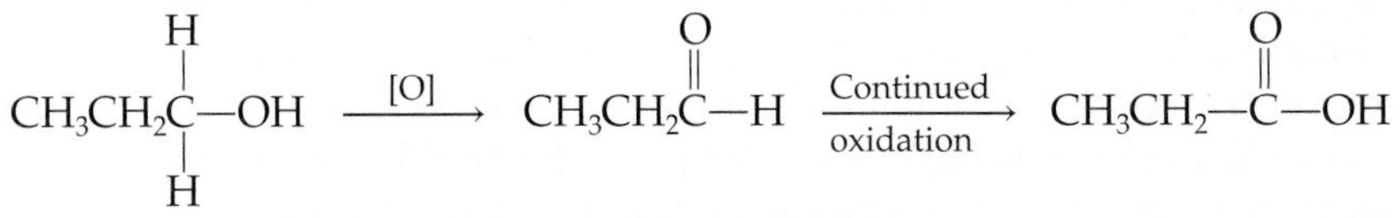

$$CH_3CH_2\overset{\overset{\displaystyle H}{|}}{\underset{\underset{\displaystyle H}{|}}{C}}-OH \xrightarrow{[O]} CH_3CH_2\overset{\overset{O}{\|}}{C}-H \xrightarrow[\text{oxidation}]{\text{Continued}} CH_3CH_2-\overset{\overset{O}{\|}}{C}-OH$$

1-Propanol   Propanal   Propanoic acid
(propyl alcohol)   (propionaldehyde)   (propionic acid)

Practice Problem 14.4

Write equations showing the synthesis of (a) ethanoic acid, (b) butanoic acid, and (c) octanoic acid by oxidation of the corresponding primary alcohol.

▶ For Further Practice: **Questions 14.49 and 14.50.**

LEARNING GOAL

5 Write equations representing acid-base reactions of carboxylic acids.

Acid-Base Reactions

The carboxylic acids behave as acids because they are proton donors. They are weak acids that dissociate to form a carboxylate anion and a hydronium ion, as shown in the following equation:

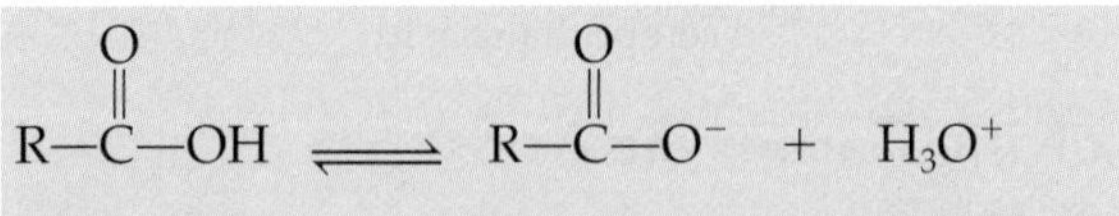

$$R-\overset{\overset{O}{\|}}{C}-OH \rightleftharpoons R-\overset{\overset{O}{\|}}{C}-O^- \;+\; H_3O^+$$

Carboxylic   Carboxylate   Hydronium
acid   anion   ion

The properties of weak acids are described in Sections 8.1 and 8.2.

Carboxylic acids are weak acids because they dissociate only slightly in solution. The majority of the acid remains in solution in the undissociated form. Typically, less than 5% of the acid is ionized (approximately five carboxylate ions to every ninety-five carboxylic acid molecules).

When strong bases are added to a carboxylic acid, neutralization occurs. The acid protons are removed by the OH^- to form water and the carboxylate ion. The position of the equilibrium is shifted to the right, owing to removal of H^+. This is an example of LeChatelier's principle.

LeChatelier's principle is described in Section 7.4.

$$R-\overset{O}{\overset{\|}{C}}-OH + NaOH \longrightarrow R-\overset{O}{\overset{\|}{C}}-O^-Na^+ + H_2O$$

Carboxylic acid | Strong base | Carboxylic acid salt | Water

The following examples show the neutralization of ethanoic (acetic) acid and benzoic acid in solutions of the strong base NaOH.

The carboxylate anion and the cation of the base form the carboxylic acid salt.

$$CH_3\overset{O}{\overset{\|}{C}}-OH + NaOH \longrightarrow CH_3\overset{O}{\overset{\|}{C}}-O^-Na^+ + H-O-H$$

Ethanoic acid (acetic acid) | Sodium hydroxide (strong base) | Sodium ethanoate (sodium acetate) | Water

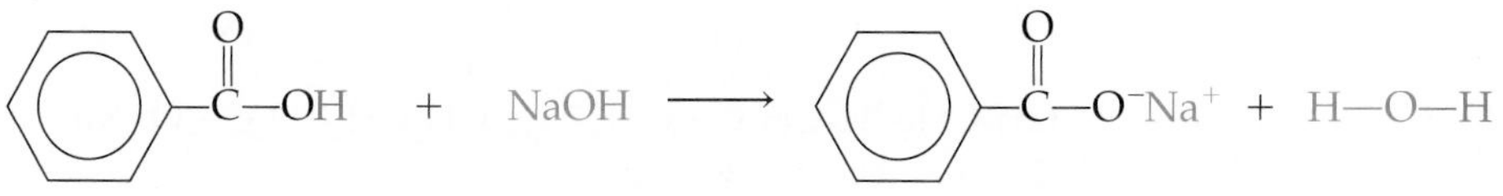

Benzoic acid | Sodium hydroxide (strong base) | Sodium benzoate | Water

Sodium benzoate is commonly used as a food preservative.

The product of the neutralization of a carboxylic acid with a strong base is a *carboxylic acid salt.* Carboxylic acid salts are ionic substances. As a result, they are very soluble in water. The long-chain carboxylic acid salts (fatty acid salts) are called *soaps.*

Soaps are made by a process called saponification, which is the base-catalyzed hydrolysis of an ester. This is described in detail in Section 14.2.

EXAMPLE 14.5 Writing an Equation to Show the Neutralization of a Carboxylic Acid by a Strong Base

LEARNING GOAL

5 Write equations representing acid-base reactions of carboxylic acids.

Write an equation showing the neutralization of propanoic acid by sodium hydroxide.

Solution

The protons of the acid are removed by the OH^- of the base. This produces water. The cation of the base—in this case, sodium ion—forms the salt of the carboxylic acid.

$$CH_3CH_2-\overset{O}{\overset{\|}{C}}-OH + NaOH \longrightarrow CH_3CH_2-\overset{O}{\overset{\|}{C}}-O^-Na^+ + H_2O$$

Propanoic acid | Sodium hydroxide | Sodium salt of propanoic acid | Water

Practice Problem 14.5

Write the formula of the organic product obtained through each of the following reactions.

a. $CH_3CH_2COOH + KOH \longrightarrow ?$

b. $CH_3CH_2CH_2COOH + Ba(OH)_2 \longrightarrow ?$

c. $CH_3CH_2CH_2CH_2CH_2COOH + KOH \longrightarrow ?$

d. Benzoic acid + sodium hydroxide $\longrightarrow$?

▶ For Further Practice: **Questions 14.52a, b, and c, 14.55, and 14.56.**

In the IUPAC system, carboxylic acid salts are named according to the following set of rules:

- Determine the parent compound, the longest continuous carbon chain bearing the carboxylate anion.
- Number the chain so that the carboxylate carbon is carbon-1.
- Replace the –ic suffix of the carboxylic acid with –ate.
- Name and number substituents in the normal manner.
- This name is then preceded by the name of the appropriate cation.

The rules are precisely the same in the common system, except that the common name of the carboxylic acid is used. In this case, the –ic suffix of the common name is replaced with –ate. Substituents are numbered using Greek letters; the α-carbon is the carbon bonded to the carboxylate group. The following examples demonstrate the two systems of nomenclature:

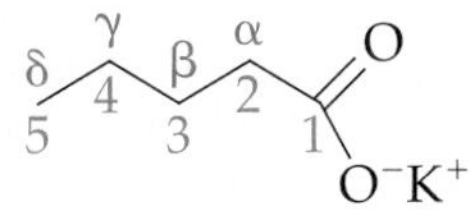

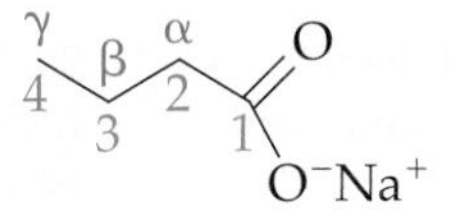

$$CH_3CH_2CH_2CH_2C-O^-K^+$$
$$CH_3CH_2CH_2C-O^-Na^+$$

	IUPAC name:	Common name:
	Potassium pentanoate	Potassium valerate
	Sodium butanoate	Sodium butyrate

EXAMPLE 14.6 Naming the Salt of a Carboxylic Acid

Write the common and IUPAC names of the salt produced in the reaction shown in Example 14.5.

Solution

$$CH_3CH_2-\overset{O}{\overset{\|}{C}}-O^-Na^+$$

IUPAC name of the parent carboxylic acid:	Propanoic acid
Replace the -ic acid ending with -ate:	Propanoate
Name of the cation of the base:	Sodium
Name of the carboxylic acid salt:	Sodium propanoate
Common name of the parent carboxylic acid:	Propionic acid
Replace the -ic acid ending with -ate:	Propionate
Name of the cation of the base:	Sodium
Name of the carboxylic acid salt:	Sodium propionate

Practice Problem 14.6

Name the products of the reactions in Practice Problem 14.5.

▶ For Further Practice: **Questions 14.57 and 14.58.**

Esterification

Carboxylic acids react with alcohols to form esters and water according to the general reaction:

$$R^1-\overset{\overset{\textstyle O}{\|}}{C}-OH + R^2OH \xrightleftharpoons{Acid} R^1-\overset{\overset{\textstyle O}{\|}}{C}-OR^2 + H_2O$$

Carboxylic   Alcohol   Ester   Water
acid

The details of these reactions will be examined in Section 14.2.

LEARNING GOAL

6 Write equations representing the preparation of an ester.

14.2 Esters

Structure and Physical Properties

Esters have the following general structure:

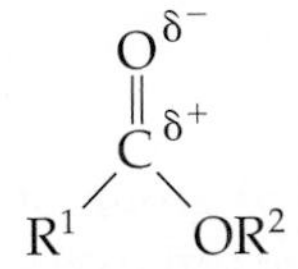

$$\underset{R^1 \qquad OR^2}{\overset{\overset{\textstyle O^{\delta-}}{\|}}{\underset{}{C^{\delta+}}}}$$

Structure of an ester

LEARNING GOAL

7 Write structures and describe the physical properties of esters.

As noted in the structure, the carbonyl group of the ester is polar and can participate in dipole-dipole attractions. However the carbonyl group is flanked by hydrocarbon chains (R^1 and R^2 in the preceding structure). Because of these hydrophobic chains, the polarity of esters is comparable to that of aldehydes and ketones. In fact, esters boil at approximately the same temperatures as aldehydes or ketones of comparable molar mass.

See A Human Perspective: The Chemistry of Flavor and Fragrance later in this chapter.

Like aldehydes and ketones, esters can form hydrogen bonds with water molecules. As a result, smaller esters are somewhat soluble in water.

Esters have pleasant aromas. Many are found in natural foodstuffs; banana oil (3-methylbutyl ethanoate; common name, isoamyl acetate), pineapples (ethyl butanoate; common name, ethyl butyrate), and raspberries (isobutyl methanoate; common name, isobutyl formate) are but a few examples.

Nomenclature

Esters are **carboxylic acid derivatives**, organic compounds derived from carboxylic acids. As shown in the general reaction equation above, they are formed from the reaction of a carboxylic acid with an alcohol, and both of these reactants are reflected in the naming of the ester.

LEARNING GOAL

8 Determine the common and IUPAC names of esters.

The simplest method to name an ester begins by recognizing the carboxylic acid and alcohol from which it is composed. This can be done by inspection of the structure:

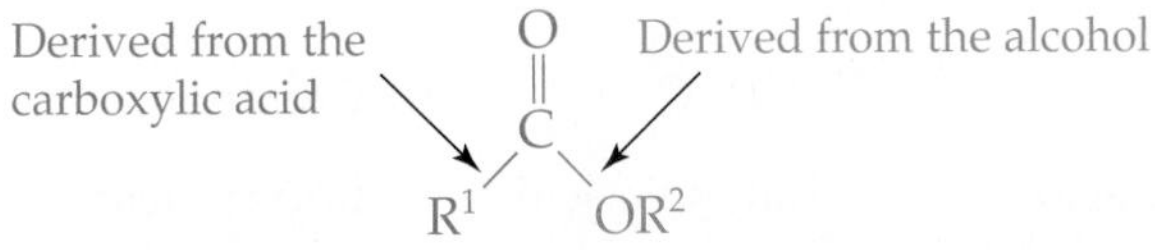

Derived from the   Derived from the alcohol
carboxylic acid

$$\underset{R^1 \qquad OR^2}{\overset{\overset{\textstyle O}{\|}}{C}}$$

The first part of the name of an ester is the alkyl (or aryl) group of the alcohol.

- Identify the name of the parent alcohol, and write the name of the corresponding alkyl or aryl group. For example, in both the IUPAC and common systems, the alkyl group of methanol (methyl alcohol) is the methyl group.

The second part of the name is derived from the carboxylic acid.

- Identify the parent carboxylic acid, and replace the *–ic* suffix with *–ate*. Thus, in the IUPAC system, ethanoic acid becomes ethanoate. In the common system, this carboxylic acid is acetic acid. The name becomes acetate.

In the following equation, we see the reaction in which ethanoic acid (acetic acid) reacts with methanol (methyl alcohol) to produce methyl ethanoate (methyl acetate).

$$\underset{\substack{\text{Ethan}oic\ acid \\ \text{(acetic acid)}}}{\text{CH}_3\overset{\overset{\text{O}}{\|}}{\text{C}}\text{—OH}} + \underset{\substack{\text{Methanol} \\ \text{(methyl alcohol)}}}{\text{CH}_3\text{OH}} \xrightleftharpoons{\text{H}^+,\ \text{heat}} \underset{\substack{\text{Methyl ethan}oate \\ \text{(methyl acetate)}}}{\text{CH}_3\overset{\overset{\text{O}}{\|}}{\text{C}}\text{—OCH}_3} + \text{H}_2\text{O}$$

Naming esters is analogous to naming the salts of carboxylic acids. Consider the following comparison:

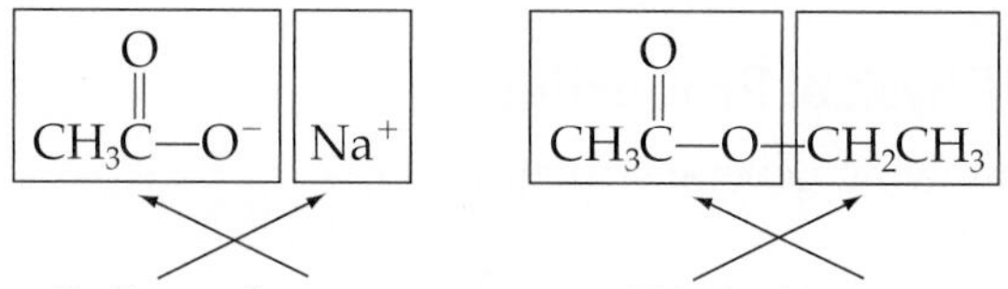

$$\underset{\substack{\text{Sodium ethanoate} \\ \text{(sodium acetate)}}}{\boxed{\text{CH}_3\overset{\overset{\text{O}}{\|}}{\text{C}}\text{—O}^-}\ \boxed{\text{Na}^+}} \qquad \underset{\substack{\text{Ethyl ethanoate} \\ \text{(ethyl acetate)}}}{\boxed{\text{CH}_3\overset{\overset{\text{O}}{\|}}{\text{C}}\text{—O}}\boxed{\text{CH}_2\text{CH}_3}}$$

As shown in this example, the alkyl group of the alcohol, rather than Na^+, has displaced the acidic hydrogen of the carboxylic acid.

EXAMPLE 14.7   **Naming Esters Using the IUPAC and Common Nomenclature Systems**

LEARNING GOAL

8 Determine the common and IUPAC names of esters.

a. The molecule shown below contributes to the flavor of pineapple. Write the IUPAC and common names for this ester.

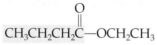

Solution

	IUPAC	Common
IUPAC and common names of parent carboxylic acid:	butanoic acid	butyric acid
Replace the *-ic acid* ending of the carboxylic acid with *-ate*:	butanoate	butyrate
Name of the alkyl portion of the alcohol:	ethyl	ethyl
IUPAC and common names of the ester:	Ethyl butanoate	Ethyl butyrate

b. The molecule shown below is associated with the characteristic flavor of apricots. Write the common and IUPAC names of this ester.

$$\text{CH}_3\text{CH}_2\text{CH}_2\overset{\overset{\text{O}}{\|}}{\text{C}}\text{—OCH}_2\text{CH}_2\text{CH}_2\text{CH}_2\text{CH}_3$$

Solution

	IUPAC	Common
IUPAC and common names of parent carboxylic acid:	butanoic acid	butyric acid
Replace the *-ic acid* ending of the carboxylic acid with *-ate*:	butanoate	butyrate
Name of the alkyl portion of the alcohol:	pentyl	pentyl
IUPAC and common names of the ester:	Pentyl butanoate	Pentyl butyrate

Practice Problem 14.7

Name each of the following esters using both the IUPAC and common nomenclature systems.

a.
$$CH_3CH_2CH_2\overset{\overset{\displaystyle O}{\|}}{C}-OCH_2CH_2CH_3$$

c.
$$CH_3\overset{\overset{\displaystyle O}{\|}}{C}-OCH_2CH_2CH_3$$

b.
$$CH_3CH_2CH_2\overset{\overset{\displaystyle O}{\|}}{C}-OCH_2CH_3$$

d.
$$CH_3CH_2\overset{\overset{\displaystyle O}{\|}}{C}-OCH_2CH_2CH_2CH_3$$

▶ For Further Practice: **Questions 14.63, 14.65, and 14.66.**

Reactions Involving Esters

Preparation of Esters

The conversion of a carboxylic acid to an ester requires heat and is catalyzed by a trace of acid (H^+). When esters are prepared directly from a carboxylic acid and an alcohol, a water molecule is lost, as in the reaction:

LEARNING GOAL

6 Write equations representing the preparation of an ester.

$$R^1\overset{\overset{\displaystyle O}{\|}}{-C}-OH + R^2OH \underset{}{\overset{H^+, \text{ heat}}{\rightleftharpoons}} R^1\overset{\overset{\displaystyle O}{\|}}{-C}-OR^2 + H_2O$$

Carboxylic acid   Alcohol   Ester   Water

$$CH_3CH_2\overset{\overset{\displaystyle O}{\|}}{C}-OH + CH_3OH \overset{H^+, \text{ heat}}{\rightleftharpoons} CH_3CH_2\overset{\overset{\displaystyle O}{\|}}{C}-OCH_3 + H-O-H$$

Propanoic acid   Methanol   Methyl propanoate
(propionic acid)   (methyl alcohol)   (methyl propionate)

Esterification is a *condensation* reaction, so called because a water molecule is removed during the reaction.

Esterification is reversible. The direction of the reaction is determined by the conditions chosen. Excess alcohol favors ester formation. The carboxylic acid is favored when excess water is present.

EXAMPLE 14.8 | **Writing Equations Representing Esterification Reactions**

LEARNING GOAL

6 Write equations representing the preparation of an ester.

Write an equation showing the esterification reactions that would produce ethyl butanoate and propyl ethanoate.

Solution

The name, ethyl butanoate, tells us that the alcohol used in the reaction is ethanol and the carboxylic acid is butanoic acid. We must remember that a trace of acid and heat are required for the reaction and that the reaction is reversible. With this information, we can write the following equation representing the reaction:

$$CH_3CH_2CH_2\overset{\overset{\displaystyle O}{\|}}{C}-OH + CH_3CH_2OH \overset{H^+, \text{ heat}}{\rightleftharpoons} CH_3CH_2CH_2\overset{\overset{\displaystyle O}{\|}}{C}-OCH_2CH_3 + H_2O$$

Butanoic acid   Ethanol   Ethyl butanoate
(butyric acid)   (ethyl butyrate)

Continued…

Similarly, the name propyl ethanoate reveals that the alcohol used in this reaction is 1-propanol and the carboxylic acid must be ethanoic acid. Knowing that we must indicate that the reaction is reversible and that heat and a trace of acid are required, we can write the following equation:

$$CH_3\overset{\overset{\displaystyle O}{\|}}{C}{-}OH + CH_3CH_2CH_2OH \underset{\longleftarrow}{\overset{H^+, \text{ heat}}{\longrightarrow}} CH_3\overset{\overset{\displaystyle O}{\|}}{C}{-}OCH_2CH_2CH_3 + H_2O$$

Ethanoic acid   1-Propanol   Propyl ethanoate
(acetic acid)   (propyl alcohol)   (propyl acetate)

Practice Problem 14.8

Write an equation showing the esterification reactions that would produce (a) butyl ethanoate and (b) ethyl propanoate.

▶ For Further Practice: **Questions 14.73 and 14.74.**

| **EXAMPLE 14.9** | **Designing the Synthesis of an Ester** |

LEARNING GOAL

Design the synthesis of ethyl propanoate from organic alcohols.

6 Write equations representing the preparation of an ester.

Solution

The ease with which alcohols are oxidized to aldehydes, ketones, or carboxylic acids (depending on the alcohol you start with and the conditions you employ), coupled with the ready availability of alcohols, provides the pathway necessary to many successful synthetic transformations. For example, let's develop a method for synthesizing ethyl propanoate, using an oxidizing agent and limiting yourself to organic alcohols that contain three or fewer carbon atoms:

$$CH_3CH_2\overset{\overset{\displaystyle O}{\|}}{C}{-}O{-}CH_2CH_3$$

Ethyl propanoate
(ethyl propionate)

Ethyl propanoate can be made from propanoic acid and ethanol:

$$CH_3CH_2\overset{\overset{\displaystyle O}{\|}}{C}{-}OH + CH_3CH_2OH \underset{\longleftarrow}{\overset{H^+, \text{ heat}}{\longrightarrow}} CH_3CH_2\overset{\overset{\displaystyle O}{\|}}{C}{-}O{-}CH_2CH_3 + H_2O$$

Propanoic acid   Ethanol   Ethyl propanoate
(propionic acid)   (ethyl alcohol)   (ethyl propionate)

Ethanol is a two-carbon alcohol that is an allowed starting material, but propanoic acid is not. Can we now make propanoic acid from an alcohol of three or fewer carbons? Yes!

$$CH_3CH_2CH_2OH \overset{[O]}{\longrightarrow} CH_3CH_2\overset{\overset{\displaystyle O}{\|}}{C}{-}H \overset{[O]}{\longrightarrow} CH_3CH_2\overset{\overset{\displaystyle O}{\|}}{C}{-}OH$$

1-Propanol   Propanal   Propanoic acid
(propyl alcohol)   (propionaldehyde)   (propionic acid)

1-Propanol is a three-carbon alcohol, an allowed starting material. The synthesis is now complete. By beginning with ethanol and 1-propanol, ethyl propanoate can be synthesized easily by the reaction shown in the equation above.

Practice Problem 14.9

Design the synthesis of (a) methyl butanoate and (b) propyl methanoate from organic alcohols.

▶ For Further Practice: **Questions 14.75 and 14.76.**

The Chemistry of Flavor and Fragrance

Carboxylic acids are often foul smelling. For instance, butyric acid is one of the worst smelling compounds imaginable—the smell of rancid butter.

$$CH_3CH_2CH_2C\overset{O}{\overset{\|}{—}}OH$$

Butanoic acid
(butyric acid)

Butyric acid is also a product of fermentation reactions carried out by *Clostridium perfringens.* This organism is the most common cause of gas gangrene. Butyric acid contributes to the notable foul smell accompanying this infection.

By forming esters of butyric acid, a chemist can generate compounds with pleasant smells. Ethyl butyrate is the essence of pineapple oil.

Volatile esters are often pleasant in both aroma and flavor. Natural fruit flavors are complex mixtures of many esters and other organic compounds. Chemists can isolate these mixtures and identify the chemical components. With this information, they are able to synthesize artificial fruit flavors, using just a few of the esters found in the natural fruit. As a result, the artificial flavors rarely have the full-bodied flavor of nature's original blend.

For Further Understanding

▶ Draw the structure of methyl salicylate, which is found in oil of wintergreen.

▶ Write an equation for the synthesis of each of the esters shown in this Perspective.

Pineapple

$$CH_3CH_2CH_2C\overset{O}{\overset{\|}{—}}OCH_2CH_3$$

Ethyl butanoate
(ethyl butyrate)

Raspberries

$$H—C\overset{O}{\overset{\|}{—}}OCH_2CHCH_3 \atop \qquad\qquad CH_3$$

Isobutyl methanoate
(isobutyl formate)

Bananas

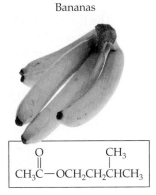

$$CH_3C\overset{O}{\overset{\|}{—}}OCH_2CH_2CHCH_3 \atop \qquad\qquad\qquad CH_3$$

3-Methylbutyl ethanoate
(isoamyl acetate)

Apples

$$CH_3CH_2CH_2C\overset{O}{\overset{\|}{—}}OCH_3$$

Methyl butanoate
(methyl butyrate)

Oranges

$$CH_3C\overset{O}{\overset{\|}{—}}OCH_2CH_2CH_2CH_2CH_2CH_2CH_2CH_3$$

Octyl ethanoate
(octyl acetate)

Apricots

$$CH_3CH_2CH_2C\overset{O}{\overset{\|}{—}}OCH_2CH_2CH_2CH_2CH_3$$

Pentyl butanoate
(pentyl butyrate)

Strawberries

$$CH_3CH_2CH_2C\overset{O}{\overset{\|}{—}}SCH_3$$

Methyl thiobutanoate
(methyl thiobutyrate)
(a thioester in which sulfur
replaces oxygen)

LEARNING GOAL

9 Write equations representing the hydrolysis of an ester.

Hydrolysis of Esters

Hydrolysis is a reaction in which a bond is broken by the addition of a water molecule. Esters undergo hydrolysis reactions in water, as shown in the general reaction:

$$\underset{\text{Ester}}{R^1-\overset{\displaystyle O}{\overset{\|}{C}}-OR^2} + \underset{\text{Water}}{H_2O} \underset{}{\overset{H^+,\ heat}{\rightleftharpoons}} \underset{\substack{\text{Carboxylic} \\ \text{acid}}}{R^1-\overset{\displaystyle O}{\overset{\|}{C}}-OH} + \underset{\text{Alcohol}}{R^2OH}$$

This reaction requires heat. A small amount of acid (H$^+$) may be added to catalyze the reaction, as in the following example:

$$\underset{\substack{\text{Propyl propanoate} \\ \text{(propyl propionate)}}}{CH_3CH_2\overset{\displaystyle O}{\overset{\|}{C}}-OCH_2CH_2CH_3} + H_2O \overset{H^+,\ heat}{\rightleftharpoons} \underset{\substack{\text{Propanoic acid} \\ \text{(propionic acid)}}}{CH_3CH_2\overset{\displaystyle O}{\overset{\|}{C}}-OH} + \underset{\substack{\text{1-Propanol} \\ \text{(propyl alcohol)}}}{CH_3CH_2CH_2OH}$$

LEARNING GOAL

10 Define the term *saponification*, and describe how soap works in the emulsification of grease and oil.

The hydrolysis of an ester can also be catalyzed by the presence of a strong base. The base-catalyzed hydrolysis of an ester is called **saponification.** This reaction is represented in the following general equation:

$$\underset{\text{Ester}}{R^1-\overset{\displaystyle O}{\overset{\|}{C}}-OR^2} + \underset{\text{Water}}{H_2O} \overset{OH^-,\ heat}{\longrightarrow} \underset{\substack{\text{Carboxylic} \\ \text{acid anion}}}{R^1-\overset{\displaystyle O}{\overset{\|}{C}}-O^-} + \underset{\text{Alcohol}}{R^2OH}$$

As we saw in Section 14.1, carboxylic acids are neutralized in the presence of a strong base (NaOH or KOH) to produce the carboxylic acid salt. Thus, under the conditions of base-catalyzed hydrolysis of an ester, the carboxylic acid cannot exist. Instead, the reaction yields the salt of the carboxylic acid having the cation of the basic catalyst.

$$\underset{\substack{\text{Butyl ethanoate} \\ \text{(butyl acetate)}}}{CH_3\overset{\displaystyle O}{\overset{\|}{C}}-OCH_2CH_2CH_2CH_3} \overset{NaOH,\ heat}{\longrightarrow} \underset{\substack{\text{Sodium ethanoate} \\ \text{(sodium acetate)}}}{CH_3\overset{\displaystyle O}{\overset{\|}{C}}-O^-Na^+} + \underset{\substack{\text{1-Butanol} \\ \text{(butyl alcohol)}}}{CH_3CH_2CH_2CH_2OH}$$

If a strong acid, such as HCl, is added to the reaction mixture, the strong base is neutralized and the carboxylic acid is formed.

$$\underset{\substack{\text{Sodium ethanoate} \\ \text{(sodium acetate)}}}{CH_3\overset{\displaystyle O}{\overset{\|}{C}}-O^-Na^+} + HCl \longrightarrow \underset{\substack{\text{Ethanoic acid} \\ \text{(acetic acid)}}}{CH_3\overset{\displaystyle O}{\overset{\|}{C}}-OH} + NaCl$$

Question 14.7 Complete each of the following reactions by drawing the structure of the missing product(s).

a. $\underset{O}{\overset{\parallel}{CH_3C}}-OCH_2CH_2CH_3 + H_2O \overset{H^+, heat}{\rightleftharpoons} ?$

b. $CH_3CH_2CH_2CH_2CH_2\overset{O}{\overset{\parallel}{C}}-OCH_2CH_2CH_3 + H_2O \overset{KOH, heat}{\longrightarrow} ?$

c. $CH_3CH_2CH_2CH_2\overset{O}{\overset{\parallel}{C}}-OCH_3 + H_2O \overset{NaOH, heat}{\longrightarrow} ?$

d. $CH_3CH_2CH_2CH_2CH_2\overset{O}{\overset{\parallel}{C}}-OCHCH_2CH_2CH_3 + H_2O \overset{H^+, heat}{\rightleftharpoons} ?$
$\qquad\qquad\qquad\qquad\qquad\overset{|}{CH_3}$

Question 14.8 Use the IUPAC Nomenclature System to name each of the products in Question 14.7.

Fats and oils are triesters of the alcohol glycerol. When they are hydrolyzed by saponification, the products are **soaps,** which are the salts of long-chain carboxylic acids (fatty acid salts). According to Roman legend, soap was discovered by washerwomen following a heavy rain on Mons Sapo ("Mount Soap"). An important sacrificial altar was located on the mountain. The rain mixed with the remains of previous animal sacrifices—wood ash and animal fat—at the base of the altar. Thus, the three substances required to make soap accidentally came together— water, fat, and alkali (potassium carbonate and potassium hydroxide, called *potash,* leached from the wood ash). The soap mixture flowed down the mountain and into the Tiber River, where the washerwomen quickly realized its value.

We still use the old Roman recipe to make soap from water, a strong base, and natural fats and oils obtained from animals or plants. The carbon chain length of the fatty acid salts governs the solubility of a soap. The lower-molar-mass carboxylic acid salts (up to twelve carbons) have greater solubility in water and give a lather containing large bubbles. The higher-molar-mass carboxylic acid salts (fourteen to twenty carbons) are much less soluble in water and produce a lather with fine bubbles. The nature of the cation also affects the solubility of the soap. In general, the potassium salts of carboxylic acids are more soluble in water than the sodium salts. The synthesis of a soap is shown in Figure 14.3.

The role of soap in the removal of soil and grease is best understood by considering the functional groups in soap molecules and understanding how they interact with oil and water. A soap is a long-chain fatty acid salt. The long hydrocarbon

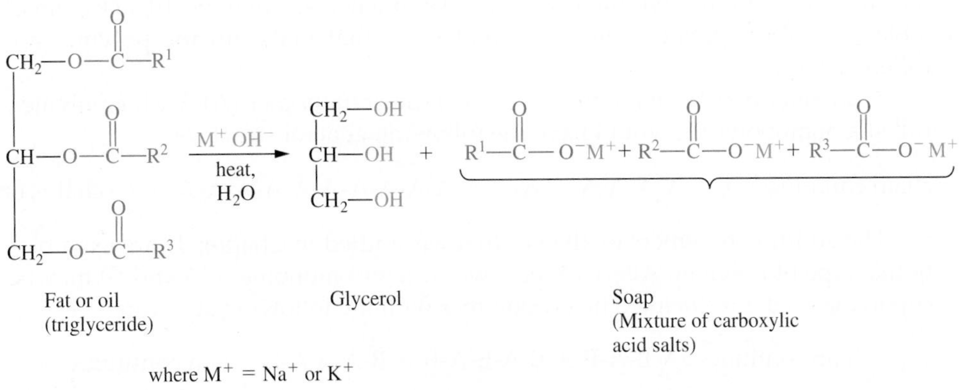

Triesters of glycerol are more commonly referred to as triglycerides. We know them as solid fats, generally from animal sources, and liquid oils, typically from plants. We will study triglycerides in detail in Section 17.3.

Figure 14.3 Saponification is the base-catalyzed hydrolysis of a glycerol triester.

chain, or "tail," resembles an alkane and dissolves in nonpolar compounds due to London dispersion forces. However, it is repelled by water and is described as *hydrophobic*, which means "water-fearing."

The carboxylate end of the molecule is highly polar and has the ability to hydrogen bond to water molecules. This region of the soap is described as *hydrophilic*, which means "water-loving."

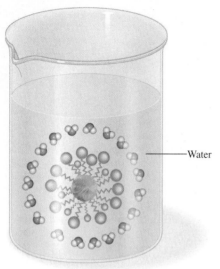

—Water

Figure 14.4 Simplified view of the action of a soap. The wiggly lines represent the long, continuous carbon chains of each soap molecule. Particles of oil and grease are surrounded by soap molecules to form a micelle.

Fatty acid tail

Hydrophobic region: repelled by water and dissolves in nonpolar substances such as oil

Carboxylate group

Hydrophilic region: dissolves in water

When soap is dissolved in water, the carboxylate end dissolves, but the long fatty acid tails are repelled by the water. When grease or oil is present in the mixture, the fatty acid tails dissolve in these nonpolar substances. The result is the formation of tiny spheres in which the carboxylate ends of the soap form the outside of the sphere and interact with water. The fatty acid tails, along with the dissolved grease or oil, are embedded within the center of the spheres. These tiny spheres are called *micelles*, as represented here in three dimensions and in Figure 14.4:

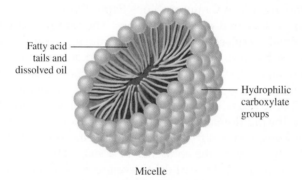

Fatty acid tails and dissolved oil

Hydrophilic carboxylate groups

Micelle

Micelles repel one another because they are surrounded on the surface by the negatively charged carboxylate ions. Mechanical action (for example, scrubbing or tumbling in a washing machine) causes oil or grease to be surrounded by soap molecules and broken into small droplets so that relatively small micelles are formed. These small micelles are then washed away.

Condensation Polymers

As we saw in Chapter 11, *polymers* are macromolecules, very large molecules. They result from the combination of many smaller molecules, usually in a repeating pattern, to give molecules whose molar mass may be 10,000 grams/mole (g/mol) or greater. The small molecules that make up the polymer are called *monomers*.

A polymer may be made from a single type of monomer (A). Such a polymer, called a homopolymer, would have the following general structure:

chain continues~A-A-A-A-A-A-A-A-A-A-A-A-A-A-A-A-A-A-A-A-A-A~chain continues

The addition polymers of alkenes that we studied in Chapter 11 are examples of this type of polymer. Alternatively, two different monomers (A and B) may be copolymerized, producing a heteropolymer with the following structure:

chain continues~A-B-A-B-A-B-A-B-A-B-A-B-A-B-A-B~chain continues

A Human Perspective

Detergents

Although soaps have served us for centuries as excellent cleansers, they do have some drawbacks. One of these is that they are salts of weak acids and thus may be converted into free fatty acids in the presence of weak acid:

$$CH_3(CH_2)_{16}\overset{\overset{\displaystyle O}{\|}}{C}\text{-O}^-\,Na^+ \ + \ HCl \longrightarrow$$

Soap

$$CH_3(CH_2)_{16}\overset{\overset{\displaystyle O}{\|}}{C}\text{-OH} \ + \ Na^+ \ + \ Cl^-$$

Free fatty acid

Free fatty acids are much less water-soluble than the sodium or potassium salts, and they tend to precipitate as soap scum. In short, they are no longer effective as soaps because the precipitate cannot emulsify the grease and dirt. Soaps are also ineffective in "hard" water, which is water having relatively high levels of calcium, magnesium, or iron, causing the following reaction:

$$2CH_3(CH_2)_{16}\overset{\overset{\displaystyle O}{\|}}{C}\text{-O}^-\,Na^+ \ + \ Ca^{2+} \longrightarrow$$

Soap (sodium salt)
(water-soluble)

$$[CH_3(CH_2)_{16}\overset{\overset{\displaystyle O}{\|}}{C}\text{-O}^-]_2\,Ca^{2+} + 2Na^+$$

Soap (calcium salt)
(water-insoluble)

The calcium and magnesium salts of the carboxylic acids are much less water-soluble than the sodium salts, and as a result precipitate, leaving rings in bathtubs and sinks. The precipitated soaps also leave a film on hair that can make it dull and may cause laundry to become gray. The solution to these problems with soap has been the detergent.

Detergents were not developed in response to the soap scum problems, but rather in response to a shortage of the natural fats (animal and plant) during World Wars I and II. Without these fats, soaps could not be made and an effective alternative cleaning agent was needed. The answer was detergents.

While soaps clean through emulsifying action, detergents are surfactants. Surfactants lower the surface tension of water so that the molecules are more likely to interact with grease and oil and less likely to interact with other water molecules. The structure of a detergent is very similar to that of a soap: There is a long hydrocarbon chain that is hydrophobic and a highly polar or charged end of the molecule. There are three common types of detergents: anionic, cationic, and nonionic. As the names suggest, these differ in the nature of the polar or charged ends of the molecules.

Anionic detergents have a negatively charged terminus, as you see in the structures of sodium dodecylsulfate (SDS) and sodium dodecylbenzenesulfonate shown here. SDS is primarily used in laundry detergents, but is also found in shampoo, bubble bath, toothpaste, and shaving foams. In higher concentrations, it is used in floor cleaners, engine degreasers, and car wash soaps. Like SDS, sodium dodecylbenzenesulfonate is largely used in laundry detergents.

Anionic Detergents

Sodium dodecylbenzensesulfonate

Sodium dodecylsulfate

Cationic detergents have a positively charged head group. In addition to being effective cleansing agents, cationic detergents have also been found to be effective antiseptics. Trimethylhexadecylammonium bromide is a cationic detergent found in topical antiseptics because it has been shown to be an effective agent against both bacteria and fungi.

Cationic Detergent

Trimethylhexadecylammonium bromide

Cationic detergents are most often found in shampoos, but are also used as fabric softeners. When added following the wash cycle, the cationic detergent neutralizes the residual charge of the anionic detergent, thereby reducing static cling.

Nonionic detergents have no charge at all on the molecule. As a result, they do not react with hard water ions, making them an excellent choice for toilet bowl cleaners to avoid unsightly buildup and to ensure that the detergent is fully active. Nonionic detergents are also very good at breaking up grease and oils and tend to foam less than ionic detergents. These properties make them very useful as dishwashing detergents. They are also used in a mixture with anionic detergents in formulations of laundry detergents.

Continued...

505

Nonionic Detergent

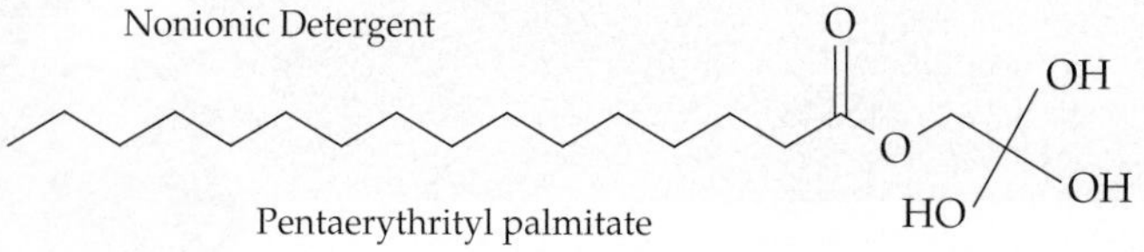

Pentaerythrityl palmitate

Whether brushing our teeth, shampooing our hair, washing the car, or doing the laundry, we find ourselves relying on the action of detergents many times through the course of the day. Our quality of life is much improved by this discovery born of wartime shortages of animal and plant fats!

For Further Understanding

▸ Compare and contrast the ways in which soaps and detergents work in the removal of dirt and grease. What is the role of agitation in the process?

▸ Below is one possible structure of an alkyl polyglucoside, the newest generation of environmentally friendly or "green" surfactants. They are synthesized in a reaction between glucose from corn and fatty alcohols from coconut or palm oil. Is this an anionic, cationic, or nonionic detergent? Explain your answer.

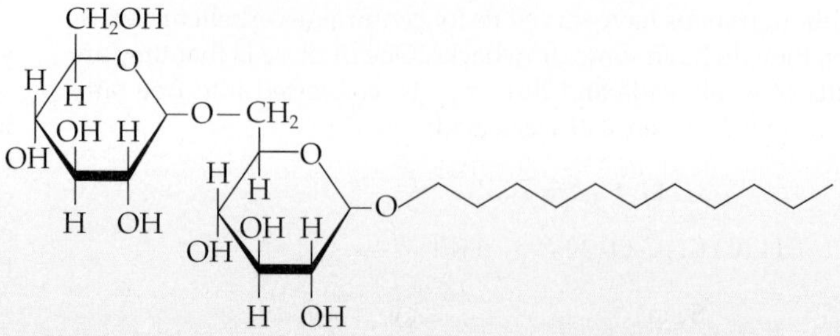

Polyesters are heteropolymers. They are also known as condensation polymers. **Condensation polymers** are formed by the polymerization of monomers in a reaction that forms a small molecule such as water or an alcohol. Polyesters are synthesized by reacting a dicarboxylic acid and a dialcohol (diol). Notice that each of the combining molecules has two reactive functional groups, highlighted in red here:

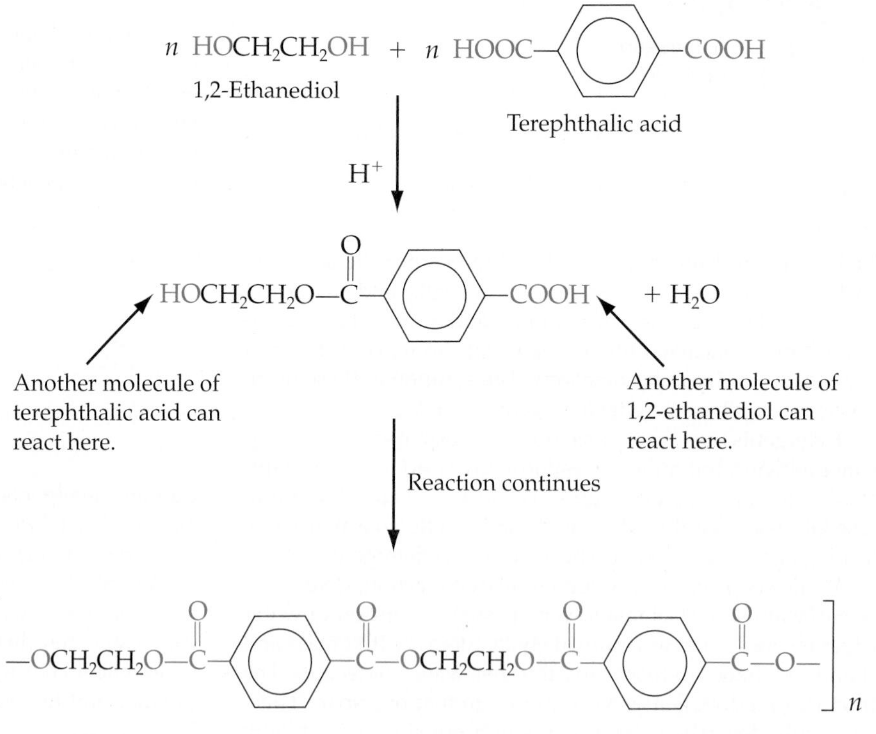

n HOCH$_2$CH$_2$OH　+　n HOOC—⬡—COOH

1,2-Ethanediol

Terephthalic acid

H$^+$

HOCH$_2$CH$_2$O—C(=O)—⬡—COOH　+ H$_2$O

Another molecule of terephthalic acid can react here.

Another molecule of 1,2-ethanediol can react here.

Reaction continues

Polyethylene terephthalate
PETE

Each time a pair of molecules reacts using one functional group from each, a new molecule is formed that still has two reactive groups. The product formed in this reaction is polyethylene terephthalate, or PETE.

When formed as fibers, polyesters are used to make fabric for clothing. These polyesters were trendy in the 1970s, during the "disco" period, but lost their popularity soon thereafter. Polyester fabrics, and a number of other synthetic polymers used in clothing, have become even more fashionable since the introduction of microfiber technology. The synthetic polymers are extruded into fibers that are only half the diameter of fine silk fibers. When these fibers are used to create fabrics, the result is a fabric that drapes freely yet retains its shape. These fabrics are generally lightweight, wrinkle resistant, and remarkably strong.

Polyester can be formed into a film called Mylar. These films, coated with aluminum foil, are used to make balloons that remain inflated for long periods. They are also used as the base for recording tapes and photographic film.

PETE can be used to make shatterproof plastic bottles, such as those used for soft drinks. However, these bottles cannot be recycled and reused directly because they cannot withstand the high temperatures required to sterilize them. PETE can't be used for any foods, such as jellies, that must be packaged at high temperatures. For these uses, a new plastic, PEN, or polyethylene naphthalate, is used.

Biodegradable plastics made of both homopolymers and heteropolymers are discussed in Green Chemistry: Garbage Bags from Potato Peels? found on page 491.

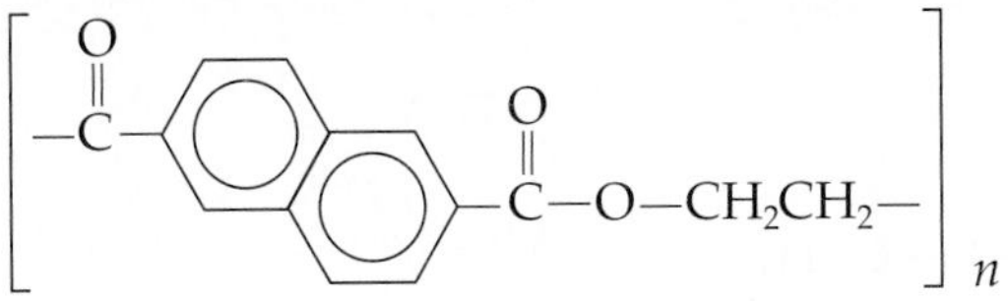

Naphthalate group   Ethylene group

14.3 Acid Chlorides and Acid Anhydrides

LEARNING GOAL

11 Determine the common and IUPAC names of acid chlorides.

Acid Chlorides

Acid chlorides are carboxylic acid derivatives having the general formula

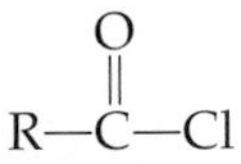

They are named by replacing the *-ic acid* ending of the common name with *-yl chloride*, or the *-oic acid* ending of the IUPAC name of the carboxylic acid with *-oyl chloride*. For example,

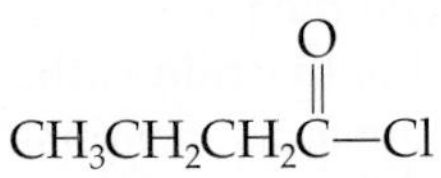

Butanoyl chloride
(butyryl chloride)

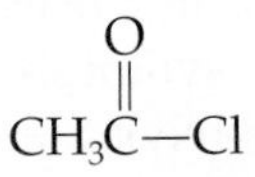

Ethanoyl chloride
(acetyl chloride)

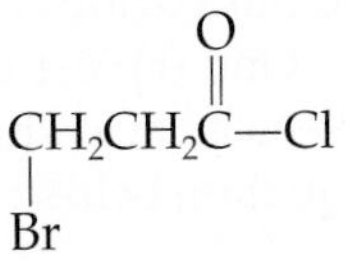

3-Bromopropanoyl chloride
(β-bromopropionyl chloride)

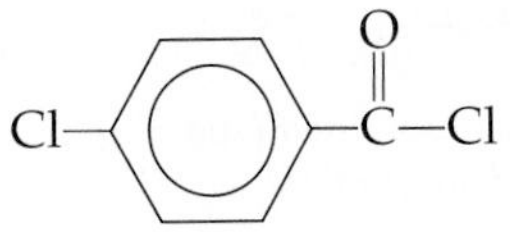

4-Chlorobenzoyl chloride
(*p*-chlorobenzoyl chloride)

Acid chlorides are noxious, irritating chemicals and must be handled with great care. They are slightly polar and boil at approximately the same temperature as the corresponding aldehyde or ketone of comparable molar mass. They react violently with water and therefore cannot be dissolved in that solvent. Acid chlorides have little commercial value other than their utility in the synthesis of esters and amides, two of the other carboxylic acid derivatives.

Acid Anhydrides

Acid anhydrides are molecules with the following general formula:

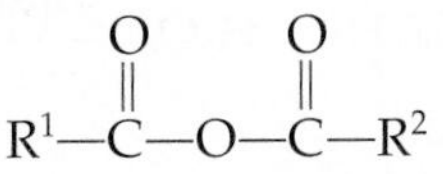

The name of the family is really quite fitting. The structure of an acid anhydride reveals that acid anhydrides are actually two carboxylic acid molecules with a water molecule removed. The word *anhydride* means "without water."

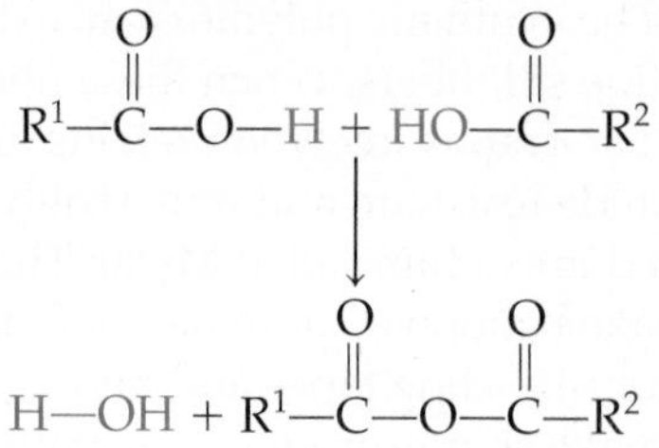

$$
\underset{\text{O}}{R^1-\overset{\displaystyle O}{\overset{\|}{C}}-O-H} + HO-\overset{\displaystyle O}{\overset{\|}{C}}-R^2
$$

$$
H-OH + R^1-\overset{\displaystyle O}{\overset{\|}{C}}-O-\overset{\displaystyle O}{\overset{\|}{C}}-R^2
$$

LEARNING GOAL

12 Determine the common and IUPAC names of acid anhydrides.

Acid anhydrides are classified as *symmetrical* if both acyl groups are the same. Symmetrical acid anhydrides are named by replacing the *acid* ending of the carboxylic acid with the word *anhydride*. For example,

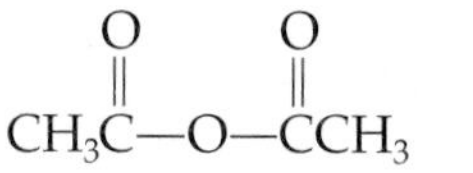

$$
CH_3\overset{\displaystyle O}{\overset{\|}{C}}-O-\overset{\displaystyle O}{\overset{\|}{C}}CH_3
$$

Ethanoic anhydride
(acetic anhydride)

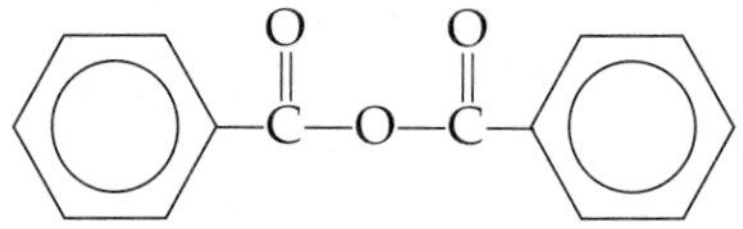

Benzoic anhydride

Unsymmetrical anhydrides are those having two different acyl groups. They are named by arranging the names of the two parent carboxylic acids and following them with the word *anhydride*. The names of the carboxylic acids may be arranged by size or alphabetically. For example:

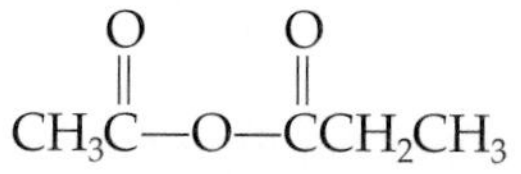

$$
CH_3\overset{\displaystyle O}{\overset{\|}{C}}-O-\overset{\displaystyle O}{\overset{\|}{C}}CH_2CH_3
$$

Ethanoic propanoic anhydride
(acetic propionic anhydride)

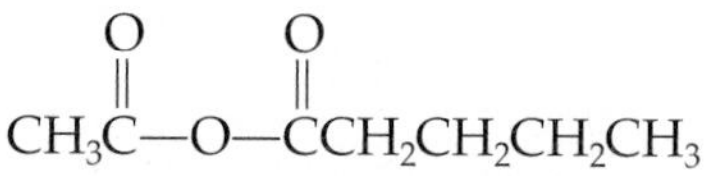

$$
CH_3\overset{\displaystyle O}{\overset{\|}{C}}-O-\overset{\displaystyle O}{\overset{\|}{C}}CH_2CH_2CH_2CH_3
$$

Ethanoic pentanoic anhydride
(acetic valeric anhydride)

LEARNING GOAL

13 Write equations representing the synthesis of acid anhydrides.

Most acid anhydrides cannot be formed in a reaction between the parent carboxylic acids. One typical pathway for the synthesis of an acid anhydride is the reaction between an acid chloride and a carboxylate anion. This general reaction is seen in the equation below:

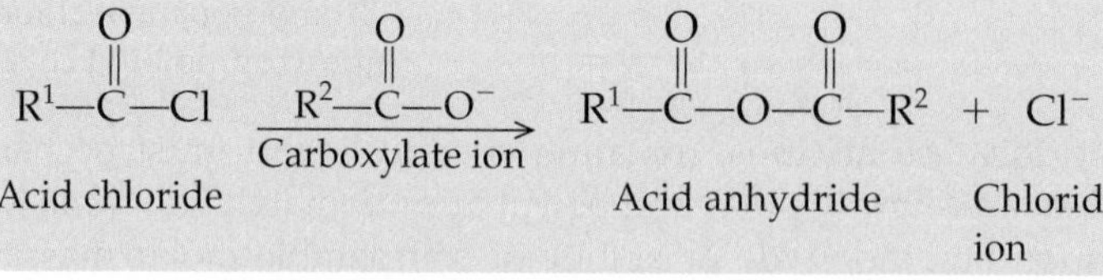

$$
\underset{\text{Acid chloride}}{R^1-\overset{\displaystyle O}{\overset{\|}{C}}-Cl} \quad \underset{\text{Carboxylate ion}}{R^2-\overset{\displaystyle O}{\overset{\|}{C}}-O^-} \longrightarrow \underset{\text{Acid anhydride}}{R^1-\overset{\displaystyle O}{\overset{\|}{C}}-O-\overset{\displaystyle O}{\overset{\|}{C}}-R^2} + \underset{\substack{\text{Chloride} \\ \text{ion}}}{Cl^-}
$$

Acid anhydrides readily undergo hydrolysis. The rate of the hydrolysis reaction may be increased by the addition of a trace of acid or hydroxide base to the solution.

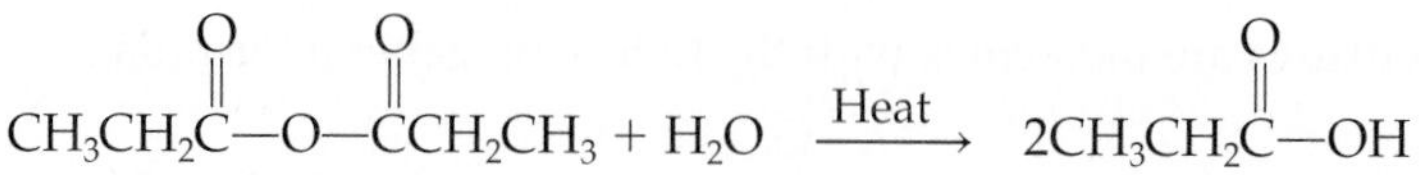

$$
\underset{\substack{\text{Propanoic anhydride} \\ \text{(propionic anhydride)}}}{CH_3CH_2\overset{\displaystyle O}{\overset{\|}{C}}-O-\overset{\displaystyle O}{\overset{\|}{C}}CH_2CH_3} + H_2O \xrightarrow{\text{Heat}} \underset{\substack{\text{Propanoic acid} \\ \text{(propionic acid)}}}{2CH_3CH_2\overset{\displaystyle O}{\overset{\|}{C}}-OH}
$$

EXAMPLE 14.10 | **Writing Equations Representing the Synthesis of Acid Anhydrides**

LEARNING GOAL

13 Write equations representing the synthesis of acid anhydrides.

Write an equation representing the synthesis of propanoic anhydride.

Solution

Propanoic anhydride can be synthesized in a reaction between propanoyl chloride and the propanoate anion. This gives us the following equation:

$$CH_3CH_2\overset{O}{\underset{\|}{C}}\!-\!Cl \xrightarrow{CH_3CH_2\overset{O}{\underset{\|}{C}}\!-\!O^-} CH_3CH_2\overset{O}{\underset{\|}{C}}\!-\!O\!-\!\overset{O}{\underset{\|}{C}}CH_2CH_3 + Cl^-$$

Propanoate ion

Propanoyl   Propanoic anhydride   Chloride
chloride   ion

Practice Problem 14.10

Write equations representing the synthesis of (a) butanoic anhydride and (b) hexanoic anhydride.

▶ For Further Practice: **Questions 14.93 and 14.94.**

EXAMPLE 14.11 | **Naming Acid Anhydrides**

LEARNING GOAL

12 Determine the common and IUPAC names of acid anhydrides.

Write the IUPAC and common names for each of the following acid anhydrides.

$$CH_3CH_2CH_2\overset{O}{\underset{\|}{C}}\!-\!O\!-\!\overset{O}{\underset{\|}{C}}CH_2CH_2CH_3$$

Solution

This is a symmetrical acid anhydride. The IUPAC name of the four-carbon parent carboxylic acid is butanoic acid (common name butyric acid). To name the anhydride, simply replace the word *acid* with the word *anhydride*. The IUPAC name of this compound is butanoic anhydride (common name butyric anhydride).

$$CH_3\overset{O}{\underset{\|}{C}}\!-\!O\!-\!\overset{O}{\underset{\|}{C}}CH_2CH_2CH_2CH_2CH_3$$

Solution

This is an unsymmetrical anhydride. The IUPAC names of the two parent carboxylic acids are ethanoic acid (two-carbon) and hexanoic acid (six-carbon). To name an unsymmetrical anhydride, the term *anhydride* is preceded by the names of the two parent acids. The IUPAC name of this compound is ethanoic hexanoic anhydride. The common names of the two parent carboxylic acids are acetic acid and caproic acid. Thus, the common name of this compound is acetic caproic anhydride.

Practice Problem 14.11

Write the common and IUPAC names for each of the following acid anhydrides.

a. $CH_3CH_2CH_2\overset{O}{\underset{\|}{C}}\!-\!O\!-\!\overset{O}{\underset{\|}{C}}CH_2CH_2CH_2CH_2CH_3$

c. $CH_3CH_2CH_2CH_2\overset{O}{\underset{\|}{C}}\!-\!O\!-\!\overset{O}{\underset{\|}{C}}CH_2CH_3$

b. $CH_3\overset{O}{\underset{\|}{C}}\!-\!O\!-\!\overset{O}{\underset{\|}{C}}CH_2CH_2CH_2CH_3$

d. $CH_3CH_2\overset{O}{\underset{\|}{C}}\!-\!O\!-\!\overset{O}{\underset{\|}{C}}CH_3$

▶ For Further Practice: **Questions 14.89 and 14.90.**

Acid anhydrides can also react with an alcohol. This reaction produces an ester and a carboxylic acid. This is an example of an acyl group transfer reaction. The **acyl group** of a carboxylic acid derivative has the following structure:

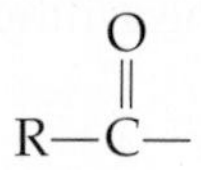

$$R-\overset{\overset{\displaystyle O}{\|}}{C}-$$

The following general equation represents the acyl group transfer reaction between an alcohol and an acid anhydride.

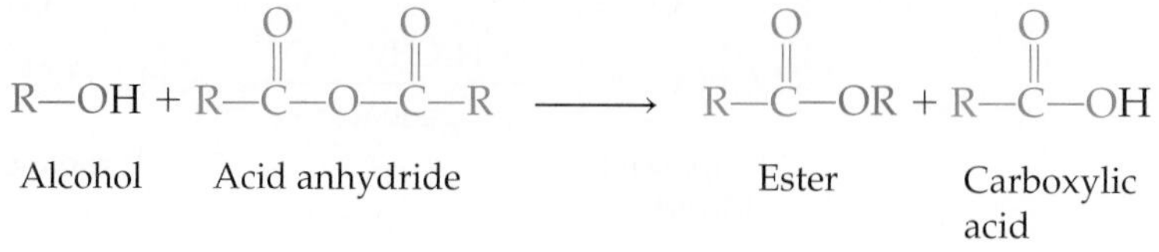

$$R-OH + R-\overset{\overset{\displaystyle O}{\|}}{C}-O-\overset{\overset{\displaystyle O}{\|}}{C}-R \longrightarrow R-\overset{\overset{\displaystyle O}{\|}}{C}-OR + R-\overset{\overset{\displaystyle O}{\|}}{C}-OH$$

     Alcohol   Acid anhydride   Ester   Carboxylic acid

The acyl group of the acid anhydride is transferred to the oxygen of the alcohol in this reaction. The alcohol and anhydride reactants and ester product are described below. The carboxylic acid product is omitted.

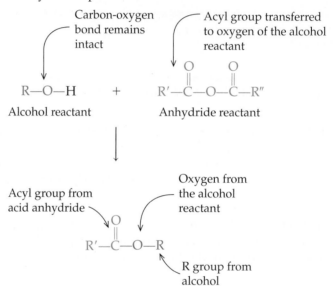

Other acyl group donors include thioesters and esters. As we will see in the final section of this chapter, acyl group transfer reactions are very important in nature, particularly in the pathways responsible for breakdown of food molecules and harvesting cellular energy.

Question 14.9 Write an equation showing the synthesis of each of the following acid anhydrides. Provide the IUPAC names of the acid chloride and carboxylate anion reactants and the acid anhydride product.

a. CH₃CHCH₂C—O—CCH₂CHCH₃ b. H—C—O—C—CH₃

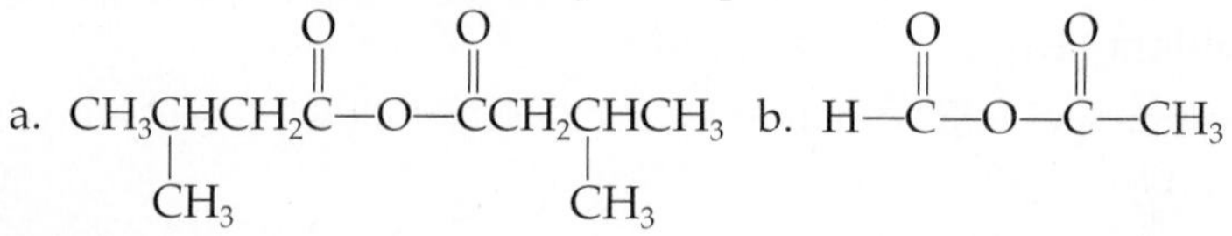

Question 14.10 Write an equation showing the synthesis of each of the following acid anhydrides. Provide the common names of the acid chloride and carboxylate anion reactants and the acid anhydride products.

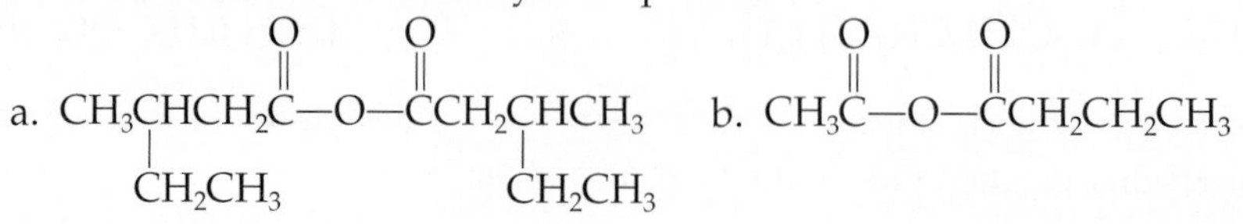

a. CH₃CHCH₂C—O—CCH₂CHCH₃ b. CH₃C—O—CCH₂CH₂CH₃

14.4 Nature's High-Energy Compounds: Phosphoesters and Thioesters

An alcohol can react with phosphoric acid to produce a phosphate ester, or **phosphoester**, as in

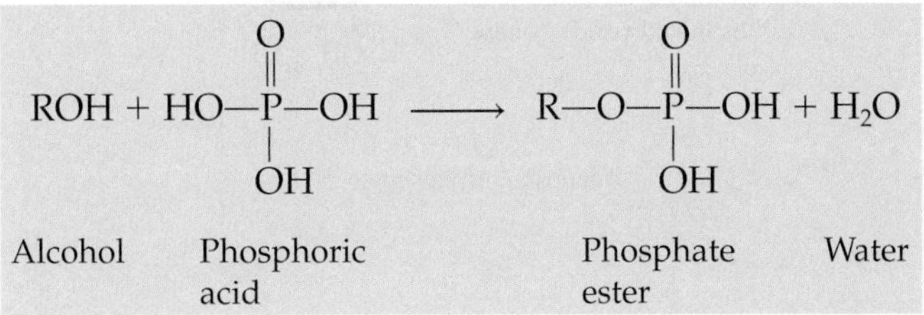

| Alcohol | Phosphoric acid | | Phosphate ester | Water |

LEARNING GOAL

14 Discuss the significance of thioesters and phosphoesters in biological systems.

Phosphoesters of simple sugars or monosaccharides are very important in the energy-harvesting biochemical pathways that provide energy for all life functions. One such pathway is *glycolysis*. This pathway is the first stage in the breakdown of sugars. The first reaction in this pathway is the formation of a phosphoester of the six-carbon sugar, glucose. The product of this reaction is glucose-6-phosphate:

The many phosphorylated intermediates in the metabolism of sugars will be discussed in Chapter 21.

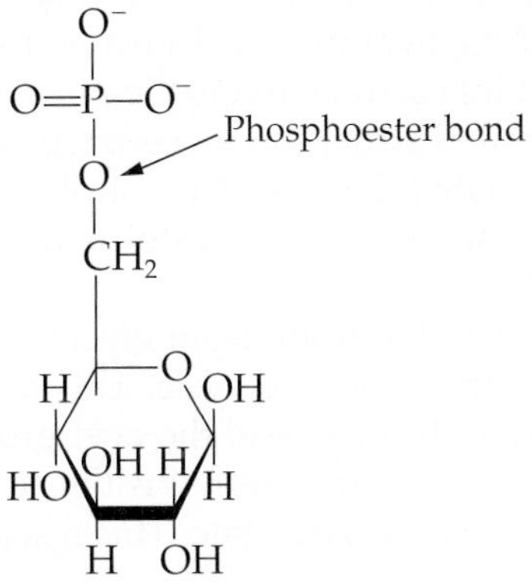

Glucose-6-phosphate

The word glycolysis comes from two Greek words that mean "splitting sugars" (glykos, "sweet," and lysis, "to split"). In this pathway, the six-carbon sugar glucose is split, and then oxidized, to produce two three-carbon molecules, called pyruvate.

When two phosphate groups react with one another, a water molecule is lost. Because water is lost, the resulting bond is called a **phosphoanhydride** bond.

| Phosphate ester | Phosphate group | | Phosphoanhydride bond |

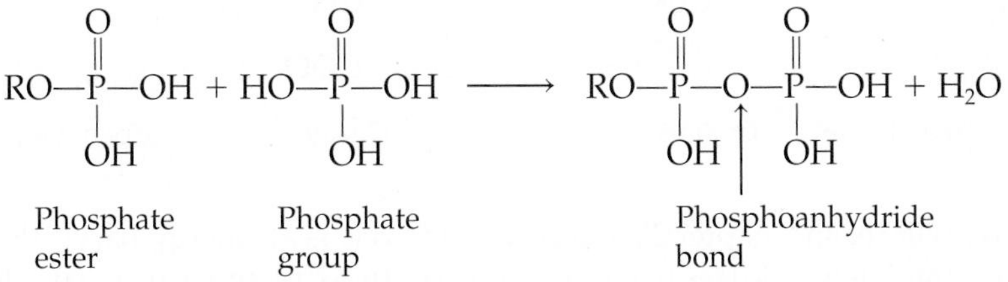

Adenosine triphosphate (ATP) is the universal energy currency for all living organisms. Much of the energy harvested in cellular metabolic reactions is stored in ATP molecules. The ATP molecule consists of a nitrogenous base (adenine) and a phosphate ester of the five-carbon sugar ribose (Figure 14.5). The triphosphate group attached to ribose is made up of three phosphate groups bonded to one another by phosphoanhydride bonds. These phosphoanhydride bonds are termed high-energy bonds because energy is released when they are broken. Thus, the metabolic energy stored in these high-energy bonds can be released for use in energy-requiring reactions.

The functions and properties of ATP in energy metabolism are discussed in Section 21.1.

Figure 14.5 The structure of adenosine triphosphate (ATP).

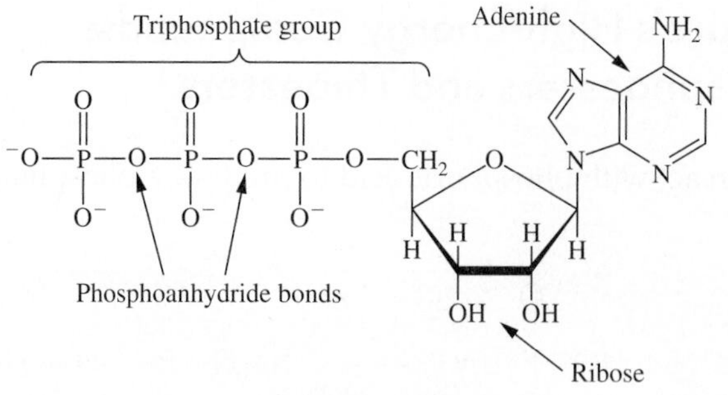

Adenosine triphosphate ATP

Thiols are described in Section 12.8.

Cellular enzymes can carry out a reaction between a thiol and a carboxylic acid to produce a **thioester:**

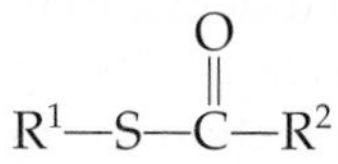

$$R^1{-}S{-}\overset{\overset{\displaystyle O}{\|}}{C}{-}R^2$$

Thioester

β-Oxidation is the pathway for the breakdown of fatty acids. Like glycolysis, it is an energy-harvesting pathway.

The reactions that produce thioesters are essential in energy-harvesting pathways as a means of "activating" acyl groups for subsequent breakdown reactions. The complex thiol coenzyme A is the most important acyl group activator in the cell. The detailed structure of coenzyme A appears in Section 12.8, but it is generally abbreviated CoA—SH to emphasize the importance of the sulfhydryl group. The most common thioester is the acetyl ester, called **acetyl coenzyme A** (acetyl CoA).

The acyl group of a carboxylic acid is named by replacing the *-oic acid* or *-ic* suffix with *-yl*. For instance, the acyl group of acetic acid is the acetyl group:

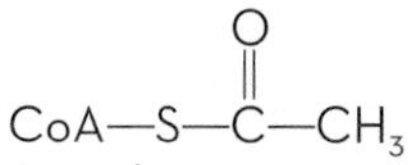

$$CoA{-}S{-}\overset{\overset{\displaystyle O}{\|}}{C}{-}CH_3$$
Acetyl coenzyme A
(acetyl CoA)

Acetyl CoA carries the acetyl group from glycolysis or β-oxidation of a fatty acid to an intermediate of the citric acid cycle. This reaction is an example of an acyl group transfer reaction. In this case, the acyl group donor is a thioester— acetyl CoA. The acyl group being transferred is the acetyl group, which is transferred to the carbonyl carbon of oxaloacetate. This reaction follows:

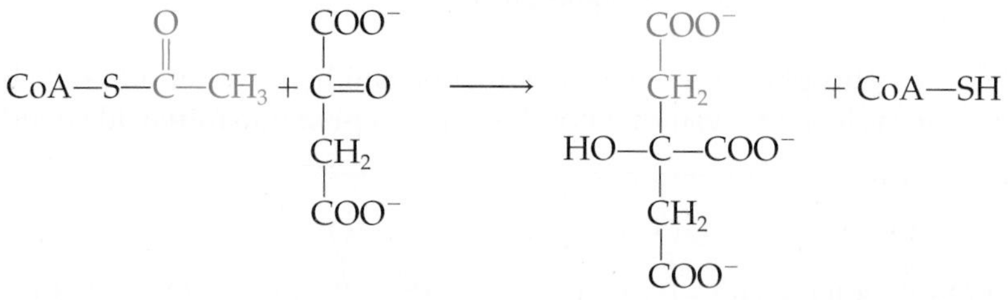

Acetyl CoA   Oxaloacetate   Citrate   Coenzyme A

Glycolysis, β-oxidation, and the citric acid cycle are cellular energy-harvesting pathways that we will study in Chapters 21, 22, and 23.

As we will see in Chapter 22, the citric acid cycle is an energy-harvesting pathway that completely oxidizes the acetyl group to two CO_2 molecules. The electrons that are harvested in the process are used to produce large amounts of ATP. Coenzyme A also serves to activate the acyl group of fatty acids during β-oxidation, the pathway by which fatty acids are oxidized to produce ATP.

A Medical Perspective

Esters for Appetite Control

Imagine a simple food additive that could reduce hunger and weight gain and, perhaps, support a weight loss regimen. Studies carried out at Imperial College London's Department of Medicine and the University of Glasgow in Scotland suggest that the simple molecule propionate could do just that.

Propionic acid is an oily liquid with a disagreeable, pungent odor. So they could not use propionic acid directly in their study. In this chapter, we have seen that the carboxylic acids with unpleasant tastes and aromas can be converted into esters that have remarkably different properties. In fact, many of the molecules that we associate with the pleasant flavors and scents of fruits are esters. So the team decided to make an ester of propionic acid and the polysaccharide inulin. This allowed them to deliver larger quantities of propionate to their treatment groups. In the gut, the ester bond is broken and the propionate is released.

Inulin is a plant polysaccharide that has been used to lower triglycerides, promote weight loss, and treat constipation. It is also used as a food additive to improve taste. Some studies suggest that inulin enhances the growth of bifidobacteria and lactobacteria, two groups of bacteria that have been associated with a healthy colon.

The compound that they developed, called inulin-propionate ester (IPE), was used to treat colon cells in tissue culture. They found that these cells were stimulated to produce two appetite-suppressing hormones called peptide YY and glucagon-like protein 1. This finding encouraged them to try IPE as an appetite-suppressant in humans. In a small study involving 20 participants, those that had a premeal dose of IPE ate 14% less than the participants who were given inulin alone.

In their most recent studies, they recruited 60 overweight individuals for a 24-week study. Half were given IPE powder to add to their food, and half were given inulin alone. At the end of the study, none of the IPE participants exhibited more than a 5% weight gain—compared to a 17% weight gain with the participants using inulin alone.

Many more and larger studies will be required to confirm these results and to determine optimal dosages of IPE, if it continues to show promise. However, the research shows that an understanding of the chemistry of molecules that have biological activity can allow the design of molecules with great potential for humankind.

For Further Study

▸ Examine the structure of inulin. With that scaffold, draw some of the many potential esters that could be produced.

▸ Why might IPE, with its chemical combination of both propionic acid and inulin, have many potential health benefits beyond weight control?

CHAPTER MAP

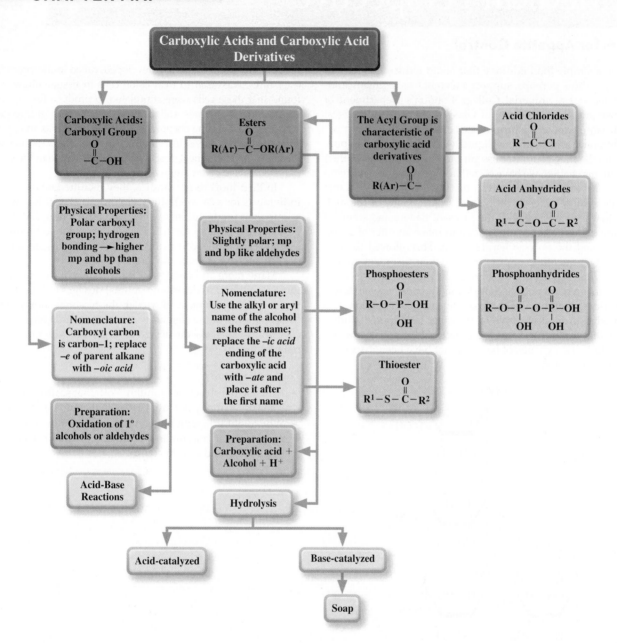

SUMMARY OF REACTIONS

Preparation of Carboxylic Acids

$$R-CH_2OH \xrightarrow{[O]} R-\overset{\overset{\displaystyle O}{\|}}{C}-H \xrightarrow{[O]} R-\overset{\overset{\displaystyle O}{\|}}{C}-OH$$

Primary alcohol   Aldehyde   Carboxylic acid

Esterification

$$R^1-\overset{\overset{\displaystyle O}{\|}}{C}-OH + R^2OH \underset{}{\overset{H^+,\,heat}{\rightleftharpoons}} R^1-\overset{\overset{\displaystyle O}{\|}}{C}-OR^2 + H_2O$$

Carboxylic   Alcohol   Ester   Water
acid

Dissociation of Carboxylic Acids

$$R-\overset{\overset{\displaystyle O}{\|}}{C}-OH \rightleftharpoons R-\overset{\overset{\displaystyle O}{\|}}{C}-O^- \ + \ H^+$$

Carboxylic acid	Carboxylate anion	Hydrogen ion

Neutralization of Carboxylic Acids

$$R-\overset{\overset{\displaystyle O}{\|}}{C}-OH + NaOH \longrightarrow R-\overset{\overset{\displaystyle O}{\|}}{C}-O^-Na^+ \ + \ H_2O$$

Carboxylic acid	Strong base	Carboxylic acid salt	Water

Synthesis of Acid Anhydrides

$$R^1-\overset{\overset{\displaystyle O}{\|}}{C}-Cl \ \xrightarrow[\substack{\text{Carboxylate}\\\text{ion}}]{R^2-\overset{\overset{\displaystyle O}{\|}}{C}-O^-} \ R^1-\overset{\overset{\displaystyle O}{\|}}{C}-O-\overset{\overset{\displaystyle O}{\|}}{C}-R^2 + Cl^-$$

Acid chloride		Acid anhydride	Chloride ion

Saponification (Base-catalyzed Hydrolysis of Esters)

$$R^1-\overset{\overset{\displaystyle O}{\|}}{C}-OR^2 + H_2O \xrightarrow{\substack{\text{NaOH,}\\\text{heat}}} R^1-\overset{\overset{\displaystyle O}{\|}}{C}-O^-Na^+ \ + \ R^2OH$$

Ester	Water	Carboxylic acid salt	Alcohol

Acid Hydrolysis of Esters

$$R^1-\overset{\overset{\displaystyle O}{\|}}{C}-OR^2 \ + H_2O \underset{}{\overset{H^+,\,heat}{\rightleftharpoons}} R^1-\overset{\overset{\displaystyle O}{\|}}{C}-OH \ + R^2OH$$

Ester	Water	Carboxylic acid	Alcohol

Formation of a Phosphoester

$$ROH + HO-\overset{\overset{\displaystyle O}{\|}}{\underset{\underset{\displaystyle OH}{|}}{P}}-OH \longrightarrow R-O-\overset{\overset{\displaystyle O}{\|}}{\underset{\underset{\displaystyle OH}{|}}{P}}-OH +H_2O$$

Alcohol	Phosphoric acid	Phosphate ester	Water

SUMMARY

14.1 Carboxylic Acids

▶ The functional group of the **carboxylic acids** is the **carboxyl group** (—COOH).

▶ Because the carboxyl group is extremely polar and carboxylic acids can form intermolecular hydrogen bonds with one another, they have higher boiling points and melting points than alcohols.

▶ The lower-molar-mass carboxylic acids are water-soluble and tend to taste sour and have unpleasant aromas.

▶ The longer-chain carboxylic acids are called **fatty acids.**

▶ In the IUPAC Nomenclature System, carboxylic acids are named by replacing the -*e* ending of the parent compound with -*oic acid*. When naming dicarboxylic acids, the -*e* is retained and the suffix -*dioic* is added.

▶ Often, common names are derived from the source of the carboxylic acid.

▶ Carboxylic acids are synthesized by the **oxidation** of primary alcohols or aldehydes.

▶ Carboxylic acids are weak acids and are neutralized by strong bases to form salts. **Soaps** are salts of long-chain carboxylic acids (fatty acids).

14.2 Esters

▶ **Esters** are **carboxylic acid derivatives** that are slightly polar and have pleasant aromas.

▶ The boiling points and melting points of esters are comparable to those of aldehydes and ketones.

▶ Esters are formed from the reaction between a carboxylic acid and an alcohol.

▶ Esters can undergo **hydrolysis** back to the parent carboxylic acid and alcohol.

▶ The base-catalyzed hydrolysis of an ester is called **saponification.**

▶ **Condensation polymers** are large molecules formed by the combination of many small molecules (monomers) that result from the joining of monomers in a reaction that also forms a small molecule, such as water or an alcohol.

14.3 Acid Chlorides and Acid Anhydrides

▶ **Acid chlorides** are noxious chemicals and are useful in the synthesis of a variety of carboxylic acid derivatives.

▶ **Acid anhydrides** are formed by the combination of an acid chloride and a carboxylate anion.

▶ Acid anhydrides can react with an alcohol to produce an ester and a carboxylic acid. This is an example of an acyl group transfer.

- **The acyl group in carboxylic acid derivatives contains the carbonyl group attached to one alkyl or one aryl group.**

14.4 Nature's High-Energy Compounds: Phosphoesters and Thioesters

- **An alcohol can react with phosphoric acid to produce a phosphate ester (phosphoester).**
- **When two phosphate groups are joined, the resulting bond is a phosphoanhydride bond.**
- **The phosphoester and phosphoanhydride bonds are important in the structure of adenosine triphosphate (ATP), the universal energy currency of all cells.**
- **Cellular enzymes carry out a reaction between a thiol and a carboxylic acid to produce a thioester. This reaction is essential for the activation of acyl groups in carbohydrate and fatty acid metabolism.**
- **Coenzyme A is the important thiol involved in these pathways, forming acetyl coenzyme A.**

ANSWERS TO PRACTICE PROBLEMS

14.1 **a.** 2,4-Dimethylpentanoic acid
b. 2,4-Dichlorobutanoic acid
c. 2,3-Dihydroxybutanoic acid
d. 2-Bromo-3-chloro-4-methylhexanoic acid

14.2 **a.** α,γ-Dimethylvaleric acid
b. α,γ-Dichlorobutyric acid
c. β,γ-Dibromovaleric acid
d. γ-Hydroxycaproic acid

14.3 **a.** *o*-Toluic acid:

b. 2,4,6-Tribromobenzoic acid:

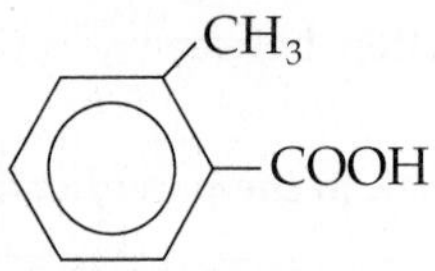

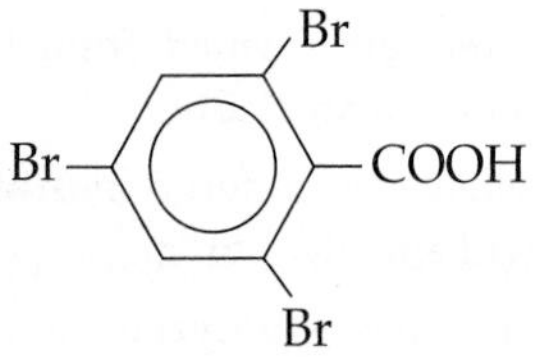

c. 2,2,2-Triphenylethanoic acid:

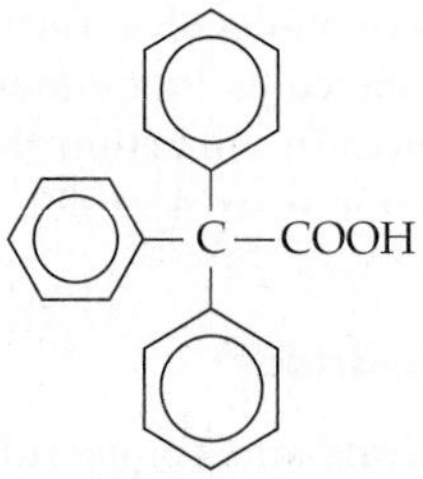

d. *p*-Toluic acid:

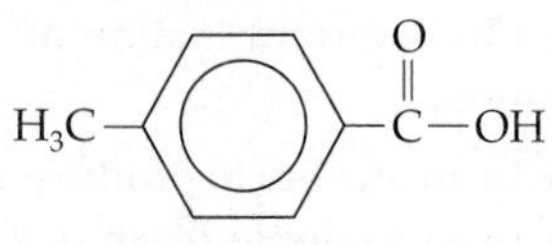

e. 3-Phenylhexanoic acid:

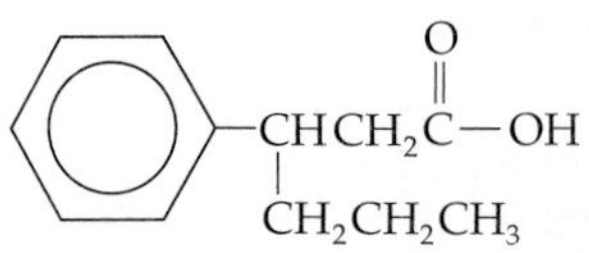

f. 3-Phenylcyclohexanecarboxylic acid:

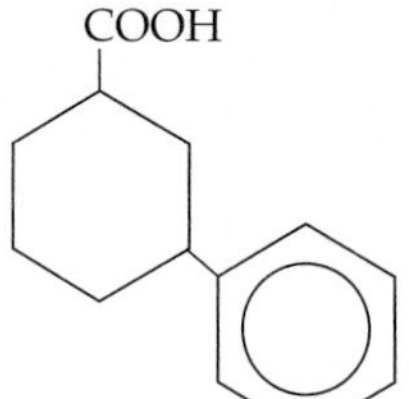

14.4 **a.** The following equation represents the synthesis of ethanoic acid from ethanol.

$$\underset{\text{Ethanol}}{CH_3CH_2OH} \xrightarrow{[O]} \underset{\text{Ethanal}}{CH_3\overset{O}{\overset{\|}{C}}-H} \xrightarrow{[O]} \underset{\text{Ethanoic acid}}{CH_3\overset{O}{\overset{\|}{C}}-OH}$$

b. The following equation represents the synthesis of butanoic acid from 1-butanol.

$$\underset{\text{1-Butanol}}{CH_3CH_2CH_2CH_2OH} \xrightarrow{[O]} \underset{\text{Butanal}}{CH_3CH_2CH_2\overset{O}{\overset{\|}{C}}-H} \xrightarrow{[O]} \underset{\text{Butanoic acid}}{CH_3CH_2CH_2\overset{O}{\overset{\|}{C}}-OH}$$

c. The following equation represents the synthesis of octanoic acid from 1-octanol.

$$\underset{\text{1-Octanol}}{CH_3(CH_2)_6CH_2OH} \xrightarrow{[O]} \underset{\text{Octanal}}{CH_3(CH_2)_6\overset{O}{\overset{\|}{C}}-H} \xrightarrow{[O]} \underset{\text{Octanoic acid}}{CH_3(CH_2)_6\overset{O}{\overset{\|}{C}}-OH}$$

14.5 **a.** $CH_3CH_2COO^-K^+$
b. $[CH_3CH_2CH_2COO^-]_2Ba^{2+}$
c. $CH_3CH_2CH_2CH_2CH_2COO^-K^+$
d.

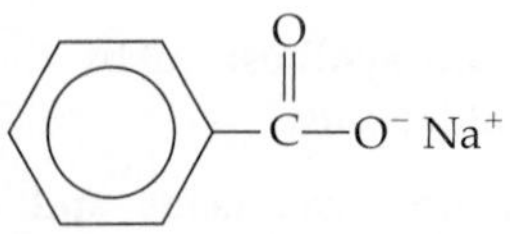

14.6 **a.** Potassium propanoate
  b. Barium butanoate
  c. Potassium hexanoate
  d. Sodium benzoate

14.7 **a.** Propyl butanoate (propyl butyrate)
  b. Ethyl butanoate (ethyl butyrate)
  c. Propyl ethanoate (propyl acetate)
  d. Butyl propanoate (butyl propionate)

14.8 **a.** The following reaction between 1-butanol and ethanoic acid produces butyl ethanoate. It requires a trace of acid and heat. It is also reversible.

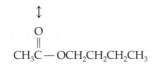

$$CH_3CH_2CH_2CH_2OH + CH_3COOH$$

$$\updownarrow$$

$$\underset{O}{\overset{O}{\parallel}}$$
$$CH_3C-OCH_2CH_2CH_2CH_3$$

  b. The following reaction between ethanol and propanoic acid produces ethyl propanoate. It requires a trace of acid and heat. It is also reversible.

$$CH_3CH_2OH + CH_3CH_2COOH$$

$$\updownarrow$$

$$\overset{O}{\parallel}$$
$$CH_3CH_2C-OCH_2CH_3$$

14.9 **a.** Methyl butanoate is made from methanol and butanoic acid.

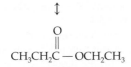

$$CH_3OH + CH_3CH_2CH_2COOH \longrightarrow CH_3CH_2CH_2\overset{O}{\overset{\parallel}{C}}OCH_3$$
Methanol   Butanoic acid   Methyl butanoate

Methanol is an allowed starting material, but butanoic acid is not. However, it can easily be produced by the oxidation of its corresponding alcohol, 1-butanol:

$$CH_3CH_2CH_2CH_2OH \xrightarrow{[O]} CH_3CH_2CH_2CHO$$
     1-Butanol   Butanal

$$\Big\downarrow {[O]}$$

$$CH_3CH_2CH_2COOH$$

Butanoic acid

  b. Propyl methanoate is made from 1-propanol and methanoic acid.

$$CH_3CH_2CH_2OH + HCOOH$$
     1-Propanol   Methanoic acid

$$\Big\downarrow$$

$$\overset{O}{\parallel}$$
$$HCOCH_2CH_2CH_3$$

Propyl methanoate

1-Propanol is an allowed starting material, but methanoic acid is not. However, it can easily be produced by the oxidation of its corresponding alcohol, methanol:

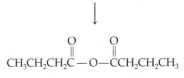

$$CH_3OH \xrightarrow{[O]} HCHO \xrightarrow{[O]} HCOOH$$

  Methanol   Methanal   Methanoic acid

14.10 **a.** The following equation represents the synthesis of butanoic anhydride:

$$CH_3CH_2CH_2\overset{O}{\overset{\parallel}{C}}O^- + CH_3CH_2CH_2\overset{O}{\overset{\parallel}{C}}-Cl$$

  Butanoate anion   Butanoyl chloride

$$\Big\downarrow$$

$$CH_3CH_2CH_2\overset{O}{\overset{\parallel}{C}}-O-\overset{O}{\overset{\parallel}{C}}CH_2CH_2CH_3$$

  Butanoic anhydride

  b. The following equation represents the synthesis of hexanoic anhydride:

$$CH_3(CH_2)_4\overset{O}{\overset{\parallel}{C}}O^- + CH_3(CH_2)_4\overset{O}{\overset{\parallel}{C}}-Cl$$

  Hexanoate anion   Hexanoyl chloride

$$\Big\downarrow$$

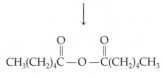

$$CH_3(CH_2)_4\overset{O}{\overset{\parallel}{C}}-O-\overset{O}{\overset{\parallel}{C}}(CH_2)_4CH_3$$

  Hexanoic anhydride

14.11 **a.** IUPAC name: Butanoic hexanoic anhydride
      Common name: Butyric caproic anhydride
  b. IUPAC name: Ethanoic pentanoic anhydride
      Common name: Acetic valeric anhydride
  c. IUPAC name: Propanoic pentanoic anhydride
      Common name: Propionic valeric anhydride
  d. IUPAC name: Ethanoic propanoic anhydride
      Common name: Acetic propionic anhydride

QUESTIONS AND PROBLEMS

Carboxylic Acids: Structure and Properties

Foundations

14.11 The functional group is largely responsible for the physical and chemical properties of the various chemical families. Explain why a carboxylic acid is more polar and has a higher boiling point than an alcohol or an aldehyde of comparable molar mass.

14.12 Explain why carboxylic acids are weak acids.

Applications

14.13 Which member of the following pairs has the higher boiling point?
  a. Pentanoic acid or pentanal
  b. 3-Pentanone or 2-pentanol
  c. 2-Pentanol or pentane

14.14 Which member of the following pairs has the higher boiling point?
  a. Ethyl propyl ether or pentanal
  b. 3-Pentanone or pentanoic acid
  c. Methanol or methanoic acid

14.15 Arrange the following from highest to lowest melting points:

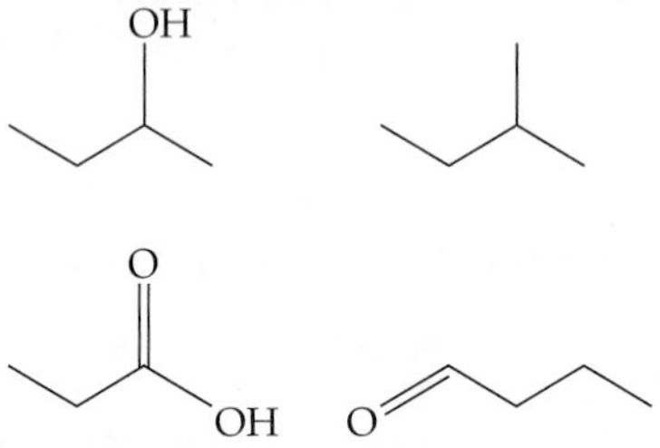

14.16 Draw the condensed formula for each of the line formulas in Question 14.15 and provide the IUPAC name for each.

14.17 Which member in each of the following pairs has the higher boiling point?
  a. Heptanoic acid or 1-heptanol
  b. Propanal or 1-propanol
  c. Methyl pentanoate or pentanoic acid
  d. 1-Butanol or butanoic acid

14.18 Which member in each of the following pairs is more soluble in water?

a. $CH_3CH_2CH_2CH_2CH_2\overset{\displaystyle O}{\overset{\|}{C}}{-}OH$ or

$$CH_3CH_2CH_2CH_2CH_2\overset{\displaystyle O}{\overset{\|}{C}}{-}O^-Na^+$$

b. $CH_3CH_2{-}O{-}CH_2CH_3$ or $CH_3CH_2\overset{\displaystyle O}{\overset{\|}{C}}{-}OCH_3$

c. $CH_3CH_2{-}O{-}CH_2CH_3$ or $CH_3CH_2CH_2CH_2CH_2CH_3$

d. Decanoic acid or ethanoic acid

14.19 Describe the properties of low-molar-mass carboxylic acids.

14.20 What are some of the biological functions of the long-chain carboxylic acids called fatty acids?

14.21 Why is citric acid added to some food products?

14.22 What is the function of lactic acid in food products? Of what significance is lactic acid in muscle metabolism?

14.23 Why is glutaric acid particularly useful in the synthesis of condensation polymers?

14.24 What is the role of octanoic acid in the control of appetite?

Carboxylic Acids: Structure and Nomenclature

Foundations

14.25 Summarize the IUPAC nomenclature rules for naming carboxylic acids.

14.26 Describe the rules for determining the common names of carboxylic acids.

Applications

14.27 Adipic acid occurs naturally in beets and is used as a food additive. What is the IUPAC name for adipic acid? Why do you think adipic acid is used as a food additive?

$$HOOCCH_2CH_2CH_2CH_2COOH$$

14.28 Propionic acid is a liquid fatty acid found in sweat and milk products. It is a bacterial fermentation product that gives the tangy flavor to Swiss cheese. What is the IUPAC name of propionic acid?

$$CH_3CH_2COOH$$

14.29 Write the condensed formula and the line formula for each of the following carboxylic acids:
  a. 3-Methylhexanoic acid
  b. 2-Ethyl-2-methylpentanoic acid
  c. 3-Methylcyclopentanecarboxylic acid

14.30 Write the condensed formula and the line formula for each of the following carboxylic acids:
  a. 2,3-Dibromocycloheptanecarboxylic acid
  b. 2-Butenoic acid
  c. 2,4,5-Trimethyloctanoic acid

14.31 Name each of the following carboxylic acids, using both the common and the IUPAC Nomenclature Systems:

a. $H{-}\overset{\displaystyle O}{\overset{\|}{C}}{-}OH$

b. $CH_3\overset{\displaystyle CH_3}{\underset{|}{C}}H CH_2\overset{\displaystyle O}{\overset{\|}{C}}{-}OH$

c.

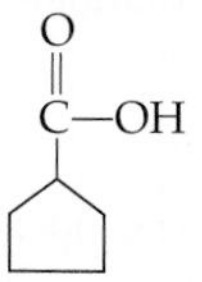

14.32 Name each of the following carboxylic acids, using both the common and IUPAC Nomenclature Systems:

a. $CH_3CH_2\overset{\displaystyle Br}{\underset{|}{C}}HCHCH_2\overset{\displaystyle O}{\overset{\|}{C}}{-}OH$
$\underset{|}{CH_3}$

b. $CH_3CH_2\overset{\displaystyle CH_2CH_3}{\underset{|}{C}}HCH_2CH_2\overset{\displaystyle O}{\overset{\|}{C}}{-}OH$

c.

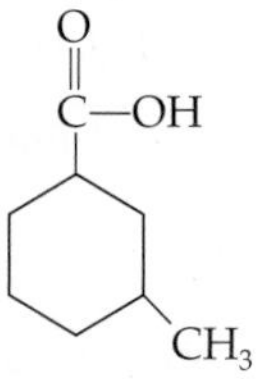

14.33 Write a complete structural formula, and determine the IUPAC name for each of the carboxylic acids of molecular formula $C_4H_8O_2$.

14.34 Write the general structure of an aldehyde, a ketone, a carboxylic acid, and an ester. What similarities exist among these structures?

14.35 Write the condensed structure of each of the following carboxylic acids:
  a. 4,4-Dimethylhexanoic acid
  b. 3-Bromo-4-methylpentanoic acid
  c. 2,3-Dinitrobenzoic acid
  d. 3-Methylcyclohexanecarboxylic acid

14.36 Use IUPAC nomenclature to write the names for each of the following carboxylic acids:

a.

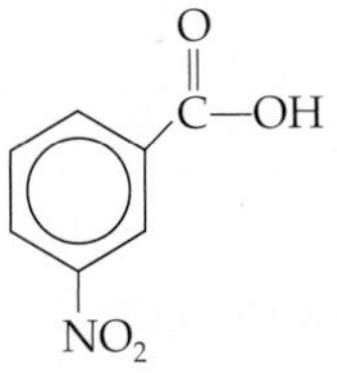

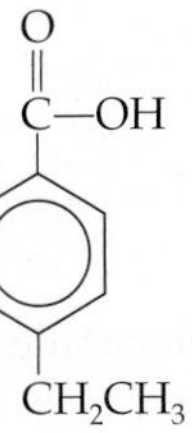

c.

b.

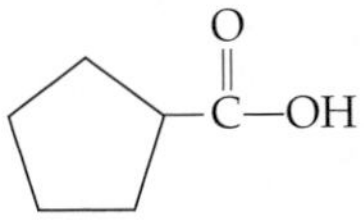

14.37 Provide the common and IUPAC names for each of the following compounds:

a. 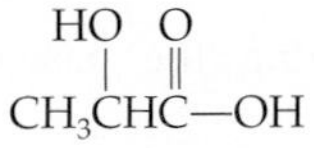 CH₃CHC—OH (with HO and O substituents shown above)

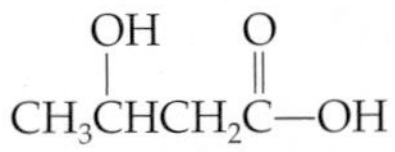

c. 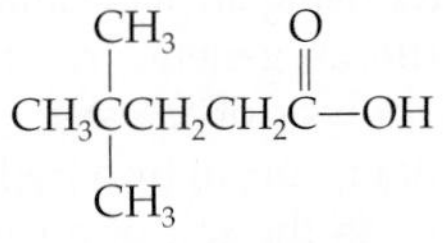 CH₃CCH₂CH₂C—OH (with CH₃ groups shown)

b. CH₃CHCH₂C—OH (with OH and O shown)

d. 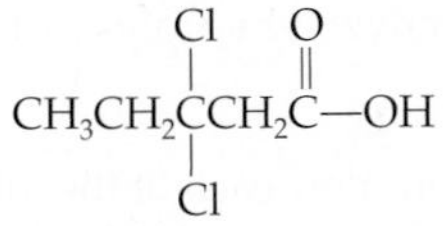 CH₃CH₂CCH₂C—OH (with Cl groups shown)

14.38 Draw the structure of each of the following carboxylic acids:
   a. β-Chlorobutyric acid
   b. α,β-Dibromovaleric acid
   c. β,γ-Dihydroxybutyric acid
   d. δ-Bromo-γ-chloro-β-methylcaproic acid

14.39 Provide the IUPAC name for each of the following aromatic carboxylic acids.

a.

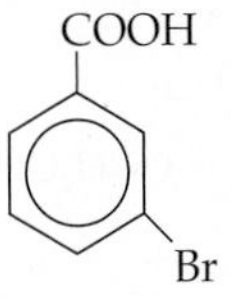

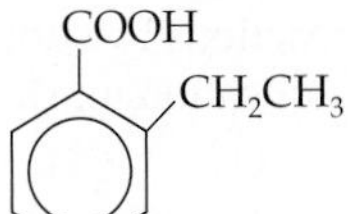

c.

b.

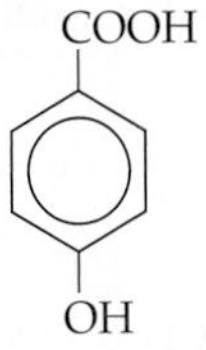

14.40 Provide the IUPAC name for each of the following aromatic carboxylic acids.

a. 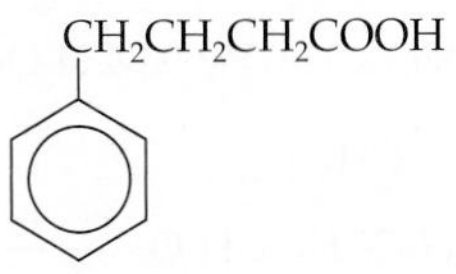 CH₂CH₂CH₂COOH

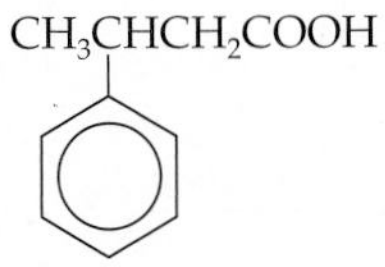

c. 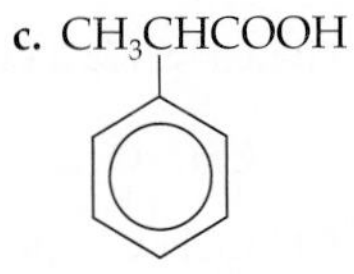 CH₃CHCOOH

b. CH₃CHCH₂COOH

Carboxylic Acids: Reactions

Foundations

14.41 Explain what is meant by oxidation in organic molecules, and provide an example of an oxidation reaction involving an aldehyde or an alcohol.

14.42 Write a general equation showing the preparation of a carboxylic acid from an alcohol.

14.43 Write a general equation showing the dissociation of a carboxylic acid in water.

14.44 Carboxylic acids are described as weak acids. To what extent do carboxylic acids generally dissociate?

14.45 What reaction occurs when a strong base is added to a carboxylic acid?

14.46 Write a general equation showing the reaction of a strong base with a carboxylic acid.

14.47 How is a soap prepared?

14.48 How do soaps assist in the removal of oil and grease from clothing?

Applications

14.49 Write an equation representing the oxidation of each of the following compounds:
   a. 1-Pentanol
   b. Butanal
   c. Butanone

14.50 Write an equation representing the oxidation of each of the following compounds:
   a. 3-Hexanol
   b. 2-Methylpentanal
   c. 3-Pentanone

14.51 Complete each of the following reactions by supplying the missing portion indicated by a question mark:

a. 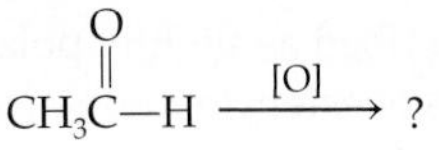 CH₃C—H $\xrightarrow{[O]}$?

b. 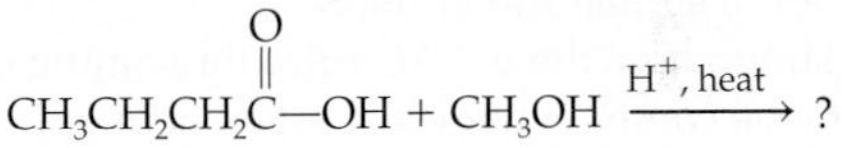 CH₃CH₂CH₂C—OH + CH₃OH $\xrightarrow{H^+, \text{heat}}$?

c.

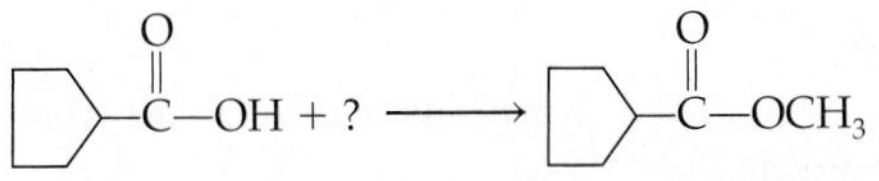

14.52 Complete each of the following reactions by supplying the missing part(s) indicated by the question mark(s):

a.

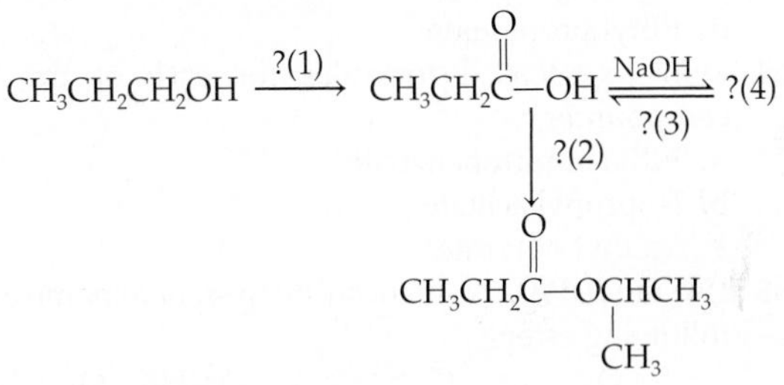

b. CH₃COOH + NaOH ⟶ ?

c. CH₃CH₂CH₂CH₂CH₂COOH + NaOH ⟶ ?

d. 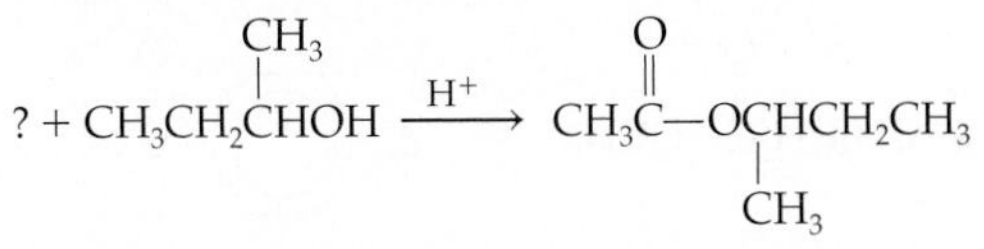 ? + CH₃CH₂CHOH $\xrightarrow{H^+}$ CH₃C—OCHCH₂CH₃ (with CH₃ groups shown)

14.53 How might $CH_3CH_2CH_2CH_2CH_2OH$ be converted to each of the following products?
   a. $CH_3CH_2CH_2CH_2CH_2CHO$
   b. $CH_3CH_2CH_2CH_2CH_2COOH$

14.54 Which of the following alcohols can be oxidized to a carboxylic acid? Name the carboxylic acid produced. For those alcohols that cannot be oxidized to a carboxylic acid, name the final product.
   a. Ethanol   **c.** 1-Propanol
   b. 2-Propanol   **d.** 3-Pentanol

14.55 Write an equation representing the neutralization of pentanoic acid with each of the following bases:
   a. NaOH
   b. KOH
   c. $Ca(OH)_2$

14.56 Write an equation representing the neutralization of each of the following carboxylic acids with KOH:
   a. 3-Chlorohexanoic acid
   b. Cyclohexanecarboxylic acid
   c. 3,4-Dimethylpentanoic acid

14.57 The calcium salt of propionic acid is added to breads as a preservative that prevents mold growth. Draw the structure of the calcium salt of propionic acid. What are the common and IUPAC names of this carboxylic acid salt?

14.58 Oxalic acid is found in the leaves of rhubarb, primarily in the form of the calcium salt. Since high levels of oxalic acid are toxic, only rhubarb stalks are used to make strawberry rhubarb pie. What is the IUPAC name of oxalic acid? Write the structure of the calcium salt of oxalic acid.

Esters: Structure, Physical Properties, and Nomenclature

Foundations

14.59 Explain why esters are described as slightly polar.

14.60 Compare the boiling points of esters to those of aldehydes or ketones of similar molar mass.

14.61 Briefly summarize the IUPAC rules for naming esters.

14.62 How are the common names of esters derived?

Applications

14.63 Draw condensed formulas for each of the following compounds:
   a. Methyl benzoate
   b. Butyl decanoate
   c. Methyl propionate
   d. Ethyl propionate

14.64 Draw condensed formulas for each of the following compounds:
   a. Ethyl *m*-nitrobenzoate
   b. Isopropyl acetate
   c. Methyl butyrate

14.65 Use the IUPAC Nomenclature System to name each of the following esters:

   a. $CH_3C(=O)-OCH_2CH_3$ **c.** $CH_3CHCH_2C(=O)-OCH_3$ (with CH_3 branch)

   b. $CH_3CH_2C(=O)-OCH_3$ **d.** 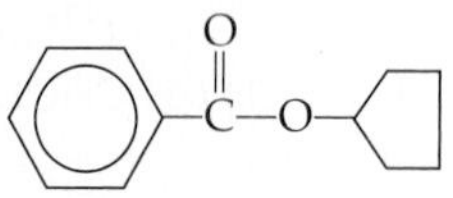 (benzene ring)—$C(=O)$—O—(cyclopentyl)

14.66 Use the IUPAC Nomenclature System to name each of the following:

   a. 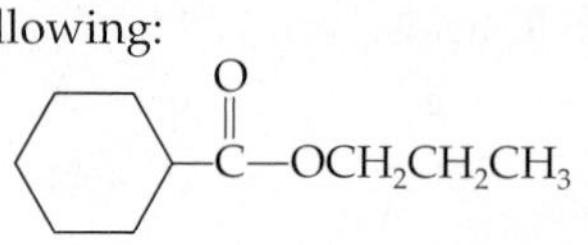 (cyclohexyl)—$C(=O)$—$OCH_2CH_2CH_3$

   b. 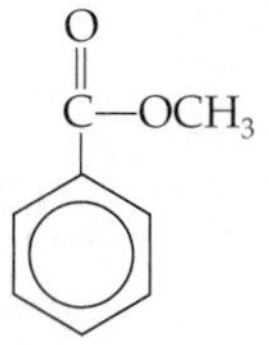 (benzene ring)—$C(=O)$—OCH_3

   c. $CH_2CHCH_2CH_2C(=O)-OCH_2CH_3$ with Br, Br on positions

Esters: Reactions

Foundations

14.67 Write a general reaction showing the preparation of an ester.

14.68 Why is preparation of an ester referred to as a condensation reaction?

14.69 Write a general reaction showing the hydrolysis of an ester using an acid catalyst.

14.70 Write a general reaction showing the base-catalyzed hydrolysis of an ester.

14.71 What is meant by a hydrolysis reaction?

14.72 Why is the salt of a carboxylic acid produced in a base-catalyzed hydrolysis of an ester?

Applications

14.73 Complete each of the following reactions by supplying the missing portion indicated with a question mark:

   a. $CH_3CH_2CH_2C(=O)-OH + CH_3CH_2OH \xrightarrow{H^+,\, heat}$?

   b. $CH_3CH_2C(=O)-OCH_2CH_3 + H_2O \xrightarrow{H^+,\, heat}$?

   c. $CH_3CHCH_2CH_2C(=O)-OH + ? \xrightarrow{H^+,\, heat}$
   (CH_3 branch)
   $CH_3CHCH_2CH_2C(=O)-OCH_2CH_2CH_3$ (CH_3 branch)

   d. $CH_3CH_2CHCH_2C(=O)-OCH_2CH_3 + H_2O \xrightarrow{OH^-,\, heat}$? (Br branch)

14.74 Complete each of the following reactions by supplying the missing portion indicated with a question mark:

   a. ? + $CH_3C(-OH)$ (with CH_3, CH_3 branches) $\xrightarrow{?}$ $CH_3CH_2C(=O)-O-C(-CH_3)$ (with CH_3, CH_3 branches)

   b. $CH_3CH_2CH_2CH_2COOH + CH_3CH_2CH_2CH_2OH \xrightarrow{H^+,\, heat}$?

   c. $CH_3CCH_2C(=O)-OCH_2CH_2CCH_3 + H_2O \xrightarrow{H^+,\, heat}$? (with CH_3, CH_3 and CH_3, CH_3 branches)

   d. $CH_3CH_2C(=O)-OCH_3 + H_2O \xrightarrow{OH^-,\, heat}$?

14.75 Design the synthesis of each of the following esters from organic alcohols.
    a. Isobutyl methanoate (raspberries)
    b. Pentyl butanoate (apricot)

14.76 Design the synthesis of each of the following esters from organic alcohols.
    a. Methyl butanoate (apples)
    b. Octyl ethanoate (oranges)

14.77 What is saponification? Give an example using specific molecules.

14.78 When the methyl ester of hexanoic acid is hydrolyzed in aqueous sodium hydroxide in the presence of heat, a homogeneous solution results. When the solution is acidified with dilute aqueous hydrochloric acid, a new product forms. What is the new product? Draw its structure.

14.79 The structure of salicylic acid is shown. If this acid reacts with methanol, the product is an ester, methyl salicylate. Methyl salicylate is known as oil of wintergreen and is often used as a flavoring agent. Draw the structure of the product of this reaction.

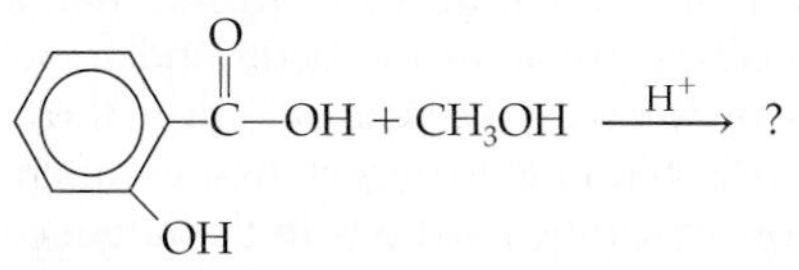

14.80 When salicylic acid reacts with acetic anhydride, one of the products is an ester, acetylsalicylic acid. Acetylsalicylic acid is the active ingredient in aspirin. Complete the following equation by drawing the structure of acetylsalicylic acid. (*Hint:* Acid anhydrides are hydrolyzed by water.)

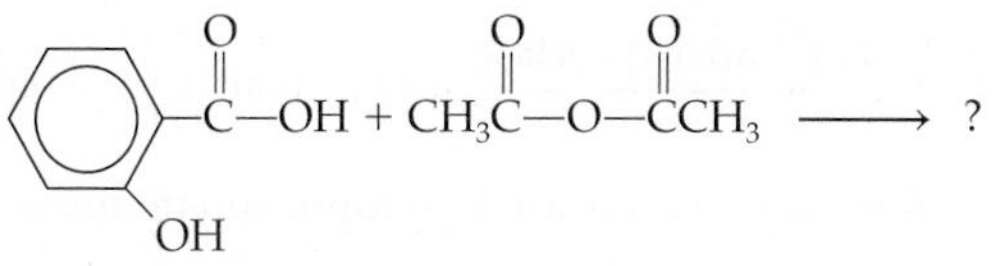

14.81 Compound A ($C_6H_{12}O_2$) reacts with water, acid, and heat to yield compound B ($C_5H_{10}O_2$) and compound C (CH_4O). Compound B is acidic. Deduce possible structures of compounds A, B, and C.

14.82 What products are formed when methyl *o*-bromobenzoate reacts with each of the following?
    a. Aqueous acid and heat
    b. Aqueous base and heat

14.83 Write an equation for the acid-catalyzed hydrolysis of each of the following esters:
    a. Propyl propanoate   **c.** Ethyl methanoate
    b. Butyl methanoate   **d.** Methyl pentanoate

14.84 Write an equation for the base-catalyzed hydrolysis of each of the following esters:
    a. Pentyl methanoate   **c.** Butyl hexanoate
    b. Hexyl propanoate   **d.** Methyl benzoate

Acid Chlorides and Acid Anhydrides

Foundations

14.85 Describe the physical properties of acid chlorides.
14.86 Describe the chemical properties of acid chlorides.
14.87 Describe the physical properties of acid anhydrides.

14.88 Write a general equation for the formation of acid anhydrides.

Applications

14.89 Write the condensed formula for each of the following compounds:
    a. Decanoic anhydride
    b. Acetic anhydride

14.90 Write the condensed formula for each of the following compounds:
    a. Valeric anhydride
    b. Benzoyl chloride

14.91 Write a condensed formula and a line formula for each of the following compounds:
    a. Octanoyl chloride
    b. Butanoyl chloride
    c. Nonanoyl chloride

14.92 Write a condensed formula and a line formula for each of the following compounds:
    a. Decanoyl chloride
    b. Heptanoyl chloride
    c. Ethanoyl chloride

14.93 Write an equation representing the synthesis of methanoic anhydride.

14.94 Write an equation representing the synthesis of octanoic anhydride.

14.95 Write an equation for the reaction of each of the following acid anhydrides with ethanol.
    a. Propanoic anhydride
    b. Ethanoic anhydride
    c. Methanoic anhydride

14.96 Write an equation for the reaction of each of the following acid anhydrides with propanol. Name each of the products using the IUPAC Nomenclature System.
    a. Butanoic anhydride
    b. Pentanoic anhydride
    c. Methanoic anhydride

Nature's High-Energy Compounds: Phosphoesters and Thioesters

14.97 By reacting phosphoric acid with an excess of ethanol, it is possible to obtain the mono-, di-, and triesters of phosphoric acid. Draw all three of these products.

14.98 What is meant by a phosphoanhydride bond?

14.99 We have described the molecule ATP as the body's energy storehouse. What do we mean by this designation? How does ATP actually store energy and provide it to the body as needed?

14.100 Write an equation for each of the following reactions:
    a. Ribose + phosphoric acid
    b. Methanol + phosphoric acid
    c. Adenosine diphosphate + phosphoric acid

14.101 Draw the thioester bond between the acetyl group and coenzyme A.

14.102 Explain the significance of thioester formation in the metabolic pathways involved in fatty acid and carbohydrate breakdown.

14.103 It is also possible to form esters of other inorganic acids such as sulfuric acid and nitric acid. One particularly noteworthy product is nitroglycerine, which is both highly unstable (explosive) and widely used in the treatment of the heart condition known as angina, a constricting pain in the chest usually resulting from coronary heart disease. In the latter case, its function is to alleviate the pain associated with angina. Nitroglycerine may be administered as a tablet (usually placed just beneath the tongue when needed) or as a salve or paste that can be applied to and absorbed through the skin. Nitroglycerine is the trinitroester of glycerol. Draw the structure of nitroglycerine, using the structure of glycerol.

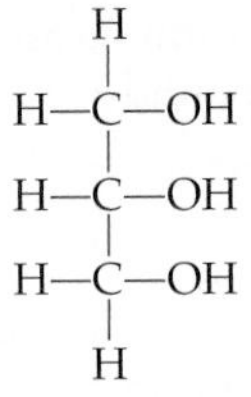

$$
\begin{array}{c}
H \\
| \\
H-C-OH \\
| \\
H-C-OH \\
| \\
H-C-OH \\
| \\
H
\end{array}
$$

Glycerol

14.104 Show the structure of the thioester that would be formed between coenzyme A and stearic acid.

CHALLENGE PROBLEMS

1. Radioactive isotopes of an element behave chemically in exactly the same manner as the nonradioactive isotopes. As a result, they can be used as tracers to investigate the details of chemical reactions. A scientist is curious about the origin of the bridging oxygen atom in an ester molecule. She has chosen to use the radioactive isotope oxygen-18 to study the following reaction:

$$
CH_3CH_2OH + CH_3\overset{\overset{\displaystyle O}{\|}}{C}-OH \xrightarrow{H^+, \text{heat}}
$$

$$
CH_3\overset{\overset{\displaystyle O}{\|}}{C}-O-CH_2CH_3 + H_2O
$$

   Design experiments using oxygen-18 that will demonstrate whether the oxygen in the water molecule came from the —OH of the alcohol or the —OH of the carboxylic acid.

2. Triglycerides are the major lipid storage form in the human body. They are formed in an esterification reaction between glycerol (1,2,3-propanetriol) and three fatty acids (long-chain carboxylic acids). Write a balanced equation for the formation of a triglyceride formed in a reaction between glycerol and three molecules of decanoic acid.

3. Chloramphenicol is a very potent, broad-spectrum antibiotic. It is reserved for life-threatening bacterial infections because it is quite toxic. It is also a very bitter-tasting chemical. As a result, children had great difficulty taking the antibiotic. A clever chemist found that the taste could be improved considerably by producing the palmitate ester. Intestinal enzymes hydrolyze the ester, producing chloramphenicol, which can then be absorbed. The following structure is the palmitate ester of chloramphenicol. Draw the structure of chloramphenicol.

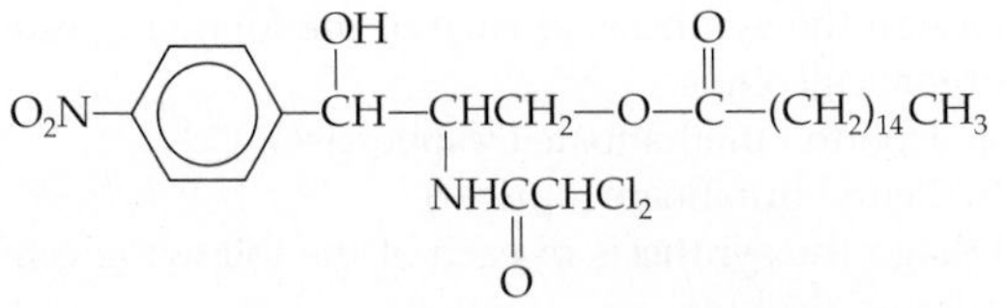

Chloramphenicol palmitate

4. Acetyl coenzyme A (acetyl CoA) can serve as a donor of acetyl groups in biochemical reactions. One such reaction is the formation of acetylcholine, an important neurotransmitter involved in nerve signal transmission at neuromuscular junctions. The structure of choline is shown below. Draw the structure of acetylcholine.

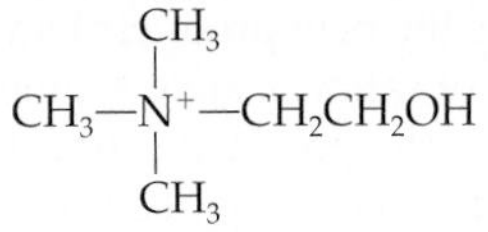

$$
\begin{array}{c}
CH_3 \\
| \\
CH_3-N^+-CH_2CH_2OH \\
| \\
CH_3
\end{array}
$$

Choline

5. Hormones are chemical messengers that are produced in a specialized tissue of the body and travel through the bloodstream to reach receptors on cells of their target tissues. This specific binding to target tissues often stimulates a cascade of enzymatic reactions in the target cells. The work of Earl Sutherland and others led to the realization that there is a *second messenger* within the target cells. Binding of the hormone to the hormone receptor in the cell membrane triggers the enzyme adenyl cyclase to produce adenosine-3′,5′-monophosphate, which is also called *cyclic AMP*, from ATP. The reaction is summarized as follows:

$$
ATP \xrightarrow{Mg^{2+}, \text{ adenyl cyclase}} \text{cyclic AMP} + PP_i + H^+
$$

   PP_i is the abbreviation for a pyrophosphate group, shown here:

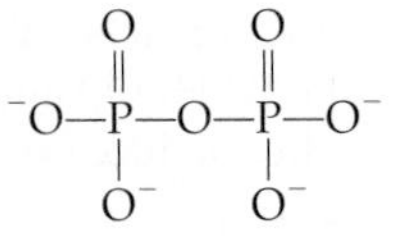

   The structure of ATP is shown here with the carbon atoms of the sugar ribose numbered according to the convention used for nucleotides:

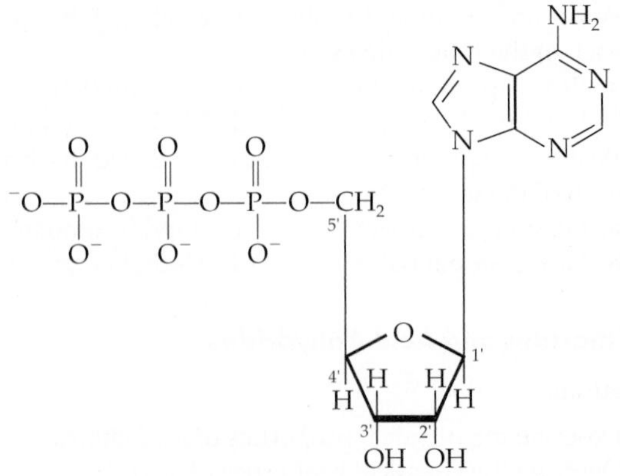

Adenosine-5′-triphosphate

   Draw the structure of adenosine-3′,5′-monophosphate.

Amines and Amides

LEARNING GOALS

1 Classify amines as primary, secondary, or tertiary.

2 Describe the physical properties of amines.

3 Draw and name simple amines using systematic and common nomenclature systems.

4 Write equations representing the synthesis of amines.

5 Write equations showing the basicity and neutralization of amines.

6 Describe the structure of quaternary ammonium salts and discuss their use as antiseptics and disinfectants.

7 Discuss the biological significance of heterocyclic amines.

8 Describe the physical properties of amides.

9 Draw the structure and write the common and IUPAC names of amides.

10 Write equations representing the preparation of amides.

11 Write equations showing the hydrolysis of amides.

12 Draw the general structure of an amino acid.

13 Draw and discuss the structure of a peptide bond.

14 Describe the function of neurotransmitters.

Ethnobotanists continue to search for medically active compounds from the rain forest.

OUTLINE

INTRODUCTION

In this chapter, we introduce an additional element into the structure of organic molecules. That element is nitrogen, the fourth most common atom in living systems. It is an important component of the structure of the nucleic acids DNA and RNA, which are the molecules that carry the genetic information for living systems. It is also essential to the structure and function of proteins, molecules that carry out the majority of the work in biological systems (enzymes), protect us from infection (antibodies), and are structural components of the cell and body.

One class of organic molecules containing nitrogen is the amines. Amines are characterized by the presence of an amino group (—NH$_2$).

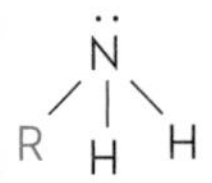

$$\overset{\displaystyle ..}{\underset{R\quad H\quad H}{N}}$$

General structure of an amine

The nitrogen atom of the amino group may have one or more of its hydrogen atoms replaced by an organic group. General structures of these types of amines are shown below:

$$\overset{\displaystyle ..}{\underset{R\quad R^1\quad H}{N}} \qquad \overset{\displaystyle ..}{\underset{R\quad R^1\quad R^2}{N}}$$

Amines are very common in biological systems and exhibit important physiological activity. Consider histamine. Histamine contributes to the inflammatory response that causes the symptoms of colds and allergies, including swollen mucous membranes, congestion, and excessive nasal secretions. We take antihistamines to help relieve these symptoms. Ephedrine is an antihistamine that has been extracted from the leaves of the ma-huang plant in China for over 2000 years. Found in over-the-counter cold medications, this antihistamine helps to shrink swollen mucous membranes and reduce nasal secretions. The structures of histamine and ephedrine are shown below.

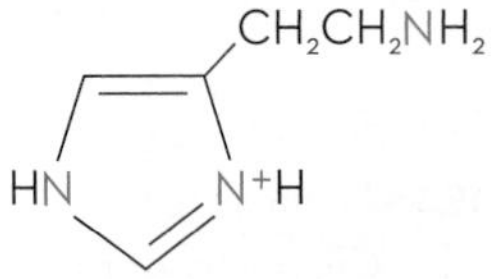

Histamine

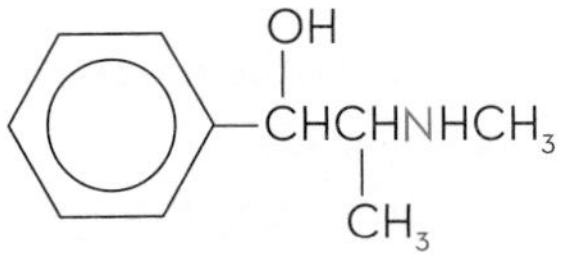

Ephedrine

Ephedra
(Ma Huang)

Ephedra, from the ma-huang plant, has been sold to promote weight loss and boost energy. What antihistamines are found in ephedra?

The other group of nitrogen-containing organic compounds we will investigate in this chapter is the amides. Amides are the products of a reaction between an amine and a carboxylic acid derivative. They have the following general structure:

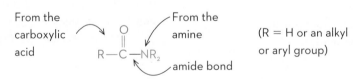

From the carboxylic acid

From the amine

amide bond

(R = H or an alkyl or aryl group)

General structure of an amide

Amino acids are the subunits from which proteins are built. They are characterized by the presence of both an amino group and a carboxyl group. When amino acids are bonded to one another to produce a protein chain, the amino group of one amino acid reacts with the carboxyl group of another amino acid. The amide bond that results is called a *peptide bond*.

In this chapter, we will explore the chemistry of the organic molecules that contain nitrogen. In upcoming chapters, we will investigate the structure and properties of the nitrogen-containing biological molecules.

The general structure of an amino acid is

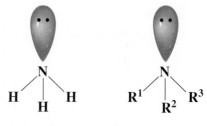

The amino group is highlighted in red and the carboxyl group in blue. Amino acids are the basic subunits of all proteins.

15.1 Amines

Structure and Physical Properties

Amines are organic derivatives of ammonia and, like ammonia, they are basic. In fact, amines are the most important type of organic base found in nature. We can think of them as substituted ammonia molecules in which one, two, or three of the ammonia hydrogens have been replaced by an organic group:

Ammonia → An amine

R substitutes for H

The structures drawn above and in Figure 15.1 reveal that, like ammonia, amines are pyramidal. The nitrogen atom has three groups bonded to it and has a non-bonding pair of electrons.

Amines are classified according to the number of alkyl or aryl groups attached to the nitrogen. A **primary (1°) amine** has one organic group attached to the nitrogen atom. A **secondary (2°) amine** has two organic groups bonded to the nitrogen atom, and a **tertiary (3°) amine** has three organic groups attached to the nitrogen atom:

The geometry of ammonia is described in Section 3.4.

| Ammonia | 1° amine (primary amine) | 2° amine (secondary amine) | 3° amine (tertiary amine) |

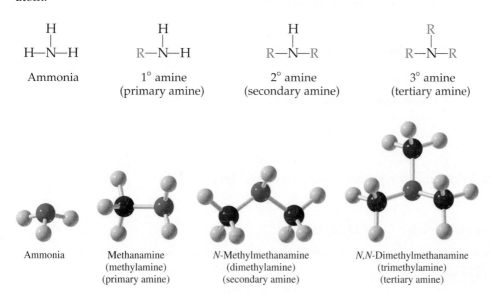

Ammonia
Methanamine (methylamine) (primary amine)
N-Methylmethanamine (dimethylamine) (secondary amine)
N,*N*-Dimethylmethanamine (trimethylamine) (tertiary amine)

Ammonia   Amine

Figure 15.1 The trigonal pyramidal structure of amines. Note the similarities in structure between an amine and the ammonia molecule.

The nitrogen atom is more electronegative than the hydrogen atoms in amines. As a result, the N—H bond is polar. In addition, the nitrogen atom contains an unshared pair of electrons. As a result, hydrogen bonding between amine molecules or between amine molecules and water can occur (Figure 15.2).

Hydrogen bonding is described in Section 5.2.

Figure 15.2 Hydrogen bonding (a) in methylamine and (b) between methylamine and water. Dotted lines represent hydrogen bonds.

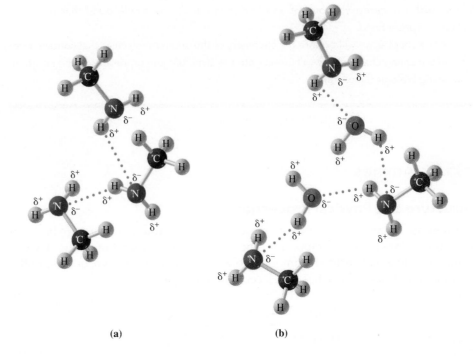

(a)   (b)

EXAMPLE 15.1   **Classifying Amines as Primary, Secondary, or Tertiary**

Classify each of the following compounds as a primary, secondary, or tertiary amine.

LEARNING GOAL

1 Classify amines as primary, secondary, or tertiary.

Solution

Compare the structure of the amine with that of ammonia.

$$\underset{\overset{|}{H-N-H}}{\overset{CH_3}{}} \qquad \underset{\overset{|}{H-N-H}}{\overset{H}{}} \qquad 1° \text{ amine: one hydrogen replaced}$$

$$\underset{\overset{|}{CH_3-N-H}}{\overset{CH_3}{}} \qquad \underset{\overset{|}{H-N-H}}{\overset{H}{}} \qquad 2° \text{ amine: two hydrogens replaced}$$

$$\underset{\overset{|}{CH_3-N-CH_3}}{\overset{CH_3}{}} \qquad \underset{\overset{|}{H-N-H}}{\overset{H}{}} \qquad 3° \text{ amine: three hydrogens replaced}$$

Practice Problem 15.1

Determine whether each of the following amines is primary, secondary, or tertiary.

$$\text{a. } \underset{\overset{|}{CH_3CH_2NCH_3}}{\overset{CH_2CH_3}{}} \qquad \text{b. } CH_3CH_2CH_2NH_2 \qquad \text{c. } \underset{\overset{|}{CH_3NCH_3}}{\overset{H}{}}$$

▶ For Further Practice: **Questions 15.29 and 15.30.**

Question 15.1   Refer to Figure 15.2 and draw a similar figure showing the hydrogen bonding that occurs between water and a 2° amine.

Question 15.2   Refer to Figure 15.2 and draw hydrogen bonding between two primary amines.

TABLE 15.1 Boiling Points of Amines

Systematic Name	Common Name	Structure	Boiling Point (°C)
	Ammonia	NH_3	−33.4
Methanamine	Methylamine	CH_3NH_2	−6.3
N-Methylmethanamine	Dimethylamine	$(CH_3)_2NH$	7.4
N,N-Dimethylmethanamine	Trimethylamine	$(CH_3)_3N$	2.9
Ethanamine	Ethylamine	$CH_3CH_2NH_2$	16.6
Propanamine	Propylamine	$CH_3CH_2CH_2NH_2$	48.7
Butanamine	Butylamine	$CH_3CH_2CH_2CH_2NH_2$	77.8

The ability of primary and secondary amines to form N—H···N hydrogen bonds is reflected in their boiling points (Table 15.1). Primary amines have boiling points well above those of alkanes of similar molar mass. Because the nitrogen atom is less electronegative than the oxygen atom, the N—H bond is less polar than the O—H bond. For this reason, amines have considerably lower boiling points than alcohols of comparable molar mass. These trends can be seen in the following examples:

LEARNING GOAL

2 Describe the physical properties of amines.

$CH_3CH_2CH_3$
Propane
M.M. = 44 g/mol
b.p. = −42.2°C

$CH_3CH_2NH_2$
Ethanamine
M.M. = 45 g/mol
b.p. = 16.6°C

CH_3CH_2OH
Ethanol
M.M. = 46 g/mol
b.p. = 78.5°C

The boiling points of primary amines increase as the carbon chain length increases because the London dispersion forces between molecules increase as the size of the molecule increases.

The boiling points of secondary amines are somewhat lower than those of primary amines. This is seen in a comparison of dimethylamine (a secondary amine) and its structural isomer ethylamine (a primary amine).

CH_3NHCH_3
Dimethylamine
M.M. = 45 g/mol
b.p. = 7.4°C

$CH_3CH_2NH_2$
Ethylamine
M.M. = 45 g/mol
b.p. = 16.6°C

Secondary amines are still able to hydrogen bond to other secondary amines. But, the nitrogen atom is now in the middle of the carbon chain, rather than at the end, and this decreases the polarity of the N—H bond. Since the N—H bond of a secondary amine is less polar than that of a primary amine, the dipole-dipole attractions between secondary amine molecules are lower. This is reflected in the lower boiling points.

Because tertiary amines do not have an N—H bond, they cannot form intermolecular hydrogen bonds with other tertiary amines. As a result, they have lower boiling points than either primary or secondary amines. This is seen in a comparison of the boiling points of several amines all having the same molar mass. First consider the boiling points of propanamine (a 1° amine) and N,N-dimethylmethanamine (a 3° amine), shown below. The boiling point of the tertiary amine (2.9°C) is significantly lower than that of the 1° amine (48.7°C). As we would expect, *N*-methylethanamine, a 2° amine, has a boiling point that is slightly lower than propanamine (1°) and much higher than *N*,*N*-dimethylmethanamine (3°). Clearly the inability of trimethylamine molecules to form intermolecular hydrogen bonds results in a much lower boiling point.

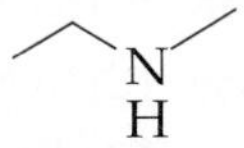

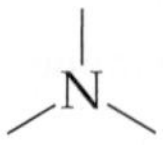

$CH_3CH_2CH_2—NH_2$

Propanamine
(propylamine)
M.M. = 59 g/mol
b.p. = 48.7°C

$CH_3CH_2—\overset{\overset{\displaystyle H}{|}}{N}—CH_3$

N-Methylethanamine
(ethylmethylamine)
M.M. = 59 g/mol
b.p. = 36.7°C

$CH_3—\overset{\overset{\displaystyle CH_3}{|}}{N}—CH_3$

N,*N*-Dimethylmethanamine
(trimethylamine)
M.M. = 59 g/mol
b.p. = 2.9°C

The intermolecular hydrogen bonds formed by primary and secondary amines are not as strong as the hydrogen bonds formed by alcohols because nitrogen is not as electronegative as oxygen. For this reason, primary and secondary amines have lower boiling points than alcohols (Table 15.2).

All amines can form intermolecular hydrogen bonds with water (O—H···N). As a result, small amines (six or fewer carbons) are soluble in water. As we have noted for other families of organic molecules, water solubility decreases as the length of the hydrocarbon (hydrophobic) portion of the molecule increases.

TABLE 15.2 Comparison of the Boiling Points of Selected Alcohols and Amines

Name	Molar Mass (g/mol)	Boiling Point (°C)
Methanol	32	64.5
Methanamine	31	−6.3
Ethanol	46	78.5
Ethanamine	45	16.6
Propanol	60	97.2
Propanamine	59	48.7

EXAMPLE 15.2 **Predicting the Physical Properties of Amines**

LEARNING GOAL

Which member of each of the following pairs of molecules has the higher boiling point?

2 Describe the physical properties of amines.

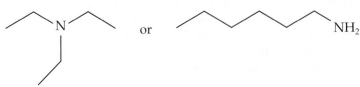

Solution

The molecule on the right, hexanamine, has a higher boiling point than the molecule on the left, *N*,*N*-diethylethanamine (triethylamine). Triethylamine is a tertiary amine; therefore, it has no N—H bond and cannot form intermolecular hydrogen bonds with other triethylamine molecules.

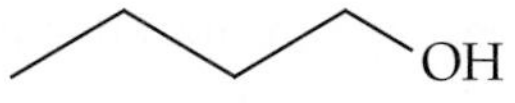

Solution

The molecule on the left, 1-butanol, has a higher boiling point than the molecule on the right, 1-butanamine. Nitrogen is not as electronegative as oxygen, thus the hydroxyl group is more polar than the amino group and forms stronger hydrogen bonds.

Practice Problem 15.2

Which compound in each of the following pairs would you predict to have a higher boiling point? Explain your reasoning.

a. Methanol or methylamine

b. Dimethylamine or water

c. Methylamine or ethylamine

d. Propylamine or butane

▶ For Further Practice: **Questions 15.18 and 15.19.**

Nomenclature

In systematic nomenclature, primary amines are named according to the following rules:

LEARNING GOAL

3 Draw and name simple amines using systematic and common nomenclature systems.

- Determine the name of the *parent compound*, the longest continuous carbon chain containing the amine group.
- Replace the *-e* ending of the alkane chain with *-amine*. Following this pattern, the alkane becomes an alkan*amine*; for instance, etha*ne* becomes ethan*amine*.
- Number the parent chain to give the carbon bearing the amine group the lowest possible number.
- Name and number all substituents, and add them as prefixes to the "alkanamine" name.

For instance,

$$CH_3{-}NH_2 \qquad CH_3CH_2CH_2{-}NH_2 \qquad \underset{\overset{|}{NH_2}}{CH_3CH_2CH_2CHCH_3}$$

Methanamine   1-Propanamine   2-Pentanamine

Secondary amines have two alkyl groups attached to the nitrogen atom.

- Determine which is the longer of the two groups; this is the parent compound.
- Name the parent compound as an alkanamine. For example, propa*ne* becomes propan*amine*.
- Name the second alkyl group as an *N*-alkyl group. For instance, a methyl group would be named *N*-methyl.
- Place the *N*-alkyl group name as a prefix to the parent amine, as shown in the following examples:

$$CH_3{-}NH{-}CH_2CH_2CH_3 \qquad\qquad CH_3CH_2{-}NH{-}CH_2CH_2CH_2CH_2CH_2CH_3$$

Methyl group   Parent chain propane   Ethyl group   Parent chain hexane
becomes *N*-methyl   becomes 1-propanamine   becomes *N*-ethyl   becomes 1-hexanamine

         N-Methyl-1-propanamine   *N*-Ethyl-1-hexanamine

Tertiary amines have three alkyl groups attached to the nitrogen atom.

- Determine which is the longest alkyl group; this is the parent compound.
- Name the parent compound as an alkanamine.
- Name the other two alkyl groups as *N*-alkyl groups, and use these names as a prefix in front of the name of the parent alkanamine.
- If the two alkyl groups are different, for instance, an ethyl and a methyl group, the two *N*-alkyl names are listed sequentially; in the example to the right below, the prefix name would be *N*-ethyl-*N*-methyl.
- If the two groups are identical, perhaps two methyl groups, the name of the prefix is written as *N*,*N*-dimethyl (below, left).
- Write the prefix names before the name of the parent amine, as shown in the following examples:

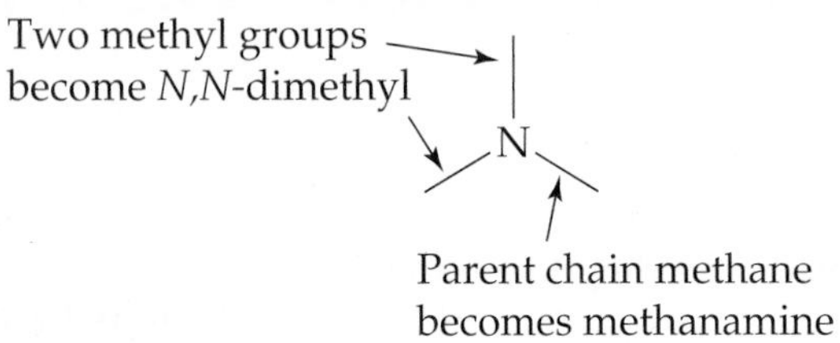

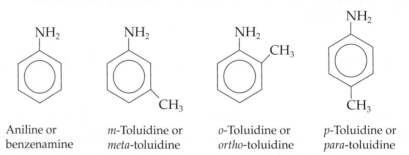

N,*N*-Dimethylmethanamine 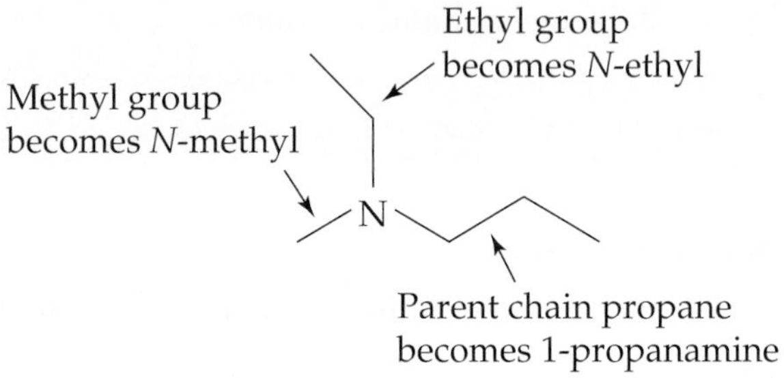   *N*-Ethyl-*N*-methyl-1-propanamine

Aromatic amines have a nitrogen atom bonded directly to the benzene ring. Several aromatic amines have special names that have also been approved for use by IUPAC. For example, the amine of benzene is given the name *aniline*. The systematic name for aniline is *benzenamine*.

| Aniline or benzenamine | *m*-Toluidine or *meta*-toluidine | *o*-Toluidine or *ortho*-toluidine | *p*-Toluidine or *para*-toluidine |

If additional groups are attached to the nitrogen of an aromatic amine, they are indicated with the letter *N*- followed by the name of the group.

EXAMPLE 15.3   **Writing the Systematic Name for an Amine**

a. Determine the systematic name for the following amine, which is used by the German cockroach as a pheromone.

$$CH_3-NH-CH_3$$

Solution

Parent compound: methane (becomes methanamine)
Additional group on N: methyl (becomes *N*-methyl)
Name: *N*-Methylmethanamine

Aniline was first isolated from the blue dye indigo, a product of the indigo plant. It is the starting material in the synthesis of hundreds of dyes. Is aniline a primary, secondary, or tertiary amine?

LEARNING GOAL

3 Draw and name simple amines using systematic and common nomenclature systems.

b. Name the following amine.

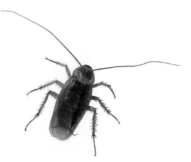

$$\text{} -NH-CH_2CH_2CH_3$$

Solution

Parent compound: cyclohexane (becomes cyclohexanamine)
Additional group on N: propyl (becomes *N*-propyl)
Name: *N*-Propylcyclohexanamine

N-methylmethanamine is used by the German cockroach as a communication pheromone. What is the common name of this amine?

Practice Problem 15.3

What is the systematic name for each of the following amines?

a. $CH_3CH_2-\underset{\underset{CH_2CH_2CH_3}{|}}{N}-CH_3$ c. $CH_3CH_2CH_2-\underset{\underset{CH_2CH_3}{|}}{N}-CH_2CH_3$

b. $CH_3CH_2\underset{\underset{NH_2}{|}}{CH}CH_2CH_2CH_3$

▶ For Further Practice: **Questions 15.21, 15.27, and 15.28.**

Common names are often used for the simple amines. The common names of the alkyl groups bonded to the amine nitrogen are followed by the ending *-amine*. Each group is listed alphabetically in one continuous word followed by the suffix *-amine*:

Tables 10.5 and 10.6 provide the names of the most common linear and branched alkyl groups.

CH_3-NH_2 $CH_3-NH-CH_3$ $CH_3-\underset{\underset{CH_3}{|}}{\overset{\overset{CH_3}{|}}{N}}-CH_3$

Methylamine   Dimethylamine   Trimethylamine

$CH_3CH_2-NH_2$ $CH_3CH_2-NH-CH_3$

Ethylamine   Ethylmethylamine

Table 15.3 compares these systems of nomenclature for a number of simple amines.

TABLE 15.3 Systematic and Common Names of Amines

Compound	Systematic Name	Common Name
R—NH₂	Alkan*amine*	Alkyl*amine*
CH₃—NH₂	Methanamine	Methylamine
CH₃CH₂—NH₂	Ethanamine	Ethylamine
CH₃CH₂CH₂—NH₂	1-Propanamine	Propylamine
CH₃—NH—CH₃	*N*-Methylmethanamine	Dimethylamine
CH₃—NH—CH₂CH₃	*N*-Methylethanamine	Ethylmethylamine
$CH_3-\underset{\underset{CH_3}{|}}{\overset{\overset{CH_3}{|}}{N}}-CH_3$	*N,N*-Dimethylmethanamine	Trimethylamine

Question 15.3   Use the structure of aniline provided and draw the condensed formula for each of the following amines.
    a. *N*-Methylaniline   c. *N*-Ethylaniline
    b. *N,N*-Dimethylaniline   d. *N*-Isopropylaniline

Question 15.4   Name each of the following amines using the systematic and common nomenclature systems.

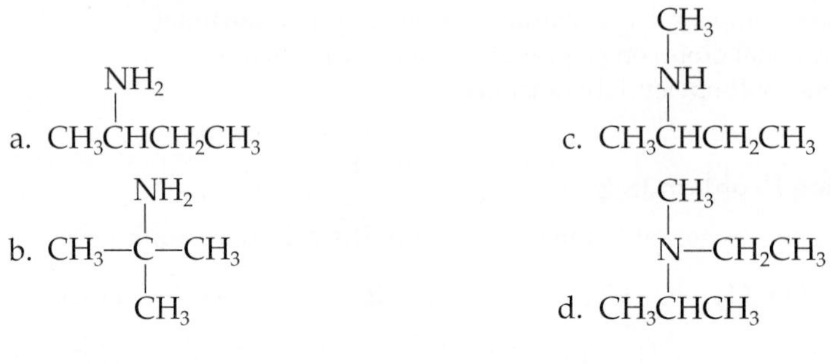

                                              CH₃
                                              |
              NH₂   NH
              | |
a. CH₃CHCH₂CH₃   c. CH₃CHCH₂CH₃

              NH₂   CH₃
              | |
b. CH₃—C—CH₃   N—CH₂CH₃
              | |
              CH₃   d. CH₃CHCH₃

Question 15.5   Draw the condensed formula for each of the following compounds.
    a. 2-Propanamine   d. 2-Methyl-2-pentanamine
    b. 3-Octanamine   e. 4-Chloro-5-iodo-1-nonanamine
    c. *N*-Ethyl-2-heptanamine   f. *N,N*-Diethyl-1-pentanamine

Question 15.6   Draw the condensed formula for each of the following compounds.
    a. Diethylmethylamine   d. Triisopropylamine
    b. 4-Methylpentylamine   e. Methyl-*t*-butylamine
    c. *N*-Methylaniline   f. Ethylhexylamine

Medically Important Amines

Although amines play many different roles in our day-to-day lives, one important use is in medicine. A host of drugs derived from amines is responsible for improving the quality of life, whereas others, such as cocaine and heroin, are highly addictive.

Amphetamines, such as benzedrine and methedrine, stimulate the central nervous system. They elevate blood pressure and pulse rate and are often used to decrease fatigue. Medically, they have been used to treat depression and epilepsy. Amphetamines have also been prescribed as diet pills because they decrease the appetite. Their use is controlled by federal law because excess use of amphetamines can cause paranoia and mental illness.

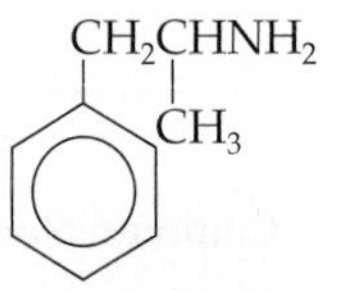

1-Phenyl-2-propanamine
(Amphetamine)

Benzedrine

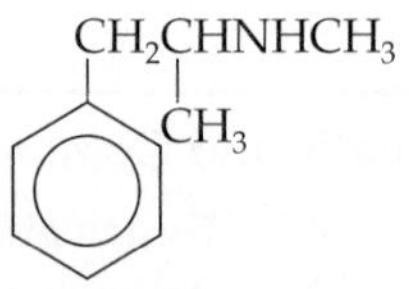

N-Methyl-1-phenyl-2-propanamine
(Methamphetamine)

Methedrine

Many of the medicinal amines are *analgesics* (pain relievers) or *anesthetics* (pain blockers). Novocaine and related compounds, for example, are used as local anesthetics. Demerol is a very strong pain reliever.

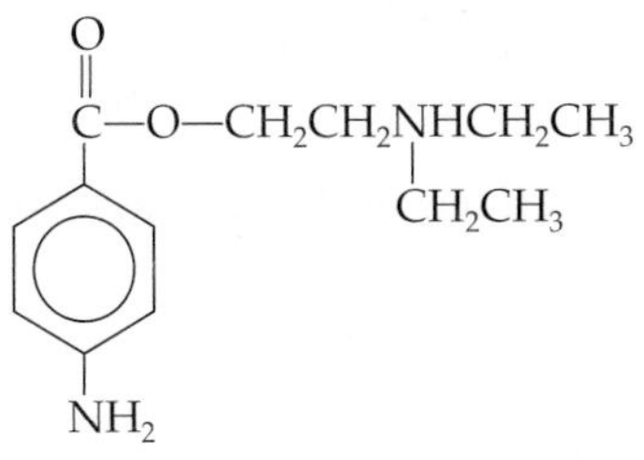

Novocaine

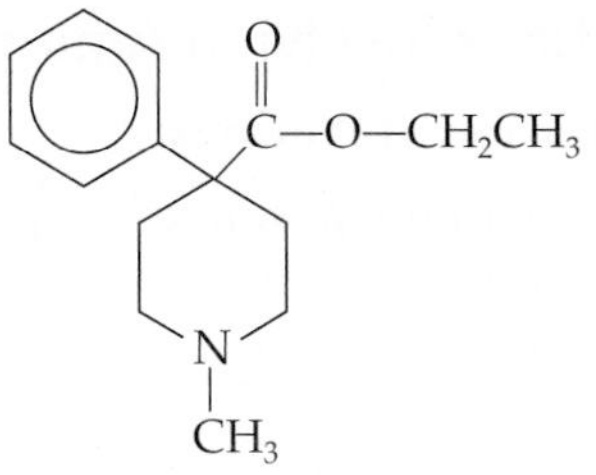

Demerol

Ephedrine, its stereoisomer pseudoephedrine, and phenylephrine (also called neosynephrine) are used as decongestants in cough syrups and nasal sprays. By shrinking the membranes that line the nasal passages, they relieve the symptoms of congestion and a stuffy nose. These compounds are very closely related to L-dopa and dopamine, which are key compounds in the function of the central nervous system.

L-Dopa, dopamine, and other key neurotransmitters are described in detail in Section 15.5.

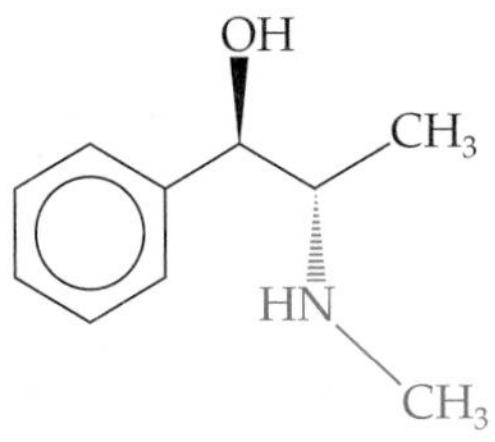

Ephedrine

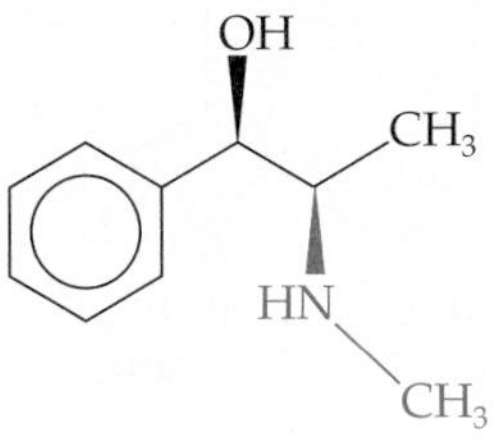

Pseudoephedrine

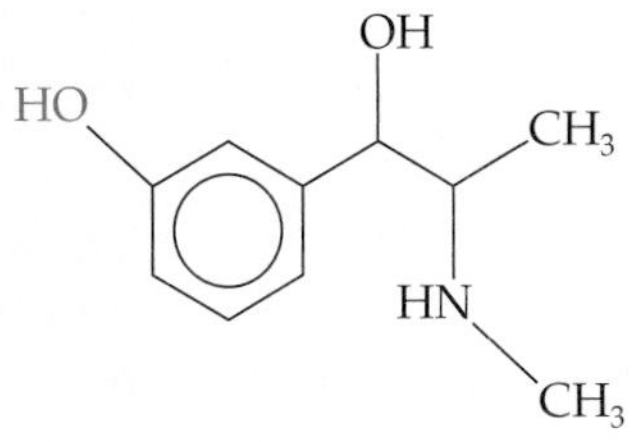

Phenylephrine (neosynephrine)

Recently, many states have restricted the sales of products containing ephedrine and pseudoephedrine, and many drugstore chains have moved these products behind the counter. The reason for these precautions is that either ephedrine or pseudoephedrine can be used as the starting material in the synthesis of methamphetamines. In response to this problem, pharmaceutical companies are replacing ephedrine and pseudoephedrine in these decongestants with phenylephrine, which cannot be used as a reactant in the synthesis of methamphetamine.

Methamphetamine use and abuse are discussed in Chemistry at the Crime Scene: Methamphetamine, found on page 536.

Ephedrine and pseudoephedrine are also the primary active ingredients in ephedra, a plant found in the deserts of central Asia. Ephedra is used as a stimulant in a variety of products that are sold over-the-counter as aids to boost energy, promote weight loss, and enhance athletic performance. In 2004, the Food and Drug Administration (FDA) banned the use of ephedra in these over-the-counter formulations after reviewing 16,000 reports of adverse side effects including nervousness, heart irregularities, seizures, heart attacks, and 80 deaths, including that of Baltimore Orioles pitcher Steve Bechler, age 23.

The *sulfa drugs,* the first chemicals used to fight bacterial infections, are synthesized from amines.

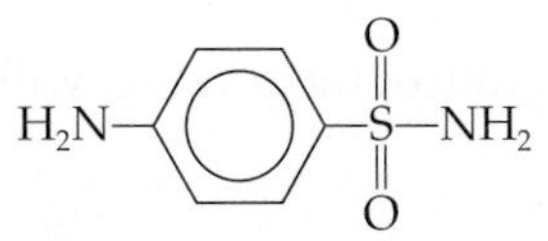

Sulfanilamide—a sulfa drug

Reactions Involving Amines

Preparation of Amines

In the laboratory, amines are prepared by the reduction of amides and nitro compounds.

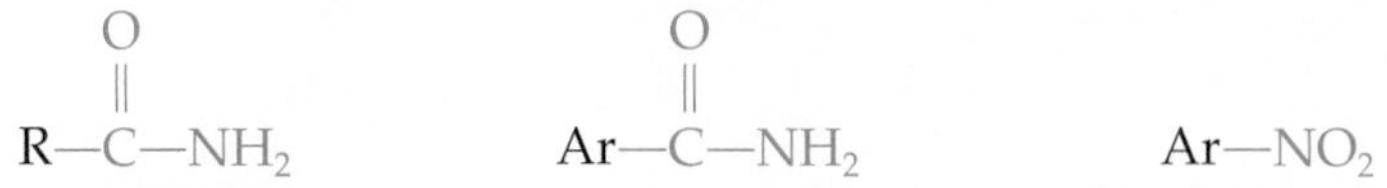

$$
\underset{\text{Examples of amides}}{R-\overset{\overset{\displaystyle O}{\|}}{C}-NH_2 \qquad\qquad Ar-\overset{\overset{\displaystyle O}{\|}}{C}-NH_2} \qquad\qquad \underset{\text{A nitro compound}}{Ar-NO_2}
$$

As we will see in Section 15.3, amides are neutral nitrogen compounds that produce an amine and a carboxylic acid when hydrolyzed. Nitro compounds are prepared by the nitration of an aromatic compound.

Primary amines are readily produced by reduction of a nitro compound, as in the following reaction:

In this reaction, the nitro compound is nitrobenzene and the product is aniline.

Amides may also be reduced to produce primary, secondary, or tertiary amines.

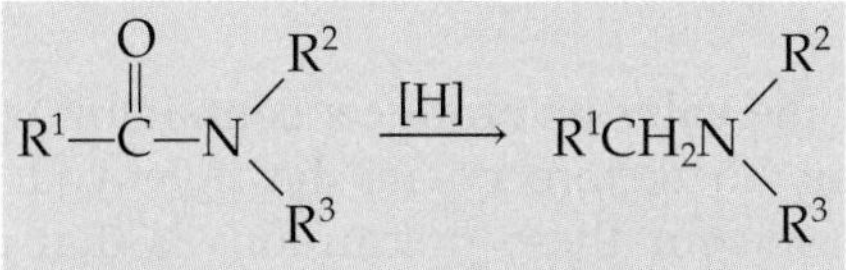

$$
\underset{\text{Amide}}{R^1-\overset{\overset{\displaystyle O}{\|}}{C}-N\!\!\begin{array}{c}R^2\\[2pt]\diagdown\\[-6pt]R^3\end{array}} \xrightarrow{\ [H]\ } \underset{\text{Amine}}{R^1CH_2N\!\!\begin{array}{c}R^2\\[2pt]\diagdown\\[-6pt]R^3\end{array}}
$$

$(R^2$ and R^3 may be a hydrogen atom or an organic group.)

If R^2 and R^3 are hydrogen atoms, the product will be a primary amine:

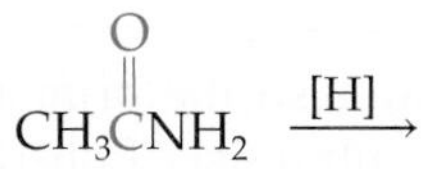

$$
\underset{\text{Ethanamide}}{CH_3\overset{\overset{\displaystyle O}{\|}}{C}NH_2} \xrightarrow{\ [H]\ } \underset{\substack{\text{Ethanamine}\\ \text{(ethylamine)}}}{CH_3CH_2NH_2}
$$

If either R^2 or R^3 is an organic group, the product will be a secondary amine:

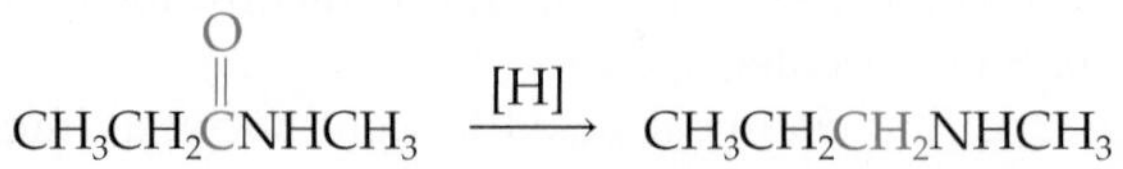

$$
\underset{\text{N-Methylpropanamide}}{CH_3CH_2\overset{\overset{\displaystyle O}{\|}}{C}NHCH_3} \xrightarrow{\ [H]\ } \underset{\substack{\text{N-Methyl-1-propanamine}\\ \text{(methylpropylamine)}}}{CH_3CH_2CH_2NHCH_3}
$$

If both R^2 and R^3 are organic groups, the product will be a tertiary amine:

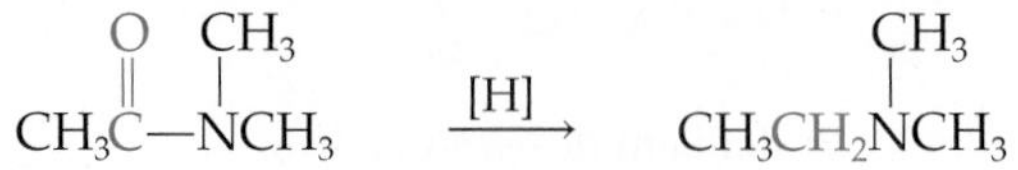

$$
\underset{\text{N,N-Dimethylethanamide}}{CH_3\overset{\overset{\displaystyle O}{\|}}{C}-N\overset{\overset{\displaystyle CH_3}{|}}{C}H_3} \xrightarrow{\ [H]\ } \underset{\text{N,N-Dimethylethanamine}}{CH_3CH_2N\overset{\overset{\displaystyle CH_3}{|}}{C}H_3}
$$

LEARNING GOAL

4 Write equations representing the synthesis of amines.

We are using the general symbol [H] to represent any reducing agent just as we used [O] to represent an oxidizing agent in previous chapters. Several different reducing agents may be used to effect the changes shown here; for example, metallic iron and acid may be used to reduce aromatic nitro compounds, and LiAlH$_4$ in ether reduces amides.

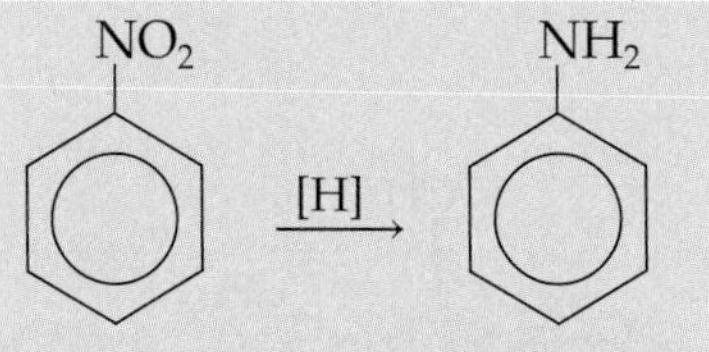

Basicity

Amines behave as weak bases, accepting H^+ when dissolved in water. The non-bonding pair (lone pair) of electrons of the nitrogen atom can be shared with a proton (H^+) from a water molecule, producing an **alkylammonium ion.** Hydroxide ions are also formed, so the resulting solution is basic.

LEARNING GOAL

5 Write equations showing the basicity and neutralization of amines.

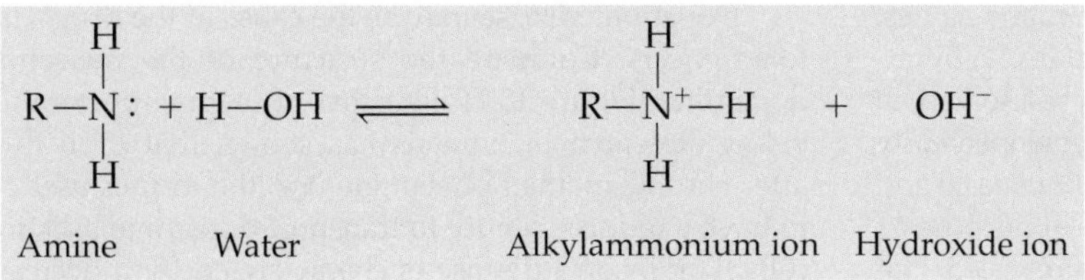

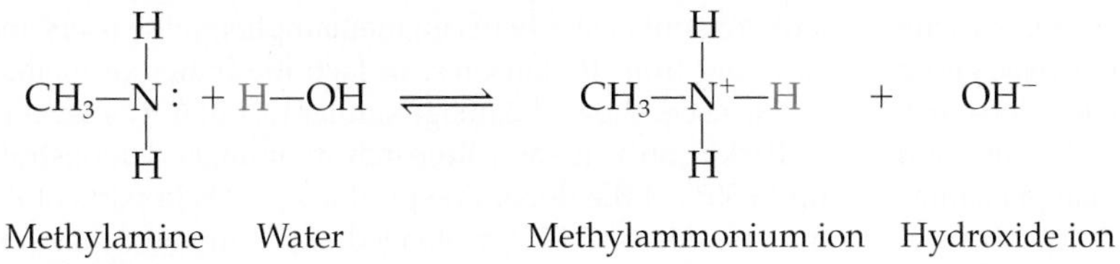

$$
\begin{array}{ccc}
\underset{\text{Amine}}{\overset{\displaystyle H}{\underset{\displaystyle H}{R-N:}}} + \underset{\text{Water}}{H-OH} & \rightleftharpoons & \underset{\text{Alkylammonium ion}}{\overset{\displaystyle H}{\underset{\displaystyle H}{R-N^+-H}}} + \underset{\text{Hydroxide ion}}{OH^-}
\end{array}
$$

$$
\begin{array}{ccc}
\underset{\text{Methylamine}}{\overset{\displaystyle H}{\underset{\displaystyle H}{CH_3-N:}}} + \underset{\text{Water}}{H-OH} & \rightleftharpoons & \underset{\text{Methylammonium ion}}{\overset{\displaystyle H}{\underset{\displaystyle H}{CH_3-N^+-H}}} + \underset{\text{Hydroxide ion}}{OH^-}
\end{array}
$$

Neutralization

Because amines are bases, they react with acids to form alkylammonium salts.

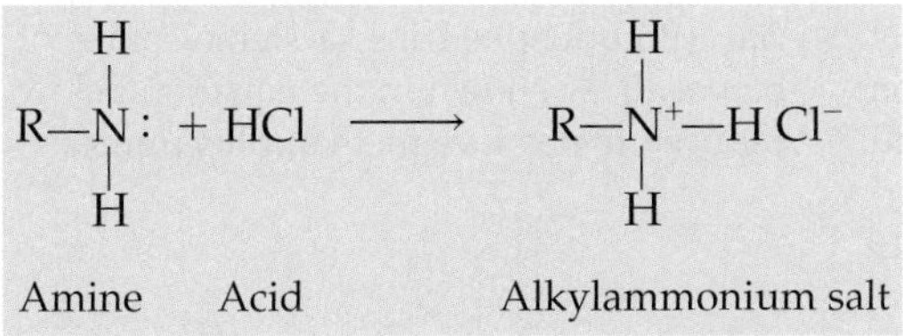

$$
\begin{array}{ccc}
\underset{\text{Amine}}{\overset{\displaystyle H}{\underset{\displaystyle H}{R-N:}}} + \underset{\text{Acid}}{HCl} & \longrightarrow & \underset{\text{Alkylammonium salt}}{\overset{\displaystyle H}{\underset{\displaystyle H}{R-N^+-H\ Cl^-}}}
\end{array}
$$

The reaction of methylamine with hydrochloric acid shown is typical of these reactions. The product is an alkylammonium salt, methylammonium chloride.

Recall that the reaction of an acid and a base gives a salt (Section 8.3).

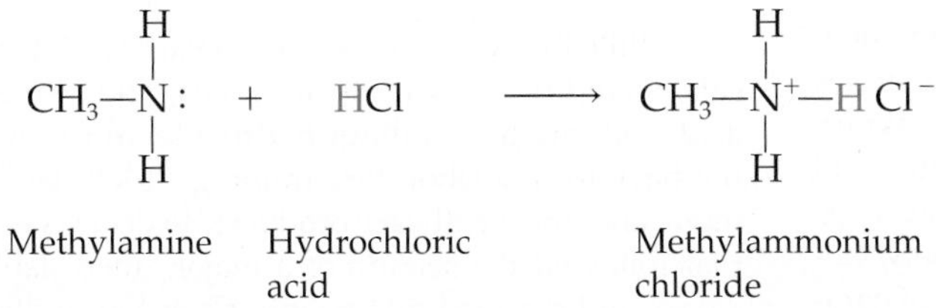

$$
\begin{array}{ccc}
\underset{\text{Methylamine}}{\overset{\displaystyle H}{\underset{\displaystyle H}{CH_3-N:}}} + \underset{\substack{\text{Hydrochloric}\\\text{acid}}}{HCl} & \longrightarrow & \underset{\substack{\text{Methylammonium}\\\text{chloride}}}{\overset{\displaystyle H}{\underset{\displaystyle H}{CH_3-N^+-H\ Cl^-}}}
\end{array}
$$

Alkylammonium salts are named by replacing the suffix *-amine* with *ammonium.* This is then followed by the name of the anion, as shown in the following examples:

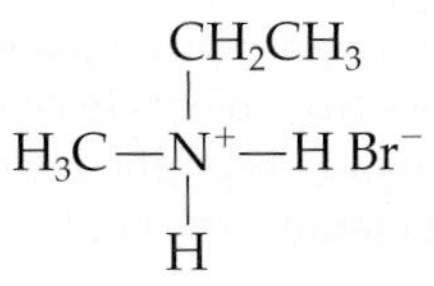

$$
\overset{\displaystyle CH_2CH_3}{\underset{\displaystyle H}{H_3C-N^+-H\ Br^-}}
$$

The ethylmethylamine salt is
ethylmethylammonium bromide

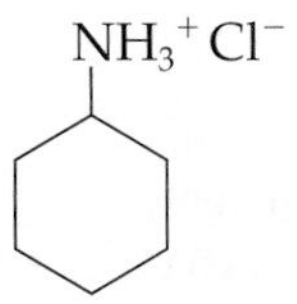

The cyclohexylamine salt is
cyclohexylammonium chloride

A variety of important drugs are amines. They are usually administered as alkylammonium salts because the salts are ionic, and therefore are much more soluble in aqueous solutions and in body fluids.

Chemistry at the Crime Scene

Methamphetamine

Methamphetamine is an addictive drug known by many names, including "speed," "crystal," "crank," "ice," and "glass." A bitter-tasting, odorless, crystalline powder, it is easily dissolved in either water or alcohol. Methamphetamine was developed early in the twentieth century and used as a decongestant in nasal and bronchial inhalers. Now it is rarely used for medical purposes.

A 2002 Health and Human Services survey revealed that twelve million Americans age twelve and older had used methamphetamine. Use of methamphetamine was once associated with white, male, blue-collar workers, but a much more diverse group now uses the drug. Although it is still used by people in jobs such as long-distance trucking that require long hours and mental and physical alertness, it is disturbing that use of methamphetamine is becoming increasingly associated with sexual activity, teenagers attending "raves," homeless people, and runaway youths.

Methamphetamine can be smoked, taken orally, snorted, or injected, depending on the form of the drug, and it alters the user's mood differently depending on how it is taken. Smoking or injecting intravenously results in a "flash" or intense rush that lasts only a few minutes (min). This may be followed by a high that lasts several hours (h). Snorting and oral ingestion result in a euphoria lasting 3 to 5 min in the case of snorting and up to 20 min in the case of oral ingestion. Because the pleasurable effects are so short-lived, methamphetamine users tend to binge to try to sustain the high.

Both the intense rush and the longer-lasting euphoria are thought to result from a release of dopamine and norepinephrine into regions of the brain that control feelings of pleasure. Once inside nerve cells (neurons), methamphetamine causes the release of dopamine and norepinephrine. At the same time, it inhibits enzymes that would normally destroy these two neurotransmitters, and the excess is transported out of neurons, causing the sensations of pleasure and euphoria. The excess norepinephrine may be responsible for the increased attention and decreased fatigue associated with methamphetamine use.

Symptoms of long-term use include addiction, anxiety, violent behavior, confusion, as well as psychotic symptoms of paranoia, hallucinations, and delusions. In severe cases, paranoia causes homicidal and/or suicidal feelings. Methamphetamine also increases heart rate and blood pressure. It can cause strokes, which result from irreversible damage to blood vessels in the brain, as well as respiratory problems, irregular heartbeat, and extreme anorexia. In extreme situations, it can cause cardiovascular collapse and death.

No physical symptoms accompany withdrawal from the drug, but psychological symptoms such as depression, anxiety, aggression, and intense craving are common. Of greatest concern is the brain damage that occurs in long-term users.

Dopamine release may be the cause of the drug's long-term toxic effects. Compare the structure of the neurotransmitter dopamine (Figure 15.7) with that of methamphetamine shown below. Research in humans has shown that even three years after chronic methamphetamine use, the former user continues to have a reduced ability to transport dopamine back into nerve cells. Parkinson's disease is characterized by a decrease in the dopamine-producing neurons in the brain; so it was logical to look for similarities between methamphetamine users and those suffering from Parkinson's. In fact, the brains of methamphetamine users showed damage similar to, but not as severe as, that in Parkinson's disease. Research in animals demonstrated that up to 50% of the dopamine-producing cells in parts of the brain may be destroyed by prolonged exposure, and that serotonin-containing neurons may sustain even worse damage.

Methamphetamine use continues to rise, in part because it is easily synthesized, or "cooked," using "recipes" that are available from many sources, including the Internet. Ephedrine, an over-the-counter decongestant drug, is the starting material. (Pseudoephedrine, a stereoisomer of ephedrine, can also be used.) As shown in the equation below, ephedrine is simply reduced to produce methamphetamine.

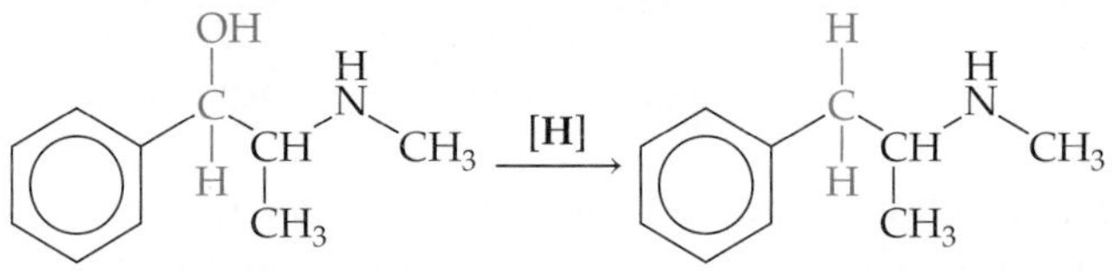

Ephedrine $\xrightarrow{[\text{H}]}$ Methamphetamine

While the synthesis involves a variety of dangerous chemicals, including anhydrous ammonia, anhydrous hydrochloric acid, sodium, and sodium hydroxide, most "meth cooks" do not have formal laboratory training. "Meth lab" fires are common, and the synthesis produces toxic wastes. The cleanup that followed the seizure of a major "meth lab" in 2003 took 8 days and required fifty people. Over four million pounds (lb) of toxic soil and 133 drums of hazardous waste were removed from the site, at a cost of $226,000.

For Further Understanding

▶ Compare the structures of methamphetamine and dopamine. Develop a hypothesis to explain why dopamine receptors also bind to and transport methamphetamine into neurons. (*Hint:* Receptors are proteins in cell membranes that have a pocket into which a specific molecule can fit.)

▶ Explain why methamphetamine is soluble in alcohols and in water.

Question 15.7 Complete each of the following reactions by supplying the missing product(s).

a. 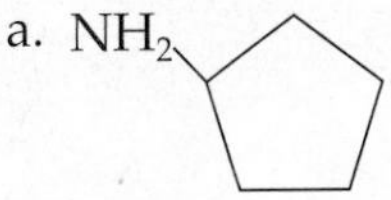 NH$_2$ [cyclopentane structure] + HBr ⟶ ?

b. $CH_3CH_2NHCH_3 + H_2O \longrightarrow$?
c. $CH_3NH_2 + H_2O \longrightarrow$?

Question 15.8 Complete each of the following reactions by supplying the missing product(s).

a. $CH_3NH_2 + HI \longrightarrow$?
b. $CH_3CH_2NH_2 + HBr \longrightarrow$?
c. $(CH_3CH_2)_2NH + HCl \longrightarrow$?

Alkylammonium salts can neutralize hydroxide ions. In this reaction, water is formed and the protonated amine cation is converted into an amine.

$$
\begin{array}{ccccc}
\overset{\displaystyle H}{\underset{\displaystyle H}{R-N^+-H}} & + & OH^- & \longrightarrow & \overset{\displaystyle H}{\underset{\displaystyle H}{R-N:}} + H-OH
\end{array}
$$

Alkylammonium salt   Hydroxide ion   Amine   Water

Thus, by adding a strong acid to a water-insoluble amine, a water-soluble alkylammonium salt can be formed. The salt can just as easily be converted back to an amine by the addition of a strong base.

The local anesthetic novocaine, which is often used in dentistry and for minor surgery, is injected as an amine salt. See Medically Important Amines earlier in this section.

Neutralization of an amine with a strong acid produces an alkylammonium salt

Neutralization of the alkylammonium salt with a strong base restores the amine

$$CH_3(CH_2)_{12}-NH_2 + HCl \longrightarrow CH_3(CH_2)_{12}-NH_3{}^+Cl^- + NaOH \longrightarrow CH_3(CH_2)_{12}-NH_2$$

Insoluble in water   Soluble in water   Insoluble in water

The different forms of cocaine provide examples of these reactions. "Crack" cocaine is an amine and a base (see Figure 15.4 and the structure below) and is generally found in the form of relatively large crystals (Figure 15.3a) that may vary in color from white to dark brown or black. When this amine reacts with HCl, the product is cocaine hydrochloride:

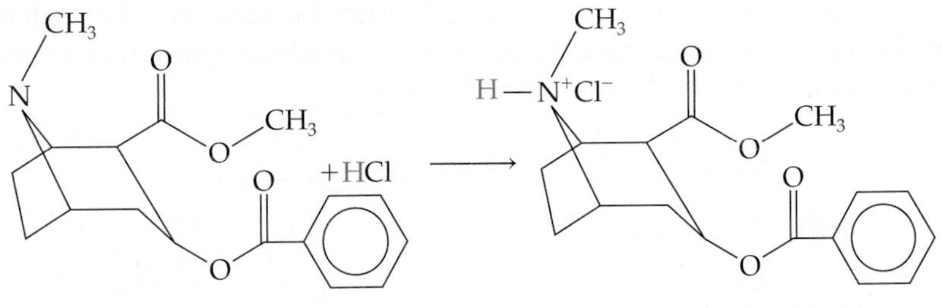

[structure of "Crack" cocaine with CH$_3$, N, O, CH$_3$, O groups] +HCl ⟶ [structure of Cocaine hydrochloride with CH$_3$, H—N$^+$Cl$^-$, O, CH$_3$, O groups]

"Crack" cocaine   Cocaine hydrochloride
(a base)   (a salt)

This salt of cocaine is a powder (Figure 15.3b) and is soluble in water. Since it is a powder, it can be snorted, and because it is water-soluble, it dissolves in the fluids

(a)

(b)

Figure 15.3 (a) Crack cocaine is a non-water-soluble base with a low melting point and a crystalline structure.
(b) Powdered cocaine is a water-soluble salt of cocaine base.
(c) Cocaine is extracted from the leaves of the coca plant.

(c)

of the nasal mucous membranes and is absorbed into the bloodstream. This is a common form of cocaine because it is the direct product of the preparation from coca leaves (Figure 15.3c). A coca paste is made from the leaves and is mixed with HCl and water. After additional processing, the product is the salt of cocaine.

Cocaine hydrochloride salt can be converted into its base form by a process called "free basing." Although the chemistry is simple, the process is dangerous because it requires highly flammable solvents. This pure cocaine is not soluble in water. It has a relatively low melting point, however, and can be smoked. Crack, so called because of the crackling noise it makes when smoked, is absorbed into the body more quickly than the snorted powder and results in a more immediate high.

Quaternary Ammonium Salts

LEARNING GOAL

6 Describe the structure of quaternary ammonium salts and discuss their use as antiseptics and disinfectants.

Quaternary ammonium salts are ammonium salts that have four organic groups bonded to the nitrogen and, as a result, do not have an unshared pair of electrons. They have the following general structure:

$$R_4N^+X^- \quad (R = \text{any alkyl or aryl group};$$
$$X^- = \text{a halide anion, most commonly } Cl^-)$$

Quaternary ammonium salts that have a very long carbon chain, sometimes called "quats," are used as disinfectants and antiseptics because they have detergent activity. Two popular quats are benzalkonium chloride (Zephiran) and cetylpyridinium chloride, found in the mouthwash Scope.

Benzalkonium
chloride

Cetylpyridinium
chloride

Phospholipids and biological membranes are discussed in Sections 17.3 and 17.6.

Choline is an important quaternary ammonium salt in the body. It is part of the hydrophilic "head" of the membrane phospholipid lecithin. Choline is also a precursor for the synthesis of the neurotransmitter acetylcholine.

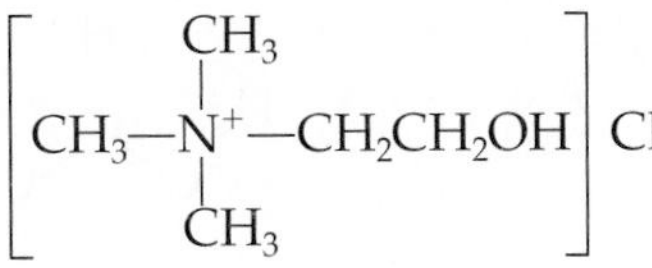

CH_3 structure, Choline

$$\begin{bmatrix} CH_3 \\ | \\ CH_3-N^+-CH_2CH_2OH \\ | \\ CH_3 \end{bmatrix} Cl^-$$

Choline

The function of acetylcholine is described in greater detail in Section 15.5.

15.2 Heterocyclic Amines

Heterocyclic amines are cyclic compounds that have at least one nitrogen atom in the ring structure. The structures and common names of several heterocyclic amines important in nature are shown here. They are represented by their structural formulas and by abbreviated line formulas.

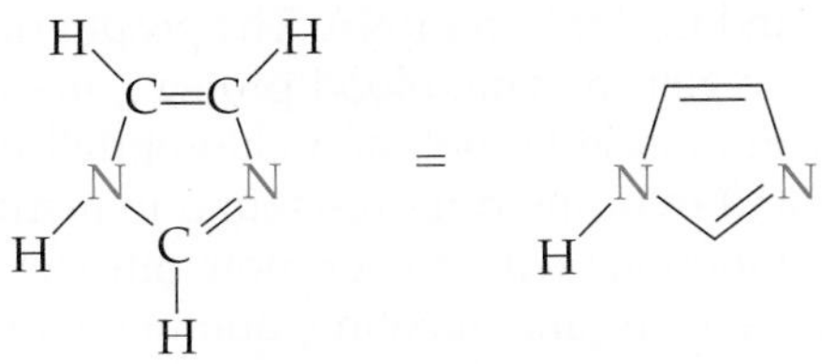

Imidazole

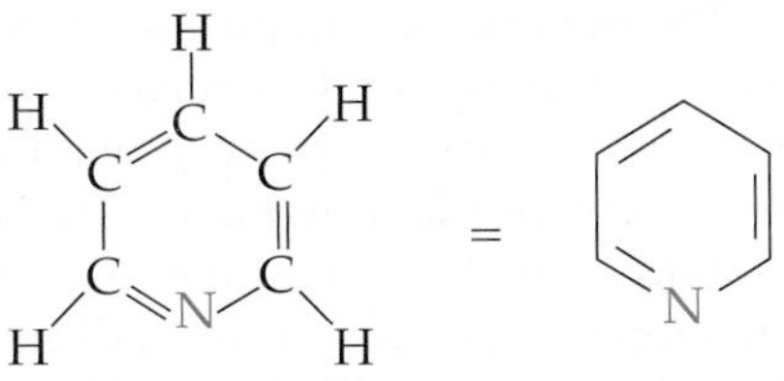

Pyridine

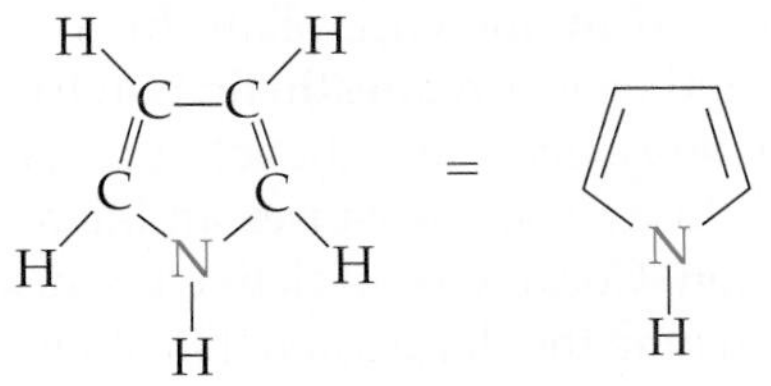

Pyrrole

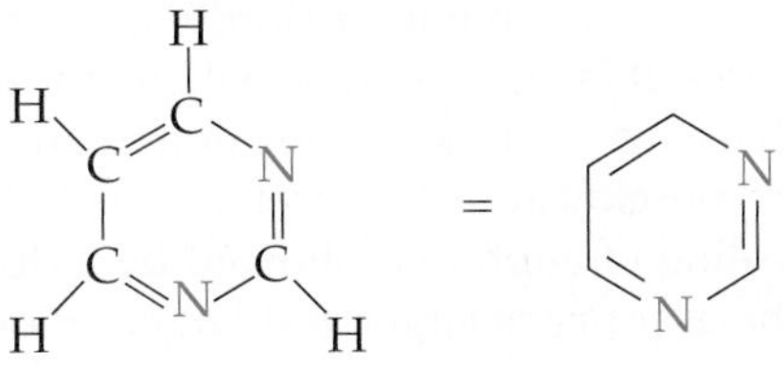

Pyrimidine

The heterocyclic amines shown below are examples of fused ring structures. Each ring pair shares two carbon atoms with the other. Thus, two fused rings share one or more common bonds as part of their ring backbones. The fused ring structures of purine, indole, and porphyrin, are shown as structural formulas and as line diagrams:

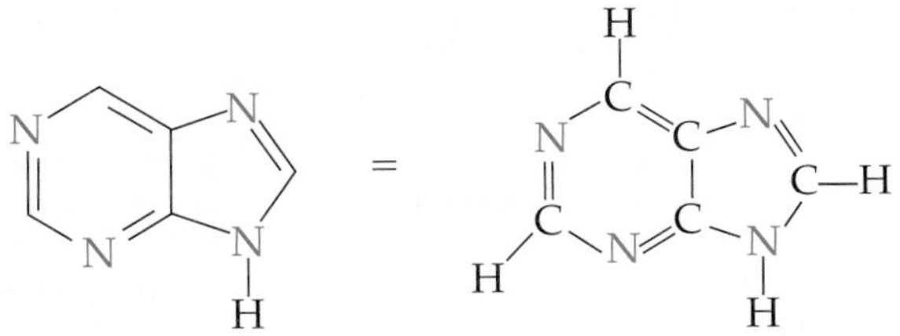

Purine

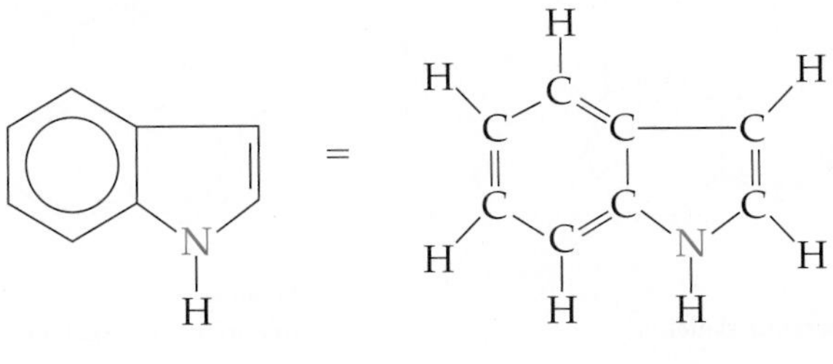

Indole

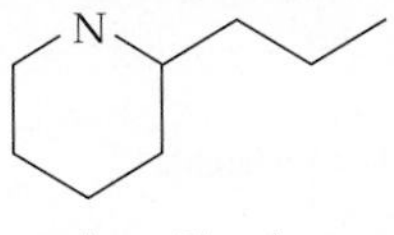

Coniine is produced by the poison hemlock plant. It is a neurotoxin that causes respiratory paralysis. It is the poison that was used to kill the Greek philosopher Socrates in 399 BC. To what class of molecules does coniine belong?

M⁺ = metal ion

Porphyrin

The structures of purines and pyrimidines are presented in Section 20.1.

The structure of the heme group found in hemoglobin and myoglobin is presented in Section 18.8.

The pyrimidine and purine rings are found in DNA and RNA. The porphyrin ring structure is found in hemoglobin (an oxygen-carrying blood protein), myoglobin (an oxygen-carrying protein found in muscle tissue), and chlorophyll (a photosynthetic plant pigment). The indole and pyridine rings are found in many **alkaloids**, which are naturally occurring compounds with one or more nitrogen-containing heterocyclic rings. Alkaloids include cocaine, nicotine, quinine, morphine, heroin, and LSD (Figure 15.4).

Lysergic acid diethylamide (LSD) is a hallucinogenic compound that may cause severe mental disorders. Cocaine is produced by the coca plant. In small doses, it is used as an anesthetic for the sinuses and eyes. An **anesthetic** is a drug that causes a lack of sensation in any part of the body (local anesthetic) or causes unconsciousness (general anesthetic). In higher doses, cocaine causes an intense feeling of euphoria followed by a deep depression. Cocaine is addictive because the user needs larger and larger amounts to overcome the depression. Nicotine is

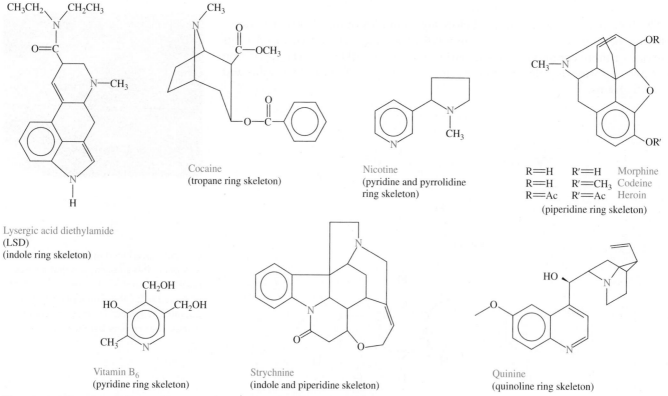

Lysergic acid diethylamide (LSD) (indole ring skeleton)

Cocaine (tropane ring skeleton)

Nicotine (pyridine and pyrrolidine ring skeleton)

R=H	R'=H	Morphine
R=H	R'=CH₃	Codeine
R=Ac	R'=Ac	Heroin

(piperidine ring skeleton)

Vitamin B₆ (pyridine ring skeleton)

Strychnine (indole and piperidine skeleton)

Quinine (quinoline ring skeleton)

Figure 15.4 Structures of several heterocyclic amines with biological activity.

one of the simplest heterocyclic amines and appears to be the addictive component of cigarette smoke.

Morphine was the first alkaloid to be isolated from the sap of the opium poppy. Morphine is a strong **analgesic**, a drug that acts as a pain killer. However, it is a powerful and addictive narcotic. Codeine, also produced by the opium poppy, is a less powerful analgesic than morphine, but it is one of the most effective cough suppressants known. Heroin is produced in the laboratory by adding two acetyl groups to morphine. It was initially made in the hopes of producing a compound with the benefits of morphine but lacking the addictive qualities. However, heroin is even more addictive than morphine.

Strychnine is found in the seeds of an Asiatic tree. It is extremely toxic and was commonly used as a rat poison at one time. Quinine, isolated from the bark of South American trees, was the first effective treatment for malaria. Vitamin B_6 is one of the water-soluble vitamins required by the body.

15.3 Amides

Amides are the products formed in a reaction between a carboxylic acid derivative and ammonia or an amine. The general structure of an amide is shown here.

From a carboxylic acid   From an amine

$$(Ar) \; R-\overset{\overset{\displaystyle O}{\|}}{C}-NH_2$$

Ethanamide

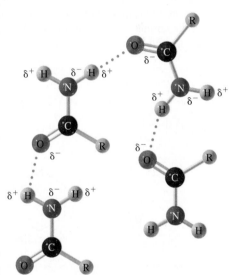

Figure 15.5 Hydrogen bonding in amides.

The amide group is composed of two portions: the carbonyl group from a carboxylic acid and the amino group from ammonia or an amine. The bond between the carbonyl carbon and the nitrogen of the amine or ammonia is called the **amide bond**.

Structure and Physical Properties

Most amides are solids at room temperature. They have very high boiling points, and the simpler ones are quite soluble in water. Both of these properties are a result of strong intermolecular hydrogen bonding between the N—H bond of one amide and the C=O group of a second amide, as shown in Figure 15.5.

Unlike amines, amides are not bases (proton acceptors). The reason is that the highly electronegative oxygen atom of the carbonyl group causes a very strong attraction between the lone pair of nitrogen electrons and the carbonyl group. As a result, the unshared pair of electrons cannot "hold" a proton.

Because of the attraction of the carbonyl group for the lone pair of nitrogen electrons, the structure of the C—N bond of an amide is a *resonance hybrid*. In the structures below, lines are used to represent pairs of electrons.

LEARNING GOAL

8 Describe the physical properties of amides.

Resonance hybrids are discussed in Section 3.4.

Kitchen Chemistry

Browning Reactions and Flavor: The Maillard Reaction

Since earliest times, humans have enjoyed the delightful flavors and aromas of meat roasted over a hot fire or seared in a hot pan. The same chemical reaction that produces the aroma of roasted meat is also the reason that bread crust is brown and more flavorful than the inside of the bread, and is responsible for the wonderful smell of roasted coffee and chocolate. The reaction is called the Maillard reaction in honor of Louis Camille Maillard, a French physician who first described it in 1910.

The reaction begins with a carbohydrate molecule, generally glucose or fructose, which binds to an amino acid in a protein chain. This reaction involves a carbonyl group of the sugar and the amino group of the amino acid. The initial product is unstable and is converted into an intermediate that further reacts with other compounds in the food to create a wide variety of aromatic heterocyclic compounds, including heterocyclic amines. Several of the simpler products of the reactions are shown here:

A thiophene

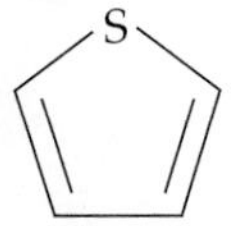

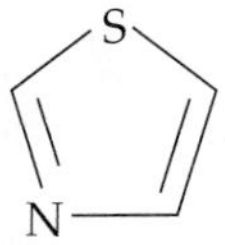

A thiazole

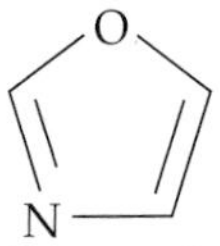

An oxazole

A pyrrole

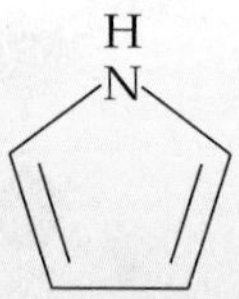

A pyridine

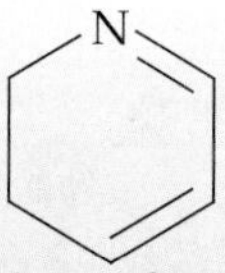

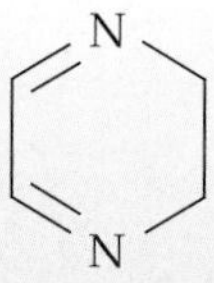

A pyrazine

The savory flavors are attributed to the peptides and amino acids in the food. Oxazoles lend a floral note to the flavor. Thiophenes and thiazoles are distinctive meaty and onion flavors. Pyrazines are in part responsible for the aroma of chocolate, and pyridines and pyrazines are associated with the flavors of green vegetables.

The Maillard reaction only occurs at approximately 250°F, which is the reason that meats cooked by boiling or steaming are typically pale and don't have the flavor profile of meats that are seared on a hot surface like a sauté pan or on a grill. Meats that are cooked in oil or in a hot oven also undergo the browning reactions because the surface of the meat quickly dehydrates and rises to the temperature of the oven.

Because foods are chemically complex, we may never fully understand the products of the Maillard reaction, but that will not keep us from enjoying the flavors and aromas that they produce.

For Further Understanding

▶ Many stew recipes begin by browning the meat, vegetables, and flour before adding any water. How will this influence the characteristics of the stew?

▶ Microwave ovens are very useful for heating leftovers or steaming vegetables, but are generally not used to prepare hamburgers or roasts. Explain the reason for this in terms of what you have just learned of the Maillard reaction.

Nomenclature

LEARNING GOAL

9 Draw the structure and write the common and IUPAC names of amides.

Nomenclature of carboxylic acids is described in Section 14.1.

The common and IUPAC names of the amides are derived from the common and IUPAC names of the carboxylic acids from which they were made. Begin by identifying the carboxylic acid parent compound. Now, remove the *-ic acid* ending of the common name or the *-oic acid* ending of the IUPAC name of the carboxylic acid, and replace it with the ending *-amide*. Several examples of the common and IUPAC nomenclature are provided in Table 15.4 and in the following structures:

Ethan*oic acid* → Ethan*amide*
or
Acet*ic acid* → Acet*amide*

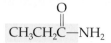

Propan*oic acid* → Propan*amide*
or
Propion*ic acid* → Propion*amide*

TABLE 15.4 IUPAC and Common Names of Simple Amides

Compound	IUPAC Name	Common Name
$R-\overset{\overset{\displaystyle O}{\|\|}}{C}-NH_2$	Alkan*amide* (-*amide* replaces the -*oic acid* ending of the IUPAC name of carboxylic acid)	Alkan*amide* (-*amide* replaces the -*ic acid* ending of the common name of carboxylic acid)
$H-\overset{\overset{\displaystyle O}{\|\|}}{C}-NH_2$	Methanamide	Formamide
$CH_3-\overset{\overset{\displaystyle O}{\|\|}}{C}-NH_2$	Ethanamide	Acetamide
$CH_3CH_2-\overset{\overset{\displaystyle O}{\|\|}}{C}-NH_2$	Propanamide	Propionamide
$H-\overset{\overset{\displaystyle O}{\|\|}}{C}-NHCH_3$	N-Methylmethanamide	N-Methylformamide
$CH_3-\overset{\overset{\displaystyle O}{\|\|}}{C}-NHCH_3$	N-Methylethanamide	N-Methylacetamide

Substituents on the nitrogen are placed as prefixes and are indicated by N-followed by the name of the substituent. There are no spaces between the prefix and the amide name. For example:

$$CH_3CH_2\overset{\overset{\displaystyle O}{\|\|}}{C}-NH-CH_3$$

N-Methylpropanamide

$$CH_3CH_2CH_2CH_2CH_2\overset{\overset{\displaystyle O}{\|\|}}{C}-NH-CH_2CH_2CH_3$$

N-Propylhexanamide

Medically Important Amides

Barbiturates, often called "downers," are derived from amides and are used as sedatives. They are also used as anticonvulsants for epileptics and for people suffering from a variety of brain disorders that manifest themselves in neurosis, anxiety, and tension.

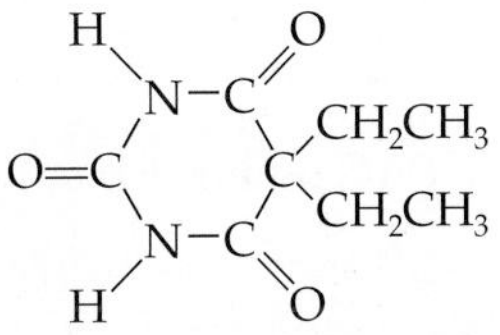

Barbital—a barbiturate

Phenacetin and acetaminophen are also amides. Acetaminophen is an aromatic amide that is commonly used in place of aspirin, particularly by people who are allergic to aspirin or who suffer stomach bleeding from the use of aspirin. It was first synthesized in 1893 and is the active ingredient in Tylenol and Datril. Like aspirin, acetaminophen relieves pain and reduces fever. However, unlike aspirin, it is not an anti-inflammatory drug.

Phenacetin was synthesized in 1887 and used as an analgesic for almost a century. Its structure and properties are similar to those of acetaminophen. However,

LEARNING GOAL

9 Draw the structure and write the common and IUPAC names of amides.

EXAMPLE 15.4 Naming Amides Using Common and IUPAC Nomenclature Systems

a. Name the following amide using both the IUPAC and common systems.

$$CH_3CH_2CH_2\overset{\overset{\displaystyle O}{\|}}{C}-NH-CH_2CH_2CH_3$$

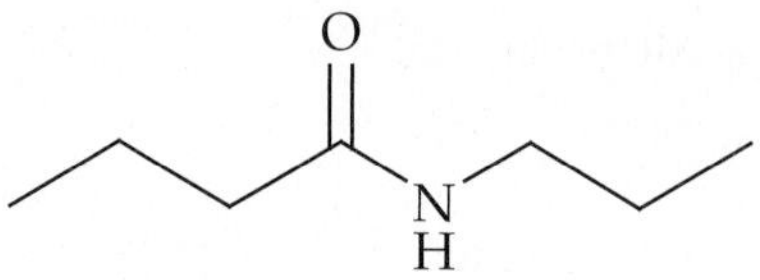

Solution

The names of amides are derived from the common and IUPAC names of the carboxylic acids from which they were made.

	IUPAC	Common
Parent carboxylic acid:	Butanoic acid (becomes butanamide)	Butyric acid (becomes butyramide)
Group on N:	propyl	propyl
Name:	*N*-Propylbutanamide	*N*-Propylbutyramide

b. Name the following amide using the IUPAC and common systems.

$$CH_3CH_2\overset{\overset{\displaystyle O}{\|}}{C}-NH-CH_2CH_2CH_2CH_2CH_3$$

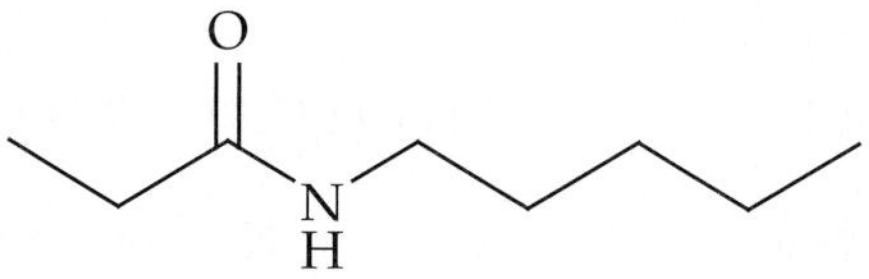

Solution

Parent carboxylic acid:	Propanoic acid (becomes propanamide)	Propionic acid (becomes propionamide)
Group on N:	pentyl	pentyl
Name:	*N*-Pentylpropanamide	*N*-Pentylpropionamide

Practice Problem 15.4

Provide the common and IUPAC names for each of the following amides.

a. $CH_3CH_2CH_2CH_2\overset{\overset{\displaystyle O}{\|}}{C}-NH-CH_2CH_2CH_2CH_2CH_3$

b. $CH_3CH_2CH_2CH_2CH_2\overset{\overset{\displaystyle O}{\|}}{C}-NH-CH_2CH_2CH_2CH_3$

▸ For Further Practice: **Questions 15.53, 15.54, and 15.57.**

it was banned by the U.S. Food and Drug Administration (FDA) in 1983 because of the kidney damage and blood disorders it causes.

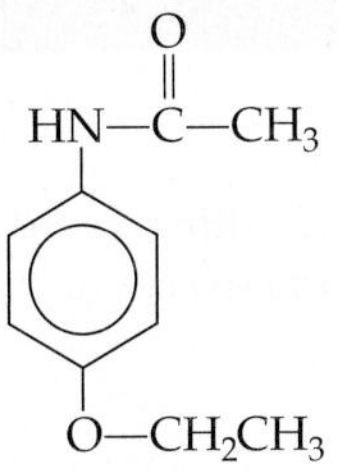

Phenacetin

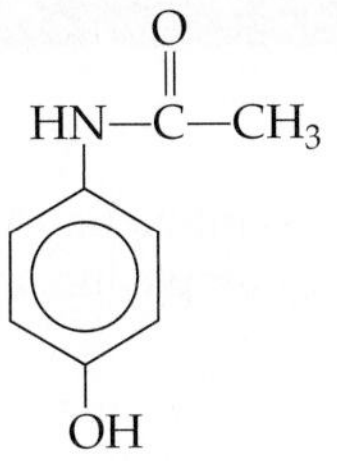

Acetaminophen

Reactions Involving Amides

Preparation of Amides

Amides may be prepared in a reaction between a carboxylic acid derivative and an amine or ammonia. The carboxylic acid derivative can be either an acid chloride or an acid anhydride.

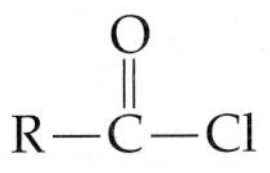

Acid chloride

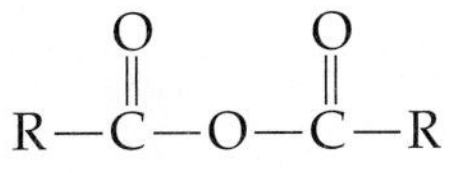

Acid anhydride

LEARNING GOAL

10 Write equations representing the preparation of amides.

Acid chlorides react rapidly with either ammonia or an amine, as shown in the following general equation:

$$R{-}\overset{\overset{\displaystyle O}{\|}}{C}{-}Cl + 2NH_3 \longrightarrow R{-}\overset{\overset{\displaystyle O}{\|}}{C}{-}NH_2 + \overset{+}{N}H_4Cl^-$$

Acid chloride; Ammonia or amine; Amide; Ammonium chloride or alkylammonium chloride

Note that two molar equivalents of ammonia or amine are required in this reaction and that this is an acyl group transfer reaction. The **acyl group**

$$R{-}\overset{\overset{\displaystyle O}{\|}}{C}{-}$$

of the acid chloride is transferred from the Cl atom to the N atom of one of the ammonia or amine molecules. The second ammonia (or amine) reacts with the HCl formed in the transfer reaction to produce ammonium chloride or alkylammonium chloride. The reaction between ethanoyl chloride and ammonia to produce ethanamide and ammonium chloride is an example of an acyl group transfer reaction. The acyl group of ethanoyl chloride (in red) is transferred to the nitrogen of the ammonia molecule (in blue).

$$H_3C{-}\overset{\overset{\displaystyle O}{\|}}{C}{-}Cl + 2NH_3 \longrightarrow H_3C{-}\overset{\overset{\displaystyle O}{\|}}{C}{-}NH_2 + \overset{+}{N}H_4Cl^-$$

Ethanoyl chloride; Ammonia; Ethanamide, A primary amide; Ammonium chloride

A Medical Perspective

Semisynthetic Penicillins

The antibacterial properties of penicillin were discovered by Alexander Fleming in 1929. These natural penicillins produced by several species of the mold *Penicillium* had a number of drawbacks. They were effective only against a type of bacteria referred to as Gram positive because of a staining reaction based on their cell wall structure. They were also very susceptible to destruction by bacterial enzymes called β-lactamases, and some were destroyed by stomach acid and had to be administered by injection.

To overcome these problems, chemists have produced semisynthetic penicillins by modifying the core structure. The core of penicillins is 6-aminopenicillanic acid, which consists of a thiazolidine ring fused to a β-lactam ring. In addition, there is an R group bonded via an amide bond to the core structure.

6-Aminopenicillanic acid

The β-lactam ring confers the antimicrobial properties. However, the R group determines the degree of antibacterial activity, the pharmacological properties, including the types of bacteria against which it is active, and the degree of resistance to the β-lactamases exhibited by any particular penicillin antibiotic. These are the properties that must be modified to produce penicillins that are acid resistant, effective with a broad spectrum of bacteria, and β-lactamase resistant.

Chemists simply remove the natural R group by cleaving the amide bond with an enzyme called an amidase. They then replace the R group and test the properties of the "new"

antibiotic. Among the resulting semisynthetic penicillins are ampicillin, methicillin, and oxacillin.

Ampicillin

Methicillin

Oxacillin

For Further Understanding

▸ Using the Internet and other resources, investigate and describe the properties of some new penicillins and the bacteria against which they are effective.

▸ Why does changing the R group of a penicillin result in altered chemical and physiological properties?

The product of a reaction between a carboxylic acid and ammonia is a **primary amide.** A primary amide is one in which the nitrogen is bonded to only one carbon, the carbonyl carbon.

The reaction between butanoyl chloride and methanamine to produce *N*-methylbutanamide is another example of an acyl group transfer reaction.

$$CH_3CH_2CH_2\overset{\displaystyle O}{\overset{\|}{C}}-Cl + 2CH_3NH_2 \longrightarrow CH_3CH_2CH_2\overset{\displaystyle O}{\overset{\|}{C}}-NH-CH_3 + CH_3\overset{+}{N}H_3Cl^-$$

| Butanoyl chloride | Methanamine | *N*-Methylbutanamide A secondary amide | Methylammonium chloride |

Methanamine is a primary amine. The product of a reaction between an acid chloride and a primary amine is a **secondary (2°) amide.** A secondary amide is one in which there are two carbons atoms bonded to the nitrogen. One is the carbonyl carbon from the acid chloride and the second is from the primary amine reactant.

The reaction of a secondary amine with an acid chloride forms a **tertiary (3°) amide.** In a tertiary amide, the nitrogen is bonded to three carbon atoms. One is the carbonyl carbon from the acid chloride, and the other two are from the secondary amine reactant. This is another example of an acyl group transfer reaction.

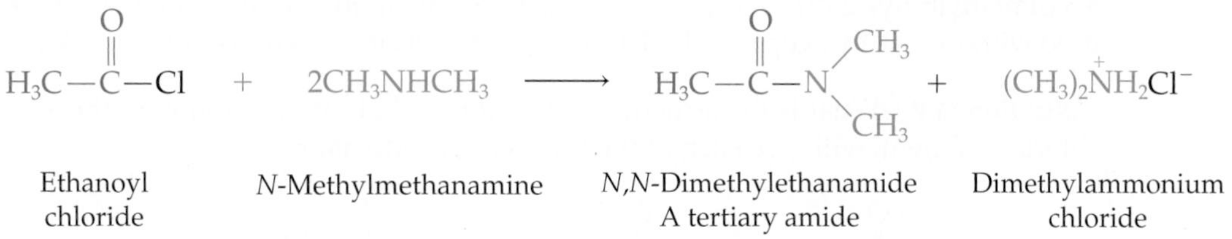

$$H_3C-\overset{\overset{\textstyle O}{\|}}{C}-Cl \quad + \quad 2CH_3NHCH_3 \quad \longrightarrow \quad H_3C-\overset{\overset{\textstyle O}{\|}}{C}-N\overset{\diagup CH_3}{\diagdown CH_3} \quad + \quad (CH_3)_2\overset{+}{N}H_2Cl^-$$

| Ethanoyl chloride | N-Methylmethanamine | N,N-Dimethylethanamide A tertiary amide | Dimethylammonium chloride |

The reaction between an amine and an acid anhydride is also an acyl group transfer. The general equation for the synthesis of an amide in the reaction between an acid anhydride and ammonia or an amine is

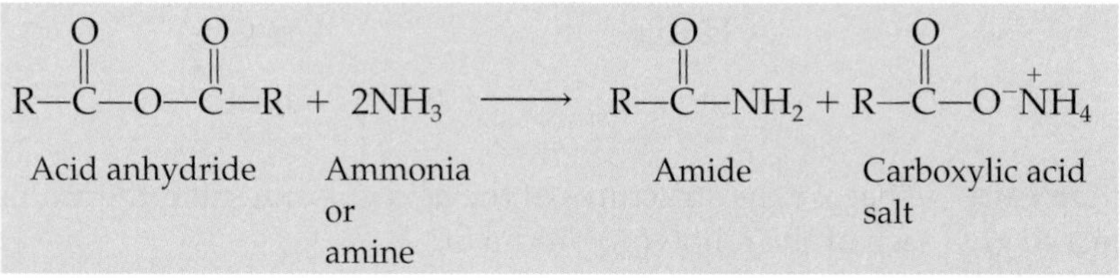

$$R-\overset{\overset{\textstyle O}{\|}}{C}-O-\overset{\overset{\textstyle O}{\|}}{C}-R + 2NH_3 \quad \longrightarrow \quad R-\overset{\overset{\textstyle O}{\|}}{C}-NH_2 + R-\overset{\overset{\textstyle O}{\|}}{C}-O^-\overset{+}{N}H_4$$

| Acid anhydride | Ammonia or amine | Amide | Carboxylic acid salt |

When subjected to heat, the carboxylic acid salt loses a water molecule to produce a second amide molecule.

A well-known commercial amide is the artificial sweetener aspartame or NutraSweet. Although the name suggests it is a sugar, it is not a sugar at all. In fact, it is the methyl ester of a molecule composed of two amino acids, aspartic acid and phenylalanine, joined by an amide bond (Figure 15.6a).

Packages of aspartame carry the warning: "Phenylketonurics: Contains Phenylalanine." Digestion of aspartame and heating to high temperatures during cooking break both the ester bond and the amide bond, which releases the amino acid phenylalanine. People with the genetic disorder phenylketonuria (PKU) cannot metabolize this amino acid. As a result, it builds up to toxic levels that can cause mental retardation in an infant born with the condition. This no longer

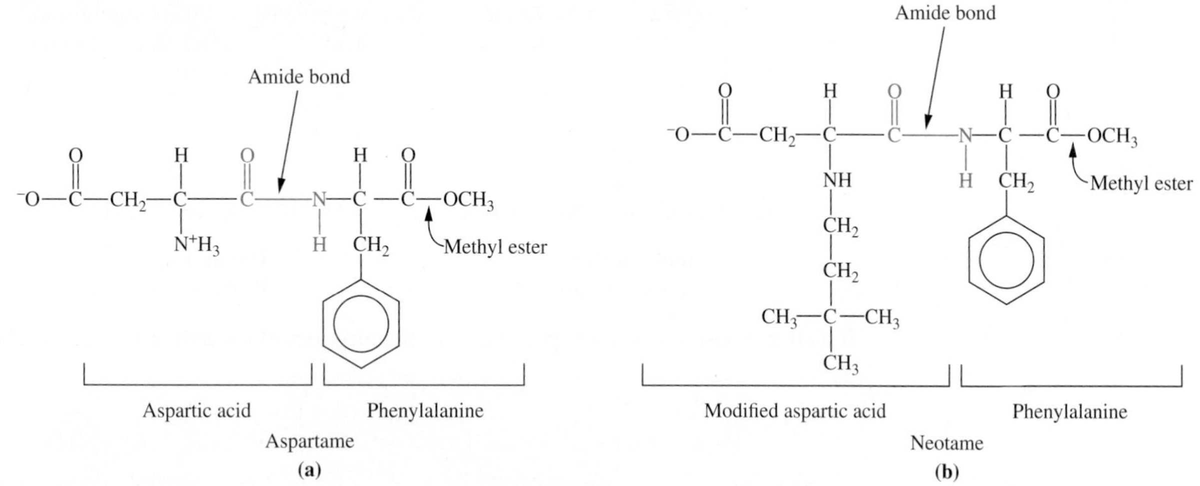

> Amino acids have both a carboxyl group and an amino group and are discussed in detail in Sections 15.4 and 18.1.

Figure 15.6 The amide bond. (a) NutraSweet, the dipeptide aspartame, is a molecule composed of two amino acids joined by an amide (peptide) bond. (b) The sweetener Neotame, is also a dipeptide. One of the amino acids has been modified so that neotame is safe for use by phenylketonurics.

occurs because every child is tested for PKU at the time of birth and each child with the disorder is treated with a diet that limits the amount of phenylalanine to only the amount required for normal growth.

In July 2002, the FDA approved a new artificial sweetener that is related to aspartame. Called neotame, it has the same core structure as aspartame, but a 3,3-dimethylbutyl group has been added to the aspartic acid (Figure 15.6b). Digestion and heating still cause breakage of the ester bond, but the bulky 3,3-dimethylbutyl group blocks the breakage of the amide bond. Neotame can be used without risk by people with PKU and also retains its sweetness during cooking.

Question 15.9   What is the structure of the amine that, on reaction with the acid chlorides shown, will give each of the following products?

$$\text{a. } ? + CH_3\overset{\displaystyle O}{\overset{\|}{C}}-Cl \longrightarrow CH_3\overset{\displaystyle O}{\overset{\|}{C}}NHCH_3 + CH_3\overset{+}{N}H_3\ Cl^-$$

$$\text{b. } ? + CH_3CH_2CH_2CH_2\underset{\underset{\displaystyle CH_2CH_3}{|}}{CH}\overset{\displaystyle O}{\overset{\|}{C}}-Cl \longrightarrow (CH_3)_2N\overset{\displaystyle O}{\overset{\|}{C}}\underset{\underset{\displaystyle CH_2CH_3}{|}}{CH}CH_2CH_2CH_2CH_3$$

$$+ (CH_3)_2\overset{+}{N}H_2\ Cl^-$$

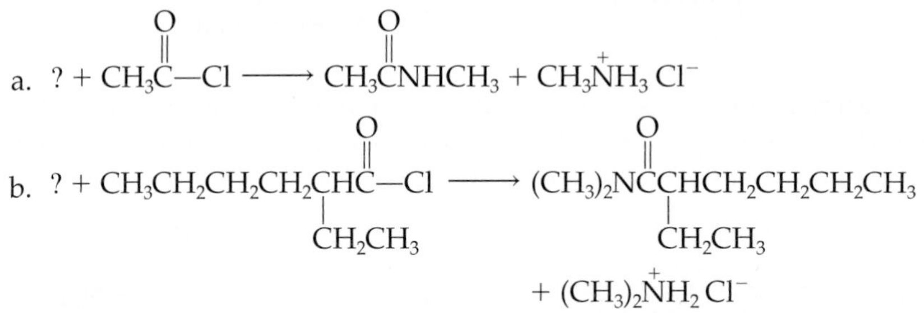

Question 15.10   What are the structures of the acid chlorides and the amines that will react to give each of the following products?
   a. N-Ethylhexanamide
   b. N-Propylbutanamide

LEARNING GOAL

11   Write equations showing the hydrolysis of amides.

Hydrolysis of Amides

Hydrolysis is a reaction in which chemical bonds are broken by the addition of water. It is very difficult to hydrolyze the amide bond. In fact, the reaction requires heat and the presence of either a strong acid or a strong base.

Hydrolysis of an amide in the presence of a strong acid produces a carboxylic acid and either an alkylammonium ion or ammonium ion, as seen in the following general equation.

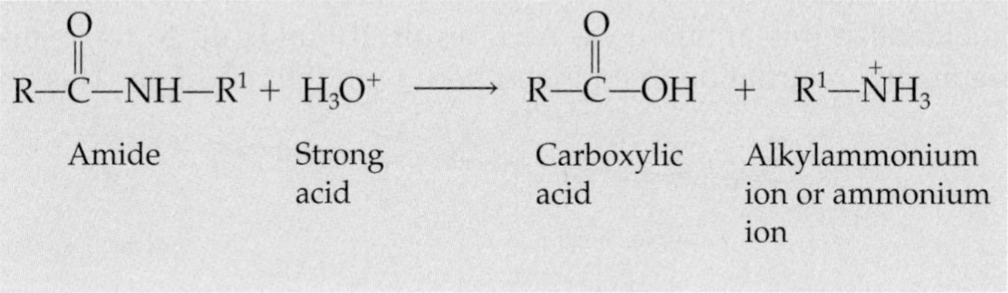

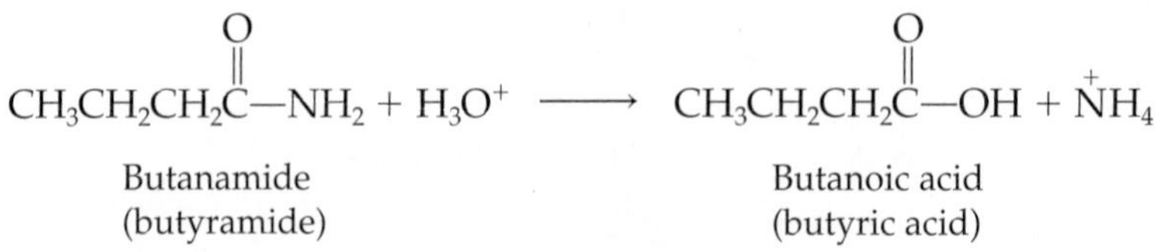

$$R\overset{\displaystyle O}{\overset{\|}{C}}-NH-R^1 + H_3O^+ \longrightarrow R\overset{\displaystyle O}{\overset{\|}{C}}-OH + R^1-\overset{+}{N}H_3$$

| Amide | Strong acid | Carboxylic acid | Alkylammonium ion or ammonium ion |

$$CH_3CH_2CH_2\overset{\displaystyle O}{\overset{\|}{C}}-NH_2 + H_3O^+ \longrightarrow CH_3CH_2CH_2\overset{\displaystyle O}{\overset{\|}{C}}-OH + \overset{+}{N}H_4$$

Butanamide (butyramide)   Butanoic acid (butyric acid)

If a strong base is used, the products are the amine and the salt of the carboxylic acid:

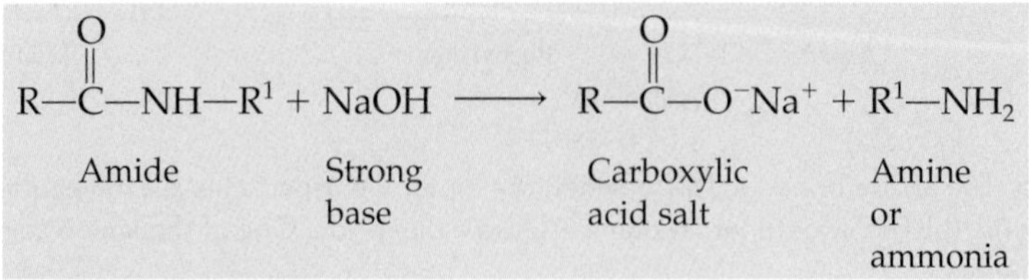

$$R\overset{\displaystyle O}{\overset{\|}{C}}-NH-R^1 + NaOH \longrightarrow R\overset{\displaystyle O}{\overset{\|}{C}}-O^-Na^+ + R^1-NH_2$$

| Amide | Strong base | Carboxylic acid salt | Amine or ammonia |

$$CH_3CH_2\overset{\overset{\displaystyle O}{\|}}{C}-NHCH_3 + NaOH \longrightarrow CH_3CH_2\overset{\overset{\displaystyle O}{\|}}{C}-O^-Na^+ + CH_3NH_2$$

N-Methylpropanamide   Sodium propanoate   Methanamine
(*N*-methylpropionamide)   (sodium propionate)   (methylamine)

15.4 A Preview of Amino Acids, Proteins, and Protein Synthesis

In Chapter 18, we will describe the structure of proteins, the molecules that carry out the majority of the biological processes essential to life. Proteins are polymers of **amino acids.** As the name suggests, amino acids have two essential functional groups, an amino group ($-NH_2$) and a carboxyl group ($-COOH$). Typically, amino acids have the following general structure:

$$H_2N-\overset{\overset{\displaystyle H}{|}}{\underset{\underset{\displaystyle R}{|}}{C}}-COOH$$

(R may be a hydrogen atom or an organic group.)

The amide bond that forms between the carboxyl group of one amino acid and the amino group of another is called the **peptide bond.**

The peptide bond is an amide bond.

$$H_3\overset{+}{N}-\overset{\overset{\displaystyle H}{|}}{\underset{\underset{\displaystyle R}{|}}{C}}-\overset{\overset{\displaystyle O}{\|}}{C}-N-\overset{\overset{\displaystyle H}{|}}{\underset{\underset{\displaystyle R}{|}}{C}}-\overset{\overset{\displaystyle O}{\|}}{C}-O^-$$

The joining of amino acids by amide bonds produces small *peptides* and larger *proteins.* Because protein structure and function are essential for life processes, it is fortunate indeed that the amide bonds that hold them together are not easily hydrolyzed at physiological pH and temperature.

The process of protein synthesis in the cell mimics amide formation in the laboratory; it involves acyl group transfer. There are several important differences between the chemistry in the laboratory and the chemistry in the cell. During protein synthesis, the **aminoacyl group** of the amino acid is transferred, rather than the acyl group of a carboxylic acid. In addition, the aminoacyl group is not transferred from a carboxylic acid derivative; it is transferred from a special carrier molecule called a **transfer RNA (tRNA).** When the aminoacyl group is covalently bonded to a tRNA, the resulting structure is called an *aminoacyl tRNA:*

Aminoacyl group

$$H_2N-\overset{\overset{\displaystyle H}{|}}{\underset{\underset{\displaystyle R}{|}}{C}}-\overset{\overset{\displaystyle O}{\|}}{C}- \text{transfer RNA}$$

The aminoacyl group of the aminoacyl tRNA is transferred to the amino group nitrogen to form a peptide bond. The transfer RNA is recycled by binding to another of the same kind of aminoacyl group.

More than 100 kinds of proteins, nucleotides, and RNA molecules participate in the incredibly intricate process of protein synthesis. In Chapter 18, we will study protein structure and learn about the many functions of proteins in the life of the cell. In Chapter 20, we will study the details of protein synthesis to see how these aminoacyl transfer reactions make us the individuals that we are.

LEARNING GOAL

12 Draw the general structure of an amino acid.

LEARNING GOAL

13 Draw and discuss the structure of a peptide bond.

In the cell the amino group is usually protonated and the carboxyl group is usually ionized to the carboxylate anion. In the future, we will represent an amino acid in the following way:

$$H_3\overset{+}{N}-\overset{\overset{\displaystyle H}{|}}{\underset{\underset{\displaystyle R}{|}}{C}}-COO^-$$

15.5   Neurotransmitters

Neurotransmitters are chemicals that carry messages, or signals, from a nerve cell to a target cell, which may be another nerve cell or a muscle cell. Neurotransmitters are classified as being *excitatory,* stimulating their target cell, or *inhibitory,* decreasing activity of the target cell. One feature shared by the neurotransmitters is that they are all nitrogen-containing compounds. Some of them have rather complex structures and one, nitric oxide (NO), consists of only two atoms. Several important neurotransmitters are discussed in the following sections.

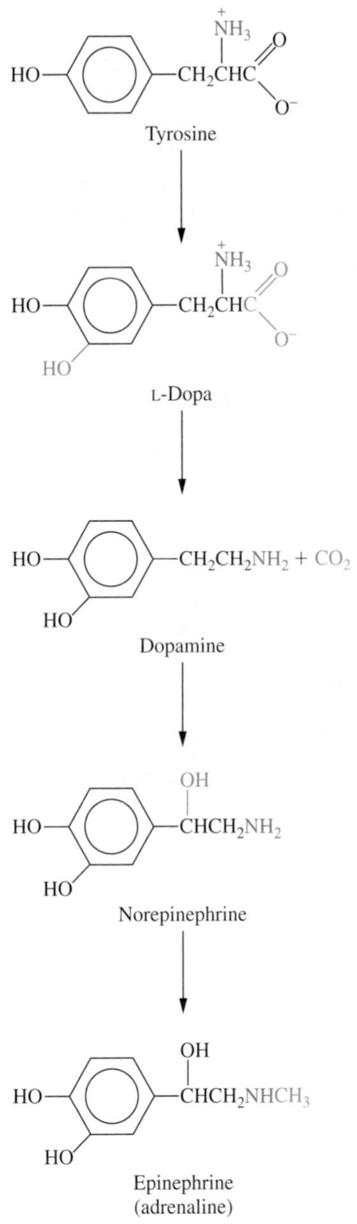

Catecholamines

All of the catecholamine neurotransmitters, including dopamine, epinephrine, and norepinephrine, are synthesized from the amino acid tyrosine (Figure 15.7). *Dopamine* is critical to good health. A deficiency in this neurotransmitter, for example, results in Parkinson's disease, a disorder characterized by tremors, monotonous speech, loss of memory and problem-solving ability, and loss of motor function. It would seem logical to treat Parkinson's disease with dopamine. Unfortunately, dopamine cannot cross the blood-brain barrier to enter brain cells. As a result, L-dopa, which is converted to dopamine in brain cells, is used to treat this disorder.

Just as too little dopamine causes Parkinson's disease, an excess is associated with schizophrenia. Dopamine also appears to play a role in addictive behavior. In proper amounts, it causes a pleasant, satisfied feeling. The greater the amount of dopamine, the more intense the sensation, the "high." Several drugs have been shown to increase the levels of dopamine. Among these are cocaine, heroin, amphetamines, alcohol, and nicotine. Marijuana also causes an increase in brain dopamine, raising the possibility that it, too, has the potential to produce addiction.

Both *epinephrine* (adrenaline) and *norepinephrine* are involved in the "fight or flight" response. Epinephrine stimulates the breakdown of glycogen to produce glucose, which is then metabolized to provide energy for the body. Norepinephrine is involved with the central nervous system in the stimulation of other glands and the constriction of blood vessels. All of these responses prepare the body to meet the stressful situation.

Serotonin

Serotonin is synthesized from the amino acid tryptophan (Figure 15.8). A deficiency of serotonin has been associated with depression. It is also thought to be involved in bulimia and anorexia nervosa, as well as the carbohydrate-cravings that characterize seasonal affective disorder (SAD), a depression caused by a decrease in daylight during autumn and winter.

Serotonin also affects the perception of pain, thermoregulation, and sleep. There are those who believe that a glass of warm milk will help you fall asleep. We have all noticed how sleepy we become after that big Thanksgiving turkey

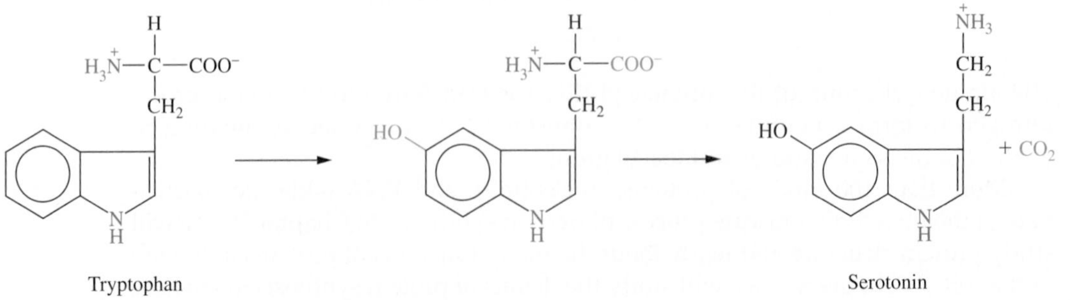

Figure 15.7 The pathway for synthesis of dopamine, epinephrine, and norepinephrine.

Figure 15.8 Synthesis of serotonin from the amino acid tryptophan.

A Medical Perspective

Opiate Biosynthesis and the Mutant Poppy

Hippocrates, the "father of medicine," left us the first record of the therapeutic use of opium (460 BC). Although not recorded, it is probably true that the addictive properties of opium were recognized soon thereafter!

The opium poppy (*Papaver somniferum*) is cultivated, legally and illegally, in many parts of the world. The flowers vary in color from white to deep red, but it is the seed pod that is sought after. In the seed pod is a milky fluid that contains morphine and codeine, and a small amount of an opioid called thebaine. The juice is extracted from the unripe seed pods and dried, and the opium alkaloids are extracted and purified.

In the legal pharmaceutical world, morphine and codeine are used to ease pain and spasmodic coughing. Thebaine is used as a reactant in pharmaceutical synthesis to produce a number of synthetic opioid compounds with a variety of biological effects. These include the analgesics oxycodone (brand name OxyContin), oxymorphone, and nalbuphine; naloxone, which is used to treat opioid overdosage; naltrexone, which is useful in helping people with narcotic or alcohol addictions to remain drug free; and buprenorphine, which is useful in the treatment of opiate addiction because it prevents withdrawal symptoms.

Approximately 40% of the world's legal opium poppies are grown on the Australian island state of Tasmania. This is big business, and the industry has developed an active research program to study the biochemical pathway for the synthesis of morphine and codeine. That pathway begins with the amino acid tyrosine, the same amino acid that is the initial reactant for the synthesis of dopamine and epinephrine (Section 15.5).

$$
\begin{array}{c}
\text{7 steps} \quad\quad \text{6 steps} \quad\quad \text{3 steps} \quad \text{1 step} \\
\text{Tyrosine} \rightarrow \text{Reticuline} \rightarrow \text{Thebaine} \rightarrow \text{Codeine} \rightarrow \text{Morphine}
\end{array}
$$

Through seven chemical reactions, tyrosine is converted to reticuline. Another six reactions convert reticuline to thebaine. Three chemical modifications convert thebaine to codeine, which undergoes an ester hydrolysis to produce morphine.

In the course of their studies, researchers produced a mutant strain of poppy that cannot make morphine or codeine, but does produce high levels of thebaine. "Norman," for "No Morphine," has been the most common strain of poppy grown in Tasmania since 1997. The mutation that causes Norman to produce high levels of thebaine is an alteration in one of the enzymes that catalyzes the conversion of thebaine to codeine. Since it can't be converted into codeine, large amounts of thebaine accumulate in the seed pods.

Synthetic opioids, such as naloxone and buprenorphine and the others mentioned above, have become much more important commercially than codeine and morphine. The economic value of Norman is that it produces large amounts of the starting material for the synthesis of these synthetic opioids, as well as the experimental synthesis of new drugs with unknown potential.

Consider the drug formulation Suboxone, a combination of buprenorphine and naloxone, approved for use in the United States in 2003 and produced by the British company Reckitt

(a)   **(b)**

(a) Opium poppies are the source of morphine and codeine.
(b) The sap of an opium poppy is white. The sap of the no-morphine mutant poppy is red.

Naloxone

many steps

Thebaine

many steps

Buprenorphine

Benckiser. This combination of synthetic opiates calms the addict's craving for opiates and yet poses little risk of being abused.

Buprenorphine or "bupe" works by binding to the same receptors in the brain to which heroin binds; but the drug is only

Continued...

551

a partial heroin agonist, so there is no high. As a result, buprenorphine is much less addictive than drugs such as methadone, leaves patients much more clearheaded, and makes it easier for them to withdraw from the drug after a few months.

Because addicts don't get high from Suboxone, it is much less likely than methadone to be stolen and sold illegally. It has the added advantage that its effects are longer lasting than those of methadone, so addicts need only one pill every 2 or 3 days. In addition, because naloxone is an opioid antagonist, it causes instant withdrawal symptoms if an addict tries to inject Suboxone for a high. As a result, Suboxone can be given to addicts to be taken at home, rather than being dispensed only at clinics, as methadone is. With all of these features, Suboxone begins to sound like a miracle drug; but as with any addiction treatment, it will only help those who want to quit and are willing to work with counselors and support groups to resolve the underlying problems that caused the addiction in the first place.

For Further Understanding

▸ In 1998, researchers in England reported that some individuals who had eaten poppy seed rolls or cake tested positive in an opiate drug screen. Using Internet or other resources, investigate the "poppy seed defense" and suggest guidelines for opiate testing that would protect the innocent.

▸ OxyContin is the brand name for a formulation of oxycodone in a timed-release tablet. It is prescribed to provide up to 12 h of relief from chronic pain. Recently, OxyContin has become a commonly abused drug and is thought to be responsible for a number of deaths. Abusers crush the timed-release tablets and ingest or snort the drug, which results in a rapid and powerful high often compared to the euphoria experienced from taking heroin. Use the Internet or other resources to explain why abuse of this prescription medication has overshadowed heroin use in some areas. Consider ways to prevent such abuse.

LEARNING GOAL

14 Describe the function of neurotransmitters.

dinner. Both milk protein and turkey are exceptionally high in tryptophan, the precursor of serotonin!

Prozac (fluoxetine), one of the antidepressant drugs, is one of the most widely prescribed drugs in the United States.

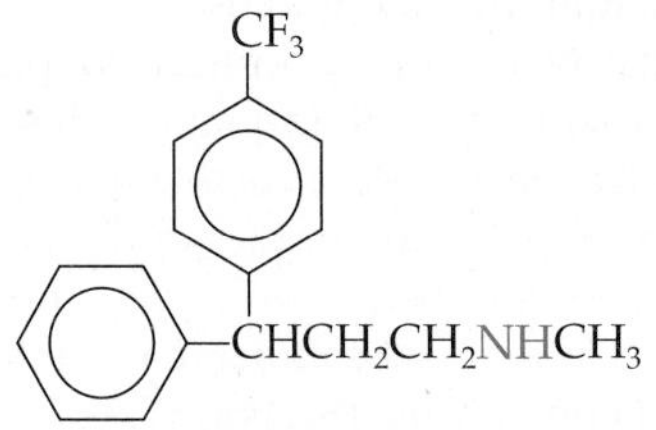

Prozac (fluoxetine)

It is a member of a class of drugs called selective serotonin reuptake inhibitors (SSRI). By inhibiting the reuptake, Prozac effectively increases the level of serotonin, relieving the symptoms of depression.

Histamine

Histamine is a neurotransmitter that is synthesized in many tissues by removing the carboxyl group from the amino acid histidine (Figure 15.9). It has many, often annoying, physiological roles. Histamine is a vasodilator that is released during the allergic response. It causes the itchy skin rash associated with poison ivy or insect bites. It also promotes the red, watery eyes and respiratory symptoms of hay fever.

Many antihistamines are available to counteract the symptoms of histamine release. These act by competing with histamine for binding to target cells. If histamine cannot bind to these target cells, the allergic response stops.

Benadryl is an antihistamine that is available as an ointment to inhibit the itchy rash response to allergens. It is also available as an oral medication to block the symptoms of systemic allergies. You need only visit the "colds and allergies" aisle of your grocery store to find dozens of medications containing antihistamines.

Histamine also stimulates secretion of stomach acid. When this response occurs frequently, the result can be chronic heartburn. The reflux of stomach acid into the esophagus can result in erosion of tissue and ulceration. The excess

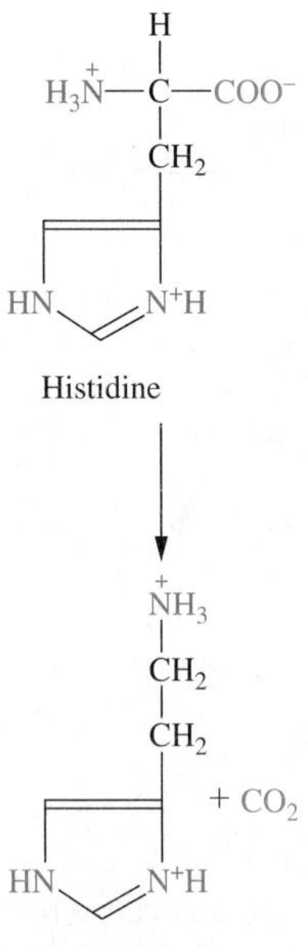

Figure 15.9 Synthesis of histamine from the amino acid histidine.

stomach acid may also contribute to development of stomach ulcers. The drug marketed as Tagamet (cimetidine) has proven to be an effective inhibitor of this histamine response, providing relief from chronic heartburn.

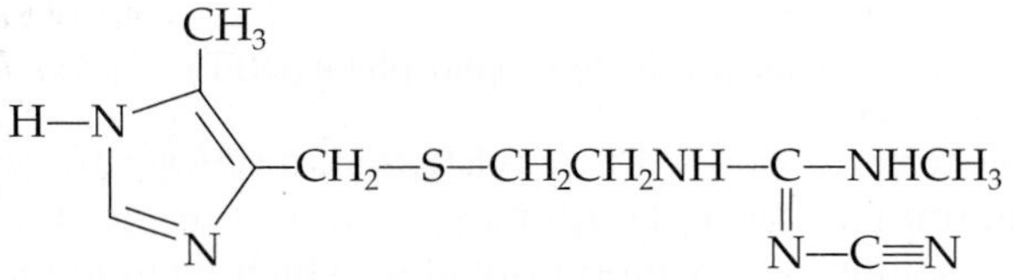

$$CH_3$$
H—N—N ... —CH$_2$—S—CH$_2$CH$_2$NH—C—NHCH$_3$
N   ‖
                               N—C≡N

Tagamet (cimetidine)

γ-Aminobutyric Acid and Glycine

γ-Aminobutyric acid (GABA) is produced by removal of a carboxyl group from the amino acid glutamate (Figure 15.10). Both GABA and the amino acid *glycine*

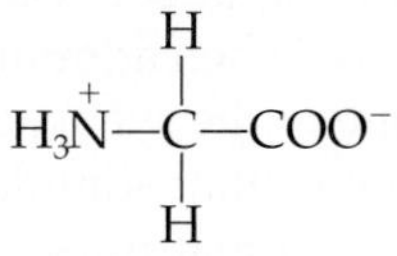

$$H_3\overset{+}{N}—\underset{H}{\overset{H}{C}}—COO^-$$

Glycine

are inhibitory neurotransmitters acting in the central nervous system. One class of tranquilizers, the benzodiazopines, relieves aggressive behavior and anxiety. These drugs have been shown to enhance the inhibitory activity of GABA, suggesting one of the roles played by this neurotransmitter.

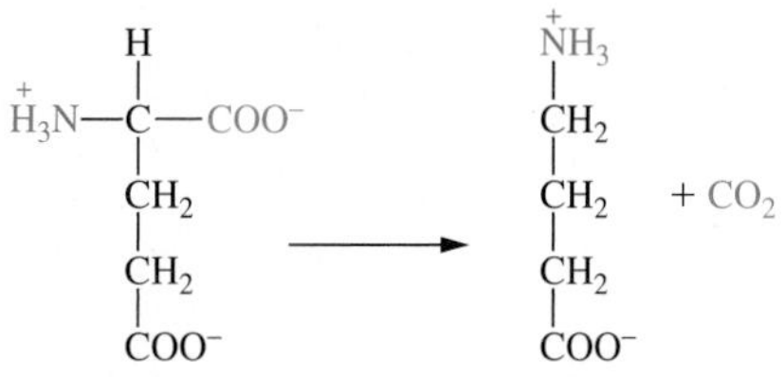

Acetylcholine

Acetylcholine is a neurotransmitter that functions at the neuromuscular junction, carrying signals from the nerve to the muscle. It is synthesized in a reaction between the quaternary ammonium ion choline and acetyl coenzyme A (Figure 15.11). When it is released from the nerve cell, acetylcholine binds to receptors on the surface of muscle cells. This binding stimulates the muscle cell to contract. Acetylcholine is then broken down to choline and acetate ion.

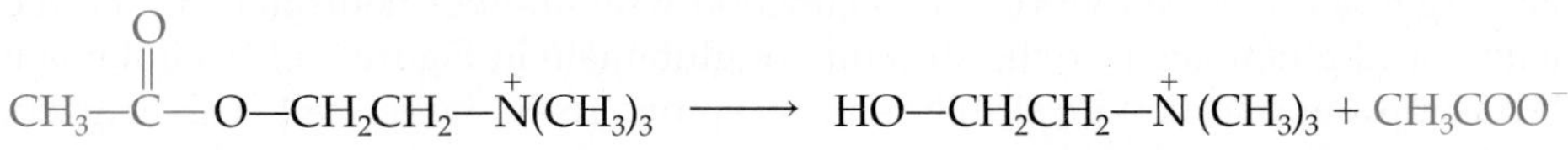

$$CH_3—\overset{O}{\overset{‖}{C}}—O—CH_2CH_2—\overset{+}{N}(CH_3)_3 \longrightarrow HO—CH_2CH_2—\overset{+}{N}(CH_3)_3 + CH_3COO^-$$

Acetylcholine   Choline   Acetate

These molecules are essentially recycled. They are taken up by the nerve cell where they are used to resynthesize acetylcholine, which is stored in the nerve cell until it is needed.

Nicotine is an agonist of acetylcholine. An agonist is a compound that binds to the receptor for another compound and causes or enhances the biological response. By binding to acetylcholine receptors, nicotine causes the sense of alertness and calm many smokers experience. Nerve cells that respond to nicotine may also signal nerve cells that produce dopamine. As noted above, the dopamine may be responsible for the addictive property of nicotine.

Inhibitors of acetylcholinesterase, the enzyme that catalyzes the breakdown of acetylcholine, are used both as poisons and as drugs. Among the most important poisons of acetylcholinesterase are a class of compounds known as organophosphates. One of these is *diisopropyl fluorophosphate* (DIFP). This molecule forms a covalently bonded intermediate with the enzyme, irreversibly inhibiting its activity.

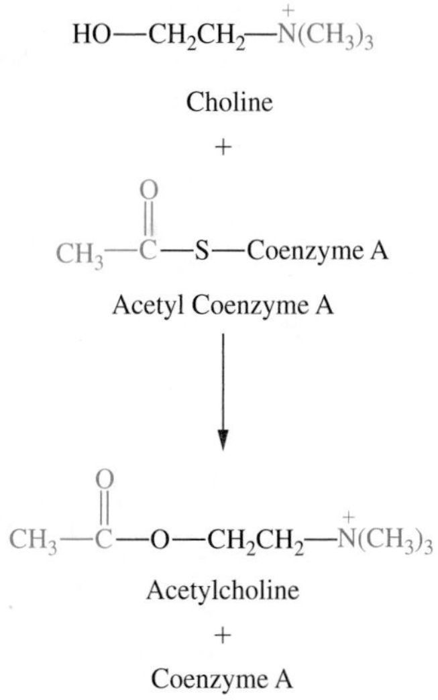

$$H_3\overset{+}{N}—\underset{|}{\overset{|}{C}}—COO^- \longrightarrow \overset{+}{N}H_3 \\ CH_2 \qquad CH_2 \\ CH_2 \qquad CH_2 \quad + CO_2 \\ CH_2 \qquad CH_2 \\ COO^- \qquad COO^-$$

Glutamate   γ-Aminobutyric acid

Figure 15.10 Synthesis of GABA from the amino acid glutamate.

$$HO—CH_2CH_2—\overset{+}{N}(CH_3)_3$$

Choline

+

$$CH_3—\overset{O}{\overset{‖}{C}}—S—Coenzyme A$$

Acetyl Coenzyme A

$$CH_3—\overset{O}{\overset{‖}{C}}—O—CH_2CH_2—\overset{+}{N}(CH_3)_3$$

Acetylcholine

+

Coenzyme A

Figure 15.11 Synthesis of acetylcholine.

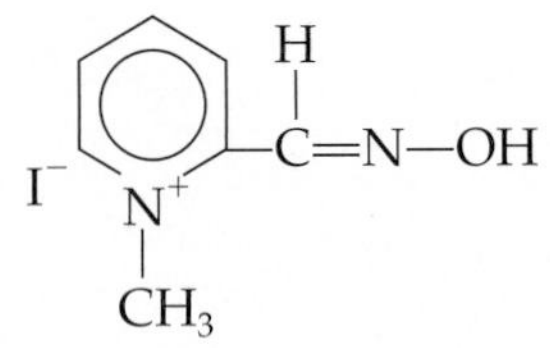

$$(CH_3)_2CH-O-\overset{\overset{\displaystyle O}{\|}}{\underset{\underset{\displaystyle F}{|}}{P}}-O-CH(CH_3)_2$$

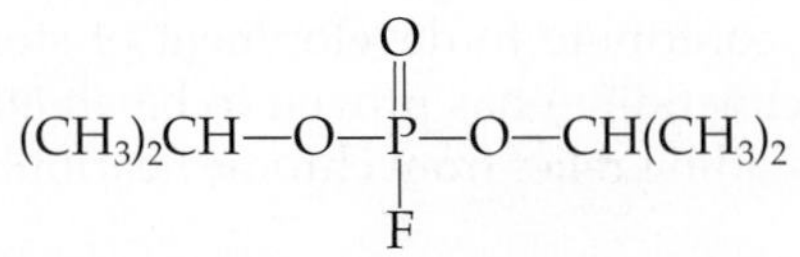

Diisopropyl fluorophosphate (DIFP)

Pyridine aldoxime methiodide

Thus, it is unable to break down the acetylcholine, and nerve transmission continues, resulting in muscle spasm. Death may occur as a result of laryngeal spasm. Antidotes for poisoning by organophosphates, which include many insecticides and nerve gases, have been developed. The antidotes work by reversing the effects of the inhibitor. One of these antidotes is *pyridine aldoxime methiodide* (PAM). This molecule displaces the organophosphate group from the active site of the enzyme, alleviating the effects of the poison.

Succinylcholine is a competitive inhibitor of acetylcholine that is used as a muscle relaxant in surgical procedures. Competitive inhibition occurs because the two molecules have structures so similar that both can bind to the acetylcholine receptor (compare the structures below). When administered to a patient, there is more succinylcholine than acetylcholine in the synapse, and thus more succinylcholine binding to the receptor. Because it cannot stimulate muscle contraction, succinylcholine causes muscles to relax. Normal muscle contraction resumes when the drug is no longer administered.

$$CH_3\overset{\overset{\displaystyle O}{\|}}{C}-O-CH_2CH_2-\overset{+}{N}(CH_3)_3$$

Acetylcholine

$$(CH_3)_3\overset{+}{N}-CH_2CH_2-O-\overset{\overset{\displaystyle O}{\|}}{C}CH_2CH_2\overset{\overset{\displaystyle O}{\|}}{C}-O-CH_2CH_2-\overset{+}{N}(CH_3)_3$$

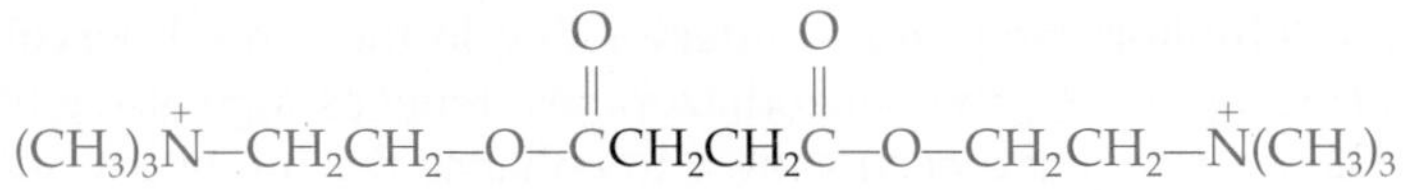

Succinylcholine

Acetylcholine nerve transmission is discussed in further detail in Chemistry at the Crime Scene: Enzymes, Nerve Agents, and Poisoning in Chapter 19.

Nitric Oxide and Glutamate

Nitric oxide (NO) is an amazing little molecule that has been shown to have many physiological functions. Among these is its ability to act as a neurotransmitter. NO is synthesized in many areas of the brain from the amino acid arginine. Research has suggested that NO works in conjunction with another neurotransmitter, the amino acid glutamate (see the structure of glutamate in Figure 15.10). Glutamate released from one nerve cell binds to receptors on its target cell. This triggers the target cell to produce NO, which then diffuses back to the original nerve cell. The NO signals the cell to release more glutamate, thus stimulating this neural pathway even further. This is a kind of positive feedback loop. It is thought that this NO-glutamate mechanism is involved in learning and the formation of memories.

CHAPTER MAP

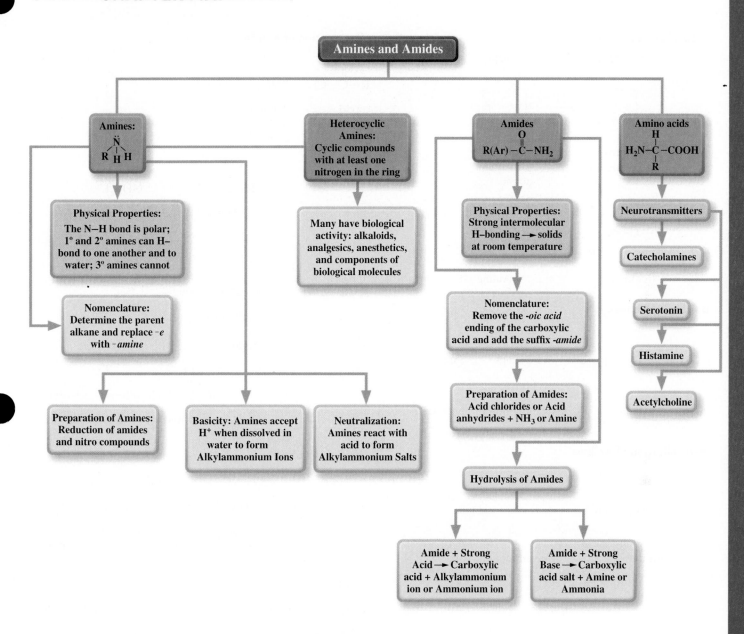

Amines and Amides

Amines:
$$\overset{\ddot{N}}{\underset{R \; H \; H}{|}}$$

Heterocyclic Amines:
Cyclic compounds with at least one nitrogen in the ring

Amides
$$\overset{O}{\underset{R(Ar)-C-NH_2}{\parallel}}$$

Amino acids
$$\overset{H}{\underset{R}{H_2N-\overset{|}{\underset{|}{C}}-COOH}}$$

Physical Properties:
The N–H bond is polar; 1º and 2º amines can H–bond to one another and to water; 3º amines cannot

Many have biological activity: alkaloids, analgesics, anesthetics, and components of biological molecules

Physical Properties:
Strong intermolecular H–bonding → solids at room temperature

Neurotransmitters

Nomenclature:
Determine the parent alkane and replace -*e* with -*amine*

Nomenclature:
Remove the -*oic acid* ending of the carboxylic acid and add the suffix -*amide*

Catecholamines

Serotonin

Preparation of Amines:
Reduction of amides and nitro compounds

Basicity: Amines accept H⁺ when dissolved in water to form Alkylammonium Ions

Neutralization:
Amines react with acid to form Alkylammonium Salts

Preparation of Amides:
Acid chlorides or Acid anhydrides + NH₃ or Amine

Histamine

Acetylcholine

Hydrolysis of Amides

Amide + Strong Acid → Carboxylic acid + Alkylammonium ion or Ammonium ion

Amide + Strong Base → Carboxylic acid salt + Amine or Ammonia

SUMMARY OF REACTIONS

Preparation of Amines

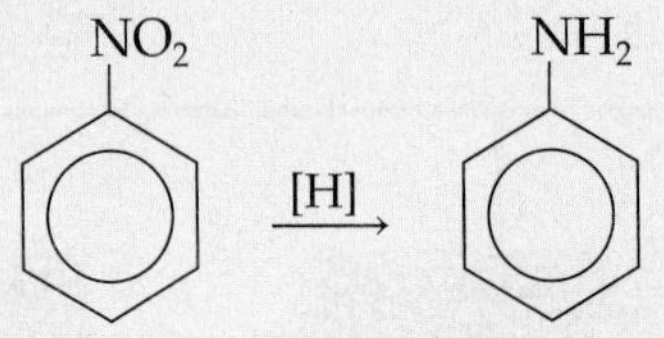

A nitro compound → An aromatic primary amine (via [H])

$$R^1-\overset{\overset{\displaystyle O}{\|}}{C}-N\overset{\diagup R^2}{\diagdown R^3} \xrightarrow{[H]} R^1CH_2N\overset{\diagup R^2}{\diagdown R^3}$$

Amide   Amine

Basicity of Amines

$$R-NH_2 + H-OH \rightleftharpoons R-\overset{\overset{\displaystyle H}{|}}{\underset{\underset{\displaystyle H}{|}}{N^+}}-H \ + \ OH^-$$

Amine   Water   Alkylammonium   Hydroxide
                         ion   ion

Neutralization of Amines

$$R-NH_2 + HCl \longrightarrow R-\overset{\overset{\displaystyle H}{|}}{\underset{\underset{\displaystyle H}{|}}{N^+}}-H\ Cl^-$$

Amine   Acid   Alkylammonium salt

Preparation of Amides

$$R-\overset{\overset{\displaystyle O}{\|}}{C}-Cl \ + \ 2NH_3 \longrightarrow$$

Acid   Ammonia
chloride   or
             amine

$$R-\overset{\overset{\displaystyle O}{\|}}{C}-NH_2 \ + \ \overset{+}{N}H_4Cl^-$$

Amide   Ammonium chloride
               or
               alkylammonium chloride

$$R-\overset{\overset{\displaystyle O}{\|}}{C}-O-\overset{\overset{\displaystyle O}{\|}}{C}-R \ + \ 2NH_3 \longrightarrow$$

Acid anhydride   Ammonia
                  or
                  amine

$$R-\overset{\overset{\displaystyle O}{\|}}{C}-NH_2 + R-\overset{\overset{\displaystyle O}{\|}}{C}-O^- \overset{+}{N}H_4$$

Amide   Carboxylic acid
               salt

Hydrolysis of Amides

$$R-\overset{\overset{\displaystyle O}{\|}}{C}-NH-R^1 + H_3O^+ \longrightarrow R-\overset{\overset{\displaystyle O}{\|}}{C}-OH + R^1-\overset{+}{N}H_3$$

Amide   Strong   Carboxylic   Alkyl-
         acid   acid   ammonium
                                 ion

$$R-\overset{\overset{\displaystyle O}{\|}}{C}-NH-R^1 + NaOH \longrightarrow R-\overset{\overset{\displaystyle O}{\|}}{C}-O^-Na^+ + R^1-NH_2$$

Amide   Strong   Carboxylic   Amine
         base   acid salt   or
                                 ammonia

SUMMARY

15.1 Amines

▶ **Amines** are a family of organic compounds that contain an amino group or substituted amino group.

▶ A **primary amine** has the general formula RNH_2; a **secondary amine** has the general formula R_2NH; and a **tertiary amine** has the general formula R_3N.

▶ In the systematic nomenclature system, amines are named as *alkanamines*.

▶ In the common system, they are named as *alkylamines*.

- Amines behave as weak bases, forming **alkylammonium ions** in water and alkylammonium salts when they react with acids.
- **Quaternary ammonium salts** are ammonium salts that have four organic groups bonded to the nitrogen atom.

15.2 Heterocyclic Amines

- **Heterocyclic amines** are cyclic compounds having at least one nitrogen atom in the ring structure.
- **Alkaloids** are natural plant products that contain at least one heterocyclic ring.
- Many alkaloids have powerful biological effects.
- Cocaine is an example of an **anesthetic,** which is a drug that causes a lack of sensation in any part of the body or which causes unconsciousness.
- Morphine is a strong **analgesic,** a drug that acts as a painkiller.

15.3 Amides

- Amides are formed in a reaction between a carboxylic acid derivative (an acid chloride or an acid anhydride) and an amine or ammonia.
  - The reaction between an acid chloride and ammonia produces a **primary (1°) amide.**
  - The reaction between an acid chloride and primary amine produces a **secondary (2°) amide.**
  - The reaction between an acid chloride and a secondary amine produces a **tertiary (3°) amide.**
- The **amide bond** is the bond between the carbonyl carbon of the **acyl group** and the nitrogen of the amine.
- In the IUPAC Nomenclature System, amides are named by replacing the *-oic acid* ending of the carboxylic acid with the *-amide* ending.
- In the common system of nomenclature, the *-ic acid* ending of the carboxylic acid is replaced by the *-amide* ending.
- Hydrolysis of an amide in the presence of a strong acid produces a carboxylic acid and an alkylammonium ion or ammonium ion; in the presence of a strong base, the products are an amine and a carboxylic acid salt.

15.4 A Preview of Amino Acids, Proteins, and Protein Synthesis

- Proteins are polymers of **amino acids** joined to one another by amide bonds called **peptide bonds.**
- During protein synthesis, the **aminoacyl group** of one amino acid is transferred from a carrier molecule called **transfer RNA (tRNA)** to the amino group nitrogen of another amino acid.

15.5 Neurotransmitters

- **Neurotransmitters** are chemicals that carry messages, or signals, from a nerve cell to a target cell, which may be another nerve cell or a muscle cell.

- Neurotransmitters may be inhibitory or excitatory, and all are nitrogen-containing compounds.
- The catecholamines include dopamine, norepinephrine, and epinephrine.
  - Too little dopamine results in Parkinson's disease. Too much is associated with schizophrenia.
  - Dopamine is also associated with addictive behavior.
- A deficiency of serotonin is associated with depression and eating disorders. Serotonin is involved in pain perception, regulation of body temperature, and sleep.
- Histamine contributes to allergy symptoms.
  - Antihistamines block histamine and provide relief from allergies.
- γ-Aminobutyric acid (GABA) and glycine are inhibitory neurotransmitters.
  - It is believed that GABA is involved in the control of aggressive behavior.
- Acetylcholine is a neurotransmitter that functions at the neuromuscular junction, carrying signals from the nerve to the muscle.
- Nitric oxide and glutamate function in a positive feedback loop that is thought to be involved in learning and the formation of memories.

ANSWERS TO PRACTICE PROBLEMS

15.1 a. Tertiary amine
  b. Primary amine
  c. Secondary amine

15.2 a. Methanol would have a higher boiling point than methylamine. The intermolecular hydrogen bonds between alcohol molecules will be stronger than those between two amines because oxygen is more electronegative than nitrogen.
  b. Water would have a higher boiling point than dimethylamine. The intermolecular hydrogen bonds between water molecules will be stronger than those between two amines because oxygen is more electronegative than nitrogen.
  c. Ethylamine will have a higher boiling point than methylamine because it has a higher molar mass.
  d. Propylamine will have a higher boiling point than butane because propylamine molecules can form intermolecular hydrogen bonds, and the nonpolar butane cannot do so.

15.3 a. *N*-Ethyl-*N*-methyl-1-propanamine
  b. 3-Hexanamine
  c. *N,N*-Diethyl-1-propanamine

15.4 a. IUPAC name: *N*-Pentylpentanamide
    Common name: *N*-Pentylvaleramide
  b. IUPAC name: *N*-Butylhexanamide
    Common name: *N*-Butylcaproamide

QUESTIONS AND PROBLEMS

Amines

Foundations

15.11 Compare the boiling points of amines, alkanes, and alcohols of the same molar mass. Explain these differences in boiling points.

15.12 Describe the water solubility of amines in relation to their carbon chain length.

15.13 Describe the systematic rules for naming amines.

15.14 How are the common names of amines derived?

15.15 Describe the physiological effects of amphetamines.

15.16 Define the terms *analgesic* and *anesthetic,* and list some amines that have these activities.

Applications

15.17 For each pair of compounds, predict which would have greater solubility in water. Explain your reasoning.
   a. Hexane or 1-pentanamine
   b. Cyclopentane or 2-butanamine

15.18 For each pair of compounds, predict which would have the higher boiling point. Explain your reasoning.
   a. Propanamine or propanol
   b. Propane or ethanamine
   c. Methanamine or water
   d. Propylmethylamine or pentane

15.19 Explain why a tertiary amine such as triethylamine has a significantly lower boiling point than its primary amine isomer, 1-hexanamine.

15.20 Draw a diagram to illustrate your answer to Question 15.19.

15.21 Use systematic nomenclature to name each of the following amines:

   a. $CH_3CH_2CHNH_2$
           |
           CH_3
   b. $CH_3CH_2CH_2CHCH_2CH_3$
                    |
                    NH_2
   c.
   d. $(CH_3)_3C{-}NH_2$

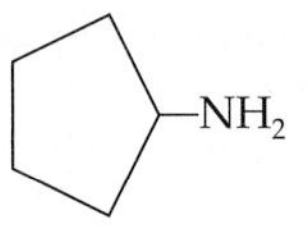

15.22 Use systematic and common nomenclature to name each of the following amines:

   a. $CH_3CH_2CH_2CH_2CH_2CH_2CH_2CH_2NH_2$

   b. $Cl{-}$$-NH_2$

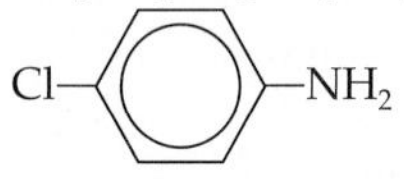

   c. $CH_3CHCH_2CH_3$
                |
                NH_2

   d. $CH_3NCH_2CH_3$
               |
               CH_3

15.23 Draw the condensed formula and line formula of each of the following compounds:
   a. Diethylamine   **d.** 3-Bromo-2-pentanamine
   b. Butylamine   **e.** Triphenylamine
   c. 3-Decanamine

15.24 Draw the condensed formula and line formula of each of the following compounds:
   a. *N,N*-Dipropylaniline
   b. Cyclohexanamine
   c. 2-Bromocyclopentanamine
   d. Tetraethylammonium iodide
   e. 3-Bromobenzenamine

15.25 Draw the condensed formula for each of the following compounds:
   a. 3-Hexanamine
   b. Hexylpentylamine
   c. Cyclobutanamine
   d. 2-Methylcyclopentanamine
   e. Triethylammonium chloride

15.26 Draw the condensed formula for each of the following compounds:
   a. 2,3-Dibromoaniline
   b. 2-Octanamine
   c. 2-Chloro-2-pentanamine
   d. *N,N*-Diethylpentanamine
   e. Diisopropylamine

15.27 Draw condensed formulas for the eight isomeric amines that have the molecular formula $C_4H_{11}N$. Name each of the isomers using the systematic names and determine whether each isomer is a 1°, 2°, or 3° amine.

15.28 Draw all of the isomeric amines of molecular formula C_3H_9N. Name each of the isomers using the systematic names, and determine whether each isomer is a primary, secondary, or tertiary amine.

15.29 Classify each of the following amines as 1°, 2°, or 3°:
   a. Cyclohexanamine
   b. Dibutylamine
   c. 2-Methyl-2-heptanamine
   d. Tripentylamine

15.30 Classify each of the following amines as primary, secondary, or tertiary:
   a. Benzenamine   **d.** Tripropylamine
   b. *N*-Ethyl-2-pentanamine   **e.** *m*-Chloroaniline
   c. Ethylmethylamine

15.31 Write an equation to show a reaction that would produce each of the following products:

   a.

   b.

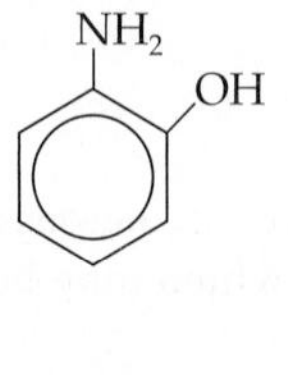

   c.

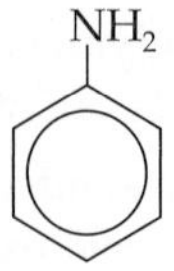

   d.

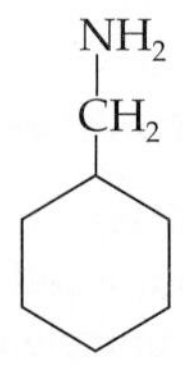

15.32 Write an equation to show a reaction that would produce each of the following amines:
  a. 1-Pentanamine
  b. *N,N*-Dimethylethanamine
  c. *N*-Ethylpropanamine

15.33 Complete each of the following equations by supplying the missing reactant or product indicated by a question mark:

$$\text{a. } CH_3\overset{\displaystyle CH_3}{\underset{\displaystyle H}{\overset{|}{\underset{|}{N}}}}H + ? \longrightarrow CH_3\overset{\displaystyle CH_3}{\underset{\displaystyle H}{\overset{|}{\underset{|}{N^+}}}}H + OH^-$$

$$\text{b. } CH_3CH_2\overset{\displaystyle CH_3}{\underset{\displaystyle CH_2CH_3}{\overset{|}{\underset{|}{N}}}} + ? \longrightarrow CH_3CH_2\overset{\displaystyle CH_3}{\underset{\displaystyle CH_2CH_3}{\overset{|}{\underset{|}{N^+}}}}H\ Br^-$$

c. $CH_3CH_2CH_2NH_2 + H_2O \longrightarrow ? + OH^-$

$$\text{d. } CH_3CH_2\overset{\displaystyle CH_2CH_3}{\underset{}{\overset{|}{N}}}H + HCl \longrightarrow ?$$

15.34 Complete each of the following equations by supplying the missing reactant or product indicated by a question mark:

a. $CH_3CH_2NH_2 + H_2O \longrightarrow ? + OH^-$

$$\text{b. } ? + HCl \longrightarrow CH_3CH_2CH_2\overset{\displaystyle CH_2CH_2CH_3}{\underset{\displaystyle H}{\overset{|}{\underset{|}{N^+}}}}H\ Cl^-$$

$$\text{c. } CH_3\overset{\displaystyle CH_3}{\underset{\displaystyle CH_3}{\overset{|}{\underset{|}{CH}}}}NH + H_2O \longrightarrow ? + ?$$

d. $NH_3 + HBr \longrightarrow ?$

15.35 Using condensed formulas, write an equation for the reaction that would produce each of the following amines:
  a. Hexanamine
  b. *N*-Methylbutanamine
  c. *N,N*-Dimethylpropanamine

15.36 Write an equation for the reaction that would produce each of the following amines:
  a. Octanamine
  b. *N*-Methylpropanamine
  c. *N,N*-Diethylpentanamine

15.37 Briefly explain why the lower-molar-mass amines (fewer than five carbons) exhibit appreciable solubility in water.

15.38 Why is the salt of an amine appreciably more soluble in water than the amine from which it was formed?

15.39 Most drugs containing amine groups are not administered as the amine but rather as the ammonium salt. Can you suggest a reason why?

15.40 Why does aspirin upset the stomach, whereas acetaminophen (Tylenol) does not?

15.41 Putrescine and cadaverine are two odoriferous amines that are produced by decaying flesh. Putrescine is 1,4-butanediamine, and cadaverine is 1,5-pentanediamine. Draw the structures of these two compounds.

15.42 How would you quickly convert an alkylammonium salt into a water-insoluble amine? Explain the rationale for your answer.

Heterocyclic Amines

15.43 Indole and pyridine rings are found in alkaloids.
  a. Sketch each ring.
  b. Name one compound containing each of the ring structures and indicate its use.

15.44 What is an alkaloid?

15.45 List some heterocyclic amines that are used in medicine.

15.46 Distinguish between the terms *analgesic* and *anesthetic*.

Amides

Foundations

15.47 Why do amides have very high boiling points?

15.48 Describe the water solubility of amides in relation to their carbon chain length.

15.49 Explain the IUPAC nomenclature rules for naming amides.

15.50 How are the common names of amides derived?

15.51 Describe the physiological effects of barbiturates.

15.52 Why is acetaminophen often recommended in place of aspirin?

Applications

15.53 Use the IUPAC and common systems of nomenclature to name the following amides:

$$\text{a. } CH_3CH_2\overset{\displaystyle O}{\overset{\|}{C}}NH_2$$

$$\text{b. } CH_3CH_2CH_2CH_2\overset{\displaystyle O}{\overset{\|}{C}}NH_2$$

$$\text{c. } CH_3\overset{\displaystyle O}{\overset{\|}{C}}N(CH_3)_2$$

15.54 Use the IUPAC Nomenclature System to name each of the following amides:

$$\text{a. } CH_3CH_2\overset{}{\underset{\displaystyle Br}{\overset{}{\underset{|}{CH}}}}CH_2\overset{\displaystyle O}{\overset{\|}{C}}NH_2 \qquad \text{c. } CH_3\overset{}{\underset{\displaystyle CH_3}{\overset{}{\underset{|}{CH}}}}\overset{\displaystyle O}{\overset{\|}{C}}NH_2$$

b.

15.55 Draw the condensed formula for each of the following amides:
  a. Propanamide   **c.** 2,3-Diethylpentanamide
  b. *N,N*-Diethylbutanamide **d.** *N*-Methylhexanamide

15.56 Draw the condensed formula for each of the following amides:
  a. *N*-Propylbutanamide   **c.** *N*-Methylpropanamide
  b. *N*-Butyloctanamide   **d.** *N*-Isopropylhexanamide

15.57 Draw the condensed formula and line formula of each of the following amides:
  a. Ethanamide
  b. *N*-Methylpropanamide
  c. *N,N*-Diethylbenzenamide
  d. 3-Bromo-4-methylhexanamide
  e. *N,N*-Dimethylacetamide

15.58 Draw the condensed formula and line formula of each of the following amides:
  a. Acetamide
  b. 4-Methylpentanamide
  c. N,N-Dimethylpropanamide
  d. Formamide
  e. N-Ethylpropionamide

15.59 The active ingredient in many insect repellents is N,N-diethyl-m-toluamide (DEET). Draw the structure of this compound. Which carboxylic acid and amine would be released by hydrolysis of this compound?

15.60 When an acid anhydride and an amine are combined, an amide is formed. This approach may be used to synthesize acetaminophen, the active ingredient in Tylenol. Using the reactants provided here, draw the structure of the amide product, acetaminophen:

$$CH_3\overset{O}{\overset{||}{C}}-O-\overset{O}{\overset{||}{C}}CH_3 + 2H_2N-\underset{}{\bigcirc}-OH$$

15.61 Explain why amides are neutral in the acid-base sense.

15.62 The amide bond is stabilized by resonance. Draw the contributing resonance forms of the amide bond.

15.63 Lidocaine is often used as a local anesthetic. For medicinal purposes, it is often used in the form of its hydrochloride salt because the salt is water-soluble. In the structure of lidocaine hydrochloride shown, locate the amide functional group.

Lidocaine hydrochloride

15.64 Locate the amine functional group in the structure of lidocaine. Is lidocaine a primary, secondary, or tertiary amine?

15.65 The antibiotic penicillin BT contains functional groups discussed in this chapter. In the structure of penicillin BT shown, locate and name as many functional groups as you can.

Penicillin BT

15.66 The structure of saccharin, an artificial sweetener, is shown. Circle the amide group.

Saccharin

15.67 Complete each of the following equations by supplying the missing reactant(s) or product(s) indicated by a question mark. Provide the systematic name for all the reactants and products.

a. $CH_3\overset{O}{\overset{||}{C}}NHCH_3 + H_3O^+ \longrightarrow ? + ?$

b. $? + H_3O^+ \longrightarrow CH_3CH_2CH_2\overset{O}{\overset{||}{C}}-OH + CH_3\overset{+}{N}H_3$

c. $CH_3\underset{CH_3}{\underset{|}{CH}}CH_2\overset{O}{\overset{||}{C}}NHCH_2CH_3 + ? \longrightarrow$

$CH_3\underset{CH_3}{\underset{|}{CH}}CH_2\overset{O}{\overset{||}{C}}-OH + ?$

15.68 Complete each of the following by supplying the missing reagents. Draw the structures of each of the reactants and products.
  a. N-Methylpropanamide + ? ⟶ propanoic acid + ?
  b. N,N-Dimethylacetamide + strong acid ⟶ ? + ?
  c. Formamide + strong acid ⟶ ? + ?

15.69 Complete each of the following equations by supplying the missing reactant(s) or product(s) indicated by a question mark.

a. $? + 2CH_3CH_2CH_2NH_2 \longrightarrow$

$CH_3CH_2CH_2NH\overset{O}{\overset{||}{C}}CH_2CH_3 +$

$CH_3CH_2\overset{O}{\overset{||}{C}}-O^-\overset{+}{N}H_3-CH_2CH_2CH_3$

b. $CH_3CH_2\overset{O}{\overset{||}{C}}-Cl + 2NH_3 \longrightarrow ? + ?$

c. $? + ? \longrightarrow CH_3CH_2CH_2\overset{O}{\overset{||}{C}}NHCH_2CH_3 +$

$CH_3CH_2-\overset{+}{N}H_3Cl^-$

15.70 Write two equations for the synthesis of each of the following amides. In one equation, use an acid chloride as a reactant. In the second equation, use an acid anhydride.
  a. Ethanamide
  b. N-Propylpentanamide
  c. Propionamide

15.71 In addition to HCl, what is the product of the reaction of ammonia with an acid chloride? Draw the structure of that product and describe its features.

15.72 In addition to HCl, what is the product of the reaction of a primary amine with an acid chloride? Draw the structure of that product and describe its features.

15.73 In addition to HCl, what is the product of the reaction of a secondary amine with an acid chloride? Draw the structure of that product and describe its features.

15.74 Write general equations for the synthesis of a primary, secondary, and tertiary amide.

A Preview of Amino Acids, Proteins, and Protein Synthesis

Foundations

15.75 Draw the general structure of an amino acid.

15.76 What is the name of the amide bond formed between two amino acids?

Applications

15.77 The amino acid glycine has a hydrogen atom as its R group, and the amino acid alanine has a methyl group. Draw these two amino acids.

15.78 Draw a dipeptide composed of glycine and alanine. Begin by drawing glycine with its amino group on the left. Circle the amide bond.

15.79 Draw the amino acid alanine (see Question 15.77). Place a star by the chiral carbon. (*Hint:* A chiral carbon is one that is bonded to four different groups or atoms.)

15.80 Does glycine have a chiral carbon? Explain your reasoning.

15.81 Describe acyl group transfer.

15.82 Describe the relationship between acyl group transfer and the process of protein synthesis.

Neurotransmitters

Foundations

15.83 Define the term *neurotransmitter*.

15.84 What are the two general classes of neurotransmitters? What distinguishes them from one another?

Applications

15.85
  a. What symptoms result from a deficiency of dopamine?
  b. What is the name of this condition?
  c. What symptoms result from an excess of dopamine?

15.86 What is the starting material in the synthesis of dopamine, epinephrine, and norepinephrine?

15.87 Explain the connection between addictive behavior and dopamine.

15.88 Why is L-dopa used to treat Parkinson's disease rather than dopamine?

15.89 What is the function of epinephrine?

15.90 What is the function of norepinephrine?

15.91 What is the starting material from which serotonin is made?

15.92 What symptoms are associated with a deficiency of serotonin?

15.93 What physiological processes are affected by serotonin?

15.94 How does Prozac relieve the symptoms of depression?

15.95 What are the physiological roles of histamine?

15.96 How do antihistamines function to control the allergic response?

15.97 What type of neurotransmitters are γ-aminobutyric acid and glycine?

15.98 Explain the evidence for a relationship between γ-aminobutyric acid and aggressive behavior.

15.99 Explain the function of acetylcholine at the neuromuscular junction.

15.100 Explain why organophosphates are considered to be poisons.

15.101 How does pyridine aldoxime methiodide function as an antidote for organophosphate poisoning?

15.102 Explain the mechanism by which glutamate and NO may function to promote development of memories and learning.

CHALLENGE PROBLEMS

1. Histamine is made and stored in blood cells called *mast cells*. Mast cells are involved in the allergic response. Release of histamine in response to an allergen causes dilation of capillaries. This, in turn, allows fluid to leak out of the capillary, resulting in local swelling. It also causes an increase in the volume of the vascular system. If this increase is great enough, a severe drop in blood pressure may cause shock. Histamine is produced by decarboxylation (removal of the carboxylate group as CO_2) of the amino acid histidine shown below. Draw the structure of histamine.

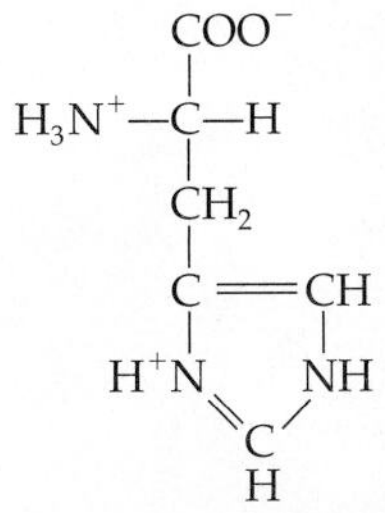

2. Carnitine tablets are sold in health food stores. It is claimed that carnitine will enhance the breakdown of body fat. Carnitine is a tertiary amine found in mitochondria, cell organelles in which food molecules are completely oxidized and ATP is produced. Carnitine is involved in transporting the acyl groups of fatty acids from the cytoplasm into the mitochondria. The fatty acyl group is transferred from a fatty acyl CoA molecule and esterified to carnitine. Inside the mitochondria, the reaction is reversed and the fatty acid is completely oxidized. The structure of carnitine is shown here:

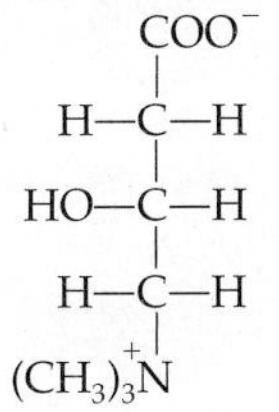

Draw the acyl carnitine molecule that is formed by esterification of palmitic acid with carnitine.

3. The amino acid proline has a structure that is unusual among amino acids. Compare the general structure of an amino acid with that of proline, shown here:

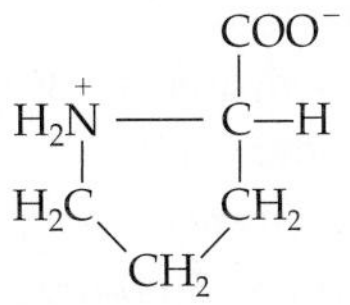

What is the major difference between proline and the other amino acids? Draw the structure of a dipeptide in which the amino group of proline forms a peptide bond with the carboxyl group of alanine.

4. Bulletproof vests are made of the polymer called Kevlar. It is produced by the copolymerization of the following two monomers:

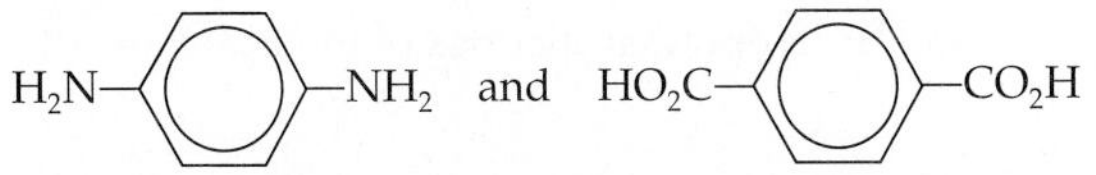

Draw the structure of a portion of the Kevlar polymer.

16

Carbohydrates

Would "looking-glass milk" be nutritious?

LEARNING GOALS

1. Explain the difference between complex and simple carbohydrates, and know the amounts of each recommended in the daily diet.

2. Apply the systems of classifying and naming monosaccharides according to the functional group and number of carbons in the chain.

3. Determine whether a molecule has a chiral center.

4. Explain stereoisomerism.

5. Identify monosaccharides as either D- or L-.

6. Draw and name the common monosaccharides using structural formulas.

7. Given the linear structure of a monosaccharide, draw the Haworth projection of its α- and β-cyclic forms and vice versa.

8. By inspection of the structure, predict whether a sugar is a reducing or a nonreducing sugar.

9. Discuss the use of the Benedict's reagent to measure the level of glucose in urine.

10. Draw and name the common disaccharides, and discuss their significance in biological systems.

11. Describe the difference between galactosemia and lactose intolerance.

12. Discuss the structural, chemical, and biochemical properties of starch, glycogen, and cellulose.

OUTLINE

INTRODUCTION

In his children's story *Through the Looking Glass*, Lewis Carroll's heroine Alice wonders whether "looking-glass milk" would be good to drink. As we will see in this chapter, many biological molecules, such as the sugars, exist as two stereoisomers, *enantiomers*, that are mirror images of one another. Because two mirror-image forms occur, it is rather remarkable that in our bodies, and in most of the biological world, only one of the two is found. For instance, the common sugars are members of the D-family, whereas all the common amino acids that make up our proteins are members of the L-family. It is not too surprising, then, that the enzymes in our bodies that break down the sugars and proteins we eat are *stereospecific*; that is, they recognize only one mirror-image isomer. Knowing this, we can make an educated guess that "looking-glass milk" could not be digested by our enzymes and therefore would not be a good source of food for us. It is even possible that it might be toxic to us!

Pharmaceutical chemists are becoming more and more concerned with the stereochemical purity of the drugs we take. Consider a few examples. In 1960, the drug thalidomide was commonly prescribed in Europe as a sedative. However, during that year, hundreds of women who took thalidomide during pregnancy gave birth to babies with severe birth defects. Thalidomide, it turned out, was a mixture of two enantiomers. One is a sedative; the other is a teratogen, a chemical that causes birth defects.

One of the common side effects of taking antihistamines for colds or allergies is drowsiness. Again, this is the result of the fact that antihistamines are mixtures of enantiomers. One causes drowsiness; the other is a good decongestant.

One enantiomer of the compound carvone is associated with the smell of spearmint; the other produces the aroma of caraway seeds or dill. One mirror-image form of limonene smells like lemons; the other has the aroma of oranges.

The pain reliever ibuprofen is currently sold as a mixture of enantiomers, but one is a much more effective analgesic than the other.

Taste, smell, and the biological effects of drugs in the body all depend on the stereochemical form of compounds and their interactions with cellular enzymes or receptors. As a result, chemists are actively working to devise methods of separating the isomers in pure form. Alternatively, methods of conducting stereospecific syntheses that produce only one stereoisomer are being sought. By preparing pure stereoisomers, the biological activity of a compound can be much more carefully controlled. This will lead to safer medications.

In this chapter, we will begin our study of stereochemistry, the spatial arrangement of atoms in molecules, with the carbohydrates. Later, we will examine the stereochemistry of the amino acids that make up our proteins and consider the stereochemical specificity of the metabolic reactions that are essential to life.

16.1 Types of Carbohydrates

We begin our study of biochemistry with the carbohydrates. Carbohydrates are produced in plants by photosynthesis (Figure 16.1). Natural carbohydrate sources such as grains and cereals, breads, sugarcane, fruits, milk, and honey are an important source of energy for animals. **Carbohydrates** include simple sugars as well as long polymers of these simple sugars, for instance potato starch, and a variety of molecules of intermediate size. The simple sugar glucose, $C_6H_{12}O_6$, is the primary energy source for the brain and nervous system and can be used by many other tissues. When "burned" by cells for energy, each gram (g) of carbohydrate releases approximately 4 kilocalories (kcal) of energy.

LEARNING GOAL

1 Explain the difference between complex and simple carbohydrates, and know the amounts of each recommended in the daily diet.

A kilocalorie is the same as the calorie (Cal) referred to in the "count-your-calories" books and on nutrition labels.

A healthy diet contains both complex carbohydrates, such as starches and cellulose, and simple sugars, such as fructose and sucrose (Figure 16.2). However, the quantity of simple sugars, especially sucrose, should be minimized because large quantities of sucrose in the diet promote obesity and tooth decay.

Complex carbohydrates are better for us than the simple sugars. Starch, found in rice, potatoes, breads, and cereals, is an excellent energy source. In addition, the complex carbohydrates, such as cellulose, provide us with an important supply of dietary fiber.

See A Medical Perspective: Tooth Decay and Simple Sugars on page 566.

It is hard to determine exactly what percentage of the daily diet *should* consist of carbohydrates. The *actual* percentage varies widely throughout the world, from 80% in the Far East, where rice is the main component of the diet, to 40–50% in the United States. Currently, it is recommended that 45–65% of the calories in the diet should come from carbohydrates, and the World Health Organization recommends that no more than 5% of the daily caloric intake should be sucrose.

Question 16.1   What is the current recommendation for the amount of carbohydrates that should be included in the diet? Of the daily intake of carbohydrates, what percentage should be simple sugar?

Question 16.2   Distinguish between simple and complex carbohydrates. What are some sources of complex carbohydrates?

Monosaccharides such as glucose and fructose are the simplest carbohydrates because they contain a single (*mono-*) sugar (*saccharide*) unit. **Disaccharides,** including sucrose and lactose, consist of two monosaccharide units joined through bridging oxygen atoms. Such a bond is called a **glycosidic bond. Oligosaccharides** consist of three to ten monosaccharide units joined by glycosidic bonds. The largest and most complex carbohydrates are the **polysaccharides,** which are long, often highly branched, chains of monosaccharides. Starch, glycogen, and cellulose are all examples of polysaccharides.

Figure 16.1 Carbohydrates are produced by plants such as this potato in the process of photosynthesis, which uses the energy of sunlight to produce hexoses from CO_2 and H_2O.

Figure 16.2 Carbohydrates from a variety of foods are an essential component of the diet.

16.2 Monosaccharides

Monosaccharides are composed of carbon, hydrogen, and oxygen. They can be classified on the basis of the functional groups they contain. A monosaccharide with a ketone (carbonyl) group is a **ketose.** In a ketose, the carbonyl group is located on carbon-2. If an aldehyde (carbonyl) group is present, it is called an **aldose.** In an aldose, the carbonyl group is located on carbon-1. Sometimes monosaccharides are called *polyhydroxyaldehydes* or *polyhydroxyketones* because they also contain many hydroxyl groups.

The importance of phosphorylated sugars in metabolic reactions is discussed in Sections 14.4 and 21.3.

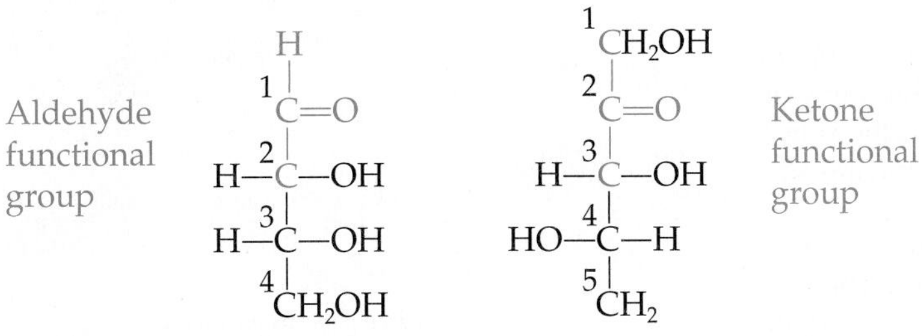

Aldehyde functional group — An aldose

Ketone functional group — A ketose

LEARNING GOAL

2 Apply the systems of classifying and naming monosaccharides according to the functional group and number of carbons in the chain.

Another system of classification tells us the number of carbon atoms in the main skeleton. A three-carbon monosaccharide is a *triose,* a four-carbon sugar is a *tetrose,* a five-carbon sugar is a *pentose,* a six-carbon sugar is a *hexose,* and so on. Combining the two classification systems gives even more information about the structure and composition of a sugar. For example, an aldotetrose is a four-carbon sugar that is also an aldehyde.

In addition to these general classification schemes, each monosaccharide has a unique name. These names are shown in blue for the following structures. Because the monosaccharides can exist in several different stereoisomers, it is important to provide the complete name. Thus, the complete names of the following structures are D-glyceraldehyde, D-glucose, and D-fructose. These names tell us that the structure represents one particular sugar and also identifies the sugar as one of two possible stereoisomers (D- or L-).

The simplest aldose is D-glyceraldehyde:

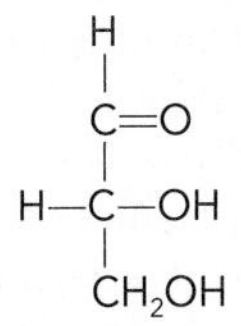

The simplest ketose is dihydroxyacetone:

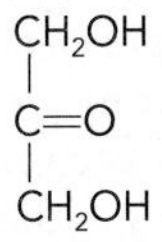

Stereoisomers are described in detail in the next section.

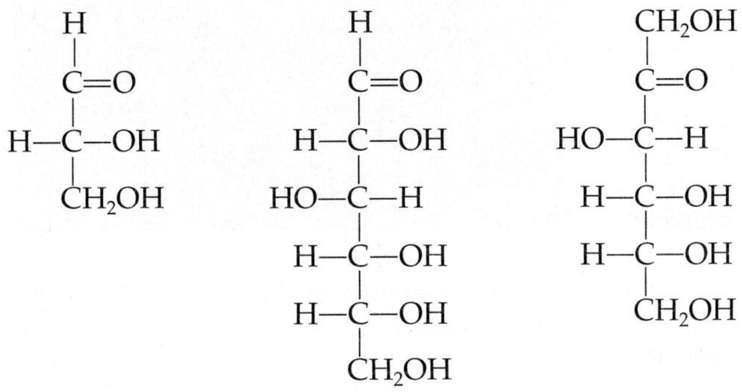

Aldose
Triose
Aldotriose
D-Glyceraldehyde

Aldose
Hexose
Aldohexose
D-Glucose

Ketose
Hexose
Ketohexose
D-Fructose

Question 16.3 What is the structural difference between an aldose and a ketose?

Question 16.4 Explain the difference between:
a. A ketohexose and an aldohexose
b. A triose and a pentose

A Medical Perspective

Tooth Decay and Simple Sugars

How many times have you heard the lecture from parents or your dentist about brushing your teeth after a sugary snack? Annoying as this lecture might be, it is based on sound scientific data that demonstrate that the cause of tooth decay is plaque and acid formed by the bacterium *Streptococcus mutans* using sucrose as its substrate.

Saliva is teeming with bacteria in concentrations up to 100 million (10^8) per milliliter (mL) of saliva! Within minutes after you brush your teeth, sticky glycoproteins in the saliva adhere to tooth surfaces. Then millions of oral bacteria immediately bind to this surface.

Although many oral bacteria stick to the tooth surface, as the diagram shows, only *S. mutans* causes cavities. The reason for this is that this organism alone can make the enzyme *glucosyl transferase*. This enzyme acts only on the disaccharide sucrose, breaking it down into glucose and fructose. The glucose is immediately added to a growing polysaccharide called *dextran*, the glue that allows the bacteria to adhere to the tooth surface, contributing to the formation of plaque.

Now the bacteria embedded in the dextran take in the fructose and use it in the lactic acid fermentation. The lactic acid that is produced lowers the pH on the tooth surface and begins to dissolve calcium from the tooth enamel. Even though we produce about 1 liter (L) of saliva each day, the acid cannot be washed away from the tooth surface because the dextran plaque is not permeable to saliva.

So what can we do to prevent tooth decay? Of course, brushing after each meal and flossing regularly reduce plaque buildup. Eating a diet rich in calcium also helps build strong tooth enamel. Foods rich in complex carbohydrates, such as fruits and vegetables, help prevent cavities in two ways. Glucosyl transferase can't use complex carbohydrates in its cavity-causing chemistry, and eating fruits and vegetables helps to mechanically remove plaque.

Perhaps the most effective way to prevent tooth decay is to avoid sucrose-containing snacks between meals. Studies have shown that eating sucrose-rich foods doesn't cause much tooth decay if followed immediately by brushing. However, even small amounts of sugar eaten between meals actively promote cavity formation.

For Further Understanding

▸ It has been suggested that tooth decay could be prevented by a vaccine that would rid the mouth of *Streptococcus mutans*. Explain this from the point of view of the chemical reactions that are described above.

▸ What steps could you take following a sugary snack to help prevent tooth decay, even when it is not possible to brush your teeth?

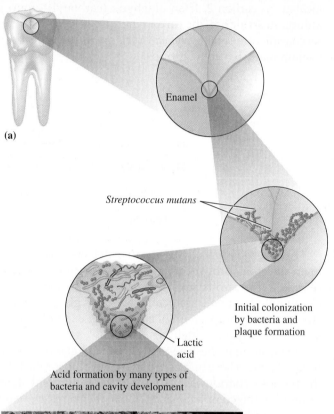

Enamel

Streptococcus mutans

Initial colonization by bacteria and plaque formation

Lactic acid

Acid formation by many types of bacteria and cavity development

(a)

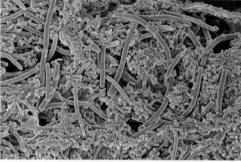

(b)

(a) The complex process of tooth decay. (b) Electron micrograph of dental plaque.

16.3 Stereoisomers and Stereochemistry

Stereoisomers

The prefixes D- and L- found in the complete name of a monosaccharide are used to identify one of two possible isomeric forms called **stereoisomers.** By definition, each member of a pair of stereoisomers must have the same molecular formula and the same bonding pattern. How then do stereoisomers of the D-family differ from those of the L-family? D- and L-isomers differ in the spatial arrangements of atoms in the molecule.

    Stereochemistry is the study of the different spatial arrangements of atoms. A general example of a pair of stereoisomers is shown in Figure 16.3. In this example, the general molecule C-abcd is formed from the bonding of a central carbon to four different groups: a, b, c, and d. This results in two possible ways to arrange the groups, rather than one. Each isomer is bonded together through the exact *same* bonding pattern, yet the two molecules are *not* identical. If they were identical, they would be superimposable. This means you can place the two molecules on top of one another and every atom and every bond of the two lie in the same space. If they cannot be superimposed, they are stereoisomers. These two stereoisomers have a mirror-image relationship that is analogous to the mirror-image relationship of the left and right hands (see Figure 16.3b).

    Two stereoisomers that are nonsuperimposable mirror images of one another are called a pair of **enantiomers.** Molecules that can exist in enantiomeric forms are called **chiral molecules.** The term simply means that as a result of different three-dimensional arrangements of atoms, the molecule can exist in two mirror-image forms.

The structures and designations of D- and L-glyceraldehyde are defined by convention. In fact, the D- and L-terminology is generally applied only to carbohydrates and amino acids. For organic molecules, the D- and L-convention has been replaced by a system that provides the absolute configuration of a chiral carbon. This system is called the (R) and (S) system.

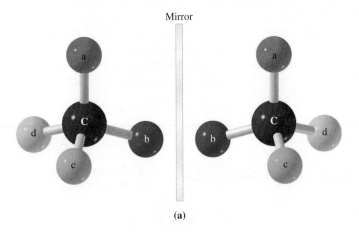

(a)

Nonsuperimposable mirror images: enantiomers

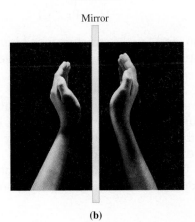

(b)

Figure 16.3 (a) A pair of enantiomers for the general molecule C-abcd. (b) Mirror-image right and left hands.

Most oxidized end

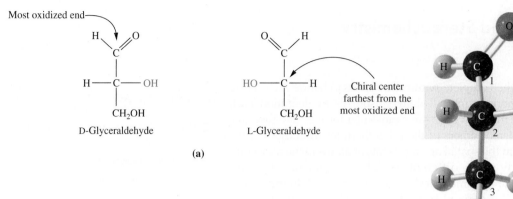

(a)

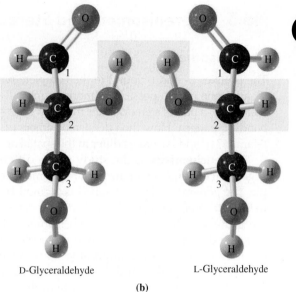

D-Glyceraldehyde   L-Glyceraldehyde

(b)

Figure 16.4 (a) Structural formulas of D- and L-glyceraldehyde. The end of the molecule with the carbonyl group is the most oxidized end. The D- or L-configuration of a monosaccharide is determined by the orientation of the functional groups attached to the chiral carbon farthest from the oxidized end. In the D-enantiomer, the —OH is to the right. In the L-enantiomer, the —OH is to the left. (b) A three-dimensional representation of D- and L-glyceraldehyde.

For any pair of nonsuperimposable mirror-image carbohydrates or amino acids, one is always designated D- and the other L-.

A carbon atom that has four *different* groups bonded to it is called a **chiral carbon** atom. Any molecule containing a chiral carbon is a chiral molecule and will exist as a pair of enantiomers. Consider the simplest chiral carbohydrate, **glyceraldehyde,** which is shown in Figure 16.4. Note that the second carbon is bonded to four different groups. It is therefore a chiral carbon. As a result, we can draw two enantiomers of glyceraldehyde that are nonsuperimposable mirror images of one another. Larger biological molecules typically have more than one chiral carbon. Example 16.1 demonstrates how to identify chiral carbons.

EXAMPLE 16.1 | **Identifying a Chiral Compound**

LEARNING GOAL

Determine which of the following compounds is chiral and which is achiral (not chiral).

3 Determine whether a molecule has a chiral center.

a.

$$H-\underset{\underset{OH}{|}}{\overset{\overset{H}{|}}{C}}_{1}-\underset{\underset{H}{|}}{\overset{\overset{OH}{|}}{C}}_{2}-\underset{\underset{Cl}{|}}{\overset{\overset{Br}{|}}{C}}_{3}-\underset{\underset{H}{|}}{\overset{\overset{H}{|}}{C}}_{4}-H$$

b.

$$H-\underset{\underset{H}{|}}{\overset{\overset{H}{|}}{C}}_{1}-\underset{\underset{Br}{|}}{\overset{\overset{Br}{|}}{C}}_{2}-\underset{\underset{Cl}{|}}{\overset{\overset{Cl}{|}}{C}}_{3}-H$$

Remember that a molecule is chiral if any of the carbon atoms in the structure is a chiral carbon. A carbon is chiral if it is bonded to four *different* groups. By simply inspecting the four groups bonded to each carbon, we can determine whether it is chiral.

a. Carbon-1 is bonded to 2 H and therefore cannot be chiral. Carbon-2 is bonded to four different groups: OH, H, CH$_2$OH, and CBrClCH$_3$. Carbon-3 is bonded to four different groups: Br, Cl, CH$_3$, and CHOHCH$_2$OH. Carbon-4 is bonded to 3 H and therefore cannot be chiral. Thus, this structure is chiral because it contains two chiral carbons. One chiral carbon is sufficient for the molecule to be chiral.
b. Carbon-1 is bonded to 3 H. Carbon-2 is bonded to 2 Br. Carbon-3 is bonded to 2 Cl. None of these carbon atoms is bonded to four different atoms. Thus, this molecule does not contain any chiral carbons and is not chiral.

Practice Problem 16.1

Place an asterisk (*) to denote each chiral carbon in the following structures.

a.

Br   Cl

b.   F   F
      Cl
      Cl   F
         Br

c.   OH   O

                H
      OH   OH

▶ For Further Practice: **Questions 16.41 and 16.42.**

Rotation of Plane-Polarized Light

Stereoisomers can be distinguished from one another by their different optical properties. Each member of a pair of stereoisomers will rotate plane-polarized light in a different direction.

As we learned in Chapter 2, white light is a form of electromagnetic radiation that consists of many different wavelengths (colors) vibrating in *planes* that are all perpendicular to the direction of the light beam. To measure optical properties of enantiomers, scientists use special light sources to produce *monochromatic light*; that is, light of a single wavelength. The monochromatic light is passed through a polarizing material, like a Polaroid lens, so that only waves in one plane can pass through. The light that emerges from the lens is **plane-polarized light** (Figure 16.5), which consists of light waves in only one plane.

Applying these principles, scientists have developed the *polarimeter* to measure the ability of a compound to change the angle of the plane of plane-polarized light (see Figure 16.5). The polarimeter allows us to measure the specific rotation of a compound; that is, the degree to which it rotates plane-polarized light. The ability to rotate plane-polarized light is called *optical activity*.

Some compounds rotate light in a clockwise direction. These are said to be *dextrorotatory* and are designated by a plus sign (+) before the specific rotation value. Other substances rotate light in a counterclockwise direction. These are called *levorotatory* and are indicated by a minus sign (−) before the specific rotation value.

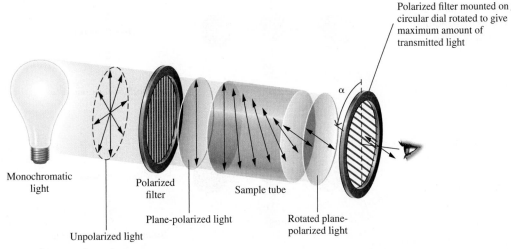

Figure 16.5 Schematic drawing of a polarimeter.

The Relationship between Molecular Structure and Optical Activity

In 1848, Louis Pasteur was the first to see a relationship between the structure of a compound and the effect of that compound on plane-polarized light. In his studies of winemaking, Pasteur noticed that salts of tartaric acid were formed as a by-product. It is a tribute to his extraordinary powers of observation that he noticed that two types of crystals were formed and that they were mirror images of one another. Using a magnifying glass and forceps, Pasteur separated the left-handed and right-handed crystals into separate piles. When he measured the optical activity of each of the mirror-image forms and of the original mixed sample, he obtained the following results:

- Both of the mirror-image crystals were optically active. In fact, the specific rotation produced by each was identical in magnitude but of opposite sign.
- A solution of the original mixture of crystals was optically inactive because the rotation caused by the two enantiomers (equal, but in opposite directions) canceled one another out.

Although Pasteur's work opened the door to understanding the relationship between structure and optical activity, it was not until 1874 that the Dutch chemist van't Hoff and the French chemist LeBel independently came up with a basis for the observed optical activity: tetrahedral carbon atoms bonded to four different atoms or groups of atoms. Thus, two enantiomers, which are identical to one another in all other chemical and physical properties, will rotate plane-polarized light to the same degree, but in opposite directions.

Fischer Projection Formulas

Emil Fischer devised a simple way to represent the structure of stereoisomers. The **Fischer Projection** is a two-dimensional drawing of a molecule that shows a chiral carbon at the intersection of two lines. The horizontal lines represent bonds projecting out of the page, and the vertical lines represent bonds that project into the page. Figure 16.6 demonstrates how to draw the Fischer Projections for the stereoisomers of bromochlorofluoromethane. In Figure 16.6a, the two isomers are represented using ball-and-stick models. The molecules are reinterpreted using the wedge-and-dash representations in Figure 16.6b. In the Fischer Projections shown in Figure 16.6c, the point at which two lines cross represents the chiral carbon. Horizontal lines replace the solid wedges, indicating that the bonds are projecting

Figure 16.6 Drawing a Fischer Projection. (a) The ball-and-stick models for the stereoisomers of bromochlorofluoromethane. (b) The wedge-and-dash and (c) Fischer Projections of these molecules.

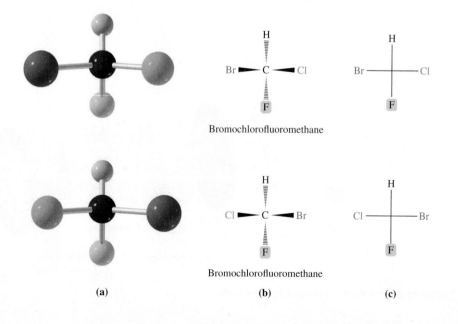

(a)   (b)   (c)

toward the reader. Vertical lines replace the dashed wedges, indicating that the bonds are projecting away from the reader. For sugars, the aldehyde or ketone group, the most oxidized carbon, is always represented at the "top." Example 16.2 will guide you through the steps for drawing a Fischer Projection.

EXAMPLE 16.2   **Drawing Fischer Projections for a Sugar**

LEARNING GOAL

5 Identify monosaccharides as either D- or L-.

Draw the Fischer Projections for the stereoisomers of glyceraldehyde.

Solution

Review the structures of the two stereoisomers of glyceraldehyde (Figure 16.4b). The ball-and-stick models can be represented using three-dimensional wedge drawings. Remember that, for sugars, the most oxidized carbon (the aldehyde or ketone group) is always drawn at the top of the structure. Here, we show the aldehyde in a condensed form, —CHO.

D-Glyceraldehyde   L-Glyceraldehyde

Remember that in the wedge diagram, the solid wedges represent bonds directed toward the reader. The dashed wedges represent bonds directed away from the reader and into the page. In these molecules, the center carbon is the only chiral carbon in the structure. To convert these wedge representations to a Fischer Projection, simply use a horizontal line in place of each solid wedge and use a vertical line to represent each dashed wedge. The chiral carbon is represented by the point at which the vertical and horizontal lines cross, as shown below.

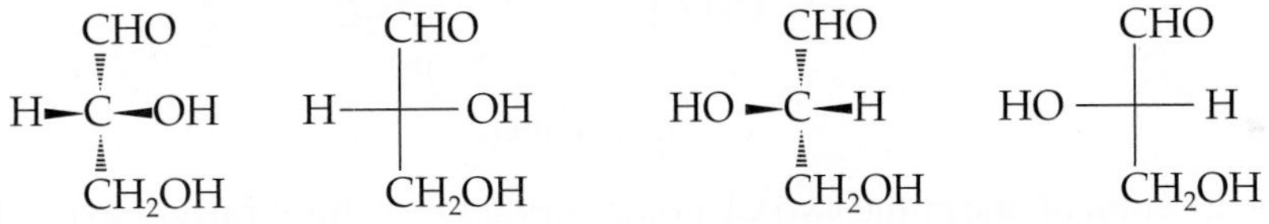

D-Glyceraldehyde   L-Glyceraldehyde

Practice Problem 16.2

Draw Fischer Projections for each of the following molecules and for their mirror images.

a. 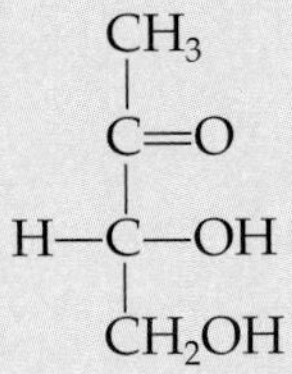
```
    CH₃
    |
    C=O
    |
H—C—OH
    |
   CH₂OH
```

b.
```
     H
     |
     C=O
     |
 H—C—OH
     |
 H—C—OH
     |
HO—C—H
     |
    CH₂OH
```

c.
```
   CH₂OH
    |
    C=O
    |
HO—C—H
    |
 H—C—OH
    |
 H—C—OH
    |
   CH₂OH
```

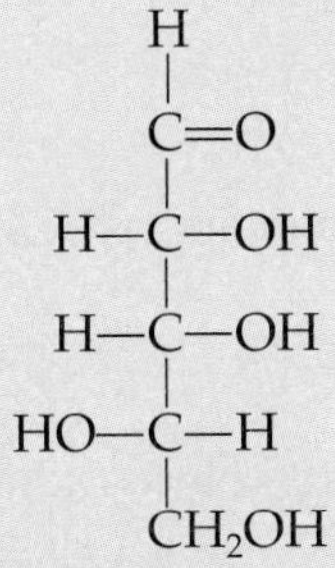

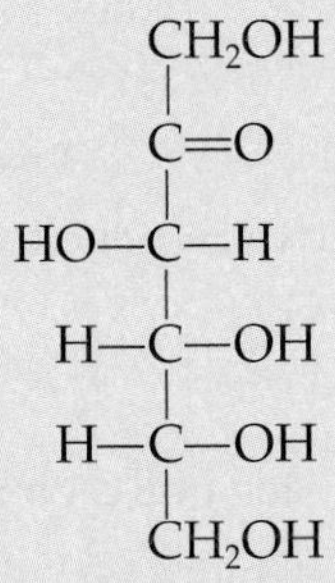

▶ For Further Practice: **Questions 16.47 and 16.48.**

Racemic Mixtures

When Louis Pasteur measured the specific rotation of the mixture of tartaric acid salt crystals, he observed that it was optically inactive. The reason was that the mixture contained equal amounts of the (+) enantiomer and the (−) enantiomer. A mixture of equal amounts of a pair of enantiomers is called a **racemic mixture,** or simply a *racemate*. The prefix ($\pm$) is used to designate a racemic mixture. In this situation, the specific rotation is zero because the rotation caused by one enantiomer is canceled by the opposite rotation caused by the mirror-image enantiomer.

Diastereomers

So far, we have looked only at molecules containing a single chiral carbon. In this case, only two enantiomers are possible. However, it is quite common to find molecules with two or more chiral carbons. For a molecule of n chiral carbons, the maximum possible number of different configurations is 2^n. Note that this formula predicts the *maximum* number of configurations. As we will see, there may actually be fewer.

EXAMPLE 16.3 | **Drawing Stereoisomers for Compounds with More than One Chiral Carbon**

LEARNING GOALS

Draw all the possible stereoisomers of 2,3,4-trichlorobutanal.

3 Determine whether a molecule has a chiral center.

Solution

4 Explain stereoisomerism.

1. There are two chiral carbons in this molecule, C-2 and C-3. Thus, there are 2^2 or 4 possible stereoisomers.
2. There are two possible configurations for each of the chiral carbons (Cl on the left or on the right). Begin by drawing an isomer with both Cl atoms on the right (a). Now draw the mirror image (b). You have now generated the first pair of enantiomers, (a) and (b).

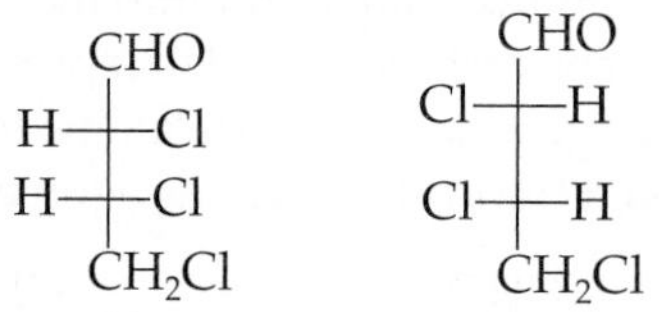

Enantiomers

3. Next, change the location of one of the two Cl atoms bonded to a chiral carbon to produce another possible isomer (c). Finally, draw the mirror image of (c) to produce the second set of enantiomers, (c) and (d).

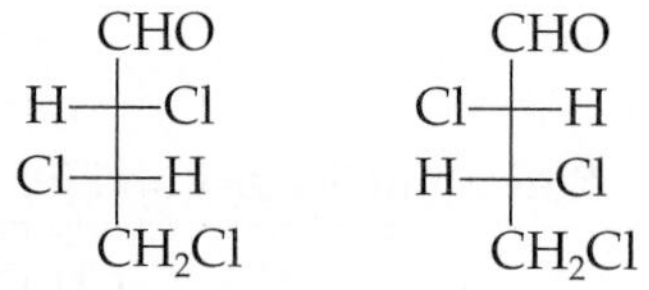

Enantiomers

4. By this systematic procedure, we have drawn the four possible isomers of 2,3,4-trichlorobutanal.

Practice Problem 16.3

Draw all of the possible stereoisomers of each of the compounds listed below. Indicate which are enantiomers.

  a. 2-Bromo-3-chlorobutane   b. 2-Chloro-3-fluoropentane   c. 2,3,4-Tribromopentanal

▶ For Further Practice: **Questions 16.49 and 16.50.**

In Example 16.3, structures (a) and (b) are clearly enantiomers, as are (c) and (d). But how do we describe the relationship between structures (a) and (c) or any of the pairs of stereoisomers that are *not* enantiomers? The term **diastereomers** is used to describe a pair of stereoisomers having two or more chiral centers and that are not enantiomers.

Although enantiomers differ from one another only in the direction of rotation of plane-polarized light, diastereomers are different in their chemical and physical properties.

Meso Compounds

As mentioned previously, the maximum number of configurations for a molecule with two chiral carbons is 2^2, or 4. However, if each of the two chiral carbons is bonded to the same four nonidentical groups, fewer than four stereoisomers exist. The example of tartaric acid, studied by Pasteur, helps to explain this phenomenon.

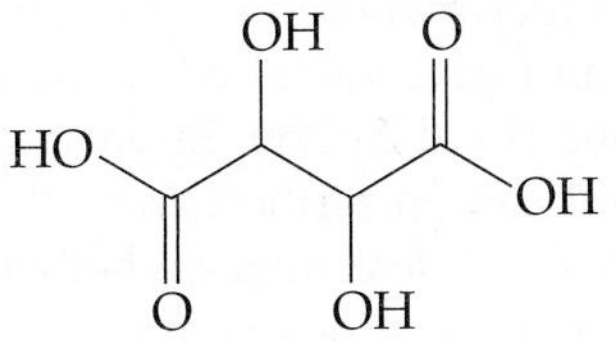

Tartaric acid

Using the steps shown in Example 16.3, we can identify two chiral carbons (2 and 3) and can draw the following four structures:

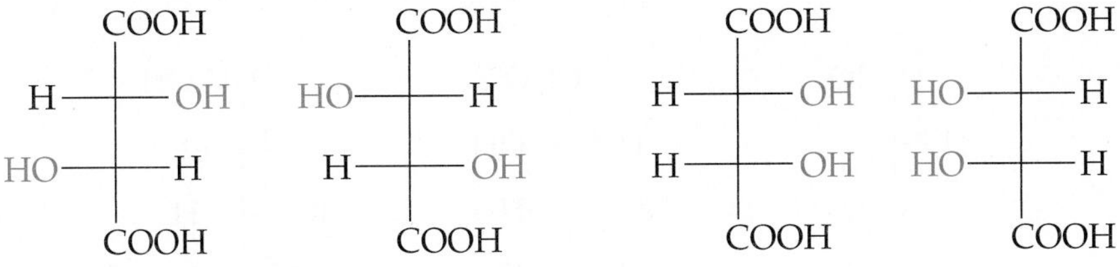

| (a) | (b) | (c) | (d) |

Enantiomers   Identical

Careful examination of structures (a) and (b) reveals that they are nonsuperimposable mirror images. They are enantiomers. Structures (c) and (d) are also mirror images, but they are identical. Structure (d) can simply be rotated 180° to produce structure (c).

Another feature of the structure represented in (c) and (d) is that it has an internal plane of symmetry. If you draw a line between the two chiral carbons (carbon-2 and carbon-3), it is clear that the top half of the molecule is the mirror image of the bottom half.

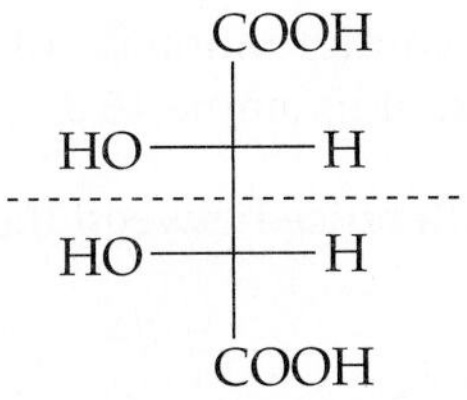

Because of this internal plane of symmetry, this compound is optically inactive; it does not rotate plane-polarized light. The rotation of plane-polarized light by chiral carbon-2 is canceled by the opposite rotation of plane-polarized light caused by carbon-3. This is a **meso compound,** a compound with two or more chiral carbon atoms and an internal plane of symmetry that causes it to be optically inactive. The molecule above is meso-tartaric acid.

The D- and L- System of Nomenclature

In 1891, Emil Fischer devised a nomenclature system that would allow scientists to distinguish between enantiomers. Fischer knew that the two enantiomers of glyceraldehyde rotated plane-polarized light in opposite directions, but he did not have the sophisticated tools needed to make an absolute connection between the structure and the direction of rotation of plane-polarized light. He simply decided that the (+) enantiomer would be the one with the hydroxyl group of the chiral carbon on the right. He called this D-glyceraldehyde. The enantiomer that rotated plane-polarized light in the (−) or levorotatory direction, he called L-glyceraldehyde (see Figure 16.4).

While specific rotation is an experimental value that must be measured, the D- and L-designations of all other monosaccharides are determined by comparison of their structures with D- and L-glyceraldehyde. Sugars with more than three carbons will have more than one chiral carbon. *By convention*, it is the position of the hydroxyl group on the chiral carbon farthest from the carbonyl group (the most oxidized end of the molecule) that determines whether a monosaccharide is in the D- or L-configuration. One way to be sure you are considering the correct chiral carbon is to number the carbon chain, giving the carbonyl group the lowest possible number. It is the chiral carbon with the highest number that is used to determine the D- or L-configuration. If the —OH group is on the right, the molecule is in the D-configuration. If the —OH group is on the left, the molecule is in the L-configuration. Almost all carbohydrates in living systems are members of the D-family.

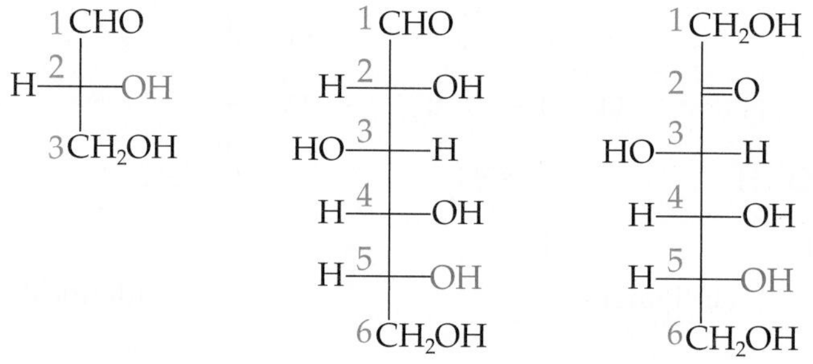

> It was not until 1952 that researchers were able to demonstrate that Fischer had guessed correctly when he proposed the structures of the (+) and (−) enantiomers of glyceraldehyde.

D-Glyceraldehyde   D-Glucose   D-Fructose

Question 16.5   Place an asterisk beside each chiral carbon in the Fischer Projections you drew for Practice Problem 16.2 at the end of Example 16.2.

Question 16.6   In the Fischer Projections you drew for Practice Problem 16.2 at the end of Example 16.2, indicate which bonds project toward you and which project into the page.

Question 16.7   Determine the configuration (D- or L-) for each of the molecules in Practice Problem 16.2 at the end of Example 16.2.

Question 16.8   Explain the difference between the D- and L-designation and the (+) and (−) designation.

16.4 Biologically Important Monosaccharides

LEARNING GOAL

6 Draw and name the common monosaccharides using structural formulas.

Monosaccharides, the simplest carbohydrates, have backbones of from three to seven carbons. There are many monosaccharides, but we will focus on those that are most common in biological systems. These include the five- and six-carbon sugars: glucose, fructose, galactose, ribose, and deoxyribose.

Glucose

Glucose is the most important sugar in the human body. It is found in numerous foods and has several common names, including dextrose, grape sugar, and blood sugar. Glucose is broken down in glycolysis and other pathways to release energy for body functions.

The concentration of glucose in the blood is critical to normal body function. As a result, it is carefully controlled by the hormones insulin and glucagon. Normal blood glucose levels are 100–120 mg glucose/100 mL blood, with the highest concentrations appearing after a meal. Insulin stimulates the uptake of the excess glucose by most cells of the body, and after 1 to 2 hours (h), levels return to normal. If blood glucose concentrations drop too low, the individual feels lightheaded and shaky. When this happens, glucagon stimulates the liver to release glucose into the blood, reestablishing normal levels. We will take a closer look at this delicate balancing act in Section 23.6.

The molecular formula of glucose, an aldohexose, is $C_6H_{12}O_6$. The structure of glucose is shown in Figure 16.7, and the method used to draw this structure is described in Example 16.4.

Why do diabetics need to use a blood glucose monitor like the one shown here?

EXAMPLE 16.4 | **Drawing the Structure of a Monosaccharide**

LEARNING GOAL

Draw the structure and Fischer Projection for D-glucose.

6 Draw and name the common monosaccharides using structural formulas.

Solution

Glucose is an aldohexose.

Step 1. Draw six carbons in a straight vertical line; each carbon is separated from the ones above and below it by a bond:

$$
\begin{array}{ll}
1 & C \\
  & | \\
2 & C \\
  & | \\
3 & C \\
  & | \\
4 & C \\
  & | \\
5 & C \\
  & | \\
6 & C \\
\end{array}
$$

Step 2. The most highly oxidized carbon is, by convention, drawn as the uppermost carbon (carbon-1). In this case, carbon-1 is an aldehyde carbon:

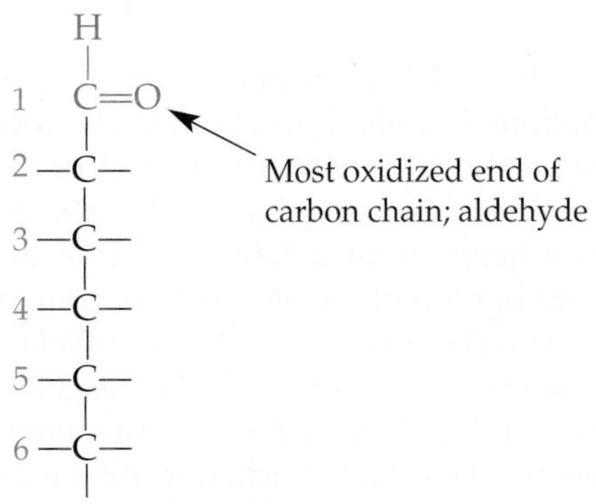

$$
\begin{array}{ll}
 & H \\
 & | \\
1 & C{=}O \longleftarrow \text{Most oxidized end of} \\
 & | \qquad\qquad \text{carbon chain; aldehyde} \\
2 & {-}C{-} \\
 & | \\
3 & {-}C{-} \\
 & | \\
4 & {-}C{-} \\
 & | \\
5 & {-}C{-} \\
 & | \\
6 & {-}C{-} \\
 & | \\
\end{array}
$$

Continued…

Step 3. The atoms are added to the next to the last carbon atom, at the bottom of the chain, to give either the D- or L-configuration as desired. Remember, when the —OH group is to the right, you have D-glucose. When in doubt, compare your structure to D-glyceraldehyde!

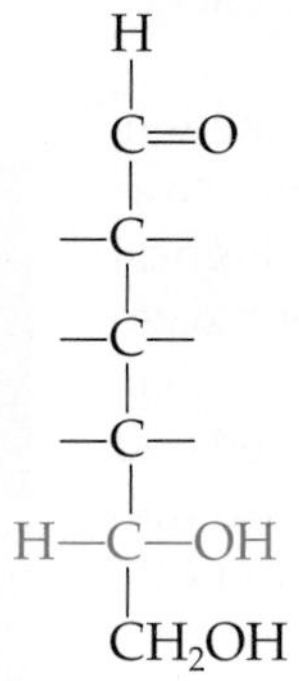

D-Isomer 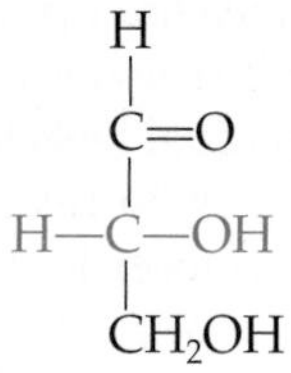   D-Glyceraldehyde

Compare chiral carbons farthest from the carbonyl group

Step 4. All the remaining atoms are then added to give the desired carbohydrate. For example, you would draw the following structure for D-glucose.

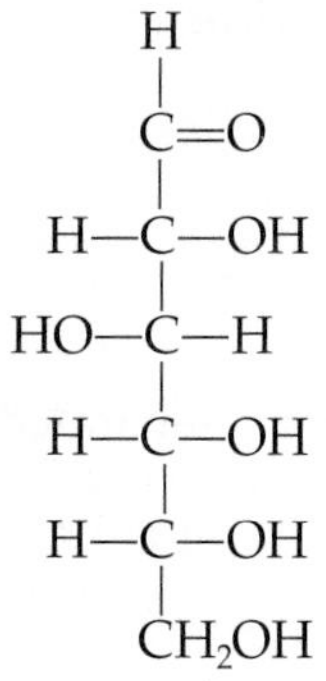

D-Glucose

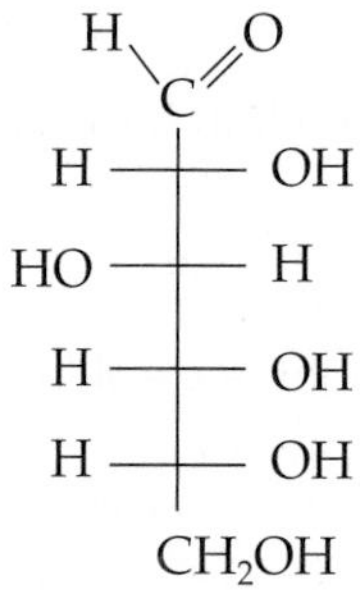

D-Glucose
(Fischer Projection)

The positions for the hydrogen atoms and the hydroxyl groups on the remaining carbons must be learned for each sugar.

Practice Problem 16.4

Draw the structures of D-ribose and L-ribose. (Information on the structure of D-ribose is found later in this chapter.)

▶ For Further Practice: **Questions 16.9 and 16.10.**

Hemiacetal structure,

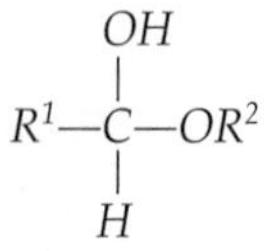

$$R^1—\underset{\underset{H}{|}}{\overset{\overset{OH}{|}}{C}}—OR^2$$

and formation are described in Section 13.4.

In actuality, the open-chain form of glucose is present in very small concentrations in cells. It exists in cyclic form under physiological conditions because the carbonyl group at C-1 of glucose reacts with the hydroxyl group at C-5 to give a six-member ring. In the discussion of aldehydes, we noted that the reaction between an aldehyde and an alcohol yields a **hemiacetal**. When the carbonyl group of the aldehyde portion of the glucose molecule reacts with the C-5 hydroxyl group, the product is a cyclic *intramolecular hemiacetal*. For D-glucose, two isomers can be formed in this reaction (see Figure 16.7). These isomers are called α- and β-D-glucose. The two isomers formed differ from one another in the location of the —OH attached to the hemiacetal carbon, C-1. Such isomers, differing in the arrangement of bonds around the hemiacetal carbon, are called **anomers.**

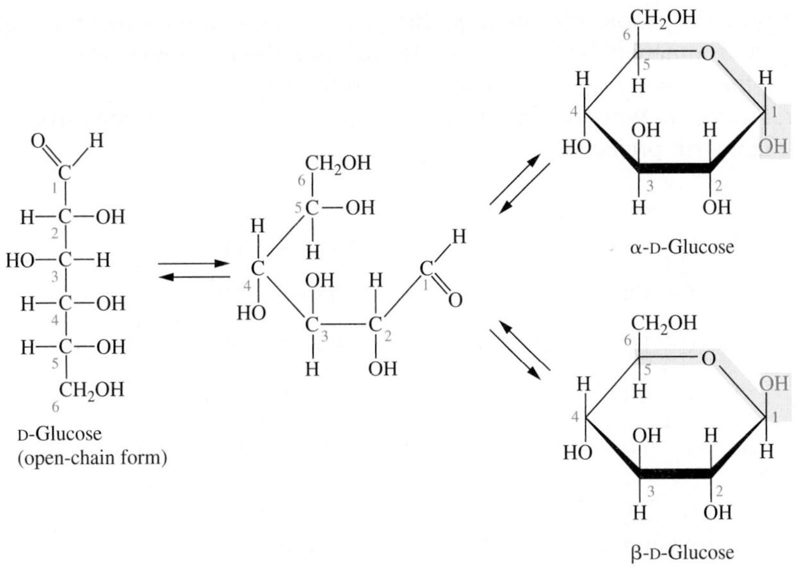

Figure 16.7 Cyclization of glucose to give α- and β-D-glucose. Note that the carbonyl carbon (C-1) becomes a chiral center in this process, yielding the α- and β-forms of glucose. This chiral center is called the anomeric carbon. The hemiacetal is highlighted in yellow.

In the α-anomers, the C-1 (*anomeric carbon*) hydroxyl group is below the ring, and in the β-anomers, the C-1 hydroxyl group is above the ring. Like the stereoisomers discussed previously, the α and β forms can be distinguished from one another because they rotate plane-polarized light differently.

Question 16.9 Draw the structure of D-galactose. (Information on the structure of D-galactose is found later in this chapter.)

Question 16.10 Draw the structure of L-galactose.

In Figure 16.7, a new type of structural formula, called a **Haworth projection,** is presented. Although on first inspection it appears complicated, it is quite simple to derive a Haworth projection from a structural formula, as Example 16.5 shows.

EXAMPLE 16.5

Drawing the Haworth Projection of a Monosaccharide from the Structural Formula

Draw the Haworth projections of α- and β-D-glucose.

Solution

1. Before attempting to draw a Haworth projection, look at the first steps of ring formation shown here:

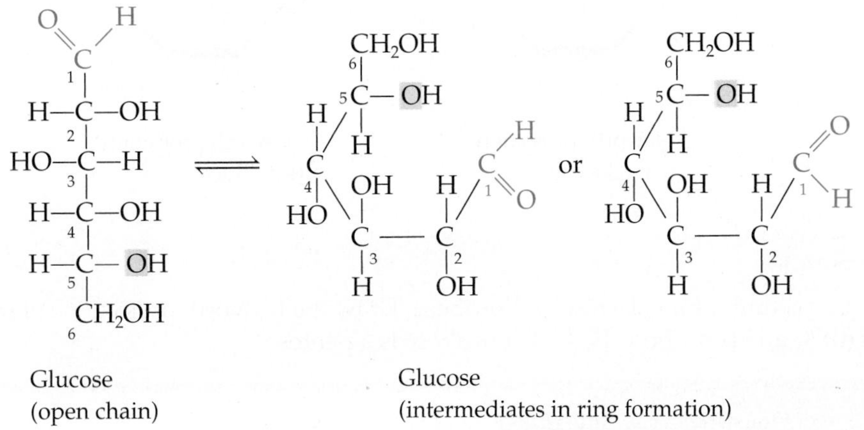

Glucose
(open chain)

Glucose
(intermediates in ring formation)

LEARNING GOAL

7 Given the linear structure of a monosaccharide, draw the Haworth projection of its α- and β-cyclic forms and vice versa.

Continued…

Try to imagine that you are seeing the preceding molecules in three dimensions. Some of the substituent groups on the molecule will be above the ring, and some will be beneath it. The question then becomes: How do you determine which groups to place above the ring and which to place beneath the ring?

2. Look at the two-dimensional structural formula. Note the groups (drawn in blue) to the left of the carbon chain. These are placed above the ring in the Haworth projection.

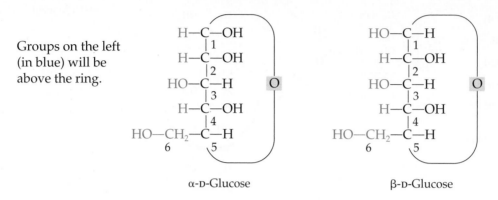

Groups on the left (in blue) will be above the ring.

α-D-Glucose   β-D-Glucose

3. Now note the groups (drawn in red) to the right of the carbon chain. These will be located beneath the carbon ring in the Haworth projection.

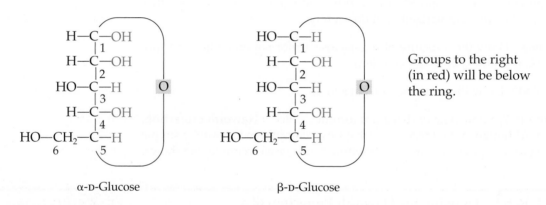

Groups to the right (in red) will be below the ring.

α-D-Glucose   β-D-Glucose

4. Thus, in the Haworth projection of the cyclic form of any D-sugar the —CH₂OH group is always "up." When the —OH group at C-1 is also "up," *cis* to the —CH₂OH group, the sugar is β-D-glucose. When the —OH group at C-1 is "down," *trans* to the —CH₂OH group, the sugar is α-D-glucose.

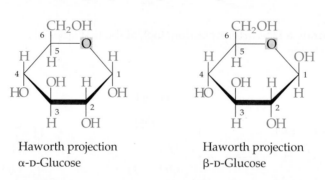

Haworth projection   Haworth projection
α-D-Glucose   β-D-Glucose

Practice Problem 16.5

Refer to the linear structures of D-galactose and D-ribose. Draw the Haworth projections of (a) α- and β-D-galactose and of (b) α- and β-D-ribose. Note that D-ribose is a pentose.

▶ For Further Practice: **Questions 16.57 and 16.58.**

Fructose

Fructose, also called levulose and fruit sugar, is the sweetest of all sugars. It is found in large amounts in honey, corn syrup, and sweet fruits. The structure of fructose is similar to that of glucose. When there is a —CH$_2$OH group instead of a —CHO group at carbon-1 and a —C=O group instead of CHOH at carbon-2, the sugar is a ketose. In this case, it is D-fructose.

Cyclization of fructose produces α- and β-D-fructose:

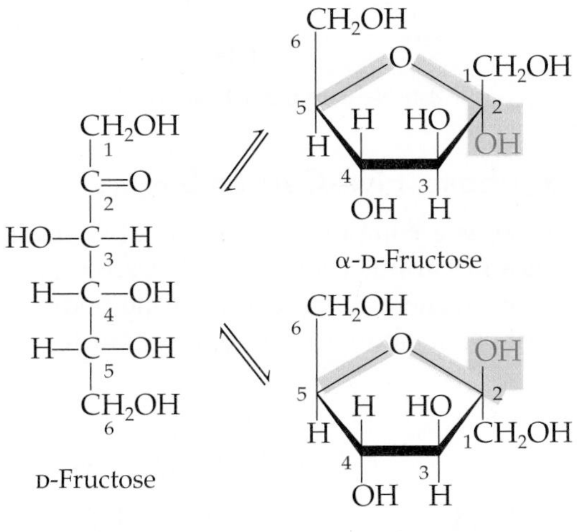

Fructose is often called fruit sugar because it contributes sweetness to ripe fruits, such as these peaches. Is fructose an aldose or a ketose?

In the equation above, the hemiacetal is highlighted. Fructose forms a five-member ring structure.

Galactose

Another important aldohexose is **galactose.** It is a diastereomer of glucose. The linear structure of D-galactose and the Haworth projections of α-D-galactose and β-D-galactose are shown here:

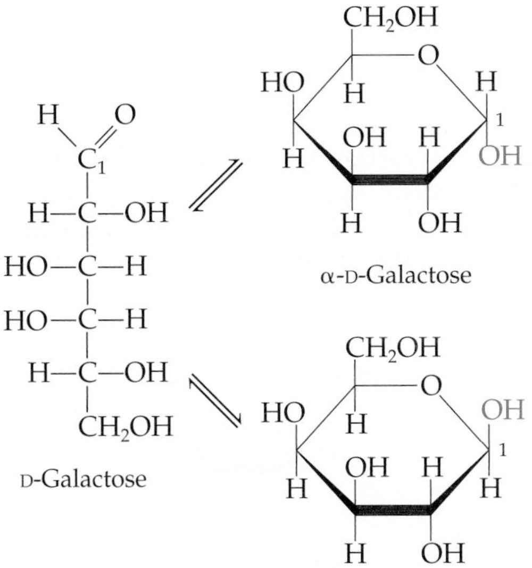

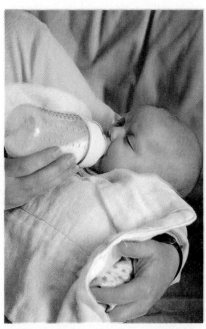

Galactose is one of the components of lactose, or milk sugar. Read about galactosemia in Section 16.5 and describe the symptoms and treatment for this genetic disorder.

Galactose is found in biological systems as a component of the disaccharide lactose, or milk sugar. This is the principal sugar found in the milk of most mammals. β-D-Galactose and a modified form, β-D-N-acetylgalactosamine, are also components of the blood group antigens.

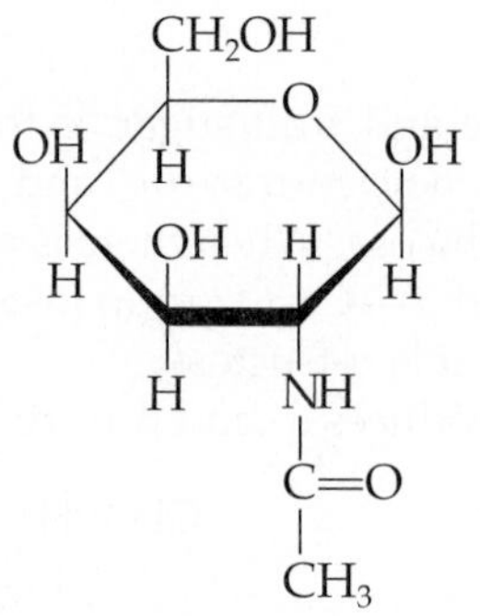

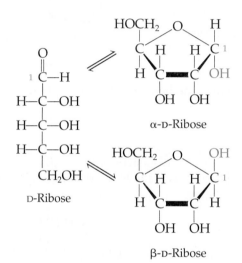

β-D-*N*-Acetylgalactosamine

Ribose and Deoxyribose, Five-Carbon Sugars

Ribose, an aldopentose, is a component of many biologically important molecules, including RNA and various coenzymes that are required by many of the enzymes that carry out biochemical reactions in the body. The structure of the five-carbon sugar D-ribose is shown in its open-chain form and in the α- and β-cyclic forms.

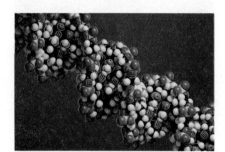

β-D-2-Deoxyribose is one of the components of the sugar-phosphate backbone of the DNA molecule. How does this molecule differ from β-D-ribose?

D-Ribose α-D-Ribose β-D-Ribose

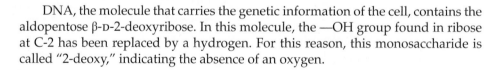

DNA, the molecule that carries the genetic information of the cell, contains the aldopentose β-D-2-deoxyribose. In this molecule, the —OH group found in ribose at C-2 has been replaced by a hydrogen. For this reason, this monosaccharide is called "2-deoxy," indicating the absence of an oxygen.

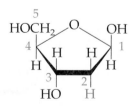

β-D-2-Deoxyribose

LEARNING GOALS

8 By inspection of the structure, predict whether a sugar is a reducing or a non-reducing sugar.

9 Discuss the use of the Benedict's reagent to measure the level of glucose in urine.

Reducing Sugars

The aldehyde group of aldoses is readily oxidized by the Benedict's reagent. Recall that the **Benedict's reagent** is a basic buffer solution that contains Cu^{2+} ions. The Cu^{2+} ions are reduced to Cu^{+} ions, which, in basic solution, precipitate

Kitchen Chemistry

The Chemistry of Caramels

Of all the treats at the holidays, my grandmother's caramels were my favorite. Unwrapped from their waxed paper, they melted in your mouth. It seemed mystical that she could mix sugar, corn syrup, butter, and cream into a very pale and unappetizing liquid and heat it to 248°F (120°C), watching it very carefully for the color change that indicated that the chemical reactions were occurring exactly as they should. I know that she never thought of herself as a chemist when she made the many holiday candies that delighted the family. But a chemist she was, carrying out reactions that are still poorly understood.

When sucrose is heated, its molecules begin to break apart into glucose and fructose. This destruction begins a series of reactions that convert the liquid sucrose, which itself is odorless, colorless, and cloyingly sweet, into literally hundreds of compounds. Some of these products are small molecules that enhance the flavor and aroma of the candy; others are little-understood polymers, caramelans, caramelens, and caramelins, that give the creamy, soft texture.

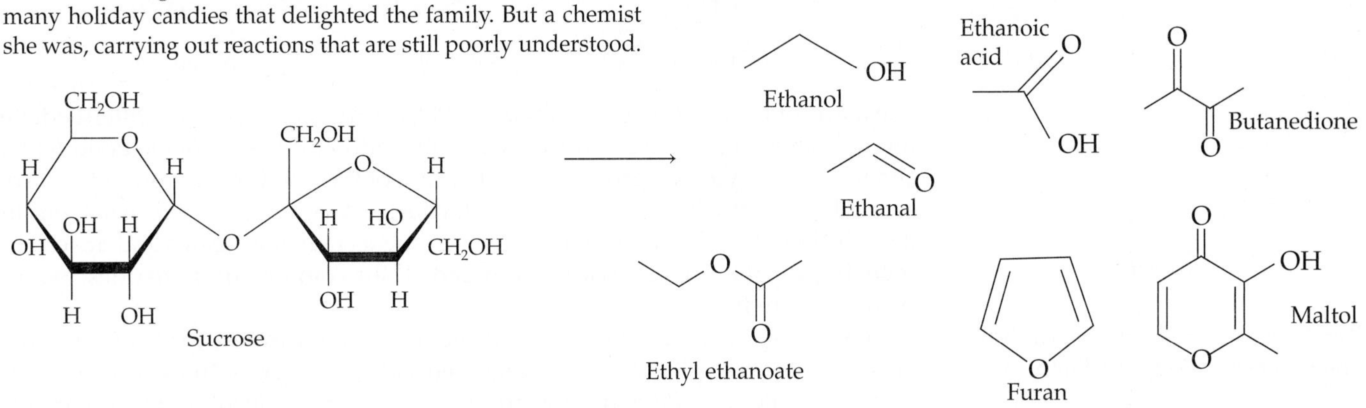

+ caramelans ($C_{24}H_{36}O_{18}$) + caramelens ($C_{36}H_{50}O_{25}$) + caramelins ($C_{125}H_{188}O_{80}$)

Of course, the sweet flavor is due to the sucrose and other sugars in the caramels. But notice that there are even slightly acidic and bitter notes that round out the flavor. As you can see in the diagram, the sour flavor is contributed by the acids produced, including acetic acid in the "equation" above. The fruity flavors are added by esters, and there is a sherry-like flavor that ethanol brings to the party. Butanedione adds a butterscotch flavor and aroma, and furans add a nutty accent. Maltol brings the aroma that we associate with caramels.

While these products have been identified, many others have not. In fact, the reactions that cause this amazing change are poorly understood. Nonetheless, we can be grateful for the delightful treats that are the result.

For Further Understanding

▸ The temperature must be very carefully controlled during candy making. Explain this in terms of chemical reactions.
▸ List some other foods and beverages that owe their color and flavor to caramelization.

---

as brick-red Cu_2O. The aldehyde group of the aldose is oxidized to a carboxylic acid, which undergoes an acid-base reaction to produce a carboxylate anion.

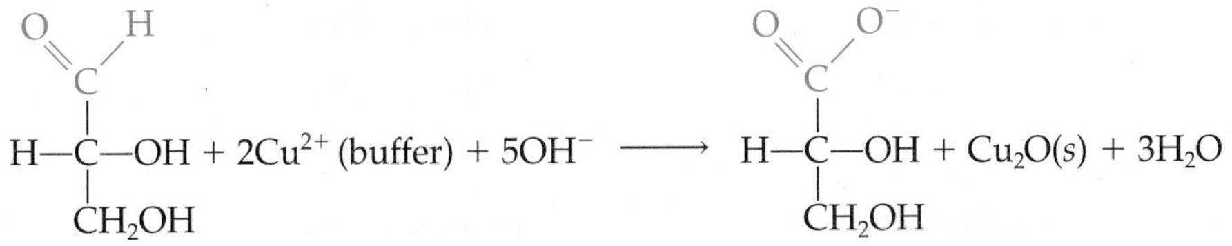

$$H-\underset{\underset{CH_2OH}{|}}{\overset{\overset{\displaystyle O \diagdown \quad H}{\diagup\diagdown}}{C}}-OH + 2Cu^{2+} \text{(buffer)} + 5OH^- \longrightarrow H-\underset{\underset{CH_2OH}{|}}{\overset{\overset{\displaystyle O \diagdown \quad O^-}{\diagup\diagdown}}{C}}-OH + Cu_2O(s) + 3H_2O$$

Although ketones generally are not easily oxidized, ketoses are an exception to that rule. Because of the —OH group on the carbon next to the carbonyl group, ketoses can be converted to aldoses, under basic conditions, via an *enediol reaction*:

$$
\begin{array}{ccc}
\mathrm{CH_2OH} & \mathrm{HO-C-H} & \mathrm{H} \\
| & \| & | \\
\mathrm{C=O} & \mathrm{C-OH} & \mathrm{C=O} \\
| & | & | \\
\mathrm{HO-C-H} & \mathrm{HO-C-H} & \mathrm{H-C-OH} \\
| & | & | \\
\mathrm{H-C-OH} \rightleftharpoons & \mathrm{H-C-OH} \rightleftharpoons & \mathrm{HO-C-H} \\
| & | & | \\
\mathrm{H-C-OH} & \mathrm{H-C-OH} & \mathrm{H-C-OH} \\
| & | & | \\
\mathrm{CH_2OH} & \mathrm{CH_2OH} & \mathrm{H-C-OH} \\
& & | \\
& & \mathrm{CH_2OH} \\
\text{D-Fructose} & \text{Enediol} & \text{D-Glucose}
\end{array}
$$

The name of the enediol reaction is derived from the structure of the intermediate through which the ketose is converted to the aldose: It has a double bond (ene), and it has two hydroxyl groups (diol). Because of this enediol reaction, ketoses are also able to react with Benedict's reagent, which is basic. Because the metal ions in the solution are reduced, the sugars are serving as reducing agents and are called **reducing sugars.** All monosaccharides and all the common disaccharides, except sucrose, are reducing sugars.

See A Medical Perspective: Diabetes Mellitus and Ketone Bodies in Chapter 23.

For many years, the Benedict's reagent was used to test for *glucosuria*, the presence of excess glucose in the urine. Individuals suffering from *Type I insulin-dependent diabetes mellitus* do not produce the hormone insulin, which controls the uptake of glucose from the blood. When the blood glucose level rises above 160 mg/100 mL, the kidney is unable to reabsorb the excess, and glucose is found in the urine. Although the level of blood glucose could be controlled by the injection of insulin, urine glucose levels were monitored to ensure that the amount of insulin injected was correct. The Benedict's reagent was a useful tool because the amount of Cu_2O formed, and hence the degree of color change in the reaction, is directly proportional to the amount of reducing sugar in the urine. A brick-red color indicates a very high concentration of glucose in the urine. Yellow, green, and blue-green solutions indicate decreasing amounts of glucose in the urine, and a blue solution indicates an insignificant concentration.

Use of Benedict's reagent to test urine glucose levels has been replaced by chemical tests that provide more accurate results. The most common technology is based on a test strip that is impregnated with the enzyme glucose oxidase and other agents that will cause a measurable color change. In one such kit, the compounds that result in color development include the enzyme peroxidase, a compound called orthotolidine, and a yellow dye. When a drop of urine is placed on the strip, the glucose oxidase catalyzes the conversion of glucose into gluconic acid and hydrogen peroxide.

$$
\begin{array}{ccc}
\mathrm{O} & & \mathrm{O} \\
\| & & \| \\
\mathrm{C-H} & & \mathrm{C-OH} \\
| & & | \\
\mathrm{H-C-OH} & \xrightarrow[\text{oxidase}]{\text{Glucose}} & \mathrm{H-C-OH} \\
| & & | \\
\mathrm{HO-C-H} \quad + \mathrm{O_2} & & \mathrm{HO-C-H} \quad + \mathrm{H_2O_2} \\
| & & | \\
\mathrm{H-C-OH} & & \mathrm{H-C-OH} \\
| & & | \\
\mathrm{H-C-OH} & & \mathrm{H-C-OH} \\
| & & | \\
\mathrm{CH_2OH} & & \mathrm{CH_2OH} \\
\text{D-Glucose} & & \text{D-Gluconic acid}
\end{array}
$$

The enzyme peroxidase catalyzes a reaction between the hydrogen peroxide and orthotolidine. This produces a blue product. The yellow dye on the test strip simply serves to "dilute" the blue end product, thereby allowing greater accuracy of the test over a wider range of glucose concentrations. The test strip remains yellow if there is no glucose in the sample. It will vary from a pale green to a dark blue, depending on the concentration of glucose in the urine sample.

Frequently, doctors recommend that diabetics monitor their *blood* glucose levels multiple times each day because this provides a more accurate indication of how well the diabetic is controlling his or her diet. Many small, inexpensive glucose meters are available that couple the oxidation of glucose by glucose oxidase with an appropriate color change system. As with the urine test, the intensity of the color change is proportional to the amount of glucose in the blood. A photometer within the device reads the color change and displays the glucose concentration. An even newer technology uses a device that detects the electrical charge generated by the oxidation of glucose. In this case, it is the amount of electrical charge that is proportional to the glucose concentration.

> Actually, glucose oxidase can only oxidize β-D-glucose. However, in the blood there is an equilibrium mixture of the α and β anomers of glucose. Fortunately, α-D-glucose is very quickly converted to β-D-glucose.

16.5 Biologically Important Disaccharides

Disaccharides consist of two monosaccharides joined through an "oxygen bridge." In biological systems, monosaccharides exist in the cyclic form and, as we have seen, they are actually hemiacetals. Recall that when a hemiacetal reacts with an alcohol, the product is an *acetal*. In the case of disaccharides, the alcohol comes from a second monosaccharide. The acetals formed are given the general name *glycosides*, and the carbon-oxygen bonds are called *glycosidic bonds*.

Glycosidic bond formation is nonspecific; that is, it can occur between a hemiacetal and any of the hydroxyl groups on the second monosaccharide. However, in biological systems, we commonly see only particular disaccharides, such as maltose (Figure 16.8), lactose (Figure 16.10), or sucrose (Figure 16.11). These specific disaccharides are produced in cells because the reactions are catalyzed by enzymes. Each enzyme catalyzes the synthesis of one specific disaccharide, ensuring that one particular pair of hydroxyl groups on the reacting monosaccharides participates in glycosidic bond formation.

LEARNING GOAL

10 Draw and name the common disaccharides, and discuss their significance in biological systems.

Maltose

If an α-D-glucose and a second glucose are linked, as shown in Figure 16.8, the disaccharide is **maltose,** or malt sugar. This is one of the intermediates in the hydrolysis of starch. Because the C-1 hydroxyl group of α-D-glucose is attached to C-4 of another glucose molecule, the disaccharide is linked by an $\alpha(1 \rightarrow 4)$ glycosidic bond.

Maltose is a reducing sugar. Any disaccharide that has a hemiacetal hydroxyl group (a free —OH group at C-1) is a reducing sugar. This is because the cyclic structure can open at this position to form a free aldehyde. Disaccharides that do not contain a hemiacetal group on C-1 do not react with the Benedict's reagent and are called **nonreducing sugars.**

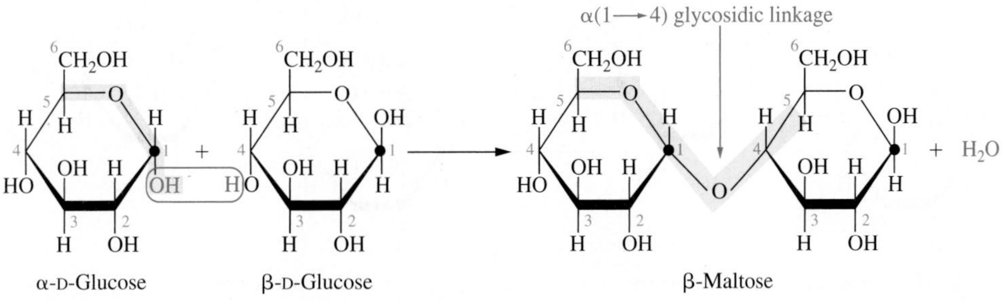

Figure 16.8 Glycosidic bond formed between the C-1 hemiacetal hydroxyl group of α-D-glucose and the C-4 alcohol hydroxyl group of β-D-glucose. The disaccharide is called β-maltose because the hydroxyl group at the reducing end of the disaccharide has the β-configuration. The hemiacetal of α-D-glucose and the acetal in β-maltose are highlighted in β yellow.

Lactose

Milk sugar, or **lactose,** is a disaccharide made up of one molecule of β-D-galactose and one of either α- or β-D-glucose. Galactose differs from glucose only in the configuration of the hydroxyl group at C-4 (Figure 16.9). In the cyclic form of glucose, the C-4 hydroxyl group is "down," and in galactose it is "up." In lactose, the C-1 hemiacetal hydroxyl group of β-D-galactose is bonded to the C-4 alcohol hydroxyl group of either an α- or β-D-glucose. The bond between the two monosaccharides is therefore a β(1 → 4) glycosidic bond (Figure 16.10).

Lactose is the principal sugar in the milk of most mammals. To be used by the body as an energy source, lactose must be hydrolyzed to produce glucose and galactose. Note that this is simply the reverse of the reaction shown in Figure 16.10. Glucose liberated by the hydrolysis of lactose is used directly in the energy-harvesting reactions of glycolysis. However, a series of reactions is necessary to convert galactose into a phosphorylated form of glucose that can be used in cellular metabolic reactions. In humans, the genetic disease **galactosemia** is caused by the absence of one or more of the enzymes needed for this conversion. A toxic compound formed from galactose accumulates in people who suffer from galactosemia. If the condition is not treated, galactosemia leads to severe mental disabilities, cataracts, and early death. However, the effects of this disease can be avoided entirely by providing galactosemic infants with a diet that does not contain galactose. Such a diet, of course, cannot contain lactose and therefore must contain no milk or milk products.

Many adults, and some children, are unable to hydrolyze lactose because they do not make the enzyme *lactase.* This condition, which affects 20% of the population of the United States, is known as **lactose intolerance.** Undigested lactose remains in the intestinal tract and causes cramping and diarrhea that can eventually lead to dehydration. Some of the lactose is metabolized by intestinal bacteria that release organic acids and CO_2 gas into the intestines, causing further discomfort. Lactose intolerance is unpleasant, but its effects can be avoided by a diet that excludes milk and milk products. Alternatively, the enzyme that hydrolyzes lactose is available in tablet form. When ingested with dairy products, it breaks down the lactose, preventing symptoms.

LEARNING GOAL

11   Describe the difference between galactosemia and lactose intolerance.

Glycolysis is discussed in Chapter 21.

Both galactosemia and lactose intolerance are treated by removing milk and milk products from the diet. Explain the difference between these two conditions.

Figure 16.9 Comparison of the cyclic forms of glucose and galactose. Note that galactose is identical to glucose except in the position of the C-4 hydroxyl group.

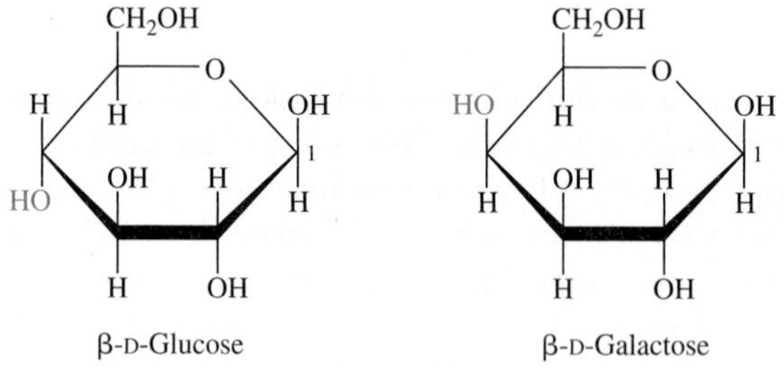

β-D-Glucose   β-D-Galactose

Figure 16.10 Glycosidic bond formed between the C-1 hemiacetal hydroxyl group of β-D-galactose and the C-4 hydroxyl group of β-D-glucose. The disaccharide is called β-lactose because the hydroxyl group at the reducing end of the disaccharide has the β-configuration.

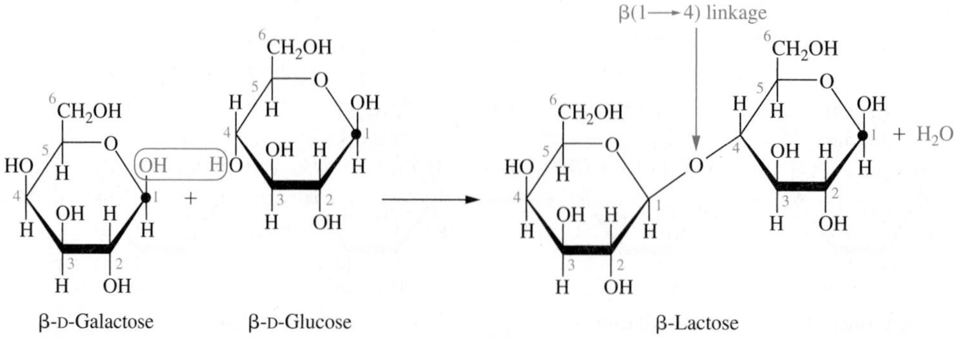

β(1 → 4) linkage

β-D-Galactose   β-D-Glucose   β-Lactose

A Medical Perspective

Human Milk Oligosaccharides

Science and society have taken human breast milk for granted since the beginning of time. We know it is a source of nutrition for the newborn, consisting of fats, proteins, and sugars in a ratio of about 1:3:7. We also know that there are antibodies and immune cells in breast milk that protect the newborn from a variety of infectious diseases. Only recently have scientists recognized that there are hundreds, or perhaps even thousands, of so-called bioactive molecules that have a profound impact on the health of the newborn.

In the late 1800s, researchers made the observation that breast-fed infants had a higher survival rate than those that were bottle-fed. They also observed that the breast-fed infants had a different complement of bacteria in their feces. This is a reflection of the infant gut microbiome, which is the collection of bacteria that reside in the colon. Much later, in the 1950s, it was recognized that there is a collection of unusual oligosaccharides in mother's milk; these were called human milk oligosaccharides or HMOs. HMOs are not destroyed by the low pH in the stomach or digested by pancreatic enzymes, and they are not absorbed by intestinal cells. Some have referred to HMOs as the "fiber" of human milk and, like fiber, they travel through the small intestine and into the colon.

After identifying nearly 200 HMOs, researchers fed them to a variety of bacteria, expecting that many species would be able to use them as an energy source. To their surprise, only one bacterium, *Bifidobacterium longum biovar infantis,* was able to grow well! While scientists are not yet sure how *B. longum bv infantis* gets into the infant's GI tract, they do know that the HMOs in human milk allow the organism to quickly multiply until it makes up 90% of the infant's microbiome.

DNA sequencing has demonstrated that *B. longum bv infantis* has the genes for every enzyme needed to break down HMOs, suggesting that the bacterium has been co-evolving with humans for a very long time. All of this co-evolution has resulted in an impressive battery of mechanisms that safeguard the health of the newborn. Because *B. longum bv infantis* digests and uses the HMOs so efficiently, it out-competes, and starves, disease-causing intestinal microorganisms. In addition, it produces a variety of short-chain fatty acids that promote the growth of other beneficial intestinal bacteria.

The HMOs also have a direct protective role in the infant gut. It has been found that they have a structure very similar to oligosaccharides that are found on the surface of the cells of the infant's GI tract. Many GI pathogens attach to these oligosaccharides on infant gut cells and in this way initiate infection and intestinal damage. Because of the presence of HMOs in the child's gut, the pathogens attach to the soluble HMOs rather than cell surfaces. They are then removed harmlessly in the feces.

Laboratories are beginning to develop supplements of probiotics (bacteria) and prebiotics (nutrients, including HMOs) in hopes that they may be useful in treating premature babies that develop a deadly disease called necrotizing enterocolitis (NEC). Studies in Canada have demonstrated that a probiotic containing *B. longum bv infantis* substantially reduced the incidence of NEC. It is hoped that continued research will provide effective supplements and allow us to learn more about the amazing protective features of human milk.

For Further Understanding

▸ Explain the protective function of HMOs against intestinal pathogens in terms of a competitive inhibition.

▸ Why have HMOs been called the "fiber" of human breast milk?

Sucrose

Sucrose is also called table sugar, cane sugar, or beet sugar. Sucrose is an important carbohydrate in plants. It is water-soluble and can easily be transported through the circulatory system of the plant. It cannot be synthesized by animals.

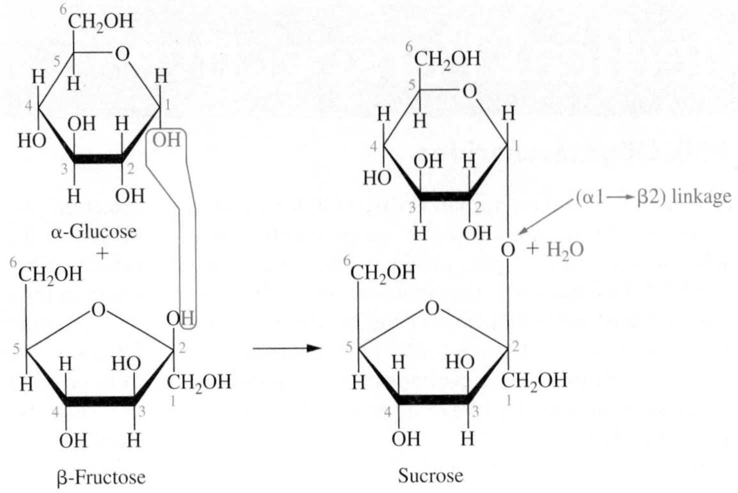

Figure 16.11 Glycosidic bond formed between the C-1 hemiacetal hydroxyl of α-D-glucose and the C-2 hemiacetal hydroxyl of β-D-fructose. This bond is called an (α1 → β2) glycosidic linkage. The disaccharide formed in this reaction is sucrose.

α-Glucose
+
β-Fructose

(α1 → β2) linkage

O + H₂O

Sucrose

High concentrations of sucrose produce a high osmotic pressure, which inhibits the growth of microorganisms, so it is used as a preservative. Of course, it is also widely used as a sweetener. In fact, it is estimated that the average American consumes 100–125 pounds (lb) of sucrose each year. It has been suggested that sucrose in the diet is undesirable because it represents a source of empty calories; that is, it contains no vitamins or minerals. However, a recent study has determined that there is a significant relationship between excess sugar consumption and an increased risk for cardiovascular disease–related mortality. In addition, the link between sucrose in the diet and dental caries, or cavities, has been scientifically verified (see A Medical Perspective: Tooth Decay and Simple Sugars on p. 566).

Sucrose is a disaccharide of α-D-glucose joined to β-D-fructose (Figure 16.11). The glycosidic linkage between α-D-glucose and β-D-fructose is quite different from those that we have examined for lactose and maltose. This bond involves the anomeric carbons of *both* sugars! This bond is called an (α1 → β2) glycosidic linkage, since it involves the C-1 hemiacetal hydroxyl group on the anomeric carbon of glucose and the C-2 hemiacetal hydroxyl group on the anomeric carbon of fructose (noted in red in Figure 16.11). Because the (α1 → β2) glycosidic bond joins both anomeric carbons, there is no hemiacetal group. As a result, the ring cannot open to the linear form and sucrose will not react with Benedict's reagent. Thus, sucrose is not a reducing sugar.

LEARNING GOAL

12 Discuss the structural, chemical, and biochemical properties of starch, glycogen, and cellulose.

A polymer (Section 11.5) is a large molecule made up of many small units, the monomers, held together by chemical bonds.

16.6 Polysaccharides

Starch

Many carbohydrates found in nature are large polymers of glucose. Thus, a polysaccharide is a large polymer composed of many monosaccharide units (the monomers) joined in one or more chains. **Homopolysaccharides** are those composed of a single monosaccharide. **Heteropolysaccharides** are those made up of two or more different monosaccharides. (See A Medical Perspective: Monosaccharide Derivatives and Heteropolysaccharides of Medical Interest on page 589.)

Plants have the ability to use the energy of sunlight to produce monosaccharides, principally glucose, from CO_2 and H_2O. Although sucrose is the major transport form of sugar in the plant, starch (a homopolysaccharide) is the principal storage form in most plants. These plants store glucose in starch granules. Nearly all plant cells contain some starch granules, but in some seeds, such as corn, as much as 80% of the cell's dry weight is starch.

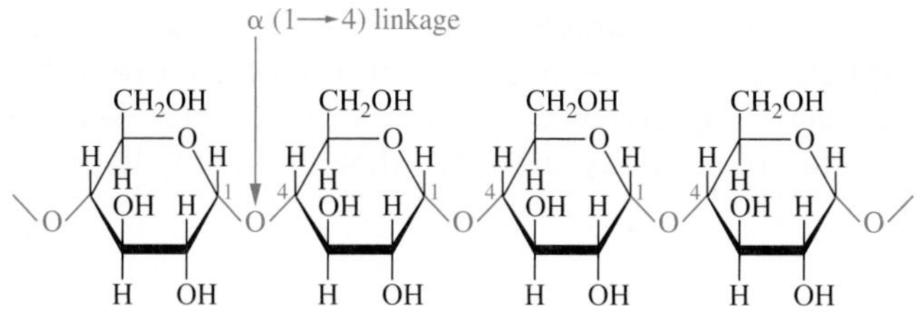

α (1 ⟶ 4) linkage

(a)

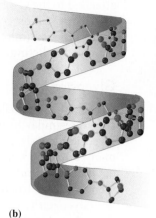

(b)

Figure 16.12 Structure of amylose. (a) A linear chain of α-D-glucose joined in α(1 → 4) glycosidic linkage makes up the primary structure of amylose. (b) Owing to hydrogen bonding, the amylose chain forms a left-handed helix that contains six glucose units per turn.

Starch is a heterogeneous material composed of the glucose polymers **amylose** and **amylopectin.** Amylose, which accounts for about 20% of the starch of a plant cell, is a linear polymer of α-D-glucose molecules connected by glycosidic bonds between C-1 of one glucose molecule and C-4 of a second glucose. Thus, the glucose units in amylose are joined by α(1 → 4) glycosidic bonds. A single chain can contain up to 4000 glucose units. Amylose coils up into a helix that repeats every six glucose units. The structure of amylose is shown in Figure 16.12.

Amylose is degraded by two types of enzymes (Figure 16.13). They are produced in the pancreas, from which they are secreted into the small intestine, and the salivary glands, from which they are secreted into the saliva. α-*Amylase* cleaves the glycosidic bonds of amylose chains at random along the chain, producing shorter polysaccharide chains. The enzyme β-*amylase* sequentially cleaves the disaccharide maltose from the reducing end of the amylose chain. The maltose is hydrolyzed into glucose by the enzyme *maltase*. The glucose is quickly absorbed by intestinal cells and used by the cells of the body as a source of energy.

Enzymes are proteins that serve as biological catalysts. They speed up biochemical reactions so that life processes can function. α- and β-Amylases are called α(1 → 4) glycosidases because they cleave α(1 → 4) glycosidic bonds.

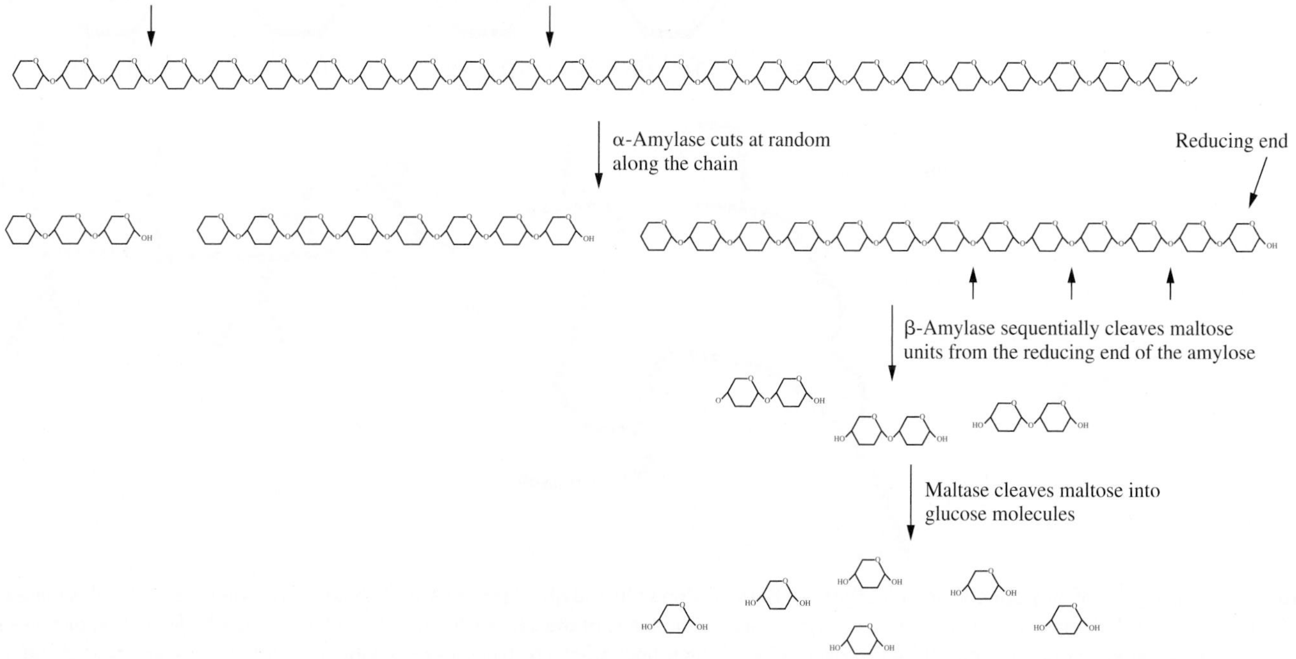

α-Amylase cuts at random along the chain

Reducing end

β-Amylase sequentially cleaves maltose units from the reducing end of the amylose

Maltase cleaves maltose into glucose molecules

Figure 16.13 The action of α-amylase, β-amylase, and maltase on amylose.

Potatoes contain large amounts of starch. Describe the composition of this starch.

Amylopectin is a highly branched amylose in which the branches are attached to the C-6 hydroxyl groups by $\alpha(1 \rightarrow 6)$ glycosidic bonds (Figure 16.14). The main chains consist of $\alpha(1 \rightarrow 4)$ glycosidic bonds. Each branch contains 20–25 glucose units, and there are so many branches that the main chain can scarcely be distinguished.

Glycogen

Glycogen is the major glucose storage molecule in animals. The structure of glycogen is similar to that of amylopectin. The "main chain" is linked by $\alpha(1 \rightarrow 4)$ glycosidic bonds, and it has numerous $\alpha(1 \rightarrow 6)$ glycosidic bonds, which provide many branch points along the chain. Glycogen differs from amylopectin only by having more and shorter branches. Otherwise, the two molecules are virtually identical. The structure of glycogen is shown in Figure 16.14.

Glycogen is stored in the liver and skeletal muscle. Glycogen synthesis and degradation in the liver are carefully regulated. As we will see in Section 21.7, these two processes are intimately involved in keeping blood glucose levels constant.

Cellulose

The most abundant polysaccharide, indeed the most abundant organic molecule in the world, is **cellulose,** a polymer of β-D-glucose units linked by $\beta(1 \rightarrow 4)$

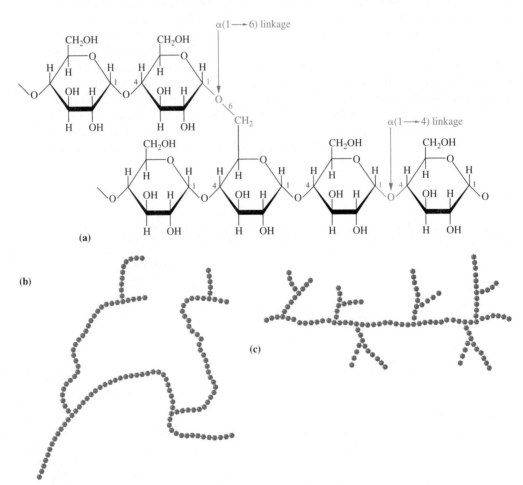

Figure 16.14 Structure of amylopectin and glycogen. (a) Both amylopectin and glycogen consist of chains of α-D-glucose molecules joined in $\alpha(1 \rightarrow 4)$ glycosidic linkages. Branching from these chains are other chains of the same structure. Branching occurs by formation of $\alpha(1 \rightarrow 6)$ glycosidic bonds between glucose units. (b) A representation of the branched-chain structure of amylopectin. (c) A representation of the branched-chain structure of glycogen. Glycogen differs from amylopectin only in that the branches are shorter and there are more of them.

A Medical Perspective

Monosaccharide Derivatives and Heteropolysaccharides of Medical Interest

Many of the carbohydrates with important functions in the human body are either derivatives of simple monosaccharides or are complex polymers of monosaccharide derivatives. One type of monosaccharide derivatives, the uronates, is formed when the terminal —CH_2OH group of a monosaccharide is oxidized to a carboxylate group. α-D-Glucuronate is a uronate of glucose (see right).

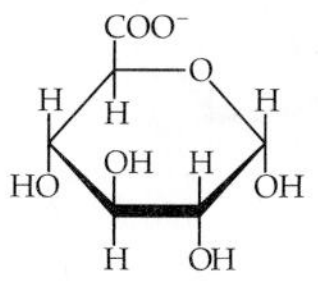

In liver cells, α-D-glucuronate is bonded to hydrophobic molecules, such as steroids, to increase their solubility in water. When bonded to the modified sugar, steroids are more readily removed from the body.

Amino sugars are a second important group of monosaccharide derivatives. In amino sugars, one of the hydroxyl groups (usually on carbon-2) is replaced by an amino group. Often, these are found in complex oligosaccharides that are attached to cellular proteins and lipids. The most common amino sugars, D-glucosamine and D-galactosamine, are often found in the N-acetyl form. N-acetylglucosamine is a component of bacterial cell walls and N-acetylgalactosamine is a component of the ABO blood group antigens.

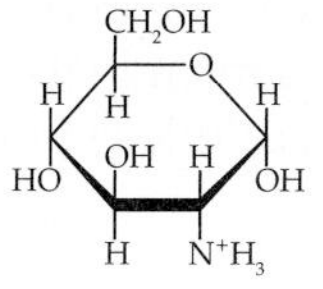

α-D-Glucosamine 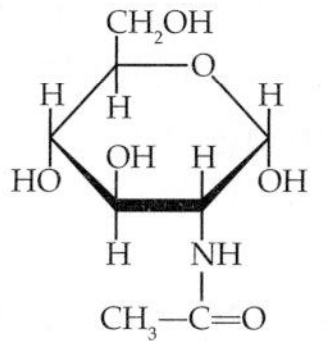   α-D-N-Acetylglucosamine

Heteropolysaccharides are long-chain polymers that contain more than one type of monosaccharide, many of which are amino sugars. These *glycosaminoglycans* include chondroitin sulfate, hyaluronic acid, and heparin. Hyaluronic acid is abundant in the fluid of joints and in the vitreous humor of the eye. Chondroitin sulfate is an important component of cartilage, and heparin has an anticoagulant function. The structures of the repeat units of these polymers are shown below.

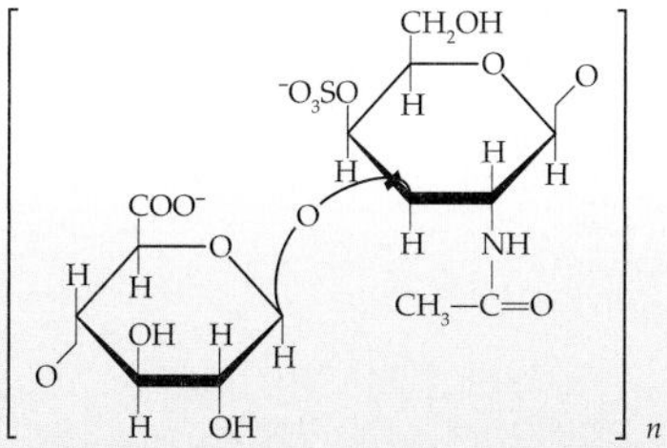

Repeat unit of chondroitin sulfate

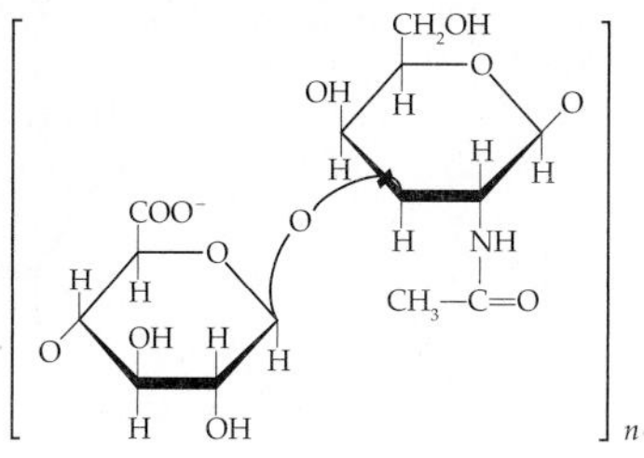

α-D-Glucuronate

Repeat unit of hyaluronic acid

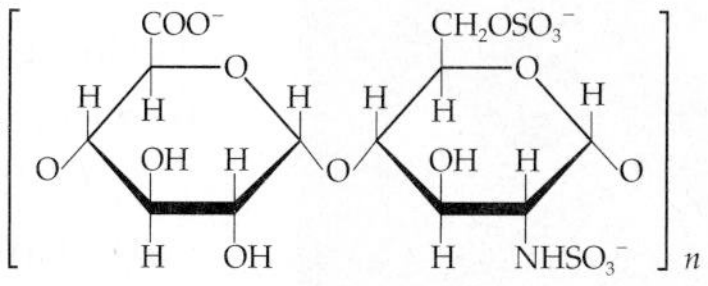

Repeat unit of heparin

Two of these molecules have been studied as potential treatments for osteoarthritis, a painful, degenerative disease of the joints. The amino sugar D-glucosamine is thought to stimulate the production of collagen. Collagen is one of the main components of articular cartilage, which is the shock-absorbing cushion within the joints. With aging, some of the D-glucosamine is lost, leading to a reduced cartilage layer and to the onset and progression of arthritis. It has been suggested that ingestion of D-glucosamine can actually "jump-start" production of cartilage and help repair eroded cartilage in arthritic joints.

It has also been suggested that chondroitin sulfate can protect existing cartilage from premature breakdown. It absorbs large amounts of water, which is thought to facilitate diffusion of nutrients into the cartilage, providing precursors for the synthesis of new cartilage. The increased fluid also acts as a shock absorber.

Capsules containing D-glucosamine and chondroitin sulfate are available over the counter, and many sufferers of osteoarthritis prefer to take this nutritional supplement as an alternative to any nonsteroidal anti-inflammatory drug (NSAID), such as ibuprofen. Although NSAIDs can reduce inflammation and pain, long-term use of NSAIDs can result in stomach ulcers, damage to auditory nerves, and kidney damage.

For Further Understanding

▶ In Chapter 15, we learned that nonsteroidal anti-inflammatory drugs (NSAIDs), such as ibuprofen, are analgesics used to treat pain, such as that associated with osteoarthritis. Why do many people prefer to treat osteoarthritis with D-glucosamine and chondroitin sulfate rather than NSAIDs?

▶ Explain why attaching a molecule such as α-D-glucuronate to a steroid molecule would increase its water solubility.

Figure 16.15 The structure of cellulose.

β(1 → 4) glycosidic bond

Vegetables contribute fiber to our diet.
What carbohydrate provides this fiber?

glycosidic bonds (Figure 16.15). A molecule of cellulose typically contains about 3000 glucose units, but the largest known cellulose, produced by the alga *Valonia*, contains 26,000 glucose molecules.

Cellulose is a structural component of the plant cell wall. The unbranched structure of the cellulose polymer and the β(1 → 4) glycosidic linkages allow cellulose molecules to form long, straight chains of parallel cellulose molecules called *fibrils*. These fibrils are quite rigid and are held together tightly by hydrogen bonds; thus, it is not surprising that cellulose is a cell wall structural element.

In contrast to glycogen, amylose, and amylopectin, cellulose *cannot* be digested by humans. The reason is that we cannot synthesize the enzyme *cellulase*, which can hydrolyze the β(1 → 4) glycosidic linkages of the cellulose polymer. Indeed, only a few animals, such as termites, cows, and goats, are able to digest cellulose. These animals have, within their digestive tracts, microorganisms that produce the enzyme cellulase. The sugars released by this microbial digestion can then be absorbed and used by these animals. In humans, cellulose from fruits and vegetables serves as fiber in the diet.

Question 16.11 What chemical reactions are catalyzed by α-amylase and β-amylase?

Question 16.12 What is the function of cellulose in the human diet? How does this relate to the structure of cellulose?

CHAPTER MAP

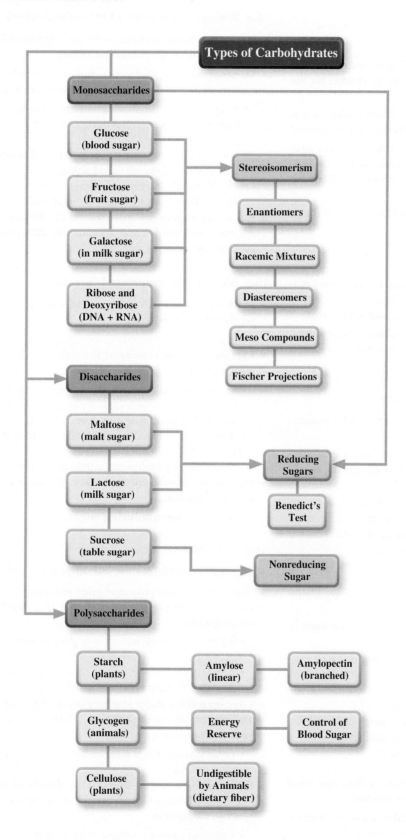

SUMMARY

16.1 Types of Carbohydrates

▶ **Carbohydrates** are found in a wide variety of naturally occurring substances and serve as principal energy sources for the body.

▶ Dietary carbohydrates include complex carbohydrates, such as starch in potatoes, and simple carbohydrates, such as sucrose.

▶ Carbohydrates are classified as **monosaccharides** (one saccharide or sugar unit), **disaccharides** (two sugar units), **oligosaccharides** (three to ten sugar units), and **polysaccharides** (many sugar units).

▶ Individual monosaccharides are joined to others through **glycosidic bonds.**

16.2 Monosaccharides

▶ Monosaccharides that have an aldehyde as their most oxidized functional group are **aldoses,** and those having a ketone group as their most oxidized functional group are **ketoses.**

▶ Monosaccharides are classified as **trioses, tetroses, pentoses, hexoses,** and so forth, depending on the number of carbon atoms in the molecule.

16.3 Stereoisomers and Stereochemistry

▶ **Stereochemistry** is the study of the different spatial arrangements of atoms.

▶ **Stereoisomers** of monosaccharides exist because of the presence of **chiral carbon** atoms. Those that are nonsuperimposable mirror images of one another are called **enantiomers.** Molecules that exist as enantiomers are **chiral molecules.**

  • Stereoisomers of monosaccharides are classified as D- or L- based on the arrangement of atoms on the chiral carbon farthest from the aldehyde or ketone group.
  • If this —OH is on the right of the molecule, the stereoisomer is of the D-family.
  • If this —OH is on the left of the molecule, the stereoisomer is of the L-family.

▶ **Each pair of stereoisomers rotates plane-polarized light in opposite directions.**

  • A polarimeter is used to measure the direction of rotation of **plane-polarized light.**
  • Compounds that rotate light in a clockwise direction are termed *dextrorotatory* and are designated (+).
  • Compounds that rotate light in a counterclockwise direction are termed *levorotatory* and are designated (−).
  • A mixture of equal amounts of a pair of enantiomers is a **racemic mixture.**

▶ **Diastereomers** are stereoisomers with more than one chiral center that are not mirror images of one another.

▶ **Meso compounds** have two chiral carbons and an internal plane of symmetry. As a result, they are achiral.

▶ **A Fischer Projection** is a two-dimensional drawing of a molecule that shows a chiral carbon at the intersection of two lines.

  • Horizontal lines represent bonds projecting out of the page.
  • Vertical lines represent bonds that project into the page.

16.4 Biologically Important Monosaccharides

▶ Important monosaccharides include **glyceraldehyde, glucose, galactose, fructose,** and **ribose.**

▶ Monosaccharides containing five or six carbon atoms can exist as five- or six-member rings. These are **hemiacetals.**

  • Ring formation produces a new chiral carbon at the original carbonyl carbon, which is designated either α or β depending on the orientation of the groups.
  • Isomers differing in the arrangement of bonds around the hemiacetal carbon are called **anomers.**
  • The **Haworth projection** is used to represent the orientation of substituents around a cyclic sugar molecule.

▶ **Reducing sugars** are oxidized by the **Benedict's reagent.** All monosaccharides and all common disaccharides, except sucrose, are reducing sugars.

16.5 Biologically Important Disaccharides

▶ **Maltose** is a disaccharide formed from α-D-glucose and a second glucose molecule.

  • It is formed in the hydrolysis of starch.

▶ **Lactose** is a disaccharide of β-D-galactose bonded β(1 → 4) with D-glucose.

  • In **galactosemia,** defective metabolism of galactose leads to accumulation of a toxic by-product. Symptoms can be avoided by excluding milk from the diet.
  • About 20% of the U.S. population has **lactose intolerance.** Caused by the inability to digest lactose, the symptoms can be avoided by excluding milk from the diet.

▶ **Sucrose** is a dimer composed of α-D-glucose bonded (α1 → β2) with β-D-fructose.

  • Sucrose is a **nonreducing sugar.**

16.6 Polysaccharides

▶ **Homopolysaccharides** are made up of a single monosaccharide.

▶ Starch, the storage polysaccharide of many plants, is a homopolysaccharide of glucose. Starch is 20% **amylose** and 80% **amylopectin.**

  • Amylose is a polymer of α-D-glucose **monomers** bonded α(1 → 4).
  • Amylopectin has a main chain like amylose and has branches that are joined α(1 → 6) to the main chain.

▶ **Glycogen** is the major storage polysaccharide of animal cells and resembles amylopectin. It differs by having more and shorter branches.

  • Liver glycogen is a reserve that is used to regulate blood glucose levels.

▶ **Cellulose** is a major structural molecule of plants. It is a β(1 → 4) polymer of D-glucose that may contain thousands of glucose monomers.

  • Cellulose cannot be digested by animals.

▶ A **heteropolysaccharide** is a polysaccharide composed of two or more different monosaccharides.

ANSWERS TO PRACTICE PROBLEMS

16.1 a.

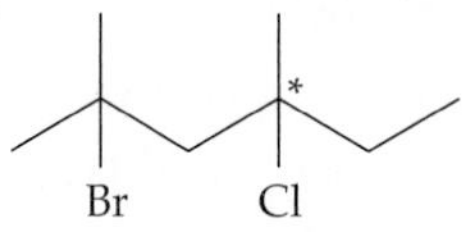

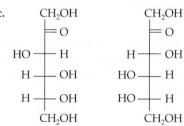

b. F   F
        | Cl |
        |* |* |
       Cl   Br F

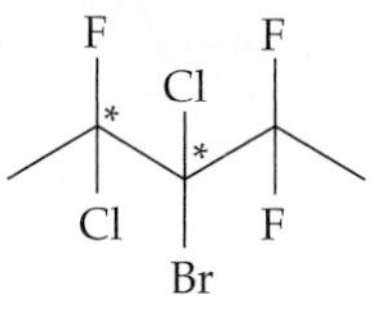

c. OH   O
        |* *|
             H
       OH   OH

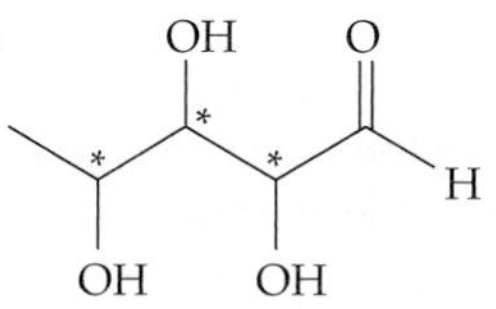

16.2 a.

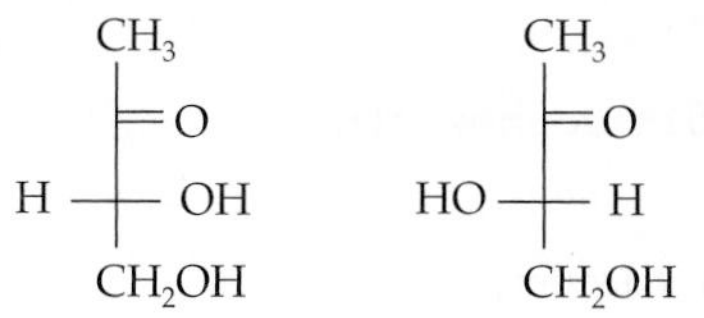

CH₃   CH₃
‖O   ‖O
H ——— OH   HO ——— H
CH₂OH   CH₂OH

b.

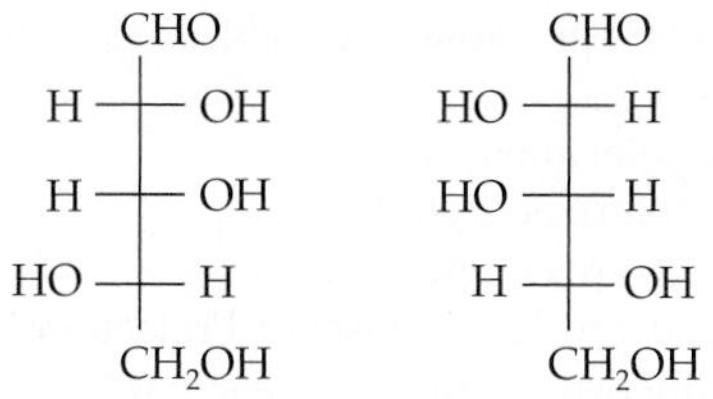

CHO   CHO
H ——— OH   HO ——— H
H ——— OH   HO ——— H
HO ——— H   H ——— OH
CH₂OH   CH₂OH

c.

CH₂OH   CH₂OH
‖O   ‖O
HO ——— H   H ——— OH
H ——— OH   HO ——— H
H ——— OH   HO ——— H
CH₂OH   CH₂OH

16.3 a.

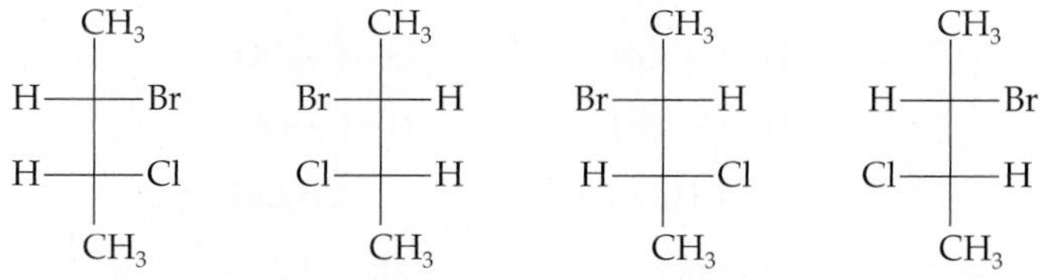

CH₃   CH₃   CH₃   CH₃
H——Br   Br——H   Br——H   H——Br
H——Cl   Cl——H   H——Cl   Cl——H
CH₃   CH₃   CH₃   CH₃
A   **B**   **C**   **D**

There are two chiral carbons in this molecule. Thus, there are four possible stereoisomers. There are two possible configurations for each chiral carbon. Compounds A and B are mirror images. Compounds C and D are also mirror images.

b.

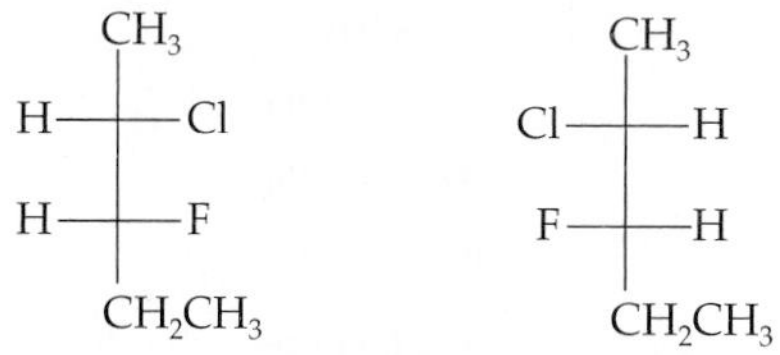

CH₃   CH₃
H——Cl   Cl——H
H——F   F——H
CH₂CH₃   CH₂CH₃
A   **B**

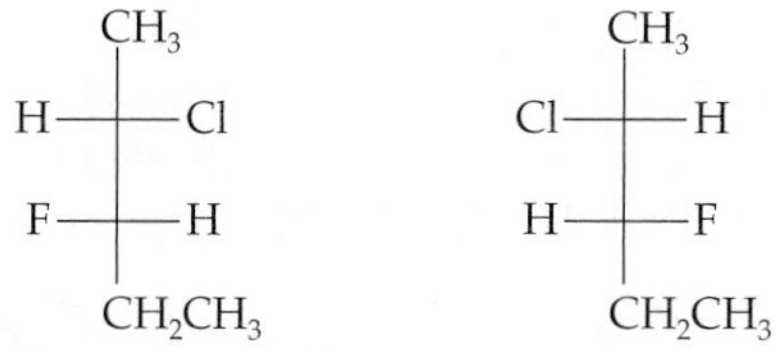

CH₃   CH₃
H——Cl   Cl——H
F——H   H——F
CH₂CH₃   CH₂CH₃
C   **D**

There are two chiral carbons in this molecule. Thus, there are four possible stereoisomers. There are two possible configurations for each chiral carbon. Compounds A and B are mirror images. Compounds C and D are also mirror images.

c.

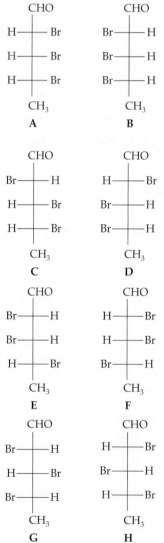

CHO   CHO
H——Br   Br——H
H——Br   Br——H
H——Br   Br——H
CH₃   CH₃
A   **B**

CHO   CHO
Br——H   H——Br
H——Br   Br——H
H——Br   Br——H
CH₃   CH₃
C   **D**

CHO   CHO
Br——H   H——Br
Br——H   H——Br
H——Br   Br——H
CH₃   CH₃
E   **F**

CHO   CHO
Br——H   H——Br
H——Br   Br——H
Br——H   H——Br
CH₃   CH₃
G   **H**

There are three chiral carbons in this molecule. Thus, there are eight possible stereoisomers. There are two possible configurations for each chiral carbon. Pairs of enantiomers include: A and B, C and D, E and F, and G and H.

16.4

D-Ribose

	CHO
H	OH
H	OH
H	OH
	CH₂OH

L-Ribose

	CHO
H	OH
H	OH
HO	H
	CH₂OH

16.5 a.

α-D-Galactose   β-D-Galactose

b.

α-D-Ribose   β-D-Ribose

QUESTIONS AND PROBLEMS

Types of Carbohydrates

Foundations

16.13 What is the difference between a monosaccharide and a disaccharide?

16.14 What is a polysaccharide?

Applications

16.15 Read the labels on some of the foods in your kitchen, and see how many products you can find that list one or more carbohydrates among the ingredients in the package. Make a list of these compounds, and attempt to classify them as monosaccharides, disaccharides, or polysaccharides.

16.16 Some disaccharides are often referred to by their common names. What are the chemical names of (a) milk sugar, (b) beet sugar, and (c) cane sugar?

16.17 How many kcal of energy are released when 1 g of carbohydrate is "burned" or oxidized?

16.18 List some natural sources of carbohydrates.

16.19 Draw and provide the names of an aldohexose and a ketohexose.

16.20 Draw and provide the name of an aldotriose.

Monosaccharides

Foundations

16.21 Define the term *aldose*.

16.22 Define the term *ketose*.

16.23 What is a tetrose?

16.24 What is a hexose?

16.25 What is a ketopentose?

16.26 What is an aldotriose?

Applications

16.27 Identify each of the following sugars.

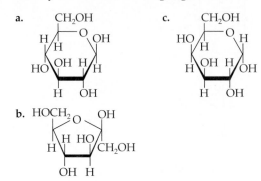

16.28 Draw the open-chain form of the sugars in Question 16.27.

16.29 Draw all of the different possible aldotrioses of molecular formula $C_3H_6O_3$.

16.30 Draw all of the different possible aldotetroses of molecular formula $C_4H_8O_4$.

Stereoisomers and Stereochemistry

Foundations

16.31 Define the term *stereoisomer*.

16.32 Define the term *enantiomer*.

16.33 Define the term *chiral carbon*.

16.34 Draw an aldotetrose. Note each chiral carbon with an asterisk (*).

16.35 Explain how a polarimeter works.

16.36 What is plane-polarized light?

16.37 What is a Fischer Projection?

16.38 How would you produce a Fischer Projection beginning with a three-dimensional model of a sugar?

16.39 Define the term *diastereomer*.

16.40 Define the term *meso compound*.

Applications

16.41 Sorbitol and mannitol are six-carbon sugar alcohols commonly used as sugar substitutes because they have fewer calories than sucrose. Indicate the chiral carbons in these molecules with an asterisk (*).

Sorbitol

	CH₂OH	
H—	C	—OH
HO—	C	—H
H—	C	—OH
H—	C	—OH
	CH₂OH	

Mannitol

	CH₂OH	
HO—	C	—H
HO—	C	—H
H—	C	—OH
H—	C	—OH
	CH₂OH	

16.42 Erythritol is a four-carbon sugar alcohol and xylitol is a five-carbon sugar alcohol. Like sorbitol and mannitol, they are used as sugar substitutes. Xylitol has the additional benefit of reducing dental cavities and promoting remineralization of the teeth and is often found in sugarless gums. Indicate the chiral carbons in these molecules with an asterisk (*).

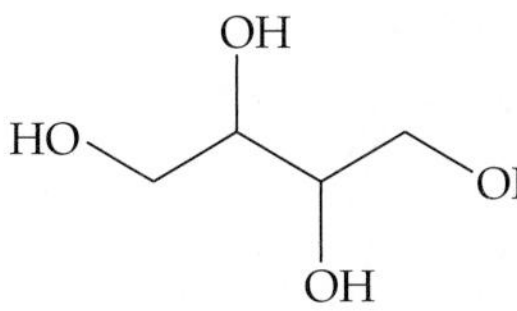

Erythritol 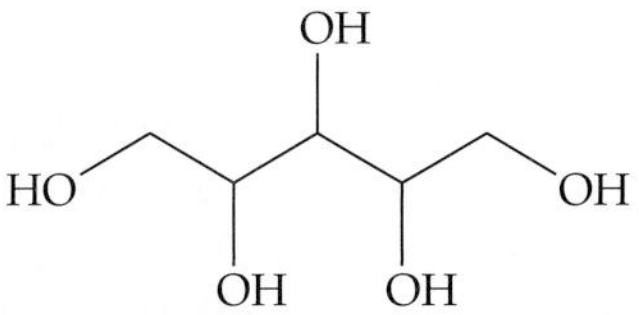   Xylitol

16.43 Is there any difference between dextrose and D-glucose?

16.44 The linear structure of D-glucose is shown in Figure 16.7. Draw its mirror image.

16.45 How are D- and L-glyceraldehyde related?

16.46 Determine whether each of the following is a D- or L-sugar:

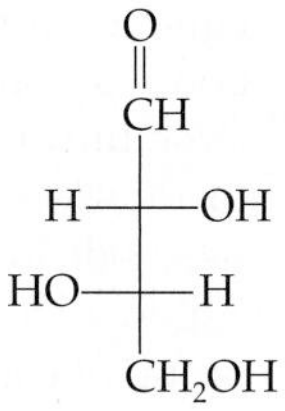

16.47 Draw a Fischer Projection formula for each of the following compounds. Indicate each of the chiral carbons with an asterisk (*).

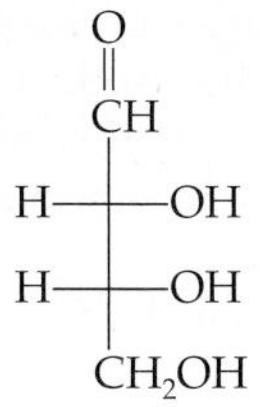

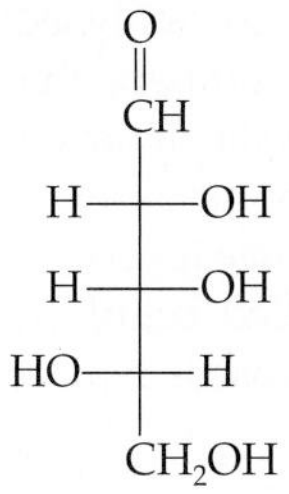

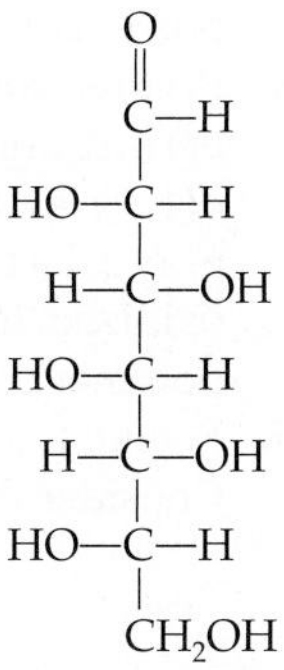

16.48 Draw a Fischer Projection formula for each of the following compounds. Indicate each of the chiral carbons with an asterisk (*).

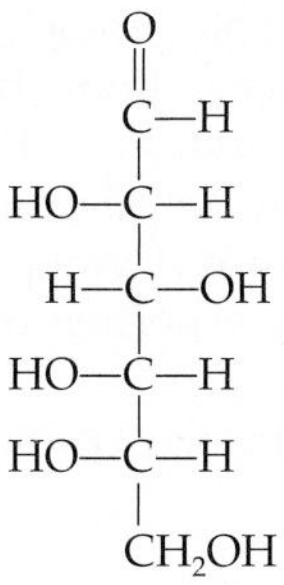

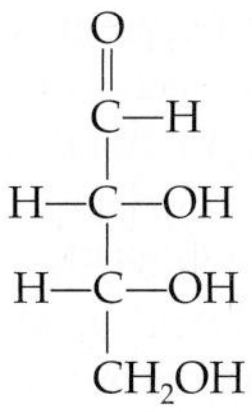

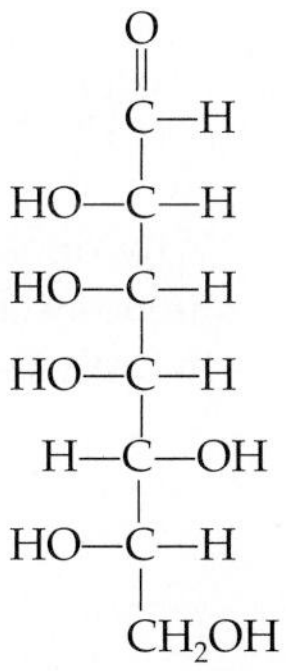

16.49 Draw all the possible stereoisomers of each of the following compounds, and indicate which are enantiomers, diastereomers, or meso compounds.

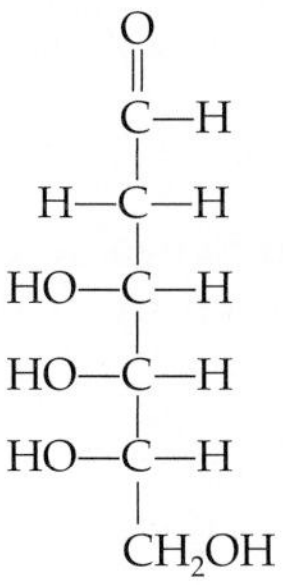

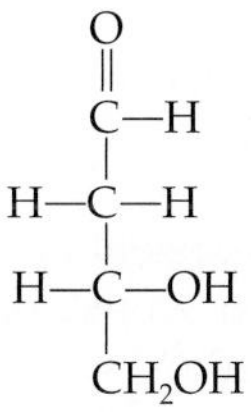

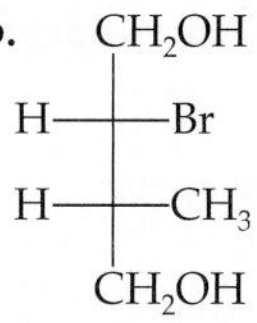

16.50 Draw all the possible stereoisomers of each of the following compounds, and indicate which are enantiomers, diastereomers, or meso compounds.

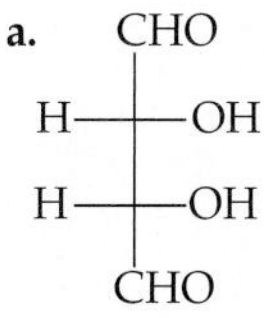

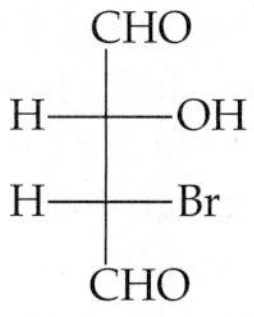

16.51 Draw all the possible stereoisomers of each of the following compounds, and indicate which are enantiomers, diastereomers, or meso compounds.

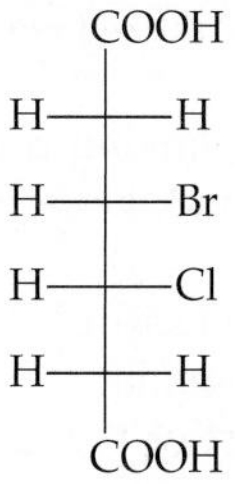

a. $CH_3CH_2CHOHCHOHCH_2CH_3$

b. $CH_2OHCCHFCHClCCH_2OH$ (with two C=O groups)

16.52 Draw all the possible stereoisomers of each of the following compounds, and indicate which are enantiomers, diastereomers, or meso compounds.

a. $CH_3CCHOHCHOHCCH_3$ (with two C=O groups)

b. $CH_2OHCHOHCHCICH_2OH$

Biologically Important Monosaccharides

Foundations

16.53 Define the term *anomer*.

16.54 What is a Haworth projection?

16.55 What is a hemiacetal?

16.56 Explain why the cyclization of D-glucose forms a hemiacetal.

Applications

16.57 Why does cyclization of D-glucose give two isomers, α- and β-D-glucose?

16.58 Draw the structure of the open-chain form of D-fructose, and show how it cyclizes to form α- and β-D-fructose.

16.59 Which of the following would give a positive Benedict's test?
a. Sucrose   c. β-Maltose
b. Glycogen   d. α-Lactose

16.60 Why was the Benedict's reagent useful for determining the amount of glucose in the urine?

16.61 Describe what is meant by a pair of enantiomers. Draw an example of a pair of enantiomers.

16.62 What is a chiral carbon atom?

16.63 When discussing sugars, what do we mean by an intramolecular hemiacetal?

16.64 Explain why ketoses can be oxidized in the Benedict's test, in contrast to ketones which cannot.

Biologically Important Disaccharides

Foundations

16.65 Define the term *disaccharide*.
16.66 What is an acetal?
16.67 What is a glycosidic bond?
16.68 Why are glycosidic bonds acetals?

Applications

16.69 Maltose is a disaccharide isolated from amylose that consists of two glucose units linked by an $\alpha(1 \rightarrow 4)$ bond. Draw the structure of this molecule.
16.70 Sucrose is a disaccharide formed by linking α-D-glucose and β-D-fructose by an $(\alpha1 \rightarrow \beta2)$ bond. Draw the structure of this disaccharide.
16.71 What is the major biological source of lactose?
16.72 What metabolic defect causes galactosemia?
16.73 What simple treatment prevents most of the ill effects of galactosemia?
16.74 What are the major physiological effects of galactosemia?
16.75 What is lactose intolerance?
16.76 What is the difference between lactose intolerance and galactosemia?

Polysaccharides

Foundations

16.77 What is a polymer?
16.78 What form of sugar is used as the major transport sugar in a plant?
16.79 What is the major storage form of sugar in a plant?
16.80 What is the major structural form of sugar in a plant?
16.81 What is a homopolysaccharide?
16.82 What is a heteropolysaccharide?
16.83 List some examples of homopolysaccharides.
16.84 List some examples of heteropolysaccharides. (*Hint:* Refer to A Medical Perspective: Monosaccharide Derivatives and Heteropolysaccharides of Medical Interest.)

Applications

16.85 What is the difference between the structure of cellulose and the structure of amylose?
16.86 How does the structure of amylose differ from that of amylopectin and glycogen?
16.87 What is the major physiological purpose of glycogen?
16.88 Where in the body do you find glycogen stored?
16.89 Where are α-amylase and β-amylase produced?
16.90 Where do α-amylase and β-amylase carry out their enzymatic functions?

CHALLENGE PROBLEMS

1. The six-member glucose ring structure is not a flat ring. Like cyclohexane, it can exist in the chair conformation. Build models of the chair conformation of α- and β-D-glucose. Draw each of these structures. Which would you predict to be the more stable isomer? Explain your reasoning.

2. The following is the structure of salicin, a bitter-tasting compound found in the bark of willow trees:

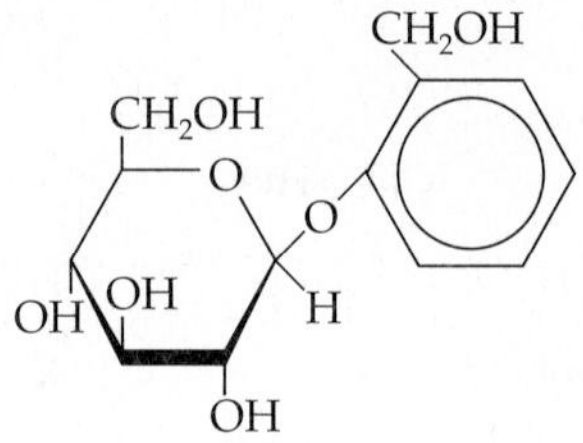

Salicin

The aromatic ring portion of this structure is quite insoluble in water. How would forming a glycosidic bond between the aromatic ring and β-D-glucose alter the solubility? Explain your answer.

3. Ancient peoples used salicin to reduce fevers. Write an equation for the acid-catalyzed hydrolysis of the glycosidic bond of salicin. Compare the aromatic product with the structure of acetylsalicylic acid (aspirin). Use this information to develop a hypothesis explaining why ancient peoples used salicin to reduce fevers.

4. Chitin is a modified cellulose in which the C-2 hydroxyl group of each glucose is replaced by

$$-NHCCH_3$$
with a $=O$ above the C

This nitrogen-containing polysaccharide makes up the shells of lobsters, crabs, and the exoskeletons of insects. Draw a portion of a chitin polymer consisting of four monomers.

5. Pectins are polysaccharides obtained from fruits and berries and used to thicken jellies and jams. Pectins are $\alpha(1 \rightarrow 4)$ linked D-galacturonic acid. D-Galacturonic acid is D-galactose in which the C-6 hydroxyl group has been oxidized to a carboxyl group. Draw a portion of a pectin polymer consisting of four monomers.

6. Peonin is a red pigment found in the petals of peony flowers. Consider the structure of peonin:

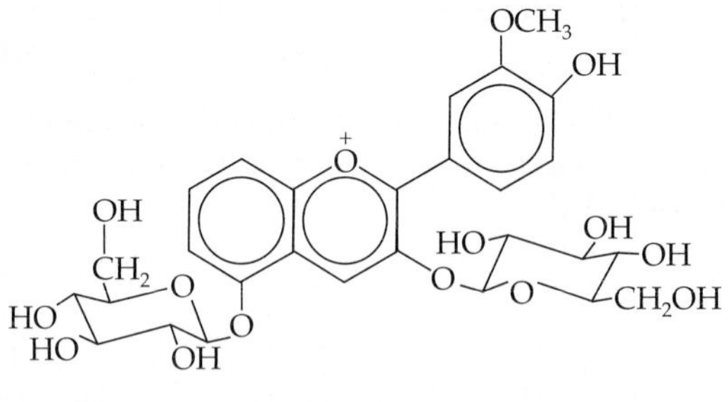

Why do you think peonin is bonded to two hexoses? What monosaccharide(s) would be produced by acid-catalyzed hydrolysis of peonin?

17

Lipids and Their Functions in Biochemical Systems

LEARNING GOALS

1. Discuss the physical and chemical properties and biological functions of each of the types of lipids.
2. Write the structures of saturated and unsaturated fatty acids.
3. Compare and contrast the structures and properties of saturated and unsaturated fatty acids.
4. Describe the functions of prostaglandins.
5. Discuss the mechanism by which aspirin reduces pain.
6. Write equations representing the reactions that fatty acids and glycerides undergo.
7. Draw the structure of a phospholipid and discuss its amphipathic nature.
8. Discuss the general classes of sphingolipids and their functions.
9. Draw the structure of the steroid nucleus, and discuss the functions of steroid hormones.
10. Describe the function of lipoproteins in triglyceride and cholesterol transport in the body.
11. Draw the structure of the cell membrane and discuss its functions.

The foxglove plant is a source of digitoxin and other cardiotonic steroids. Read A Medical Perspective: Steroids and the Treatment of Heart Disease in this chapter, and explain why the ingestion of foxglove can have deadly consequences.

OUTLINE

INTRODUCTION

Lipids seem to be the most controversial group of biological molecules, particularly in the fields of medicine and nutrition. We are concerned about the use of anabolic steroids by athletes. Although these hormones increase muscle mass and enhance performance, we are just beginning to understand the damage they cause to the body.

We worry about what types of dietary fat we should consume. We hear frequently about the amounts of saturated fats and cholesterol in our diets because a strong correlation has been found between these lipids and heart disease. Large quantities of dietary saturated fats may also predispose an individual to colon, esophageal, stomach, and breast cancers. As a result, we are advised to reduce our intake of cholesterol and saturated fats.

In this chapter, we will study the diverse collection of molecules referred to as lipids. We will see that triglycerides are both a dietary source of energy and a (sometimes unwanted) storage form of energy. Other lipids serve as structural components of the cell; for instance, phospholipids and cholesterol are components of the membranes around each of our cells. Some of the chemical messengers of our bodies are lipids. These include the steroid hormones and the hormonelike prostaglandins. Even some of the vitamins that are required in our diet are lipids, and any diet that is completely fat-free will result in deficiencies of these vitamins.

This quick tour through a few of the potential hazards of lipids and their essential roles in our bodies makes it easy to understand why lipids are so controversial and why so much literature has been published about the lipids in our diets. But, in fact, standards of fat intake have not been experimentally determined. The most recent U.S. Dietary Guidelines recommend that our daily fat intake be 20–35% of our daily caloric intake. Fewer than 10% of the calories we consume should be saturated fats, and cholesterol intake should be less than 300 milligrams/day. Overall, the intake of fats and oils that are high in saturated fats or in *trans*-fatty acids should be limited.

17.1   Biological Functions of Lipids

The term **lipids** actually refers to a collection of organic molecules of varying chemical composition. They are grouped together on the basis of their solubility in nonpolar solvents. Lipids may be subdivided into four main types:

- *Fatty acids* (saturated and unsaturated)
- *Glycerides* (glycerol-containing lipids)
- *Nonglyceride lipids* (sphingolipids, steroids, waxes)
- *Complex lipids* (lipoproteins)

In this chapter, we examine the structure, properties, chemical reactions, and biological functions of each of the lipid groups shown in Figure 17.1.

As a result of differences in their structures, lipids serve many different functions in the human body. The following brief list will give you an idea of the importance of lipids in biological processes:

- *Energy source.* Like carbohydrates, lipids are an excellent source of energy for the body. When oxidized, each gram of fat releases 9 kilocalories (kcal) of energy, or more than twice the energy released by oxidation of a gram of carbohydrate.
- *Energy storage.* Most of the energy stored in the body is in the form of lipids (triglycerides). Stored in fat cells called *adipocytes,* these fats are a particularly rich source of energy for the body.
- *Cell membrane structural components.* Phosphoglycerides, sphingolipids, and steroids make up the basic structure of all cell membranes. These membranes control the flow of molecules into and out of cells and allow cell-to-cell communication.

A Medical Perspective

Lifesaving Lipids

In the intensive-care nursery, the premature infant struggles for life. Born three and a half months early, the baby weighs only 1.6 pounds (lb), and the lungs labor to provide enough oxygen to keep the tiny body alive. Premature infants often have respiratory difficulties because they have not yet begun to produce *pulmonary surfactant.*

Pulmonary surfactant is a combination of phospholipids and proteins that reduces surface tension in the alveoli of the lungs. (Alveoli are the small, thin-walled air sacs in the lungs.) This allows efficient gas exchange across the membranes of the alveolar cells; oxygen can more easily diffuse from the air into the tissues, and carbon dioxide can easily diffuse from the tissues into the air. Without pulmonary surfactant, gas exchange in the lungs is very poor.

Pulmonary surfactant is not produced until early in the sixth month of pregnancy. Premature babies born before they have begun secretion of natural surfactant suffer from *respiratory distress syndrome (RDS)*, which is caused by the severe difficulty they have obtaining enough oxygen from the air that they breathe.

Until recently, RDS was a major cause of death among premature infants, but now a lifesaving treatment is available. A fine aerosol of an artificial surfactant is administered directly into the trachea. The Glaxo-Wellcome Company product EXO-SURF Neonatal contains the phospholipid lecithin to reduce surface tension; 1-hexadecanol, which spreads the lecithin; and a polymer called *tyloxapol,* which disperses both the lecithin and the 1-hexadecanol.

Artificial pulmonary surfactant therapy has dramatically reduced premature infant death caused by RDS and appears to have reduced overall mortality for all babies born weighing less than 700 grams (g) (about 1.5 lb). Advances such as this have come about as a result of research on the makeup of body tissues and secretions in both healthy and diseased individuals. Often, such basic research provides the information needed to develop effective therapies.

For Further Understanding

▶ Draw the structure of 1-hexadecanol.
▶ Draw the structure of lecithin. Explain how lecithin reduces surface tension.

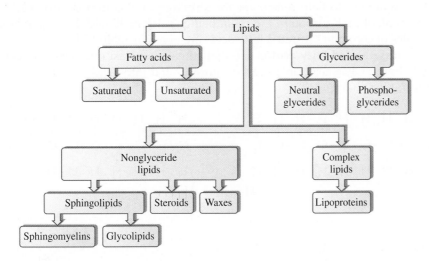

Figure 17.1 Types of lipids.

- *Hormones.* The steroid hormones are critical chemical messengers that allow tissues of the body to communicate with one another. The hormonelike prostaglandins exert strong biological effects on both the cells that produce them and other cells of the body.
- *Vitamins.* The lipid-soluble vitamins, A, D, E, and K, play a major role in the regulation of several critical biological processes, including blood clotting and vision.
- *Vitamin absorption.* Dietary fat serves as a carrier of the lipid-soluble vitamins. All are transported into cells of the small intestine in association with fat molecules. Therefore, a diet that is too low in fat (less than 20% of calories) can result in a deficiency of these four vitamins.

- *Protection.* Fats serve as a shock absorber, or protective layer, for the vital organs. About 4% of the total body fat is reserved for this critical function.
- *Insulation.* Fat stored beneath the skin (subcutaneous fat) serves to insulate the body from extremes of cold temperatures.

17.2 Fatty Acids

Structure and Properties

LEARNING GOAL

2 Write the structures of saturated and unsaturated fatty acids.

Fatty acids are long-chain monocarboxylic acids. As a consequence of their biosynthesis, fatty acids generally contain an *even number* of carbon atoms. The general formula for a **saturated fatty acid** is $CH_3(CH_2)_nCOOH$, in which n in biological systems is an even number. Recall that —COOH is a representation of the carboxyl group. If $n = 16$, the result is an 18-carbon saturated fatty acid, stearic acid, having the following structural formula:

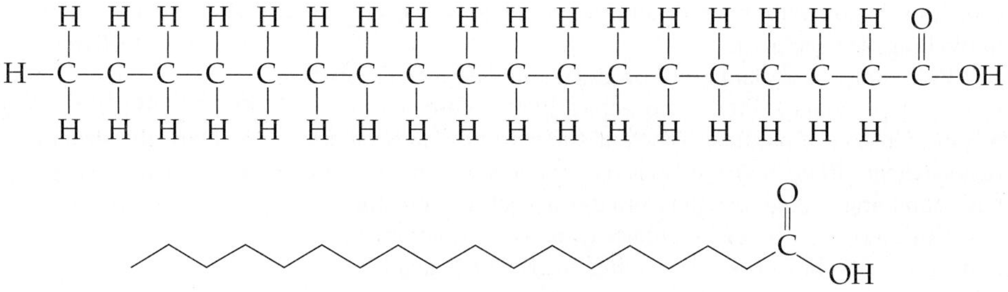

The saturated fatty acids may be thought of as derivatives of alkanes, the saturated hydrocarbons described in Chapter 10.

Note that each of the carbons in the chain is bonded to the maximum number of hydrogen atoms. To help remember the structure of a saturated fatty acid, you might think of each carbon in the chain being "saturated" with hydrogen atoms. Examples of common saturated fatty acids are given in Table 17.1.

An **unsaturated fatty acid** is one that contains at least one carbon-to-carbon double bond. Oleic acid, an 18-carbon unsaturated fatty acid, has the following structural formula:

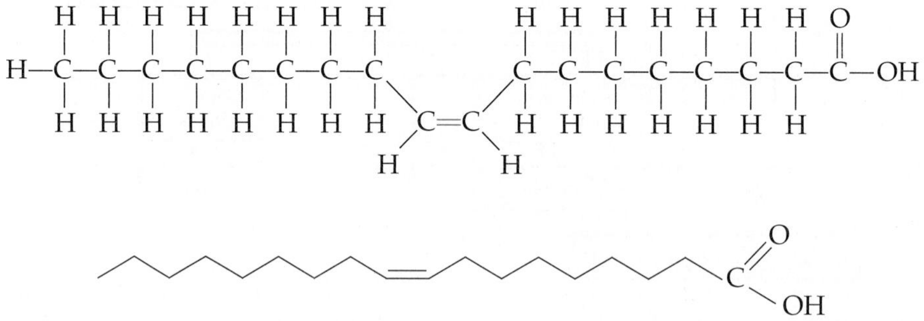

The unsaturated fatty acids may be thought of as derivatives of the alkenes, the unsaturated hydrocarbons discussed in Chapter 11.

A discussion of trans-fatty acids is found in Section 11.3.

Because of the double bonds, the carbon atoms involved in these bonds are not "saturated" with hydrogen atoms. The double bonds found in almost all naturally occurring unsaturated fatty acids are in the *cis* configuration. In addition, the double bonds are not randomly located in the hydrocarbon chain. Both the placement and the geometric configuration of the double bonds are dictated by the enzymes that catalyze the biosynthesis of unsaturated fatty acids. Examples of common unsaturated fatty acids are also given in Table 17.1. The similarities and differences between saturated and unsaturated fatty acids are described in Table 17.2.

Examination of Table 17.1 and Figure 17.2 reveals several interesting and important points about the physical properties of fatty acids.

EXAMPLE 17.1	Writing the Structural Formula of an Unsaturated Fatty Acid

Draw the structural formula for palmitoleic acid.

LEARNING GOAL

2 Write the structures of saturated and unsaturated fatty acids.

Solution

The IUPAC name of palmitoleic acid is *cis*-9-hexadecenoic acid. The name tells us that this is a 16-carbon fatty acid having a carbon-to-carbon double bond between carbons 9 and 10. The name also reveals that this is the *cis* isomer.

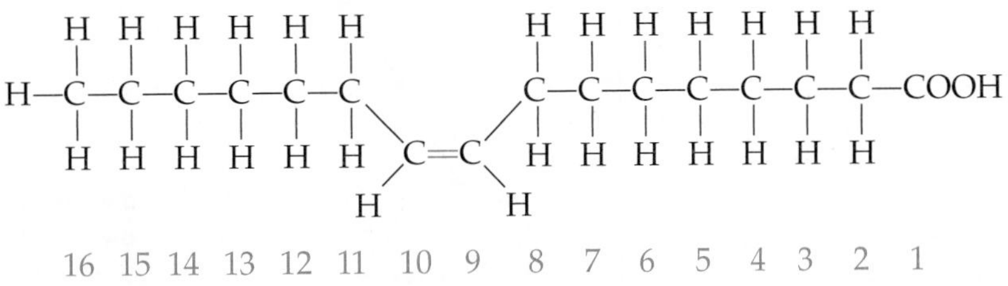

16 15 14 13 12 11 10 9   8   7   6   5   4   3   2   1

Practice Problem 17.1

Draw the line formulas for (a) oleic acid and (b) linoleic acid.

▶ For Further Practice: **Questions 17.1 a and c and 17.28.**

TABLE 17.1 Common Saturated and Unsaturated Fatty Acids

Common Saturated Fatty Acids

Common Name	IUPAC Name	Melting Point (°C)	Molar Mass	Condensed Formula
Capric	Decanoic	32	172.26	$CH_3(CH_2)_8COOH$
Lauric	Dodecanoic	44	200.32	$CH_3(CH_2)_{10}COOH$
Myristic	Tetradecanoic	54	228.37	$CH_3(CH_2)_{12}COOH$
Palmitic	Hexadecanoic	63	256.42	$CH_3(CH_2)_{14}COOH$
Stearic	Octadecanoic	70	284.48	$CH_3(CH_2)_{16}COOH$
Arachidic	Eicosanoic	77	312.53	$CH_3(CH_2)_{18}COOH$

Common Unsaturated Fatty Acids

Common Name	IUPAC Name	Melting Point (°C)	Molar Mass	Number of Double Bonds	Position of Double Bond(s)
Palmitoleic	*cis*-9-Hexadecenoic	0	254.41	1	9
Oleic	*cis*-9-Octadecenoic	16	282.46	1	9
Linoleic	*cis,cis*-9,12-Octadecadienoic	5	280.45	2	9, 12
Linolenic	All *cis*-9,12,15-Octadecatrienoic	−11	278.43	3	9, 12, 15
Arachidonic	All *cis*-5,8,11,14-Eicosatetraenoic	−50	304.47	4	5, 8, 11, 14

Condensed Formula

Palmitoleic	$CH_3(CH_2)_5CH = CH(CH_2)_7COOH$
Oleic	$CH_3(CH_2)_7CH = CH(CH_2)_7COOH$
Linoleic	$CH_3(CH_2)_4CH = CH-CH_2-CH = CH(CH_2)_7COOH$
Linolenic	$CH_3CH_2CH = CH-CH_2-CH = CH-CH_2-CH = CH(CH_2)_7COOH$
Arachidonic	$CH_3(CH_2)_4CH = CH-CH_2-CH = CH-CH_2-CH = CH-CH_2-CH = CH-(CH_2)_3COOH$

Figure 17.2 The melting points of fatty acids. Melting points of both saturated and unsaturated fatty acids increase as the number of carbon atoms in the chain increases. The melting points of unsaturated fatty acids are lower than those of the corresponding saturated fatty acid with the same number of carbon atoms. Also, as the number of double bonds in the chain increases, the melting points decrease.

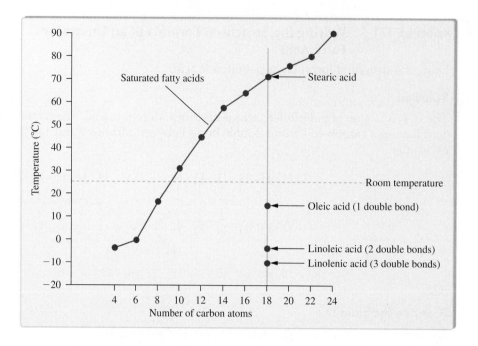

Explain why the olive oil in the photo above is liquid at room temperature but the beef fat is solid.

TABLE 17.2 Similarities and Differences Between Saturated and Unsaturated Fatty Acids

Property	Saturated Fatty Acid	Unsaturated Fatty Acid
Chemical composition	Carbon, hydrogen, oxygen	Carbon, hydrogen, oxygen
Chemical structure	Hydrocarbon chain with a terminal carboxyl group	Hydrocarbon chain with a terminal carboxyl group
Carbon-carbon bonds within the hydrocarbon chain	Only C—C single bonds	At least one C—C double bond
Hydrocarbon chains are characteristic of what group of hydrocarbons	Alkanes	Alkenes
"Shape" of hydrocarbon chain	Linear, fully extended	Bend in carbon chain at site of C—C double bond
Physical state at room temperature	Solid	Liquid
Melting point for two fatty acids of the same hydrocarbon chain length	Higher	Lower
Relationship between melting point and chain length	Longer chain length, higher melting point	Longer chain length, higher melting point

LEARNING GOAL

3 Compare and contrast the structures and properties of saturated and unsaturated fatty acids.

The relationship between alkane chain length and melting point is described in Section 10.2.

- The melting points of saturated fatty acids increase with increasing carbon number due to London dispersion forces, as is the case with alkanes. Saturated fatty acids containing ten or more carbons are solids at room temperature.
- The melting point of a saturated fatty acid is greater than that of an unsaturated fatty acid of the same chain length. The reason is that saturated fatty acid chains tend to be fully extended and to stack in a regular structure, thereby causing increased intermolecular London dispersion force attraction. Introduction of a *cis* double bond into the hydrocarbon chain produces

a rigid 30° bend. Such "kinked" molecules cannot stack in an organized arrangement and thus have lower intermolecular attractions and lower melting points.

- As in the case for saturated fatty acids, the melting points of unsaturated fatty acids increase with increasing hydrocarbon chain length.

The relationship between alkene chain length and melting point is described in Section 11.1.

Question 17.1 Draw formulas for each of the following fatty acids:
a. Oleic acid
c. Linoleic acid
b. Lauric acid
d. Stearic acid

Question 17.2 What is the IUPAC name for each of the fatty acids in Question 17.1? (*Hint:* Review the naming of carboxylic acids in Section 14.1, and Table 17.1.)

Omega-3 Fatty Acids

In 2002, the American Heart Association (AHA) issued dietary guidelines that recommend that we include at least two servings of "oily" fish in our diet each week. Among the fish recommended are salmon, albacore tuna, sardines, lake trout, and mackerel. The reason for this recommendation is that these fish contain high levels of two omega-3 fatty acids called eicosapentaenoic acid (EPA) and docosahexaenoic acid (DHA). The AHA further recommended a third omega-3 fatty acid, α-linolenic acid, which is found in flax seed, soybeans, and canola, as well as in oil made from these plants.

The name of this group of fatty acids arises from the position of the double bond nearest the terminal *methyl group* of the molecule. In these fatty acids, it is the third carbon from the end, designated omega (ω), which is the location of the double bond.

All *cis*-9,12,15-Octadecatrienoic acid
(α-Linolenic acid or ALA)

All *cis*-5,8,11,14,17-Eicosapentaenoic acid
(EPA)

All *cis*-4,7,10,13,16,19-Docosahexaenoic acid
(DHA)

The American Heart Association recommends that we include two meals of fish, such as salmon, in our diets each week. However, there is concern about the amount of heavy metals and other environmental contaminants in wild-caught fish. Go online to investigate the reasons that these contaminants accumulate in these fish.

The reason for this dietary recommendation was research that supported the idea that omega-3 fatty acids reduce the risk of cardiovascular disease by decreasing blood clot formation, blood triglyceride levels, and growth of atherosclerotic plaque. Because of these effects, arterial health improved and blood pressure decreased, as did the risk of sudden death and heart arrhythmias.

In some cases, the reason for the effect can be understood. For instance, EPA is a precursor for the synthesis of prostacyclin, which inhibits clumping of platelets and thus reduces clot formation. DHA is one of the major fatty acids in the phospholipids of sperm and brain cells, as well as in the retina. It has also been shown to reduce triglyceride levels, although the mechanism is not understood.

Linolenic acid is an essential fatty acid and must be acquired through the diet. Although it, too, seems to reduce the incidence of cardiovascular disease, it is not clear whether it acts alone or because it is the precursor of DHA and EPA.

Linoleic acid is also an essential fatty acid, required for the synthesis of arachidonic acid, the precursor for many prostaglandins. These two fatty acids are termed omega-6 fatty acids because the first double bond in the molecule is six carbon atoms from the methyl (ω) end of the molecule.

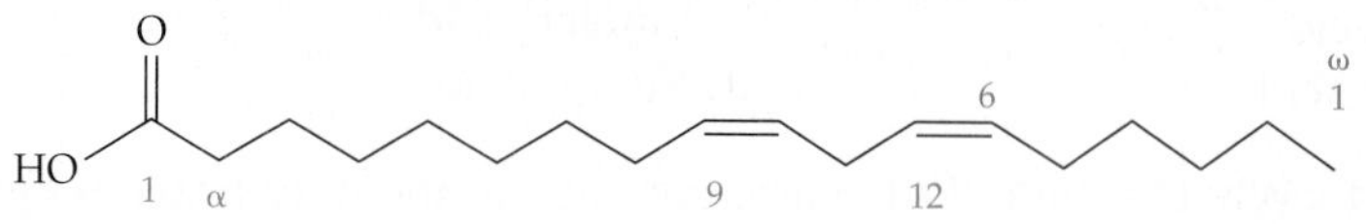

cis,cis-9,12-Octadecadienoic acid
(Linoleic acid)

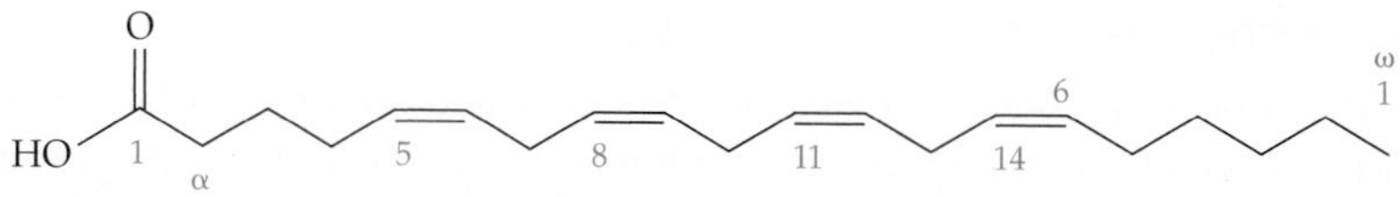

All cis-5,8,11,14-Eicosatetraenoic acid
(Arachidonic acid)

LEARNING GOAL

4 Describe the functions of prostaglandins.

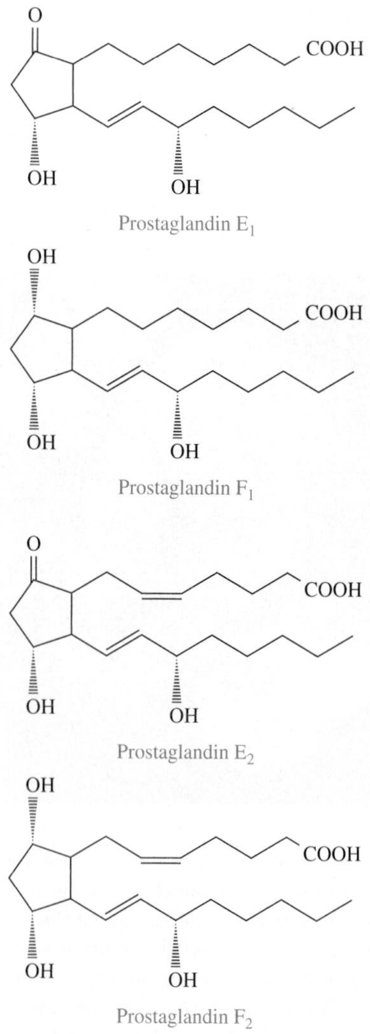

It is intriguing to note that the omega-3 fatty acids are precursors of prostaglandins that exhibit anti-inflammatory effects, and the omega-6 fatty acids are precursors of prostaglandins that have inflammatory effects. This has led researchers to suggest that the amount of omega-6 fatty acids in our diets should not exceed 2–4 times the amount of omega-3. In fact, in the United States, the diet often contains 10–30 times more omega-6 fatty acids than omega-3 fatty acids. Thus, changing the ratio of the two in the diet could increase the levels of anti-inflammatory prostaglandins and reduce the level of inflammatory prostaglandins. To encourage this dietary change, the National Institutes of Health have issued recommended daily intakes of four of these fatty acids: 650 mg/day of EPA and DHA, 2.22 g/day of α-linolenic acid, and 4.44 g/day of linoleic acid.

Eicosanoids: Prostaglandins, Leukotrienes, and Thromboxanes

Some of the unsaturated fatty acids containing more than one double bond cannot be synthesized by the body. For many years, it has been known that linolenic acid, also called α-linolenic acid to distinguish it from isomeric forms, and linoleic acid, called the **essential fatty acids**, are necessary for specific biochemical functions and must be supplied in the diet (see Table 17.1). The function of linoleic acid became clear in the 1960s when it was discovered that linoleic acid is required for the biosynthesis of **arachidonic acid**, the precursor of a class of hormonelike molecules known as **eicosanoids**. The name is derived from the Greek word *eikos*, meaning "twenty," because they are all derivatives of 20-carbon fatty acids. The eicosanoids include three groups of structurally related compounds: the prostaglandins, the leukotrienes, and the thromboxanes.

Prostaglandins are extremely potent biological molecules with hormonelike activity. They got their name because they were originally isolated from seminal fluid produced in the prostate gland. More recently, they also have been isolated from most animal tissues. Prostaglandins are unsaturated carboxylic acids consisting of a 20-carbon skeleton that contains a five-carbon ring.

Several general classes of prostaglandins are grouped under the designations A, B, E, and F, among others. The nomenclature of prostaglandins is based on the arrangement of the carbon skeleton and the number and orientation of double bonds, hydroxyl groups, and ketone groups. For example, in the name PGF$_2$, PG stands for prostaglandin, F indicates a particular group of prostaglandins with a hydroxyl group bonded to carbon-9, and 2 indicates that there are two carbon-carbon double bonds in the compound. The examples in Figure 17.3 illustrate the general structure of prostaglandins and the current nomenclature system.

Prostaglandin E$_1$

Prostaglandin F$_1$

Prostaglandin E$_2$

Prostaglandin F$_2$

Figure 17.3 The structures of four prostaglandins.

Prostaglandins are made in most tissues, and exert their biological effects both on the cells that produce them and on other cells in the immediate vicinity. Because the prostaglandins and the closely related leukotrienes and thromboxanes affect so many body processes and because they often cause opposing effects in different tissues, it can be difficult to keep track of their many regulatory functions. The following is a brief summary of some of the biological processes that are thought to be regulated by the prostaglandins, leukotrienes, and thromboxanes.

- **Blood clotting.** Blood clots form when a blood vessel is damaged, yet such clotting along the walls of undamaged vessels could result in heart attack or stroke. *Thromboxane A_2* (Figure 17.4) is produced by platelets in the blood and stimulates constriction of the blood vessels and aggregation of the platelets. Conversely, PGI_2 (prostacyclin) is produced by the cells lining the blood vessels and has precisely the opposite effect of thromboxane A_2. Prostacyclin inhibits platelet aggregation and causes dilation of blood vessels and thus prevents the untimely production of blood clots.

- **The inflammatory response.** The inflammatory response is another of the body's protective mechanisms. When tissue is damaged by mechanical injury, burns, or invasion by microorganisms, a variety of white blood cells descend on the damaged site to try to minimize the tissue destruction. The result of this response is swelling, redness, fever, and pain. Prostaglandins are thought to promote certain aspects of the inflammatory response, especially pain and fever. Drugs such as aspirin block prostaglandin synthesis and help to relieve the symptoms.

- **Reproductive system.** PGE_2 stimulates smooth muscle contraction, particularly uterine contractions. An increase in the level of prostaglandins has been noted immediately before the onset of labor. PGE_2 has also been used to induce second trimester abortions. There is strong evidence that dysmenorrhea (painful menstruation) suffered by many women may be the result of an excess of two prostaglandins. Indeed, drugs such as ibuprofen that inhibit prostaglandin synthesis have been approved by the Food and Drug Administration (FDA) and are found to provide relief from these symptoms.

- **Gastrointestinal tract.** Prostaglandins have been shown to both inhibit the secretion of acid and increase the secretion of a protective mucus layer into the stomach. In this way, prostaglandins help to protect the stomach lining. Because aspirin inhibits prostaglandin synthesis, prolonged use may actually encourage stomach ulcers by inhibiting the formation of the normal protective mucus layer, while simultaneously allowing increased secretion of stomach acid.

- **Kidneys.** Prostaglandins produced in the kidneys cause the renal blood vessels to dilate. The greater flow of blood through the kidney results in increased water and electrolyte excretion.

- **Respiratory tract.** Eicosanoids produced by certain white blood cells, the *leukotrienes* (see Figure 17.4), promote the constriction of the bronchi associated with asthma. Other prostaglandins promote bronchodilation.

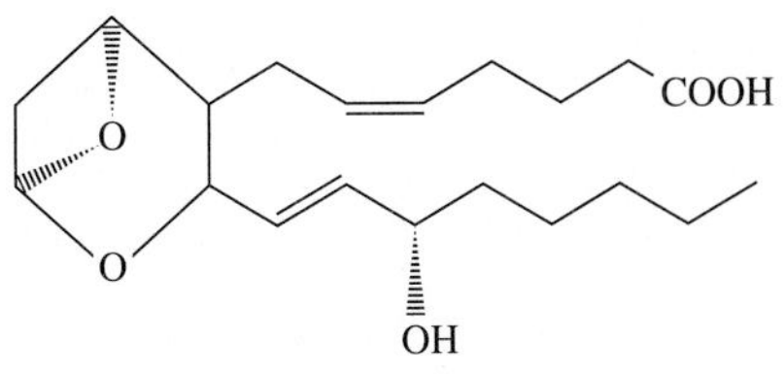

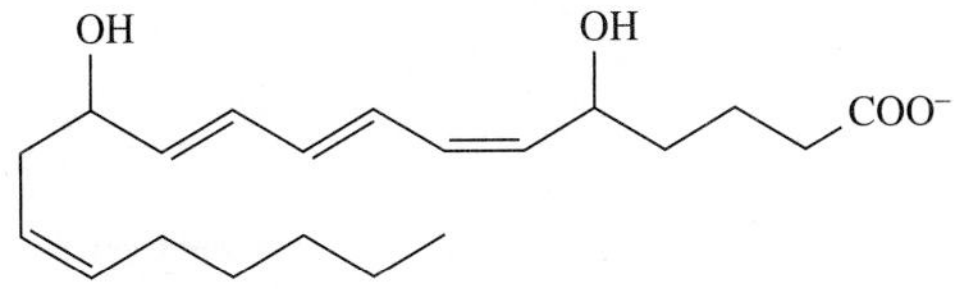

As this brief survey suggests, the prostaglandins have numerous, often antagonistic effects. Although they do not fit the formal definition of a hormone (a substance produced in a specialized tissue and transported by the circulatory system to target tissues *elsewhere* in the body), the prostaglandins are clearly strong biological regulators with far-reaching effects.

As mentioned, prostaglandins stimulate the inflammatory response and, as a result, are partially responsible for the cascade of events that cause pain. Aspirin has long been known to alleviate such pain, and we now know that it does so by inhibiting the synthesis of prostaglandins (Figure 17.5).

The first two steps of prostaglandin synthesis (Figure 17.6), the release of arachidonic acid from the membrane and its conversion to PGH_2 by the enzyme cyclooxygenase, occur in all tissues that are able to produce prostaglandins.

Figure 17.4 The structures of thromboxane A_2 and leukotriene B_4.

A hormone is a chemical signal that is produced by a specialized tissue and is carried by the bloodstream to send a message to target tissues. Eicosanoids are referred to as hormonelike because they affect the cells that produce them, as well as other target tissues.

LEARNING GOAL

5 Discuss the mechanism by which aspirin reduces pain.

Figure 17.5 Aspirin inhibits the synthesis of prostaglandins by acetylating the enzyme cyclooxygenase. The acetylated enzyme is no longer functional.

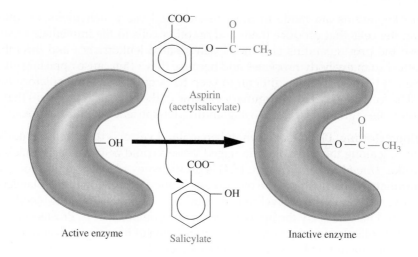

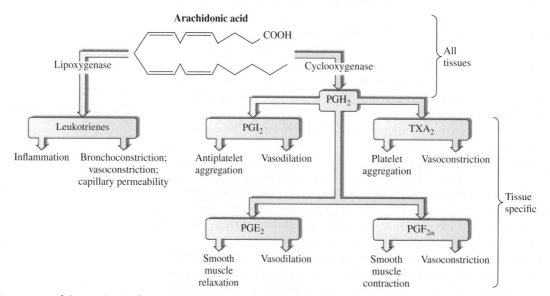

Figure 17.6 A summary of the synthesis of several prostaglandins from arachidonic acid.

The conversion of PGH_2 into the other biologically active forms is tissue-specific and requires the appropriate enzymes, which are found only in certain tissues.

Aspirin works by inhibiting the cyclooxygenase, which catalyzes the first step in the pathway leading from arachidonic acid to PGH_2. The acetyl group of aspirin becomes covalently bound to the enzyme, thereby inactivating it (Figure 17.5). Because the reaction catalyzed by cyclooxygenase occurs in all cells, aspirin effectively inhibits synthesis of all of the prostaglandins.

17.3   Glycerides

Neutral Glycerides

An ester is formed in the reaction between an alcohol and a carboxylic acid. See also Section 14.2.

Glycerides are lipid esters that contain the glycerol molecule and fatty acids. They may be subdivided into two classes: neutral glycerides and phosphoglycerides. Neutral glycerides are nonionic and nonpolar. Phosphoglyceride molecules have a polar region, the phosphoryl group, in addition to the nonpolar fatty acid tails. The structures of each of these types of glycerides are critical to their function.

The esterification of glycerol with a fatty acid produces a **neutral glyceride**. Esterification may occur at one, two, or all three positions, producing **monoglycerides**, **diglycerides**, or **triglycerides**. You will also see these referred to as *mono-*, *di-*, or *triacylglycerols*.

EXAMPLE 17.2 **Writing an Equation for the Synthesis of a Monoglyceride**

LEARNING GOAL

Write a general equation for the esterification of glycerol and one fatty acid.

6 Write equations representing the reactions that fatty acids and glycerides undergo.

Solution

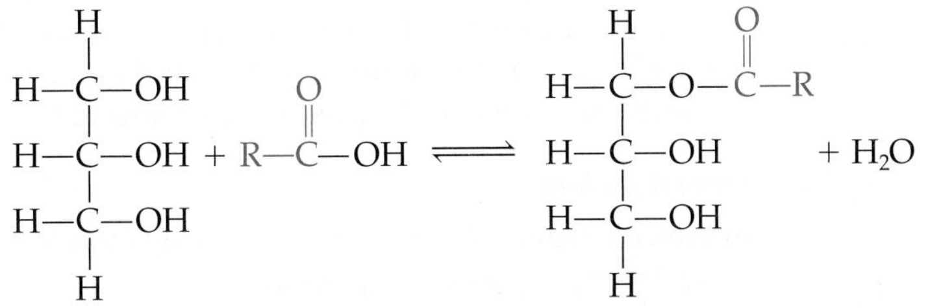

Glycerol   Fatty acid   Monoglyceride   Water

Practice Problem 17.2

Write equations for the following esterification reactions.

a. Glycerol with two molecules of stearic acid
b. Glycerol with one molecule of myristic acid

▶ For Further Practice: **Questions 17.41 and 17.42.**

Although monoglycerides and diglycerides are present in nature, the most important neutral glycerides are the triglycerides, the major component of fat cells. The triglyceride consists of a glycerol backbone (shown in black) joined to three fatty acid units through ester bonds (shown in red). The formation of a triglyceride is shown in the following equation:

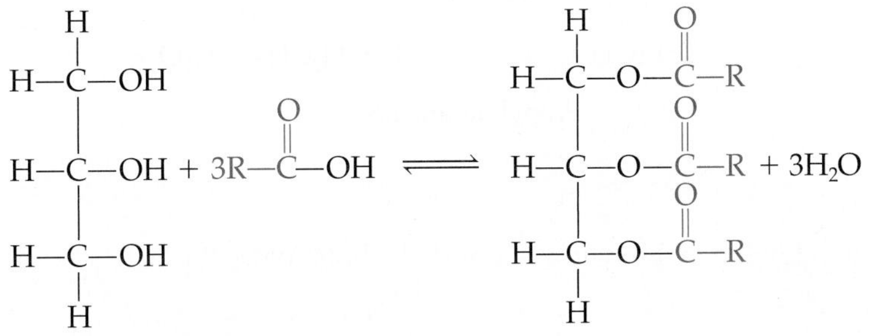

Glycerol   Fatty acids   Triglyceride   Water

Because there are no charges (+ or −) on these molecules, they are called *neutral glycerides*. These long molecules readily stack with one another and constitute the majority of the lipids stored in the body's fat cells.

The principal function of triglycerides in biochemical systems is the storage of energy. If more energy-rich nutrients are consumed than are required for metabolic processes, much of the excess is converted to neutral glycerides and stored as triglycerides in fat cells of *adipose tissue*. When energy is needed, the triglycerides are metabolized by the body, and energy is released. For this reason, exercise, along with moderate reduction in caloric intake, is recommended for overweight individuals. Exercise, an energy-demanding process, increases the rate of metabolism of fats and results in weight loss.

Lipid metabolism is discussed in Chapter 23.

See A Human Perspective: Losing Those Unwanted Pounds of Adipose Tissue in Chapter 23.

Chemical Reactions of Fatty Acids and Glycerides

The reactions of fatty acids are identical to those of short-chain carboxylic acids. The major reactions of fatty acids include the following:

- *Esterification:* The reaction between the carboxyl group of a fatty acid and the hydroxyl group of an alcohol.
- *Addition at the double bond:* This generally involves the addition of hydrogen (H_2) to the double bond of an unsaturated fatty acid.

The major reactions of glycerides are

- *Acid hydrolysis:* Breaking the ester bond of a glyceride by the addition of a water molecule in the presence of a strong acid.
- *Saponification:* Breaking the ester bond of a glyceride by the addition of a water molecule in the presence of a strong base.

Esterification

In **esterification**, fatty acids react with alcohols to form esters and water according to the following general equation:

$$R^1{-}\overset{\overset{\displaystyle O}{\|}}{C}{-}OH + R^2OH \xrightarrow{H^+,\ heat} R^1{-}\overset{\overset{\displaystyle O}{\|}}{C}{-}OR^2 + H{-}OH$$

Fatty acid   Alcohol   Ester   Water

EXAMPLE 17.3   **Writing Equations Representing the Esterification of Fatty Acids**

Write an equation representing the esterification of capric acid with propyl alcohol, and write the IUPAC name of each of the organic reactants and products.

Solution

$$CH_3(CH_2)_8{-}\overset{\overset{\displaystyle O}{\|}}{C}{-}OH + CH_3CH_2CH_2OH \xrightarrow{H^+,\ heat}$$

Decanoic acid   1-Propanol

$$CH_3(CH_2)_8{-}\overset{\overset{\displaystyle O}{\|}}{C}{-}O{-}CH_2CH_2CH_3 + H_2O$$

Propyl decanoate

Practice Problem 17.3

Write an equation for the esterification of the following carboxylic acids and alcohols. Write the IUPAC names for all of the organic reactants and products.

    a. Lauric acid and ethyl alcohol
    b. Palmitic acid and 1-pentanol

▶ For Further Practice: **Questions 17.31 and 17.32.**

Reaction at the Double Bond (Unsaturated Fatty Acids)

Hydrogenation is an example of an addition reaction. The following is a typical example of the addition of hydrogen to the double bonds of a fatty acid:

$$CH_3(CH_2)_4CH{=}CHCH_2CH{=}CH(CH_2)_7COOH \xrightarrow{2H_2,\ Ni} CH_3(CH_2)_{16}COOH$$

Linoleic acid   Stearic acid

<table>
<tr><td>EXAMPLE 17.4</td><td>Writing Equations for the Hydrogenation
of a Fatty Acid</td></tr>
</table>

Write an equation representing the hydrogenation of oleic acid and write the IUPAC name of each of the organic reactants and products.

Solution

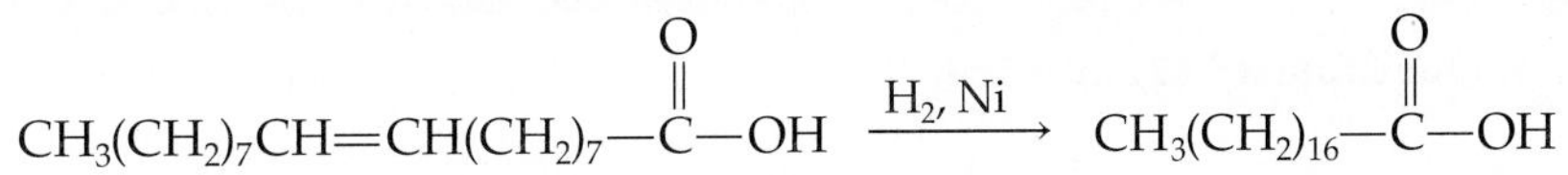

$$CH_3(CH_2)_7CH{=}CH(CH_2)_7{-}\overset{\overset{\displaystyle O}{\|}}{C}{-}OH \xrightarrow{H_2,\ Ni} CH_3(CH_2)_{16}{-}\overset{\overset{\displaystyle O}{\|}}{C}{-}OH$$

cis-9-Octadecenoic acid   Octadecanoic acid

Practice Problem 17.4

Write balanced equations showing the hydrogenation of (a) *cis*-9-hexadecenoic acid and (b) arachidonic acid.

▶ For Further Practice: **Questions 17.37 and 17.38.**

Hydrogenation is used in the food industry to convert polyunsaturated vegetable oils into saturated solid fats. *Partial hydrogenation* is carried out to add hydrogen to some, but not all, double bonds in polyunsaturated oils. In this way, liquid vegetable oils are converted into solid form. Crisco is one example of a hydrogenated vegetable oil.

Margarine is also produced by partial hydrogenation of vegetable oils, such as corn oil or soybean oil. The extent of hydrogenation is carefully controlled so that the solid fat will be spreadable and have the consistency of butter when eaten. If too many double bonds were hydrogenated, the resulting product would have the undesirable consistency of animal fat. Artificial color is added to the product, and it may be mixed with milk to produce a butterlike appearance and flavor.

Hydrogenation of vegetable oils produces a mixture of *cis* and *trans* unsaturated fatty acids. The *trans* unsaturated fatty acids are thought to contribute to atherosclerosis (hardening of the arteries).

Acid Hydrolysis

Recall that hydrolysis is the reverse of esterification, producing fatty acids from esters. In acid hydrolysis of an ester, the products are a carboxylic acid (fatty acid, in this case) and an alcohol:

Acid hydrolysis is discussed in Section 14.2.

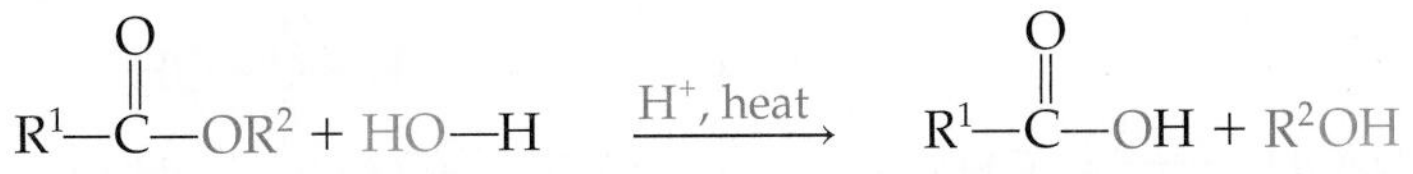

$$R^1{-}\overset{\overset{\displaystyle O}{\|}}{C}{-}OR^2 + HO{-}H \xrightarrow{H^+,\ heat} R^1{-}\overset{\overset{\displaystyle O}{\|}}{C}{-}OH + R^2OH$$

Ester   Water   Fatty acid   Alcohol

<table>
<tr><td>EXAMPLE 17.5</td><td>Writing Equations Representing the Acid Hydrolysis
of a Monoglyceride</td></tr>
</table>

Write an equation representing the acid hydrolysis of a monoglyceride composed of one decanoic acid molecule esterified to glycerol.

Solution

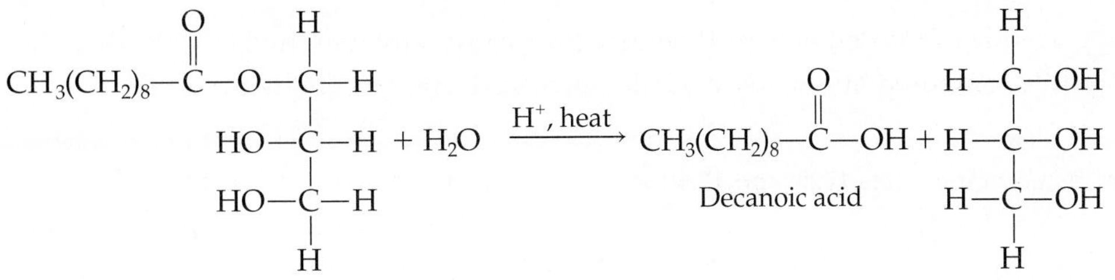

$$
CH_3(CH_2)_8{-}\overset{\overset{\displaystyle O}{\|}}{C}{-}O{-}\overset{\overset{\displaystyle H}{|}}{\underset{|}{C}}{-}H
$$

$$
\begin{array}{c}
HO{-}\overset{|}{\underset{|}{C}}{-}H \\
HO{-}\overset{|}{\underset{|}{C}}{-}H \\
\overset{|}{H}
\end{array}
+ H_2O \xrightarrow{H^+,\ heat}
CH_3(CH_2)_8{-}\overset{\overset{\displaystyle O}{\|}}{C}{-}OH +
\begin{array}{c}
H \\
H{-}\overset{|}{\underset{|}{C}}{-}OH \\
H{-}\overset{|}{\underset{|}{C}}{-}OH \\
H{-}\overset{|}{\underset{|}{C}}{-}OH \\
H
\end{array}
$$

Decanoic acid   1,2,3-Propanetriol
(glycerol)

Continued…

Practice Problem 17.5

Write a complete equation for the acid hydrolysis of the following monoglycerides.

a. Tetradecanoic acid esterified to glycerol

b. Dodecanoic acid esterified to glycerol

▶ For Further Practice: **Questions 17.33 and 17.34.**

Saponification is described in Section 14.2.

Saponification

Saponification is the base-catalyzed hydrolysis of an ester. In saponification, the products are a carboxylic acid salt (fatty acid salt, in this case) and an alcohol:

$$R^1-\overset{\overset{\displaystyle O}{\|}}{C}-OR^2 + NaOH \longrightarrow R^1-\overset{\overset{\displaystyle O}{\|}}{C}-O^-Na^+ + R^2OH$$

Ester   Base   Salt   Alcohol

The role of soaps in the removal of dirt and grease is described in Section 14.2.

Examples of micelles are shown in Figures 14.4 and 23.1.

The long-chain carboxylic acid salt or fatty acid salt that is the product of this reaction is a soap. Because soaps have a long uncharged hydrocarbon tail and a negatively charged terminus (the carboxylate group), they form micelles that dissolve oil and dirt particles. Thus, the dirt is emulsified and broken into small particles, and can be rinsed away.

EXAMPLE 17.6 **Writing Equations Representing the Base-Catalyzed Hydrolysis of a Monoglyceride**

LEARNING GOAL

Write an equation representing the base-catalyzed hydrolysis of a monoglyceride composed of dodecanoic acid and glycerol.

6 Write equations representing the reactions that fatty acids and glycerides undergo.

Solution

$$
\begin{array}{c}
CH_3(CH_2)_{10}-\overset{\overset{\displaystyle O}{\|}}{C}-O-\overset{\overset{\displaystyle H}{|}}{\underset{|}{C}}-H \\
HO-\overset{|}{\underset{|}{C}}-H + NaOH \\
HO-\overset{|}{\underset{|}{C}}-H \\
H
\end{array}
\longrightarrow
CH_3(CH_2)_{10}-\overset{\overset{\displaystyle O}{\|}}{C}-O^-Na^+ \;+\;
\begin{array}{c}
H \\
H-\overset{|}{\underset{|}{C}}-OH \\
H-\overset{|}{\underset{|}{C}}-OH \\
H-\overset{|}{\underset{|}{C}}-OH \\
H
\end{array}
$$

Sodium dodecanoate

1,2,3-Propanetriol
(glycerol)

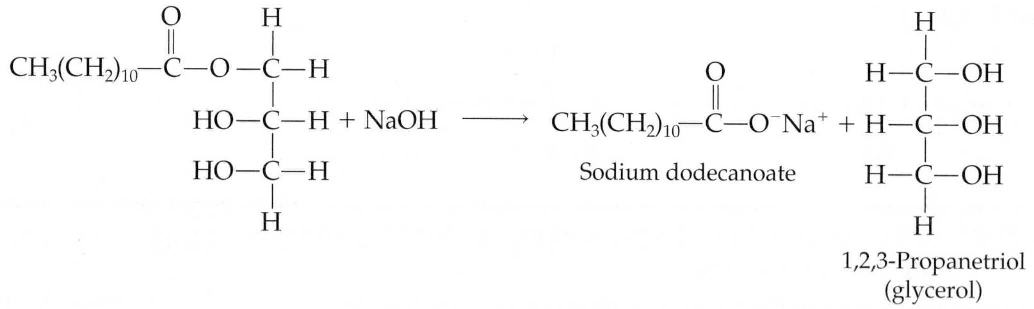

Practice Problem 17.6

Write a complete equation for the following reactions.

a. A monoglyceride composed of hexadecanoic acid and glycerol with KOH

b. A diglyceride composed of octanoic acid, decanoic acid, and glycerol with NaOH

▶ For Further Practice: **Questions 17.39 and 17.40.**

Problems can arise when "hard" water is used for cleaning because the high concentrations of Ca^{2+} and Mg^{2+} in such water cause fatty acid salts to precipitate.

Not only does this interfere with the emulsifying action of the soap, it also leaves a hard scum on the surface of sinks and tubs.

$$2R-\overset{\overset{\textstyle O}{\|}}{C}-O^- + Ca^{2+} \longrightarrow (R-\overset{\overset{\textstyle O}{\|}}{C}-O^-)_2Ca^{2+}(s)$$

Question 17.3 Write the complete equation for the esterification of myristic acid and ethyl alcohol. Write the IUPAC name for each of the organic reactants and products.

Question 17.4 Write the complete equation for the esterification of arachidic acid and ethyl alcohol. Write the IUPAC name for each of the organic reactants and products.

Question 17.5 Write a complete equation for the acid hydrolysis of pentyl butyrate. Write the IUPAC name for each of the organic reactants and products.

Question 17.6 Write a complete equation for the acid hydrolysis of butyl acetate. Write the IUPAC name for each of the organic reactants and products.

Question 17.7 Write a complete equation for the reaction of butyl acetate and KOH. Write the IUPAC name for each of the organic reactants and products.

Question 17.8 Write a complete equation for the reaction of methyl butyrate and NaOH. Write the IUPAC name for each of the organic reactants and products.

Question 17.9 Write a balanced equation for the hydrogenation of linolenic acid.

Question 17.10 Write a balanced equation for the hydrogenation of 2-hexenoic acid.

Phosphoglycerides

Phospholipids are a group of lipids that are phosphate esters. The presence of the phosphoryl group results in a molecule with a polar head (the phosphoryl group) and a nonpolar tail (the alkyl chain of the fatty acid). Because the phosphoryl group ionizes in solution, a charged lipid results.

The most abundant membrane lipids are derived from glycerol-3-phosphate and are known as **phosphoglycerides** or *phosphoacylglycerols*. Phosphoglycerides contain acyl groups derived from long-chain fatty acids at C-1 and C-2 of glycerol-3-phosphate. At C-3, the phosphoryl group is joined to glycerol by a phosphoester bond. The simplest phosphoglyceride contains a free phosphoryl group and is known as a **phosphatidate** (Figure 17.7). When the phosphoryl group is attached to another hydrophilic molecule, a more complex phosphoglyceride is formed. For example, *phosphatidylcholine* (*lecithin*) and *phosphatidylethanolamine* (*cephalin*) are found in the membranes of most cells (Figure 17.7).

Lecithin possesses a polar "head" and a nonpolar "tail." Thus, it is an *amphipathic* molecule. This structure is similar to that of soap and detergent molecules, discussed earlier. The ionic "head" is hydrophilic and interacts with water molecules, whereas the nonpolar "tail" is hydrophobic and interacts with nonpolar molecules. This amphipathic nature is central to the structure and function of cell membranes.

In addition to being a component of cell membranes, lecithin is the major phospholipid in pulmonary surfactant. It is also found in egg yolks and soybeans and is used as an emulsifying agent in ice cream. An **emulsifying agent** aids in the suspension of triglycerides in water. The amphipathic lecithin serves as a bridge, holding together the highly polar water molecules and the nonpolar triglycerides. Emulsification occurs because the hydrophilic head of lecithin dissolves in water and its hydrophobic tail dissolves in the triglycerides.

LEARNING GOAL

7 Draw the structure of a phospholipid and discuss its amphipathic nature.

Phosphoesters are described in Section 14.4.

See A Medical Perspective: Lifesaving Lipids at the beginning of this chapter.

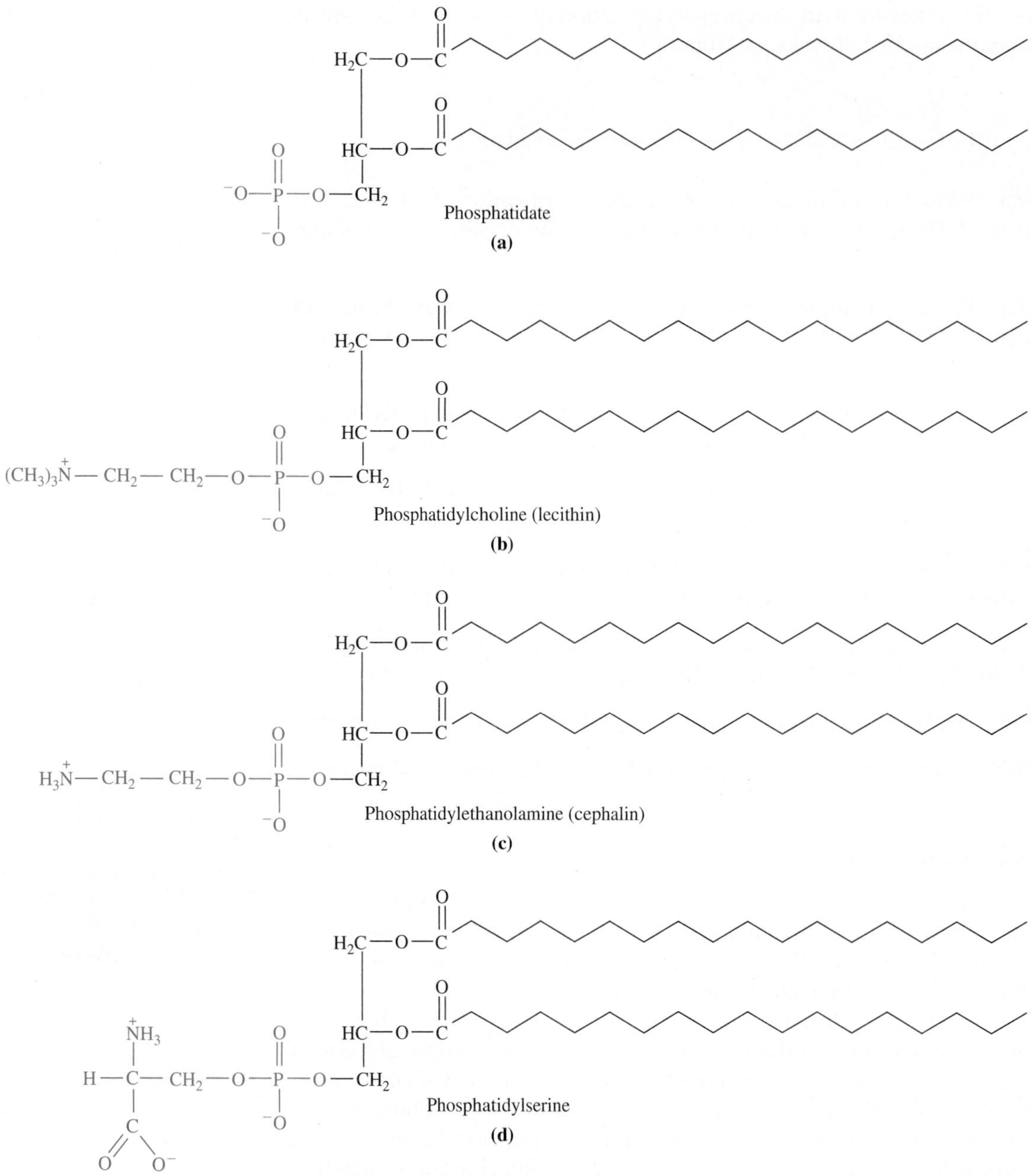

Figure 17.7 The structures of (a) phosphatidate and the common membrane phospholipids, (b) phosphatidylcholine (lecithin), (c) phosphatidylethanolamine (cephalin), and (d) phosphatidylserine.

Cephalin is similar in general structure to lecithin; the amine group bonded to the phosphoryl group is the only difference.

Question 17.11 Using condensed formulas, draw the mono-, di-, and triglycerides that would result from the esterification of glycerol with each of the following fatty acids.

    a. Oleic acid

    b. Capric acid

Question 17.12 Using condensed formulas, draw the mono-, di-, and triglycerides that would result from the esterification of glycerol with each of the following fatty acids.

    a. Palmitic acid

    b. Lauric acid

Chemistry at the Crime Scene

Adipocere and Mummies of Soap

One November evening in 1911, widower Patrick Higgins stepped into his local pub in Abercorn, Scotland. To the surprise of his drinking companions, he did not have his two young boys with him. Neighbors knew that the boys had been a great burden on Patrick since the death of his wife, so they believed Patrick's story that he left William, age 6, and John, age 4, with two women in Edinburgh who had offered to adopt the boys.

More than 18 months had passed when an object was seen floating in the Hopetoun Quarry, an unused, flooded quarry near town. When the object was fished out, it was obvious that it was the body of a young boy; the rescuers were stunned to find another small body tied to the first by a rope. How were these bodies preserved after such a long time and why did they float? The answer is that their bodies had almost completely turned into adipocere, or more simply, soap.

Forensic scientists are trying to understand the nature of the reaction that creates *adipocere*, the technical term for the yellowish-white, greasy, waxlike substance that results from the saponification of fatty tissue. Some researchers hope that this information may allow determination of the postmortem interval (length of time since death). Others simply value the process because it helps preserve the body so well that even after long periods, it can be easily recognized and any wounds or injuries can be observed.

It is known that adipocere is produced when body fat is hydrolyzed (water is needed) to release fatty acids. Because the fatty acids lower the pH in the tissues, they inhibit many of the bacteria that would begin the process of decay. Certain other bacteria, particularly *Clostridium welchii*, an organism that cannot grow in the presence of oxygen, is known to speed up the formation of adipocere in moist, warm, anaerobic (oxygenless) environments. Adipocere forms first in subcutaneous tissues, including the cheeks, breasts, and buttocks. Given appropriate warmth and damp conditions, it may be seen as early as 3 to 4 weeks after death, but more commonly it is not observed until 5 to 6 months after death.

Adipocere formation in John and William Higgins was so extensive that their former neighbors had no trouble recognizing them. At the postmortem, another advantage of adipocere formation became obvious—it had preserved the stomach contents of the boys! From this, the coroner learned that the boys had eaten Scotch broth about an hour before they died. Investigators were able to find the woman who had given the broth to the boys and, from her testimony, learned that she had fed them on the last day they were seen in the village. Clearly, their father had lied about the adoption by two Edinburgh women! In under one and one-half hours, a jury convicted the father of murdering his sons and he was hanged in October 1913.

For Further Understanding

▶ Adipocere is the technical term for "soap mummification." It comes from the Latin words *adipis* or fat, as in adipose tissue, and *cera*, which means wax. Draw a triglyceride composed of the fatty acids myristic acid, stearic acid, and oleic acid. Write a balanced equation showing a possible reaction that would lead to the formation of adipocere.

▶ Forensic scientists are studying adipocere formation as a possible source of information to determine the postmortem interval (length of time since death) of bodies of murder or accident victims. Among the factors being studied are the type of soil, including pH, moisture, temperature, and presence or absence of lime. How might each of these factors influence the rate of adipocere formation and hence the determination of the postmortem interval?

17.4 Nonglyceride Lipids

Sphingolipids

Sphingolipids are lipids that are not derived from glycerol. Like phospholipids, sphingolipids are amphipathic, having a polar head group and two nonpolar hydrocarbon tails, one of which is derived from a fatty acid. Sphingolipids are structural components of cellular membranes. They are derived from sphingosine, a long-chain, nitrogen-containing (amino) alcohol:

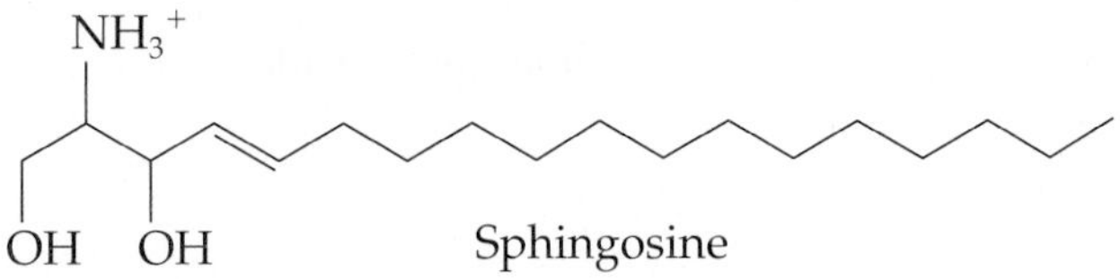

LEARNING GOAL

8 Discuss the general classes of sphingolipids and their functions.

Sphingosine

The sphingolipids include the sphingomyelins and the glycosphingolipids. The **sphingomyelins** are the only class of sphingolipids that are also phospholipids. In sphingomyelin, the acyl group of the fatty acid is bonded to sphingosine through an amide bond.

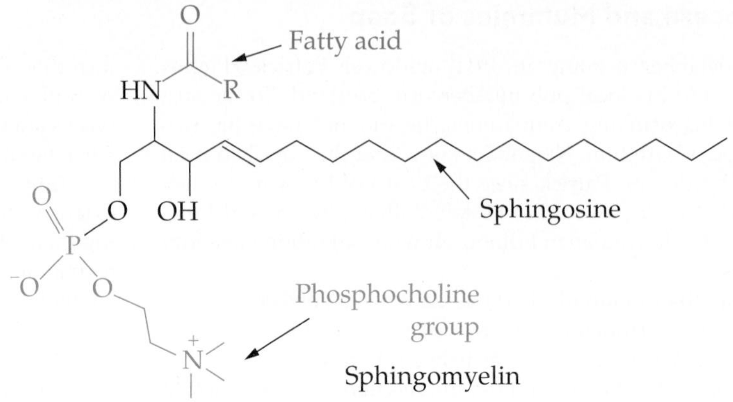

Sphingomyelin

Sphingomyelins are located throughout the body, but are particularly important structural lipid components of nerve cell membranes. They are found in abundance in the myelin sheath that surrounds and insulates cells of the central nervous system. In humans, about 25% of the lipids of the myelin sheath are sphingomyelins. Their role is essential to proper cerebral function and nerve transmission.

Glycosphingolipids, or *glycolipids,* include the cerebrosides, sulfatides, and gangliosides and are built on a ceramide backbone structure, which is a fatty acid amide derivative of sphingosine:

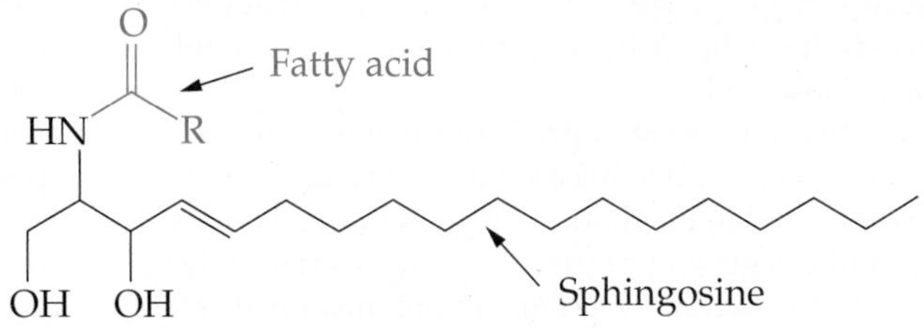

Ceramide

The *cerebrosides* are characterized by the presence of a single monosaccharide head group. Two common cerebrosides are glucocerebroside, found in the membranes of macrophages (cells that protect the body by ingesting and destroying foreign microorganisms) and galactocerebroside, found almost exclusively in the membranes of brain cells. Glucocerebroside consists of ceramide bonded to the hexose glucose; galactocerebroside consists of ceramide joined to the monosaccharide galactose.

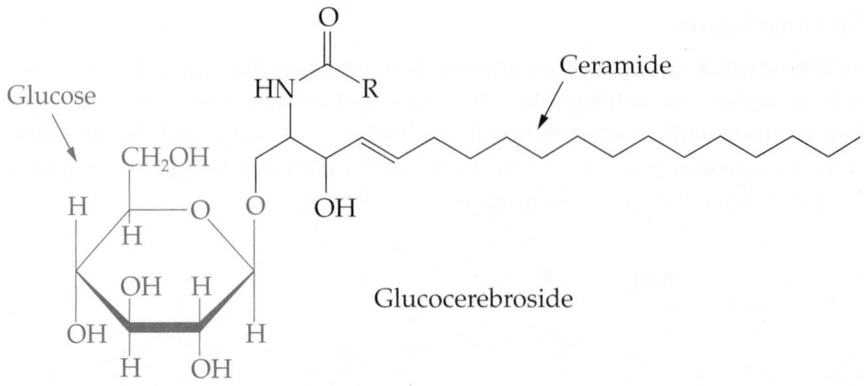

Glucocerebroside

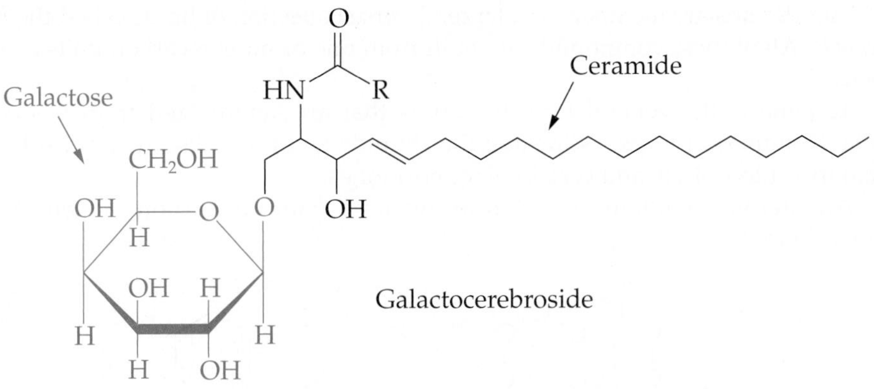

Galactocerebroside

Sulfatides are derivatives of galactocerebroside that contain a sulfate group. Notice that they carry a negative charge at physiological pH.

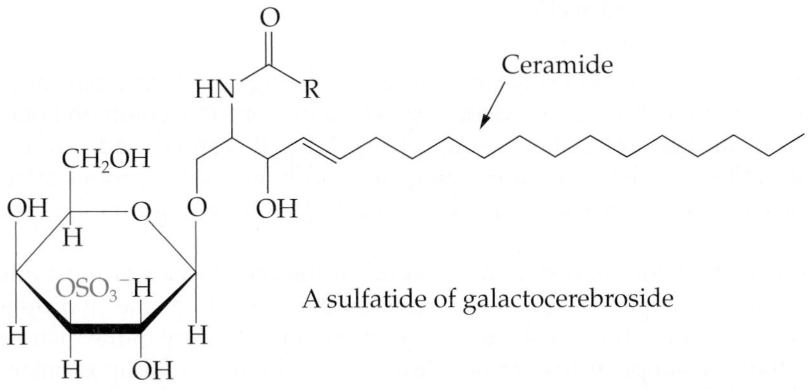

A sulfatide of galactocerebroside

Gangliosides are glycolipids that possess oligosaccharide groups rather than a single monosaccharide. These oligosaccharide groups always contain one or more molecules of *N*-acetylneuraminic acid (sialic acid). First isolated from membranes of nerve tissue, gangliosides are found in most tissues of the body.

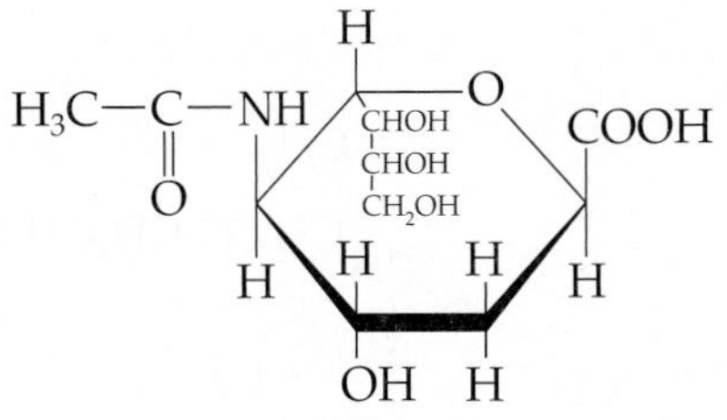

N-Acetylneuraminic acid (sialic acid)

LEARNING GOAL

9 Draw the structure of the steroid nucleus, and discuss the functions of steroid hormones.

Steroids

Steroids are a naturally occurring family of organic molecules of biochemical and medical interest. A great deal of controversy has surrounded various steroids. We worry about the amount of cholesterol in the diet and the possible health effects. We are concerned about the use of anabolic steroids by athletes wishing to build muscle mass and improve their performance. However, members of this family of molecules derived from cholesterol have many important functions in the body. The bile salts that aid in the emulsification and digestion of lipids are steroid molecules, as are the sex hormones testosterone and estrone.

Lipid digestion is described in Section 23.1.

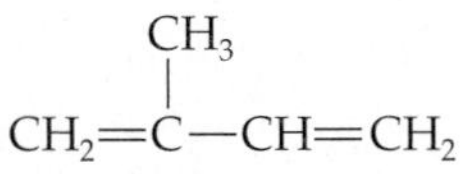

$$CH_2\!=\!\overset{\overset{\displaystyle CH_3}{|}}{C}\!-\!CH\!=\!CH_2$$

Isoprene

The steroids are members of a large, diverse collection of lipids called the *isoprenoids*. All of these compounds are built from one or more 5-carbon units called *isoprene*.

Terpene is the general term for lipids that are synthesized from isoprene units. Examples of terpenes include the steroids and bile salts, the lipid-soluble vitamins, chlorophyll, and certain plant hormones.

All steroids contain the steroid nucleus (steroid carbon skeleton) as shown here:

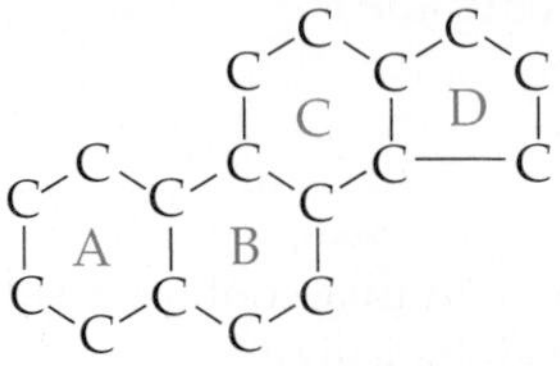

Carbon skeleton of   Steroid nucleus
the steroid nucleus

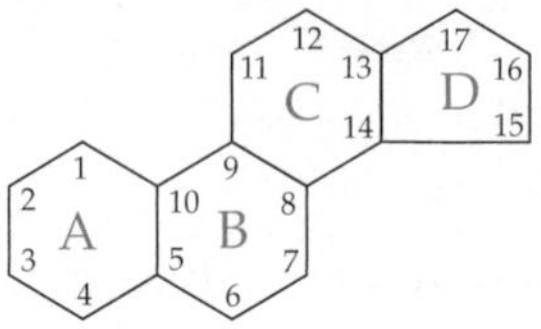

The steroid carbon skeleton consists of four fused rings. Each ring pair has two carbons in common. Thus, two fused rings share one or more common bonds as part of their ring backbones. For example, rings A and B, B and C, and C and D are all fused in the preceding structure. Many steroids have methyl groups attached to carbons 10 and 13, as well as an alkyl, alcohol, or ketone group attached to carbon-17.

Cholesterol, a common steroid, is found in the membranes of most animal cells. It is an amphipathic molecule and is readily soluble in the hydrophobic region of membranes. It is involved in regulation of the fluidity of the membrane as a result of the nonpolar fused ring. However, the hydroxyl group is polar and functions like the polar heads of sphingolipids and phospholipids. There is a strong correlation between the concentration of cholesterol found in the blood plasma and heart disease, particularly **atherosclerosis** (hardening of the arteries). Cholesterol, in combination with other substances, contributes to a narrowing of the artery passageway. As narrowing increases, more pressure is necessary to ensure adequate blood flow, and high blood pressure (*hypertension*) develops. Hypertension is also linked to heart disease.

This breakfast is high in cholesterol and saturated fats. Why do nutritionists recommend that we limit the amount of such foods in our diets?

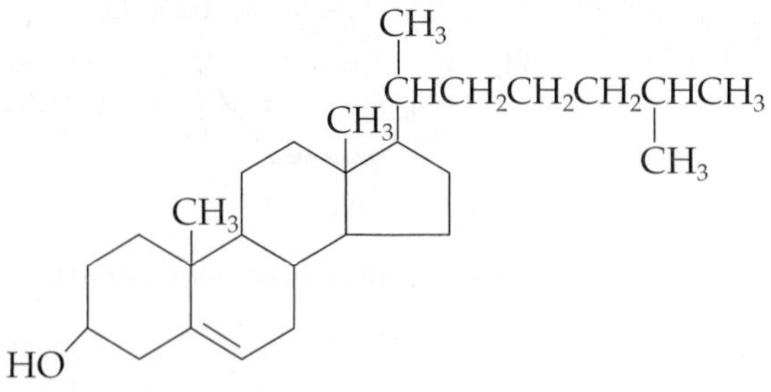

Bile salts are described in greater detail in Section 23.1.

Cholesterol

Egg yolks contain a high concentration of cholesterol, as do many dairy products and animal fats. As a result, it has been recommended that the amounts of these products in the diet be regulated to moderate the dietary intake of cholesterol.

Bile salts are amphipathic derivatives of cholesterol that are synthesized in the liver and stored in the gallbladder. The principal bile salts in humans are cholate and chenodeoxycholate.

A Medical Perspective

Disorders of Sphingolipid Metabolism

There are a number of human genetic disorders that are caused by a deficiency in one of the enzymes responsible for the breakdown of sphingolipids. In general, the symptoms are caused by the accumulation of abnormally large amounts of these lipids within particular cells. It is interesting to note that three of these diseases, Niemann-Pick disease, Gaucher's disease, and Tay-Sachs disease are found much more frequently among Ashkenazi Jews of Northern European heritage than among other ethnic groups.

Of the four subtypes of Niemann-Pick disease, type A is the most severe. It is inherited as a recessive disorder (that is, a defective copy of the gene must be inherited from each parent) that results in an absence of the enzyme sphingomyelinase. The absence of this enzyme causes the storage of large amounts of sphingomyelin and cholesterol in the brain, bone marrow, liver, and spleen.

Symptoms may begin when a baby is only a few months old. The parents may notice a delay in motor development and/or problems with feeding. Although the infants may develop some motor skills, they quickly begin to regress as they lose muscle strength and tone, as well as vision and hearing. The disease progresses rapidly and the children typically die within the first few years of life.

Tay-Sachs disease is a lipid storage disease caused by an absence of the enzyme hexosaminidase, which functions in ganglioside metabolism. As a result of the enzyme deficiency, the ganglioside accumulates in the cells of the brain, causing neurological deterioration. Like Niemann-Pick disease, it is an autosomal recessive genetic trait that becomes apparent in the first few months of the life of an infant and rapidly progresses to death within a few years. Symptoms include listlessness, irritability, seizures, paralysis, loss of muscle tone and function, blindness, deafness, and delayed mental and social skills.

Gaucher's disease is an autosomal recessive genetic disorder resulting in a deficiency of the enzyme glucocerebrosidase. In the normal situation, this enzyme breaks down glucocerebroside, which is an intermediate in the synthesis and degradation of complex glycosphingolipids found in cellular membranes. In Gaucher's disease, glucocerebroside builds up in macrophages found in the liver, spleen, and bone marrow. These cells become engorged with excess lipid and displace healthy, normal cells in bone marrow. The symptoms of Gaucher's disease include severe anemia, thrombocytopenia (reduction in the number of platelets), and hepatosplenomegaly (enlargement of the spleen and liver). There can also be skeletal problems including bone deterioration and secondary fractures.

Fabry's disease is an X-linked inherited disorder caused by the deficiency of the enzyme α-galactosidase A. This disease afflicts as many as 50,000 people worldwide. Typically, symptoms, including pain in the fingers and toes and a red rash around the waist, begin to appear when individuals reach their early twenties. A preliminary diagnosis can be confirmed by determining the concentration of the enzyme α-galactosidase A. Patients with Fabry's disease have an increased risk of kidney and heart disease, and a reduced life expectancy. Because this is an X-linked disorder, it is more common among males than females.

For Further Understanding

▸ A defect in the enzyme sphingomyelinase is the cause of Niemann-Pick disease. Write a chemical equation to represent the reaction catalyzed by the enzyme sphingomyelinase, the cleavage of sphingomyelin to produce phosphorylcholine and ceramide. Show the structural formulas for the reactant and products of this reaction.

▸ A defect in the enzyme glucocerebrosidase is the cause of Gaucher's disease. Write a chemical equation to represent the reaction catalyzed by the enzyme glucocerebrosidase, the cleavage of glucocerebroside to produce glucose and ceramide. Show the structural formulas for the reactant and products of this reaction.

Cholate   Chenodeoxycholate

Bile salts are emulsifying agents whose polar hydroxyl groups interact with water and whose hydrophobic regions bind to lipids. Following a meal, bile flows from the gallbladder to the duodenum (the uppermost region of the small intestine). Here, the bile salts emulsify dietary fats into small droplets

A Medical Perspective

Steroids and the Treatment of Heart Disease

The foxglove plant (*Digitalis purpurea*) is an herb that produces one of the most powerful known stimulants of heart muscle. The active ingredients of the foxglove plant (digitalis) are the so-called cardiac glycosides, or *cardiotonic steroids,* which include digitoxin, digosin, and gitalin.

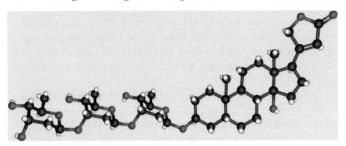

The structure of digitoxin, one of the cardiotonic steroids produced by the foxglove plant.

These drugs are used clinically in the treatment of congestive heart failure, which results when the heart is not beating with strong, efficient strokes. When the blood is not propelled through the cardiovascular system efficiently, fluid builds up in the lungs and lower extremities (edema). The major symptoms of congestive heart failure are an enlarged heart, weakness, edema, shortness of breath, and fluid accumulation in the lungs.

This condition was originally described in 1785 by a physician, William Withering, who found a peasant woman whose folk medicine was famous as a treatment for chronic heart problems. Her potion contained a mixture of more than twenty herbs, but Dr. Withering, a botanist as well as physician, quickly discovered that foxglove was the active ingredient in the mixture. Withering used *Digitalis purpurea* successfully to treat congestive heart failure and even described some cautions in its use.

The cardiotonic steroids are extremely strong heart stimulants. A dose as low as 1 milligram (mg) increases the stroke volume of the heart (volume of blood per contraction), increases the strength of the contraction, and reduces the heart rate. When the heart is pumping more efficiently because of stimulation by digitalis, the edema disappears.

Digitalis can be used to control congestive heart failure, but the dose must be carefully determined and monitored because

Digitalis purpurea, the foxglove plant.

the therapeutic dose is close to the dose that causes toxicity. The symptoms that result from high body levels of cardiotonic steroids include vomiting, blurred vision and lightheadedness, increased water loss, convulsions, and death. Only a physician can determine the initial dose and maintenance schedule for an individual to control congestive heart failure and yet avoid the toxic side effects.

For Further Understanding

▶ Foxglove is a perennial plant; that is, a plant that will grow back each year for at least 3 years. Occasionally, foxglove first-year growth has been mistaken for comfrey, another plant with medical applications. Greeks and Romans used comfrey to treat wounds and to stop heavy bleeding, as well as for bronchial problems. Explain why the use of foxglove in place of comfrey might have fatal consequences.

▶ Drugs such as digitalis are referred to as *cardiac glycosides* and as *cardiotonic steroids.* Explain why both these names are valid.

---

that can be more readily digested by lipases (lipid digesting enzymes) also found in the small intestine.

Steroids play a role in the reproductive cycle. In a series of chemical reactions, cholesterol is converted to the steroid *progesterone,* the most important hormone associated with pregnancy. Produced in the ovaries and in the placenta, progesterone is responsible for both the successful initiation and the successful completion of pregnancy. It prepares the lining of the uterus to accept the fertilized egg. Once the egg is attached, progesterone is involved in the development of the fetus and plays a role in the suppression of further ovulation during pregnancy.

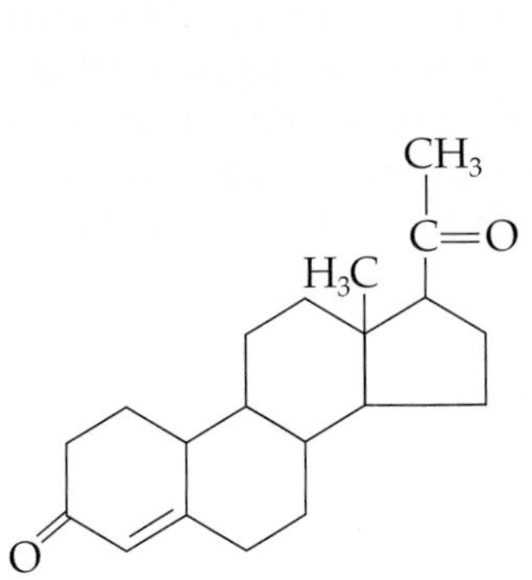

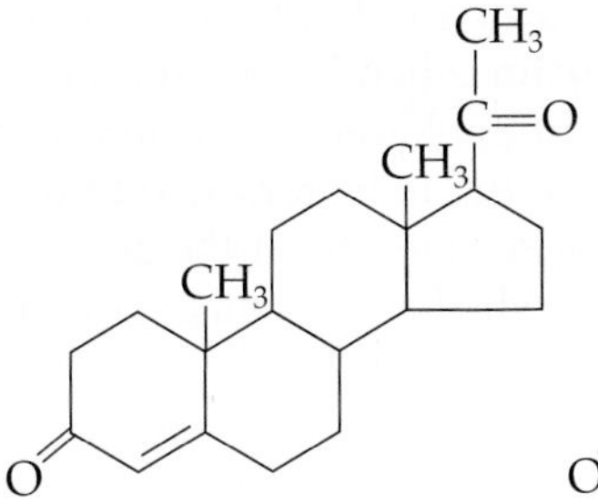

CH₃
|
C=O

H₃C

O

Progesterone

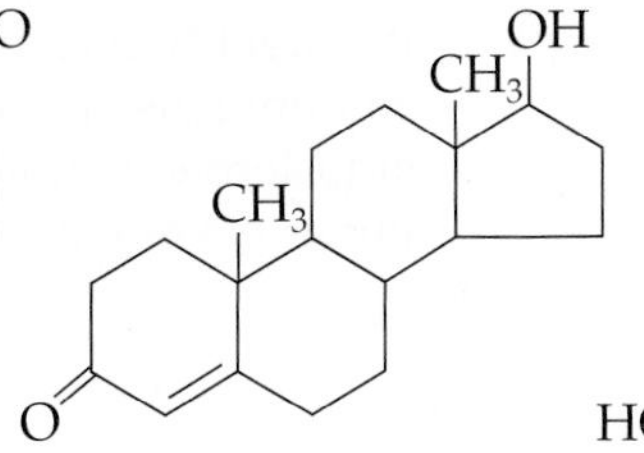

Testosterone

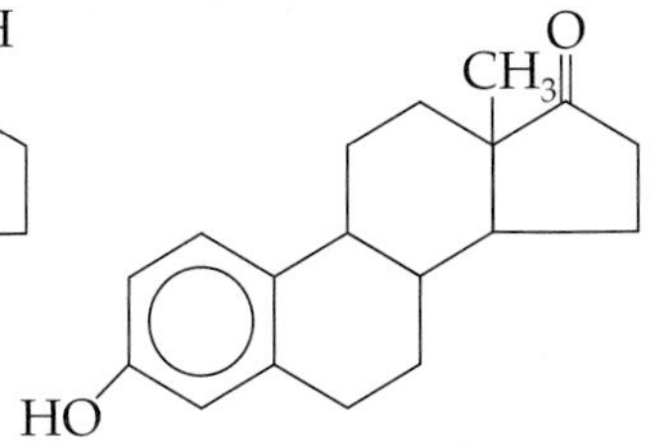

Estrone

Testosterone, a male sex hormone found in the testes, and *estrone*, a female sex hormone, are both produced by the chemical modification of progesterone. These hormones are involved in the development of male and female sex characteristics.

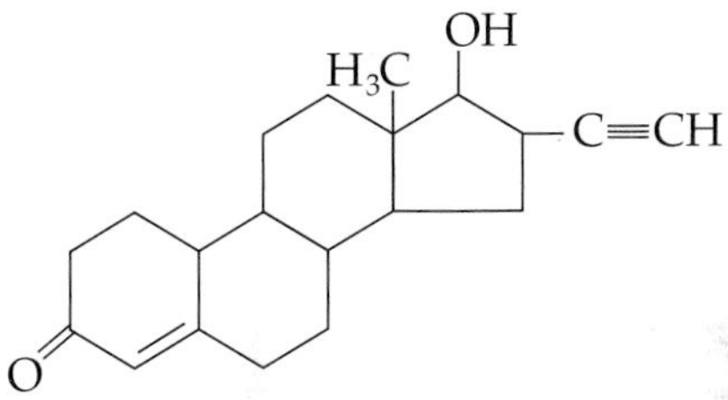

Many steroids, including progesterone, have played important roles in the development of birth control agents. 19-Norprogesterone was one of the first synthetic birth control agents. It is approximately ten times as effective as progesterone in providing birth control. However, its utility was severely limited because this compound could not be administered orally and had to be taken by injection. A related compound, norlutin (chemical name: 17-α-ethynyl-19-nortestosterone), was found to provide both the strength and the effectiveness of 19-norprogesterone and could be taken orally.

Norlutin

Currently "combination" oral contraceptives are prescribed most frequently. These include a progesterone and an estrogen. These newer products confer better contraceptive protection than either agent administered individually. They are also used to regulate menstruation in patients with heavy menstrual bleeding. First investigated in the late 1950s and approved by the FDA in 1961, there are at least thirty combination pills currently available. In addition, a transdermal patch for the treatment of postmenopausal osteoporosis is being investigated.

All of these compounds act by inducing a false pregnancy, which prevents ovulation. When oral contraception is discontinued, ovulation usually returns within three menstrual cycles. Although there have been problems associated with "the pill," it appears to be an effective and safe method of family planning for much of the population.

Cortisone is a steroid important to the proper regulation of a number of biochemical processes. For example, it is involved in the metabolism of carbohydrates. Cortisone is also used in the treatment of rheumatoid arthritis, asthma, gastrointestinal disorders, many skin conditions, and a variety of other diseases. However, treatment with cortisone is not without risk. Some of the possible side effects of cortisone therapy include fluid retention, sodium retention, and potassium loss that can lead to congestive heart failure. Other side effects include muscle weakness, osteoporosis, gastrointestinal upsets including peptic ulcers, and neurological symptoms, including vertigo, headaches, and convulsions.

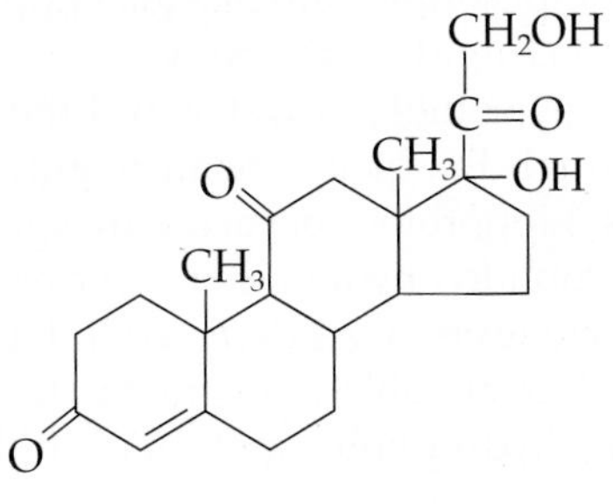

Newer birth control pill formulations include both a progesterone and an estrogen. How do these steroids prevent pregnancy?

CH₂OH
|
C=O

Cortisone

Aldosterone is a steroid hormone produced by the adrenal cortex and secreted into the bloodstream when blood sodium ion levels are too low. Upon reaching its target tissues in the kidney, aldosterone activates a set of reactions that cause sodium ions and water to be returned to the blood. If sodium levels are elevated, aldosterone is not secreted from the adrenal cortex and the sodium ions filtered out of the blood by the kidney will be excreted.

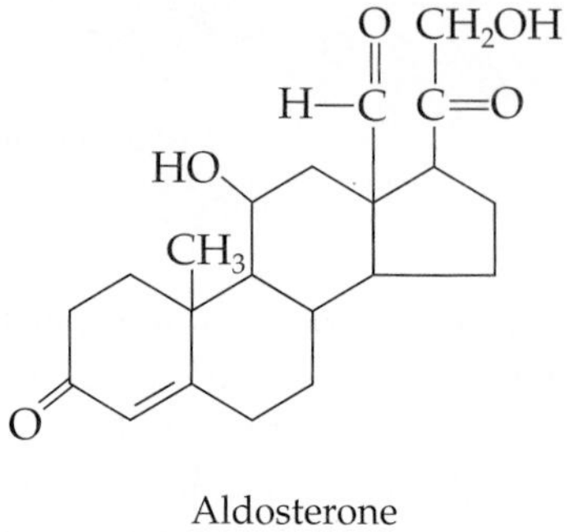

Aldosterone

Question 17.13   Draw the structure of the steroid nucleus. Note the locations of the A, B, C, and D steroid rings.

Question 17.14   What is meant by the term *fused ring?*

Waxes

Waxes are derived from many different sources and have a variety of chemical compositions, depending on the source. Paraffin wax, for example, is composed of a mixture of solid hydrocarbons (usually straight-chain compounds). The natural waxes generally are composed of a long-chain fatty acid esterified to a long-chain alcohol. Because the long hydrocarbon tails are extremely hydrophobic, waxes are completely insoluble in water. Waxes are also solid at room temperature, owing to their high molar masses. Two examples of waxes are myricyl palmitate, a major component of beeswax, and whale oil (spermaceti wax), from the head of the sperm whale, which is composed of cetyl palmitate.

Naturally occurring waxes have a variety of uses. Lanolin, which serves as a protective coating for hair and skin, is used in skin creams and ointments. Carnauba wax is used in automobile polish. Whale oil was once used as a fuel, in ointments, and in candles. However, synthetic waxes have replaced whale oil to a large extent, because of efforts to ban the hunting of whales.

$$CH_3(CH_2)_{14}-\overset{\overset{\displaystyle O}{\|}}{C}-O-(CH_2)_{29}CH_3$$

Myricyl palmitate
(beeswax)

$$CH_3(CH_2)_{14}-\overset{\overset{\displaystyle O}{\|}}{C}-O-(CH_2)_{15}CH_3$$

Cetyl palmitate
(whale oil)

17.5   Complex Lipids

Complex lipids are lipids that are bonded to other types of molecules. The most common and important complex lipids are plasma lipoproteins, which are responsible for the transport of other lipids in the body.

Lipids are only sparingly soluble in water, and the movement of lipids from one organ to another through the bloodstream requires a transport system that uses **plasma lipoproteins**. Lipoprotein particles are spheres that consist of a core of hydrophobic lipids surrounded by an outer layer or shell of amphipathic proteins, phospholipids, and cholesterol. As depicted in Figure 17.8, the amphipathic molecules of the outer surface are able to interact with the aqueous environment of the bloodstream and the hydrophobic molecules in the interior of the lipoprotein particle.

LEARNING GOAL

10   Describe the function of lipoproteins in triglyceride and cholesterol transport in the body.

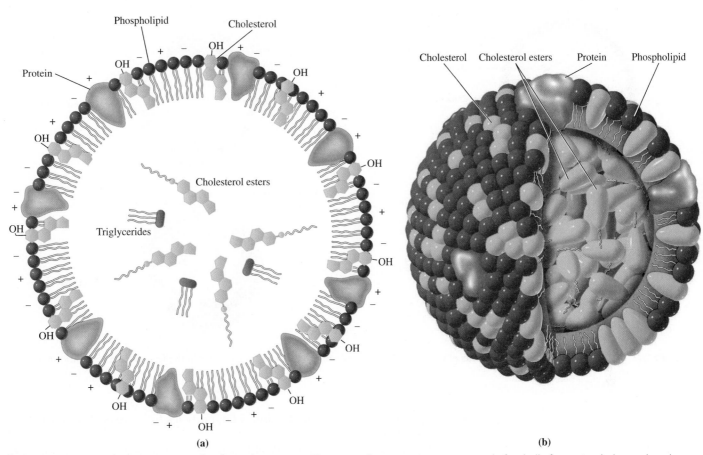

(a)   (b)

Figure 17.8 A model for the structure of a plasma lipoprotein. The various lipoproteins are composed of a shell of protein, cholesterol, and phospholipids surrounding more hydrophobic molecules such as triglycerides or cholesterol esters (cholesterol esterified to fatty acids). (a) Cross section, (b) three-dimensional view.

There are four major classes of human plasma lipoproteins:

- **Chylomicrons,** which have a density of less than 0.95 g/mL, carry dietary triglycerides from the intestine to other tissues. The remaining lipoproteins are classified by their densities.
- **Very low density lipoproteins (VLDL)** have a density of 0.95–1.019 g/mL. They bind triglycerides synthesized in the liver and carry them to adipose and other tissues for storage.
- **Low-density lipoproteins (LDL)** are characterized by a density of 1.019–1.063 g/mL. They carry cholesterol from the liver to peripheral tissues and help regulate cholesterol levels in those tissues. These are richest in cholesterol, frequently carrying 40% of the plasma cholesterol.
- **High-density lipoproteins (HDL)** have a density of 1.063–1.210 g/mL. They are bound to plasma cholesterol; however, they transport cholesterol from peripheral tissues to the liver.

A summary of the composition of each of the plasma lipoproteins is presented in Figure 17.9.

Chylomicrons are aggregates of triglycerides and protein that transport dietary triglycerides to cells throughout the body. Not all lipids in the blood are derived directly from the diet. Triglycerides and cholesterol are also synthesized in the liver and transported through the blood in lipoprotein packages. Triglycerides are assembled into VLDL particles that carry the energy-rich lipid molecules either to tissues requiring an energy source or to adipose tissue for storage.

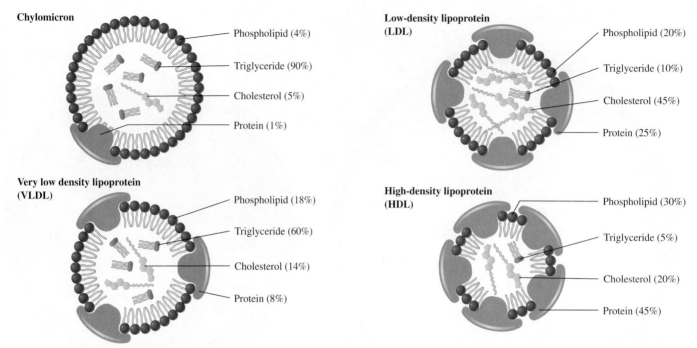

Chylomicron

Phospholipid (4%)

Triglyceride (90%)

Cholesterol (5%)

Protein (1%)

Low-density lipoprotein (LDL)

Phospholipid (20%)

Triglyceride (10%)

Cholesterol (45%)

Protein (25%)

Very low density lipoprotein (VLDL)

Phospholipid (18%)

Triglyceride (60%)

Cholesterol (14%)

Protein (8%)

High-density lipoprotein (HDL)

Phospholipid (30%)

Triglyceride (5%)

Cholesterol (20%)

Protein (45%)

Figure 17.9 A summary of the relative amounts of cholesterol, phospholipid, protein, and triglycerides in the four classes of lipoproteins.

Similarly, cholesterol is assembled into LDL particles for transport from the liver to peripheral tissues.

Entry of LDL particles into the cell is dependent on a specific recognition event and binding between the LDL particle and a protein receptor embedded within the membrane. Low-density lipoprotein receptors (LDL receptors) are found in the membranes of cells outside the liver and are responsible for the uptake of cholesterol by the cells of various tissues. LDL (lipoprotein bound to cholesterol) binds specifically to the LDL receptor, and the complex is taken into the cell by a process called *receptor-mediated endocytosis* (Figure 17.10). The membrane begins to be pulled into the cell at the site of the LDL receptor complexes. This draws the entire LDL particle into the cell. Eventually, the portion of the membrane surrounding the LDL particles pinches away from the cell membrane and forms a membrane around the LDL particles. As we will see in Section 17.6, membranes are fluid and readily flow. Thus, they can form a vesicle or endosome containing the LDL particles.

Cellular digestive organelles known as *lysosomes* fuse with the endosomes. This fusion is accomplished when the membranes of the endosome and the lysosome flow together to create one larger membrane-bound body or vesicle. Hydrolytic enzymes from the lysosome then digest the entire complex to release cholesterol into the cytoplasm of the cell. There, cholesterol inhibits its own biosynthesis and activates an enzyme that stores cholesterol in cholesterol ester droplets. High concentrations of cholesterol inside the cell also inhibit the synthesis of LDL receptors to ensure that the cell will not take up too much cholesterol. People who have a genetic defect in the gene coding for the LDL receptor do not take up as much cholesterol. As a result, they accumulate LDL cholesterol in the plasma. This excess plasma cholesterol is then deposited on the artery walls, causing atherosclerosis. This disease is called *hypercholesterolemia*.

Liver lipoprotein receptors enable large amounts of cholesterol to be removed from the blood, thus ensuring low concentrations of cholesterol in the blood plasma. Other factors being equal, the person with the most lipoprotein receptors will be the least vulnerable to a high-cholesterol diet and have the least likelihood of developing atherosclerosis.

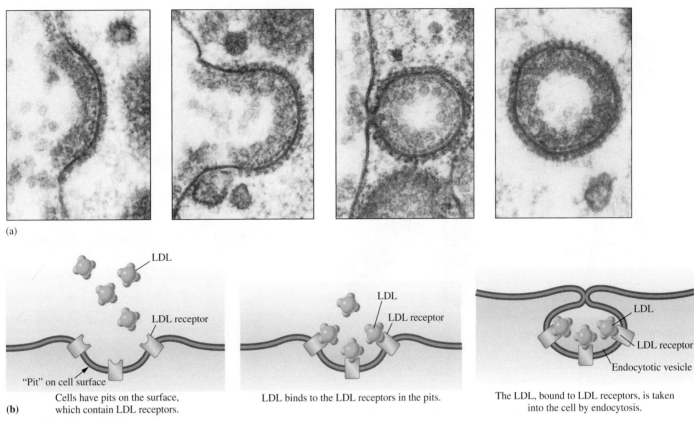

(a)

LDL

LDL receptor

"Pit" on cell surface

(b) Cells have pits on the surface, which contain LDL receptors.

LDL

LDL receptor

LDL binds to the LDL receptors in the pits.

LDL

LDL receptor

Endocytotic vesicle

The LDL, bound to LDL receptors, is taken into the cell by endocytosis.

Figure 17.10 Receptor-mediated endocytosis. (a) Electron micrographs of the process of receptor-mediated endocytosis. (b) Summary of the events of receptor-mediated endocytosis of LDL.

There is also evidence that high levels of HDL in the blood help reduce the incidence of atherosclerosis, perhaps because HDL carries cholesterol from the peripheral tissues back to the liver. In the liver, some of the cholesterol is used for bile synthesis and secreted into the intestines, from which it is excreted.

A final correlation has been made between diet and atherosclerosis. People whose diet is high in saturated fats tend to have high levels of cholesterol in the blood. Although the relationship between saturated fatty acids and cholesterol metabolism is unclear, it is known that a diet rich in unsaturated fats results in decreased cholesterol levels. In fact, the use of unsaturated fat in the diet results in a decrease in the level of LDL and an increase in the level of HDL. With the positive correlation between heart disease and high cholesterol levels, the current dietary recommendations include a diet that is low in fat and the substitution of unsaturated fats (vegetable oils) for saturated fats (animal fats).

> Recently, an inflammatory protein, the C-reactive protein (CRP), has been implicated in atherosclerosis. A test for the level of this protein in the blood is being suggested as a way to predict the risk of heart attack. A high-sensitivity CRP test is now widely available.

Question 17.15 What is the mechanism of uptake of cholesterol from plasma?

Question 17.16 What is the role of lysosomes in the metabolism of plasma lipoproteins?

17.6 The Structure of Biological Membranes

Biological membranes are *lipid bilayers* in which the hydrophobic hydrocarbon tails are packed in the center of the bilayer and the ionic head groups are exposed on the surface to interact with water (Figure 17.11). The hydrocarbon tails of membrane phospholipids provide a thin shell of nonpolar material that prevents the mixing of molecules on either side. The nonpolar tails of membrane

LEARNING GOAL

11 Draw the structure of the cell membrane and discuss its functions.

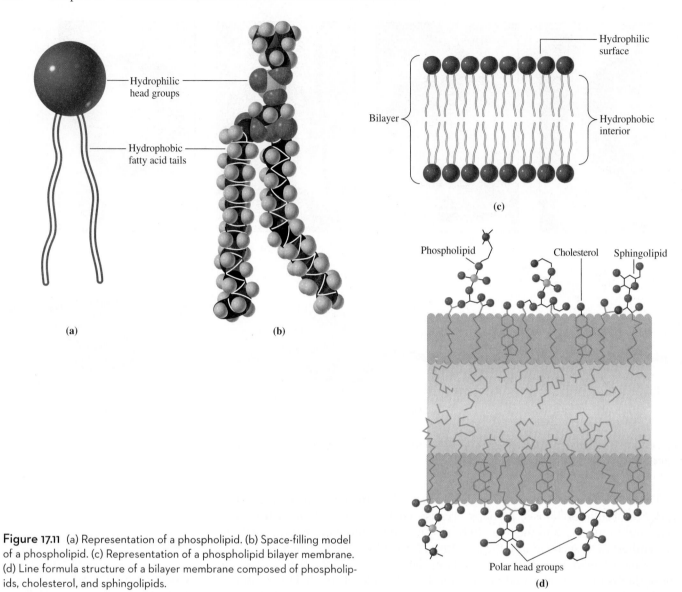

Figure 17.11 (a) Representation of a phospholipid. (b) Space-filling model of a phospholipid. (c) Representation of a phospholipid bilayer membrane. (d) Line formula structure of a bilayer membrane composed of phospholipids, cholesterol, and sphingolipids.

phospholipids thus provide a barrier between the interior of the cell and its surroundings. The polar heads of lipids are exposed to water, and they are highly solvated.

The two layers of the phospholipid bilayer membrane are not identical in composition. For instance, in human red blood cells, approximately 80% of the phospholipids in the outer layer of the membrane are phosphatidylcholine and sphingomyelin; whereas phosphatidylethanolamine and phosphatidylserine make up approximately 80% of the inner layer. In addition, carbohydrate groups are found attached only to those phospholipids found on the outer layer of a membrane. These *glycolipids* (lipids bonded to carbohydrate groups) participate in receptor and recognition functions.

Fluid Mosaic Structure of Biological Membranes

As we have just noted, membranes are not static; they are composed of molecules in motion. The fluidity of biological membranes is determined by the proportions of saturated and unsaturated fatty acid groups in the membrane phospholipids. About half of the fatty acids that are isolated from membrane lipids from all sources are unsaturated.

The unsaturated fatty acid tails of the phospholipids contribute to membrane fluidity because of the bends introduced into the hydrocarbon chain by the double bonds. Because of these "kinks," the fatty acid tails do not pack together tightly.

We also find that the percentage of unsaturated fatty acid groups in membrane lipids is inversely proportional to the temperature of the environment. Bacteria, for example, have different ratios of saturated and unsaturated fatty acids in their membrane lipids, depending on the temperatures of their surroundings. For instance, the membranes of bacteria that grow in the Arctic Ocean have high levels of unsaturated fatty acids so that their membranes remain fluid even at these frigid temperatures. Conversely, the organisms that live in the hot springs of Yellowstone National Park, with temperatures near the boiling point of water, have membranes with high levels of saturated fatty acids. This flexibility in fatty acid content enables the bacteria to maintain the same membrane fluidity over a temperature range of almost 100°C.

The bacteria growing in this hot spring in Yellowstone National Park are called thermophiles because they live at temperatures approaching the boiling point of water. What type of fatty acids do you think will be found in their membrane phospholipids?

Generally, the body temperatures of mammals are quite constant, and the fatty acid composition of their membrane lipids is therefore usually very uniform. One interesting exception is the reindeer. Much of the year, the reindeer must travel through ice and snow. Thus, the hooves and lower legs must function at much colder temperatures than the rest of the body. Because of this, the percentage of unsaturation in the membranes varies along the length of the reindeer leg. We find that the proportion of unsaturated fatty acids increases closer to the hoof, permitting the membranes to function in the low temperatures of ice and snow to which the lower leg is exposed.

Thus, membranes are fluid, regardless of the environmental temperature conditions. In fact, it has been estimated that membranes have the consistency of olive oil.

Although the hydrophobic barrier created by the fluid lipid bilayer is an important feature of membranes, the proteins embedded within the lipid bilayer are equally important and are responsible for critical cellular functions. The presence of these membrane proteins was revealed by an electron microscopic technique called *freeze-fracture*. Cells are frozen to very cold temperatures and then fractured with a very fine diamond knife. Some of the cells are fractured between the two layers of the lipid bilayer. When viewed with the electron microscope, the membrane appeared to be a mosaic, studded with proteins. Because of the fluidity of membranes and the appearance of the proteins seen by electron microscopy, our concept of membrane structure is called the **fluid mosaic model** (Figure 17.12).

Some of the observed proteins, called **peripheral membrane proteins,** are bound only to one of the surfaces of the membrane by interactions between ionic head groups of the membrane lipids and ionic amino acids on the surface of the peripheral protein. Other membrane proteins, called **transmembrane proteins,** are embedded within the membrane and extend completely through it, being exposed both inside and outside the cell.

Just as the phospholipid composition of the membrane is asymmetric, so too is the orientation of transmembrane proteins. Each transmembrane protein has hydrophobic regions that associate with the fatty acid tails of membrane phospholipids. Each also has a unique hydrophilic domain that is always found associated with the outer layer of the membrane and is located on the outside of the cell. This region of the protein typically has oligosaccharides covalently attached. Hence, these proteins are *glycoproteins.* Similarly, each transmembrane protein has a second hydrophilic domain that is always found associated with the inner layer of the membrane and projects into the cytoplasm of the cell. Typically, this region of the transmembrane protein is attached to filaments of the cytoplasmic skeleton.

Membranes are dynamic structures. The mobility of proteins embedded in biological membranes was studied by labeling certain proteins in human and

A Medical Perspective

Liposome Delivery Systems

Liposomes were discovered by Dr. Alec Bangham in 1961. During his studies on phospholipids and blood clotting, he found that if he mixed phospholipids and water, tiny phospholipid bilayer sacs, called liposomes, would form spontaneously. Since that first observation, liposomes have been developed as efficient delivery systems for everything from antitumor and antiviral drugs, to the hair-loss therapy minoxidil!

If a drug is included in the solution during formation of liposomes, the phospholipids will form a sac around the solution. In this way, the drug becomes encapsulated within the phospholipid sphere. These liposomes can be injected intravenously or applied to body surfaces. Sometimes scientists include hydrophilic molecules in the surface of the liposome. This increases the length of time that they will remain in circulation in the bloodstream. These so-called stealth liposomes are being used to carry anticancer drugs, such as doxorubicin and mitoxantrone. Liposomes are also being used as carriers for the antiviral drugs, such as AZT and ddC, that are used to treat human immunodeficiency virus (HIV) infection.

A clever trick to help target the drug-carrying liposome is to include an antibody on the surface of the liposome. These antibodies are proteins designed to bind specifically to the surface of a tumor cell. Upon attaching to the surface of the tumor cell, the liposome "membrane" fuses with the cell membrane. In this way, the deadly chemicals are delivered only to those cells targeted for destruction. This helps to avoid many of the unpleasant side effects of chemotherapy treatment that occur when normal healthy cells are killed by the drug.

Another application of liposomes is in the cosmetics industry. Liposomes can be formed that encapsulate a vitamin, herbal agent, or other nutritional element. When applied to the skin, the liposomes pass easily through the outer layer of dead skin, delivering their contents to the living skin cells beneath. As with the pharmaceutical liposomes, these liposomes, sometimes called *cosmeceuticals,* fuse with skin cells. Thus, they directly deliver the beneficial cosmetic agent directly to the cells that can benefit the most.

Since their accidental discovery 50 years ago, much has been learned about the formation of liposomes and ways to engineer them for more efficient delivery of their contents. This is another example of the marriage of serendipity, an accidental discovery, with scientific research and technological application. As the development of new types of liposomes continues, we can expect that even more ways will be found to improve the human condition.

For Further Understanding

▸ From what you know of the structure of phospholipids, explain the molecular interactions that cause liposomes to form.

▸ Could you use liposome technology to deliver a hydrophobic drug? Explain your answer.

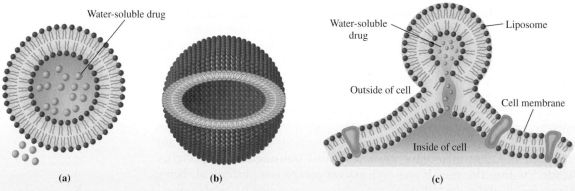

(a) Cross section of a liposome, (b) three-dimensional view of a liposome, and (c) liposome fusing with cell membrane.

mouse cell membranes with red and green fluorescent dyes. The human and mouse cells were fused; in other words, special techniques were used to cause the membranes of the mouse and human cells to flow together to create a single cell. The new cell was observed through a special ultraviolet or fluorescence microscope. The red and green patches were localized within regions of their original cell membranes when the experiment began. Forty minutes later, the color patches were uniformly distributed in the fused cellular membrane (Figure 17.13). This experiment suggests that we can think of the fluid mosaic membrane as an ocean filled with mobile, floating icebergs.

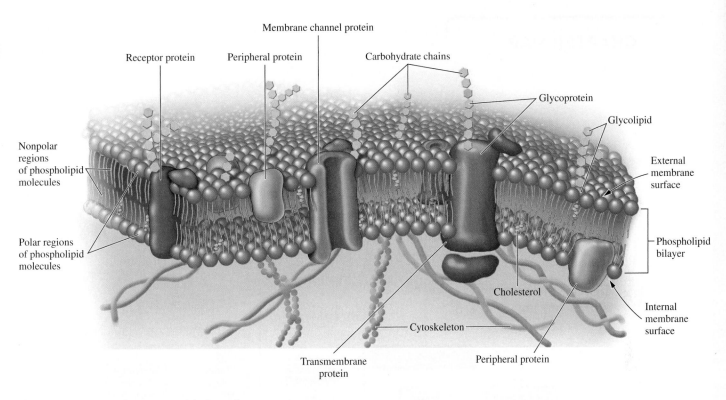

Figure 17.12 The fluid mosaic model of membrane structure.

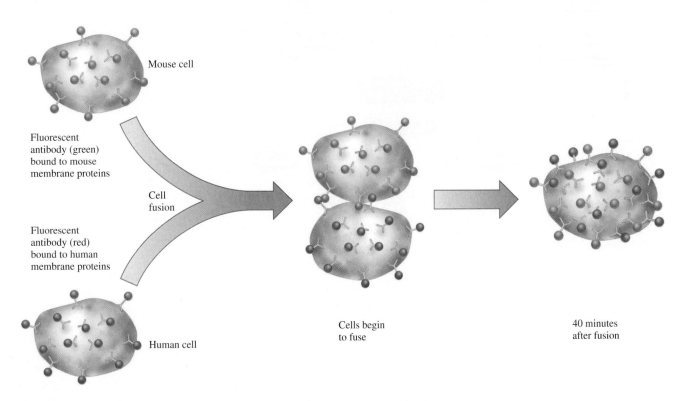

Figure 17.13 Demonstration that membranes are fluid and that proteins move freely in the plane of the lipid bilayer.

CHAPTER MAP

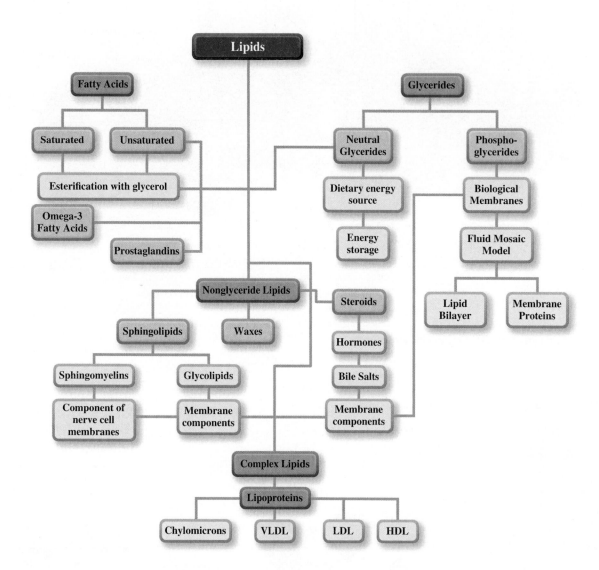

SUMMARY

17.1 Biological Functions of Lipids

▶ **Lipids** are organic molecules characterized by their solubility in nonpolar solvents.

▶ Lipids serve many functions in the body, including energy storage, protection of organs, insulation, absorption of vitamins, energy sources, and hormones.

17.2 Fatty Acids

▶ **Fatty acids** may be **saturated** or **unsaturated** carboxylic acids containing 12–24 carbon atoms.

▶ Fatty acids with even numbers of carbon atoms occur most frequently in nature.

▶ Omega-3 fatty acids are important components of a healthy diet.

▶ **Essential fatty acids** are those that must be provided in the diet. Linoleic and linolenic acids are essential fatty acids.

▶ Linoleic acid is needed for the synthesis of **arachidonic acid,** which is the precursor of the **eicosanoids.**

▶ Among the eicosanoids are the **prostaglandins,** thromboxanes, and leukotrienes, all of which have important physiological effects.

17.3 Glycerides

▶ **Glycerides** are the most abundant lipids. The **neutral glycerides** are esters of one (**monoglycerides**), two (**diglycerides**), or three (**triglycerides**) fatty acids and glycerol.

▶ The reactions of fatty acids are identical to those of carboxylic acids: **esterification, hydrogenation,** and production by acid hydrolysis of esters. The base-catalyzed hydrolysis of a glyceride is **saponification.**

▶ **Phospholipids** are important components of biological membranes.

▶ **Phosphoglycerides** are esters derived from glycerol-3-phosphate and two fatty acids.

▶ The simplest phosphoglyceride is **phosphatidate.**

▶ Because of their amphipathic nature, phospholipids are excellent **emulsifying agents.**

17.4 Nonglyceride Lipids

▶ Nonglyceride lipids include the **sphingolipids, steroids,** and **waxes.**

▶ **Sphingomyelin** is a component of the myelin sheath around cells of the central nervous system.

▶ **Steroids** are important for many biochemical functions. They are **terpenes** because they are synthesized from isoprene units.

  • **Cholesterol** is a membrane component. However, high concentrations in the blood can contribute to **atherosclerosis.**

  • Testosterone, progesterone, and estrone are sex hormones.
  • Cortisone is an anti-inflammatory steroid that is important in the regulation of biological pathways.

17.5 Complex Lipids

▶ **Plasma lipoproteins** are **complex lipids** that transport other lipids through the bloodstream.

▶ **Chylomicrons** carry dietary triglycerides from the intestine to other tissues.

▶ **Very low density lipoproteins** carry triglycerides synthesized in the liver to other tissues for storage.

▶ **Low-density lipoproteins** carry cholesterol to peripheral tissues and help regulate cholesterol levels.

▶ **High-density lipoproteins** transport cholesterol from peripheral tissues to the liver.

17.6 The Structure of Biological Membranes

▶ The **fluid mosaic model** of membrane structure pictures biological membranes that are composed of lipid bilayers in which proteins are embedded.

▶ Membrane lipids contain polar head groups and nonpolar hydrocarbon tails.

▶ Membrane proteins may be embedded within the membrane (**transmembrane proteins**) or may lie on one surface of the membrane (**peripheral membrane proteins**).

ANSWERS TO PRACTICE PROBLEMS

17.1 a. $CH_3(CH_2)_7CH=CH(CH_2)_7COOH$

Oleic acid

b. $CH_3(CH_2)_4CH=CH-CH_2-CH=CH(CH_2)_7COOH$

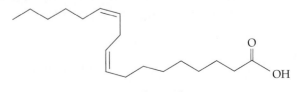

Linoleic acid

17.2 a.

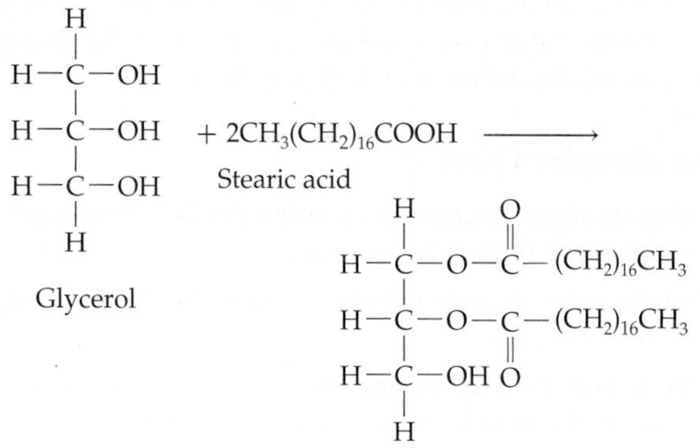

$$\underset{\text{Glycerol}}{\begin{array}{c} H \\ | \\ H-C-OH \\ | \\ H-C-OH \\ | \\ H-C-OH \\ | \\ H \end{array}} + 2CH_3(CH_2)_{16}COOH \underset{\text{Stearic acid}}{} \longrightarrow$$

$$\begin{array}{c} H \quad\quad O \\ | \quad\quad\; \| \\ H-C-O-C-(CH_2)_{16}CH_3 \\ | \\ H-C-O-C-(CH_2)_{16}CH_3 \\ | \quad\quad\quad\;\; \| \\ H-C-OH \quad O \\ | \\ H \end{array}$$

b.

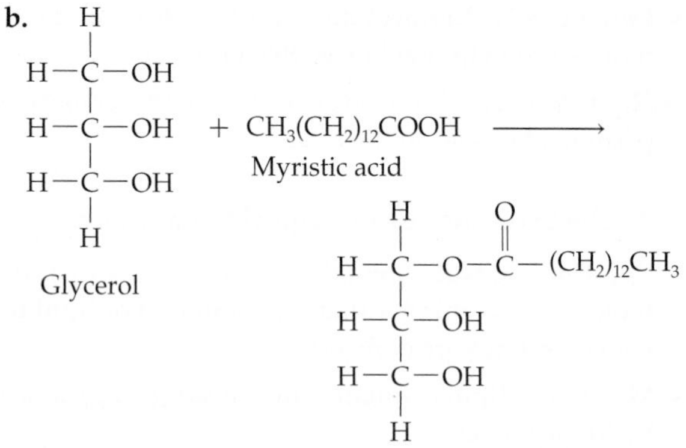

$$\underset{\text{Glycerol}}{\begin{array}{c} H \\ | \\ H-C-OH \\ | \\ H-C-OH \\ | \\ H-C-OH \\ | \\ H \end{array}} + CH_3(CH_2)_{12}COOH \underset{\text{Myristic acid}}{} \longrightarrow$$

$$\begin{array}{c} H \quad\quad O \\ | \quad\quad\; \| \\ H-C-O-C-(CH_2)_{12}CH_3 \\ | \\ H-C-OH \\ | \\ H-C-OH \\ | \\ H \end{array}$$

17.3 a. $\underset{\substack{\text{Dodecanoic acid}\\ \text{Lauric acid}}}{CH_3(CH_2)_{10}COOH} + \underset{\text{Ethanol}}{CH_3CH_2OH} \xrightarrow{H^+,\,\text{heat}}$

$$\underset{\text{Ethyl dodecanoate}}{CH_3(CH_2)_{10}\overset{\displaystyle O}{\overset{\displaystyle \|}{C}}OCH_2CH_3}$$

b. $\underset{\substack{\text{Hexadecanoic acid}\\ \text{Palmitic acid}}}{CH_3(CH_2)_{14}COOH} + \underset{\text{1-Pentanol}}{CH_3(CH_2)_3CH_2OH} \xrightarrow{H^+,\,\text{heat}}$

$$\underset{\text{Pentyl hexadecanoate}}{CH_3(CH_2)_{14}\overset{\displaystyle O}{\overset{\displaystyle \|}{C}}O(CH_2)_4CH_3}$$

17.4 a. $\underset{\textit{cis}\text{-9-Hexadecenoic acid}}{CH_3(CH_2)_5CH{=}CH(CH_2)_7COOH} + H_2$

$$\downarrow Ni$$

$$\underset{\text{Hexadecanoic acid}}{CH_3(CH_2)_{14}COOH}$$

b. $CH_3(CH_2)_4CH{=}CH-CH_2-CH{=}CH-CH_2-CH{=}$
$CH-CH_2-CH{=}CH-(CH_2)_3COOH + 4H_2$

$$\downarrow Ni$$

$$CH_3(CH_2)_{18}COOH$$

17.5 a.

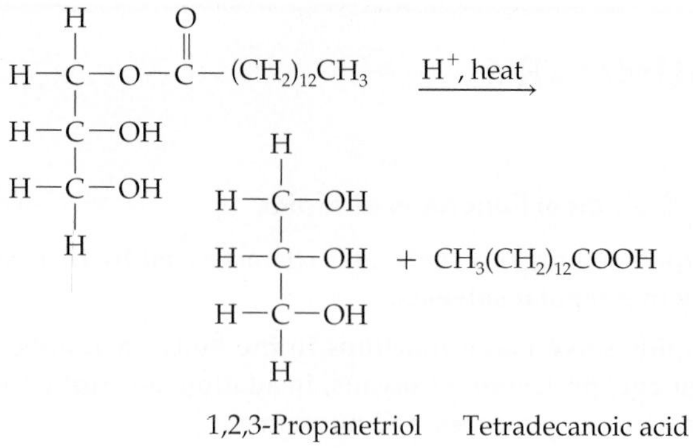

$$\begin{array}{c} H \quad\quad O \\ | \quad\quad\; \| \\ H-C-O-C-(CH_2)_{12}CH_3 \\ | \\ H-C-OH \\ | \\ H-C-OH \\ | \\ H \end{array} \xrightarrow{H^+,\,\text{heat}}$$

$$\underset{\text{1,2,3-Propanetriol}\quad\text{Tetradecanoic acid}}{\begin{array}{c} H \\ | \\ H-C-OH \\ | \\ H-C-OH \\ | \\ H-C-OH \\ | \\ H \end{array} + CH_3(CH_2)_{12}COOH}$$

b.

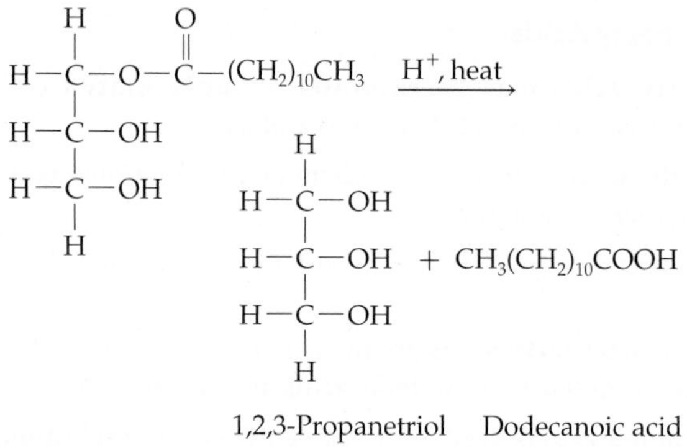

$$\begin{array}{c} H \quad\quad O \\ | \quad\quad\; \| \\ H-C-O-C-(CH_2)_{10}CH_3 \\ | \\ H-C-OH \\ | \\ H-C-OH \\ | \\ H \end{array} \xrightarrow{H^+,\,\text{heat}}$$

$$\underset{\text{1,2,3-Propanetriol}\quad\text{Dodecanoic acid}}{\begin{array}{c} H \\ | \\ H-C-OH \\ | \\ H-C-OH \\ | \\ H-C-OH \\ | \\ H \end{array} + CH_3(CH_2)_{10}COOH}$$

17.6 a.

$$\begin{array}{c} H \quad\quad O \\ | \quad\quad\; \| \\ H-C-O-C-(CH_2)_{14}CH_3 \\ | \\ H-C-OH \\ | \\ H-C-OH \\ | \\ H \end{array} + KOH \longrightarrow$$

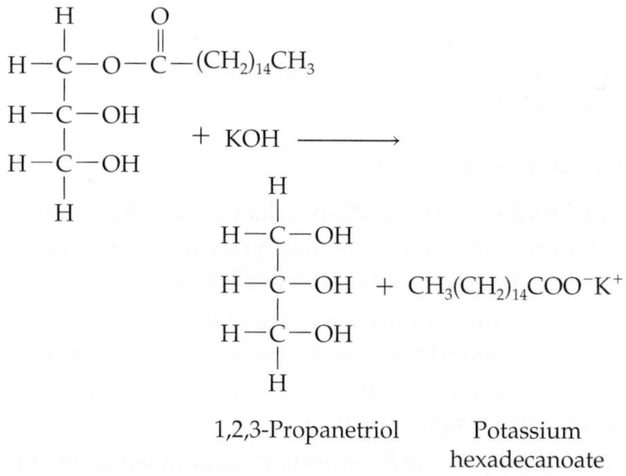

$$\underset{\substack{\text{1,2,3-Propanetriol}\quad\quad\text{Potassium}\\ \text{hexadecanoate}}}{\begin{array}{c} H \\ | \\ H-C-OH \\ | \\ H-C-OH \\ | \\ H-C-OH \\ | \\ H \end{array} + CH_3(CH_2)_{14}COO^-K^+}$$

b.

$$\begin{array}{c} H \quad\quad O \\ | \quad\quad\; \| \\ H-C-O-C-(CH_2)_6CH_3 \\ | \quad\quad\; O \\ | \quad\quad\; \| \\ H-C-O-C-(CH_2)_8CH_3 \\ | \\ H-C-OH \\ | \\ H \end{array} + 2\,NaOH \longrightarrow$$

$$\underset{\text{1,2,3-Propanetriol}\quad\text{Sodium decanoate}}{\begin{array}{c} H \\ | \\ H-C-OH \\ | \\ H-C-OH \\ | \\ H-C-OH \\ | \\ H \end{array}} \begin{array}{c} O \\ \| \\ + CH_3(CH_2)_6CO^-Na^+ \\ \text{Sodium octanoate} \\ O \\ \| \\ + CH_3(CH_2)_8CO^-Na^+ \end{array}$$

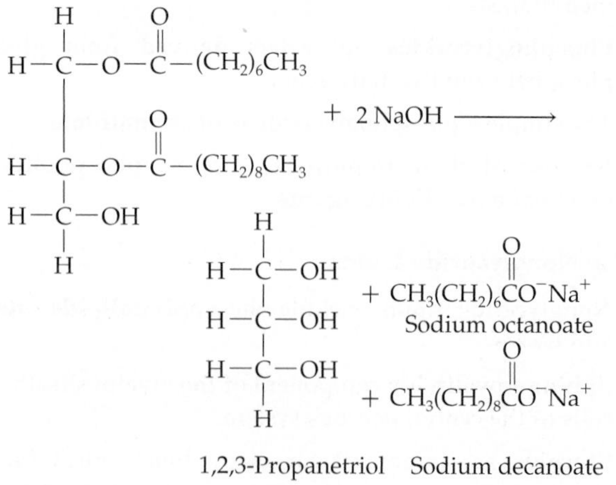

QUESTIONS AND PROBLEMS

Biological Functions of Lipids

Foundations

17.17 List the four main groups of lipids.
17.18 List the biological functions of lipids.

Applications

17.19 In terms of solubility, explain why a diet that contains no lipids can lead to a deficiency of the lipid-soluble vitamins.
17.20 Why are lipids (triglycerides) such an efficient molecule for the storage of energy in the body?

Fatty Acids

Foundations

17.21 What is the difference between a saturated and an unsaturated fatty acid?
17.22 Write the structures for a saturated and an unsaturated fatty acid.
17.23 As the length of the hydrocarbon chain of saturated fatty acids increases, what is the effect on the melting point?
17.24 As the number of carbon-carbon double bonds in fatty acids increases, what is the effect on the melting point?

Applications

17.25 Explain the relationship between fatty acid chain length and melting points that you described in the answer to Question 17.23.
17.26 Explain the relationship you described in the answer to Question 17.24 for the effect of the number of carbon-carbon double bonds in fatty acids on their melting points.
17.27 Draw the structures of each of the following fatty acids:
 a. Decanoic acid
 b. Stearic acid
17.28 Draw the structures of each of the following fatty acids:
 a. *trans*-5-Decenoic acid   **b.** *cis*-5-Decenoic acid
17.29 What are the common and IUPAC names of each of the following fatty acids?
 a. $C_{15}H_{31}COOH$
 b. $C_{11}H_{23}COOH$
17.30 What are the common and IUPAC names of each of the following fatty acids?
 a. $CH_3(CH_2)_5CH=CH(CH_2)_7COOH$
 b. $CH_3(CH_2)_7CH=CH(CH_2)_7COOH$
17.31 Write an equation for the esterification of glycerol with three molecules of myristic acid.
17.32 Write an equation for the esterification of glycerol with three molecules of palmitic acid.
17.33 Write an equation for the acid hydrolysis of a triglyceride containing three oleic acid molecules.
17.34 Write an equation for the acid hydrolysis of a triglyceride containing three capric acid molecules.
17.35 Write equations for the reactions of octanoic acid and stearic acid with KOH.
17.36 Write equations for the reactions of lauric acid and linoleic acid with KOH.
17.37 Using line formulas, write an equation for the hydrogenation of all *cis*-5,8,11,14,17-eicosapentaenoic acid.

17.38 Using line formulas, write an equation for the hydrogenation of all *cis*-4,7,10,13,16,19-docosahexaenoic acid.
17.39 Write an equation for the base-catalyzed hydrolysis of a triglyceride containing a molecule of lauric acid, a molecule of stearic acid, and a molecule of capric acid.
17.40 Write an equation for the base-catalyzed hydrolysis of a triglyceride containing a molecule of palmitoleic acid, a molecule of oleic acid, and a molecule of palmitic acid.
17.41 Write an equation for the esterification of glycerol with a molecule of hexadecanoic acid, a molecule of dodecanoic acid, and a molecule of decanoic acid.
17.42 Write an equation for the esterification of glycerol with a molecule of capric acid, a molecule of oleic acid, and a molecule of stearic acid.
17.43 What is the function of the essential fatty acids?
17.44 What molecules are formed from arachidonic acid?
17.45 What is the biochemical basis for the effectiveness of aspirin in decreasing the inflammatory response?
17.46 What is the role of prostaglandins in the inflammatory response?
17.47 List four effects of prostaglandins.
17.48 What are the functions of thromboxane A_2 and leukotrienes?
17.49 What do the terms *omega-3* and *omega-6* indicate about structures of the fatty acids in those classifications?
17.50 What foods are good sources of EPA and DHA?
17.51 Summarize the health benefits associated with omega-3 fatty acids.
17.52 List some foods that are good sources of α-linolenic acid.
17.53 Explain the relationship between increased levels of omega-3 fatty acids and a decreased risk of cardiovascular disease.
17.54 Do you think that a diet higher in omega-3 fatty acids would be an effective treatment for the symptoms of arthritis? Defend your answer.
17.55 Explain the logic behind decreasing the ratio of omega-6 to omega-3 fatty acids in the diet.
17.56 What is the recommendation of the National Institutes of Health for intake of DHA, EPA, linoleic acid, and linolenic acid?

Glycerides

Foundations

17.57 Define the term *glyceride*.
17.58 Define the term *phosphatidate*.
17.59 What are emulsifying agents and what are their practical uses?
17.60 Why are triglycerides also referred to as triacylglycerols?

Applications

17.61 What do you predict the physical state would be of a triglyceride with three saturated fatty acid tails? Explain your reasoning.
17.62 What do you predict the physical state would be of a triglyceride with three unsaturated fatty acid tails? Explain your reasoning.
17.63 Draw the structure of the triglyceride molecule formed by esterification at C-1, C-2, and C-3 with hexadecanoic acid, *trans*-9-hexadecenoic acid, and *cis*-9-hexadecenoic acid, respectively.

17.64 Draw one possible structure of a triglyceride that contains the three fatty acids capric acid, lauric acid, and arachidonic acid.

17.65 Draw the structure of the phosphatidate formed between glycerol-3-phosphate that is esterified at C-1 and C-2 with capric and lauric acids, respectively.

17.66 Draw the structure of a lecithin molecule in which the fatty acyl groups are derived from arachidic acid.

17.67 What are the structural differences between triglycerides (triacylglycerols) and phospholipids?

17.68 How are the structural differences between triglycerides and phospholipids reflected in their different biological functions?

Nonglyceride Lipids

Foundations

17.69 Define the term *sphingolipid.*

17.70 What are the two major types of sphingolipids?

17.71 Define the term *glycosphingolipid.*

17.72 Distinguish among the three types of glycosphingolipids, cerebrosides, sulfatides, and gangliosides.

Applications

17.73 What is the biological function of sphingomyelin?

17.74 Why are sphingomyelins amphipathic?

17.75 What is the role of cholesterol in biological membranes?

17.76 How does cholesterol contribute to atherosclerosis?

17.77 What are the biological functions of progesterone, testosterone, and estrone?

17.78 How has our understanding of the steroid sex hormones contributed to the development of oral contraceptives?

17.79 What is the medical application of cortisone?

17.80 What are the possible side effects of cortisone treatment?

17.81 A wax found in beeswax is myricyl palmitate. What fatty acid and what alcohol are used to form this compound?

17.82 A wax found in the head of sperm whales is cetyl palmitate. What fatty acid and what alcohol are used to form this compound?

17.83 What are isoprenoids?

17.84 What is a terpene?

17.85 List some important biological molecules that are terpenes.

17.86 Draw the five-carbon isoprene unit.

Complex Lipids

Foundations

17.87 What are the four major types of plasma lipoproteins?

17.88 What is the function of each of the four types of plasma lipoproteins?

Applications

17.89 There is a single, unique structure for the cholesterol molecule. What is meant by the terms *good* and *bad* cholesterol?

17.90 Distinguish among the four plasma lipoproteins in terms of their composition and their function.

17.91 What is the relationship between atherosclerosis and high blood pressure?

17.92 How is LDL taken into cells?

17.93 How does a genetic defect in the LDL receptor contribute to atherosclerosis?

17.94 What is the correlation between saturated fats in the diet and atherosclerosis?

The Structure of Biological Membranes

Foundations

17.95 What is the basic structure of a biological membrane?

17.96 Describe the fluid mosaic model of membrane structure.

17.97 Describe peripheral membrane proteins.

17.98 Describe transmembrane proteins and list some of their functions.

Applications

17.99 What is the major effect of cholesterol on the properties of biological membranes?

17.100 Why do the hydrocarbon tails of membrane phospholipids provide a barrier between the inside and outside of the cell?

17.101 What experimental observation shows that proteins diffuse within the lipid bilayers of biological membranes?

17.102 Why don't proteins turn around in biological membranes like revolving doors?

17.103 How will the properties of a biological membrane change if the fatty acid tails of the phospholipids are converted from saturated to unsaturated chains?

17.104 What is the function of unsaturation in the hydrocarbon tails of membrane lipids?

CHALLENGE PROBLEMS

1. Olestra is a fat substitute that provides no calories, yet has all the properties of a naturally occurring fat. It has a creamy, tongue-pleasing consistency. Unlike other fat substitutes, olestra can withstand heating. Thus, it can be used to prepare foods such as potato chips and crackers. Olestra is a sucrose polyester and is produced by esterification of six, seven, or eight fatty acids to molecules of sucrose. Draw the structure of one such molecule having eight stearic acid acyl groups attached.

2. Liposomes can be made by vigorously mixing phospholipids (like phosphatidylcholine) in water. When the mixture is allowed to settle, spherical vesicles form that are surrounded by a phospholipid bilayer "membrane." Pharmaceutical chemists are trying to develop liposomes as a targeted drug delivery system. By adding the drug of choice to the mixture described above, liposomes form around the solution of the drug. Specific proteins can be incorporated into the mixture that will end up within the phospholipid bilayers of the liposomes. These proteins are able to bind to targets on the surface of particular kinds of cells in the body. Explain why injection of liposome encapsulated pharmaceuticals might be a good drug delivery system.

3. "Cholesterol is bad and should be eliminated from the diet." Do you agree or disagree? Defend your answer.

4. Why would a phospholipid such as lecithin be a good emulsifying agent for ice cream?

5. When a plant becomes cold-adapted, the composition of the membranes changes. What changes in fatty acid and cholesterol composition would you predict? Explain your reasoning.

18

Protein Structure and Function

LEARNING GOALS

1 List the functions of proteins.

2 Draw the general structure of an amino acid, and classify amino acids based on their R groups.

3 Describe the primary structure of proteins, and draw the structure of the peptide bond.

4 Draw the structures of small peptides and name them.

5 Describe the types of secondary structure of a protein.

6 Discuss the forces that maintain secondary structure.

7 Describe the structure and functions of fibrous proteins.

8 Describe the tertiary and quaternary structure of a protein.

9 List the R group interactions that maintain protein conformation.

10 List examples of proteins that require prosthetic groups, and explain the way in which they function.

11 Discuss the importance of the three-dimensional structure of a protein to its function.

12 Describe the roles of hemoglobin and myoglobin.

13 Describe how extremes of pH and temperature cause denaturation of proteins.

14 Explain the difference between essential and nonessential amino acids.

Silk fibers are harvested from the cocoons of silkworms.

OUTLINE

INTRODUCTION

In the 1800s, Johannes Mulder came up with the name **protein,** a term derived from a Greek word that means "of first importance." Indeed, proteins are a very important class of food molecules because they provide an organism not only with carbon and hydrogen, but also with nitrogen and sulfur. These latter two elements are unavailable from fats and carbohydrates, the other major classes of food molecules.

In addition to their dietary importance, the proteins are the most abundant macromolecules in the cell, and have a wide variety of biological functions. **Enzymes** are biological catalysts and most of them are proteins. Reactions that would take days or weeks or require extremely high temperatures without enzymes are completed in an instant. **Defense proteins** include **antibodies** (also called *immunoglobulins*), which are specific protein molecules produced by specialized cells of the immune system in response to foreign **antigens.** These foreign invaders include bacteria and viruses that infect the body. Each antibody has regions that precisely fit and bind to a single antigen, helping to destroy it or remove it from the body. **Transport proteins** carry materials from one place to another in the body. The protein *transferrin* transports iron from the liver to the bone marrow, where it is used to synthesize the heme group for hemoglobin. The proteins *hemoglobin* and *myoglobin* are responsible for transport and storage of oxygen in higher organisms, respectively. **Regulatory proteins** control many aspects of cell function, including metabolism and reproduction. We can function only within a limited set of conditions. For life to exist, body temperature, the pH of the blood, and blood glucose levels must be carefully regulated. Many of the hormones that regulate body function, such as *insulin* and *glucagon*, are proteins. **Structural proteins** provide mechanical support to large animals and provide them with their outer coverings. Our hair and fingernails are largely composed of the protein *keratin*. Other proteins provide mechanical strength for our bones, tendons, and skin. Without such support, large, multicellular organisms like ourselves could not exist. **Movement proteins** are necessary for all forms of movement. Our muscles, including that most important muscle, the heart, contract and expand through the interaction of *actin* and *myosin* proteins. Sperm can swim because they have long flagella made up of proteins. **Nutrient proteins** serve as sources of amino acids for embryos or infants. Egg *albumin* and *casein* in milk are examples of nutrient storage proteins.

18.1   Protein Building Blocks: The α-Amino Acids

Structure of Amino Acids

The proteins of the body are made up of some combination of twenty different subunits called α-**amino acids.** The general structure of an α-amino acid is shown in Figure 18.1. Nineteen of the twenty amino acids that are commonly isolated from proteins have this same general structure; they are primary amines on the α-carbon. The remaining amino acid, proline, is a secondary amine.

Notice that the α-carbon in the general structure is attached to a carboxylate group (a carboxyl group that has lost a proton, $—COO^-$) and a protonated amino group (an amino group that has gained a proton, $—N^+H_3$). At pH 7, a condition required for life functions, you will not find amino acids in which the carboxylate group is protonated ($—COOH$) and the amino group is unprotonated ($—NH_2$). Under these conditions, the carboxyl group is in the conjugate base form ($—COO^-$), and the amino group is in its conjugate acid form ($—N^+H_3$). Any neutral molecule with equal numbers of positive and negative charges is called a *zwitterion*. Thus, amino acids in water exist as dipolar ions called zwitterions.

Conjugate acids and bases are described in detail in Section 8.1.

The α-carbon of each amino acid is also bonded to a hydrogen atom and a side chain, or R group. In a protein, the R groups interact with one another through a variety of weak attractive forces. These interactions participate in folding the protein chain into a precise three-dimensional shape that determines its ultimate function. They also serve to maintain that three-dimensional conformation.

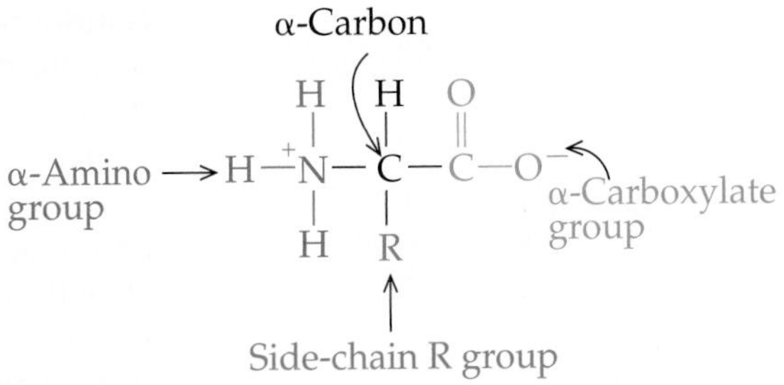

α-Carbon

α-Amino → group

α-Carboxylate group

Side-chain R group

Figure 18.1 General structure of an α-amino acid. All amino acids isolated from proteins, with the exception of proline, have this general structure.

Stereoisomers of Amino Acids

The α-carbon is attached to four different groups in all amino acids except glycine. The α-carbon of most α-amino acids is therefore chiral, allowing mirror-image forms, enantiomers, to exist. Glycine has two hydrogen atoms attached to the α-carbon and is the only amino acid commonly found in proteins that is not chiral.

The L-configuration of α-amino acids is isolated from proteins. The D-L notation is very similar to that discussed for carbohydrates, but instead of the —OH group we use the —N^+H_3 group to determine which is D- and which is L- (Figure 18.2). In Figure 18.2a, we see a comparison of D- and L-glyceraldehyde with D- and L-alanine. Notice that the most oxidized end of the molecule, the carbonyl group of glyceraldehyde or carboxyl group of alanine, is drawn at the top of the molecule. In the D-isomer of glyceraldehyde, the —OH group is on the right. Similarly, in the D-isomer of alanine, the —N^+H_3 is on the right. In the L-isomers of the two

Stereochemistry is discussed in Section 16.3.

Figure 18.2 (a) Structures of D- and L-glyceraldehyde and their relationship to D- and L-alanine. (b) Ball-and-stick models of D- and L-alanine.

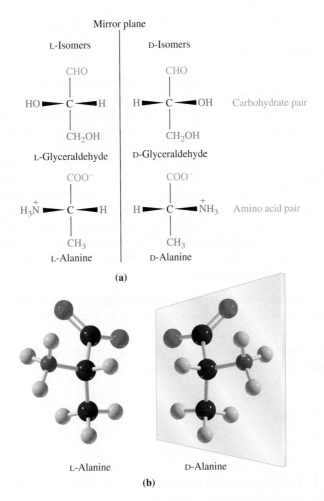

Mirror plane

L-Isomers   D-Isomers

CHO   CHO

HO—C—H   H—C—OH   Carbohydrate pair

CH$_2$OH   CH$_2$OH

L-Glyceraldehyde   D-Glyceraldehyde

COO$^-$ COO$^-$

H$_3$$^+$N—C—H   H—C—$^+NH_3$ Amino acid pair

CH$_3$ CH$_3$

L-Alanine   D-Alanine

(a)

L-Alanine   D-Alanine

(b)

compounds, the —OH and —N$^+$H$_3$ groups are on the left. By this comparison with the enantiomers of glyceraldehyde, we can define the D- and L-enantiomers of the amino acids. Figure 18.2b shows ball-and-stick models of the D- and L-isomers of alanine.

In Chapter 16, we learned that almost all of the monosaccharides found in nature are in the D-family. Just the opposite is true of the α-amino acids. Almost all of the α-amino acids isolated from proteins in nature are members of the L-family. In other words, the orientation of the four groups around the chiral carbon of these α-amino acids resembles the orientation of the four groups around the chiral carbon of L-glyceraldehyde.

Classes of Amino Acids

Because all amino acids have a carboxyl group and an amino group, all differences between amino acids depend upon their side-chain R groups. The amino acids are grouped in Figure 18.3 according to the polarity of their side chains.

The side chains of some amino acids are nonpolar. They prefer contact with one another over contact with water and are said to be **hydrophobic** ("water-fearing") **amino acids.** They are generally found buried in the interior of proteins, where they can associate with one another and remain isolated from water. Nine amino acids fall into this category: alanine, valine, leucine, isoleucine, proline, glycine, methionine, phenylalanine, and tryptophan. The R group of proline is unique; it is actually bonded to the α-amino group, forming a secondary amine.

The side chains of the remaining amino acids are polar. Because they are attracted to polar water molecules, they are said to be **hydrophilic** ("water-loving") **amino acids.** The hydrophilic side chains are often found on the surfaces of proteins. The polar amino acids can be subdivided into three classes.

- *Polar, neutral amino acids* have R groups that have a high affinity for water but are not ionic at pH 7. Serine, threonine, tyrosine, cysteine, asparagine, and glutamine fall into this category. Most of these amino acids associate with one another by hydrogen bonding, but cysteine molecules form disulfide bonds with one another, as we will discuss in Section 18.5.
- *Negatively charged amino acids* have ionized carboxyl groups in their side chains. At pH 7, these amino acids have a net charge of −1. Aspartate and glutamate are the two amino acids in this category. They are acidic amino acids because ionization of the carboxylic acid releases a proton.
- *Positively charged amino acids.* At pH 7, lysine, arginine, and histidine have a net positive charge because their side chains contain positive groups. These amino groups are basic because the side chain reacts with water, picking up a proton and releasing a hydroxide anion.

The names of the amino acids can be abbreviated by a three-letter code and by a one-letter code. These abbreviations are shown in Table 18.1.

> The hydrophobic interaction between nonpolar R groups is one of the forces that helps maintain the proper three-dimensional shape of a protein.

> Hydrogen bonding (Section 5.2) is another weak interaction that helps maintain the proper three-dimensional structure of a protein. The positively and negatively charged amino acids within a protein interact with one another to form ionic bridges that also help to keep the protein chain folded in a precise way.

Question 18.1   Write the one-letter and three-letter abbreviations, and draw the structure of each of the following amino acids.
   a. Glycine
   b. Proline
   c. Threonine
   d. Aspartate
   e. Lysine

Question 18.2   Indicate whether the side chains of each of the amino acids listed in Question 18.1 is polar, nonpolar, basic, or acidic.

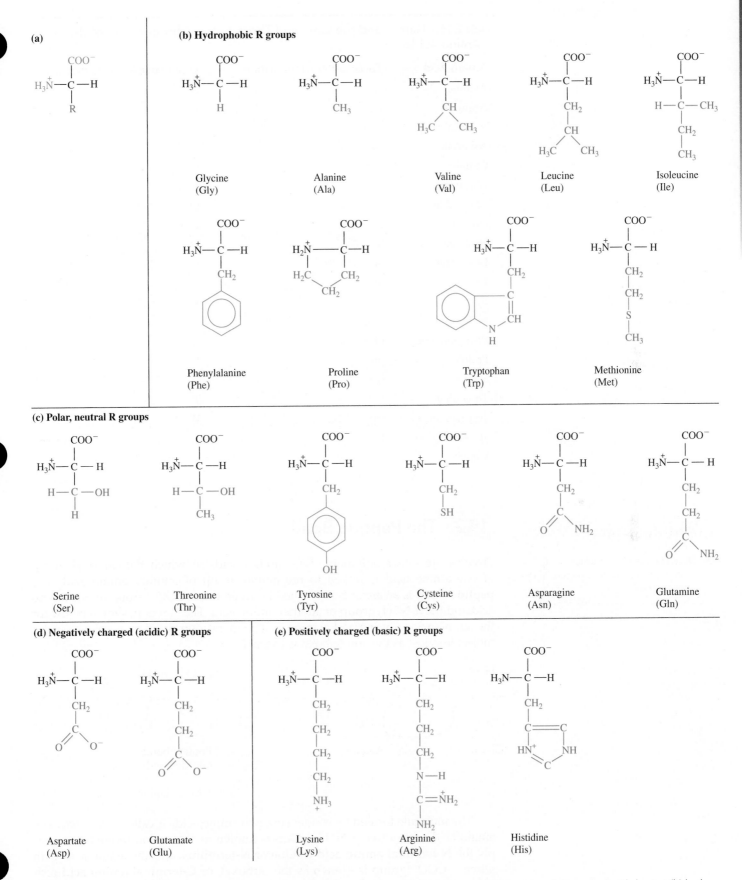

Figure 18.3 Structures of the amino acids at pH 7.0. (a) The general structure of an amino acid. Structures of the amino acids having (b) hydrophobic; (c) polar, neutral; (d) negatively charged; and (e) positively charged R groups.

TABLE 18.1 Names and the One- and Three-Letter Abbreviations of the α-Amino Acids

Amino Acid	Three-Letter Abbreviation	One-Letter Abbreviation
Alanine	ala	A
Arginine	arg	R
Asparagine	asn	N
Aspartate	asp	D
Cysteine	cys	C
Glutamate	glu	E
Glutamine	gln	Q
Glycine	gly	G
Histidine	his	H
Isoleucine	ile	I
Leucine	leu	L
Lysine	lys	K
Methionine	met	M
Phenylalanine	phe	F
Proline	pro	P
Serine	ser	S
Threonine	thr	T
Tryptophan	trp	W
Tyrosine	tyr	Y
Valine	val	V

LEARNING GOAL

3 Describe the primary structure of proteins, and draw the structure of the peptide bond.

18.2 The Peptide Bond

Proteins are linear polymers of L-α-amino acids in which the carboxyl group of one amino acid is linked to the amino group of another amino acid. The **peptide bond** is an *amide bond* formed between the —COO⁻ group of one amino acid and the α-N⁺H₃ group of another amino acid. The reaction, shown below for the amino acids glycine and alanine, is a condensation reaction, because a water molecule is lost as the amide bond is formed.

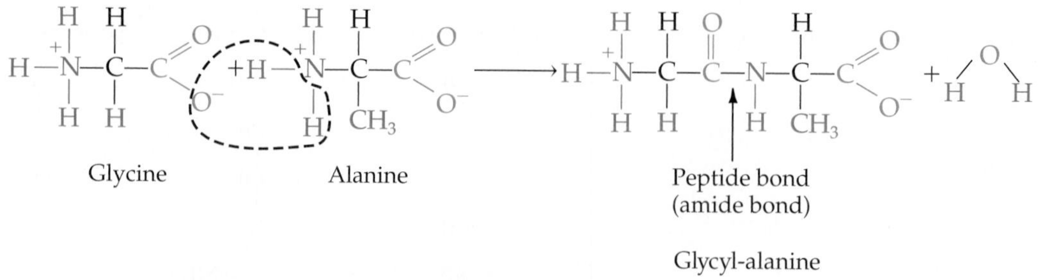

The molecule formed by condensing two amino acids is called a *dipeptide*. The amino acid with a free α-N⁺H₃ group is known as the amino terminal, or simply the **N-terminal amino acid** residue or **N-terminus,** and the amino acid with a free —COO⁻ group is known as the carboxyl, or **C-terminal amino acid** residue or **C-terminus.** Structures of proteins are conventionally written with their N-terminal amino acid on the left.

The number of amino acids in small peptides is indicated by the prefixes *di-* (two units), *tri-* (three units), *tetra-* (four units), and so forth. Peptides are named as derivatives of the C-terminal amino acid, which receives its entire name. For all other amino acids, the ending *-ine* is changed to *-yl*. Thus, the dipeptide in the example on the previous page has glycine as its *N*-terminal amino acid, as indicated by its full name, glycyl-alanine.

The dipeptide formed from alanine and glycine that has alanine as its *N*-terminal amino acid is named alanyl-glycine. These two dipeptides have the same amino acid composition, but different amino acid sequences.

To understand why the N-terminus is placed first and the C-terminus is placed last, we need to look at the process of protein synthesis. As we will see in Section 20.6, the N-terminus is the first amino acid residue of the protein. It forms a peptide bond involving its carboxyl group and the amino group of the second amino acid residue in the protein. Thus, a free amino group literally projects from the "left" end of the protein. Similarly, the C-terminal amino acid is the last amino acid residue added to the protein during protein synthesis. Because the peptide bond is formed between the amino group of this amino acid residue and the carboxyl group of the previous amino acid residue, a free carboxyl group projects from the "right" end of the protein chain.

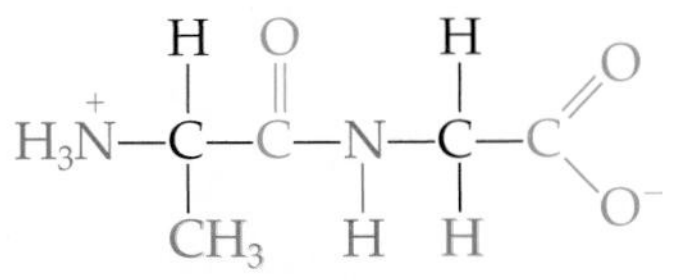

Alanyl-glycine
(ala-gly)

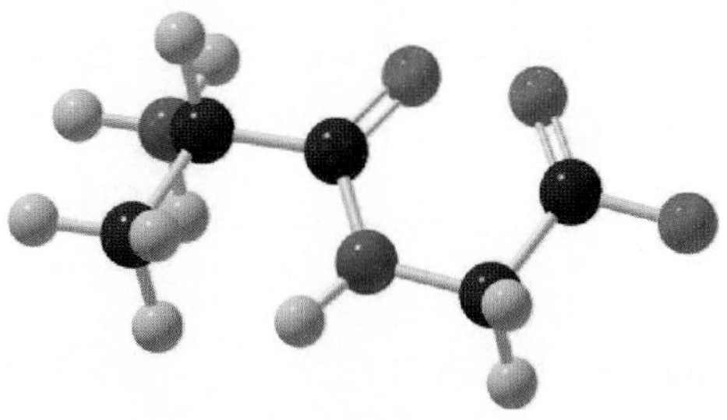

Alanyl-glycine

The structures of small peptides can easily be drawn with practice if certain rules are followed. First, note that the backbone of the peptide contains the repeating sequence

N—C—C—N—C—C—N—C—C
2 1   2 1   2 1

in which N is the α-amino group, carbon-2 is the α-carbon, and carbon-1 is the carboxyl group. Carbon-2 is always bonded to a hydrogen atom and to the R group side chain that is unique to each amino acid. Continue drawing as outlined in Example 18.1.

EXAMPLE 18.1 | **Writing the Structure of a Tripeptide**

Draw the structure of the tripeptide alanyl-glycyl-valine.

LEARNING GOAL

4 Draw the structures of small peptides and name them.

Solution

Step 1. Write the backbone for a tripeptide. It will contain three sets of three atoms, or nine atoms in all. Remember that the N-terminal amino acid is written to the left.

N—C—C   N—C—C   N—C—C
Set 1   Set 2   Set 3

Step 2. Add oxygens to the carboxyl carbons and hydrogens to the amino nitrogens:

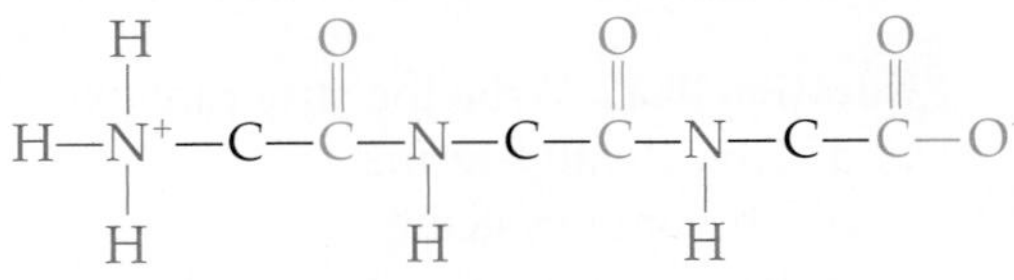

Continued...

Step 3. Add hydrogens to the α-carbons:

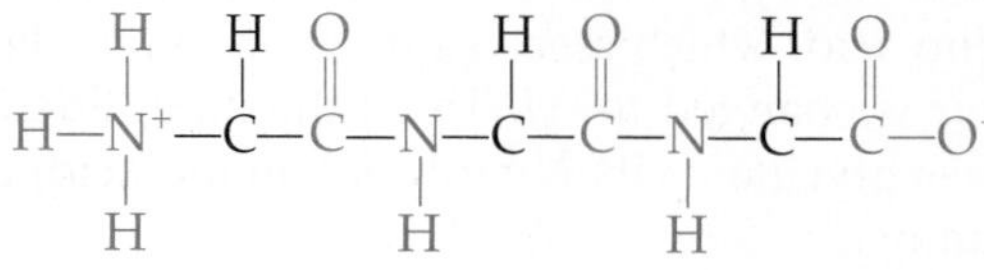

$$H{-}N^+{-}C{-}C{-}N{-}C{-}C{-}N{-}C{-}C{-}O^-$$

Step 4. Add the side chains. In this example (ala-gly-val) they are, from left to right, —CH₃, —H, and —CH(CH₃)₂:

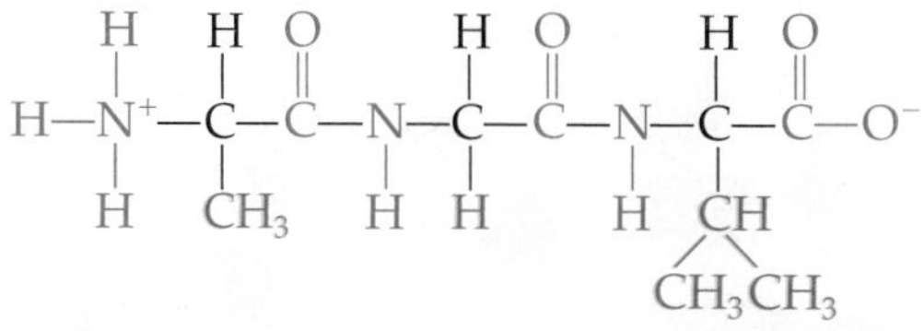

$$H{-}N^+{-}C{-}C{-}N{-}C{-}C{-}N{-}C{-}C{-}O^-$$

CH₃ CH₃

Practice Problem 18.1

Write the structure of each of the following peptides at pH 7.

a. Methionyl-leucyl-cysteine   b. Tyrosyl-seryl-histidine   c. Arginyl-isoleucyl-glutamine

▶ For Further Practice: **Questions 18.37 and 18.38.**

At first, it appears logical to think that a long polymer of amino acids would undergo constant change in conformation because of free rotation around the —N—C—C— single bonds of the peptide backbone. In reality, this is not the case. An explanation for this phenomenon resulted from the early X-ray diffraction studies of Linus Pauling. By interpreting the pattern formed when X-rays were diffracted by a crystal of pure protein, Pauling discovered that peptide bonds are both planar (flat) and rigid and that the N—C bonds are shorter than expected. What did all of this mean? Pauling concluded that the peptide bond has a partially double bond character because it exhibits resonance.

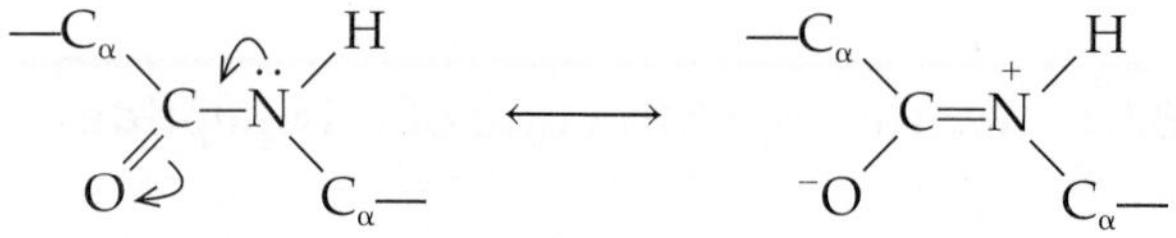

This means that there is free rotation around only two of the three single bonds of the peptide backbone (Figure 18.4a), which limits the number of possible conformations for any peptide. A second feature of the rigid peptide bond is that the R groups on adjacent amino acids are on opposite sides of the extended peptide chain (Figure 18.4b).

Question 18.3 Write the structure of each of the following peptides at pH 7:
a. Alanyl-phenylalanine
b. Lysyl-alanine
c. Phenylalanyl-tyrosyl-leucine

Question 18.4 Write the structure of each of the following peptides at pH 7:
a. Glycyl-valyl-serine
b. Threonyl-cysteine
c. Isoleucyl-methionyl-aspartate

The New Protein

Dietary trends come and go. Should we eat a diet that is low in fat and high in carbohydrates or follow the Mediterranean diet in which up to a third of the calories are fats, such as olive oil? Should we become vegetarians or remove all animal products from our diet and become vegans? Perhaps we should revert to our caveperson roots and follow the Paleo diet.

Currently, one of the most popular trends is the low-carbohydrate, high-protein diet. A 2008 study published in the *American Journal of Clinical Nutrition* reported that an increase in protein in the diet helped people feel full more quickly and stay full for a longer period. It also reported that subjects burned more calories as heat and some were found to build or maintain more lean muscle instead of storing excess energy as fat. The ability to lose weight through a low-carbohydrate, high-protein diet has been the claim of Dr. Robert Atkins since the publication of his first book, *The Atkins Diet*, in 1972. The high-protein diet has become so popular that we now see breakfast cereals and breads, foods that we have traditionally thought of as sources of carbohydrates, that have been developed in high-protein options and included in the diet!

Another changing trend is the type of protein that we eat. In the past, a high-protein diet would have brought images of steaks, pork chops, or bacon and eggs. But concerns about heart disease and obesity have caused a shift in our choices of proteins. Increasingly we are turning away from red meats and selecting chicken or fish instead. In fact, about 70% of Millennials, born between 1980 and 2000, report that they eat meat alternatives several times a week.

The protein of choice to produce meat substitutes is soy protein. Soy protein is inexpensive and is the only vegetable protein that provides all of the essential amino acids. It is even considered to be heart-healthy, since it lowers harmful LDL cholesterol while maintaining beneficial HDL cholesterol. It is also taste-neutral and can be flavored to prepare either savory or sweet products.

The traditional source of soy protein is tofu. It is prepared by coagulating soy milk and then pressing the curds into solid blocks. Depending on the way in which the blocks are formed, the tofu may be soft, firm, or extra firm. Like the purified protein, the versatile tofu will work as well in sweet dishes, such as sesame tofu, as in hot and spicy dishes, such as Hunan tofu.

Modern technology has allowed the food industry to form soy protein into all kinds of interesting textures that mimic many other foods in our diet. It can be formed into small crisps that have the shape and texture of Rice Krispies. These are found in a variety of snack bars that are composed of clusters of the crisps found in many flavors. Nuts, dried fruit, and chocolate are used in the bars to provide a variety of tempting, high-protein snacks. Soy protein can be extruded into a form that nearly perfectly mimics a variety

of meats. Some of the available versions include burgers, hot dogs, ground meat, chicken patties, and even barbecue ribs. Soy protein also finds its way into crackers, cookies, breakfast cereals, and waffles.

With all these meatless options, it has become a simple matter to swap out a few meat meals each week with healthier meat-substitute options and to enjoy flavors and textures similar to those of the animal protein options.

For Further Understanding

▶ Soy protein is a complete protein. What does that mean in terms of its amino acid composition?

▶ Soy protein is produced by a process called isoelectric precipitation. First, it is dissolved in an alkaline solution. Impurities, such as fiber, are removed. Then the pH of the solution is lowered and the protein precipitates out of the solution. It can then be spray dried for use in a variety of products. What features of the amino acids of soy protein are responsible for this behavior under alkaline and acidic conditions?

Figure 18.4 (a) There is free rotation around only two of the three single bonds of a peptide backbone. (b) This model of an eight amino acid peptide shows that the R groups on adjacent amino acids are on opposite sides of the chain because of the rigid peptide bond.

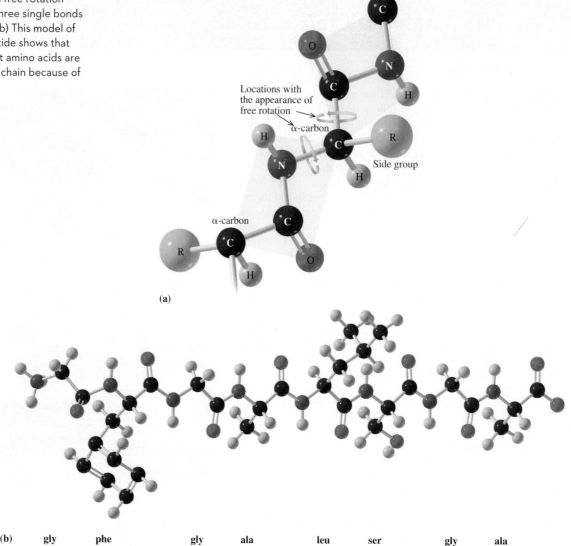

(a)

(b)   gly   phe   gly   ala   leu   ser   gly   ala

18.3   The Primary Structure of Proteins

LEARNING GOAL

3   Describe the primary structure of proteins, and draw the structure of the peptide bond.

The genetic code and the process of protein synthesis are described in Sections 20.5 and 20.6.

The **primary structure** of a protein is the amino acid sequence of the protein chain. It results from the covalent bonding between the amino acid residues in the chain (peptide bonds). The primary structures of proteins are translations of information contained in genes. Each protein has a different primary structure with different amino acid residues in different places along the chain. This sequence of amino acid residues is dictated by the sequence of the gene.

Ultimately, it is the primary structure of a protein that will determine its biologically active form. The interactions among the R groups of the amino acids in the protein chain depend on the location of those R groups along the chain. These interactions will govern how the protein chain folds, which, in turn, dictates its final three-dimensional structure and its biological function.

18.4   The Secondary Structure of Proteins

LEARNING GOAL

5   Describe the types of secondary structure of a protein.

Regions of the primary sequence of a protein, the chain of covalently linked amino acids, fold into regularly repeating structures that resemble designs in a tapestry. These repeating structures define the **secondary structure** of the protein. The secondary structure is the result of hydrogen bonding between the amide hydrogens and carbonyl oxygens of the peptide bonds. Many hydrogen bonds are needed to maintain the secondary structure and thereby the overall structure of the protein. Different regions of a protein chain may have different types of secondary structure. Some regions of a protein chain may have a random or nonregular structure; however, the two most common types of secondary structure are the α-helix and the β-pleated sheet because they maximize hydrogen bonding in the backbone.

α-Helix

The most common type of secondary structure is a coiled, helical conformation known as the **α-helix** (Figure 18.5). The α-helix has several important features.

- Every amide hydrogen and carbonyl oxygen associated with the peptide backbone is involved in a hydrogen bond when the chain coils into an α-helix. These hydrogen bonds lock the α-helix into place.
- Every carbonyl oxygen is hydrogen-bonded to an amide hydrogen four amino acids away in the chain.
- The hydrogen bonds of the α-helix are parallel to the long axis of the helix (see Figure 18.5).

LEARNING GOAL

6 Discuss the forces that maintain secondary structure.

In the photo above, some of the children have straight hair and some have curly hair. Knowing that the primary structure of a protein is dictated by the sequence of a gene, in this case a keratin gene, develop a hypothesis to explain hair curliness.

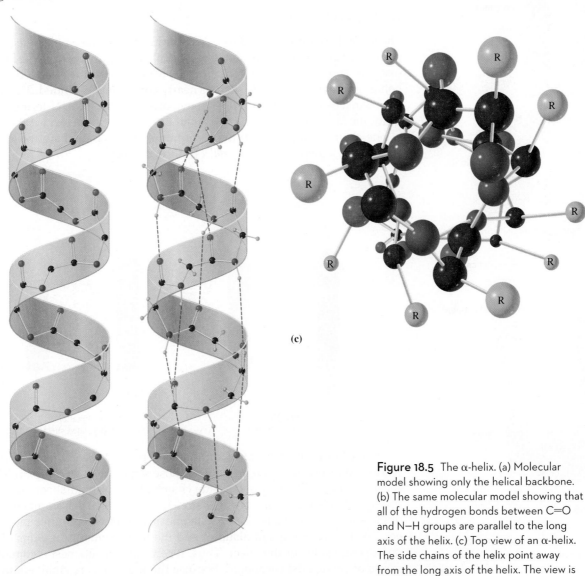

(c)

(a)   **(b)**

Figure 18.5 The α-helix. (a) Molecular model showing only the helical backbone. (b) The same molecular model showing that all of the hydrogen bonds between C=O and N—H groups are parallel to the long axis of the helix. (c) Top view of an α-helix. The side chains of the helix point away from the long axis of the helix. The view is into the barrel of the helix.

7 Describe the structure and functions of fibrous proteins.

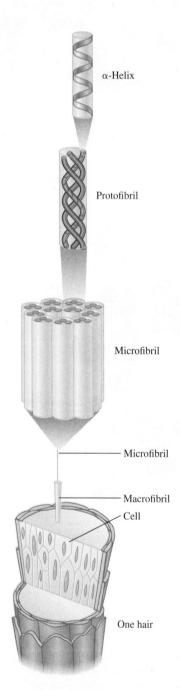

Figure 18.6 Structure of the α-keratins. These proteins are assemblies of triple-helical protofibrils that are assembled in an array known as a *microfibril*. These in turn are assembled into macrofibrils. Hair is a collection of macrofibrils and hair cells.

- The polypeptide chain in an α-helix is right-handed. It is oriented like a normal screw. If you turn a screw clockwise, it goes into the wall; turned counterclockwise, it comes out of the wall.
- The repeat distance of the helix, or its pitch, is 5.4 angstroms (Å), and there are 3.6 amino acids per turn of the helix.

Fibrous proteins are structural proteins arranged in fibers or sheets that have only one type of secondary structure. The **α-keratins** are fibrous proteins that form the covering (hair, wool, nails, hooves, and fur) of most land animals. Human hair provides a typical example of the structure of the α-keratins. The proteins of hair consist almost exclusively of polypeptide chains coiled up into α-helices. A single α-helix is coiled in a bundle with two other helices to give a three-stranded superstructure called a *protofibril* that is part of an array known as a *microfibril* (Figure 18.6). These structures, which resemble "molecular pigtails," possess great mechanical strength, and they are virtually insoluble in water.

The major structural property of a coiled coil superstructure of α-helices is its great mechanical strength. This property is applied very efficiently in both the fibrous proteins of skin and those of muscle. As you can imagine, these proteins must be very strong to carry out their functions of mechanical support and muscle contraction.

β-Pleated Sheet

The second common secondary structure in proteins resembles the pleated folds of drapery and is known as the **β-pleated sheet** (Figure 18.7a). All of the carbonyl oxygens and amide hydrogens in a β-pleated sheet are involved in hydrogen bonds, and the polypeptide chain is nearly completely extended. The polypeptide chains in a β-pleated sheet can have two orientations. If the N-termini are head to head, the structure is known as a *parallel* β-pleated sheet. And if the N-terminus of one chain is aligned with the C-terminus of a second chain (head to tail), the structure is known as an *antiparallel* β-pleated sheet.

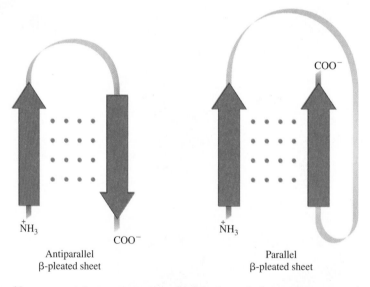

Antiparallel
β-pleated sheet

Parallel
β-pleated sheet

Some fibrous proteins are composed of β-pleated sheets. For example, the silkworm produces *silk fibroin,* a protein whose structure is an antiparallel β-pleated sheet (Figure 18.7). The polypeptide chains of a β-pleated sheet are almost completely extended, and silk does not stretch easily. Glycine accounts for nearly half of the amino acids of silk fibroin. Alanine and serine account for most of the others. The methyl groups of alanines and the hydroxymethyl groups of serines lie on opposite sides of the sheet. Thus, the stacked sheets nestle comfortably, like sheets of corrugated cardboard, because the R groups are small enough to allow the stacked-sheet superstructure.

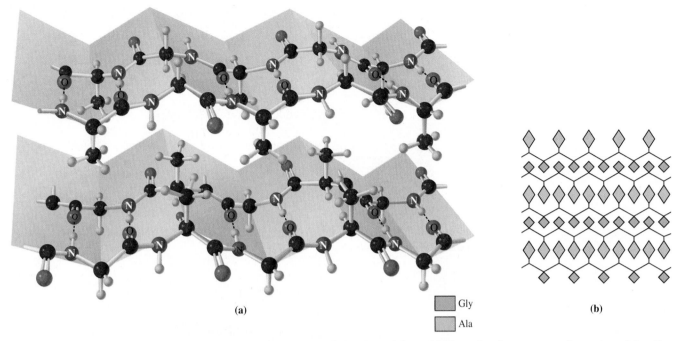

Gly
Ala

(a)   **(b)**

Figure 18.7 The structure of silk fibroin is almost entirely antiparallel β-pleated sheet. (a) The molecular structure of a portion of the silk fibroin protein. (b) A schematic representation of the antiparallel β-pleated sheet with the nestled R groups.

18.5 The Tertiary Structure of Proteins

Most fibrous proteins, such as silk, collagen, and the α-keratins, are almost completely insoluble in water. (Our skin would do us very little good if it dissolved in the rain.) The majority of cellular proteins, however, are soluble in the cell cytoplasm. Soluble proteins are usually **globular proteins**. Globular proteins have three-dimensional structures called the **tertiary structure** of the protein, which are distinct from their secondary structure. Tertiary structure refers to the three-dimensional shape of the entire peptide chain. The regions of secondary structure, α-helix and β-pleated sheet, further fold on themselves to achieve the tertiary structure.

We have seen that the forces that maintain the secondary structure of a protein are hydrogen bonds between the amide hydrogen and the carbonyl oxygen of the peptide bond. What are the forces that maintain the tertiary structure of a protein? The globular tertiary structure forms spontaneously and is maintained as a result of interactions among the side chains, the R groups, of the amino acids. The structure is maintained by the following molecular interactions:

- van der Waals forces (London dispersion forces and dipole-dipole attractions) between the hydrophobic R groups
- Hydrogen bonds between the polar R groups
- Ionic bonds (salt bridges) between the oppositely charged R groups
- Covalent bonds between the thiol-containing amino acid residues. Two of the polar cysteines can be oxidized to a dimeric amino acid called *cystine* (Figure 18.8). The disulfide bond of cystine can be a cross-link between different proteins, or it can tie two segments within a protein together.

The bonds that maintain the tertiary structure of proteins are shown in Figure 18.9. The importance of these bonds becomes clear when we realize that it is the tertiary structure of the protein that defines its biological function. Most of the time, nonpolar side chains of amino acid residues are buried,

LEARNING GOAL

8 Describe the tertiary and quaternary structure of a protein.

LEARNING GOAL

9 List the R group interactions that maintain protein conformation.

Thiols were discussed in Section 12.8.

In the next section, we will see that some proteins have an additional level of structure, quaternary structure, that also influences function.

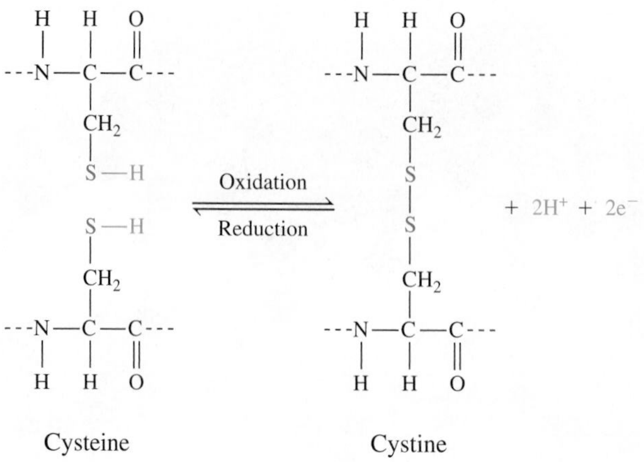

Figure 18.8 Oxidation of two cysteines to give the dimer cystine. This reaction occurs in cells and is readily reversible.

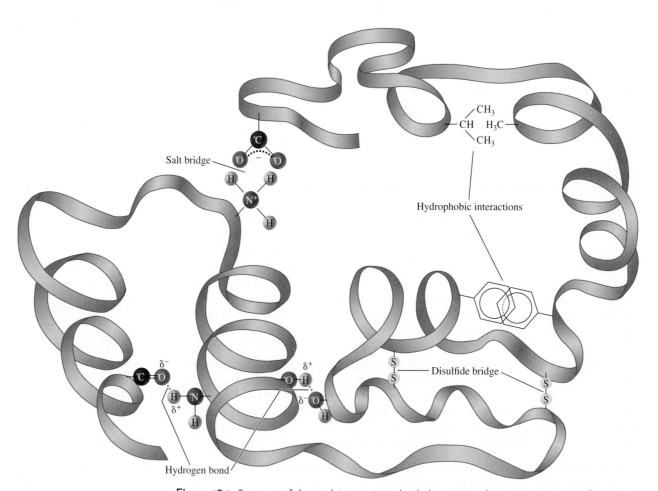

Figure 18.9 Summary of the weak interactions that help maintain the tertiary structure of a protein.

closely packed, in the interior of a globular protein, out of contact with water. Polar and charged side chain amino acid residues lie on the surfaces of globular proteins. Globular proteins are extremely compact. The tertiary structure can contain regions of α-helix and regions of β-pleated sheet. "Hinge" regions of random coil connect regions of α-helix and β-pleated sheet. Because of its cyclic structure, proline disrupts an α-helix. As a result, proline is often found in these hinge regions.

A Medical Perspective

Collagen, Cosmetic Procedures, and Clinical Applications

Collagen is the most abundant protein in the human body, making up about one-third of the total protein content. It provides mechanical strength to bone, tendon, skin, and blood vessels. Collagen fibers in bone provide a scaffolding around which *hydroxyapatite* (a calcium phosphate polymer) crystals are arranged. Skin contains loosely woven collagen fibers that can expand in all directions. The corneas of the eyes are composed of collagen. As we consider these tissues, we realize that they have quite different properties, ranging from tensile strength (tendons) and flexibility (blood vessels) to transparency (cornea).

How could such diverse structures be composed of a single protein? The answer lies in the fact that collagen is actually a family of twenty genetically distinct, but closely related proteins. Although the differences in the amino acid sequence of these different collagen proteins allow them to carry out a variety of functions in the body, they all have a similar three-dimensional structure. Collagen is composed of three left-handed polypeptide helices that are twisted around one another to form a "superhelix" called a *triple helix*. Each of the individual peptide chains of collagen is a left-handed helix, but they are wrapped around one another in the right-handed sense.

Every third amino acid in the collagen chain is glycine. It is important to the structure because the triple-stranded helix forms as a result of interchain hydrogen bonding involving glycine. Thus, every third amino acid on one strand is in very close contact with the other two strands. Glycine has another advantage: It is the only amino acid with an R group small enough for the space allowed by the triple-stranded structure.

Collagen injections have been used in cosmetic procedures to add fullness to lips or minimize the appearance of wrinkled or sunken facial skin. This procedure is no longer very common because of the relatively high frequency of allergic reactions to the bovine collagen used in the injections. Popular alternatives to collagen injections include the patient's own fat and hyaluronic acid (see also A Medical Perspective: Monosaccharide Derivatives and Heteropolysaccharides of Medical Interest in Chapter 16). Another alternative, although an expensive one, is the injection of recombinant human collagen. This reduces the incidence of allergic reactions.

Collagen is also used in the preparation of artificial skin for severe burn patients. The collagen is used in combination with silicones, glycosaminoglycans (such as hyaluronic acid), growth factors, and human fibroblasts, which are the most common type of cell in connective tissue and which promote wound healing.

Collagen even finds its way into our diet. When partially hydrolyzed, the three polypeptide strands separate from one another and then curl up into globular random coils. The product is gelatin, found most notably in gelatin desserts such as JELL-O. However, gelatin is found in many other foods, as well as in dietary supplements that claim to improve fingernail and skin condition.

Structure of the
collagen triple helix

Two unusual, hydroxylated amino acids account for nearly one-fourth of the amino acids in collagen. These amino acids are 4-hydroxyproline and 5-hydroxylysine.

4-Hydroxyproline   5-Hydroxylysine

Structures of 4-hydroxyproline and 5-hydroxylysine, two amino acids found only in collagen

These amino acids are an important component of the structure of collagen because they form covalent cross-linkages between adjacent molecules within the triple strand. They can also participate in interstrand hydrogen bonding to further strengthen the structure.

When collagen is synthesized, the amino acids proline and lysine are incorporated into the chain of amino acids. These are later modified by two enzymes to form 4-hydroxyproline and 5-hydroxylysine. Both of these enzymes require vitamin C to carry out these reactions. In fact, this is the major known

Continued…

physiological function of vitamin C. Without hydroxylation, hydrogen bonds cannot form and the triple helix is weak, resulting in fragile blood vessels.

OH
|
O   O   C—CH₂OH
|
H

HO   OH

Vitamin C
(ascorbic acid)

People who are deprived of vitamin C, as were sailors on long voyages before the eighteenth century, develop *scurvy*, a disease of collagen metabolism. The symptoms of scurvy include skin lesions, fragile blood vessels, and bleeding gums. The British Navy provided the antidote to scurvy by including limes, which are rich in vitamin C, in the diets of its sailors. The epithet *limey*, a slang term for *British*, entered the English language as a result.

For Further Understanding

▸ What feature of glycine is responsible for its importance in the hydrogen bonding that maintains the helical structure of collagen?

▸ Collagen may have great tensile strength (as in tendons) and flexibility (as in skin and blood vessels), and may even be transparent (cornea of the eye). Propose a hypothesis to explain the biochemical differences between different forms of collagen that give rise to such different properties.

18.6   The Quaternary Structure of Proteins

LEARNING GOAL

8 Describe the tertiary and quaternary structure of a protein.

LEARNING GOAL

10 List examples of proteins that require prosthetic groups, and explain the way in which they function.

LEARNING GOALS

3 Describe the primary structure of proteins, and draw the structure of the peptide bond.

5 Describe the types of secondary structure of a protein.

6 Discuss the forces that maintain secondary structure.

7 Describe the structure and functions of fibrous proteins.

8 Describe the tertiary and quaternary structure of a protein.

9 List the R group interactions that maintain protein conformation.

For many proteins, the functional form is not composed of a single peptide but is rather an aggregate of smaller globular peptides. For instance, the protein hemoglobin is composed of four individual globular peptide subunits: two identical α-subunits and two identical β-subunits. Only when the four peptides are bound to one another is the protein molecule functional. The association of several polypeptides to produce a functional protein defines the **quaternary structure** of a protein.

The forces that hold the quaternary structure of a protein are the same as those that hold the tertiary structure. These include van der Waals forces between hydrophobic R groups, hydrogen bonds between polar R groups, ionic bridges between oppositely charged R groups, and disulfide bridges.

In some cases, quaternary structure of a functional protein involves binding to a nonprotein group. This additional group is called a **prosthetic group.** For example, many of the receptor proteins on cell surfaces are **glycoproteins.** These are proteins with sugar groups covalently attached. Each of the subunits of hemoglobin is bound to an iron-containing heme group. The heme group is a large, unsaturated organic cyclic amine with an iron ion coordinated within it. As in the case of hemoglobin, the prosthetic group often determines the function of a protein. For instance, in hemoglobin it is the iron-containing heme groups that have the ability to bind reversibly to oxygen.

Question 18.5   Describe the four levels of protein structure.

Question 18.6   What are the weak interactions that maintain the tertiary structure of a protein?

18.7   An Overview of Protein Structure and Function

Let's summarize the various types of protein structure and their relationship to one another (Figure 18.10).

• *Primary Structure:* The primary structure of the protein is the amino acid sequence of the protein. The primary structure results from the formation of covalent peptide bonds between amino acids. Peptide bonds are amide bonds formed between the carboxylate group of one amino acid and the amino group of another.

Figure 18.10 Summary of the four levels of protein structure, using hemoglobin as an example.

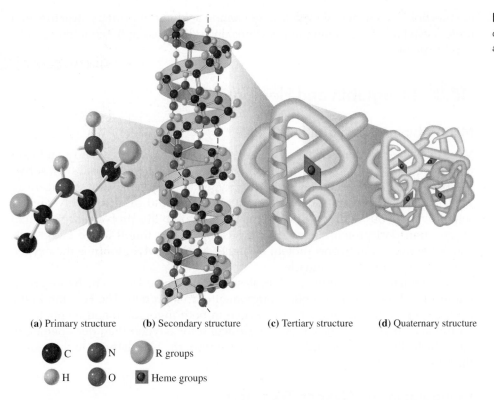

(a) Primary structure   **(b)** Secondary structure   **(c)** Tertiary structure   **(d)** Quaternary structure

C   N   R groups

H   O   Heme groups

- *Secondary Structure:* As the protein chain grows, numerous opportunities for noncovalent interactions in the backbone of the polypeptide chain become available. These cause regions of the chain to fold and orient themselves in a variety of conformational arrangements. The secondary level of structure includes the α-helix and the β-pleated sheet, which are the result of hydrogen bonding between the amide hydrogens and carbonyl oxygens of the peptide bonds. Different regions of the chain may be involved in different types of secondary structure arrangements; some regions might be α-helix and others might be a β-pleated sheet.
- *Tertiary Structure:* When we discuss tertiary structure, we are interested in the overall folding of the entire chain. In other words, we are concerned with the further folding of the secondary structure. Are the two ends of the chain close together or far apart? What general shape is involved? Both noncovalent interactions between the R groups of the amino acids and covalent disulfide bridges play a role in determining the tertiary structure. The noncovalent interactions include hydrogen bonding, ionic bonding, and van der Waals forces (London dispersion forces and dipole-dipole attractions).
- *Quaternary Structure:* Like tertiary structure, quaternary structure is concerned with the topological, spatial arrangements of two or more peptide chains with respect to each other. How is one chain oriented with respect to another? What is the overall shape of the final functional protein?

The quaternary structure is maintained by the same forces that are responsible for the tertiary structure. It is the tertiary and quaternary structures of the protein that ultimately define its function. Some have a fibrous structure with great mechanical strength. These make up the major structural components of the cell and the organism. Often they are also responsible for the movement of the organism. Others fold into globular shapes. Most of the transport proteins, regulatory proteins, and enzymes are globular proteins. The very precise three-dimensional structure of each of these proteins allows each to carry out its very specific function in the body. As we will see with the example of sickle cell hemoglobin in the

LEARNING GOAL

11 Discuss the importance of the three-dimensional structure of a protein to its function.

The three-dimensional structure of a protein is the feature that allows it to carry out its specific biological function. However, we must always remember that it is the primary structure, the order of the R groups, that determines how the protein will fold and what the ultimate shape will be.

next section, the change of even a single amino acid in the primary structure of a protein can have far-reaching implications, including loss of function that can be life-threatening.

18.8 Myoglobin and Hemoglobin

Myoglobin and Oxygen Storage

Most of the cells of our bodies are buried in the interior of the body and cannot directly get food molecules or eliminate waste. The circulatory system solves this problem by delivering nutrients and oxygen to body cells and carrying away wastes. Our cells require a steady supply of oxygen, but oxygen is only slightly soluble in aqueous solutions. To overcome this solubility problem, we have an oxygen transport protein, **hemoglobin.** Hemoglobin is found in red blood cells and is the oxygen transport protein of higher animals. **Myoglobin** is the oxygen storage protein of skeletal muscle.

The structure of myoglobin (Mb) is shown in Figure 18.11. The **heme group** (Figure 18.12) is also an essential component of this protein. The Fe^{2+} ion in the heme group is the binding site for oxygen in both myoglobin and hemoglobin. Fortunately, myoglobin has a greater attraction for oxygen than does hemoglobin, which allows efficient transfer of oxygen from the bloodstream to the cells of the body.

Hemoglobin and Oxygen Transport

Hemoglobin (Hb) is a tetramer composed of four polypeptide subunits: two α-subunits and two β-subunits (Figure 18.13). Because each subunit of hemoglobin contains a heme group, a hemoglobin molecule can bind four molecules of oxygen:

$$\text{Hb} + 4O_2 \longrightarrow \text{Hb}(O_2)_4$$

Deoxyhemoglobin   Oxyhemoglobin

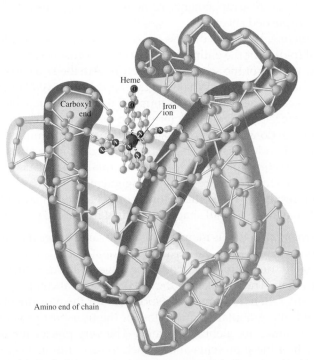

Figure 18.11 Myoglobin. The heme group has an iron ion to which oxygen binds.

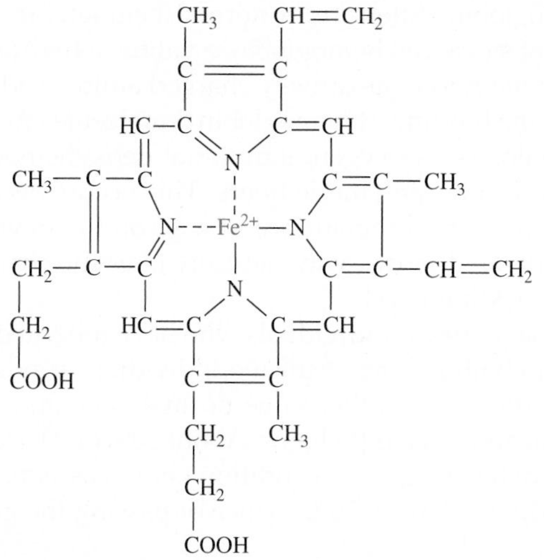

Figure 18.12 Structure of the heme prosthetic group, which binds to myoglobin and hemoglobin.

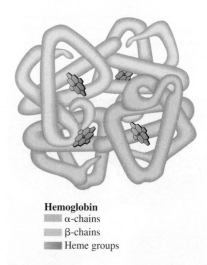

Hemoglobin
- α-chains
- β-chains
- Heme groups

Figure 18.13 Structure of hemoglobin. The protein contains four subunits, designated α and β. The α- and β-subunits face each other across a central cavity. Each subunit in the tetramer contains a heme group that binds oxygen.

The oxygenation of hemoglobin in the lungs and the transfer of oxygen from hemoglobin to myoglobin in the tissues are very complex processes. We begin our investigation of these events with the inhalation of a breath of air.

The oxygenation of hemoglobin in the lungs is greatly favored by differences in the oxygen partial pressure (pO_2) in the lungs and in the blood. The pO_2 in the air in the lungs is approximately 100 millimeters of mercury (mm Hg); the pO_2 in oxygen-depleted blood is only about 40 mm Hg. Oxygen diffuses from the region of high pO_2 in the lungs to the region of low pO_2 in the blood. There it enters red blood cells and binds to the Fe^{2+} ions of the heme groups of deoxyhemoglobin, forming oxyhemoglobin. This binding actually helps bring more O_2 into the blood.

Oxygen Transport from Mother to Fetus

A fetus receives its oxygen from its mother by simple diffusion across the placenta. If both the fetus and the mother had the same type of hemoglobin, this transfer process would not be efficient, because the hemoglobin of the fetus and the mother would have the same affinity for oxygen. The fetus, however, has a unique type of hemoglobin, called *fetal hemoglobin*. This unique hemoglobin molecule has a greater affinity for oxygen than does the mother's hemoglobin. Oxygen is therefore efficiently transported, via the circulatory system, from the lungs of the mother to the fetus. The biosynthesis of fetal hemoglobin stops shortly after birth when the genes encoding fetal hemoglobin are switched "off" and the genes coding for adult hemoglobin are switched "on."

Question 18.7 Why is oxygen efficiently transferred from hemoglobin in the blood to myoglobin in the muscles?

Question 18.8 How is oxygen efficiently transferred from mother to fetus?

Sickle Cell Anemia

Sickle cell anemia is a human genetic disease that first appeared in tropical west and central Africa. It afflicts about 0.4% of African Americans. These individuals produce a mutant hemoglobin known as sickle cell hemoglobin (Hb S). Sickle cell anemia receives its name from the sickled appearance of the red blood cells that form in this condition (Figure 18.14). The sickled cells are unable to pass through the small capillaries of the circulatory system, and circulation is hindered. This results in damage to many organs, especially bone and kidney, and can lead to death at an early age.

The genetic basis of this alteration is discussed in Chapter 20.

When hemoglobin is carrying O_2, it is called oxyhemoglobin. When it is not bound to O_2, it is called deoxyhemoglobin.

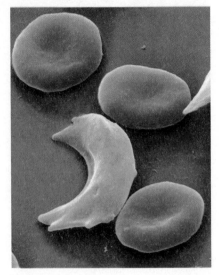

Figure 18.14 Scanning electron micrographs of normal and sickled red blood cells.

Sickle cell hemoglobin differs from normal hemoglobin by a single amino acid. In the β-chain of sickle cell hemoglobin, a valine (a hydrophobic amino acid) has replaced a glutamic acid (a negatively charged amino acid). This substitution provides a basis for the binding of hemoglobin S molecules to one another. When oxyhemoglobin S unloads its oxygen, individual deoxyhemoglobin S molecules bind to one another as long polymeric fibers. This occurs because the valine fits into a hydrophobic pocket on the surface of a second deoxyhemoglobin S molecule. The fibers generated in this way radically alter the shape of the red blood cell, resulting in the sickling effect.

Sickle cell anemia occurs in individuals who have inherited the gene for sickle cell hemoglobin from both parents. Afflicted individuals produce 90–100% defective β-chains. Individuals who inherit one normal gene and one defective gene produce both normal and altered β-chains. About 10% of African Americans carry a single copy of the defective gene, a condition known as *sickle cell trait*. Although not severely affected, they have a 50% chance of passing the gene to each of their children.

An interesting relationship exists between sickle cell trait and resistance to malaria. In some parts of Africa, up to 20% of the population has sickle cell trait. In those same parts of Africa, one of the leading causes of death is malaria. The presence of sickle cell trait is linked to an increased resistance to malaria because the malarial parasite cannot feed efficiently on sickled red blood cells. People who have sickle cell disease die young; those without sickle cell trait have a high probability of succumbing to malaria. Occupying the middle ground, people who have sickle cell trait do not suffer much from sickle cell anemia and simultaneously resist deadly malaria. Because those with sickle cell trait have a greater chance of survival and reproduction, the sickle cell hemoglobin gene is maintained in the population.

18.9   Proteins in the Blood

The blood plasma of a healthy individual typically contains 60–80 grams per liter (g/L) of protein. This protein can be separated into five classes designated α through γ. The separation is based on the overall surface charge on each of the types of protein.

The most abundant protein in the blood is albumin, making up about 55% of the blood protein. Albumin contributes to the osmotic pressure of the blood simply because it is a dissolved molecule. It also serves as a nonspecific transport molecule for important metabolites that are otherwise poorly soluble in water. Among the molecules transported through the blood by albumin are bilirubin (a waste product of the breakdown of hemoglobin), Ca^{2+}, and fatty acids (organic anions).

The α-globulins ($α_1$ and $α_2$) make up 13% of the plasma proteins. They include glycoproteins (proteins with sugar groups attached), high-density lipoproteins, haptoglobin (a transport protein for free hemoglobin), ceruloplasmin (a copper transport protein), prothrombin (a protein involved in blood clotting), and very low density lipoproteins. The most abundant $α_1$-globulin is $α_1$-antitrypsin. Although the name leads us to believe that this protein inhibits a digestive enzyme, trypsin, the primary function of $α_1$-antitrypsin is the inactivation of an enzyme that causes damage in the lungs (see also A Medical Perspective: $α_1$-Antitrypsin and Familial Emphysema in Chapter 19). $α_1$-Antichymotrypsin is another inhibitor found in the bloodstream. This protein, along with amyloid proteins, is found in the amyloid

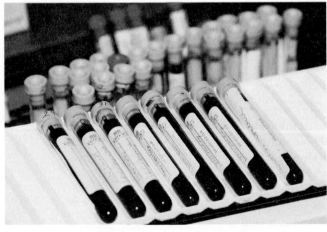

Blood samples drawn from patients

plaques characteristic of Alzheimer's disease (AD). As a result, it has been suggested that an overproduction of this protein may contribute to AD. In the blood, α_1-antichymotrypsin is also found complexed to prostate specific antigen (PSA), the protein antigen that is measured as an indicator of prostate cancer. Elevated PSA levels are observed in those with the disease. It is interesting to note that PSA is a chymotrypsin-like proteolytic enzyme.

The β-globulins represent 13% of the blood plasma proteins and include transferrin (an iron transport protein) and low-density lipoprotein. Fibrinogen, a protein involved in coagulation of blood, comprises 7% of the plasma protein. Finally, the γ-globulins, IgG, IgM, IgA, IgD, and IgE, make up the remaining 11% of the plasma proteins. The γ-globulins are synthesized by B lymphocytes, but most of the remaining plasma proteins are synthesized in the liver. In fact, a frequent hallmark of liver disease is reduced amounts of one or more of the plasma proteins.

18.10 Denaturation of Proteins

We have shown that the shape of a protein is absolutely essential to its function. We have also mentioned that life can exist only within a rather narrow range of temperature and pH. How are these two concepts related? As we will see, extremes of pH or temperature have a drastic effect on protein conformation, causing the molecules to lose their characteristic three-dimensional shape. **Denaturation** occurs when the organized structures of a globular protein, the α-helix, the β-pleated sheet, and tertiary folds become completely disorganized. However, it does not alter the primary structure. Denaturation of an α-helical protein is shown in Figure 18.15.

LEARNING GOAL

13 Describe how extremes of pH and temperature cause denaturation of proteins.

Temperature

Consider the effect of increasing temperature on a solution of proteins—for instance, egg white. At first, increasing the temperature simply increases the rate of molecular movement, the movement of the individual molecules within the solution. Then, as the temperature continues to increase, the bonds within the proteins begin to vibrate more violently. Eventually, the weak interactions, like hydrogen bonds and hydrophobic interactions, that maintain the protein structure are disrupted. The protein molecules are denatured as they lose their characteristic three-dimensional conformation and become completely disorganized. **Coagulation** occurs as the protein molecules then unfold and become entangled. At this point, they are no longer in solution; they have aggregated to become a solid and will precipitate out of

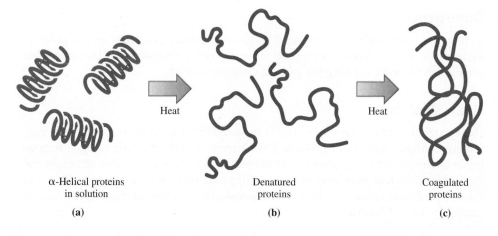

α-Helical proteins in solution   Denatured proteins   Coagulated proteins

(a)   (b)   (c)

Figure 18.15 The denaturation of proteins by heat. (a) The α-helical proteins are in solution. (b) As heat is applied, the hydrogen bonds maintaining the secondary structure are disrupted, and the protein structure becomes disorganized. The protein is denatured. (c) The denatured proteins clump together, or coagulate, and are now in an insoluble form.

Proteins in each of the foods shown here have been denatured. What agent caused denaturation in each case?

Lactate fermentation is discussed in Section 21.4.

the solution. (see Figure 18.15). The egg white began as a viscous solution of egg albumins; but when it was cooked, the proteins were denatured and coagulated to become solid.

Many of the proteins of our cells—for instance, the enzymes—are in the same kind of viscous solution within the cytoplasm. To continue to function properly, they must remain in solution and maintain the correct three-dimensional configuration. If the body temperature becomes too high, or if local regions of the body are subjected to very high temperatures, as when you touch a hot cookie sheet, cellular proteins become denatured. They lose their function, and the cell or the organism dies.

pH

Because of the R groups of the amino acids, all proteins have a characteristic electrical charge. Because every protein has a different amino acid composition, each will have a characteristic net electrical charge on its surface. The positively and negatively charged R groups on the surface of the molecule interact with ions and water molecules, and these interactions keep the protein in solution within the cytoplasm.

When the pH of the solution is changed dramatically, the acid or base will change the charge of the protein, interfering with the salt bridges and hydrogen bonds that stabilize the tertiary structure.

This is a reaction that you have probably observed in your own kitchen. When milk sits in the refrigerator for a prolonged period, the bacteria in the milk begin to grow. They use the milk sugar, lactose, as an energy source in the process of fermentation and produce lactic acid as a by-product. As the bacteria continue to grow, the concentration of lactic acid increases. The additional acid results in the protonation of exposed carboxylate groups on the surface of the dissolved milk proteins. As a result, they lose their characteristic surface charge and can no longer associate with ions and water or repel one another. Under these conditions, the proteins tend to clump together and precipitate out of solution, forming a solid curd.

Organic Solvents

Polar organic solvents, such as rubbing alcohol (2-propanol), denature proteins by disrupting hydrogen bonds within the protein, in addition to forming hydrogen bonds with the solvent, water. The nonpolar regions of these solvents interfere with hydrophobic interactions in the interior of the protein molecule, thereby disrupting the conformation. Traditionally, a 70% solution of rubbing alcohol was often used as a disinfectant or antiseptic. However, recent evidence suggests that it is not an effective agent in this capacity.

Detergents

Detergents have both a hydrophobic region (the fatty acid tail) and a polar or hydrophilic region. When detergents interact with proteins, they disrupt hydrophobic interactions, causing the protein chain to unfold.

Heavy Metals

Heavy metals such as mercury (Hg^{2+}) or lead (Pb^{2+}) may form bonds with negatively charged side chain groups. This interferes with the salt bridges formed between amino acid R groups of the protein chain, resulting in loss of conformation. Heavy metals may also bind to sulfhydryl groups of a protein. This may cause a profound change in the three-dimensional structure of the protein, accompanied by loss of function.

Egg Foams: Meringues and Soufflés

The transformation of an egg white into a bowl of snow-white fluff simply by whisking vigorously for a few minutes seems quite amazing. Somehow, the liquid egg white and the air whisked into it form a structure that appears solid and has the body to form lovely peaks. Chefs have been using this method for about 350 years to provide delightful meringues for pies and the base for sweet or savory soufflés and mousses.

This foam is really just a bowl of air bubbles, and each of the bubbles is encased in a thin film of denatured egg protein. Two physical stresses, brought about by whipping the egg whites, cause the proteins to denature. First, the drag of the whisk through the liquid creates a pulling force that unfolds and stretches out the normally globular proteins. The interface of the water and the air further causes unfolding because of the two very different environments. The denatured proteins tend to gather at the interface of the air and the water. Their hydrophilic regions remain in the water and their hydrophobic regions project into the air. In this way, a layer of denatured protein forms around the bubbles, causing them to be stable.

With time, the foam will separate. So how can we stabilize the foam for the final dish? One common way is simply to bake it. The heat of baking causes the ovalbumin, which isn't denatured by the whipping, to denature. This further adds to the wall of protein surrounding the air bubbles. At the same time, much of the water evaporates from the foam. This further aids the transformation of the egg foam into a permanent solid form—a meringue.

French chefs in the eighteenth century noticed that the use of a copper bowl to create egg foams created a more stable foam. We now know that the reason for this is that copper forms an extremely tight bond to —SH groups. This prevents disulfide bond formation which, in turn, keeps the proteins from binding to one another too tightly. When these stronger disulfide bonds form, the protein meshwork around the bubbles tends to become heavy and to collapse the fragile structure. Of course, copper bowls are expensive and difficult to care for. It turns out that the addition of a bit of acid will do the same thing; the addition of 1/8 teaspoon (tsp) of cream of tartar or ½ tsp of lemon juice per egg at the beginning will also inhibit disulfide bond formation and promote a more stable foam.

Soufflés are simply a savory or sweet mixture or batter into which egg foam is folded. When placed in the oven, they puff up into delightfully light dishes that please both the eye and the palate. The principle governing the behavior of a soufflé is Charles's law. As you learned much earlier in this book, the volume of a gas is proportional to its temperature. When the soufflé is baked, the gas trapped within the bubbles expands as the temperature rises. Although the cooked mixture gives some stability to the risen soufflé, Charles's law determines the behavior of the soufflé when it is taken from the oven and served. The result—the soufflé will fall.

Both the rise and the fall of a soufflé depend on the ingredients in the mixture and the details of the way in which it is baked. Modern chefs sometimes use unusual apparatus to prepare their soufflés. The area of molecular gastronomy has combined the chemistry lab with the kitchen and the results are sometimes astounding. Hervé This, a French physical chemist and father of the molecular gastronomy movement, describes baking a soufflé in a vacuum to provide the highest, lightest soufflé yet served!

For Further Understanding

- Why does the use of a vacuum lead to a lighter soufflé?
- The presence of egg yolks or detergents makes it impossible to create an egg foam. Explain this observation.

Mechanical Stress

Stirring, whipping, or shaking can disrupt the weak interactions that maintain protein conformation. The drag of the whisk or beater creates a pulling force that unfolds and stretches the proteins. This is part of the reason that whipping egg whites produces a stiff meringue. For more information on the preparation of egg foams and their use in the kitchen, see Kitchen Chemistry: Egg Foams: Meringues and Soufflés.

Question 18.9 How does high temperature denature proteins?

Question 18.10 How does extremely low pH cause proteins to coagulate?

A Medical Perspective

Medications from Venoms

Some scientists are traveling to the ends of the earth to find exotic and venomous creatures—and all in the hopes of discovering new drugs and medications. This is not a totally new idea. There are actually a number of drugs on the market that have been developed to mimic the action of protein venoms. Captopril has been used since 1981 to treat hypertension and, in some cases, congestive heart failure. It was designed to mimic a peptide found in the venom of the lancehead viper, a South American snake. Scientists determined that the biological activity resided in the sulfhydryl group of the peptide, and so the relatively simple molecule became a very effective addition to our arsenal of hypertensive medications.

Captopril

A 39-amino-acid peptide called Byetta has been demonstrated to help control the blood glucose levels of type 2 diabetics. Byetta, which is injected subcutaneously twice a day, is a synthetic version of a hormone found in the saliva of the Gila monster. It is thought that the peptide enhances the glucose-dependent secretion of insulin by the β-cells of the pancreas.

Prialt, an unusual pain medication, is based on a toxin found in cone snails. It is a hydrophilic peptide and is not effective when administered orally or intravenously. It must be introduced directly into the cerebrospinal fluid. As a result, it is reserved for the treatment of severe, chronic pain that is not managed by traditional methods.

Other venom-based drugs are currently in the development stage. A peptide based on the venom of a sea anemone is being studied for the promise that it may combat autoimmune disorders. Another peptide originally isolated from tarantula venom is being studied in the hope that it will be effective against muscular dystrophy.

For many years, we considered these venom peptides only from the point of view of the potentially deadly reactions of those who are bitten. Indeed, much of the study still involves the development of anti-venom that can be more widely available, especially in developing countries. The recognition that the many bio-active molecules in venom represent a potential treasure trove of new medications is sparking intensive research into methods for identifying and studying these proteins.

For Further Understanding

▶ Byetta is a 39-amino-acid peptide amide with the following structure: HGEGTFTSDLSKQMEEEAVRLFIEWLKNGGP SSGAPPPS-NH$_2$. Write out the amino acid sequence of this peptide using the names of the amino acids. Draw the structure of the N-terminus.

▶ The development of Captopril has been considered a case of "biopiracy," because the Brazilian tribe that initially used the venom for poison arrow tips have not received any benefits from the development of the drug. Go onto the Internet and read about the efforts to protect fair access to the genetic resources of the environment and the indigenous peoples who have discovered them.

18.11 Dietary Protein and Protein Digestion

LEARNING GOAL

14 Explain the difference between essential and nonessential amino acids.

Proteins, as well as carbohydrates and fats, are an energy source in the diet. As do carbohydrates and fats, proteins serve several dietary purposes. They can be oxidized to provide energy. In addition, the amino acids liberated by the hydrolysis of proteins are used directly in biosynthesis. The protein synthetic machinery of the cell can incorporate amino acids, released by the digestion of dietary protein, directly into new cellular proteins. Amino acids are also used in the biosynthesis of a large number of important molecules called the *nitrogen compounds*. This group includes some hormones, the heme groups of hemoglobin and myoglobin, and the nitrogen-containing bases found in DNA and RNA.

Digestion of dietary protein begins in the stomach. The stomach enzyme *pepsin* begins the digestion by hydrolyzing some of the peptide bonds of the protein. This breaks the protein down into smaller peptides.

Production of pepsin and other proteolytic digestive enzymes must be carefully controlled because the active enzymes would digest and destroy the cell that produces them. Thus, pepsin is actually synthesized and secreted in an inactive form called *pepsinogen*. Pepsinogen has an additional forty-two amino acids in its primary structure. These are removed in the stomach to produce active pepsin.

The inactive form of a proteolytic enzyme is called a proenzyme. These are discussed in Section 19.9.

Protein digestion continues in the small intestine where the enzymes trypsin, chymotrypsin, elastase, and others catalyze the hydrolysis of peptide bonds at different sites in the protein. For instance, chymotrypsin cleaves peptide bonds on the carbonyl side of aromatic amino acids and trypsin cleaves peptide bonds on the carbonyl side of basic amino acids. Together these proteolytic enzymes degrade large dietary proteins into amino acids that can be absorbed by cells of the small intestine.

The specificity of proteolytic enzymes is described in Section 19.11.

Amino acids can be divided into two major nutritional classes. **Essential amino acids** are those that cannot be synthesized by the body and are required in the diet. **Nonessential amino acids** are those amino acids that can be synthesized by the body and need not be included in the diet. Table 18.2 lists the essential and nonessential amino acids.

See Section 19.11 and Figure 19.13 for a more detailed picture of the action of digestive proteases.

Proteins are also classified as *complete* or *incomplete*. Protein derived from animal sources is generally **complete protein.** That is, it provides all of the essential and nonessential amino acids in approximately the correct amounts for biosynthesis. In contrast, protein derived from vegetable sources is generally **incomplete protein** because it lacks a sufficient amount of one or more essential amino acids. In fact, soy protein is the only known complete vegetable protein. See A Human Perspective: The New Protein found earlier in this chapter. People who want to maintain a strictly vegetarian diet or for whom animal protein is often not

TABLE 18.2 The Essential and Nonessential Amino Acids

Essential Amino Acids	Nonessential Amino Acids
Isoleucine	Alanine
Leucine	Arginine[1]
Lysine	Asparagine
Methionine	Aspartate
Phenylalanine	Cysteine[2]
Threonine	Glutamate
Tryptophan	Glutamine
Valine	Glycine
	Histidine[1]
	Proline
	Serine
	Tyrosine[2]

[1]Histidine and arginine are essential amino acids for infants but not for healthy adults.
[2]Cysteine and tyrosine are considered to be semiessential amino acids. They are required by premature infants and adults who are ill.

available have the problem that most high-protein vegetables do not have all of the essential amino acids needed to ensure a sufficient daily intake. For example, the major protein of beans contains abundant lysine and tryptophan but very little methionine, whereas corn contains considerable methionine but very little tryptophan or lysine. A mixture of corn and beans, however, satisfies both requirements. This combination, called *succotash,* was a staple of the diet of Native Americans for centuries.

Eating a few vegetarian meals each week can provide all the required amino acids and simultaneously help reduce the amount of saturated fats in the diet. Many ethnic foods apply the principle of mixing protein sources. Mexican foods such as tortillas and refried beans, Cajun dishes of spicy beans and rice, Indian cuisine of rice and lentils, and even the traditional American peanut butter sandwich are examples of ways to mix foods to provide complete protein.

Question 18.11   Why must vegetable sources of protein be mixed to provide an adequate diet?

Question 18.12   What are some common sources of dietary protein?

CHAPTER MAP

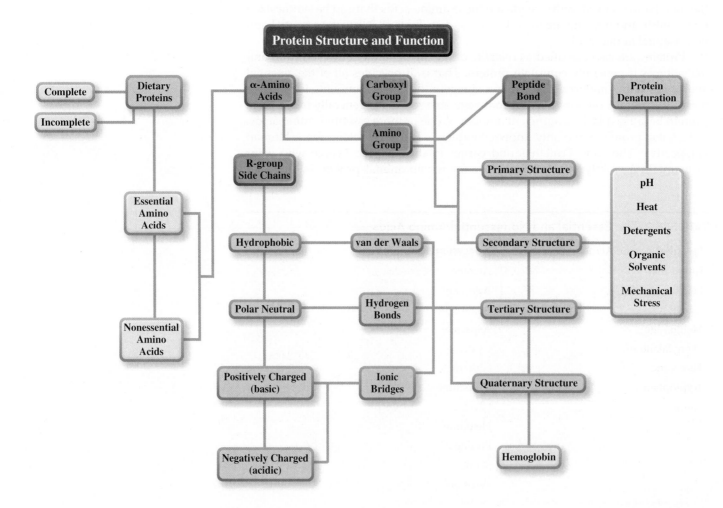

SUMMARY

Introduction

▶ **Proteins** have many functions in the body.
- **Enzymes** are biological catalysts.
- **Defense proteins,** including **antibodies,** are produced in response to foreign **antigens.**
- **Transport proteins** carry materials throughout the body.
- **Regulatory proteins** control many aspects of cellular function.
- **Movement proteins** such as actin and myosin in muscle cells are needed for all types of motion.
- **Nutrient proteins** such as egg albumin and casein in milk, serve as amino acid sources for embryos or infants.
- **Structural proteins** provide mechanical support.

18.1 Protein Building Blocks: The α-Amino Acids

▶ Proteins are made up of twenty different α-**amino acids,** each having an α-carboxylate group and an α-amino group.

▶ Amino acids differ from one another in their side-chain R groups. Differences in polarity distinguish different groups of amino acids.
- Some amino acid side chains are nonpolar. These are **hydrophobic amino acids.**
- Some amino acids have polar or charged side chains. These are **hydrophilic amino acids.**

▶ All amino acids except glycine are chiral molecules.

▶ Naturally occurring amino acids are generally L-amino acids.

18.2 The Peptide Bond

▶ Amino acids are joined by **peptide bonds** to produce peptides and proteins.

▶ The peptide bond is an amide bond formed between the α-carboxylate group of one amino acid and the α-amino group of another.

▶ The peptide bond is planar and relatively rigid.

▶ In a peptide chain, the amino acid with a free carboxylate group is called the **C-terminal amino acid** residue **(C-terminus);** the amino acid with a free amino group is called the **N-terminal amino acid** residue **(N-terminus).**

18.3 The Primary Structure of Proteins

▶ Proteins are linear polymers of amino acids.

▶ The linear sequence of amino acids defines the **primary structure** of the peptide.

18.4 The Secondary Structure of Proteins

▶ The **secondary structure** of a protein is the folding of regions of the primary sequence into an α-**helix** or a β-**pleated sheet.**

▶ These two structures are maintained by hydrogen bonds between the amide nitrogen and the carbonyl oxygen of the peptide bond.

▶ Some structural proteins, such as the α-**keratins,** are entirely composed of α-helix.

▶ Some **fibrous proteins,** such as silk fibroin, are composed of β-pleated sheets.

18.5 The Tertiary Structure of Proteins

▶ **Globular proteins** have varying amounts of α-helix or β-pleated sheet folded into higher levels of structure called **tertiary structure.** Tertiary structure refers to the three-dimensional folding of the entire protein chain.

▶ The tertiary structure of a protein is maintained by attractive forces between the R groups of amino acids, including the following:
- Hydrophobic interactions
- Hydrogen bonds
- Ionic bridges
- Disulfide bonds

18.6 The Quaternary Structure of Proteins

▶ Proteins composed of more than one peptide are said to have **quaternary structure.**

▶ Attractive forces between R groups hold the peptide subunits together.

▶ Some proteins require an additional nonprotein **prosthetic group** in order to carry out their functions.

▶ **Glycoproteins** have covalently bonded sugar groups.

18.7 An Overview of Protein Structure and Function

▶ The primary structure of a protein dictates the way in which it folds into secondary and tertiary levels of structure and whether it will associate with other protein subunits, producing quaternary structure.

▶ An amino acid change in the protein can have drastic effects on protein folding and thus on the ability of the protein to function.

18.8 Myoglobin and Hemoglobin

▶ **Myoglobin,** the oxygen storage protein of skeletal muscle, has a prosthetic group called the **heme group,** which is the site of oxygen binding.

▶ **Hemoglobin,** which transports oxygen from the lungs to the tissues, consists of four peptides, each of which has a heme group.

▶ Myoglobin has a greater affinity for oxygen than hemoglobin; so oxygen is efficiently transferred from hemoglobin in the blood to myoglobin in the tissues.

▶ Fetal hemoglobin has a greater affinity for oxygen than maternal hemoglobin, allowing efficient transfer of oxygen across the placenta from mother to fetus.

▶ **Sickle cell anemia** is caused by a mutant hemoglobin gene.

18.9 Proteins in the Blood

▶ Blood plasma contains 60–80 g/L of protein.

▶ These proteins can be separated into five classes designated α through γ.

▶ Albumin is the most abundant protein in the blood (55%).

▶ The α-globulins and β-globulins each make up 13% of the blood proteins.

▶ Fibrinogen comprises 7% of the blood proteins.

▶ The remaining proteins in the blood are the γ-globulins.

18.10 Denaturation of Proteins

▶ Heat disrupts the hydrogen bonds and hydrophobic interactions that maintain protein structure, causing the protein to unfold and lose its organized structure. This is **denaturation.**

▶ **Coagulation,** or clumping, occurs when the protein chains unfold and become entangled. When this happens, the protein is no longer water-soluble.

▶ Changes in pH interfere with ionic bridges and hydrogen bonds that stabilize and maintain tertiary structure.

18.11 Dietary Protein and Protein Digestion

▶ **Essential amino acids** must be acquired in the diet.

▶ **Nonessential amino acids** can be synthesized by the body.

▶ **Complete proteins** contain all the essential and nonessential amino acids.

▶ **Incomplete proteins** are missing one or more essential amino acids.

▶ Protein digestion begins in the stomach, where proteins are degraded by pepsin.

▶ Further digestion occurs in the small intestine by enzymes such as trypsin and chymotrypsin.

ANSWERS TO PRACTICE PROBLEMS

18.1 a. Methionyl-leucyl-cysteine

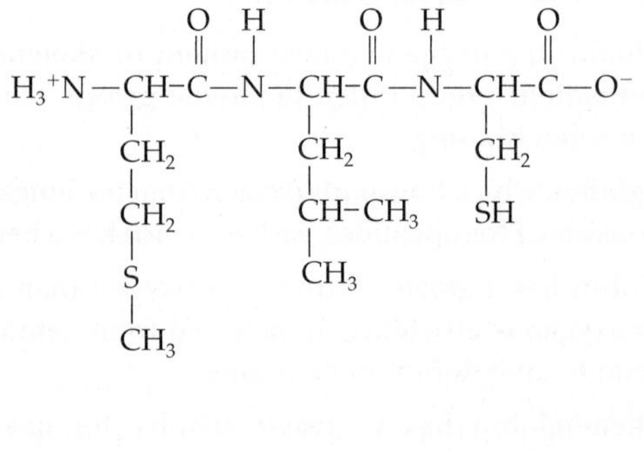

b. Tyrosyl-seryl-histidine

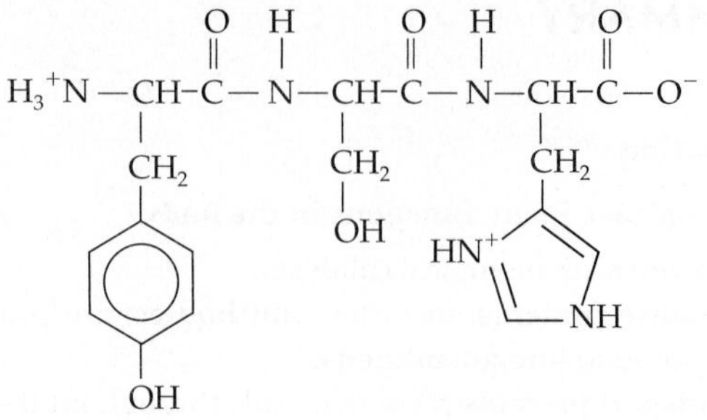

c. Arginyl-isoleucyl-glutamine

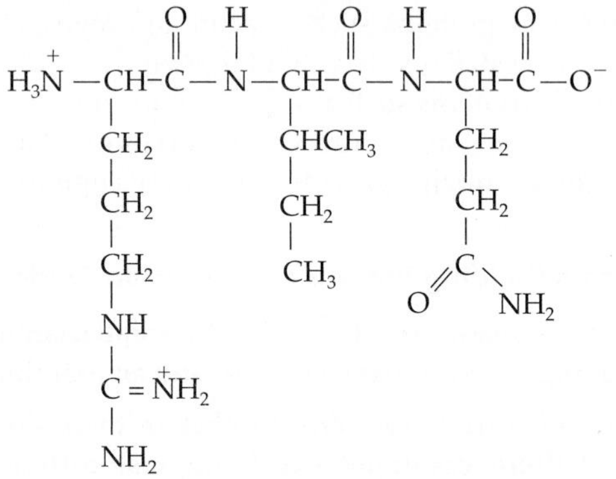

QUESTIONS AND PROBLEMS

Introduction to Protein Functions

Foundations

18.13 Define the term *enzyme.*

18.14 Define the term *antibody.*

18.15 What is a transport protein?

18.16 What are the functions of structural proteins?

Applications

18.17 Of what significance are enzymes in the cell?

18.18 How do antibodies protect us against infection?

18.19 List two transport proteins and describe their significance to the organism.

18.20 What is the function of regulatory proteins?

18.21 Provide two examples of nutrient proteins.

18.22 Provide two examples of proteins that are required for movement.

Protein Building Blocks: The α-Amino Acids

Foundations

18.23 Describe the basic general structure of an L-α-amino acid, and draw its structure.

18.24 Draw the D- and L-isomers of valine. Which would you expect to find in nature?

18.25 What is a zwitterion?

18.26 Why are amino acids zwitterions at pH 7.0?

Applications

18.27 What is a chiral carbon?

18.28 Why are all of the α-amino acids except glycine chiral?

18.29 What is the importance of the R groups of the amino acids?

18.30 Describe the classification of the R groups of the amino acids, and provide an example of each class.

18.31 Write the structures of the six amino acids that have polar, neutral side chains.

18.32 Write the structures of the positively charged amino acids. Indicate whether you would expect to find each on the surface or buried in a globular protein.

The Peptide Bond

Foundations

18.33 Define the term *peptide bond*.

18.34 What type of bond is the peptide bond? Explain why the peptide bond is rigid.

18.35 What observations led Linus Pauling and his colleagues to hypothesize that the peptide bond exists as a resonance hybrid?

18.36 Draw the resonance hybrids that represent the peptide bond.

Applications

18.37 Write the structure of each of the following peptides:
   a. Phe-val-tyr
   b. Ala-glu-cys
   c. Asn-leu-gly

18.38 Write the structure of each of the following peptides:
   a. Lys-trp-pro
   b. Gln-ser-his
   c. Arg-met-asp

The Primary Structure of Proteins

Foundations

18.39 Define the *primary structure* of a protein.

18.40 What type of bond joins the amino acids to one another in the primary structure of a protein?

Applications

18.41 How does the primary structure of a protein determine its three-dimensional shape and biological function?

18.42 Explain the relationship between the primary structure of a protein and the gene for that protein.

18.43 Write a balanced equation showing peptide bond formation between leucine and arginine.

18.44 Write a balanced equation showing peptide bond formation between threonine and aspartate.

The Secondary Structure of Proteins

Foundations

18.45 Define the secondary structure of a protein.

18.46 What are the two most common types of secondary structure?

Applications

18.47 What type of secondary structure is characteristic of:
   a. The α-keratins?
   b. Silk fibroin?

18.48 Describe the forces that maintain the two types of secondary structure: α-helix and β-pleated sheet.

18.49 Define fibrous proteins.

18.50 What is the relationship between the structure of fibrous proteins and their functions?

18.51 Describe a parallel β-pleated sheet.

18.52 Compare a parallel β-pleated sheet to an antiparallel β-pleated sheet.

The Tertiary Structure of Proteins

Foundations

18.53 Define the tertiary structure of a protein.

18.54 Use examples of specific amino acids to show the variety of weak interactions that maintain tertiary protein structure.

Applications

18.55 Write the structure of the amino acid produced by the oxidation of cysteine.

18.56 What is the role of cystine in maintaining protein structure?

18.57 Explain the relationship between the secondary and tertiary protein structures.

18.58 Why is the amino acid proline often found in the random coil hinge regions of the tertiary structure?

The Quaternary Structure of Proteins

Foundations

18.59 Describe the quaternary structure of proteins.

18.60 What weak interactions are responsible for maintaining quaternary protein structure?

Applications

18.61 What is a glycoprotein?

18.62 What is a prosthetic group?

An Overview of Protein Structure and Function

Applications

18.63 Why is hydrogen bonding so important to protein structure?

18.64 Explain why α-keratins that have many disulfide bonds between adjacent polypeptide chains are much less elastic and much harder than those without disulfide bonds.

18.65 How does the structure of the peptide bond make the structure of proteins relatively rigid?

18.66 The primary structure of a protein known as histone H4, which tightly binds DNA, is identical in all mammals and differs by only one amino acid between the calf and pea seedlings. What does this extraordinary conservation of primary structure imply about the importance of that one amino acid?

18.67 What does it mean to say that the structure of proteins is genetically determined?

18.68 Explain why genetic mutations that result in the replacement of one amino acid with another can lead to the formation of a protein that cannot carry out its biological function.

Myoglobin and Hemoglobin

Foundations

18.69 What is the function of hemoglobin?

18.70 What is the function of myoglobin?

18.71 Describe the structure of hemoglobin.

18.72 Describe the structure of myoglobin.

18.73 What is the function of heme in hemoglobin and myoglobin?

18.74 Write an equation representing the binding to and release of oxygen from hemoglobin.

Applications

18.75 Carbon monoxide binds tightly to the heme groups of hemoglobin and myoglobin. How does this affinity reflect the toxicity of carbon monoxide?

18.76 The blood of the horseshoe crab is blue because of the presence of a protein called *hemocyanin*. What is the function of hemocyanin?

18.77 Why does replacement of glutamic acid with valine alter hemoglobin and ultimately result in sickle cell anemia?

18.78 How do sickled red blood cells hinder circulation?

18.79 What is the difference between sickle cell disease and sickle cell trait?

18.80 How is it possible for sickle cell trait to confer a survival benefit on the person who possesses it?

Proteins in the Blood

Foundations

18.81 What is the most abundant protein in the blood?

18.82 List the functions of several α-globulins.

Applications

18.83 Develop a hypothesis to explain why albumin in the blood can serve as a nonspecific carrier for such diverse substances as bilirubin, Ca^{2+}, and fatty acids. (*Hint:* Consider what you know about the structures of amino acid R groups.)

18.84 Fibrinogen and prothrombin are both involved in formation of blood clots when they are converted into proteolytic enzymes. However, they are normally found in the blood in an inactive form. Develop an explanation for this observation.

Denaturation of Proteins

Foundations

18.85 Define the term *denaturation*.

18.86 What is the difference between denaturation and coagulation?

Applications

18.87 Why is heat an effective means of sterilization?

18.88 As you increase the temperature of an enzyme-catalyzed reaction, the rate of the reaction initially increases. It then reaches a maximum rate and finally dramatically declines. Keeping in mind that enzymes are proteins, how do you explain these changes in reaction rate?

18.89 Yogurt is produced from milk by the action of dairy bacteria. These bacteria produce lactic acid as a by-product

of their metabolism. The pH decrease causes the milk proteins to coagulate. Why are food preservatives not required to inhibit the growth of bacteria in yogurt?

18.90 Wine is made from the juice of grapes by varieties of yeast. The yeast cells produce ethanol as a by-product of their fermentation. However, when the ethanol concentration reaches 12–13%, all the yeast die. Explain this observation.

Dietary Protein and Protein Digestion

Foundations

18.91 Define the term *essential amino acid*.

18.92 Define the term *nonessential amino acid*.

18.93 Define the term *complete protein*.

18.94 Define the term *incomplete protein*.

Applications

18.95 Write an equation representing the action of the proteolytic enzyme chymotrypsin. (*Hint:* In order to write the structure of a dipeptide that would be an appropriate reactant, you must consider what is known about where chymotrypsin cleaves a protein chain.)

18.96 Write an equation representing the action of the proteolytic enzyme trypsin. (*Hint:* In order to write the structure of a dipeptide that would be an appropriate reactant, you must consider what is known about where trypsin cleaves a protein chain.)

18.97 Why is it necessary to mix vegetable proteins to provide an adequate vegetarian diet?

18.98 Name some ethnic foods that apply the principle of mixing vegetable proteins to provide all of the essential amino acids.

18.99 Why must synthesis of digestive enzymes be carefully controlled?

18.100 What is the relationship between pepsin and pepsinogen?

CHALLENGE PROBLEMS

1. Calculate the length of an α-helical polypeptide that is twenty amino acids long. Calculate the length of a region of antiparallel β-pleated sheet that is forty amino acids long.

2. Proteins involved in transport of molecules or ions into or out of cells are found in the membranes of all cells. They are classified as transmembrane proteins because some regions are embedded within the lipid bilayer, whereas other regions protrude into the cytoplasm or outside the cell. Review the classification of amino acids based on the properties of their R groups. What type of amino acids would you expect to find in the regions of the proteins embedded within the membrane? What type of amino acids would you expect to find on the surface of the regions in the cytoplasm or that protrude outside the cell?

3. A biochemist is trying to purify the enzyme hexokinase from a bacterium that normally grows in the Arctic Ocean at 5° C. In the next lab, a graduate student is trying to purify the same protein from a bacterium that grows in the vent of a

volcano at 98° C. To maintain the structure of the protein from the Arctic bacterium, the first biochemist must carry out all her purification procedures at refrigerator temperatures. The second biochemist must perform all his experiments in a warm room incubator. In molecular terms, explain why the same kind of enzyme from organisms with different optimal temperatures for growth can have such different thermal properties.

4. The α-keratin of hair is rich in the amino acid cysteine. The location of these cysteines in the protein chain is genetically determined. As a result of the location of the cysteines in the protein, a person may have curly, wavy, or straight hair. How can the location of cysteines in α-keratin result in these different styles of hair? Propose a hypothesis to explain how a "perm" causes straight hair to become curly.

5. Calculate the number of different pentapeptides you can make in which the amino acids phenylalanine, glycine, serine, leucine, and histidine are each found. Imagine how many proteins could be made from the twenty amino acids commonly found in proteins.

19 Enzymes

A hot-spring-fed lake at Yellowstone National Park. Explain how bacteria can thrive at temperatures near the boiling point of water.

OUTLINE

LEARNING GOALS

1 Classify enzymes according to the type of reaction catalyzed and the type of specificity.

2 Give examples of the correlation between an enzyme's common name and its function.

3 Describe the effect that enzymes have on the activation energy of a reaction.

4 Explain the effect of substrate concentration on enzyme-catalyzed reactions.

5 Discuss the role of the active site and the importance of enzyme specificity.

6 Describe the difference between the lock-and-key model and the induced fit model of enzyme-substrate complex formation.

7 Discuss the roles of cofactors and coenzymes in enzyme activity.

8 Explain how pH and temperature affect the rate of an enzyme-catalyzed reaction.

9 Describe the mechanisms used by cells to regulate enzyme activity.

10 Discuss the mechanisms by which certain chemicals inhibit enzyme activity.

11 Discuss the role of the enzyme chymotrypsin and other serine proteases.

12 Provide examples of medical uses of enzymes.

INTRODUCTION

Imagine the earth about four billion years ago: It was young then, not even a billion years old. Beginning as a red-hot molten sphere, slowly the earth's surface had cooled and become solid rock. But the interior, still extremely hot, erupted through the crust, spewing hot gases and lava. Eventually these eruptions produced craggy landmasses and an atmosphere composed of gases like hydrogen, carbon dioxide, ammonia, and water vapor. As the water vapor cooled, it condensed into liquid water, forming ponds and shallow seas.

At the dawn of biological life, the surface of the earth was still very hot and covered with rocky peaks and hot shallow oceans. The atmosphere was not very inviting either—filled with noxious gases and containing no molecular oxygen. Yet this is the environment where life on our planet began.

Some scientists think that they have found bacteria—living fossils—that may be very closely related to the first inhabitants of earth. These bacteria thrive at temperatures higher than the boiling point of water. Some need only H_2, CO_2, and H_2O for their metabolic processes, and they quickly die in the presence of molecular oxygen.

But this lifestyle raises some uncomfortable questions. For instance, how do these bacteria survive at these extreme temperatures that would cook the life-forms with which we are more familiar? Researcher Mike Adams of the University of Georgia has found some of the answers. Adams and his students have studied the structure of an enzyme from one of these extraordinary bacteria. He found that the three-dimensional structure of the super-hot enzymes is held together by many more attractive forces than the structure of the low-temperature version of the same enzyme. Thus, these proteins are stable and functional even at temperatures above the boiling point of water.

In Chapter 18, we studied the structure and properties of proteins. We are now going to apply that knowledge to the study of a group of proteins that do the majority of the work for the cell. These special proteins, the **enzymes,** catalyze the biochemical reactions that break down food molecules to allow the cell to harvest energy. They also catalyze the biosynthetic reactions that produce the molecules required for cellular life. In this chapter, we will study the properties of this extraordinary group of proteins and learn how they dramatically speed up biochemical reactions.

The enzymes discussed in this chapter are proteins; however, several ribonucleic acid (RNA) molecules have been demonstrated to have the ability to catalyze biological reactions. These are called *ribozymes*.

19.1 Nomenclature and Classification

Classification of Enzymes

Enzymes may be classified according to the type of reaction that they catalyze. The six classes are as follows.

Oxidoreductases

Oxidoreductases are enzymes that catalyze oxidation–reduction (redox) reactions. *Lactate dehydrogenase* is an oxidoreductase that removes hydrogen from a molecule of lactate. Other subclasses of the oxidoreductases include oxidases and reductases.

LEARNING GOAL

1 Classify enzymes according to the type of reaction catalyzed and the type of specificity.

Recall that redox reactions involve electron transfer from one substance to another (Section 4.8). In organic and biochemistry, oxidation is typically recognized as the loss of hydrogen atoms or gain of oxygen atoms. Similarly, reduction is recognized as the loss of oxygen atoms or the gain of hydrogen atoms (Section 12.5).

$$\underset{\text{Lactate}}{\text{HO}-\underset{\underset{\text{CH}_3}{|}}{\overset{\overset{\text{COO}^-}{|}}{\text{C}}}-\text{H} + \text{NAD}^+} \xrightarrow{\text{Lactate dehydrogenase}} \underset{\text{Pyruvate}}{\overset{\overset{\text{COO}^-}{|}}{\underset{\underset{\text{CH}_3}{|}}{\text{C}}}=\text{O} + \text{NADH}}$$

The significance of phosphate group transfers in energy metabolism is discussed in Sections 21.1 and 21.3.

Transferases

Transferases are enzymes that catalyze the transfer of functional groups from one molecule to another. For example, a *transaminase* catalyzes the transfer of an amino functional group, and a *transmethylase* catalyzes the transfer of a methyl group. A *kinase* catalyzes the transfer of a phosphoryl group. Kinases play a major role in energy-harvesting processes involving ATP. Hexokinase is an enzyme that catalyzes the transfer of a phosphoryl group from adenosine triphosphate (ATP) to a molecule of glucose in the first reaction of glycolysis:

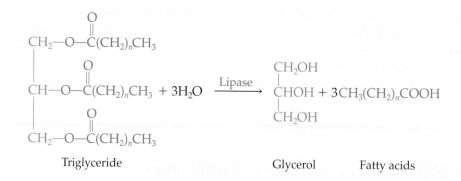

Glucose   Glucose-6-phosphate

Hydrolases

Hydrolases catalyze hydrolysis reactions; that is, the addition of a water molecule to a bond, resulting in bond breakage. These reactions are important in the digestive process. For example, glycosidases, such as α-amylase, catalyze the hydrolysis of glycosidic bonds between monosaccharides in a polysaccharide and proteases catalyze the hydrolysis of peptide bonds in proteins to release amino acids. Lipases catalyze the hydrolysis of the ester bonds in triglycerides:

These kittens are adorable, but there is nothing lovely about the aroma of an untended kitty litter box. Urease, a hydrolase, catalyzes the reaction responsible for this smell. Write an equation representing the reaction catalyzed by urease, and identify the offending molecule.

Triglyceride   Glycerol   Fatty acids

Hydrolysis of esters is described in Section 14.2. The action of lipases in digestion is discussed in Section 23.1.

Lyases

Lyases catalyze the addition of a group to a double bond or the removal of a group to form a double bond. *Fumarase* is an example of a lyase. In the citric acid cycle, fumarase catalyzes the addition of a water molecule to the double bond of the substrate fumarate. The product is malate.

The reactions of the citric acid cycle are described in Section 22.4.

Fumarate   Malate

Citrate lyase catalyzes a far more complicated reaction in which we see the removal of a group and formation of a double bond. Specifically, citrate lyase catalyzes the removal of an acetyl group from a molecule of citrate. The products of this reaction include oxaloacetate, acetyl CoA, ADP, and an inorganic phosphate group (P_i):

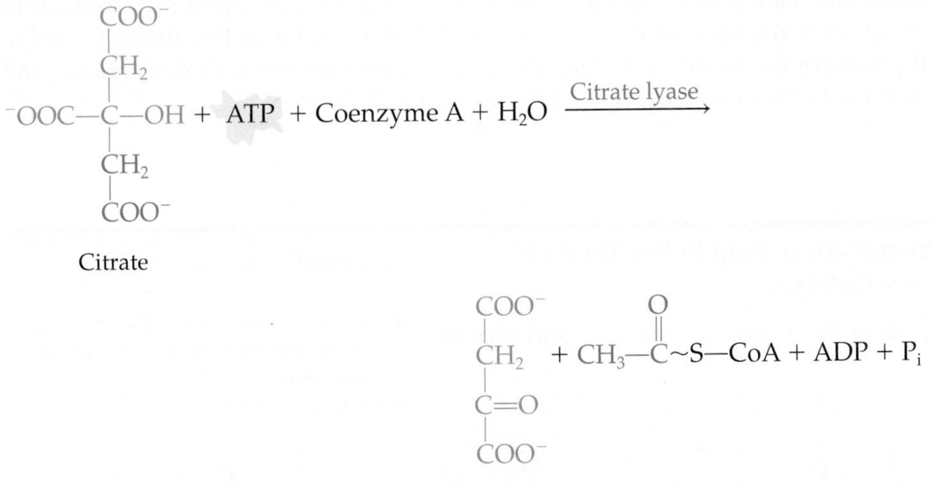

$$
\begin{array}{c}
COO^- \\
|\\
CH_2 \\
|\\
{}^-OOC-C-OH \\
|\\
CH_2 \\
|\\
COO^-
\end{array}
\; + \; ATP \; + \; \text{Coenzyme A} + H_2O \xrightarrow{\text{Citrate lyase}}
$$

Citrate

Recall that the squiggle (~) represents a high-energy bond.

$$
\begin{array}{c}
COO^- \\
|\\
CH_2 \\
|\\
C=O \\
|\\
COO^-
\end{array}
\; + \; CH_3-\overset{\displaystyle O}{\overset{\|}{C}}{\sim}S-CoA + ADP + P_i
$$

Oxaloacetate   Acetyl CoA

Isomerases

Isomerases rearrange the functional groups within a molecule and catalyze the conversion of one isomer into another. For example, *phosphoglycerate mutase* converts one structural isomer, 3-phosphoglycerate, into another, 2-phosphoglycerate:

This reaction is important in glycolysis, an energy-harvesting pathway described in Chapter 21.

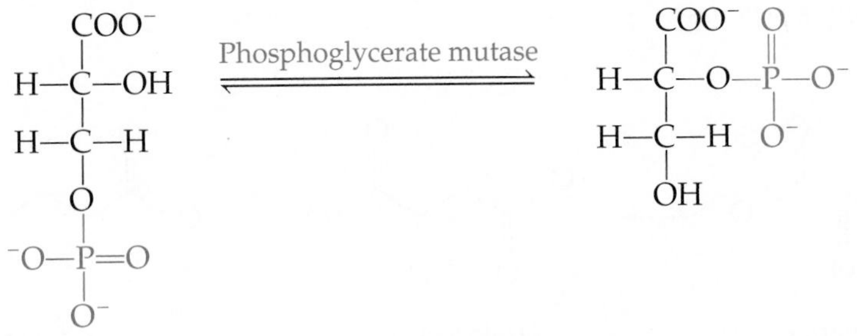

$$
\begin{array}{c}
COO^- \\
|\\
H-C-OH \\
|\\
H-C-H \\
|\\
O \\
|\\
{}^-O-P=O \\
|\\
O^-
\end{array}
\xrightleftharpoons{\text{Phosphoglycerate mutase}}
\begin{array}{c}
COO^- \\
|\\
H-C-O-\overset{\displaystyle O}{\overset{\|}{P}}-O^- \\
|\qquad\quad | \\
H-C-H \quad O^- \\
|\\
OH
\end{array}
$$

3-Phosphoglycerate   2-Phosphoglycerate

Ligases

Ligases are enzymes that catalyze a reaction in which a C—C, C—S, C—O, or C—N bond is made or broken. This is accompanied by an ATP-ADP interconversion. For example, *DNA ligase* catalyzes the joining of the hydroxyl group of a nucleotide in a DNA strand with the phosphoryl group of the adjacent nucleotide to form a phosphoester bond:

The use of DNA ligase in recombinant DNA studies is detailed in Section 20.8.

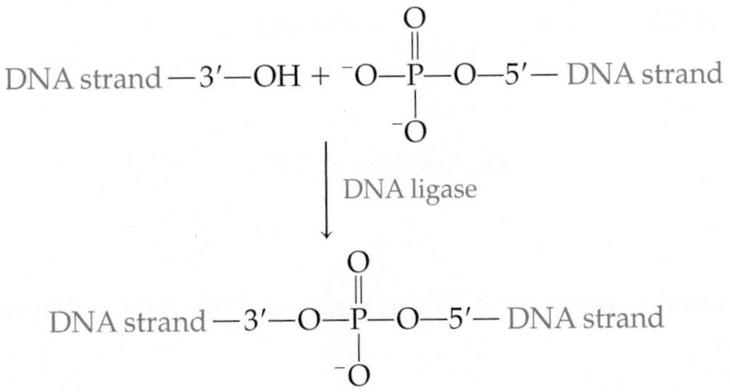

$$
\text{DNA strand}-3'-OH + {}^-O-\overset{\displaystyle O}{\overset{\|}{P}}-O-5'-\text{DNA strand}
$$
$$
\overset{|}{{}^-O}
$$

$$\Big\downarrow \text{DNA ligase}$$

$$
\text{DNA strand}-3'-O-\overset{\displaystyle O}{\overset{\|}{P}}-O-5'-\text{DNA strand}
$$
$$
\overset{|}{{}^-O}
$$

LEARNING GOAL

2 Give examples of the correlation between an enzyme's common name and its function.

Nomenclature of Enzymes

The common names for some enzymes are derived from the name of the **substrate,** the reactant that binds to the enzyme and is converted into product. In many cases, the name of the enzyme is simply derived by adding the suffix *-ase* to the name of the substrate. For instance, *urease* catalyzes the hydrolysis of urea and *lactase* catalyzes the hydrolysis of the disaccharide lactose.

EXAMPLE 19.1 Classifying Enzymes According to the Type of Reaction that They Catalyze

LEARNING GOAL

1 Classify enzymes according to the type of reaction catalyzed and the type of specificity.

Classify the enzyme that catalyzes each of the following reactions, and explain your reasoning.

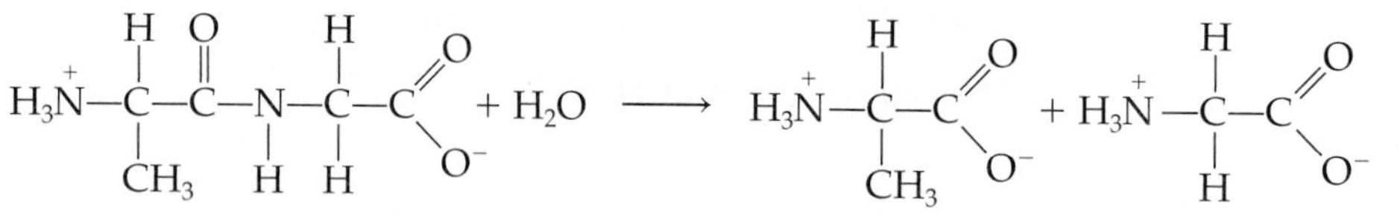

Alanyl-glycine Alanine Glycine

Solution

The reaction occurring here involves breaking a bond, in this case a peptide bond, by adding a water molecule. The enzyme is classified as a *hydrolase,* specifically a *peptidase.*

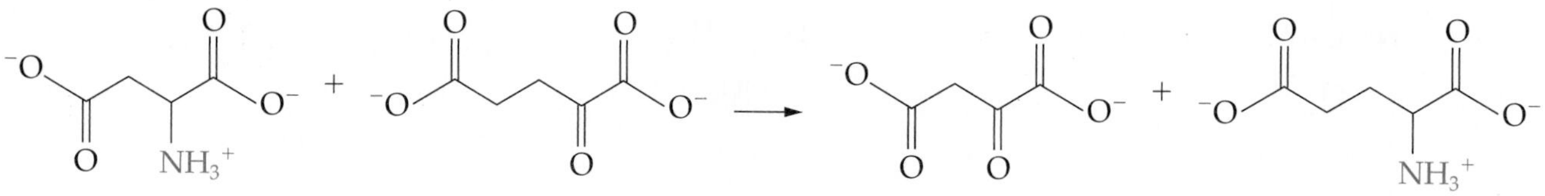

Aspartate α-Ketoglutarate Oxaloacetate Glutamate

Solution

In this reaction, an amino group is transferred from the amino acid aspartate to α-ketoglutarate, producing oxaloacetate and the amino acid glutamate. The enzyme, *aspartate transaminase,* is an example of a *transferase.*

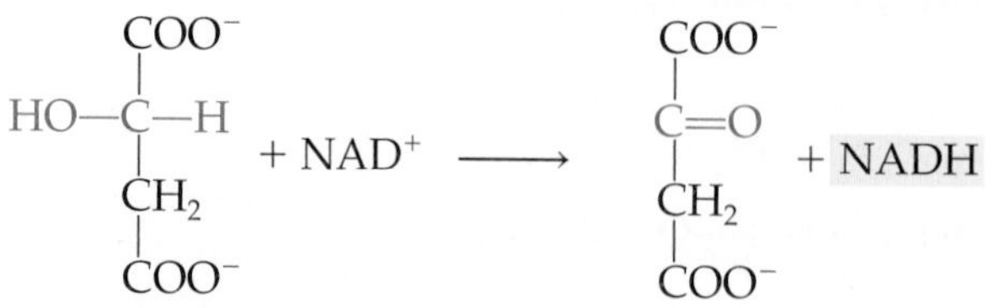

Malate Oxaloacetate

Solution

In this reaction, the reactant malate is oxidized and the coenzyme NAD^+ is reduced. The enzyme that catalyzes this reaction, *malate dehydrogenase,* is an *oxidoreductase.*

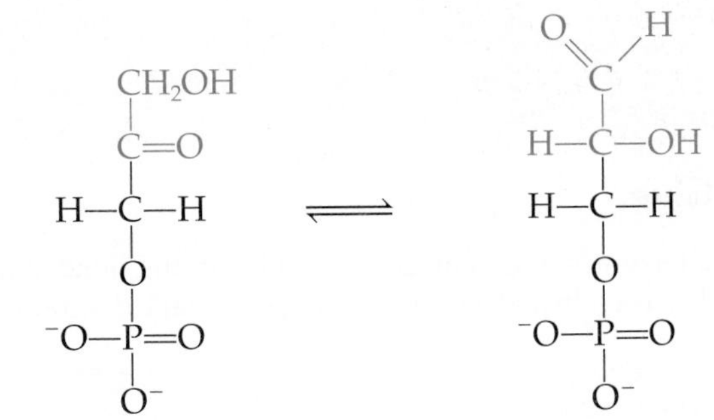

$$\text{Dihydroxyacetone phosphate} \rightleftharpoons \text{Glyceraldehyde-3-phosphate}$$

Dihydroxyacetone phosphate (left):

CH₂OH — C=O — H—C—H — O — ⁻O—P=O — O⁻

Glyceraldehyde-3-phosphate (right):

O=C—H — H—C—OH — H—C—H — O — ⁻O—P=O — O⁻

Solution

Careful inspection of the structure of the reactant and the product reveals that they each have the same number of carbon, hydrogen, oxygen, and phosphorus atoms; thus, they must be structural isomers. The enzyme must be an *isomerase*. Its name is *triose phosphate isomerase.*

Practice Problem 19.1

To which class of enzymes does each of the following belong?

a. Pyruvate kinase
b. Alanine transaminase
c. Triose phosphate isomerase
d. Pyruvate dehydrogenase
e. Lactase
f. Phosphofructokinase
g. Lipase
h. Acetoacetate decarboxylase
i. Succinate ehydrogenase

▶ For Further Practice: **Questions 19.23 and 19.25.**

Names of other enzymes reflect the type of reaction that they catalyze. *Dehydrogenases* catalyze the removal of hydrogen atoms from a substrate, while *decarboxylases* catalyze the removal of carboxyl groups. *Hydrogenases* and *carboxylases* carry out the opposite reaction, adding hydrogen atoms or carboxyl groups to their substrates.

Thus, the common name of an enzyme often tells us a great deal about the function of an enzyme. Yet other enzymes have historical names that have no relationship to either the substrates or the reactions that they catalyze. A few examples include catalase, trypsin, pepsin, and chymotrypsin. In these cases, the names of the enzymes and the reactions that they catalyze must simply be memorized.

The systematic names for enzymes tell us the substrate, the type of reaction that is catalyzed, and the name of any coenzyme that is required. For instance, the systematic name of the oxidoreductase lactate dehydrogenase is lactate: NAD oxidoreductase.

Coenzymes are molecules required by some enzymes to serve as donors or acceptors of electrons, hydrogen atoms, or other functional groups during a chemical reaction. Coenzymes are discussed in Section 19.7.

Question 19.1 Write an equation representing the reaction catalyzed by each of the enzymes listed in Practice Problem 19.1a through d at the end of Example 19.1. (*Hint:* You may need to refer to the index of this book to learn more about the substrates and the reactions that are catalyzed.)

Question 19.2 Write an equation representing the reaction catalyzed by each of the enzymes listed in Practice Problem 19.1e through i at the end of Example 19.1. (*Hint:* You may need to refer to the index of this book to learn more about the substrates and the reactions that are catalyzed.)

Question 19.3 What is the substrate for each of the following enzymes?
a. Sucrase   b. Pyruvate decarboxylase   c. Succinate dehydrogenase

Question 19.4 What chemical reaction is mediated by each of the enzymes in Question 19.3?

Kitchen Chemistry

Transglutaminase: aka Meat Glue

Transglutaminases belong to a family of enzymes that catalyze a reaction between a glutamine residue in a protein and a lysine residue in the same or another protein, as shown in the following schematic diagram:

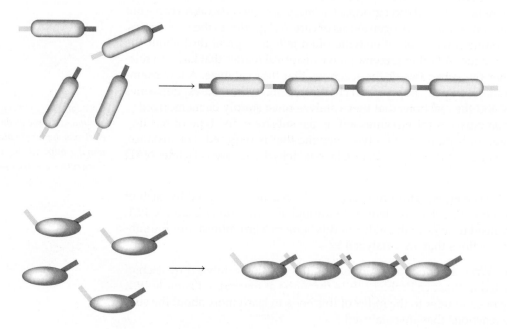

Cross-linked Proteins

Possible Cross-linking Reactions

The result of these reactions is the formation of large polymers of protein that are very tightly linked to one another. In the body, transglutaminases form large, generally insoluble protein polymers that are important to the organism as barriers. One transglutaminase is factor XIII, which is essential for the formation of blood clots. Those with a genetic deficiency of factor XIII are prone to hemorrhage, and the condition can be treated by providing the patient with the missing transglutaminase.

The great British chef Heston Blumenthal is thought to have introduced transglutaminases into the kitchen. In this country, Wylie Dufresne, executive chef and owner of wd-50 in New York City, has brought the so-called meat glue into American cuisine. Meat glue can be used to make consistent, uniform portions of meat or fish from smaller scraps. These portions will cook more evenly, be attractive on the plate, and reduce the waste of such unproductive bits. It is also used to prepare sausages without casings and to prepare unusual meat combinations, such as a chicken and beef loaf. But the creative ideas for the use of this enzyme are much more fanciful. Using gelatin as a binder, protein-based noodles can be made from peanut butter or shrimp.

Edible art, such as Adam Melonas's octopop, combines artistry and savory flavors. Octopus legs are fused with transglutaminase and cooked at a very low temperature. They are then dipped into a saffron and orange carrageenan (a linear, sulfated polysaccharide derived from red seaweed) gel and placed as a centerpiece on stalks of dill flowers.

As beautiful and delicious as the octopop is, and as many economical uses of transglutaminases there may be, there is

Octopops

growing concern about the misuse of transaminases in the food industry. Unscrupulous marketers now have the tools to form sub-standard meat cuts into "higher priced" cuts. The result could be economic loss for the consumer and potential health hazards.

For Further Understanding

▶ When forming a larger piece of meat from smaller ones, the enzyme is sprinkled on the surfaces of the meat. It is then wrapped tightly in plastic wrap and stored in the refrigerator for about 24 hours. Explain the logic of each of these steps.

▶ What health hazards might be of concern if meat is improperly treated with transglutaminase?

19.2 The Effect of Enzymes on the Activation Energy of a Reaction

How does an enzyme speed up a chemical reaction? It changes the path by which the reaction occurs, providing a lower energy route for the conversion of the substrate into the **product**, the substance that results from the enzyme-catalyzed reaction. Thus, enzymes speed up reactions by lowering the activation energy of the reaction. The energy diagrams in Figure 19.1 show that the energy difference between reactant (substrate) and product is not changed. It is only the activation energy that is reduced.

LEARNING GOAL

3 Describe the effect that enzymes have on the activation energy of a reaction.

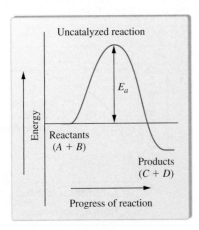

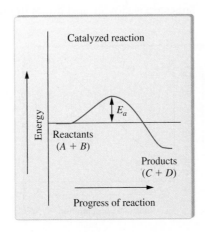

Figure 19.1 Diagram of the difference in energy between the reactants (A and B) and products (C and D) for a reaction. Enzymes cannot change this energy difference but act by lowering the activation energy (E_a) for the reaction, thereby speeding up the reaction.

(a)   (b)

Equilibrium constants are described in Section 7.4.

Recall that every chemical reaction is characterized by an equilibrium constant. Consider, for example, the simple equilibrium

$$aA \rightleftharpoons bB$$

Energy, rate, and equilibrium are described in Chapter 7.

The equilibrium constant for this reaction, K_{eq}, is defined as

$$K_{eq} = \frac{[B]^b}{[A]^a} = \frac{[\text{product}]^b}{[\text{reactant}]^a}$$

The activation energy (Section 7.3) of a reaction is the threshold energy that must be overcome to produce a chemical reaction.

This equilibrium constant is actually a reflection of the difference in energy between reactants and products—that is, a measure of the relative stabilities of the reactants and products. No matter how the chemical reaction occurs (which path it follows), the difference in energy between the reactants and the products is always the same. **For this reason, an enzyme cannot alter the equilibrium constant for the reaction that it catalyzes.** It can only provide a lower energy path for the conversion of reactant to product and, in this way, speed up the reaction (Figure 19.1).

19.3 The Effect of Substrate Concentration on Enzyme-Catalyzed Reactions

The rates of uncatalyzed chemical reactions often double every time the substrate concentration is doubled (Figure 19.2a). Therefore, as long as the substrate concentration increases, there is a direct increase in the rate of the reaction. For enzyme-catalyzed reactions, however, this is not the case. Although the rate of the reaction is initially responsive to the substrate concentration, at a certain concentration of substrate the rate of the reaction reaches a maximum value. A graph of the rate of reaction, V, versus the substrate concentration, $[S]$, is shown in Figure 19.2b. We see that the rate of the reaction initially increases rapidly as the substrate concentration is increased but that the rate levels off at a maximum value. At its maximum rate, the active sites of all the enzyme molecules are occupied by a substrate molecule. The active site is the region of the enzyme that specifically binds the substrate and catalyzes the reaction. A new molecule of substrate cannot bind to the enzyme molecule until the substrate molecule already held in the active site is converted to product and released.

From this information, we realize that an enzyme-catalyzed reaction must occur in two stages:

1. The formation of an *enzyme-substrate complex.* This binding of the substrate to the active site of the enzyme is a rapid step.
2. Conversion of substrate into product and release of the product and enzyme. This step is slower and limits the rate of the overall reaction.

Figure 19.2 Plot of the rate or velocity, *V*, of a reaction versus the concentration of substrate, [*S*], for (a) an uncatalyzed reaction and (b) an enzyme-catalyzed reaction. For an enzyme-catalyzed reaction, the rate is at a maximum when all of the enzyme molecules are bound to the substrate. Beyond this concentration of substrate, further increases in substrate concentration have no effect on the rate of the reaction.

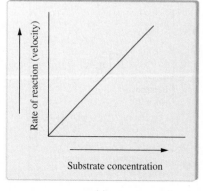

(a)

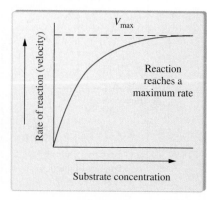

(b)

Stage 2 is called the rate-limiting step because the rate of the reaction is controlled, or limited, by the speed with which the substrate is converted into product and is released. Ultimately, the reaction rate is dependent on the amount of enzyme that is available.

19.4 The Enzyme-Substrate Complex

The following series of reversible reactions represents the steps in an enzyme-catalyzed reaction. The first step (highlighted in blue) involves the encounter of the enzyme with its substrate and the formation of an **enzyme-substrate complex.**

LEARNING GOAL

5 Discuss the role of the active site and the importance of enzyme specificity.

$$E + S \underset{\text{Step I}}{\rightleftharpoons} ES \underset{\text{Step II}}{\rightleftharpoons} ES^* \underset{\text{Step III}}{\rightleftharpoons} EP \underset{\text{Step IV}}{\rightleftharpoons} E + P$$

| Enzyme + substrate | Enzyme–substrate complex | Transition state | Enzyme–product complex | Enzyme + product |

The part of the enzyme that binds with the substrate is called the **active site.** The characteristics of the active site that are crucial to enzyme function include the following:

- Enzyme active sites are pockets or clefts in the surface of the enzyme. The amino acid R groups in the active site that are involved in catalysis are called *catalytic groups.*
- The shape of the active site is complementary to the shape of the substrate. That is, the substrate fits neatly into the active site of the enzyme.
- An enzyme attracts and holds its substrate by weak, noncovalent interactions. These include hydrogen bonding, van der Waals forces, dipole-dipole attractions, and electrostatic attractions. The amino acid R groups involved in substrate binding, and not necessarily catalysis, make up the *binding site.*
- The conformation of the active site determines the specificity of the enzyme because only the substrate that fits into the active site will be used in a reaction.

LEARNING GOAL

6 Describe the difference between the lock-and-key model and the induced fit model of enzyme-substrate complex formation.

The **lock-and-key model** of enzyme activity, shown in Figure 19.3a, was devised by Emil Fischer in 1894. At that time, it was thought that the substrate simply snapped into place like a piece of a jigsaw puzzle or a key into a lock.

Today we know that proteins are flexible molecules. This led Daniel E. Koshland, Jr., to propose a more sophisticated model of the way enzymes and substrates interact. This model, proposed in 1958, is called the **induced fit model** (Figure 19.3b). In this model, the active site of the enzyme is not a rigid pocket into which the substrate fits precisely; rather, it is a flexible pocket that *approximates* the shape of the substrate. When the substrate enters the pocket, the active site "molds" itself around the substrate. This produces the perfect enzyme-substrate "fit."

The overall shape of a protein is maintained by many weak interactions. At any time, a few of these weak interactions may be broken by heat energy or a local change in pH. If only a few bonds are broken, they will re-form very quickly. The overall result is that there is a brief change in the shape of the enzyme. Thus, the protein or enzyme can be viewed as a flexible molecule, changing shape slightly in response to minor local changes.

Question 19.5 Compare the lock-and-key and induced fit models of enzyme-substrate binding.

Question 19.6 What is the relationship between an enzyme active site and its substrate?

Figure 19.3 (a) The lock-and-key model of enzyme-substrate binding assumes that the enzyme active site has a rigid structure that is precisely complementary in shape and charge distribution to the substrate. (b) The induced fit model of enzyme-substrate binding. As the enzyme binds to the substrate, the shape of the active site conforms precisely to the shape of the substrate. The shape of the substrate may also change.

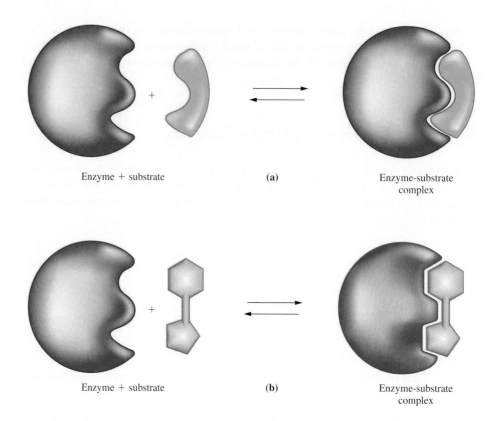

Enzyme + substrate   **(a)**   Enzyme-substrate complex

Enzyme + substrate   **(b)**   Enzyme-substrate complex

19.5   Specificity of the Enzyme-Substrate Complex

For an enzyme-substrate interaction to occur, the surfaces of the enzyme and substrate must be complementary. It is this requirement for a specific fit that determines whether an enzyme will bind to a particular substrate and carry out a chemical reaction.

Enzyme specificity is the ability of an enzyme to bind only one, or a very few, substrates. To illustrate the specificity of enzymes, consider the following reactions.

The enzyme urease catalyzes the hydrolysis of urea to carbon dioxide and ammonia as follows:

$$H_2N-\overset{\overset{\displaystyle O}{\|}}{C}-NH_2 + H_2O \xrightarrow{\text{Urease}} CO_2 + 2NH_3$$

Urea

Methylurea, in contrast, though structurally similar to urea, is not affected by urease:

$$H_2N-\overset{\overset{\displaystyle O}{\|}}{C}-NHCH_3 + H_2O \xrightarrow{\text{Urease}} \text{no reaction}$$

Methylurea

Not all enzymes exhibit the same degree of specificity. Four classes of enzyme specificity have been observed.

- **Absolute specificity:** An enzyme that catalyzes the reaction of only one substrate has absolute specificity. *Aminoacyl tRNA synthetases* exhibit absolute specificity. Each must attach the correct amino acid to the correct transfer RNA molecule. If the wrong amino acid is attached to the transfer RNA, it may be mistakenly added to a peptide chain, producing a nonfunctional protein.

LEARNING GOAL

5 Discuss the role of the active site and the importance of enzyme specificity.

Aminoacyl tRNA synthetases are discussed in Section 20.6. Aminoacyl group transfer reactions were described in Section 15.4.

- **Group specificity:** An enzyme that catalyzes reactions involving similar molecules containing the same functional group has group specificity. *Hexokinase* is a group-specific enzyme that catalyzes the addition of a phosphoryl group to the hexose sugar glucose in the first step of glycolysis. Hexokinase can also add a phosphoryl group to several other six-carbon sugars.
- **Linkage specificity:** An enzyme that catalyzes the formation or breakage of only certain bonds in a molecule has linkage specificity. *Proteases,* such as trypsin, chymotrypsin, and elastase, are enzymes that selectively hydrolyze peptide bonds. Thus, these enzymes are linkage specific.
- **Stereochemical specificity**: An enzyme that can distinguish one enantiomer from the other has stereochemical specificity. Most of the enzymes of the human body show stereochemical specificity. Because we use only D-sugars and L-amino acids, the enzymes involved in digestion and metabolism recognize only those particular stereoisomers.

Hexokinase activity is described in Section 21.3.

Proteolytic enzymes are discussed in Section 19.11.

19.6   The Transition State and Product Formation

How does enzyme-substrate binding result in a faster chemical reaction? To answer this question, we must once again look at the steps of an enzyme-catalyzed reaction, focusing on the steps highlighted in blue:

$$E + S \underset{\text{Step I}}{\rightleftharpoons} ES \underset{\text{Step II}}{\rightleftharpoons} ES^* \underset{\text{Step III}}{\rightleftharpoons} EP \underset{\text{Step IV}}{\rightleftharpoons} E + P$$

| Enzyme + substrate | Enzyme– substrate complex | Transition state | Enzyme– product complex | Enzyme + product |

After the formation of the enzyme-substrate complex in Step I, the flexible enzyme and substrate interact, changing the substrate into a configuration that is no longer energetically stable (Step II). This is the **transition state,** a state in which the substrate is in an intermediate form, having features of both the substrate and the product. This state favors conversion of the substrate into product (Step III) and release of the product (Step IV). Notice that the enzyme is completely unchanged by these events.

What kinds of transition state changes might occur in the substrate that would make a reaction proceed more rapidly?

1. The enzyme might put "stress" on a bond and thereby promote bond breakage, as in the example of sucrase. The enzyme catalyzes the hydrolysis of sucrose into glucose and fructose. The formation of the enzyme-substrate complex (Figures 19.4a and 19.4b) results in a change in the shape of the enzyme. This, in turn, may stretch or distort the bond between glucose and fructose, weakening the bond and allowing it to be broken much more easily than in the absence of the enzyme (Figure 19.4c through e).
2. An enzyme may facilitate a reaction by bringing two reactants close to one another and in the proper orientation for reaction to occur. If we look at the reaction between glucose and fructose to produce sucrose (Figure 19.5a), we see that each of the sugars has five hydroxyl groups that could undergo condensation to produce a disaccharide. By random molecular collision, there is a one in twenty-five chance that the two molecules will collide in the proper orientation to produce sucrose. The probability is actually much less because most molecular collisions won't have enough energy to overcome the energy of activation. The enzyme can bring the two molecules close together in the correct alignment (Figure 19.5b), forming the transition state and greatly speeding up the reaction.

Figure 19.4 Bond breakage is facilitated by the enzyme as a result of stress on a bond. (a, b) The enzyme-substrate complex is formed. (c) In the transition state, the enzyme changes shape and thereby puts stress on the glycosidic bond holding the two monosaccharides together. This lowers the energy of activation of this reaction. (d, e) The bond is broken, and the products are released.

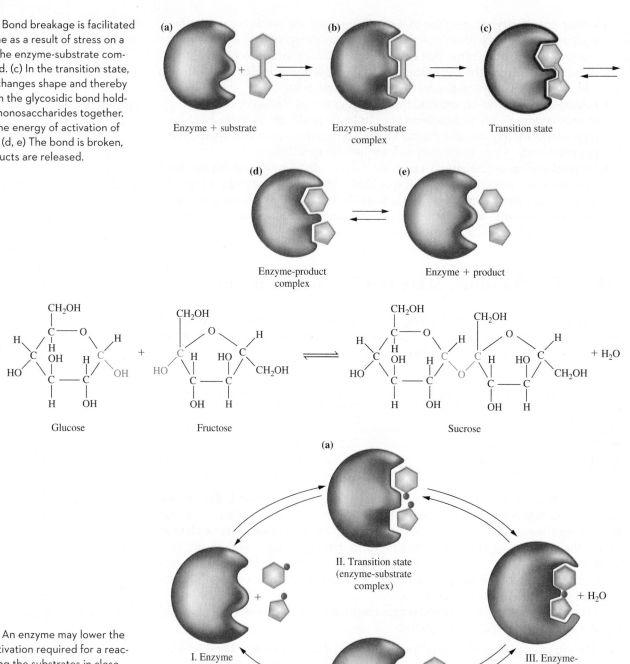

Glucose   Fructose   Sucrose

Figure 19.5 An enzyme may lower the energy of activation required for a reaction by holding the substrates in close proximity and in the correct orientation. (a) A condensation reaction between glucose and fructose to produce sucrose. (b) The enzyme-substrate complex forms, bringing the two monosaccharides together with the correct hydroxyl groups extended toward one another.

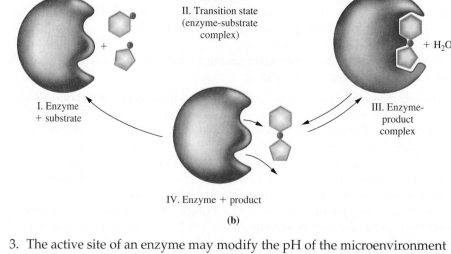

3. The active site of an enzyme may modify the pH of the microenvironment surrounding the substrate. For instance, it may serve as a donor or an acceptor of H^+. This would cause a change in the pH in the vicinity of the substrate without disturbing the normal pH elsewhere in the cell.

Question 19.7   Summarize three ways in which an enzyme might lower the energy of activation of a reaction.

Question 19.8   What is the transition state in an enzyme-catalyzed reaction?

A Medical Perspective

HIV Protease Inhibitors and Pharmaceutical Drug Design

In 1981, the Centers for Disease Control in Atlanta, Georgia, recognized a new disease syndrome, acquired immune deficiency syndrome (AIDS). The syndrome is characterized by an impaired immune system, a variety of opportunistic infections and cancer, and brain damage that results in dementia. It soon became apparent that the disease was being transmitted by blood and blood products, as well as by sexual contact.

The earliest drugs that proved effective in the treatment of human immunodeficiency virus (HIV) infections all inhibited replication of the genetic material of the virus. While these treatments were initially effective, prolonging the lives of many, it was not long before viral mutants resistant to these drugs began to appear. Clearly, a new approach was needed.

In 1989, a group of scientists revealed the three-dimensional structure of the HIV protease. This structure is shown in the accompanying figure. This enzyme is necessary for viral replication because the virus has an unusual strategy for making all of its proteins. Rather than make each protein individually, it makes large "polyproteins" that must then be cut by the HIV protease to form the final proteins required for viral replication.

Since scientists realized that this enzyme was essential for HIV replication, they decided to engineer a substance that would inhibit the enzyme by binding irreversibly to the active site, in essence plugging it up. The challenge, then, was to design a molecule that would be the plug. Researchers knew the primary structure (amino acid sequence) of the HIV protease from earlier nucleic acid sequencing studies. By 1989, they also had a very complete picture of the three-dimensional nature of the molecule, which they had obtained by X-ray crystallography.

Putting all of this information into a sophisticated computer modeling program, they could look at the protease from any angle. They could see the location of each of the R groups of each of the amino acids in the active site. This kind of information allowed the scientists to design molecules that would be complementary to the shape and charge distribution of the enzyme active site—in other words, structural analogs of the normal protease substrate. It was not long before the scientists had produced several candidates for the HIV protease inhibitor.

But, there are many tests that a drug candidate must pass before it can be introduced into the market as safe and effective. Scientists had to show that the candidate drugs would bind effectively to the HIV protease and block its function, thereby inhibiting virus replication. Properties such as the solubility, the efficiency of absorption by the body, the period of activity in the body, and the toxicity of the drug candidates all had to be determined.

There are currently ten protease inhibitors available to combat HIV infection. *Saquinavir* was the first HIV protease inhibitor to reach the market. Introduced in 1995, it is effective

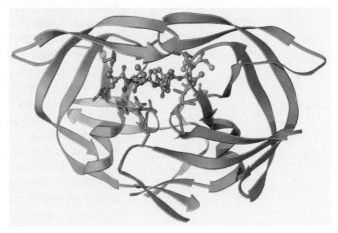

The human immunodeficiency virus protease

against both HIV-1 and HIV-2 and is generally well tolerated by the patient. *Ritonavir* came to market in 1996. Because it causes significant gastrointestinal side effects and can inhibit cellular metabolism, it is used only in combination with other protease inhibitors. *Indinavir* also reached the market in 1996 and was designed to have greater potency by the addition of a terminal phenyl group. *Nelfinavir* was introduced in 1997 and was the first protease inhibitor to be prescribed for pediatric AIDS patients. *Amprenavir* became available in 1999 and was useful because of its greater oral bioavailability. However, it was supplanted in 2004 by a similar drug, *fosamprenavir,* which was demonstrated to have even better oral bioavailability, thereby reducing the number of pills the patient had to take. *Lopinavir* was first marketed in 2000 and is valuable because it reduces the probability of the virus becoming drug resistant. *Atazanavir* was introduced in 2003 and has an even better resistance profile than lopinavir. *Tipranavir*, available since 2005, has very broad antiviral activity against strains of the virus that are resistant to other protease inhibitors. *Darunavir* reached the market in 2006 and is recommended for adults and adolescents. The availability of these ten HIV protease inhibitors is a testament both to the urgent need for HIV treatments and to the technology available to attack the problem.

For Further Understanding

▶ What particular concerns would you have, as a medical researcher, about administering a drug that is a protease inhibitor?

▶ Often, a protease inhibitor is prescribed along with an inhibitor of replication of the viral genetic material. Develop a hypothesis to explain this strategy.

19.7 Cofactors and Coenzymes

Some enzymes require an additional nonprotein *prosthetic group* to function. In this case, the protein portion is called the **apoenzyme,** and the nonprotein group is called the **cofactor.** Together they form the active enzyme called the **holoenzyme.** Cofactors may be metal ions, organic compounds, or organometallic compounds, and must be bound to the enzyme to maintain the correct shape of the active site (Figure 19.6). Thus, these enzymes are only active when the cofactor is bound to them.

Other enzymes require the temporary binding of a **coenzyme.** Such binding is generally mediated by weak interactions like hydrogen bonds. Coenzymes are organic molecules that serve as carriers of electrons or chemical groups. In chemical reactions, they may either donate groups to the substrate or accept groups that are removed from the substrate. The example in Figure 19.7 shows a coenzyme accepting a functional group from one substrate and donating it to the second substrate in a reaction catalyzed by a transferase.

Often, coenzymes contain modified vitamins as part of their structure. A **vitamin** is an organic substance that is required in the diet in only small amounts. Of the water-soluble vitamins, only vitamin C has not been associated with a coenzyme. Table 19.1 is a summary of some coenzymes and the water-soluble vitamins from which they are made.

Nicotinamide adenine dinucleotide (NAD^+), shown in Figure 19.8, is an example of a coenzyme that is of critical importance in the oxidation reactions of the cellular energy-harvesting processes. NAD^+ can accept a hydride ion, a hydrogen atom with two electrons, from the substrate of these reactions. The substrate is oxidized, and the portion of NAD^+ that is derived from the vitamin *niacin* is reduced to produce NADH. Also shown in Figure 19.8 is the hydride ion carrier $NADP^+$ and the hydrogen atom carrier FAD. Both are used in the oxidation-reduction reactions that harvest energy for the cell. Unlike NADH and $FADH_2$, NADPH serves as "reducing power" for the cell by donating hydride ions in biochemical reactions. Like NAD^+, $NADP^+$ is derived from niacin. FAD is made from the vitamin *riboflavin*.

Figure 19.6 (a) The apoenzyme is unable to bind to its substrate. (b) When the required cofactor—in this case, a copper ion, Cu^{2+}—is available, it binds to the apoenzyme. Now the active site takes on the correct configuration, the enzyme-substrate complex forms, and the reaction occurs.

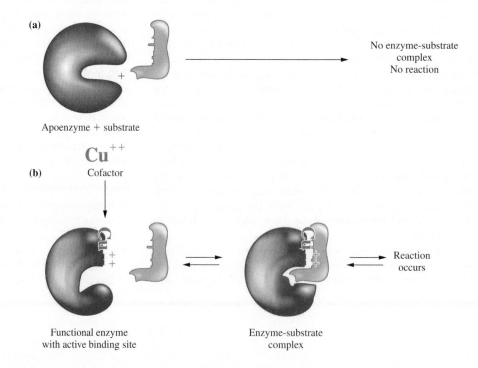

(a)

Apoenzyme + substrate

No enzyme-substrate complex
No reaction

Cu^{++}

(b)   Cofactor

Functional enzyme
with active binding site

Enzyme-substrate
complex

Reaction
occurs

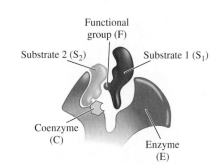

Figure 19.7 Some enzymes require a coenzyme to facilitate the reaction.

1. An enzyme with a coenzyme positioned to react with two substrates.

Functional group (F)

Substrate 2 (S₂)   Substrate 1 (S₁)

Coenzyme (C)

Enzyme (E)

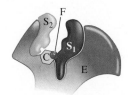

2. Coenzyme picks up a functional group from substrate 1.

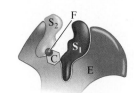

3. Coenzyme transfers the functional group to substrate 2.

4. Products are released from enzyme.

TABLE 19.1 The Water-Soluble Vitamins and Their Coenzymes

Vitamin	Coenzyme	Function
Thiamine (B₁)	Thiamine pyrophosphate	Decarboxylation reactions
Riboflavin (B₂)	Flavin mononucleotide (FMN)	Carrier of H atoms
	Flavin adenine dinucleotide (FAD)	
Niacin (B₃)	Nicotinamide adenine dinucleotide (NAD⁺)	Carrier of hydride ions
	Nicotinamide adenine dinucleotide phosphate (NADP⁺)	
Pyridoxine (B₆)	Pyridoxal phosphate	Carriers of amino and carboxyl groups
	Pyridoxamine phosphate	
Cyanocobalamin (B₁₂)	Deoxyadenosyl cobalamin	Coenzyme in amino acid metabolism
Folic acid	Tetrahydrofolic acid	Coenzyme for 1-C transfer
Pantothenic acid	Coenzyme A	Acyl group carrier
Biotin	Biocytin	Coenzyme in CO₂ fixation
Ascorbic acid	Unknown	Hydroxylation of proline and lysine in collagen

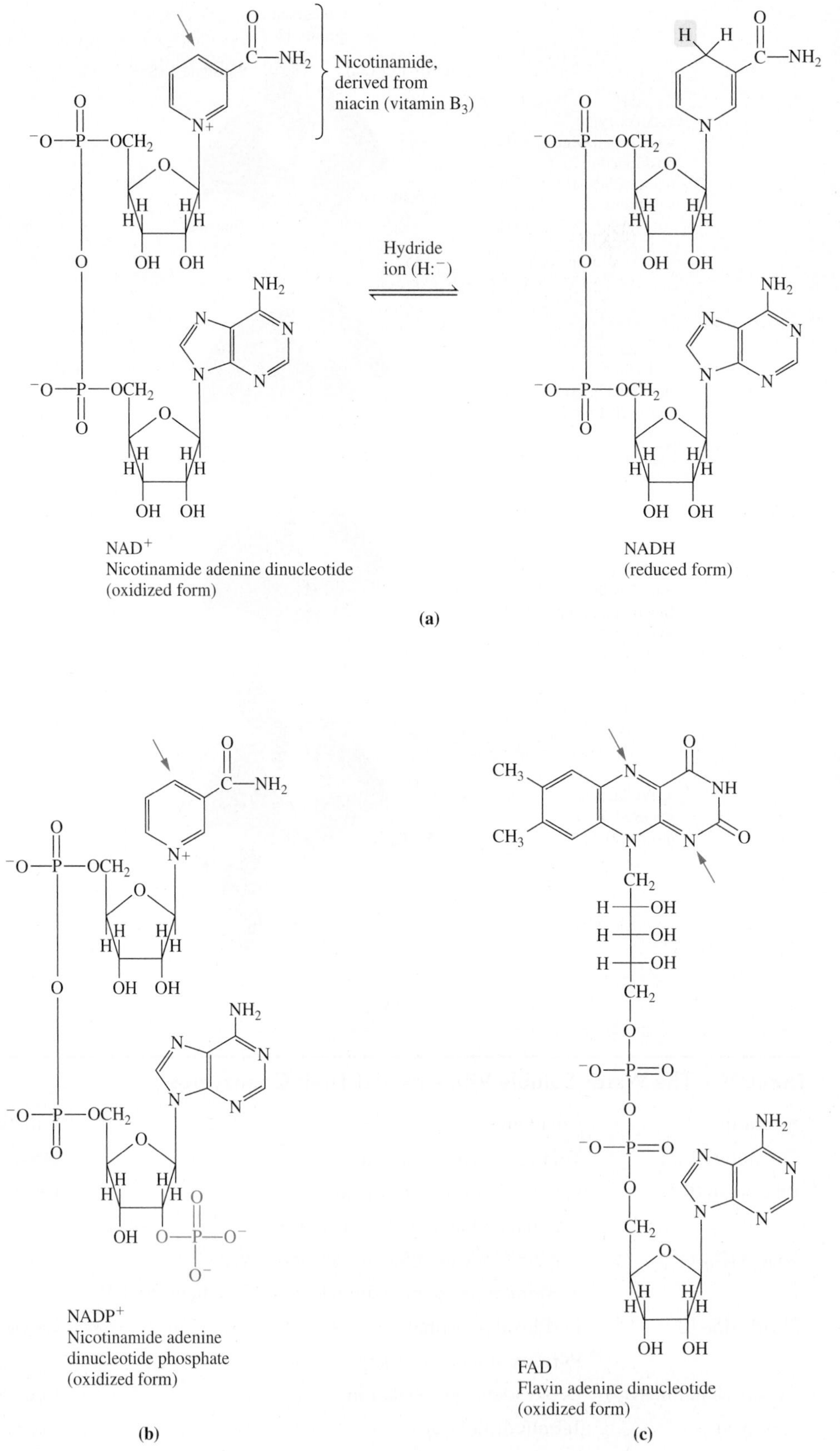

Figure 19.8 The structure of three coenzymes. (a) The oxidized and reduced forms of nicotinamide adenine dinucleotide. (b) The oxidized form of the closely related hydride ion carrier, nicotinamide adenine dinucleotide phosphate (NADP$^+$), which accepts hydride ions at the same position as NAD$^+$ (colored arrow). (c) The oxidized form of flavin adenine dinucleotide (FAD) accepts hydrogen atoms at the positions indicated by the colored arrows.

NAD$^+$
Nicotinamide adenine dinucleotide
(oxidized form)

NADH
(reduced form)

(a)

NADP$^+$
Nicotinamide adenine
dinucleotide phosphate
(oxidized form)

(b)

FAD
Flavin adenine dinucleotide
(oxidized form)

(c)

Question 19.9   Why does the body require water-soluble vitamins?

Question 19.10   What are the coenzymes formed from each of the following vitamins? What are the functions of each of these coenzymes?
   a. Pantothenic acid
   b. Niacin
   c. Riboflavin

19.8   Environmental Effects

Effect of pH

Most enzymes are active only within a very narrow pH range. The cytoplasm of the cell has a pH of 7, and most enzymes function best at this pH. A plot of reaction rate versus pH for a typical enzyme is shown in Figure 19.9.

The pH at which an enzyme functions optimally is called the **pH optimum.** Making the solution more basic or more acidic sharply decreases the rate of the reaction. These pH changes alter the degree of ionization of amino acid R groups in the protein, as well as the extent to which they can hydrogen bond. This causes the enzyme to lose its biologically active configuration; it becomes *denatured.* Less drastic changes in the R groups of an enzyme active site can also destroy the ability to form the enzyme-substrate complex.

Some environments within the body must function at a pH far from 7. For instance, the pH of the stomach is approximately 2 as a result of the secretion of hydrochloric acid by cells of the stomach lining. The proteolytic digestive enzyme *pepsin* must effectively degrade proteins at this extreme pH. In the case of pepsin, the enzyme has evolved an amino acid sequence that can maintain a stable tertiary structure at pH 2 and is most active in the hydrolysis of peptides that have been denatured by very low pH. Thus, pepsin has a pH optimum of 2.

In a similar fashion, another proteolytic enzyme, *trypsin,* functions under the conditions of higher pH found in the intestine. Both pepsin and trypsin cleave peptide bonds by virtually identical mechanisms, yet their amino acid sequences have evolved so that they are stable and active in very different environments.

The body has used the adaptation of enzymes to different environments to protect itself against one of its own destructive defense mechanisms. Within the cytoplasm of a cell are organelles called *lysosomes.* Christian de Duve, who discovered lysosomes in 1956, called them "suicide bags" because they are membrane-bound vesicles containing about fifty different kinds of hydrolases that degrade large biological molecules into small molecules.

If the hydrolytic enzymes of the lysosome were accidentally released into the cytoplasm of the cell, the result would be the destruction of cellular macromolecules and death of the cell. Because of this danger, the cell invests a great deal of energy in maintaining the integrity of the lysosomal membranes. An additional protective mechanism relies on the fact that lysosomal enzymes function optimally at an acid pH (pH 4.8). Should some of these enzymes leak out of the lysosome or should a lysosome accidentally rupture, the cytoplasmic pH of 7.0–7.3 renders them inactive.

Effect of Temperature

The enzymes in our cells are rapidly destroyed if the temperature of their environment rises much above 37°C, but they remain stable at much lower temperatures. This is why enzymes used for clinical assays are stored in refrigerators or freezers before use. Figure 19.10 shows the effects of temperature on enzyme-catalyzed and uncatalyzed reactions. The rate of the uncatalyzed reaction steadily increases with increasing temperature because more collisions occur with sufficient energy to overcome the energy barrier for the reaction. The rate of an enzyme-catalyzed

LEARNING GOAL

8 Explain how pH and temperature affect the rate of an enzyme-catalyzed reaction.

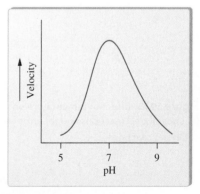

Figure 19.9 Effect of pH on the rate of an enzyme-catalyzed reaction. This enzyme functions most efficiently at pH 7. The rate of the reaction falls rapidly as the solution is made either more acidic or more basic.

A Medical Perspective

α₁-Antitrypsin and Familial Emphysema

Nearly two million people in the United States suffer from emphysema. Emphysema is a respiratory disease caused by destruction of the alveoli, the tiny, elastic air sacs of the lung. This damage results from the irreversible destruction of a protein called *elastin*, which is needed for the strength and flexibility of the walls of the alveoli. When elastin is destroyed, the small air passages in the lungs, called *bronchioles*, become narrower or may even collapse. This severely limits the flow of air into and out of the lung, causing respiratory distress, and in extreme conditions, death.

Some people have a genetic predisposition to emphysema. This is called *familial emphysema*. These individuals have a genetic defect in the gene that encodes the human plasma protein α₁-antitrypsin. As the name suggests, α₁-antitrypsin is an inhibitor of the proteolytic enzyme trypsin. But, as we have seen in this chapter, trypsin is just one member of a large family of proteolytic enzymes called the *serine proteases*. In the case of the α₁-antitrypsin activity in the lung, it is the inhibition of the enzyme elastase that is the critical event.

Elastase damages or destroys elastin, which in turn promotes the development of emphysema. People with normal levels of α₁-antitrypsin are protected from familial emphysema because their α₁-antitrypsin inhibits elastase and, thus, protects the elastin. The result is healthy alveoli in the lungs. However, individuals with a genetic predisposition to emphysema have very low levels of α₁-antitrypsin. This is due to a mutation that causes a single amino acid substitution in the protein chain. Because elastase in the lungs is not effectively controlled, severe lung damage characteristic of emphysema occurs.

Emphysema is also caused by cigarette smoking. Is there a link between these two forms of emphysema? The answer is yes; research has revealed that components of cigarette smoke cause the oxidation of a methionine near the amino terminus of α₁-antitrypsin. This chemical damage destroys α₁-antitrypsin activity. There are enzymes in the lung that reduce the methionine, converting it back to its original chemical form and restoring α₁-antitrypsin activity. However, it is obvious that over a long period, smoking seriously reduces the level of α₁-antitrypsin activity. The accumulated lung damage results in emphysema in many chronic smokers.

At the current time, the standard treatment of emphysema is the use of inhaled oxygen. Studies have shown that intravenous infusion of α₁-antitrypsin isolated from human blood is both safe and effective. However, the level of α₁-antitrypsin in the blood must be maintained by repeated administration.

The α₁-antitrypsin gene has been cloned. In experiments with sheep, it was shown that the protein remains stable when administered as an aerosol. It is still functional after it has passed through the pulmonary epithelium. This research offers hope of an effective treatment for this frightful disease.

For Further Understanding

▸ Draw the structure of methionine, and write an equation showing the reversible oxidation of this amino acid.

▸ Develop a hypothesis to explain why an excess of elastase causes emphysema. What is the role of elastase in this disease?

reaction also increases with modest increases in temperature because there are increasing numbers of collisions between the enzyme and the substrate. At the **temperature optimum,** the enzyme is functioning optimally and the rate of the reaction is maximal. Above the temperature optimum, increasing temperature begins to increase the vibrational energy of the bonds within the enzyme. Eventually, so many bonds and weak interactions are disrupted that the enzyme becomes denatured, and the reaction stops.

Figure 19.10 Effect of temperature on (a) uncatalyzed reactions and (b) enzyme-catalyzed reactions.

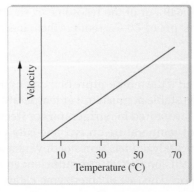

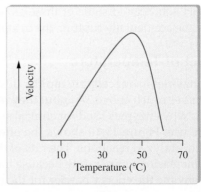

(a)

(b)

Because heating enzymes and other proteins destroys their three-dimensional structure, and hence their activity, a cell cannot survive very high temperatures. Thus, heat is an effective means of sterilizing medical instruments and solutions for transfusion or clinical tests. Although instruments can be sterilized by dry heat (160°C) applied for at least 2 hours (h) in a dry air oven, autoclaving is a quicker, more reliable procedure. The autoclave works on the principle of the pressure cooker. Air is pumped out of the chamber, and steam under pressure is pumped into the chamber until a pressure of 2 atmospheres (atm) is achieved. The pressure causes the temperature of the steam, which would be 100°C at atmospheric pressure, to rise to 121°C. Within 20 minutes (min), all the bacteria and viruses are killed. This is the most effective means of destroying the very heat-resistant endospores that are formed by many bacteria of clinical interest. These bacteria include the genera *Bacillus* and *Clostridium,* which are responsible for such unpleasant and deadly diseases as anthrax, gas gangrene, tetanus, and botulism food poisoning.

See also the Introduction to this chapter.

However, not all enzymes are inactivated by heating, even to rather high temperatures. Certain bacteria live in such out-of-the-way places as coal slag heaps, which are actually burning. Others live in deep vents on the ocean floor where temperatures and pressures are extremely high. Still others grow in the hot springs of Yellowstone National Park, where they thrive at temperatures near the boiling point of water. These organisms, along with their enzymes, survive under such incredible conditions because the amino acid sequences of their proteins dictate structures that are stable at such seemingly impossible temperature extremes.

Question 19.11 How does a decrease in pH alter the activity of an enzyme?

Question 19.12 Heating is an effective mechanism for killing bacteria on surgical instruments. How does elevated temperature result in cellular death?

19.9 Regulation of Enzyme Activity

Enzyme activity is often regulated by the cell. Often the reason for this is to conserve energy. If the cell runs out of chemical energy, it will die; therefore, many mechanisms exist to conserve cellular energy. For instance, it is a great waste of energy to produce an enzyme if the substrate is not available. Similarly, if the product of an enzyme-catalyzed reaction is present in excess, it is a waste of energy for the enzyme to continue producing more of the unwanted product.

The simplest mechanism of enzyme regulation is to produce the enzyme only when the substrate is present. This mechanism is used by bacteria to regulate the enzymes needed to break down various sugars to yield ATP for cellular work. The bacteria have no control over their environment or over what food sources, if any, might be available. It would be an enormous waste of energy to produce all of the enzymes needed to break down all the possible sugars. Thus, the bacteria save energy by producing the enzymes only when a specific sugar substrate is available. Other mechanisms for regulating enzyme activity include use of allosteric enzymes, feedback inhibition, production of proenzymes, and protein modification. Let's take a look at these regulatory mechanisms in some detail.

LEARNING GOAL

9 Describe the mechanisms used by cells to regulate enzyme activity.

Allosteric Enzymes

One type of enzyme regulation involves enzymes that have more than a single binding site. These enzymes, called **allosteric enzymes,** have active sites that can be altered by the binding of small molecules called *effector molecules.* As shown in Figure 19.11, the effector binding alters the shape of the active site of the enzyme. In **negative allosterism,** effector binding converts the active site to an inactive configuration. In **positive allosterism,** effector binding converts the active site to an active configuration. In either case, binding of the effector molecule regulates enzyme activity by determining whether it will be active or inactive.

Allosteric means "other forms."

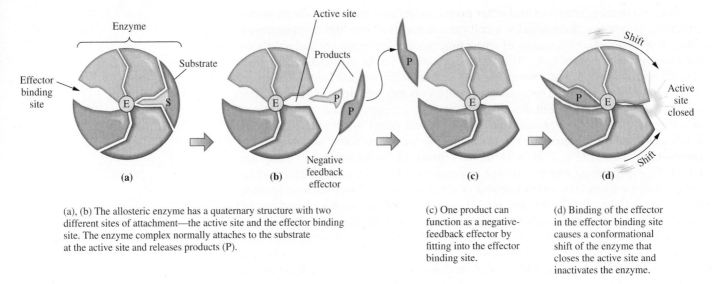

(a), (b) The allosteric enzyme has a quaternary structure with two different sites of attachment—the active site and the effector binding site. The enzyme complex normally attaches to the substrate at the active site and releases products (P).

(c) One product can function as a negative-feedback effector by fitting into the effector binding site.

(d) Binding of the effector in the effector binding site causes a conformational shift of the enzyme that closes the active site and inactivates the enzyme.

Figure 19.11 A mechanism of negative allosterism. This is an example of feedback inhibition.

An example of allosterism is found in glycolysis, which is the first stage of the breakdown of carbohydrates to produce ATP energy for the cell. This pathway must be responsive to the demands of the body. When more energy is required, the reactions of the pathway should occur more quickly, producing more ATP. However, if the energy demand is low, the reactions should slow down.

The third reaction in glycolysis is the transfer of a phosphoryl group from an ATP molecule to a molecule of fructose-6-phosphate. This reaction, shown here, is catalyzed by an enzyme called *phosphofructokinase:*

Fructose-6-phosphate + ATP   →(Phosphofructokinase)→   Fructose-1,6-bisphosphate + ADP

Phosphofructokinase activity is sensitive to both positive and negative allosterism. For instance, when ATP is present in abundance, it is a signal that the body has sufficient energy, and the pathway should slow down. ATP is a negative allosteric effector of phosphofructokinase, inhibiting the activity of the enzyme. Conversely, an abundance of AMP, which is a precursor of ATP, is evidence that the body needs to make ATP. When AMP binds to an effector binding site on phosphofructokinase, enzyme activity is increased, speeding up the reaction and the entire pathway. Thus, AMP is a positive allosteric effector of the enzyme.

Feedback Inhibition

Allosteric enzymes are the basis for **feedback inhibition** of biochemical pathways. This system functions on the same principle as the thermostat on your furnace. You set the thermostat at 70°F; the furnace turns on and produces heat until the sensor in the thermostat registers a room temperature of 70°F. It then signals the furnace to shut off.

Feedback inhibition usually regulates pathways of enzymes involved in the synthesis of a biological molecule. Such a pathway can be shown schematically as follows:

$$A \xrightarrow{E_1} B \xrightarrow{E_2} C \xrightarrow{E_3} D \xrightarrow{E_4} E \xrightarrow{E_5} F$$

In this pathway, the starting material, A, is converted to B by the enzyme E_1. Enzyme E_2 immediately converts B to C, and so on until the final product, F, has been synthesized. If F is no longer needed, it is a waste of cellular energy to continue to produce it.

To avoid this waste of energy, the cell uses feedback inhibition, in which the product can shut off the entire pathway for its own synthesis. In this example, the product, F, acts as a negative allosteric effector on one of the early enzymes of the pathway, E_1. When F is present in excess, it binds to the effector-binding site, causing the active site to close so that it cannot bind to substrate A. Thus, A is not converted to B and the entire pathway ceases to operate. The product, F, has turned off all the steps involved in its own synthesis, just as the heat produced by the furnace is ultimately responsible for turning off the furnace itself.

When the concentration of F drops, it will dissociate from the effector binding site. When this occurs, the enzyme is once again active. Thus, feedback inhibition is an effective metabolic on-off switch.

Proenzymes

Another means of regulating enzyme activity involves the production of the enzyme in an inactive form called a **proenzyme.** The proenzyme is converted by proteolysis (hydrolysis of the protein) to the active form when it has reached the site of its activity. This strategy is used with enzymes, like the digestive enzymes, that could destroy the cells in which they are made. An example of this strategy is seen with the enzyme pepsin, a proteolytic enzyme that acts in the stomach. The cells that produce pepsin actually produce an inactive proenzyme, called *pepsinogen*. Pepsinogen has an additional forty-two amino acids. In the presence of stomach acid and previously activated pepsin, the extra forty-two amino acids are cleaved off, and the proenzyme is transformed into the active enzyme. Table 19.2 lists several other proenzymes and the enzymes that convert them to active form.

Protein Modification

Protein modification is another mechanism that the cell can use to turn an enzyme on or off. This is a process in which a chemical group is covalently added to or removed from the protein. This covalent modification either activates the enzyme or turns it off.

The most common type of protein modification is phosphorylation or dephosphorylation of an enzyme. Typically, the phosphoryl group is added to (or removed from) the R group of a serine, tyrosine, or threonine in the protein chain of the enzyme. Notice that these three amino acids have a free —OH in their R group, which serves as the site for the addition of the phosphoryl group.

TABLE 19.2 Proenzymes of the Digestive Tract

Proenzyme	Activator	Enzyme
Proelastase	Trypsin	Elastase
Trypsinogen	Trypsin	Trypsin
Chymotrypsinogen A	Trypsin + chymotrypsin	Chymotrypsin
Pepsinogen	Acid pH + pepsin	Pepsin
Procarboxypeptidases	Trypsin	Carboxypeptidase A, Carboxypeptidase B

The covalent modification of an enzyme's structure is catalyzed by other enzymes. *Protein kinases* add phosphoryl groups to a target enzyme, while *phosphatases* remove them. For some enzymes, it is the phosphorylated form that is active. For instance, in adipose tissue, phosphorylation activates the enzyme triacylglycerol lipase, an enzyme that breaks triglycerides down to fatty acids and glycerol. Glycogen phosphorylase, an enzyme involved in the breakdown of glycogen, is also activated by the addition of a phosphoryl group. However, for some enzymes, phosphorylation inactivates the enzyme. This is true for glycogen synthase, an enzyme involved in the synthesis of glycogen. When this enzyme is phosphorylated, it becomes inactive.

The convenient aspect of this type of regulation is the reversibility. An enzyme can quickly be turned on or off in response to environmental or physiological conditions.

19.10   Inhibition of Enzyme Activity

LEARNING GOAL

10 Discuss the mechanisms by which certain chemicals inhibit enzyme activity.

Many chemicals can bind to enzymes and either eliminate or drastically reduce their catalytic ability. These chemicals, called *enzyme inhibitors*, have been used for hundreds of years. When she poisoned her victims with arsenic, Lucretia Borgia was unaware that it was binding to the thiol groups of cysteine amino acids in the proteins of her victims and thus interfering with the formation of disulfide bonds needed to stabilize the tertiary structure of enzymes. However, she was well aware of the deadly toxicity of heavy metal salts like arsenic and mercury. When you take penicillin for a bacterial infection, you are taking another enzyme inhibitor. Penicillin inhibits several enzymes that are involved in the synthesis of bacterial cell walls.

Enzyme inhibitors are classified on the basis of whether the inhibition is reversible or irreversible, competitive or noncompetitive. Reversibility deals with whether the inhibitor will eventually dissociate from the enzyme, releasing it in the active form. Competition refers to whether the inhibitor is a structural analog, or look-alike, of the natural substrate. If so, the inhibitor and substrate will compete for the enzyme active site.

Irreversible Inhibitors

Irreversible enzyme inhibitors, such as arsenic, usually bind very tightly, sometimes even covalently, to the enzyme. This generally involves binding of the inhibitor to one of the R groups of an amino acid in the active site. Inhibitor binding may block the active site binding groups so that the enzyme-substrate complex cannot form. Alternatively, an inhibitor may interfere with the catalytic groups of the active site, thereby effectively eliminating catalysis. Irreversible inhibitors, which include snake venoms and nerve gases, generally inhibit many different enzymes.

Reversible, Competitive Inhibitors

Reversible, competitive enzyme inhibitors are often referred to as **structural analogs;** that is, they are molecules that resemble the structure and charge distribution of the natural substrate for a particular enzyme. Because of this resemblance, the inhibitor can occupy the enzyme active site. However, no reaction can occur, and enzyme activity is inhibited (Figure 19.12). This inhibition is competitive because the inhibitor and the substrate compete for binding to the enzyme active site. Thus, the degree of inhibition depends on their relative concentrations. If the inhibitor is in excess or binds more strongly to the active site, it will occupy the active site more frequently, and enzyme activity will be greatly decreased. On the other hand, if the natural substrate is present in excess, it will more frequently occupy the active site, and there will be little inhibition.

Chemistry at the Crime Scene

Enzymes, Nerve Agents, and Poisoning

The transmission of nerve impulses at the *neuromuscular junction* involves many steps, one of which is the activity of a critical enzyme, called *acetylcholinesterase,* which catalyzes the hydrolysis of the chemical messenger, *acetylcholine,* that initiated the nerve impulse. The need for this enzyme activity becomes clear when we consider the events that begin with a message from the nerve cell and end in the appropriate response by the muscle cell. Acetylcholine is a *neurotransmitter,* that is, a chemical messenger that transmits a message from the nerve cell to the muscle cell. Acetylcholine is stored in membrane-bound bags, called *synaptic vesicles,* in the nerve cell ending.

Acetylcholinesterase comes into play in the following way. The arrival of a nerve impulse at the end plate of the nerve axon causes an influx of Ca^{2+}. This causes the acetylcholine-containing vesicles to migrate to the nerve cell membrane that is in contact with the muscle cell. This is called the *presynaptic membrane.* The vesicles fuse with the presynaptic membrane and release the neurotransmitter. The acetylcholine then diffuses across the *nerve synapse* (the space between the nerve and muscle cells) and binds to the acetylcholine receptor protein (R) in the *postsynaptic membrane* of the muscle cell. This receptor then opens pores in the membrane through which Na^+ and K^+ ions flow into and out of the cell, respectively. This generates the nerve impulse and causes the muscle to contract. If acetylcholine remains at the neuromuscular junction, it will continue to stimulate the muscle contraction. To stop this continued stimulation, acetylcholine is hydrolyzed, and hence, destroyed by acetylcholinesterase. When this happens, nerve stimulation ceases.

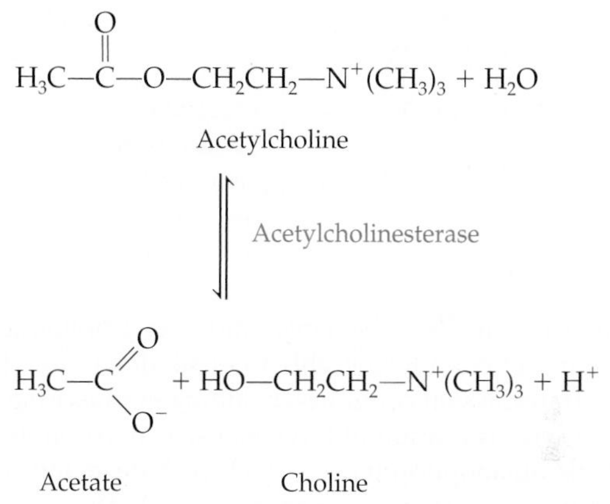

$$H_3C-\overset{\overset{O}{\|}}{C}-O-CH_2CH_2-N^+(CH_3)_3 + H_2O$$

Acetylcholine

Acetylcholinesterase

$$H_3C-\overset{\overset{O}{\diagup}}{\underset{O^-}{C}} + HO-CH_2CH_2-N^+(CH_3)_3 + H^+$$

Acetate   Choline

Inhibitors of acetylcholinesterase are used both as poisons and as drugs. Among the most important inhibitors of acetylcholinesterase are a class of compounds known as *organophosphates.* One of these is the nerve agent Sarin (isopropylmethylfluorophosphate). Sarin forms a covalently bonded intermediate with the active site of acetylcholinesterase. Thus, it acts as an irreversible, noncompetitive inhibitor.

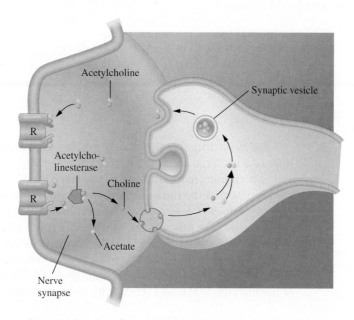

Schematic diagram of the synapse at the neuromuscular junction

Continued…

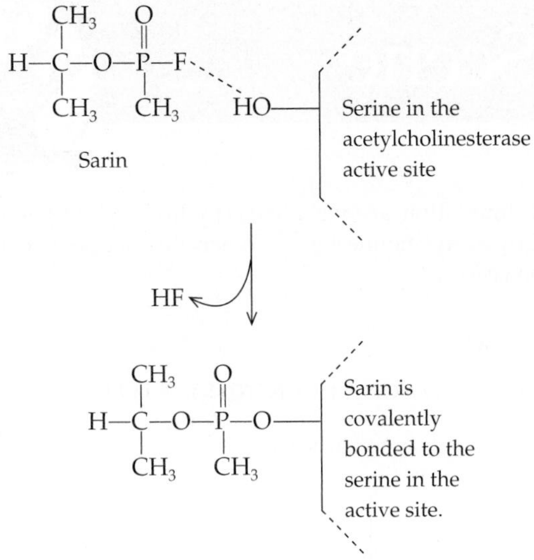

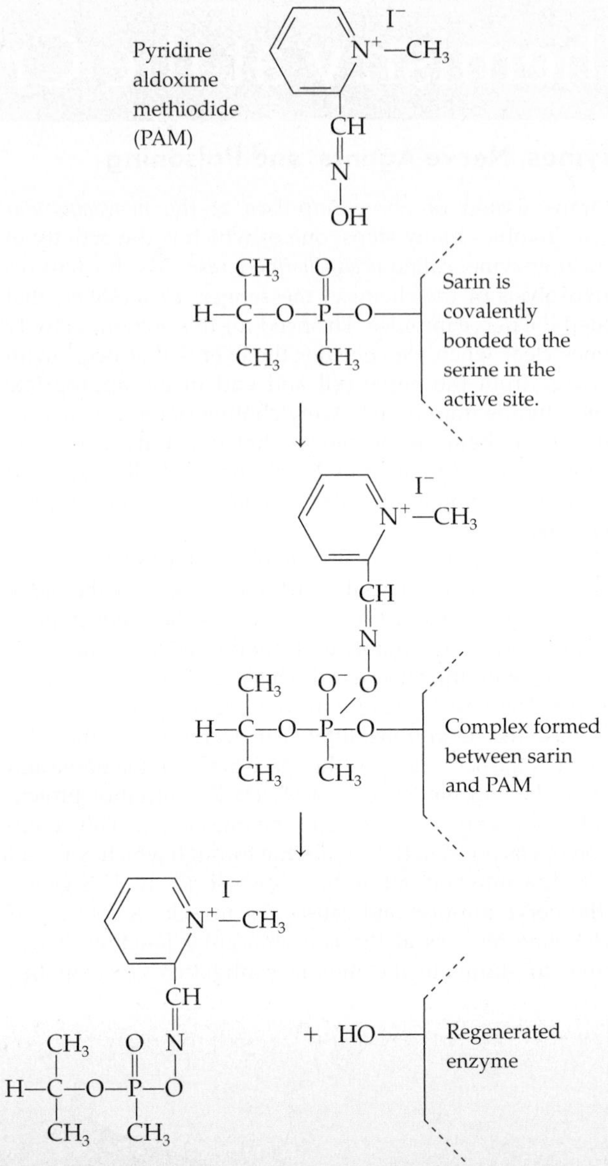

Sarin

Serine in the acetylcholinesterase active site

HF

Sarin is covalently bonded to the serine in the active site.

Pyridine aldoxime methiodide (PAM)

Sarin is covalently bonded to the serine in the active site.

Complex formed between sarin and PAM

+ HO— Regenerated enzyme

The covalent intermediate is stable, and acetylcholinesterase is therefore inactive, no longer able to break down acetylcholine. Nerve transmission continues, resulting in muscle spasm. Death may occur as a result of laryngeal spasm. Antidotes for poisoning by organophosphates, which include many insecticides and nerve gases, have been developed. The antidotes work by reversing the effects of the inhibitor. One of these antidotes is known as *PAM*, an acronym for *pyridine aldoxime methiodide*. This molecule displaces the organophosphate group from the active site of the enzyme, alleviating the effects of the poison.

For Further Understanding

▶ Botulinum toxin inhibits release of neurotransmitters from the presynaptic membrane. What symptoms do you predict would result from this?

▶ Why must Na^+ and K^+ enter and exit the cell through a protein channel?

The sulfa drugs, the first antimicrobics to be discovered, are **competitive inhibitors** of a bacterial enzyme needed for the synthesis of the vitamin folic acid. *Folic acid* is a vitamin required for the transfer of methyl groups in the biosynthesis of methionine and the nitrogenous bases required to make DNA and RNA. Humans cannot synthesize folic acid and must obtain it from the diet. Bacteria, on the other hand, must make folic acid because they cannot take it in from the environment.

para-Aminobenzoic acid (PABA) is the substrate for an early step in folic acid synthesis. The sulfa drugs, the prototype of which was discovered in the 1930s by Gerhard Domagk, are structural analogs of PABA and thus competitive inhibitors of the enzyme that uses PABA as its normal substrate.

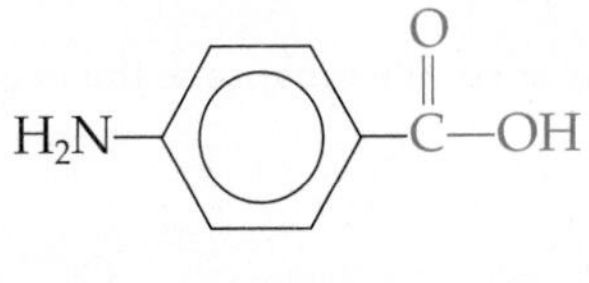

In addition to the folic acid supplied in our diets, we obtain folic acid from our intestinal bacteria.

p-Aminobenzoic acid

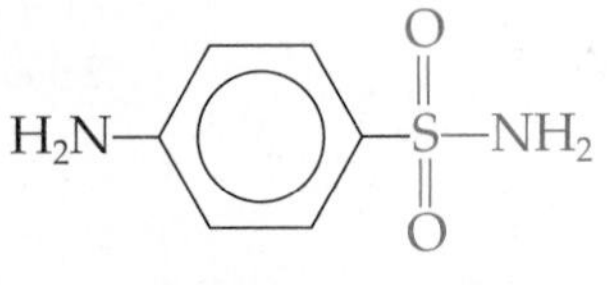

Sulfanilamide

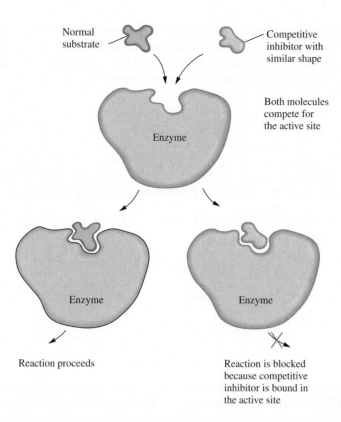

Figure 19.12 Competitive inhibition.

Normal substrate

Competitive inhibitor with similar shape

Both molecules compete for the active site

Enzyme

Enzyme

Enzyme

Reaction proceeds

Reaction is blocked because competitive inhibitor is bound in the active site

If the correct substrate (PABA) is bound by the enzyme, the reaction occurs, and the bacterium lives. However, if the sulfa drug is present in excess over PABA, it binds more frequently to the active site of the enzyme. No folic acid will be produced, and the bacterial cell will die.

Because we obtain our folic acid from our diets, sulfa drugs do not harm us. However, bacteria are selectively killed. Luckily, we can capitalize on this property for the treatment of bacterial infections, and as a result, sulfa drugs have saved countless lives. Although bacterial infection was the major cause of death before the discovery of sulfa drugs and other antibiotics, death caused by bacterial infection is relatively rare at present.

Question 19.13 Why are irreversible inhibitors considered to be poisons?

Question 19.14 Explain the difference between an irreversible inhibitor and a reversible, competitive inhibitor.

Question 19.15 What is a structural analog?

Question 19.16 How can structural analogs serve as enzyme inhibitors?

19.11 Proteolytic Enzymes

Proteolytic enzymes break the peptide bonds that maintain the primary protein structure. *Chymotrypsin,* for example, is an enzyme that hydrolyzes dietary proteins in the small intestine. It acts specifically at peptide bonds on the carbonyl side of the peptide bond. The C-terminal amino acids of the peptides released by bond cleavage are methionine, tyrosine, tryptophan, and phenylalanine. The specificity of chymotrypsin depends upon the presence of a *hydrophobic pocket,* a cluster of hydrophobic amino acids brought together by the three-dimensional folding of the protein chain. The flat aromatic side chains of certain amino acids (tyrosine, tryptophan, phenylalanine) slide into this pocket, providing the binding specificity required for catalysis at the catalytic site (Figure 19.13).

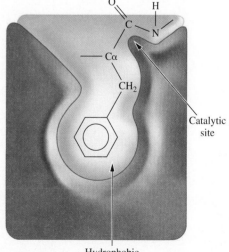

Hydrophobic pocket

Figure 19.13 The specificity of chymotrypsin is determined by a hydrophobic pocket that holds the aromatic side chain of the substrate. This brings the peptide bond to be cleaved into the catalytic domain of the active site.

Catalytic site

11 Discuss the role of the enzyme chymotrypsin and other serine proteases.

How can we determine which bond is cleaved by a protease such as chymotrypsin? To know which bond is cleaved, we must write out the sequence of amino acids in the region of the peptide that is being cleaved. This can be determined experimentally by amino acid sequencing techniques. Remember that the N-terminal amino acid is written to the left and the C-terminal amino acid to the right. Consider a protein having within it the sequence —Ala-Phe-Gly—. A reaction is set up in which the enzyme, chymotrypsin, is mixed with the protein substrate. After the reaction has occurred, the products are purified, and their amino acid sequences are determined. Experiments of this sort show that chymotrypsin cleaves the bond between phenylalanine and glycine, which is the peptide bond on the carbonyl side of amino acids having an aromatic side chain.

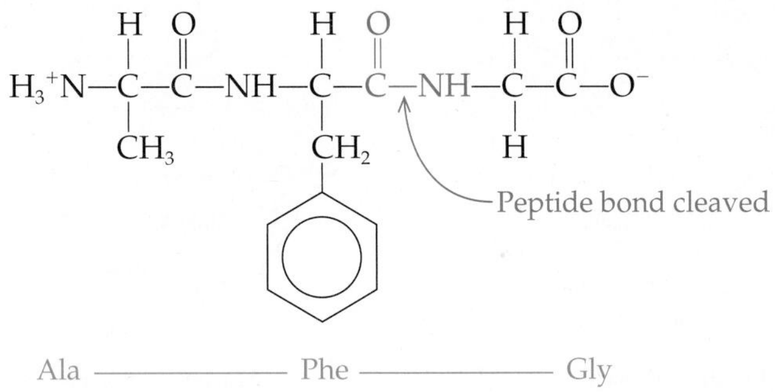

Ala ——————— Phe ——————— Gly

The **pancreatic serine proteases** trypsin, chymotrypsin, and elastase all hydrolyze peptide bonds. As the name suggests, they are produced in the pancreas and subsequently transported to the small intestine. These enzymes are the result of *divergent evolution* in which a single ancestral gene was first duplicated. Then each copy evolved individually. They have similar primary structures, similar tertiary structures, and virtually identical mechanisms of action. However, as a result of evolution, these enzymes all have different specificities:

- Chymotrypsin cleaves peptide bonds on the carbonyl side of aromatic amino acids and large, hydrophobic amino acids such as methionine.
- Trypsin cleaves peptide bonds on the carbonyl side of basic amino acids.
- Elastase cleaves peptide bonds on the carbonyl side of glycine and alanine.

These enzymes are called *serine proteases* because they have the amino acid serine in the catalytic region of the active site that is essential for hydrolysis of the peptide bond.

These enzymes have different pockets for the side chains of their substrates; *different keys fit different locks.* This difference manifests itself in the substrate specificity. For example, the binding pocket of trypsin is long, narrow, and negatively charged to accommodate lysine or arginine R groups. Yet although the binding pockets have undergone divergent evolution, the catalytic sites have remained unchanged, and the mechanism of proteolytic action is the same for all the serine proteases. In each case, the mechanism involves a serine R group.

Question 19.17   Draw the structural formulas of the following peptides, and show which bond would be cleaved by chymotrypsin.
   a. ala-phe-ala
   b. tyr-ala-tyr

Question 19.18   Draw the structural formulas of the following peptides, and show which bond would be cleaved by chymotrypsin.
   a. trp-val-gly
   b. phe-ala-pro

Question 19.19   Draw the structural formula of the peptide val-phe-ala-gly-leu. Which bond would be cleaved if this peptide were reacted with chymotrypsin? With elastase?

Question 19.20   Draw the structural formula of the peptide trp-val-lys-ala-ser. Show which bonds would be cleaved by trypsin, chymotrypsin, and elastase.

19.12 Uses of Enzymes in Medicine

Analysis of blood serum for levels (concentrations) of certain enzymes can provide a wealth of information about a patient's medical condition. Often, such tests are used to confirm a preliminary diagnosis based on the disease symptoms or clinical picture. These tests, called *enzyme assays,* are very precise and specific because they are based on the specificity of the enzyme-substrate complex.

Acute myocardial infarction (AMI) occurs when the blood supply to the heart muscle is blocked for an extended time. If this lack of blood supply, called *ischemia,* is prolonged, the myocardium suffers irreversible cell damage and muscle death, or infarction. When this happens, the concentration of cardiac enzymes in the blood rises dramatically as the dead cells release their contents into the bloodstream.

Three cardiac biomarkers have become the primary tools used to assess myocardial disease and suspected AMI. These are myoglobin, creatine kinase-MB (CK-MB), and cardiac troponin I. Of these three, only troponin is cardiac specific. In fact, it is so reliable that the American College of Cardiology has stated that any elevation of troponin is "abnormal and represents cardiac injury."

Myoglobin is the smallest of these three proteins and diffuses most rapidly through the vascular system. Thus, it is the first cardiac biomarker to appear, becoming elevated as early as 30 min after onset of chest pain. Myoglobin has another benefit in following a myocardial infarction. It is rapidly cleared from the body by the kidneys, returning to normal levels within 16 to 36 h after a heart attack. If the physician sees this decline in myoglobin levels, followed by a subsequent rise, it is an indication that the patient has had a second myocardial infarction.

Creatine kinase-MB is one of the most important cardiac biomarkers, even though it is found primarily in muscle and the brain. Levels typically rise 3 to 8 h after chest pains begin. Within another 48 to 72 h, the CK-MB levels return to normal. As a result, like myoglobin, CK-MB can also be used to diagnose a second AMI.

The physician also has enzymes available to treat a heart attack patient. Most AMIs are the result of a *thrombus,* or clot, within a coronary blood vessel. The clot restricts blood flow to the heart muscle. One technique that shows promise for treatment following a coronary thrombosis, a heart attack caused by the formation of a clot, is destruction of the clot by intravenous or intracoronary injection of an enzyme called *streptokinase.* This enzyme, formerly purified from the pathogenic bacterium *Streptococcus pyogenes* but now available through recombinant DNA techniques, catalyzes the production of the proteolytic enzyme plasmin from its proenzyme, plasminogen. Plasmin can degrade a fibrin clot into subunits. This has the effect of dissolving the clot that is responsible for restricted blood flow to the heart, but there is an additional protective function as well. The subunits produced by plasmin degradation of fibrin clots are able to inhibit further clot formation by inhibiting thrombin.

Recombinant DNA technology has provided medical science with yet another, perhaps more promising, clot-dissolving enzyme. *Tissue-type plasminogen activator (TPA)* is a proteolytic enzyme that occurs naturally in the body as a part of the anti-clotting mechanisms. TPA converts the proenzyme, plasminogen, into the active enzyme, plasmin. Injection of TPA within 2 h of the initial chest pain can significantly improve the circulation to the heart and greatly improve the patient's chances of survival.

Elevated blood serum concentrations of the enzymes amylase and lipase are indications of pancreatitis, an inflammation of the pancreas. Liver diseases such as cirrhosis and hepatitis result in elevated levels of alanine aminotransferase/serum glutamate–pyruvate transaminase (ALT/SGPT) and aspartate aminotransferase/serum glutamate–oxaloacetate transaminase (AST/SGOT) in blood serum. In fact, these two enzymes also increase in concentration following a heart attack, but the physician can differentiate between these two conditions by considering the relative increase in the two enzymes. If ALT/SGPT is elevated to a greater extent than AST/SGOT, it can be concluded that the problem is liver dysfunction.

Enzymes are also used as analytical reagents in the clinical laboratory owing to their specificity. They often selectively react with one substance of interest, producing a product that is easily measured. An example of this is the clinical analysis of urea in blood. The measurement of urea levels in blood is difficult because of the complexity of blood. However, if urea is converted to ammonia using the enzyme urease, the ammonia becomes an *indicator* of urea, because it is produced from urea, and it is easily measured. This test, called the *blood urea nitrogen (BUN) test,* is useful in the diagnosis of kidney malfunction and serves as one example of the utility of enzymes in clinical chemistry.

See A Medical Perspective: Disorders of Sphingolipid Metabolism in Chapter 17.

Enzyme replacement therapy can also be used in the treatment of certain diseases. One such disease, Gaucher's disease, is a genetic disorder resulting in a deficiency of the enzyme *glucocerebrosidase.* In the normal situation, this enzyme breaks down a glycolipid called *glucocerebroside,* which is an intermediate in the synthesis and degradation of complex glycosphingolipids found in cellular membranes. Glucocerebrosidase is found in the lysosomes, where it hydrolyzes glucocerebroside into glucose and ceramide.

Glucocerebroside

Glucocerebrosidase

Glucose   Ceramide

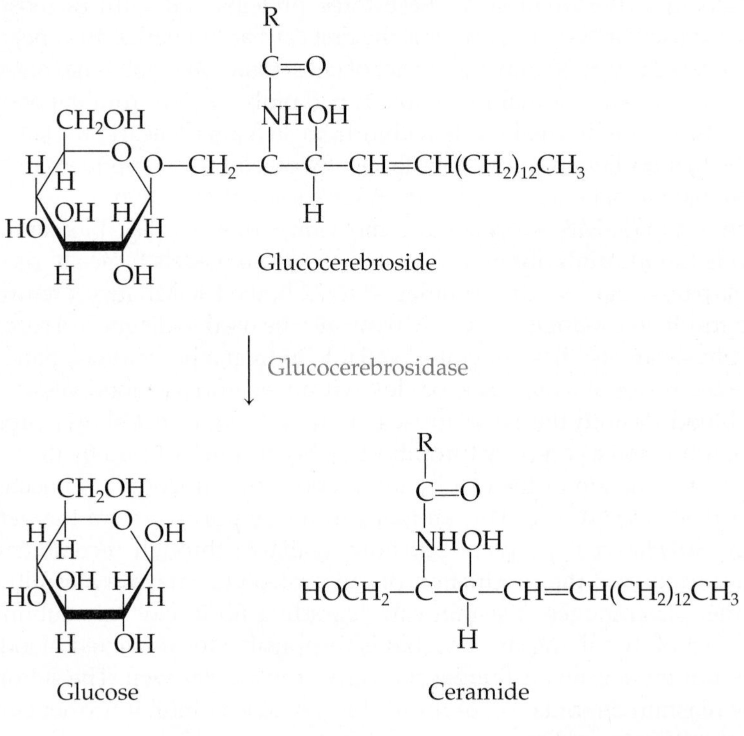

In Gaucher's disease, the enzyme is not present and glucocerebroside builds up in macrophages found in the liver, spleen, and bone marrow. These cells become engorged with excess lipid that cannot be metabolized and then displace healthy, normal cells in bone marrow. The symptoms of Gaucher's disease include severe anemia, thrombocytopenia (reduction in the number of platelets), and hepatosplenomegaly (enlargement of the spleen and liver). There can also be skeletal problems, including bone deterioration and secondary fractures.

Recombinant DNA technology has been used by the Genzyme Corporation to produce the human lysosomal enzyme β-glucocerebrosidase. Given the trade name *Cerezyme,* the enzyme hydrolyzes glucocerebroside into glucose and ceramide so that the products can be metabolized normally. Patients receive Cerezyme intravenously over the course of 1 to 2 h. The dosage and treatment schedule can be tailored to the individual. The results of testing are very encouraging. Patients experience improved red blood cell and platelet counts and reduced hepatosplenomegaly.

CHAPTER MAP

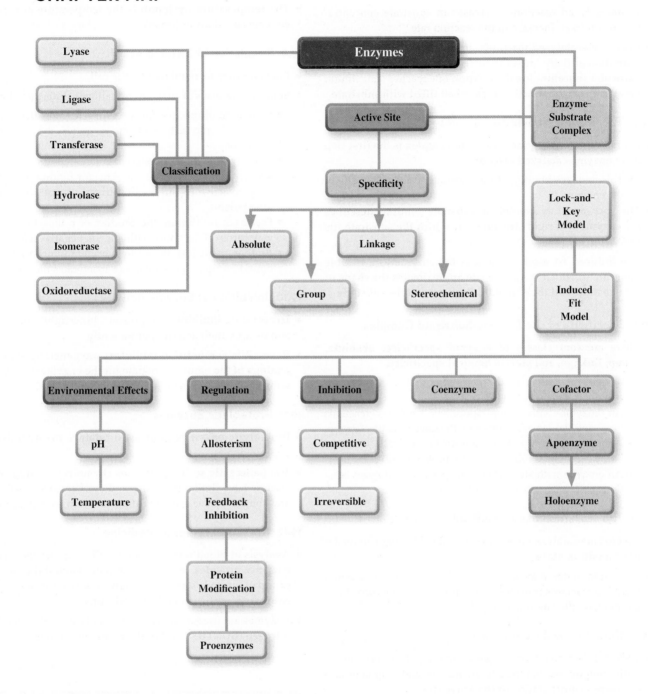

SUMMARY

19.1 Nomenclature and Classification

▶ **Enzymes** are most frequently named by using the common system of nomenclature.

- Common names are useful because they are often derived from the name of the substrate and/or the type of reaction catalyzed by the enzyme.

▶ Enzymes convert one or more **substrates** into one or more **products**.

▶ There are six classes of enzymes, categorized according to function: **oxidoreductases, transferases, hydrolases, lyases, isomerases,** and **ligases.**

19.2 The Effect of Enzymes on the Activation Energy of a Reaction

▶ Enzymes are biological catalysts that lower the activation energy of a reaction but do not alter the equilibrium constant.

19.3 The Effect of Substrate Concentration on Enzyme-Catalyzed Reactions

▶ In uncatalyzed reactions, increases in substrate concentration result in an increase in the reaction rate.

▶ In enzyme-catalyzed reactions, an increase in substrate concentration initially causes an increase in reaction rate, but at a particular concentration the reaction rate reaches a maximum because the enzyme active sites are all filled with substrate.

19.4 The Enzyme-Substrate Complex

▶ Formation of an **enzyme-substrate complex** is the first step of an enzyme-catalyzed reaction.

  • This involves binding of the substrate to the **active site** of the enzyme.

▶ The **lock-and-key model** of substrate binding describes the enzyme as a rigid structure into which the substrate fits precisely.

▶ The **induced fit model** describes the enzyme as a flexible molecule in which the active site approximates the shape of the substrate and then "molds" itself around the substrate.

19.5 Specificity of the Enzyme-Substrate Complex

▶ There are four classes of **enzyme specificity: absolute, group, linkage,** and **stereochemical specificity.**

  • An enzyme with absolute specificity catalyzes the reaction of only a single substrate.
  • An enzyme with group specificity catalyzes reactions involving similar substrates with the same functional group.
  • An enzyme with linkage specificity catalyzes reactions involving similar substrates with the same kind of bond.
  • An enzyme with stereochemical specificity catalyzes reactions involving only one enantiomer.

19.6 The Transition State and Product Formation

▶ An enzyme-catalyzed reaction is mediated through an unstable **transition state.**

▶ Transition states may involve putting "stress" on a bond, bringing reactants into close proximity and in the correct orientation, or altering the local pH.

19.7 Cofactors and Coenzymes

▶ **Cofactors** are metal ions, organic compounds, or organometallic compounds that bind to an enzyme and help maintain the correct configuration of the active site.

  • The protein portion is the **apoenzyme;** the protein portion bound to the cofactor is the **holoenzyme.**

▶ **Coenzymes** are organic groups that bind transiently to an enzyme during the reaction and that accept or donate chemical groups.

  • Coenzymes often contain modified **vitamins** as part of their structure.

19.8 Environmental Effects

▶ Enzymes are sensitive to pH and temperature and are quickly inactivated by extremes of pH or high temperature.

▶ The pH at which an enzyme functions optimally is the **pH optimum.**

▶ The **temperature optimum** is the temperature at which an enzyme functions optimally.

19.9 Regulation of Enzyme Activity

▶ Enzymes may be regulated by the cell.

▶ Some mechanisms of enzyme regulation include the following:

  • Formation of inactive forms or **proenzymes** that are later converted into active enzymes under the appropriate conditions.
  • Allosterism: **Allosteric enzymes** have an effector binding site, as well as an active site. Effector binding renders the enzyme active **(positive allosterism)** or inactive **(negative allosterism).**
  • **Feedback inhibition:** the product of a biosynthetic pathway turns off the entire pathway via negative allosterism.
  • **Protein modification:** adding or removing a covalently bound group either activates or inactivates the enzyme.

19.10 Inhibition of Enzyme Activity

▶ **Irreversible inhibitors,** or poisons, bind tightly to enzymes and destroy their activity permanently.

▶ **Reversible competitive inhibitors** are generally **structural analogs** of the natural substrate for the enzyme that compete with the natural substrate for binding to the active side.

19.11 Proteolytic Enzymes

▶ **Proteolytic enzymes** (proteases) catalyze the hydrolysis of peptide bonds.

▶ The **pancreatic serine proteases** chymotrypsin, trypsin, and elastase have similar structures and mechanisms of action and are thought to have evolved from a common ancestral protease.

19.12 Uses of Enzymes in Medicine

▶ Analysis of blood serum for unusually high levels of certain enzymes provides valuable information about the condition of a patient and is used to diagnose heart attack, liver disease, pancreatitis, and other conditions.

▶ Enzymes are used as analytical reagents, as in the blood urea nitrogen (BUN) test, and in the treatment of disease.

ANSWERS TO PRACTICE PROBLEMS

19.1 a. Pyruvate kinase is a transferase.
  b. Alanine transaminase is a transferase.
  c. Triose phosphate isomerase is an isomerase.
  d. Pyruvate dehydrogenase is an oxidoreductase.
  e. Lactase is a hydrolase.
  f. Phosphofructokinase is a transferase.
  g. Lipase is a hydrolase.
  h. Acetoacetate decarboxylase is a transferase.
  i. Succinate dehydrogenase is an oxidoreductase.

QUESTIONS AND PROBLEMS

Nomenclature and Classification

Foundations

19.21 How are the common names of enzymes often derived?

19.22 What is the most common characteristic used to classify enzymes?

Applications

19.23 Match each of the following substrates with its corresponding enzyme:

1.	Urea	a.	Lipase
2.	Hydrogen peroxide	b.	Glucose-6-phosphatase
3.	Lipid	c.	Peroxidase
4.	Aspartic acid	d.	Sucrase
5.	Glucose-6-phosphate	e.	Urease
6.	Sucrose	f.	Aspartase

19.24 Give a systematic name for the enzyme that would act on each of the following substrates:

a.	Alanine	d.	Ribose
b.	Citrate	e.	Methylamine
c.	Ampicillin		

19.25 Describe the function implied by the name of each of the following enzymes:
  a. Citrate decarboxylase
  b. Adenosine diphosphate phosphorylase
  c. Oxalate reductase
  d. Nitrite oxidase
  e. *cis-trans* Isomerase

19.26 List the six classes of enzymes based on the type of reaction catalyzed. Briefly describe the function of each class, and provide an example of each.

The Effect of Enzymes on the Activation Energy of a Reaction

Foundations

19.27 Define the term *substrate.*

19.28 Define the term *product.*

Applications

19.29 What is the activation energy of a reaction?

19.30 What is the effect of an enzyme on the activation energy of a reaction?

19.31 Write and explain the equation for the equilibrium constant of an enzyme-mediated reaction. Does the enzyme alter the K_{eq}?

19.32 If an enzyme does not alter the equilibrium constant of a reaction, how does it speed up the reaction?

The Effect of Substrate Concentration on Enzyme-Catalyzed Reactions

Foundations

19.33 What is the effect of doubling the substrate concentration on the rate of a chemical reaction?

19.34 Why doesn't the rate of an enzyme-catalyzed reaction increase indefinitely when the substrate concentration is made very large?

Applications

19.35 What is meant by the term *rate-limiting step?*

19.36 How does the rate-limiting step influence an enzyme-catalyzed reaction?

19.37 Draw a graph that describes the effect of increasing the concentration of the substrate on the rate of an enzyme-catalyzed reaction.

19.38 What does a graph of enzyme activity versus substrate concentration tell us about the nature of enzyme-catalyzed reactions?

The Enzyme-Substrate Complex

Foundations

19.39 Define the term *enzyme-substrate complex.*

19.40 Define the term *active site.*

19.41 What are catalytic groups of an enzyme active site?

19.42 What is the binding site of an enzyme active site?

Applications

19.43 Name three major properties of enzyme active sites.

19.44 If enzyme active sites are small, why are enzymes so large?

19.45 What is the lock-and-key model of enzyme-substrate binding?

19.46 Why is the induced fit model of enzyme-substrate binding a much more accurate model than the lock-and-key model?

Specificity of the Enzyme-Substrate Complex

Foundations

19.47 Define the term *enzyme specificity.*

19.48 What region of an enzyme is responsible for its specificity?

19.49 What is meant by the term *group specificity?*

19.50 What is meant by the term *linkage specificity?*

19.51 What is meant by the term *absolute specificity?*

19.52 What is meant by the term *stereochemical specificity?*

Applications

19.53 Provide an example of an enzyme with group specificity and explain the advantage of group specificity for that particular enzyme.

19.54 Provide an example of an enzyme with linkage specificity and explain the advantage of linkage specificity for that particular enzyme.

19.55 Provide an example of an enzyme with absolute specificity and explain the advantage of absolute specificity for that particular enzyme.

19.56 Provide an example of an enzyme with stereochemical specificity and explain the advantage of stereochemical specificity for that particular enzyme.

The Transition State and Product Formation

Foundations

19.57 Outline the four general stages in an enzyme-catalyzed reaction.

19.58 Describe the transition state.

Applications

19.59 What types of transition states might be envisioned that would decrease the energy of activation of a reaction?

19.60 If an enzyme catalyzed a reaction by modifying the local pH, what kind of amino acid R groups would you expect to find in the active site?

Cofactors and Coenzymes

Foundations

19.61 What is the role of a cofactor in enzyme activity?
19.62 How does a coenzyme function in an enzyme-catalyzed reaction?

Applications

19.63 List each of the vitamins found in modified form in a coenzyme and list the coenzymes.
19.64 List the functions of each of the coenzymes. What classes of enzymes would require these coenzymes?

Environmental Effects

Foundations

19.65 Define the temperature optimum for an enzyme.
19.66 Define the optimum pH for enzyme activity.
19.67 How will each of the following changes in conditions alter the rate of an enzyme-catalyzed reaction?
   a. Decreasing the temperature from 37°C to 10°C
   b. Increasing the pH of the solution from 7 to 11
   c. Heating the enzyme from 37°C to 100°C
19.68 Why does an enzyme lose activity when the pH is drastically changed from optimum pH?

Applications

19.69 High temperature is an effective mechanism for killing bacteria on surgical instruments. How does high temperature result in cellular death?
19.70 An increase in temperature will increase the rate of a reaction if a nonenzymatic catalyst is used; however, an increase in temperature will eventually *decrease* the rate of a reaction when an enzyme catalyst is used. Explain the apparent contradiction of these two statements.
19.71 What is the function of the lysosome?
19.72 Of what significance is it that lysosomal enzymes have a pH optimum of 4.8?
19.73 Why are enzymes that are used for clinical assays in hospitals stored in refrigerators?
19.74 Why do extremes of pH inactivate enzymes?

Regulation of Enzyme Activity

Foundations

19.75 **a.** Why is it important for cells to regulate the level of enzyme activity?
   b. Why must synthesis of digestive enzymes be carefully controlled?
19.76 What is an allosteric enzyme?
19.77 What is the difference between positive and negative allosterism?
19.78 **a.** Define feedback inhibition.
   b. Describe the role of allosteric enzymes in feedback inhibition.
   c. Is this positive or negative allosterism?
19.79 What is a proenzyme?

19.80 Three proenzymes that are involved in digestion of proteins in the stomach and intestines are pepsinogen, chymotrypsinogen, and trypsinogen. What is the advantage of producing these enzymes as inactive peptides?

Applications

19.81 The blood clotting mechanism consists of a set of proenzymes that act in a cascade that results in formation of a blood clot. Develop a hypothesis to explain the value of this mechanism for blood clotting.
19.82 What is the benefit for an enzyme such as triacylglycerol lipase to be regulated by covalent modification, in this case phosphorylation?

Inhibition of Enzyme Activity

Foundations

19.83 Define *competitive enzyme inhibition.*
19.84 How do the sulfa drugs selectively kill bacteria while causing no harm to humans?
19.85 Describe the structure of a structural analog.
19.86 How can structural analogs serve as enzyme inhibitors?
19.87 Define *irreversible enzyme inhibition.*
19.88 Why are irreversible enzyme inhibitors often called *poisons?*

Applications

19.89 Suppose that a certain drug company manufactured a compound that had nearly the same structure as a substrate for a certain enzyme but that could not be acted upon chemically by the enzyme. What type of interaction would the compound have with the enzyme?
19.90 The addition of phenylthiourea to a preparation of the enzyme polyphenoloxidase completely inhibits the activity of the enzyme.
   a. Knowing that phenylthiourea binds all copper ions, what conclusion can you draw about whether polyphenoloxidase requires a cofactor?
   b. What kind of inhibitor is phenylthiourea?

Proteolytic Enzymes

Foundations

19.91 What is the function of a proteolytic enzyme?
19.92 Where are the proteolytic enzymes pepsin and trypsin formed? Where do they carry out their function?

Applications

19.93 What do the similar structures of chymotrypsin, trypsin, and elastase suggest about their evolutionary relationship?
19.94 What properties are shared by chymotrypsin, trypsin, and elastase?
19.95 Draw the complete structural formula for the peptide tyr-lys-ala-phe. Show which bond would be broken when this peptide is reacted with chymotrypsin.
19.96 Repeat Question 19.95 for the peptide trp-pro-gly-tyr.
19.97 The sequence of a peptide that contains ten amino acids is as follows:

   ala-gly-val-leu-trp-lys-ser-phe-arg-pro

   Which peptide bond(s) are cleaved by elastase, trypsin, and chymotrypsin?
19.98 What structural features of trypsin, chymotrypsin, and elastase account for their different specificities?

Uses of Enzymes in Medicine

Foundations

19.99 How are blood serum levels of certain enzymes used in medical diagnosis?

19.100 How are enzymes used in medical treatment? Provide an example.

Applications

19.101 List the enzymes whose levels are elevated in blood serum following a myocardial infarction.

19.102 List the enzymes whose levels are elevated as a result of hepatitis or cirrhosis of the liver.

19.103 How is urease used in the diagnosis of kidney malfunction?

19.104 What medical condition is indicated by elevated blood serum levels of amylase and lipase?

CHALLENGE PROBLEMS

1. Ethylene glycol is a poison that causes about fifty deaths a year in the United States. Treating people who have drunk ethylene glycol with massive doses of ethanol can save their lives. Suggest a reason for the effect of ethanol.

2. Generally speaking, feedback inhibition involves regulation of the first step in a pathway. Consider the following hypothetical pathway:

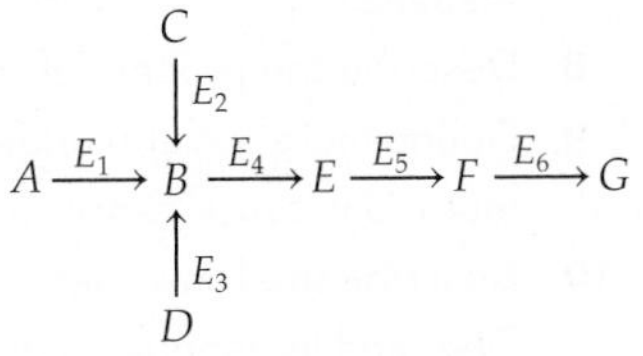

$$C$$
$$\downarrow E_2$$
$$A \xrightarrow{E_1} B \xrightarrow{E_4} E \xrightarrow{E_5} F \xrightarrow{E_6} G$$
$$\uparrow E_3$$
$$D$$

Which step in this pathway do you think should be regulated? Explain your reasoning.

3. In an amplification cascade, each step greatly increases the amount of substrate available for the next step, so that a very large amount of the final product is made. Consider the following hypothetical amplification cascade:

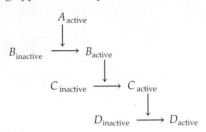

If each active enzyme in the pathway converts 100 molecules of its substrate to active form, how many molecules of D will be produced if the pathway begins with one molecule of A?

4. L-1-(p-toluenesulfonyl)-amido-2-phenylethylchloromethyl ketone (TPCK, shown below) inhibits chymotrypsin, but not trypsin. Propose a hypothesis to explain this observation.

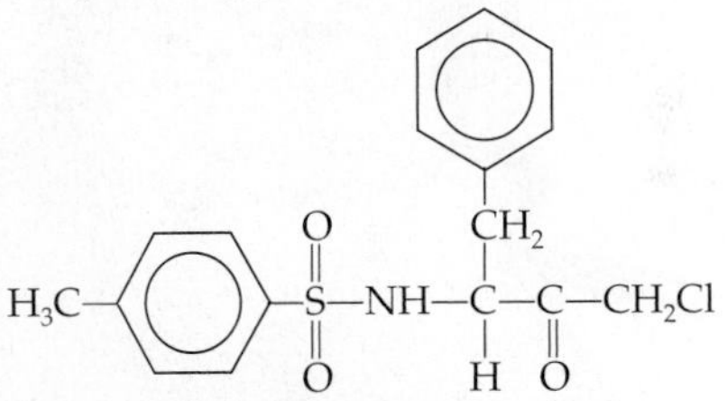

$$H_3C \text{—}\bigcirc\text{—}\overset{O}{\underset{O}{S}}\text{—NH—}\overset{CH_2}{\underset{H}{C}}\text{—}\overset{}{\underset{O}{C}}\text{—}CH_2Cl$$

5. A graduate student is trying to make a "map" of a short peptide so that she can eventually determine the amino acid sequence. She digested the peptide with several proteases and determined the sizes of the resultant digestion products.

Enzyme	M.M. of Digestion Products
Trypsin	2000, 3000
Chymotrypsin	500, 1000, 3500
Elastase	500, 1000, 1500, 2000

Suggest experiments that would allow the student to map the order of the enzyme digestion sites along the peptide.

20 Introduction to Molecular Genetics

These twin girls are identical. Explain why this is so.

LEARNING GOALS

1 Draw the general structure of DNA and RNA nucleotides.

2 Describe the structure of DNA and compare it with RNA.

3 Explain DNA replication.

4 List three classes of RNA molecules and describe their functions.

5 Explain the process of transcription.

6 List and explain the three types of post-transcriptional modifications of eukaryotic mRNA.

7 Describe the essential elements of the genetic code, and develop a "feel" for its elegance.

8 Describe the process of translation.

9 Define mutation and understand how mutations cause cancer and cell death.

10 Describe the tools used in the study of DNA and in genetic engineering.

11 Describe the process of polymerase chain reaction, and discuss potential uses of the process.

12 Discuss strategies for genome analysis and DNA sequencing.

INTRODUCTION

Look around at the students in your chemistry class. They all share many traits: upright stance, a head with two eyes, a nose, and a mouth facing forward, one ear on each side of the head, and so on. You would have no difficulty listing the similarities that define you and your classmates as *Homo sapiens*.

As you look more closely at the individuals, you begin to notice many differences. Eye color, hair color, skin color, the shape of the nose, height, body build: all these traits, and many more, show amazing variety from one person to the next. Even within one family, in which the similarities may be more pronounced, each individual has a unique appearance. In fact, only identical twins look exactly alike—well, most of the time.

The molecule responsible for all these similarities and differences is deoxyribonucleic acid (DNA). Tightly wound up in structures called *chromosomes* in the nucleus of the cell, DNA carries the genetic code to produce the thousands of different proteins that make us who we are. These proteins include enzymes that are responsible for production of the pigment melanin. The more melanin we are genetically programmed to make, the darker our hair, eyes, and skin will be. Others are structural proteins. The gene for α-keratin that makes up hair determines whether our hair will be wavy, straight, or curly. Thousands of genes carry the genetic information for thousands of proteins that dictate our form and, some believe, our behavior.

Genetic traits are passed from one generation to the next. When a sperm fertilizes an ovum, a zygote is created from a single set of maternal chromosomes and a single set of paternal chromosomes. As this fertilized egg divides, each daughter cell will receive one copy of each of these chromosomes. The genes on these chromosomes will direct fetal development from that fertilized cell to a newborn with all the characteristics we recognize as human.

In this chapter, we will explore the structure of DNA and the molecular events that translate the genetic information of a gene into the structure of a protein.

20.1 The Structure of the Nucleotide

Even before the philosopher Aristotle observed that "like begets like," humans were curious about the way in which family likenesses are passed from one generation to the next. In the 1860s Gregor Mendel combined astute observations, careful experimental design, and mathematical analysis to explain inheritance. Presented at a meeting in 1865 and published in 1866, Mendel's brilliant work was largely ignored by a scientific community that simply could not understand it.

At about the same time (1869), Friedrich Miescher discovered a substance in the nuclei of white blood cells recovered from pus. Chemical analysis of this substance, which he called *nuclein,* revealed that it contained 14% nitrogen and 3% phosphorus in addition to carbon, hydrogen, and oxygen. As microscopes improved in the last decades of the nineteenth century, biologists were able to peer into the nuclei of cells. They observed structures, later called *chromosomes,* which seemed to play a critical role in the process of cell division. Interestingly, egg and sperm cells were observed to have only half the chromosomes of the cells that produced them.

Chemical analysis of chromosomes indicated that they were composed of both protein and nuclein. But which of these molecules represented the genetic material? Most were convinced that the answer to this question was protein. The reasoning was that the genetic material must have a structure that would allow it to encode the enormous variation seen in the biological world. Both nuclein and protein were known to be polymers. However, proteins were polymers of twenty different subunits, the amino acids. Based on the results of Phoebus Levene, working with Emil Fischer and Albrecht Kossel, nuclein was composed of only four subunits. It appeared to lack the complexity required of a molecule responsible for the great diversity seen among plants, animals, and microbes.

In 1950, the genetic information was demonstrated to be nuclein, now called *deoxyribonucleic acid,* or DNA. In 1953, just over 60 years ago, James Watson and Francis Crick published a paper describing the structure of the DNA molecule.

Chemical Composition of DNA and RNA

Two types of nucleic acids are important to the cell. The first is **deoxyribonucleic acid (DNA),** which carries all of the genetic information for an organism. The second type is **ribonucleic acid (RNA),** which is responsible for interpreting the genetic information into proteins that will carry out the essential cellular functions.

The components of these nucleic acids include a five-carbon sugar, phosphate, and four heterocyclic amines called *nitrogenous bases.* The sugar in DNA is 2′-deoxyribose and the sugar in RNA is ribose. These differ only in the absence of a hydroxyl group at the carbon-2 position of 2′-deoxyribose (Figure 20.1). The nitrogenous bases are divided into two families known as **pyrimidines** and **purines** (Figure 20.1). The pyrimidine bases in DNA are cytosine and thymine. The pyrimidines found in RNA are cytosine and uracil (Figure 20.1). Notice that these three pyrimidines differ from one another only in the positioning of certain functional groups around the ring.

The major purines of both DNA and RNA are adenine and guanine (Figure 20.1). As with the pyrimidines, the purines differ from one another only in the location of functional groups around the ring.

Nucleosides

Nucleosides are produced by the combination of a sugar, either ribose (in RNA) or 2′-deoxyribose (in DNA), with a purine or a pyrimidine base. Because there are two cyclic molecules in a nucleoside, we need an easy way to describe the ring atoms of each. For this reason the ring atoms of the sugar are designated with a prime to distinguish them from atoms in the base (Figure 20.2). The covalent bond between the sugar and the base is called a β-*N-glycosidic linkage.* The general structures of a purine nucleoside and a pyrimidine nucleoside are shown in Figure 20.2.

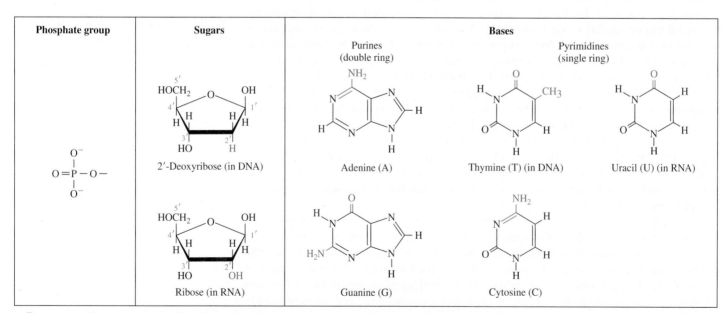

Figure 20.1 The components of nucleic acids include phosphate groups, the five-carbon sugars ribose and deoxyribose, and purine and pyrimidine nitrogenous bases. The ring positions of the sugars are designated with primes (′) to distinguish them from the ring positions of the bases.

N-1 of pyrimidines and N-9 of purines participate in the glycosidic bonds of nucleosides. The nucleosides formed with ribose and adenine or guanine are called adenosine or guanosine, respectively. If 2'-deoxyribose is the sugar in these nucleosides, they are called 2'-deoxyadenosine and 2'-deoxyguanosine. The ribonucleosides formed from cytosine and uracil are called cytidine and uridine, respectively. The deoxyribonucleosides of cytosine and thymine are called 2'-deoxycytidine and thymidine, respectively. No prefix is needed for thymidine because it is found only in DNA.

Nucleotide Structure

From the work of Watson and Crick, as well as that of Miescher, Levene, and many others, we now know that deoxyribonucleic acid (DNA) and ribonucleic acid (RNA) are long polymers of **nucleotides.** Every nucleotide is composed of a nitrogenous base, a five-carbon sugar, and at least one phosphoryl group.

Each nucleotide consists of either ribose or deoxyribose, one of the five nitrogenous bases, and one or more phosphoryl groups (Figure 20.3). A nucleotide with the sugar ribose is a **ribonucleotide,** and one having the sugar 2'-deoxyribose is a **deoxyribonucleotide** (Figure 20.1).

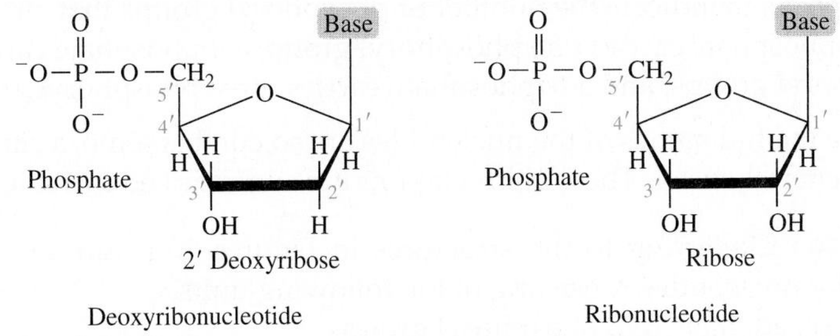

The covalent bond between the sugar and the phosphoryl group is a phosphoester bond formed by a condensation reaction between the 5'-OH of the sugar and an —OH of the phosphoryl group. As noted above, the bond between the base and the sugar is a β-N-glycosidic linkage that joins the 1'-carbon of the sugar and a nitrogen atom of the base (N-9 of purines and N-1 of pyrimidines).

To name a nucleotide (Figure 20.3), simply begin with the name of the nitrogenous base, and apply the following simple rules:

- Remove the -ine ending, and replace it with either -osine for purines or -idine for pyrimidines. Uracil is the one exception to this rule. In this case, the -acil ending is replaced with -idine, producing the name *uridine*.

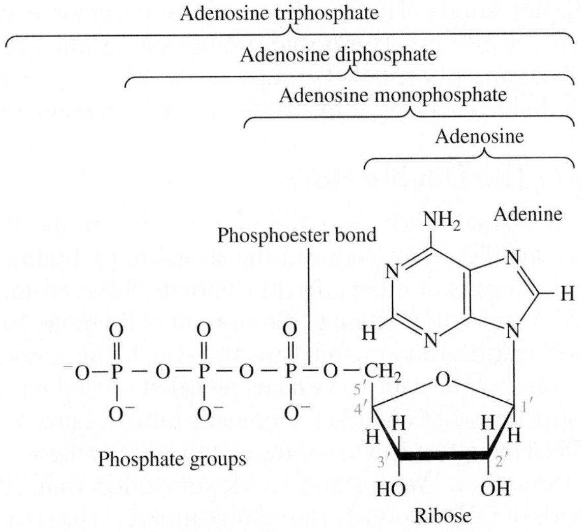

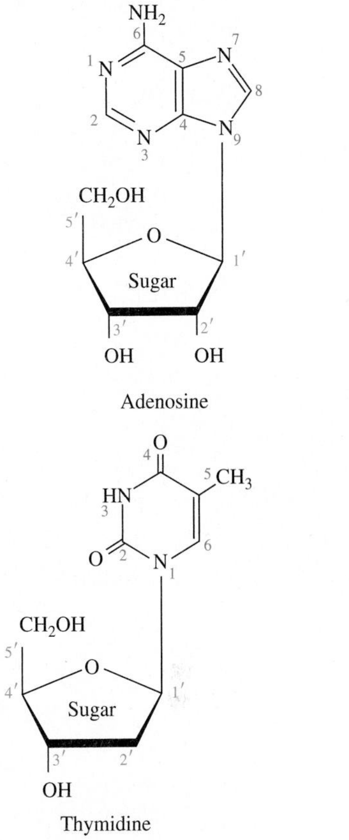

Figure 20.2 General structures of a purine and a pyrimidine nucleoside. Notice that the N-glycosidic linkage involves the 1' carbon of the sugar and either the N-1 of the pyrimidine or N-9 of the purine.

Figure 20.3 The structure of the ribonucleotide adenosine triphosphate.

TABLE 20.1 Names and Abbreviations of the Ribonucleotides and Deoxyribonucleotides Containing Adenine

Nucleotide	Abbreviation
Deoxyadenosine monophosphate	dAMP
Deoxyadenosine diphosphate	dADP
Deoxyadenosine triphosphate	dATP
Adenosine monophosphate	AMP
Adenosine diphosphate	ADP
Adenosine triphosphate	ATP

- Nucleotides with the sugar ribose are ribonucleotides, and those having the sugar 2'-deoxyribose are deoxyribonucleotides. For a deoxyribonucleotide, the prefix *deoxy-* is placed before the modified nitrogenous base name. No prefix is required for ribonucleotides, or for thymidine, which is found only in DNA.
- Add a prefix to indicate the number of phosphoryl groups that are attached. A *mono*phosphate carries one phosphoryl group; a *di*phosphate carries two phosphoryl groups; and a *tri*phosphate carries three phosphoryl groups.

Because the full names of the nucleotides are so cumbersome, a simple abbreviation is generally used. These abbreviations are summarized in Table 20.1.

Question 20.1   Referring to the structures in Figures 20.1 and 20.3, draw the structures for nucleotides consisting of the following units.
    a. Ribose, adenine, two phosphoryl groups
    b. 2'-Deoxyribose, guanine, three phosphoryl groups

Question 20.2   Referring to the structures in Figures 20.1 and 20.3, draw the structures for nucleotides consisting of the following units.
    a. 2'-Deoxyribose, thymine, one phosphoryl group
    b. Ribose, cytosine, three phosphoryl groups
    c. Ribose, uracil, one phosphoryl group

20.2   The Structure of DNA and RNA

LEARNING GOAL

2   Describe the structure of DNA and compare it with RNA.

A single strand of DNA is a polymer of nucleotides bonded to one another by 3'–5' phosphodiester bonds. The backbone of the polymer is called the *sugar-phosphate backbone* because it is composed of alternating units of the five-carbon sugar 2'-deoxyribose and phosphoryl groups in phosphodiester linkage. A nitrogenous base is bonded to each sugar by an *N*-glycosidic linkage (Figure 20.4).

DNA Structure: The Double Helix

James Watson and Francis Crick were the first to describe the three-dimensional structure of DNA, in 1953. They deduced the structure by building models based on the experimental results of others. Irwin Chargaff observed that the amount of adenine in any DNA molecule is equal to the amount of thymine. Similarly, he found that the amounts of cytosine and guanine are also equal. The X-ray diffraction studies of Rosalind Franklin and Maurice Wilkens revealed several repeat distances that characterize the structure of DNA: 0.34 nanometers (nm), 3.4 nm, and 2 nm. (Look at the structure of DNA in Figure 20.5 to see the significance of these measurements.)

With this information, Watson and Crick concluded that DNA is a **double helix** of two strands of DNA wound around one another. The structure of the double helix is often compared to a spiral staircase. The sugar-phosphate backbones

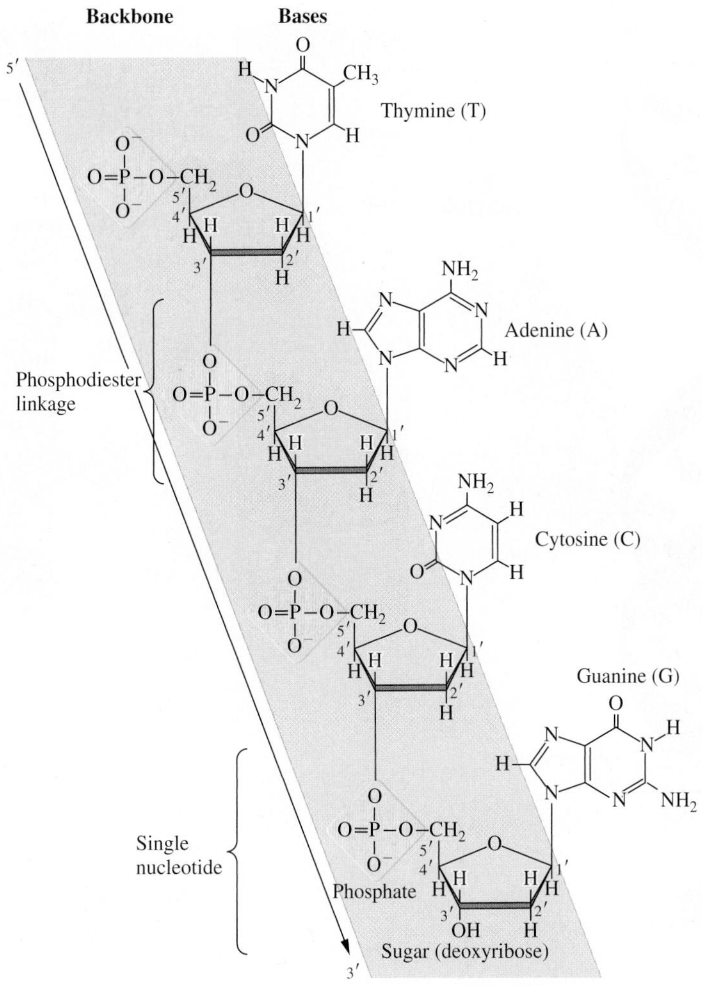

Backbone   **Bases**

Figure 20.4 The covalent, primary structure of DNA.

of the two strands of DNA spiral around the outside of the helix like the handrails on a spiral staircase. The nitrogenous bases extend into the center at right angles to the axis of the helix. You can imagine the nitrogenous bases forming the steps of the staircase. The structure of this elegant molecule is shown in Figure 20.5.

One noncovalent attraction that helps maintain the double helix structure is hydrogen bonding between the nitrogenous bases in the center of the helix. Adenine forms two hydrogen bonds with thymine, and cytosine forms three hydrogen bonds with guanine (Figure 20.5). These are called **base pairs.** The two strands of DNA are **complementary strands** because the sequence of bases on one automatically determines the sequence of bases on the other. When there is an adenine on one strand, there will always be a thymine in the same location on the opposite strand.

The diameter of the double helix is 2.0 nm. This is dictated by the dimensions of the purine-pyrimidine base pairs. The helix completes one turn every ten base pairs. One complete turn is 3.4 nm. Thus, each base pair advances the helix by 0.34 nm. These dimensions explain the repeat distances observed by X-ray diffraction.

One last important feature of the DNA double helix is that the two strands are **antiparallel strands,** as this example shows:

Base-pairing explains Chargaff's observation that the amount of adenine always equals the amount of thymine and the amount of guanine always equals the amount of cytosine for any DNA sample.

```
5'  P—S—P—S—P—S—P—S—P—S—P—S—OH 3'
        |     |     |     |     |     |
        A     T     G     C     G     A
        :     :     :     :     :     :
        T     A     C     G     C     T
        |     |     |     |     |     |
3' OH—S—P—S—P—S—P—S—P—S—P—S—P 5'
```

Key Features

- Two strands of DNA form a right-handed double helix.

- The bases in opposite strands hydrogen bond according to the AT/GC rule.

- The two strands are antiparallel with regard to their 5' to 3' directionality.

- There are ~10.0 nucleotides in each strand per complete 360° turn of the helix.

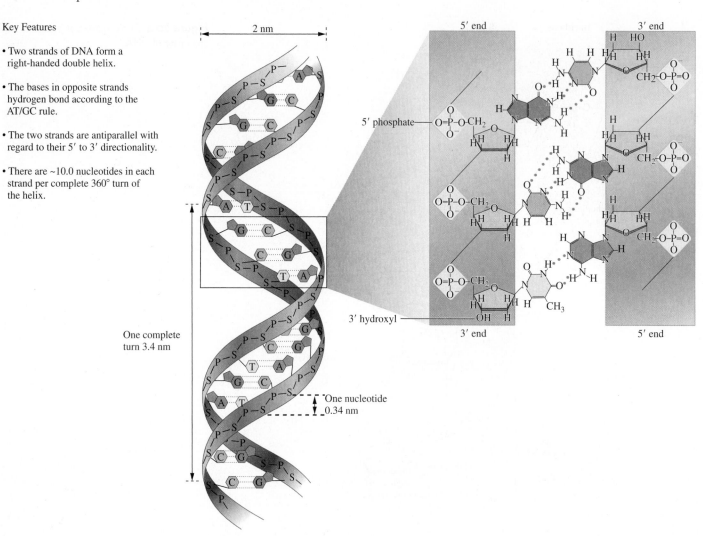

Figure 20.5 Schematic ribbon diagram of the DNA double helix showing the dimensions of the DNA molecule and the antiparallel orientation of the two strands.

In other words, the two strands of the helix run in opposite directions (see Figure 20.5). Only when the two strands are antiparallel can the base pairs form the hydrogen bonds that hold the two strands together.

Chromosomes

Chromosomes are pieces of DNA that carry the genetic instructions, or genes, of an organism. Organisms such as the prokaryotes have only a single chromosome and its structure is relatively simple. Others, the eukaryotes, have many chromosomes, each of which has many different levels of structure. The complete set of genetic information in all the chromosomes of an organism is called the **genome.**

Prokaryotes are organisms with a simple cellular structure in which there is no true nucleus surrounded by a nuclear membrane and there are no true membrane-bound organelles. This group includes all of the bacteria. In these organisms, the chromosome is a circular DNA molecule that is supercoiled, which means that the helix is coiled on itself. The supercoiled DNA molecule is attached to a complex of proteins at roughly forty sites along its length, forming a series of loops. This structure, called the nucleoid, can be seen in Figure 20.6.

Eukaryotes are organisms that have cells containing a true nucleus enclosed by a nuclear membrane. They also have a variety of membrane-bound

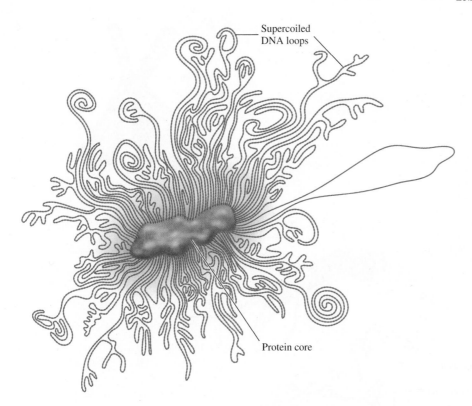

Figure 20.6 Structure of a bacterial nucleoid. The nucleoid is made up of the supercoiled, circular chromosome attached to a protein core.

Supercoiled
DNA loops

Protein core

organelles that segregate different cellular functions into different compartments. As an example, the reactions of aerobic respiration are located within the mitochondria.

All animals, plants, and fungi are eukaryotes. The number and size of the chromosomes of eukaryotes vary from one species to the next. For instance, humans have twenty-three pairs of chromosomes, while the adder's tongue fern has 631 pairs of chromosomes. But the chromosome structure is the same for all those organisms that have been studied.

Eukaryotic chromosomes are very complex structures (Figure 20.7). The first level of structure is the **nucleosome,** which consists of a strand of DNA wrapped around a small disk made up of histone proteins. At this level, the DNA looks like beads along a string. The string of beads then coils into a larger structure called the *condensed fiber.* This complex of DNA and protein is termed *chromatin* and makes up the eukaryotic chromosomes. The full complexities of the eukaryotic chromosome are not yet understood, but there are probably many such levels of coiled structures.

Some human genetic disorders are characterized by unusual chromosome numbers. For instance, Down syndrome is characterized by an extra copy of chromosome 21. The presence of this additional chromosome causes the traits associated with Down syndrome, including varying degrees of mental challenges, a flattened face, and short stature. The presence of an additional chromosome 18 causes Edward syndrome and an extra chromosome 13 causes Patau syndrome. Both of these are extremely rare and result in extreme mental and physical defects and early death. The presence of extra copies of the sex chromosomes, X or Y, is not lethal. Males with two X chromosomes and one Y suffer from Klinefelter syndrome and show sexual immaturity and breast development. Males with an extra Y chromosome are unusually tall, as are women with three X chromosomes. A woman with only a single X chromosome experiences Turner syndrome, including short stature, a webbed neck, and sexual immaturity. All other abnormalities in chromosome number are thought to be lethal to the fetus. In fact, it is thought that 50% of all miscarriages are the result of abnormal chromosome numbers.

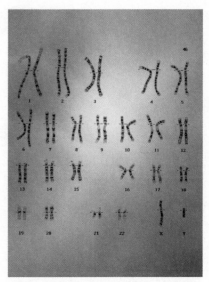

This karyotype shows the twenty-three pairs of chromosomes of humans. What type of genetic disorders could be identified by karyotyping?

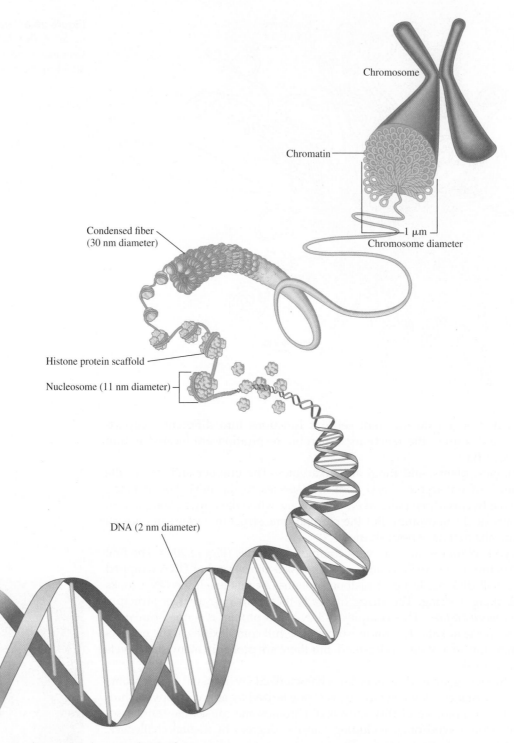

Figure 20.7 The eukaryotic chromosome has many levels of structure.

RNA Structure

The sugar-phosphate backbone of RNA consists of ribonucleotides, also linked by 3′–5′ phosphodiester bonds. These phosphodiester bonds are identical to those found in DNA. However, RNA molecules differ from DNA molecules in three basic properties.

- RNA molecules are usually *single-stranded*.
- The sugar-phosphate backbone of RNA consists of *ribonucleotides* linked by 3′–5′ phosphodiester bonds. Thus, the sugar *ribose* is found in place of 2′-deoxyribose.
- The nitrogenous base *uracil* (U) replaces thymine (T).

Molecular Genetics and Detection of Human Genetic Disorders

It is estimated that 3–5% of the human population suffers from a serious genetic defect. That's 350 million people! But what if genetic disease could be detected and "cured"? Two new technologies, *gene therapy* and *preimplantation diagnosis,* may help us realize this dream.

For a couple with a history of genetic disorders in the family, pregnancy is a time of anxiety. Through *genetic counseling* these couples can learn the probability that their child has the defect. For several hundred genetic disorders, the uncertainty can be eliminated. *Amniocentesis* (removal of 10–20 mL of fluid from the sac around the fetus) and *chorionic villus sampling* (removal of cells from a fetal membrane) are two procedures that are used to obtain fetal cells for genetic testing. Fetal cells are cultured and tested by enzyme assays and DNA tests to look for genetic disorders. *Noninvasive prenatal testing* (NIPT) takes advantage of the fact that a small amount of the baby's DNA crosses into the mother's bloodstream. Analyzing the DNA in a sample of the mother's blood allows physicians to check whether the fetus has a higher chance of some disorders. Results are available in a week or two. Since NIPT is a screening method, amniocentesis or chorionic villus sampling may be recommended to confirm the result. If a genetic defect is diagnosed, the parents must make a difficult decision: to abort the fetus or to carry the child to term and deal with the effects of the genetic disorder.

The power of modern molecular genetics is obvious in our ability to find a "bad" gene from just a few cells. But scientists have developed an even more impressive way to test for genetic defects before the embryo implants into the uterine lining. This technique, called *preimplantation diagnosis,* involves fertilizing a human egg and allowing the resulting zygote to divide in a sterile petri dish. When the zygote consists of eight to sixteen cells, *one* cell is removed for genetic testing. Only genetically normal embryos are implanted in the mother. Thus, the genetic defects that we can detect could be eliminated from the population by preimplantation diagnosis because only a zygote with "good" genes is used.

Gene therapy is a second way in which genetic disorders may one day be eliminated. Foreign genes, including one for growth hormone, have been introduced into fertilized mouse eggs and the zygotes implanted in female mice. The baby mice born with the foreign growth hormone gene were about three times larger than their normal littermates! One day, this kind of technology may be used to introduce normal genes into human fertilized eggs carrying a defective gene, thereby replacing the defective gene with a normal one.

In this chapter, we will examine the molecules that carry and express our genetic information, DNA and RNA. Only by understanding the structure and function of these molecules have we been able to develop the amazing array of genetic tools that currently exists. We hope that as we continue to learn more about human genetics, we will be able to detect and one day correct most of the known genetic disorders.

For Further Understanding

▶ Go online to investigate the concerns that limit the use of gene therapy in humans.

▶ What is the advantage of preimplantation diagnosis?

Although RNA molecules are single-stranded, base pairing between uracil and adenine and between guanine and cytosine can still occur. We will show the importance of this property as we examine the way in which RNA molecules are involved in the expression of the genetic information in DNA.

20.3 DNA Replication

DNA must be replicated before a cell divides so that each daughter cell inherits a copy of each gene. A cell that is missing a critical gene will die, just as an individual with a genetic disorder, a defect in an important gene, may die early in life. Thus, it is essential that the process of DNA replication produces an absolutely accurate copy of the original genetic information. If mistakes are made in critical genes, the result may be lethal mutations.

The structure of the DNA molecule suggested the mechanism for its accurate replication. Since adenine can base pair only with thymine and cytosine with guanine, Watson and Crick first suggested that an enzyme could "read" the nitrogenous bases on one strand of a DNA molecule and add complementary bases to a strand of DNA being synthesized. The product of this mechanism would be a new DNA molecule in which one strand is the original, or parent, strand and the

LEARNING GOAL

3 Explain DNA replication.

Isotopes are atoms of the same element having the same number of protons but different numbers of neutrons and, therefore, different mass numbers.

second strand is a newly synthesized, or daughter, strand. This mode of DNA replication is called **semiconservative replication** (Figure 20.8).

Experimental evidence for this mechanism of DNA replication was provided by an experiment designed by Matthew Meselson and Franklin Stahl in 1958. *Escherichia coli* cells were grown in a medium in which $^{15}NH_4^+$ was the sole nitrogen source. ^{15}N is a nonradioactive, heavy isotope of nitrogen. Thus, growing the cells in this medium resulted in all of the cellular DNA containing this heavy isotope.

The cells containing only $^{15}NH_4^+$ were then added to a medium containing only the abundant isotope of nitrogen, $^{14}NH_4^+$, and were allowed to grow for one cycle of cell division. When the daughter DNA molecules were isolated and analyzed, it was found that each was made up of one strand of "heavy" DNA, the parental strand, and one strand of "light" DNA, the new daughter strand. After a second round of cell division, half of the isolated DNA contained no ^{15}N and half contained a 50/50 mixture of ^{14}N and ^{15}N-labeled DNA (Figure 20.9). This demonstrated conclusively that each parental strand of the DNA molecule serves as the template for the synthesis of a daughter strand and that each newly synthesized DNA molecule is composed of one parental strand and one newly synthesized daughter strand.

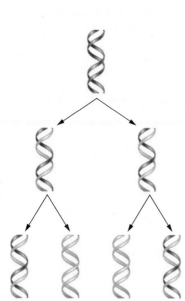

Figure 20.9 Representation of the Meselson and Stahl experiment. The DNA from cells grown in medium containing $^{15}NH_4^+$ is shown in purple. After a single generation in medium containing $^{14}NH_4^+$, the daughter DNA molecules have one ^{15}N-labeled parent strand and one ^{14}N-labeled daughter strand (blue). After a second generation in $^{14}NH_4^+$ containing medium, there are equal numbers of $^{14}N/^{15}N$ DNA molecules and $^{14}N/^{14}N$ DNA molecules.

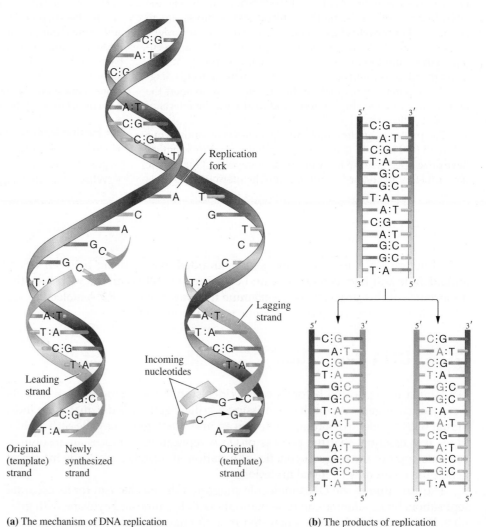

(a) The mechanism of DNA replication

(b) The products of replication

Figure 20.8 In semiconservative DNA replication, each parent strand serves as a template for the synthesis of a new daughter strand.

Bacterial DNA Replication

The bacterial chromosome is a circular molecule of DNA made up of about three million nucleotides. DNA replication begins at a unique sequence on the circular chromosome known as the **replication origin** (Figure 20.10). Replication occurs bidirectionally at the rate of about 500 new nucleotides every second! The point at which the new deoxyribonucleotide is added to the growing daughter strand is called the **replication fork** (Figure 20.10). It is here that the DNA has been opened to allow binding of the various proteins and enzymes responsible for DNA replication. Since DNA synthesis occurs bidirectionally, there are two replication forks moving in opposite directions. Replication is complete when the replication forks collide approximately halfway around the circular chromosome.

The first step in DNA replication (Figure 20.11) is the separation of the strands of DNA. The protein *helicase* does this by breaking the hydrogen bonds between the base pairs. This, in turn, causes supercoiling of the molecule. This stress is relieved by the enzyme *topoisomerase*, which travels along the DNA ahead of the replication fork. At this point, *single-strand binding protein* binds to the separated strands, preventing them from coming back together. In the next step, the enzyme *primase* catalyzes the synthesis of a small piece of RNA (ten to twelve nucleotides) called an *RNA primer* that serves to "prime" the process of DNA replication.

Now the enzyme **DNA polymerase III** "reads" each parental strand, also called the *template*, and catalyzes the polymerization of a complementary daughter strand. Deoxyribonucleotide triphosphate molecules are the precursors for DNA replication (Figure 20.12). In this reaction, a pyrophosphate group is released as a phosphoester bond is formed between the 5′-phosphoryl group of the nucleotide being added to the chain and the 3′-OH of the nucleotide on the daughter strand. This is called 5′ to 3′ synthesis.

One complicating factor in the process of DNA replication is the fact that the two strands of DNA are antiparallel to one another. DNA polymerase III can only catalyze DNA chain elongation in the 5′ to 3′ direction, yet the replication fork proceeds in one direction, while both strands are replicated simultaneously. Another complication is the need for an RNA primer to serve as the starting point for DNA replication. As a result of these two obstacles, there are different mechanisms for replication of the two strands. One strand, called the **leading strand,** is replicated continuously. The opposite strand, called the **lagging strand,** is replicated discontinuously.

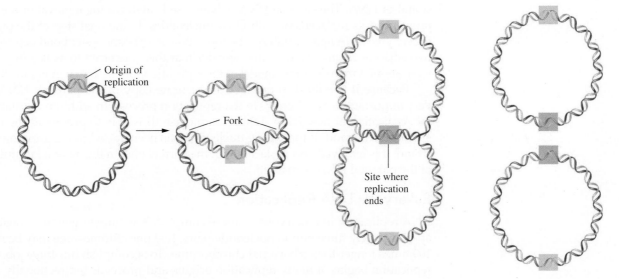

Figure 20.10 Bacterial chromosome replication.

Functions of key proteins involved with DNA replication

- **DNA helicase** breaks the hydrogen bonds between the DNA strands.

- **Topoisomerase** alleviates positive supercoiling.

- **Single-strand binding proteins** keep the parental strands apart.

- **Primase** synthesizes an RNA primer.

- **DNA polymerase III** synthesizes a daughter strand of DNA.

- **DNA polymerase I** excises the RNA primers and fills in with DNA.

- **DNA ligase** covalently links the DNA fragments together.

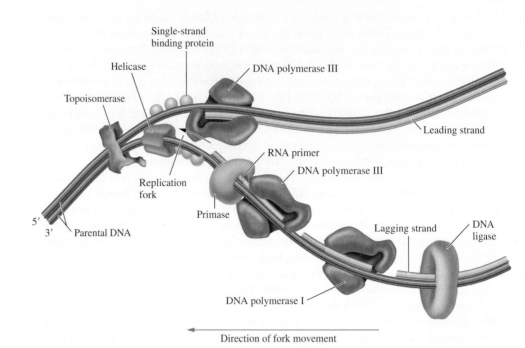

Figure 20.11 Because the two strands of DNA are antiparallel and DNA polymerase can only catalyze 5′ ⟶ 3′ replication, only one of the two DNA strands (top strand) can be read continuously to produce a daughter strand. The other must be synthesized in segments that are extended away from the direction of movement of the replication fork (bottom strand). These discontinuous segments are later covalently joined together by DNA ligase.

The two mechanisms are shown in Figures 20.11 and 20.12. For the leading strand, a single RNA primer is produced at the replication origin and DNA polymerase III continuously catalyzes the addition of nucleotides in the 5′ to 3′ direction, beginning with addition of the first nucleotide to the RNA primer.

On the lagging strand, many RNA primers are produced as the replication fork proceeds along the molecule. DNA polymerase III catalyzes DNA chain elongation from each of these primers. When the new strand "bumps" into a previous one, synthesis stops at that site. Meanwhile, at the replication fork, a new primer is being synthesized by primase. The final steps of synthesis on the lagging strand involve removal of the primers, repair of the gaps, and sealing of the fragments into an intact strand of DNA. The enzyme DNA polymerase I catalyzes the removal of the RNA primer and its replacement with DNA nucleotides. In the final step of the process, the enzyme DNA ligase catalyzes the formation of a phosphoester bond between the two adjacent fragments. It is little wonder that this is referred to as lagging strand replication! A more accurate model of the replication fork is shown in Figure 20.13.

Because it is critical to produce an accurate copy of the parental DNA, it is very important to avoid errors in the replication process. In addition to catalyzing the replication of new DNA, DNA polymerase III is able to proofread the newly synthesized strand. If the wrong nucleotide has been added to the growing DNA strand, it is removed and replaced with the correct one. In this way, a faithful copy of the parental DNA is ensured.

Eukaryotic DNA Replication

DNA replication in eukaryotes is more complex. The human genome consists of approximately three billion nucleotide pairs. Just one chromosome may be nearly 100 times longer than a bacterial chromosome. To accomplish this huge job, DNA replication begins at many replication origins and proceeds bidirectionally along each chromosome.

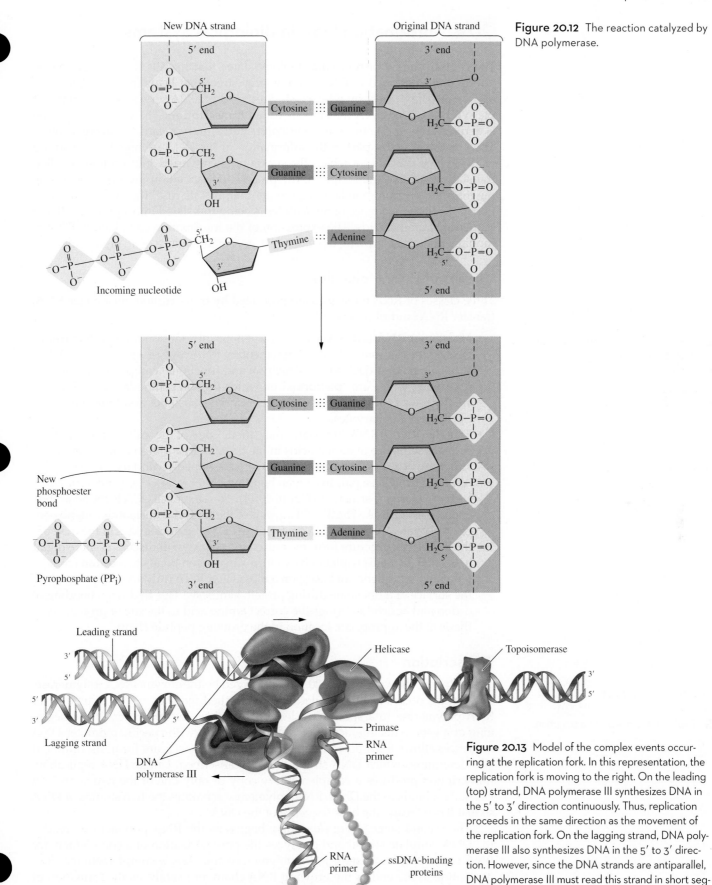

Figure 20.12 The reaction catalyzed by DNA polymerase.

Figure 20.13 Model of the complex events occurring at the replication fork. In this representation, the replication fork is moving to the right. On the leading (top) strand, DNA polymerase III synthesizes DNA in the 5′ to 3′ direction continuously. Thus, replication proceeds in the same direction as the movement of the replication fork. On the lagging strand, DNA polymerase III also synthesizes DNA in the 5′ to 3′ direction. However, since the DNA strands are antiparallel, DNA polymerase III must read this strand in short segments (discontinuously) and in the opposite direction of the movement of the replication fork.

20.4 Information Flow in Biological Systems

The **central dogma** of molecular biology states that in cells the flow of genetic information contained in DNA is a one-way street that leads from DNA to RNA to protein. The process by which a single strand of DNA serves as a template for the synthesis of an RNA molecule is called **transcription.** The word *transcription* is derived from the Latin word *transcribere* and simply means "to make a copy." Thus, in this process, part of the information in the DNA is copied into a strand of RNA. The process by which the message is converted into protein is called **translation.** Unlike transcription, the process of translation involves converting the information from one language to another. In this case, the genetic information in the linear sequence of nucleotides is being translated into a protein, a linear sequence of amino acids. The expression of the information contained in DNA is fundamental to the growth, development, and maintenance of all organisms.

Classes of RNA Molecules

Three classes of RNA molecules are produced by transcription: messenger RNA, transfer RNA, and ribosomal RNA.

LEARNING GOAL

4 List three classes of RNA molecules and describe their functions.

- **Messenger RNA (mRNA)** carries the genetic information for a protein from DNA to the ribosomes. It is a complementary RNA copy of a gene on the DNA.
- **Ribosomal RNA (rRNA)** is a structural and functional component of the ribosomes, which are "platforms" on which protein synthesis occurs. There are three types of rRNA molecules in bacterial ribosomes and four in the ribosomes of eukaryotes.
- **Transfer RNA (tRNA)** translates the genetic code of the mRNA into the primary sequence of amino acids in the protein. In addition to the primary structure, tRNA molecules have a cloverleaf-shaped secondary structure resulting from base pair hydrogen bonding (A—U and G—C) and a roughly L-shaped tertiary structure (Figure 20.14). The sequence CCA is found at the 3' end of the tRNA. The 3'—OH group of the terminal nucleotide, adenosine, can be covalently attached to an amino acid. Three nucleotides at the base of the cloverleaf structure form the **anticodon.** As we will discuss in more detail in Section 20.6, this triplet of bases forms hydrogen bonds to a **codon** (complementary sequence of bases) on a messenger RNA (mRNA) molecule on the surface of a ribosome during protein synthesis. This hydrogen bonding of codon and anticodon brings the correct amino acid to the site of protein synthesis at the appropriate location in the growing peptide chain.

Transcription

LEARNING GOAL

5 Explain the process of transcription.

Transcription, shown in Figure 20.15, is catalyzed by the enzyme **RNA polymerase.** The process occurs in three stages. The first, called *initiation,* involves binding of RNA polymerase to a specific nucleotide sequence, the **promoter,** at the beginning of a gene. This interaction of RNA polymerase with specific promoter DNA sequences allows RNA polymerase to recognize the start point for transcription. It also determines which DNA strand will be transcribed. Unlike DNA replication, transcription produces a complementary copy of only one of the two strands of DNA. As it binds to the DNA, RNA polymerase separates the two strands of DNA so that it can "read" the base sequence of the DNA.

The second stage, *chain elongation,* begins as the RNA polymerase "reads" the DNA template strand and catalyzes the polymerization of a complementary RNA copy. With each step, RNA polymerase transfers a complementary ribonucleotide to the end of the growing RNA chain and catalyzes the formation of a 3'–5' phosphodiester bond between the 5' phosphoryl group of the incoming ribonucleotide and the 3' hydroxyl group of the last ribonucleotide of the growing RNA chain.

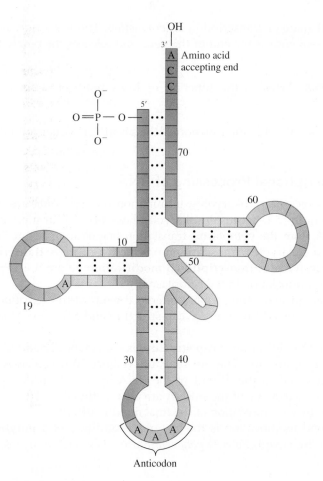

Figure 20.14 Structure of tRNA. The primary structure of a tRNA is the linear sequence of ribonucleotides. Here we see the hydrogen-bonded secondary structure of a tRNA, showing the three loops and the amino acid accepting end.

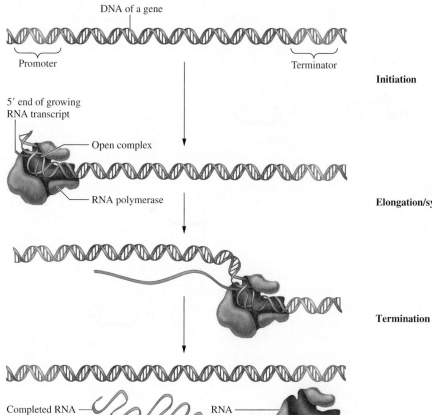

Figure 20.15 The stages of transcription.

The final stage of transcription is *termination*. The RNA polymerase finds a termination sequence at the end of the gene and releases the newly formed RNA molecule.

Question 20.3   What is the function of RNA polymerase in the process of transcription?

Question 20.4   What is the function of the promoter sequence in the process of transcription?

Post-transcriptional Processing of RNA

In bacteria, which are prokaryotes, termination releases a mature mRNA ready for translation. In fact, because prokaryotes have no nuclear membrane separating the DNA from the cytoplasm, translation begins long before the mRNA is completed. In eukaryotes, transcription produces a **primary transcript** that must undergo extensive **post-transcriptional modification** before it is exported out of the nucleus for translation in the cytoplasm.

Eukaryotic primary transcripts undergo three post-transcriptional modifications. These are the addition of a 5′ cap structure and a 3′ poly(A) tail, and RNA splicing.

In the first modification, a **cap structure** is enzymatically added to the 5′ end of the primary transcript. The cap structure (Figure 20.16) consists of 7-methyl-guanosine attached to the 5′ end of the RNA by a 5′–5′ triphosphate bridge. The first two nucleotides of the mRNA are also methylated. The cap structure is required for efficient translation of the final mature mRNA.

The second modification is the enzymatic addition of a **poly(A) tail** to the 3′ end of the transcript. *Poly(A) polymerase* uses ATP and catalyzes the stepwise

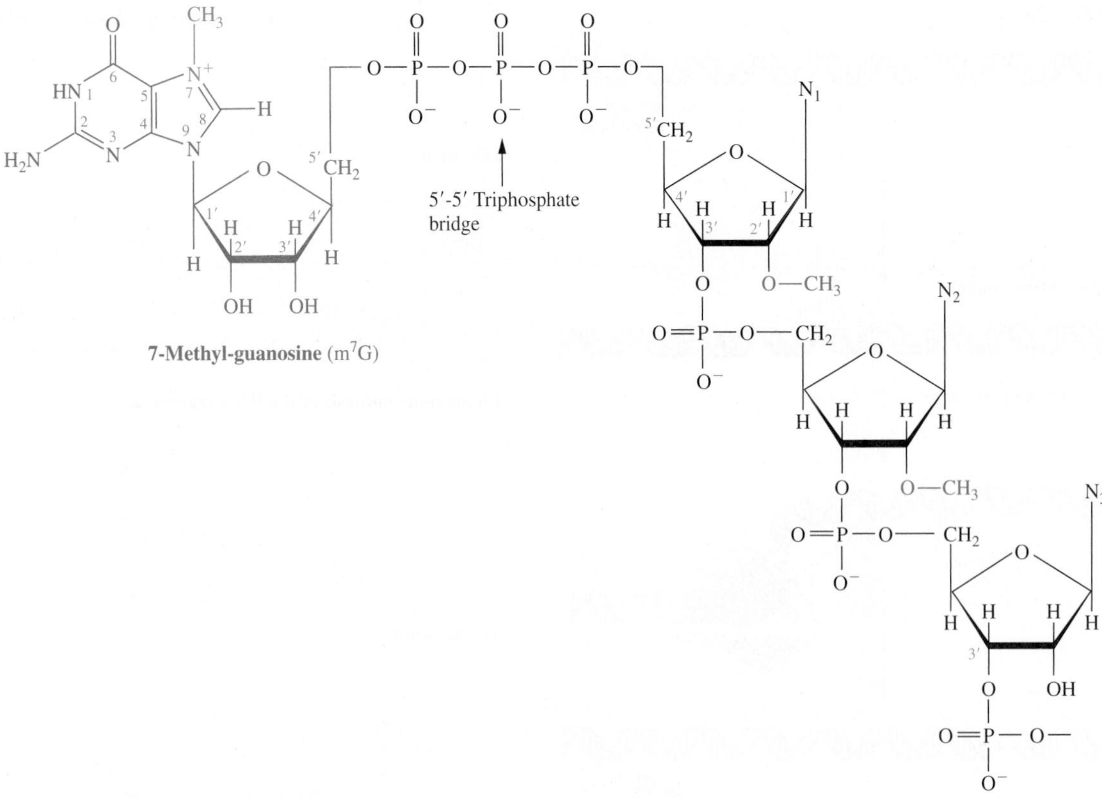

LEARNING GOAL

6 List and explain the three types of post-transcriptional modifications of eukaryotic mRNA.

Figure 20.16 The 5′-methylated cap structure of eukaryotic mRNA. N_1, N_2, and N_3 represent any of the four nitrogenous bases adenine, guanine, cytosine, or uracil.

polymerization of 100 to 200 adenosine nucleotides on the 3' end of the RNA. The poly(A) tail protects the 3' end of the mRNA from enzymatic degradation and thus prolongs the lifetime of the mRNA.

The third modification, **RNA splicing,** involves the removal of portions of the primary transcript that are not protein coding. Bacterial genes are continuous; all the nucleotide sequences of the gene are found in the mRNA. However, study of the gene structure of eukaryotes revealed a fascinating difference. Eukaryotic genes are discontinuous; there are *extra* DNA sequences within these genes that do not encode any amino acid sequences for the protein. These sequences are called *intervening sequences* or **introns.** The primary transcript contains both the introns and the protein coding sequences, called **exons.** The presence of introns in the mRNA would make it impossible for the process of translation to synthesize the correct protein. Therefore, they must be removed, which is done by the process of RNA splicing.

As you can imagine, RNA splicing must be very precise. If too much, or too little, RNA is removed, the mRNA will not carry the correct code for the protein. Thus, there are "signals" in the DNA to mark the boundaries of the introns. The sequence GpU is always found at the intron's 5' boundary and the sequence ApG is found at the 3' boundary.

Recognition of the splice boundaries and stabilization of the splicing complex requires the assistance of particles called *spliceosomes.* Spliceosomes are composed of a variety of *small nuclear ribonucleoproteins* (snRNPs, read "snurps"). Each snRNP consists of a small RNA and associated proteins. The RNA components of different snRNPs are complementary to different sequences involved in splicing. By hydrogen bonding to a splice boundary or intron sequences, the snRNPs recognize and bring together the sequences involved in the splicing reactions.

One of the first eukaryotic genes shown to contain introns was the gene for the β subunit of adult hemoglobin (Figure 20.17). On the DNA, the gene for β-hemoglobin is 1200 nucleotides long, but only 438 nucleotides carry the genetic information for protein. The remaining sequences are found in two introns of 116 and 646 nucleotides that are removed by splicing before translation. It is interesting that the larger intron is longer than the final β-hemoglobin mRNA! In the genes that have been studied, introns have been found to range in size from 50 to 20,000 nucleotides in length, and there may be many throughout a gene. Thus, a typical human gene might be ten to thirty times longer than the final mRNA.

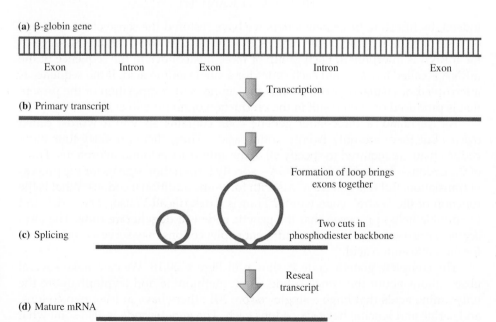

(a) β-globin gene

Exon   Intron   Exon   Intron   Exon

Transcription

(b) Primary transcript

Formation of loop brings exons together

Two cuts in phosphodiester backbone

(c) Splicing

Reseal transcript

(d) Mature mRNA

Figure 20.17 Schematic diagram of mRNA splicing. (a) The β-globin gene contains protein coding exons, as well as noncoding sequences called *introns.* (b) The primary transcript of the DNA carries both the introns and the exons. (c) The introns are looped out, the phosphodiester backbone of the mRNA is cut twice, and the pieces are joined together. (d) The final mature mRNA now carries only the coding sequences (exons) of the gene.

20.5 The Genetic Code

The mRNA carries the genetic code for a protein. But what is the nature of this code? In 1954, George Gamow proposed that because there are only four "letters" in the DNA alphabet (A, T, G, and C) and because there are twenty amino acids, the genetic code must contain words made of at least three letters taken from the four letters in the DNA alphabet. How did he come to this conclusion? He reasoned that a code of two-letter words constructed from any combination of the four letters has a "vocabulary" of only sixteen words (4^2). In other words, there are only sixteen different ways to put A, T, C, and G together two bases at a time (AA, AT, AC, AG, TT, TA, etc.). That is not enough to encode all twenty amino acids. A code of four-letter words gives 256 words (4^4), far more than are needed. A code of three-letter words, however, has a possible vocabulary of sixty-four words (4^3), sufficient to encode the twenty amino acids but not too excessive.

A series of elegant experiments proved that Gamow was correct by demonstrating that the genetic code is, indeed, a triplet code. Mutations were introduced into the DNA of a bacterial virus. These mutations inserted (or deleted) one, two, or three nucleotides into a gene. The researchers then looked for the protein encoded by that gene. When one or two nucleotides were inserted, no protein was produced. However, when a third base was inserted, the sense of the mRNA was restored, and the protein was made. You can imagine this experiment by using a sentence composed of only three-letter words. For instance,

THE CAT RAN OUT

What happens to the "sense" of the sentence if we insert one letter?

THE FCA TRA NOU T

The reading frame of the sentence has been altered, and the sentence is now nonsense. Can we now restore the sense of the sentence by inserting a second letter?

THE FAC ATR ANO UT

No, we have not restored the sense of the sentence. Once again, we have altered the reading frame, but because our code has only three-letter words, the sentence is still nonsense. If we now insert a third letter, it should restore the correct reading frame:

THE FAT CAT RAN OUT

Indeed, by inserting three new letters we have restored the sense of the message by restoring the reading frame. This is exactly the way in which the message of the mRNA is interpreted. Each group of three nucleotides in the sequence of the mRNA is called a *codon*, and each codes for a single amino acid. If the sequence is interrupted or changed, it can change the amino acid composition of the protein that is produced or even result in the production of no protein at all.

As we noted, a three-letter genetic code contains sixty-four words, called *codons*, but there are only twenty amino acids. Thus, there are forty-four more codons than are required to specify all of the amino acids found in proteins. Three of the codons—UAA, UAG, and UGA—specify termination signals for the process of translation. But this still leaves us with forty-one additional codons. What is the function of the "extra" code words? Francis Crick (recall Watson and Crick and the double helix) proposed that the genetic code is a **degenerate code.** The term *degenerate* is used to indicate that different triplet codons may serve as code words for the same amino acid.

The complete genetic code is shown in Figure 20.18. We can make several observations about the genetic code. First, methionine and tryptophan are the only amino acids that have a single codon. All others have at least two codons, and serine and leucine have six codons each. The genetic code is also somewhat

FIRST BASE	SECOND BASE				THIRD BASE
	U	C	A	G	
U	UUU Phenylalanine	UCU Serine	UAU Tyrosine	UGU Cysteine	U
	UUC Phenylalanine	UCC Serine	UAC Tyrosine	UGC Cysteine	C
	UUA Leucine	UCA Serine	UAA STOP	UGA STOP	A
	UUG Leucine	UCG Serine	UAG STOP	UGG Tryptophan	G
C	CUU Leucine	CCU Proline	CAU Histidine	CGU Arginine	U
	CUC Leucine	CCC Proline	CAC Histidine	CGC Arginine	C
	CUA Leucine	CCA Proline	CAA Glutamine	CGA Arginine	A
	CUG Leucine	CCG Proline	CAG Glutamine	CGG Arginine	G
A	AUU Isoleucine	ACU Threonine	AAU Asparagine	AGU Serine	U
	AUC Isoleucine	ACC Threonine	AAC Asparagine	AGC Serine	C
	AUA Isoleucine	ACA Threonine	AAA Lysine	AGA Arginine	A
	AUG (START) Methionine	ACG Threonine	AAG Lysine	AGG Arginine	G
G	GUU Valine	GCU Alanine	GAU Aspartic acid	GGU Glycine	U
	GUC Valine	GCC Alanine	GAC Aspartic acid	GGC Glycine	C
	GUA Valine	GCA Alanine	GAA Glutamic acid	GGA Glycine	A
	GUG Valine	GCG Alanine	GAG Glutamic acid	GGG Glycine	G

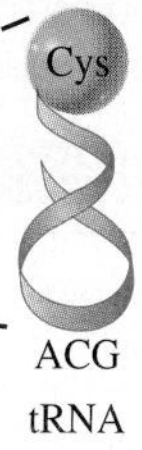

Cys
ACG
tRNA

Asn
UUG
tRNA

Figure 20.18 The genetic code. The table shows the possible codons found in mRNA. To read the universal biological language from this chart, find the first base in the column on the left, the second base from the row across the top, and the third base from the column to the right. This will direct you to one of the sixty-four squares in the matrix. Within that square, you will find the codon and the amino acid that it specifies. In the cell, this message is decoded by tRNA molecules like those shown to the right of the table.

mutation-resistant. For those amino acids that have multiple codons, the first two bases are often identical and thus identify the amino acid, and only the third position is variable. Mutations—changes in the nucleotide sequence—in the third position therefore often have no effect on the amino acid that is incorporated into a protein.

Question 20.5 Why is the genetic code said to be degenerate?

Question 20.6 Why is the genetic code said to be mutation-resistant?

20.6 Protein Synthesis

The process of protein synthesis is called *translation*. It involves translating the genetic information from the sequence of nucleotides into the sequence of amino acids in the primary structure of a protein. Figure 20.19 shows the relationship through which the nucleotide sequence of a DNA molecule is transcribed into

LEARNING GOAL

8 Describe the process of translation.

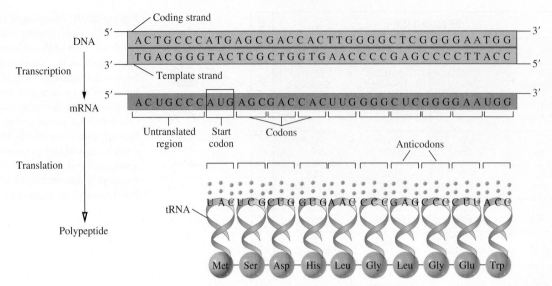

Figure 20.19 Messenger RNA (mRNA) is an RNA copy of one strand of a gene in the DNA. Each codon on the mRNA that specifies a particular amino acid is recognized by the complementary anticodon on a transfer RNA (tRNA).

a complementary sequence of ribonucleotides, the mRNA molecule. Each mRNA has a short untranslated region followed by the sequences that carry the information for the order of the amino acids in the protein that will be produced in the process of translation. That genetic information is the sequence of codons along the mRNA. The decoding process is carried out by tRNA molecules.

Translation is carried out on **ribosomes,** which are complexes of ribosomal RNA (rRNA) and proteins. Each ribosome is made up of two subunits: a small and a large ribosomal subunit (Figure 20.20a). In eukaryotic cells, the small ribosomal subunit contains one rRNA molecule and thirty-three different ribosomal proteins, and the large subunit contains three rRNA molecules and about forty-nine different proteins.

Protein synthesis involves the simultaneous action of many ribosomes on a single mRNA molecule. These complexes of many ribosomes along a single mRNA are known as *polyribosomes* or **polysomes** (Figure 20.20b). Each ribosome

Figure 20.20 Structure of the ribosome. (a) The large and small subunits form the functional complex in association with an mRNA molecule. (b) A polyribosome translating the mRNA for a β-globin chain of hemoglobin.

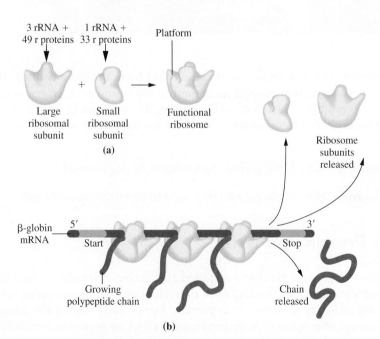

synthesizes one copy of the protein molecule encoded by the mRNA. Thus, many copies of a protein are simultaneously produced.

The Role of Transfer RNA

The codons of mRNA must be read if the genetic message is to be translated into protein. The molecule that decodes the information in the mRNA molecule into the primary structure of a protein is transfer RNA (tRNA). To decode the genetic message into the primary sequence of a protein, the tRNA must faithfully perform two functions.

First, the tRNA must covalently bind one, and only one, specific amino acid. There is at least one transfer RNA for each amino acid. All tRNA molecules have the sequence CCA at their 3' ends. This is the site where the amino acid will be covalently attached to the tRNA molecule. Each tRNA is specifically recognized by the active site of an enzyme called an **aminoacyl tRNA synthetase.** This enzyme also recognizes the correct amino acid and covalently links the amino acid to the 3' end of the tRNA molecule (Figure 20.21). The resulting structure is called an **aminoacyl tRNA.** The covalently bound amino acid will be transferred from the tRNA to a growing polypeptide chain during protein synthesis.

Second, the tRNA must be able to recognize the appropriate codon on the mRNA that calls for that amino acid. This is mediated through a sequence of three bases called the *anticodon,* which is located at the bottom of the tRNA cloverleaf (refer to Figure 20.14). The anticodon sequence for each tRNA is complementary to the codon on the mRNA that specifies a particular amino acid. As you can see in Figure 20.19, the anticodon-codon complementary hydrogen bonding will bring the correct amino acid to the site of protein synthesis.

Question 20.7   How are codons related to anticodons?

Question 20.8   If the sequence of a codon on the mRNA is 5'-AUG-3', what will the sequence of the anticodon be? Remember that the hydrogen bonding rules require antiparallel strands. It is easiest to write the anticodon first 3' $\longrightarrow$ 5' and then reverse it to the 5' $\longrightarrow$ 3' order.

The Process of Translation

Initiation

The first stage of protein synthesis is *initiation.* Proteins called **initiation factors** assist in the formation of a translation complex composed of an mRNA molecule, the small and large ribosomal subunits, and the initiator tRNA. This initiator tRNA recognizes the codon AUG and carries the amino acid methionine.

The ribosome has two sites for binding tRNA molecules. The first site, called the **peptidyl tRNA binding site (P-site),** holds the peptidyl tRNA, the growing peptide bound to a tRNA molecule. The second site, called the **aminoacyl tRNA binding site (A-site),** holds the aminoacyl tRNA carrying the next amino acid to be added to the peptide chain. Each of the tRNA molecules is hydrogen bonded to the mRNA molecule by codon-anticodon complementarity. The entire complex is further stabilized by the fact that the mRNA is also bound to the ribosome. Figure 20.22a shows the series of events that result in the formation of the initiation complex. The initiator methionyl tRNA occupies the P-site in this complex.

Chain Elongation

The second stage of translation is *chain elongation.* This occurs in three steps that are repeated until protein synthesis is complete. We enter the action after a

LEARNING GOAL

8 Describe the process of translation.

Figure 20.21 Aminoacyl tRNA synthetase binds the amino acid in one region of the active site and the appropriate tRNA in another. The acylation reaction occurs and the aminoacyl tRNA is released.

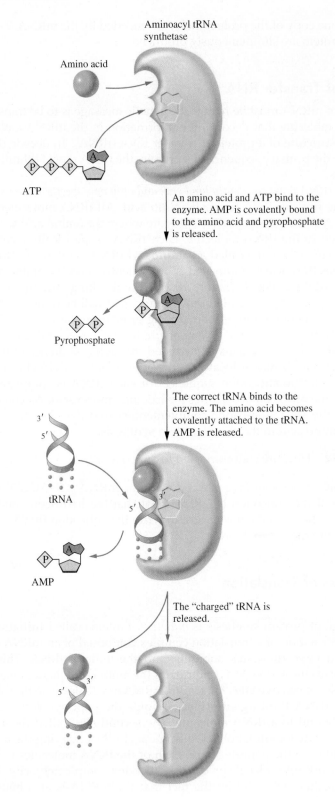

Aminoacyl tRNA synthetase

Amino acid

ATP

An amino acid and ATP bind to the enzyme. AMP is covalently bound to the amino acid and pyrophosphate is released.

Pyrophosphate

The correct tRNA binds to the enzyme. The amino acid becomes covalently attached to the tRNA. AMP is released.

3′

5′

tRNA

AMP

The "charged" tRNA is released.

5′   3′

tetrapeptide has already been assembled, and a peptidyl tRNA occupies the P-site (Figure 20.22b).

The first event is the binding of an aminoacyl-tRNA molecule to the empty A-site. Next, peptide bond formation occurs. This is catalyzed by an enzyme on the ribosome called *peptidyl transferase*. Now the peptide chain is shifted to the tRNA

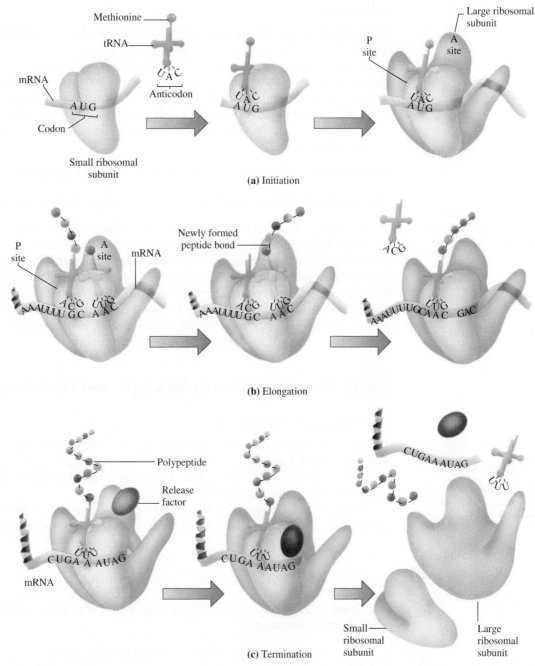

Figure 20.22 (a) Formation of an initiation complex sets protein synthesis in motion. The mRNA and proteins called *initiation factors* bind to the small ribosomal subunit. Next, a charged methionyl tRNA molecule binds, and finally, the initiation factors are released, and the large subunit binds. (b) The elongation phase of protein synthesis involves addition of new amino acids to the C-terminus of the growing peptide. An aminoacyl tRNA molecule binds at the empty A-site, and the peptide bond is formed. The uncharged tRNA molecule is released, and the peptidyl tRNA is shifted to the P-site as the ribosome moves along the mRNA. (c) Termination of protein synthesis occurs when a release factor binds the stop codon on mRNA. This leads to the hydrolysis of the ester bond linking the peptide to the peptidyl tRNA molecule in the P-site. The ribosome then dissociates into its two subunits, releasing the mRNA and the newly synthesized peptide.

that occupies the A-site. Finally, the tRNA in the P-site falls away, and the ribosome changes positions so that the next codon on the mRNA occupies the A-site. This movement of the ribosome is called **translocation**. The process shifts the new peptidyl tRNA from the A-site to the P-site. The chain elongation stage of translation requires the hydrolysis of GTP to GDP and P_i. Several **elongation factors** are also involved in this process.

Recent evidence indicates that the peptidyl transferase is a catalytic region of the 28S ribosomal RNA.

Termination

The last stage of translation is *termination.* There are three **termination codons**—UAA, UAG, and UGA—for which there are no corresponding tRNA molecules. When one of these "stop" codons is encountered, translation is terminated. A **release factor** binds the empty A-site. The peptidyl transferase that had previously catalyzed peptide bond formation hydrolyzes the ester bond between the peptidyl tRNA and the last amino acid of the newly synthesized protein (Figure 20.22c). At this point, the tRNA, the newly synthesized peptide, and the two ribosomal subunits are released.

> **Question 20.9**   What is the function of the ribosomal P-site in protein synthesis?

> **Question 20.10**   What is the function of the ribosomal A-site in protein synthesis?

The peptide that is released following translation is not necessarily in its final functional form. In some cases, the peptide is proteolytically cleaved before it becomes functional. Synthesis of digestive enzymes uses this strategy. Sometimes the protein must associate with other peptides to form a functional protein, as in the case of hemoglobin. Cellular enzymes add carbohydrate or lipid groups to some proteins, especially those that will end up on the cell surface. These final modifications are specific for particular proteins and, like the sequence of the protein itself, are directed by the cellular genetic information.

Post-translational proteolytic cleavage of digestive enzymes is discussed in Section 19.11.

The quaternary structure of hemoglobin is described in Section 18.8.

20.7   Mutation, Ultraviolet Light, and DNA Repair

The Nature of Mutations

LEARNING GOAL

9 Define mutation and understand how mutations cause cancer and cell death.

Changes can occur in the nucleotide sequence of a DNA molecule. Such a genetic change is called a **mutation.** Mutations can arise from mistakes made by DNA polymerase during DNA replication. They also result from the action of chemicals, called **mutagens,** that damage the DNA.

Mutations are classified by the kind of change that occurs in the DNA. The substitution of a single nucleotide for another is called a **point mutation:**

ATG**G**ACTTC:	normal DNA sequence
ATG**C**ACTTC:	point mutation

Sometimes a single nucleotide or even large sections of DNA are lost. These are called **deletion mutations:**

ATG**GAC**TTC:	normal DNA sequence
ATGTTC:	deletion mutation

Occasionally, one or more nucleotides are added to a DNA sequence. These are called **insertion mutations:**

ATGGACTTC:	normal DNA sequence
ATG**CTC**GACTTC:	insertion mutation

The Results of Mutations

Some mutations are **silent mutations;** that is, they cause no change in the protein. Often, however, a mutation has a negative effect on the health of the organism. The effect of a mutation depends on how it alters the genetic code for a protein. Consider the two codons for glutamic acid: GAA and GAG. A point mutation that alters the third nucleotide of GA**A** to GA**G** will still result in the incorporation of glutamic acid at the correct position in the protein. Similarly, a GAG to GAA mutation will also be silent.

Many mutations are not silent. There are approximately four thousand human genetic disorders that result from such mutations. These occur because the mutation in the DNA changes the codon and results in incorporation of the wrong amino acid into the protein. This causes the protein to be nonfunctional or to function improperly.

Consider the human genetic disorder sickle cell anemia. In the normal β-chain of hemoglobin, the sixth amino acid is glutamic acid. In the β-chain of sickle cell hemoglobin, the sixth amino acid is valine. How did this amino acid substitution arise? The answer lies in examination of the codons for glutamic acid and valine:

Glutamic acid:	GAA or GAG
Valine:	GUG, GUC, GUA, or GUU

A point mutation of A ⟶ U in the second nucleotide changes some codons for glutamic acid into codons for valine:

$$GAA \longrightarrow GUA$$
$$GAG \longrightarrow GUG$$

Glutamic acid codon   Valine codon

This mutation in a single codon leads to the change in amino acid sequence at position 6 in the β-chain of human hemoglobin from glutamic acid to valine. The result of this seemingly minor change is sickle cell anemia in individuals who inherit two copies of the mutant gene.

Question 20.11 The sequence of a gene on the mRNA is normally AUGCCC-GACUUU. A point mutation in the gene results in the mRNA sequence AUGCGC-GACUUU. What are the amino acid sequences of the normal and mutant proteins? Would you expect this to be a silent mutation?

Question 20.12 The sequence of a gene on the mRNA is normally AUGCCC-GACUUU. A point mutation in the gene results in the mRNA sequence AUGCCG-GACUUU. What are the amino acid sequences of the normal and mutant proteins? Would you expect this to be a silent mutation?

Mutagens and Carcinogens

Any chemical that causes a change in the DNA sequence is called a *mutagen*. Often, mutagens are also **carcinogens,** cancer-causing chemicals. Most cancers result from mutations in a single normal cell. These mutations result in the loss of normal growth control, causing the abnormal cell to proliferate. If that growth is not controlled or destroyed, it will result in the death of the individual. We are exposed to many carcinogens in the course of our lives. Sometimes we are exposed to a carcinogen by accident, but in some cases it is by choice. Cigarette smoke has about three thousand chemical components and several are potent mutagens. As a result, people who smoke have a much greater chance of developing lung cancer than those who don't.

Ultraviolet Light Damage and DNA Repair

Ultraviolet (UV) light is another agent that causes damage to DNA. Absorption of UV light by DNA causes adjacent pyrimidine bases to become covalently linked. The product is called a **pyrimidine dimer.** As a result of pyrimidine dimer formation, there is no hydrogen bonding between these pyrimidine molecules and the complementary bases on the other DNA strand. This stretch of DNA cannot be replicated or transcribed!

A Medical Perspective

Epigenomics

In Section 20.2, we learned about the organization of the chromatin in eukaryotic cells in the form of chromosomes. We also saw the way in which the genetic information in the cell is transcribed and translated into the proteins that carry out the work of the cell and serve as structural elements. But we have not considered the way in which the expression of the genes in the cell is controlled.

There are about 25,000 protein-coding genes in a mammalian cell. Many of these are needed in all cell types; for instance, the genes that code for the enzymes of glycolysis and the citric acid cycle. But many of those genes are not needed for all cells. Only cells of the immune system need to produce antibody molecules, and only cells of the hair follicles need to produce the keratin protein that makes up hair.

There are many mechanisms involved in the control of gene expression in eukaryotic cells. We know that there are regions of the DNA called *enhancers* that bind to different types of *transcription factors* that control the rate of transcription of a particular gene. But more recently, scientists have learned that there are other types of chemical markers that control whether a region of the genome will be active transcriptionally or will be silent. The study of epigenomics is shedding light on the way in which simple chemical modifications of the DNA and the histone proteins determine whether genes will be transcribed, or not.

The first type of modification that was discovered is the methylation of regions of the DNA. The enzymes that catalyze this reaction are called *DNA methyltransferases*. The most common methylation is found on carbon-5 of cytosine found next to a guanine (CpG).

5-Methyl-2'-deoxycytidine

The methylation pattern of the DNA varies from cell to cell and from organism to organism and even changes over the course of development. This reflects the needs of the cell for the products of certain genes. Scientists have found that transcription of the genes of the DNA regions that are highly methylated is repressed. Those regions that are unmethylated are transcriptionally active.

DNA methylation modifies gene expression through a multistep process involving *chromatin remodeling.* Enzymes in the nucleus are stimulated to chemically modify the histone proteins. Several types of modifications are known. Two chemical modifications known to repress transcription are methylation and acetylation of histone proteins. Paradoxically, methylation of histone proteins has been implicated in both the activation and the repression of transcription. The effect is dependent on which amino acid in the histone is methylated!

Recently, scientists from a number of laboratories around the world have published epigenomic maps from 111 human cell types, including kidney, pancreas, brain, and muscle. They also studied cancer cells and cells associated with Alzheimer's disease (AD). One of the intriguing outcomes of these studies was the possible identification of the cells that cause AD. It had always been assumed that neurons were the source of the disease. The epigenomic study suggests that the genes associated with AD are not active in neurons, but are active in cells called microglia, which are immune cells found in the brain. They also observed epigenomic differences between normal melanocytes (pigment-producing cells) and those from a melanoma.

While it is too early to understand these differences completely, it appears that the study of the epigenome may provide medical science with new tools to combat diseases such as Alzheimer's and cancer.

For Further Study

▶ Develop a hypothesis to explain why DNA methylation and chromatin remodeling might change the rate of transcription of a gene.
▶ Develop a hypothesis to explain why chromatin remodeling changes during embryonic development.

Pyrimidine dimers were originally called thymine dimers because thymine is more commonly involved in these reactions than cytosine.

Bacteria such as *Escherichia coli* have four different mechanisms to repair UV light damage. However, even a repair process can make a mistake. Mutations occur when the UV damage repair system makes an error and causes a change in the nucleotide sequence of the DNA.

In medicine, the pyrimidine dimerization reaction is used to advantage in hospitals where germicidal (UV) light is used to kill bacteria in the air and on environmental surfaces, such as in a vacant operating room. This cell death is caused by pyrimidine dimer formation on a massive scale. The repair systems of the bacteria are overwhelmed, and the cells die.

Of course, the same type of pyrimidine dimer formation can occur in human cells as well. Lying out in the sun all day to acquire a fashionable tan exposes the skin to large amounts of UV light. This damages the skin by formation of many pyrimidine dimers. Exposure to high levels of UV from sunlight or tanning booths has been linked to a rising incidence of skin cancer in human populations.

Consequences of Defects in DNA Repair

The human repair system for pyrimidine dimers is quite complex, requiring at least five enzymes. The first step in repair of the pyrimidine dimer is the cleavage of the sugar-phosphate backbone of the DNA near the site of the damage. The enzyme that performs this is called a *repair endonuclease*. If the gene encoding this enzyme is defective, pyrimidine dimers cannot be repaired. The accumulation of mutations combined with a simultaneous decrease in the efficiency of DNA repair mechanisms leads to an increased incidence of cancer. For example, a mutation in the repair endonuclease gene, or in other genes in the repair pathway, results in the genetic skin disorder called *xeroderma pigmentosum*. People who suffer from xeroderma pigmentosum are extremely sensitive to the ultraviolet rays of sunlight and develop multiple skin cancers, usually before the age of twenty.

20.8 Recombinant DNA

Tools Used in the Study of DNA

Many of the techniques and tools used in recombinant DNA studies were developed or discovered during basic studies on bacterial DNA replication and gene expression. These include many enzymes that catalyze reactions of DNA molecules, cloning vectors, and hybridization techniques.

Restriction Enzymes

Restriction enzymes, often called *restriction endonucleases,* are bacterial enzymes that "cut" the sugar-phosphate backbone of DNA molecules at specific nucleotide sequences. The first of these enzymes to be purified and studied was called EcoR1. The name is derived from the genus and species name of the bacteria from which it was isolated, in this case *Escherichia coli,* or *E. coli.* The following is the specific nucleotide sequence recognized by EcoR1:

$$5'\text{------------- GAATTC ------------- } 3'$$
$$3'\text{------------- CTTAAG ------------- } 5'$$

When EcoR1 cuts the DNA at this site, it does so in a staggered fashion. Specifically, it cuts between the G and the first A on both strands. Cutting produces two DNA fragments with the following structure:

$$5'\text{--------------- G} \qquad \text{AATTC --------------- } 3'$$
$$3'\text{--------------- CTTAA} \qquad \text{G --------------- } 5'$$

These staggered termini are called *sticky ends* because they can reassociate with one another by hydrogen bonding. This is a property of the DNA fragments generated by restriction enzymes that is very important to gene cloning.

Examples of other restriction enzymes and their specific recognition sequences are listed in Table 20.2. The sites on the sugar-phosphate backbone that are cut by the enzymes are indicated by slashes.

These enzymes are used to digest large DNA molecules into smaller fragments of specific size. Because a restriction enzyme always cuts at the same site, DNA from a particular individual generates a reproducible set of DNA fragments. This is convenient for the study or cloning of DNA from any source.

This mother is applying sunscreen to her daughter to shield her from UV radiation. Explain the kind of damage that UV light can cause and what the potential long-term effects may be.

LEARNING GOAL

10 Describe the tools used in the study of DNA and in genetic engineering.

TABLE 20.2 Common Restriction Enzymes and Their Recognition Sequences

Restriction Enzyme	Recognition Sequence
BamHI	5'-G/GATCC-3'
	3'-CCTAG/G-5'
HindIII	5'-A/AGCTT-3'
	3'-TTCGA/A-5'
SalI	5'-G/TCGAC-3'
	3'-CAGCT/G-5'
BglII	5'-A/GATCT-3'
	3'-TCTAG/A-5'
PstI	5'-CTGCA/G-3'
	3'-G/ACGTC-5'

DNA Cloning Vectors

DNA cloning experiments allow us to isolate single copies of a gene and then produce billions of copies. To produce multiple copies of a gene, it may be joined to a **cloning vector.** A cloning vector is a piece of DNA having its own replication origin so that it can be replicated inside a host cell. Often the bacterium *E. coli* serves as the host cell in which the vector carrying the cloned DNA is replicated in abundance.

There are two major kinds of cloning vectors. The first are bacterial virus or phage vectors. These are bacterial viruses that have been genetically altered to allow the addition of cloned DNA fragments. These viruses have all the genes required to replicate 100 to 200 copies of the virus (and cloned fragment) per infected cell.

The second commonly used vector is a plasmid vector. Plasmids are extra pieces of circular DNA found in most kinds of bacteria. The plasmids that are used as cloning vectors often contain antibiotic resistance genes that are useful in the selection of cells containing a plasmid.

Each plasmid has its own replication origin to allow efficient DNA replication in the bacterial host cell. Most plasmid vectors also have a *selectable marker,* often a gene for resistance to an antibiotic. Finally, plasmid vectors have a gene that has several restriction enzyme sites useful for cloning. The valuable feature of this gene is that it is inactivated when a DNA fragment has been cloned into it. Thus, cells containing a plasmid carrying a cloned DNA fragment can be recognized by their ability to grow in the presence of antibiotic and by loss of function of the gene into which the cloned DNA has been inserted.

Genetic Engineering

Now that we have assembled most of the tools needed for a cloning experiment, we must decide which gene to clone. The example that we will use is the cloning of the β-globin genes for normal and sickle cell hemoglobin. DNA from an individual with normal hemoglobin is digested with a restriction enzyme. This is the target DNA. The vector DNA must be digested with the same enzyme (Figure 20.23). In our example, the restriction enzyme cuts within the *lacZ* gene, which codes for the enzyme β-galactosidase.

The digested vector and target DNA are mixed together under conditions that encourage the sticky ends of the target and vector DNA to hybridize with one another. The sticky ends are then covalently linked by the enzyme DNA ligase. This enzyme catalyzes the formation of phosphoester bonds between the two pieces of DNA.

Now the recombinant DNA molecules are introduced into bacterial cells by a process called transformation. Next, the cells of the transformation mixture are

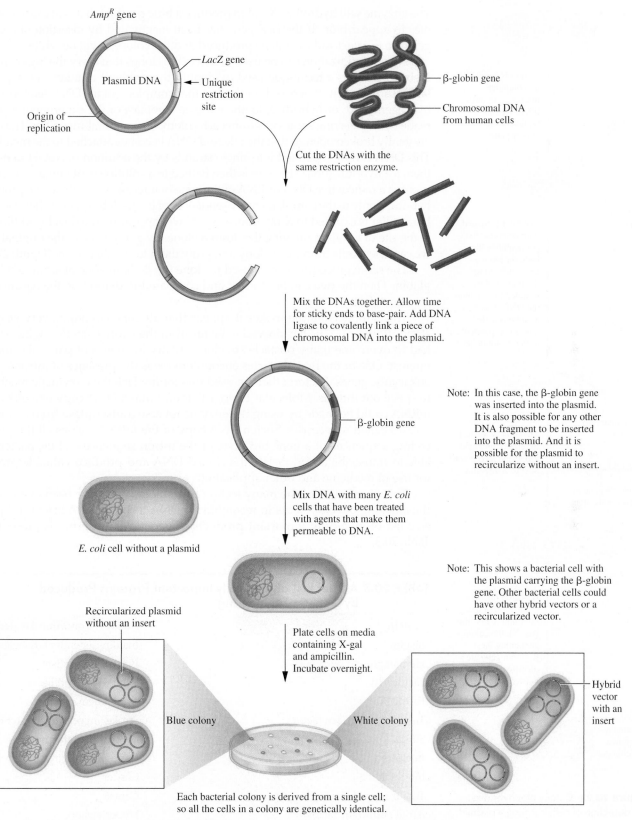

Figure 20.23 Cloning of eukaryotic DNA into a plasmid cloning vector.

plated on a solid nutrient agar medium containing the antibiotic ampicillin and the β-galactosidase substrate X-gal (5-bromo-4-chloro-3-indolyl-β-D-galactoside). Only those cells containing the antibiotic resistance gene will survive and grow into bacterial colonies. Cells with an intact *lacZ* gene will produce the enzyme β-galactosidase.

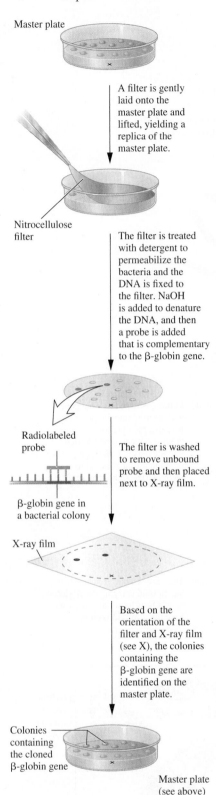

Master plate

A filter is gently laid onto the master plate and lifted, yielding a replica of the master plate.

Nitrocellulose filter

The filter is treated with detergent to permeabilize the bacteria and the DNA is fixed to the filter. NaOH is added to denature the DNA, and then a probe is added that is complementary to the β-globin gene.

Radiolabeled probe

The filter is washed to remove unbound probe and then placed next to X-ray film.

β-globin gene in a bacterial colony

X-ray film

Based on the orientation of the filter and X-ray film (see X), the colonies containing the β-globin gene are identified on the master plate.

Colonies containing the cloned β-globin gene

Master plate (see above)

Figure 20.24 Colony blot hybridization for detection of cells carrying a plasmid clone of the β-chain gene of hemoglobin.

The enzyme will hydrolyze X-gal to produce a blue product that will cause the colonies to appear blue. If the *lacZ* gene has been inactivated by insertion of a cloned gene, no β-galactosidase will be produced and the colonies will be white.

Now hybridization can be used to detect the clones that carry the β-globin gene. **Hybridization** is a technique used to identify the presence of a gene on a particular DNA fragment. It is based on the fact that complementary DNA sequences will hydrogen bond, or hybridize, to one another. A replica of the experimental plate is made by transferring some cells from each colony onto a membrane filter. These cells are gently broken open so that the released DNA becomes attached to the membrane. This DNA is now separated into single strands by the addition of NaOH to prepare them for hybridization. The filter is then bathed in a solution containing a probe. The probe is a radioactive DNA or RNA of the gene of interest. The radioactive probe will hybridize only to the complementary sequences of the β-globin gene. When the membrane filter is exposed to X-ray film, a "spot" will appear on the developed film only at the site of a colony carrying the desired clone. By going back to the original plate, we can select cells from that colony and grow them for further study (Figure 20.24).

The same procedure can be used to clone the β-chain gene of sickle cell hemoglobin. Then the two can be studied and compared to determine the nature of the genetic defect.

This simple example makes it appear that all gene cloning is very easy and straightforward. This has proved to be far from the truth. Genetic engineers have had to overcome many obstacles to clone eukaryotic genes of particular medical interest. One of the first obstacles encountered was the presence of introns within eukaryotic genes. Bacteria that are used for cloning lack the enzymatic machinery to splice out introns. Molecular biologists found that a DNA copy of a eukaryotic mRNA could be made by using the enzyme reverse transcriptase from a family of viruses called *retroviruses*. Such a DNA copy of the mRNA carries all the protein-coding sequences of a gene but none of the intron sequences. Thus, bacteria are able to transcribe and translate the cloned DNA and produce valuable products for use in medicine and other applications.

This is only one of the many technical problems that have been overcome by the amazing developments in recombinant DNA technology. A brief but impressive list of medically important products of genetic engineering is presented in Table 20.3.

TABLE 20.3 A Brief List of Medically Important Proteins Produced by Genetic Engineering

Protein	Medical Condition Treated
Insulin	Insulin-dependent diabetes
Human growth hormone	Pituitary dwarfism
Factor VIII	Type A hemophilia
Factor IX	Type B hemophilia
Tissue plasminogen factor	Stroke, myocardial infarction
Streptokinase	Myocardial infarction
Interferon	Cancer, some virus infections
Interleukin-2	Cancer
Tumor necrosis factor	Cancer
Atrial natriuretic factor	Hypertension
Erythropoietin	Anemia
Thymosin α-1	Stimulate immune system
Hepatitis B virus (HBV) vaccine	Prevent HBV viral hepatitis
Influenza vaccine	Prevent influenza infection

20.9   Polymerase Chain Reaction

A bacterium originally isolated from a hot spring in Yellowstone National Park provides the key to a powerful molecular tool for the study of DNA. Polymerase chain reaction (PCR) allows scientists to produce unlimited amounts of any gene of interest and the bacterium *Thermus aquaticus* produces a heat-stable DNA polymerase (Taq polymerase) that allows the process to work.

The human genome consists of approximately three billion base pairs of DNA. But suppose you are interested in studying only one gene, perhaps the gene responsible for muscular dystrophy or cystic fibrosis. It's like looking for a needle in a haystack. Using PCR, a scientist can make millions of copies of the gene of interest, while ignoring the thousands of other genes on human chromosomes.

The secret to this specificity is the synthesis of a DNA primer, a short piece of single-stranded DNA that will specifically hybridize to the beginning of a particular gene. DNA polymerases require a primer for initiation of DNA synthesis because they act by adding new nucleotides to the 3′—OH of the last nucleotide of the primer.

To perform PCR, a small amount of DNA is mixed with Taq polymerase, the primer, and the four DNA nucleotide triphosphates. The mixture is then placed in an instrument called a *thermocycler*. The temperature in the thermocycler is raised to 94–96°C for several minutes to separate the two strands of DNA. Because the Taq polymerase is heat-stable, it is not denatured by these temperatures. The temperature is then dropped to 50–56°C to allow the primers to hybridize to the target DNA. Finally, the temperature is raised to 72°C to allow Taq polymerase to act, reading the template DNA strand and polymerizing a daughter strand extended from the primer. At the end of this step, the amount of the gene has doubled (Figure 20.25).

Now the three steps are repeated. With each cycle, the amount of the gene is doubled. Theoretically, after thirty cycles, you have one billion times more DNA than you started with!

PCR can be used in genetic screening to detect the gene responsible for muscular dystrophy. It can also be used to diagnose disease. For instance, it can be used to amplify small amounts of HIV in the blood. It can also be used by forensic scientists to amplify DNA from a single hair follicle or a tiny drop of blood at a crime scene.

20.10   The Human Genome Project

In 1990, the Department of Energy and the National Institutes of Health began the Human Genome Project (HGP), a multinational project that would extend into the next millennium. The goals of the HGP were to identify all of the genes in human DNA and to sequence the entire three billion nucleotide pairs of the genome. In order to accomplish these goals, enormous computer databases had to be developed to store the information, and computer software had to be designed to analyze it.

Initially, the HGP planned to complete the work by the year 2005. However, as a result of technological advances made by those in the project, a working draft of the human genome was published in February 2001 and the successful completion of the project was announced on April 14, 2003.

Genetic Strategies for Genome Analysis

The strategy for the HGP was rather straightforward. In order to determine the DNA sequence of the human genome, genomic libraries had to be produced. A *genomic library* is a set of clones representing the entire genome. The DNA

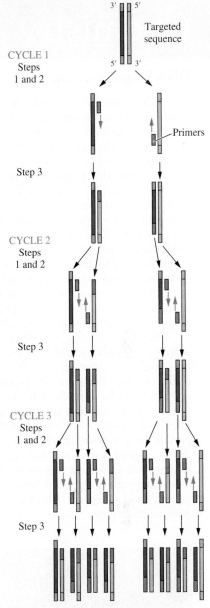

Figure 20.25 Polymerase chain reaction.

LEARNING GOAL

11   Describe the process of polymerase chain reaction, and discuss potential uses of the process.

LEARNING GOAL

12   Discuss strategies for genome analysis and DNA sequencing.

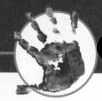

Chemistry at the Crime Scene

DNA Fingerprinting

Four U.S. Army helicopters swept over the field of illicit coca plants (*Erythroxylum* spp.) growing in a mountainous region of northern Colombia. When the soldiers were certain that the fields were unguarded, a fifth helicopter landed. From it emerged a team of researchers from the Agricultural Research Service of the U.S. Department of Agriculture (ARS-USDA). Quickly the scientists gathered leaves from mature plants, as well as from seedlings growing in a coca nursery, and returned to the helicopter with their valuable samples. With a final sweep over the field, the Army helicopters sprayed herbicides to kill the coca plants.

From 1997 to 2001, this scene was repeated in regions of Colombia known to have the highest coca production. The reason for these collections was to study the genetic diversity of the coca plants being grown for the illegal production of cocaine. The tool selected for this study was DNA fingerprinting.

DNA fingerprinting was developed in the 1980s by Alec Jeffries of the University of Leicester in England. The idea grew out of basic molecular genetic studies of the human genome. Scientists observed that some DNA sequences varied greatly from one person to the next. Such hypervariable regions are made up of variable numbers of repeats of short DNA sequences. They are located at many sites on different chromosomes. Each person has a different number of repeats and when his or her DNA is digested with restriction enzymes, a unique set of DNA fragments is generated. Jeffries invented DNA fingerprinting by developing a set of DNA probes that detect these variable number tandem repeats (VNTRs) when used in hybridization with Southern blots.

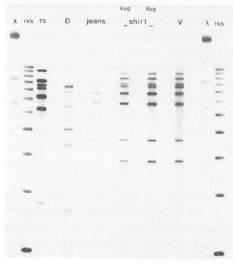

An example of a DNA fingerprint used in a criminal case. The DNA sample designated *V* is that of the victim, and the sample designated *D* is that of the defendant. The samples labeled *jeans* and *shirt* were taken from the clothing of the defendant. The DNA bands from the defendant's clothing clearly match the DNA bands of the victim, providing evidence of the guilt of the defendant.

Although several variations of DNA fingerprinting exist, the basic technique is quite simple. DNA is digested with restriction enzymes, producing a set of DNA fragments. These are separated by electrophoresis through an agarose gel. The DNA fragments are then transferred to membrane filters and hybridized with the radioactive probe DNA. The bands that hybridize the radioactive probe are visualized by exposing the membrane to X-ray film and developing a "picture" of the gel. The result is what Jeffries calls a *DNA fingerprint*, a set of twenty-five to sixty DNA bands that are unique to an individual.

DNA fingerprinting is now routinely used for paternity testing, testing for certain genetic disorders, and identification of the dead in cases where no other identification is available. DNA fingerprints are used as evidence in criminal cases involving rape and murder. In such cases, the evidence may be little more than a hair with an intact follicle on the clothing of the victim.

Less widely known is the use of DNA fingerprinting to study genetic diversity in natural populations of plants and animals. The greater the genetic diversity, the healthier the population is likely to be. Populations with low genetic diversity face a far higher probability of extinction under adverse conditions. Customs officials have used DNA fingerprinting to determine whether confiscated elephant tusks were taken illegally from an endangered population of elephants or were obtained from a legally harvested population.

Coca nursery next to a mature field in Colombia

In the case of the coca plants, ARS wanted to know whether the drug cartels were developing improved strains that might be hardier or more pest resistant or that have a higher concentration of cocaine. Their conclusions, which you can read in *Phytochemistry* (64: 187–197, 2003), were that the drug cartels have introduced significant genetic modification into coca plants in Colombia in the last two decades. In addition, some of these new variants, those producing the highest levels of cocaine, have been transplanted to other regions of the country. All of this indicates that the cocaine agribusiness is thriving.

For Further Understanding

▶ As this sampling of applications suggests, DNA fingerprinting has become an invaluable tool in law enforcement, medicine, and basic research. What other applications of this technology can you think of?

▶ Do some research on the development of DNA fingerprinting as a research and forensics tool. What is the probability that two individuals will have the same DNA fingerprint? How are these probabilities determined?

sequences of each of these clones could then be determined. Of course, once the sequence of each of these clones is determined, there is no way to know how they are arranged along the chromosomes.

A second technique, called *chromosome walking,* provides both DNA sequence information and a method for identifying the DNA sequences next to it on the chromosome. This method requires clones that are overlapping. To accomplish this, libraries of clones are made using many different restriction enzymes. The DNA sequence of a fragment is determined. Then, that information is used to develop a probe for any clones in the library that are overlapping. Each time a DNA fragment is sequenced, the information is used to identify overlapping clones. This process continues, allowing scientists to walk along the chromosome in two directions until the entire sequence is cloned, mapped, and sequenced.

DNA Sequencing

The method of DNA sequencing used is based on a technique developed by Frederick Sanger. A cloned piece of DNA is separated into its two strands. Each of these will serve as a template strand to carry out DNA replication in test tubes. A primer strand is also needed. This is a short piece of DNA that will hybridize to the template strand. The primer is the starting point for addition of new nucleotides during DNA synthesis.

The DNA is then placed in four test tubes with all of the enzymes and nucleotides required for DNA synthesis. In addition, each tube contains an unusual nucleotide, called a dideoxynucleotide. These nucleotides differ from the standard nucleotides by having a hydrogen atom at the 3′ position of the deoxyribose, rather than a hydroxyl group. When a dideoxynucleotide is incorporated into a growing DNA chain, it acts as a chain terminator. Because it does not have a 3′-hydroxyl group, no phosphoester bond can be formed with another nucleotide and no further polymerization can occur.

Each of the four tubes containing the DNA, enzymes, and an excess of the nucleotides required for replication will also have a small amount of one of the four dideoxynucleotides. In the tube that receives dideoxyadenosine triphosphate (ddA), for example, DNA synthesis will begin. As replication proceeds, either the standard nucleotide or ddA will be incorporated into the growing strand. Since the standard nucleotide is present in excess, the dideoxynucleotide will be incorporated infrequently and randomly. This produces a family of DNA fragments that terminate at the location of one of the deoxyadenosines in the molecule.

The same reaction is done with each of the dideoxynucleotides. The DNA fragments are then separated by gel electrophoresis on a DNA sequencing gel. The four reactions are placed in four wells, side by side, on the gel. Following

A Medical Perspective

A Genetic Approach to Familial Emphysema

Familial emphysema is a human genetic disease resulting from the inability to produce the protein α_1-antitrypsin. See also A Medical Perspective: α_1-Antitrypsin and Familial Emphysema, in Chapter 19. In individuals who have inherited one or two copies of the normal α_1-antitrypsin gene, this serum protein protects the lungs from the enzyme elastase. Normally, elastase fights bacteria and helps in the destruction and removal of dead lung tissue. However, the enzyme can also cause lung damage. By inhibiting elastase, α_1-antitrypsin prevents lung damage. Individuals who have inherited two defective α_1-antitrypsin genes do not produce this protein and suffer from familial, or A1AD, emphysema. In the absence of α_1-antitrypsin, the elastase and other proteases cause the severe lung damage characteristic of emphysema.

A1AD is the second most common genetic disorder in Caucasians. It is estimated that there are 100,000 sufferers in the United States and that one in five Americans carries the gene. The disorder, discovered in 1963, is often misdiagnosed as asthma or chronic obstructive pulmonary disease. In fact, it is estimated that fewer than 5% of the sufferers are diagnosed with A1AD.

The α_1-antitrypsin gene has been cloned. Early experiments with sheep showed that the protein remains stable when administered as an aerosol and remains functional after it has passed through the pulmonary epithelium. This research offers hope of an effective treatment for this disease.

The current treatment involves weekly IV injections of α_1-antitrypsin. The supply of the protein, purified from human plasma that has been demonstrated to be virus-free, is rather limited. Thus, the injections are expensive. In addition, they are painful. These two factors cause some sufferers to refuse the treatment.

Dr. Terry Flotte and his colleagues at the University of Florida have cloned the gene for α_1-antitrypsin into the DNA of adeno-associated virus (AAV). This virus is an ideal vector for human gene replacement therapy because it replicates only in cells that are not dividing and it does not stimulate a strong immune or inflammatory response. The researchers injected the virus carrying the cloned α_1-antitrypsin gene into the muscle tissue of mice, then tested for the level of α_1-antitrypsin in the blood. The results were very promising. Effective levels of α_1-antitrypsin were produced in the muscle cells of the mice and secreted into the bloodstream. Furthermore, the level of α_1-antitrypsin remained at therapeutic levels for more than four months. In the past 10 years, three clinical trials in humans have been completed and a fourth is ongoing. In 2009, Dr. Flotte and his colleagues at the University of Massachusetts Medical School and the University of Florida reported that three patients treated with the recombinant AAV were able to produce α_1-antitrypsin for up to 1 year. Although the levels of enzyme produced were not considered to be therapeutic, this clinical trial demonstrated that introduction of a functioning gene can result in production of the protein. Preliminary results of an ongoing Phase 2 clinical trial suggest that therapeutic concentrations of α_1-antitrypsin can be achieved. The research team is optimistic that by modifying the design of the recombinant AAV or the method of delivery into the body, they will be able to develop an effective therapy for familial emphysema.

For Further Understanding

▸ Of the three treatments described in this perspective, which do you think has the highest probability of success in the long term? Defend your answer.

▸ Explain the lung damage that results from A1AD.

---

electrophoresis, the DNA sequence can be read directly from the gel, as shown in Figure 20.26.

When chain termination DNA sequencing was first done, radioactive isotopes were used to label the DNA strands. However, new technology has resulted in automated systems that employ dideoxynucleotides that are labeled with fluorescent dyes, a different color for each dideoxynucleotide. Because each reaction (A, G, C, and T) will be a different color, all the reactions can be done in a single reaction mixture and the products separated on a single lane of a sequencing gel. A computer then "reads" the gel by distinguishing the color of each DNA band. The sequence information is directly stored into a databank for later analysis.

There is currently a massive amount of DNA sequence information available on the Internet. A 2014 article in *The Scientist* estimated that nearly 18,000 bacterial genomes had been sequenced, along with over 350 fungal genomes, nearly 100 insect genomes, and thousands of human genomes. The new generation DNA sequencing, based on the types of principles we have studied, employs advanced

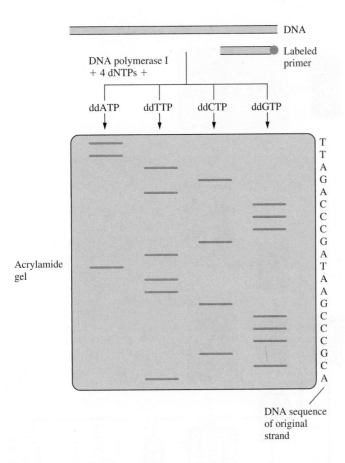

Figure 20.26 DNA sequencing by chain termination requires a template DNA strand and a radioactive primer. These are placed into each of four reaction mixtures that contain DNA polymerase, the four DNA nucleotides (dATP, dCTP, dGTP, and TTP), as well as one of the four dideoxynucleotides. Following the reaction, the products are separated on a DNA sequencing gel. The sequence is read from an autoradiograph of the gel.

technology that allows low-cost, high-throughput sequencing that is generating the astounding body of genetic information.

It has allowed researchers to compare the sequences of normal genes and their mutant counterparts. It has also allowed the study of ancient DNA, answering questions about the differences, and similarities, between modern humans and our Neanderthal relatives from 130,000 years ago.

This immense collection of DNA information has given rise to an entirely new branch of science. The field of **bioinformatics** brings together the disciplines of computer science, mathematics, statistics, DNA technology, and engineering to devise methods and software tools for organizing, understanding, analyzing, and applying the knowledge we gain from these DNA sequences.

CHAPTER MAP

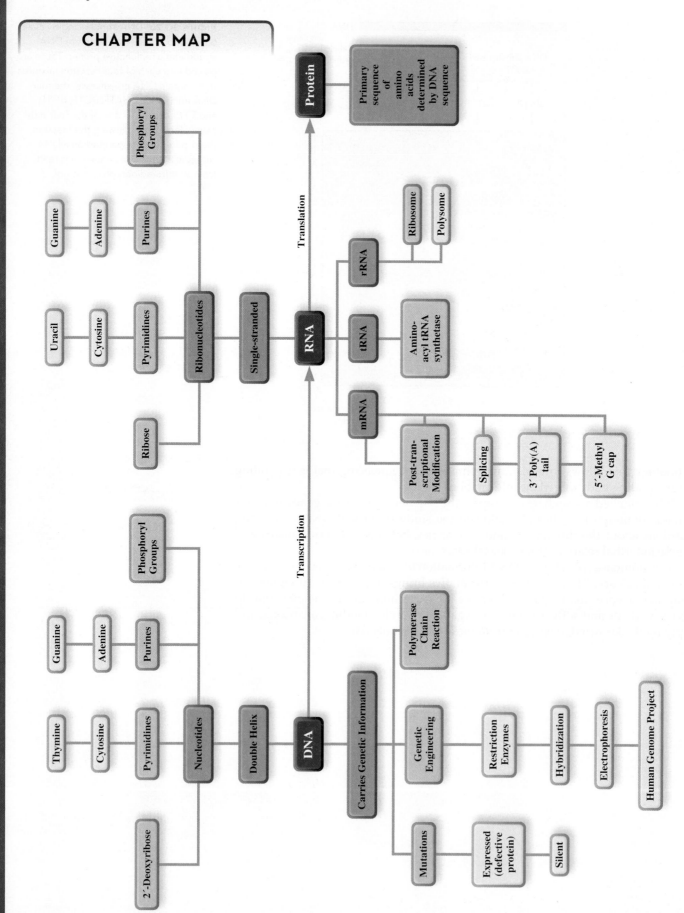

SUMMARY

20.1 The Structure of the Nucleotide

▶ **Deoxyribonucleic acid** (DNA) and **ribonucleic acid** (RNA) are polymers of **nucleotides.**

▶ **Nucleosides** are produced by the combination of a sugar, either ribose in RNA or 2′-deoxyribose in DNA, and a nitrogenous base.

▶ Nucleotides are composed of a nucleoside bonded to one, two, or three phosphoryl groups.

▶ There are two types of nitrogenous bases:
  - **Purines:** adenine and guanine
  - **Pyrimidines:** cytosine, thymine, and uracil

▶ **Deoxyribonucleotides** are the subunits of DNA.

▶ **Ribonucleotides** are the subunits of RNA.

20.2 The Structure of DNA and RNA

▶ Nucleotides are joined by 3′–5′ phosphodiester bonds in both DNA and RNA.

▶ DNA is a **double helix,** two strands of nucleotide polymers wound around one another with the sugar-phosphate backbone on the outside and complementary pairs of bases extended into the center of the helix.

▶ **Base pairs** are held together by hydrogen bonds.
  - Adenine base pairs with thymine.
  - Cytosine base pairs with guanine.

▶ The two **complementary strands** of DNA are **antiparallel** to one another.

▶ RNA is single-stranded.

▶ **Prokaryotes** are organisms with a simple cellular structure and in which there is no true membrane-bound nucleus and no true membrane-bound organelles.

▶ **Eukaryotes** are organisms that have cells containing a true nucleus with a nuclear membrane and a variety of organelles that segregate a variety of cellular functions from one another.
  - Eukaryotic **chromosomes** have a complex structure with a first-level structure called the **nucleosome.**

▶ The **genome** is the complete set of genetic information of an organism.

20.3 DNA Replication

▶ DNA replication involves synthesis of a faithful copy of the DNA molecule.
  - It begins at a **replication origin** and proceeds at the **replication fork.**

▶ DNA replication is **semiconservative;** each daughter molecule consists of one parental strand and one newly synthesized strand.

▶ **DNA polymerase III** "reads" each parental strand and synthesizes the complementary daughter strand according to the rules of base pairing.

  - The **leading strand** is replicated continuously.
  - The opposite **lagging strand** is replicated discontinuously.

20.4 Information Flow in Biological Systems

▶ The **central dogma** states that the flow of biological information in cells is DNA → RNA → protein.

▶ There are three classes of RNA: **messenger RNA** (mRNA), **transfer RNA** (tRNA), and **ribosomal RNA** (rRNA).

▶ **Transcription** is the process by which RNA is synthesized and occurs in three stages: initiation, elongation, and termination.

▶ **RNA polymerase** catalyzes transcription beginning at the **promoter** of the gene.

▶ Eukaryotic genes contain **introns,** sequences that do not encode protein.
  - Introns are removed from the **primary transcript** by RNA **splicing.**
  - The final RNA contains only the protein coding sequences called **exons.**

▶ Eukaryotic mRNA undergoes further **post-transcriptional modification,** including addition of a **5′ cap structure** and a **3′ poly(A) tail.**

20.5 The Genetic Code

▶ The genetic code is a triplet code, and each code word is called a **codon.**

▶ There are sixty-four codons in the genetic code.
  - There are three termination codons: UAA, UAG, and UGA.
  - The remaining sixty-one specify an amino acid.

▶ Most amino acids have several codons, causing the genetic code to be called **degenerate.**

20.6 Protein Synthesis

▶ The process of protein synthesis is called **translation.**

▶ The codons on the mRNA are decoded by tRNA; the **anticodon** on the tRNA is complementary to a codon on the mRNA, and the tRNA is covalently linked to the correct amino acid.
  - The tRNA bonded to the correct amino acid is called an **aminoacyl tRNA.**
  - An **aminoacyl tRNA synthetase** catalyzes bond formation between the tRNA and the amino acid.

▶ Hydrogen bonding between the codon and anticodon brings the correct amino acid to the site of protein synthesis.

▶ The stages of translation are initiation, chain elongation, and termination.
  - **Initiation factors** and **elongation factors** facilitate initiation and chain elongation.
  - **Translocation** is the movement of the ribosome along the mRNA during chain elongation.
  - A **termination codon** signals the end of the translation process.
  - A **release factor** causes peptidyl transferase to hydrolyze the bond between the peptide and the peptidyl tRNA, releasing the completed peptide.

▶ Protein synthesis occurs on **ribosomes. The aminoacyl tRNA binding site** of the ribosome holds the aminoacyl tRNA.

The **peptidyl tRNA binding site** of the ribosome holds the tRNA bonded to the growing peptide chain.

▶ Many ribosomes together translating a single mRNA constitute a **polysome.**

20.7 Mutation, Ultraviolet Light, and DNA Repair

▶ Any change in a DNA sequence is a **mutation.**

- Mutations may be **silent mutations** if they do not cause any change in the function of the protein encoded by the gene.
- Mutations that destroy or damage the function of a protein may have effects that range from a mild genetic disorder to death.

▶ Agents that cause mutations are called **mutagens.**

- Mutagens may be **carcinogens,** cancer-causing chemicals.

▶ Mutations are classified by the type of change that occurs in the DNA sequence. They may be **point mutations, deletion mutations,** or **insertion mutations.**

▶ Ultraviolet light causes the formation of **pyrimidine dimers.**

- Errors in the repair of pyrimidine dimers can cause UV-induced mutations.
- Germicidal lamps emit UV light that kills bacteria on environmental surfaces.
- UV light damage to the skin can cause skin cancer.

20.8 Recombinant DNA

▶ **Restriction enzymes, hybridization,** and **cloning vectors** are all tools required for genetic engineering.

▶ Cloning a DNA fragment involves digestion of the target and the vector DNA with a restriction enzyme.

- DNA ligase joins the target and vector DNA covalently.
- The recombinant DNA molecules are introduced into bacteria by transformation.
- The desired clones are located by using antibiotic selection and hybridization.

▶ Many eukaryotic genes have been cloned for the purpose of producing medically important proteins.

20.9 Polymerase Chain Reaction

▶ Using a heat-stable DNA polymerase from the bacterium *Thermus aquaticus* and specific DNA primers, polymerase chain reaction (PCR) allows the amplification of specific DNA fragments that are present in small quantities.

▶ PCR is useful in genetic screening, diagnosis of viral or bacterial disease, and in forensic science.

20.10 The Human Genome Project

▶ The Human Genome Project has identified and mapped the genes of the human genome and determined the complete DNA sequence of each of the chromosomes.

▶ DNA libraries were generated and the DNA sequences determined.

▶ Chromosome walking was used to map the sequences along the chromosomes.

▶ DNA sequencing involves reactions in which DNA polymerase copies specific DNA sequences.

- Nucleotide analogs that cause chain termination (dideoxynucleotides) are incorporated randomly into the growing DNA chain.
- This generates a family of DNA fragments that differ in size by one nucleotide.
- DNA sequencing gels separate these fragments and provide DNA sequence data.

▶ **Bioinformatics** is an interdisciplinary field that uses computer information sciences and DNA technology to develop methods to understand, analyze, and apply DNA sequence information.

QUESTIONS AND PROBLEMS

The Structure of the Nucleotide

Foundations

20.13 What is a heterocyclic amine?

20.14 What components of nucleic acids are heterocyclic amines?

Applications

20.15 Draw the structure of the purine ring, and indicate the nitrogen that is bonded to sugars in nucleotides.

20.16 a. Draw the ring structure of the pyrimidines.
    b. In a nucleotide, which nitrogen atom of pyrimidine rings is bonded to the sugar?

20.17 ATP is the universal energy currency of the cell. What components make up the ATP nucleotide?

20.18 One of the energy-harvesting steps of the citric acid cycle results in the production of GTP. What is the structure of the GTP nucleotide?

The Structure of DNA and RNA

Foundations

20.19 The two strands of a DNA molecule are antiparallel. What is meant by this description?

20.20 List three differences between DNA and RNA.

Applications

20.21 What is the significance of the following repeat distances in the structure of the DNA molecule: 0.34 nm, 3.4 nm, and 2 nm?

20.22 Except for the functional groups attached to the rings, the nitrogenous bases are largely flat, hydrophobic molecules. Explain why the arrangement of the purines and pyrimidines found in DNA molecules is very stable.

20.23 Draw the adenine-thymine base pair and indicate the hydrogen bonds that link them.

20.24 Draw the guanine-cytosine base pair and indicate the hydrogen bonds that link them.

20.25 Write the structure that results when deoxycytosine-5′-monophosphate is linked by a 3′ → 5′ phosphodiester bond to thymidine-5′-monophosphate.

20.26 Write the structure that results when adenosine-5′-monophosphate is linked by a 3′ → 5′ phosphodiester bond to uridine-5′-monophosphate.

20.27 Describe the structure of the prokaryotic chromosome.

20.28 Describe the structure of the eukaryotic chromosome.

DNA Replication

Foundations

20.29 What is meant by semiconservative DNA replication?

20.30 Draw a diagram illustrating semiconservative DNA replication.

Applications

20.31 What are the two primary functions of DNA polymerase III?

20.32 **a.** Why is DNA polymerase said to be template-directed?
b. Why is DNA replication a self-correcting process?

20.33 If a DNA strand had the nucleotide sequence

5′-ATGCCCGAGCTGATTGATCAGA-3′

what would the sequence of the complementary daughter strand be?

20.34 If the sequence of a double-stranded DNA molecule is

5′-CATAAGTCGAGACCGTTACTCACTACTGGAC-3′
      | | | | | | | | | | | | | | | | | | | | | | | | | | | | | | |
3′-GTATTCAGCTCTGGCAATGAGTGATGACCTG-5′

what would the sequence of the two daughter DNA molecules be after DNA replication? Indicate which strands are newly synthesized and which are parental.

20.35 What is the replication origin of a DNA molecule?

20.36 What is occurring at the replication fork?

20.37 What is the function of the enzyme helicase?

20.38 What is the function of the enzyme primase?

20.39 What role does the RNA primer play in DNA replication?

20.40 Explain the differences between leading strand and lagging strand replication.

Information Flow in Biological Systems

Foundations

20.41 What is the central dogma of molecular biology?

20.42 What are the roles of DNA, RNA, and protein in information flow in biological systems?

20.43 On what molecule is the anticodon found?

20.44 On what molecule is the codon found?

Applications

20.45 If a gene had the nucleotide sequence

5′-TACGGGCATAGGCCTTAAAGCTAGCTT-3′

what would the sequence of the mRNA be?

20.46 If an RNA strand has the nucleotide sequence

5′-AUGCCAUAACGAUACCCAGUC-3′

what was the sequence of the DNA strand that was transcribed?

20.47 What is meant by the term *RNA splicing?*

20.48 The following is the unspliced transcript of a eukaryotic gene:

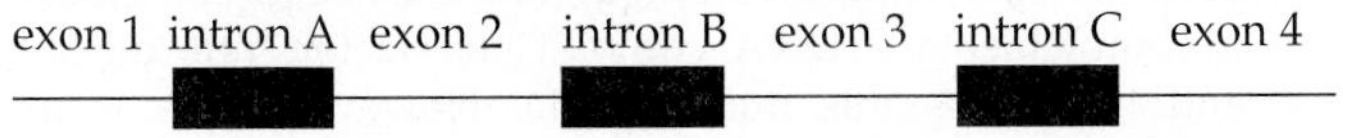

exon 1 intron A exon 2   intron B   exon 3   intron C   exon 4

What would the structure of the final mature mRNA look like, and which of the above sequences would be found in the mature mRNA?

20.49 List the three classes of RNA molecules.

20.50 What is the function of each of the classes of RNA molecules?

20.51 What is the function of the spliceosome?

20.52 What are snRNPs? How do they facilitate RNA splicing?

20.53 What is a poly(A) tail?

20.54 What is the purpose of the poly(A) tail on eukaryotic mRNA?

20.55 What is the cap structure?

20.56 What is the function of the cap structure on eukaryotic mRNA?

The Genetic Code

Foundations

20.57 How many codons constitute the genetic code?

20.58 What is meant by a triplet code?

20.59 What is meant by the reading frame of a gene?

20.60 What happens to the reading frame of a gene if a nucleotide is deleted?

Applications

20.61 Which two amino acids are encoded by only one codon?

20.62 Which amino acids are encoded by six codons?

20.63 An essential gene has the codon 5′-UUU-3′ in a critical position. If this codon is mutated to the sequence 5′-UUA-3′, what is the expected consequence for the cell?

20.64 An essential gene has the codon 5′-UUA-3′ in a critical position. If this codon is mutated to the sequence 5′-UUG-3′, what is the expected consequence for the cell?

Protein Synthesis

Foundations

20.65 What is the function of ribosomes?

20.66 What are the two tRNA binding sites on the ribosome?

Applications

20.67 Write one of the possible mRNA sequences that encodes the following peptide: Ala-gly-leu-cys-met-trp-tyr-ser-ile-gly.

20.68 Why are there several alternative mRNA sequences that could encode the same peptide?

20.69 Explain how a change in the sequence of nucleotides of a gene, a mutation, may alter the sequence of amino acids in the protein encoded by that gene.

20.70 What peptide sequence would be formed from the mRNA 5′-AUGUGUAGUGACCAACCGAUUUCACUGUGA-3′?

The following diagram shows the reaction that produces an aminoacyl tRNA, in this case methionyl tRNA. Use this diagram to answer Questions 20.71 and 20.72.

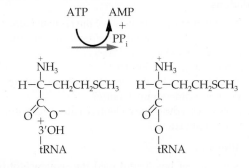

The amino acyl linkage is formed between the 3′—OH of the tRNA and the carboxylate group of the amino acid methionine.

20.71 By what type of bond is an amino acid linked to a tRNA molecule in an aminoacyl tRNA molecule?

20.72 Draw the structure of an alanine residue bound to the 3' position of adenine at the 3' end of alanyl tRNA.

Mutation, Ultraviolet Light, and DNA Repair

Foundations

20.73 Define the term *point mutation*.

20.74 What are deletion and insertion mutations?

Applications

20.75 Why are some mutations silent?

20.76 Which is more likely to be a silent mutation, a point mutation or a deletion mutation? Explain your reasoning.

20.77 What damage does UV light cause in DNA, and how does this lead to mutations?

20.78 Explain why UV lights are effective germicides on environmental surfaces.

20.79 What is a carcinogen? Why are carcinogens also mutagens?

20.80 **a.** What causes the genetic disease xeroderma pigmentosum?
   b. Why are people who suffer from xeroderma pigmentosum prone to cancer?

Recombinant DNA

Foundations

20.81 What is a restriction enzyme?

20.82 Of what value are restriction enzymes in recombinant DNA research?

20.83 What is a selectable marker?

20.84 What is a cloning vector?

Applications

20.85 Name three products of recombinant DNA that are of value in the field of medicine.

20.86 **a.** What is the ultimate goal of genetic engineering?
   b. What ethical issues does this goal raise?

Polymerase Chain Reaction

20.87 After twelve cycles of polymerase chain reaction, how many copies of target DNA would you have for each original molecule in the mixture?

20.88 How is the polymerase chain reaction applied in forensic science?

The Human Genome Project

Foundations

20.89 What were the major goals of the Human Genome Project?

20.90 What are the potential benefits of the information gained in the Human Genome Project?

20.91 What is a genome library?

20.92 What is meant by the term *chromosome walking?*

20.93 What is a dideoxynucleotide?

20.94 How does a dideoxynucleotide cause chain termination in DNA replication?

Applications

20.95 A researcher has determined the sequence of the following pieces of DNA. Using this sequence information, map the location of these pieces relative to one another.

   a. 5'-AGCTCCTGATTTCATACAGTTTCTACTACCTACTA-3'
   b. 5'-AGACATTCTATCTACCTAGACTATGTTCAGAA-3'
   c. 5'-TTCAGAACTCATTCAGACCTACTACTATACCTTGGGAGCTCCT-3'
   d. 5'-ACCTACTAGACTATACTACTACTAAGGGGACTATTCCAGACTT-3'

20.96 Draw a DNA sequencing gel that would represent the sequence shown below. Be sure to label which lanes of the gel represent each of the four dideoxynucleotides in the chain termination reaction mixture.

5'-GACTATCCTAG-3'

CHALLENGE PROBLEMS

1. It has been suggested that the triplet genetic code evolved from a two-nucleotide code. Perhaps there were fewer amino acids in the ancient proteins. Examine the genetic code in Figure 20.18. What features of the code support this hypothesis?

2. The strands of DNA can be separated by heating the DNA sample. The input heat energy breaks the hydrogen bonds between base pairs, allowing the strands to separate from one another. Suppose that you are given two DNA samples. One has a G + C content of 70% and the other has a G + C content of 45%. Which of these samples will require a higher temperature to separate the strands? Explain your answer.

3. A mutation produces a tRNA with a new anticodon. Originally, the anticodon was 5'-CCA-3'; the mutant anticodon is 5'-UCA-3'. What effect will this mutant tRNA have on cellular translation?

4. You have just cloned an EcoR1 fragment that is 1650 base pairs (bp) and contains the gene for the hormone leptin. Your first job is to prepare a restriction enzyme map of the recombinant plasmid. You know that you have cloned into a plasmid vector that is 805 bp and that has only one EcoR1 site (the one into which you cloned). There are no other restriction enzyme sites in the plasmid. The following table shows the restriction enzymes used and the DNA fragment sizes that resulted. Draw a map of the circular recombinant plasmid and a representation of the gel from which the fragment sizes were obtained.

Restriction Enzymes	DNA Fragment Sizes (bp)
EcoR1	805, 1650
EcoR1 + BamHI	450, 805, 1200
EcoR1 + SalI	200, 805, 1450
BamHI + SalI	200, 250, 805, 1200

5. A scientist is interested in cloning the gene for blood-clotting factor VIII into bacteria so that large amounts of the protein can be produced to treat hemophiliacs. Knowing that bacterial cells cannot carry out RNA splicing, she clones a complementary DNA copy of the factor VIII mRNA and introduces this into bacteria. However, there is no transcription of the cloned factor VIII gene. How could the scientist engineer the gene so that the bacterial cell RNA polymerase will transcribe it?

Carbohydrate Metabolism

LEARNING GOALS

1 Discuss the importance of ATP in cellular energy transfer processes.

2 Describe the three stages of catabolism of dietary proteins, carbohydrates, and lipids.

3 Discuss glycolysis in terms of its two major segments.

4 Looking at an equation representing any of the chemical reactions that occur in glycolysis, describe the kind of reaction that is occurring and the significance of that reaction to the pathway.

5 Describe the mechanism of regulation of the rate of glycolysis. Discuss particular examples of that regulation.

6 Discuss the practical and metabolic roles of fermentation reactions.

7 List several products of the pentose phosphate pathway that are required for biosynthesis.

8 Compare glycolysis and gluconeogenesis.

9 Summarize the regulation of blood glucose levels by glycogenesis and glycogenolysis.

All activity requires a source of energy.

OUTLINE

INTRODUCTION

When you awoke this morning, a flood of chemicals called *neurotransmitters* was sent from cell to cell in your nervous system. As these chemical signals accumulated, you gradually became aware of your surroundings. Chemical signals from your nerves to your muscles propelled you out of your warm bed to prepare for your day.

For breakfast, you had a glass of milk, two eggs, and buttered toast, thus providing your body with needed molecules in the form of carbohydrates, proteins, lipids, vitamins, and minerals. As you ran out the door, enzymes in your digestive tract were dismantling the macromolecules of your breakfast. Other enzymes in your cells were busy converting the chemical energy of food molecules into adenosine triphosphate (ATP), the universal energy currency of all cells.

Cells need a ready supply of cellular energy for the many cellular functions that support these activities. They need energy for active transport, to move molecules between the environment and the cell. Energy is needed for biosynthesis of all types of molecules, including the neurotransmitters that helped you awake this morning. Finally, energy is needed for mechanical work, including the muscle contractions that allowed you to get out of bed, have breakfast, and run out the door. Other examples of energy-requiring processes are found in Table 21.1.

Our diet includes three major sources of energy: carbohydrates, fats, and proteins. Each of these types of large biological molecules must be broken down into its basic subunits—simple sugars, fatty acids and glycerol, and amino acids—before they can be taken into the cell and used to produce cellular energy. Of these classes of food molecules, carbohydrates are the most readily used. The pathway for the first stage of carbohydrate breakdown is called *glycolysis*. We find the same pathway in organisms as different as the simple bacterium and humans.

Recall that the potential energy of a compound is the bond energy of that compound.

In this chapter, we are going to examine the steps of this ancient energy-harvesting pathway. We will see that it is responsible for the capture of some of the bond energy of carbohydrates and the storage of that energy in the molecular form of adenosine triphosphate (ATP). Glycolysis actually releases and stores very little (2.2%) of the potential energy of glucose, but the pathway also serves as a source of biosynthetic building blocks. It also modifies the carbohydrates in such a way that other pathways are able to release as much as 40% of the potential energy.

21.1   ATP: The Cellular Energy Currency

LEARNING GOAL

1   Discuss the importance of ATP in cellular energy transfer processes.

Catabolism is the set of metabolic pathways that break down complex macromolecules into simpler ones and, in the process, harvest part of their potential energy for use by the cell. One of those energy-requiring functions is **anabolism,** or biosynthesis; others are described in Table 21.1.

With a series of enzymes, biochemical pathways in the cell carry out a step-by-step oxidation of the sugar glucose and other fuel molecules. Small amounts of energy are released at several points in the pathway, and that energy is harvested and saved in the bonds of a molecule that has been called the *universal energy currency.* This molecule is **adenosine triphosphate (ATP).**

ATP serves as a "go-between" molecule that couples the *exergonic* (energy releasing) reactions of catabolism and the endergonic (energy requiring) reactions of anabolism. To understand how this molecule harvests the energy and releases it for energy-requiring reactions, we must take a look at the structure of this amazing molecule (Figure 21.1). ATP is a **nucleotide,** which means it is a molecule composed of a nitrogenous base; a five-carbon sugar; and one, two, or three phosphoryl groups.

TABLE 21.1 The Types of Cellular Work that Require Energy

Biosynthesis: Synthesis of Metabolic Intermediates and Macromolecules
Synthesis of glucose from CO_2 and H_2O in the process of photosynthesis in plants
Synthesis of amino acids
Synthesis of nucleotides
Synthesis of lipids
Protein synthesis from amino acids
Synthesis of nucleic acids
Synthesis of organelles and membranes
Active Transport: Movement of Ions and Molecules
Transport of H^+ to maintain constant pH
Transport of food molecules into the cell
Transport of K^+ and Na^+ into and out of nerve cells for transmission of nerve impulses
Secretion of HCl from parietal cells into the stomach
Transport of waste from the blood into the urine in the kidneys
Transport of amino acids and most hexose sugars into the blood from the intestine
Accumulation of calcium ions in the mitochondria
Motility
Contraction and flexion of muscle cells
Separation of chromosomes during cell division
Ability of sperm to swim via flagella
Movement of foreign substances out of the respiratory tract by cilia on the epithelial lining of the trachea
Translocation of eggs into the fallopian tubes by cilia in the female reproductive tract

In ATP, a phosphoester bond joins the first phosphoryl group to the five-carbon sugar ribose. The next two phosphoryl groups are joined to one another by phosphoanhydride bonds (Figure 21.1). Recall that the phosphoanhydride bond is a *high-energy bond*. When it is broken or hydrolyzed, a large amount of energy is released. When the phosphoanhydride bond of ATP is broken, the energy released can be used for cellular work. These high-energy bonds are indicated as squiggles (~) in Figure 21.1.

Nature's high-energy bonds, including phosphoanhydride and phosphoester bonds, are discussed in Section 14.4.

Figure 21.1 The structure of the universal energy currency, ATP.

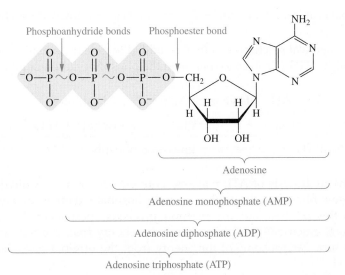

Adenosine

Adenosine monophosphate (AMP)

Adenosine diphosphate (ADP)

Adenosine triphosphate (ATP)

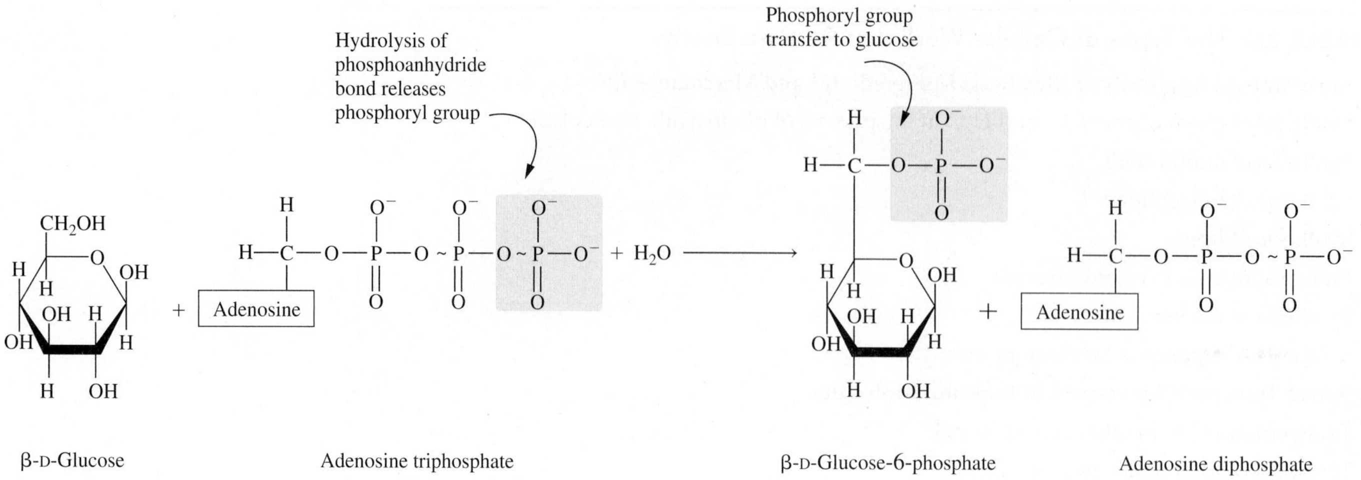

Figure 21.2 The hydrolysis of the phosphoanhydride bond of ATP releases inorganic phosphate and energy. In this coupled reaction catalyzed by an enzyme, the phosphoryl group and some of the released energy are transferred to β-D-glucose.

See Sections 14.4 and 20.1 for further information on the structure of ATP and hydrolysis of the phosphoanhydride bonds.

Hydrolysis of ATP yields adenosine diphosphate (ADP), an inorganic phosphate group (P_i), and energy (Figure 21.2). The energy released by this hydrolysis of ATP is then used to drive biological processes; for instance, the phosphorylation of glucose or fructose.

An example of the way in which the energy of ATP is used can be seen in the first step of glycolysis, the anaerobic degradation of glucose to harvest chemical energy. The first step involves the transfer of a phosphoryl group, $—PO_3^{2-}$, from ATP to the C-6 hydroxyl group of glucose (Figure 21.2). This reaction is catalyzed by the enzyme hexokinase.

This reaction can be dissected to reveal the role of ATP as a source of energy. Although this is a coupled reaction, we can think of it as a two-step process. The first step is the hydrolysis of ATP to ADP and phosphate, abbreviated P_i. This is an exergonic reaction that *releases* about 7 kcal/mol of energy:

$$ATP + H_2O \longrightarrow ADP + P_i + 7 \text{ kcal/mol}$$

The second step, the synthesis of glucose-6-phosphate from glucose and phosphate, is an endergonic reaction that *requires* 3.0 kcal/mol:

$$3.0 \text{ kcal/mol} + glucose + P_i \longrightarrow glucose\text{-}6\text{-}phosphate + H_2O$$

These two chemical reactions can then be added to give the equation showing the way in which ATP hydrolysis is *coupled* to the phosphorylation of glucose:

$$ATP + H_2O \longrightarrow ADP + P_i + 7 \text{ kcal/mol}$$
$$\underline{3.0 \text{ kcal/mol} + glucose + P_i \longrightarrow glucose\text{-}6\text{-}phosphate + H_2O}$$
$$\text{Net: } ATP + glucose \longrightarrow glucose\text{-}6\text{-}phosphate + ADP + 4 \text{ kcal/mol}$$

Because the hydrolysis of ATP releases more energy than is required to synthesize glucose-6-phosphate from glucose and phosphate, there is an overall energy release in this process and the reaction proceeds spontaneously to the right. The product, glucose-6-phosphate, has more energy than the reactant, glucose, because it now carries some of the energy from the original phosphoanhydride bond of ATP.

The primary function of all catabolic pathways is to harvest the chemical energy of fuel molecules and to store that energy by the production of ATP. This continuous production of ATP is what provides the stored potential energy that is used to power most cellular functions.

Question 21.1   Why is ATP called the universal energy currency?

Question 21.2   List five biological activities that require ATP.

21.2   Overview of Catabolic Processes

Carbohydrates, fats, and proteins can all be degraded to release energy, but carbohydrates are the most readily used energy source. When we eat a meal, we are eating quantities of these nutrients that will provide the energy required for life processes. We can organize catabolic processes into three stages, which are summarized in Figure 21.3.

LEARNING GOAL

2 Describe the three stages of catabolism of dietary proteins, carbohydrates, and lipids.

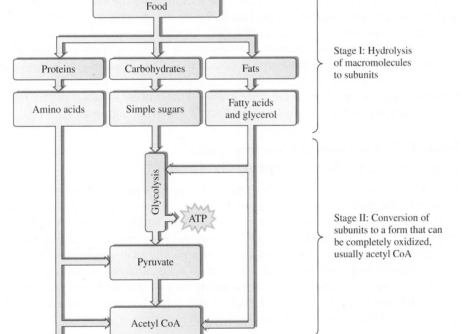

Figure 21.3 The three stages of the conversion of food into cellular energy in the form of ATP.

Describe the path of the carbohydrates, lipids, and proteins in this meal from digestion through the biochemical energy-harvesting reactions.

Stage I: Hydrolysis of Dietary Macromolecules into Small Subunits

The purpose of the first stage of catabolism is to degrade large food molecules into their component subunits. These subunits—simple sugars, amino acids, fatty acids, and glycerol—are then taken into the cells of the body for use as an energy source.

Polysaccharides are hydrolyzed to monosaccharides. Salivary amylase begins the hydrolysis of starch in the mouth. In the small intestine, pancreatic amylase further hydrolyzes the starch into maltose. Maltase catalyzes the hydrolysis of maltose, producing two glucose molecules. Similarly, sucrose is hydrolyzed to glucose and fructose by the enzyme sucrase, and lactose (milk sugar) is degraded into glucose and galactose by the enzyme lactase in the small intestine. The monosaccharides are taken up by the epithelial cells of the intestine in an energy-requiring process called *active transport.*

In the laboratory, a strong acid or base and high temperatures are required for hydrolysis of amide bonds (Section 15.3). However, this reaction proceeds quickly under physiological conditions when catalyzed by enzymes (Section 19.11).

The digestion of proteins begins in the stomach, where the low pH denatures the proteins so that they are more easily hydrolyzed by the enzyme pepsin. They are further degraded in the small intestine by trypsin, chymotrypsin, elastase, and other proteases. The products of protein digestion—amino acids and short oligopeptides—are taken up by the cells lining the intestine. This uptake also involves an active transport mechanism.

The digestion and transport of fats are considered in greater detail in Chapter 23.

Digestion of fats does not begin until the food reaches the small intestine. Fats arrive in the duodenum, the first portion of the small intestine, in the form of large fat globules. Bile salts produced by the liver break these up into an emulsion of tiny fat droplets. Because the small droplets have a greater surface area, the lipids are now more accessible to the action of pancreatic lipase. This enzyme hydrolyzes the fats into fatty acids and glycerol, which are taken up by intestinal cells by a transport process that does not require energy. This process is called *passive transport.* A summary of these hydrolysis reactions is shown in Figure 21.4.

Stage II: Conversion of Monomers into a Form that Can Be Completely Oxidized

The citric acid cycle is considered in detail in Section 22.4.

The monosaccharides, amino acids, fatty acids, and glycerol must now be assimilated into the pathways of energy metabolism. The two major pathways are glycolysis and the citric acid cycle (see Figure 21.3). Sugars usually enter the glycolysis pathway in the form of glucose or fructose. They are eventually converted to acetyl CoA, which is a form that can be completely oxidized in the citric acid cycle. Amino groups are removed from amino acids, and the remaining carbon skeletons enter the catabolic processes at many steps. Fatty acids are converted to acetyl CoA and enter the citric acid cycle in that form. Glycerol, produced by the hydrolysis of fats, enters energy metabolism via glycolysis.

Stage III: The Complete Oxidation of Nutrients and the Production of ATP

Oxidative phosphorylation is described in Section 22.6.

Acetyl CoA carries acetyl groups, two-carbon remnants of the nutrients, to the citric acid cycle. Acetyl CoA enters the cycle, and electrons and hydrogen atoms are harvested during the complete oxidation of the acetyl group to CO_2. Coenzyme A is released (recycled) to carry additional acetyl groups to the pathway. The electrons and hydrogen atoms that are harvested are used in the process of *oxidative phosphorylation* to produce ATP.

Question 21.3 Briefly describe the three stages of catabolism.

Question 21.4 Discuss the digestion of dietary carbohydrates, lipids, and proteins.

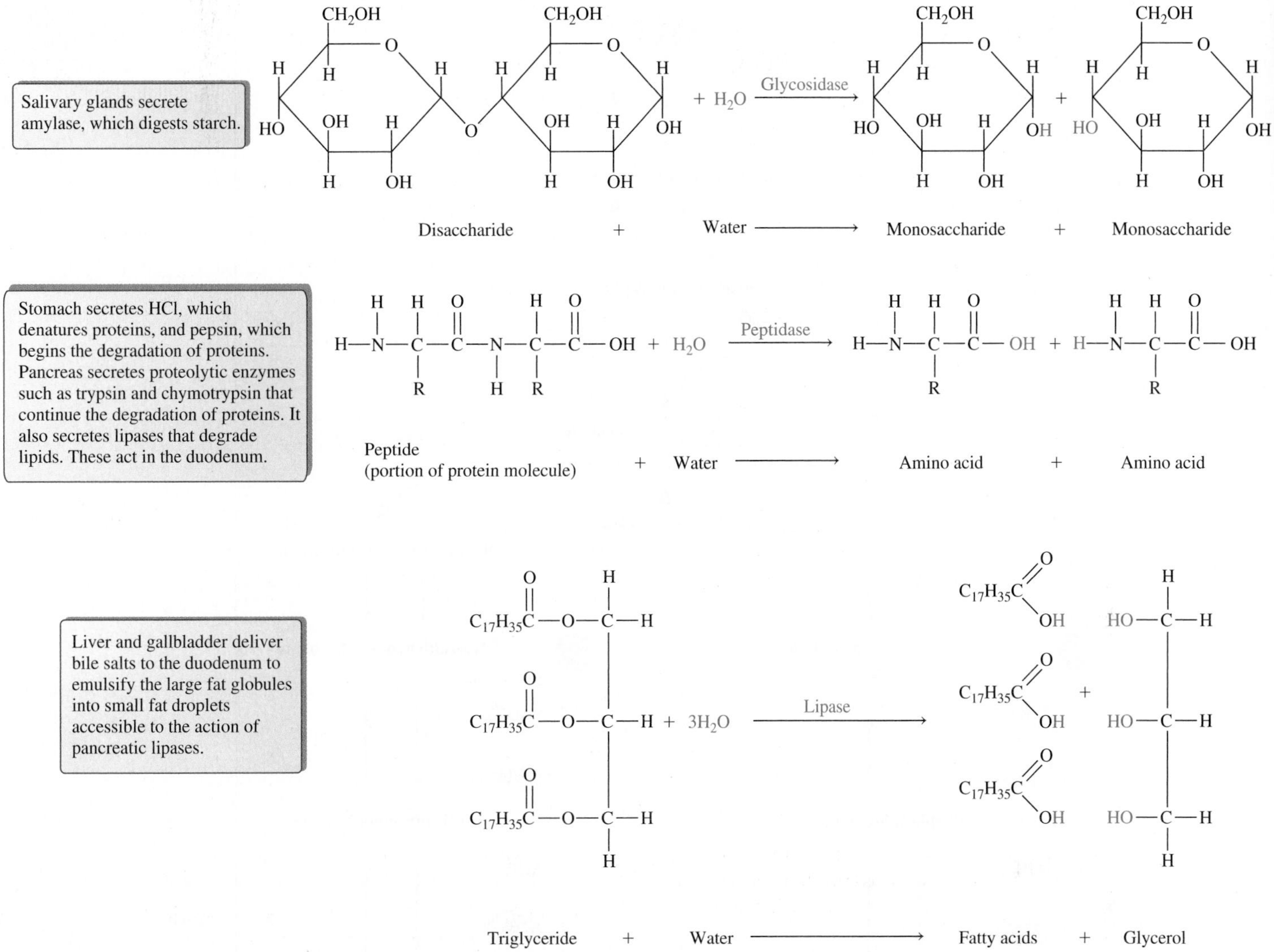

Salivary glands secrete amylase, which digests starch.

Disaccharide + Water ⟶ Monosaccharide + Monosaccharide

Stomach secretes HCl, which denatures proteins, and pepsin, which begins the degradation of proteins. Pancreas secretes proteolytic enzymes such as trypsin and chymotrypsin that continue the degradation of proteins. It also secretes lipases that degrade lipids. These act in the duodenum.

Peptide (portion of protein molecule) + Water ⟶ Amino acid + Amino acid

Liver and gallbladder deliver bile salts to the duodenum to emulsify the large fat globules into small fat droplets accessible to the action of pancreatic lipases.

Triglyceride + Water ⟶ Fatty acids + Glycerol

Figure 21.4 A summary of the hydrolysis reactions of carbohydrates, proteins, and fats, and their locations in the digestive tract.

21.3 Glycolysis

An Overview

Glycolysis, also known as the Embden-Meyerhof Pathway, is a pathway for carbohydrate catabolism that begins with the substrate D-glucose. The pathway evolved at a time when the earth's atmosphere was *anaerobic;* no free oxygen was available. As a result, glycolysis requires no oxygen; it is an anaerobic process. Further, it must have evolved in very simple, single-celled organisms that lacked complex organelles, much like bacteria. As a result, glycolysis is a process carried out by enzymes that are free in the cytoplasm. To this day, glycolysis remains an anaerobic process carried out by cytoplasmic enzymes, even in cells as complex as our own.

The ten steps of glycolysis, catalyzed by ten enzymes, are outlined in Figure 21.5. The first reactions of glycolysis involve an energy investment. ATP molecules are hydrolyzed, energy is released, and phosphoryl groups are added to the hexose sugars. In the remaining steps of glycolysis, energy is harvested to produce a net gain of ATP.

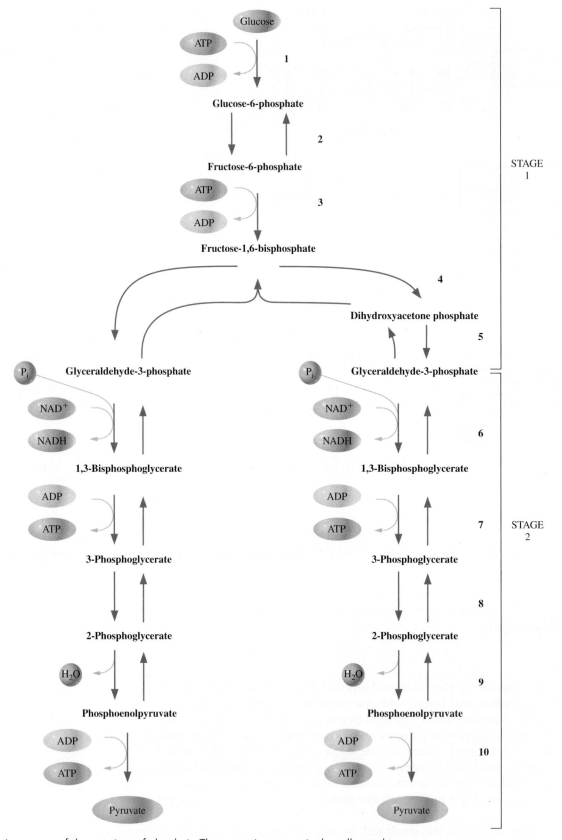

Figure 21.5 A summary of the reactions of glycolysis. These reactions occur in the cell cytoplasm.

The three major products of glycolysis are seen in Figure 21.5. These are chemical energy in the form of ATP, chemical energy in the form of NADH, and two three-carbon pyruvate molecules. Each of these products is considered below:

- **Chemical energy as ATP.** Four ATP molecules are formed by the process of **substrate-level phosphorylation.** In this process, a high-energy phosphoryl group from one of the substrates is transferred to ADP to form ATP. The two substrates involved in these transfer reactions are 1,3-bisphosphoglycerate and phosphoenolpyruvate (see Figure 21.5, steps 7 and 10). Although four ATP molecules are produced during glycolysis, the *net* gain is only two ATP molecules because two ATP molecules are used early in glycolysis (Figure 21.5, steps 1 and 3).

- **Chemical energy in the form of reduced NAD$^+$, NADH. Nicotinamide adenine dinucleotide (NAD$^+$)** is a coenzyme derived from the vitamin niacin. The reduced form of NAD$^+$, NADH, carries hydride anions, hydrogen atoms with two electrons (H:$^-$), removed during the oxidation of glyceraldehyde-3-phosphate (see Figure 21.5, step 6). Under aerobic conditions, the electrons and hydrogen atom are transported from the cytoplasm into the mitochondria. Here they enter an electron transport system for the generation of ATP by **oxidative phosphorylation.** Under anaerobic conditions, NADH is used as a source of electrons in fermentation reactions.

The structure of NAD$^+$ and the way it functions as a hydride anion carrier are shown in Figure 19.8 and described in Section 19.7.

Aerobic Environment

NAD$^+$ regenerated when H:$^-$ is donated to an electron transport system

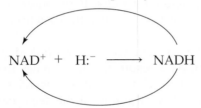

$$NAD^+ \ + \ H:^- \longrightarrow NADH$$

Anaerobic Environment

NAD$^+$ regenerated when H:$^-$ is donated in a fermentation reaction

- **Two pyruvate molecules.** At the end of glycolysis, the six-carbon glucose molecule has been converted into two three-carbon pyruvate molecules. The fate of the pyruvate also depends on whether the reactions are occurring in the presence or absence of oxygen. Under aerobic conditions, it is used to produce acetyl CoA destined for the citric acid cycle and complete oxidation. Under anaerobic conditions it is used as an electron acceptor in fermentation reactions.

Both pyruvate and NADH must be used in some way so that glycolysis can continue to function and produce ATP. The buildup of pyruvate would cause glycolysis to stop, thereby stopping the production of ATP. If NADH were to build up, there would be no NAD$^+$ available for step 6, in which glyceraldehyde-3-phosphate is oxidized and NAD$^+$ is reduced (accepts the hydride anion). This would stop glycolysis, and thus NADH must be reoxidized so that glycolysis can continue to produce ATP for the cell.

Biological Effects of Genetic Disorders of Glycolysis

Before we look at each of the reactions of glycolysis in detail, let's consider the symptoms that arise if a person has a genetic defect in one of the enzymes.

Symptoms include muscle myopathy [muscle *(myo)* and disorder *(pathy)*], which involves damage to the muscle as a result of the inability to extract energy from food molecules. In its mildest form, myopathy can cause exercise intolerance, which is the onset of fatigue when exercising. In more severe forms, it can cause muscle breakdown (rhabdomyolysis), in which the muscle cells, starved of energy, begin to die. Another symptom of rhabdomyolysis is the release of myoglobin into the blood and eventually into the urine. This condition, called *myoglobinuria,* results in urine that is the color of cola soft drinks, and may even damage the kidneys.

Myopathy may include severe muscle pain. Patients often describe it as a cramp, but it is not a cramp since the muscle is not able to contract because of the lack of energy. Rather, the pain is caused by cell death and tissue damage that result from an inability to produce enough ATP.

Another symptom is hemolytic anemia (anemia that results from the lysis of red blood cells). Red blood cells are completely dependent on glycolysis for their ATP. A defect in one of the enzymes of glycolysis results in insufficient ATP and resultant cell death.

Tarui's disease is caused by a deficiency of phosphofructokinase. Although this is not a sex-linked disorder, the great majority of sufferers are males (nine males to one female). The disorder is most frequently found in U.S. Ashkenazi Jews and Italian families. Onset of symptoms typically occurs between the ages of twenty and forty, although some severe cases have been reported in infants and young children. Patients experiencing the late-onset form of Tarui's disease typically experienced exercise intolerance when they were younger. Vigorous exercise results in myoglobinuria and severe muscle pain. Meals high in carbohydrates worsen the exercise intolerance. Early-onset disease is often associated with respiratory failure, cardiomyopathy (heart muscle disease), seizures, and cortical blindness.

Phosphoglycerate kinase deficiency is a sex-linked genetic disorder (located on the X chromosome). As a result, far more males than females suffer from this disease. There are many clinical features associated with this deficiency, although only rarely are they all found in the same patient. These symptoms range from mental challenge and seizures to a slowly progressive myopathy and hemolytic anemia.

Phosphoglycerate mutase deficiency has been mapped on chromosome 7. The disorder is found predominantly in U.S African American, Italian, and Japanese families. The clinical features include exercise intolerance, muscle pain, and myoglobinuria following more intense exercise.

These are just three of the disorders associated with deficiencies of the enzymes of glycolysis, but they make it clear that the pathway is critical to our health. The most extreme deficiency of one of these enzymes will cause death of the fetus.

LEARNING GOAL

3 Discuss glycolysis in terms of its two major segments.

LEARNING GOAL

4 Looking at an equation representing any of the chemical reactions that occur in glycolysis, describe the kind of reaction that is occurring and the significance of that reaction to the pathway.

Reactions of Glycolysis

Glycolysis can be divided into two major segments. The first set of steps, reactions 1–5, is the investment of ATP energy. The second major segment involves the remaining reactions of the pathway (6–10), those that result in a net energy yield.

Reaction 1

The substrate, glucose, is phosphorylated by the enzyme *hexokinase* in a coupled phosphorylation reaction. The source of the phosphoryl group is ATP. At first, this reaction seems contrary to the overall purpose of catabolism, the *production* of ATP. The expenditure of ATP in these early reactions must be thought of as an "investment." The cell actually goes into energy "debt" in these early reactions, but this is absolutely necessary to get the pathway started.

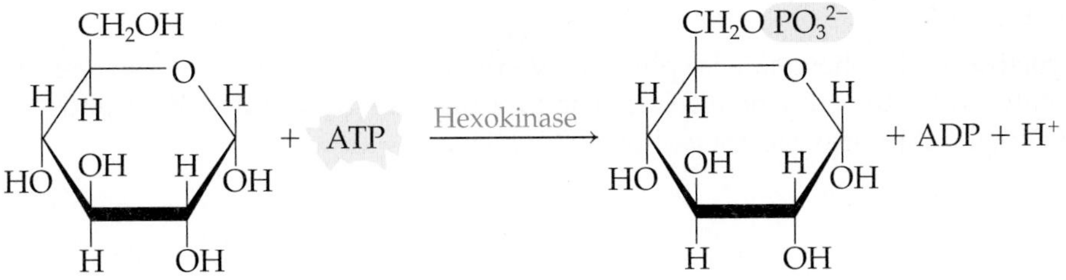

Glucose

Glucose-6-phosphate

The enzyme name can tell us a lot about the reaction (see Section 19.1). The suffix -*kinase* tells us that the enzyme is a transferase that will transfer a phosphoryl group, in this case from an ATP molecule to the substrate. The prefix *hexo-* gives us a hint that the substrate is a six-carbon sugar. Hexokinase predominantly phosphorylates the six-carbon sugar glucose.

Reaction 2

The glucose-6-phosphate formed in the first reaction is rearranged to produce the structural isomer fructose-6-phosphate. The enzyme *phosphoglucose isomerase* catalyzes this isomerization. The result is that the C-1 carbon of the six-carbon sugar is exposed; it is no longer part of the ring structure. Examination of the open-chain structures reveals that this isomerization converts an aldose into a ketose.

The enzyme name, phosphoglucose isomerase, provides clues to the reaction that is being catalyzed (Section 19.1). *Isomerase* tells us that the enzyme will catalyze the interconversion of one isomer into another. *Phosphoglucose* suggests that the substrate is a phosphorylated form of glucose.

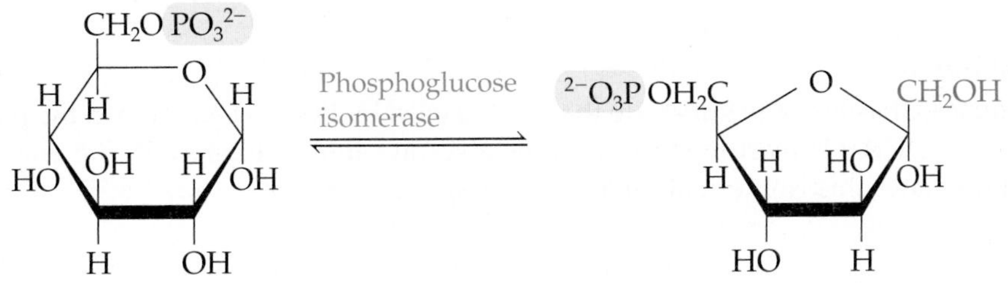

Glucose-6-phosphate

Fructose-6-phosphate

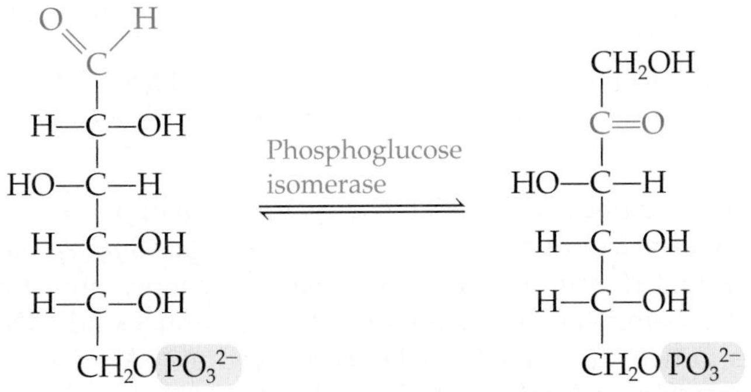

Glucose-6-phosphate
(an aldose)

Fructose-6-phosphate
(a ketose)

Reaction 3

A second energy "investment" is catalyzed by the enzyme *phosphofructokinase*. A phosphoanhydride bond in ATP is hydrolyzed, and a phosphoester linkage between the phosphoryl group and the C-1 hydroxyl group of fructose-6-phosphate is formed. The product is fructose-1,6-bisphosphate.

The suffix -*kinase* in the name of the enzyme tells us that this is a coupled reaction: ATP is hydrolyzed and a phosphoryl group is transferred to another molecule. The prefix *phosphofructo-* tells us the other molecule is a phosphorylated form of fructose.

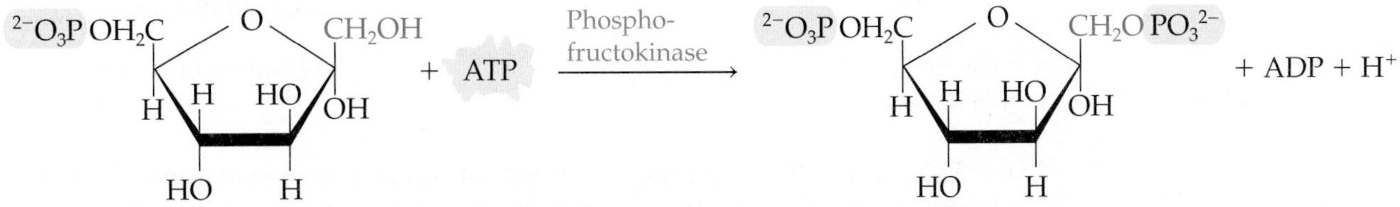

Fructose-6-phosphate

Fructose-1,6-bisphosphate

Reaction 4

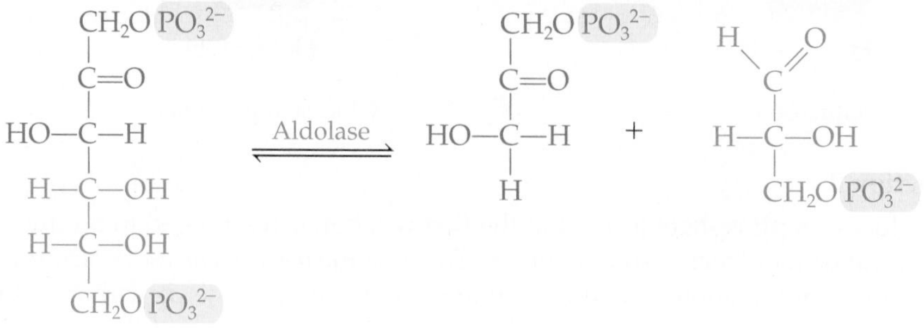

Fructose-1,6-bisphosphate is split into two three-carbon intermediates in a reaction catalyzed by the enzyme *aldolase*. The products are glyceraldehyde-3-phosphate (G3P) and dihydroxyacetone phosphate (DHAP).

Fructose-   Dihydroxyacetone   Glyceraldehyde-
1,6-bisphosphate   phosphate   3-phosphate

Reaction 5

Because G3P is the only substrate that can be used by the next enzyme in the pathway, the DHAP is rearranged to become a second molecule of G3P. The enzyme that mediates this isomerization is *triose phosphate isomerase*.

The enzyme name hints that two isomers of a phosphorylated three-carbon sugar are going to be interconverted (Section 19.1). The ketone dihydroxyacetone phosphate and its isomeric aldehyde, glyceraldehyde-3-phosphate are interconverted through an enediol intermediate.

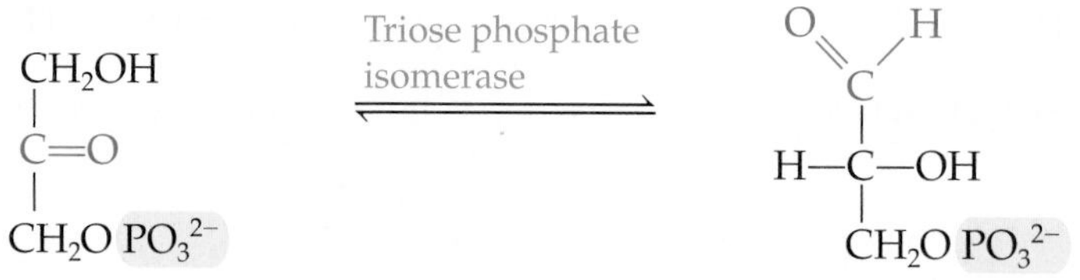

Dihydroxyacetone phosphate   Glyceraldehyde-3-phosphate

Reaction 6

In this reaction, the aldehyde glyceraldehyde-3-phosphate is oxidized to a carboxylic acid in a reaction catalyzed by *glyceraldehyde-3-phosphate dehydrogenase*. This is the first step in glycolysis that harvests energy, and it involves the reduction of the coenzyme nicotinamide adenine dinucleotide (NAD^+). This reaction occurs in two steps. First, NAD^+ is reduced to NADH as the aldehyde group of glyceraldehyde-3-phosphate is oxidized to a carboxyl group. Second, an inorganic phosphate group is transferred to the carboxyl group to give 1,3-bisphosphoglycerate. Notice that the new bond is denoted with a squiggle (~), indicating that this is a high-energy bond.

The name glyceraldehyde-3-phosphate dehydrogenase tells us that the substrate glyceraldehyde-3-phosphate is going to be oxidized. In this reaction, we see that the aldehyde group has been oxidized to a carboxylate group (Section 13.4).

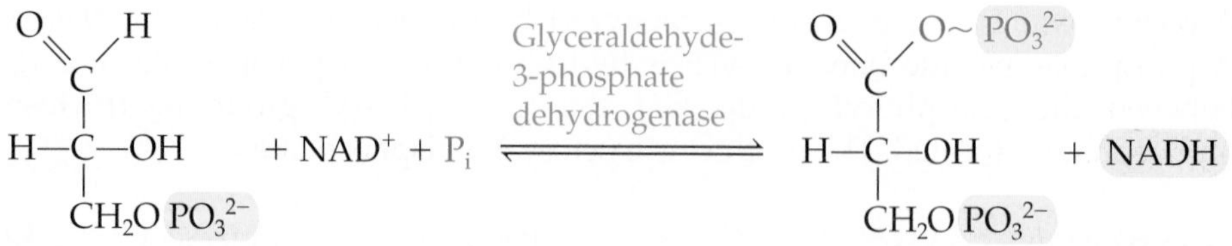

Glyceraldehyde-   1,3-Bisphosphoglycerate
3-phosphate

This, and all remaining reactions of glycolysis, occur twice for each glucose because each glucose has been converted into two molecules of glyceraldehyde-3-phosphate.

Reaction 7

In this reaction, energy is harvested in the form of *ATP*. The enzyme *phosphoglycerate kinase* catalyzes the transfer of the phosphoryl group of 1,3-bisphosphoglycerate to ADP. This is the first substrate-level phosphorylation of glycolysis, and it produces ATP and 3-phosphoglycerate. It is a coupled reaction in which the high-energy bond is hydrolyzed and the energy released is used to drive the synthesis of ATP.

Once again, the enzyme name reveals a great deal about the reaction. The suffix *-kinase* tells us that a phosphoryl group will be transferred. In this case, a phosphoester bond in the substrate 1,3-bisphosphoglycerate is hydrolyzed and ADP is phosphorylated. Note that this is a reversible reaction.

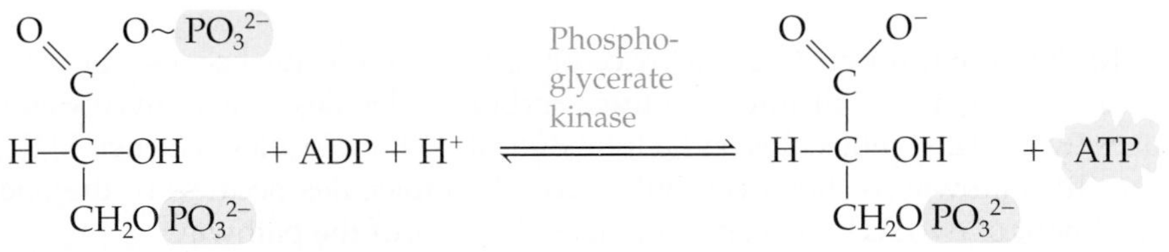

1,3-Bisphosphoglycerate   3-Phosphoglycerate

Reaction 8

3-Phosphoglycerate is isomerized to produce 2-phosphoglycerate in a reaction catalyzed by the enzyme *phosphoglycerate mutase*. The phosphoryl group attached to the third carbon of 3-phosphoglycerate is transferred to the second carbon.

The suffix *-mutase* indicates another type of isomerase. Notice that the chemical formulas of the substrate and reactant are the same. The only difference is in the location of the phosphoryl group.

3-Phosphoglycerate   2-Phosphoglycerate

Reaction 9

In this step, the enzyme *enolase* catalyzes the dehydration (removal of a water molecule) of 2-phosphoglycerate. The energy-rich product is phosphoenolpyruvate, the highest energy phosphorylated compound in our metabolism.

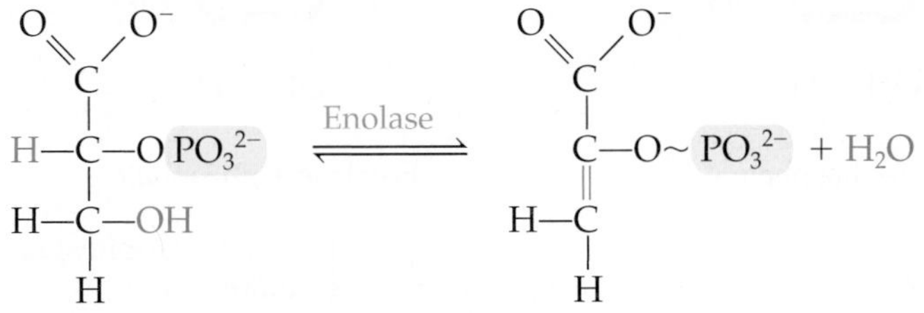

2-Phosphoglycerate   Phosphoenolpyruvate

Reaction 10

The final substrate-level phosphorylation in the pathway is catalyzed by *pyruvate kinase*. Phosphoenolpyruvate serves as a donor of the phosphoryl group that is transferred to ADP to produce ATP. This is another coupled reaction in which hydrolysis of the phosphoester bond in phosphoenolpyruvate provides energy for the formation of the phosphoanhydride bond of ATP. The final product of glycolysis is pyruvate.

The enzyme name indicates that a phosphoryl group will be transferred (kinase) and that the product will be pyruvate.

$$\underset{\text{Phosphoenolpyruvate}}{\overset{\displaystyle \begin{array}{c} O \diagup \overset{\displaystyle \nwarrow}{C} \diagdown O^- \\ \mid \\ C-O\sim PO_3^{2-} \\ \parallel \\ H-C \\ \mid \\ H \end{array}}{}} + ADP + H^+ \xrightarrow{\text{Pyruvate kinase}} \underset{\text{Pyruvate}}{\overset{\displaystyle \begin{array}{c} O \diagup \overset{\displaystyle \nwarrow}{C} \diagdown O^- \\ \mid \\ C=O \\ \mid \\ CH_3 \end{array}}{}} + ATP$$

Reactions 6 through 10 occur twice per glucose molecule because the starting six-carbon sugar is split into two three-carbon molecules. Thus, glycolysis produces two NADH molecules and a total of four ATP molecules. The net ATP gain from this pathway is, however, only two ATP molecules because of the energy investment of two ATP molecules in the early steps of the pathway.

Entry of Fructose into Glycolysis

Depending on the tissue, fructose enters glycolysis in different ways. In the muscle, where hexokinase is abundant, the enzyme phosphorylates fructose to fructose-6-phosphate, which directly enters glycolysis. There is much less hexokinase in the liver, but fructokinase is present. *Fructokinase* phosphorylates fructose to produce fructose-1-phosphate. This product is cleaved into dihydroxyacetone phosphate (DHAP) and glyceraldehyde by the enzyme *fructose-1-phosphate aldolase*. The glyceraldehyde is phosphorylated by *triose kinase* to produce glyceraldehyde-3-phosphate (G3P). The DHAP and G3P enter glycolysis directly.

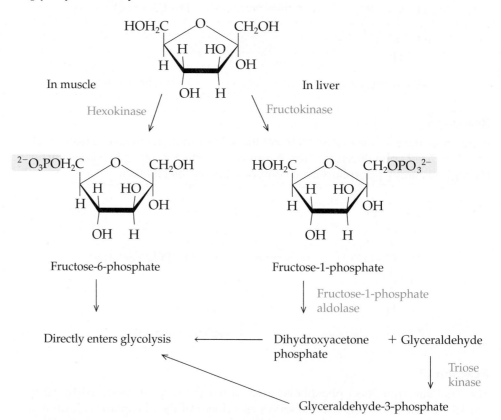

Question 21.5   What is substrate-level phosphorylation?

Question 21.6   What are the major products of glycolysis?

A Medical Perspective

High Fructose Corn Syrup

A controversy over the health risks of high fructose corn syrup (HFCS) has been raging in recent years. This sweetener was introduced into the food market forty years ago. Some have made the correlation that the increase in the use of HFCS parallels the increase in obesity in the U.S. population. In 1970, approximately 15% of the population was obese. That level currently stands at 33% of the U.S. population.

Until recently, there were no data to suggest that the observed correlation was valid. But in 2010, Dr. Bart Hoebel and his colleagues at Princeton University carried out two experiments in rats that suggest that there is a relationship between HFCS in the diet and obesity.

In the first experiment, one set of rats was given water sweetened with sucrose at a concentration typical of a soft drink. A second set of rats was provided water sweetened with HFCS at a concentration half that found in soft drinks. Both sets of rats were provided their standard diet of rat chow. Dr. Hoebel's group observed that the rats provided water sweetened with HFCS gained much more weight than those given water with sucrose.

In a long-term study of the effects of HFCS, the group monitored weight gain, blood triglycerides, and body fat over a period of 6 months. In this study, rats on rat chow only were compared with rats on a high HFCS diet. Their results were startling. Not only did the HFCS rats gain 48% more weight, they exhibited significant increases in blood triglycerides and deposition of abdominal fat. These are conditions that, in humans, are associated with metabolic syndrome which, in turn, is associated with coronary artery disease, high blood pressure, cancer, and diabetes.

Both HFCS and sucrose contain glucose and fructose; so what differences might cause these results? The primary difference is that HFCS contains a higher concentration of fructose than sucrose (55% compared to 50%). Could this increased amount of fructose be the cause of the symptoms?

The way the body metabolizes HFCS compared to sucrose may provide some clues. First, sucrose is a disaccharide that must be hydrolyzed before it can be absorbed by cells of the body. HFCS is a mixture of the monosaccharides glucose and fructose; these can be immediately absorbed. In addition,

unlike glucose, fructose does not stimulate an increase in insulin levels. Since insulin controls the release of leptin, a hormone that signals the satiety center of the brain, thus reducing hunger, a reduced amount of leptin would promote over-eating and weight gain.

The last observation is that fructose enters glycolysis by a different path than glucose. As we saw in the text, fructose is phosphorylated to fructose-1-phosphate in the liver. This is directly cleaved by fructose-1-phosphate aldolase to produce dihydroxyacetone phosphate (DHAP) and glyceraldehye. The DHAP is quickly isomerized and the glyceraldehyde is phosphorylated; both reactions produce glyceraldehyde-3-phosphate. This leads to the production of acetyl CoA via reactions that are unregulated by the normal regulatory steps of glycolysis. The reactions catalyzed by hexokinase and phosphofructokinase, the two earliest regulatory enzymes in glycolysis, are bypassed in fructose metabolism. The final regulatory step, catalyzed by pyruvate kinase, is mediated by fructose-1,6,-bisphosphate, which is reduced when there is an excess of fructose in the cell. The result, some argue, is that high levels of acetyl CoA produced by this unregulated pathway promote the synthesis of fats that are then deposited in adipose tissue.

While Dr. Hoebel's experiments involved rats, the work of Dr. Kathleen Page of the University of Southern California has shown that fructose has interesting effects in humans. She and her team demonstrated that fructose *activates* the hypothalamus, which is the part of the brain that regulates hunger. Glucose, on the other hand, *suppresses* the hypothalamus, thereby triggering a feeling of being full. All of these results suggest that we need to learn more about the metabolism of fructose and the presence of HFCS in the diet.

For Further Understanding

▸ A correlation has been found between the increase of HFCS in the diet and an increase in obesity in the U.S. population. Why is this not proof that a relationship between the two exists?

▸ Discuss the difficulties of proving the effects of foods such as HFCS in humans.

Question 21.7 Describe an overview of the reactions of glycolysis.

Question 21.8 How do the names of the first three enzymes of the glycolytic pathway relate to the reactions they catalyze?

Regulation of Glycolysis

Energy-harvesting pathways, such as glycolysis, are responsive to the energy needs of the cell. Reactions of the pathway speed up when there is a demand for ATP. They slow down when there is abundant ATP to meet the energy requirements of the cell.

LEARNING GOAL

5 Describe the mechanism of regulation of the rate of glycolysis. Discuss particular examples of that regulation.

There are additional mechanisms that regulate the rate of glycolysis, but we will focus on those that involve allosteric enzymes (Section 19.9).

One of the major mechanisms for the control of the rate of glycolysis is the use of *allosteric enzymes*. In addition to the active site, which binds the substrate, allosteric enzymes have an effector binding site, which binds a chemical signal that alters the rate at which the enzyme catalyzes the reaction. Effector binding may increase (positive allosterism) or decrease the rate of reaction (negative allosterism).

The chemical signals, or effectors, that indicate the energy needs of the cell include molecules such as ATP. When the ATP concentration is high, the cell must have sufficient energy. Similarly, ADP and AMP, which are precursors of ATP, are indicators that the cell is in need of ATP. In fact, all of these molecules are allosteric effectors that alter the rate of irreversible reactions catalyzed by enzymes in the glycolytic pathway.

The enzyme hexokinase, which catalyzes the phosphorylation of glucose, is allosterically inhibited by the product of the reaction it catalyzes, glucose-6-phosphate. A buildup of this product indicates that the reactions of glycolysis are not proceeding at a rapid rate, presumably because the cell has enough energy.

Phosphofructokinase, the enzyme that catalyzes the third reaction in glycolysis, is a key regulatory enzyme in the pathway. ATP is an allosteric inhibitor of phosphofructokinase, whereas AMP and ADP are allosteric activators. Another allosteric inhibitor of phosphofructokinase is citrate. As we will see in the next chapter, citrate is the first intermediate in the citric acid cycle, a pathway that results in the complete oxidation of the pyruvate. A high concentration of citrate signals that sufficient substrate is entering the citric acid cycle. The inhibition of phosphofructokinase by citrate is an example of *feedback inhibition*: the product, citrate, allosterically inhibits the activity of an enzyme early in the pathway.

The last enzyme in glycolysis, pyruvate kinase, is also subject to allosteric regulation. In this case, fructose-1,6-bisphosphate, the product of the reaction catalyzed by phosphofructokinase, is the allosteric activator. Thus, activation of phosphofructokinase results in the activation of pyruvate kinase. This is an example of *feedforward activation* because the product of an earlier reaction causes activation of an enzyme later in the pathway.

LEARNING GOAL

6 Discuss the practical and metabolic roles of fermentation reactions.

Aerobic respiration is discussed in Chapter 22.

When you exercise beyond the ability of your heart and lungs to provide sufficient oxygen to muscle, the lactate fermentation kicks in. Write an equation representing the reaction catalyzed by lactate dehydrogenase, and explain how this reaction enables muscle to continue working.

21.4 Fermentations

In the overview of glycolysis, we noted that the end product pyruvate must be used up and the NADH must be reoxidized in order for glycolysis to continue. If the cell is functioning under aerobic conditions, NADH will be reoxidized, and pyruvate will be completely oxidized by aerobic respiration. Under anaerobic conditions, however, different types of fermentation reactions accomplish these purposes. **Fermentations** are catabolic reactions that occur with no net oxidation. Pyruvate or an organic compound produced from pyruvate is reduced as NADH is oxidized.

Lactate Fermentation

Lactate fermentation is familiar to anyone who has performed strenuous exercise. If you exercise so hard that your lungs and circulatory system can't deliver enough oxygen to the working muscles, your aerobic (oxygen-requiring) energy-harvesting pathways are not able to supply enough ATP to your muscles. But the muscles still demand energy. Under these anaerobic conditions, lactate fermentation begins. In this reaction, the enzyme *lactate dehydrogenase* reduces pyruvate to lactate. NADH is the reducing agent for this process (Figure 21.6). As pyruvate is reduced, NADH is oxidized, and NAD^+ is again available, permitting glycolysis to continue.

The lactate produced in the working muscle passes into the blood. Eventually, if strenuous exercise is continued, the concentration of lactate becomes so high

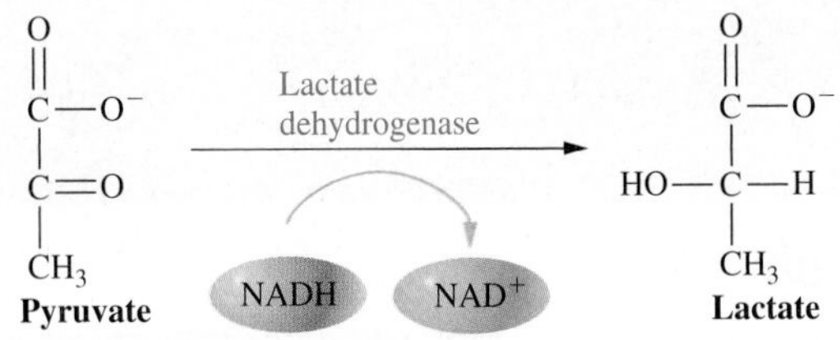

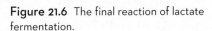

Figure 21.6 The final reaction of lactate fermentation.

that this fermentation can no longer continue. Glycolysis, and thus ATP production, stops. The muscle, deprived of energy, can no longer function. This point of exhaustion is called the **anaerobic threshold.**

A variety of bacteria are able to carry out lactate fermentation under anaerobic conditions. This is of great importance in the dairy industry, because these organisms are used to produce yogurt and some cheeses. The tangy flavor of yogurt is contributed by the lactate produced by these bacteria. Unfortunately, similar organisms also cause milk to spoil.

As we saw in A Medical Perspective: Tooth Decay and Simple Sugars (Chapter 16), the lactate produced by oral bacteria is responsible for the gradual removal of calcium from tooth enamel and the resulting dental cavities.

Alcohol Fermentation

Alcohol fermentation has been appreciated, if not understood, since the dawn of civilization. The fermentation process itself was discovered by Louis Pasteur during his studies of the chemistry of wine making and "diseases of wines." Under anaerobic conditions, yeast ferment the sugars produced by fruit and grains. The pyruvate produced by glycolysis undergoes two reactions of the alcohol fermentation (Figure 21.7):

These applications and other fermentations are described in A Human Perspective: Fermentations: The Good, the Bad, and the Ugly.

- *Pyruvate decarboxylase* removes CO_2 from the pyruvate, producing ethanal (acetaldehyde).
- *Alcohol dehydrogenase* catalyzes the reduction of ethanal to ethanol and, more importantly, the oxidation of NADH to NAD^+.

The regeneration of NAD^+ allows glycolysis to continue, just as in the case of lactate fermentation.

The two products of alcohol fermentation, then, are ethanol and CO_2. We take advantage of this fermentation in the production of wines and other alcoholic beverages and in the process of bread making.

Question 21.9 How is the alcohol fermentation in yeast similar to lactate production in skeletal muscle?

Question 21.10 Why must pyruvate be used and NADH be reoxidized so that glycolysis can continue?

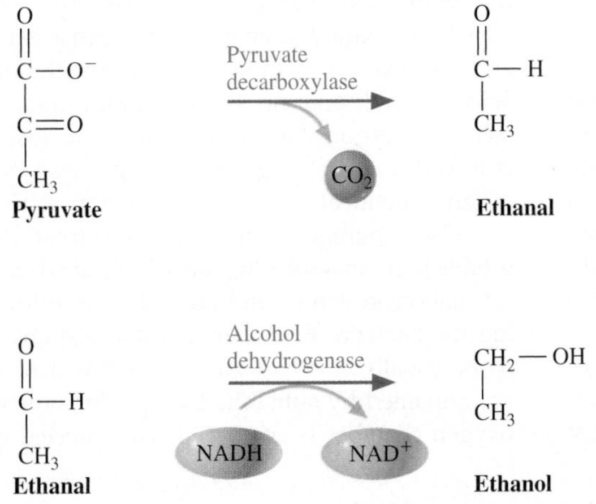

Figure 21.7 The final two reactions of alcohol fermentation.

A Human Perspective

Fermentations: The Good, the Bad, and the Ugly

In this chapter, we have seen that fermentation is an anaerobic, cytoplasmic process that allows continued ATP generation by glycolysis. ATP production can continue because the pyruvate produced by the pathway is utilized in the fermentation and because NAD^+ is regenerated.

The stable end products of alcohol fermentation are CO_2 and ethanol. These have been used by humankind in a variety of ways, including the production of alcoholic beverages, bread making, and alternative fuel sources.

If alcohol fermentation is carried out by using fruit juices in a vented vat, the CO_2 will escape, and the result will be a still wine (not bubbly). But conditions must remain anaerobic; otherwise, fermentation will stop, and aerobic energy-harvesting reactions will ruin the wine. Fortunately for vintners (wine makers), when a vat is fermenting actively, enough CO_2 is produced to create a layer that keeps the oxygen-containing air away from the fermenting juice, thus maintaining an anaerobic atmosphere.

Now suppose we want to make a sparkling wine, such as champagne. To do this, we simply have to trap the CO_2 produced. In this case, the fermentation proceeds in a sealed bottle, a very strong bottle. Both the fermentation products, CO_2 and ethanol, accumulate. Under pressure within the sealed bottle the CO_2 remains in solution. When the top is "popped," the pressure is released, and the CO_2 comes out of solution in the form of bubbles.

In either case, the fermentation continues until the alcohol concentration reaches 12–13%. At that point, the yeast "stews in its own juices"! That is, 12–13% ethanol kills the yeast cells that produce it. This points out a last generalization about fermentations. The stable fermentation end product, whether it is lactate or ethanol, eventually accumulates to a concentration that is toxic to the organism. Muscle fatigue is the early effect of lactate buildup in the working muscle. In the same way, continued accumulation of the fermentation product can lead to concentrations that are fatal if there is no means of getting rid of the toxic product or of getting away from it. For single-celled organisms, the result is generally death. Our bodies have evolved in such a way that lactate buildup contributes to muscle fatigue that causes the exerciser to stop the exercise. Then the lactate is removed from the blood and converted to glucose by the process of gluconeogenesis.

Another application of alcohol fermentation is the use of yeast in bread making. When we mix the water, sugar, and dried yeast, the yeast cells begin to grow and carry out the process of fermentation. This mixture is then added to the flour, milk, shortening, and salt, and the dough is placed in a warm place to rise. The yeast continues to grow and ferment the sugar, producing CO_2 that causes the bread to rise. Of course, when we bake the bread, the yeast cells are killed, and the ethanol evaporates, but we are left with a light and airy loaf of bread.

Today, alcohol produced by fermentation is being used as an alternative fuel to replace the use of some fossil fuels. Geneticists and bioengineers are trying to develop strains of yeast

The production of bread, wine, and cheese depends on fermentation processes.

that can survive higher alcohol concentrations and thus convert more of the sugar of corn and other grains into alcohol.

Bacteria perform a variety of other fermentations. The propionibacteria produce propionic acid and CO_2. The acid gives Swiss cheese its characteristic flavor, and the CO_2 gas produces the characteristic holes in the cheese. Other bacteria, the clostridia, perform a fermentation that is responsible in part for the horrible symptoms of gas gangrene. When these bacteria are inadvertently introduced into deep tissues by a puncture wound, they find a nice anaerobic environment in which to grow. In fact, these organisms are *obligate anaerobes;* that is, they are killed by even a small amount of oxygen. As they grow, they perform a fermentation called the *butyric acid, butanol, acetone fermentation.* This results in the formation of CO_2, the gas associated with gas gangrene. The CO_2 infiltrates the local tissues and helps to maintain an anaerobic environment because oxygen from the local blood supply cannot enter the area of the wound. Now able to grow well, these bacteria produce a variety of toxins and enzymes that cause extensive tissue death and necrosis. In addition, the fermentation produces acetic acid, ethanol, acetone, isopropanol, butanol, and butyric acid (which is responsible, along with the necrosis, for the characteristic foul smell of gas gangrene). Certainly, the presence of these organic chemicals in the wound enhances tissue death.

Gas gangrene is very difficult to treat. Because the bacteria establish an anaerobic region of cell death and cut off the local circulation, systemic antibiotics do not infiltrate the wound and kill the bacteria. Even our immune response is stymied. Treatment usually involves surgical removal of the necrotic tissue accompanied by antibiotic therapy. In some cases, a hyperbaric oxygen chamber is employed. The infected extremity is placed in an environment with a very high partial pressure of oxygen. The oxygen forced into the tissues is poisonous to the bacteria, and they die.

These are but a few examples of the fermentations that have an effect on humans. Regardless of the specific chemical reactions, all fermentations share the following traits:

- They use pyruvate produced in glycolysis.
- They reoxidize the NADH produced in glycolysis.
- They are self-limiting because the accumulated stable fermentation end product eventually kills the cell that produces it.

For Further Understanding

▸ Write condensed structural formulas for each of the fermentation products made by clostridia in gas gangrene. Identify the functional groups and provide the IUPAC name for each.

▸ Explain the importance of utilizing pyruvate and reoxidizing NADH to the ability of a cell to continue producing ATP.

21.5 The Pentose Phosphate Pathway

The **pentose phosphate pathway** is an alternative pathway for glucose oxidation. It provides the cell with energy in the form of NADPH, which is the reducing agent required for many biosynthetic pathways.

The details of the pentose phosphate pathway will not be covered in this text. But an overview of the key reactions will allow us to understand the importance of the pathway (Figure 21.8).

$$\text{glucose-6-phosphate} + 2\text{NADP}^+ + \text{H}_2\text{O} \longrightarrow$$

$$\text{ribulose-5-phosphate} + 2\text{NADPH} + \text{CO}_2$$

The pentose phosphate pathway provides several molecules that are important in biosynthesis. The first is reducing power in the form of NADPH. It also provides sugar phosphates that are required for biosynthesis. For instance, *ribose-5-phosphate* is used for the synthesis of nucleotides such as ATP. The four-carbon sugar phosphate, *erythrose-4-phosphate,* is produced in the third stage of the pentose phosphate pathway (not shown in Figure 21.8). It is a precursor of the amino acids phenylalanine, tyrosine, and tryptophan.

The pentose phosphate pathway is most active in tissues involved in cholesterol and fatty acid biosynthesis. These two processes require abundant NADPH. Thus the liver, which is the site of cholesterol synthesis and a major site for fatty acid biosynthesis, and adipose (fat) tissue, where active fatty acid synthesis also occurs, have very high levels of pentose phosphate pathway enzymes.

LEARNING GOAL

7 List several products of the pentose phosphate pathway that are required for biosynthesis.

The pathway for fatty acid biosynthesis is discussed in Section 23.4.

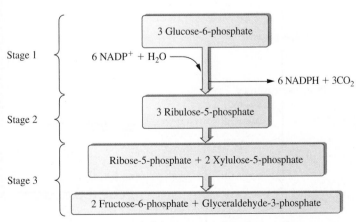

Figure 21.8 Summary of the major stages of the pentose phosphate pathway.

8 Compare glycolysis and gluconeogenesis.

Under extreme conditions of starvation the brain eventually switches to the use of ketone bodies. Ketone bodies are produced, under certain circumstances, from the breakdown of lipids (Section 23.3).

21.6 Gluconeogenesis: The Synthesis of Glucose

Under normal conditions, we have enough glucose to satisfy our needs. However, under some conditions the body must make glucose. This is necessary following strenuous exercise, to replenish the liver and muscle supplies of glycogen. It also occurs during starvation, so that the body can maintain adequate blood glucose levels to supply the brain cells and red blood cells. Under normal conditions, these two tissues use only glucose for energy.

Glucose is produced by the process of **gluconeogenesis** (*gleuko*, Greek *sweet*; *neo*, Latin *new*; *genesis*, Latin *produce*), the production of glucose from noncarbohydrate starting materials (Figure 21.9). Gluconeogenesis, an anabolic pathway, occurs primarily in the liver. Lactate, all the amino acids except leucine and lysine, and glycerol from fats can all be used to make glucose. However, the amino acids and glycerol are generally used only under starvation conditions.

At first glance, gluconeogenesis appears to be simply the reverse of glycolysis (compare Figures 21.9 and 21.5), because the intermediates of the two pathways are identical. But this is not the case, because steps 1, 3, and 10 of glycolysis

Figure 21.9 Comparison of the reactions of glycolysis and gluconeogenesis.

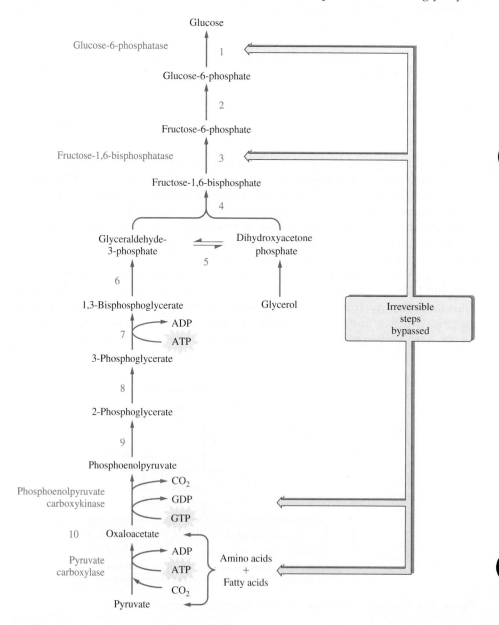

are irreversible, and therefore the reverse reactions must be carried out by other enzymes. In step 1 of glycolysis, hexokinase catalyzes the phosphorylation of glucose. In gluconeogenesis, the dephosphorylation of glucose-6-phosphate is carried out by the enzyme *glucose-6-phosphatase,* which is found in the liver but not in muscle. Similarly, reaction 3, the phosphorylation of fructose-6-phosphate catalyzed by phosphofructokinase, is irreversible. That step is bypassed in gluconeogenesis by using the enzyme *fructose-1,6-bisphosphatase.* Finally, the phosphorylation of ADP catalyzed by pyruvate kinase, step 10 of glycolysis, cannot be reversed. The conversion of pyruvate to phosphoenolpyruvate actually involves two enzymes and some unusual reactions. First, the enzyme *pyruvate carboxylase* adds CO_2 to pyruvate. The product is the four-carbon compound oxaloacetate. Then *phosphoenolpyruvate carboxykinase* removes the CO_2 and adds a phosphoryl group. The donor of the phosphoryl group in this unusual reaction is **guanosine triphosphate (GTP).** This is a nucleotide like ATP, except that the nitrogenous base is guanine.

If glycolysis and gluconeogenesis were not regulated in some fashion, the two pathways would occur simultaneously, with the disastrous effect that nothing would get done. Three convenient sites for this regulation are the three bypass reactions. Step 3 of glycolysis is catalyzed by the enzyme phosphofructokinase. This enzyme is stimulated by high concentrations of AMP, ADP, and inorganic phosphate, signals that the cell needs energy. When the enzyme is active, glycolysis proceeds. On the other hand, when ATP is plentiful, phosphofructokinase is inhibited, and fructose-1,6-bisphosphatase is stimulated. The net result is that in times of energy excess (high concentrations of ATP), gluconeogenesis will occur.

The conversion of lactate into glucose is important in mammals. As the muscles work, they produce lactate, which is converted back to glucose in the liver. The glucose is transported into the blood and from there back to the muscle. In the muscle, it can be catabolized to produce ATP, or it can be used to replenish the muscle stores of glycogen. This cyclic process between the liver and skeletal muscles is called the **Cori Cycle** and is shown in Figure 21.10. Through this cycle, gluconeogenesis produces enough glucose to restore the depleted muscle glycogen reservoir within 48 hours (h).

Question 21.11 What are the major differences between gluconeogenesis and glycolysis?

Question 21.12 What do the three irreversible reactions of glycolysis have in common?

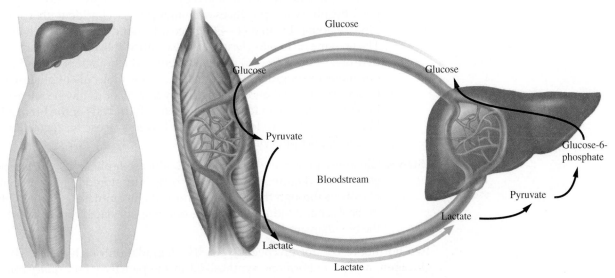

Figure 21.10 The Cori Cycle.

21.7 Glycogen Synthesis and Degradation

9 Summarize the regulation of blood glucose levels by glycogenesis and glycogenolysis.

Marathon runners often carbo-load in the days before a race. The goal is to build stores of muscle glycogen. Carbo-loading involves reduced exercise the week before the race, along with a diet that is as high as 70% carbohydrate. Explain how this helps build the runner's endurance.

Glucose is the sole source of energy of mammalian red blood cells and the major source of energy for the brain. Neither red blood cells nor the brain can store glucose; thus, a constant supply must be available as blood glucose. This is provided by dietary glucose and by the production of glucose either by gluconeogenesis or by **glycogenolysis,** the degradation of glycogen. Glycogen is a long, branched-chain polymer of glucose. Stored in the liver and skeletal muscles, it is the principal storage form of glucose.

The total amount of glucose in the blood of a 70-kilogram (kg) (approximately 150-pound) adult is about 20 grams (g), but the brain alone consumes 5–6 g of glucose per h. Breakdown of glycogen in the liver mobilizes the glucose when hormonal signals register a need for increased levels of blood glucose. Skeletal muscle also contains substantial stores of glycogen, which provide energy for rapid muscle contraction. However, this glycogen is not able to contribute to blood glucose.

The Structure of Glycogen

Glycogen is a highly branched glucose polymer in which the "main chain" is linked by $\alpha\,(1 \rightarrow 4)$ glycosidic bonds. The polymer also has numerous $\alpha\,(1 \rightarrow 6)$ glycosidic bonds, which provide many branch points along the chain (Figure 21.11). **Glycogen granules** with a diameter of 10–40 nanometers (nm) are found in the cytoplasm of liver and muscle cells. These granules exist in complexes with the enzymes that are responsible for glycogen synthesis and degradation.

Glycogenolysis: Glycogen Degradation

The steps in glycogenolysis, or glycogen degradation, are summarized as follows.

Step 1. The enzyme glycogen phosphorylase catalyzes *phosphorolysis* of a glucose at one end of a glycogen polymer (Figure 21.11). The reaction involves the displacement of a glucose unit of glycogen by a phosphate group. As a result of phosphorolysis, glucose-1-phosphate is produced without using ATP as the phosphoryl group donor.

Step 2. Glycogen contains many branches bound to the $\alpha\,(1 \rightarrow 4)$ backbone by $\alpha\,(1 \rightarrow 6)$ glycosidic bonds. These branches must be removed to allow the complete degradation of glycogen. The extensive action of glycogen phosphorylase produces a smaller polysaccharide with a single glucose bound by an $\alpha\,(1 \rightarrow 6)$ glycosidic bond to the main chain. The enzyme $\alpha\,(1 \rightarrow 6)$ *glycosidase,* also called the *debranching enzyme,* hydrolyzes the $\alpha\,(1 \rightarrow 6)$ glycosidic bond at a branch point and frees one molecule of glucose (Figure 21.12). This molecule of glucose can be phosphorylated and utilized in glycolysis, or it may be released into the bloodstream for use elsewhere. Hydrolysis of the branch bond liberates another stretch of $\alpha\,(1 \rightarrow 4)$-linked glucose for the action of glycogen phosphorylase.

Step 3. Glucose-1-phosphate is converted to glucose-6-phosphate by *phosphoglucomutase* (Figure 21.13). Glucose originally stored in glycogen enters glycolysis through the action of phosphoglucomutase. Alternatively, in the liver and kidneys it may be dephosphorylated for transport into the bloodstream.

Two hormones control glycogenolysis, the degradation of glycogen. These are **glucagon,** a peptide hormone synthesized in the pancreas, and *epinephrine,* produced in the adrenal glands. Glucagon is released from the pancreas in response

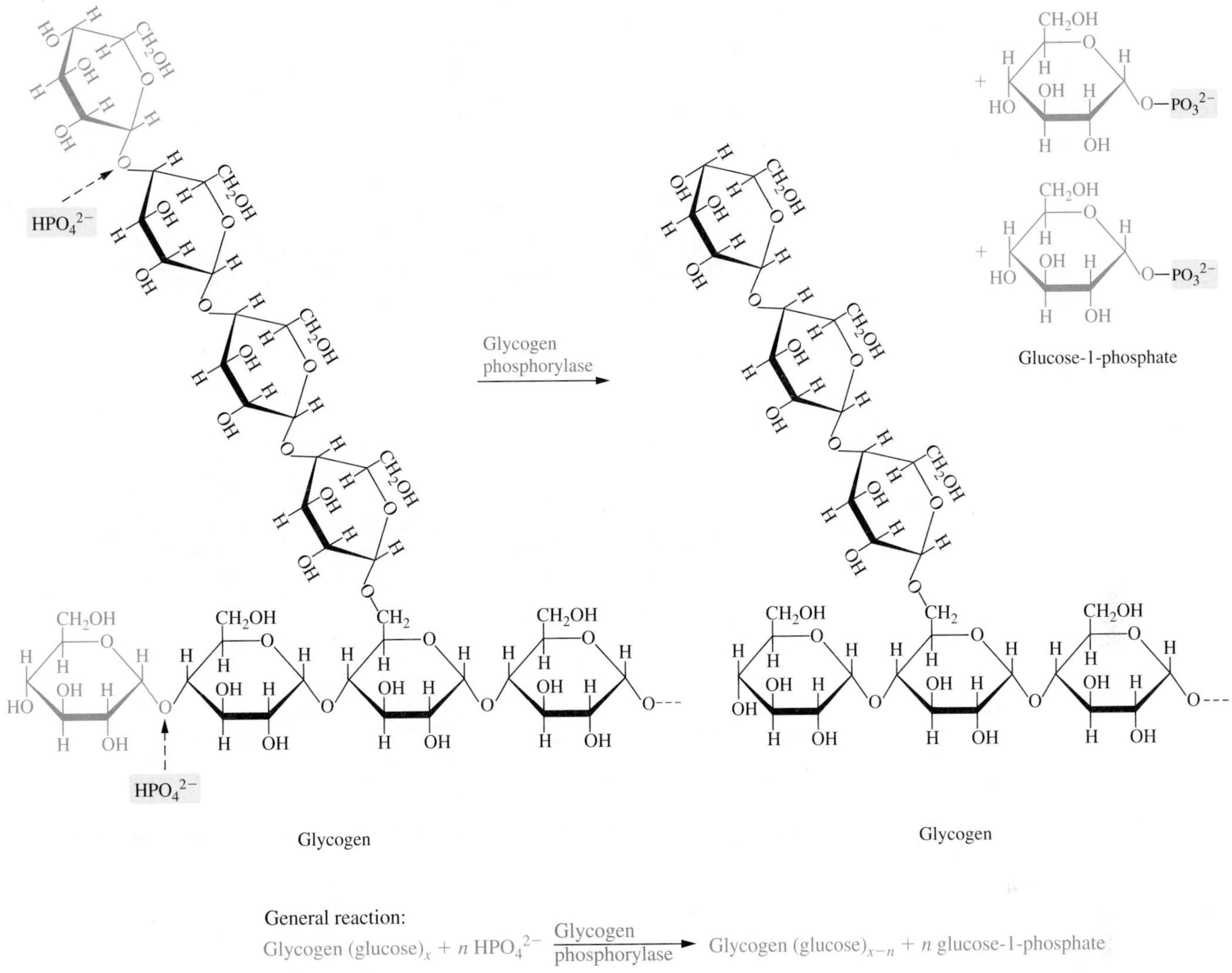

General reaction:

$$\text{Glycogen (glucose)}_x + n\ \text{HPO}_4^{2-} \xrightarrow[\text{phosphorylase}]{\text{Glycogen}} \text{Glycogen (glucose)}_{x-n} + n\ \text{glucose-1-phosphate}$$

Figure 21.11 The action of glycogen phosphorylase in glycogenolysis.

to low blood glucose, and epinephrine is released from the adrenal glands in response to a threat or a stress. Both situations require an increase in blood glucose, and both hormones function by altering the activity of two enzymes, glycogen phosphorylase and glycogen synthase. *Glycogen phosphorylase* is involved in glycogen degradation and is activated; *glycogen synthase* is involved in glycogen synthesis and is inactivated.

Question 21.13 Explain the role of glycogen phosphorylase in glycogenolysis.

Question 21.14 How does the action of glycogen phosphorylase and phosphoglucomutase result in an energy savings for the cell if the product, glucose-6-phosphate, is used directly in glycolysis?

Glycogenesis: Glycogen Synthesis

The hormone **insulin,** produced by the pancreas in response to high blood glucose levels, stimulates the synthesis of glycogen, **glycogenesis.** Insulin is perhaps one of the most influential hormones in the body because it directly alters the metabolism and uptake of glucose in all but a few cells.

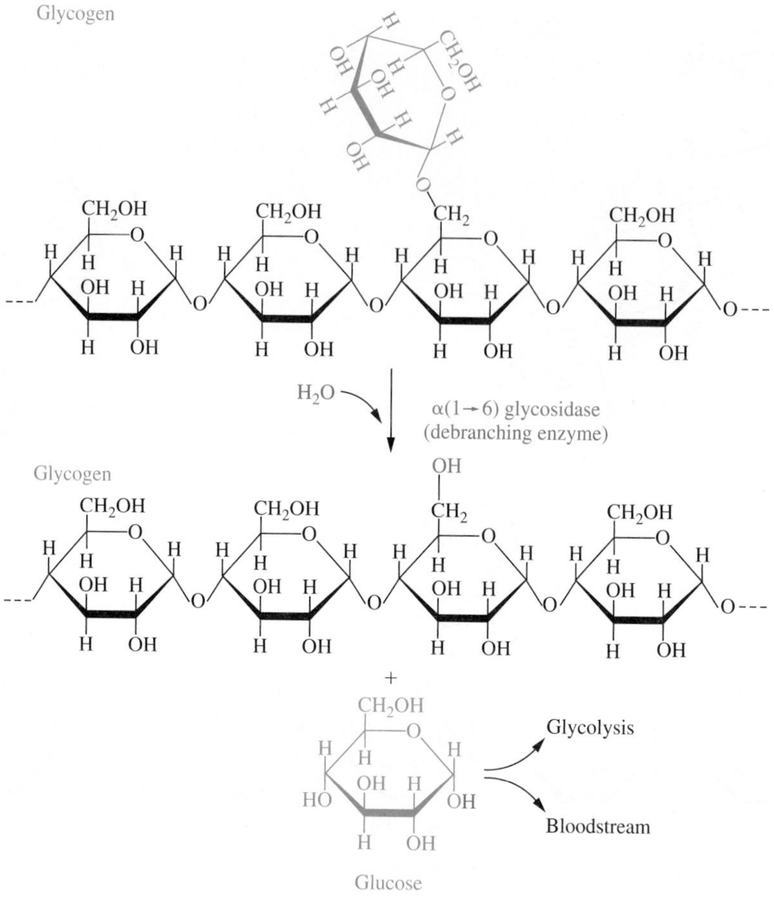

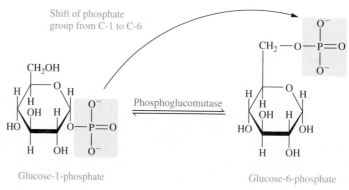

Figure 21.12 The action of α(1→6) glycosidase (debranching enzyme) in glycogen degradation.

Figure 21.13 The action of phosphoglucomutase in glycogen degradation.

When blood glucose rises, as after a meal, the beta cells of the pancreas secrete insulin. It immediately accelerates the uptake of glucose by all the cells of the body except the brain and certain blood cells. In these cells, the uptake of glucose is insulin-independent. The increased uptake of glucose is especially marked in the liver, heart, skeletal muscle, and adipose tissue.

In the liver, insulin promotes glycogen synthesis and storage by inhibiting glycogen phosphorylase, thus inhibiting glycogen degradation. It also stimulates glycogen synthase and glucokinase, two enzymes that are involved in glycogen synthesis.

Although glycogenesis and glycogenolysis share some reactions in common, the two pathways are not simply the reverse of one another. Glycogenesis involves some very unusual reactions, which we will now examine in detail.

The first reaction of glycogen synthesis in the liver traps glucose within the cell by phosphorylating it. In this reaction, catalyzed by the enzyme *glucokinase*, ATP serves as a phosphoryl donor, and glucose-6-phosphate is formed:

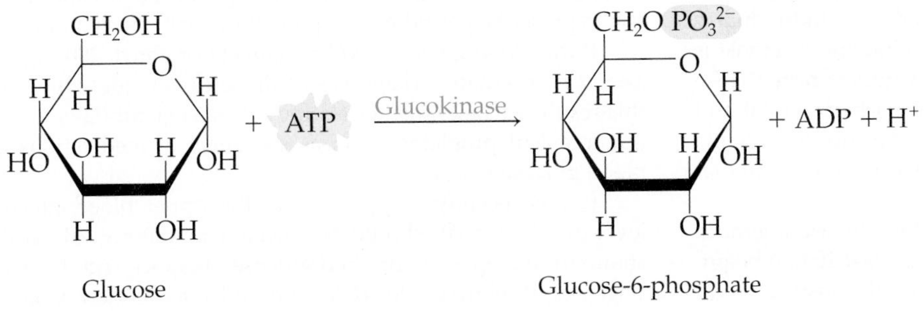

Glucose   Glucose-6-phosphate

The second reaction of glycogenesis is the reverse of one of the reactions of glycogenolysis. The glucose-6-phosphate formed in the first step is isomerized to glucose-1-phosphate. The enzyme that catalyzes this step is phosphoglucomutase:

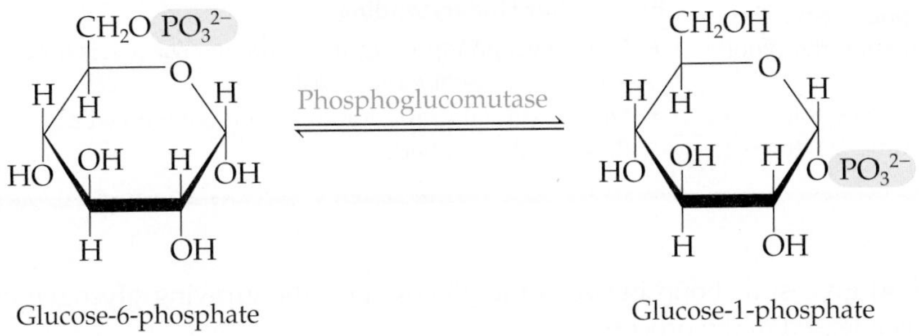

Glucose-6-phosphate   Glucose-1-phosphate

The glucose-1-phosphate must now be activated before it can be added to the growing glycogen chain. The high-energy compound that accomplishes this is the nucleotide **uridine triphosphate (UTP).** In this reaction, mediated by the enzyme *pyrophosphorylase*, the C-1 phosphoryl group of glucose is linked to the α-phosphoryl group of UTP to produce UDP-glucose:

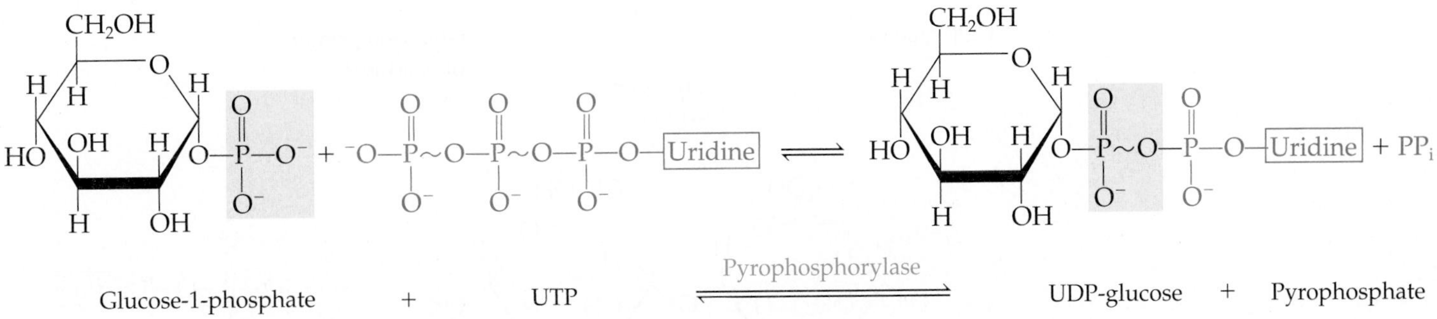

Glucose-1-phosphate   +   UTP   UDP-glucose   +   Pyrophosphate

This is accompanied by the release of a pyrophosphate group (PP_i). The structure of UDP-glucose is shown in Figure 21.14.

The UDP-glucose can now be used to extend glycogen chains. The enzyme glycogen synthase breaks the phosphoester linkage of UDP-glucose and forms an

A Medical Perspective

Diagnosing Diabetes

When diagnosing diabetes, doctors take many factors and symptoms into consideration. However, there are two primary tests that are performed to determine whether an individual is properly regulating blood glucose levels. First and foremost is the fasting blood glucose test. A person who has fasted since midnight should have a blood glucose level between 70 and 110 milligrams per deciliter (mg/dL) in the morning. If the level is 140 mg/dL on at least two occasions, a diagnosis of diabetes is generally made.

The second commonly used test is the glucose tolerance test. For this test, the subject must fast for at least 10 h. A beginning blood sample is drawn to determine the fasting blood glucose level. This will serve as the background level for the test. The subject ingests 50–100 g of glucose (40 g/m² body surface), and the blood glucose level is measured at 30 minutes (min), and at 1, 2, and 3 h after ingesting the glucose.

A graph is made of the blood glucose levels over time. For a person who does not have diabetes, the curve will show a peak of blood glucose at approximately 1 h. There will be a reduction in the level, and perhaps a slight hypoglycemia (low blood glucose level) over the next hour. Thereafter, the blood glucose level stabilizes at normal levels.

An individual is said to have impaired glucose tolerance if the blood glucose level remains between 140 and 200 mg/dL

2 h after ingestion of the glucose solution. This suggests that there is a risk of the individual developing diabetes and a reason to prescribe periodic testing to allow early intervention.

If the blood glucose level remains at or above 200 mg/dL after 2 h, a tentative diagnosis of diabetes is made. However, this result warrants further testing on subsequent days to rule out transient problems, such as the effect of medications on blood glucose levels.

It was recently suggested that the upper blood glucose level of 200 mg/dL should be lowered to 180 mg/dL as the standard to diagnose impaired glucose tolerance and diabetes. This would allow earlier detection and intervention. Considering the grave nature of long-term diabetic complications, it is thought to be very beneficial to begin treatment at an early stage to maintain constant blood glucose levels. For more information on diabetes, see A Medical Perspective: Diabetes Mellitus and Ketone Bodies, in Chapter 23.

For Further Understanding

▶ Draw a graph representing blood glucose levels for a normal glucose tolerance test.

▶ Draw a similar graph for an individual who would be diagnosed as diabetic.

$\alpha(1 \rightarrow 4)$ glycosidic bond between the glucose and the growing glycogen chain. UDP is released in the process.

UDP-glucose   +   Glycogen primer (*n* residues)

Glycogen synthase

Glycogen (*n* + 1 residues)   +   UDP

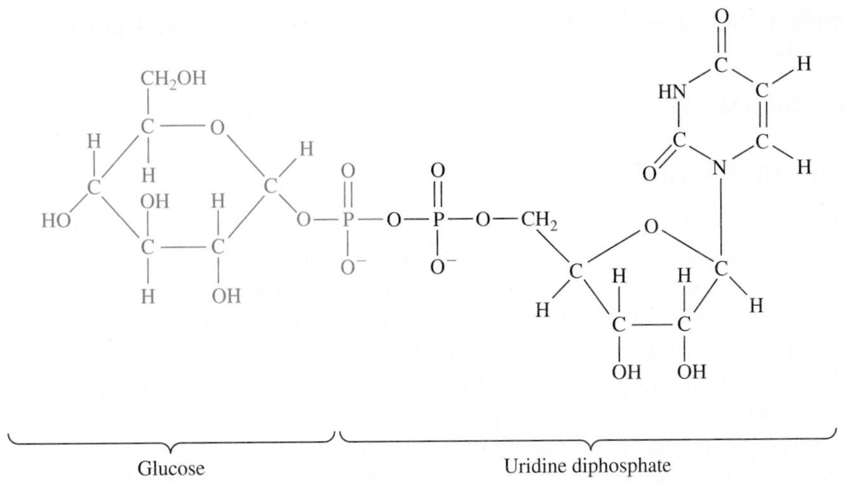

Figure 21.14 The structure of UDP-glucose.

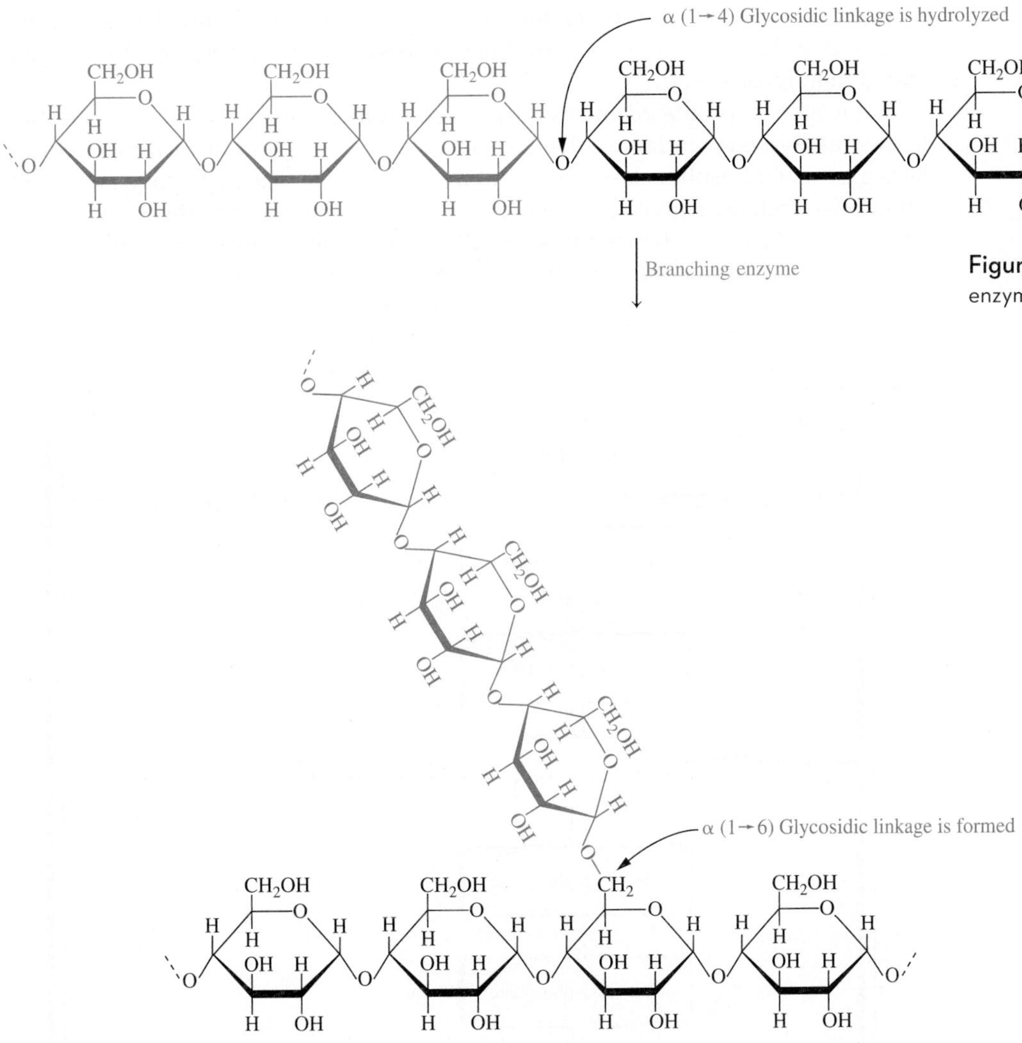

Figure 21.15 The action of the branching enzyme in glycogen synthesis.

Finally, we must introduce the $\alpha(1 \rightarrow 6)$ glycosidic linkages to form the branches. The branches are quite important to proper glycogen utilization. As Figure 21.15 shows, the *branching enzyme* removes a section of the linear $\alpha(1 \rightarrow 4)$ linked glycogen and reattaches it in $\alpha(1 \rightarrow 6)$ glycosidic linkage elsewhere in the chain.

Question 21.15   Describe the way in which glucokinase traps glucose inside liver cells.

Question 21.16   Describe the reaction catalyzed by the branching enzyme.

Compatibility of Glycogenesis and Glycogenolysis

As was the case with glycolysis and gluconeogenesis, it would be futile for the cell to carry out glycogen synthesis and degradation simultaneously. The results achieved by the action of one pathway would be undone by the other. This problem is avoided by a series of hormonal controls that activate the enzymes of one pathway while inactivating the enzymes of the other pathway.

When the blood glucose level is too high, a condition known as **hyperglycemia,** insulin stimulates the uptake of glucose via a transport mechanism. It further stimulates the trapping of the glucose by the elevated activity of glucokinase. Finally, it activates glycogen synthase, the last enzyme in the synthesis of glycogen chains. To further accelerate storage, insulin *inhibits* the first enzyme in glycogen degradation, glycogen phosphorylase. The net effect, seen in Figure 21.16, is that glucose is removed from the bloodstream and converted into glycogen in the liver. When the glycogen stores are filled, excess glucose is converted to fat and stored in adipose tissue.

Glucagon is produced in response to low blood glucose levels, a condition known as **hypoglycemia,** and has an effect opposite to that of insulin. It stimulates glycogen phosphorylase, which catalyzes the first stage of glycogen degradation. This accelerates glycogenolysis and release of glucose into the bloodstream. The effect is further enhanced because glucagon inhibits glycogen synthase. The opposing effects of insulin and glucagon are summarized in Figure 21.16.

Figure 21.16 The opposing effects of the hormones insulin and glucagon on glycogen metabolism.

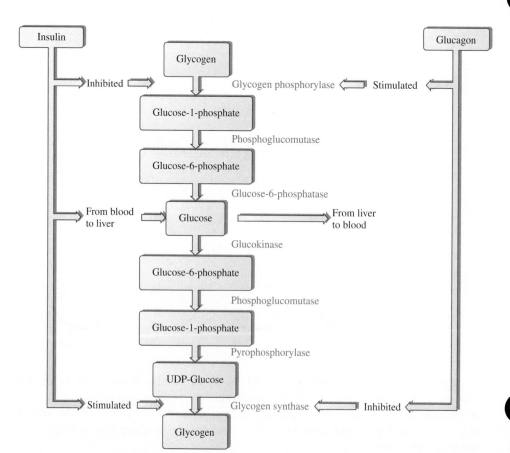

Glycogen Storage Diseases

Glycogen metabolism is important for the proper function of many aspects of cellular metabolism. Many diseases of glycogen metabolism have been discovered. Generally, these are diseases that result in the excessive accumulation of glycogen in the liver, muscle, and tubules of the kidneys. Often they are caused by defects in one of the enzymes involved in the degradation of glycogen.

One example is an inherited defect of glycogen metabolism known as *von Gierke's disease*. This disease results from a defective gene for glucose-6-phosphatase, which catalyzes the final step of gluconeogenesis and glycogenolysis. People who lack glucose-6-phosphatase cannot convert glucose-6-phosphate to glucose. As we have seen, the liver is the primary source of blood glucose, and much of this glucose is produced by gluconeogenesis. Glucose-6-phosphate, unlike glucose, cannot cross the cell membrane, and the liver of a person suffering from von Gierke's disease cannot provide him or her with glucose. The blood sugar level falls precipitously low between meals. In addition, the lack of glucose-6-phosphatase also affects glycogen metabolism. Because glucose-6-phosphatase is absent, the supply of glucose-6-phosphate in the liver is large. This glucose-6-phosphate can also be converted to glycogen. A person suffering from von Gierke's disease has a massively enlarged liver as a result of enormously increased stores of glycogen.

Defects in other enzymes of glycogen metabolism also exist. *Cori's disease* is caused by a genetic defect in the debranching enzyme. As a result, individuals who have this disease cannot completely degrade glycogen and thus use their glycogen stores very inefficiently.

On the other side of the coin, *Andersen's disease* results from a genetic defect in the branching enzyme. Individuals who have this disease produce very long, unbranched glycogen chains. This genetic disorder results in decreased efficiency of glycogen storage.

A final example of a glycogen storage disease is *McArdle's disease*. In this syndrome, the muscle cells lack the enzyme glycogen phosphorylase and cannot degrade glycogen to glucose. Individuals who have this disease have little tolerance for physical exercise because their muscles cannot provide enough glucose for the necessary energy-harvesting processes. It is interesting to note that the liver enzyme glycogen phosphorylase is perfectly normal, and these people respond appropriately with a rise of blood glucose levels under the influence of glucagon or epinephrine.

For Further Understanding

▶ Write equations showing the reactions catalyzed by the enzymes that are defective in each of the genetic disorders described in this perspective.

▶ There are different forms of glycogen phosphorylase, one found in the liver and the other in skeletal muscle. Discuss the differences you would expect between a defect in the muscle enzyme and a defect in the liver enzyme.

This elegant system of hormonal control ensures that the reactions involved in glycogen degradation and synthesis do not compete with one another. In this way, they provide glucose when the blood level is too low, and they cause the storage of glucose in times of excess.

Question 21.17 Explain how glucagon affects the synthesis and degradation of glycogen.

Question 21.18 How does insulin affect the storage and degradation of glycogen?

CHAPTER MAP

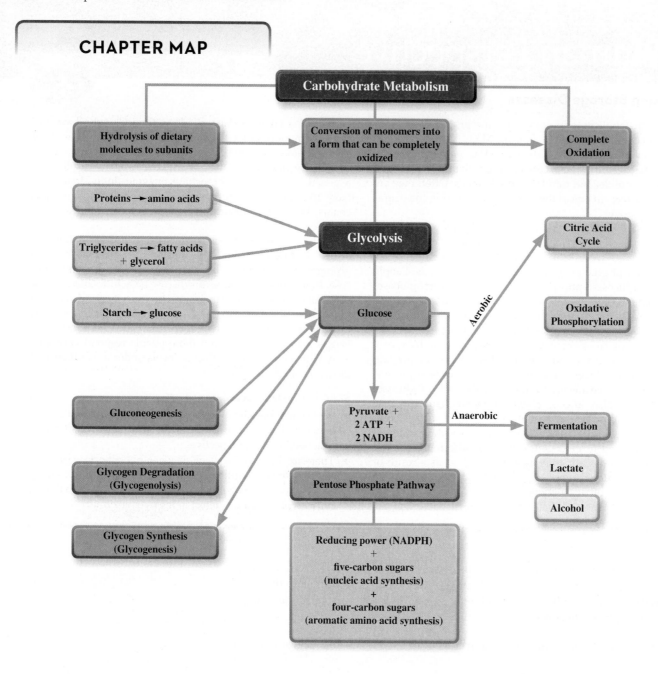

SUMMARY

21.1 ATP: The Cellular Energy Currency

▶ **Adenosine triphosphate (ATP)** is a **nucleotide** composed of adenine, the sugar ribose, and a triphosphate group.

▶ The energy released by the hydrolysis of the phosphoanhydride bond between the second and third phosphoryl groups provides the energy for most cellular work.

▶ The energy harvested during the degradation of fuel molecules, **catabolism,** is stored in ATP.

▶ Biosynthetic reactions, **anabolism,** utilize ATP as an energy source.

21.2 Overview of Catabolic Processes

▶ The body needs a supply of ATP to carry out life processes.

▶ To provide this ATP, we consume a variety of energy-rich food molecules: carbohydrates, lipids, and proteins.

▶ In digestion, these food molecules are degraded into smaller molecules that are absorbed by our cells: monosaccharides, fatty acids and glycerol, and amino acids.

▶ Through catabolic reactions, these molecules are used to produce ATP.

21.3 Glycolysis

▶ **Glycolysis** is the pathway for glucose catabolism that leads to pyruvate. Glycolysis:
  • is anaerobic
  • occurs in the cytoplasm of the cell

- produces a net harvest of two ATP and two NADH [formed from the coenzyme nicotinamide adenine dinucleotide (NAD^+)]
▶ The ATP is produced by **substrate-level phosphorylation.**
▶ Under aerobic conditions, the electrons carried by NADH are used to produce ATP by **oxidative phosphorylation.**
▶ The rate of glycolysis responds to the energy demands of the cell.
  - This regulation of the rate of glycolysis occurs through the allosteric enzymes hexokinase, phosphofructokinase, and pyruvate kinase.

21.4 Fermentations

▶ **Fermentations** are catabolic reactions that occur with no net oxidation.
▶ Under anaerobic conditions, the NADH produced by glycolysis is used to reduce pyruvate.
▶ The lactate fermentation reduces pyruvate to produce lactate.
  - This occurs in working muscles when there is not enough oxygen to provide ATP through aerobic catabolic reactions.
  - If lactate builds up to a high enough concentration, it can inhibit glycolysis and thus ATP production. At this point, the **anaerobic threshold,** there is not enough energy for muscles to continue to function.
▶ The alcohol fermentation reduces pyruvate to ethanol.
  - The alcohol fermentation is used to produce wines and other alcoholic beverages.
  - The alcohol fermentation is also used in bread making.

21.5 The Pentose Phosphate Pathway

▶ The **pentose phosphate pathway** is an alternative pathway for glucose degradation and is particularly abundant in the liver and adipose tissue.
▶ Products of the pathway include:
  - NADPH, a reducing agent for biosynthetic reactions.
  - Ribose-5-phosphate for nucleotide synthesis.
  - Erythrose-4-phosphate for biosynthesis of several amino acids (tryptophan, tyrosine, and phenylalanine).

21.6 Gluconeogenesis: The Synthesis of Glucose

▶ Gluconeogenesis is the pathway for glucose synthesis from noncarbohydrate starting materials.
  - It occurs in the mammalian liver.
  - Starting materials include lactate and all amino acids except lysine and leucine, and glycerol.
▶ Gluconeogenesis is not simply the reverse of glycolysis.
  - Three steps in glycolysis in which ATP is produced or consumed are bypassed in gluconeogenesis by using other enzymes.
  - All the other enzymes in the two pathways are the same.
  - **Guanosine triphosphate (GTP)** is a phosphoryl group donor in a reaction that converts oxaloacetate into phosphoenolpyruvate. The oxaloacetate is produced by the carboxylation of pyruvate.

▶ The **Cori Cycle** is a metabolic pathway in which lactate produced by working muscle is taken up by the liver and converted to glucose by gluconeogenesis.

21.7 Glycogen Synthesis and Degradation

▶ **Glycogen** is a long, branched-chain polymer of glucose.
  - It is stored in **glycogen granules** that consist of glycogen and the enzymes required for glycogen synthesis and degradation.
  - It is found in the liver and muscle.
▶ **Glycogenolysis** is the pathway by which the glycogen polymer is broken down into individual monomers (glucose molecules).
▶ **Glycogenesis** is the metabolic pathway by which the polymer glycogen is synthesized from glucose molecules.
  - Uridine triphosphate (UTP) is a ribonucleotide involved in glycogenesis.
▶ Glycogen synthesis and degradation are under hormonal control.
  - **Insulin** inhibits glycogen degradation and stimulates glycogen synthesis when blood glucose levels are too high **(hyperglycemia).**
  - **Glucagon** inhibits glycogen synthesis and stimulates glycogen degradation when blood glucose levels are too low **(hypoglycemia).**

QUESTIONS AND PROBLEMS

ATP: The Cellular Energy Currency

Foundations

21.19 What molecule is primarily responsible for conserving the energy released in catabolism?

21.20 Describe the structure of ATP.

Applications

21.21 Write a reaction showing the hydrolysis of the terminal phosphoanhydride bond of ATP.

21.22 What is meant by the term *high-energy bond*?

21.23 What is meant by a coupled reaction?

21.24 Compare and contrast anabolism and catabolism in terms of their roles in metabolism and their relationship to ATP.

Overview of Catabolic Processes

Foundations

21.25 What is the most readily used energy source in the diet?

21.26 What is a hydrolysis reaction?

Applications

21.27 Write an equation showing the hydrolysis of maltose.

21.28 Write an equation showing the hydrolysis of sucrose.

21.29 Write a balanced equation showing the hydrolysis of the following peptide: phe-ala-glu-met-lys.

21.30 Describe the stages of protein digestion, including the location of each.

21.31 Write an equation showing the hydrolysis of a triglyceride consisting of glycerol, oleic acid, linoleic acid, and stearic acid.

21.32 How are fatty acids taken up into the cell?

21.33 Write an equation showing the hydrolysis of the dipeptide alanyl-leucine.

21.34 How are amino acids transported into the cell?

Glycolysis

Foundations

21.35 Define glycolysis and describe its role in cellular metabolism.

21.36 What are the end products of glycolysis?

21.37 Why does glycolysis require a supply of NAD^+ to function?

21.38 Why must the NADH produced in glycolysis be reoxidized to NAD^+?

21.39 What is the net energy yield of ATP in glycolysis?

21.40 How many molecules of ATP are produced by substrate-level phosphorylation during glycolysis?

21.41 Explain how muscle is able to carry out rapid contraction for prolonged periods even though its supply of ATP is sufficient only for a fraction of a second of rapid contraction.

21.42 Where in the muscle cell does glycolysis occur?

21.43 Write the balanced chemical equation for glycolysis.

21.44 Write a chemical equation for the transfer of a phosphoryl group from ATP to fructose-6-phosphate.

21.45 Match each of the following enzymes with the reaction that it catalyzes.
  a. Phosphoglucose isomerase
  b. Phosphofructokinase
  c. Triose phosphate isomerase
  d. Aldolase
  e. Hexokinase
  f. Enolase
  g. Glyceraldehyde-3-phosphate dehydrogenase
  h. Phosphoglycerate kinase
  i. Pyruvate kinase
  j. Phosphoglycerate mutase

  1. Phosphorylation of glucose
  2. Phosphorylation of fructose-6-phosphate
  3. Dephosphorylation of pyruvate
  4. Conversion of fructose-1,6-bisphosphate to dihydroxyacetone phosphate and glyceraldehyde-3-phosphate
  5. Phosphorylation and oxidation of glyceraldehyde-3-phosphate to produce 1,3-bisphosphoglycerate and NADH
  6. Conversion of dihydroxyacetone phosphate into glyceraldehyde-3-phosphate
  7. Isomerization of glucose-6-phosphate into fructose-6-phosphate
  8. Isomerization of 3-phosphoglycerate into 2-phosphoglycerate
  9. Dehydration of 2-phosphoglycerate to produce phosphoenolpyruvate
  10. Substrate-level phosphorylation involving transfer of a phosphoryl group from 1,3-bisphosphoglycerate to ADP

21.46 Match each of the following enzymes with the appropriate class of enzymes that it represents. (*Hint:* An enzyme classification may be used more than once or not at all.)
  a. Phosphoglucose isomerase
  b. Phosphofructokinase
  c. Triose phosphate isomerase
  d. Aldolase
  e. Hexokinase
  f. Enolase
  g. Glyceraldehyde-3-phosphate dehydrogenase
  h. Phosphoglycerate kinase
  i. Pyruvate kinase
  j. Phosphoglycerate mutase

  1. Transferase
  2. Oxidoreductase
  3. Kinase
  4. Hydrolase
  5. Lyase
  6. Isomerase

Applications

21.47 Describe the symptoms associated with a genetic deficiency of an enzyme in the glycolysis pathway.

21.48 Why are red blood cells particularly susceptible to a deficiency of an enzyme in the glycolysis pathway?

21.49 What is the cause of myoglobinuria?

21.50 Describe exercise intolerance and the cause of the condition.

Examine the following pair of equations and use them to answer Questions 21.51–21.54.

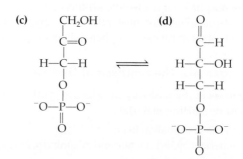

21.51 What type of enzyme would catalyze each of these reactions?

21.52 To which family of organic molecules do a and d belong? To which family of organic molecules do b and c belong?

21.53 What is the name of the type of intermediate formed in each of these reactions?

21.54 Draw the intermediate that would be formed in each of these reactions.

21.55 When an enzyme has the term *kinase* in the name, what type of reaction do you expect it to catalyze?

21.56 What features do the reactions catalyzed by hexokinase and phosphofructokinase share in common?

21.57 What is the role of NAD^+ in a biochemical oxidation reaction?

21.58 Write the equation for the reaction catalyzed by glyceraldehyde-3-phosphate dehydrogenase. Highlight the chemical changes that show this to be an oxidation reaction.

21.59 What is the importance of the regulation of glycolysis?

21.60 Explain the role of allosteric enzymes in control of glycolysis.

21.61 What molecules serve as allosteric effectors of phosphofructokinase?

21.62 What molecule serves as an allosteric inhibitor of hexokinase?

21.63 Explain the role of citrate in the feedback inhibition of glycolysis.

21.64 Explain the feedforward activation mechanism that results in the activation of pyruvate kinase.

Fermentations

Foundations

21.65 Write a balanced chemical equation for the conversion of acetaldehyde to ethanol.

21.66 Write a balanced chemical equation for the conversion of pyruvate to lactate.

Applications

21.67 After running a 100-meter (m) dash, a sprinter had a high concentration of muscle lactate. What process is responsible for production of lactate?

21.68 If the muscle of an organism had no lactate dehydrogenase, could anaerobic glycolysis occur in those muscle cells? Explain your answer.

21.69 What food products are the result of lactate fermentation?

21.70 Explain the value of alcohol fermentation in bread making.

21.71 What enzyme catalyzes the reduction of pyruvate to lactate?

21.72 What enzymes catalyze the conversion of pyruvate to ethanol and carbon dioxide?

21.73 A child was brought to the doctor's office suffering from a strange set of symptoms. When the child exercised hard, she became giddy and behaved as though drunk. What do you think is the metabolic basis of these symptoms?

21.74 A family started a batch of wine by adding yeast to grape juice and placing the mixture in a sealed bottle. Two weeks later, the bottle exploded. What metabolic reactions—and specifically, what product of those reactions—caused the bottle to explode?

The Pentose Phosphate Pathway

21.75 Of what value are the ribose-5-phosphate and erythrose-4-phosphate that are produced in the pentose phosphate pathway?

21.76 Of what value is the NADPH that is produced in the pentose phosphate pathway?

Gluconeogenesis: The Synthesis of Glucose

21.77 Define gluconeogenesis and describe its role in metabolism.

21.78 What is the role of guanosine triphosphate in gluconeogenesis?

21.79 What organ is primarily responsible for gluconeogenesis?

21.80 What is the physiological function of gluconeogenesis?

21.81 Lactate can be converted to glucose by gluconeogenesis. To what metabolic intermediate must lactate be converted so that it can be a substrate for the enzymes of gluconeogenesis?

21.82 L-Alanine can be converted to pyruvate. Can L-alanine also be converted to glucose? Explain your answer.

21.83 Explain why gluconeogenesis is not simply the reversal of glycolysis.

21.84 In step 10 of glycolysis, phosphoenolpyruvate is converted to pyruvate, and ATP is produced by substrate-level phosphorylation. How is this reaction bypassed in gluconeogenesis?

21.85 Which steps in the glycolysis pathway are irreversible?

21.86 What enzymatic reactions of gluconeogenesis bypass the irreversible steps of glycolysis?

Glycogen Synthesis and Degradation

Foundations

21.87 What organs are primarily responsible for maintaining the proper blood glucose level?

21.88 Why must the blood glucose level be carefully regulated?

21.89 What does the term *hypoglycemia* mean?

21.90 What does the term *hyperglycemia* mean?

Applications

21.91 a. What enzymes involved in glycogen metabolism are stimulated by insulin?
b. What effect does this have on glycogen metabolism?
c. What effect does this have on blood glucose levels?

21.92 a. What enzyme in glycogen metabolism is stimulated by glucagon?
b. What effect does this have on glycogen metabolism?
c. What effect does this have on blood glucose levels?

21.93 Explain how a defect in glycogen metabolism can cause hypoglycemia.

21.94 What defects of glycogen metabolism would lead to a large increase in the concentration of liver glycogen?

21.95 Write a "word" equation showing the reaction catalyzed by glycogen phosphorylase.

21.96 Describe the function of the debranching enzyme in glycogen degradation.

21.97 Write a balanced equation for the reaction catalyzed by phosphoglucomutase. What is the role of this enzyme in glycogen degradation? What is the role of this enzyme in glycogen synthesis?

21.98 Draw the structure of UDP-glucose and describe its role in glycogen synthesis.

21.99 Write a balanced equation for the reaction catalyzed by glucokinase. What is the function of this enzyme in glycogen synthesis?

21.100 Write a "word" equation showing the reaction catalyzed by glycogen synthase.

CHALLENGE PROBLEMS

1. An enzyme that hydrolyzes ATP (an ATPase) bound to the plasma membrane of certain tumor cells has an abnormally high activity. How will this activity affect the rate of glycolysis?

2. Explain why no net oxidation occurs during anaerobic glycolysis followed by lactate fermentation.

3. A certain person was found to have a defect in glycogen metabolism. The liver of this person could (a) make glucose-6-phosphate from lactate and (b) synthesize glucose-6-phosphate from glycogen but (c) could not synthesize glycogen from glucose-6-phosphate. What enzyme is defective?

4. A scientist added phosphate labeled with radioactive phosphorus (^{32}P) to a bacterial culture growing anaerobically (without O_2). She then purified all the compounds produced during glycolysis. Look carefully at the steps of the pathway. Predict which of the intermediates of the pathway would be the first one to contain radioactive phosphate. On which carbon of this compound would you expect to find the radioactive phosphate?

5. A 2-month-old baby was brought to the hospital suffering from seizures. He deteriorated progressively over time, showing psychomotor retardation. Blood tests revealed a high concentration of lactate and pyruvate. Although blood levels of alanine were high, they did not stimulate gluconeogenesis. The doctor measured the activity of pyruvate carboxylase in the baby and found it to be only 1% of the normal level. What reaction is catalyzed by pyruvate carboxylase? How could this deficiency cause the baby's symptoms and test results?

Aerobic Respiration and Energy Production

LEARNING GOALS

1 Name the regions of the mitochondria and the function of each region.

2 Describe the reaction that results in the conversion of pyruvate to acetyl CoA, describing the location of the reaction and the components of the pyruvate dehydrogenase complex.

3 Summarize the reactions of aerobic respiration.

4 Looking at an equation representing any of the chemical reactions that occur in the citric acid cycle, describe the kind of reaction that is occurring and the significance of that reaction to the pathway.

5 Explain the mechanisms for the control of the citric acid cycle.

6 Describe the process of oxidative phosphorylation.

7 Describe the conversion of amino acids to molecules that can enter the citric acid cycle.

8 Explain the importance of the urea cycle, and describe its essential steps.

9 Discuss the cause and effect of hyperammonemia.

10 Summarize the role of the citric acid cycle in catabolism and anabolism.

Rock climbing demands a great deal of energy.

OUTLINE

INTRODUCTION

In this chapter, we will be studying the amazing, intricate set of reactions that allow us to completely degrade fuel molecules such as sugars and amino acids. These oxygen-requiring reactions, called *aerobic respiration*, occur in cellular organelles called *mitochondria*.

We are used to thinking of the organelles as a collection of membrane-bound structures that are synthesized under the direction of the genetic information in the nucleus of the cell. Not so with the mitochondria. These organelles have their own genetic information and are able to make some of their own proteins. They grow and multiply in a way very similar to simple bacteria. This, along with other information on the structure and activities of mitochondria, has led researchers to conclude that the mitochondria are actually the descendants of bacteria captured by eukaryotic cells millions of years ago.

We inherit all of our mitochondria from our mothers and, like the mitochondria themselves, some genetic diseases of energy metabolism are maternally inherited. One such disease, Leber's hereditary optic neuropathy (LHON), causes blindness and heart problems. People with LHON have a reduced ability to make ATP. As a result, sensitive tissues that demand a great deal of energy eventually die. LHON sufferers eventually lose their sight because the optic nerve dies from lack of energy.

Researchers have identified and cloned a mutant mitochondrial gene that is responsible for LHON. The defect is a mutant form of *NADH dehydrogenase*, a huge, complex enzyme that accepts electrons from NADH and sends them on through an electron transport system. Passage of electrons through the electron transport system allows the synthesis of ATP. If NADH dehydrogenase is defective, passage of electrons through the electron transport system is less efficient, and less ATP is made. In LHON sufferers, the result is eventual blindness.

In this chapter and the next, we will study some of the important biochemical reactions that occur in the mitochondria. Here, the final oxidations of carbohydrates, lipids, and proteins occur. Here, also, the electrons harvested in these oxidation reactions are used to make ATP. In these remarkably efficient reactions, nearly 40% of the potential energy of glucose is stored as ATP.

22.1 The Mitochondria

Mitochondria are football-shaped organelles that are roughly the size of a bacterial cell. They are surrounded by an **outer mitochondrial membrane** and an **inner mitochondrial membrane** (Figure 22.1). The space between the two membranes is the **intermembrane space**, and the space inside of the inner membrane is the **matrix space**. The enzymes of the citric acid cycle, of the β-oxidation pathway for the breakdown of fatty acids, and for the degradation of amino acids are all found in the mitochondrial matrix space.

Structure and Function

The outer mitochondrial membrane has many small pores through which small molecules (less than 10,000 g/mol) can pass. Thus, the small molecules to be oxidized for the production of ATP can easily enter the mitochondrial intermembrane space.

The inner membrane is highly folded to create a large surface area. The folded membranes are known as **cristae.** The inner mitochondrial membrane is almost completely impermeable to most substances. For this reason, it has many transport proteins to bring particular fuel molecules into the matrix space. Also embedded

---

LEARNING GOAL

1 Name the regions of the mitochondria and the function of each region.

An organelle is a compartment within the cytoplasm that has a specialized function.

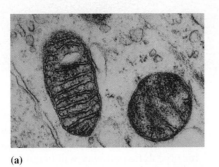

Figure 22.1 Structure of the mitochondrion. (a) Electron micrograph of mitochondria. (b) Schematic drawing of the mitochondrion.

(a)

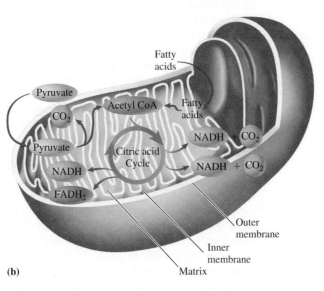

(b)

within the inner mitochondrial membrane are the protein electron carriers of the *electron transport system*, and *ATP synthase*. ATP synthase is a large complex of many proteins that catalyzes the synthesis of ATP.

Origin of the Mitochondria

Not only are mitochondria roughly the size of bacteria, they have several other features that have led researchers to suspect they may once have been free-living bacteria that were "captured" by eukaryotic cells. They have their own genetic information (DNA). They also make their own ribosomes that are very similar to those of bacteria. These ribosomes allow the mitochondria to synthesize some of their own proteins. Finally, mitochondria are actually self-replicating; they grow in size and divide to produce new mitochondria. All of these characteristics suggest that the mitochondria that produce the majority of the ATP for our cells evolved from bacteria "captured" perhaps as long as 1.5×10^9 years ago.

As we saw in Chapter 20, ribosomes are complexes of protein and RNA that serve as small platforms for protein synthesis.

Question 22.1   What is the function of the mitochondria?

Question 22.2   How do the mitochondria differ from the other components of eukaryotic cells?

Question 22.3   Draw a schematic diagram of a mitochondrion and label the parts of this organelle.

Question 22.4   Describe the evidence that suggests that mitochondria evolved from free-living bacteria.

A Human Perspective

Exercise and Energy Metabolism

The Olympic sprinters get set in the blocks. The gun goes off, and roughly 10 seconds (s) later the 100-meter (m) dash is over. Elsewhere, the marathoners line up. They will run 26 miles (mi) and 385 yards (yd) in a little over 2 hours (h). Both sports involve running, but they utilize very different sources of energy.

Let's look at the sprinter first. The immediate source of energy for the sprinter is stored ATP. But the quantity of stored ATP is very small, only about 3 ounces (oz). This allows the sprinter to run as fast as he or she can for about 3 s. Obviously, another source of stored energy must be tapped, and that energy store is *creatine phosphate:*

The structure of creatine phosphate

Creatine phosphate, stored in the muscle, donates its high-energy phosphate to ADP to produce new supplies of ATP.

This will keep our runner in motion for another 5 or 6 s before the store of creatine phosphate is also depleted. This is almost enough energy to finish the 100-m dash, but in reality, all the runners are slowing down, owing to energy depletion, and the winner is the sprinter who is slowing down the least!

Consider a longer race, the 400-m or the 800-m. These runners run at maximum capacity for much longer. When they have depleted their ATP and creatine phosphate stores, they must synthesize more ATP. Of course, the cells have been making ATP all the time, but now the demand for energy is much greater. To supply this increased demand, the anaerobic energy-generating reactions (glycolysis and lactate fermentation, Chapter 21) and aerobic processes (citric acid cycle and oxidative phosphorylation) begin to function much more rapidly. Often, however, these athletes are running so strenuously that they cannot provide enough oxygen to the exercising muscle to allow oxidative phosphorylation to function efficiently. When this happens, the muscles must rely on glycolysis and lactate fermentation to provide *most* of the energy requirement. The chemical by-product of these anaerobic processes, lactate, builds up in the muscle and diffuses into the bloodstream. However, the concentration of lactate inevitably builds up in the working muscle and causes muscle fatigue and, eventually, muscle failure. Thus, exercise that depends primarily on anaerobic ATP production cannot continue for very long.

The marathoner presents us with a different scenario. This runner will deplete his or her stores of ATP and creatine phosphate as quickly as a short-distance runner. The anaerobic glycolytic pathway will begin to degrade glucose provided by the blood at a more rapid rate, as will the citric acid cycle and oxidative phosphorylation. The major difference in ATP production between the long-distance runner and the short- or middle-distance runner is that the muscles of the long-distance runner derive almost all their energy through aerobic pathways. These individuals continue to run long distances at a pace that allows them to supply virtually all the oxygen needed by the exercising muscle. In fact, only aerobic pathways can provide a constant supply of ATP for exercise that goes on for hours. Theoretically, under such conditions our runner could run indefinitely, utilizing first his or her stored glycogen and eventually stored lipids. Of course, in reality, other factors such as dehydration and fatigue place limits on the athlete's ability to continue.

From this we can conclude that long-distance runners must have a great capacity to produce ATP aerobically, in the mitochondria, whereas short- and middle-distance runners need a great capacity to produce energy anaerobically, in the cytoplasm of the muscle cells. It is interesting to note that the muscles of these runners reflect these diverse needs.

When researchers examine muscle tissue that has been surgically removed, they find two predominant types of muscle fibers. *Fast-twitch muscle fibers* are large, relatively plump, pale cells. They have only a few mitochondria but contain a large reserve of glycogen and high concentrations of the enzymes needed for glycolysis and lactate fermentation. These muscle fibers fatigue rather quickly because fermentation is inefficient, quickly depleting the cell's glycogen store and causing the accumulation of lactate.

Slow-twitch muscle fiber cells are about half the diameter of fast-twitch muscle cells and are red. The red color is a result of the high concentrations of myoglobin in these cells. Recall that myoglobin stores oxygen for the cell (Section 18.8) and facilitates rapid diffusion of oxygen throughout the cell. In addition, slow-twitch muscle fiber cells are packed with mitochondria. With this abundance of oxygen and mitochondria, these cells have the capacity for extended ATP production via aerobic pathways—ideal for endurance sports like marathon racing.

It is not surprising, then, that researchers have found that the muscles of sprinters have many more fast-twitch muscle fibers and those of endurance athletes have many more slow-twitch muscle fibers. One question that many researchers

Phosphoryl group transfer from creatine phosphate to ADP is catalyzed by the enzyme creatine kinase.

The sprinter relies on fast-twitch muscle fibers.

The marathon runner largely uses slow-twitch muscle fibers.

are trying to answer is whether the type of muscle fibers an individual has is a function of genetic makeup or training. Is a marathon runner born to be a long-distance runner, or are his or her abilities due to the type of training the runner undergoes? There is no doubt that the training regimen for an endurance runner does indeed increase the number of slow-twitch muscle fibers and that of a sprinter increases the number of fast-twitch muscle fibers. But there is intriguing new evidence to suggest that the muscles of endurance athletes have a greater proportion of slow-twitch muscle fibers before they ever begin training. It appears that some of us truly were born to run.

For Further Understanding

▸ It has been said that the winner of the 100-m race is the one who is slowing down the least. Explain this observation in terms of energy-harvesting pathways.

▸ Design an experiment to safely test whether the type of muscle fibers a runner has are the result of training or genetic makeup.

22.2 Conversion of Pyruvate to Acetyl CoA

As we saw in Chapter 21, under *anaerobic* conditions, glucose is broken down into two pyruvate molecules that are then converted to a stable fermentation product. This limited degradation of glucose releases very little of the potential energy of glucose. Under *aerobic* conditions, the cells can use oxygen and completely oxidize glucose to CO_2 in a metabolic pathway called the *citric acid cycle.*

This pathway is often referred to as the *Krebs cycle* in honor of Sir Hans Krebs, who worked out the steps of this cyclic pathway from his own experimental data and that of other researchers. It is also called the *tricarboxylic acid (TCA) cycle* because several of the early intermediates in the pathway have three carboxylate groups.

Once pyruvate enters the mitochondria, it must be converted to a two-carbon acetyl group. This acetyl group must be "activated" to enter the reactions of the citric acid cycle. Activation occurs when the acetyl group is bonded to the thiol group of coenzyme A. **Coenzyme A** is a large thiol derived from ATP and the vitamin pantothenic acid (Figure 22.2). It is an acceptor of acetyl groups (in red in Figure 22.2), which are bonded to it through a high-energy thioester bond. The acetyl coenzyme A **(acetyl CoA)** formed is the "activated" form of the acetyl group.

Figure 22.3 shows us the reaction that converts pyruvate to acetyl CoA. First, pyruvate is decarboxylated, which means it loses a carboxyl group that is released as CO_2. Next it is oxidized, and the hydride anion ($H:^-$) that is removed is accepted by NAD^+. Finally, the remaining acetyl group, $CH_3CO—$, is linked to coenzyme A by a thioester bond. This very complex reaction is carried out by three enzymes and five coenzymes that are organized together in a single bundle

LEARNING GOAL

2 Describe the reaction that results in the conversion of pyruvate to acetyl CoA, describing the location of the reaction and the components of the pyruvate dehydrogenase complex.

Coenzyme A is described in Sections 12.8 and 14.4. Thioester bonds are discussed in Section 14.4.

In Section 19.7, we learned that the hydride anion is a hydrogen atom with two electrons.

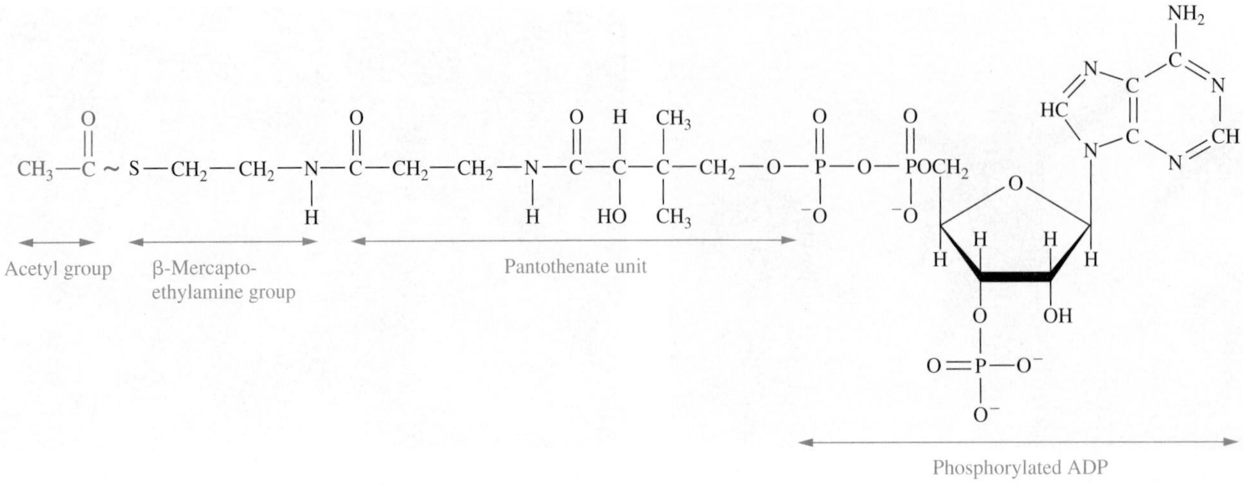

Acetyl coenzyme A
(acetyl CoA)

Figure 22.2 The structure of acetyl CoA. The bond between the acetyl group and coenzyme A is a high-energy thioester bond.

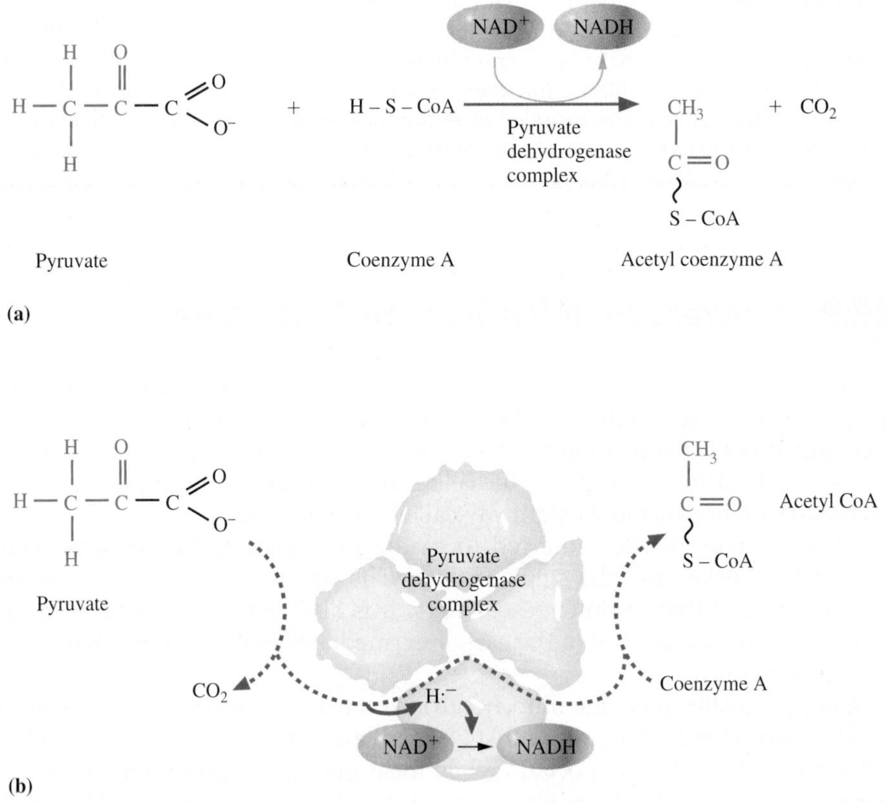

Figure 22.3 The decarboxylation and oxidation of pyruvate to produce acetyl CoA. (a) The overall reaction in which CO_2 and an H:$^-$ are removed from pyruvate and the remaining acetyl group is attached to coenzyme A. This requires the concerted action of three enzymes and five coenzymes. (b) The pyruvate dehydrogenase complex that carries out this reaction is actually a cluster of enzymes and coenzymes. The substrate is passed from one enzyme to the next as the reaction occurs.

called the **pyruvate dehydrogenase complex** (see Figure 22.3). This organization allows the substrate to be passed from one enzyme to the next as each chemical reaction occurs. A schematic representation of this "disassembly line" is shown in Figure 22.3b.

This single reaction requires four coenzymes made from four different vitamins, in addition to the coenzyme lipoamide. These are thiamine pyrophosphate, derived from thiamine (Vitamin B_1); FAD, derived from riboflavin (Vitamin B_2); NAD$^+$, derived from niacin; and coenzyme A, derived from pantothenic acid. Obviously, a deficiency in any of these vitamins would seriously reduce the

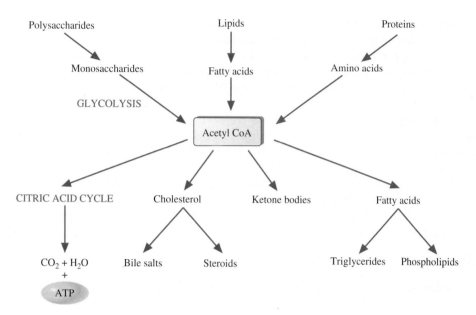

Figure 22.4 The central role of acetyl CoA in cellular metabolism.

amount of acetyl CoA that our cells could produce. This, in turn, would limit the amount of ATP that the body could make and would contribute to vitamin-deficiency diseases. Fortunately, a well-balanced diet provides an adequate supply of these and other vitamins.

In Figure 22.4, we see that acetyl CoA is a central character in cellular metabolism. It is produced by the degradation of glucose, fatty acids, and some amino acids. The major function of acetyl CoA in energy-harvesting pathways is to carry the acetyl group to the citric acid cycle, in which it will be used to produce large amounts of ATP. In addition to these catabolic duties, the acetyl group of acetyl CoA can also be used for *anabolic* or biosynthetic reactions to produce cholesterol and fatty acids. It is through this intermediate, acetyl CoA, that all the energy sources (fats, proteins, and carbohydrates) are interconvertible.

Question 22.5 What vitamins are required for acetyl CoA production from pyruvate?

Question 22.6 What is the major role of coenzyme A in catabolic reactions?

22.3 An Overview of Aerobic Respiration

Aerobic respiration is the oxygen-requiring breakdown of food molecules and production of ATP. The different steps of aerobic respiration occur in different compartments of the mitochondria.

The enzymes for the citric acid cycle are found in the mitochondrial matrix space. The first enzyme catalyzes a reaction that joins the acetyl group of acetyl CoA (two carbons) to a four-carbon molecule (oxaloacetate) to produce citrate (six carbons). The remaining enzymes catalyze a series of rearrangements, decarboxylations (removal of CO_2), and oxidation-reduction reactions. The eventual products of this cyclic pathway are two CO_2 molecules and oxaloacetate—the molecule we began with.

At several steps in the citric acid cycle, a substrate is oxidized. In three of these steps, a pair of electrons is transferred from the substrate to NAD^+, producing NADH (three NADH molecules per turn of the cycle). At another step, a pair of electrons is transferred from a substrate to FAD, producing $FADH_2$ (one $FADH_2$ molecule per turn of the cycle).

LEARNING GOAL

3 Summarize the reactions of aerobic respiration.

Remember (Section 19.7) that it is really the hydride anion with its pair of electrons ($H:^-$) that is transferred to NAD^+ to produce NADH. Similarly, a pair of hydrogen atoms (and thus two electrons) are transferred to FAD to produce $FADH_2$.

The electrons are passed from NADH or FADH$_2$, through an electron transport system located in the inner mitochondrial membrane, and finally to the terminal electron acceptor, molecular oxygen (O$_2$). The transfer of electrons through the electron transport system causes protons (H$^+$) to be pumped from the mitochondrial matrix into the intermembrane compartment. The result is a high-energy proton reservoir.

In the final step, the energy of the H$^+$ reservoir is used to make ATP. This last step is carried out by the enzyme complex ATP synthase. As protons flow back into the mitochondrial matrix through a pore in the ATP synthase complex, the enzyme catalyzes the synthesis of ATP.

This long, involved process is called *oxidative phosphorylation*, because the energy of electrons from the *oxidation* of substrates in the citric acid cycle is used to *phosphorylate* ADP and produce ATP. The details of each of these steps will be examined in upcoming sections.

Question 22.7 What is meant by the term *oxidative phosphorylation?*

Question 22.8 What does the term *aerobic respiration* mean?

22.4 The Citric Acid Cycle (The Krebs Cycle)

Biological Effects of Disorders of the Citric Acid Cycle

In Chapter 21, we saw that deficiencies of the enzymes involved in glycolysis cause debilitating conditions and, in severe form, result in death. This is also true of deficiencies of enzymes in the citric acid cycle (Table 22.1). A number of these deficiencies and the mutations that cause them have been studied and, up to this date, no treatment has been found for any of them. As with glycolysis, these genetic deficiencies emphasize the importance of the citric acid cycle to the life of the organism.

Mutations in the fumarase gene cause encephalopathy, which is a syndrome with a variety of neurological symptoms ranging from subtle personality changes to psychosis, lethargy, involuntary muscle spasms, tremors, and seizures. A newborn with a fumarase deficiency may exhibit muscle weakness (hypotonia) and poor feeding. As the child ages, the neurological symptoms become more severe and include severe developmental delay, brain deformation, psychomotor deficits, and seizures. Many children with this deficiency die in infancy or childhood.

A deficiency of α-ketoglutarate dehydrogenase is characterized by chronic lactic acidosis and progressive encephalopathy and hypotonia. The life expectancy of a child born with this deficiency is 2 to 3 years, with death resulting from neurological deterioration.

TABLE 22.1 Some Citric Acid Cycle Enzyme Deficiencies and the Associated Disorders

Enzyme Deficiency	Disorder
Fumarase	Early encephalopathy, seizures, and muscular hypotonia
α-Ketoglutarate dehydrogenase	Hypotonia, severe encephalopathy, psychotic behavior
Succinate dehydrogenase	Leigh disease (subacute necrotizing encephalomyelopathy), paraganglioma
Aconitase	Friedreich ataxia

A variety of mutations in the succinate dehydrogenase gene have been identified. One of these, SdhA, has been associated with Leigh disease, a disorder that generally affects children between 3 months and 2 years of age. It results in loss of motor skills and eventual death. Other mutations lead to tumors called *paragangliomas*. These are typically found in the head and neck regions and, depending on the nature of the mutation, may be malignant or benign.

One manifestation of aconitase deficiency is characterized by myopathy and exercise intolerance. In fact, physical exertion may prove fatal for some patients as a result of circulatory shock. A second manifestation is Friedreich ataxia. Typically, symptoms appear in children between the ages of five and fifteen and include muscle weakness, loss of coordination, impaired hearing or speech, and heart disorders.

Reactions of the Citric Acid Cycle

The **citric acid cycle** is the final stage of the breakdown of carbohydrates, fats, and amino acids released from dietary proteins (Figure 22.5). To understand this important cycle, let's follow the fate of the acetyl group of an acetyl CoA as it passes through the citric acid cycle. The numbered reactions listed below correspond to the steps in the citric acid cycle that are summarized in Figure 22.5.

LEARNING GOAL

4 Looking at an equation representing any of the chemical reactions that occur in the citric acid cycle, describe the kind of reaction that is occurring and the significance of that reaction to the pathway.

Reaction 1. This is a condensation reaction between the acetyl group of acetyl CoA and oxaloacetate. It is catalyzed by the enzyme *citrate synthase.* The product formed is citrate, a six-carbon molecule with three carboxylate anions:

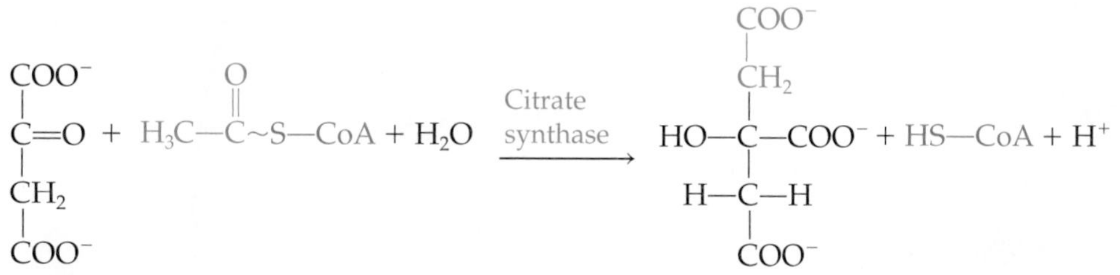

Oxaloacetate   Acetyl CoA   Citrate   Coenzyme A

Reaction 2. The enzyme *aconitase* catalyzes the dehydration of citrate, producing *cis*-aconitate. The same enzyme, aconitase, then catalyzes addition of a water molecule to the *cis*-aconitate, converting it to isocitrate. The net effect of these two steps is the isomerization of citrate to isocitrate:

Notice that the conversion of citrate to cis-aconitate is a biological example of the dehydration of an alcohol to produce an alkene (Section 12.4). The conversion of cis-aconitate to isocitrate is a biochemical example of the hydration of an alkene to produce an alcohol (Sections 11.5 and 12.4).

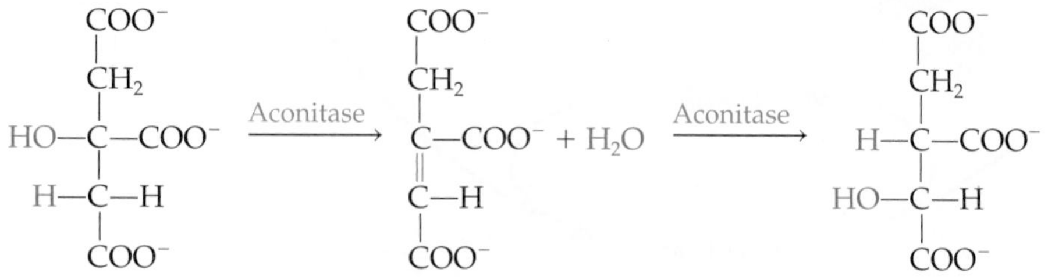

Citrate   *cis*-Aconitate   Isocitrate

Reaction 3. The first oxidative step of the citric acid cycle is catalyzed by *isocitrate dehydrogenase.* It is a complex reaction in which three things happen:
   a. the hydroxyl group of isocitrate is oxidized to a ketone,
   b. carbon dioxide is released, and
   c. NAD^+ is reduced to NADH.

The oxidation of a secondary alcohol produces a ketone (Sections 12.4 and 13.4).

Figure 22.5 The reactions of the citric acid cycle.

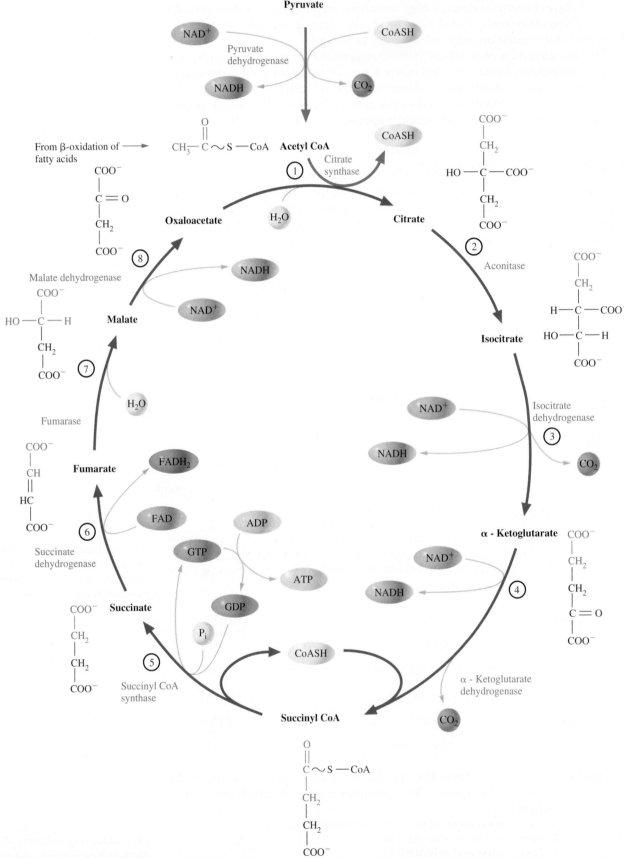

This is an oxidative decarboxylation reaction, and the product is α-ketoglutarate:

The structure of NAD⁺ and its reduction to NADH are shown in Figure 19.8.

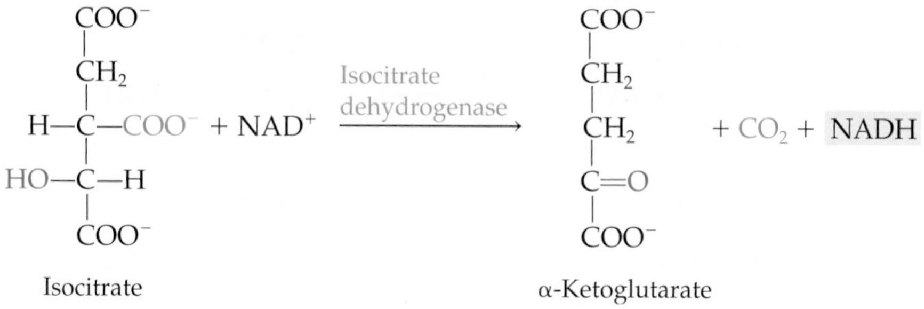

Isocitrate → α-Ketoglutarate

Remember, in organic (and thus biochemical) reactions, oxidation can be recognized as a gain of oxygen or loss of hydrogen (Section 12.5).

Reaction 4. Coenzyme A enters the picture again as the *α-ketoglutarate dehydrogenase* complex carries out a complex series of reactions similar to those catalyzed by the pyruvate dehydrogenase complex. The same coenzymes are required and, once again, three chemical events occur:

a. α-ketoglutarate loses a carboxylate group as CO_2,
b. it is oxidized and NAD^+ is reduced to NADH, and
c. coenzyme A combines with the product, succinate, to form succinyl CoA. The bond formed between succinate and coenzyme A is a high-energy thioester bond.

The pyruvate dehydrogenase complex was described in Section 22.2 and shown in Figure 22.3.

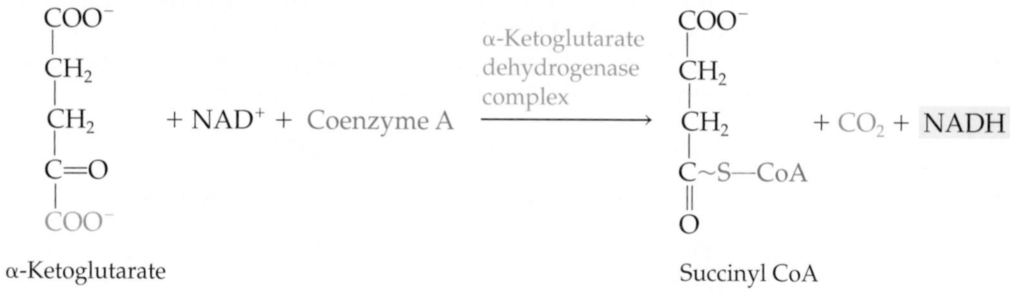

α-Ketoglutarate → Succinyl CoA

Reaction 5. Succinyl CoA is converted to succinate in this step, which once more is chemically very involved. The enzyme *succinyl CoA synthase* catalyzes a coupled reaction in which the high-energy thioester bond of succinyl CoA is hydrolyzed and an inorganic phosphate group is added to GDP to make GTP:

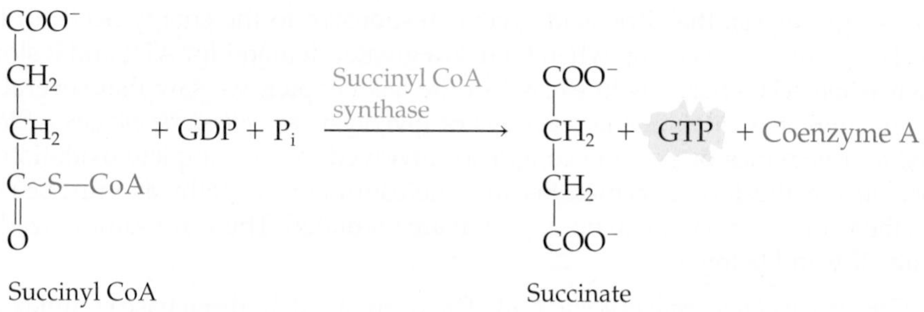

Succinyl CoA → Succinate

Another enzyme, *dinucleotide diphosphokinase,* then catalyzes the transfer of a phosphoryl group from GTP to ADP to make ATP:

$$GTP + ADP \xrightarrow{\text{Dinucleotide diphosphokinase}} GDP + ATP$$

The structure of FAD was shown in Figure 19.8.

Reaction 6. *Succinate dehydrogenase* then catalyzes the oxidation of succinate to fumarate in the next step. The oxidizing agent, *flavin adenine dinucleotide (FAD)*, is reduced in this step:

$$
\begin{array}{c}
\text{COO}^- \\
|\\
\text{CH}_2 \\
|\\
\text{CH}_2 \\
|\\
\text{COO}^-
\end{array}
+ \text{FAD}
\quad\xrightarrow{\text{Succinate dehydrogenase}}\quad
\begin{array}{c}
\text{COO}^- \\
|\\
\text{C—H} \\
\|\\
\text{H—C} \\
|\\
\text{COO}^-
\end{array}
+ \textbf{FADH}_2
$$

Succinate   Fumarate

We studied hydrogenation of alkenes to produce alkanes in Section 11.5. This is simply the reverse.

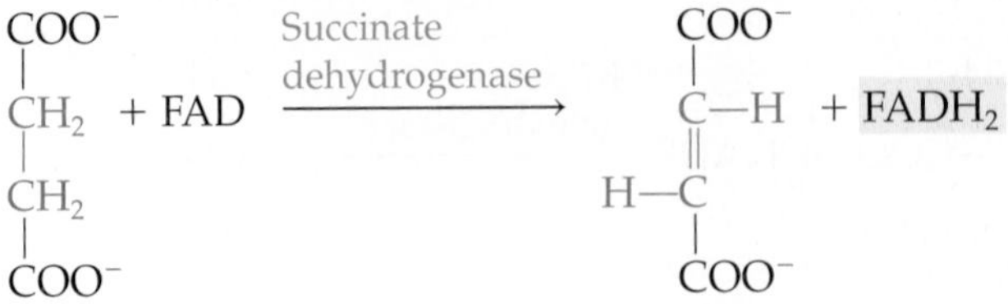

Reaction 7. Addition of H_2O to the double bond of fumarate gives malate. The enzyme *fumarase* catalyzes this reaction:

This reaction is a biological example of the hydration of an alkene to produce an alcohol (Sections 11.5 and 12.4).

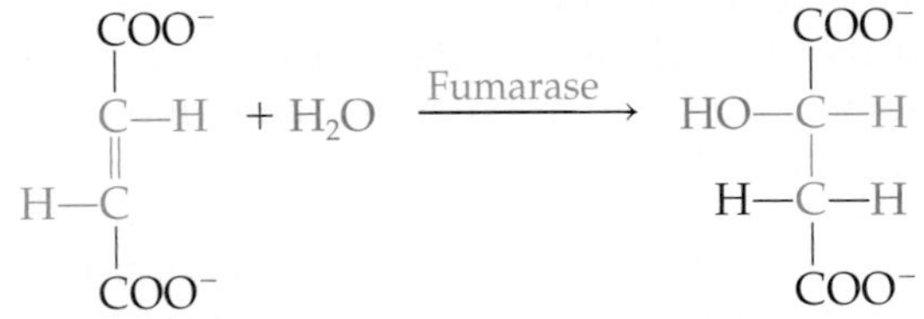

$$
\begin{array}{c}
\text{COO}^- \\
|\\
\text{C—H} \\
\|\\
\text{H—C} \\
|\\
\text{COO}^-
\end{array}
+ \text{H}_2\text{O}
\quad\xrightarrow{\text{Fumarase}}\quad
\begin{array}{c}
\text{COO}^- \\
|\\
\text{HO—C—H} \\
|\\
\text{H—C—H} \\
|\\
\text{COO}^-
\end{array}
$$

Fumarate   Malate

This reaction is a biochemical example of the oxidation of a secondary alcohol to a ketone, which we studied in Sections 12.4 and 13.4.

Reaction 8. In the final step of the citric acid cycle, *malate dehydrogenase* catalyzes the reduction of NAD^+ to NADH and the oxidation of malate to oxaloacetate. Because the citric acid cycle "began" with the addition of an acetyl group to oxaloacetate, we have come full circle.

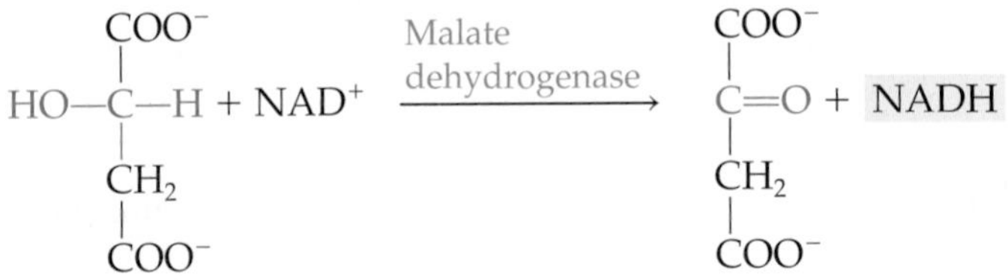

$$
\begin{array}{c}
\text{COO}^- \\
|\\
\text{HO—C—H} \\
|\\
\text{CH}_2 \\
|\\
\text{COO}^-
\end{array}
+ \text{NAD}^+
\quad\xrightarrow{\text{Malate dehydrogenase}}\quad
\begin{array}{c}
\text{COO}^- \\
|\\
\text{C=O} \\
|\\
\text{CH}_2 \\
|\\
\text{COO}^-
\end{array}
+ \textbf{NADH}
$$

Malate   Oxaloacetate

22.5   Control of the Citric Acid Cycle

LEARNING GOAL

5 Explain the mechanisms for the control of the citric acid cycle.

Just like glycolysis, the citric acid cycle is responsive to the energy needs of the cell. The pathway speeds up when there is a greater demand for ATP, and it slows down when ATP energy is in excess. In the last chapter, we saw that several of the enzymes that catalyze the reactions of glycolysis are *allosteric enzymes*. Similarly, four enzymes or enzyme complexes involved in the complete oxidation of pyruvate are allosteric enzymes. Because the control of the pathway must be precise, there are several enzymatic steps that are regulated. These are summarized in Figure 22.6 and below:

1. *Conversion of pyruvate to acetyl CoA.* The pyruvate dehydrogenase complex is inhibited by high concentrations of ATP, acetyl CoA, and NADH. Of course, the presence of these compounds in abundance signals that the cell has an adequate supply of energy, and thus energy metabolism is slowed.

2. *Synthesis of citrate from oxaloacetate and acetyl CoA.* The enzyme citrate synthase is an allosteric enzyme. In this case, the negative effector is ATP. Again, this is logical because an excess of ATP indicates that the cell has an abundance of energy.

3. *Oxidation and decarboxylation of isocitrate to α-ketoglutarate.* Isocitrate dehydrogenase is also an allosteric enzyme; however, the enzyme is controlled by the positive allosteric effector, ADP. ADP is a signal that the levels of ATP must be low, and therefore the rate of the citric acid cycle should be increased. Interestingly, isocitrate dehydrogenase is also *inhibited* by high levels of NADH and ATP.

4. *Conversion of α-ketoglutarate to succinyl CoA.* The α-ketoglutarate dehydrogenase complex is inhibited by high levels of the products of the reactions that it catalyzes, namely, NADH and succinyl CoA. It is further inhibited by high concentrations of ATP.

Allosteric enzymes bind to effectors, such as ATP or ADP, that alter the shape of the enzyme active site, either stimulating the rate of the reaction (positive allosterism) or inhibiting the reaction (negative allosterism). For more detail, see Section 19.9.

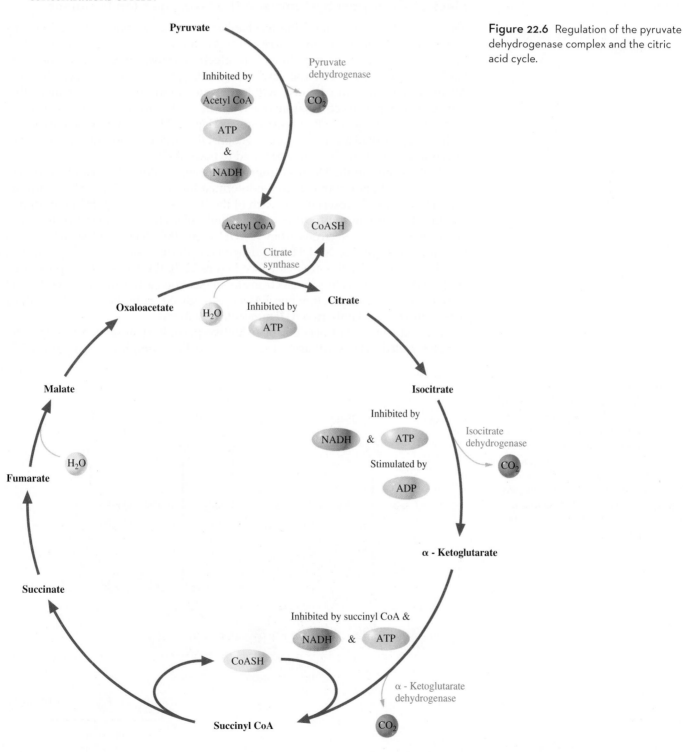

Figure 22.6 Regulation of the pyruvate dehydrogenase complex and the citric acid cycle.

22.6 Oxidative Phosphorylation

The electrons carried by NADH can be used to produce three ATP molecules, and those carried by $FADH_2$ can be used to produce two ATP molecules. We turn now to the process by which the energy of electrons carried by these coenzymes is converted to ATP energy. It is a series of reactions called **oxidative phosphorylation,** which couples the *oxidation* of NADH and $FADH_2$ to the *phosphorylation* of ADP to generate ATP.

Electron Transport Systems and the Hydrogen Ion Gradient

Before we try to understand the mechanism of oxidative phosphorylation, let's first look at the molecules that carry out this complex process. Embedded within the mitochondrial inner membrane are **electron transport systems.** These are made up of a series of electron carriers, including coenzymes and cytochromes. All these molecules are located within the membrane in an arrangement that allows them to pass electrons from one to the next. This array of electron carriers is called the *respiratory electron transport system* (Figure 22.7). As you would expect in such sequential oxidation-reduction reactions, the electrons lose some energy with each transfer. Some of this energy is used to make ATP.

At three sites in the electron transport system, protons (H^+) can be pumped from the mitochondrial matrix to the intermembrane space. These H^+ contribute to a high-energy H^+ reservoir. At each of the three sites, enough H^+ are pumped into the H^+ reservoir to produce one ATP molecule. The first site is NADH dehydrogenase. Because electrons from NADH enter the electron transport system by being transferred to NADH dehydrogenase, all three sites actively pump H^+, and three ATP molecules are made (see Figure 22.7). $FADH_2$ is a less "powerful" electron donor. It transfers its electrons to an electron carrier that follows NADH dehydrogenase. As a result, when $FADH_2$ is oxidized, only the second and third sites pump H^+, and only two ATP molecules are made.

The last component needed for oxidative phosphorylation is a multiprotein complex called **ATP synthase,** also called the **F_0F_1 complex** (see Figure 22.7).

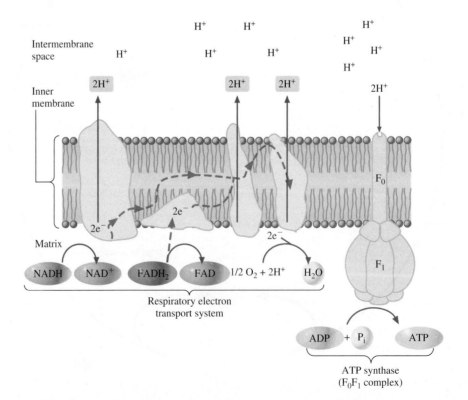

Figure 22.7 Electrons flow from NADH to molecular oxygen through a series of electron carriers embedded in the inner mitochondrial membrane. Protons are pumped from the mitochondrial matrix space into the intermembrane space. This results in a hydrogen ion reservoir in the intermembrane space. As protons pass through the channel in ATP synthase, their energy is used to phosphorylate ADP and produce ATP.

The F_0 portion of the molecule is a channel through which H^+ pass. It spans the inner mitochondrial membrane, as shown in Figure 22.7. The F_1 part of the molecule is an enzyme that catalyzes the phosphorylation of ADP to produce ATP.

ATP Synthase and the Production of ATP

How does all this complicated machinery actually function? NADH carries electrons, originally from glucose, to the first carrier of the electron transport system, NADH dehydrogenase (see Figure 22.7). There, NADH is oxidized to NAD^+, which returns to the site of the citric acid cycle to be reduced again. As the dashed red line shows, the pair of electrons is passed to the next electron carrier, and H^+ are pumped to the intermembrane compartment. The electrons are passed sequentially through the electron transport system, and at two additional sites, H^+ from the matrix are pumped into the intermembrane compartment. With each transfer the electrons lose some of their potential energy. It is this energy that is used to transport H^+ across the inner mitochondrial membrane and into the H^+ reservoir. As mentioned earlier, $FADH_2$ donates its electrons to a carrier of lower energy and fewer H^+ are pumped into the reservoir.

Finally, the electrons arrive at the last carrier. They now have too little energy to accomplish any more work, but they *must* be donated to some final electron acceptor so that the electron transport system can continue to function. In aerobic organisms, the **terminal electron acceptor** is molecular oxygen, O_2, and the product is water.

As the electron transport system continues to function, a high concentration of protons builds up in the intermembrane space. This creates an H^+ gradient across the inner mitochondrial membrane. Such a gradient is an enormous energy source, like water stored behind a dam. The mitochondria make use of the potential energy of the gradient to synthesize ATP energy.

ATP synthase harvests the energy of this gradient by making ATP. H^+ pass through the F_0 channel back into the matrix. This causes F_1 to become an active enzyme that catalyzes the phosphorylation of ADP to produce ATP. In this way, the energy of the H^+ reservoir is harvested to make ATP.

The importance of keeping the electron transport system functioning becomes obvious when we consider what occurs in cyanide poisoning. Cyanide binds to the heme group iron of cytochrome oxidase, one of the electron carriers in the electron transport system, instantly stopping electron transfers and causing death within minutes!

Question 22.9 Write a balanced chemical equation for the reduction of NAD^+.

Question 22.10 Write a balanced chemical equation for the reduction of FAD.

Summary of the Energy Yield

Now that we have studied the reactions of the citric acid cycle and oxidative phosphorylation, we can calculate the total energy yield, in ATP, that is produced from a single glucose molecule.

One turn of the citric acid cycle results in the production of two CO_2 molecules, three NADH molecules, one $FADH_2$ molecule, and one ATP molecule. Oxidative phosphorylation yields three ATP molecules per NADH molecule and two ATP molecules per $FADH_2$ molecule. The only exception to these energy yields is the NADH produced in the cytoplasm during glycolysis. Oxidative phosphorylation yields only two ATP molecules per cytoplasmic NADH molecule. The reason for this is that energy must be expended to shuttle electrons from NADH in the cytoplasm to $FADH_2$ in the mitochondrion.

Knowing this information and keeping in mind that two turns of the citric acid cycle are required, we can sum up the total energy yield from the complete oxidation of one glucose molecule, as shown in Example 22.1.

Aerobic metabolism is very much more efficient than anaerobic metabolism. The abundant energy harvested by aerobic metabolism has had enormous consequences for the biological world. Much of the energy released by the oxidation of fuels is not lost as heat but conserved in the form of ATP. Organisms that possess abundant energy have evolved into multicellular organisms and developed specialized functions. As a consequence of their energy requirements, all multicellular organisms are aerobic.

Babies with Three Parents?

It has been estimated that nearly 800 children are born each year to women at risk for mitochondrial genetic disorders. In addition to Leber's hereditary optic neuropathy, described in the introduction to this chapter, there are a number of other mitochondrial disorders that affect the brain, heart, and skeletal muscles in varying degrees of severity. Symptoms range from poor growth, visual or auditory problems, mental disabilities, diabetes, and neurological disorders of varying severity.

A technique has been developed that could ensure that a woman at risk of having a child with a mitochondrial disorder would be able to have a child born with normal mitochondria. As we saw in this chapter, children inherit all their mitochondria from the mother. The technique of producing a three-parent baby involves taking the ovum of a woman with normal mitochondria and removing the nucleus of the cell. The nucleus is then replaced with the nucleus of the ovum of the prospective mother. That new ovum is then fertilized with sperm from the father.

Genetically, these children would have the nuclear DNA, and all their genes, from the biological parents. Only the cytoplasm of the ovum and the healthy mitochondria would come from the donor mother. In 2015, the Parliament of the United Kingdom approved this "three-parent" process. As a result, the first child conceived by this process could be born in 2016.

As is to be expected, there is quite a bit of controversy surrounding this procedure. Those in the research community who have been developing the technology report that they have investigated safety concerns and are convinced that it is not dangerous. Many are convinced that this is an opportunity to allow couples at risk of having children with mitochondrial disorders to have babies that are healthy and can expect to have full lives.

For Further Understanding

▸ Go on the Internet and investigate the variety of types of mitochondrial disorders.

▸ One concern with this technology is that it will alter the mitochondrial genome of all future generations. Do you find this to be a reasonable objection? Support your answer with arguments based on your knowledge of the genetics of human reproduction and the mitochondria.

EXAMPLE 22.1 Determining the Yield of ATP from Aerobic Respiration

Calculate the number of ATP produced by the complete oxidation of one molecule of glucose.

Solution

Glycolysis:

Substrate-level phosphorylation	2 ATP
2 NADH × 2 ATP/cytoplasmic NADH	4 ATP

Conversion of 2 pyruvate molecules to 2 acetyl CoA molecules:

2 NADH × 3 ATP/NADH	6 ATP

Citric acid cycle (two turns):

2 GTP × 1 ATP/GTP	2 ATP
6 NADH × 3 ATP/NADH	18 ATP
2 FADH$_2$ × 2 ATP/FADH$_2$	4 ATP
	36 ATP

This represents an energy harvest of about 40% of the potential energy of glucose.

Practice Problem 22.1

Calculate the number of ATP produced by the complete oxidation of pyruvate.

▸ For Further Practice: **Questions 22.41, 22.42, and 22.44.**

22.7 The Degradation of Amino Acids

Carbohydrates are not our only source of energy. As we saw in Chapter 21, dietary protein is digested to amino acids that can also be used as an energy source, although this is not their major metabolic function. Most of the amino acids used for energy come from the diet. In fact, it is only under starvation conditions, when stored glycogen has been depleted, that the body begins to burn its own protein, for instance from muscle, as a fuel.

The fate of the mixture of amino acids provided by digestion of protein depends upon a balance between the need for amino acids for biosynthesis and the need for cellular energy. Only those amino acids that are not needed for protein synthesis are eventually converted into citric acid cycle intermediates and used as fuel.

The degradation of amino acids occurs primarily in the liver and takes place in two stages.

- Stage one is the removal of the α-amino group.
- Stage two is the degradation of the amino acid carbon skeleton.

In land mammals, the amino group generally ends up in urea, which is excreted in the urine. The carbon skeletons can be converted into a variety of compounds, including citric acid cycle intermediates, pyruvate, acetyl CoA, or acetoacetyl CoA. The degradation of the carbon skeletons is summarized in Figure 22.8. Deamination reactions and the fate of the carbon skeletons of amino acids are the focus of this section.

Removal of α-Amino Groups: Transamination

The first stage of amino acid degradation, the removal of the α-amino group, is usually accomplished by a **transamination** reaction. **Transaminases** catalyze the transfer of the α-amino group from an α-amino acid to an α-keto acid:

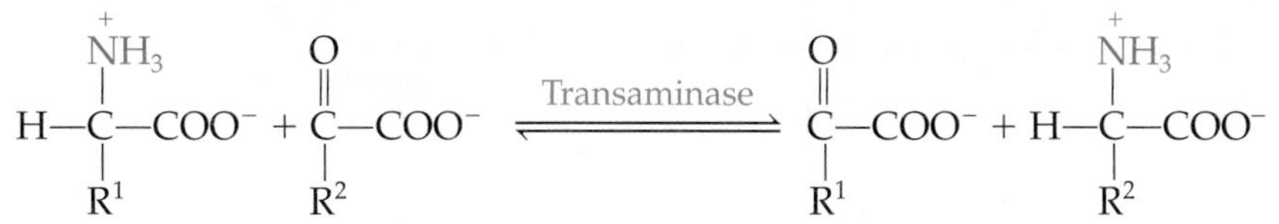

| Donor amino acid | Acceptor keto acid | α-Keto acid of amino acid | New amino acid |

The α-amino group of a great many amino acids is transferred to α-ketoglutarate to produce the amino acid glutamate and a new keto acid. This glutamate family of transaminases is especially important because the α-keto acid corresponding to glutamate is α-ketoglutarate, a citric acid cycle intermediate. The glutamate transaminases thus provide a direct link between amino acid degradation and the citric acid cycle.

Aspartate transaminase catalyzes the transfer of the α-amino group of aspartate to α-ketoglutarate, producing oxaloacetate and glutamate:

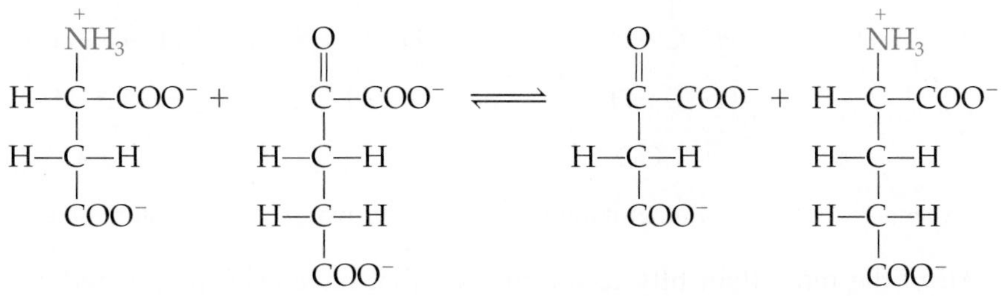

Aspartate α-Ketoglutarate Oxaloacetate Glutamate

LEARNING GOAL

7 Describe the conversion of amino acids to molecules that can enter the citric acid cycle.

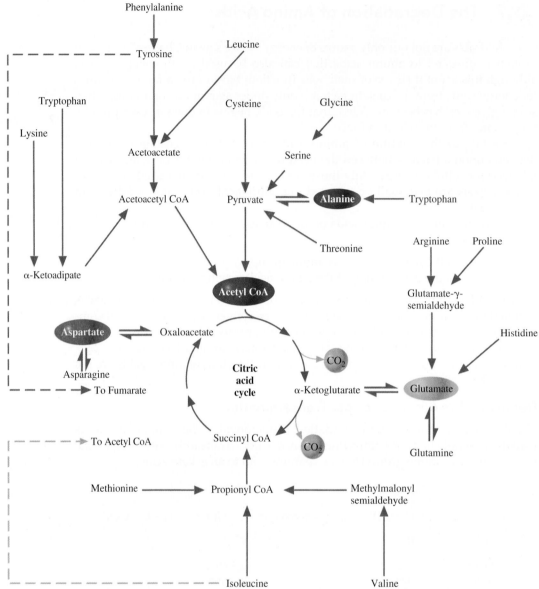

Figure 22.8 The carbon skeletons of amino acids can be converted to citric acid cycle intermediates and completely oxidized to produce ATP energy.

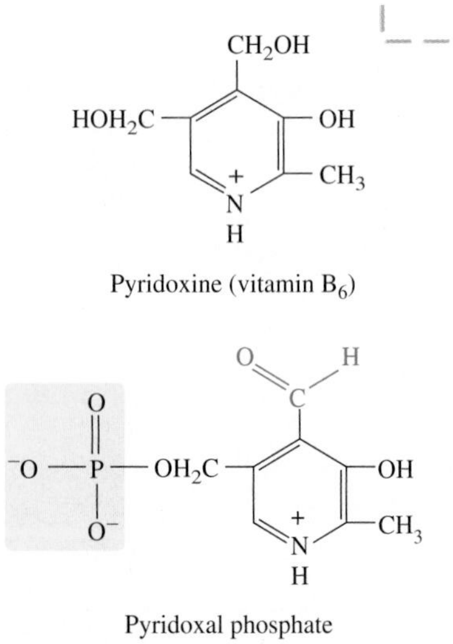

Pyridoxine (vitamin B_6)

Pyridoxal phosphate

Figure 22.9 The structure of pyridoxal phosphate, the coenzyme required for all transamination reactions, and pyridoxine, vitamin B_6, the vitamin from which it is derived.

Another important transaminase in mammalian tissues is *alanine transaminase*, which catalyzes the transfer of the α-amino group of alanine to α-ketoglutarate and produces pyruvate and glutamate:

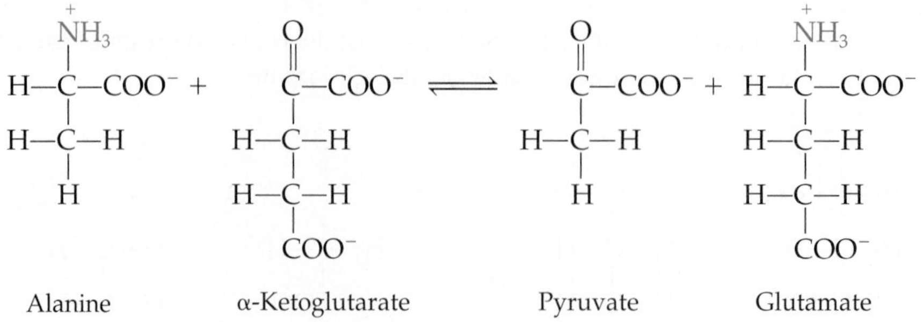

Alanine   α-Ketoglutarate   Pyruvate   Glutamate

All of the more than fifty transaminases that have been discovered require the coenzyme **pyridoxal phosphate.** This coenzyme is derived from vitamin B_6 (pyridoxine, Figure 22.9).

   The transamination reactions appear to be a simple transfer, but in reality, the reaction is much more complex. The transaminase binds the amino acid (aspartate in Figure 22.10a) in its active site. Then, the α-amino group of aspartate is transferred to pyridoxal phosphate, producing pyridoxamine phosphate and oxaloacetate (Figure 22.10b). The amino group is then transferred to an α-keto acid; in this case, α-ketoglutarate (Figure 22.10c), to produce the amino acid glutamate (Figure 22.10d). Next, we will examine the fate of the amino group that has been transferred to α-ketoglutarate to produce glutamate.

Question 22.11   What is the role of pyridoxal phosphate in transamination reactions?

Question 22.12   What is the function of a transaminase?

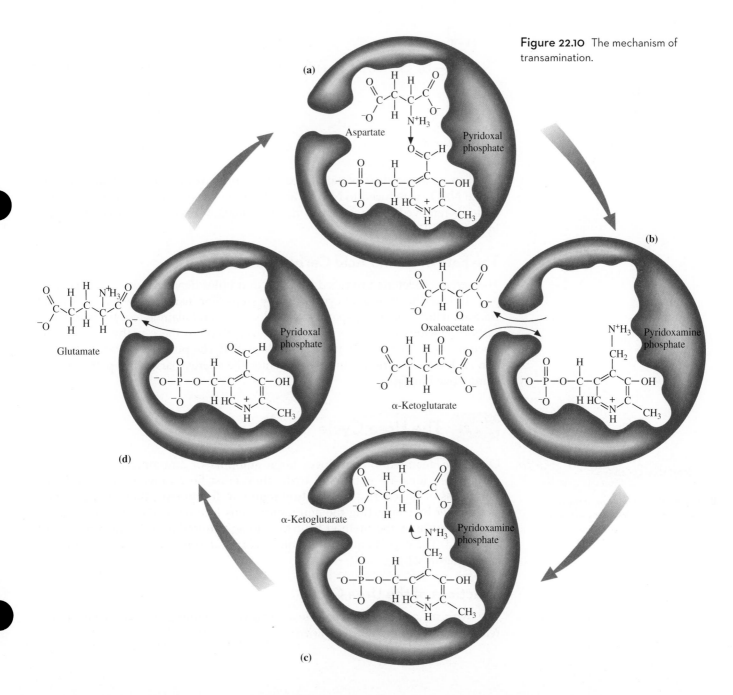

Figure 22.10 The mechanism of transamination.

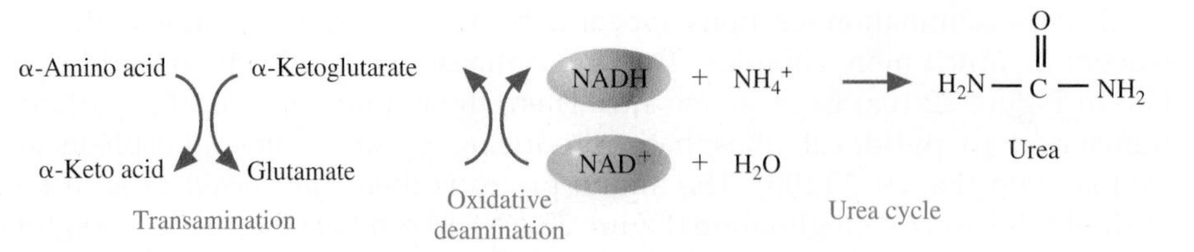

Figure 22.11 Summary of the deamination of an α-amino acid and the fate of the ammonium ion (NH_4^+).

Removal of α-Amino Groups: Oxidative Deamination

In the next stage of amino acid degradation, ammonium ion is liberated from the glutamate formed by the transaminase. This breakdown of glutamate, catalyzed by the enzyme *glutamate dehydrogenase,* occurs as follows:

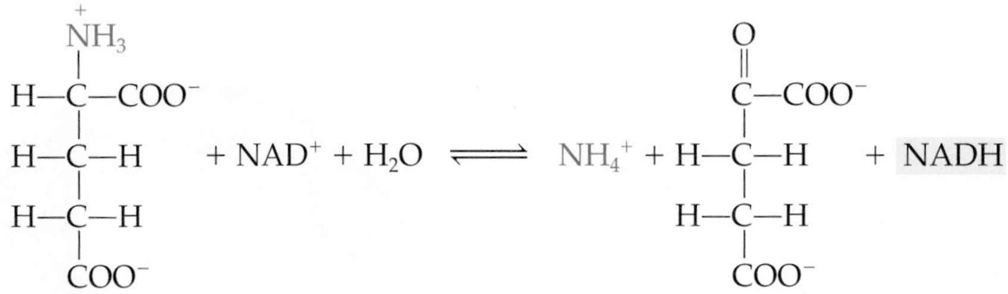

Glutamate α-Ketoglutarate

This is an example of an **oxidative deamination,** an oxidation-reduction process in which NAD^+ is reduced to NADH and the amino acid is deaminated (the amino group is removed). A summary of the deamination reactions described is shown in Figure 22.11.

The Fate of Amino Acid Carbon Skeletons

The carbon skeletons produced by these and other deamination reactions enter glycolysis or the citric acid cycle at many steps. For instance, we have seen that transamination converts aspartate to oxaloacetate and alanine to pyruvate. These carbon skeletons will be completely oxidized through these pathways, and the electrons harvested will be used to produce ATP. The positions at which the carbon skeletons of various amino acids enter the energy-harvesting pathways are summarized in Figure 22.8.

22.8 The Urea Cycle

LEARNING GOAL

8 Explain the importance of the urea cycle, and describe its essential steps.

Oxidative deamination produces large amounts of ammonium ion. Because ammonium ions are extremely toxic, they must be removed from the body, regardless of the energy expenditure required. In humans, they are detoxified in the liver by converting the ammonium ions into urea. This pathway, called the **urea cycle,** is the method by which toxic ammonium ions are kept out of the blood. The excess ammonium ions incorporated in urea are excreted in the urine (Figure 22.12).

Reactions of the Urea Cycle

The five reactions of the urea cycle are shown in Figure 22.12, and details of the reactions are summarized as follows.

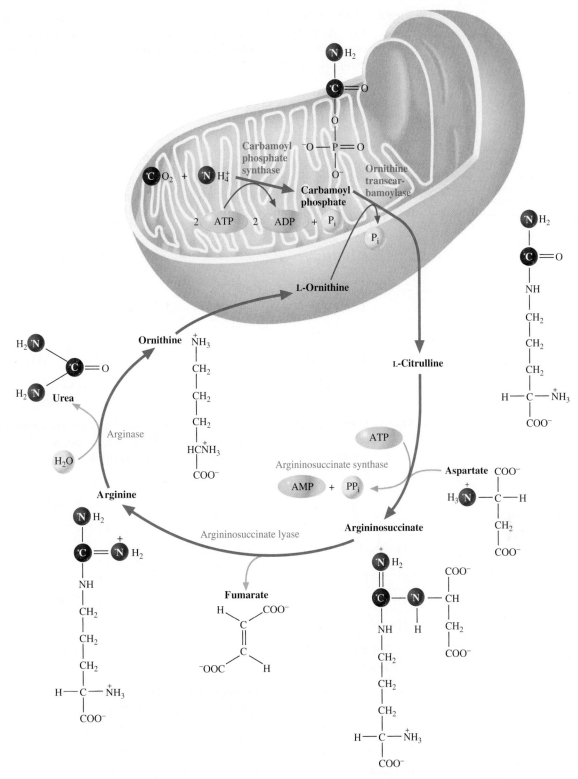

Figure 22.12 The urea cycle converts ammonium ions into urea, which is less toxic. The intracellular locations of the reactions are indicated. Citrulline, formed in the reaction between ornithine and carbamoyl phosphate, is transported out of the mitochondrion and into the cytoplasm. Ornithine, a substrate for the formation of citrulline, is transported from the cytoplasm into the mitochondrion.

Step 1. The first step of the cycle is a reaction in which CO_2 and NH_4^+ form carbamoyl phosphate. This reaction, which also requires ATP and H_2O, occurs in the mitochondria and is catalyzed by the enzyme *carbamoyl phosphate synthase.*

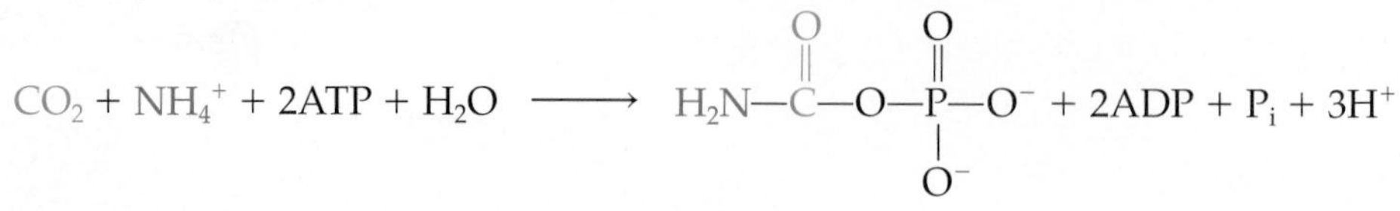

$$CO_2 + NH_4^+ + 2ATP + H_2O \longrightarrow H_2N-\overset{\overset{O}{\|}}{C}-O-\overset{\overset{O}{\|}}{\underset{\underset{O^-}{|}}{P}}-O^- + 2ADP + P_i + 3H^+$$

Carbamoyl phosphate

The urea cycle involves several unusual amino acids that are not found in polypeptides.

Step 2. The carbamoyl phosphate now condenses with the amino acid ornithine to produce the amino acid citrulline. This reaction also occurs in the mitochondria and is catalyzed by the enzyme *ornithine transcarbamoylase.*

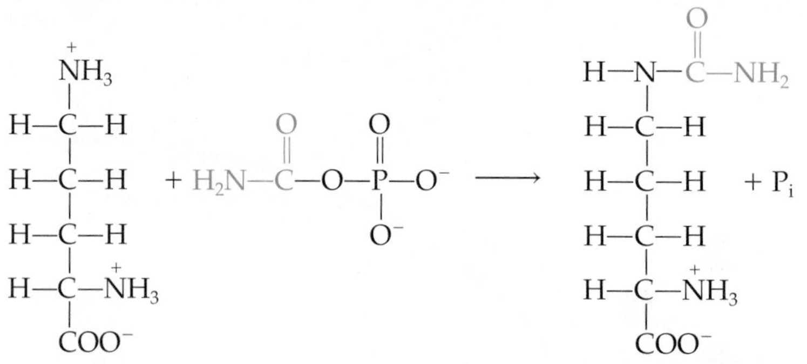

Ornithine   Carbamoyl phosphate   Citrulline

The abbreviation PP$_i$ represents the pyrophosphate group, which consists of two phosphate groups joined by a phosphoanhydride bond:

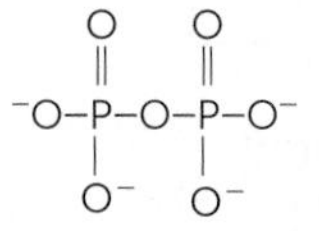

$$^-O-\overset{\overset{O}{\|}}{\underset{\underset{O^-}{|}}{P}}-O-\overset{\overset{O}{\|}}{\underset{\underset{O^-}{|}}{P}}-O^-$$

Step 3. Citrulline is transported into the cytoplasm and now condenses with aspartate to produce argininosuccinate. This reaction, which requires energy released by the hydrolysis of ATP, is catalyzed by the enzyme *argininosuccinate synthase.*

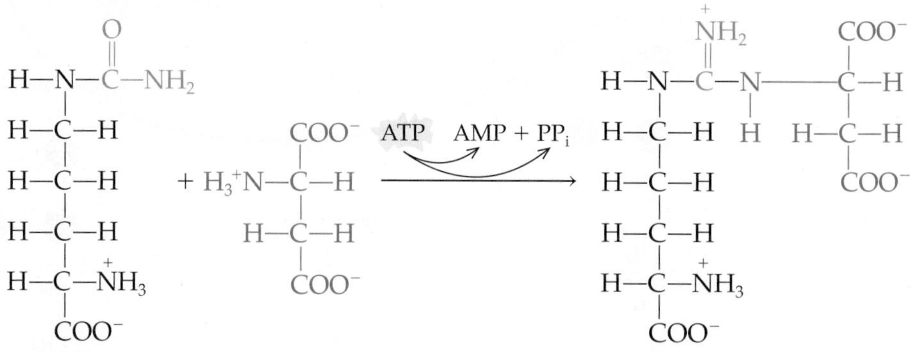

Citrulline   Aspartate   Argininosuccinate

Step 4. Now the argininosuccinate is cleaved to produce the amino acid arginine and the citric acid cycle intermediate fumarate. This reaction is catalyzed by the enzyme *argininosuccinate lyase.*

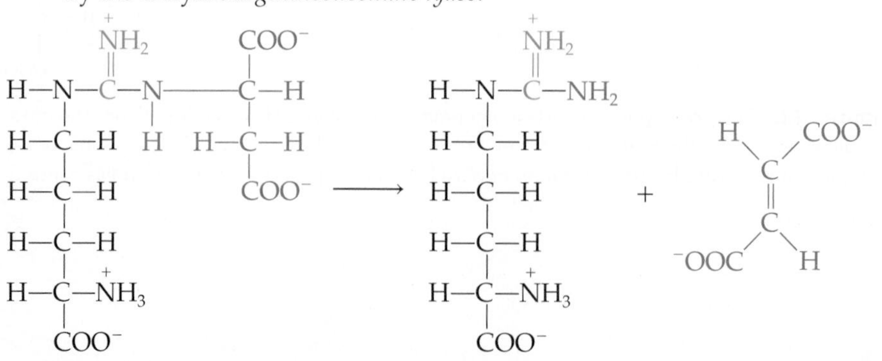

Argininosuccinate   Arginine   Fumarate

A Medical Perspective

Pyruvate Carboxylase Deficiency

Pyruvate carboxylase is the enzyme that converts pyruvate to oxaloacetate.

$$\text{Pyruvate} + CO_2 + ATP + H_2O \longrightarrow$$
$$\text{Oxaloacetate} + ADP + 2H^+$$

This reaction is important because it provides oxaloacetate for the citric acid cycle when the supplies have run low because of the demands of biosynthesis. It is also the enzyme that catalyzes the first step in gluconeogenesis, the pathway that provides the body with needed glucose in times of starvation or periods of exercise that deplete glycogen stores. But somehow these descriptions don't fill us with a sense of the importance of this enzyme and its jobs. It is not until we investigate a case study of a child born with pyruvate carboxylase deficiency that we see the full impact of this enzyme.

Pyruvate carboxylase deficiency is found in about 1 in 250,000 births; however, there is an increased incidence in native North American Indians who speak the Algonquin dialect and in the French. There are two types of genetic disorders that have been described. In the neonatal form of the disease, there is a complete absence of the enzyme. Symptoms are apparent at birth, and the child is born with brain abnormalities. In the infantile form, the patient develops symptoms early in infancy. Again, it is neurological symptoms that draw attention to the condition. The infants do not develop mental or psychomotor skills. They may develop seizures and/or respiratory depression. In both cases, it is the brain that suffers the greatest damage. In fact, this is the case in most of the disorders that reduce energy metabolism because the brain has such high energy requirements.

Biochemically, patients exhibit quite a variety of symptoms. They show acidosis (low blood pH) due to accumulations of lactate and extremely high pyruvate concentrations in the blood. Blood levels of alanine are also high, and large doses of alanine do not stimulate gluconeogenesis. Furthermore, a patient's cells accumulate lipid.

We can understand each of these symptoms by considering the pathways affected by the absence of this single enzyme. Lactic acidosis results from the fact that the body must rely on glycolysis and lactate fermentation for most of its energy needs. Alanine levels are high because it isn't being transaminated to pyruvate efficiently, because pyruvate levels are so high. In addition, alanine can't be converted to glucose by gluconeogenesis. Although the excess alanine is taken up by the liver and converted to pyruvate, the pyruvate can't be converted to glucose. Lipids accumulate because a great deal of pyruvate is converted to acetyl CoA. However, the acetyl CoA is not used to produce citrate as a result of the absence of oxaloacetate. So, the acetyl CoA is thus used to synthesize fatty acids, which are stored as triglycerides.

Dietary intervention has been tried. One such regimen is to supplement the diet with aspartic acid and glutamic acid. The theory behind this treatment is as complex as the many symptoms of the disorder. Both amino acids can be aminated (amino groups added) in non-nervous tissue. This produces asparagine and glutamine, both of which are able to cross the blood-brain barrier. Glutamine is deaminated to glutamate, which is then transaminated to α-ketoglutarate, indirectly replenishing oxaloacetate. Asparagine can be deaminated to aspartate, which can be converted to oxaloacetate. This serves as a second supply of oxaloacetate. To date, these attempts at dietary intervention have not proved successful. Perhaps in time research will provide the tools for enzyme replacement therapy or gene therapy that could alleviate the symptoms.

For Further Understanding

▶ Write an equation showing the reaction catalyzed by pyruvate carboxylase using structural formulas for pyruvate and oxaloacetate.

▶ Supplementing the diet with asparagine and glutamine was tried as a treatment for pyruvate carboxylase deficiency. Write equations showing the reactions that convert these amino acids into citric acid cycle intermediates.

Step 5. Finally, arginine is hydrolyzed to generate urea, to be excreted, and ornithine, the original reactant in the cycle. *Arginase* is the enzyme that catalyzes this reaction.

| Arginine | Water | | Urea | | Ornithine |

Note that one of the amino groups in urea is derived from the ammonium ion and the second is derived from the amino acid aspartate.

LEARNING GOAL

9 Discuss the cause and effect of
hyperammonemia.

There are genetically transmitted diseases that result from a deficiency of one of the enzymes of the urea cycle. The importance of the urea cycle is apparent when we consider the terrible symptoms suffered by afflicted individuals. A deficiency of urea cycle enzymes causes an elevation of the concentration of NH_4^+, a condition known as **hyperammonemia.** If there is a complete deficiency of one of the enzymes of the urea cycle, the result is death in early infancy. If there is a partial deficiency of one of the enzymes of the urea cycle, the result may be mental challenge, convulsions, and vomiting. In these milder forms of hyperammonemia, a low-protein diet leads to a lower concentration of NH_4^+ in blood and less severe clinical symptoms.

Question 22.13   What is the purpose of the urea cycle?

Question 22.14   Where do the reactions of the urea cycle occur?

22.9   Overview of Anabolism: The Citric Acid Cycle as a Source of Biosynthetic Intermediates

LEARNING GOAL

10 Summarize the role of the citric acid
cycle in catabolism and anabolism.

So far, we have talked about the citric acid cycle only as an energy-harvesting mechanism. We have seen that dietary carbohydrates and amino acids enter the pathway at various stages and are oxidized to generate NADH and $FADH_2$, which, by means of oxidative phosphorylation, are used to make ATP.

However, the role of the citric acid cycle in cellular metabolism involves more than just **catabolism.** It plays a key role in **anabolism,** or biosynthesis, as well. Figure 22.13 shows the central role of glycolysis and the citric acid cycle as energy-harvesting reactions, as well as their role as a source of biosynthetic precursors.

As you may already suspect from the fact that amino acids can be converted into citric acid cycle intermediates, these same citric acid cycle intermediates can also be used as starting materials for the synthesis of amino acids. Oxaloacetate provides the carbon skeleton for the one-step synthesis of the amino acid aspartate by the transamination reaction:

$$\text{oxaloacetate} + \text{glutamate} \rightleftharpoons \text{aspartate} + \alpha\text{-ketoglutarate}$$

Aside from providing aspartate for protein synthesis, this reaction provides aspartate for the urea cycle.

Asparagine is made from aspartate by the amination reaction

$$\text{aspartate} + NH_4^+ + \text{ATP} \longrightarrow \text{asparagine} + \text{AMP} + PP_i + H^+$$

The nine amino acids not shown in Figure 22.13 (histidine, isoleucine, leucine, lysine, methionine, phenylalanine, threonine, tryptophan, and valine) are called the essential amino acids (Section 18.11) because they cannot be synthesized by humans. Arginine is an essential amino acid for infants and adults under physical stress.

α-Ketoglutarate serves as the starting carbon chain for the family of amino acids, including glutamate, glutamine, proline, and arginine. Glutamate is especially important because it serves as the donor of the α-amino group of almost all other amino acids. It is synthesized from NH_4^+ and α-ketoglutarate in a reaction mediated by glutamate dehydrogenase. This is the reverse of the reaction shown in Figure 22.11 and previously described. In this case, the coenzyme that serves as the reducing agent is NADPH.

$$NH_4^+ + \alpha\text{-ketoglutarate} + \text{NADPH} \rightleftharpoons \text{L-glutamate} + NADP^+ + H_2O$$

Glutamine, proline, and arginine are synthesized from glutamate.

Examination of Figure 22.13 reveals that serine, glycine, and cysteine are synthesized from 3-phosphoglycerate; alanine is synthesized from pyruvate; and tyrosine is produced from phosphoenolpyruvate and the four-carbon sugar erythrose-4-phosphate, which, in turn, is synthesized from glucose-6-phosphate in the pentose phosphate pathway. In addition to the amino acid precursors, glycolysis and the citric acid cycle also provide precursors for lipids and the nitrogenous

Actually, in humans tyrosine is made from the essential amino acid phenylalanine.

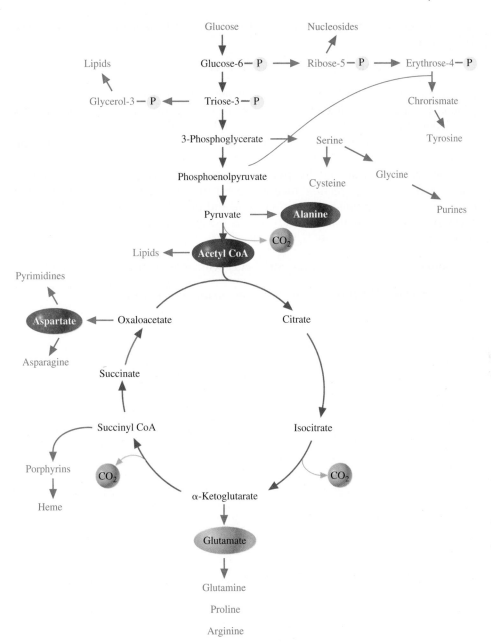

Figure 22.13 Glycolysis, the pentose phosphate pathway, and the citric acid cycle also provide a variety of precursors for the biosynthesis of amino acids, nitrogenous bases, and porphyrins.

bases required to make DNA, the molecule that carries the genetic information. They also generate precursors for heme, the prosthetic group that is required for hemoglobin, myoglobin, and the cytochromes.

Clearly, the reactions of glycolysis and the citric acid cycle are central to both anabolic and catabolic cellular activities. Metabolic pathways that function in both anabolism and catabolism are called **amphibolic pathways.** Consider for a moment the difficulties that the dual nature of these pathways could present to the cell. When the cell is actively growing, there is a great demand for biosynthetic precursors to build new cell structures. A close look at Figure 22.13 shows us that periods of active cell growth and biosynthesis may deplete the supply of citric acid cycle intermediates. The problem is, the processes of growth and biosynthesis also require a great deal of ATP!

The solution to this problem is to have an alternative pathway for oxaloacetate synthesis that can produce enough oxaloacetate to supply the anabolic and catabolic requirements of the cell. Although bacteria and plants have several mechanisms, the only way that mammalian cells can produce more oxaloacetate is by the

Carboxylation of pyruvate during gluco-neogenesis is discussed in Section 21.6.

See also A Medical Perspective: Pyruvate Carboxylase Deficiency earlier in this chapter.

carboxylation of pyruvate, a reaction that is also important in gluconeogenesis. This reaction is

$$\text{pyruvate} + CO_2 + ATP \longrightarrow \text{oxaloacetate} + ADP + P_i$$

The enzyme that catalyzes this reaction is *pyruvate carboxylase.* It is a conjugated protein having as its covalently linked prosthetic group the vitamin *biotin.* This enzyme is "turned on" by high levels of acetyl CoA, a signal that the cell requires high levels of the citric acid cycle intermediates, particularly oxaloacetate, the beginning substrate.

The reaction catalyzed by pyruvate carboxylase is called an **anaplerotic reaction.** The term *anaplerotic* means "to fill up." Indeed, this critical enzyme must constantly replenish the oxaloacetate and thus, indirectly, all the citric acid cycle intermediates that are withdrawn as biosynthetic precursors for the reactions summarized in Figure 22.13.

Question 22.15   Explain how the citric acid cycle serves as an amphibolic pathway.

Question 22.16   What is the function of an anaplerotic reaction?

CHAPTER MAP

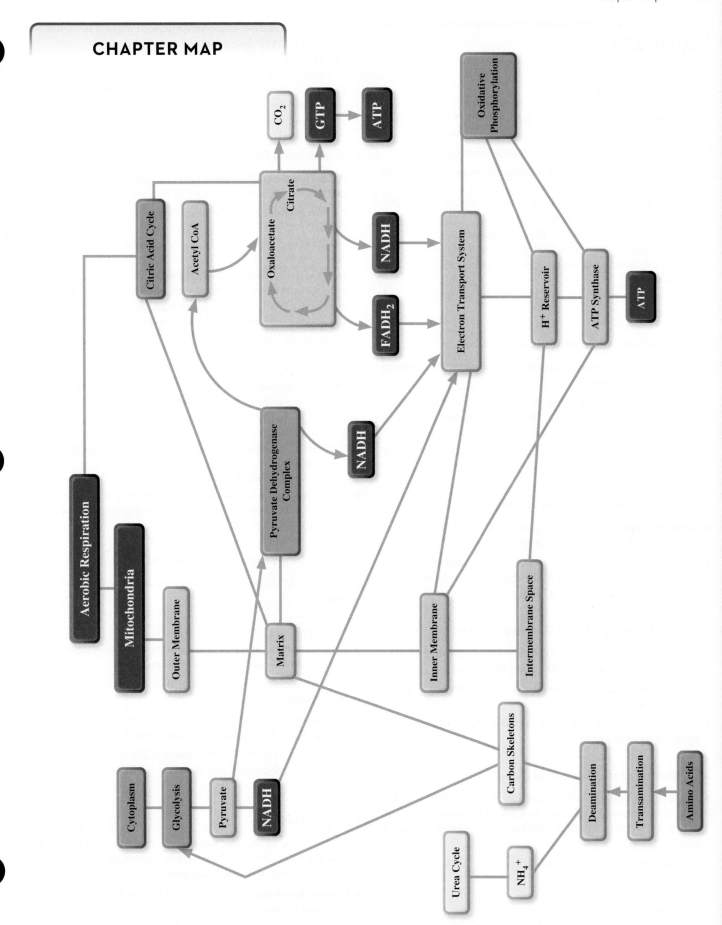

SUMMARY

22.1 The Mitochondria

▶ **Mitochondria** are aerobic cell organelles that are responsible for most of the ATP production in eukaryotic cells.

▶ Mitochondria are enclosed by a double membrane.

- The **outer mitochondrial membrane** permits low-molar-mass molecules to pass through.
- The **inner mitochondrial membrane** is almost completely impermeable to most molecules.
  - It is the site of oxidative phosphorylation.
  - It is highly folded into **cristae** to increase the surface area.
- The **intermembrane space** is the site of the high-energy H^+ reservoir.
- The **matrix space** contains the enzymes of the citric acid cycle.

22.2 Conversion of Pyruvate to Acetyl CoA

▶ Under aerobic conditions, pyruvate is oxidized by the **pyruvate dehydrogenase complex.**

▶ In this reaction, **coenzyme A** bonds to an acetyl group through a thioester bond to produce acetyl coenzyme A or **acetyl CoA.**

▶ Acetyl CoA is a central molecule in both anabolism and catabolism.

22.3 An Overview of Aerobic Respiration

▶ **Aerobic respiration** is the oxygen-requiring degradation of food molecules and production of ATP.

▶ Oxidative phosphorylation is the process that uses high-energy electrons harvested by oxidation of substrates of the citric acid cycle to produce ATP.

22.4 The Citric Acid Cycle (The Krebs Cycle)

▶ The **citric acid cycle** is the final pathway for the degradation of carbohydrates, amino acids, and fatty acids.

▶ This pathway carries out the complete oxidation of carbon skeletons of food molecules.

▶ The enzymes of the citric acid cycle are found in the matrix space of the mitochondria.

22.5 Control of the Citric Acid Cycle

▶ Because the rate of ATP production by the cell must vary with the amount of available oxygen and the energy requirements of the body, the citric acid cycle is regulated at several steps. These are:

- Conversion of pyruvate to acetyl CoA (pyruvate dehydrogenase)
- Synthesis of citrate from oxaloacetate and acetyl CoA (citrate synthase)

- Oxidation and decarboxylation of isocitrate to α-ketoglutarate (isocitrate dehydrogenase)
- Conversion of α-ketoglutarate to succinyl CoA (α-ketoglutarate dehydrogenase)

▶ This regulation allows the cells to produce more ATP when it is needed for activity and less ATP when the body is at rest and the requirement for ATP is lower.

22.6 Oxidative Phosphorylation

▶ **Oxidative phosphorylation** is the process by which NADH and $FADH_2$ are oxidized and ATP is produced.

- Two molecules of ATP are produced when one $FADH_2$ is oxidized.
- Three molecules of ATP are produced when one NADH is oxidized.

▶ The **electron transport system** is the series of electron transport proteins embedded within the inner mitochondrial membrane that accepts high-energy electrons from NADH and $FADH_2$ and transfers them in stepwise fashion to O_2.

- At three sites in the electron transport system, H^+ are pumped into the high-energy H^+ reservoir in the intermembrane space.
- **ATP synthase**, or the F_0F_1 **complex**, is a protein complex in the inner mitochondrial membrane.
  - H^+ pass through the F_0 portion of the complex and return to the matrix space.
  - The F_1 portion catalyzes the phosphorylation of ADP to form ATP.
- The **terminal electron acceptor** is molecular oxygen (O_2).

▶ The complete oxidation of a glucose molecule by glycolysis, the citric acid cycle, and oxidative phosphorylation produces thirty-six ATP molecules.

22.7 The Degradation of Amino Acids

▶ Amino acid degradation occurs in the liver and has two stages.

- Removal of the α-amino group
- Degradation of the carbon skeleton

▶ Removal of the α-amino group occurs in a **transamination** reaction and is catalyzed by a **transaminase.**

▶ All transaminases require the coenzyme **pyridoxal phosphate.**

▶ **Oxidative deamination** is the reaction by which the glutamate produced in transamination is deaminated.

▶ The carbon skeletons of the amino acids enter a number of reactions of glycolysis and the citric acid cycle and are completely oxidized.

22.8 The Urea Cycle

▶ In the **urea cycle,** the toxic ammonium ions released by deamination of amino acids are incorporated into urea, which is excreted through the urine.

▶ **Hyperammonemia** is a genetic condition that results from a mutation of one of the enzymes of the urea cycle.

22.9 Overview of Anabolism: The Citric Acid Cycle as a Source of Biosynthetic Intermediates

▶ The reactions of the citric acid cycle have key roles in both **anabolism** and **catabolism.**

▶ Amino acids, lipids, and nitrogenous bases for nucleic acids all can be synthesized from many intermediates in the citric acid cycle and glycolysis.

▶ Metabolic pathways that function in both anabolism and catabolism are called **amphibolic pathways.**

▶ An **anaplerotic reaction** replenishes a substrate needed for a biochemical reaction.

ANSWERS TO PRACTICE PROBLEMS

22.1 The ATP yield from one molecule of pyruvate can be summarized as follows:

Conversion of pyruvate to acetyl CoA

$$1 \text{ NADH} \times 3 \text{ ATP/NADH} = 3 \text{ ATP}$$

Citric Acid Cycle

$1 \text{ GTP} \times 1 \text{ ATP/GTP}$	$= 1 \text{ ATP}$
$3 \text{ NADH} \times 3 \text{ ATP/NADH}$	$= 9 \text{ ATP}$
$1 \text{ FADH}_2 \times 2 \text{ ATP/FADH}_2$	$= \underline{2 \text{ ATP}}$
Total	$= 15 \text{ ATP}$

QUESTIONS AND PROBLEMS

The Mitochondria

Foundations

22.17 Define the term *mitochondrion.*
22.18 Define the term *cristae.*

Applications

22.19 What is the function of the intermembrane compartment of the mitochondria?
22.20 What biochemical processes occur in the matrix space of the mitochondria?
22.21 In what important way do the inner and outer mitochondrial membranes differ?
22.22 What kinds of proteins are found in the inner mitochondrial membrane?

Conversion of Pyruvate to Acetyl CoA

Foundations

22.23 What is coenzyme A?
22.24 What is the role of coenzyme A in the reaction catalyzed by pyruvate dehydrogenase?
22.25 In the reaction catalyzed by pyruvate dehydrogenase, pyruvate is decarboxylated. What is meant by the term *decarboxylation?*

22.26 In the reaction catalyzed by pyruvate dehydrogenase, pyruvate is also oxidized. What substance is reduced when pyruvate is oxidized? What is the product of that reduction reaction?

Applications

22.27 Under what metabolic conditions is pyruvate converted to acetyl CoA?
22.28 Write a chemical equation for the production of acetyl CoA from pyruvate. Under what conditions does this reaction occur?
22.29 How could a deficiency of riboflavin, thiamine, niacin, or pantothenic acid reduce the amount of ATP the body can produce?
22.30 In what form are the vitamins riboflavin, thiamine, niacin, and pantothenic acid needed by the pyruvate dehydrogenase complex?

The Citric Acid Cycle (The Krebs Cycle)

Foundations

22.31 The reaction catalyzed by citrate synthase is a condensation reaction. Define the term *condensation.*
22.32 The pair of reactions catalyzed by aconitase results in the conversion of isocitrate to its isomer citrate. What are isomers?
22.33 What general type of reaction is occurring in the conversion of isocitrate to α-ketoglutarate?
22.34 Write the equation for the conversion of isocitrate to α-ketoglutarate, and circle the chemical change that reveals the type of reaction that is occurring.
22.35 The reaction catalyzed by succinate dehydrogenase is a dehydrogenation reaction. What is meant by the term *dehydrogenation reaction?*
22.36 The reaction catalyzed by fumarase is an example of the hydration of an alkene to produce an alcohol. Write the equation for this reaction. What is meant by the term *hydration reaction?*
22.37 Match each of the following enzymes with the class of enzyme to which it belongs. (*Hint:* An enzyme classification may be used more than once or not at all.)

a. Citrate synthase	**1.** Transferase
b. Aconitase	**2.** Oxidoreductase
c. Isocitrate dehydrogenase	**3.** Kinase
d. α-Ketoglutarate dehydrogenase	**4.** Hydrolase
e. Succinyl CoA synthase	**5.** Lyase
f. Succinate dehydrogenase	**6.** Isomerase
g. Fumarase	
h. Malate dehydrogenase	

22.38 Describe the reaction catalyzed by each of the enzymes listed in Question 22.37.
22.39 Is the following statement true or false? If false, rewrite the statement to make it accurate. Acetyl CoA transfers an acetyl group from pyruvate to citrate.
22.40 Are the following statements true or false? If false, rewrite the statements to make them accurate. Glycolysis and the citric acid cycle are aerobic processes. These anabolic processes occur in the mitochondria and the cytoplasm, respectively.
22.41 How many ions of NAD^+ are reduced to molecules of NADH during one turn of the citric acid cycle?

22.42 How many molecules of FAD are converted to $FADH_2$ during one turn of the citric acid cycle?

22.43 What is the net yield of ATP for anaerobic glycolysis?

22.44 How many molecules of ATP are produced by the complete degradation of glucose via glycolysis, the citric acid cycle, and oxidative phosphorylation?

22.45 What is the function of acetyl CoA in the citric acid cycle?

22.46 What is the function of oxaloacetate in the citric acid cycle?

22.47 GTP is formed in one step of the citric acid cycle. How is this GTP converted into ATP?

22.48 What is the chemical meaning of the term *decarboxylation?* Give an example of a decarboxylase.

Applications

22.49 Fumarase converts fumarate to malate. Explain this reaction in terms of the chemistry of alcohols and alkenes.

22.50 The enzyme aconitase catalyzes the isomerization of citrate into isocitrate. Discuss the two reactions catalyzed by aconitase in terms of the chemistry of alcohols and alkenes.

22.51 A bacterial culture is given ^{14}C-labeled pyruvate as its sole source of carbon and energy. The following is the structure of the radiolabeled pyruvate.

$$*CH_3-\overset{\overset{O}{\|}}{C}-\overset{\overset{O}{\|}}{C}-O^-$$

Follow the fate of the radioactive carbon through the reactions of the citric acid cycle.

22.52 A bacterial culture is given ^{14}C-labeled pyruvate as its sole source of carbon and energy. The following is the structure of the radiolabeled pyruvate.

$$CH_3-\overset{\overset{O}{\|}}{C^*}-\overset{\overset{O}{\|}}{C}-O^-$$

Follow the fate of the radioactive carbon through the reactions of the citric acid cycle.

22.53 In the oxidation of malate to oxaloacetate, what is the structural evidence that an oxidation reaction has occurred? What functional groups are involved?

22.54 In the oxidation of succinate to fumarate, what is the structural evidence that an oxidation reaction has occurred? What functional groups are involved?

22.55 To what class of enzymes does dinucleotide diphosphokinase belong? Explain your answer.

22.56 To what class of enzymes does succinate dehydrogenase belong? Explain your answer.

22.57 Explain why mutations of the citric acid cycle enzymes frequently appear first in the central nervous system.

22.58 Why would a deficiency of α-ketoglutarate dehydrogenase cause chronic lactic acidosis?

22.59 Explain why deficiencies of citric acid cycle enzymes cause hypotonia.

22.60 Why is myoglobinuria associated with genetic disorders of the enzymes of the citric acid cycle?

Control of the Citric Acid Cycle

Foundations

22.61 Define the term *allosteric enzyme.*
22.62 Define the term *effector.*

22.63 Why are allosteric enzymes an efficient means to regulate a biochemical pathway?

22.64 Why are ADP and ATP efficient effector molecules for allosteric enzymes that regulate a biochemical pathway such as the citric acid cycle?

Applications

22.65 What four allosteric enzymes or enzyme complexes are responsible for the regulation of the citric acid cycle?

22.66 Which of the four allosteric enzymes or enzyme complexes in the citric acid cycle are under negative allosteric control? Which are under positive allosteric control?

22.67 What is the importance of the regulation of the citric acid cycle?

22.68 Explain the role of allosteric enzymes in control of the citric acid cycle.

22.69 What molecule serves as a signal to increase the rate of the reactions of the citric acid cycle?

22.70 What molecules serve as signals to decrease the rate of the reactions of the citric acid cycle?

Oxidative Phosphorylation

Foundations

22.71 Define the term *electron transport system.*
22.72 What is the terminal electron acceptor in aerobic respiration?

Applications

22.73 How many molecules of ATP are produced when one molecule of NADH is oxidized by oxidative phosphorylation?

22.74 How many molecules of ATP are produced when one molecule of $FADH_2$ is oxidized by oxidative phosphorylation?

22.75 What is the source of energy for the synthesis of ATP in mitochondria?

22.76 What is the name of the enzyme that catalyzes ATP synthesis in mitochondria?

22.77 What is the function of the electron transport systems of the mitochondria?

22.78 What is the cellular location of the electron transport systems?

22.79 **a.** Compare the number of molecules of ATP produced by glycolysis to the number of ATP molecules produced by oxidation of glucose by aerobic respiration.
b. Which pathway produces more ATP? Explain.

22.80 At which steps in the citric acid cycle do oxidation-reduction reactions occur?

The Degradation of Amino Acids

Foundations

22.81 What chemical transformation is carried out by transaminases?

22.82 Write a chemical equation for the transfer of an amino group from alanine to α-ketoglutarate, catalyzed by a transaminase.

22.83 Why is the glutamate family of transaminases so important?

22.84 What biochemical reaction is catalyzed by glutamate dehydrogenase?

Applications

22.85 Into which citric acid cycle intermediate is each of the following amino acids converted?

- **a.** Alanine
- **b.** Glutamate
- **c.** Aspartate
- **d.** Phenylalanine
- **e.** Threonine
- **f.** Arginine

22.86 What is the net ATP yield for degradation of each of the amino acids listed in Question 22.85?

22.87 Explain the mechanism of transamination.

22.88 What is the role of vitamin B_6 in transamination?

The Urea Cycle

22.89 What metabolic condition is produced if the urea cycle does not function properly?

22.90 What is hyperammonemia? How are mild forms of this disease treated?

22.91 The structure of urea is

- **a.** What substances are the sources of each of the amino groups in the urea molecule?
- **b.** What substance is the source of the carbonyl group?

22.92 What is the energy source used for the urea cycle?

Overview of Anabolism: The Citric Acid Cycle as a Source of Biosynthetic Intermediates

Foundations

22.93 Define the term *anabolism.*

22.94 Define the term *catabolism.*

Applications

22.95 From which citric acid cycle intermediate is the amino acid glutamate synthesized?

22.96 What amino acids are synthesized from α-ketoglutarate?

22.97 What is the role of the citric acid cycle in biosynthesis?

22.98 How are citric acid cycle intermediates replenished when they are in demand for biosynthesis?

22.99 What is meant by the term *essential amino acid?*

22.100 What are the nine essential amino acids?

22.101 Write a balanced equation for the reaction catalyzed by pyruvate carboxylase.

22.102 How does the reaction described in Question 22.101 allow the citric acid cycle to fulfill its roles in both catabolism and anabolism?

CHALLENGE PROBLEMS

1. A 1-month-old baby boy was brought to the hospital showing severely delayed development and cerebral atrophy. Blood tests showed high levels of lactate and pyruvate. By 3 months of age, very high levels of succinate and fumarate were found in the urine. Fumarase activity was absent in the liver and muscle tissue. The baby died at 5 months of age. This was the first reported case of fumarase deficiency, and the defect was recognized too late for effective therapy to be administered. What reaction is catalyzed by fumarase? How would a deficiency of this mitochondrial enzyme account for the baby's symptoms and test results?

2. A certain bacterium can grow with ethanol as its only source of energy and carbon. Propose a pathway to describe how ethanol can enter a pathway that would allow ATP production and synthesis of precursors for biosynthesis.

3. Fluoroacetate has been used as a rat poison and can be fatal when eaten by humans. Patients with fluoroacetate poisoning accumulate citrate and fluorocitrate within the cells. What enzyme is inhibited by fluoroacetate? Explain your reasoning.

4. The pyruvate dehydrogenase complex is activated by removal of a phosphoryl group from *pyruvate dehydrogenase*. This reaction is catalyzed by the enzyme pyruvate dehydrogenase phosphate phosphatase. A baby is born with a defect in this enzyme. What effects would this defect have on the rate of each of the following pathways: aerobic respiration, glycolysis, lactate fermentation? Explain your reasoning.

5. Pyruvate dehydrogenase phosphate phosphatase is stimulated by Ca^{2+}. In muscles, the Ca^{2+} concentration increases dramatically during muscle contraction. How would the elevated Ca^{2+} concentration affect the rate of glycolysis and the citric acid cycle?

6. Liver contains high levels of nucleic acids. When excess nucleic acids are degraded, ribose-5-phosphate is one of the degradation products that accumulate in the cell. Can this substance be used as a source of energy? What pathway would be used?

7. In birds, arginine is an essential amino acid. Can birds produce urea as a means of removing ammonium ions from the blood? Explain your reasoning.

23

Fatty Acid Metabolism

Olive oil, prepared by pressing ripe olives, is a healthy alternative in our diet.

INTRODUCTION

Triglycerides are our most concentrated energy reserve, yielding 9 kilocalories per gram (kcal/g) when completely oxidized. Compare this with the energy yield for carbohydrates, which is only 4 kcal/g. Lipid metabolism in mammals is extremely complex. There is great concern about the epidemic of obesity that is afflicting the United States (see A Medical Perspective: Obesity: A Genetic Disorder? later in this chapter). No matter the source of dietary calories, when excess calories are ingested, our bodies store them as triglycerides in adipose tissue.

For some animals, large lipid stores are essential for survival. Animals such as bears that may hibernate for as many as 7 months rely solely on their fat reserves as the source of metabolic energy. The ruby-throated hummingbird migrates from New England to the West Indies in the fall of the year. As the time of migration nears, the hummingbirds accumulate as much as 40% of their body weight as triglycerides. This provides them with the energy to fly non-stop at speeds of nearly 30 miles per hour (mph) for as long as 60 hours (h).

The metabolism of fatty acids and lipids revolves around the fate of acetyl CoA. We saw in Chapter 22 that, under aerobic conditions, pyruvate is converted to acetyl CoA, which feeds into the citric acid cycle. Fatty acids are also degraded to acetyl CoA and oxidized by the citric acid cycle. Moreover, acetyl CoA is itself the starting material for the biosynthesis of fatty acids, cholesterol, and steroid hormones. Acetyl CoA is thus a key intermediary in lipid metabolism.

23.1 Lipid Metabolism in Animals

Digestion and Absorption of Dietary Triglycerides

Triglycerides, also called *triacylglycerols,* are fatty acid esters of glycerol and are highly hydrophobic ("water fearing"). Because of this, they must be processed before they can be digested, absorbed, and metabolized. Because processing of dietary lipids occurs in the small intestine, the water-soluble **lipases,** enzymes that hydrolyze triglycerides, which are found in the stomach and in the saliva are not very effective. In fact, most dietary fat arrives in the duodenum, the first part of the small intestine, in the form of fat globules. These fat globules stimulate the secretion of bile from the gallbladder. **Bile** is composed of micelles of lecithin, cholesterol, protein, bile salts, inorganic ions, and bile pigments. **Micelles** (Figure 23.1) are aggregations of amphipathic molecules, which are molecules having a polar region and a nonpolar region. The nonpolar (hydrophobic or "water fearing") ends of bile salts tend to bunch together when placed in water. The hydrophilic ("water loving") regions of these molecules interact with water. Bile salts are made in the liver and stored in the gallbladder, awaiting the stimulus to be secreted into the duodenum. The major bile salts in humans are cholate and chenodeoxycholate (Figure 23.2).

Cholesterol is almost completely insoluble in water, but the conversion of cholesterol to bile salts creates *detergents* whose polar heads make them soluble in the aqueous phase and whose hydrophobic tails bind triglycerides. After a meal is eaten, bile flows through the common bile duct into the duodenum, where bile salts emulsify the fat globules into tiny droplets. This increases the surface area of the lipid molecules, allowing them to be more easily hydrolyzed by lipases (Figure 23.3).

Much of the lipid in these droplets is in the form of **triglycerides,** which are fatty acid esters of glycerol. A protein called **colipase** binds to the surface of the lipid droplets and helps pancreatic lipases stick to the surface and hydrolyze the ester bonds between the glycerol and fatty acids of the triglycerides (Figure 23.4). In this process, two of the three fatty acids are liberated, and the monoglycerides and free fatty acids produced mix freely with the micelles of bile. These micelles

LEARNING GOAL

1 Summarize the digestion and storage of lipids.

See Sections 14.2 and 17.2 for a discussion of micelles.

Triglycerides are described in Section 17.3.

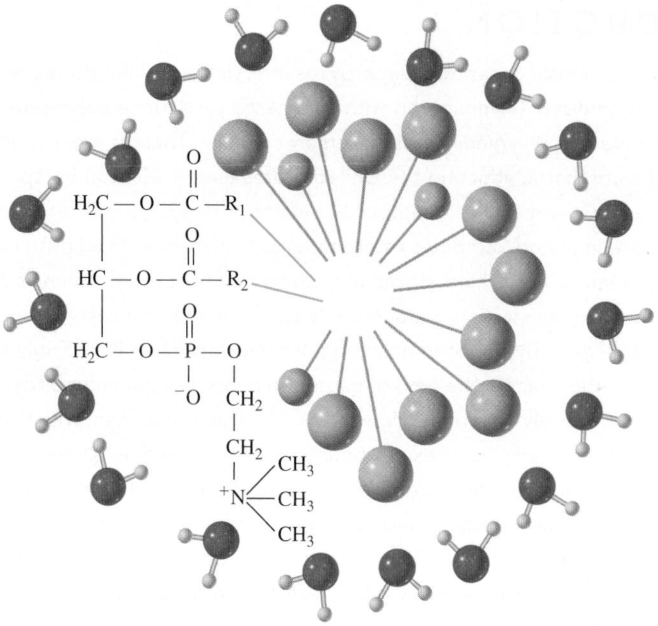

Figure 23.1 The structure of a micelle formed from the phospholipid lecithin. The straight lines represent the long hydrophobic fatty acid tails, and the spheres represent the hydrophilic heads of the phospholipid.

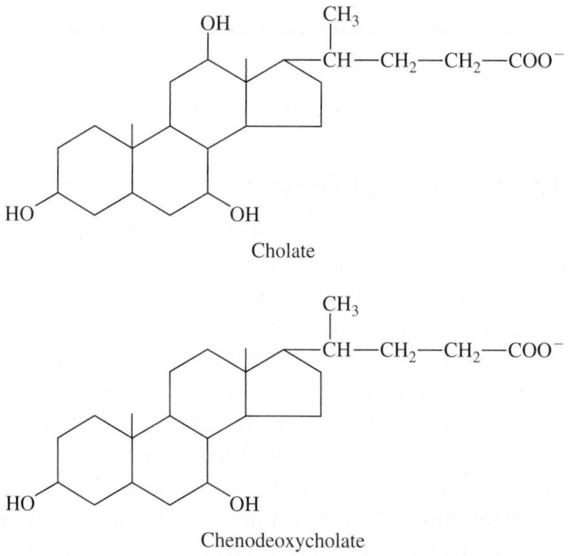

Figure 23.2 Structures of the most common bile salts in human bile: cholate and chenodeoxycholate.

Cholate

Chenodeoxycholate

are readily absorbed through the membranes of the intestinal epithelial cells (Figure 23.3).

Surprisingly, the monoglycerides and fatty acids are then reassembled into triglycerides that are combined with protein to produce the class of plasma lipoproteins called **chylomicrons** (Figure 23.3). These collections of lipid and protein are secreted into small lymphatic vessels and eventually arrive in the bloodstream. In the bloodstream, the triglycerides are once again hydrolyzed to produce glycerol and free fatty acids that are then absorbed by the cells. If the body needs energy, these molecules are degraded to produce ATP. If the body does not need energy, these energy-rich molecules are stored.

Plasma lipoproteins are described in Section 17.5.

Lipid Storage

Fatty acids are stored in the form of triglycerides. Most of the body's triglyceride molecules are stored as fat droplets in the cytoplasm of **adipocytes** (fat cells) that make up **adipose tissue.** Each adipocyte contains a large fat droplet

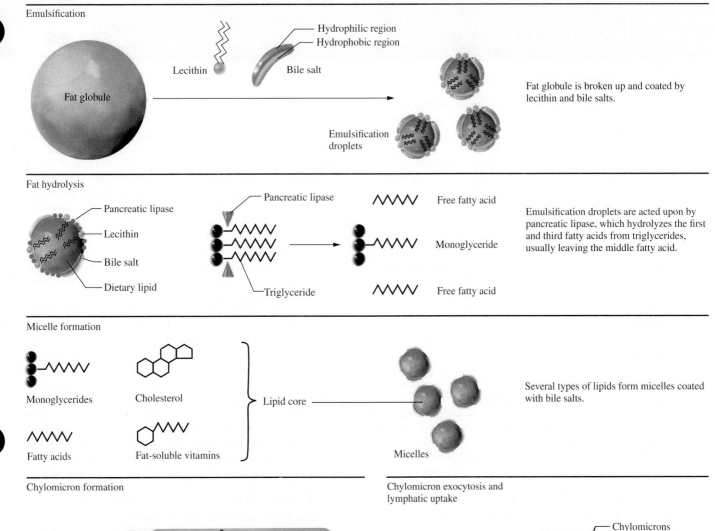

Emulsification

Fat globule is broken up and coated by lecithin and bile salts.

Fat hydrolysis

Emulsification droplets are acted upon by pancreatic lipase, which hydrolyzes the first and third fatty acids from triglycerides, usually leaving the middle fatty acid.

Micelle formation

Several types of lipids form micelles coated with bile salts.

Chylomicron formation

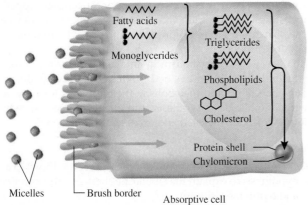

Intestinal cells absorb lipids from micelles, resynthesize triglycerides, and package triglycerides, cholesterol, and phospholipids into protein-coated chylomicrons.

Chylomicron exocytosis and lymphatic uptake

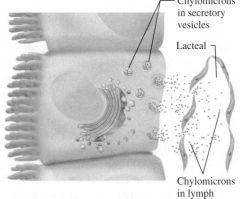

Golgi complex packages chylomicrons into secretory vesicles; chylomicrons are released from the basal cell membrane by exocytosis and enter the lacteal (lymphatic capillary).

Figure 23.3 Stages of lipid digestion in the intestinal tract.

that accounts for nearly the entire volume of the cell. Other cells, such as those of cardiac muscle, contain a few small fat droplets. In these cells, the fat droplets are surrounded by mitochondria. When the cells need energy, triglycerides are hydrolyzed to release fatty acids that are transported into the matrix space

A Medical Perspective

Obesity: A Genetic Disorder?

Approximately a third of all Americans are obese; that is, they are more than 20% overweight. One million are morbidly obese; they carry so much extra weight that it threatens their health. Many obese people simply eat too much and exercise too little, but others actually gain weight even though they eat fewer calories than people of normal weight. This observation led many researchers to the hypothesis that obesity in some people is a genetic disorder.

This hypothesis was supported by the 1950 discovery of an obesity mutation in mice. Selective breeding produced a strain of genetically obese mice from the original mutant mouse. The hypothesis was further strengthened by the results of experiments performed in the 1970s by Douglas Coleman. Coleman connected the circulatory systems of a genetically obese mouse and a normal mouse. The obese mouse started eating less and lost weight. Coleman concluded that there was a substance in the blood of normal mice that signals the brain to decrease the appetite. Obese mice, he hypothesized, can't produce this "satiety factor," and thus they continue to eat and gain weight.

In 1987, Jeffrey Friedman assembled a team of researchers to map and then clone the obesity gene that was responsible for appetite control. In 1994, after 7 years of intense effort, the scientists achieved their goal, but they still had to demonstrate that the protein encoded by the cloned obesity gene did, indeed, have a metabolic effect. The gene was modified to be compatible with the genetic system of bacteria so that they could be used to manufacture the protein. When the engineered gene was then introduced into bacteria, they produced an abundance of the protein product. The protein was then purified in preparation for animal testing.

The researchers calculated that a normal mouse has about 12.5 milligrams (mg) of the protein in its blood. They injected that amount into each of ten mice that were so fat they couldn't squeeze into the feeding tunnels used for normal mice. The day after the first injection, graduate student Jeff Halaas observed that the mice had eaten less food. Injections were given daily, and each day the obese mice ate less. After 2 weeks of treatment, each of the ten mice had lost about 30% of its weight. In addition, the mice had become more active and their metabolisms had speeded up.

When normal mice underwent similar treatment, their body fat fell from 12.2% to 0.67%, which meant that these mice had no extra fat tissue. The 0.67% of their body weight represented by fat was accounted for by the membranes that surround each of the cells of their bodies! Because of the dramatic

results, Friedman and his colleagues called the protein *leptin*, from the Greek word *leptos*, meaning slender.

The leptin protein is a hormone that functions as a signal in a metabolic thermostat. Fat cells produce leptin and secrete it into the bloodstream. As a result, the leptin concentration in a normal person is proportional to the amount of body fat. The blood concentration of the hormone is monitored by the hypothalamus, a region known to control appetite and set metabolic rates. When the concentration reaches a certain level, it triggers the hypothalamus to suppress the appetite. If no leptin or only small amounts of it are produced, the hypothalamus "thinks" that the individual has too little body fat or is starving. Under these circumstances it does not send a signal to suppress hunger and the individual continues to eat.

The human leptin gene also has been cloned and shown to correct genetic obesity in mice. Unfortunately, the dramatic results achieved with mice were *not* observed with humans. Why? It seems that nearly all of the obese volunteers already produced an abundance of leptin. In fact, fewer than ten people have been found, to date, who do not produce leptin and many obese people have very high levels of leptin in the blood. It appears that, as with type 2 diabetes and insulin, these people are no longer sensitive to the leptin produced by their fat cells.

Ghrelin is another hormone that influences appetite. Produced in the stomach, ghrelin stimulates appetite. As you would predict, the level of ghrelin is high before a meal and decreases following a meal. Many obese people have high levels of ghrelin and therefore experience constant hunger.

Obestatin is another hormone that has been discovered to influence body weight. This hormone decreases appetite. Interestingly, both ghrelin and obestatin are encoded in a single gene. When the protein is produced, it is cleaved into the two peptide hormones.

Clearly lipid metabolism in animals is a complex process and is not yet fully understood. In this chapter, we will study other aspects of lipid metabolism: the pathways for fatty acid degradation and biosynthesis and the processes by which dietary lipids are digested and excess lipids are stored.

For Further Understanding

▶ Go online to learn the cause of type 2 diabetes. Develop a hypothesis to explain the similarities between this condition and leptin insensitivity.

▶ Design an experiment to demonstrate the opposing effects of ghrelin and obestatin.

of the mitochondria. There the fatty acids are completely oxidized, and ATP is produced.

The fatty acids provided by the hydrolysis of triglycerides are a very rich energy source for the body. The complete oxidation of fatty acids releases much more energy than the oxidation of a comparable amount of glycogen.

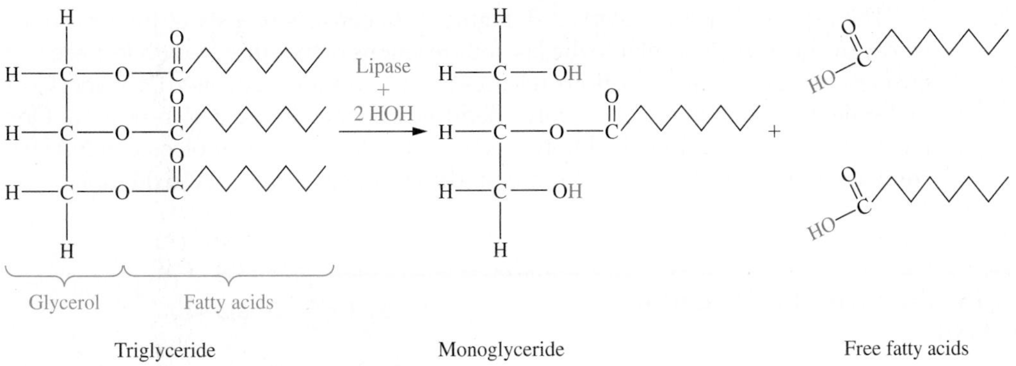

Figure 23.4 The action of pancreatic lipase in the hydrolysis of dietary lipids.

Question 23.1 How do bile salts aid in the digestion of dietary lipids?

Question 23.2 Why must dietary lipids be processed before enzymatic digestion can be effective?

23.2 Fatty Acid Degradation

An Overview of Fatty Acid Degradation

Early in the twentieth century, a very clever experiment was done to determine how fatty acids are degraded. Recall from Chapter 9 that radioactive elements can be incorporated into biological molecules and followed through the body. A German biochemist, Franz Knoop, devised a similar kind of labeling experiment long before radioactive tracers were available. Knoop fed dogs fatty acids in which the usual terminal methyl group had a phenyl group attached to it. Such molecules are called ω-labeled (omega-labeled) fatty acids (Figure 23.5). When he isolated the metabolized fatty acids from the urine of the dogs, he found that phenyl acetate was formed when the fatty acid had an even number of carbon atoms in the chain. But benzoate was formed when the fatty acid had an odd number of carbon atoms. Knoop interpreted these data to mean that the degradation of fatty acids occurs by the removal of two-carbon acetate groups from the carboxyl end of the fatty acid. We now know that the two-carbon fragments produced by the degradation of fatty acids are not acetate, but acetyl CoA. The pathway for the breakdown of fatty acids into acetyl CoA is called β-**oxidation.**

LEARNING GOAL

2 Describe the degradation of fatty acids by β-oxidation.

LEARNING GOAL

3 Explain the role of acetyl CoA in fatty acid metabolism.

This pathway is called β-oxidation because it involves the stepwise oxidation of the β-carbon of the fatty acid.

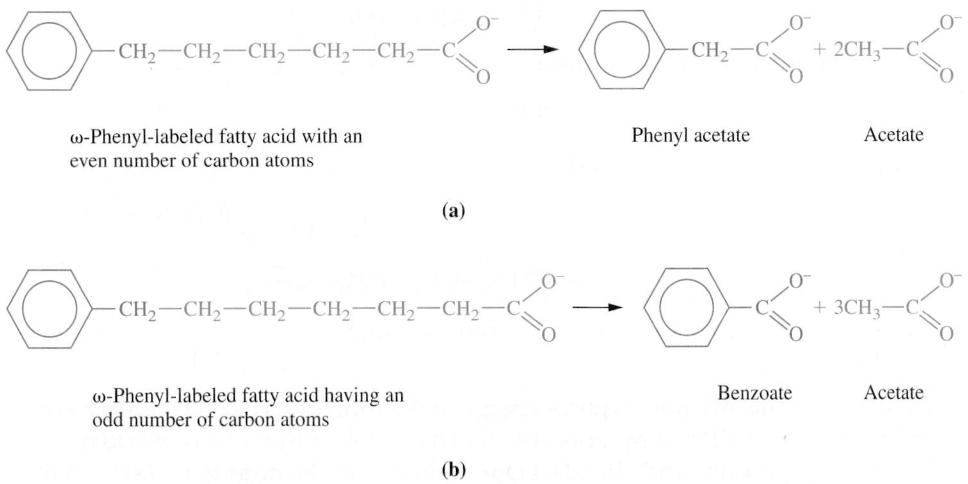

ω-Phenyl-labeled fatty acid with an even number of carbon atoms

Phenyl acetate   Acetate

(a)

ω-Phenyl-labeled fatty acid having an odd number of carbon atoms

Benzoate   Acetate

(b)

Figure 23.5 The last carbon of the chain is called the ω-carbon (omega-carbon), so the attached phenyl group is an ω-phenyl group. Oxidation of ω-phenyl-labeled fatty acids occurs two carbons at a time. (a) Fatty acids having an even number of carbon atoms are degraded to phenyl acetate and "acetate." (b) Oxidation of ω-phenyl-labeled fatty acids that contain an odd number of carbon atoms yields benzoate and "acetate."

Review Section 22.6 for the ATP yields that result from oxidation of FADH₂ and NADH.

The β-oxidation cycle (steps 2–5, Figure 23.6) consists of a set of four reactions whose overall form is similar to the last four reactions of the citric acid cycle. Each trip through the sequence of reactions releases acetyl CoA and returns a fatty acyl CoA molecule that has two fewer carbons. Reduced coenzymes are also produced. One molecule of $FADH_2$, equivalent to two ATP molecules, and one molecule of NADH, equivalent to three ATP molecules, are produced for each cycle of β-oxidation.

EXAMPLE 23.1   **Predicting the Products of β-Oxidation of a Fatty Acid**

LEARNING GOAL

2 Describe the degradation of fatty acids by β-oxidation.

What products would be produced by the β-oxidation of 10-phenyldecanoic acid?

Solution

This ten-carbon fatty acid would be broken down into four acetyl CoA molecules and one phenyl acetate molecule. Because four cycles through β-oxidation are required to break down a ten-carbon fatty acid, four NADH molecules and four $FADH_2$ molecules would also be produced.

Practice Problem 23.1

What products would be formed by β-oxidation of each of the following fatty acids?

   a. 9-Phenylnonanoic acid   c. 7-Phenylheptanoic acid
   b. 8-Phenyloctanoic acid   d. 12-Phenyldodecanoic acid

▶ Further Practice: **Questions 23.41 and 23.42.**

The Reactions of β-Oxidation

LEARNING GOAL

2 Describe the degradation of fatty acids by β-oxidation.

The enzymes that catalyze the β-oxidation of fatty acids are located in the matrix space of the mitochondria. Special transport mechanisms are required to bring fatty acid molecules into the mitochondrial matrix. Once inside, the fatty acids are degraded by the reactions of β-oxidation. As we will see, these reactions interact with oxidative phosphorylation and the citric acid cycle to produce ATP.

Reaction 1. The first step is an *activation* reaction that results in the production of a fatty acyl CoA molecule. A thioester bond is formed between coenzyme A and the fatty acid:

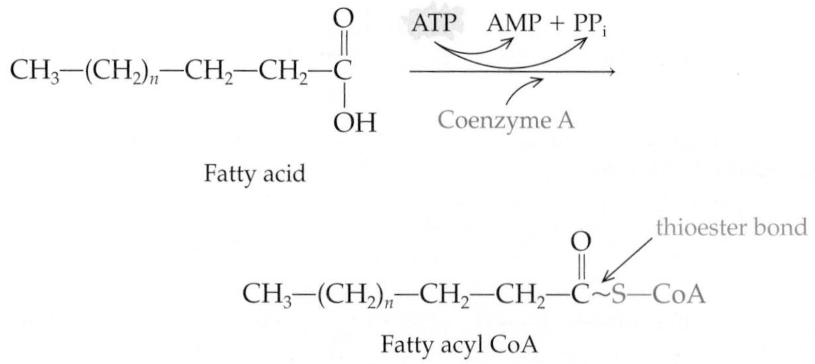

$$CH_3-(CH_2)_n-CH_2-CH_2-\underset{\underset{OH}{|}}{\overset{\overset{O}{\|}}{C}} \quad \xrightarrow[\text{Coenzyme A}]{\text{ATP}\quad\text{AMP} + PP_i}$$

Fatty acid

$$CH_3-(CH_2)_n-CH_2-CH_2-\overset{\overset{O}{\|}}{C}\!\sim\!S-CoA$$

thioester bond

Fatty acyl CoA

This reaction requires energy in the form of ATP, which is cleaved to AMP and pyrophosphate. This involves hydrolysis of two phosphoanhydride bonds. Here again we see the need to invest a small

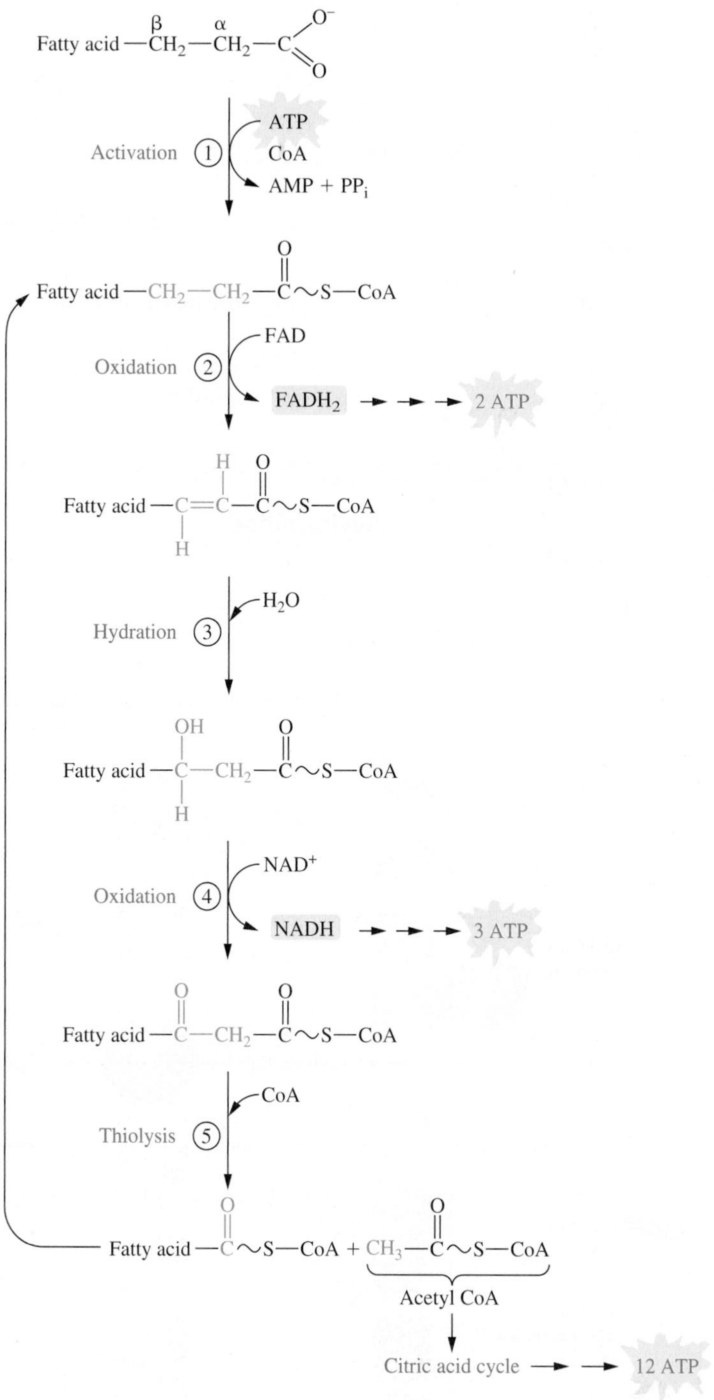

Figure 23.6 The reactions in β-oxidation of fatty acids.

amount of energy so that a much greater amount of energy can be harvested later in the pathway. Coenzyme A is also required for this step. The product, a fatty acyl CoA, has a *high-energy* thioester bond between the fatty acid and coenzyme A. *Acyl-CoA ligase*, which catalyzes this reaction, is located in the outer membrane of the mitochondria. The mechanism that brings the fatty acyl CoA into the mitochondrial matrix involves a carrier molecule called L-*carnitine*. The first step, catalyzed by the enzyme *carnitine acyltransferase I*, is the transfer of the fatty acyl group to carnitine, producing acylcarnitine and coenzyme A (Figure 23.7a). Next, a carrier protein located in the mitochondrial inner membrane transfers the acylcarnitine into

Acyl group transfer reactions are described in Section 14.4.

Figure 23.7 (a) The reaction catalyzed by carnitine acyltransferase I. (b) The transport of fatty acids into the mitochondrial matrix.

(a)

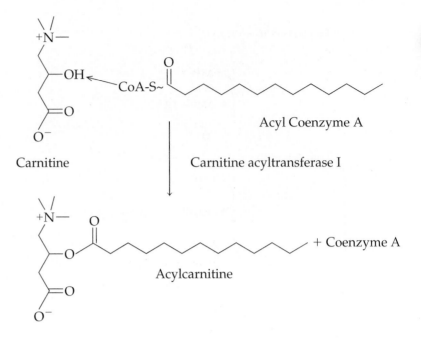

Carnitine

Acyl Coenzyme A

Carnitine acyltransferase I

Acylcarnitine

+ Coenzyme A

(b)

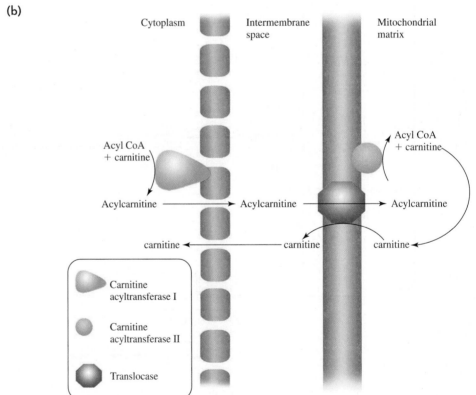

the mitochondrial matrix. There *carnitine acyltransferase II* catalyzes the regeneration of fatty acyl CoA, which now becomes involved in the remaining reactions of β-oxidation (Figure 23.7b).

Reaction 2. The next reaction is an *oxidation* reaction that removes a pair of hydrogen atoms from the fatty acid. These are used to reduce FAD to produce $FADH_2$. This *dehydrogenation* reaction is catalyzed by

Carnitine: The Fat Mover

Carnitine is a quaternary ammonium compound synthesized in the liver and kidneys from the amino acids lysine and methionine. This synthesis requires vitamin C. Carnitine can also be obtained through the diet. Red meats are an excellent source of carnitine, but it can be found also in nuts, legumes, broccoli, fruits, and cereals.

Carnitine

L-Carnitine, the active enantiomer of carnitine, has been sold as a weight-loss supplement. The idea behind this is that additional carnitine will transport more fatty acids to the site of β-oxidation and in that way promote weight loss. Although the idea is clever, there has never been any scientific evidence to demonstrate that the claim of enhanced weight loss is true. However, a research study in 2007 did find evidence that regular supplements of L-carnitine in the elderly improved energy metabolism and neurotransmitter function in the brain. Further study will be required to fully understand the use of L-carnitine as a nutritional supplement.

Physicians have observed L-carnitine deficiency in children. Systemic primary carnitine deficiency is caused by a mutation in the membrane transport system that brings L-carnitine into the cell. As a result, β-oxidation of fatty acids is defective. The disorder produces a variety of symptoms that present in infancy or childhood. Among these is metabolic decompensation between 3 months and 2 years of age. Decompensation is the inability of the heart to maintain adequate blood circulation, which results in edema and labored breathing. Other symptoms that are more episodic include hypoglycemia, lethargy, irritability, and an enlarged liver. Some children exhibit myopathy of the heart and skeletal muscles between the ages of 2 and 4 years.

If diagnosed before there is irreversible organ damage, the symptoms can be treated with L-carnitine taken orally. A dose of 100–400 mg/kg/day brings about improvement of metabolic decompensation, as well as improved cardiac and skeletal muscle function. Thereafter, the condition can be managed through oral administration of L-carnitine.

For Further Understanding

▸ Design an experiment that might demonstrate whether L-carnitine promotes weight loss.

▸ Explain why cardiac and skeletal muscles are particularly susceptible to damage as a result of primary carnitine deficiency.

---

the enzyme *acyl-CoA dehydrogenase* and results in the formation of a carbon-carbon double bond:

$$CH_3-(CH_2)_n-CH_2-CH_2-\overset{\overset{O}{\|}}{C}\sim S-CoA \xrightarrow{\hspace{1cm}} $$

FAD → FADH₂

$$CH_3-(CH_2)_n-\overset{H}{\underset{H}{C}}=\overset{}{C}-\overset{\overset{O}{\|}}{C}\sim S-CoA$$

Oxidative phosphorylation yields two ATP molecules for each molecule of FADH₂ produced by this oxidation-reduction reaction.

Reaction 3. The third reaction involves the *hydration* of the double bond produced in reaction 2. As a result, the β-carbon is hydroxylated. This reaction is catalyzed by the enzyme *enoyl-CoA hydrase*.

$$CH_3-(CH_2)_n-\overset{H}{\underset{H}{C}}=\overset{}{C}-\overset{\overset{O}{\|}}{C}\sim S-CoA \xrightarrow{H_2O}$$

$$CH_3-(CH_2)_n-\overset{OH}{\underset{H}{C}}-CH_2-\overset{\overset{O}{\|}}{C}\sim S-CoA$$

Reaction 4. In this *oxidation* reaction the hydroxyl group of the β-carbon is now dehydrogenated. NAD⁺ is reduced to form NADH that is subsequently used to produce three ATP molecules by oxidative phosphorylation. L-β-*Hydroxyacyl-CoA dehydrogenase* catalyzes this reaction.

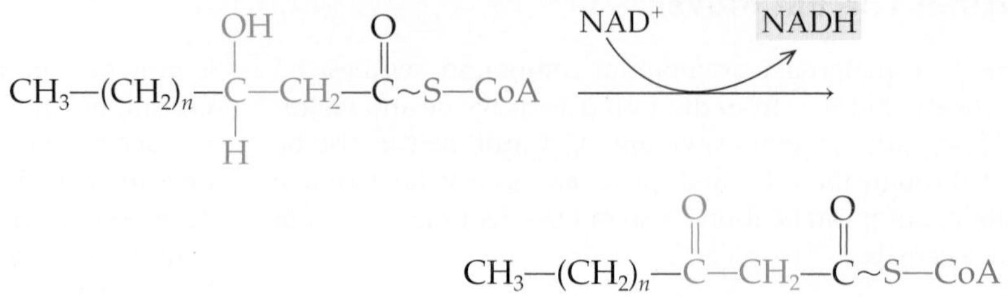

$$CH_3-(CH_2)_n-\underset{\underset{H}{|}}{\overset{\overset{OH}{|}}{C}}-CH_2-\overset{O}{\overset{||}{C}}\sim S-CoA \xrightarrow[]{NAD^+ \quad NADH}$$

$$CH_3-(CH_2)_n-\overset{O}{\overset{||}{C}}-CH_2-\overset{O}{\overset{||}{C}}\sim S-CoA$$

Reaction 5. The final reaction, catalyzed by the enzyme *thiolase,* is the cleavage that releases acetyl CoA. This is accomplished by *thiolysis,* attack of a molecule of coenzyme A on the β-carbon. The result is the release of acetyl CoA and a fatty acyl CoA that is two carbons shorter than the beginning fatty acid:

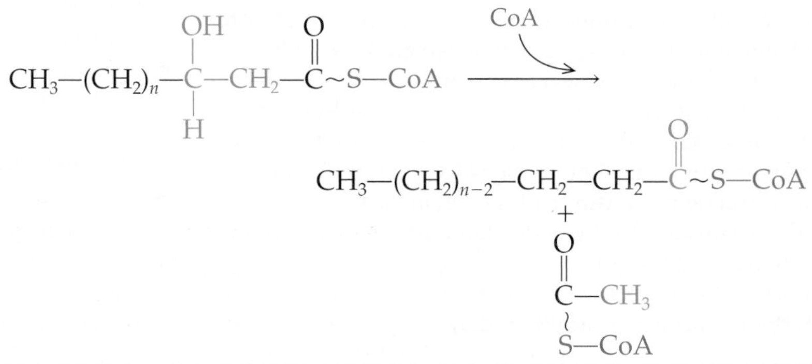

$$CH_3-(CH_2)_n-\underset{\underset{H}{|}}{\overset{\overset{OH}{|}}{C}}-CH_2-\overset{O}{\overset{||}{C}}\sim S-CoA \xrightarrow{\quad CoA \quad}$$

$$CH_3-(CH_2)_{n-2}-CH_2-CH_2-\overset{O}{\overset{||}{C}}\sim S-CoA$$
$$+$$
$$\overset{O}{\overset{||}{C}}-CH_3$$
$$S-CoA$$

The shortened fatty acyl CoA is further oxidized by cycling through reactions 2–5 until the fatty acid carbon chain is completely degraded to acetyl CoA. The acetyl CoA produced by β-oxidation of fatty acids then enters the reactions of the citric acid cycle. Of course, this eventually results in the production of 12 ATP molecules per molecule of acetyl CoA released during β-oxidation.

As an example of the energy yield from β-oxidation, the balance sheet for ATP production when the sixteen-carbon fatty acid palmitic acid is degraded by β-oxidation is summarized in Figure 23.8. Complete oxidation of palmitate results in production of 129 molecules of ATP, *three and one half times more energy than results from the complete oxidation of an equivalent amount of glucose.*

EXAMPLE 23.2	**Calculating the Amount of ATP Produced in Complete Oxidation of a Fatty Acid**

LEARNING GOAL

How many molecules of ATP are produced in the complete oxidation of stearic acid, an eighteen-carbon saturated fatty acid?

2 Describe the degradation of fatty acids by β-oxidation.

Solution

Step 1 (activation)	−2 ATP
Steps 2–5:	
8 FADH₂ × 2 ATP/FADH₂	16 ATP
8 NADH × 3 ATP/NADH	24 ATP

9 acetyl CoA (to citric acid cycle):

9×1 GTP $\times 1$ ATP/GTP	9 ATP
9×3 NADH $\times 3$ ATP/NADH	81 ATP
9×1 FADH$_2$ $\times 2$ ATP/FADH$_2$	18 ATP
	146 ATP

Practice Problem 23.2

Write out the sequence of steps for β-oxidation of butyryl CoA. What is the energy yield from the complete degradation of butyryl CoA via β-oxidation, the citric acid cycle, and oxidative phosphorylation?

▶ For Further Practice: **Questions 23.43 and 23.44.**

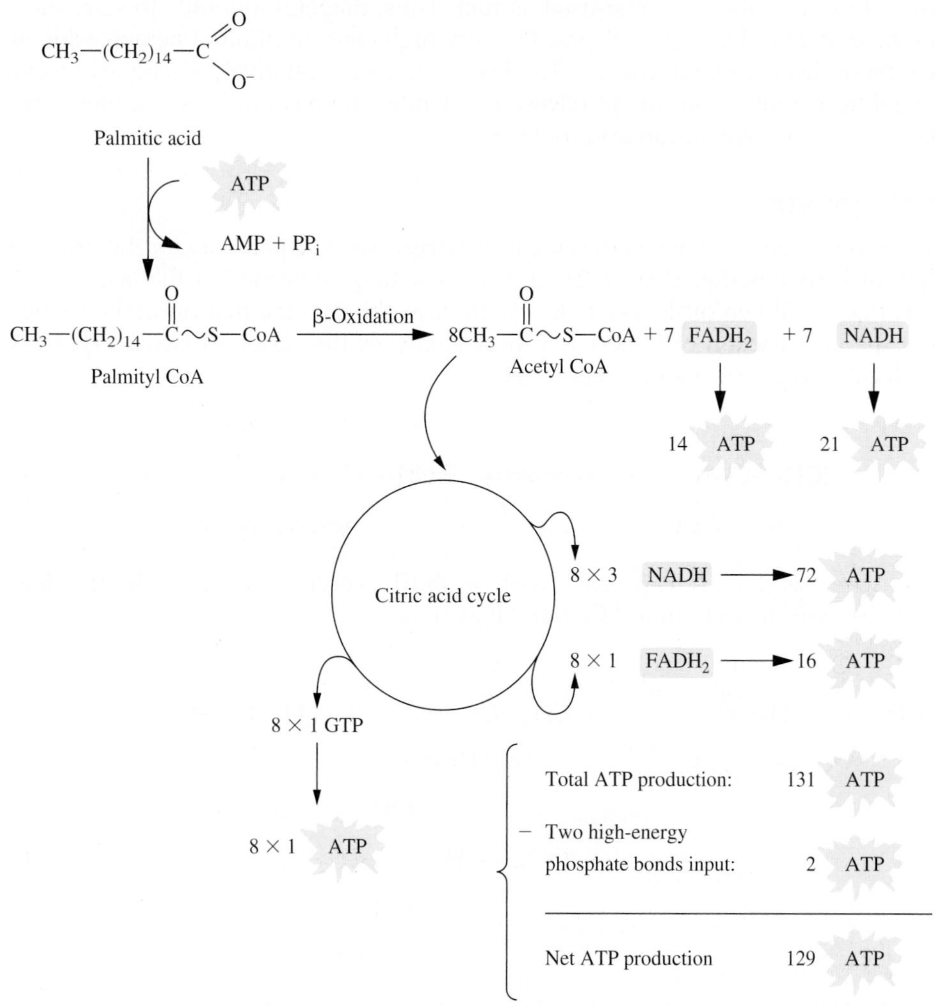

Figure 23.8 Complete oxidation of palmitic acid yields 129 molecules of ATP. Note that the activation step is considered to be an expenditure of two high-energy phosphoanhydride bonds because ATP is hydrolyzed to AMP + PP$_i$.

23.3 Ketone Bodies

For the acetyl CoA produced by the β-oxidation of fatty acids to efficiently enter the citric acid cycle, there must be an adequate supply of oxaloacetate. If glycolysis and β-oxidation are occurring at the same rate, there will be a steady supply of pyruvate (from glycolysis) that can be converted to oxaloacetate. But what happens

LEARNING GOAL

4 Understand the role of ketone body production in β-oxidation.

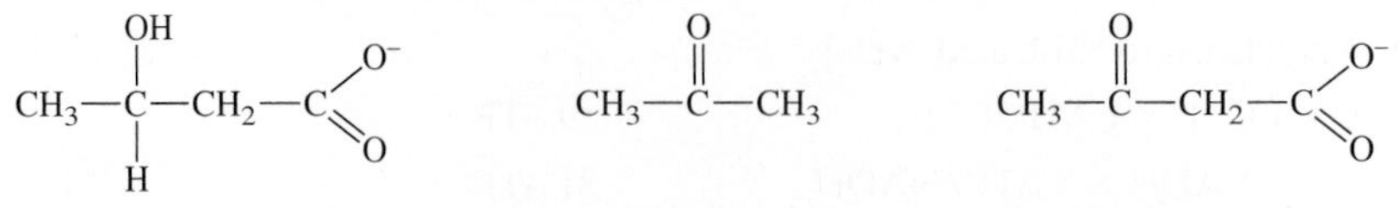

Figure 23.9 Structures of ketone bodies.

$$CH_3-\underset{\underset{H}{|}}{\overset{\overset{OH}{|}}{C}}-CH_2-\overset{O}{\underset{O^-}{C}} \qquad CH_3-\overset{O}{\overset{\|}{C}}-CH_3 \qquad CH_3-\overset{O}{\overset{\|}{C}}-CH_2-\overset{O}{\underset{O^-}{C}}$$

β-Hydroxybutyrate   Acetone   Acetoacetate

See Section 22.9 for a review of the reactions that provide oxaloacetate.

Diabetes mellitus is a disease characterized by the appearance of glucose in the urine as a result of high blood glucose levels. The disease is often caused by the inability to produce the hormone insulin.

if the supply of oxaloacetate is too low to allow all of the acetyl CoA to enter the citric acid cycle? Under these conditions, acetyl CoA is converted to the so-called **ketone bodies:** β-hydroxybutyrate, acetone, and acetoacetate (Figure 23.9).

Ketosis

Ketosis, abnormally high levels of blood ketone bodies, is a situation that arises under some pathological conditions, such as starvation, a diet that is extremely low in carbohydrates (as with the high-protein diets), or uncontrolled **diabetes mellitus.** The carbohydrate intake of a diabetic is normal, but the carbohydrates cannot get into the cell to be used as fuel. Thus, diabetes amounts to starvation in the midst of plenty. In diabetes, the very high concentration of ketone acids in the blood leads to **ketoacidosis.** The ketone acids are relatively strong acids and therefore readily dissociate to release H$^+$. Under these conditions, the blood pH becomes acidic, which can lead to death.

Ketogenesis

The production of ketone bodies is called *ketogenesis.* The pathway for the production of ketone bodies (Figure 23.10) begins with a "reversal" of the last step of β-oxidation. When oxaloacetate levels are low, the enzyme that normally carries out the last reaction of β-oxidation now catalyzes the fusion of two acetyl CoA molecules to produce acetoacetyl CoA:

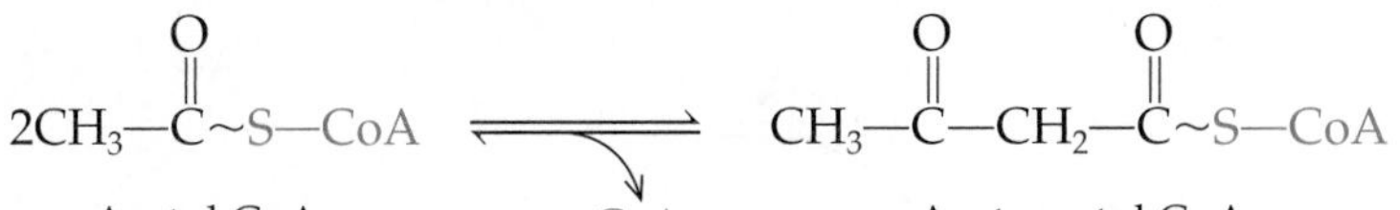

$$2CH_3-\overset{O}{\overset{\|}{C}}{\sim}S-CoA \rightleftharpoons CH_3-\overset{O}{\overset{\|}{C}}-CH_2-\overset{O}{\overset{\|}{C}}{\sim}S-CoA$$

Acetyl CoA   CoA   Acetoacetyl CoA

Acetoacetyl CoA can react with a third acetyl CoA molecule to yield β-hydroxy-β-methylglutaryl CoA (HMG-CoA):

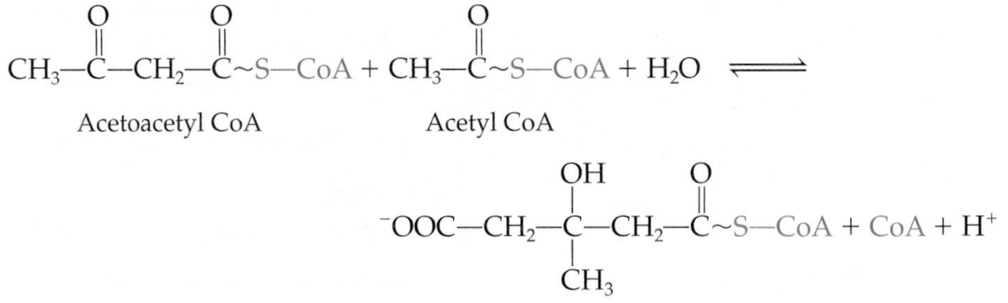

$$CH_3-\overset{O}{\overset{\|}{C}}-CH_2-\overset{O}{\overset{\|}{C}}{\sim}S-CoA + CH_3-\overset{O}{\overset{\|}{C}}{\sim}S-CoA + H_2O \rightleftharpoons$$

Acetoacetyl CoA   Acetyl CoA

$$^-OOC-CH_2-\underset{\underset{CH_3}{|}}{\overset{\overset{OH}{|}}{C}}-CH_2-\overset{O}{\overset{\|}{C}}{\sim}S-CoA + CoA + H^+$$

HMG-CoA

If HMG-CoA were formed in the cytoplasm, it would serve as a precursor for cholesterol biosynthesis. But ketogenesis, like β-oxidation, occurs in the mitochondrial matrix, and here HMG-CoA is cleaved to yield acetoacetate and acetyl CoA:

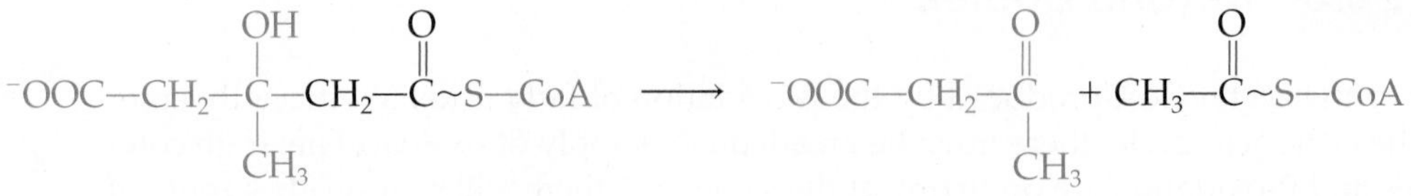

$$^-OOC-CH_2-\underset{\underset{CH_3}{|}}{\overset{\overset{OH}{|}}{C}}-CH_2-\overset{O}{\overset{\|}{C}}{\sim}S-CoA \longrightarrow {}^-OOC-CH_2-\underset{\underset{CH_3}{|}}{\overset{O}{\overset{\|}{C}}} + CH_3-\overset{O}{\overset{\|}{C}}{\sim}S-CoA$$

HMG-CoA   Acetoacetate   Acetyl CoA

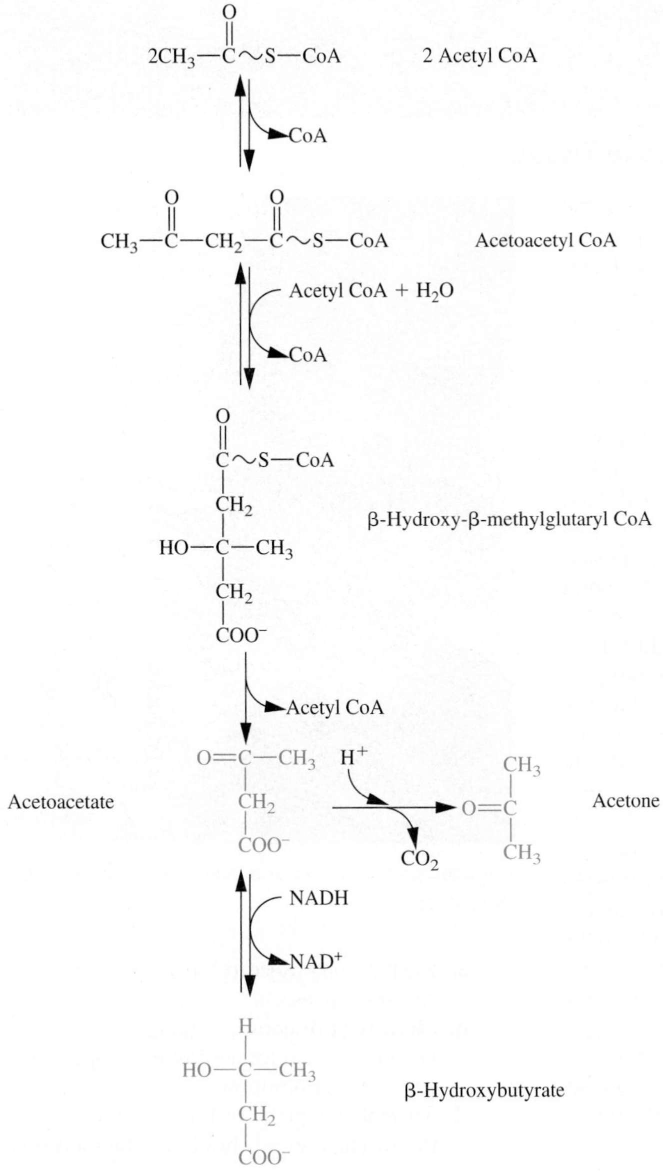

Figure 23.10 Summary of the reactions involved in ketogenesis.

In very small amounts, acetoacetate spontaneously loses carbon dioxide to give acetone. This is the reaction that causes the "acetone breath" that is often associated with uncontrolled diabetes mellitus.

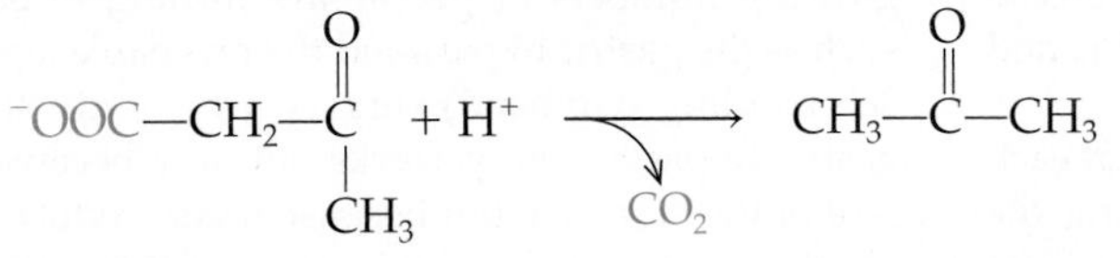

Acetoacetate   Acetone

More frequently, it undergoes NADH-dependent reduction to produce β-hydroxybutyrate:

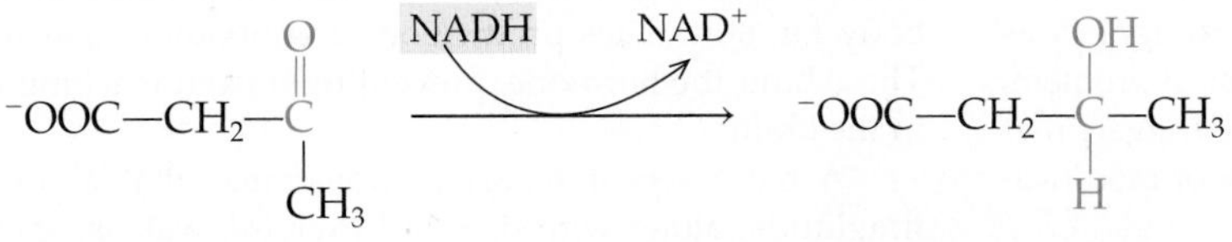

Acetoacetate   β-Hydroxybutyrate

A Human Perspective

Losing Those Unwanted Pounds of Adipose Tissue

Weight, or overweight, is a topic of great concern to the American populace. A glance through almost any popular magazine quickly informs us that by today's standards, "beautiful" is synonymous with "thin." The models in all these magazines are extremely thin, and there are literally dozens of ads for weight-loss programs. Americans spend millions of dollars each year trying to attain this slim ideal of the fashion models.

Studies have revealed that this slim ideal is often below a desirable, healthy body weight. In fact, the suggested weight for a 6-foot (ft) tall male between 18 and 39 years of age is 179 pounds (lb). For a 5'6" female in the same age range, the desired weight is 142 lb. For a 5'1" female, 126 lb is recommended. Just as being too thin can cause health problems, so too can obesity.

What is obesity, and does it have disadvantages beyond aesthetics? An individual is considered to be obese if his or her body weight is more than 20% above the ideal weight for his or her height. Being overweight carries with it a wide range of physical problems, including elevated blood cholesterol levels; high blood pressure; an increased incidence of diabetes, cancer, and heart disease; and an increased probability of early death. It often causes psychological problems as well, such as guilt and low self-esteem.

Many factors may contribute to obesity. These include genetic factors, a sedentary lifestyle, and a preference for high-calorie, high-fat foods. However, the real concern is how to lose weight. How can we lose weight wisely and safely and keep the weight off for the rest of our lives? The prevalence and financial success of the quick-weight-loss programs suggest that the majority of people want a program that is rapid and effortless. Unfortunately, most programs that promise dramatic weight reduction with little effort are usually ineffective or, worse, unsafe. The truth is that weight loss and management are best obtained by a program involving three elements.

1. *Reduced caloric intake.* One pound of body fat is equivalent to 3500 Calories (Cal). So if you want to lose 2 lb each week, a reasonable goal, you must reduce your caloric intake by 1000 Cal per day. Remember that diets recommending fewer than 1200 Cal per day are difficult to maintain because they are not very satisfying and may be unsafe because they don't provide all the required vitamins and minerals.
2. *Exercise.* Increase energy expenditures by 200–400 Cal each day. You may choose walking, running, or mowing the lawn; the type of activity doesn't matter, as long as you get moving. Exercise has additional benefits. It increases cardiovascular fitness, provides a psychological lift, and may increase the base rate at which you burn calories after exercise is finished.
3. *Behavior modification.* For some people, overweight is as much a psychological problem as it is a physical problem, and half the battle is learning to recognize the triggers that cause overeating. Several principles of behavior modification have been found to be very helpful.

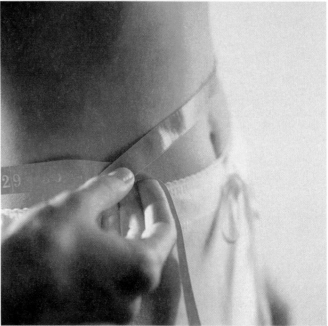

Reduced caloric intake and exercise are the keys to permanent weight loss.

    a. Keep a diary. Record the amount of foods eaten and the circumstances.
    b. Identify your eating triggers.
    c. Develop a plan for avoiding or coping with your trigger situations or emotions.
    d. Set realistic goals, and reward yourself when you reach them. The reward should not be food related.

Traditionally, there has been no "quick fix" for safe, effective weight control. A commitment has to be made to modify existing diet and exercise habits. Most important, those habits have to be avoided forever and replaced by new, healthier behaviors and attitudes.

Frustrated by attempts to modify their diet and exercise, growing numbers of people are turning to bariatric surgery, such as the gastric bypass and the reversible laparoscopic stomach banding (lap-band) surgery. One study suggests that the gastric bypass surgery works not only because it reduces the size of the stomach, but because it also reduces the amount of ghrelin produced. Ghrelin is a hormone produced in the stomach that stimulates appetite.

Another interesting approach to the problem has been the development of an antiobesity vaccine. The vaccine, which slowed weight gain in rats and reduced the amount of stored body fat, stimulates production of antibodies against ghrelin. These bind the hormone, preventing it from reaching its target in the brain.

A more recent report demonstrated that a drug called liraglutide, along with diet and exercise, was associated with clinically significant weight loss in overweight or obese patients. Liraglutide mimics the action of glucagon-like peptide-1, a hormone produced in the intestine that decreases food intake by increasing satiety in the brain. It is hoped that greater understanding of the complex systems that control appetite will lead to additional effective drugs to combat obesity and the many health issues it causes.

For Further Understanding

▶ In terms of the energy-harvesting reactions we have studied in Chapters 22 and 23, explain how reduced caloric intake and an increase in activity level contribute to weight loss.

▶ If you increased your energy expenditure by 200 Cal per day and did not change your eating habits, how long would it take you to lose 10 lb?

Acetoacetate and β-hydroxybutyrate are produced primarily in the liver. These metabolites diffuse into the blood and are circulated to other tissues, where they may be reconverted to acetyl CoA and used to produce ATP. In fact, the heart muscle derives most of its metabolic energy from the oxidation of ketone bodies, not from the oxidation of glucose. Other tissues that are best adapted to the use of glucose will increasingly rely on ketone bodies for energy when glucose becomes unavailable or limited. This is particularly true of the brain.

Question 23.3   What conditions lead to excess production of ketone bodies?

Question 23.4   What is the cause of the characteristic "acetone breath" that is associated with uncontrolled diabetes mellitus?

23.4 Fatty Acid Synthesis

All organisms possess the ability to synthesize fatty acids. In humans, the excess acetyl CoA produced by carbohydrate degradation is used to make fatty acids that are then stored as triglycerides.

A Comparison of Fatty Acid Synthesis and Degradation

On first examination, fatty acid synthesis appears to be simply the reverse of β-oxidation. Specifically, the fatty acid chain is constructed by the sequential addition of two-carbon acetyl groups (Figure 23.11). Although the chemistry of fatty acid synthesis and breakdown are similar, there are several major differences between β-oxidation and fatty acid biosynthesis. These are summarized as follows.

- **Intracellular location.** The enzymes responsible for fatty acid biosynthesis are located in the cytoplasm of the cell, whereas those responsible for β-oxidation of fatty acids are in the mitochondria.
- **Acyl group carriers.** The activated intermediates of fatty acid biosynthesis are bound to a carrier molecule called the **acyl carrier protein (ACP)**. In β-oxidation, the acyl group carrier was coenzyme A. However, there are important similarities between these two carriers. Both contain the **phosphopantetheine** group, which is made from the vitamin pantothenic acid. In both cases, the fatty acyl group is bound by a thioester bond to the phosphopantetheine group.

LEARNING GOAL

5 Compare β-oxidation of fatty acids and fatty acid biosynthesis.

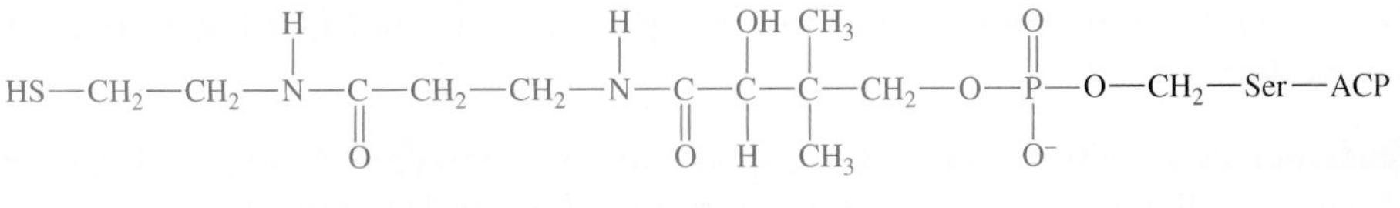

Phosphopantetheine prosthetic group of ACP

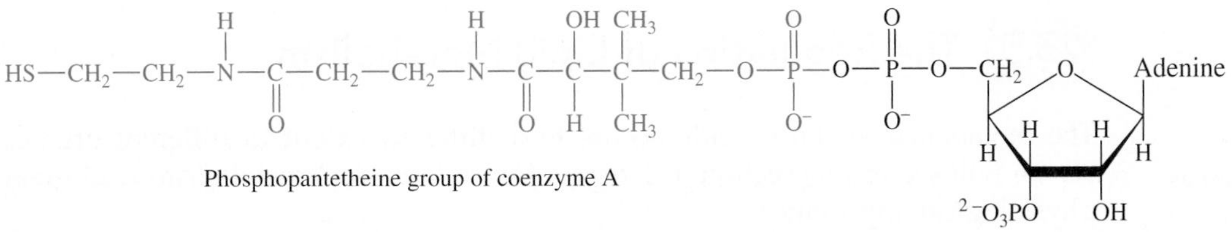

Phosphopantetheine group of coenzyme A

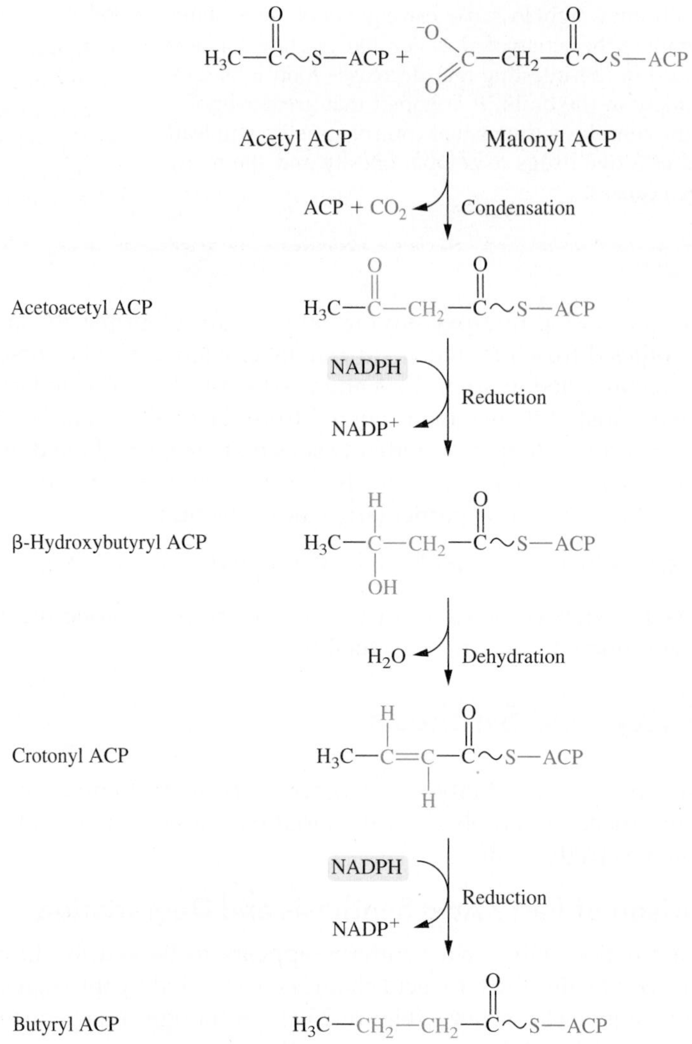

Figure 23.11 Summary of fatty acid synthesis. Malonyl ACP is produced in two reactions: carboxylation of acetyl CoA to produce malonyl CoA and transfer of the malonyl acyl group from malonyl CoA to ACP.

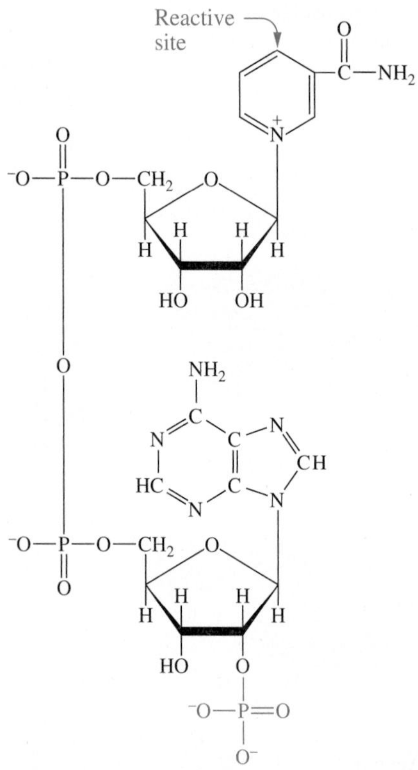

Acetyl ACP   Malonyl ACP

ACP + CO$_2$ ⟵ Condensation

Acetoacetyl ACP

β-Hydroxybutyryl ACP

NADPH ⟶ Reduction ⟵ NADP$^+$

H$_2$O ⟵ Dehydration

Crotonyl ACP

NADPH ⟶ Reduction ⟵ NADP$^+$

Butyryl ACP

Figure 23.12 Structure of NADPH. The phosphate group shown in red is the structural feature that distinguishes NADPH from NADH.

- **Enzymes involved.** Fatty acid biosynthesis is carried out by a multienzyme complex known as *fatty acid synthase.* The enzymes responsible for β-oxidation are not physically associated in such complexes.
- **Electron carriers.** NADH and FADH$_2$ are produced in β-oxidation of fatty acids. However, the reducing agent for fatty acid synthesis is NADPH. NADH and NADPH differ only by the presence of a phosphate group bound to the ribose ring of NADPH (Figure 23.12). The enzymes that use these coenzymes, however, are easily able to distinguish them on this basis.

Question 23.5 List the four major differences between β-oxidation and fatty acid biosynthesis that reveal that the two processes are not just the reverse of one another.

Question 23.6 What chemical group is part of coenzyme A and acyl carrier protein and allows both molecules to form thioester bonds to fatty acids?

23.5 The Regulation of Lipid Metabolism

The metabolism of fatty acids occurs to a different extent in different organs. As we will see in this section, the regulation of fatty acid metabolism is of great physiological importance.

A Medical Perspective

Diabetes Mellitus and Ketone Bodies

More than one person, found unconscious on the streets of some metropolis, has been carted to jail only to die of complications arising from uncontrolled diabetes mellitus. Others are fortunate enough to arrive in hospital emergency rooms. A quick test for diabetes mellitus–induced coma is the odor of acetone on the breath of the afflicted person. Acetone is one of several metabolites produced by diabetics that are known collectively as *ketone bodies.*

The term *diabetes* was used by the ancient Greeks to designate diseases in which excess urine is produced. Two thousand years later, in the eighteenth century, the urine of certain individuals was found to contain sugar, and the name *diabetes mellitus* (Latin: *mellitus,* sweetened with honey) was given to this disease. People suffering from diabetes mellitus waste away as they excrete large amounts of sugar-containing urine.

The cause of insulin-dependent diabetes mellitus is inadequate production of insulin by the body. Insulin is secreted in response to high blood glucose levels. It binds to the membrane receptor protein on its target cells. Binding increases the rate of transport of glucose across the membrane and stimulates glycogen synthesis, lipid biosynthesis, and protein synthesis. As a result, the blood glucose level is reduced. Clearly, the inability to produce sufficient insulin seriously impairs the body's ability to regulate metabolism.

Individuals suffering from diabetes mellitus do not produce enough insulin to properly regulate blood glucose levels. This generally results from the destruction of the β-cells of the islets of Langerhans. One theory to explain the mysterious disappearance of these cells is that a virus infection stimulates the immune system to produce antibodies that cause the destruction of the β-cells.

In the absence of insulin, the uptake of glucose into the tissues is not stimulated, and a great deal of glucose is eliminated in the urine. Without insulin, then, adipose cells are unable to take up the glucose required to synthesize triglycerides. As a result, the rate of fat hydrolysis is much greater than the rate of fat resynthesis, and large quantities of free fatty acids are

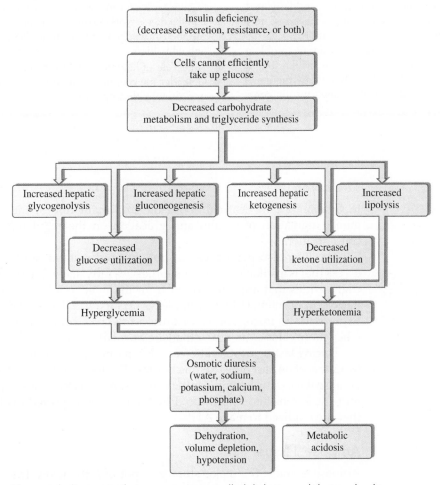

The metabolic events that occur in uncontrolled diabetes and that can lead to coma and death.

Continued…

liberated into the bloodstream. Because glucose is not being efficiently taken into cells, carbohydrate metabolism slows, and there is an increase in the rate of lipid catabolism. In the liver, this lipid catabolism results in the production of ketone bodies: acetone, acetoacetate, and β-hydroxybutyrate.

A similar situation can develop from improper eating, fasting, or dieting—any situation in which the body is not provided with sufficient energy in the form of carbohydrates. These ketone bodies cannot all be oxidized by the citric acid cycle, which is limited by the supply of oxaloacetate. The acetone concentration in blood rises to levels so high that acetone can be detected in the breath of untreated diabetics. The elevated concentration of ketones in the blood can overwhelm the buffering capacity of the blood, resulting in ketoacidosis. Ketones, too, will be excreted through the kidney. In fact, the presence of excess ketones in the urine can raise the osmotic concentration of the urine so that it behaves as an "osmotic diuretic," causing the excretion of enormous amounts of water. As a result, the patient may become severely dehydrated. In extreme cases, the combination of dehydration and ketoacidosis may lead to coma and death.

It has been observed that diabetics also have a higher than normal level of glucagon in the blood. As we have seen, glucagon stimulates lipid catabolism and ketogenesis. It may be that the symptoms previously described result from both the deficiency of insulin and the elevated glucagon levels. The absence of insulin may cause the elevated blood glucose and fatty acid levels, whereas the glucagon, by stimulating ketogenesis, may be responsible for the ketoacidosis and dehydration.

There is no cure for diabetes. However, when the problem is the result of the inability to produce active insulin, blood glucose levels can be controlled moderately well by the injection of human insulin produced from the cloned insulin gene.

Unfortunately, one or even a few injections of insulin each day cannot mimic the precise control of blood glucose accomplished by the pancreas.

As a result, diabetics suffer progressive tissue degeneration that leads to early death. One primary cause of this degeneration is atherosclerosis, the deposition of plaque on the walls of blood vessels. This causes a high frequency of strokes, heart attack, and gangrene of the feet and lower extremities, often necessitating amputation. Kidney failure causes the death of about 20% of diabetics under 40 years of age, and diabetic retinopathy (various kinds of damage to the retina of the eye) ranks fourth among the leading causes of blindness in the United States. Nerves are also damaged, resulting in neuropathies that can cause pain or numbness, particularly of the feet.

There is no doubt that insulin injections prolong the life of diabetics, but only the presence of a fully functioning pancreas can allow a diabetic to live a life free of the complications noted here. At present, pancreas transplants do not have a good track record. Only about 50% of the transplants are functioning after 1 year. It is hoped that improved transplantation techniques will be developed so that diabetics can live a normal life span, free of debilitating disease.

For Further Understanding

▶ The Atkins' low-carbohydrate diet recommends that dieters test their urine for the presence of ketone bodies as an indicator that the diet is working. In terms of lipid and carbohydrate metabolism, explain why ketone bodies are being produced and why this is an indication that the diet is working.

▶ An excess of ketone bodies in the blood causes ketoacidosis. Consider the chemical structure of the ketone bodies and explain why they are acids.

The Liver

In Chapter 21, we learned about the important role of the liver and the hormonal regulation of insulin and glucagon on the regulation of blood glucose concentration.

The liver also plays a central role in lipid metabolism (Figure 23.13). When excess fuel is available, the liver synthesizes fatty acids. These are used to produce triglycerides that are transported from the liver to adipose tissues by very low density lipoprotein (VLDL) complexes. In fact, VLDL complexes provide adipose tissue with its major source of fatty acids. This transport is particularly active when more calories are eaten than are burned!

During fasting or starvation conditions, however, the liver converts fatty acids to acetoacetate and other ketone bodies. The liver cannot use these ketone bodies because it lacks an enzyme for the conversion of acetoacetate to acetyl CoA. Therefore, the ketone bodies produced by the liver are exported to other organs where they are oxidized to make ATP.

Adipose Tissue

Adipose tissue is the major storage depot of fatty acids. Triglycerides produced by the liver are transported through the bloodstream as components of VLDL complexes. The triglycerides are hydrolyzed by the same lipases that act on chylomicrons, and the fatty acids are absorbed by adipose tissue. The synthesis of triglycerides in adipose tissue requires glycerol-3-phosphate, which is produced

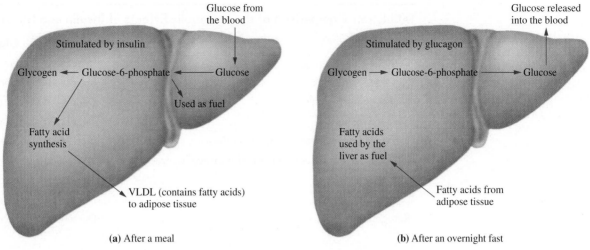

Figure 23.13 The liver controls the concentration of blood glucose.

from glucose by glycolysis. Thus, adipose cells must have a ready source of glucose to synthesize and store triglycerides.

Triglycerides are constantly being hydrolyzed and resynthesized in the cells of adipose tissue. Lipases that are under hormonal control determine the rate of hydrolysis of triglycerides into fatty acids and glycerol. If glucose is in limited supply, there will not be sufficient glycerol-3-phosphate for the resynthesis of triglycerides, and the fatty acids and glycerol are exported to the liver for further processing (Figure 23.14).

Muscle Tissue

The energy demand of *resting* muscle is generally supplied by the β-oxidation of fatty acids. The heart muscle actually prefers ketone bodies over glucose. *Working* muscle, however, obtains energy by degradation of its own supply of glycogen.

The Brain

Under normal conditions, the brain uses glucose as its sole source of metabolic energy. When the body is in the resting state, about 60% of the free glucose of the body is used by the brain. Starvation depletes glycogen stores, and the amount of glucose available to the brain drops sharply. The ketone bodies acetoacetate and β-hydroxybutyrate are then used by the brain as an alternative energy source. Fatty acids are transported in the blood in complexes with proteins and cannot cross the blood-brain barrier to be used by brain cells as an energy source. But ketone bodies, which have a free carboxylate group, are soluble in blood and can enter the brain.

Question 23.7   Under what conditions does the liver synthesize fatty acids and triglycerides?

Question 23.8   What is the role of VLDL in triglyceride metabolism?

23.6   The Effects of Insulin and Glucagon on Cellular Metabolism

The hormone **insulin** is produced by the β-cells of the islets of Langerhans in the pancreas. It is secreted from these cells in response to an increase in the blood glucose level. Insulin lowers the concentration of blood glucose by causing a number of changes in metabolism (Table 23.1).

The simplest way to lower blood glucose levels is to stimulate storage of glucose, both as glycogen and as triglycerides. *Insulin therefore activates biosynthetic processes and inhibits catabolic processes.*

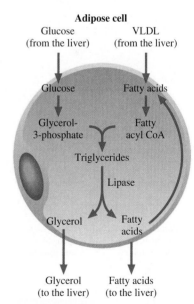

Glycerol-3-phosphate

Figure 23.14 Synthesis and degradation of triglycerides in adipose tissue.

LEARNING GOAL

7 Summarize the antagonistic effects of glucagon and insulin.

TABLE 23.1 Comparison of the Metabolic Effects of Insulin and Glucagon

Actions	Insulin	Glucagon
Cellular glucose transport	Increased	No effect
Glycogen synthesis	Increased	Decreased
Glycogenolysis in liver	Decreased	Increased
Gluconeogenesis	Decreased	Increased
Amino acid uptake and protein synthesis	Increased	No effect
Inhibition of amino acid release and protein degradation	Decreased	No effect
Lipogenesis	Increased	No effect
Lipolysis	Decreased	Increased
Ketogenesis	Decreased	Increased

The effect of insulin on glycogen metabolism is described in Section 21.7.

Insulin acts only on those cells, known as *target cells,* that possess a specific insulin receptor protein in their plasma membranes. The major target cells for insulin are liver, adipose, and muscle cells.

The blood glucose level is normally about 10 millimolar (mM). However, a substantial meal increases the concentration of blood glucose considerably and stimulates insulin secretion. Subsequent binding of insulin to the plasma membrane insulin receptor increases the rate of transport of glucose across the membrane and into cells.

Insulin exerts a variety of effects on all aspects of cellular metabolism:

- **Carbohydrate metabolism.** Insulin stimulates glycogen synthesis. At the same time, it inhibits glycogenolysis and gluconeogenesis. The overall result of these activities is the storage of excess glucose.
- **Protein metabolism.** Insulin stimulates the transport and uptake of amino acids, as well as the incorporation of amino acids into proteins.
- **Lipid metabolism.** Insulin stimulates the uptake of glucose by adipose cells, as well as the synthesis and storage of triglycerides. As we have seen, storage of lipids requires a source of glucose, and insulin helps the process by increasing the available glucose. At the same time, insulin inhibits the breakdown of stored triglycerides.

As you may have already guessed, insulin is only part of the overall regulation of cellular metabolism in the body. A second hormone, **glucagon,** is secreted by the α-cells of the islets of Langerhans in response to decreased blood glucose levels. The effects of glucagon, generally the opposite of the effects of insulin, are summarized in Table 23.1. Although it has no direct effect on glucose uptake, glucagon inhibits glycogen synthesis and stimulates glycogenolysis and gluconeogenesis. It also stimulates the breakdown of fats and ketogenesis.

The antagonistic effects of these two hormones, seen in Figure 23.15, are critical for the maintenance of adequate blood glucose levels. During fasting, low blood glucose levels stimulate production of glucagon, which increases blood glucose by stimulating the breakdown of glycogen and the production of glucose by gluconeogenesis. This ensures a ready supply of glucose for the tissues, especially the brain. On the other hand, when blood glucose levels are too high, insulin is secreted. It stimulates the removal of the excess glucose by enhancing uptake and inducing pathways for storage.

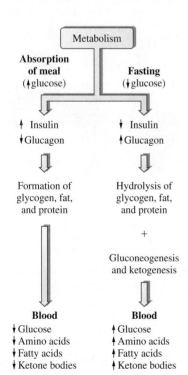

Figure 23.15 A summary of the antagonistic effects of insulin and glucagon.

Question 23.9 Summarize the effects of the hormone insulin on carbohydrate, lipid, and amino acid metabolism.

Question 23.10 Summarize the effects of the hormone glucagon on carbohydrate and lipid metabolism.

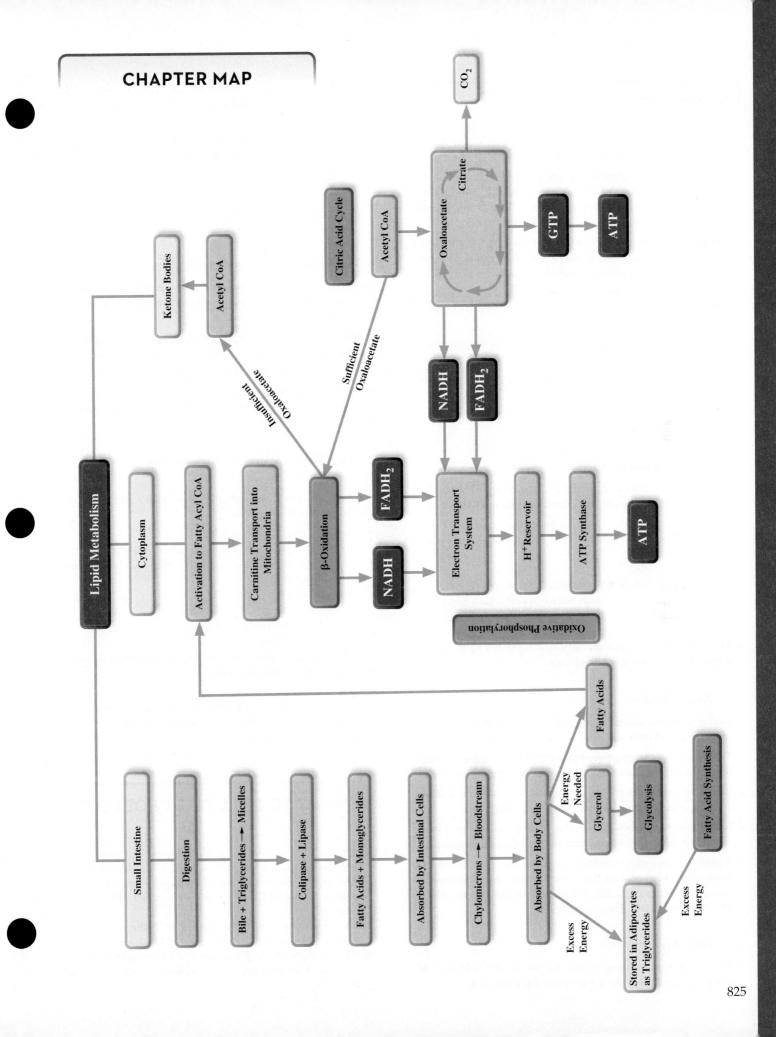

CHAPTER MAP

825

SUMMARY

23.1 Lipid Metabolism in Animals

▶ Dietary lipids **(triglycerides)** are emulsified into tiny fat droplets in the intestine by the action of **bile.** Bile is composed of **micelles** consisting of lecithin, cholesterol, protein, bile salts, inorganic ions, and bile pigments.

▶ **Colipase** binds to the surface of these fat droplets and helps pancreatic **lipase** stick to the surface so that it can catalyze the hydrolysis of triglycerides into monoglycerides and fatty acids.

▶ These are absorbed by the intestinal epithelial cells, reassembled into triglycerides, and packaged with proteins to form **chylomicrons.**

▶ Chylomicrons are transported to the cells of the body through the bloodstream.

▶ Fatty acids are stored as triglycerides (triacylglycerols) in fat droplets in the cytoplasm of **adipocytes.**

▶ **Adipose tissue** is composed of adipocytes and is a means of energy storage in the body.

23.2 Fatty Acid Degradation

▶ Fatty acids are degraded to acetyl CoA in the mitochondria by the **β-oxidation** pathway, which consists of five steps:

- Production of a fatty acyl CoA molecule
- Oxidation of the fatty acid by an FAD-dependent dehydrogenase
- Hydration
- Oxidation by an NAD^+-dependent dehydrogenase
- Cleavage of the chain with release of acetyl CoA and a fatty acyl CoA that is two carbons shorter than the beginning fatty acid.

▶ The last four reactions are repeated until the fatty acid is completely degraded to acetyl CoA.

23.3 Ketone Bodies

▶ Under some conditions, fatty acid degradation occurs more rapidly than glycolysis.

▶ This results in a large amount of acetyl CoA from β-oxidation, but an insufficient amount of oxaloacetate from glycolysis.

▶ In this situation, acetyl CoA is converted to **ketone bodies:** acetone, acetoacetate, and β-hydroxybutyrate.

▶ **Ketosis,** an abnormally high level of ketone bodies in the blood, may result from starvation or uncontrolled **diabetes mellitus.**

▶ A high level of ketone bodies in the blood causes **ketoacidosis** because the ketone acids β-hydroxybutyrate and acetoacetate are relatively strong acids. When they dissociate, the blood pH may drop to life-threatening levels.

23.4 Fatty Acid Synthesis

▶ Fatty acid biosynthesis occurs by sequential addition of acetyl groups, but is not the reverse of β-oxidation.

▶ Although the chemical reactions are similar, fatty acid synthesis differs from β-oxidation in the following ways:

- It occurs in the cytoplasm.
- It utilizes **acyl carrier protein (ACP)** rather than coenzyme A, both of which have the **phosphopantetheine** group as the functional part of the molecule that binds to the acetyl or acyl group.
- It utilizes NADPH, rather than NADH.
- It is carried out by a multienzyme complex called *fatty acid synthase.*

23.5 The Regulation of Lipid Metabolism

▶ The liver synthesizes fatty acids when energy is in excess. These are transported to adipose tissue by very low density lipoprotein (VLDL) complexes.

▶ In adipose tissue, the major storage depot for fatty acids, triglycerides are constantly hydrolyzed and resynthesized.

▶ Muscle can oxidize glucose, fatty acids, and ketone bodies.

▶ The brain uses only glucose as a fuel except in prolonged starvation or fasting, when it will use ketone bodies as fuel.

23.6 The Effects of Insulin and Glucagon on Cellular Metabolism

▶ **Insulin** activates biosynthetic processes and inhibits catabolic processes as follows:

- Stimulates glycogen synthesis
- Inhibits glycogen degradation and gluconeogenesis
- Stimulates transport and uptake of amino acids and their incorporation into proteins
- Stimulates uptake of glucose by adipose cells, as well as synthesis and storage of triglycerides
- Inhibits breakdown of stored triglycerides
- Inhibits ketogenesis

▶ The effects of **glucagon** are generally the opposite of those of insulin in the following ways:

- Inhibits glycogen synthesis
- Stimulates glycogen breakdown and gluconeogenesis
- Stimulates the breakdown of stored triglycerides
- Stimulates ketogenesis

ANSWERS TO PRACTICE PROBLEMS

23.1 a. The products of the β-oxidation of 9-phenylnonanoic acid are 4 acetyl CoA, 1 benzoate, 4 NADH, and 4 $FADH_2$.

b. The products of the β-oxidation of 8-phenyloctanoic acid are 3 acetyl CoA, 1 phenyl acetate, 3 NADH, and 3 $FADH_2$.

c. The products of the β-oxidation of 7-phenylheptanoic acid are 3 acetyl CoA, 1 benzoate, 3 NADH, and 3 $FADH_2$.

d. The products of the β-oxidation of 12-phenyldodecanoic acid are 5 acetyl CoA, 1 phenyl acetate, 5 NADH, and 5 $FADH_2$.

23.2 The following equations show the steps for the β-oxidation of butyryl CoA:

$$CH_3CH_2CH_2-\overset{\overset{\displaystyle O}{\|}}{C}-S-CoA$$

$$\Big\downarrow \curvearrowright FAD$$

$$CH_3CH=CH-\overset{\overset{\displaystyle O}{\|}}{C}-S-CoA + FADH_2$$

$$\Big\downarrow \curvearrowright H_2O$$

$$\overset{\overset{\displaystyle OH}{|}}{CH_3CH}-CH_2-\overset{\overset{\displaystyle O}{\|}}{C}-S-CoA$$

$$\Big\downarrow \curvearrowright NAD^+$$

$$CH_3-\overset{\overset{\displaystyle O}{\|}}{C}-CH_2-\overset{\overset{\displaystyle O}{\|}}{C}-S-CoA + NADH$$

$$\Big\downarrow \curvearrowright \text{Coenzyme A}$$

$$2\ CH_3-\overset{\overset{\displaystyle O}{\|}}{C}-S-CoA$$

The energy yield from the complete degradation of butyryl CoA via β-oxidation, the citric acid cycle, and oxidative phosphorylation is summarized below:

Butyryl CoA has already been activated, so no ATP investment is needed.

β-Oxidation:

1 FADH₂ × 2 ATP/FADH₂	2 ATP
1 NADH × 3 ATP/NADH	3 ATP

Citric acid cycle (2 acetyl CoA):

6 NADH × 3 ATP/NADH	18 ATP
2 FADH₂ × 2 ATP/FADH₂	4 ATP
2 GTP × 1 ATP/GTP	2 ATP
	29 ATP

QUESTIONS AND PROBLEMS

Lipid Metabolism in Animals

Foundations

23.11 Draw the structures of the bile salts cholate and chenodeoxycholate.

23.12 To what class of lipids do the bile salts belong?

23.13 Define the term *micelle*.

23.14 In Figure 23.1, a micelle composed of the phospholipid lecithin is shown. Why is lecithin a good molecule for the formation of micelles?

23.15 Define the term *triglyceride*.

23.16 Draw the structure of a triglyceride composed of glycerol, palmitoleic acid, linolenic acid, and oleic acid.

23.17 Describe the structure of chylomicrons.

23.18 Why is colipase needed for lipid digestion?

Applications

23.19 What is the major storage form of fatty acids?

23.20 What tissue is the major storage depot for lipids?

23.21 What is the outstanding structural feature of an adipocyte?

23.22 What is the major metabolic function of adipose tissue?

23.23 What is the general reaction catalyzed by lipases?

23.24 Why are the lipases that are found in saliva and in the stomach not very effective at digesting triglycerides?

23.25 List three major biological molecules for which acetyl CoA is a precursor.

23.26 Why are triglycerides more efficient energy-storage molecules than glycogen?

23.27 What is the function of chylomicrons?

23.28 **a.** What are very low density lipoproteins?
  b. Compare the function of VLDLs with that of chylomicrons.

23.29 What is the function of the bile salts in the digestion of dietary lipids?

23.30 What is the function of colipase in the digestion of dietary lipids?

23.31 Describe the stages of lipid digestion.

23.32 Describe the transport of lipids digested in the lumen of the intestines to the cells of the body.

Fatty Acid Degradation

Foundations

23.33 What is the energy source for the activation of a fatty acid in preparation for β-oxidation?

23.34 Which bond in fatty acyl CoA is a high-energy bond?

23.35 What is carnitine?

23.36 Explain the mechanism by which a fatty acyl group is brought into the mitochondrial matrix.

23.37 Explain why the reaction catalyzed by acyl-CoA dehydrogenase is an example of an oxidation reaction.

23.38 What is the reactant that is oxidized in the reaction catalyzed by acyl-CoA dehydrogenase? What is the reactant that is reduced in this reaction?

23.39 What is the product of the hydration of an alkene?

23.40 Which reaction in β-oxidation is a hydration reaction? What is the name of the enzyme that catalyzes this reaction? Write an equation representing this reaction.

Applications

23.41 What products are formed when the ω-phenyl-labeled carboxylic acid 14-phenyltetradecanoic acid is degraded by β-oxidation?

23.42 What products are formed when the ω-phenyl-labeled carboxylic acid 5-phenylpentanoic acid is degraded by β-oxidation?

23.43 Calculate the number of ATP molecules produced by complete β-oxidation of the fourteen-carbon saturated fatty acid tetradecanoic acid (common name: myristic acid).

23.44 **a.** Write the sequence of steps that would be followed for one round of β-oxidation of hexanoic acid.
  b. Calculate the number of ATP molecules produced by complete β-oxidation of hexanoic acid.

23.45 Calculate the number of ATP molecules produced by complete β-oxidation of lauric acid.

23.46 Calculate the number of ATP molecules produced by the complete β-oxidation of eicosanoic acid.

23.47 What is the fate of the acetyl CoA produced by β-oxidation?

23.48 How many ATP molecules are produced from each acetyl CoA molecule generated in β-oxidation that enters the citric acid cycle?

Ketone Bodies

Foundations

23.49 What are ketone bodies?

23.50 What are the chemical properties of ketone bodies?

23.51 What is ketosis?

23.52 Define *ketoacidosis*.

23.53 In what part of the cell does ketogenesis occur? Be specific.

23.54 What would be the fate of HMG-CoA produced in ketogenesis if it were produced in the cell cytoplasm?

Applications

23.55 Draw the structures of acetoacetate and β-hydroxybutyrate.

23.56 Describe the relationship between the formation of ketone bodies and β-oxidation.

23.57 Why do uncontrolled diabetics produce large amounts of ketone bodies?

23.58 How does the presence of ketone bodies in the blood lead to ketoacidosis?

23.59 When does the heart use ketone bodies?

23.60 When does the brain use ketone bodies?

Fatty Acid Synthesis

Foundations

23.61 Where in the cell does fatty acid biosynthesis occur?

23.62 What is the acyl group carrier in fatty acid biosynthesis?

23.63 Draw the structure of NADPH.

23.64 What is the function of NADPH in fatty acid biosynthesis?

Applications

23.65 **a.** What is the role of the phosphopantetheine group in fatty acid biosynthesis?
   b. From what molecule is phosphopantetheine made?

23.66 What molecules involved in fatty acid degradation and fatty acid biosynthesis contain the phosphopantetheine group?

23.67 How does the structure of fatty acid synthase differ from that of the enzymes that carry out β-oxidation?

23.68 In what cellular compartments do fatty acid biosynthesis and β-oxidation occur?

The Regulation of Lipid Metabolism

Foundations

23.69 Under what conditions does the liver synthesize ketone bodies?

23.70 Under what conditions does the brain use ketone bodies as a source of energy?

23.71 What pathway provides the majority of the ATP for *working* muscle?

23.72 How are the fatty acids synthesized in the liver transported to adipose tissue?

Applications

23.73 Which pathway provides the majority of the ATP for *resting* muscle?

23.74 Why is the liver unable to utilize ketone bodies as an energy source?

23.75 Why is the brain able to use ketone bodies under starvation conditions?

23.76 How are triglyceride synthesis and degradation regulated in adipose tissue?

23.77 What are the major fuels of the heart, brain, and liver?

23.78 Why can't the brain use fatty acids as fuel?

23.79 Briefly describe triglyceride metabolism in an adipocyte.

23.80 What is the source of the glycerol molecule that is used in the synthesis of triglycerides?

The Effects of Insulin and Glucagon on Cellular Metabolism

Foundations

23.81 In general, what is the effect of insulin on catabolic and anabolic or biosynthetic processes?

23.82 What is the trigger that causes insulin to be secreted into the bloodstream?

23.83 What is meant by the term *target cell?*

23.84 What are the primary target cells of insulin?

23.85 What is the trigger that causes glucagon to be secreted into the bloodstream?

23.86 What are the primary target cells of glucagon?

Applications

23.87 Where is insulin produced?

23.88 Where is glucagon produced?

23.89 How does insulin affect carbohydrate metabolism?

23.90 How does glucagon affect carbohydrate metabolism?

23.91 How does insulin affect lipid metabolism?

23.92 How does glucagon affect lipid metabolism?

23.93 Explain the importance of the antagonistic effects of insulin and glucagon.

23.94 How does the presence of the insulin receptor on the surface of a cell identify that cell as a target cell?

CHALLENGE PROBLEMS

1. Suppose that fatty acids were degraded by sequential oxidation of the α-carbon. What product(s) would Knoop have obtained with fatty acids with even numbers of carbon atoms? What product(s) would he have obtained with fatty acids with odd numbers of carbon atoms?

2. Oil-eating bacteria can oxidize long-chain alkanes. In the first step of the pathway, the enzyme monooxygenase catalyzes a reaction that converts the long-chain alkane into a primary alcohol. Data from research studies indicate that three more reactions are required to allow the primary alcohol to enter the β-oxidation pathway. Propose a pathway that would convert the long-chain alcohol into a product that could enter the β-oxidation pathway.

3. A young woman sought the advice of her physician because she was 30 lb overweight. The excess weight was in the form of triglycerides carried in adipose tissue. Yet when the woman described her diet, it became obvious that she actually ate very moderate amounts of fatty foods. Most of her caloric intake was in the form of carbohydrates. This included candy, cake, beer, and soft drinks. Explain how

the excess calories consumed in the form of carbohydrates ended up being stored as triglycerides in adipose tissue.

4. Olestra is a fat substitute that provides no calories, yet has a creamy, tongue-pleasing consistency. Because it can withstand heating, it can be used to prepare foods such as potato chips and crackers. Olestra is a sucrose polyester produced by esterification of six, seven, or eight fatty acids to molecules of sucrose. Develop a hypothesis to explain why olestra is not a source of dietary calories.

5. Carnitine is a tertiary amine found in mitochondria that is involved in transporting the acyl groups of fatty acids from the cytoplasm into the mitochondria. The fatty acyl group is transferred from a fatty acyl CoA molecule and esterified to carnitine. Inside the mitochondria, the reaction is reversed and the fatty acid enters the β-oxidation pathway.

A 17-year-old male went to a university medical center complaining of fatigue and poor exercise tolerance. Muscle biopsies revealed droplets of triglycerides in his muscle cells. Biochemical analysis showed that he had only one-fifth of the normal amount of carnitine in his muscle cells.

What effect will carnitine deficiency have on β-oxidation? What effect will carnitine deficiency have on glucose metabolism?

6. Acetyl CoA carboxylase catalyzes the formation of malonyl CoA from acetyl CoA and the bicarbonate anion, a reaction that requires the hydrolysis of ATP. Write a balanced equation showing this reaction.

The reaction catalyzed by acetyl CoA carboxylase is the rate-limiting step in fatty acid biosynthesis. The malonyl group is transferred from coenzyme A to acyl carrier protein; similarly, the acetyl group is transferred from coenzyme A to acyl carrier protein. This provides the two beginning substrates of fatty acid biosynthesis shown in Figure 23.11.

Consider the following case study. A baby boy was brought to the emergency room with severe respiratory distress. Examination revealed muscle pathology, poor growth, and severe brain damage. A liver biopsy revealed that the child didn't make acetyl CoA carboxylase. What metabolic pathway is defective in this child? How is this defect related to the respiratory distress suffered by the baby?

Glossary

A

absolute specificity (19.5) the property of an enzyme that allows it to bind and catalyze the reaction of only one substrate

accuracy (1.4) the nearness of an experimental value to the true value

acetal (13.4) the family of organic compounds formed via the reaction of two molecules of alcohol with an aldehyde or a ketone in the presence of an acid catalyst

acetyl coenzyme A (acetyl CoA) (14.4, 22.2) a molecule composed of coenzyme A and an acetyl group; the intermediate that provides acetyl groups for complete oxidation by aerobic respiration

acid anhydride (14.3) the product formed by the combination of an acid chloride and a carboxylate ion; structurally they are two carboxylic acids with a water molecule removed:

$$(Ar)R-\overset{\overset{O}{\|}}{C}-O-\overset{\overset{O}{\|}}{C}-R(Ar)$$

acid-base reaction (4.7) reaction that involves the transfer of a hydrogen ion (H^+) from one reactant to another

acid chloride (14.3) member of the family of organic compounds with the general formula

$$(Ar)R-\overset{\overset{O}{\|}}{C}-Cl$$

activated complex (7.3) the arrangement of atoms at the top of the potential energy barrier as a reaction proceeds

activation energy (7.3) the threshold energy that must be overcome to produce a chemical reaction

active site (19.4) the cleft in the surface of an enzyme that is the site of substrate binding

acyl carrier protein (ACP) (23.4) the protein that forms a thioester linkage with fatty acids during fatty acid synthesis

acyl group (14.3, 15.3) the functional group found in carboxylic acid derivatives that contains the carbonyl group attached to one alkyl or aryl group:

$$(Ar)R-\overset{\overset{O}{\|}}{C}-$$

addition polymer (11.5) polymers prepared by the sequential addition of a monomer

addition reaction (11.5, 13.4) a reaction in which two molecules add together to form a new molecule; often involves the addition of one molecule to a double or triple bond in an unsaturated molecule;

e.g., the addition of alcohol to an aldehyde or ketone to form a hemiacetal

adenosine triphosphate (ATP) (14.4, 21.1) a nucleotide composed of the purine adenine, the sugar ribose, and three phosphoryl groups; the primary energy storage and transport molecule used by the cells in cellular metabolism

adipocyte (23.1) a fat cell

adipose tissue (23.1) fatty tissue that stores most of the body lipids

aerobic respiration (22.3) the oxygen-requiring degradation of food molecules and production of ATP

alcohol (12.1) an organic compound that contains a hydroxyl group (—OH) attached to an alkyl group

aldehyde (13.1) a class of organic molecules characterized by a carbonyl group; the carbonyl carbon is bonded to a hydrogen atom and to another hydrogen or an alkyl or aryl group. Aldehydes have the following general structure:

$$(Ar)-\overset{\overset{O}{\|}}{C}-H \qquad R-\overset{\overset{O}{\|}}{C}-H$$

aldose (16.2) a sugar that contains an aldehyde (carbonyl) group

aliphatic hydrocarbon (10.1) any member of the alkanes, alkenes, and alkynes or the substituted alkanes, alkenes, and alkynes

alkali metal (2.4) an element within Group IA (1) of the periodic table

alkaline earth metal (2.4) an element within Group IIA (2) of the periodic table

alkaloid (15.2) a class of naturally occurring compounds that contain one or more nitrogen heterocyclic rings; many of the alkaloids have medicinal and other physiological effects

alkane (10.2) a hydrocarbon that contains only carbon and hydrogen and is bonded together through carbon-hydrogen and carbon-carbon single bonds; a saturated hydrocarbon with the general molecular formula C_nH_{2n+2}

alkene (11.1) a hydrocarbon that contains one or more carbon-carbon double bond; an unsaturated hydrocarbon with the general formula C_nH_{2n}

alkyl group (10.2) a hydrocarbon group that results from the removal of one hydrogen from the original hydrocarbon (e.g., methyl, —CH₃; ethyl, —CH₂CH₃)

alkyl halide (10.5) a substituted hydrocarbon with the general structure R—X, in which R— represents any alkyl group and X = a halogen (F—, Cl—, Br—, or I—)

alkylammonium ion (15.1) the ion formed when the lone pair of electrons of the

nitrogen atom of an amine is shared with a proton (H^+) from a water molecule

alkyne (11.1) a hydrocarbon that contains one or more carbon-carbon triple bond; an unsaturated hydrocarbon with the general formula C_nH_{2n-2}

allosteric enzyme (19.9) an enzyme that has an effector binding site and an active site; effector binding changes the shape of the active site, rendering it either active or inactive

alpha particle (9.1) a particle consisting of two protons and two neutrons; the alpha particle is identical to a helium nucleus

amide bond (15.3) the bond between the carbonyl carbon of a carboxylic acid and the amino nitrogen of an amine

amides (15.3) the family of organic compounds formed by the reaction between a carboxylic acid derivative and an amine and characterized by the amide group

amines (15.1) the family of organic molecules with the general formula RNH_2, R_2NH, or R_3N (R— can represent either an alkyl or aryl group); they may be viewed as substituted ammonia molecules in which one or more of the ammonia hydrogens has been substituted by a more complex organic group

α-amino acid (15.4, 18.1) the subunits of proteins composed of an α-carbon bonded to a carboxylate group, a protonated amino group, a hydrogen atom, and a variable R group

aminoacyl group (15.4) the functional group that is characteristic of an amino acid; the aminoacyl group has the following general structure:

$$H_3\overset{+}{N}-\overset{\overset{H}{|}}{\underset{R}{C}}-\overset{\overset{O}{\|}}{C}-$$

aminoacyl tRNA (20.6) the transfer RNA covalently linked to the correct amino acid

aminoacyl tRNA binding site of ribosome (A-site) (20.6) a pocket on the surface of a ribosome that holds the aminoacyl tRNA during translation

aminoacyl tRNA synthetase (20.6) an enzyme that recognizes one tRNA and covalently links the appropriate amino acid to it

amorphous solid (5.3) a solid with no organized, regular structure

amphibolic pathway (22.9) a metabolic pathway that functions in both anabolism and catabolism

amphiprotic (8.1) a substance that can behave either as a Brønsted acid or a Brønsted base

amylopectin (16.6)   a highly branched form of amylose; the branches are attached to the C-6 hydroxyl by $\alpha(1 \rightarrow 6)$ glycosidic linkage; a component of starch

amylose (16.6)   a linear polymer of α-D-glucose molecules bonded in $\alpha(1 \rightarrow 4)$ glycosidic linkage that is a component of starch; a polysaccharide storage form

anabolism (21.1, 22.9)   all of the cellular energy-requiring biosynthetic pathways

anaerobic threshold (21.4)   the point at which the level of lactate in the exercising muscle inhibits glycolysis, and the muscle, deprived of energy, ceases to function

analgesic (15.2)   any drug that acts as a painkiller, e.g., aspirin, acetaminophen

anaplerotic reaction (22.9)   a reaction that replenishes a substrate needed for a biochemical pathway

anesthetic (15.2)   a drug that causes a lack of sensation in part of the body (local anesthetic) or causes unconsciousness (general anesthetic)

anion (2.6)   a negatively charged atom or group of atoms

anode (4.8)   the positively charged electrode in an electrical cell

anomers (16.4)   isomers of cyclic monosaccharides that differ from one another in the arrangement of bonds around the hemiacetal carbon

antibodies (18: Intro)   immunoglobulins; specific glycoproteins produced by cells of the immune system in response to invasion by infectious agents

anticodon (20.6)   a sequence of three ribonucleotides on a tRNA that are complementary to a codon on the mRNA; codon-anticodon binding results in delivery of the correct amino acid to the site of protein synthesis

antigen (18: Intro)   any substance that is able to stimulate the immune system; generally a protein or large carbohydrate

antiparallel strands (20.2)   a term describing the polarities of the two strands of the DNA double helix; on one strand, the sugar-phosphate backbone advances in the $5' \rightarrow 3'$ direction; on the opposite, complementary strand, the sugar-phosphate backbone advances in the $3' \rightarrow 5'$ direction

apoenzyme (19.7)   the protein portion of an enzyme that requires a cofactor to function in catalysis

aqueous solution (6.1)   any solution in which the solvent is water

arachidonic acid (17.2)   a fatty acid derived from linoleic acid; the precursor of the prostaglandins

aromatic hydrocarbon (10.1, 11.6)   an organic compound that contains the benzene ring or a derivative of the benzene ring

Arrhenius acid (8.1)   a substance that dissociates to produce H^+

Arrhenius base (8.1)   a substance that dissociates to produce OH^-

Arrhenius theory (8.1)   a theory that describes an acid as a substance that dissociates to produce H^+, and a base as a substance that dissociates to produce OH^-

artificial radioactivity (9.5)   radiation that results from the conversion of a stable nucleus to another, unstable nucleus

atherosclerosis (17.4)   deposition of excess plasma cholesterol and other lipids and proteins on the walls of arteries, resulting in decreased artery diameter and increased blood pressure

atom (2.1)   the smallest unit of an element that retains the properties of that element

atomic mass (2.1, 4.1)   the mass of an atom expressed in atomic mass units

atomic mass unit (4.1) $1/12$ of the mass of a ^{12}C atom, equivalent to 1.661×10^{-24} g

atomic number (2.1)   the number of protons in the nucleus of an atom; it is a characteristic identifier of an element

atomic orbital (2.3, 2.5)   a specific region of space where an electron may be found

ATP synthase (22.6)   a multiprotein complex within the inner mitochondrial membrane that uses the energy of the proton (H^+) gradient to produce ATP

autoionization (8.1)   also known as *self-ionization*, the reaction of a substance, such as water, with itself to produce a positive and a negative ion

Avogadro's law (5.1)   a law that states that the volume is directly proportional to the number of moles of gas particles, assuming that the pressure and temperature are constant

Avogadro's number (4.1) 6.022×10^{23} particles of matter contained in 1 mol of a substance

B

background radiation (9.6)   the radiation that emanates from natural sources

barometer (5.1)   a device for measuring pressure

base pair (20.2)   a hydrogen-bonded pair of nitrogenous bases within the DNA double helix; the standard base pairs always involve a purine and a pyrimidine; in particular, adenine always base pairs with thymine and cytosine with guanine

Becquerel (9.7)   amount of radioactive material that produces 1 atomic disintegration per second

Benedict's reagent (16.4)   a buffered solution of Cu^{2+} ions that can be used to test for reducing sugars or to distinguish between aldehydes and ketones

Benedict's test (13.4)   a test used to determine the presence of reducing sugars or to distinguish between aldehydes and ketones; it requires a buffered solution of Cu^{2+} ions that are reduced to Cu^+, which precipitates as brick-red Cu_2O

bent (3.4)   a planar molecule with bond angles other than 180°

beta particle (9.1)   an electron formed in the nucleus by the conversion of a neutron into a proton

bile (23.1)   micelles of lecithin, cholesterol, bile salts, protein, inorganic ions, and bile pigments that aid in lipid digestion by emulsifying fat droplets

binding energy (9.3)   the energy required to break down the nucleus into its component parts

bioinformatics (20.10)   an interdisciplinary field that uses computer information sciences and DNA technology to devise methods for understanding, analyzing, and applying DNA sequence information

boat conformation (10.4)   a form of a six-member cycloalkane that resembles a rowboat. It is less stable than the chair conformation because the hydrogen atoms are not perfectly staggered

boiling point (3.3)   the temperature at which the vapor pressure of a liquid is equal to the atmospheric pressure

bond energy (3.4)   the amount of energy necessary to break a chemical bond

Boyle's law (5.1)   a law stating that the volume of a gas varies inversely with the pressure exerted if the temperature and number of moles of gas are constant

breeder reactor (9.4)   a nuclear reactor that produces its own fuel in the process of providing electrical energy

Brønsted-Lowry acid (8.1)   a proton donor

Brønsted-Lowry base (8.1)   a proton acceptor

buffer capacity (8.4)   a measure of the ability of a solution to resist large changes in pH when a strong acid or strong base is added

buffer solution (8.4)   a solution containing a weak acid or base and its salt (the conjugate base or acid) that is resistant to large changes in pH upon addition of strong acids or bases

buret (8.3)   a device calibrated to deliver accurately known volumes of liquid, as in a titration

C

C-terminal amino acid (18.2)   the amino acid in a peptide that has a free α-CO_2^- group; the last amino acid in a peptide

calorimetry (7.2)   the measurement of heat energy changes during a chemical reaction

cap structure (20.4)   a 7-methylguanosine unit covalently bonded to the 5' end of a mRNA by a 5'–5' triphosphate bridge

carbinol carbon (12.1)   that carbon in an alcohol to which the hydroxyl group is attached

carbohydrate (16.1)   generally sugars and polymers of sugars; the primary source of energy for the cell

carbonyl group (13: Intro)   the functional group that contains a carbon-oxygen double bond: —C═O; the functional group found in aldehydes and ketones

carboxyl group (14.1) the —COOH functional group; the functional group found in carboxylic acids

carboxylic acid (14.1) a member of the family of organic compounds that contain the —COOH functional group

carboxylic acid derivative (14.2) any of several families of organic compounds, including the esters and amides, that are derived from carboxylic acids and have the general formula

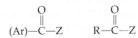

$Z =$ —OR or OAr for the esters, and
$Z =$ —NH$_2$ for the amides

carcinogen (20.7) any chemical or physical agent that causes mutations in the DNA that lead to uncontrolled cell growth or cancer

catabolism (21.1, 22.9) the degradation of fuel molecules and production of ATP for cellular functions

catalyst (7.3) any substance that increases the rate of a chemical reaction (by lowering the activation energy of the reaction) and that is not destroyed in the course of the reaction

cathode (4.8) the negatively charged electrode in an electrical cell

cation (2.6) a positively charged atom or group of atoms

cellulose (16.6) a polymer of β-D-glucose linked by β(1 → 4) glycosidic bonds

central dogma (20.4) a statement of the directional transfer of the genetic information in cells: DNA → RNA → Protein

chain reaction (9.4) the process in a fission reactor that involves neutron production and causes subsequent reactions accompanied by the production of more neutrons in a continuing process

chair conformation (10.4) the most stable conformation for a six-member cycloalkane; so-called for its resemblance to a lawn chair

Charles's law (5.1) a law stating that the volume of a gas is directly proportional to the temperature of the gas, assuming that the pressure and number of moles of the gas are constant

chemical bond (3.1) the attractive force holding two atomic nuclei together in a chemical compound

chemical change (1.2) a process in which one or more atoms of a substance is rearranged, removed, replaced, or added to produce a new substance

chemical equation (4.3) a record of chemical change, showing the conversion of reactants to products

chemical equilibrium (7.4) the state of a reaction in which the rates of the forward and reverse reactions are equal

chemical formula (4.2) the representation of a compound or ion in which elemental symbols represent types of atoms, and subscripts show the relative numbers of atoms

chemical properties (1.2) characteristics of a substance that relate to the substance's participation in a chemical reaction

chemical reaction (1.2) a process in which atoms are rearranged to produce new combinations

chemistry (1.1) the study of matter, its chemical and physical properties, the chemical and physical changes it undergoes, and the energy changes that accompany these processes

chiral carbon (16.3) a carbon atom bonded to four different atoms or groups of atoms

chiral molecule (16.3) molecule capable of existing in mirror-image forms

cholesterol (17.4) a twenty-seven-carbon steroid ring structure that serves as the precursor of the steroid hormones

chromosome (20.2) a piece of DNA that carries the genetic instructions, or genes, of an organism

chylomicron (17.5, 23.1) a plasma lipoprotein (aggregate of protein and triglycerides) that carries triglycerides from the intestine to all body tissues via the bloodstream

cis-trans isomers (10.3) isomers that differ from one another in the placement of substituents on a double bond or ring

citric acid cycle (22.4) a cyclic biochemical pathway that is the final stage of degradation of carbohydrates, fats, and amino acids. It results in the complete oxidation of acetyl groups derived from these dietary fuels

cloning vector (20.8) a DNA molecule that can carry a cloned DNA fragment into a cell and that has a replication origin that allows the DNA to be replicated abundantly within the host cell

coagulation (18.10) the process by which proteins in solution are denatured and aggregate with one another to produce a solid

codon (20.5) a group of three ribonucleotides on the mRNA that specifies the addition of a specific amino acid onto the growing peptide chain

coenzyme (19.7) an organic group required by some enzymes; it generally serves as a donor or acceptor of electrons or a functional group in a reaction

coenzyme A (22.2) a molecule derived from ATP and the vitamin pantothenic acid; coenzyme A functions in the transfer of acetyl groups in lipid and carbohydrate metabolism

cofactor (19.7) an inorganic group, usually a metal ion, that must be bound to an apoenzyme to maintain the correct configuration of the active site

colipase (23.1) a protein that aids in lipid digestion by binding to the surface of lipid droplets and facilitating binding of pancreatic lipase

colligative property (6.4) property of a solution that is dependent only on the concentration of solute particles

colloidal suspension (6.1) a heterogeneous mixture of solute particles in a solvent; distribution of solute particles is not uniform because of the size of the particles

combination reaction (4.3) a reaction in which two substances join to form another substance

combined gas law (5.1) an equation that describes the behavior of a gas when volume, pressure, and temperature may change simultaneously

combustion (4.8) the oxidation of hydrocarbons by burning in the presence of air to produce carbon dioxide and water

competitive inhibitor (19.10) a structural analog; a molecule that has a structure very similar to the natural substrate of an enzyme, competes with the natural substrate for binding to the enzyme active site, and inhibits the reaction

complementary strands (20.2) the opposite strands of the double helix which are hydrogen-bonded to one another such that adenine and thymine or guanine and cytosine are always paired

complete protein (18.11) a protein source that contains all the essential and nonessential amino acids

complex lipid (17.5) a lipid bonded to other types of molecules

compound (1.2) a substance that is characterized by constant composition and that can be chemically broken down into elements

concentration (1.6, 6.2) a measure of the quantity of a substance contained in a specified volume of solution

concentration gradient (6.4) region where concentration decreases over distance

condensation (5.2) the conversion of a gas to a liquid

condensation polymer (14.2) a polymer, which is a large molecule formed by the combination of many small molecules (monomers) that results from the joining of monomers in a reaction that forms a small molecule, such as water or an alcohol

condensed formula (10.2) a structural formula showing all of the atoms in a molecule and placing them in a sequential arrangement that details which atoms are bonded to each other; the bonds themselves are not shown

conformations, conformers (10.4) discrete, distinct isomeric structures that may be converted, one to the other, by rotation about the bonds in the molecule

conjugate acid (8.1) substance that has one more proton than the base from which it is derived

conjugate acid-base pair (8.1) two species related to each other through the gain or loss of a proton

conjugate base (8.1) substance that has one fewer proton than the acid from which it is derived

constitutional isomers (10.2)   two molecules having the same molecular formulas, but different chemical structures

Cori Cycle (21.6)   a metabolic pathway in which the lactate produced by working muscle is taken up by cells in the liver and converted back to glucose by gluconeogenesis

corrosion (4.8)   the unwanted oxidation of a metal

covalent bonding (3.1)   a pair of electrons shared between two atoms

covalent solid (5.3)   a collection of atoms held together by covalent bonds

cristae (22.1)   the folds of the inner membrane of the mitochondria

crystal lattice (3.1)   a unit of a solid characterized by a regular arrangement of components

crystalline solid (5.3)   a solid having a regular repeating atomic structure

curie (9.7)   the quantity of radioactive material that produces 3.7×10^{10} nuclear disintegrations per second

cycloalkane (10.3)   a cyclic alkane; a saturated hydrocarbon that has the general formula C_nH_{2n}

D

Dalton's law (5.1)   also called the law of partial pressures; states that the total pressure exerted by a gas mixture is the sum of the partial pressures of the component gases

data (1.1)   facts resulting from an experiment

decomposition reaction (4.3)   the breakdown of a substance into two or more substances

defense proteins (18: Intro)   proteins that defend the body against infectious diseases; antibodies are defense proteins

degenerate code (20.5)   a term used to describe the fact that several triplet codons may be used to specify a single amino acid in the genetic code

dehydration (of alcohols) (12.4)   a reaction that involves the loss of a water molecule, in this case the loss of water from an alcohol and the simultaneous formation of an alkene

deletion mutation (20.7)   a mutation that results in the loss of one or more nucleotides from a DNA sequence

denaturation (18.10)   the process by which the organized structure of a protein is disrupted, resulting in a completely disorganized, nonfunctional form of the protein

density (1.6)   mass per unit volume of a substance

deoxyribonucleic acid (DNA) (20.1)   the nucleic acid molecule that carries all of the genetic information of an organism; the DNA molecule is a double helix composed of two strands, each of which is composed of phosphate groups, deoxyribose, and the nitrogenous bases thymine, cytosine, adenine, and guanine

deoxyribonucleotide (20.1)   a nucleoside phosphate or nucleotide composed of a nitrogenous base in β-N-glycosidic linkage to the 1′ carbon of the sugar 2′-deoxyribose and with one, two, or three phosphoryl groups esterified at the hydroxyl of the 5′ carbon

deviation (1.4)   the amount of variation present in a set of replicate measurements

diabetes mellitus (23.3)   a disease caused by the production of insufficient levels of insulin and characterized by the appearance of very high levels of glucose in the blood and urine

dialysis (6.5)   the removal of waste material via transport across a membrane

diastereomers (16.3)   stereoisomers with at least two chiral carbons that are not mirror images of one another

diffusion (6.4)   net movement of solute or solvent molecules from a region of high concentration to a region of low concentration

diglyceride (17.3)   the product of esterification of glycerol at two positions

dipole-dipole interactions (5.2)   attractive forces between polar molecules

disaccharide (16.1)   a sugar composed of two monosaccharides joined through an oxygen atom bridge

dissociation (3.3)   production of positive and negative ions when an ionic compound dissolves in water

disulfide (12.8)   an organic compound that contains a disulfide group (—S—S—)

DNA polymerase III (20.3)   the enzyme that catalyzes the polymerization of daughter DNA strands using the parental strand as a template

double bond (3.4)   a bond in which two pairs of electrons are shared by two atoms

double helix (20.2)   the spiral staircase–like structure of the DNA molecule characterized by two sugar-phosphate backbones wound around the outside and nitrogenous bases extending into the center

double-replacement reaction (4.3)   a chemical change in which cations and anions "exchange partners"

dynamic equilibrium (7.4)   the state that exists when the rate of change in the concentration of products and reactants is equal, resulting in no net concentration change

E

eicosanoid (17.2)   any of the derivatives of twenty-carbon fatty acids, including the prostaglandins, leukotrienes, and thromboxanes

electrolysis (4.8)   an electrochemical process that uses electrical energy to cause nonspontaneous oxidation-reduction reactions to occur

electrolyte (3.3, 6.1)   a material that dissolves in water to produce a solution that conducts an electrical current

electromagnetic radiation (2.3)   energy that is propagated as waves at the speed of light

electron (2.1)   a negatively charged particle outside of the nucleus of an atom

electron affinity (2.7)   the energy released when an electron is added to an isolated atom

electron configuration (2.5)   the arrangement of electrons around the nucleus of an atom, an ion, or a collection of nuclei of a molecule

electron density (2.3)   the probability of finding the electron in a particular location

electron transport system (22.6)   the series of electron transport proteins embedded in the inner mitochondrial membrane that accept high-energy electrons from NADH and $FADH_2$ and transfer them in stepwise fashion to molecular oxygen (O_2)

electronegativity (3.1)   a measure of the tendency of an atom in a molecule to attract shared electrons

element (1.2)   a substance that cannot be decomposed into simpler substances by chemical or physical means

elimination reaction (12.4)   a reaction in which a molecule loses atoms or ions from its structure

elongation factor (20.6)   proteins that facilitate the elongation phase of translation

emulsifying agent (17.3)   a bipolar molecule that aids in the suspension of fats in water

enantiomers (16.3)   stereoisomers that are nonsuperimposable mirror images of one another

endothermic reaction (7.1)   a chemical or physical change in which energy is absorbed

energy (1.1)   the ability to do work

energy level (2.3)   one of numerous atomic regions where electrons may be found

enol (13.4)   a tautomer containing a carbon-carbon double bond and a hydroxyl group

enthalpy (7.1)   a term that represents heat energy

entropy (7.1)   a measure of randomness or disorder

enzyme (18: Intro, 19: Intro)   a protein that serves as a biological catalyst

enzyme specificity (19.5)   the ability of an enzyme to bind to only one, or a very few, substrates and thus catalyze only a single reaction

enzyme-substrate complex (19.4)   a molecular aggregate formed when the substrate binds to the active site of the enzyme

equilibrium constant (7.4)   number equal to the ratio of the equilibrium concentrations of products to the equilibrium concentrations of reactants, each raised to the power corresponding to its coefficient in the balanced equation

equivalence point (8.3)   the situation in which reactants have been mixed in the molar ratio corresponding to the balanced equation

error (1.4)   the difference between the true value and the experimental value for data or results

essential amino acid (18.11)   an amino acid that cannot be synthesized by the body and must therefore be supplied by the diet

essential fatty acids (17.2)   the fatty acids linolenic and linoleic acids that must be supplied in the diet because they cannot be synthesized by the body

ester (14.2)   a carboxylic acid derivative formed by the reaction of a carboxylic acid and an alcohol. Esters have the following general formula:

$$(Ar)R\overset{\overset{\displaystyle O}{\|}}{-C-}OR(Ar)$$

esterification (17.2)   the formation of an ester in the reaction of a carboxylic acid and an alcohol

ether (12.7)   an organic compound that contains two alkyl and/or aryl groups attached to an oxygen atom; R—O—R, Ar—O—R, and Ar—O—Ar

eukaryote (20.2)   an organism having cells containing a true nucleus enclosed by a nuclear membrane and having a variety of membrane-bound organelles that segregate different cellular functions into different compartments

evaporation (5.2)   the conversion of a liquid to a gas below the boiling point of the liquid

excited state (2.3)   an electronic state of an atom when energy has been adsorbed by the ground state atom and one or more electrons are promoted into a higher energy level

exon (20.4)   protein-coding sequences of a gene found on the final mature mRNA

exothermic reaction (7.1)   a chemical or physical change that releases energy

extensive property (1.2)   a property of a substance that depends on the quantity of the substance

F

F$_0$ F$_1$ complex (22.6)   an alternative term for ATP synthase, the multiprotein complex in the inner mitochondrial membrane that uses the energy of the proton gradient to produce ATP

fatty acid (14.1, 17.2)   any member of the family of continuous-chain carboxylic acids that generally contain four to twenty carbon atoms; the most concentrated source of energy used by the cell

feedback inhibition (19.9)   the process whereby excess product of a biosynthetic pathway turns off the entire pathway for its own synthesis

fermentation (12.3, 21.4)   anaerobic (in the absence of oxygen) catabolic reactions that occur with no net oxidation. Pyruvate or an organic compound produced from pyruvate is reduced as NADH is oxidized

fibrous protein (18.4)   a protein composed of peptides arranged in long sheets or fibers

Fischer Projection (16.3)   a two-dimensional drawing of a molecule that shows a chiral carbon at the intersection of two lines and horizontal lines representing bonds projecting out of the page, and vertical lines representing bonds that project into the page

fission (9.4)   the splitting of heavy nuclei into lighter nuclei accompanied by the release of large quantities of energy

fluid mosaic model (17.6)   the model of membrane structure that describes the fluid nature of the lipid bilayer and the presence of numerous proteins embedded within the membrane

formula (3.2)   the representation of the fundamental compound unit using chemical symbols and numerical subscripts

formula mass (4.2)   the mass of a formula unit of a compound relative to a standard (carbon-12)

formula unit (4.2)   the smallest collection of atoms from which the formula of a compound can be established

free energy (7.1)   the combined contribution of entropy and enthalpy for a chemical reaction

fructose (16.4)   a ketohexose that is also called levulose and fruit sugar; the sweetest of all sugars, abundant in honey and fruits

fuel value (7.2)   the amount of energy derived from a given mass of material

functional group (10.1)   an atom (or group of atoms and their bonds) that imparts specific chemical and physical properties to a molecule

fusion (9.4)   the joining of light nuclei to form heavier nuclei, accompanied by the release of large amounts of energy

G

galactose (16.4)   an aldohexose that is a component of lactose (milk sugar)

galactosemia (16.5)   a human genetic disease caused by the inability to convert galactose to a phosphorylated form of glucose (glucose-1-phosphate) that can be used in cellular metabolic reactions

gamma ray (9.1)   a high-energy emission from nuclear processes, traveling at the speed of light; the high-energy region of the electromagnetic spectrum

gaseous state (1.2)   a physical state of matter characterized by a lack of fixed shape or volume and ease of compressibility

genome (20.2)   the complete set of genetic information in all the chromosomes of an organism

geometric isomer (10.3, 11.3)   an isomer that differs from another isomer in the placement of substituents on a double bond or a ring

globular protein (18.5)   a protein composed of polypeptide chains that are tightly folded into a compact spherical shape

glucagon (21.7, 23.6)   a peptide hormone synthesized by the α-cells of the islets of Langerhans in the pancreas and secreted in response to low blood glucose levels; glucagon promotes glycogenolysis and gluconeogenesis and thereby increases the concentration of blood glucose

gluconeogenesis (21.6)   the synthesis of glucose from noncarbohydrate precursors

glucose (16.4)   an aldohexose, the most abundant monosaccharide; it is a component of many disaccharides, such as lactose and sucrose, and of polysaccharides, such as cellulose, starch, and glycogen

glyceraldehyde (16.4)   an aldotriose that is the simplest carbohydrate; phosphorylated forms of glyceraldehyde are important intermediates in cellular metabolic reactions

glyceride (17.3)   a lipid that contains glycerol

glycogen (16.6, 21.7)   a long, branched polymer of glucose stored in liver and muscles of animals; it consists of a linear backbone of α-D-glucose in $\alpha(1 \rightarrow 4)$ linkage, with numerous short branches attached to the C-6 hydroxyl group by $\alpha(1 \rightarrow 6)$ linkage

glycogenesis (21.7)   the metabolic pathway that results in the addition of glucose to growing glycogen polymers when blood glucose levels are high

glycogen granule (21.7)   a core of glycogen surrounded by enzymes responsible for glycogen synthesis and degradation

glycogenolysis (21.7)   the biochemical pathway that results in the removal of glucose molecules from glycogen polymers when blood glucose levels are low

glycolysis (21.3)   the enzymatic pathway that converts a glucose molecule into two molecules of pyruvate; this anaerobic process generates a net energy yield of two molecules of ATP and two molecules of NADH

glycoprotein (18.6)   a protein bonded to sugar groups

glycosidic bond (16.1)   the bond between the hydroxyl group of the C-1 carbon of one sugar and a hydroxyl group of another sugar

gray (9.7)   the absorption of 1 joule of energy by 1 kg of matter

ground state (2.3)   the electronic state of an atom in which all of the electrons are in the lowest possible energy levels

group (2.4)   any one of eighteen vertical columns of elements; often referred to as a *family*

group specificity (19.5)   an enzyme that catalyzes reactions involving similar substrate molecules having the same functional groups

guanosine triphosphate (GTP) (21.6)   a nucleotide composed of the purine guanine, the sugar ribose, and three phosphoryl groups

H

half-life ($t_{1/2}$) (9.3, 9.6)   the length of time required for one-half of the initial mass of an isotope to decay to products

halogen (2.4)   an element found in Group VIIA (17) of the periodic table

halogenation (10.5, 11.5)   a reaction in which one of the C—H bonds of a hydrocarbon is replaced with a C—X bond (X = Br or Cl generally)

Haworth projection (16.4)   a means of representing the orientation of substituent groups around a cyclic sugar molecule

heat (7.1)   energy transferred between a system and its surroundings due to a temperature difference between system and surroundings

α-helix (18.4) a right-handed coiled secondary structure maintained by hydrogen bonds between the amide hydrogen of one amino acid and the carbonyl oxygen of an amino acid four residues away

heme group (18.8) the chemical group found in hemoglobin and myoglobin that is responsible for the ability to carry oxygen

hemiacetal (13.4, 16.4) the family of organic compounds formed via the reaction of one molecule of alcohol with an aldehyde or a ketone in the presence of an acid catalyst

hemoglobin (18.8) the major protein component of red blood cells; the function of this red, iron-containing protein is transport of oxygen

Henderson-Hasselbalch equation (8.4) an equation for calculating the pH of a buffer system:

$$pH = pK_a + \log \frac{[\text{conjugate base}]}{[\text{weak acid}]}$$

Henry's law (6.1) a law stating that the number of moles of a gas dissolved in a liquid at a given temperature is proportional to the partial pressure of the gas

heterocyclic amine (15.2) a heterocyclic compound that contains nitrogen in at least one position in the ring skeleton

heterocyclic aromatic compound (11.7) cyclic aromatic compound having at least one atom other than carbon in the structure of the aromatic ring

heterogeneous mixture (1.2) a mixture of two or more substances characterized by nonuniform composition

heteropolysaccharide (16.6) a polysaccharide composed of two or more different monosaccharides

hexose (16.2) a six-carbon monosaccharide

high-density lipoprotein (HDL) (17.5) a plasma lipoprotein that transports cholesterol from peripheral tissue to the liver

holoenzyme (19.7) an active enzyme consisting of an apoenzyme bound to a cofactor

homogeneous mixture (1.2) a mixture of two or more substances characterized by uniform composition

homopolysaccharide (16.6) a polysaccharide composed of identical monosaccharides

hybridization (20.8) a technique for identifying DNA or RNA sequences that is based on specific hydrogen bonding between a radioactive probe and complementary DNA or RNA sequences

hydrate (4.2) any substance that has water molecules incorporated in its structure

hydration (11.5, 12.4) a reaction in which water is added to a molecule, e.g., the addition of water to an alkene to form an alcohol

hydrocarbon (10.1) a compound composed solely of the elements carbon and hydrogen

hydrogen bonding (5.2) the attractive force between a hydrogen atom covalently bonded to a small, highly electronegative atom and another atom containing an unshared pair of electrons

hydrogenation (11.5, 13.4, 17.2) a reaction in which hydrogen (H_2) is added to a double or a triple bond

hydrohalogenation (11.5) the addition of a hydrohalogen (HCl, HBr, or HI) to an unsaturated bond

hydrolase (19.1) an enzyme that catalyzes hydrolysis reactions

hydrolysis (14.2) a chemical change that involves the reaction of a molecule with water; the process by which molecules are broken into their constituents by addition of water

hydronium ion (8.1) a protonated water molecule, H_3O^+

hydrophilic amino acid (18.1) "water loving"; a polar or ionic amino acid that has a high affinity for water

hydrophobic amino acid (18.1) "water fearing"; a nonpolar amino acid that prefers contact with other nonpolar molecules over contact with water

hydroxide ion (8.1) the anion consisting of one oxygen atom and one hydrogen atom ($-OH^-$)

hydroxyl group (12.1) the —OH functional group that is characteristic of alcohols

hyperammonemia (22.8) a genetic defect in one of the enzymes of the urea cycle that results in toxic or even fatal elevation of the concentration of ammonium ions in the body

hyperglycemia (21.7) blood glucose levels that are higher than normal

hypertonic solution (6.4) the more concentrated solution of two separated by a semipermeable membrane

hypoglycemia (21.7) blood glucose levels that are lower than normal

hypothesis (1.1) an attempt to explain observations in a commonsense way

hypotonic solution (6.4) the more dilute solution of two separated by a semipermeable membrane

I

ideal gas (5.1) a gas in which the particles do not interact and the volume of the individual gas particles is assumed to be negligible

ideal gas law (5.1) a law stating that for an ideal gas the product of pressure and volume is proportional to the product of the number of moles of the gas and its temperature; the proportionality constant for an ideal gas is symbolized R

incomplete protein (18.11) a protein source that does not contain all the essential and nonessential amino acids

indicator (8.3) a solute that shows some condition of a solution (such as acidity or basicity) by its color

induced fit model (19.4) the theory of enzyme-substrate binding that assumes that the enzyme is a flexible molecule and that both the substrate and the enzyme change their shapes to accommodate one another as the enzyme-substrate complex forms

initiation factors (20.6) proteins that are required for formation of the translation initiation complex, which is composed of the large and small ribosomal subunits, the mRNA, and the initiator tRNA, methionyl tRNA

inner mitochondrial membrane (22.1) the highly folded, impermeable membrane within the mitochondrion that is the location of the electron transport system and ATP synthase

insertion mutation (20.7) a mutation that results in the addition of one or more nucleotides to a DNA sequence

insulin (21.7, 23.6) a hormone released from the pancreas in response to high blood glucose levels; insulin stimulates glycogenesis, fat storage, and cellular uptake and storage of glucose from the blood

intensive property (1.2) a property of a substance that is independent of the quantity of the substance

intermembrane space (22.1) the region between the outer and inner mitochondrial membranes, which is the location of the proton (H^+) reservoir that drives ATP synthesis

intermolecular force (3.5) any attractive force that occurs between molecules

intramolecular force (3.5) any attractive force that occurs within molecules

intron (20.4) a noncoding sequence within a eukaryotic gene that must be removed from the primary transcript to produce a functional mRNA

ion (2.6) an electrically charged particle formed by the gain or loss of electrons

ionic bonding (3.1) an electrostatic attractive force between ions resulting from electron transfer

ionic solid (5.3) a solid composed of positive and negative ions in a regular three-dimensional crystalline arrangement

ionization energy (2.7) the energy needed to remove an electron from an atom in the gas phase

ionizing radiation (9.1, 9.5) radiation that is sufficiently high in energy to cause ion formation upon impact with an atom

ion product constant for water (8.1) the product of the hydronium and hydroxide ion concentrations in pure water at a specified temperature; at 25°C, it has a value of 1.0×10^{-14}

irreversible enzyme inhibitor (19.10) a chemical that binds strongly to the R groups of an amino acid in the active site and eliminates enzyme activity

isoelectronic (2.6) atoms, ions, and molecules containing the same number of electrons

isomerase (19.1) an enzyme that catalyzes the conversion of one isomer to another

isomers (3.4) molecules having the same molecular formula but different chemical structures

isotonic solution (6.4) a solution that has the same solute concentration as another solution with which it is being compared; a solution that has the same osmotic pressure as a solution existing within a cell

isotope (2.1)   atom of the same element that differs in mass because it contains different numbers of neutrons

IUPAC Nomenclature System (10.2)   the International Union of Pure and Applied Chemistry (IUPAC) standard, universal system for the nomenclature of organic compounds

K

α-keratin (18.4)   a member of the family of fibrous proteins that form the covering of most land animals; major components of fur, skin, beaks, and nails

ketoacidosis (23.3)   a drop in the pH of the blood caused by elevated levels of ketone bodies

ketone (13.1)   a family of organic molecules characterized by a carbonyl group; the carbonyl carbon is bonded to two alkyl groups, two aryl groups, or one alkyl and one aryl group; ketones have the following general structures:

$$\begin{array}{ccc} O & O & O \\ \| & \| & \| \\ R-C-R & R-C-(Ar) & (Ar)-C-(Ar) \end{array}$$

ketone bodies (23.3)   acetone, acetoacetone, and β-hydroxybutyrate produced from fatty acids in the liver via acetyl CoA

ketose (16.2)   a sugar that contains a ketone (carbonyl) group

ketosis (23.3)   an abnormal rise in the level of ketone bodies in the blood

kinetic energy (1.6)   the energy resulting from motion of an object [kinetic energy $= 1/2(\text{mass})(\text{velocity})^2$]

kinetic molecular theory (5.1)   the fundamental model of particle behavior in the gas phase

kinetics (7.3)   the study of rates of chemical reactions

L

lactose (16.5)   a disaccharide composed of β-D-galactose and either α- or β-D-glucose in β(1 ⟶ 4) glycosidic linkage; milk sugar

lactose intolerance (16.5)   the inability to produce the digestive enzyme lactase, which degrades lactose to galactose and glucose

lagging strand (20.3)   in DNA replication, the strand that is synthesized discontinuously from numerous RNA primers

law of conservation of mass (4.3)   a law stating that, in chemical change, matter cannot be created or destroyed

leading strand (20.3)   in DNA replication, the strand that is synthesized continuously from a single RNA primer

LeChatelier's principle (7.4)   a law stating that when a system at equilibrium is disturbed, the equilibrium shifts in the direction that minimizes the disturbance

lethal dose (LD$_{50}$) (9.7)   the quantity of toxic material (such as radiation) that causes the death of 50% of a population of an organism

Lewis structure (3.1)   representation of a molecule (or polyatomic ion) that shows valence electron arrangement among the atoms in a molecule (or polyatomic ion)

Lewis symbol (3.1)   representation of an atom (or ion) using the atomic symbol (for the nucleus and core electrons) and dots to represent valence electrons

ligase (19.1)   an enzyme that catalyzes the joining of two molecules

linear (3.4)   the structure of a molecule in which the bond angle(s) about the central atom(s) is (are) 180°

line formula (10.2)   the simplest representation of a molecule, in which it is assumed that there is a carbon atom at any location where two or more lines intersect, there is a carbon at the end of any line, and each carbon is bonded to the correct number of hydrogen atoms

linkage specificity (19.5)   the property of an enzyme that allows it to catalyze reactions involving only one kind of bond in the substrate molecule

lipase (23.1)   an enzyme that hydrolyzes the ester linkage between glycerol and the fatty acids of triglycerides

lipid (17.1)   a member of the group of biological molecules of varying composition that are classified together on the basis of their solubility in nonpolar solvents

liquid state (1.2)   a physical state of matter characterized by a fixed volume and the absence of a fixed shape

lock-and-key model (19.4)   the theory of enzyme-substrate binding that depicts enzymes as inflexible molecules; the substrate fits into the rigid active site in the same way a key fits into a lock

London dispersion forces (5.2)   weak attractive forces between molecules that result from short-lived dipoles that occur because of the continuous movement of electrons in the molecules

lone pair (3.1)   an electron pair that is not involved in bonding

low-density lipoprotein (LDL) (17.5)   a plasma lipoprotein that carries cholesterol to peripheral tissues and helps to regulate cholesterol levels in those tissues

lyase (19.1)   an enzyme that catalyzes a reaction involving double bonds

M

maltose (16.5)   a disaccharide composed of α-D-glucose and a second glucose molecule in α(1 ⟶ 4) glycosidic linkage

Markovnikov's rule (11.5)   the rule stating that a hydrogen atom, adding to a carbon-carbon double bond, will add to the carbon having the larger number of hydrogens attached to it

mass (1.3)   a quantity of matter

mass/mass percent [% (m/m)] (6.2)   the concentration of a solution expressed as a ratio of mass of solute to mass of solution multiplied by 100%

mass number (2.1)   the sum of the number of protons and neutrons in an atom

mass/volume percent [% (m/V)] (6.2)   the concentration of a solution expressed as a ratio of grams of solute to milliliters of solution multiplied by 100%

matrix space (22.1)   the region of the mitochondrion within the inner membrane; the location of the enzymes that carry out the reactions of the citric acid cycle and β-oxidation of fatty acids

matter (1.1)   anything that has mass and occupies space

melting point (3.3, 5.2)   the temperature at which a solid converts to a liquid

meso compound (16.3)   a special case of stereoisomers that occurs when a molecule has two chiral carbons and an internal plane of symmetry; these molecules are achiral because they have a plane of symmetry within the molecule

messenger RNA (20.4)   an RNA species produced by transcription and that specifies the amino acid sequence for a protein

metal (2.4)   an element located on the left side of the periodic table (left of the "staircase" boundary)

metallic solid (5.3)   a solid composed of metal atoms held together by metallic bonds

metalloid (2.4)   an element along the "staircase" boundary between metals and nonmetals; metalloids exhibit both metallic and nonmetallic properties

metastable isotope (9.3)   an isotope that will give up some energy to produce a more stable form of the same isotope

micelle (23.1)   an aggregation of molecules having nonpolar and polar regions; the nonpolar regions of the molecules aggregate, leaving the polar regions facing the surrounding water

mitochondria (22.1)   the cellular "power plants" in which the reactions of the citric acid cycle, β-oxidation of fatty acids, the electron transport system, and ATP synthase function to produce ATP

mixture (1.2)   a material composed of two or more substances

molality (6.4)   the number of moles of solute per kilogram of solvent

molarity (6.3)   the number of moles of solute per liter of solution

molar mass (4.1, 4.2)   the mass in grams of 1 mol of a substance

molar volume (5.1)   the volume occupied by 1 mol of a substance

mole (4.1)   the amount of substance containing Avogadro's number of particles

molecular formula (10.2)   a formula that provides the atoms and number of each type of atom in a molecule but gives no information regarding the bonding pattern involved in the structure of the molecule

molecular solid (5.3)   a solid in which the molecules are held together by dipole-dipole and London dispersion forces (van der Waals forces)

molecule (3.1)   a unit in which the atoms of two or more elements are held together by chemical bonds

monatomic ion (3.2)   an ion formed by electron gain or loss from a single atom

monoglyceride (17.3)   the product of the esterification of glycerol at one position

monomer (11.5)   the individual molecules from which a polymer is formed

monosaccharide (16.1)   the simplest type of carbohydrate consisting of a single saccharide unit

movement protein (18: Intro)   a protein involved in any aspect of movement in an organism, for instance, actin and myosin in muscle tissue and flagellin that composes bacterial flagella

mutagen (20.7)   any chemical or physical agent that causes changes in the nucleotide sequence of a gene

mutation (20.7)   any change in the nucleotide sequence of a gene

myoglobin (18.8)   the oxygen storage protein found in muscle

N

N-terminal amino acid (18.2)   the amino acid in a peptide that has a free α-N^+H_3 group; the first amino acid of a peptide

natural radioactivity (9.5)   the spontaneous decay of a nucleus to produce high-energy particles or rays

negative allosterism (19.9)   effector binding inactivates the active site of an allosteric enzyme

neurotransmitter (15.5)   a chemical that carries a message, or signal, from a nerve cell to a target cell

neutral glyceride (17.3)   the product of the esterification of glycerol at one, two, or three positions

neutralization (4.7, 8.3)   the reaction between an acid and a base

neutron (2.1)   an uncharged particle, with the same mass as the proton, in the nucleus of an atom

nicotinamide adenine dinucleotide (NAD$^+$) (21.3)   a molecule synthesized from the vitamin niacin and the nucleotide ATP and that serves as a carrier of hydride anions; a coenzyme that is an oxidizing agent used in a variety of metabolic processes

noble gas (2.4)   elements in Group VIIIA (18) of the periodic table

nomenclature (3.2)   a system for naming chemical compounds

nonelectrolyte (3.3, 6.1)   a substance that, when dissolved in water, produces a solution that does not conduct an electrical current

nonessential amino acid (18.11)   any amino acid that can be synthesized by the body

nonmetal (2.4)   an element located on the right side of the periodic table (right of the "staircase" boundary)

nonreducing sugar (16.5)   a sugar that cannot be oxidized by Benedict's or Tollens' reagent

normal boiling point (5.2)   the temperature at which a substance will boil at 1 atm of pressure

nuclear equation (9.2)   a balanced equation accounting for the products and reactants in a nuclear reaction

nuclear imaging (9.5)   the generation of images of components of the body (organs, tissues) using techniques based on the measurement of radiation

nuclear medicine (9.5)   a field of medicine that uses radioisotopes for diagnostic and therapeutic purposes

nuclear reactor (9.4)   a device for conversion of nuclear energy into electrical energy

nucleoside (20.1)   a molecule composed of a nitrogenous base and a five-carbon sugar

nucleosome (20.2)   the first level of chromosome structure consisting of a strand of DNA wrapped around a small disk of histone proteins

nucleotide (20.1, 21.1)   a molecule composed of a nitrogenous base, a five-carbon sugar, and one, two, or three phosphoryl groups

nucleus (2.1)   the small, dense center of positive charge in the atom

nuclide (9.1)   any atom characterized by an atomic number and a mass number

nutrient protein (18: Intro)   a protein that serves as a source of amino acids for embryos or infants

nutritional Calorie (7.2)   equivalent to 1 kilocalorie (1000 calories); also known as a large Calorie

O

octet rule (2.6)   a rule predicting that atoms form the most stable molecules or ions when they are surrounded by eight electrons in their highest occupied energy level

oligosaccharide (16.1)   an intermediate-sized carbohydrate composed of from three to ten monosaccharides

osmolarity (6.4)   the molarity of particles in solution

osmosis (6.4)   net flow of a solvent across a semipermeable membrane in response to a concentration gradient

osmotic pressure (6.4)   the net force with which water enters a solution through a semipermeable membrane; alternatively, the pressure required to stop net transfer of solvent across a semipermeable membrane

outer mitochondrial membrane (22.1)   the membrane that surrounds the mitochondrion and separates it from the contents of the cytoplasm; it is highly permeable to small "food" molecules

β-oxidation (23.2)   the biochemical pathway that results in the oxidation of fatty acids and the production of acetyl CoA

oxidation (4.8, 12.4, 13.4, 14.1)   a loss of electrons; in organic compounds, it may be recognized as a loss of hydrogen atoms or the gain of oxygen

oxidation-reduction reaction (4.8)   also called *redox reaction*, a reaction involving the transfer of one or more electrons from one reactant to another

oxidative deamination (22.7)   an oxidation-reduction reaction in which NAD^+ is reduced and the amino acid is deaminated

oxidative phosphorylation (21.3, 22.6)   production of ATP using the energy of electrons harvested during biological oxidation-reduction reactions

oxidizing agent (4.8)   a substance that oxidizes, or removes electrons from, another substance; the oxidizing agent is reduced in the process

oxidoreductase (19.1)   an enzyme that catalyzes an oxidation-reduction reaction

P

pancreatic serine proteases (19.11)   a family of proteolytic enzymes, including trypsin, chymotrypsin, and elastase, that arose by divergent evolution

parent compound or parent chain (10.2)   in the IUPAC Nomenclature System, the parent compound is the longest carbon-carbon chain containing the principal functional group in the molecule that is being named

partial pressure (5.1)   the pressure exerted by one component of a gas mixture

parts per million (6.2)   number of parts of solute in one million parts of solvent

parts per thousand (6.2)   number of parts of solute per thousand parts of solvent

pentose (16.2)   a five-carbon monosaccharide

pentose phosphate pathway (21.5)   an alternative pathway for glucose degradation that provides the cell with reducing power in the form of NADPH

peptide bond (15.4, 18.2)   the amide bond between two amino acids in a peptide chain

peptidyl tRNA binding site of ribosome (P-site) (20.6)   a pocket on the surface of the ribosome that holds the tRNA bound to the growing peptide chain

percent yield (4.9)   the ratio of the actual and theoretical yields of a chemical reaction multiplied by 100%

period (2.4)   any one of seven horizontal rows of elements in the periodic table

periodic law (2.4)   a law stating that properties of elements are periodic functions of their atomic numbers (Note that Mendeleev's original statement was based on atomic masses.)

peripheral membrane protein (17.6)   a protein bound to either the inner or the outer surface of a membrane

phenol (12.6)   an organic compound that contains a hydroxyl group (—OH) attached to a benzene ring

phenyl group (11.6)   a benzene ring that has had a hydrogen atom removed, C_6H_5—

pH optimum (19.8)   the pH at which an enzyme catalyzes the reaction at maximum efficiency

phosphatidate (17.3)   a molecule of glycerol with fatty acids esterified to C-1 and C-2 of glycerol and a free phosphoryl group esterified at C-3

phosphoanhydride (14.4)   the bond formed when two phosphate groups react with one another and a water molecule is lost

phosphoester (14.4)   the product of the reaction between phosphoric acid and an alcohol

phosphoglyceride (17.3)   a molecule with fatty acids esterified at the C-1 and C-2 positions of glycerol and a phosphoryl group esterified at the C-3 position

phospholipid (17.3)   a lipid containing a phosphoryl group

phosphopantetheine (23.4)   the portion of coenzyme A and the acyl carrier protein that is derived from the vitamin pantothenic acid

photon (2.3)   a particle of light

pH scale (8.2)   a numerical representation of acidity or basicity of a solution; $pH = -\log[H_3O^+]$

physical change (1.2)   a change in the form of a substance but not in its chemical composition; no chemical bonds are broken in a physical change

physical equilibrium (7.4)   occurs between two phases of the same substance

physical property (1.2)   a characteristic of a substance that can be observed without the substance undergoing change (examples include color, density, melting and boiling points)

plane-polarized light (16.3)   light in which all of the light waves are vibrating in the same, parallel direction

plasma lipoprotein (17.5)   a complex composed of lipid and protein that is responsible for the transport of lipids throughout the body

β-pleated sheet (18.4)   a common secondary structure of a peptide chain that resembles the pleats of an Oriental fan

point mutation (20.7)   the substitution of one nucleotide pair for another within a gene

polar covalent bond (3.1)   a covalent bond in which the electrons are not equally shared

poly(A) tail (20.4)   a tract of 100–200 adenosine monophosphate units covalently attached to the 3′ end of eukaryotic messenger RNA molecules

polyatomic ion (3.2)   an ion containing a number of atoms

polymer (11.5)   a very large molecule formed by the combination of many small molecules (called *monomers*) (e.g., polyamides, nylons)

polyprotic substance (8.3)   a substance that can accept or donate more than one proton per molecule

polysaccharide (16.1)   a large, complex carbohydrate composed of long chains of monosaccharides

polysome (20.6)   complexes of many ribosomes all simultaneously translating a single mRNA

positive allosterism (19.9)   effector binding activates the active site of an allosteric enzyme

positron (9.1)   particle that has the same mass as an electron but opposite (+) charge

post-transcriptional modification (20.4)   alterations of the primary transcripts produced in eukaryotic cells; these include addition of a poly(A) tail to the 3′ end of the mRNA, addition of the cap structure to the 5′ end of the mRNA, and RNA splicing

potential energy (1.6)   stored energy or energy caused by position or composition

precipitate (4.5, 6.1)   an insoluble substance formed and separated from a solution

precision (1.4)   the degree of agreement among replicate measurements of the same quantity

pressure (5.1)   a force per unit area

primary (1°) alcohol (12.1)   an alcohol with the general formula RCH_2OH

primary (1°) amide (15.3)   an amide produced in a reaction between a carboxylic acid and ammonia and having only one carbon, the carbonyl carbon, bonded to the nitrogen. They have the following general structure:

primary (1°) amine (15.1)   an amine with the general formula RNH_2

primary (1°) carbon (10.2)   a carbon atom that is bonded to only one other carbon atom

primary structure (of a protein) (18.3)   the linear sequence of amino acids in a protein chain determined by the genetic information of the gene for each protein

primary transcript (20.4)   the RNA product of transcription in eukaryotic cells, before post-transcriptional modifications are carried out

principal energy level (2.5)   a region where electrons may be found; has integral values $n = 1$, $n = 2$, and so forth

product (4.3, 19.1)   the chemical species that results from a chemical reaction and that appears on the right side of a chemical equation

proenzyme (19.9)   the inactive form of a proteolytic enzyme

prokaryote (20.2)   an organism with simple cellular structure in which there is no true nucleus enclosed by a nuclear membrane and there are no true membrane-bound organelles in the cytoplasm

promoter (20.4)   the sequence of nucleotides immediately before a gene that is recognized by the RNA polymerase and signals the start point and direction of transcription

properties (1.2)   characteristics of matter

prostaglandins (17.2)   a family of hormonelike substances derived from the twenty-carbon fatty acid, arachidonic acid; produced by many cells of the body, they regulate many body functions

prosthetic group (18.6)   the nonprotein portion of a protein that is essential to the biological activity of the protein; often a complex organic compound

protein (18: Intro)   a macromolecule whose primary structure is a linear sequence of α-amino acids and whose final structure results from folding of the chain into a specific three-dimensional structure; proteins serve as catalysts, structural components, and nutritional elements for the cell

protein modification (19.9)   a means of enzyme regulation in which a chemical group is covalently added to or removed from a protein. The chemical modification either turns the enzyme on or turns it off

proteolytic enzyme (19.11)   an enzyme that hydrolyzes the peptide bonds between amino acids in a protein chain

proton (2.1)   a positively charged particle in the nucleus of an atom

pure substance (1.2)   a substance with constant composition

purine (20.1)   a family of nitrogenous bases (heterocyclic amines) that are components of DNA and RNA and consist of a six-sided ring fused to a five-sided ring; the common purines in nucleic acids are adenine and guanine

pyridoxal phosphate (22.7)   a coenzyme derived from vitamin B_6 that is required for all transamination reactions

pyrimidine (20.1)   a family of nitrogenous bases (heterocyclic amines) that are components of nucleic acids and consist of a single six-sided ring; the common pyrimidines of DNA are cytosine and thymine; the common pyrimidines of RNA are cytosine and uracil

pyrimidine dimer (20.7)   UV light–induced covalent bonding of two adjacent pyrimidine bases in a strand of DNA

pyruvate dehydrogenase complex (22.2)   a complex of all the enzymes and coenzymes required for the synthesis of CO_2 and acetyl CoA from pyruvate

Q

quaternary ammonium salt (15.1)   an amine salt with the general formula $R_4N^+A^-$ (in which R— can be an alkyl or aryl group or a hydrogen atom and A^- can be any anion)

quaternary (4°) carbon (10.2)   a carbon atom that is bonded to four other carbon atoms

quaternary structure (of a protein) (18.6)   aggregation of more than one folded peptide chain to yield a functional protein

R

racemic mixture (16.3)   a mixture of equal amounts of a pair of enantiomers

rad (9.7)   abbreviation for *radiation absorbed dose*, the absorption of 2.4×10^{-3} calories of energy per kilogram of absorbing tissue

radioactivity (9.1)   the process by which atoms emit high-energy particles or rays; the spontaneous decomposition of a nucleus to produce a different nucleus

radiocarbon dating (9.3)   the estimation of the age of objects through measurement of isotopic ratios of carbon

Raoult's law (6.4)   a law stating that the vapor pressure of a component is equal to its mole fraction times the vapor pressure of the pure component

rate constant (7.3)   the proportionality constant that relates the rate of a reaction and the concentration of reactants

rate law (7.3)   expresses the rate of a reaction in terms of reactant concentration and a rate constant

rate (of chemical reaction) (7.3)   the change in concentration of a reactant or product per unit time

reactant (4.3)   starting material for a chemical reaction, appearing on the left side of a chemical equation

reaction order (7.3)   the exponent of each concentration term in the rate equation

reducing agent (4.8)   a substance that reduces, or donates electrons to, another substance; the reducing agent is itself oxidized in the process

reducing sugar (16.4)   a sugar that can be oxidized by Benedict's or Tollens' reagents; includes all monosaccharides and most disaccharides

reduction (4.8, 12.4)   the gain of electrons; in organic compounds it may be recognized by a gain of hydrogen or loss of oxygen

regulatory proteins (18: Intro)   proteins that control cell functions such as metabolism and reproduction

release factor (20.6)   a protein that binds to the termination codon in the empty A-site of the ribosome and causes the peptidyl transferase to hydrolyze the bond between the peptide and the peptidyl tRNA

rem (9.7)   abbreviation for *roentgen equivalent for man*, the product of rad and RBE

replication fork (20.3)   the point at which new nucleotides are added to the growing daughter DNA strand

replication origin (20.3)   the region of a DNA molecule where DNA replication always begins

representative element (2.4)   member of the groups of the periodic table designated as A

resonance (3.4)   a condition that occurs when more than one valid Lewis structure can be written for a particular molecule

restriction enzyme (20.8)   a bacterial enzyme that recognizes specific nucleotide sequences on a DNA molecule and cuts the sugar-phosphate backbone of the DNA at or near that site

result (1.1)   the outcome of a designed experiment, often determined from individual bits of data

reversible, competitive enzyme inhibitor (19.10)   a chemical that resembles the structure and charge distribution of the natural substrate and competes with it for the active site of an enzyme

reversible reaction (7.4)   a reaction that will proceed in either direction, reactants to products or products to reactants

ribonucleic acid (RNA) (20.1)   single-stranded nucleic acid molecules that are composed of phosphoryl groups, ribose, and the nitrogenous bases uracil, cytosine, adenine, and guanine

ribonucleotide (20.1)   a ribonucleoside phosphate or nucleotide composed of a nitrogenous base in β-*N*-glycosidic linkage to the 1′ carbon of the sugar ribose and with one, two, or three phosphoryl groups esterified at the hydroxyl of the 5′ carbon of the ribose

ribose (16.4)   a five-carbon monosaccharide that is a component of RNA and many coenzymes

ribosomal RNA (rRNA) (20.4)   the RNA species that are structural and functional components of the small and large ribosomal subunits

ribosome (20.6)   an organelle composed of a large and a small subunit, each of which is made up of ribosomal RNA and proteins; the platform on which translation occurs and that carries the enzymatic activity that forms peptide bonds

RNA polymerase (20.4)   the enzyme that catalyzes the synthesis of RNA molecules using DNA as the template

RNA splicing (20.4)   removal of portions of the primary transcript that do not encode protein sequences

roentgen (9.7)   the dose of radiation producing 2.1×10^9 ions in 1 cm^3 of air at 0°C and 1 atm of pressure

S

saccharide (16.1)   a sugar molecule

saponification (14.2, 17.3)   a reaction in which a soap is produced; more generally, the hydrolysis of an ester by an aqueous base

saturated fatty acid (17.2)   a long-chain monocarboxylic acid in which each carbon of the chain is bonded to the maximum number of hydrogen atoms

saturated hydrocarbon (10.1)   an alkane; a hydrocarbon that contains only carbon and hydrogen bonded together through carbon-hydrogen and carbon-carbon single bonds

saturated solution (6.1)   one in which undissolved solute is in equilibrium with the solution

scientific law (1.1)   a summary of a large quantity of information

scientific method (1.1)   the process of studying our surroundings that is based on experimentation

scientific notation (1.4)   a system used to represent numbers as powers of ten

secondary (2°) alcohol (12.1)   an alcohol with the general formula R_2CHOH

secondary (2°) amide (15.3)   an amide produced in a reaction between an acid chloride and a primary amine and having the following general structure:

$$R^1 - \overset{\overset{\textstyle O}{\|}}{C} - \overset{\overset{\textstyle R^2}{|}}{\underset{\underset{\textstyle H}{|}}{N}}$$

secondary (2°) amine (15.1)   an amine with the general formula R_2NH

secondary (2°) carbon (10.2)   a carbon atom that is bonded to two other carbon atoms

secondary structure (of a protein) (18.4)   folding of the primary structure of a protein into an α-helix or a β-pleated sheet; folding is maintained by hydrogen bonds between the amide hydrogen and the carbonyl oxygen of the peptide bond

selectively permeable membrane (6.4)   a membrane that restricts diffusion of some ions and molecules (based on size and charge) across the membrane

self-ionization (8.1)   transfer of a proton from one water molecule to another

semiconservative DNA replication (20.3)   DNA polymerase "reads" each parental strand of DNA and produces a complementary daughter strand; thus, all newly synthesized DNA molecules consist of one parental and one daughter strand

semipermeable membrane (6.4)   a membrane permeable to the solvent but not the solute; a material that allows the transport of certain substances from one side of the membrane to the other

shielding (9.6)   material used to provide protection from radiation

sickle cell anemia (18.8)   a human genetic disorder resulting from inheriting mutant hemoglobin genes from both parents

sievert (9.7)   the biological effect that results when 1 Gy of radiation energy is absorbed by human tissue

significant figures (1.4)   all digits in a number known with certainty and the first uncertain digit

silent mutation (20.7)   a mutation that changes the sequence of the DNA but does not alter the amino acid sequence of the protein encoded by the DNA

single bond (3.4)   a bond in which one pair of electrons is shared by two atoms

single-replacement reaction (4.3)   also called *substitution reaction*, one in which one atom in a molecule is displaced by another

soap (14.1)   any of a variety of the alkali metal salts of fatty acids

solid state (1.2)   a physical state of matter characterized by its rigidity and fixed volume and shape

solubility (3.5, 6.1)   the amount of a substance that will dissolve in a given volume of solvent at a specified temperature

solute (6.1)   a component of a solution that is present in lesser quantity than the solvent

solution (6.1)   a homogeneous (uniform) mixture of two or more substances

solvent (6.1)   the solution component that is present in the largest quantity

specific gravity (1.6)   the ratio of the density of a substance to the density of water at 4°C or any specified temperature

specific heat (7.2)   the quantity of heat (calories) required to raise the temperature of 1 g of a substance 1 degree Celsius

spectroscopy (2.3)   the measurement of intensity and energy of electromagnetic radiation

speed of light (2.3) 2.99×10^8 m/s in a vacuum

sphingolipid (17.4)   a phospholipid that is derived from the amino alcohol sphingosine rather than from glycerol

sphingomyelin (17.4)   a sphingolipid found in abundance in the myelin sheath that surrounds and insulates cells of the central nervous system

standard solution (8.3)   a solution whose concentration is accurately known

standard temperature and pressure (STP) (5.1)   defined as 273 K and 1 atm

stereochemical specificity (19.5)   the property of an enzyme that allows it to catalyze reactions involving only one enantiomer of the substrate

stereochemistry (16.3)   the study of the spatial arrangement of atoms in a molecule

stereoisomers (10.3, 16.3)   a pair of molecules having the same structural formula and bonding pattern but differing in the arrangement of the atoms in space

steroid (17.4)   a lipid derived from cholesterol and composed of one five-sided ring and three six-sided rings; the steroids include sex hormones and anti-inflammatory compounds

structural analog (19.10)   a chemical having a structure and charge distribution very similar to those of a natural enzyme substrate

structural formula (10.2)   a formula showing all of the atoms in a molecule and exhibiting all bonds as lines

structural isomers (10.2)   molecules having the same molecular formula but different chemical structures

structural protein (18: Intro)   a protein that provides mechanical support for large plants and animals

sublevel (2.5)   a set of equal-energy orbitals within a principal energy level

sublimation (5.3)   a process whereby some molecules in the solid state convert directly to the gaseous state

substituted hydrocarbon (10.1)   a hydrocarbon in which one or more hydrogen atoms is replaced by another atom or group of atoms

substitution reaction (10.5, 11.6)   a reaction that results in the replacement of one group for another

substrate (19.1)   the reactant in a chemical reaction that binds to an enzyme active site and is converted to product

substrate-level phosphorylation (21.3)   the production of ATP by the transfer of a phosphoryl group from the substrate of a reaction to ADP

sucrose (16.5)   a disaccharide composed of α-D-glucose and β-D-fructose in (α1 ⟶ β2) glycosidic linkage; table sugar

supersaturated solution (6.1)   a solution that is more concentrated than a saturated solution (Note that such a solution is not at equilibrium.)

surface tension (5.2)   a measure of the strength of the attractive forces at the surface of a liquid

surfactant (5.2)   a substance that decreases the surface tension of a liquid

surroundings (7.1)   the universe outside of the system

suspension (6.1)   a heterogeneous mixture of particles; the suspended particles are larger than those found in a colloidal suspension

system (7.1)   the process under study

T

tautomers (13.4)   structural isomers that differ from one another in the placement of a hydrogen atom and a double bond

temperature (1.6)   a measure of the relative "hotness" or "coldness" of an object

temperature optimum (19.8)   the temperature at which an enzyme functions optimally and the rate of reaction is maximal

terminal electron acceptor (22.6)   the final electron acceptor in an electron transport system that removes the low-energy electrons from the system; in aerobic organisms, the terminal electron acceptor is molecular oxygen

termination codon (20.6)   a triplet of ribonucleotides with no corresponding anticodon on a tRNA; as a result, translation will end because there is no amino acid to transfer to the peptide chain

terpene (17.4)   the general term for lipids that are synthesized from isoprene units; the terpenes include steroids, bile salts, lipid-soluble vitamins, and chlorophyll

tertiary (3°) alcohol (12.1)   an alcohol with the general formula R_3COH

tertiary (3°) amide (15.3)   an amide produced in a reaction between a secondary amine and an acid chloride and having the following general structure:

$$
\begin{array}{c}
\quad\; O \\
\quad\; \| \\
\quad\; C \\
R^1 \diagup \;\; \diagdown N \diagup R^2 \\
\qquad\;\; | \\
\qquad\;\; R^3
\end{array}
$$

tertiary (3°) amine (15.1)   an amine with the general formula R_3N

tertiary (3°) carbon (10.2)   a carbon atom that is bonded to three other carbon atoms

tertiary structure (of a protein) (18.5)   the globular, three-dimensional structure of a protein that results from folding

the regions of secondary structure; this folding occurs spontaneously as a result of interactions of the side chains or R groups of the amino acids

tetrahedral structure (3.4)   a molecule consisting of four groups attached to a central atom that occupy the four corners of an imagined regular tetrahedron

tetrose (16.2)   a four-carbon monosaccharide

theoretical yield (4.9)   the maximum amount of product that can be produced from a given amount of reactant

theory (1.1)   a hypothesis supported by extensive testing that explains and predicts facts

thermodynamics (7.1)   the branch of science that deals with the relationship between energies of systems, work, and heat

thioester (14.4)   the product of a reaction between a thiol and a carboxylic acid

thiol (12.8)   an organic compound that contains a thiol group (—SH)

titration (8.3)   the process of adding a solution from a buret to a sample until a reaction is complete, at which time the volume is accurately measured and the concentration of the sample is calculated

Tollens' test (13.4)   a test reagent (silver nitrate in ammonium hydroxide) used to distinguish aldehydes and ketones; also called the Tollens' silver mirror test

tracer (9.5)   a radioisotope that is rapidly and selectively transmitted to the part of the body for which diagnosis is desired

transaminase (22.7)   an enzyme that catalyzes the transfer of an amino group from one molecule to another

transamination (22.7)   a reaction in which an amino group is transferred from one molecule to another

transcription (20.4)   the synthesis of RNA from a DNA template

transferase (19.1)   an enzyme that catalyzes the transfer of a functional group from one molecule to another

transfer RNA (tRNA) (15.4, 20.4)   small RNAs that bind to a specific amino acid at the 3' end and mediate its addition at the appropriate site in a growing peptide chain; accomplished by recognition of the correct codon on the mRNA by the complementary anticodon on the tRNA

transition element (2.4)   any element located between Groups IIA (2) and IIIA (13) in the long periods of the periodic table

transition state (19.6)   the unstable intermediate in catalysis in which the enzyme has altered the form of the substrate so that it now shares properties of both the substrate and the product

translation (20.6)   the synthesis of a protein from the genetic code carried on the mRNA

translocation (20.6)   movement of the ribosome along the mRNA during translation

transmembrane protein (17.6)   a protein that is embedded within a membrane and crosses the lipid bilayer, protruding from the membrane both inside and outside the cell

transport protein (18: Intro) a protein that transports materials across the cell membrane or throughout the body

triglyceride (17.3, 23.1) triacylglycerol; a molecule composed of glycerol esterified to three fatty acids

trigonal planar (3.4) a molecular geometry in which a central atom is bonded to three atoms that lie at the vertices of an equilateral triangle. All atoms lie within one plane and all bond angles are 120°

trigonal pyramidal (3.4) a nonplanar structure involving three groups bonded to a central atom in which each group is equidistant from the central atom

triose (16.2) a three-carbon monosaccharide

triple bond (3.4) a bond in which three pairs of electrons are shared by two atoms

true solution (6.1) a homogeneous mixture with uniform properties throughout

U

unit (1.3) a determinate quantity (of length, time, etc.) that has been adopted as a standard of measurement

unsaturated fatty acid (17.2) a long-chain monocarboxylic acid having at least one carbon-to-carbon double bond

unsaturated hydrocarbon (10.1, 11: Intro) a hydrocarbon containing at least one multiple (double or triple) bond

urea cycle (22.8) a cyclic series of reactions that detoxifies ammonium ions by incorporating them into urea, which is excreted from the body

uridine triphosphate (UTP) (21.7) a nucleotide composed of the pyrimidine uracil, the sugar ribose, and three phosphoryl groups and that serves as a carrier of glucose-1-phosphate in glycogenesis

V

valence electron (2.6) electron in the outermost shell (principal quantum level) of an atom

valence-shell electron-pair repulsion (VSEPR) theory (3.4) a model that predicts molecular geometry using the premise that electron pairs will arrange themselves as far apart as possible, to minimize electron repulsion

van der Waals forces (5.2) a general term for intermolecular forces that include dipole-dipole and London dispersion forces

vapor pressure lowering (6.4) the decrease in the tendency of a liquid to become a gas when a solute is added

vapor pressure of a liquid (5.2) the pressure exerted by the vapor at the surface of a liquid at equilibrium

very low density lipoprotein (VLDL) (17.5) a plasma lipoprotein that binds triglycerides synthesized by the liver and carries them to adipose tissue for storage

viscosity (5.2) a measure of the resistance to flow of a substance at constant temperature

vitamin (19.7) an organic substance that is required in the diet in small amounts; water-soluble vitamins are used in the synthesis of coenzymes required for the function of cellular enzymes; lipid-soluble vitamins are involved in calcium metabolism, vision, and blood clotting

voltaic cell (4.8) an electrochemical cell that converts chemical energy into electrical energy

W

wax (17.4) a collection of lipids that are generally considered to be esters of long-chain alcohols

weight (1.3) the force exerted on an object by gravity

Z

Zaitsev's rule (12.4) states that in an elimination reaction, the alkene with the greatest number of alkyl groups on the double-bonded carbon (the more highly substituted alkene) is the major product of the reaction

Answers to Odd-Numbered Questions and Problems

Chapter 1

1.1 Homogeneous mixture

1.3 a. Physical property
b. Chemical property
c. Physical property

1.5 a. Extensive property
b. Intensive property

1.7
a. Three   d. Two
b. Three   e. Three
c. Four   f. Two

1.9
a. 2.4×10^{-3} d. 6.73×10^{5}
b. 1.80×10^{-2} e. 7.2420×10^{1}
c. 2.24×10^{2} f. 8.3×10^{-1}

1.11
a. 61.4   d. 63.7
b. 6.17   e. 8.77
c. 6.65×10^{-2}

1.13
a. 8.09
b. 5.9
c. 20.19

1.15 8.84×10^{-4}

1.17
a. 51   d. 8.6×10^{-1}
b. 8.0×10^{1} e. 4.63×10^{8}
c. 5.80×10^{1} f. 2.0×10^{-3}

1.19 2.49×10^{3} J

1.21 Chemistry is the study of matter, its chemical and physical properties, the chemical and physical changes it undergoes, and the energy changes that accompany those processes.

When wood is burned, energy is released while its matter undergoes chemical and physical changes.

1.23 Experiments are necessary to test hypotheses and theories. Experimental results that are derived from measured or observed data may or may not support the hypotheses and theories.

1.25 You need the cost per gallon, the average miles per gallon, and the distance between New York City and Washington, D.C.

1.27 According to the model, methane has one carbon atom and four hydrogen atoms per molecule. The model also provides the spatial relationship of the atoms, and the location and number of bonds that hold the unit together.

1.29 A hypothesis is essentially an "educated guess." A theory is a hypothesis supported by extensive experimentation; it can explain and predict new facts.

1.31 Many examples exist; one could involve the time it takes to get to work: hypothesis—taking the "back roads" is faster than choosing the crowded interstate highway. The experiment would involve driving each path several times, starting at the same time, timing each trip, and calculating an average time for each path.

1.33 A theory. The word "potential" indicates the scientific status of the statement.

1.35 Freeze the contents of two beakers, one containing pure water, and the other containing salt dissolved in water. Slowly warm each container and measure the temperature of each as they convert from the solid to the liquid state. This temperature is the melting point; hence, the freezing point. Note that the melting and freezing point of a solution are the same temperature.

1.37 The gaseous state, the liquid state, and the solid state are the three states of matter.

1.39 A pure substance has constant composition with only a single substance, whereas a mixture is composed of two or more substances.

1.41 An intensive property is a property of matter that is independent of the quantity of the substance. A substance's boiling point is an example of an intensive property.

1.43 Mixtures are composed of two or more substances. A homogeneous mixture has uniform composition, while a heterogeneous mixture has non-uniform composition.

1.45 Physical properties are characteristics of a substance that can be observed without the substance undergoing a change in chemical composition. Chemical properties can be observed only through chemical reactions that result in a change in chemical composition of the substance.

1.47 a. Gas
b. Solid
c. Solid

1.49

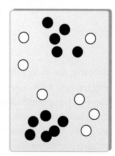

1.51 a. Chemical reaction
b. Physical change
c. Physical change

1.53 a. Physical property
b. Chemical property

1.55
a. Mixture   c. Mixture
b. Pure substance   d. Pure substance

1.57
a. Heterogeneous   c. Homogeneous
b. Homogeneous   d. Homogeneous

1.59 The state of matter represented in the diagram is the gaseous state. The diagram represents a homogeneous mixture.

1.61 a. Surface area and mass
b. Color and shape

1.63 Mass describes the quantity of matter in an object.

1.65 Length is the distance between two points.

1.67 A liter (L) is the volume occupied by 1000 g of water at 4°C.

1.69 mm < m < km

1.71 23.92

1.73 **a.** Precision is a measure of the agreement of replicate results.
b. Accuracy is the degree of agreement between the true value and measured value.

1.75 **a.** 3   **d.** 4
b. 3   **e.** 4
c. 3   **f.** 3

1.77 **a.** 3.87×10^{-3} **d.** 24.3
b. 5.20×10^{-2} **e.** 2.40×10^{2}
c. 2.62×10^{-3} **f.** 2.41

1.79 **a.** 1.5×10^{4} **d.** 1139.42
b. 2.41×10^{-1} **e.** 7.21×10^{3}
c. 5.99

1.81 **a.** 1.23×10^{1} **e.** 9.2×10^{7}
b. 5.69×10^{-2} **f.** 5.280×10^{-3}
c. -1.527×10^{3} **g.** 1.279×10^{0}
d. 7.89×10^{-7} **h.** -5.3177×10^{2}

1.83 These measurements have high levels of precision and accuracy.

1.85 Unlike the English system, the metric system is systematic. Since it is composed of a set of units that are related to each other decimally, it is simpler to convert one unit to another.

1.87 **a.** k, 10^{3} **c.** μ, 10^{-6}
b. c, 10^{-2}

1.89 $\dfrac{1\ \text{ft}}{12\ \text{in}}$ and $\dfrac{12\ \text{in}}{1\ \text{ft}}$

1.91 **a.** 32 oz   **d.** 9.1×10^{5} mg
b. 1.0×10^{-3} t   **e.** 9.1×10^{1} dag
c. 9.1×10^{2} g

1.93 **a.** 6.6×10^{-3} lb   **d.** 3.0×10^{2} cg
b. 1.1×10^{-1} oz   **e.** 3.0×10^{3} mg
c. 3.0×10^{-3} kg

1.95 15.0 mg

1.97 13.4 m²

1.99 4 L

1.101 101°F

1.103 5 cm is shorter than 5 in.

1.105 5.0 μg is smaller than 5.0 mg.

1.107 **a.** First, the boundary measurements that are in feet should be converted to kilometers. Using the unit relationships provided in the textbook, feet can first be converted to yards. Then, yards can be converted to meters, and meters can be converted to kilometers. The boundary measurements that are in meters also need to be converted to kilometers. Finally, after all of the lengths have the same units (km), they can be added together to determine the circumference of the property.
b. 0.470 km

1.109 Celsius, Fahrenheit, and Kelvin

1.111 Energy may be categorized as either kinetic energy, the energy of motion, or potential energy, the energy of position.

1.113 **a.** Extensive property   **c.** Intensive property
b. Extensive property

1.115 **a.** 10.0°C   **b.** 283.2 K

1.117 **a.** 293.2 K   **b.** 68.0°F

1.119 3×10^{4} J

1.121 6.00 g/mL

1.123 6.20×10^{2} L

1.125 1.08×10^{3} g

1.127 5110 g salt

1.129 Lead has the lowest density and platinum has the greatest density.

1.131 12.6 mL

1.133 1.03 g/mL

1.135 0.789

1.137 1.05

Chapter 2

2.1 **a.** 35 protons, 35 electrons, 44 neutrons
b. 35 protons, 35 electrons, 46 neutrons
c. 26 protons, 26 electrons, 30 neutrons

2.3 Electron density is the probability that an electron will be found in a particular region of an atomic orbital.

2.5 **a.** Na, metal   **c.** Mn, metal
b. Ra, metal   **d.** Mg, metal

2.7 **a.** Zr (zirconium)   **c.** Cr (chromium)
b. 22.99 amu   **d.** At (astatine)

2.9 a. Helium, atomic number = 2, mass = 4.009 amu
b. Fluorine, atomic number = 9, mass = 19.00 amu
c. Manganese, atomic number = 25, mass = 54.94 amu

2.11 a. 8 protons, 10 electrons
b. 12 protons, 10 electrons
c. 26 protons, 23 electrons

2.13 **a.** $1s^{2}\, 2s^{2}\, 2p^{6}\, 3s^{2}\, 3p^{6}$ [Ar]
b. $1s^{2}\, 2s^{2}\, 2p^{6}\, 3s^{2}\, 3p$ [Ar]
c. $1s^{2}\, 2s^{2}\, 2p^{6}\, 3s^{2}\, 3p^{6}\, 4s^{2}\, 3d^{10}\, 4p^{6}$ [Kr]
d. $1s^{2}\, 2s^{2}\, 2p^{6}\, 3s^{2}\, 3p^{6}\, 4s^{2}\, 3d^{10}\, 4p^{6}$ [Kr]

2.15 **a.** K^{+} and Ar are isoelectronic.
b. Sr^{2+} and Kr are isoelectronic.
c. S^{2-} and Ar are isoelectronic.
d. Mg^{2+} and Ne are isoelectronic.
e. P^{3-} and Ar are isoelectronic.
f. Be^{2+} and He are isoelectronic.

2.17 **a.** (Smallest) F, N, Be (largest)
b. (Lowest) Be, N, F (highest)
c. (Lowest) Be, N, F, (highest)

2.19 The mass number is equal to the sum of the number of protons and neutrons in an atom.
The atomic mass is the weighted average of the masses of isotopes of an element in amu.

2.21 **a.** Neutrons   **c.** Protons, neutrons
b. Protons   **d.** Nucleus, negative

2.23 Isotopes of an element have different numbers of neutrons. They have similar chemical behavior.

2.25 **a.** True   **c.** True
b. True

2.27 **a.** 56 protons, 80 neutrons, 56 electrons
b. 84 protons, 125 neutrons, 84 electrons
c. 48 protons, 65 neutrons, 48 electrons

2.29 $^{19}_{9}\text{F}$

2.31 **a.** All isotopes of Rn have 86 protons.
b. 134 neutrons

2.33 **a.** 34   **b.** 46

2.35 **a.** $^{1}_{1}\text{H}$ **b.** $^{14}_{6}\text{C}$

2.37 63.55 amu

2.39 • All matter consists of tiny particles called atoms.
• Atoms cannot be created, divided, destroyed, or converted to any other type of atom.
• All atoms of a particular element have identical properties.
• Atoms of different elements have different properties.

- Atoms combine in simple whole-number ratios.
- Chemical change involves joining, separating, or rearranging atoms.

2.41 Our understanding of the nucleus is based on the gold foil experiment performed by Geiger and interpreted by Rutherford. In this experiment, Geiger bombarded a piece of gold foil with alpha particles, and observed that some alpha particles passed straight through the foil, others were deflected and some simply bounced back. This led Rutherford to propose that the atom consisted of a small, dense nucleus (alpha particles bounced back), surrounded by a cloud of electrons (some alpha particles were deflected). The size of the nucleus is small when compared to the volume of the atom (alpha particles were able to pass through the foil).

2.43
  a. Dalton—developed the law of multiple proportions; determined the relative atomic weights of the elements known at that time; developed the first scientific atomic theory.
  b. Crookes—developed the cathode ray tube and discovered "cathode rays"; characterized electron properties.
  c. Chadwick—demonstrated the existence of the neutron.
  d. Goldstein—identified positive charge in the atom.

2.45 A cathode ray is the negatively charged particle formed in a cathode ray tube. It was characterized as an electron, with a mass of nearly zero and a charge of 1^-.

2.47 Electrons surround the nucleus in a diffuse region. Electrons are negatively charged, and electrons are very low in mass in contrast to protons and neutrons.

2.49 It was believed that protons and electrons were uniformly distributed throughout the atom.

2.51 Spectroscopy is the measurement of intensity and energy of electromagnetic radiation.

2.53 Electromagnetic radiation, or light, travels in waves from its source. Each wavelength of light has its own characteristic energy.

2.55 False.

2.57 Infrared radiation has greater energy than microwave radiation. Energy is inversely proportional to the wavelength. Since infrared radiation has shorter wavelengths than microwave radiation, it has more energy than microwave radiation.

2.59 When electrical energy is applied to a sample of hydrogen gas, the electrons in lower orbits are "excited" to higher orbits. As they fall back down into lower orbits, they release an amount of energy equal to the amount of energy they absorbed to jump to the higher orbit. This energy may be released in the form of light, and the wavelength is proportional to the energy difference. This produces a line spectrum that is characteristic of hydrogen.

2.61 According to Bohr, Planck, and others, electrons exist only in certain allowed regions, quantum levels, outside of the nucleus.

2.63
- Electrons are found in orbits at discrete distances from the nucleus.
- The orbits are quantized—they are of discrete energies.
- Electrons can only be found in these orbits, never in between (they are able to jump instantaneously from orbit to orbit).
- Electrons can undergo transitions—if an electron absorbs energy, it will jump to a higher orbit; when the electron falls back to a lower orbit, it will release energy.

2.65 Bohr's atomic model was the first to successfully account for electronic properties of atoms—specifically, the interaction of atoms and light (spectroscopy).

2.67
  a. atomic number 11, atomic mass 22.99, sodium
  b. atomic number 19, atomic mass 39.10, potassium
  c. atomic number 12, atomic mass 24.31, magnesium
  d. atomic number 5, atomic mass 10.81, boron

2.69 Group IA (or 1) is known collectively as the alkali metals and consists of Li, Na, K, Rb, Cs, and Fr.

2.71 Group VIIA (or 17) is known collectively as the halogens and consists of fluorine, chlorine, bromine, iodine, and astatine.

2.73
  a. Na, Ni, Al   **c.** Ar
  b. Na, Al

2.75

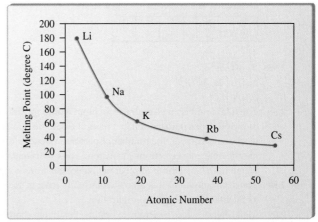

According to periodic law, the physical and chemical properties of the elements are periodic functions of their atomic numbers. As atomic number increases, the melting point decreases.

2.77 A principal energy level is designated $n = 1, 2, 3$, and so forth. It is similar to Bohr's orbit in concept. A sublevel is a part of a principal energy level and is designated $s, p, d,$ and f.

2.79 See Figure 2.11 for a sketch of the s atomic orbital. The s orbital represents the probability of finding an electron in a region of space surrounding the nucleus.

2.81 $2\,e^-$ for $n = 1$
$8\,e^-$ for $n = 2$
$18\,e^-$ for $n = 3$

2.83 According to the Pauli exclusion principle, each orbital can hold up to two electrons with the electrons spinning in opposite directions (paired). Therefore, since the d sublevel has five orbitals, it can contain a maximum of ten electrons.

2.85
  a. $1s^2\,2s^2\,2p^6\,3s^2\,3p^1$
  b. $1s^2\,2s^2\,2p^6\,3s^1$
  c. $1s^2\,2s^2\,2p^6\,3s^2\,3p^6\,4s^2\,3d^1$

2.87
  a. $1s^2\,2s^2\,2p^1$

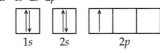

  b. $1s^2\,2s^2\,2p^6\,3s^2\,3p^4$

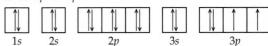

  c. $1s^2\,2s^2\,2p^6\,3s^2\,3p^6$

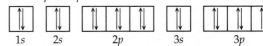

2.89 **a.** Not possible; $n = 1$ level can have only s-level orbitals.
  b. Possible; the electron configuration that is shown represents the carbon atom.
  c. Not possible; cannot have two identical orbitals ($2s^2$).
  d. Not possible; cannot have three electrons in an s orbital ($2s^3$)

2.91 Diagram A is incorrect.

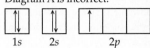

1s   2s   2p

Diagram B is correct.

Diagram C is incorrect.

1s   2s   2p

2.93 **a.** [Kr] $5s^2\,4d^2$
  b. [Ar] $4s^2\,3d^{10}\,4p^5$
  c. [Ar] $4s^1$

2.95 The total number of electrons in an atom would include all of the electrons in the atom. In a neutral atom, the number of electrons is equal to the number of protons. Valence electrons are the outermost electrons in an atom. For a representative element, the maximum number of valence electrons is eight.

2.97 The octet rule states that atoms will usually react in such a way as to obtain a noble gas configuration.

2.99 Metals tend to lose electrons to become positively charged cations.

2.101

Atom	Total electrons	Valence electrons	Principal energy level number
a. H	1	1	1
b. Na	11	1	3
c. B	5	3	2
d. F	9	7	2
e. Ne	10	8	2
f. He	2	2	1

2.103 **a.** 17 protons, 18 electrons
  b. 20 protons, 18 electrons
  c. 26 protons, 24 electrons

2.105 **a.** Two   **c.** One
  b. One   **d.** Two

2.107 **a.** Li$^+$ **c.** S^{2-}
  b. Ca^{2+}

2.109 **a.** Isoelectronic
  b. Isoelectronic

2.111 **a.** $1s^2\,2s^2\,2p^6\,3s^2\,3p^6\,4s^2\,3d^{10}\,4p^6\,5s^2\,4d^{10}\,5p^6$ [Xe]
  b. $1s^2\,2s^2\,2p^6\,3s^2\,3p^6\,4s^2\,3d^{10}\,4p^6\,5s^2\,4d^{10}\,5p^6$ [Xe]
  c. $1s^2\,2s^2\,2p^6\,3s^2\,3p^6\,4s^2\,3d^{10}\,4p^6$ [Kr]
  d. $1s^2\,2s^2\,2p^6$ [Ne]

2.113 Atomic size decreases from left to right across a period in the periodic table.

2.115 Ionization energy is the energy required to remove an electron from an isolated atom.

2.117 Na + ionization energy $\longrightarrow$ Na$^+$ + e$^-$

2.119 **a.** (Smallest) F, O, N (Largest)
  b. (Smallest) Li, K, Cs (Largest)
  c. (Smallest) Cl, Br, I (Largest)
  d. (Smallest) Be, Mg, Ra (Largest)

2.121 **a.** (Smallest) N, O, F (Largest)
  b. (Smallest) Cs, K, Li (Largest)

2.123 **a.** (Largest) Li, Na, K (Smallest)
  b. (Largest) Te, Sn, Sr (Smallest)

2.125 **a.** A positive ion is always smaller than its parent atom because the positive charge of the nucleus is shared among fewer electrons in the ion. As a result, each electron is pulled closer to the nucleus and the volume of the ion decreases.
  b. The fluoride ion has a completed octet of electrons and an electron configuration resembling its nearest noble gas. The electron affinity of fluorine is very high; therefore, it is energetically favorable for the fluorine atom to gain the electron.

2.127 Cl$^-$ is larger because it has a smaller nuclear (positive) charge.

Chapter 3

3.1 **a.** Potassium cyanide
  b. Magnesium sulfide
  c. Magnesium acetate

3.3 **a.** The bonded nuclei are closer together when a double bond exists, in comparison to a single bond.
  b. The bond strength increases as the bond order increases. Therefore, a double bond is stronger than a single bond. The distance of separation and bond strength are inversely related.

3.5 **a.** H—P̈—H   P̈ (H, H, H)   trigonal pyramidal
            |
            H

  b. H—Si̇—H   Si (H, H, H)   tetrahedral
           |
           H

3.7 **a.** The bond is polar.   **c.** The bond is nonpolar.
            $\overset{+\longrightarrow}{\text{S}\,-\,\text{O}}$

  b. The bond is polar.   **d.** The bond is polar.
            $\overset{+\longrightarrow}{\text{C}\equiv\text{N}}$ $\overset{+\longrightarrow}{\text{I}\,-\,\text{Cl}}$

3.9 **a.** Nonpolar   **c.** Polar
  b. Polar   **d.** Nonpolar

3.11 **a.** H$_2$O   **c.** NH$_3$
  b. CO   **d.** ICl

3.13 **a.** H·   **c.** ·Si̇·
  b. He: **d.** ·N̈·

3.15 **a.** Li$^+$ **b.** Mg^{2+} **c.** $\left[\,:\ddot{\text{Cl}}:\,\right]^-$ **d.** $\left[\,:\ddot{\text{P}}:\,\right]^{3-}$

3.17 Covalent bonding involves a sharing of electrons between atoms to complete the octet of electrons for each atom participating in the bond. Ionic bonding involves a transfer of one or more electrons from one atom to another. An ionic bond is the electrostatic force between the resulting anion and a cation.

3.19 Electronegativity values increase as we proceed left to right and bottom to top of the table. The most electronegative elements are located in the upper right corner of the periodic table.

3.21 **a.** Ionic   **c.** Nonpolar covalent
  b. Polar covalent   **d.** Polar covalent

3.23 **a.** Li· + :B̈r· $\longrightarrow$ Li$^+$ + $\left[\,:\ddot{\text{Br}}:\,\right]^-$

  b. ·Mg· + 2:C̈l· $\longrightarrow$ Mg^{2+} + 2$\left[\,:\ddot{\text{Cl}}:\,\right]^-$

  c. ·P̈· + 3H· $\longrightarrow$ H—P̈—H
                                      |
                                      H

3.25 He has two valence electrons (electron configuration $1s^2$) and a complete N = 1 level. It has a stable electron configuration, with no tendency to gain or lose electrons, and satisfies the octet rule (2 e^- for period 1). Hence, it is nonreactive.
He:

3.27 MgS

3.29 **a.** Sodium ion
b. Copper(I) ion (or cuprous ion)
c. Magnesium ion

3.31 **a.** Bicarbonate ion
b. Hydronium ion
c. Carbonate ion

3.33 **a.** K^+ **b.** Ni^{2+}

3.35 **a.** SO_4^{2-} **b.** NO_3^-

3.37 **a.** Al_2O_3 **b.** Li_2S

3.39 **a.** Magnesium chloride
b. Aluminum chloride
c. Copper (II) nitrate

3.41 **a.** NaCl   **b.** $MgBr_2$

3.43 **a.** AgCN   **b.** NH_4Cl

3.45 **a.** CuO   **b.** Fe_2O_3

3.47 **a.** $NaNO_3$ **b.** $Mg(NO_3)_2$

3.49 **a.** NH_4I **b.** $(NH_4)_2SO_4$

3.51 **a.** Nitrogen dioxide   **c.** Sulfur trioxide
b. Selenium trioxide

3.53 **a.** SiO_2 **b.** SO_2

3.55 Ionic solid-state compounds exist in regular, repeating, three-dimensional structures; the crystal lattice. The crystal lattice is made up of positive and negative ions. Solid-state covalent compounds are made up of molecules that may be arranged in a regular crystalline pattern or in an irregular (amorphous) structure.

3.57 The boiling points of ionic solids are generally much higher than those of covalent solids.

3.59 KCl would be expected to exist as a solid at room temperature; it is an ionic compound, and ionic compounds are characterized by high melting points.

3.61 Water will have a higher boiling point. Water is a polar molecule with strong intermolecular attractive forces, whereas carbon tetrachloride is a nonpolar molecule with weak intermolecular attractive forces. More energy—hence, a higher temperature is required to overcome the attractive forces among the water molecules.

3.63 Yes, $MgCl_2$ in water forms an electrolytic solution. $MgCl_2$ is an ionic solid that dissociates in water, and the resulting solution is capable of conducting a current of electricity.

3.65 The least electronegative atom will usually be the central atom. The central atom is often the element in the compound for which there is only one atom.

3.67 For polyatomic cations, subtract one electron for each unit of positive charge.

3.69 Single bond < double bond < triple bond
(Lowest energy)   (Highest energy)

3.71 C_5H_{12}. As the size of the hydrocarbon increases, the number of possible isomers increases.

3.73 Resonance can occur when more than one valid Lewis structure can be written for a molecule. Each individual structure which can be drawn is a resonance form. The true nature of the structure for the molecule is the resonance hybrid, which consists of the "average" of the resonance forms.

3.75 120°

3.77 True. A molecule containing only nonpolar bonds must be a nonpolar molecule.

3.79 **a.** $:\!\ddot{C}l\!-\!N\!-\!\ddot{C}l\!:$ **c.** $:\!\ddot{S}\!=\!C\!=\!\ddot{S}\!:$
          $\overset{|}{\underset{}{:\!\ddot{C}l\!:}}$

b. $H\!-\!\overset{H}{\underset{H}{C}}\!-\!\ddot{O}\!-\!H$ **d.** $:\!\ddot{C}l\!-\!\overset{:\ddot{C}l:}{\underset{H}{C}}\!-\!H$

3.81 $H\!-\!\overset{H}{\underset{H}{C}}\!-\!\overset{\ddot{O}:}{C}\!\diagdown\! H$

3.83 $H\!-\!\overset{H}{\underset{H}{C}}\!-\!\overset{\ddot{O}}{\underset{H}{C}}\!-\!\overset{H}{\underset{H}{C}}\!-\!H$

3.85 $[:\!N\!\equiv\!O:]^+$

3.87 $[:\!\ddot{O}\!-\!H]^-$

3.89 $\left[H\!-\!\overset{H}{\underset{H}{C}}\!-\!\overset{:\ddot{O}:}{C}\!-\!\ddot{O}\!: \right]^- \rightleftharpoons \left[H\!-\!\overset{H}{\underset{H}{C}}\!-\!\overset{:\ddot{O}:}{C}\!=\!\ddot{O} \right]^-$

3.91 **a.** $:\!C\!\equiv\!O:$ **b.**   **c.** $\ddot{O}\!=\!C\!=\!\ddot{O}$

3.93 $:\!\ddot{C}l\!-\!Be\!-\!\ddot{C}l\!:$

3.95
$$:\!\ddot{F}\!: $$
$$:\!\ddot{F}\diagdown\!\underset{Se}{|}\!\diagup\!\ddot{F}\!:$$
$$:\!\ddot{F}\diagup\!\underset{}{|}\!\diagdown\!\ddot{F}\!:$$
$$:\!\ddot{F}\!:$$

3.97 **a.** Bent   **b.** Trigonal planar

$:\!\ddot{O}\!-\!\overset{}{\underset{\diagdown \ddot{O}:}{S}}$

$\overset{\ddot{O}}{\underset{:\ddot{O}\diagup \overset{S}{} \diagdown \ddot{O}:}{\|}}$

3.99 **a.** Polar   **b.** Nonpolar

3.101 Both a. CO_2 and c. CF_4

3.103 A molecule containing no polar bonds *must* be nonpolar. A molecule containing polar bonds may or may not itself be polar. It depends upon the number and arrangement of the bonds.

3.105 **a.** $:\!\ddot{C}l\!-\!\overset{N}{\underset{:\ddot{C}l:}{}}\!-\!\ddot{C}l\!:$ **c.** $:\!\ddot{S}\!=\!C\!=\!\ddot{S}\!:$
          Linear
Trigonal   nonpolar
pyramidal   not water soluble
polar
water soluble

b. $H\!-\!\overset{H}{\underset{H}{C}}\!-\!\ddot{O}\!-\!H$ **d.** $H\!-\!\overset{H}{\underset{:\ddot{C}l:}{C}}\!-\!\ddot{C}l\!:$

Tetrahedral around C   Tetrahedral
bent around O   polar
polar, water soluble   water soluble

3.107 Polar compounds have strong intermolecular attractive forces. Higher temperatures are needed to overcome these forces and convert the solid to a liquid; hence, we predict higher melting points for polar compounds when compared to nonpolar compounds.

3.109 Yes

3.111 **a.** NH_3 **b.** CF_4 **c.** NaCl

Chapter 4

4.1 3.33×10^{-10} g Hg

4.3 Formula mass = 194.20 amu
Molar mass = 194.20 g/mol

4.5 **a.** DR   **c.** DR
b. SR   **d.** D

4.7 The black precipitate is CuS. $Cu^{2+}(aq) + S^{2-}(aq) \longrightarrow CuS(s)$

4.9 $Ca \rightarrow Ca^{2+} + 2e^-$ (oxidation ½ reaction)
$S + 2e^- \rightarrow S^{2-}$ (reduction ½ reaction)
$Ca + S \rightarrow CaS$ (complete reaction)

4.11 Oxidizing agent: S
Reducing agent: Ca
Substance oxidized: Ca
Substance reduced: S

4.13 **a.** In order for $Cr^{3+}(aq)$ to form $Cr(s)$ and the electrode to be electro-plated, Cr^{3+} must gain 3 electrons and undergo reduction. Therefore, the electrode must have a negative charge.
b. Cathode
c. $Cr^{3+}(aq) + 3e^- \rightarrow Cr(s)$

4.15 **a.** 47.9 g NO_2 **b.** 52.2%

4.17 **a.** 200.6 amu   **c.** 24.31 amu
b. 83.80 amu

4.19 **a.** 28.09 g/mol   **c.** 74.92 g/mol
b. 107.9 g/mol

4.21 39.95 g

4.23 6.0×10^{19} carbon atoms

4.25 1.7×10^{-22} mol As

4.27 40.36 g Ne

4.29 4.00 g/mol

4.31 **a.** 5.00 mol He   **c.** 4.2×10^{-2} mol Cl_2
b. 1.7 mol Na

4.33 1.62×10^3 g Ag

4.35 8.37×10^{22} Ag atoms

4.37 A molecule is a single unit composed of atoms joined by covalent bonds. An ion pair is composed of positive and negatively charged ions joined by electrostatic attraction, the ionic bond. The ion pairs, unlike the molecule, do not form single units; the electrostatic charge is directed to other ions in a crystal lattice as well.

4.39 **a.** 58.44 amu and 58.44 g/mol
b. 142.04 amu and 142.04 g/mol
c. 357.49 amu and 357.49 g/mol

4.41 32.00 amu and 32.00 g/mol

4.43 249.70 amu and 249.70 g/mol

4.45 **a.** 0.257 mol NaCl   **b.** 0.106 mol Na_2SO_4

4.47 **a.** 18.02 g H_2O **c.** 40.0 g He
b. 116.9 g NaCl   **d.** 2.02×10^2 g H_2

4.49 **a.** 1.60 g CH_4 **c.** 4.00 g NaOH
b. 10.0 g $CaCO_3$ **d.** 9.81 g H_2SO_4

4.51 **a.** 0.420 mol KBr   **c.** 6.57×10^{-1} mol CS_2
b. 0.415 mol $MgSO_4$ **d.** 2.14×10^{-1} mol $Al_2(CO_3)_3$

4.53 The ultimate basis for a balanced chemical equation is the law of conservation of mass. No mass may be gained or lost in a chemical reaction, and the chemical equation must reflect this fact.

4.55 A reactant is the starting material for a chemical reaction. Reactants are found on the left side of the reaction arrow.

4.57 Heat is necessary for the reaction to occur.

4.59 **a.** D   **c.** C
b. DR   **d.** SR

4.61 The subscript tells us the number of atoms or ions contained in one unit of the compound.

4.63 If we change the subscript, we change the identity of the compound.

4.65 **a.** $2C_2H_6(g) + 7O_2(g) \rightarrow 4CO_2(g) + 6H_2O(g)$
b. $6K_2O(s) + P_4O_{10}(s) \rightarrow 4K_3PO_4(s)$
c. $MgBr_2(aq) + H_2SO_4(aq) \rightarrow 2HBr(g) + MgSO_4(aq)$
d. $C_2H_5OH(l) + 3O_2(g) \rightarrow 2CO_2(g) + 3H_2O(g)$

4.67 **a.** $Ca(s) + F_2(g) \rightarrow CaF_2(s)$
b. $2Mg(s) + O_2(g) \rightarrow 2MgO(s)$
c. $3H_2(g) + N_2(g) \rightarrow 2NH_3(g)$

4.69 **a.** $2C_4H_{10}(g) + 13O_2(g) \rightarrow 10H_2O(g) + 8CO_2(g)$
b. $Au_2S_3(s) + 3H_2(g) \rightarrow 2Au(s) + 3H_2S(g)$
c. $Al(OH)_3(s) + 3HCl(aq) \rightarrow AlCl_3(aq) + 3H_2O(l)$
d. $(NH_4)_2Cr_2O_7(s) \rightarrow Cr_2O_3(s) + N_2(g) + 4H_2O(g)$

4.71 **a.** $N_2(g) + 3H_2(g) \rightarrow 2NH_3(g)$
b. $HCl(aq) + NaOH(aq) \rightarrow NaCl(aq) + H_2O(l)$
c. $C_6H_{12}O_6(s) + 6O_2(g) \rightarrow 6H_2O(l) + 6CO_2(g)$
d. $Na_2CO_3(s) \rightarrow Na_2O(s) + CO_2(g)$

4.73 **a.** Na_2SO_4 will not form a precipitate.
b. $BaSO_4$ will form a precipitate.
c. $BaCO_3$ will form a precipitate.
d. K_2CO_3 will not form a precipitate.

4.75 Yes. PbI_2

4.77 Yes. $CaCO_3$

4.79 An ionic equation shows all reactants and products as free ions unless they are precipitates. Ions that appear on both sides of the equation do not appear in the net ionic equation. The net ionic equation only shows the chemical species that actually undergo change.

4.81 $Ag^+(aq) + Br^-(aq) \rightarrow AgBr(s)$

4.83 An acid loses a hydrogen cation.

4.85 HCN is the acid, and KOH is the base.

4.87 The species oxidized *loses* electrons.

4.89 During an oxidation-reduction reaction the species *oxidized* is the reducing agent.

4.91
$$\underset{\substack{\text{substance reduced} \\ \text{oxidizing agent}}}{Cl_2} \quad + \quad \underset{\substack{\text{substance oxidized} \\ \text{reducing agent}}}{2KI} \quad \rightarrow 2KCl + I_2$$

4.93 $2I^- \rightarrow I_2 + 2e^-$ (oxidation ½ reaction)
$Cl_2 + 2e^- \rightarrow 2Cl^-$ (reduction ½ reaction)

4.95 An oxidation-reduction reaction must take place to produce electron flow in a voltaic cell.

4.97 Storage battery

4.99 The coefficients represent the relative number of moles of product(s) and reactant(s).

4.101 27.7 g B_2H_6

4.103 0.658 mol $CrCl_3$

4.105 **a.** $N_2(g) + 3H_2(g) \rightarrow 2NH_3(g)$
b. Three moles of H_2 will react with one mole of N_2.
c. One mole of N_2 will produce two moles of the product NH_3.
d. 1.50 mol H_2
e. 17.0 g NH_3

4.107 **a.** 149.21 g/mol **c.** 32.00 g O
   b. 1.20×10^{24} O atoms **d.** 10.7 g O
4.109 7.39 g O_2
4.111 6.77×10^4 g CO_2
4.113 70.6 g $C_{10}H_{22}$
4.115 9.13×10^2 g N_2
4.117 92.6%
4.119 6.85×10^2 g N_2

Chapter 5

5.1 **a.** 0.954 atm **c.** 0.730 atm
   b. 0.382 atm **d.** 6.5 atm
5.3 2.91 atm
5.5 Radon (Rn) is a collection of atoms (recall that all atoms are inherently nonpolar), and nitrogen dioxide (NO_2) molecules are polar. Since nonpolar molecules are only weakly attracted to each other, they exhibit more ideal gas behavior.
5.7 Molecules with complex structures, which do not "slide" smoothly past each other, and polar molecules tend to have higher viscosities.
5.9 Evaporation is the conversion of a liquid to a gas at a temperature lower than the boiling point of the liquid. Condensation is the conversion of a gas to a liquid at a temperature lower than the boiling point of the liquid.
5.11 $CO_2 < CH_3Cl < CH_3OH$. Only CH_3OH exhibits London dispersion forces, dipole-dipole interactions, and hydrogen bonding. Hence, CH_3OH has the strongest intermolecular forces and, therefore, the highest boiling point.
5.13 **a.** Ionic solids generally have high melting points and a tendency to be hard and brittle.
   b. Table salt (NaCl) and calcium chloride ($CaCl_2$)
5.15 A monometer can be used to measure O_2 gas pressure in terms of the height of a column of liquid (mercury, for example) that is supported by the force exerted on the surface of the liquid by the O_2 gas being measured.
5.17 **a.** 1.24 atm **c.** 0.197 atm
   b. 0.0954 atm **d.** 1.23 kPa
5.19 **a.** 10.4 psi **c.** 15 psi
   b. 3.00 psi **d.** 22.1 psi
5.21 In all cases, gas particles are much farther apart than similar particles in the liquid or solid state. In most cases, particles in the liquid state are, on average, farther apart than those in the solid state. Water is the exception; liquid water's molecules are closer together than they are in the solid state.
5.23 Gases are easily compressed simply because there is a great deal of space between particles; they can be pushed closer together (compressed) because the space is available.
5.25 Gas particles are in continuous, random motion. They are free (minimal attractive forces between particles) to roam, up to the boundary of their container.
5.27 Gases exhibit more ideal behavior at low pressures. At low pressures, gas particles are more widely separated and therefore the attractive forces between particles are less. The ideal gas model assumes negligible attractive forces between gas particles.
5.29 The kinetic molecular theory states that the average kinetic energy of the gas particles increases as the temperature increases. Kinetic energy is proportional to (velocity)^2. Therefore, as the temperature increases, the gas particle velocity increases and the rate of mixing increases as well.

5.31 Boyle's law states that the volume of a gas varies inversely with the gas pressure if the temperature and the number of moles of gas are held constant.
5.33 Volume will decrease according to Boyle's law. Volume is inversely proportional to the pressure exerted on the gas.
5.35 1 atm
5.37 5 L·atm
5.39 5.23 atm
5.41 Charles's law states that the volume of a gas varies directly with the absolute temperature if pressure and number of moles of gas are constant.
5.43 The Kelvin scale is the only scale that is directly proportional to molecular motion, and it is this motion that determines the physical properties of gases.
5.45 No. The volume is proportional to the temperature in K, not Celsius.
5.47 0.96 L
5.49 1.51 L
5.51 120°F
5.53 • Volume and temperature are *directly* proportional; increasing T increases V.
   • Volume and pressure are *inversely* proportional; decreasing P increases V.
   Therefore, both variables work together to *increase* the volume.
5.55 $V_f = \dfrac{P_i V_i T_f}{P_f T_i}$
5.57 1.82×10^{-2} L
5.59 1.5 atm
5.61 Avogadro's law states that equal volumes of any ideal gas contain the same number of moles if measured at constant temperature and pressure.
5.63 6.00 L
5.65 No. One mole of an ideal gas will occupy exactly 22.4 L; however, there is no completely ideal gas, and careful measurement will show different volumes for gases exhibiting varying degrees of ideality.
5.67 Standard temperature is 273 K.
5.69 0.80 mol
5.71 35.2 L
5.73 1.25 g/L
5.75 0.276 mol
5.77 5.94×10^{-2} L
5.79 172°C
5.81 9.08×10^3 L
5.83 Dalton's law states that the total pressure of a mixture of gases is the sum of the partial pressures of the component gases.
5.85 0.74 atm
5.87 0.29 atm
5.89 Limitations to the ideal gas model arise from interactive forces that are present between the individual atoms or molecules of a gas. These interactive forces are present in gases composed of polar molecules. The forces increase as the temperature of the gas decreases or the pressure of the gas increases.
5.91 When temperature increases, the attractive forces present in gases decrease. CO behaves more ideally at 50 K than at 5 K.
5.93 Intermolecular forces in liquids are considerably stronger than intermolecular forces in gases. Particles are, on average, much closer together in liquids and the strength of attraction decreases as the distance of separation increases.
5.95 The vapor pressure of a liquid increases as the temperature of the liquid increases.

5.97 Viscosity is the resistance to flow caused by intermolecular attractive forces. Complex molecules may become entangled and not slide smoothly across one another.

5.99 All molecules exhibit London dispersion forces. This is because electrons are in constant motion in all molecules.

5.101 Only methanol exhibits hydrogen bonding. Methanol has an oxygen atom bonded to a hydrogen atom, a necessary condition for hydrogen bonding.

5.103 Propylene glycol

5.105 Solids are essentially incompressible because the average distance of separation among particles in the solid state is small. There is literally no space for the particles to crowd closer together.

5.107 **a.** High melting temperature, brittle
b. High melting temperature, hard

5.109 Beryllium. Metallic solids are good electrical conductors. Carbon forms covalent solids that are poor electrical conductors.

5.111 Mercury. Mercury is a liquid at room temperature, whereas chromium is a solid at room temperature. Liquids have higher vapor pressures than solids.

Chapter 6

6.1 A chemical analysis must be performed in order to determine the identity of all components, a qualitative analysis. If only one component is found, it is a pure substance; two or more components indicates a true solution.

6.3 After the container of soft drink is opened, CO_2 diffuses into the surrounding atmosphere; consequently, the partial pressure of CO_2 over the soft drink decreases and the equilibrium

$$CO_2(g) \rightleftharpoons CO_2(aq)$$

shifts to the left, lowering the concentration of CO_2 in the soft drink.

6.5 0.19 M

6.7 0.125 mol HCl

6.9 Pure water

6.11 4.0×10^{-2} mol particles/L KCl, 3×10^{-1} mol particles/L glucose

6.13 0.110 mol/L

6.15 A solution is described as clear if it efficiently transmits light, showing no evidence of suspended particles. The solution does not have to be colorless to meet these conditions.

6.17 **a.** Electrolyte   **b.** Nonelectrolyte   **c.** Electrolyte

6.19 A true solution contains more than one substance, with the particles having a diameter less than 1×10^{-9} m. Particles with diameters of 1×10^{-9} m to 2×10^{-7} m are colloids. A suspension contains particles much larger than 2×10^{-7} m.

6.21 A saturated solution is one in which undissolved solute is in equilibrium with the solution. A supersaturated solution is a solution that is more concentrated than a saturated solution.

6.23 A colloidal dispersion of albumin is not completely homogenous. The colloid particles scatter light (Tyndall effect). Saline solution is completely homogenous, and the dissolved NaCl ions do not scatter light.

6.25 CCl_4 is more likely to form a solution in benzene (C_6H_6). The rule "like dissolves like" suggests that CCl_4 is soluble in benzene because both CCl_4 and benzene are nonpolar.

6.27 Stream temperature is much lower in early spring than mid-August. Henry's law predicts that the concentration of dissolved oxygen in the stream is greater at lower water temperatures. Trout require oxygen and thrive in early spring.

6.29 0.033 mol/L

6.31 **a.** 6.60% $C_6H_{12}O_6$ **b.** 2.00% NaCl

6.33 **a.** 10.0% ethanol   **b.** 5.00% ethanol

6.35 **a.** 21.0% NaCl   **b.** 3.75% NaCl

6.37 19.5% KNO_3

6.39 1.00 g sugar

6.41 **a.** 2.25 g NaCl   **b.** 3.13 g $NaC_2H_3O_2$

6.43 2.0×10^{-3} ppt

6.45 0.04% (m/m) solution is more concentrated

6.47 0.50 M

6.49 0.900 M

6.51 Laboratory managers often purchase concentrated solutions for practical reasons such as economy and conservation of storage space.

6.53 **a.** 1.46 g NaCl   **b.** 9.00 g $C_6H_{12}O_6$

6.55 158 g glucose

6.57 0.266 L

6.59 50.0 mL

6.61 20.0 M

6.63 $5.00 \times 10^{-2} M$

6.65 A colligative property is a solution property that depends on the concentration of solute particles rather than the identity of the particles.

6.67 Salt is an ionic substance that dissociates in water to produce positive and negative ions. These ions (or particles) lower the freezing point of water. If the concentration of salt particles is large, the freezing point may be depressed below the surrounding temperature, and the ice would melt.

6.69 Raoult's law states that when a solute is added to a solvent, the vapor pressure of the solvent decreases in proportion to the concentration of the solute.

6.71 One mole of $CaCl_2$ produces three moles of particles in solution, whereas one mole of NaCl produces two moles of particles in solution. Therefore, a one molar $CaCl_2$ solution contains a greater number of particles than a one molar NaCl solution and will produce a greater freezing-point depression.

6.73 **a.** $-2.79°C$ **b.** $-5.58°C$

6.75 Freezing Temperature for NaCl Solution = $-1.86°C$
Freezing Temperature for Sucrose Solution = $-0.93°C$

6.77 Sucrose

6.79 A → B

6.81 No net flow

6.83 1.0×10^{-3} mol particles/L

6.85 24 atm

6.87 Hypertonic

6.89 Hypotonic

6.91 Water is often termed the "universal solvent" because it is a polar molecule and will dissolve, at least to some extent, most ionic and polar covalent compounds. The majority of our body mass is water and this water is an important part of the nutrient transport system due to its solvent properties. This is true in other animals and plants as well. Because of its ability to hydrogen bond, water has a high boiling point and a low vapor pressure. Also, water is abundant and easily purified.

6.93

$$H-\underset{\underset{H}{|}}{N} : \delta^- \delta^+ H$$
$$\underset{\underset{H}{|}}{O} : \delta^-$$

6.95 The number of particles in solution is dependent on the degree of dissociation.

6.97   In dialysis, sodium ions move from a region of high concentration to a region of low concentration. If we wish to remove (transport) sodium ions from the blood, they can move to a region of lower concentration, the dialysis solution.

6.99   The shelf life is a function of the stability of the ammonia-water solution. The ammonia can react with the water to convert to the extremely soluble and stable ammonium ion. Also, ammonia and water are polar molecules. Polar interactions, particularly hydrogen bonding, are strong and contribute to the long-term solution stability.

6.101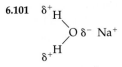

6.103 Polar; like dissolves like (H_2O is polar)

6.105 Elevated concentrations of sodium ion in the blood may cause confusion, stupor, or coma.

6.107 Elevated concentrations of sodium ion in the blood may occur whenever large amounts of water are lost. Diarrhea, diabetes, and certain high-protein diets are particularly problematic.

6.109 0.10 eq/L

6.111 **a.** 0.154 mol/L   **b.** 0.154 mol/L

6.113 4.0×10^{-2} mol/L

Chapter 7

7.1   **a.** Exothermic
        b. Exothermic

7.3   He(g). Gases have a greater degree of disorder than solids.

7.5 $\Delta G = (+) - T(-)$
         ΔG must always be positive. A positive value for ΔG indicates a nonspontaneous process.

7.7 2.7×10^3 J

7.9   Heat energy produced by the friction of striking the match provides the activation energy necessary for this combustion process.

7.11 If the enzyme catalyzed a process needed to sustain life, the substance interfering with that enzyme would be classified as a poison.

7.13 At a busy restaurant during lunchtime, approximately the same number of people will enter and exit the restaurant at any given moment. Throughout lunchtime, the number of people in the restaurant may be essentially unchanged, but the identity of the individuals in the restaurant is continually changing.

7.15 Measure the concentrations of products and reactants at a series of times until no further concentration change is observed.

7.17 Product formation

7.19 The first law of thermodynamics, the law of conservation of energy, states that the energy of the universe is constant.

7.21 An exothermic reaction is one in which energy is released during chemical change.

7.23 A fuel must release heat in the combustion (oxidation) process.

7.25 Free energy is the combined contribution of entropy and enthalpy for a chemical reaction.

7.27 Enthalpy is a measure of heat energy.

7.29 **a.** Entropy increases. Conversion of a solid to a liquid results in an increase in disorder of the substance. Solids retain their shape while liquids will flow and their shapes are determined by their container.
        b. Entropy increases. Conversion of a liquid to a gas results in an increase in disorder of the substance. Gas particles

move randomly with very weak interactions between particles, much weaker than those interactions in the liquid state.

7.31 Isopropyl alcohol quickly evaporates (liquid $\longrightarrow$ gas) after being applied to the skin. Conversion of a liquid to a gas requires heat energy. The heat energy is supplied by the skin. When this heat is lost, the skin temperature drops.

7.33 $\Delta G = (-) - T(+)$
         ΔG must always be negative; the process is spontaneous.

7.35 Fuel value is the amount of energy per gram of food.

7.37 The temperature of the water (or solution) is measured in a calorimeter. If the reaction being studied is exothermic, released energy heats the water and the temperature increases. In an endothermic reaction, heat flows from the water to the reaction and the water temperature decreases.

7.39 Joule

7.41 Double-walled containers, used in calorimeters, provide a small airspace between the part of the calorimeter (inside wall) containing the sample solution and the outside wall contacting the surroundings. This gap makes heat transfer more difficult.

7.43 1.20×10^3 cal

7.45 8×10^{-2} nutritional Calories/g substance

7.47 Decomposition of leaves and twigs to produce soil.

7.49 The activated complex is the arrangement of reactants in an unstable transition state as a chemical reaction proceeds. The activated complex must form in order to convert reactants to products.

7.51 The rate of a reaction is the change in concentration of a reactant or product per unit time. The rate constant is the proportionality constant that relates rate and concentration. The order is the exponent of each concentration term in the rate equation.

7.53 A catalyst increases the rate of a reaction without itself undergoing change.

7.55 An increase in concentration of reactants means that there are more molecules in a certain volume. The probability of collision is enhanced because they travel a shorter distance before meeting another molecule. The rate is proportional to the number of collisions per unit time.

7.57 A catalyst speeds up a chemical reaction by facilitating the formation of the activated complex, thus lowering the activation energy, the energy barrier for the reaction.

7.59 See textbook Figure 7.11.

7.61 Rate = $k[CH_4][O_2]$ Increase

7.63 Rate = $k[N_2O_4]^n$ Note: n must be experimentally determined.

7.65 The rate of the reaction increases four-fold.

7.67 LeChatelier's principle states that when a system at equilibrium is disturbed, the equilibrium shifts in the direction that minimizes the disturbance.

7.69 A physical equilibrium occurs between two phases of the same substance. A chemical equilibrium is a state of a chemical reaction in which the rates of the forward and reverse reactions are equal.

7.71 Products

7.73 **a.** False. A slow reaction may go to completion, but take a longer period of time.
        b. False. The rate of forward and reverse reactions is equal in a dynamic equilibrium situation.

7.75 (I)

7.77 A dynamic equilibrium has fixed concentrations of all reactants and products—these concentrations do not change with time. However, the process is dynamic because products and reactants are continuously being formed and consumed. The concentrations do not change because the rates of production and consumption are equal.

7.79 The position of the equilibrium addresses the question of whether products or reactants are favored. In a hypothetical equilibrium, if the equilibrium position shifts to the left, that means more reactants are formed. If the equilibrium position shifts to the right, that means more products are formed.

7.81 1. Concentration
2. Heat
3. Pressure

7.83 $K_{eq} = \dfrac{[NO_2]^2}{[N_2O_4]}$

7.85 $K_{eq} = \dfrac{[NH_3]^2}{[N_2][H_2]^3}$

7.87 $K_{eq} = \dfrac{[H_2S]^2}{[H_2]^2[S_2]}$

7.89 $7.7 \times 10^{-5}\,M$

7.91 **a.** Equilibrium shifts to the left.   **c.** No change
b. No change

7.93 **a.** PCl_3 increases   **d.** PCl_3 decreases
b. PCl_3 decreases   **e.** PCl_3 remains the same
c. PCl_3 decreases

7.95 Decrease

7.97 $K_{eq} = \dfrac{[CO][H_2]}{[H_2O]}$

7.99 False. The position of equilibrium is not affected by a catalyst, only the rate at which equilibrium is attained.

7.101 Carbon dioxide is dissolved in cola. Heating shifts the equilibrium to the right.

$$CO_2(l) \rightleftharpoons CO_2(g)$$

Since gases are less soluble at elevated temperatures, the pressure buildup from the carbon dioxide gas in the sealed bottle can lead to an explosion.

7.103 **a.** $K_{eq} = \dfrac{[SO_3]^2}{[SO_2]^2[O_2]}$ **b.** 3.0×10^2

7.105 Adding $SO_2(g)$ or $O_2(g)$ or increasing the pressure

Chapter 8

8.1 **a.** Brønsted-Lowry acid
b. Brønsted-Lowry acid
c. Brønsted-Lowry base
d. Brønsted-Lowry base

8.3 **a.** $HF(aq) + H_2O(l) \rightleftharpoons F^-(aq) + H_3O^+(aq)$
b. $C_6H_5COO^-(aq) + H_2O(l) \rightleftharpoons C_6H_5COOH(aq) + OH^-(aq)$

8.5 **a.** HF/F^- and H_3O^+/H_2O
b. H_2O/OH^- and $C_6H_5COOH/C_6H_5COO^-$

8.7 $4.0 \times 10^{-13}\,M$

8.9 $1.0 \times 10^{-11}\,M$

8.11 pH = 4.74

8.13 pH = 4.87

8.15 $CO_2 + H_2O \rightleftharpoons H_2CO_3 \rightleftharpoons H_3O^+ + HCO_3^-$
An increase in the partial pressure of CO_2 is a stress on the left side of the equilibrium. The equilibrium will shift to the right in an effort to decrease the concentration of CO_2. This will cause the molar concentration of H_2CO_3 to increase.

8.17 In Question 8.15, the equilibrium shifts to the right. Therefore, the molar concentration of H_3O^+ should increase.
In Question 8.16, the equilibrium shifts to the left. Therefore, the molar concentration of H_3O^+ should decrease.

8.19 $pH = pK_a + \log \dfrac{[HCO_3^-]}{[H_2CO_3]}$

8.21 **a.** An Arrhenius acid is a substance that dissociates, producing hydrogen ions.
b. A Brønsted-Lowry acid is a substance that behaves as a proton donor.

8.23 The Brønsted-Lowry theory provides a broader view of acid-base theory than does the Arrhenius theory. Brønsted-Lowry emphasizes the role of the solvent in the dissociation process.

8.25 **a.** Brønsted-Lowry acid   **c.** Amphiprotic
b. Brønsted-Lowry base

8.27 **a.** Brønsted-Lowry acid
b. Amphiprotic
c. Brønsted-Lowry base

8.29 **a.** $HNO_2(aq) + H_2O(l) \rightleftharpoons H_3O^+(aq) + NO_2^-(aq)$
b. $HCN(aq) + H_2O(l) \rightleftharpoons H_3O^+(aq) + CN^-(aq)$
c. $CH_3CH_2CH_2COO^-(aq) + H_2O(l) \longrightarrow CH_3CH_2CH_2COOH(aq) + OH^-(aq)$

8.31 HCN

8.33 I^-

8.35 HNO_3

8.37 CN^-

8.39 HF

8.41 **a.** HCN/CN^- and NH_4^+/NH_3
b. HCO_3^-/CO_3^{2-} and HCl/Cl^-

8.43 Concentration refers to the quantity of acid or base contained in a specified volume of solvent. Strength refers to the degree of dissociation of the acid or base.

8.45 **a.** Weak   **b.** Weak   **c.** Weak

8.47 **a.** $1.0 \times 10^{-7}\,M$ **b.** $1.0 \times 10^{-11}\,M$

8.49 **a.** $1.0 \times 10^{-10}\,M$ **b.** $1.0 \times 10^{-12}\,M$

8.51 $1.7 \times 10^{-11}\,M$

8.53 The beaker containing $0.10\,M\ CH_3COOH$. CH_3COOH is a weaker acid than HCl. Thus, it has a higher pH than HCl.

8.55 **a.** pH = 2.00   **b.** pH = 4.00

8.57 **a.** $[H_3O^+] = 1.0 \times 10^{-1}\,M$
b. $[H_3O^+] = 1.0 \times 10^{-5}\,M$

8.59 pH = 11.00

8.61 **a.** $[H_3O^+] = 5.0 \times 10^{-2}\,M$
     $[OH^-] = 2.0 \times 10^{-13}\,M$
b. $[H_3O^+] = 2.0 \times 10^{-10}\,M$
     $[OH^-] = 5.0 \times 10^{-5}\,M$

8.63 A neutralization reaction is one in which an acid and a base react to produce water and a salt (a "neutral" solution).

8.65 **a.** $[H_3O^+] = 1.0 \times 10^{-6}\,M$
     $[OH^-] = 1.0 \times 10^{-8}\,M$
b. $[H_3O^+] = 6.3 \times 10^{-6}\,M$
     $[OH^-] = 1.6 \times 10^{-9}\,M$
c. $[H_3O^+] = 1.6 \times 10^{-8}\,M$
     $[OH^-] = 6.3 \times 10^{-7}\,M$

8.67 The statement is incorrect. The pH = 3 solution is 1000 times as acidic as the pH = 6 solution because pH is a logarithmic function.

8.69 **a.** $[H_3O^+] = 1.0 \times 10^{-5}\,M$
b. $[H_3O^+] = 1.0 \times 10^{-12}\,M$
c. $[H_3O^+] = 3.2 \times 10^{-6}\,M$

8.71 **a.** pH = 6.00   **b.** pH = 8.00   **c.** pH = 3.25

8.73 pH = 3.12

8.75 pH = 10.74

8.77 4 mol HCl

8.79 An indicator is a substance that is added to a solution and changes color as the solution reaches a certain pH. It is often used in the technique of titration to determine the equivalence point.

8.81 $HNO_3(aq) + NaOH(aq) \longrightarrow H_2O(l) + NaNO_3(aq)$

8.83 $H^+(aq) + OH^-(aq) \longrightarrow H_2O(l)$ or
$H_3O^+(aq) + OH^-(aq) \longrightarrow 2H_2O(l)$

8.85 Two protons

8.87 0.1800 M

8.89 13.33 mL

8.91 Step 1. $H_2CO_3(aq) + H_2O(l) \rightleftharpoons H_3O^+(aq) + HCO_3^-(aq)$
Step 2. $HCO_3^-(aq) + H_2O(l) \rightleftharpoons H_3O^+(aq) + CO_3^{2-}(aq)$

8.93 **a.** NH_3 and NH_4Cl can form a buffer solution.
b. HNO_3 and KNO_3 cannot form a buffer solution.

8.95 HCl/NaCl is not a buffer solution. It is a strong acid and a salt. The NaOH would produce a significant pH change. On the other hand, CH_3COOH/CH_3COONa is a buffer solution because it is a weak acid and its salt. Therefore, it would resist significant pH change upon addition of a strong base.

8.97 The equilibrium reaction is:
$CO_2 + H_2O \rightleftharpoons H_2CO_3 \rightleftharpoons H_3O^+ + HCO_3^-$
A situation of high blood CO_2 levels and low pH is termed acidosis. A high concentration of CO_2 is a stress on the left side of the equilibrium. The equilibrium will shift to the right in an effort to decrease the concentration of CO_2. This will cause the molar concentration of carbonic acid (H_2CO_3) to increase.

8.99 **a.** Addition of strong acid is equivalent to adding H_3O^+. This is a stress on the right side of the equilibrium, and the equilibrium will shift to the left. Consequently, the [CH_3COOH] increases.
b. Water, in this case, is a solvent and does not appear in the equilibrium expression. Hence, it does not alter the position of the equilibrium.

8.101 [H_3O^+] = $2.32 \times 10^{-7} M$

8.103 CH_3COO^-, a conjugate base, reacts with added H_3O^+ to maintain pH.

8.105 pH = 4.74

8.107 $11.2 = \dfrac{[HCO_3^-]}{[H_2CO_3]}$

Chapter 9

9.1 X-ray, ultraviolet, visible, infrared, microwave, and radiowave

9.3 $^{144}_{60}Nd \rightarrow ^{140}_{58}Ce + ^{4}_{2}He$

9.5 $^{131}_{53}I \rightarrow ^{131}_{54}Xe + ^{0}_{-1}e$

9.7 A positron has a positive charge, and a beta particle has a negative charge.

9.9 201 hours

9.11 1/4 of the radioisotope remains after two half-lives.

9.13 Isotopes with short half-lives release their radiation rapidly. There is much more radiation per unit time observed with short half-life substances; hence, the signal is stronger and the sensitivity of the procedure is enhanced.

9.15 The rem takes into account the relative biological effect of the radiation in addition to the quantity of radiation. This provides a more meaningful estimate of potential radiation damage to human tissue.

9.17 Natural radioactivity is the spontaneous decay of a nucleus to produce high-energy particles or rays.

9.19 Two protons and two neutrons

9.21 An electron with a −1 charge.

9.23 • charge, α = +2, β = −1
• mass, α = 4 amu, β = 0.000549 amu
• velocity, α = 10% of the speed of light, β = 90% of the speed of light

9.25 Chemical reactions involve joining, separating, and rearranging atoms; valence electrons are critically involved. Nuclear reactions only involve changes in nuclear composition.

9.27 $^{4}_{2}He$

9.29 A helium atom has two electrons; an alpha particle has no electrons.

9.31 Alpha particles, beta particles, and positrons are matter; gamma radiation is pure energy. Alpha particles are large and relatively slow moving. They are the least energetic and least penetrating. Beta particles and positrons are smaller, faster, and more penetrating than alpha particles. Gamma radiation moves at the speed of light, is highly energetic, and is most penetrating.

9.33 $^{15}_{7}N$

9.35 $^{235}_{92}U$

9.37 $^{1}_{1}H$ = 0 neutrons and 1 proton
$^{2}_{1}H$ = 1 neutron and 1 proton
$^{3}_{1}H$ = 2 neutrons and 1 proton

9.39 $^{60}_{27}Co \rightarrow ^{60}_{28}Ni + ^{0}_{-1}\beta + \gamma$

9.41 $^{23}_{11}Na + ^{2}_{1}H \rightarrow ^{24}_{11}Na + ^{1}_{1}H$

9.43 $^{24}_{10}Ne \rightarrow ^{0}_{-1}\beta + ^{24}_{11}Na$

9.45 $^{140}_{55}Cs \rightarrow ^{140}_{56}Ba + ^{0}_{-1}e$

9.47 $^{209}_{83}Bi + ^{54}_{24}Cr \rightarrow ^{262}_{107}Bh + ^{1}_{0}n$

9.49 $^{27}_{12}Mg \rightarrow ^{0}_{-1}e + ^{27}_{13}Al$

9.51 $^{12}_{7}N \rightarrow ^{12}_{6}C + ^{0}_{+1}e$

9.53 $^{241}_{95}Am \rightarrow ^{4}_{2}\alpha + ^{237}_{93}Np$

9.55 • Nuclei for light atoms tend to be most stable if their neutron/proton ratio is close to 1.
• Nuclei with more than 84 protons tend to be unstable.
• Isotopes with a "magic number" of protons or neutrons (2, 8, 20, 50, 82, or 126 protons or neutrons) tend to be stable.
• Isotopes with even numbers of protons or neutrons tend to be more stable.

9.57 15 half-lives

9.59 $^{20}_{8}O$; Oxygen-20 has 20 − 8 = 12 neutrons, an n/p of 12/8, or 1.5. The n/p is probably too high for stability even though it does have a "magic number" of protons and an even number of protons and neutrons.

9.61 Chromium-48 has 48 − 24 = 24 neutrons, an n/p of 24/24, or 1.0. It also has an even number of protons and neutrons. It would probably be stable.

9.63 0.40 mg of iodine-131 remains

9.65 13 mg of iron-59 remains

9.67 Radiocarbon dating is a process used to determine the age of objects. The ratio of the masses of the stable isotope, carbon-12, and unstable isotope, carbon-14, is measured. Using this value and the half-life of carbon-14, the age of the coffin may be calculated.

9.69 Fission

9.71 **a.** The fission process involves the breaking down of large, unstable nuclei into smaller, more stable nuclei. This process releases some of the binding energy in the form of heat and/or light.

b. The heat generated during the fission process could be used to generate steam, which is then used to drive a turbine to create electricity.

9.73 $^3_1H + ^1_1H \rightarrow ^4_2He$ + energy

9.75 A "breeder" reactor creates the fuel that can be used by a conventional fission reactor during its fission process.

9.77 The reaction in a fission reactor that involves neutron production and causes subsequent reactions accompanied by the production of more neutrons in a continuing process.

9.79 High operating temperatures

9.81 Radiation therapy provides sufficient energy to destroy molecules critical to the reproduction of cancer cells.

9.83 Natural radioactivity is a spontaneous process; artificial radioactivity is nonspontaneous and results from a nuclear reaction that produces an unstable nucleus.

9.85 **a.** Technetium-99 m is used to study the heart (cardiac output, size, and shape), kidney (follow-up procedure for kidney transplant), and liver and spleen (size, shape, presence of tumors).

b. Xenon-133 is used to locate regions of reduced ventilation and presence of tumors in the lung.

9.87 $^{108}_{47}Ag + ^4_2He \rightarrow ^{112}_{49}In$

9.89 Background radiation, radiation from natural sources, is emitted by the sun as cosmic radiation, and from naturally radioactive isotopes found throughout our environment.

9.91 Level decreases

9.93 Yes, it would lead to a positive effect. Potential damage is often directly proportional to the time of exposure.

9.95 Yes, it would lead to a positive effect. The operator of the robotic device could be located far from the source, no physical contact with the source is necessary, and barriers of lead or other shielding can isolate the control and the robot.

9.97 Yes. Concrete has a higher density than wood and thus serves as a better radiation shield.

9.99 Relative biological effect is a measure of the damage to biological tissue caused by different forms of radiation.

9.101 **a.** The curie is the amount of radioactive material needed to produce 3.7×10^{10} atomic disintegrations per second.

b. The roentgen is the amount of radioactive material needed to produce 2×10^9 ion-pairs when passing through 1 cc of air at 0°C.

c. The becquerel is the amount of radioactive material needed to produce 1 atomic disintegration per second.

9.103 A film badge detects gamma radiation by darkening photographic film in proportion to the amount of radiation exposure over time. Badges are periodically collected and evaluated for their level of exposure. This mirrors the level of exposure of the personnel wearing the badges.

Chapter 10

10.1 The student could test the solubility of the substance in water and in an organic solvent, such as hexane. Solubility in hexane would suggest an organic substance; whereas solubility in water would indicate an inorganic compound. The student could also determine the melting and boiling points of the substance. If the melting and boiling points are very high, an inorganic substance would be suspected.

10.3 **a.**

b.

c.

10.5 **a.** The monobromination of propane will produce two products, as shown in the following two equations:

$$CH_3CH_2CH_3 + Br_2 \xrightarrow{\text{Light or heat}} CH_3CH_2CH_2Br + HBr$$
$$CH_3CH_2CH_3 + Br_2 \xrightarrow{\text{Light or heat}} CH_3CHBrCH_3 + HBr$$

b. The monochlorination of butane will produce two products, as shown in the following two equations:

$$CH_3CH_2CH_2CH_3 + Cl_2 \xrightarrow{\text{Light or heat}} CH_3CH_2CH_2CH_2Cl + HCl$$
$$CH_3CH_2CH_2CH_3 + Cl_2 \xrightarrow{\text{Light or heat}} CH_3CH_2CHClCH_3 + HCl$$

c. The monochlorination of cyclobutane:

d. The monobromination of pentane will produce three products as shown in the following equations:

$$CH_3CH_2CH_2CH_2CH_3 + Br_2 \xrightarrow{\text{Light or heat}} CH_3CH_2CH_2CH_2CH_2Br + HBr$$
$$CH_3CH_2CH_2CH_2CH_3 + Br_2 \xrightarrow{\text{Light or heat}} CH_3CH_2CH_2CHBrCH_3 + HBr$$
$$CH_3CH_2CH_2CH_2CH_3 + Br_2 \xrightarrow{\text{Light or heat}} CH_3CH_2CHBrCH_2CH_3 + HBr$$

10.7 The number of organic compounds is nearly limitless because carbon forms stable covalent bonds with other carbon atoms in a variety of different patterns. In addition, carbon can form stable bonds with other elements and functional groups. Finally, carbon can form double or triple bonds with other carbon atoms to produce organic molecules with different properties.

10.9 Because ionic substances often form three-dimensional crystals made up of many positive and negative ions, they generally have much higher melting and boiling points than covalent compounds.

10.11 **a.** $LiCl > H_2O > CH_4$

b. $NaCl > C_3H_8 > C_2H_6$

10.13 **a.** LiCl would be a solid; H_2O would be a liquid; and CH_4 would be a gas.

b. NaCl would be a solid; both C_3H_8 and C_2H_6 would be gases.

10.15 **a.** Water-soluble inorganic compounds

b. Inorganic compounds

c. Organic compounds

d. Inorganic compounds

e. Organic compounds

10.17 a. $C_{19}H_{40}$

b.

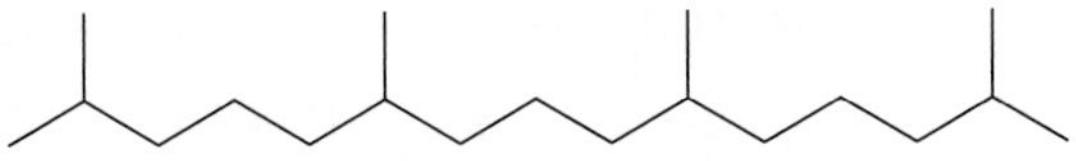

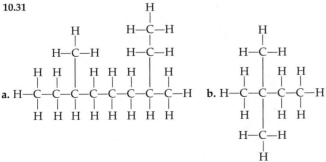

$CH(CH_3)_2(CH_2)_3CH(CH_3)(CH_2)_3CH(CH_3)(CH_2)_3CH(CH_3)_2$

c. 268.51 g/mol

10.19 a. $CH_3CH_2CH(CH_3)_2$
b. $CH_3CH_2C(CH_3)_2(CH_2)_2CH(CH_3)_2$
c. $CH_3CH_2C(CH_3)_2(CH_2)_3CH(CH_3)CH(CH_3)_2$

10.21 a.   **b.** 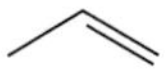   **c.**

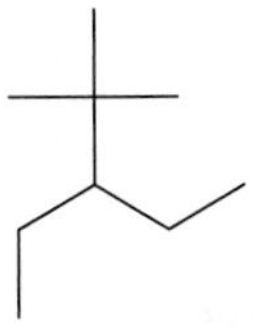

10.23 a. $(CH_3)_3CCH(CH_2CH_3)_2$
b. CH_3CHCH_2
c. $CH_3CH_2CH_3$

10.25

a.   **b.**

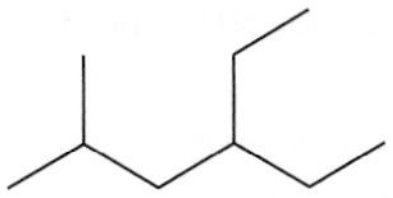

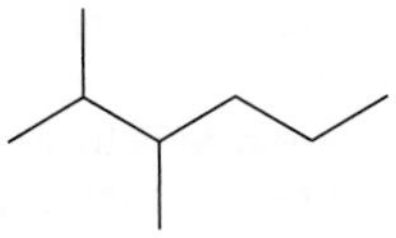

c. 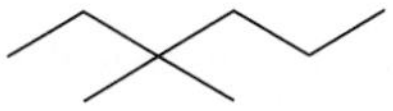   **d.**

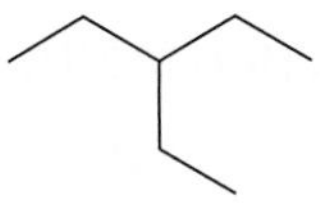

10.27 a.   Tricosane:

Pentacosane:

Heptacosane:

b. Tricosane: 324.61 g/mol
Pentacosane: 352.67 g/mol
Heptacosane: 380.72 g/mol

10.29

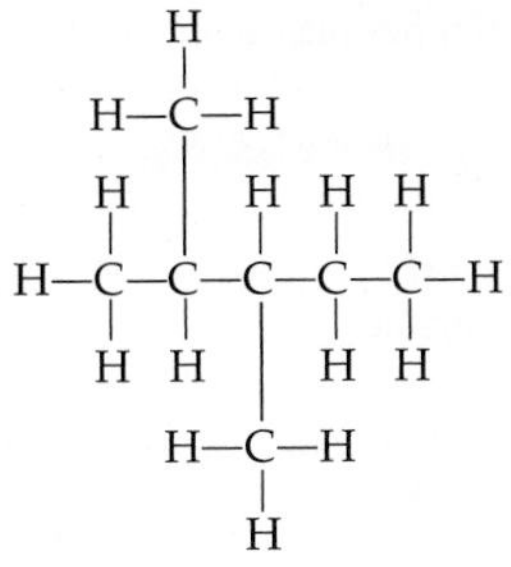

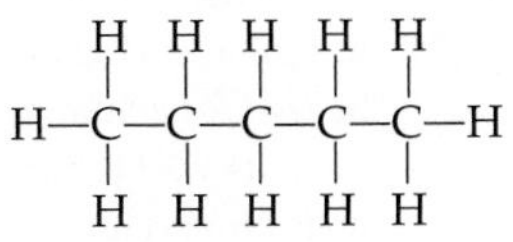

a. H—C—C—C—C—H   **b.** H—C—C—C—C—H

10.31

a. H—C—C—C—C—C—C—C—H   **b.** H—C—C—C—C—H

10.33 a. Hydroxyl group   **e.** Ester group
b. Amino group   **f.** Ether group
c. Carbonyl group   **g.** Halide
d. Carboxyl group

10.35 a. C_nH_{2n+2} **d.** C_nH_{2n}
b. C_nH_{2n-2} **e.** C_nH_{2n-2}
c. C_nH_{2n}

10.37 Alkanes have only carbon-to-carbon and carbon-to-hydrogen single bonds, as in the molecule ethane:

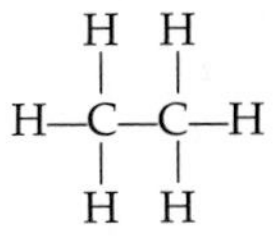

Alkenes have at least one carbon-to-carbon double bond, as in the molecule ethene:

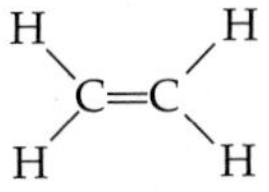

Alkynes have at least one carbon-to-carbon triple bond, as in the molecule ethyne:

$H—C≡C—H$

10.39

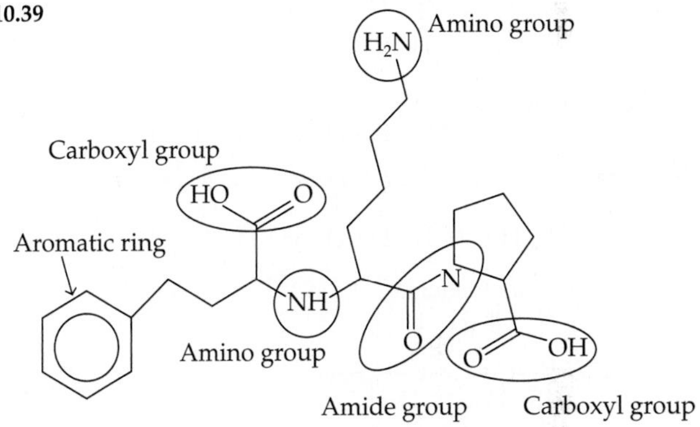

10.41

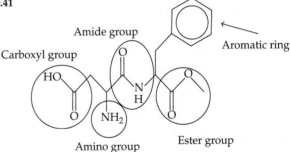

10.43 van der Waals forces are the attractive forces between neutral molecules. They include dipole-dipole attractions and London dispersion forces.

10.45 London dispersion forces result from the attraction of two molecules that experience short-lived dipoles as a result of transient shifts in the electron cloud. Larger molecules with more electrons exhibit a stronger attraction. As a result, they will have higher melting and boiling points.

10.47 Hydrocarbons are nonpolar molecules, and hence are not soluble in water.

10.49 a. Heptane > Hexane > Butane > Ethane
b. $CH_3CH_2CH_2CH_2CH_2CH_2CH_2CH_2CH_3 >$
$CH_3CH_2CH_2CH_2CH_3 > CH_3CH_2CH_3$

10.51 a. Heptane and hexane would be liquid at room temperature; butane and ethane would be gases.
  b. $CH_3CH_2CH_2CH_2CH_2CH_2CH_2CH_2CH_3$ and $CH_3CH_2CH_2CH_2CH_3$ would be liquids at room temperature; $CH_3CH_2CH_3$ would be a gas.

10.53 Nonane: $CH_3CH_2CH_2CH_2CH_2CH_2CH_2CH_2CH_3$
  Pentane: $CH_3CH_2CH_2CH_2CH_3$
  Propane: $CH_3CH_2CH_3$

10.55 a.

Br

b.

c.

10.57 a. 2,2-Dibromobutane:

H Br H H
| | | |
H—C—C—C—C—H
| | | |
H Br H H

  b. 2-Iododecane:

H I H H H H H H H H
| | | | | | | | | |
H—C—C—C—C—C—C—C—C—C—C—H
| | | | | | | | | |
H H H H H H H H H H

  c. 1,2-Dichloropentane:

H Cl H H H
| | | | |
Cl—C—C—C—C—C—H
| | | | |
H H H H H

  d. 1-Bromo-2-methylpentane:

H
|
H—C—H
|
H | H H H
| | | | |
H—C—C—C—C—C—H
| | | | |
Br H H H H

10.59 a. 3-Methylpentane
  b. 2,5-Dimethylhexane
  c. 1-Bromoheptane
  d. 1-Chloro-3-methylbutane

10.61 a. 2-Chloropropane
  b. 2-Iodobutane
  c. 2,2-Dibromopropane
  d. 1-Chloro-2-methylpropane
  e. 2-Iodo-2-methylpropane

10.63 a. The straight chain isomers of molecular formula C_4H_9Br:

H H H H
| | | |
H—C—C—C—C—Br
| | | |
H H H H

H H Br H
| | | |
H—C—C—C—C—H
| | | |
H H H H

  b. The straight chain isomers of molecular formula $C_4H_8Br_2$:

H H H Br
| | | |
H—C—C—C—C—Br
| | | |
H H H H

H H Br H
| | | |
H—C—C—C—C—Br
| | | |
H H H H

H Br H H
| | | |
H—C—C—C—C—Br
| | | |
H H H H

H H H H
| | | |
Br—C—C—C—C—Br
| | | |
H H H H

H H Br H
| | | |
H—C—C—C—C—H
| | | |
H H Br H

H Br Br H
| | | |
H—C—C—C—C—H
| | | |
H H H H

10.65 a. 2-Chlorohexane   **c.** 3-Chloropentane
  b. 1,4-Dibromobutane   **d.** 2-Methylheptane

10.67 a. The first pair of molecules are constitutional isomers: hexane and 2-methylpentane.
  b. The second pair of molecules are identical. Both are heptane.

10.69 Structures "a" and "c"

10.71 a. Incorrect: 3-Methylhexane   **c.** Incorrect: 3-Methylheptane
  b. Incorrect: 2-Methylbutane   **d.** Correct

10.73 a.

CH_3
|
$CH_3CHCH_2CHCH_3$ The name given in the problem
| is correct.
CH_3

  b.
CH_3 The correct name is
| 4-methylheptane.
$CH_3CH_2CH_2CHCH_2CH_2CH_3$

  c. $I—CH_2CH_2CH_2CH_2CH_2—I$ The name given in the problem is correct.

  d. $CH_3CH_2CH_2CH_2CH_2CHCH_2CH_2CH_3$ The correct
| name is
CH_2CH_3 4-ethylnonane.

  e. Br   Br   The name given
| | in the problem
$CH_2CH_2CH_2CH_2CH_2CCH_2CH_3$ is correct.
                    |
                    CH_3

10.75 Cycloalkanes are a family of molecules having carbon-to-carbon bonds in a ring structure.

10.77 The general formula for a cycloalkane is C_nH_{2n}.

10.79 a. Chlorocyclopropane
  b. *cis*-1,2-Dichlorocyclopropane
  c. *trans*-1,2-Dichlorocyclopropane
  d. Bromocyclopropane

10.81 a.

c.

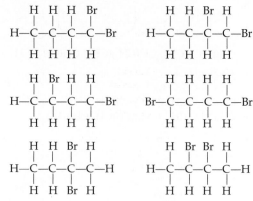

b.

d.

10.83   There are three structural isomers of dichlorocyclopropane. Two of these isomers are geometric isomers.

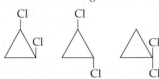

10.85   **a.** Incorrect—1,2-Dibromocyclobutane
b. Incorrect—1,2-Diethylcyclobutane
c. Correct
d. Incorrect—1,2,3-Trichlorocyclohexane

10.87   **a.** *cis*-1,3-Dibromocyclopentane

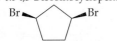

b. *trans*-1,2-Dimethylcyclobutane

c. *cis*-1,2-Dichlorocyclopropane

d. *trans*-1,4-Diethylcyclohexane

10.89   **a.** *cis*-1,2-Dibromocyclopentane
b. *trans*-1,3-Dibromocyclopentane
c. *cis*-1,2-Dimethylcyclohexane
d. *cis*-1,2-Dimethylcyclopropane

10.91   Conformational isomers are distinct isomeric structures that may be converted into one another by rotation about the bonds in the molecule.

10.93   In the chair conformation, the hydrogen atoms, and thus the electron pairs of the C—H bonds, are farther from one another. As a result, there is less electron repulsion and the structure is more stable. In the boat conformation, the electron pairs are more crowded. This causes greater electron repulsion, producing a less stable conformation.

10.95   Combustion is the oxidation of hydrocarbons by burning in the presence of air to produce carbon dioxide and water.

10.97   **a.** $C_3H_8 + 5O_2 \longrightarrow 4H_2O + 3CO_2$
b. $C_7H_{16} + 11O_2 \longrightarrow 8H_2O + 7CO_2$
c. $C_9H_{20} + 14O_2 \longrightarrow 10H_2O + 9CO_2$
d. $2C_{10}H_{22} + 31O_2 \longrightarrow 22H_2O + 20CO_2$

10.99 $2C_{16}H_{34} + 49O_2 \longrightarrow 32CO_2 + 34H_2O$

10.101   **a.** $8CO_2 + 10H_2O$
b.

$$Br—\underset{\underset{CH_3}{|}}{\overset{\overset{CH_3}{|}}{C}}—CH_3 + CH_3CHCH_2Br + 2HBr$$

c. Cl_2 and light

10.103   The following molecules are all isomers of C_6H_{14}.

$CH_3CH_2CH_2CH_2CH_2CH_3$

Hexane

$CH_3\overset{\overset{CH_3}{|}}{CH}CH_2CH_2CH_3$

2-Methylpentane

$CH_3CH_2\overset{\overset{CH_3}{|}}{CH}CH_2CH_3$

3-Methylpentane

$CH_3\overset{\overset{CH_3}{|}}{CH}\overset{\overset{CH_3}{|}}{CH}CH_3$

2,3-Dimethylbutane

$CH_3\overset{\overset{CH_3}{|}}{\underset{\underset{CH_3}{|}}{C}}CH_2CH_3$

2,2-Dimethylbutane

a. 2,3-Dimethylbutane produces only two monobrominated derivatives: 1-bromo-2,3-dimethylbutane and 2-bromo-2, 3-dimethylbutane.
b. Hexane produces three monobrominated products: 1-bromohexane, 2-bromohexane, and 3-bromohexane. 2,2-Dimethylbutane also produces three monobrominated products: 1-bromo-2,2-dimethylbutane, 2-bromo-3, 3-dimethylbutane, and 1-bromo-3,3-dimethylbutane.
c. 3-Methylpentane produces four monobrominated products: 1-bromo-3-methylpentane, 2-bromo-3-methylpentane, 3-bromo-3-methylpentane, and 1-bromo-2-ethylbutane.

10.105   The hydrocarbon is cyclooctane, and it has the molecular formula C_8H_{16}.

$$\bigcirc + 12\,O_2 \longrightarrow 8\,CO_2 + 8\,HO_2 + \text{heat energy}$$

Chapter 11

11.1   a. $CH_2BrCH_2C{\equiv}CCH_2CH_3$

b. $CH_3C{\equiv}CCH_3$

c. $ClC{\equiv}CCl$

d. $HC{\equiv}C(CH_2)_7I$

11.3   a.

cis-3-Hexene   trans-3-Hexene

b.

trans-2,3-Dibromo-2-butene   cis-2,3-Dibromo-2-butene

11.5 Molecule "c" can exist as *cis*- and *trans*-isomers because there are two different groups on each of the carbon atoms attached by the double bond.

11.7 a.

$$CH_3CH_2 \quad\quad CH_2CH_2CH_2CH_3$$
$$C=C$$
$$H \quad\quad\quad H$$

c.

$$CH_3 \quad\quad Cl$$
$$C=C$$
$$Cl \quad\quad CH_3$$

b.

$$CH_3 \quad\quad H$$
$$C=C \quad Cl$$
$$H \quad\quad CH_2CHCH_3$$

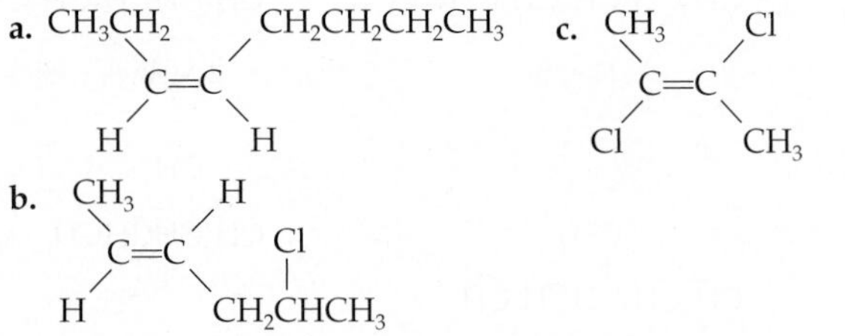

11.9 The hydrogenation of the *cis* and *trans* isomers of 2-pentene would produce the same product, pentane.

11.11 a.

$$H_3C-C\equiv C-CH_3 + 2H_2 \xrightarrow{Ni}$$

H H H H
H—C—C—C—C—H
H H H H

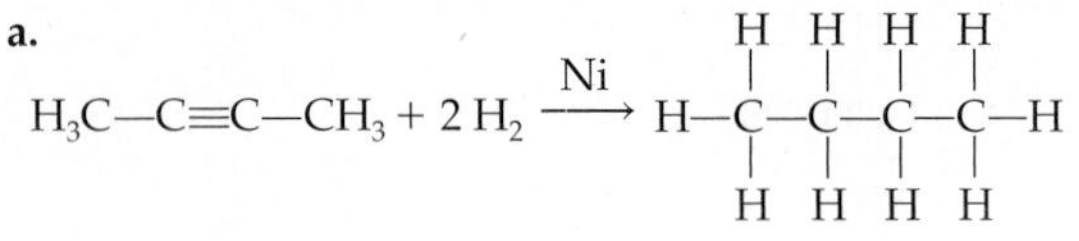

2-Butyne   Butane

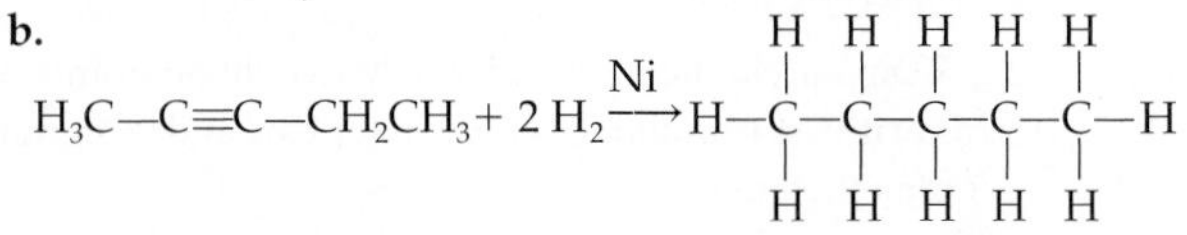

b.

$$H_3C-C\equiv C-CH_2CH_3 + 2H_2 \xrightarrow{Ni}$$

H H H H H
H—C—C—C—C—C—H
H H H H H

2-Pentyne   Pentane

11.13

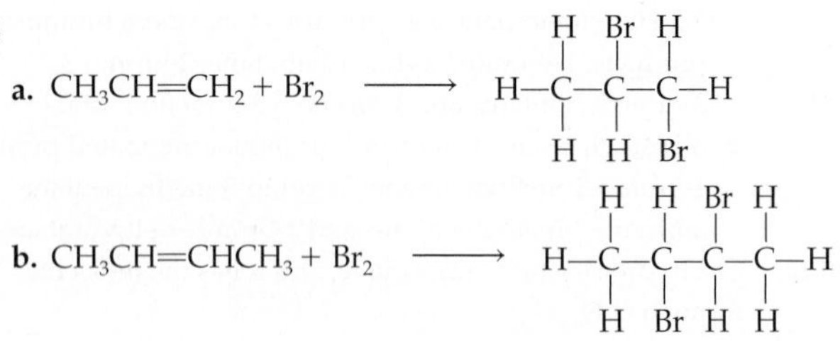

a. $CH_3CH=CH_2 + Br_2 \longrightarrow$

H Br H
H—C—C—C—H
H H Br

b. $CH_3CH=CHCH_3 + Br_2 \longrightarrow$

H H Br H
H—C—C—C—C—H
H Br H H

11.15

a. $CH_3C\equiv CCH_3 + 2Cl_2 \longrightarrow$

H Cl Cl H
H—C—C—C—C—H
H Cl Cl H

b. $CH_3C\equiv CCH_2CH_3 + 2Cl_2 \longrightarrow$

H Cl Cl H H
H—C—C—C—C—C—H
H Cl Cl H H

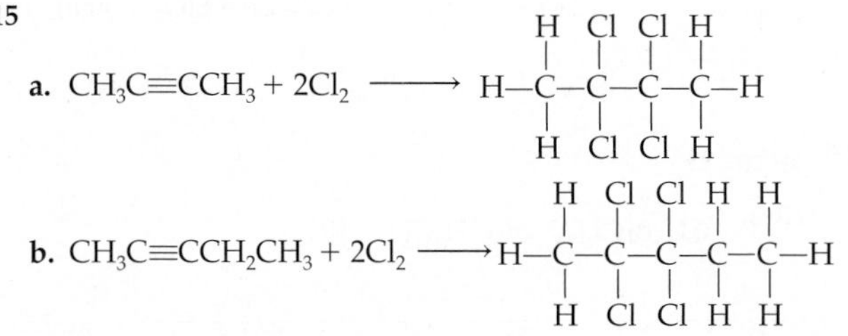

11.17 a. $CH_3CH=CHCH_3 + H_2O \xrightarrow{H+} CH_3CHOHCH_2CH_3$

b. $H_2C=CHCH_2CH_2CH(CH_3)_2 + H_2O \xrightarrow{H+}$

$CH_3CHOHCH_2CH_2CH(CH_3)_2$
(Major product)

$H_2C=CHCH_2CH_2CH(CH_3)_2 + H_2O \xrightarrow{H+}$

$CH_2OHCH_2CH_2CH_2CH(CH_3)_2$
(Minor product)

c. $CH_3CH_2CH_2CH=CHCH_2CH_3 + H_2O \xrightarrow{H+}$

$CH_3CH_2CH_2CHOHCH_2CH_2CH_3$

$CH_3CH_2CH_2CH=CHCH_2CH_3 + H_2O \xrightarrow{H+}$

$CH_3CH_2CH_2CH_2CHOHCH_2CH_3$

d. $CH_3CHClCH=CHCHClCH_3 + H_2O \xrightarrow{H+}$

$CH_3CHClCHOHCH_2CHClCH_3$
Only product

11.19 a. $H_3CC\equiv CH + H_2O \xrightarrow{H^+}$

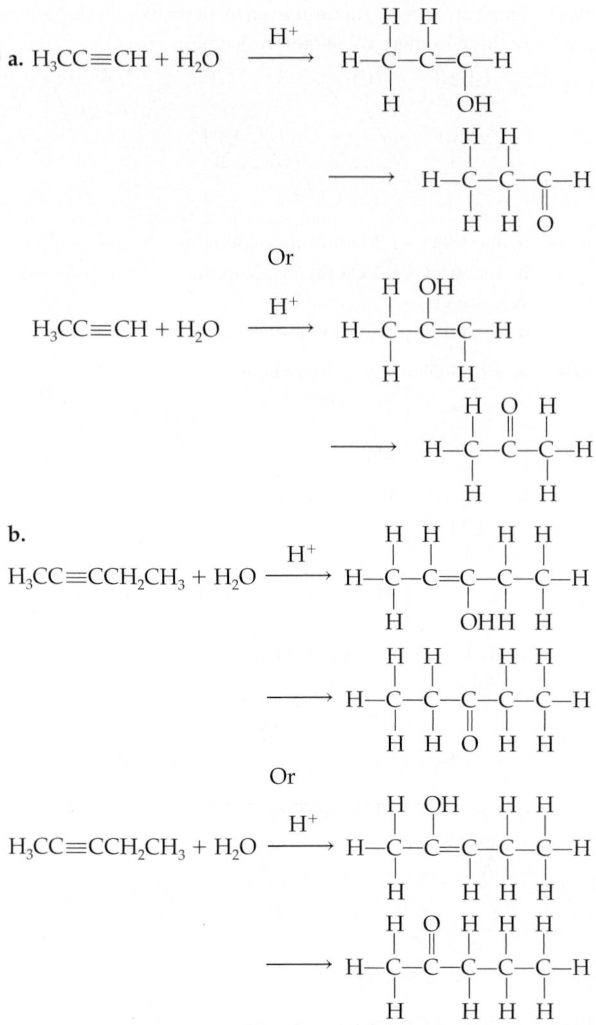

H H
H—C—C=C—H
H OH

H H
H—C—C—C—H
H H O

Or

$H_3CC\equiv CH + H_2O \xrightarrow{H^+}$

H OH
H—C—C=C—H
H H

H O H
H—C—C—C—H
H H

b. $H_3CC\equiv CCH_2CH_3 + H_2O \xrightarrow{H^+}$

H H H H
H—C—C=C—C—C—H
H OHH H

H H H H
H—C—C—C—C—C—H
H H O H H

Or

$H_3CC\equiv CCH_2CH_3 + H_2O \xrightarrow{H^+}$

H OH H H
H—C—C=C—C—C—H
H H H H

H O H H H
H—C—C—C—C—C—H
H H H H

11.21 a.

d.

b.

e.

c.

f.

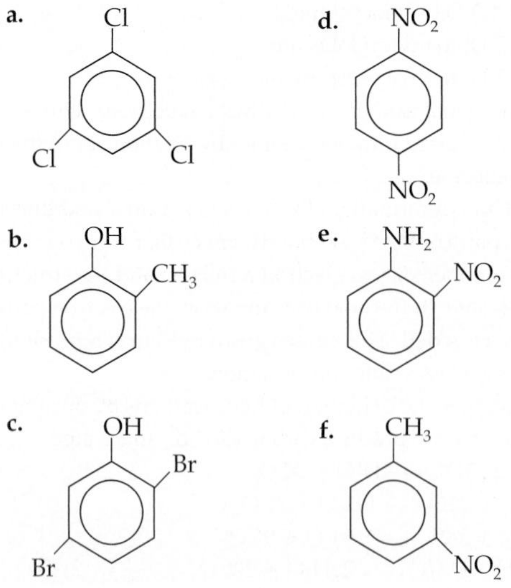

11.23 As the length of the hydrocarbon chain increases, the London dispersion forces between the molecules increase. The stronger these attractive forces between molecules are, the higher the boiling point will be.

11.25 The general formula for an alkane is C_nH_{2n+2}.
The general formula for an alkene is C_nH_{2n}.
The general formula for an alkyne is C_nH_{2n-2}.

11.27 Ethene is a planar molecule. All of the bond angles are 120°.

11.29 In alkanes, such as ethane, the four bonds around each carbon atom have tetrahedral geometry. The bond angles are 109.5°. In alkenes, such as ethene, each carbon is bonded by two single bonds and one double bond. The molecule is planar and each bond angle is approximately 120°.

11.31 Ethyne is a linear molecule. All of the bond angles are 180°.

11.33 In alkanes, such as ethane, the four bonds around each carbon atom have tetrahedral geometry. The bond angles are 109.5°. In alkenes, such as ethene, each carbon is bonded by two single bonds and one double bond. The molecule is planar and each bond angle is approximately 120°. In alkynes, such as ethyne, each carbon is bonded by one single bond and one triple bond. The molecule is linear and the bond angles are 180°.

11.35 **a.** 2-Pentyne > Propyne > Ethyne
 b. 3-Decene > 2-Butene > Ethene

11.37 Identify the longest carbon chain containing the carbon-to-carbon double or triple bond. Replace the *–ane* suffix of the alkane name with *–ene* for an alkene or *-yne* for an alkyne. Number the chain to give the lowest number to the first of the two carbons involved in the double or triple bond. Determine the name and carbon number of each substituent group and place that information as a prefix in front of the name of the parent compound.

11.39 Geometric isomers of alkenes differ from one another in the placement of substituents attached to each of the carbon atoms of the double bond. Of the pair of geometric isomers, the *cis*-isomer is the one in which identical groups are on the same side of the double bond.

11.41 **a.** H$_3$C, CH$_2$CH$_2$CH$_3$ / C=C / H$_3$C, H

 b. CH$_3$CH$_2$, H / C=C / H, CH$_2$CH$_2$CH$_3$

 c. ClCH$_2$, CH$_2$CH$_3$ / C=C / H, H

 d. (H$_3$C)$_2$CCl, CH$_2$CH$_2$CH$_3$ / C=C / H, H

 e. (H$_3$C)$_2$CH, H / C=C / H, CHBrCH(CH$_3$)CH$_2$CH$_3$

11.43 **a.** 3-Methyl-1-pentene
 b. 7-Bromo-1-heptene
 c. 5-Bromo-3-heptene
 d. 1-*t*-Butyl-4-methylcyclohexene

11.45 **a.** CH$_2$FCH$_2$CHFCH$_2$CH$_2$F
 b.
 H, H / C=C / H$_3$C, CH$_2$CH$_2$CH$_2$CH$_2$CH$_3$
 c. CH$_3$CH$_2$CH$_2$C≡CCH$_2$CH$_2$CH$_3$

11.47 **a.** 1-Heptene can only be drawn one way. Therefore, a *cis-trans* isomer does not exist.
 b. 2-Heptene can be drawn two ways. Therefore, *cis-trans* isomers do exist.
 c. 3-Heptene can be drawn two ways. Therefore, *cis-trans* isomers do exist.
 d. 2-Methyl-2-hexene can only be drawn one way. Therefore, *cis-trans* isomers do not exist.
 e. 3-Methyl-2-hexene can be drawn two ways. Therefore, *cis-trans* isomers do exist.

11.49 Alkenes b and c would not exhibit *cis-trans* isomerism.

11.51 Alkenes b and d can exist as both *cis-* and *trans-* isomers.

11.53 **a.** 1,5-Nonadiene   **c.** 2,5-Octadiene
 b. 1,4,7-Nonatriene   **d.** 4-Methyl-2,5-heptadiene

11.55
R, R / C=C / R, R + H$_2$ →(Pt, Pd, or Ni / heat or pressure)→ H R / R—C—C—R / R H

11.57
R, R / C=C / R, R + X$_2$ → X R / R—C—C—R / R X

11.59
R, R / C=C / R, R + H$_2$O →(H$^+$)→ H R / R—C—C—R / R OH

11.61 The primary difference between complete hydrogenation of an alkene and an alkyne is that 2 moles of H$_2$ are required for the complete hydrogenation of an alkyne.

11.63
 a. CH$_2$=CH(CH$_2$)$_4$CH$_3$ + H$_2$O →(H$^+$)→ CH$_3$CHOH(CH$_2$)$_4$CH$_3$ Major Product + CH$_2$OH(CH$_2$)$_5$CH$_3$ Minor Product

 b. CH$_3$CH=CH(CH$_2$)$_3$CH$_3$ + HBr → CH$_3$CH$_2$CHBr(CH$_2$)$_3$CH$_3$ + CH$_3$CHBr(CH$_2$)$_4$CH$_3$

 c. CH$_3$CH$_2$CH=CH(CH$_2$)$_2$CH$_3$ + H$_2$ →(Pt, Pd, or Ni / heat or pressure)→ CH$_3$(CH$_2$)$_5$CH$_3$

 d. CH$_3$C=CHCH$_2$CH$_2$CH$_3$ | CH$_3$ + HCl →
 Cl | CH$_3$C(CH$_2$)$_3$CH$_3$ | CH$_3$ Major Product
 + CH$_3$ | CH$_3$CHCH(CH$_2$)$_2$CH$_3$ | Cl Minor Product

11.65 **a.** H$_2$ **d.** 19O$_2$ → 12CO$_2$ + 14H$_2$O
 b. H$_2$O   **e.** Cl$_2$
 c. HBr   **f.**

11.67 a.

$$H_3CC{\equiv}CCH_3 + 2H_2 \xrightarrow[\text{heat or pressure}]{\text{Pt, Pd, or Ni}} H_3C{-}CH_2{-}CH_2{-}CH_3$$

2-Butyne

b.

$$CH_3CH_2C{\equiv}CCH_3 + 2X_2 \longrightarrow CH_3CH_2{-}CX_2{-}CX_2{-}CH_3$$

2-Pentyne

11.69 **a.** Reactant—*cis*-2-butene; Only product—butane
b. Reactant—1-butene; Major product—2-butanol
c. Reactant—2-butene; Only product—2,3-dichlorobutane
d. Reactant—1-pentene; Major product— 2-bromopentane

11.71 a. Hydrogenation of 4-chlorocyclooctene

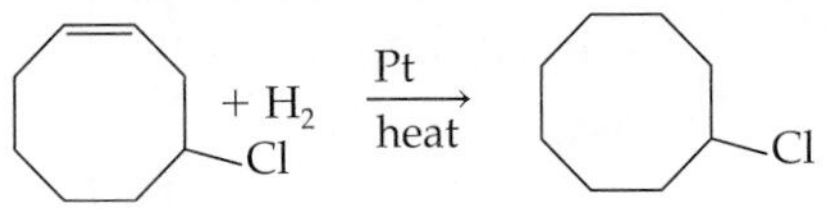

b. Halogenation of 1,3-cyclooctadiene

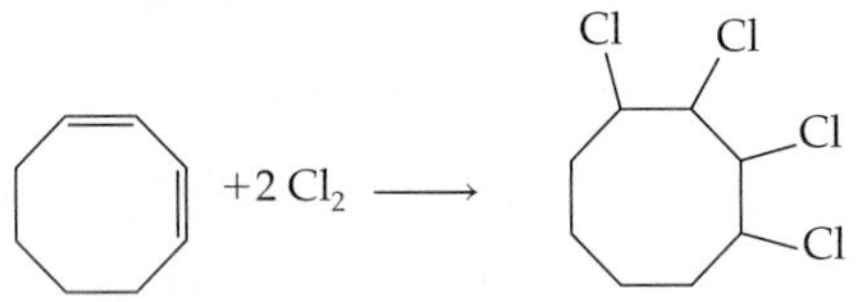

c. Hydration of 3-methylcyclobutene

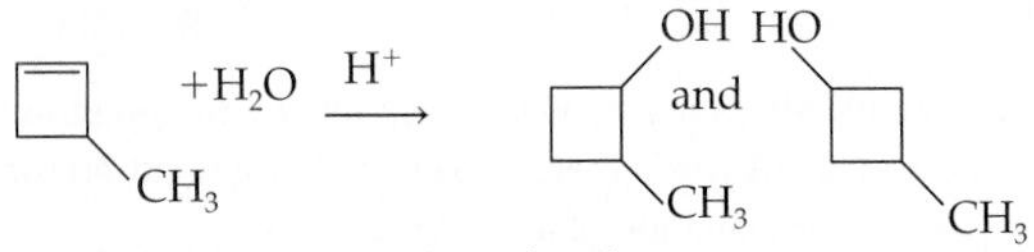

d. Hydrohalogenation of cyclopentene

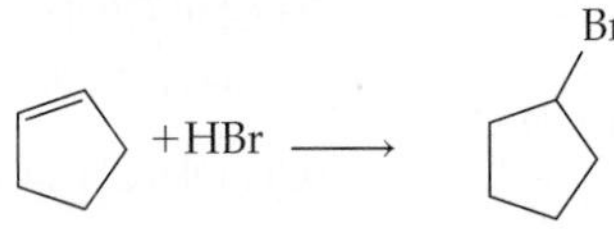

11.73 $CH_2{=}CHCH_2CH_2CH_3$, $CH_3CH{=}CHCH_2CH_3$, $CH_3C(CH_3){=}CHCH_3$, $CH_2{=}C(CH_3)CH_2CH_3$, $CH_2{=}CHCH(CH_3)CH_3$

11.75 a. $CH_3CHBrCH_2CH_3$

b. $CH_3CH_2{-}CI(CH_3){-}CH_2CH_2CH_3 + CH_3CHICH(CH_3)CH_2CH_2CH_3$

(Major product) (Minor product)

c.

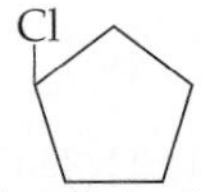

11.77 A polymer is a macromolecule composed of repeating structural units called *monomers*.

11.79 Polyvinyl chloride (PVC) is used in pipes, detergent bottles, and cleanser bottles.

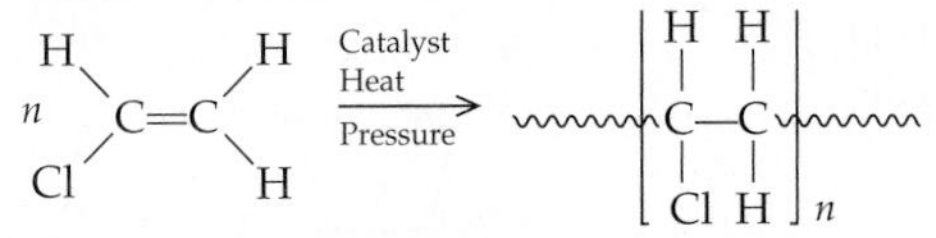

11.81 The IUPAC name for (a) is 2-pentene, for (b) is 3-bromo-1-propene, and for (c) is 3,4-dimethylcyclohexene.

a. These products will be formed in approximately equal amounts.

$$CH_3CH{=}CHCH_2CH_3 + H_2O \xrightarrow{H^+} CH_3CHOHCH_2CH_2CH_3$$

$$CH_3CH{=}CHCH_2CH_3 + H_2O \xrightarrow{H^+} CH_3CH_2CHOHCH_2CH_3$$

b. $CH_3BrCH{=}CH_2 + H_2O \xrightarrow{H^+}$

$CH_2BrCH_2CH_2OH + CH_2BrCHOHCH_3$

Minor product Major product

c. These products will be formed in approximately equal amounts.

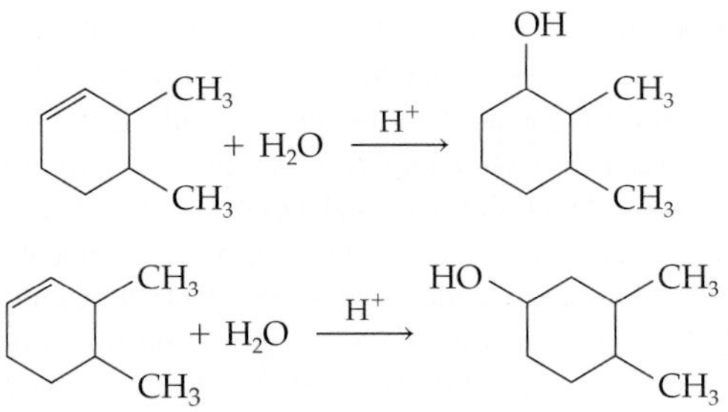

11.83 **a.** This is the minor product of this reaction.

$$H_2C{=}CHCH_2CH(CH_3)_2 + H_2O \xrightarrow{H^+} CH_2OHCH_2CH_2CH(CH_3)_2$$

b. $CH_3CH{=}CHCH_2CH_2CH_3 + HBr \xrightarrow{H^+} CH_3CH_2CHBrCH_2CH_2CH_3$

or

$CH_3CH_2CH{=}CHCH_2CH_3 + HBr \longrightarrow CH_3CH_2CHBrCH_2CH_2CH_3$

c.

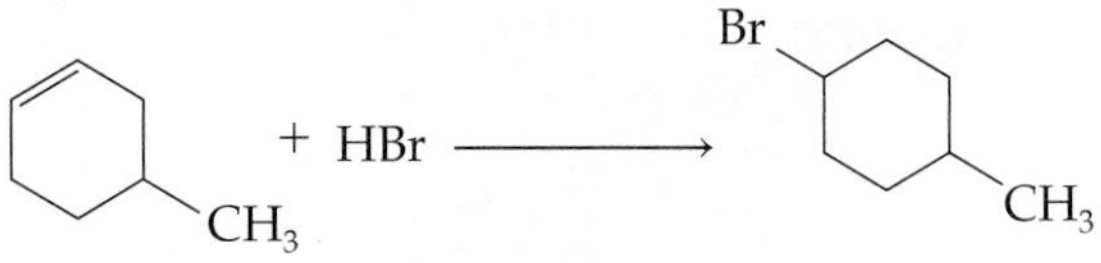

d.

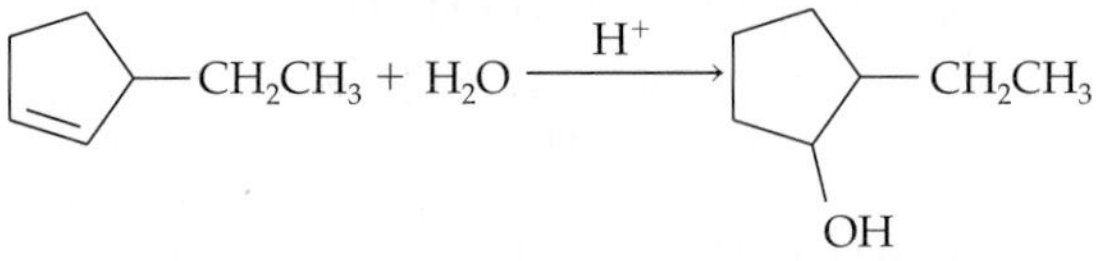

11.85 **a.**

$$CH_2{=}CHCH_2CH{=}CHCH_3 + 2H_2 \xrightarrow[\text{heat}]{\text{Pt}} CH_3(CH_2)_4CH_3$$

1,4-Hexadiene Hexane

b.

$$CH_3CH{=}CHCH{=}CHCH{=}CHCH_3 + 3H_2 \xrightarrow[\text{heat}]{\text{Ni}} CH_3(CH_2)_6CH_3$$

2,4,6-Octatriene Octane

c.

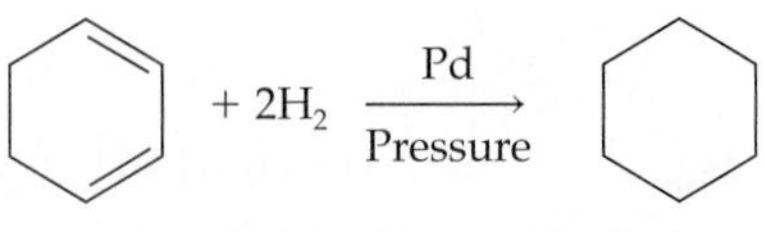

1,3-Cyclohexadiene Cyclohexane

d.

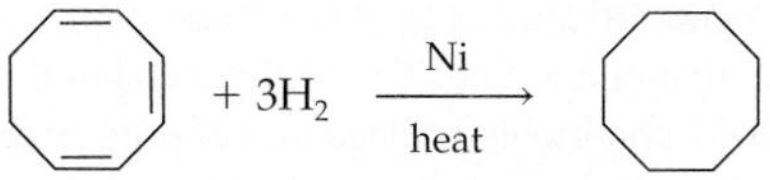

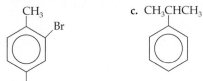

1,3,5-Cyclooctatriene   Cyclooctane

11.87   The term aromatic hydrocarbon was first used as a term to describe the pleasant-smelling resins of tropical trees.

11.89   Resonance hybrids are molecules for which more than one valid Lewis structure can be written.

11.91   **a.**

c. CH_3CHCH_3

b.

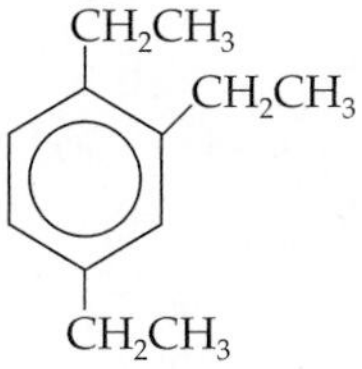

d.

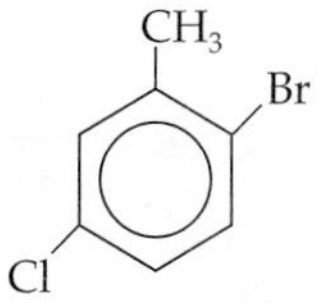

11.93   **a.**   OH

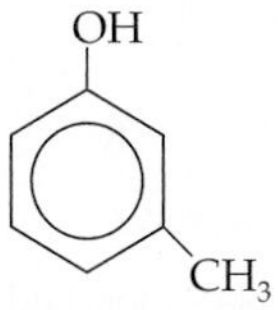

c. NO_2

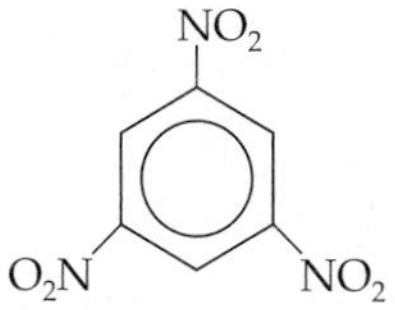

b. $CH_2CH_2CH_3$

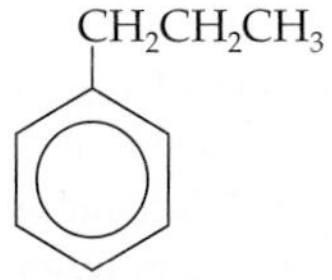

d. CH_3

11.95   Kekulé proposed that single and double carbon-carbon bonds alternate around the benzene ring. To explain why benzene does not react like other unsaturated compounds, he proposed that the double and single bonds shift positions rapidly.

11.97   An addition reaction involves addition of a molecule to a double or triple bond in an unsaturated molecule. In a substitution reaction, one chemical group replaces another.

11.99   **a.**

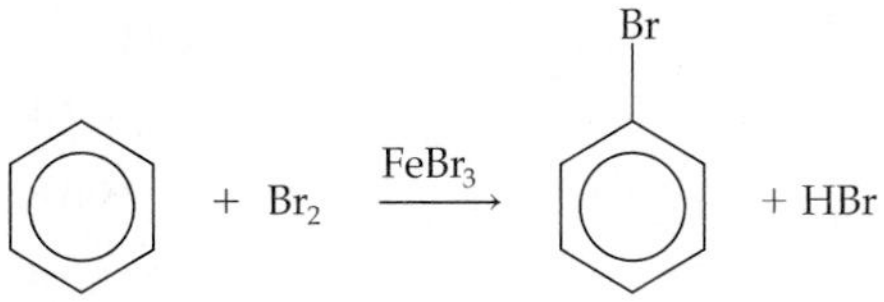

b.

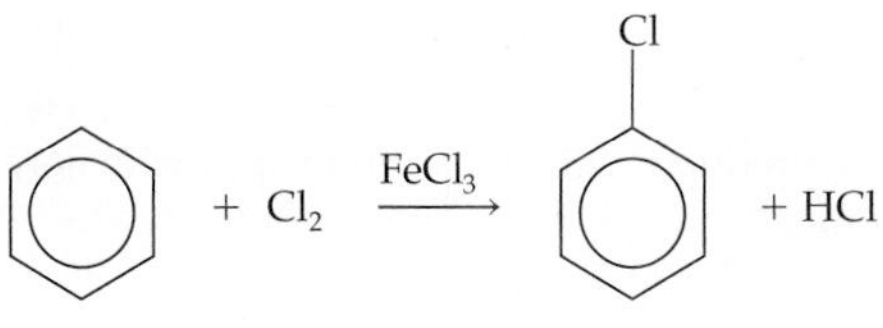

c.

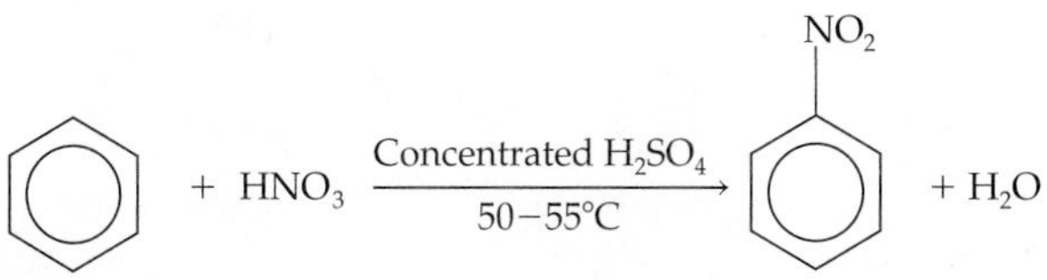

11.101

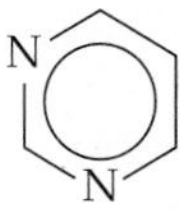

Pyrimidine

11.103

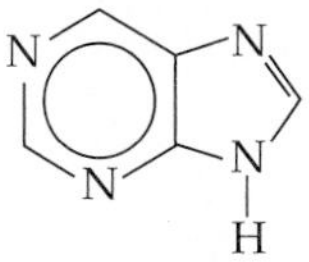

Purine

Chapter 12

12.1   **a.** 2-Methyl-1-propanol
CH_3CHCH_2OH
CH_3

b. 2-Chlorocyclopentanol

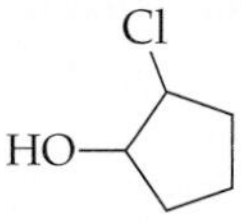

c. 2,4-Dimethylcyclohexanol

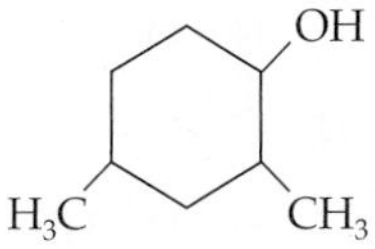

d. 2,3-Dichloro-3-hexanol
$CH_3CHCCH_2CH_2CH_3$

12.3   IUPAC name: 1-Butanol
Common name: Butyl alcohol
Primary alcohol

12.5   **a.** Ethanol is a primary alcohol. The product is ethene.
b. 2-Propanol is a secondary alcohol. The product is propene.
c. 4-Methyl-3-hexanol is a secondary alcohol. The products are 3-methyl-3-hexene and 4-methyl-2-hexene.
d. 2-Methyl-2-propanol is a tertiary alcohol. The product is 2-methyl-1-propene.

12.7   **a.** The reactant is 2-butanol and the product is butanone.
b. The reactant is 2-pentanol and the product is 2-pentanone.

12.9   Simple phenols are somewhat soluble in water because they have the polar hydroxyl group.

12.11   Ethers have much lower boiling points than alcohols because ether molecules cannot hydrogen bond to one another.

12.13   The longer the hydrocarbon tail of an alcohol becomes, the less water soluble it will be.

12.15   The carbinol carbon is the one to which the hydroxyl group is bonded.

12.17   **a.** Primary alcohol   **d.** Tertiary alcohol
b. Secondary alcohol   **e.** Tertiary alcohol
c. Tertiary alcohol

12.19   **a.** Primary alcohol   **d.** Primary alcohol
b. Secondary alcohol   **e.** Secondary alcohol
c. Primary alcohol

12.21 a. 2-Nonanol is a secondary alcohol:
$CH_3CHOHCH_2CH_2CH_2CH_2CH_2CH_3$
b. 2-Heptanol is a secondary alcohol:
$CH_3CHOHCH_2CH_2CH_2CH_2CH_3$
c. 2-Undecanol is a secondary alcohol:
$CH_3CHOHCH_2CH_2CH_2CH_2CH_2CH_2CH_2CH_3$

12.23   a < d < c < b

12.25 a. CH_3CH_2OH
b. $CH_3CH_2CH_2CH_2OH$
c. CH_3CHCH_3
　　$|$
　　OH

12.27   The IUPAC rules for the nomenclature of alcohols require you to name the parent compound, that is the longest continuous carbon chain bonded to the —OH group. Replace the —e ending of the parent alkane with –ol of the alcohol. Number the parent chain so that the carbon bearing the hydroxyl group has the lowest possible number. Name and number all other substituents. If there is more than one hydroxyl group, the –ol ending will be modified to reflect the number. If there are two —OH groups, the suffix –diol is used; if it has three —OH groups, the suffix –triol is used, etc.

12.29 a. 1,4-Hexanediol
b. 2,3-Pentanediol
c. 2-Methyl-3-pentanol

12.31 a.

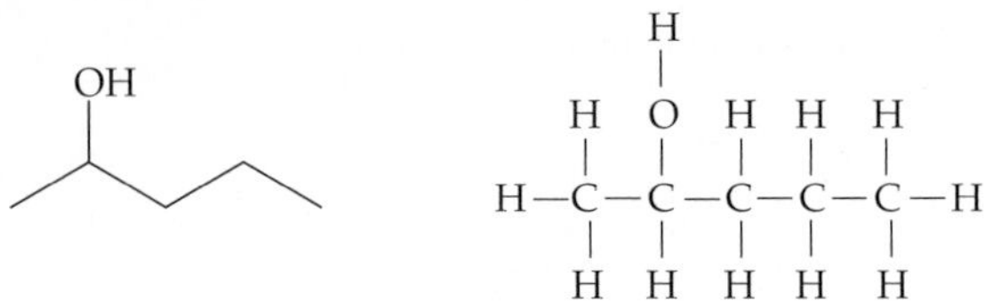

b.

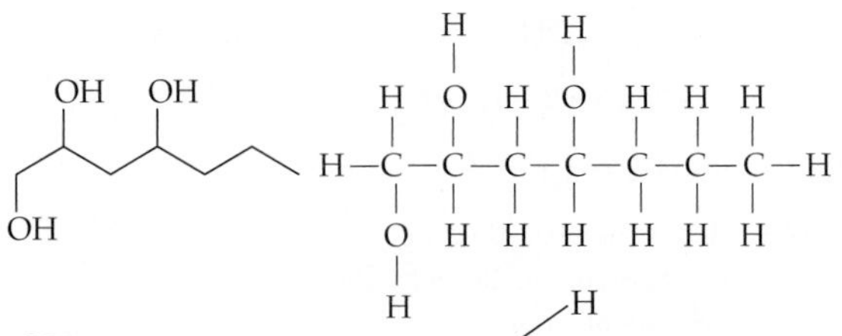

c.

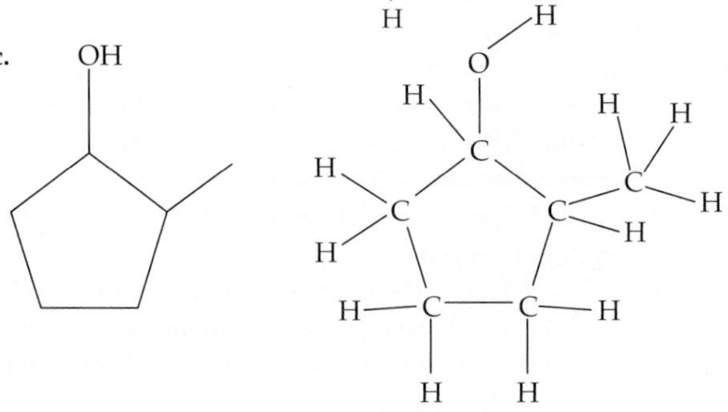

12.33 a. Cyclopentanol　　**c.** 3-Methylcyclohexanol
b. Cycloheptanol
12.35 a. Methyl alcohol　　**c.** Ethylene glycol
b. Ethyl alcohol　　**d.** Propyl alcohol
12.37 a. $CH_3CHOHCH_2CH(CH_3)CH_2CH_3$　**d.** $CH_3CHOH(CH_2)_6CH_3$
b. $CH(CH_3)_2CH_2OH$　　**e.**
c. $CH_2OH(CH_2)_3CH_2OH$

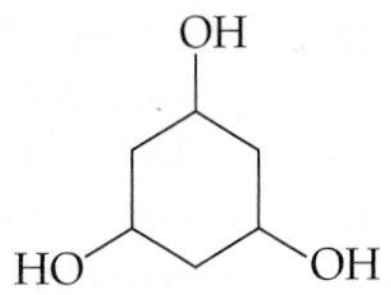

12.39   Denatured alcohol is 100% ethanol to which benzene or methanol is added. The additive makes the ethanol unfit to drink and prevents illegal use of pure ethanol.

12.41   Fermentation is the anaerobic degradation of sugar that involves no net oxidation. The alcohol fermentation, carried out by yeast, produces ethanol and carbon dioxide.

12.43   When the ethanol concentration in a fermentation reaches 12–13%, the yeast producing the ethanol are killed by it. To produce a liquor of higher alcohol concentration, the product of the original fermentation must be distilled.

12.45

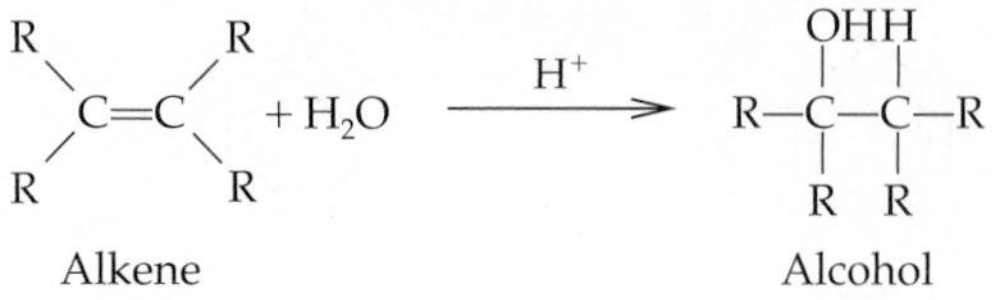

12.47

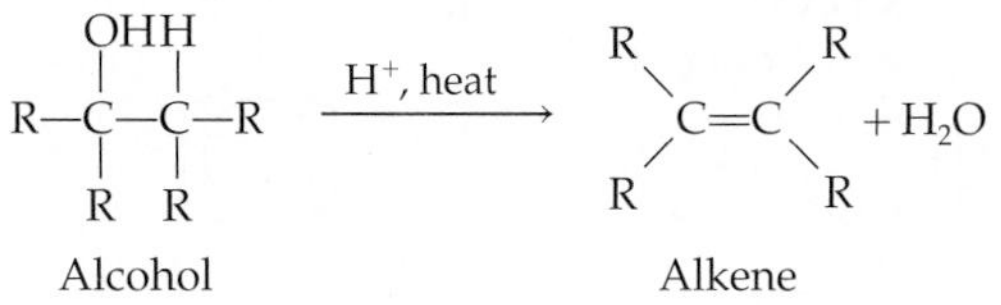

12.49

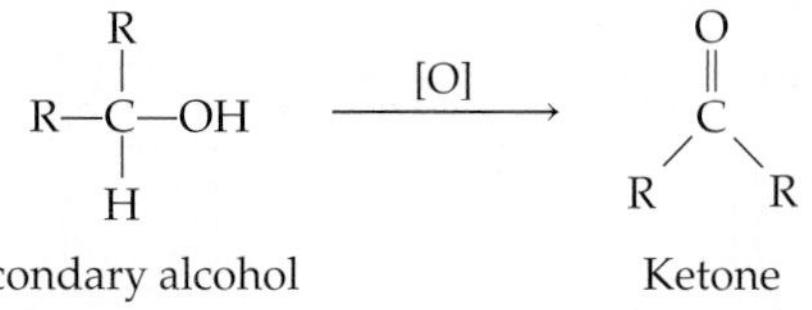

12.51 a. The predicted products are 1-hexanol (minor) and 2-hexanol (major).
b. The predicted products are 2-hexanol and 3-hexanol. These products will be formed in approximately equal amounts.
c. The predicted products are 5-methyl-3-hexanol and 2-methyl-3-hexanol. These products will be formed in approximately equal amounts.
d. The predicted products are 2,2-dimethyl-4-heptanol and 2,2-dimethyl-3-heptanol. These products will be formed in approximately equal amounts.

12.53 a.

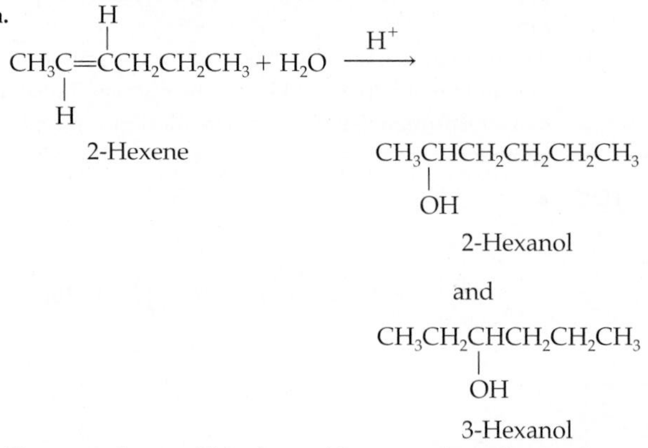

These products will be formed in approximately equal amounts.

b.

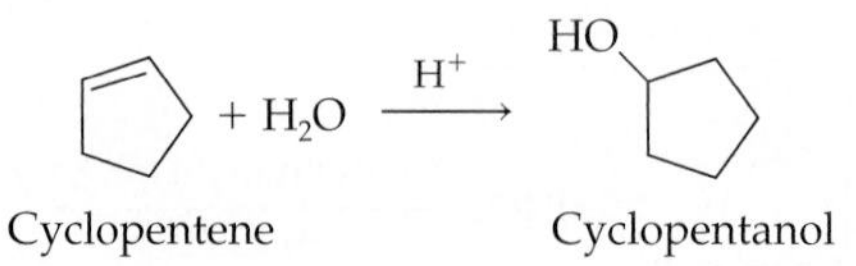

c. $CH_2{=}CHCH_2CH_2CH_2CH_2CH_2CH_3 + H_2O \xrightarrow{H^+}$

1-Octene

$\underset{\underset{OH}{|}}{CH_3CHCH_2CH_2CH_2CH_2CH_2CH_3}$

2-Octanol
(Major product)
and

$\underset{\underset{OH}{|}}{CH_2CH_2CH_2CH_2CH_2CH_2CH_2CH_3}$

1-Octanol
(Minor product)

d.

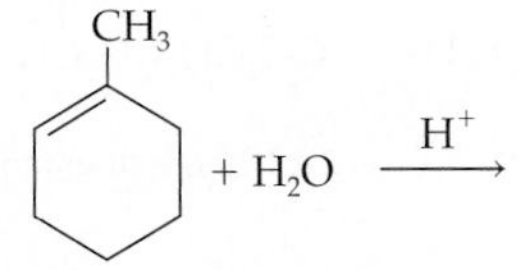

1-Methylcyclohexene

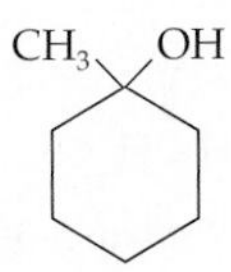

1-Methylcyclohexanol   2-Methylcyclohexanol
(Major product)   (Minor product)

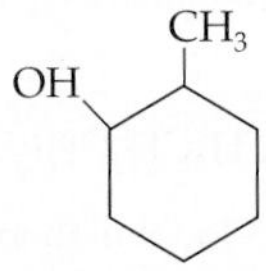

12.55 **a.** Butanone
   b. N.R.
   c. Cyclohexanone
   d. N.R.
12.57 **a.** 3-Pentanol; 3-Pentanone
   b. 1-Propanol; Propanal (Upon further oxidation, propanoic acid would be formed.)
   c. 4-Methyl-2-pentanol; 4-Methyl-2-pentanone
   d. 2-Methyl-2-butanol; N.R.
   e. 3-Phenyl-1-propanol; 3-Phenylpropanal (Upon further oxidation, 3-phenylpropanoic acid will be formed.)
12.59

$CH_3CH_2OH \xrightarrow{\text{liver enzymes}} \underset{O}{\overset{O}{CH_3-\overset{\|}{C}-H}}$

Ethanol   Ethanal

The product, ethanal, is responsible for the symptoms of a hangover.
12.61 The reaction in which a water molecule is added to 1-butene is a hydration reaction.

$CH_3CH_2CH{=}CH_2 + H_2O \xrightarrow{H^+} \underset{\underset{OH}{|}}{CH_3CH_2CHCH_3}$

1-Butene   2-Butanol

12.63

$CH_3CH{=}CH_2 \xrightarrow{H_2O,\,H^+} \underset{\underset{OH}{|}}{CH_3CHCH_3} \xrightarrow{[O]} \underset{\overset{O}{\|}}{CH_3CCH_3}$

Propene   2-Propanol   Propanone
(propylene)   (isopropanol)   (acetone)

12.65 **a.**

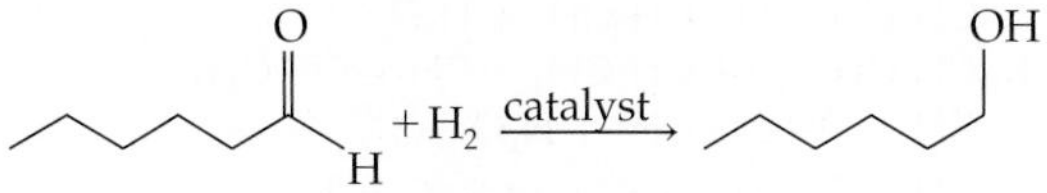

b.

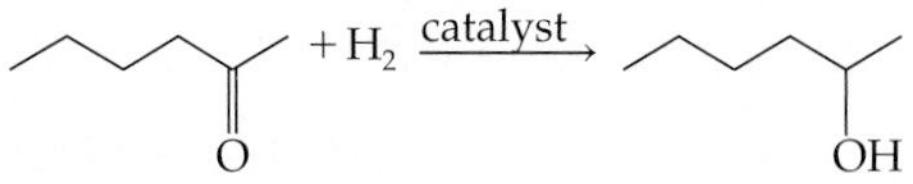

c.

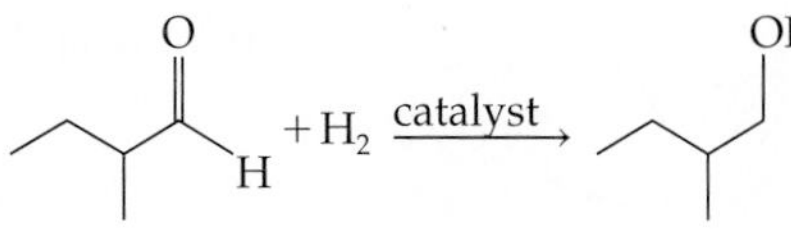

d.

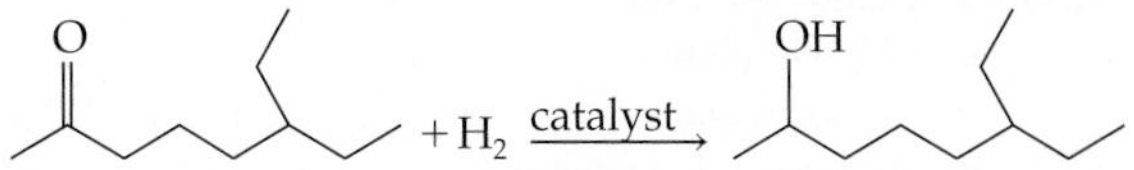

12.67 Oxidation is a loss of electrons, whereas reduction is a gain of electrons.
12.69

$CH_3CH_2CH_3 < CH_3CH_2CH_2OH < CH_3CH_2\overset{O}{\overset{\|}{C}}{-}H < CH_3CH_2\overset{O}{\overset{\|}{C}}{-}OH$

12.71 Phenols are compounds with an —OH attached to a benzene ring.

12.73   Picric acid: 2,4,6,-Trinitrotoluene:

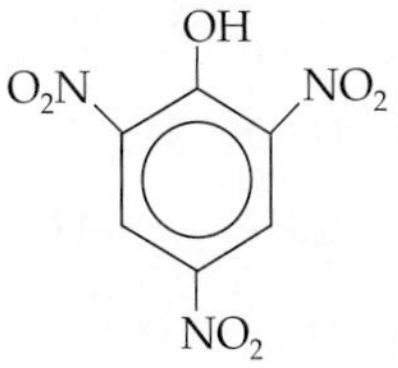

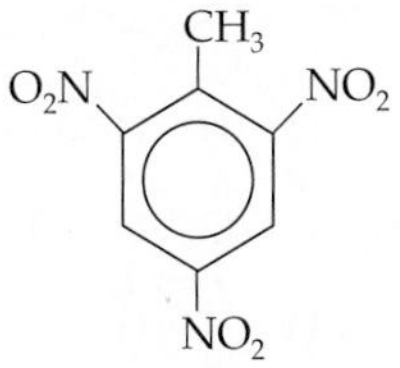

Picric acid is water-soluble because of the polar hydroxyl group that can form hydrogen bonds with water.
12.75 Hexachlorophene, hexylresorcinol, and o-phenylphenol are phenol compounds used as antiseptics or disinfectants.
12.77 Ethers have much lower boiling points than alcohols of similar molar mass, but higher boiling points than alkanes of similar molar mass. The boiling points are higher than alkanes because the R—O—R bond is polar. However, there is no —OH group, so ether molecules cannot hydrogen bond to one another. This is the reason that the boiling points are lower than alcohols of similar molar mass.
12.79 Alcohols of molecular formula $C_4H_{10}O$

$CH_3CH_2CH_2CH_2OH,$ $\underset{\underset{OH}{|}}{CH_3CHCH_2CH_3},$

$\underset{\underset{CH_3}{|}}{CH_3CHCH_2OH},$ $\underset{\underset{CH_3}{|}}{\overset{\overset{OH}{|}}{CH_3-C-CH_3}}$

Ethers of molecular formula $C_4H_{10}O$
$CH_3{-}O{-}CH_2CH_2CH_3$ $CH_3CH_2{-}O{-}CH_2CH_3$
$\underset{\underset{CH_3}{|}}{CH_3{-}O{-}CHCH_3}$

12.81 **a.** $CH_3CH_2{-}O{-}CH_2CH_3 + H_2O$
b. $CH_3CH_2{-}O{-}CH_2CH_3 + CH_3{-}O{-}CH_3 + CH_3{-}O{-}CH_2CH_3 + H_2O$
c. $CH_3{-}O{-}CH_3 + CH_3{-}O{-}\underset{\underset{CH_3}{|}}{C}HCH_3 + CH_3\underset{\underset{CH_3}{|}}{C}H{-}O{-}\underset{\underset{CH_3}{|}}{C}HCH_3 + H_2O$
d. 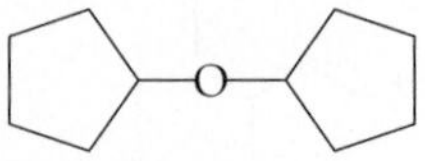 $+ H_2O$

12.83 **a.** 2-Ethoxypentane
b. 2-Methoxybutane
c. 1-Ethoxybutane
d. Methoxycyclopentane

12.85 **a.** 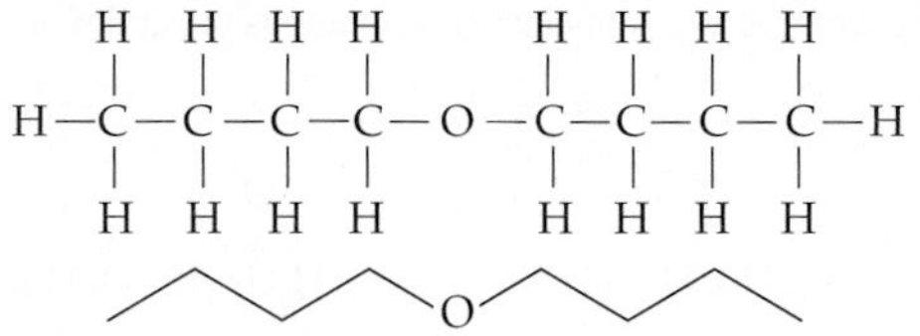

b. 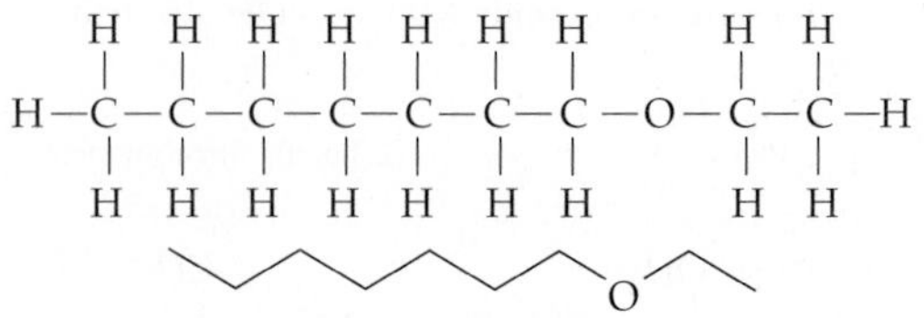

c. 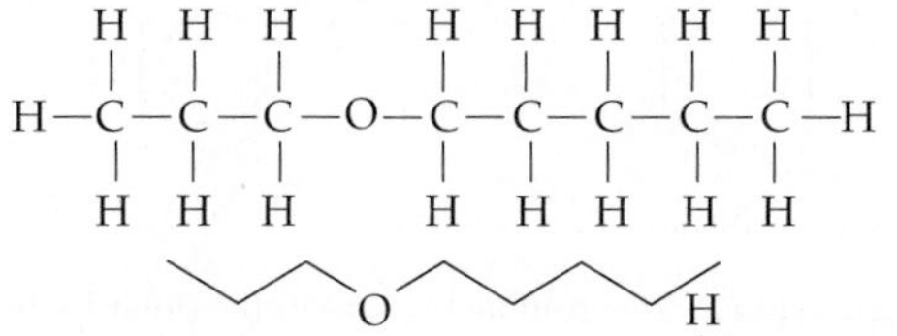

d. 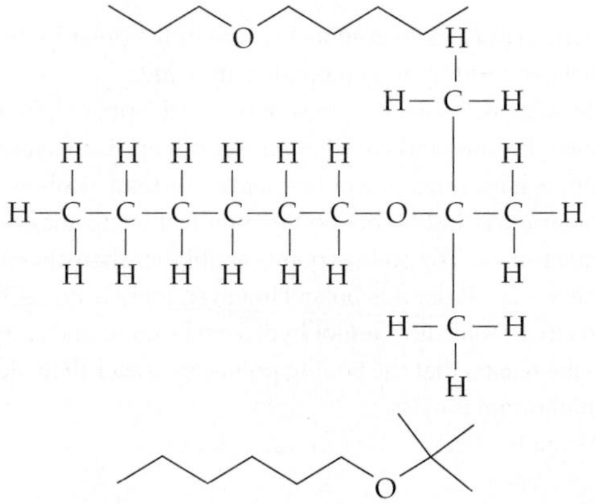

12.87 Thiols contain the sulfhydryl group (—SH). The sulfhydryl group is similar to the hydroxyl group (—OH) of alcohols, except that a sulfur atom replaces the oxygen atom.

12.89 Cystine: 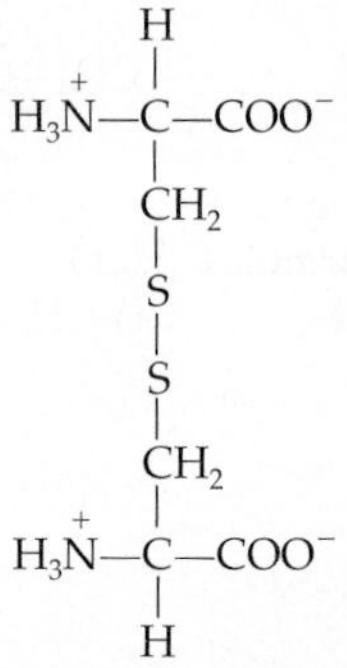

12.91 **a.** 1-Propanethiol **b.** 2-Butanethiol **c.** 2-Methyl-2-butanethiol **d.** 1,4-Cyclohexanedithiol

Chapter 13

13.1 **a.** $CH_3{-}\overset{\overset{O}{\|}}{C}{-}CH_3$ **b.** $CH_3\overset{\overset{OH}{|}}{C}HCH_2CH_2CH_3$

13.3 **a.** $CH_3CH_2\overset{\overset{O}{\|}}{C}{-}OH$ **b.** $CH_3\overset{\overset{O}{\|}}{C}{-}OH$

13.5 **a.** 2,3-Dichloropentanal
$CH_3CH_2CHClCHCl\overset{\overset{O}{\|}}{C}H$
b. 2-Bromobutanal
$CH_3CH_2CHBr\overset{\overset{O}{\|}}{C}H$
c. 4-Methylhexanal
$CH_3CH_2CH(CH_3)CH_2CH_2\overset{\overset{O}{\|}}{C}H$
d. Butanal
$CH_3CH_2CH_2\overset{\overset{O}{\|}}{C}H$
e. 2,4-Dimethylpentanal
$CH_3CH(CH_3)CH_2CH(CH_3)\overset{\overset{O}{\|}}{C}H$

13.7 **a.** 3-Iodobutanone **b.** 4-Methyl-2-octanone **c.** 3-Methylbutanone **d.** 2-Methyl-3-pentanone **e.** 2-Fluoro-3-pentanone

13.9 $CH_3{-}\overset{\overset{O}{\|}}{C}{-}H$

13.11 **a.** Reduction **b.** Reduction **c.** Reduction **d.** Oxidation **e.** Reduction

13.13 **a.** Hemiacetal **b.** Acetal **c.** Acetal **d.** Hemiacetal

13.15 As the carbon chain length increases, the compounds become less polar and more hydrocarbon-like. As a result, their solubility in water decreases.

13.17 A good solvent should dissolve a wide range of compounds. Simple ketones are considered to be universal solvents because they have both a polar carbonyl group and nonpolar side chains. As a result, they dissolve organic compounds and are also miscible in water.

13.19 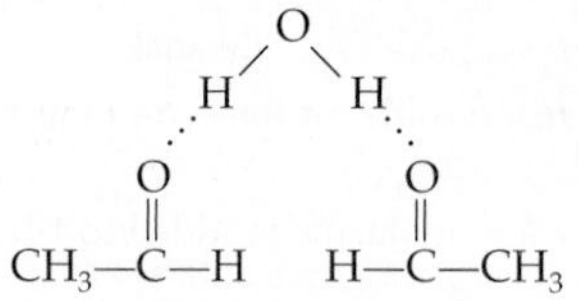

13.21 Alcohols have higher boiling points than aldehydes or ketones of comparable molar mass because alcohol molecules can form intermolecular hydrogen bonds with one another. Aldehydes and ketones cannot form intermolecular hydrogen bonds.

13.23 **a.** 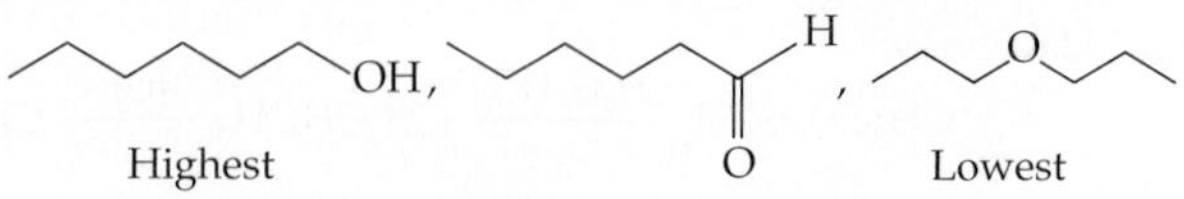
Highest Lowest

b. 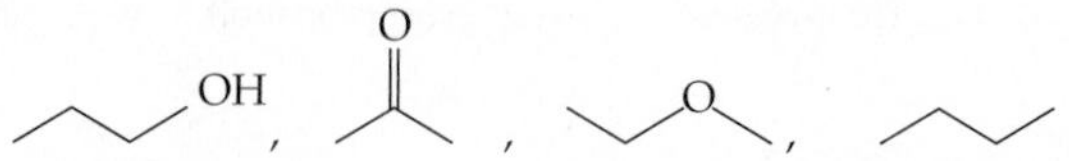
Highest Lowest

13.25 To name an aldehyde using the IUPAC Nomenclature System, identify and name the longest carbon chain containing the carbonyl group. Replace the final *-e* of the alkane name with *-al*. Number and name all substituents as usual. Remember that the carbonyl carbon is always carbon-1 and does not need to be numbered in the name of the compound.

13.27 The common names of aldehydes are derived from the same Latin roots as the corresponding carboxylic acids. For instance, methanal is formaldehyde; ethanal is acetaldehyde; propanal is propionaldehyde, *etc.*

Substituted aldehydes are named as derivatives of the straight-chain parent compound. Greek letters are used to indicate the position of substituents. The carbon nearest the carbonyl group is the α-carbon, the next is the β-carbon, and so on.

13.29 a.

CH_3CH (with O double bonded)   (acetaldehyde structure with O and H)

b.

$(CH_3)_2CHCH(CH_3)CH_2CH$ (with O)   (branched aldehyde structure)

c.

$CH_3(CH_2)_4CH(CH_2CH_3)CH$ (with O)   (branched aldehyde structure)

d.

$CH_2ClCH_2CHCl(CH_2)_3CH$ (with O)   (Cl-substituted aldehyde structure)

13.31 a. $CH_3CH(CH_3)CCH_2CH_3$ (with O)   (ketone structure)

b. $CH_3(CH_2)_2CCH_2CH_3$ (with O)   (ketone structure)

c. $CH_3(CH_2)_3C(CH_2)_3CH_3$ (with O)   (ketone structure)

d. $CH_3(CH_2)_5C(CH_2)_6CH_3$ (with O)   (ketone structure)

13.33 a. Butanone   **b.** 2-Ethylhexanal

13.35 a. 3-Nitrobenzaldehyde   **b.** 3,4-Dihydroxycyclopentanone

13.37 7-Hydroxy-3,7-dimethyloctanal

13.39 a. 4,6-Dimethyl-3-heptanone   **b.** 3,3-Dimethylcyclopentanone

13.41 a. Acetone   **d.** Propionaldehyde
    b. Ethyl methyl ketone   **e.** Methyl isopropyl ketone
    c. Acetaldehyde

13.43 a. 3-Hydroxybutanal   **b.** 2-Methylpentanal

$CH_3CHOHCH_2CH$ (with O) $CH_3(CH_2)_2CH(CH_3)CH$ (with O)

c. 4-Bromohexanal   **d.** 3-Iodopentanal

$CH_3CH_2CHBr(CH_2)_2CH$ (with O) $CH_3CH_2CHICH_2CH$ (with O)

e. 2-Hydroxy-3-methylheptanal

$CH_3(CH_2)_3CH(CH_3)CHOHCH$ (with O)

13.45 Acetone is a good solvent because it can dissolve a wide range of compounds. It has both a polar carbonyl group and nonpolar side chains. As a result, it dissolves organic compounds and is also miscible in water.

13.47 The liver

13.49 In organic molecules, oxidation may be recognized as a gain of oxygen or a loss of hydrogen. An aldehyde may be oxidized to form a carboxylic acid as in the following example in which ethanal is oxidized to produce ethanoic acid.

$$H_3C-\overset{O}{\underset{\|}{C}}-H \xrightarrow{[O]} H_3C-\overset{O}{\underset{\|}{C}}-OH$$

Ethanal   Ethanoic acid

13.51 Addition reactions of aldehydes or ketones are those in which a second molecule is added to the double bond of the carbonyl group.

13.53 The following equation represents the oxidation of an aldehyde. The product is a carboxylic acid.

$$R-\overset{O}{\underset{\|}{C}}-H \xrightarrow{[O]} R-\overset{O}{\underset{\|}{C}}-OH$$

Aldehyde   Carboxylic acid

13.55 The following general equation represents the addition of one alcohol molecule to an aldehyde:

$$R-\overset{O}{\underset{\|}{C}}-H + R'OH \underset{}{\overset{H^+}{\rightleftharpoons}} R-\overset{OH}{\underset{H}{\overset{|}{\underset{|}{C}}}}-OR'$$

Aldehyde   Hemiacetal

The following general equation represents the addition of one alcohol molecule to a ketone:

$$R-\overset{O}{\underset{\|}{C}}-R + R'OH \underset{}{\overset{H^+}{\rightleftharpoons}} R-\overset{OH}{\underset{R}{\overset{|}{\underset{|}{C}}}}-OR'$$

Ketone   Hemiacetal

13.57 a.   OH   13.57

(structure) $\xrightarrow{[O]}$ (structure)

4-Methyl-2-heptanol   4-Methyl-2-heptanone

b.

(structure) OH $\xrightarrow{[O]}$ (structure) O ... H

3,4-Dimethyl-1-pentanol   3,4-Dimethylpentanal

c.   OH

(structure) $\xrightarrow{[O]}$ (structure) O

4-Ethyl-2-heptanol   4-Ethyl-2-heptanone

d.

(structure) OH Cl Cl $\xrightarrow{[O]}$ (structure) O Cl Cl

5,7-Dichloro-3-heptanol   5,7-Dichloro-3-heptanone

13.59 $R-CH_2OH \xrightarrow{[O]} R-\overset{O}{\underset{\|}{C}}-H \xrightarrow{[O]} R-\overset{O}{\underset{\|}{C}}-OH$

Primary   Aldehyde   Carboxylic
alcohol   acid

13.61 a. Reduction reaction

$$CH_3-\overset{\overset{\displaystyle O}{\|}}{C}-H + H_2 \xrightarrow{Pt} CH_3CH_2OH$$

Ethanal   Ethanol

b. Reduction reaction

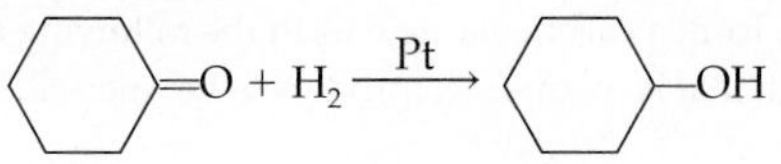

=O + H₂ →(Pt) —OH

Cyclohexanone   Cyclohexanol

c. Oxidation reaction

$$CH_3\overset{\overset{\displaystyle OH}{|}}{C}HCH_3 \xrightarrow{[O]} CH_3-\overset{\overset{\displaystyle O}{\|}}{C}-CH_3$$

2-Propanol   Propanone

13.63 a.

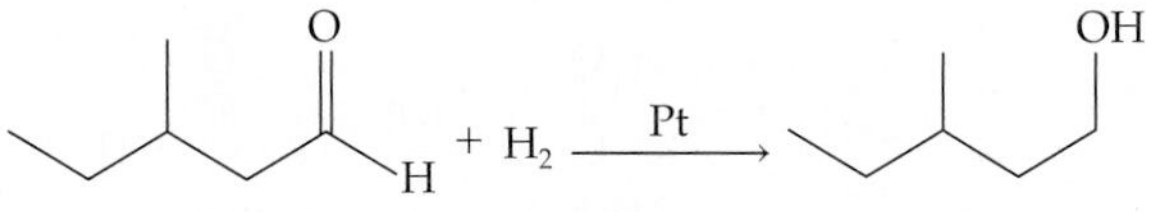

+ H₂ →(Pt)

b.

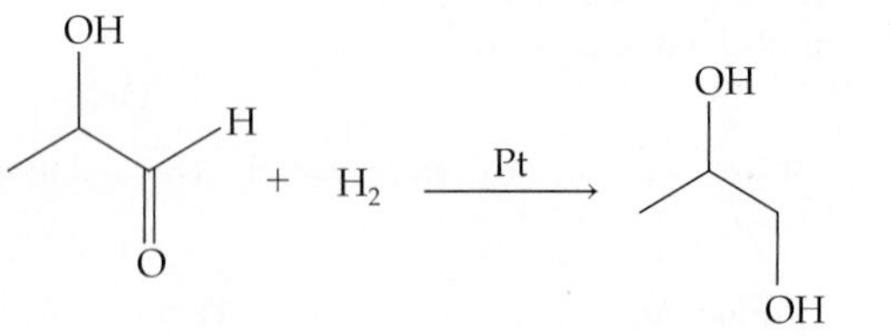

+ H₂ →(Pt)

c.

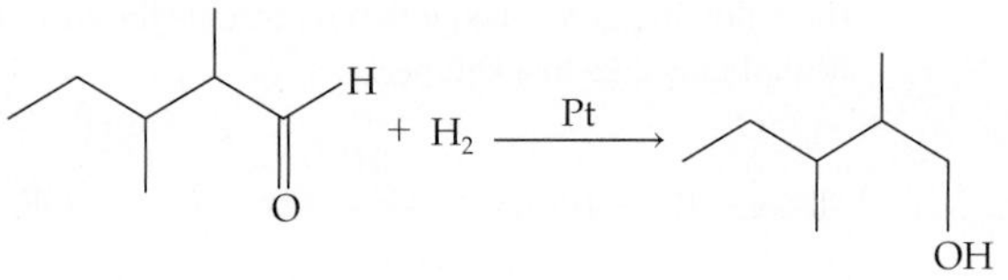

+ H₂ →(Pt)

d.

+ H₂ →(Pt)

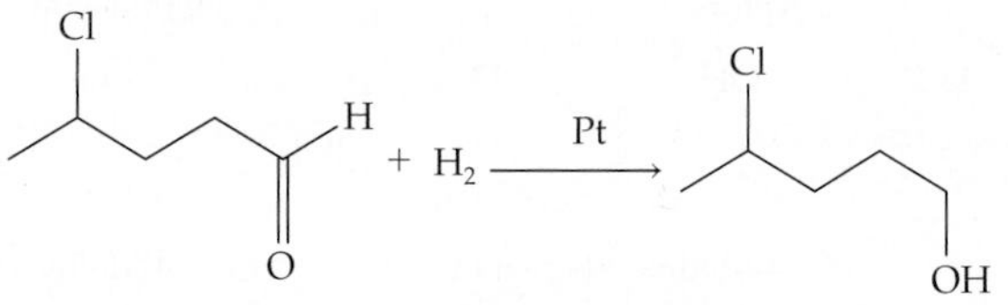

13.65 a.

$$CH_3CH_2CH_2\overset{\overset{\displaystyle O}{\|}}{C}H + H_2 \xrightarrow{Pt} CH_3CH_2CH_2CH_2OH$$

Butanal   1-Butanol

b.

$$CH_3CH_2\overset{\overset{\displaystyle}{|}}{\underset{\underset{\displaystyle CH_3}{|}}{C}}HCH_2\overset{\overset{\displaystyle O}{\|}}{C}H + H_2 \xrightarrow{Pt} CH_3CH_2\underset{\underset{\displaystyle CH_3}{|}}{C}HCH_2CH_2OH$$

3-Methylpentanal   3-Methyl-1-pentanol

c.

$$CH_3\underset{\underset{\displaystyle CH_3}{|}}{C}H\overset{\overset{\displaystyle O}{\|}}{C}H + H_2 \xrightarrow{Pt} CH_3\underset{\underset{\displaystyle CH_3}{|}}{C}HCH_2OH$$

2-Methylpropanal   2-Methyl-1-propanol

13.67 Only (c) 3-methylbutanal and (f) acetaldehyde would give a positive Tollens' test.

13.69 a.

$$CH_3CH_2\overset{\overset{\displaystyle O}{\|}}{C}H + CH_3CH_2OH \xrightarrow{H^+} CH_3CH_2-\underset{\underset{\displaystyle OCH_2CH_3}{|}}{\overset{\overset{\displaystyle OH}{|}}{C}}-H$$

b.

$$CH_3\overset{\overset{\displaystyle O}{\|}}{C}H + CH_3CH_2OH \xrightarrow{H^+} CH_3-\underset{\underset{\displaystyle OCH_2CH_3}{|}}{\overset{\overset{\displaystyle OH}{|}}{C}}-H$$

13.71 a.

$$CH_3CH_2\overset{\overset{\displaystyle O}{\|}}{C}H + 2\,CH_3OH \xrightarrow{H^+} CH_3CH_2-\underset{\underset{\displaystyle OCH_3}{|}}{\overset{\overset{\displaystyle OCH_3}{|}}{C}}-H + H_2O$$

b.

$$CH_3\overset{\overset{\displaystyle O}{\|}}{C}H + 2\,CH_3OH \xrightarrow{H^+} CH_3-\underset{\underset{\displaystyle OCH_3}{|}}{\overset{\overset{\displaystyle OCH_3}{|}}{C}}-H + H_2O$$

13.73 a.

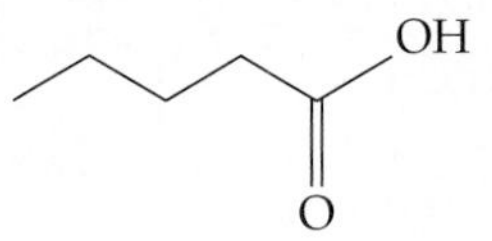

b.

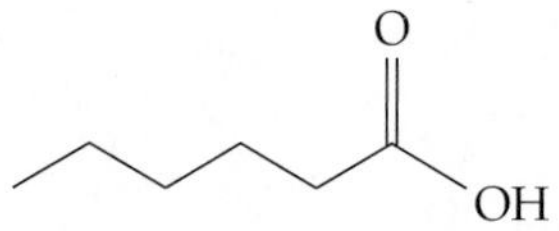

c.

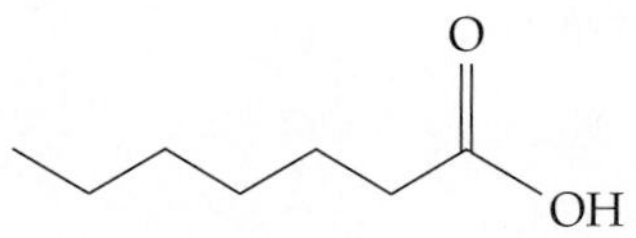

d.

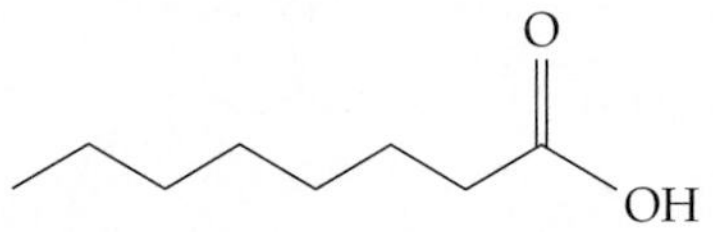

13.75 a. Methanal
b. Propanal

13.77 a. False
b. True
c. False
d. False

13.79

$$CH_3\overset{\overset{\displaystyle O}{\|}}{C}CH_3$$

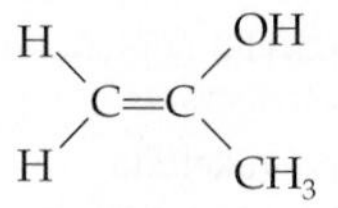

Keto form of   Enol form of
Propanone   Propanone

13.81 a.

$$CH_3CH_2CH_2-\underset{\underset{\displaystyle OCH_2CH_3}{|}}{\overset{\overset{\displaystyle OH}{|}}{C}}-CH_3$$

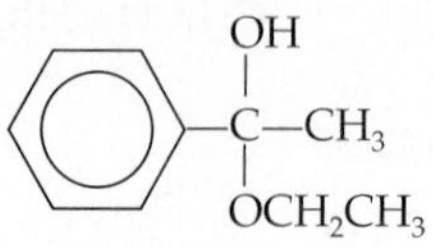

c.

b.

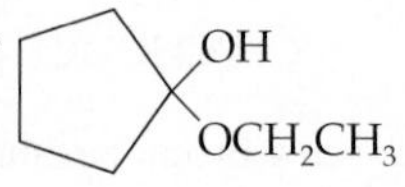

13.83 **(1)** $2CH_3CH_2OH$
(2) $KMnO_4/OH^-$
(3) $CH_3CH=CH_2$

Chapter 14

14.1 **a.** Ketone
b. Ketone
c. Alkane

14.3 The carboxyl group consists of two very polar groups, the carbonyl group and the hydroxyl group. Thus, carboxylic acids are very polar, in addition to which, they can hydrogen bond to one another. Aldehydes are polar, as a result of the carbonyl group, but cannot hydrogen bond to one another. As a result, carboxylic acids have higher boiling points than aldehydes of the same carbon chain length.

14.5 **a.** 3-Methylcyclohexanecarboxylic acid
b. 2-Ethylcyclopentanecarboxylic acid

14.7 **a.** $CH_3COOH + CH_3CH_2CH_2OH$
      Ethanoic acid   1-Propanol
b. $CH_3CH_2CH_2CH_2CH_2COO^-K^+ + CH_3CH_2CH_2OH$
      Potassium hexanoate   1-Propanol
c. $CH_3CH_2CH_2CH_2COO^-Na^+ + CH_3OH$
      Sodium pentanoate   Methanol
d. $CH_3CH_2CH_2CH_2CH_2COOH + CH_3CHCH_2CH_2CH_3$
                                                           |
                                                          OH
      Hexanoic acid   2-Pentanol

14.9 **a.**

3-Methylbutanoyl chloride

3-Methylbutanoic anhydride

b.

Methanoyl chloride   Ethanoic methanoic anhydride

14.11 Aldehydes are polar, as a result of the carbonyl group, but cannot hydrogen bond to one another. Alcohols are polar and can hydrogen bond as a result of the polar hydroxyl group. The carboxyl group of the carboxylic acids consists of both of these groups: the carbonyl group and the hydroxyl group. Thus, carboxylic acids are more polar than either aldehydes or alcohols, in addition to which, they can hydrogen bond to one another. As a result, carboxylic acids have higher boiling points than aldehydes or alcohols of comparable molar mass.

14.13 **a.** Pentanoic acid
b. 2-Pentanol
c. 2-Pentanol

14.15

Propanoic acid   2-Butanol

Butanal   2-Methylbutane

14.17 **a.** Heptanoic acid
b. 1-Propanol
c. Pentanoic acid
d. Butanoic acid

14.19 The smaller carboxylic acids are water-soluble. They have sharp, sour tastes and unpleasant aromas.

14.21 Citric acid is found naturally in citrus fruits. It is added to foods to give them a tart flavor or to act as a food preservative and antioxidant.

14.23 Glutaric acid is useful in the synthesis of condensation polymers because it has an odd number of carbons in the chain, which reduces the elasticity of the polymer.

14.25 Determine the name of the parent compound; that is, the longest carbon chain containing the carboxyl group. Change the -e ending of the alkane name to -oic acid. Number the chain so that the carboxyl carbon is carbon-1. Name and number substituents in the usual way.

14.27 The IUPAC name for adipic acid is hexanedioic acid. Adipic acid is a natural food additive that reduces spoilage by lowering the pH and thereby inhibiting the growth of bacteria and fungi.

14.29 **a.** $CH_3(CH_2)_2CH(CH_3)CH_2COOH$

b. $CH_3(CH_2)_2C(CH_3)(CH_2CH_3)COOH$

c.

14.31 **a.** IUPAC name: Methanoic acid
      Common name: Formic acid
b. IUPAC name: 3-Methylbutanoic acid
      Common name: β-Methylbutyric acid
c. IUPAC name: Cyclopentanecarboxylic acid
      Common name: Cyclovalericcarboxylic acid

14.33

Butanoic acid   Methylpropanoic acid

14.35 a.

$$CH_3CH_2\underset{\underset{CH_3}{|}}{\overset{\overset{CH_3}{|}}{C}}CH_2CH_2COOH$$

c.

A benzene ring with —COOH, and O$_2$N and NO$_2$ substituents.

b.

$$CH_3\underset{Br}{\overset{CH_3}{\underset{|}{CHCHCH_2COOH}}}$$

d.

A cyclohexane ring with —COOH and H$_3$C substituents.

14.37 a. IUPAC name: 2-Hydroxypropanoic acid
Common name: α-Hydroxypropionic acid
b. IUPAC name: 3-Hydroxybutanoic acid
Common name: β-Hydroxybutyric acid
c. IUPAC name: 4,4-Dimethylpentanoic acid
Common name: γ,γ-Dimethylvaleric acid
d. IUPAC name: 3, 3-Dichloropentanoic acid
Common name: β,β-Dichlorovaleric acid

14.39 a. 3-Bromobenzoic acid (or *meta*-bromobenzoic acid or *m*-bromobenzoic acid)
b. 2-Ethylbenzoic acid (or *ortho*-ethylbenzoic acid or *o*-ethylbenzoic acid)
c. 4-Hydroxybenzoic acid (or *para*-hydroxybenzoic acid or *p*-hydroxybenzoic acid)

14.41 In organic molecules, oxidation may be recognized as a gain of oxygen or a loss of hydrogen. An aldehyde may be oxidized to form a carboxylic acid as in the following example in which ethanal is oxidized to produce ethanoic acid.

$$H_3C-\overset{\overset{O}{\|}}{C}-H \xrightarrow{[O]} H_3C-\overset{\overset{O}{\|}}{C}-OH$$
Ethanal   Ethanoic acid

14.43 The following general equation represents the dissociation of a carboxylic acid.

$$R-\overset{\overset{O}{\|}}{C}-OH \rightleftharpoons R-\overset{\overset{O}{\|}}{C}-O^- + H^+$$

14.45 When a strong base is added to a carboxylic acid, neutralization occurs.

14.47 Soaps are made from water, a strong base, and natural fats or oils.

14.49 a.

Structures showing oxidation steps: alcohol $\xrightarrow{[O]}$ aldehyde $\xrightarrow{[O]}$ carboxylic acid.

b.

$$\xrightarrow{[O]}$$ aldehyde to carboxylic acid.

c.

$$\xrightarrow{[O]}$$ No Reaction

14.51 a. CH$_3$COOH

b.

$$CH_3CH_2CH_2-\overset{\overset{O}{\|}}{C}-O-CH_3 + H_2O$$

c. CH$_3$OH

14.53 a. The oxidation of 1-pentanol yields pentanal.
b. Continued oxidation of pentanal yields pentanoic acid.

14.55

a.

$$\text{(butanoic acid)} + NaOH \longrightarrow \text{(sodium salt } O^-Na^+) + H_2O$$

b.

$$\text{(butanoic acid)} + KOH \longrightarrow \text{(potassium salt } O^-K^+) + H_2O$$

c.

$$2\ \text{(butanoic acid)} + Ca(OH)_2 \longrightarrow \text{(calcium salt } Ca^{2+}) + 2H_2O$$

14.57 The structure of the calcium salt of propionic acid is $[CH_3CH_2\ COO^-]_2Ca^{2+}$. The common name of this salt is calcium propionate and the IUPAC name is calcium propanoate.

14.59 Esters are slightly polar as a result of the polar carbonyl group within the structure.

14.61 Esters are formed in the reaction of a carboxylic acid with an alcohol. The name is derived by using the alkyl or aryl portion of the alcohol IUPAC name as the first name. The *-ic* acid ending of the IUPAC name of the carboxylic acid is replaced with *-ate* and follows the name of the aryl or alkyl group.

14.63 a.

$$\overset{\overset{O}{\|}}{C}-OCH_3$$ (attached to benzene ring)

b.

$$CH_3CH_2CH_2CH_2CH_2CH_2CH_2CH_2CH_2-\overset{\overset{O}{\|}}{C}-O-CH_2CH_2CH_2CH_3$$

c.

$$CH_3CH_2-\overset{\overset{O}{\|}}{C}-O-CH_3$$

d.

$$CH_3CH_2-\overset{\overset{O}{\|}}{C}-O-CH_2CH_3$$

14.65 a. Ethyl ethanoate   **c.** Methyl-3-methylbutanoate
b. Methyl propanoate   **d.** Cyclopentyl benzoate

14.67 The following equation shows the general reaction for the preparation of an ester:

$$R-\overset{\overset{O}{\|}}{C}-OH + R-OH \underset{}{\overset{H^+,\ heat}{\rightleftharpoons}} R-\overset{\overset{O}{\|}}{C}-OR + H_2O$$
Carboxylic   Alcohol   Ester   Water
acid

14.69 The following equation shows the general reaction for the acid-catalyzed hydrolysis of an ester:

$$R-\overset{\overset{O}{\|}}{C}-OR + H_2O \underset{}{\overset{H^+,\ heat}{\rightleftharpoons}} R-\overset{\overset{O}{\|}}{C}-OH + R-OH$$
Ester   Water   Carboxylic   Alcohol
acid

14.71 A hydrolysis reaction is the cleavage of any bond by the addition of a water molecule.

14.73 a.

$$CH_3CH_2CH_2-\overset{\overset{\displaystyle O}{\|}}{C}-O-CH_2CH_3$$

b.

$$CH_3CH_2-\overset{\overset{\displaystyle O}{\|}}{C}-OH + CH_3CH_2OH$$

c. $CH_3CH_2CH_2OH$

d.

$$CH_3CH_2\overset{\overset{\displaystyle Br}{|}}{C}HCH_2-\overset{\overset{\displaystyle O}{\|}}{C}-O^- + CH_3CH_2OH$$

14.75 a. Isobutyl methanoate is made from isobutyl alcohol (IUPAC name 2-methyl-1-propanol) and methanoic acid.

$$CH_3\overset{\overset{\displaystyle CH_3}{|}}{C}HCH_2OH + HCOOH \rightarrow HCO\overset{\overset{\displaystyle O}{\|}}{}CH_2\overset{\overset{\displaystyle CH_3}{|}}{C}HCH_3$$

Isobutyl alcohol   Methanoic acid   Isobutyl methanoate

Isobutyl alcohol is an allowed starting material, but methanoic acid is not. However, it can easily be produced by the oxidation of its corresponding alcohol, methanol:

$$CH_3OH \overset{[O]}{\longrightarrow} HCHO \overset{[O]}{\longrightarrow} HCOOH$$

Methanol   Methanal   Methanoic acid

b. Pentyl butanoate is made from 1-pentanol and butanoic acid.

$$CH_3(CH_2)_3CH_2OH + CH_3CH_2CH_2COOH$$

1-Pentanol   Butanoic acid

$$\rightarrow CH_3CH_2CH_2\overset{\overset{\displaystyle O}{\|}}{C}OCH_2(CH_2)_3CH_3$$

Pentyl butanoate

Pentanol is an allowed starting material but butanoic acid is not. However, it can easily be produced by the oxidation of its corresponding alcohol, 1-butanol:

$$CH_3CH_2CH_2CH_2OH \overset{[O]}{\longrightarrow} CH_3CH_2CH_2CHO$$

1-Butanol   Butanal

$$\overset{[O]}{\longrightarrow} CH_3CH_2CH_2COOH$$

Butanoic acid

14.77 Saponification is a reaction in which soap is produced. More generally, it is the hydrolysis of an ester in the presence of a base. The following reaction shows the base-catalyzed hydrolysis of an ester:

$$CH_3(CH_2)_{14}-\overset{\overset{\displaystyle O}{\|}}{C}-O-CH_3 + NaOH \longrightarrow$$

$$CH_3(CH_2)_{14}-\overset{\overset{\displaystyle O}{\|}}{C}-O^- Na^+ + CH_3OH$$

14.79

Salicylic acid

Methyl salicylate

14.81 Compound A is

$$CH_3CH_2CH_2CH_2-\overset{\overset{\displaystyle O}{\|}}{C}-O-CH_3$$

Compound B is

$$CH_3CH_2CH_2CH_2-\overset{\overset{\displaystyle O}{\|}}{C}-OH$$

Compound C is CH_3OH

14.83 a.

$$CH_3CH_2-\overset{\overset{\displaystyle O}{\|}}{C}-OCH_2CH_2CH_3 \underset{}{\overset{H^+, heat}{\rightleftharpoons}} CH_3CH_2-\overset{\overset{\displaystyle O}{\|}}{C}-OH$$

Propyl propanoate   Propanoic acid

$$+ CH_3CH_2CH_2OH$$

1-Propanol

b.

$$H-\overset{\overset{\displaystyle O}{\|}}{C}-OCH_2CH_2CH_2CH_3 \underset{}{\overset{H^+, heat}{\rightleftharpoons}} H-\overset{\overset{\displaystyle O}{\|}}{C}-OH$$

Butyl methanoate   Methanoic acid

$$+ CH_3CH_2CH_2CH_2OH$$

1-Butanol

c.

$$H-\overset{\overset{\displaystyle O}{\|}}{C}-OCH_2CH_3 \underset{}{\overset{H^+, heat}{\rightleftharpoons}} H-\overset{\overset{\displaystyle O}{\|}}{C}-OH$$

Ethyl methanoate   Methanoic acid

$$+ CH_3CH_2OH$$

Ethanol

d.

$$CH_3CH_2CH_2CH_2-\overset{\overset{\displaystyle O}{\|}}{C}-OCH_3 \underset{}{\overset{H^+, heat}{\rightleftharpoons}}$$

Methyl pentanoate

$$CH_3CH_2CH_2CH_2-\overset{\overset{\displaystyle O}{\|}}{C}-OH + CH_3OH$$

Pentanoic acid   Methanol

14.85 Acid chlorides are noxious, irritating chemicals. They are slightly polar and have boiling points similar to comparable aldehydes or ketones. They cannot be dissolved in water because they react violently with it.

14.87 Acid anhydrides have much lower boiling points than carboxylic acids of comparable molar mass. They are also less soluble in water, and often react with it.

14.89 a.

$$CH_3(CH_2)_8-\overset{\overset{\displaystyle O}{\|}}{C}-O-\overset{\overset{\displaystyle O}{\|}}{C}-(CH_2)_8CH_3$$

b.

$$CH_3-\overset{\overset{\displaystyle O}{\|}}{C}-O-\overset{\overset{\displaystyle O}{\|}}{C}-CH_3$$

14.91 a.

$$CH_3(CH_2)_6\overset{\overset{\displaystyle O}{\|}}{C}Cl$$

b.

$$CH_3(CH_2)_2\overset{\overset{\displaystyle O}{\|}}{C}Cl$$

c.

$$CH_3(CH_2)_7\overset{\overset{\displaystyle O}{\|}}{C}Cl$$

14.93   The following equation represents the synthesis of methanoic anhydride:

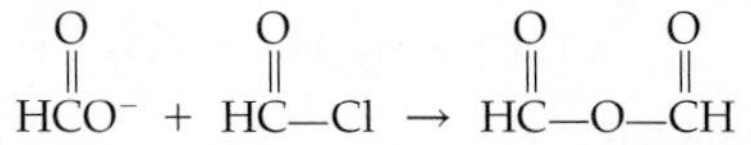

$$\underset{\substack{\text{Methanoate} \\ \text{anion}}}{HCO^-} + \underset{\substack{\text{Methanoic} \\ \text{chloride}}}{\overset{\overset{\displaystyle O}{\|}}{HC-Cl}} \longrightarrow \underset{\substack{\text{Methanoic} \\ \text{anhydride}}}{\overset{\overset{\displaystyle O}{\|}\quad\overset{\displaystyle O}{\|}}{HC-O-CH}}$$

14.95   **a.**

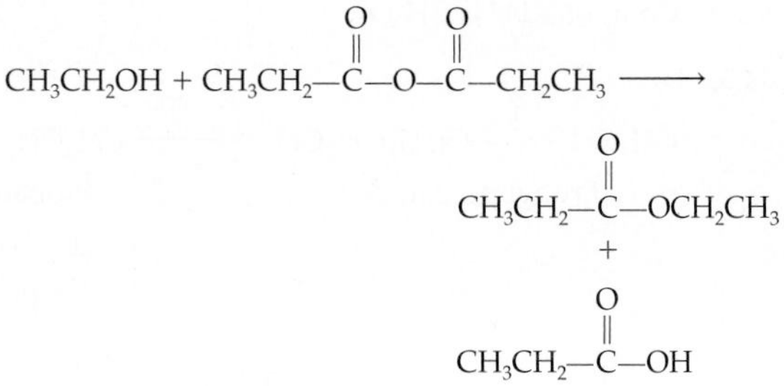

$$CH_3CH_2OH + CH_3CH_2-\overset{\overset{\displaystyle O}{\|}}{C}-O-\overset{\overset{\displaystyle O}{\|}}{C}-CH_2CH_3 \longrightarrow$$

$$CH_3CH_2-\overset{\overset{\displaystyle O}{\|}}{C}-OCH_2CH_3$$

+

$$CH_3CH_2-\overset{\overset{\displaystyle O}{\|}}{C}-OH$$

b.

$$CH_3CH_2OH + CH_3-\overset{\overset{\displaystyle O}{\|}}{C}-O-\overset{\overset{\displaystyle O}{\|}}{C}-CH_3 \longrightarrow$$

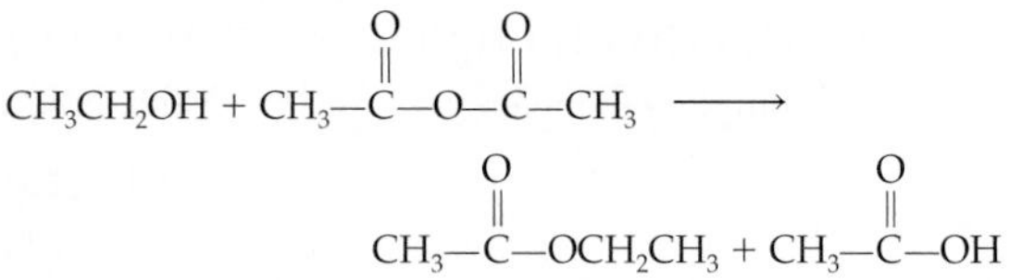

$$CH_3-\overset{\overset{\displaystyle O}{\|}}{C}-OCH_2CH_3 + CH_3-\overset{\overset{\displaystyle O}{\|}}{C}-OH$$

c.

$$CH_3CH_2OH + H-\overset{\overset{\displaystyle O}{\|}}{C}-O-\overset{\overset{\displaystyle O}{\|}}{C}-H \longrightarrow$$

$$H-\overset{\overset{\displaystyle O}{\|}}{C}-OCH_2CH_3 + H-\overset{\overset{\displaystyle O}{\|}}{C}-OH$$

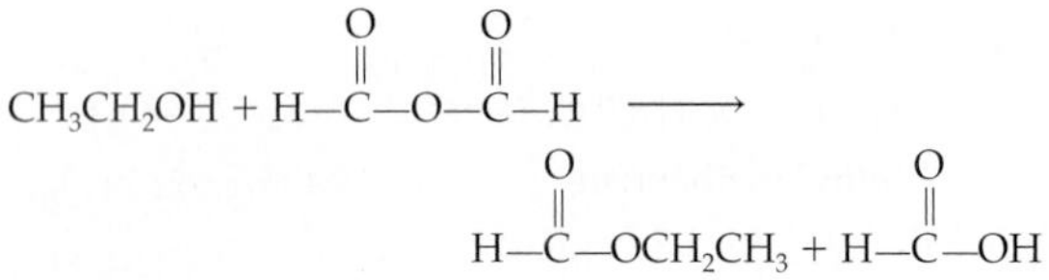

14.97   **a.** Monoester:

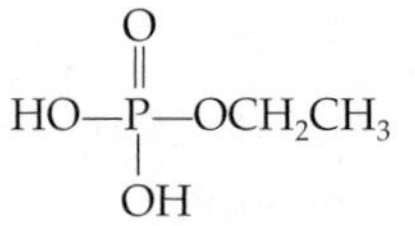

$$\underset{\overset{\displaystyle |}{OH}}{\overset{\overset{\displaystyle O}{\|}}{HO-P-OCH_2CH_3}}$$

b. Diester:

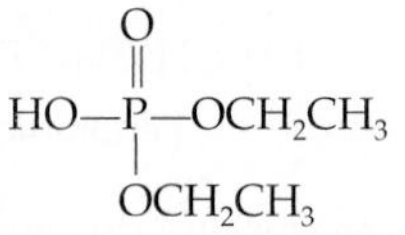

$$\underset{\overset{\displaystyle |}{OCH_2CH_3}}{\overset{\overset{\displaystyle O}{\|}}{HO-P-OCH_2CH_3}}$$

c. Triester:

$$\underset{\overset{\displaystyle |}{OCH_2CH_3}}{\overset{\overset{\displaystyle O}{\|}}{CH_3CH_2O-P-OCH_2CH_3}}$$

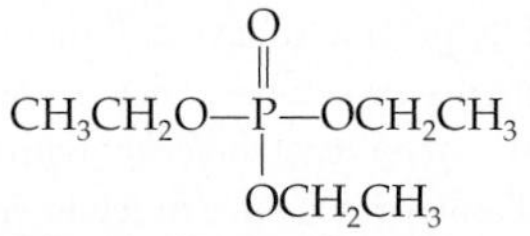

14.99   ATP is the molecule used to store the energy released in metabolic reactions. The energy is stored in the phosphoanhydride bonds between two phosphoryl groups. The energy is released when the bond is hydrolyzed. A portion of the energy can be transferred to another molecule if the phosphoryl group is transferred from ATP to the other molecule.

14.101

$$CH_3-\overset{\overset{\displaystyle O}{\|}}{C}\sim S-\text{COENZYME A}$$

The squiggle denotes a high-energy bond.

14.103

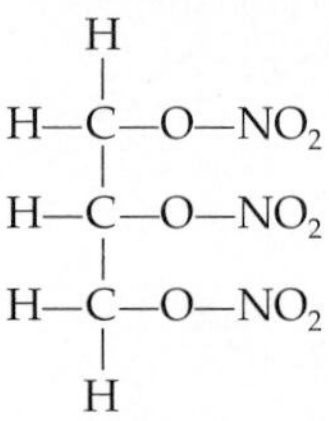

$$\begin{array}{c} H \\ | \\ H-C-O-NO_2 \\ | \\ H-C-O-NO_2 \\ | \\ H-C-O-NO_2 \\ | \\ H \end{array}$$

Chapter 15

15.1

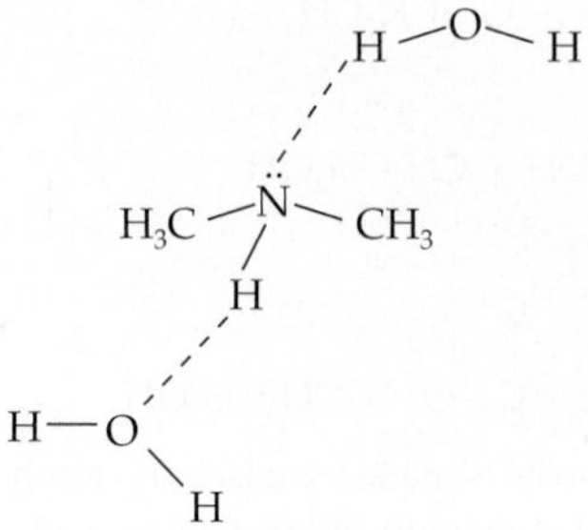

15.3   **a.**

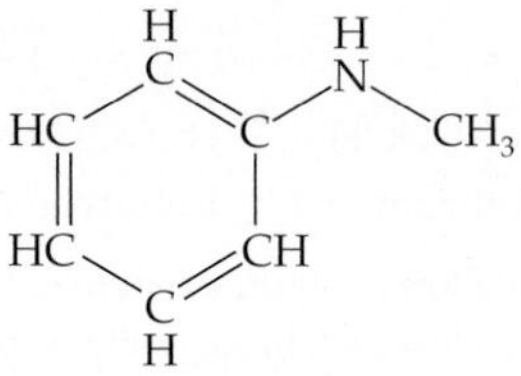

b.

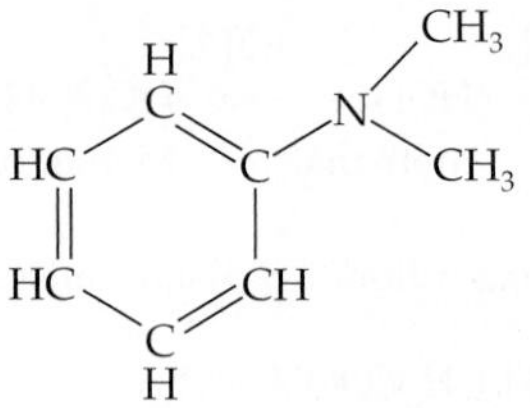

c.

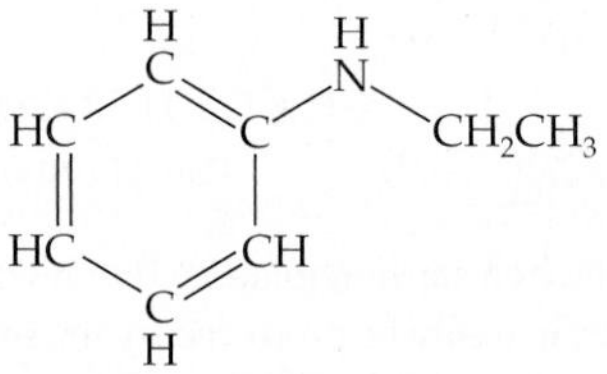

d.

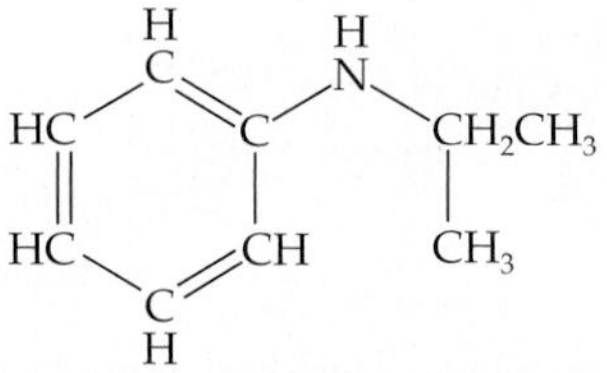

15.5   **a.**

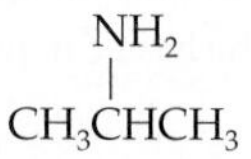

$$\underset{}{\overset{\overset{\displaystyle NH_2}{|}}{CH_3CHCH_3}}$$

b.

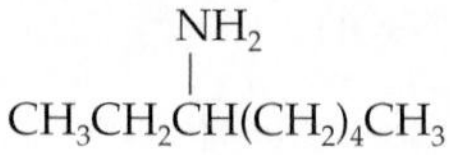

$$\underset{}{\overset{\overset{\displaystyle NH_2}{|}}{CH_3CH_2CH(CH_2)_4CH_3}}$$

c.

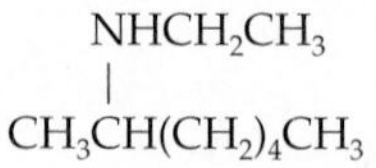

$$\underset{}{\overset{\overset{\displaystyle NHCH_2CH_3}{|}}{CH_3CH(CH_2)_4CH_3}}$$

d. $NH_2C(CH_3)_2(CH_2)_2CH_3$

e. $NH_2(CH_2)_3CHCH(CH_2)_3CH_3$
$$\underset{Cl\ \ I}{}$$

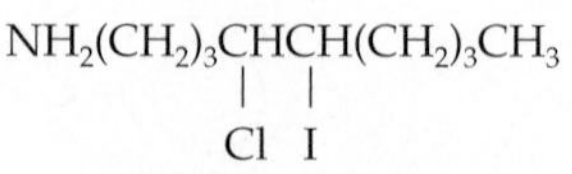

f. $(CH_3CH_2)_2N(CH_2)_4CH_3$

15.7 a. [structure: cyclopentyl-N⁺H₃Br⁻]

b.
$$CH_3CH_2—\overset{\overset{\displaystyle H}{|}}{\underset{\underset{\displaystyle H}{|}}{N^+}}—CH_3 + OH^-$$

c. $CH_3—N^+H_3 + OH^-$

15.9 a. $CH_3—NH_2$

b. $CH_3—\overset{\overset{\displaystyle CH_3}{|}}{NH}$

15.11 The nitrogen atom is more electronegative than the hydrogen atom in amines; thus, the N—H bond is polar and hydrogen bonding can occur between primary or secondary amine molecules. Thus, amines have a higher boiling point than alkanes, which are nonpolar. Because nitrogen is not as electronegative as oxygen, the N—H bond is not as polar as the O—H. As a result, intermolecular hydrogen bonds between primary and secondary amine molecules are not as strong as the hydrogen bonds between alcohol molecules. Thus, alcohols have a higher boiling point.

15.13 In systematic nomenclature, primary amines are named by determining the name of the parent compound, the longest continuous carbon chain containing the amine group. The *-e* ending of the alkane chain is replaced with *-amine.* Thus, an alkane becomes an alkanamine. The parent chain is then numbered to give the carbon bearing the amine group the lowest possible number. Finally, all substituents are named and numbered and added as prefixes to the "alkanamine" name.

15.15 Amphetamines elevate blood pressure and pulse rate. They also decrease the appetite.

15.17 a. 1-Pentamine would be more soluble in water because it has a polar amine group that can form hydrogen bonds with water molecules.

b. 2-Butamine would be more soluble in water because it has a polar amine group that can form hydrogen bonds with water molecules.

15.19 Triethylamine molecules cannot form hydrogen bonds with one another, but 1-hexanamine molecules are able to do so.

15.21 a. 2-Butanamine
b. 3-Hexanamine
c. Cyclopentanamine
d. 2-Methyl-2-propanamine

15.23 a. $CH_3CH_2NHCH_2CH_3$ [structure: secondary amine with N—H]

b. $CH_3(CH_2)_3NH_2$ [zig-zag chain with NH₂]

c. $CH_3(CH_2)_6\overset{\overset{\displaystyle NH_2}{|}}{CH}CH_2CH_3$ [zig-zag chain with NH₂]

d. $CH_3\overset{\overset{\displaystyle NH_2}{|}}{CH}CHCH_2CH_3$ with Br   [zig-zag with NH₂ and Br]

15.25 a. $CH_3CH_2CH(NH_2)(CH_2)_2CH_3$

b. $CH_3(CH_2)_5NH(CH_2)_4CH_3$

c. $H_2C—\overset{\overset{\displaystyle NH_2}{|}}{CH}$
$||$
$H_2C—CH_2$

d. [structure: cyclopentane ring with NH₂ and CH₃ substituents]
$$\overset{\displaystyle NH_2}{\underset{\displaystyle}{}}$$

e. $\overset{\displaystyle Cl^-}{CH_3CH_2N^+HCH_2CH_3}$
$\overset{|}{CH_2CH_3}$

15.27 $CH_3CH_2CH_2CH_2NH_2$ $CH_3CH_2\overset{\overset{\displaystyle}{|}}{CH}CH_3$ with NH_2

1-Butanamine   2-Butanamine
(Primary amine)   (Primary amine)

$CH_3\overset{\overset{\displaystyle}{|}}{CH}CH_2NH_2$ with CH_3 $CH_3—\overset{\overset{\displaystyle CH_3}{|}}{\underset{\underset{\displaystyle NH_2}{|}}{C}}—CH_3$

2-Methyl-1-propanamine   2-Methyl-2-propanamine
(Primary amine)   (Primary amine)

$CH_3CH_2—\overset{\overset{\displaystyle CH_3}{|}}{N}—CH_3$ $CH_3CH_2—NH—CH_2CH_3$

N,N-Dimethylethanamine   N-Ethylethanamine
(Tertiary amine)   (Secondary amine)

$\underset{\underset{\displaystyle NH—CH_3}{|}}{CH_3CHCH_3}$ $CH_3CH_2CH_2—NH—CH_3$

N-Methyl-2-propanamine   N-Methyl-1-propanamine
(Secondary amine)   (Secondary amine)

15.29 a. Primary   **c.** Primary
b. Secondary   **d.** Tertiary

15.31 a. [benzene ring with NO₂ and CH₃] $\xrightarrow{[H]}$ [benzene ring with NH₂ and CH₃]

b. [benzene ring with NO₂ and OH] $\xrightarrow{[H]}$ [benzene ring with NH₂ and OH]

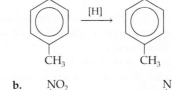

c.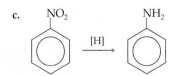

$$NO_2 \xrightarrow{[H]} NH_2$$

d.

$$\underset{\text{(cyclohexane)}}{\overset{O}{\underset{\|}{C}}-NH_2} \xrightarrow{[H]} \underset{\text{(cyclohexane)}}{H_2C-NH_2}$$

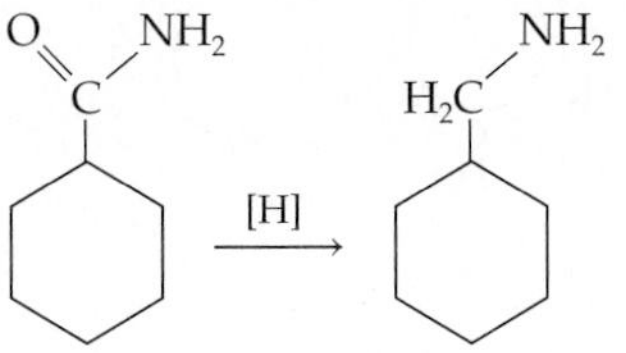

15.33 a. H_2O
   b. HBr
   c. $CH_3CH_2CH_2{-}N^+H_3$
   d. $CH_3CH_2{-}N^+H_2Cl^-$
               $\underset{CH_2CH_3}{|}$

15.35 a. $CH_3(CH_2)_4\overset{O}{\overset{\|}{C}}NH_2 \xrightarrow{[H]} CH_3(CH_2)_5NH_2$

   b. $CH_3(CH_2)_2\overset{O}{\overset{\|}{C}}NHCH_3 \xrightarrow{[H]} CH_3(CH_2)_3NHCH_3$

   c. $CH_3CH_2\overset{O}{\overset{\|}{C}}\underset{CH_3}{\underset{|}{N}}CH_3 \xrightarrow{[H]} CH_3(CH_2)_2\underset{CH_3}{\underset{|}{N}}CH_3$

15.37 Lower molar mass amines are soluble in water because the N—H bond is polar and can form hydrogen bonds with water molecules.

15.39 Drugs containing amine groups are generally administered as ammonium salts because the salt is more soluble in water and, hence, in body fluids.

15.41 Putrescine (1,4-Butanediamine):
$$\underset{NH_2}{\underset{|}{CH_2}}CH_2CH_2\underset{NH_2}{\underset{|}{CH_2}}$$

   Cadaverine (1,5-Pentanediamine):
$$\underset{NH_2}{\underset{|}{CH_2}}CH_2CH_2CH_2\underset{NH_2}{\underset{|}{CH_2}}$$

15.43 a.

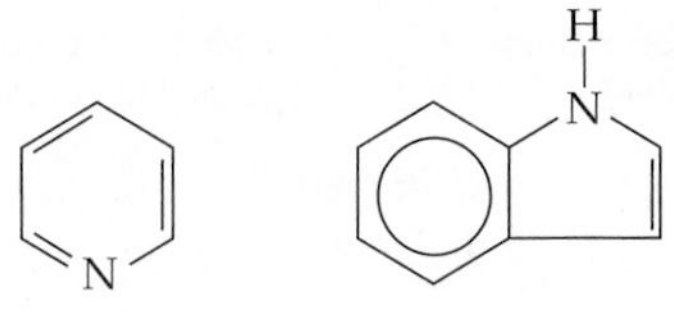

Pyridine   Indole

   b. The indole ring is found in lysergic acid diethylamide, which is a hallucinogenic drug. The pyridine ring is found in vitamin B_6, an essential water-soluble vitamin.

15.45 Morphine, codeine, quinine, and vitamin B_6

15.47 Amides have very high boiling points because the amide group consists of two very polar functional groups, the carbonyl group and the amino group. Strong intermolecular hydrogen bonding between the N—H bond of one amide and the C=O group of a second amide results in very high boiling points.

15.49 The IUPAC names of amides are derived from the IUPAC names of the carboxylic acids from which they are derived. The *-oic acid* ending of the carboxylic acid is replaced with the *-amide* ending.

15.51 Barbiturates are often called "downers" because they act as sedatives. They are sometimes used as anticonvulsants for epileptics and people suffering from other disorders that manifest as neurosis, anxiety, or tension.

15.53 a. IUPAC name: Propanamide
      Common name: Propionamide
   b. IUPAC name: Pentanamide
      Common name: Valeramide
   c. IUPAC name: *N,N*-Dimethylethanamide
      Common name: *N,N*-Dimethylacetamide

15.55 a.
$$CH_3CH_2\overset{O}{\overset{\|}{C}}NH_2$$

   b.
$$CH_3(CH_2)_2\overset{O}{\overset{\|}{C}}N(CH_2CH_3)_2$$

   c.
$$(CH_3CH_2)_2CHCH(CH_2CH_3)\overset{O}{\overset{\|}{C}}NH_2$$

   d.
$$CH_3(CH_2)_4\overset{O}{\overset{\|}{C}}NHCH_3$$

15.57 a. $CH_3{-}\overset{O}{\overset{\|}{C}}{-}NH_2,$

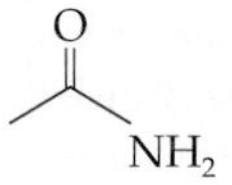

   b. $CH_3CH_2{-}\overset{O}{\overset{\|}{C}}{-}NHCH_3,$

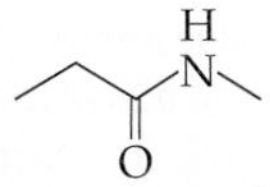

   c.

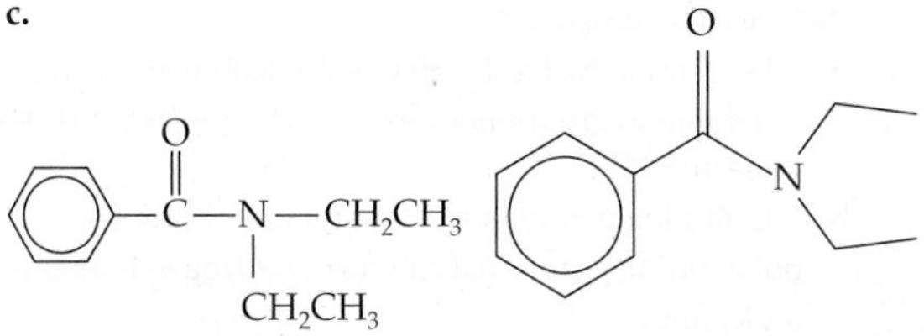

   d. $CH_3CH_2\underset{Br}{\underset{|}{CH}}\underset{CH_3}{\overset{CH_3}{\overset{|}{C}}}CH_2{-}\overset{O}{\overset{\|}{C}}{-}NH_2,$

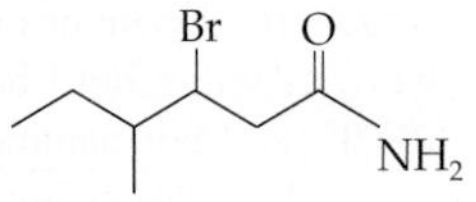

   e. $CH_3{-}\overset{O}{\overset{\|}{C}}{-}\underset{CH_3}{\underset{|}{N}}{-}CH_3,$

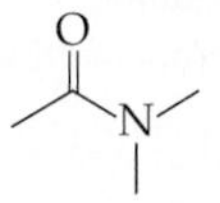

15.59 *N,N*-Diethyl-*m*-toluamide:

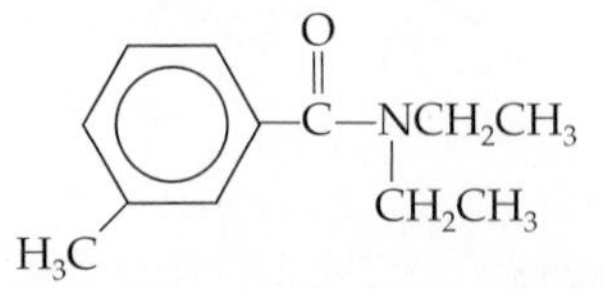

   Hydrolysis of this compound would release the carboxylic acid *m*-toluic acid and the amine *N*-ethylethanamine (diethylamine).

15.61 Amides are not proton acceptors (bases) because the highly electronegative carbonyl oxygen has a strong attraction for the nitrogen lone pair of electrons. As a result, they cannot "hold" a proton.

15.63

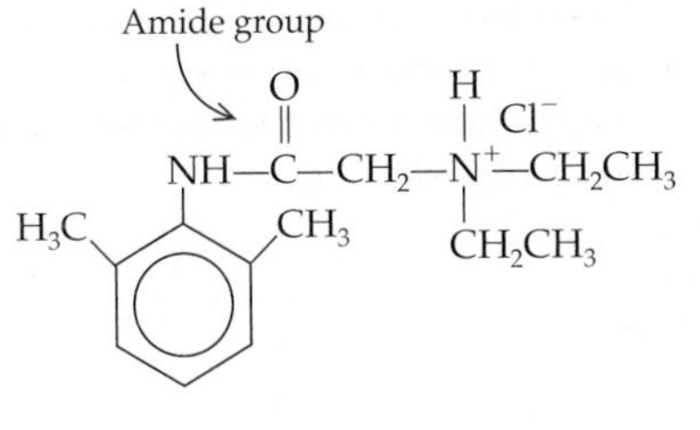

Amide group

Lidocaine hydrochloride

15.65

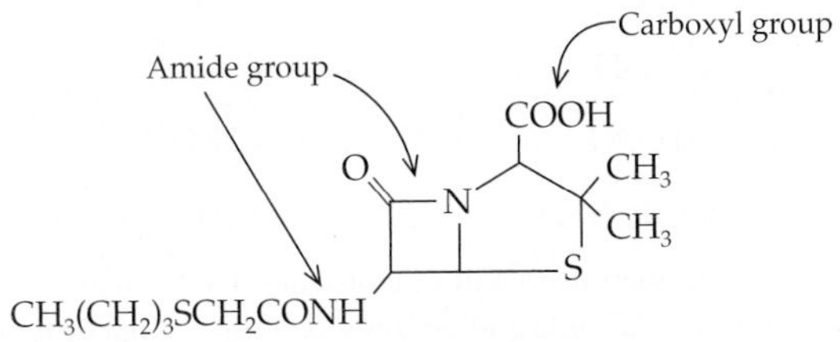

Amide group, Carboxyl group

$CH_3(CH_2)_3SCH_2CONH$

Penicillin BT

15.67

a.

$CH_3—\overset{O}{\overset{\|}{C}}—NHCH_3 + H_3O^+ \longrightarrow$

N-Methylethanamide

$CH_3COOH + CH_3N^+H_3$

Ethanoic acid   Methylammonium ion

b.

$CH_3CH_2CH_2—\overset{O}{\overset{\|}{C}}—NH—CH_3 + H_3O^+ \longrightarrow$

N-Methylbutanamide

$CH_3CH_2CH_2COOH + CH_3N^+H_3$

Butanoic acid   Methylammonium ion

c.

$CH_3\overset{CH_3}{\overset{|}{C}}HCH_2—\overset{O}{\overset{\|}{C}}—NH—CH_2CH_3 + H_3O^+ \longrightarrow$

N-Ethyl-3-methylbutanamide

$CH_3\overset{}{C}HCH_2COOH + CH_3CH_2N^+H_3$
$\quad\overset{|}{CH_3}$

3-Methylbutanoic acid   Ethylammonium ion

15.69   a.

$CH_3CH_2—\overset{O}{\overset{\|}{C}}—O—\overset{O}{\overset{\|}{C}}—CH_2CH_3$

b.

$CH_3CH_2—\overset{O}{\overset{\|}{C}}—NH_2 + NH_4^+Cl^-$

c.

$CH_3CH_2CH_2—\overset{O}{\overset{\|}{C}}—Cl + 2CH_3CH_2NH_2$

15.71   A primary (1°) amide is the product of the reaction between ammonia and an acid chloride. A primary amide has only one carbon, the carbonyl carbon, bonded to the nitrogen and has the following general structure:

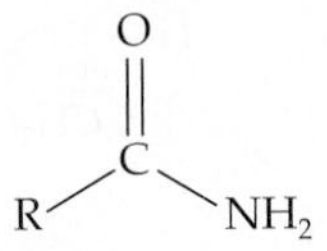

$R—\overset{O}{\overset{\|}{C}}—NH_2$

15.73   A tertiary (3°) amide is the product of a reaction between a secondary amine and an acid chloride. A tertiary amide has three

carbon atoms bonded to the nitrogen. One is the carbonyl carbon from the acid chloride and the other two are from the secondary amine reactant. The following is the general structure:

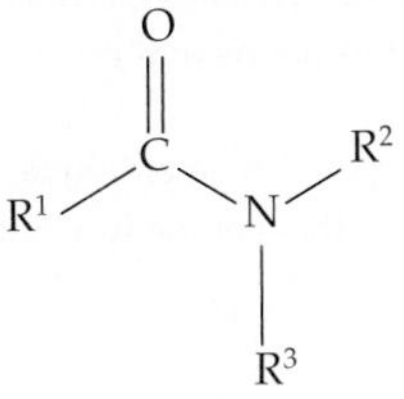

15.75

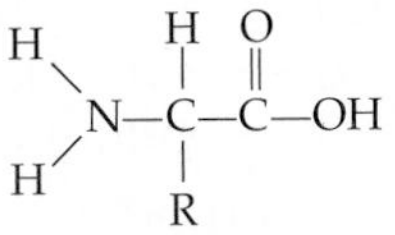

15.77   Glycine: Alanine:

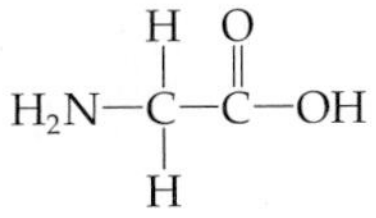

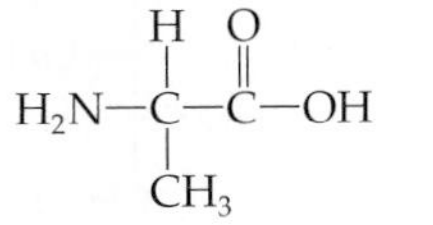

15.79

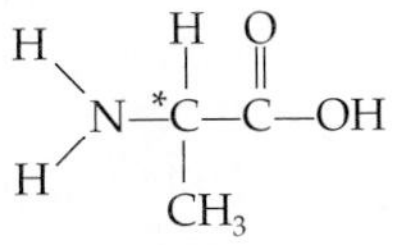

15.81   In an acyl group transfer reaction, the acyl group of an acid chloride is transferred from the Cl of the acid chloride to the N of an amine or ammonia. The product is an amide.

15.83   A chemical that carries messages or signals from a nerve to a target cell

15.85   a. Tremors, monotonous speech, loss of memory and problem-solving ability, and loss of motor function
b. Parkinson's disease
c. Schizophrenia, intense satiety sensations

15.87   In proper amounts, dopamine causes a pleasant, satisfied feeling. This feeling becomes intense as the amount of dopamine increases. Several drugs, including cocaine, heroin, amphetamines, alcohol, and nicotine increase the levels of dopamine. It is thought that the intense satiety response this brings about may contribute to addiction to these substances.

15.89   Epinephrine is a component of the flight or fight response. It stimulates glycogen breakdown to provide the body with glucose to supply the needed energy for this stress response.

15.91   The amino acid tryptophan

15.93   Perception of pain, thermoregulation, and sleep

15.95   Promotes the itchy skin rash associated with poison ivy and insect bites; the respiratory symptoms characteristic of hay fever; secretion of stomach acid

15.97   Inhibitory neurotransmitters

15.99   When acetylcholine is released from a nerve cell, it binds to receptors on the surface of muscle cells. This binding stimulates the muscle cell to contract. To stop the contraction, the acetylcholine is then broken down to choline and acetate ion. This is catalyzed by the enzyme acetylcholinesterase.

15.101   Organophosphates inactivate acetylcholinesterase by binding covalently to it. Since acetylcholine is not broken down, nerve transmission continues, resulting in muscle spasm. Pyridine aldoxime methiodide (PAM) is an antidote to organophosphate poisoning because it displaces the organophosphate, thereby allowing acetycholinesterase to function.

Chapter 16

16.1 It is currently recommended that 45–55% of the calories in the diet should be carbohydrates. Of that amount, the World Health Organization recommends that no more than 5% should be simple sugars.

16.3 An aldose is a sugar with an aldehyde functional group. A ketose is a sugar with a ketone functional group.

16.5 a.

CH₃ / =O / H—*—OH / CH₂OH   CH₃ / =O / HO—*—H / CH₂OH   c. CH₂OH / =O / HO—*—H / H—*—OH / H—*—OH / CH₂OH   CH₂OH / =O / H—*—OH / HO—*—H / HO—*—H / CH₂OH

b.

CHO / H—*—OH / H—*—OH / HO—*—H / CH₂OH   CHO / HO—*—H / HO—*—H / H—*—OH / CH₂OH

16.7 a. D-
b. L-
c. D-

16.9

CHO / H—OH / HO—H / HO—H / H—OH / CH₂OH

D-Galactose

16.11 α-Amylase and β-amylase are digestive enzymes that break down the starch amylose. α-Amylase cleaves glycosidic bonds of the amylose chain at random, producing shorter polysaccharide chains. β-Amylase sequentially cleaves maltose (a disaccharide of glucose) from the reducing end of the polysaccharide chain.

16.13 A monosaccharide is the simplest sugar and consists of a single saccharide unit. A disaccharide is made up of two monosaccharides joined covalently by a glycosidic bond.

16.15 Mashed potato flakes, rice, and corn starch contain amylose and amylopectin, both of which are polysaccharides. A candy bar contains sucrose, a disaccharide. Orange juice contains fructose, a monosaccharide. It may also contain sucrose if the label indicates that sugar has been added.

16.17 Four

16.19

O ‖ C—H / H—C—OH / HO—C—H / HO—C—H / H—C—OH / CH₂OH

D-Galactose
(An aldohexose)

CH₂OH / C=O / HO—C—H / H—C—OH / H—C—OH / CH₂OH

D-Fructose
(A ketohexose)

16.21 An aldose is a sugar that contains an aldehyde (carbonyl) group.

16.23 A tetrose is a sugar with a four-carbon backbone.

16.25 A ketopentose is a sugar with a five-carbon backbone and containing a ketone (carbonyl) group.

16.27 a. β-D-Glucose
b. β-D-Fructose
c. α-D-Galactose

16.29

O ‖ C—H / H—C—OH / CH₂OH

D-Glyceraldehyde

O ‖ C—H / HO—C—H / CH₂OH

L-Glyceraldehyde

16.31 Stereoisomers are a pair of molecules that have the same structural formula and bonding pattern but that differ in the arrangement of the atoms in space.

16.33 A chiral carbon is one that is bonded to four different chemical groups.

16.35 A polarimeter converts monochromatic light into monochromatic plane-polarized light. This plane-polarized light is passed through a sample and into an analyzer. If the sample is optically active, it will rotate the plane of the light. The degree and angle of rotation are measured by the analyzer.

16.37 A Fischer Projection is a two-dimensional drawing of a molecule that shows a chiral carbon at the intersection of two lines. Horizontal lines at the intersection represent bonds projecting out of the page, and vertical lines represent bonds that project into the page.

16.39 Diastereomers are a pair of stereoisomers that are not enantiomers.

16.41

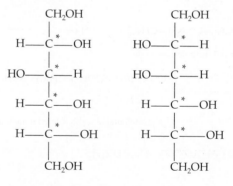

Sorbitol   Mannitol

16.43 Dextrose is a common name used for D-glucose.

16.45 D- and L-Glyceraldehyde are a pair of enantiomers; that is, they are nonsuperimposable mirror images of one another.

16.47

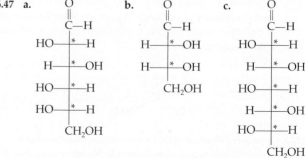

a.

O ‖ C—H / HO—*—H / H—*—OH / HO—*—H / HO—*—H / CH₂OH

b.

O ‖ C—H / H—*—OH / H—*—OH / CH₂OH

c.

O ‖ C—H / HO—*—H / H—*—OH / HO—*—H / H—*—OH / HO—*—H / CH₂OH

16.49 a.

A	B	C	D

The stereoisomers are A, C, and D. Compounds A and B are identical and they are meso compounds because they have an internal plane of symmetry. Compounds C and D are enantiomers. Compound A is a diastereomer to both compounds C and D.

b.

A	B	C	D

There are four possible stereoisomers. The two pairs of enantiomers are A with B and C with D. Compounds A and B are both diastereomers to compounds C and D.

16.51 a.

A	B	C	D

The stereoisomers are A, C, and D. Compounds A and B are identical and they are meso compounds because they have an internal plane of symmetry. Compounds C and D are enantiomers. Compound A is a diastereomer to both compounds C and D.

b.

A	B	C	D

There are four possible stereoisomers. Compounds A and B are enantiomers, and compounds C and D are enantiomers. Compounds A and B are diastereomers to both compounds C and D.

16.53 Anomers are isomers that differ in the arrangement of bonds around the hemiacetal carbon.

16.55 A hemiacetal is a member of the family of organic compounds formed in the reaction of one molecule of alcohol with an aldehyde or a ketone.

16.57 When the carbonyl group at C-1 of D-glucose reacts with the C-5 hydroxyl group, a new chiral carbon is created (C-1). In the α-isomer of the cyclic sugar, the C-1 hydroxyl group is below the ring; and in the β-isomer, the C-1 hydroxyl group is above the ring.

16.59 β-Maltose and α-lactose would give positive Benedict's tests. Glycogen would give only a weak reaction because there are fewer reducing ends for a given mass of the carbohydrate.

16.61 Enantiomers are stereoisomers that are nonsuperimposable mirror images of one another. For instance:

D-Glyceraldehyde   L-Glyceraldehyde

16.63 An aldehyde sugar forms an intramolecular hemiacetal when the carbonyl group of the monosaccharide reacts with a hydroxyl group on one of the other carbon atoms.

16.65 A disaccharide is a simple carbohydrate composed of two monosaccharides.

16.67 A glycosidic bond is the bond formed between the hydroxyl group of the C-1 carbon of one sugar and a hydroxyl group of another sugar.

16.69

β-Maltose

16.71 Milk

16.73 Eliminating milk and milk products from the diet

16.75 Lactose intolerance is the inability to produce the enzyme lactase that hydrolyzes the milk sugar lactose into its component monosaccharides: glucose and galactose.

16.77 A polymer is a very large molecule formed by the combination of many small molecules, called monomers.

16.79 Starch

16.81 Homopolysaccharides are a class of polysaccharides that are composed of a single monosaccharide.

16.83 Starch, glycogen, and cellulose are examples of homopolysaccharides. These homopolysaccharides are all made up of glucose.

16.85 The glucose units of amylose are joined by α (1 → 4) glycosidic bonds and those of cellulose are bonded together by β (1 → 4) glycosidic bonds.

16.87 Glycogen serves as a storage molecule for glucose.

16.89 The salivary glands and the pancreas

Chapter 17

17.1 **a.** $CH_3(CH_2)_7CH=CH(CH_2)_7COOH$
b. $CH_3(CH_2)_{10}COOH$
c. $CH_3(CH_2)_4CH=CH-CH_2-CH=CH(CH_2)_7COOH$
d. $CH_3(CH_2)_{16}COOH$

17.3

17.5

17.7

$$CH_3\overset{\displaystyle O}{\overset{\|}{C}}-O-CH_2(CH_2)_2CH_3 \ + \ KOH \ \rightarrow$$

Butyl acetate

$$CH_3\overset{\displaystyle O}{\overset{\|}{C}}-O^- K^+ \ + \ CH_3(CH_2)_2CH_2OH$$

Potassium   1-Butanol
acetate

17.9

$$CH_3CH_2CH=CHCH_2CH=CHCH_2CH=CH(CH_2)_7COOH \ + \ 3H_2$$

All *cis*-9,12,15-Octadecatrienoic acid

↓ Ni

$$CH_3(CH_2)_{16}COOH$$

Octadecanoic acid

17.11 a.

$$CH_3(CH_2)_7CH=CH(CH_2)_7-\overset{}{\underset{\displaystyle O}{C}}-O-CH_2$$

$$CH-OH$$

$$CH_2-OH$$

$$CH_3(CH_2)_7CH=CH(CH_2)_7-\overset{}{\underset{\displaystyle O}{C}}-O-CH_2$$

$$CH_3(CH_2)_7CH=CH(CH_2)_7-\overset{}{\underset{\displaystyle O}{C}}-O-CH$$

$$CH_2-OH$$

$$CH_3(CH_2)_7CH=CH(CH_2)_7-\overset{}{\underset{\displaystyle O}{C}}-O-CH_2$$

$$CH_3(CH_2)_7CH=CH(CH_2)_7-\overset{}{\underset{\displaystyle O}{C}}-O-CH$$

$$CH_3(CH_2)_7CH=CH(CH_2)_7-\overset{}{\underset{\displaystyle O}{C}}-O-CH_2$$

b.

$$CH_3(CH_2)_8-\overset{}{\underset{\displaystyle O}{C}}-O-CH_2$$

$$CH-OH$$

$$CH_2-OH$$

$$CH_3(CH_2)_8-\overset{}{\underset{\displaystyle O}{C}}-O-CH_2$$

$$CH_3(CH_2)_8-\overset{}{\underset{\displaystyle O}{C}}-O-CH$$

$$CH_2-OH$$

$$CH_3(CH_2)_8-\overset{}{\underset{\displaystyle O}{C}}-O-CH_2$$

$$CH_3(CH_2)_8-\overset{}{\underset{\displaystyle O}{C}}-O-CH$$

$$CH_3(CH_2)_8-\overset{}{\underset{\displaystyle O}{C}}-O-CH_2$$

17.13

Steroid nucleus

17.15 Receptor-mediated endocytosis

17.17 Fatty acids, glycerides, nonglyceride lipids, and complex lipids

17.19 Lipid-soluble vitamins are transported into cells of the small intestine in association with dietary fat molecules. Thus, a diet low in fat reduces the amount of vitamins A, D, E, and K that enters the body.

17.21 A saturated fatty acid is one in which the hydrocarbon tail has only carbon-to-carbon single bonds. An unsaturated fatty acid has at least one carbon-to-carbon double bond.

17.23 The melting points increase.

17.25 The melting points of fatty acids increase as the length of the hydrocarbon chains increase. This is because the intermolecular attractive forces, including London dispersion forces, increase as the length of the hydrocarbon chain increases.

17.27 a. Decanoic acid $CH_3(CH_2)_8COOH$
   b. Stearic acid $CH_3(CH_2)_{16}COOH$

17.29 a. IUPAC name: Hexadecanoic acid
      Common name: Palmitic acid
   b. IUPAC name: Dodecanoic acid
      Common name: Lauric acid

17.31 Esterification of glycerol with three molecules of myristic acid:

$$CH_2OH$$
$$CHOH \ + 3\ CH_3(CH_2)_{12}-\overset{\displaystyle O}{\overset{\|}{C}}-OH \longrightarrow$$
$$CH_2OH$$

$$CH_3(CH_2)_{12}-\overset{\displaystyle O}{\overset{\|}{C}}-O-CH_2$$

$$CH_3(CH_2)_{12}-\overset{\displaystyle O}{\overset{\|}{C}}-O-CH \ + 3H_2O$$

$$CH_3(CH_2)_{12}-\overset{\displaystyle O}{\overset{\|}{C}}-O-CH_2$$

17.33

$$H-\overset{\displaystyle H}{\underset{}{C}}-O-\overset{\displaystyle O}{\overset{\|}{C}}-(CH_2)_7CH=CH(CH_2)_7CH_3$$

$$H-\overset{}{\underset{}{C}}-O-\overset{\displaystyle O}{\overset{\|}{C}}-(CH_2)_7CH=CH(CH_2)_7CH_3 \ + 3H_2O$$

$$H-\overset{}{\underset{\displaystyle H}{C}}-O-\overset{\displaystyle O}{\overset{\|}{C}}-(CH_2)_7CH=CH(CH_2)_7CH_3$$

↓ H^+, heat

$$H-\overset{\displaystyle H}{\underset{}{C}}-OH \ + \ HO-\overset{\displaystyle O}{\overset{\|}{C}}-(CH_2)_7CH=CH(CH_2)_7CH_3$$

$$H-\overset{}{\underset{}{C}}-OH \ + \ HO-\overset{\displaystyle O}{\overset{\|}{C}}-(CH_2)_7CH=CH(CH_2)_7CH_3$$

$$H-\overset{}{\underset{\displaystyle H}{C}}-OH \ + \ HO-\overset{\displaystyle O}{\overset{\|}{C}}-(CH_2)_7CH=CH(CH_2)_7CH_3$$

Glycerol   3 Oleic acid molecules

17.35

$$HO-\overset{\overset{\displaystyle O}{\|}}{C}-(CH_2)_6CH_3 \xrightarrow{\text{KOH}} K^+\ {}^-O-\overset{\overset{\displaystyle O}{\|}}{C}-(CH_2)_6CH_3$$

Octanoic acid   Potassium octanoate

$$HO-\overset{\overset{\displaystyle O}{\|}}{C}-(CH_2)_{16}CH_3 \xrightarrow{\text{KOH}} K^+\ {}^-O-\overset{\overset{\displaystyle O}{\|}}{C}-(CH_2)_{16}CH_3$$

Stearic acid   Potassium stearate

17.37   This line drawing of EPA shows the bends or "kinks" introduced into the molecule by the double bonds.

All *cis*-5,8,11,14,17-Eicosapentaenoic acid (EPA)

$$+\ 5H_2 \Big| \text{Ni}$$

Eicosanoic acid

17.39

$$\text{H}-\overset{\overset{\displaystyle H}{|}}{\underset{\underset{\displaystyle H}{|}}{C}}-O-\overset{\overset{\displaystyle O}{\|}}{C}-(CH_2)_{10}CH_3 \quad \text{Lauric acid}$$

$$-O-\overset{\overset{\displaystyle O}{\|}}{C}-(CH_2)_{16}CH_3 \quad \text{Stearic acid}$$

$$-O-\overset{\overset{\displaystyle O}{\|}}{C}-(CH_2)_8CH_3 \quad \text{Capric acid}$$

$$\Big\downarrow \text{NaOH}$$

$$\text{H}-\overset{\overset{\displaystyle H}{|}}{\underset{\underset{\displaystyle H}{|}}{C}}-OH \quad + \quad Na^+{}^-O-\overset{\overset{\displaystyle O}{\|}}{C}-(CH_2)_{10}CH_3$$

$$\text{H}-\overset{|}{C}-OH \quad + \quad Na^+{}^-O-\overset{\overset{\displaystyle O}{\|}}{C}-(CH_2)_{16}CH_3$$

$$\text{H}-\overset{|}{C}-OH \quad + \quad Na^+{}^-O-\overset{\overset{\displaystyle O}{\|}}{C}-(CH_2)_8CH_3$$

17.41

$$\text{H}-\overset{\overset{\displaystyle H}{|}}{\underset{\underset{\displaystyle H}{|}}{C}}-OH \quad + \quad HO-\overset{\overset{\displaystyle O}{\|}}{C}-(CH_2)_8CH_3 \quad \text{Decanoic acid}$$

$$\text{H}-\overset{|}{C}-OH \quad + \quad HO-\overset{\overset{\displaystyle O}{\|}}{C}-(CH_2)_{10}CH_3 \quad \text{Dodecanoic acid}$$

$$\text{H}-\overset{|}{C}-OH \quad + \quad HO-\overset{\overset{\displaystyle O}{\|}}{C}-(CH_2)_{14}CH_3 \quad \text{Hexadecanoic acid}$$

Glycerol

$$\Big\downarrow H^+, \text{heat}$$

$$\text{H}-\overset{\overset{\displaystyle H}{|}}{\underset{\underset{\displaystyle H}{|}}{C}}-O-\overset{\overset{\displaystyle O}{\|}}{C}-(CH_2)_8CH_3$$

$$\text{H}-\overset{|}{C}-O-\overset{\overset{\displaystyle O}{\|}}{C}-(CH_2)_{10}CH_3 \ + \ 3H_2O$$

$$\text{H}-\overset{|}{C}-O-\overset{\overset{\displaystyle O}{\|}}{C}-(CH_2)_{14}CH_3$$

17.43   The essential fatty acid linoleic acid is required for the synthesis of arachidonic acid, a precursor for the synthesis of the prostaglandins, a group of hormonelike molecules.

17.45   Aspirin effectively decreases the inflammatory response by inhibiting the synthesis of all prostaglandins. Aspirin works by inhibiting cyclooxygenase, the first enzyme in prostaglandin biosynthesis. This inhibition results from the transfer of an acetyl group from aspirin to the enzyme. Because cyclooxygenase is found in all cells, synthesis of all prostaglandins is inhibited.

17.47   Smooth muscle contraction, enhancement of fever and swelling associated with the inflammatory response, bronchial dilation, inhibition of secretion of acid into the stomach

17.49   The name of these fatty acids arises from the position of the double bond nearest the terminal *methyl group* of the molecule. The terminal methyl group is designated omega (ω). In ω-3 fatty acids, the double bond nearest the ω methyl group is three carbons along the chain. In ω-6 fatty acids, the nearest double bond is six carbons from the end.

17.51   Omega-3 fatty acids reduce the risk of cardiovascular disease by decreasing blood clot formation, blood triglyceride levels, and growth of atherosclerotic plaque.

17.53   The decrease in blood clot formation, along with the reduced blood triglyceride levels and decreased atherosclerotic plaque result in improved arterial health. This, in turn, results in lower blood pressure and a decreased risk of sudden death and heart arrhythmias.

17.55   Omega-3 fatty acids are precursors of prostaglandins that exhibit anti-inflammatory effects. On the other hand, omega-6 fatty acids are precursors to prostaglandins that have inflammatory effects. To reduce the inflammatory response contribution to cardiovascular disease, it is logical to increase the amount of omega-3 fatty acids in the diet and to decrease the amount of omega-6 fatty acids.

17.57   A glyceride is a lipid ester that contains the glycerol molecule and from one to three fatty acids.

17.59   An emulsifying agent is a molecule that aids in the suspension of triglycerides in water. They are amphipathic molecules, such as lecithin, that serve as bridges holding together the highly polar water molecules and the nonpolar triglycerides.

17.61   A triglyceride with three saturated fatty acid tails would be a solid at room temperature. The long, straight fatty acid tails would stack with one another because of strong intermolecular and intramolecular London dispersion force attractions.

17.63

$$CH_3(CH_2)_{14}-\overset{\overset{\displaystyle O}{\|}}{C}-O-\underset{1}{CH_2}$$

$$\overset{H}{\underset{CH_3(CH_2)_4CH_2}{\diagdown}}C=C\overset{CH_2(CH_2)_6-\overset{\overset{\displaystyle O}{\|}}{C}-O-\underset{2}{CH}}{\diagup}$$

$$\overset{CH_3(CH_2)_4CH_2}{\diagdown}C=C\overset{CH_2(CH_2)_6-\overset{\overset{\displaystyle O}{\|}}{C}-O-\underset{3}{CH_2}}{\diagup}$$

17.65

$$CH_3CH_2CH_2CH_2CH_2CH_2CH_2CH_2CH_2-\overset{\overset{\displaystyle O}{\|}}{C}-O-\underset{1}{CH_2}$$

$$CH_3CH_2CH_2CH_2CH_2CH_2CH_2CH_2CH_2CH_2-\underset{\underset{\displaystyle O}{\|}}{C}-O-\underset{2}{CH}$$

$$CH_2-O-\overset{\overset{\displaystyle O}{\|}}{\underset{\underset{\displaystyle O^-}{|}}{P}}-O^-$$

17.67 Triglycerides consist of three fatty acids esterified to the three hydroxyl groups of glycerol. In phospholipids, there are only two fatty acids esterified to glycerol. A phosphoryl group is esterified (phosphoester linkage) to the third hydroxyl group.

17.69 A sphingolipid is a lipid that is not derived from glycerol, but rather from sphingosine, a long-chain, nitrogen-containing (amino) alcohol. Like phospholipids, sphingolipids are amphipathic.

17.71 A glycosphingolipid or glycolipid is a lipid that is built on a ceramide backbone structure. Ceramide is a fatty acid derivative of sphingosine.

17.73 Sphingomyelins are important structural lipid components of nerve cell membranes. They are found in the myelin sheath that surrounds and insulates cells of the central nervous system.

17.75 Cholesterol is readily soluble in the hydrophobic region of biological membranes. It is involved in regulating the fluidity of the membrane.

17.77 Progesterone is the most important hormone associated with pregnancy. Testosterone is needed for development of male secondary sexual characteristics. Estrone is required for proper development of female secondary sexual characteristics.

17.79 Cortisone is used to treat rheumatoid arthritis, asthma, gastrointestinal disorders, and many skin conditions.

17.81 Myricyl palmitate (beeswax) is made up of the fatty acid palmitic acid and the alcohol myricyl alcohol—$CH_3(CH_2)_{28}CH_2OH$.

17.83 Isoprenoids are a large, diverse collection of lipids that are synthesized from the isoprene unit:

$$CH_2=\overset{\overset{\displaystyle CH_3}{|}}{C}-CH=CH_2$$

17.85 Steroids and bile salts, lipid-soluble vitamins, certain plant hormones, and chlorophyll

17.87 Chylomicrons, high-density lipoproteins, low-density lipoproteins, and very low density lipoproteins

17.89 The terms "good" and "bad" cholesterol refer to two classes of lipoprotein complexes. The high-density lipoproteins, or HDL, are considered to be "good" cholesterol because a correlation has been made between elevated levels of HDL and a reduced incidence of atherosclerosis. Low-density lipoproteins, or LDL, are considered

to be "bad" cholesterol because evidence suggests that a high level of LDL is associated with increased risk of atherosclerosis.

17.91 Atherosclerosis results when cholesterol and other substances coat the arteries, causing a narrowing of the passageways. As the passageways become narrower, greater pressure is required to provide adequate blood flow. This results in higher blood pressure (hypertension).

17.93 If the LDL receptor is defective, it cannot function to remove cholesterol-bearing LDL particles from the blood. The excess cholesterol, along with other substances, will accumulate along the walls of the arteries, causing atherosclerosis.

17.95 The basic structure of a biological membrane is a bilayer of phospholipid molecules arranged so that the hydrophobic hydrocarbon tails are packed in the center and the hydrophilic head groups are exposed on the inner and outer surfaces.

17.97 A peripheral membrane protein is bound to only one surface of the membrane, either inside or outside the cell.

17.99 Cholesterol is freely soluble in the hydrophobic layer of a biological membrane. It moderates the fluidity of the membrane by disrupting the stacking of the fatty acid tails of membrane phospholipids.

17.101 Specific membrane proteins on human and mouse cells were labeled with red and green fluorescent dyes, respectively. The human and mouse cells were fused into single-celled hybrids and were observed using a microscope with an ultraviolet light source. The ultraviolet light caused the dyes to fluoresce. Initially, the dyes were localized in regions of the membrane representing the original human or mouse cell. Within an hour, the proteins were evenly distributed throughout the membrane of the fused cell.

17.103 If the fatty acid tails of the membrane phospholipids are converted from saturated to unsaturated, the fluidity of the membrane will increase.

Chapter 18

18.1   **a.** Glycine (gly or G):

$$H_3{}^+N-\overset{\overset{\displaystyle COO^-}{|}}{\underset{\underset{\displaystyle H}{|}}{C}}-H$$

b. Proline (pro or P):

$$\overset{\qquad\qquad COO^-}{\underset{H_2C\diagdown_{\underset{\displaystyle H_2}{C}}\diagup CH_2}{H_2{}^+N\text{———}CH}}$$

c. Threonine (thr or T):

$$\overset{\overset{\displaystyle COO^-}{|}}{\underset{\underset{\underset{\displaystyle CH_3}{|}}{\overset{\displaystyle |}{H-C-OH}}}{H_3{}^+N-C-H}}$$

d. Aspartate (asp or D):

$$\overset{\overset{\displaystyle COO^-}{|}}{\underset{\underset{\underset{\displaystyle COO^-}{|}}{\overset{\displaystyle |}{H-C-H}}}{H_3{}^+N-C-H}}$$

e. Lysine (lys or K):

$$
\begin{array}{c}
COO^- \\
| \\
H_3{}^+N-C-H \\
| \\
H-C-H \\
| \\
H-C-H \\
| \\
H-C-H \\
| \\
H-C-H \\
| \\
N^+H_3
\end{array}
$$

18.3 a. Alanyl-phenylalanine:

$$
\begin{array}{c}
\quad\; H \;\; O \;\; H \;\; H \\
\quad\; | \quad || \quad | \quad | \\
H_3{}^+N-C-C-N-C-COO^- \\
\quad\; | \qquad\qquad | \\
\quad\; CH_3 \qquad\quad CH_2
\end{array}
$$

b. Lysyl-alanine:

$$
\begin{array}{c}
\quad\; H \;\; O \;\; H \;\; H \\
\quad\; | \quad || \quad | \quad | \\
H_3{}^+N-C-C-N-C-COO^- \\
\quad\; | \qquad\qquad | \\
\quad\; CH_2 \qquad\quad CH_3 \\
\quad\; | \\
\quad\; CH_2 \\
\quad\; | \\
\quad\; CH_2 \\
\quad\; | \\
\quad\; CH_2 \\
\quad\; | \\
\quad\; N^+H_3
\end{array}
$$

c. Phenylalanyl-tyrosyl-leucine:

$$
\begin{array}{c}
\quad H \;\; O \;\; H \;\; H \;\; O \;\; H \;\; H \\
\quad | \quad || \quad | \quad | \quad || \quad | \quad | \\
H_3{}^+N-C-C-N-C-C-N-C-COO^- \\
\quad | \qquad\qquad | \qquad\qquad | \\
\quad CH_2 \qquad\quad CH_2 \qquad\; CH_2 \\
\qquad\qquad\qquad\qquad\qquad CHCH_3 \\
\qquad\qquad\qquad\qquad\qquad CH_3 \\
\\
\qquad\qquad\qquad\qquad OH
\end{array}
$$

18.5 The primary structure of a protein is the amino acid sequence of the protein chain. Regular, repeating folding of the peptide chain caused by hydrogen bonding between the amide nitrogens and carbonyl oxygens of the peptide bond is the secondary structure of a protein. The two most common types of secondary structure are the α-helix and the β-pleated sheet. Tertiary structure is the further folding of the regions of the α-helix and β-pleated sheet into a compact, spherical structure. Formation and maintenance of the tertiary structure results from weak attractions between amino acid R groups. The binding of two or more peptides to produce a functional protein defines the quaternary structure.

18.7 Oxygen is efficiently transferred from hemoglobin to myoglobin in the muscle because myoglobin has a greater affinity for oxygen.

18.9 High temperature disrupts the hydrogen bonds and other weak interactions that maintain protein structure.

18.11 Vegetables vary in amino acid composition. Most vegetables do not provide all of the amino acid requirements of the body. By eating a variety of different vegetables, all the amino acid requirements of the human body can be met.

18.13 An enzyme is a protein that serves as a biological catalyst, speeding up biological reactions.

18.15 A transport protein is a protein that transports materials across the cell membrane or throughout the body.

18.17 Enzymes speed up reactions that might take days or weeks to occur on their own. They also catalyze reactions that might require very high temperatures or harsh conditions if carried out in the laboratory. In the body, these reactions occur quickly under physiological conditions.

18.19 Transferrin is a transport protein that carries iron from the liver to the bone marrow, where it is used to produce the heme group for hemoglobin and myoglobin. Hemoglobin transports oxygen in the blood.

18.21 Egg albumin is a nutrient protein that serves as a source of protein for the developing chick. Casein is the nutrient storage protein in milk, providing protein, a source of amino acids, for mammals.

18.23 The general structure of an L-α-amino acid has a carbon in the center that is referred to as the alpha carbon. Bonded to the alpha carbon are a protonated amino group, a carboxylate group, a hydrogen atom, and a side chain, R.

$$
\begin{array}{c}
COO^- \\
| \\
H_3{}^+N-C-H \\
| \\
R
\end{array}
$$

18.25 A zwitterion is a neutral molecule with equal numbers of positive and negative charges. Under physiological conditions, amino acids are zwitterions.

18.27 A chiral carbon is one that has four different atoms or groups of atoms attached to it.

18.29 Interactions between the R groups of the amino acids in a polypeptide chain are important for the formation and maintenance of the tertiary and quaternary structures of proteins.

18.31

$$
\begin{array}{c}
COO^- \\
| \\
H_3{}^+N-C-H \\
| \\
CH_2 \\
| \\
OH
\end{array}
\qquad
\begin{array}{c}
COO^- \\
| \\
H_3{}^+N-C-H \\
| \\
H-C-OH \\
| \\
CH_3
\end{array}
\qquad
\begin{array}{c}
COO^- \\
| \\
H_3{}^+N-C-H \\
| \\
CH_2 \\
| \\
SH
\end{array}
$$

$$
\text{L-Serine} \qquad\qquad \text{L-Threonine} \qquad\qquad \text{L-Cysteine}
$$

$$
\begin{array}{c}
COO^- \\
| \\
H_3{}^+N-C-H \\
| \\
CH_2 \\
\\
\\
OH
\end{array}
\qquad
\begin{array}{c}
COO^- \\
| \\
H_3{}^+N-C-H \\
| \\
CH_2 \\
| \\
C \\
O \quad NH_2
\end{array}
\qquad
\begin{array}{c}
COO^- \\
| \\
H_3{}^+N-C-H \\
| \\
CH_2 \\
| \\
CH_2 \\
| \\
C \\
O \quad NH_2
\end{array}
$$

$$
\text{L-Tyrosine} \qquad\quad \text{L-Asparagine} \qquad\quad \text{L-Glutamine}
$$

18.33 A peptide bond is an amide bond between two amino acids in a peptide chain.

18.35 Linus Pauling and his colleagues carried out X-ray diffraction studies of protein. Interpretation of the pattern formed when X-rays were diffracted by a crystal of pure protein led Pauling to conclude that peptide bonds are both planar (flat) and rigid and that the N—C bonds are shorter than expected. In other words, they deduced that the peptide bond has a partial double bond character because it exhibits resonance. There is no free rotation about the amide bond because the carbonyl group of the amide bond has a strong attraction for the amide nitrogen lone pair of electrons.

18.37

a.

$$H_3^+N-CH-\overset{\overset{\displaystyle O}{\|}}{C}-N(H)-CH-\overset{\overset{\displaystyle O}{\|}}{C}-N(H)-CH-\overset{\overset{\displaystyle O}{\|}}{C}-O^-$$

with side chains CH_2 (benzyl, phenyl ring); $CH-CH_3$ and CH_3 (isoleucine); CH_2 (phenol ring with OH).

b.

$$H_3^+N-CH-\overset{\overset{\displaystyle O}{\|}}{C}-N(H)-CH-\overset{\overset{\displaystyle O}{\|}}{C}-N(H)-CH-\overset{\overset{\displaystyle O}{\|}}{C}-O^-$$

with side chains CH_3; $CH_2-CH_2-C=O$ with O^-; CH_2-SH.

c.

$$H_3^+N-CH-\overset{\overset{\displaystyle O}{\|}}{C}-N(H)-CH-\overset{\overset{\displaystyle O}{\|}}{C}-N(H)-CH-\overset{\overset{\displaystyle O}{\|}}{C}-O^-$$

with side chains $CH_2-C=O$ with NH_2; $CH_2-CH-CH_3$ with CH_3; H.

18.39 The primary structure of a protein is the sequence of amino acids bonded to one another by peptide bonds.

18.41 The primary structure of a protein determines its three-dimensional shape and biological function because the location of R groups along the protein chain is determined by the primary structure. The interactions among the R groups, based on their location in the chain, will govern how the protein folds. This, in turn, dictates its three-dimensional structure and biological function.

18.43

$$H_3^+N-CH-\overset{\overset{\displaystyle O}{\|}}{C}-O^- \quad H_3^+N-CH-\overset{\overset{\displaystyle O}{\|}}{C}-O^- \longrightarrow H_3^+N-CH-\overset{\overset{\displaystyle O}{\|}}{C}-N(H)-CH-\overset{\overset{\displaystyle O}{\|}}{C}-O^- + H_2O$$

with side chains: first $CH_2-CH-CH_3$ with CH_3; second $CH_2-CH_2-CH_2-NH-C=NH_2^+$ with NH_2.

18.45 The secondary structure of a protein is the folding of the primary structure into an α-helix or β-pleated sheet.

18.47 a. α-Helix
b. β-Pleated sheet

18.49 A fibrous protein is one that is composed of peptides arranged in long sheets or fibers.

18.51 A parallel β-pleated sheet is one in which the hydrogen bonded peptide chains have their amino-termini aligned head-to-head.

18.53 The tertiary structure of a protein is the globular, three-dimensional structure of a protein that results from folding the regions of secondary structure.

18.55

$$
\begin{array}{c}
COO^- \\
| \\
H_3^+N-C-H \\
| \\
CH_2 \\
| \\
S \\
| \\
S \\
| \\
CH_2 \\
| \\
H_3^+N-C-H \\
| \\
COO^-
\end{array}
$$

18.57 The tertiary structure is a level of folding of a protein chain that has already undergone secondary folding. The regions of α-helix and β-pleated sheet are folded into a globular structure.

18.59 Quaternary protein structure is the aggregation of two or more folded peptide chains to produce a functional protein.

18.61 A glycoprotein is a protein with covalently attached sugars.

18.63 Hydrogen bonding maintains the secondary structure of a protein and contributes to the stability of the tertiary and quaternary levels of structure.

18.65 The peptide bond exhibits resonance, which results in a partially double bonded character. This causes the rigidity of the peptide bond.

$$
\left[
\begin{array}{ccc}
\overset{\ddot{\ddot{O}}}{\underset{\displaystyle \|}{}} & & \overset{\ddot{O}^-}{} \\
R-C-\ddot{N}-R' & \longleftrightarrow & R-C=\overset{+}{N}-R' \\
\quad\quad | & & \quad\quad | \\
\quad\quad H & & \quad\quad H
\end{array}
\right]
$$

18.67 The code for the primary structure of a protein is carried in the genetic information (DNA).

18.69 The function of hemoglobin is to carry oxygen from the lungs to oxygen-demanding tissues throughout the body. Hemoglobin is found in red blood cells.

18.71 Hemoglobin is a protein composed of four subunits—two α-globin and two β-globin subunits. Each subunit holds a heme group, which in turn carries an Fe^{2+} ion.

18.73 The function of the heme group in hemoglobin and myoglobin is to bind to molecular oxygen.

18.75 Because carbon monoxide binds tightly to the heme groups of hemoglobin, it is not easily removed or replaced by oxygen. As a result, the effects of oxygen deprivation (suffocation) occur.

18.77 When sickle cell hemoglobin (HbS) is deoxygenated, the amino acid valine fits into a hydrophobic pocket on the surface of another HbS molecule. Many such sickle cell hemoglobin molecules polymerize into long rods that cause the red blood cell to sickle. In normal hemoglobin, glutamic acid is found in the place of the valine. This negatively charged amino acid will not "fit" into the hydrophobic pocket.

18.79 When individuals have one copy of the sickle cell gene and one copy of the normal gene, they are said to carry the sickle cell trait. These individuals will not suffer serious side effects, but may pass the trait to their offspring. Individuals with two copies of the sickle cell globin gene exhibit all the symptoms of the disease and are said to have sickle cell anemia.

18.81 Albumin

18.83 Albumin in the blood can serve as a carrier for Ca^{2+} because it contains acidic amino acids. The negative charges in the acidic amino acids can form salt bridges (or ionic bonds) with Ca^{2+}. Albumin can also serve as a carrier for fatty acids because it contains basic amino acids. The positive charges in the basic amino acids can form salt bridges with the anionic fatty acids.

18.85 Denaturation is the process by which the organized structure of a protein is disrupted, resulting in a completely disorganized, nonfunctional form of the protein.

18.87 Heat is an effective means of sterilization because it destroys the proteins of microbial life-forms, including fungi, bacteria, and viruses.

18.89 The low pH of the yogurt denatures the proteins of microbial contaminants, inhibiting their growth.

18.91 An essential amino acid is one that must be provided in the diet because it cannot be synthesized in the body.

18.93 A complete protein is one that contains all of the essential and nonessential amino acids.

18.95 Chymotrypsin catalyzes the hydrolysis of peptide bonds on the carbonyl side of aromatic amino acids.

18.97 In a vegetarian diet, vegetables are the only source of dietary protein. Because most individual vegetable sources do not provide all the needed amino acids, vegetables must be mixed to provide all the essential and nonessential amino acids in the amounts required for biosynthesis.

18.99 Synthesis of digestive enzymes must be carefully controlled because the active enzyme would digest, and thus destroy, the cell that produces it.

Chapter 19

19.1 **a.** Pyruvate kinase catalyzes the transfer of a phosphoryl group from phosphoenolpyruvate to adenosine diphosphate.

b. Alanine transaminase catalyzes the transfer of an amino group from alanine to α-ketoglutarate, producing pyruvate and glutamate.

c. Triose phosphate isomerase catalyzes the isomerization of the ketone dihydroxyacetone phosphate to the aldehyde glyceraldehyde-3-phosphate.

d. Pyruvate dehydrogenase catalyzes the oxidation and decarboxylation of pyruvate, producing acetyl coenzyme A and CO_2.

19.3 **a.** Sucrose
   b. Pyruvate
   c. Succinate

19.5 The induced fit model assumes that the enzyme is flexible. Both the enzyme and the substrate are able to change shape to form the enzyme-substrate complex. The lock-and-key model assumes that the enzyme is inflexible (the lock) and the substrate (the key) fits into a specific rigid site (the active site) on the enzyme to form the enzyme-substrate complex.

19.7 An enzyme might distort a bond, thereby catalyzing bond breakage. An enzyme could bring two reactants into close proximity and in the proper orientation for the reaction to occur. Finally, an enzyme could alter the pH of the microenvironment of the active site, thereby serving as a transient donor or acceptor of H^+.

19.9 Water-soluble vitamins are required by the body for the synthesis of coenzymes that are required for the function of a variety of enzymes.

19.11 A decrease in pH will change the degree of ionization of the R groups within a peptide chain. This disturbs the weak interactions that maintain the structure of an enzyme, which may denature the enzyme. Less drastic alterations in the charge of R groups in the active site of the enzyme can inhibit enzyme-substrate binding or destroy the catalytic ability of the active site.

19.13 Irreversible inhibitors bind very tightly, sometimes even covalently, to an R group in enzyme active sites. They generally inhibit many different enzymes. The loss of enzyme activity impairs normal cellular metabolism, resulting in the death of the cell or the individual.

19.15 A structural analog is a molecule that has a structure and charge distribution very similar to that of the natural substrate of an enzyme. Generally, they are able to bind to the enzyme active site. This inhibits enzyme activity because the normal substrate must compete with the structural analog to form an enzyme-substrate complex.

19.17 a.

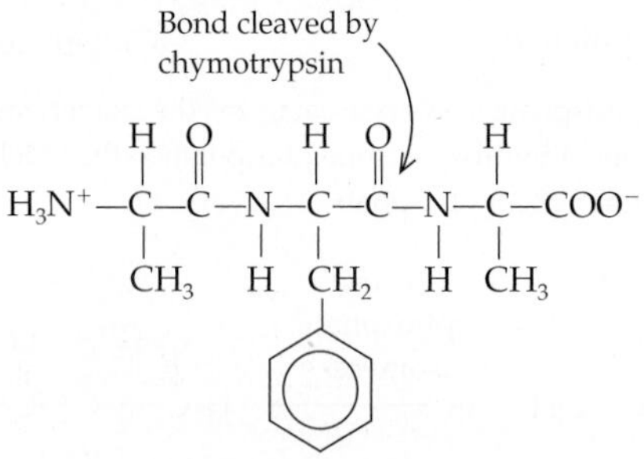

Bond cleaved by chymotrypsin

$$H_3N^+ —C—C—N—C—C—N—C—COO^-$$

ala-phe-ala

b.

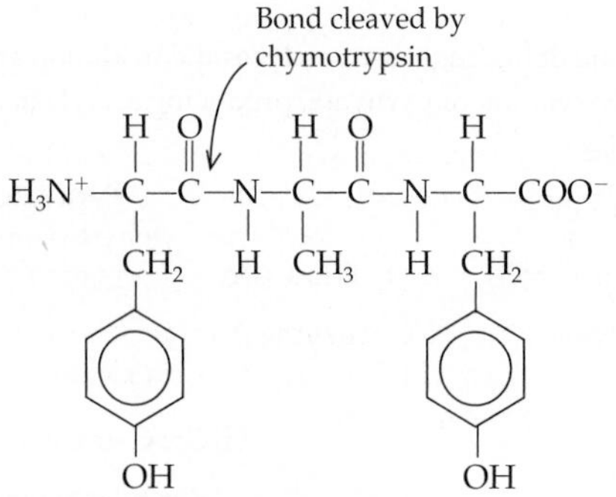

Bond cleaved by chymotrypsin

$$H_3N^+ —C—C—N—C—C—N—C—COO^-$$

tyr-ala-tyr

19.19

Chymotrypsin   Elastase   Elastase

$$H_3^+N—C—C—N—C—C—N—C—C—N—C—C—N—C—COO^-$$

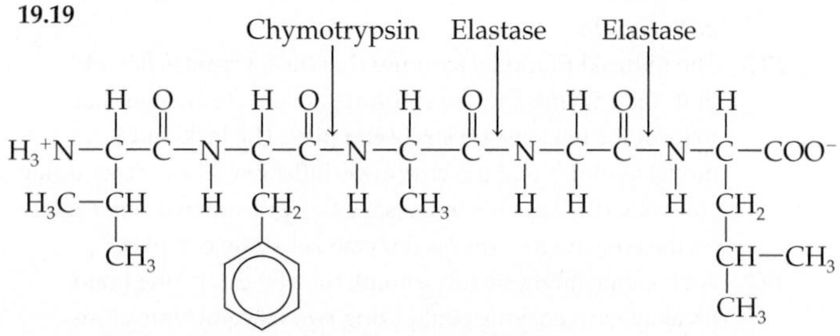

19.21 The common name of an enzyme is often derived from the name of the substrate and/or the type of reaction that it catalyzes.

19.23 1. Urease
2. Peroxidase
3. Lipase
4. Aspartase
5. Glucose-6-phosphatase
6. Sucrase

19.25 a. Citrate decarboxylase catalyzes the cleavage of a carboxyl group from citrate.
b. Adenosine diphosphate phosphorylase catalyzes the addition of a phosphate group to ADP.
c. Oxalate reductase catalyzes the reduction of oxalate.
d. Nitrite oxidase catalyzes the oxidation of nitrite.
e. *cis-trans* Isomerase catalyzes interconversion of *cis* and *trans* isomers.

19.27 A substrate is the reactant in an enzyme-catalyzed reaction that binds to the active site of the enzyme and is converted into product.

19.29 The activation energy of a reaction is the energy required for the reaction to occur.

19.31 The equilibrium constant for a chemical reaction is a reflection of the difference in energy of the reactants and products. Consider the following reaction:

$$aA + bB \longrightarrow cC + dD$$

The equilibrium constant for this reaction is:

$$K_{eq} = [D]^d[C]^c/[A]^a[B]^b = [products]/[reactants]$$

Because the difference in energy between reactants and products is the same regardless of what path the reaction takes, an enzyme does not alter the equilibrium constant of a reaction.

19.33 The rate of an uncatalyzed chemical reaction typically doubles every time the substrate concentration is doubled.

19.35 The rate-limiting step is that step in an enzyme-catalyzed reaction that is the slowest, and hence limits the speed with which the substrate can be converted into product.

19.37

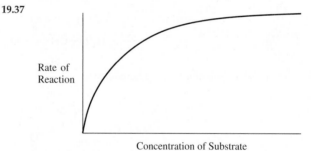

Rate of Reaction

Concentration of Substrate

19.39 The enzyme-substrate complex is the molecular aggregate formed when the substrate binds to the active site of an enzyme.

19.41 The catalytic groups of an enzyme active site are those functional groups that are involved in carrying out catalysis.

19.43 Enzyme active sites are pockets in the surface of an enzyme that include R groups involved in binding and R groups involved in catalysis. The shape of the active site is complementary to the shape of the substrate. Thus, the conformation of the active site determines the specificity of the enzyme. Enzyme-substrate binding involves weak, noncovalent interactions.

19.45 The lock-and-key model of enzyme-substrate binding was proposed by Emil Fischer in 1894. He thought that the active site was a rigid region of the enzyme into which the substrate fit perfectly. Thus, the model purports that the substrate simply snaps into place within the active site, like two pieces of a jigsaw puzzle fitting together.

19.47 Enzyme specificity is the ability of an enzyme to bind to only one, or a very few, substrates and thus catalyze only a single reaction.

19.49 Group specificity means that an enzyme catalyzes reactions involving similar molecules having the same functional group.

19.51 Absolute specificity means that an enzyme catalyzes the reaction of only one substrate.

19.53 Hexokinase has group specificity. The advantage is that the cell does not need to encode many enzymes to carry out the phosphorylation of six-carbon sugars. Hexokinase can carry out many of these reactions.

19.55 Methionyl tRNA synthetase has absolute specificity. This is the enzyme that attaches the amino acid methionine to the transfer RNA (tRNA) that will carry the amino acid to the site of protein synthesis. If the wrong amino acid were attached to the tRNA, it could be incorporated into the protein, destroying its correct three-dimensional structure and biological function.

19.57 The first step of an enzyme-catalyzed reaction is the formation of the enzyme-substrate complex. In the second step, the transition state is formed. This is the state in which the substrate assumes a form intermediate between the original substrate and the product. In step 3, the substrate is converted to product and the enzyme-product complex is formed. Step 4 involves the release of the product and regeneration of the enzyme in its original form.

19.59 In a reaction involving bond breaking, the enzyme might distort a bond, producing a transition state in which the bond is stressed. An enzyme could bring two reactants into close proximity and in the proper orientation for the reaction to occur, producing a transition state in which the proximity of the reactants facilitates bond formation. Finally, an enzyme could alter the pH of the microenvironment of the active site, thereby serving as a transient donor or acceptor of H^+.

19.61 A cofactor helps maintain the shape of the active site of an enzyme.

19.63 Thiamine (B_1) is found in the coenzyme thiamine pyrophosphate. Riboflavin (B_2) is found in both flavin mononucleotide and flavin adenine dinucleotide. Niacin (B_3) is found in both nicotinamide adenine dinucleotide and nicotinamide adenine dinucleotide phosphate. Pyridoxine (B_6) is found in both pyridoxal phosphate and pyridoxamine phosphate. Cyanocobalamin (B_{12}) is found in deoxyadenosyl cobalamin. Folic acid is found in tetrahydrofolic acid. Pantothenic acid is found in coenzyme A. Biotin is found in biocytin.

19.65 At the temperature optimum, the enzyme is functioning optimally and the rate of the reaction is maximal. Above the temperature optimum, increasing temperature begins to denature the enzyme and stop the reaction.

19.67 Each of the following answers assumes that the enzyme was purified from an organism with optimal conditions for life near 37°C, pH 7.
  a. Decreasing the temperature from 37°C to 10°C will cause the rate of an enzyme-catalyzed reaction to decrease because the frequency of collisions between enzyme and substrate will decrease as the rate of molecular movement decreases.
  b. Increasing the pH from 7 to 11 will generally cause a decrease in the rate of an enzyme-catalyzed reaction. In fact, most enzymes would be denatured by a pH of 11 and enzyme activity would cease.
  c. Heating an enzyme from 37°C to 100°C will destroy enzyme activity because the enzyme would be denatured by the extreme heat.

19.69 High temperature denatures bacterial enzymes and structural proteins. Because the life of the cell is dependent on the function of these proteins, the cell dies.

19.71 A lysosome is a membrane-bound vesicle in the cytoplasm of cells that contains approximately fifty hydrolytic enzymes. Some of the enzymes in the lysosomes can degrade proteins to amino acids, others hydrolyze polysaccharides into monosaccharides, and some degrade lipids and nucleic acids. The lysosome contains these enzymes to prevent degradation of large biological molecules that are important to maintain cell integrity.

19.73 Enzymes used for clinical assays in hospitals are typically stored at refrigerator temperatures to ensure that they are not denatured by heat. In this way, they retain their activity for long periods.

19.75 **a.** Cells regulate the level of enzyme activity to conserve energy. It is a waste of cellular energy to produce an enzyme if its substrate is not present or if its product is in excess.
  b. Production of proteolytic digestive enzymes must be carefully controlled because the active enzyme could destroy the cell that produces it. Thus, they are produced in an inactive form in the cell and are only activated at the site where they carry out digestion.

19.77 In positive allosterism, binding of the effector molecule turns the enzyme on. In negative allosterism, binding of the effector molecule turns the enzyme off.

19.79 A proenzyme is the inactive form of an enzyme that is converted to the active form at the site of its activity.

19.81 Blood clotting is a critical protective mechanism in the body, preventing excessive loss of blood following an injury. However, it can be a dangerous mechanism if it is triggered inappropriately. The resulting clot could cause a heart attack or stroke. By having a cascade of proteolytic reactions leading to the final formation of the clot, there are many steps at which the process can be regulated. This ensures that it will only be activated under the appropriate conditions.

19.83 Competitive enzyme inhibition occurs when a structural analog of the normal substrate occupies the enzyme active site so that the reaction cannot occur. The structural analog and the normal substrate compete for the active site. Thus, the rate of the reaction will depend on the relative concentrations of the two molecules.

19.85 A structural analog has a shape and charge distribution that are very similar to those of the normal substrate for an enzyme.

19.87 Irreversible inhibitors bind tightly to and block the active site of an enzyme and eliminate catalysis at the site.

19.89 The compound would be a competitive inhibitor of the enzyme.

19.91 A proteolytic enzyme catalyzes the cleavage of the peptide bond that maintains the primary protein structure.

19.93 The structural similarities among chymotrypsin, trypsin, and elastase suggest that these enzymes evolved from a single ancestral gene that was duplicated. Each copy then evolved independently.

19.95

tyr-lys-ala-phe

19.97   Elastase will cleave the peptide bonds on the carbonyl side of alanine and glycine. Trypsin will cleave the peptide bonds on the carbonyl side of lysine and arginine. Chymotrypsin will cleave the peptide bonds on the carbonyl side of tryptophan and phenylalanine.

19.99   Analysis of blood serum for levels of certain enzymes can confirm a preliminary diagnosis that was made based on disease symptoms or a clinical picture. When cells die, they release their enzymes into the bloodstream. Enzyme assays can measure amounts of certain enzymes in the blood.

19.101   Creatine kinase-MB and aspartate aminotransferase (AST/SGOT)

19.103   Urease is used in the clinical analysis of urea in blood. In a test called the blood urea nitrogen test (BUN), urea is converted to ammonia using the enzyme urease; the ammonia becomes an indicator of urea. This allows for the levels of urea to be measured. This measurement is useful in the diagnosis of kidney malfunction.

Chapter 20

20.1 a. Adenosine diphosphate:

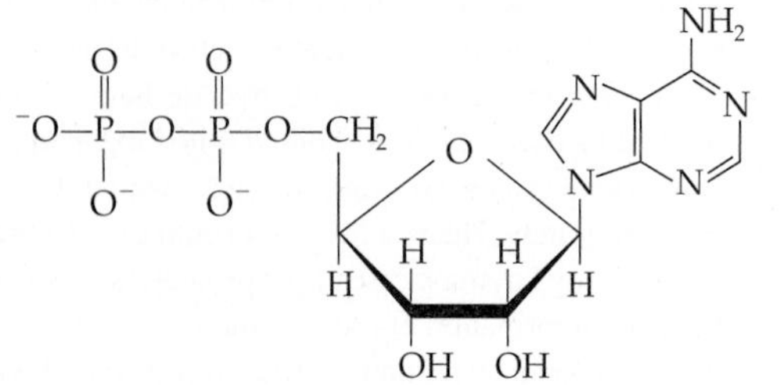

b. Deoxyguanosine triphosphate:

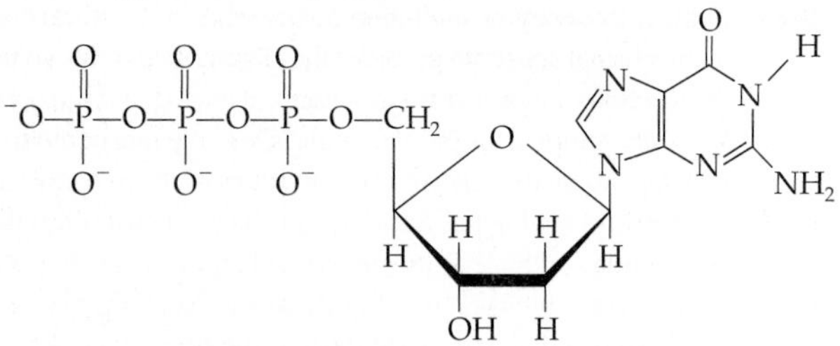

20.3   The RNA polymerase recognizes the promoter site for a gene, separates the strands of DNA, and catalyzes the polymerization of an RNA strand complementary to the DNA strand that carries the genetic code for a protein. It recognizes a termination site at the end of the gene and releases the RNA molecule.

20.5   The genetic code is said to be degenerate because several different triplet codons may serve as code words for a single amino acid.

20.7   The nitrogenous bases of the codons are complementary to those of the anticodons. As a result, they are able to hydrogen bond to one another according to the base pairing rules.

20.9   The ribosomal P-site holds the peptidyl tRNA during protein synthesis. The peptidyl tRNA is the tRNA carrying the growing peptide chain. The only exception to this is during initiation of translation when the P-site holds the initiator tRNA.

20.11   The normal mRNA sequence, AUG-CCC-GAC-UUU, would encode the peptide sequence methionine-proline-aspartate-phenylalanine. The mutant mRNA sequence, AUG-CGC-GAC-UUU, would encode the mutant peptide sequence methionine-arginine-aspartate-phenylalanine. This would not be a silent mutation because a hydrophobic amino acid (proline) has been replaced by a positively charged amino acid (arginine).

20.13   A heterocyclic amine is a compound that contains nitrogen in at least one position of the ring skeleton.

20.15   It is the N-9 of the purine that forms the N-glycosidic bond with C-1 of the five-carbon sugar. The general structure of the purine ring is shown below:

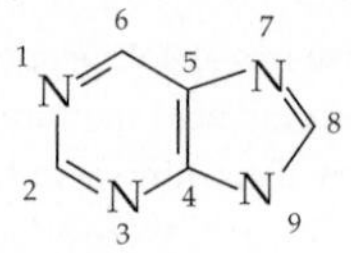

20.17   The ATP nucleotide is composed of the five-carbon sugar ribose, the purine adenine, and a triphosphate group.

20.19   The two strands of DNA in the double helix are said to be antiparallel because they run in opposite directions. One strand progresses in the $5' \rightarrow 3'$ direction, and the opposite strand progresses in the $3' \rightarrow 5'$ direction.

20.21   The DNA double helix is 2 nm in width. The nitrogenous bases are stacked at a distance of 0.34 nm from one another. One complete turn of the helix is 3.4 nm, or 10 base pairs.

20.23

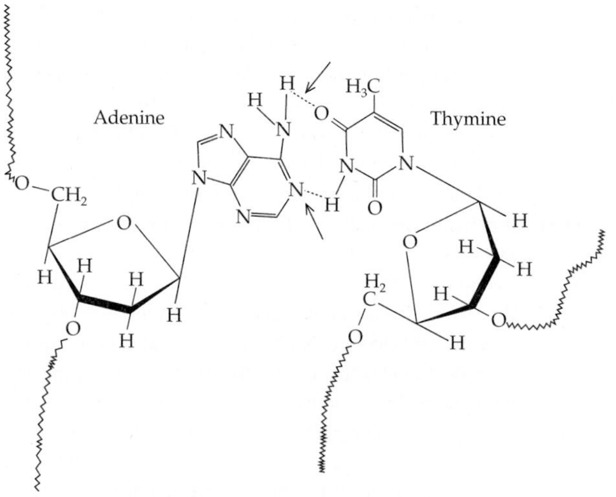

20.25

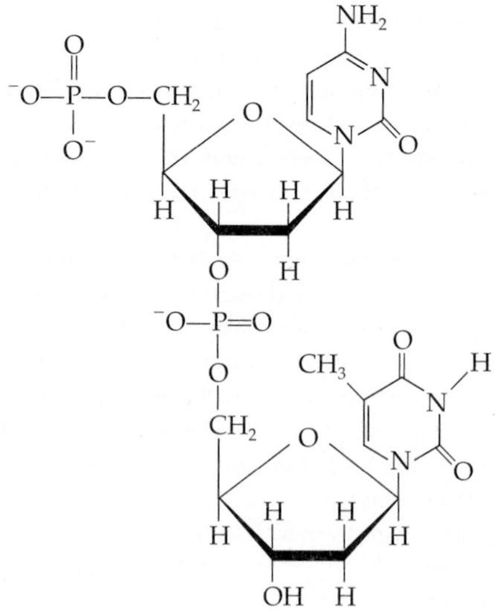

20.27   The prokaryotic chromosome is a circular DNA molecule that is supercoiled; that is, the helix is coiled on itself.

20.29   The term semiconservative DNA replication refers to the fact that each parental DNA strand serves as the template for the synthesis of a daughter strand. As a result, each of the daughter DNA molecules is made up of one strand of the original parental DNA and one strand of newly synthesized DNA.

20.31 The two primary functions of DNA polymerase III are to read a template DNA strand and catalyze the polymerization of a new daughter strand, and to proofread the newly synthesized strand and correct any errors by removing the incorrectly inserted nucleotide and adding the proper one.

20.33 3′-TACGGGCTCGACTAACTAGTCT-5′

20.35 The replication origin of a DNA molecule is the unique sequence on the DNA molecule where DNA replication begins.

20.37 The enzyme helicase separates the strands of DNA at the origin of DNA replication so that the proteins involved in replication can interact with the nitrogenous base pairs.

20.39 The RNA primer "primes" DNA replication by providing a 3′—OH which can be used by DNA polymerase III for the addition of the next nucleotide in the growing DNA chain.

20.41 DNA → RNA → Protein

20.43 Anticodons are found on transfer RNA molecules.

20.45 3′-AUGCCCGUAUCCGGAAUUUCGAUCGAA-5′

20.47 RNA splicing is the process by which the noncoding sequences (introns) of the primary transcript of a eukaryotic mRNA are removed and the protein coding sequences (exons) are spliced together.

20.49 Messenger RNA, transfer RNA, and ribosomal RNA

20.51 Spliceosomes are small ribonucleoprotein complexes that carry out RNA splicing.

20.53 The poly(A) tail is a stretch of 100–200 adenosine nucleotides polymerized onto the 3′ end of a mRNA by the enzyme poly(A) polymerase.

20.55 The cap structure is made up of the nucleotide 7-methylguanosine attached to the 5′ end of a mRNA by a 5′-5′ triphosphate bridge. Generally, the first two nucleotides of the mRNA are also methylated.

20.57 Sixty-four

20.59 The reading frame of a gene is the sequential set of triplet codons that carries the genetic code for the primary structure of a protein.

20.61 Methionine and tryptophan

20.63 The codon 5′-UUU-3′ encodes the amino acid phenylalanine. The mutant codon 5′-UUA-3′ encodes the amino acid leucine. Both leucine and phenylalanine are hydrophobic amino acids; however, leucine has a smaller R group. It is possible that the smaller R group would disrupt the structure of the protein.

20.65 The ribosomes serve as a platform on which protein synthesis can occur. They also carry the enzymatic activity that forms peptide bonds.

20.67 5′-AUG GCU GGG CUU UGU AUG UGG UAU UCU AUU GGG UAA-3′

20.69 The sequence of DNA nucleotides in a gene is transcribed to produce a complementary sequence of RNA nucleotides in a messenger RNA (mRNA). In the process of translation, the sequence of the mRNA is read sequentially in words of three nucleotides (codons) to produce a protein. Each codon calls for the addition of a particular amino acid to the growing peptide chain. If one of those codons has been altered by mutation, it may now call for the addition of the wrong amino acid to the growing peptide chain. This could result in improper folding of the protein and in loss of biological function.

20.71 An ester bond

20.73 A point mutation is the substitution of one nucleotide pair for another in a gene.

20.75 Some mutations are silent because the change in the nucleotide sequence does not alter the amino acid sequence of the protein. This can happen because there are many amino acids encoded by multiple codons.

20.77 UV light causes the formation of pyrimidine dimers, the covalent bonding of two adjacent pyrimidine bases. Mutations occur when the UV damage repair system makes an error during the repair process. This causes a change in the nucleotide sequence of the DNA.

20.79 **a.** A carcinogen is a compound that causes cancer. Cancers are caused by mutations in the genes responsible for controlling cell division.

b. Carcinogens cause DNA damage that results in changes in the nucleotide sequence of the gene. Thus, carcinogens are also mutagens.

20.81 A restriction enzyme is a bacterial enzyme that "cuts" the sugar–phosphate backbone of DNA molecules at a specific nucleotide sequence.

20.83 A selectable marker is a genetic trait that can be used to detect the presence of a plasmid in a bacterium. Many plasmids have antibiotic resistance genes as selectable markers. Bacteria containing the plasmid will be able to grow in the presence of the antibiotic; those without the plasmid will be killed.

20.85 Human insulin, interferon, human growth hormone, and human blood clotting factor VIII

20.87 4096 copies

20.89 The goals of the Human Genome Project were to identify and map all of the genes of the human genome and to determine the DNA sequences of the complete three billion nucleotide pairs.

20.91 A genome library is a set of clones that represents all of the DNA sequences in the genome of an organism.

20.93 A dideoxynucleotide is one that has hydrogen atoms rather than hydroxyl groups bonded to both the 2′ and 3′ carbons of the five-carbon sugar.

20.95 Sequences that these DNA sequences have in common are highlighted in bold.

a. 5′-**AGCTCCT**GATTTCATACAGTTTCTACT**ACCTACTA**-3′

b. 5′-AGACATTCTATCTACCTAGACTATG**TTCAGAA**-3′

c. 5′-**TTCAGAA**CTCATTCAGACCTACTACTATACCTTGG **GAGCTCCT**-3′

d. 5′-**ACCTACTA**GACTATACTACTACTAAGGGGACTATT CCAGACTT-3′

The 5′ end of sequence (a) is identical to the 3′ end of sequence (c). The 3′ end of sequence (a) is identical to the 5′ end of sequence (d). The 3′ end of sequence (b) is identical to the 5′ end of sequence (c). From 5′ to 3′, the sequences would form the following map:

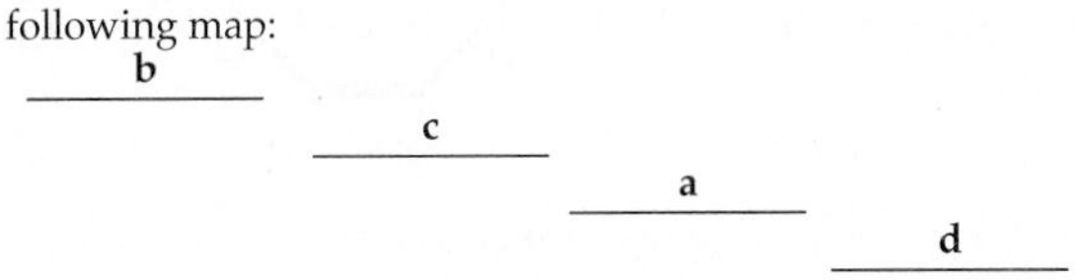

```
      b
 _____
         _____
         c
               _____
               a
                     _____
                     d
```

Chapter 21

21.1 ATP is called the universal energy currency because it is the major molecule used by all organisms to store energy.

21.3 The first stage of catabolism is the digestion (hydrolysis) of dietary macromolecules in the stomach and intestine.

In the second stage of catabolism, monosaccharides, amino acids, fatty acids, and glycerol are converted by metabolic reactions into molecules that can be completely oxidized.

In the third stage of catabolism, the two-carbon acetyl group of acetyl CoA is completely oxidized by the reactions of the citric acid cycle. The energy of the electrons harvested in these oxidation reactions is used to make ATP.

21.5 Substrate level phosphorylation is one way the cell can make ATP. In this reaction, a high-energy phosphoryl group of a substrate in the reaction is transferred to ADP to produce ATP.

21.7 Glycolysis is a pathway involving ten reactions. In reactions 1–3, energy is invested in the beginning substrate, glucose. This is done by transferring high-energy phosphoryl groups from ATP to the intermediates in the pathway. The product is fructose-1,6-bisphosphate. In the energy-harvesting reactions of glycolysis, fructose-1,6-bisphosphate is split into two three-carbon molecules that begin a series of rearrangement, oxidation-reduction, and substrate-level phosphorylation reactions that produce four ATP, two NADH, and two pyruvate molecules. Because of the investment of two ATP in the early steps of glycolysis, the net yield of ATP is two.

21.9 Both the alcohol and lactate fermentations are anaerobic reactions that use the pyruvate and re-oxidize the NADH produced in glycolysis.

21.11 Gluconeogenesis (synthesis of glucose from noncarbohydrate sources) appears to be the reverse of glycolysis (the first stage of carbohydrate degradation) because the intermediates in the two pathways are the same. However, reactions 1, 3, and 10 of glycolysis are not reversible reactions. Thus, the reverse reactions must be carried out by different enzymes.

21.13 The enzyme glycogen phosphorylase catalyzes the phosphorolysis of a glucose unit at one end of a glycogen molecule. The reaction involves the displacement of the glucose by a phosphate group. The products are glucose-1-phosphate and a glycogen molecule that is one glucose unit shorter.

21.15 Glucokinase traps glucose within the liver cell by phosphorylating it. Because the product, glucose-6-phosphate, is charged, it cannot be exported from the cell.

21.17 Glucagon indirectly stimulates glycogen phosphorylase, the first enzyme of glycogenolysis. This speeds up glycogen degradation. Glucagon also inhibits glycogen synthase, the first enzyme in glycogenesis. This inhibits glycogen synthesis.

21.19 ATP

21.21

Adenosine triphosphate

$+ H_2O \longrightarrow$

Adenosine diphosphate

Inorganic phosphate group

21.23 A coupled reaction is one that can be thought of as a two-step process. In a coupled reaction, two reactions occur simultaneously. Frequently, one of the reactions releases the energy that drives the second, energy-requiring reaction.

21.25 Carbohydrates

21.27 The following equation represents the hydrolysis of maltose:

β-Maltose

$+ H_2O$

$\longrightarrow \quad 2$

β-D-Glucose

21.29

$+ 4H_2O$

21.31 The hydrolysis of a triglyceride containing oleic acid, stearic acid, and linoleic acid is represented in the following equation:

$H-\overset{H}{\underset{H}{C}}-O-\overset{O}{C}-(CH_2)_7CH=CH(CH_2)_7CH_3$

$H-C-O-\overset{O}{C}-(CH_2)_{16}CH_3 \qquad + 3H_2O \longrightarrow$

$H-C-O-\overset{O}{C}-(CH_2)_7CH=CHCH_2CH=CH(CH_2)_4CH_3$

Glycerol + Oleic acid + Stearic acid + Linoleic acid

21.33 The hydrolysis of the dipeptide alanyl leucine is represented in the following equation:

Alanyl leucine

Alanine + Leucine

21.35 Glycolysis is the enzymatic pathway that converts a glucose molecule into two molecules of pyruvate. The pathway generates a net energy yield of two ATP and two NADH. Glycolysis is the first stage of carbohydrate catabolism.

21.37 Glycolysis requires NAD^+ for reaction 6 in which glyceraldehyde-3-phosphate dehydrogenase catalyzes the oxidation of glyceraldehyde-3-phosphate. NAD^+ is reduced.

21.39 Two ATP per glucose

21.41 Although muscle cells have enough ATP stored for only a few seconds of activity, glycolysis speeds up dramatically when there is a demand for more energy. If the cells have a sufficient supply of oxygen, aerobic respiration (the citric acid cycle and oxidative phosphorylation) will contribute large amounts of ATP. If oxygen is limited, the lactate fermentation will speed up. This will use up the pyruvate and re-oxidize the NADH produced by glycolysis and allow continued synthesis of ATP for muscle contraction.

21.43 $C_6H_{12}O_6 + 2ADP + 2P_i + 2NAD+ \rightarrow$
Glucose $2C_3H_3O_3 + 2ATP + 2NADH + 2H_2O$
                                   Pyruvate

21.45
a. 7	**f.** 9
b. 2	**g.** 5
c. 6	**h.** 10
d. 4	**i.** 3
e. 1	**j.** 8

21.47 Myopathy and hemolytic anemia are symptoms associated with a genetic defect in some of the enzymes of glycolysis. Myopathy can lead to exercise intolerance, muscle breakdown, and blood

in the urine. Tarui's disease is also caused by a deficiency in one of the enzymes of glycolysis. Its symptoms include muscle pain, exercise intolerance, respiratory failure, heart muscle disease, seizures, and blindness.

21.49 If a person is deficient in some of the enzymes of glycolysis, muscle cells may begin to die, which can lead to the release of myoglobin into the blood and the urine. This condition is called myoglobinuria, and it results in urine that is the color of cola soft drinks.

21.51 Isomerase

21.53 Enediol

21.55 A kinase transfers a phosphoryl group from one molecule to another.

21.57 NAD^+ is reduced, accepting a hydride anion.

21.59 To optimize efficiency and minimize waste, it is important that energy-harvesting pathways, such as glycolysis, respond to the energy demands of the cell. If energy in the form of ATP is abundant, there is no need for the pathway to continue at a rapid rate. When this is the case, allosteric enzymes that catalyze the reactions of the pathway are inhibited by binding to their negative effectors. Similarly, when there is a great demand for ATP, the pathway speeds up as a result of the action of allosteric enzymes binding to positive effectors.

21.61 ATP and citrate are allosteric inhibitors of phosphofructokinase, whereas AMP and ADP are allosteric activators.

21.63 Citrate, which is the first intermediate in the citric acid cycle, is an allosteric inhibitor of phosphofructokinase. The citric acid cycle is a pathway that results in the complete oxidation of the pyruvate produced by glycolysis. A high concentration of citrate signals that sufficient substrate is entering the citric acid cycle. The inhibition of phosphofructokinase by citrate is an example of feedback inhibition: the product, citrate, allosterically inhibits the activity of an enzyme early in the pathway.

21.65

Acetaldehyde → Ethanol

21.67 Lactate fermentation

21.69 Yogurt and some cheeses

21.71 Lactate dehydrogenase

21.73 This child must have the enzymes to carry out the alcohol fermentation. When the child exercised hard, there was not enough oxygen in the cells to maintain aerobic respiration. As a result, glycolysis and the alcohol fermentation were responsible for the majority of the ATP production by the child. The accumulation of alcohol (ethanol) in the child caused the symptoms of drunkenness.

21.75 The ribose-5-phosphate is used for the biosynthesis of nucleotides. The erythrose-4-phosphate is used for the biosynthesis of aromatic amino acids.

21.77 Gluconeogenesis is production of glucose from noncarbohydrate starting materials. This pathway can provide glucose when starvation or strenuous exercise leads to a depletion of glucose from the body.

21.79 The liver

21.81 Lactate is first converted to pyruvate.

21.83 Because steps 1, 3, and 10 of glycolysis are irreversible, gluconeogenesis is not simply the reverse of glycolysis. The reverse reactions must be carried out by different enzymes.

21.85 Steps 1, 3, and 10 of glycolysis are irreversible. Step 1 is the transfer of a phosphoryl group from ATP to carbon-6 of glucose and is catalyzed by hexokinase. Step 3 is the transfer of a phosphoryl group from ATP to carbon-1 of fructose-6-phosphate and is catalyzed by phosphofructokinase. Step 10 is the substrate-level phosphorylation in which a phosphoryl group is transferred from phosphoenolpyruvate to ADP and is catalyzed by pyruvate kinase.

21.87 The liver and pancreas

21.89 Hypoglycemia is the condition in which blood glucose levels are too low.

21.91 **a.** Insulin stimulates glycogen synthase, the first enzyme in glycogen synthesis. It also stimulates uptake of glucose from the bloodstream into cells and phosphorylation of glucose by the enzyme glucokinase.

   b. This traps glucose within liver cells and increases the storage of glucose in the form of glycogen.

   c. These processes decrease blood glucose levels.

21.93 Any defect in the enzymes required to degrade glycogen or export glucose from liver cells will result in a reduced ability of the liver to provide glucose at times when blood glucose levels are low. This will cause hypoglycemia.

21.95 Glycogen phosphorylase catalyzes phosphorolysis of a glucose at one end of a glycogen polymer. The reaction involves the displacement of a glucose unit of glycogen by a phosphate group. As a result, glucose-1-phosphate is produced.

$$\text{Glycogen(glucose)}_x + n\text{HPO}_4^{2-} \longrightarrow \text{Glycogen(glucose)}_{n\text{-}x} + n\text{glucose-1-phosphate}$$

21.97 In glycogen degradation, phosphoglucomutase converts glucose-1-phosphate to glucose-6-phosphate. In glycogen synthesis, phosphoglucomutase converts glucose-6-phosphate to glucose-1-phosphate.

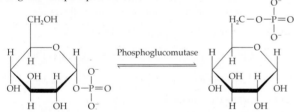

21.99 Glucokinase converts glucose to glucose-6-phosphate as the first reaction of glycogen synthesis.

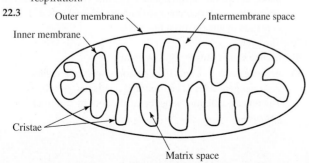

Chapter 22

22.1 Mitochondria are the organelles responsible for aerobic respiration.

22.3

Outer membrane
Intermembrane space
Inner membrane
Cristae
Matrix space

22.5 Pyruvate is converted to acetyl CoA by the pyruvate dehydrogenase complex. This huge enzyme complex requires four coenzymes, each of which is made from a different vitamin. The four coenzymes are thiamine pyrophosphate (made from thiamine), FAD (made from riboflavin), NAD⁺ (made from niacin), and coenzyme A (made from the vitamin pantothenic acid). The coenzyme lipoamide is also involved in this reaction.

22.7 Oxidative phosphorylation is the process by which the energy of electrons harvested from oxidation of a fuel molecule is used to phosphorylate ADP to produce ATP.

22.9 $\text{NAD}^+ + \text{H:}^- \longrightarrow \text{NADH}$

22.11 During transamination reactions, the α-amino group is transferred to the coenzyme pyridoxal phosphate. In the last part of the reaction, the α-amino group is transferred from pyridoxal phosphate to an α-keto acid.

22.13 The purpose of the urea cycle is to convert toxic ammonium ions to urea, which is excreted in the urine of land animals.

22.15 An amphibolic pathway is a metabolic pathway that functions both in anabolism and catabolism. The citric acid cycle is amphibolic because it has a catabolic function—it completely oxidizes the acetyl group carried by acetyl CoA to provide electrons for ATP synthesis. Because citric acid cycle intermediates are precursors for the biosynthesis of many other molecules, it also serves a function in anabolism.

22.17 The mitochondrion is an organelle that serves as the cellular power plant. The reactions of the citric acid cycle, the electron transport system, and ATP synthase function together within the mitochondrion to harvest ATP energy for the cell.

22.19 The intermembrane compartment is the location of the high-energy proton (H⁺) reservoir produced by the electron transport system. The energy of this H⁺ reservoir is used to make ATP.

22.21 The outer mitochondrial membrane is freely permeable to substances of molar mass less than 10,000 g/mol. The inner mitochondrial membrane is highly impermeable. Embedded within the inner mitochondrial membrane are the electron carriers of the electron transport system, and ATP synthase, the multisubunit enzyme that makes ATP.

22.23 Coenzyme A is a molecule derived from ATP and the vitamin pantothenic acid. It functions in the transfer of acetyl groups in lipid and carbohydrate metabolism.

22.25 Decarboxylation is a chemical reaction in which a carboxyl group is removed from a molecule.

22.27 Under aerobic conditions pyruvate is converted to acetyl CoA.

22.29 The coenzymes NAD⁺, FAD, thiamine pyrophosphate, and coenzyme A are required by the pyruvate dehydrogenase complex for the conversion of pyruvate to acetyl CoA. These coenzymes are synthesized from the vitamins niacin, riboflavin, thiamine, and pantothenic acid, respectively. If the vitamins are not available, the coenzymes will not be available and pyruvate cannot be converted to acetyl CoA. Because the complete oxidation of the acetyl group of acetyl CoA produces the vast majority of the ATP for the body, ATP production would be severely inhibited by a deficiency of any of these vitamins.

22.31 A condensation is a reaction in which aldehydes or ketones react to form larger molecules.

22.33 Oxidation reduction

22.35 A dehydrogenation reaction is an oxidation reaction in which protons and electrons are removed from a molecule.

22.37 **a.** 1　　　　　　　**e.** 4
    b. 6　　　　　　　**f.** 2
    c. 2　　　　　　　**g.** 5
    d. 2　　　　　　　**h.** 2

22.39 True

22.41 Three

22.43 Two ATP per glucose

22.45 The function of acetyl CoA in the citric acid cycle is to bring the two-carbon remnant (acetyl group) of pyruvate from glycolysis and transfer it to oxaloacetate. In this way, the acetyl group enters the citric acid cycle for the final stages of oxidation.

22.47 The high-energy phosphoryl group of the GTP is transferred to ADP to produce ATP. This reaction is catalyzed by the enzyme dinucleotide diphosphokinase.

22.49 Fumarate contains an alkene carbon-carbon double bond. Addition of water to the double bond of fumarate gives malate. The enzyme fumarase catalyzes this reaction. When water is added to the alkene double bond, one of the carbons forms a new bond to —OH, and the other carbon forms a new bond to —H. As a result, the alkene becomes an alcohol.

22.51 First, pyruvate is converted to acetyl CoA:

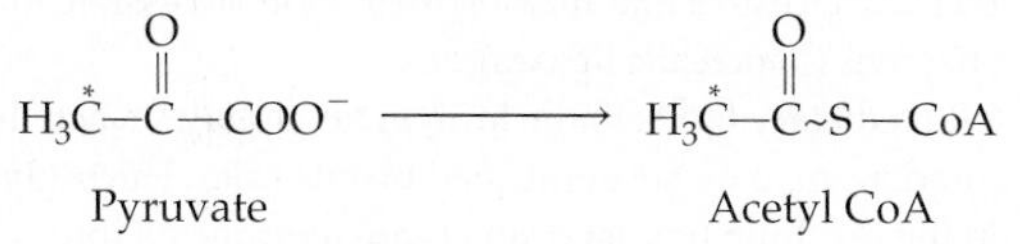

$$H_3\overset{*}{C}-\overset{O}{\underset{||}{C}}-COO^- \longrightarrow H_3\overset{*}{C}-\overset{O}{\underset{||}{C}}\sim S-CoA$$

Pyruvate → Acetyl CoA

Then, citrate is formed from oxaloacetate and the radiolabeled acetyl CoA.

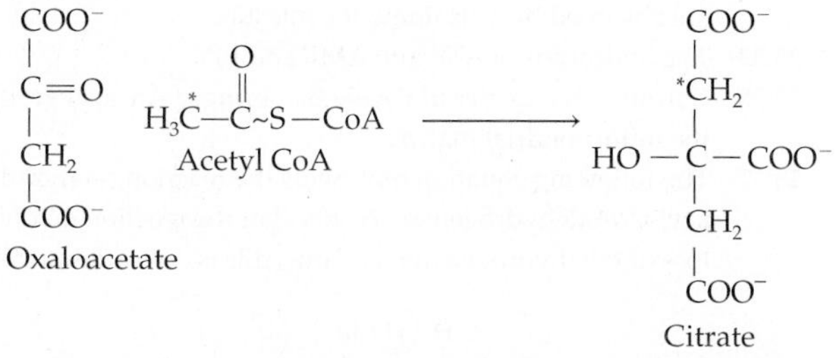

(structures: Oxaloacetate + Acetyl CoA → Citrate)

The following structures are the intermediates of the citric acid cycle. An asterisk is on the radiolabeled carbon, and circles are on the —COO⁻ groups that are released as CO_2.

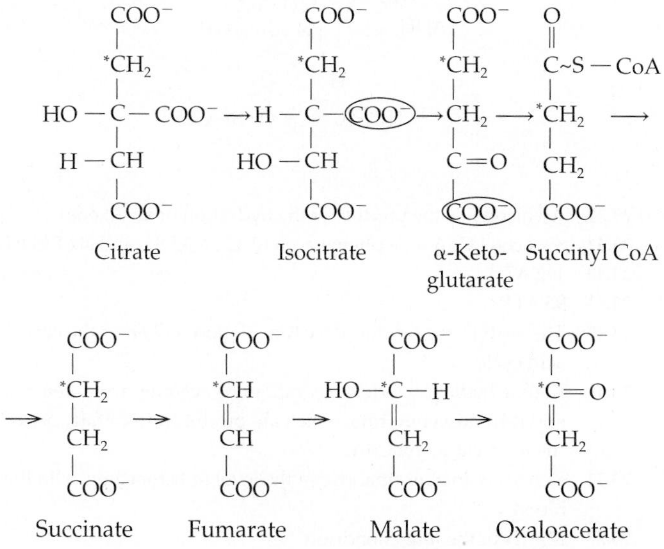

(structures: Citrate → Isocitrate → α-Ketoglutarate → Succinyl CoA)

(structures: Succinate → Fumarate → Malate → Oxaloacetate)

22.53 This reaction is an example of the oxidation of a secondary alcohol to a ketone. The two functional groups are the hydroxyl group of the alcohol and the carbonyl group of the ketone.

22.55 It is a kinase because it transfers a phosphoryl group from one molecule to another. Kinases are a specific type of transferase.

22.57 Mutations of the citric acid cycle enzymes frequently appear first in the central nervous system because of the high energy (ATP) demands of this tissue.

22.59 Deficiencies of citric acid cycle enzymes cause hypotonia because there is insufficient ATP.

22.61 An allosteric enzyme is one that has an effector binding site and an active site. Effector binding can change the shape of the active site, causing it to be active or inactive.

22.63 Allosteric enzymes are an efficient means to regulate a biochemical pathway because they bind to effectors, such as ATP or ADP, that alter the shape of the enzyme active site, either stimulating the rate of the reaction or inhibiting the reaction.

22.65 The citric acid cycle is regulated by the following four enzymes or enzyme complexes: pyruvate dehydrogenase complex, citrate synthase, isocitrate dehydrogenase, and the α-ketoglutarate dehydrogenase complex.

22.67 Energy-harvesting pathways, such as the citric acid cycle, must be responsive to the energy needs of the cell. If the energy requirements are high, as during exercise, the reactions must speed up. If energy demands are low and ATP is in excess, the reactions of the pathway slow down.

22.69 ADP

22.71 The electron transport system is a series of electron transport proteins embedded in the inner mitochondrial membrane that accept high-energy electrons from NADH and $FADH_2$ and transfer them in stepwise fashion to molecular oxygen (O_2).

22.73 Three ATP

22.75 The oxidation of a variety of fuel molecules, including carbohydrates, the carbon skeletons of amino acids, and fatty acids provides the electrons. The energy of these electrons is used to produce an H^+ reservoir. The energy of this proton reservoir is used for ATP synthesis.

22.77 The electron transport system passes electrons harvested during oxidation of fuel molecules to molecular oxygen. At three sites, protons are pumped from the mitochondrial matrix into the intermembrane compartment. Thus, the electron transport system builds the high-energy H^+ reservoir that provides energy for ATP synthesis.

22.79 **a.** Two ATP per glucose (net yield) are produced in glycolysis, whereas the complete oxidation of glucose in aerobic respiration (glycolysis, the citric acid cycle, and oxidative phosphorylation) results in the production of 36 ATP per glucose.

b. Thus, aerobic respiration harvests nearly 40% of the potential energy of glucose, and anaerobic glycolysis harvests only about 2% of the potential energy of glucose.

22.81 Transaminases transfer amino groups from amino acids to ketoacids.

22.83 The glutamate family of transaminases is very important because the ketoacid corresponding to glutamate is α-ketoglutarate, one of the citric acid cycle intermediates. This provides a link between the citric acid cycle and amino acid metabolism. These transaminases provide amino groups for amino acid synthesis and collect amino groups during catabolism of amino acids.

22.85 **a.** Pyruvate **d.** Acetyl CoA
b. α-Ketoglutarate **e.** Succinate
c. Oxaloacetate **f.** α-Ketoglutarate

22.87 Transaminase binds to the amino acid in its active site. Then, the α-amino group is transferred to pyridoxal phosphate, producing pyridoxamine phosphate. The amino group is then transferred to an α-keto acid.

22.89 Hyperammonemia

22.91 **a.** The source of one amino group of urea is the ammonium ion, and the source of the other is the α-amino group of the amino acid aspartate.

   b. The carbonyl group of urea is derived from CO_2.

22.93 Anabolism is a term used to describe all of the cellular energy-requiring biosynthetic pathways.

22.95 α-Ketoglutarate

22.97 Citric acid cycle intermediates are the starting materials for the biosynthesis of many biological molecules.

22.99 An essential amino acid is one that cannot be synthesized by the body and must be provided in the diet.

22.101

$$\underset{\text{Pyruvate}}{\overset{\displaystyle O}{\underset{\underset{CH_3}{|}}{\overset{||}{C}}-COO^-}} + CO_2 + ATP \longrightarrow \underset{\text{Oxaloacetate}}{\overset{\displaystyle O}{\underset{\underset{\underset{COO^-}{|}}{\underset{CH_2}{|}}}{\overset{||}{C}}-COO^-}} + ADP + P_i$$

Chapter 23

23.1 Because dietary lipids are hydrophobic, they arrive in the small intestine as large fat globules. The bile salts emulsify these fat globules into tiny fat droplets. This greatly increases the surface area of the lipids, allowing them to be more accessible to pancreatic lipases and thus more easily digested.

23.3 Starvation, a diet low in carbohydrates, and diabetes mellitus are conditions that lead to the production of ketone bodies.

23.5 **(1)** Fatty acid biosynthesis occurs in the cytoplasm, whereas β-oxidation occurs in the mitochondria.

   (2) The acyl group carrier in fatty acid biosynthesis is acyl carrier protein, while the acyl group carrier in β-oxidation is coenzyme A.

   (3) The seven enzymes of fatty acid biosynthesis are associated as a multienzyme complex called fatty acid synthase. The enzymes involved in β-oxidation are not physically associated with one another.

   (4) NADPH is the reducing agent used in fatty acid biosynthesis. NADH and $FADH_2$ are produced by β-oxidation.

23.7 When excess fuel is available, the liver synthesizes fatty acids and triglycerides.

23.9 Insulin stimulates uptake of glucose and amino acids by cells, glycogen and protein synthesis, and storage of lipids. It inhibits glycogenolysis, gluconeogenesis, breakdown of stored triglycerides, and ketogenesis.

23.11

Cholate

Chenodeoxycholate

23.13 A micelle is an aggregation of molecules having nonpolar and polar regions; the nonpolar regions of the molecules aggregate, leaving the polar regions facing the surrounding water.

23.15 A triglyceride is a molecule composed of glycerol esterified to three fatty acids.

23.17 Triglycerides, cholesterol, and phospholipids are packaged into a protein-coated shell to produce the class of plasma lipoproteins called chylomicrons.

23.19 Triglycerides

23.21 The large fat globule that takes up nearly the entire cytoplasm

23.23 Lipases catalyze the hydrolysis of the ester bonds of triglycerides.

23.25 Acetyl CoA is the precursor for fatty acids, several amino acids, cholesterol, and other steroids.

23.27 Chylomicrons carry dietary triglycerides from the intestine to all tissues via the bloodstream.

23.29 Bile salts are detergents that emulsify the lipids, increasing their surface area and making them more accessible to digestive enzymes (pancreatic lipases).

23.31 When dietary lipids in the form of fat globules reach the duodenum, they are emulsified by bile salts. The triglycerides in the resulting tiny fat droplets are hydrolyzed into monoglycerides and fatty acids by the action of pancreatic lipases, assisted by colipase. The monoglycerides and fatty acids are absorbed by cells lining the intestine.

23.33 The hydrolysis of ATP into AMP and PP_i

23.35 Carnitine is a carrier molecule that brings fatty acyl groups into the mitochondrial matrix.

23.37 The following equation represents the reaction catalyzed by acyl-CoA dehydrogenase. Notice that the reaction involves the loss of two hydrogen atoms. Thus, this is an oxidation reaction.

23.39 An alcohol is the product of the hydration of an alkene.

23.41 Six acetyl CoA, one phenyl acetate, six NADH, and six $FADH_2$

23.43 112 ATP

23.45 95 ATP

23.47 The acetyl CoA produced by β-oxidation will enter the citric acid cycle.

23.49 Ketone bodies include the compounds acetone, acetoacetone, and β-hydroxybutyrate, which are produced from fatty acids in the liver via acetyl CoA.

23.51 Ketosis is an abnormal rise in the level of ketone bodies in the blood.

23.53 Matrix of the mitochondrion

23.55

Acetoacetate   β-Hydroxybutyrate

23.57 In those suffering from uncontrolled diabetes, the glucose in the blood cannot get into the cells of the body. The excess glucose is excreted in the urine. Body cells degrade fatty acids because glucose is not available. β-oxidation of fatty acids yields enormous quantities of acetyl CoA, so much acetyl CoA, in fact, that it cannot all enter the citric acid cycle because there is not enough oxaloacetate available. Excess acetyl CoA is used for ketogenesis.

23.59 Ketone bodies are the preferred energy source of the heart.

23.61 Cytoplasm

23.63

23.65 **a.** The phosphopantetheine group allows formation of a high-energy thioester bond with a fatty acid.
   b. It is derived from the vitamin pantothenic acid.

23.67 Fatty acid synthase is a huge multienzyme complex consisting of the seven enzymes involved in fatty acid synthesis. It is found in the cell cytoplasm. The enzymes involved in β-oxidation are not physically associated with one another. They are free in the mitochondrial matrix space.

23.69 The liver produces ketone bodies under conditions of starvation or fasting.

23.71 Working muscle obtains most of its energy from the degradation of its supply of glycogen.

23.73 β-oxidation of fatty acids

23.75 Because ketone bodies have a free carboxylate group and are soluble in the blood, they can enter the brain and be used as an energy source.

23.77 Ketone bodies are the major fuel for the heart. Glucose is the major energy source of the brain, and the liver obtains most of its energy from the oxidation of amino acid carbon skeletons.

23.79 Fatty acids are absorbed from the bloodstream by adipocytes. Using glycerol-3-phosphate, produced as a by-product of glycolysis, triglycerides are synthesized. Triglycerides are constantly being hydrolyzed and resynthesized in adipocytes. The rates of hydrolysis and synthesis are determined by lipases that are under hormonal control.

23.81 In general, insulin stimulates anabolic processes and inhibits catabolic processes.

23.83 A target cell is one that has a receptor for a particular hormone.

23.85 Decreased blood glucose levels

23.87 In the β-cells of the islets of Langerhans in the pancreas.

23.89 Insulin stimulates the uptake of glucose from the blood into cells. It enhances glucose storage by stimulating glycogenesis and inhibiting glycogen degradation and gluconeogenesis.

23.91 Insulin stimulates synthesis and storage of triglycerides.

23.93 Insulin is secreted when blood glucose levels are high. It facilitates the uptake and storage of glucose by target cells to restore normal blood glucose levels. Glucagon is secreted when blood glucose levels are too low. It stimulates release of glucose into the blood to restore normal levels.

Credits

Photo Credits

Design Elements
A Human Perspective: © kristiansekulic/Vetta/Getty Images RF; Chemistry at the Crime Scene: © McGraw-Hill Education; A Medical Perspective: © Martin Barraud/OJOImages/Age Fotostock RF; Green Chemistry: © ViewStock/Getty Images RF, © Tetra Images/Alamy RF; Kitchen Chemistry: © Stockbrokerxtra Images/Photolibrary RF.

Front Matter
Table of Contents 1: © Purestock/SuperStock RF; 2: Image Science and Analysis Laboratory, 3: © Javier Trueba/MSF/Science Source; 4: © ImageSource/Veer RF; 5: © Royalty-Free/DigitalStock/Corbis RF; 6: © John Gillmoure/Spirit/Corbis RF; 7: © Jorg Greuel/Getty Images RF; 8: © Richard Carey/Getty Images RF; 9: © Javier Larrea/Pixtal/Age Fotostock RF; 10: © Digital Vision/PunchStock RF; 11: © Digital Vision/Getty Images RF; 12: © Photodisc/Getty Images RF; 13: © Iconotec RF; 14: © Stockbyte/Corbis RF; 15: © Parvinder Sethi; 16: © Phanie/Science Source; 17: © Steve Satushek/Getty Images; 18: © Fotohunter/Getty Images; 19: Courtesy Dr. Robert E. Shoemaker; 20: © Barbara Penoyar/Photodisc/Getty Images RF; 21: © Chris Falkenstein/Photodisc/Getty Images RF; 22: © Chris Falkenstein/Photodisc/Getty Images RF; 23: © Emilio Simion/Getty Images RF.

Chapter 1
Opener: © Purestock/SuperStock RF; Page 3 (top right): © Digital Vision/Getty Images RF; Page 5 (top right): © Graham Bell/Cardinal/Corbis RF; 1.2d: © David M. Dennis/Age Fotostock; 1.2c: © David Parker/Segate Microelectronics/SPL/Science Source; 1.2b: © Patrik Stollarz/AFP/Getty Images; 1.2a: © Adam Gault/OJO Images/Getty Images RF; 1.4c: © Danaè R. QuirkDorr, Ph.D.; 1.4a: © ImageSource RF; 1.4b: © ImageSource RF; 1.5b: © McGraw-Hill Education/Jeff Topping, photographer; 1.5c: © McGraw-Hill Education/Louis Rosenstock, photographer; 1.5a: © Corey Hochachka/DesignPics RF; 1.6a: © McGraw-Hill Education/Ken Karp, photographer; 1.6b: © McGraw-Hill Education/Ken Karp, photographer; Page 11: © McGraw-Hill Education/John Thoeming, photographer; 1.7a-c: © McGraw-Hill Education/Louis Rosenstock, photographer; 1.9a-d: © McGraw-Hill Education/Louis Rosenstock, photographer; Page 21: © Adam Gault/Getty Images RF; Page 25 (top center): © Fred Hutchinson Cancer Research Center/Susan M. Parkhurst, Ph.D., photographer; Page 25 (top right): © Maximilian Weinzierl/Alamy; Page 30: © ImageSource/Corbis RF; 1.12: © McGraw-Hill Education/Louis Rosenstock, photographer; Page 35: © Photolink/Getty Images RF.

Chapter 2
Opener: Image Science and Analysis Laboratory, NASA-Johnson Space Center; 2.1: © IBM Corporation, Almaden Research Center; 2.6: © Yoav Levy/Phototakeusa.com; 2.7a: © McGraw-Hill Education; Page 54: (top right) © Scott Camazine/Science Source; Page 54 (bottom left): © Earth Satellite Corp./SPL/Science Source; Page 56: © Photodisc/Getty Images RF; Page 59: © McGraw-Hill Education/Stephen Frisch, photographer; 2.14: © McGraw-Hill Education/Jacques Cornell, photographer.

Chapter 3
Opener: © Javier Trueba/MSF/Science Source; 3.1c: © McGraw-Hill Education/Dennis Strete, photographer; Page 99: © Comstock/PictureQuest RF; Page 102: © Pixtal/Age Fotostock RF; 3.14: © David A. Tietz/EditorialImage, LLC.

Chapter 4
Opener: © ImageSource/Veer RF; Page 129: © Royalty-Free/Corbis RF; 4.1: © McGraw-Hill Education/Louis Rosenstock, photographer; 4.3: © McGraw-Hill Education/Louis Rosenstock, photographer; Page 145: © McGraw-Hill Education/Charles D. Winters, photographer; Page 148: © McGraw-Hill Education/Stephen Frisch, photographer; Page 149: © McGraw-Hill Education/Louis Rosenstock, photographer; 4.6a-b: © McGraw-Hill Education/Stephen Frisch, photographer; Page 157: © David R .Frazier Photolibrary, Inc RF; Page 160: © Keith Eng, 2008 RF; Page 163: © Radius Images/Getty Images RF; 4.10a: © Stockbyte/Getty Images RF; 4.10b: © Dynamic Graphics Group/Getty Images RF.

Chapter 5
Opener: © Royalty-Free/DigitalStock/Corbis RF; Page 173: © Javier Larrea/Age Fotostock RF; Page 175: © Archive Holdings Inc/The Image Bank/Getty Images; 5.2a-b: © McGraw-Hill Education/Louis Rosenstock, photographer; 5.5: © Open Door/Alamy RF; Page 189: © Keith Thomas Productions/Brand X Pictures/PictureQuest RF; Page 192: © Mikael Karlsson RF; Page 195: © Photodisc/Getty Images RF.

Chapter 6
Opener: © John Gillmoure/Spirit/Corbis RF; Page 201: © Keith Eng; 6.1: © McGraw-Hill Education; Page 205: © Martin Strmiska/Alamy RF; Page 206: © Don Farrall/Photodisc/Getty Images RF; 6.4: © McGraw-Hill Education/Louis Rosenstock, photographer; Page 219: © Ingram Publishing/SuperStock RF; 6.6a: © David M. Phillips/Science Source; 6.6b: © David M. Phillips/Science Source; 6.6c: © David M. Phillips/Science Source; 6.7: © Sean Justice/Corbis RF; Page 222 (top right): Courtesy of Rita Colwell, National Science Foundation; 6.8a-b: © McGraw-Hill Education/Louis Rosenstock, photographer; Page 228: © AJ Photo/Science Source.

Chapter 7
Opener: © Jorg Greuel/Getty Images RF; Page 235: U.S. Coast Guard; Page 238: Warren Gretz/National Renewable Energy Laboratory/U.S. Department of Energy; Page 242: © McGraw-Hill Education/Jill Braaten, photographer; 7.9: © McGraw-Hill Education/Ken Karp, photographer; Page 248: © Alan Marsh/DesignPics/PunchStock RF; Page 251: © mykidsmom/Getty Images RF; 7.15: © McGraw-Hill Education/Ken Karp, photographer; 7.16: © McGraw-Hill Education/Ken Karp, photographer; Page 263: © McGraw-Hill Education/Louis Rosenstock, photographer.

Chapter 8
Opener: © Richard Carey/Getty Images RF; Page 271: © Evgeny Terentev/Getty Images RF; 8.1: © ballyscanlon/Photodisc/Getty Images RF; 8.5a: © McGraw-Hill Education/Charles D. Winters, photographer; 8.5b: © McGraw-Hill Education; Page 283: © Steve Cole/Vetta/Getty Images RF; Page 284: © Danita Delimont/Alamy RF; Page 285 (center left): © McGraw-Hill Education/Stephen Frisch, photographer; Page 285 (center): © McGraw-Hill Education/Stephen Frisch, photographer; Page 287 (center right): © McGraw-Hill Education/Louis Rosenstock, photographer; Page 287 (center left): © McGraw-Hill Education/Louis Rosenstock, photographer; Page 289: © Terry Wild Studio/McGraw-Hill Education; Page 292: © McGraw-Hill Higher Education/Auburn University Photographic Services; Page 294: © McGraw-Hill Education/Louis Rosenstock, photographer.

Chapter 9
Opener: © Javier Larrea/Pixtal/Age Fotostock RF; Page 302: © Mark Kostich/Vetta/Getty Images RF; Page 305: Space Telescope Science Institute (STScI)/NASA; Page 313: © Ingram Publishing RF; 9.3: © Gianni Tortoli/Science Source; Page 315: © Brand X Pictures/Alamy RF; 9.6: NRC File Photo/U.S. Nuclear Regulatory Commission (NRC); Page 317: © Steve Allen/Brand X Pictures/Alamy RF; Page 318: © Science Photo Library-MIRIAMMASLO/Brand X Pictures/Getty Images; 9.7: © Blair Seitz/Science Source; Page 321 (top right): © Don Carstens/Artville RF; 9.8: © Bristaol-Myers Squibb Medical Imaging; 9.9: © U.S. DoE/Mark Marten/Science Source; 9.1: © Stockbyte/Punchstock RF; 9.11: © Arthur S Aubry/Photodisc/Getty Images RF; Page 325: © Keith Eng.

Chapter 10

Opener: © Digital Vision/PunchStock RF; Page 336: IODP/NOAA/John W. Beck, Marine Laboratory Specialist (Photographer); Page 356: © Digital Stock/Corbis RF; Page 357 (center right): © Digital Stock/Corbis RF; Page 357 (top right): © C Squared Studios/Photodisc/Getty Images RF; Page 358: © Ingram Publishing/Fotosearch RF; Page 360: © Royalty-Free/DigitalStock/Corbis RF.

Chapter 11

Opener: © Digital Vision/Getty Images RF; Page 371: © Photolink/Photodisc/Getty Images RF; Page 372: (center left) © F. Schussler/PhotoLink/Getty Images RF; Page 372 (bottom left): © BananaStock/PunchStock RF; Page 377: © Martia lColomb/Photodisc/Getty Images RF; Page 378: © Steven P. Lynch; Page 384: (center right) © Andrew Dernie/Photodisc/Getty Images RF; Page 384: (top right) © Ingram Publishing/SuperStock RF; Page 384 (bottom right): © Steven P. Lynch; Page 385: © Akira Kaede/Digital Vision/Getty Images RF; Page 387: © Photodisc/Getty Images RF; 11.4: © McGraw-Hill Education/Ken Karp, photographer; Page 397: © moodboard/Corbis RF.

Chapter 12

Opener: © C Squared Studios/Photodisc/Getty Images RF; Page 416: © Barry Gregg/Corbis RF; 12.3: © PhotoLink/Getty Images RF; Page 424: © C. Sherburne/PhotoLink/Getty Images RF; 12.4: © Stan Fellerman/DigitalStock/Corbis RF; Page 435: © Food Collection/StockFood RF; 12.8: © Glow Wellness/Getty Images; 12.9: © Photodisc/Getty Images RF; Page 442: © Photodisc Collection/Getty Images RF.

Chapter 13

Opener: © Iconotec RF; Page 454 (top right): © Andrew Syred/Science Source; Page 454: (bottom left) © Víctor Suárez/Age Fotostock; Page 457: © Photodisc/Getty Images RF; Page 460: © IZA Stock/Getty Images RF; Page 460: (center left) © PhotoLink/Photodisc/Getty Images RF; Page 461: (center right) © IT Stock/Punch Stock RF; Page 461: (center left) © Goodshoot/PunchStock RF; Page 461: (bottom center) © Dynamic Graphics Group/Punch Stock RF; Page 461: (bottom right) © John Foxx Images/Imagestate Media RF; Page 461: (center) © Eisenhut and Mayer Wien/Photolibrary/Getty Images; 13.4a-c: © McGraw-Hill Education/Charles D. Winters, photographer; 13.5: © Andrew Lambert Photography/Science Source; Page 469: © Michael Grayson/Flickr RF/Getty Images RF.

Chapter 14

Opener: © Stockbyte/Corbis RF; Page 484: © peng wu/Getty Images; Page 487: © Digital Stock/Corbis RF; Page 490: © Reuters/Corbis; Page 502 (center right): © Digital Stock/Corbis RF; Page 502 (center left): © S. Alden/PhotoLink/Getty Images RF; Page 502 (center): © C Squared Studios/Photodisc/Getty Images RF; Page 502 (center): © John A. Rizzo/Photodisc/Getty

Images RF; Page 502 (bottom center): © Digital Stock/Corbis RF; Page 502 (bottom left): © Digital Stock/Corbis RF; Page 502 (bottom right): © Digital Stock/Corbis RF.

Chapter 15

Opener: © Parvinder Sethi; Page 525: © Spike Mafford/Photodisc/Getty Images RF; Page 531: © Duncan Smith/SPL/Science Source; Page 532: © Creatas/PunchStock RF; Page 539 (top left): © Digital Stock/Corbis RF; Page 539 (top center): © Digital Stock/Corbis RF; Page 539 (top right): © Hobbs/PhotoLink/Getty Images RF; Page 540: © Steven P. Lynch; Page 552 (top right): © Photodisc/PhotoLink/Getty Images RF; Page 552 (top right): © Phil Larkin, CSIRO Plant Industry.

Chapter 16

Opener: © Phanie/Science Source; 16.1: Scott Bauer/U.S. Department of Agriculture; 16.2: © Mitch Hrdlicka/Photodisc/Getty Images RF; Page 567: © Steve Gschmeissner/SPL/Getty Images RF; 16.3: © Royalty-Free/Corbis; Page 576: © Digital Stock/Corbis RF; Page 580 (top left): Keith Weller/U.S. Department of Agriculture; Page 580 (center left): © McGraw-Hill Education/Jill Braaten, photographer; Page 581: © M. Freeman/Photodisc/PhotoLink/Getty Images RF; Page 582 (center): © Food Collection/SuperStock RF; Page 582 (center left): © Food and Drink Photos/Alamy; Page 586: © JGI/Blend Images/Getty Images RF; Page 587: © Pixtal/SuperStock RF; Page 589: © John A. Rizzo/Photodisc/Getty Images RF; Page 591: © Digital Stock/Corbis RF.

Chapter 17

Opener: © Steve Satushek/Getty Images; Page 603 (center right): © Chris Shorten/Cole Group/Photodisc/Getty Image RF; Page 603 (center left): © Cole Group/Photodisc/Getty Images RF; Page 604: © Michael Lamotte/Cole Group/Photodisc/Getty Images RF; Page 617: © Photodisc/Photo Link/Getty Images RF; Page 618: © Photodisc/Getty Images RF; Page 619: © Steven P .Lynch; 17.1: © James Dennis/Phototake; Page 626: © Royalty-Free/Corbis.

Chapter 18

Opener: © Fotohunter/Getty Images; Page 642: © Pat Hastings/Getty Images; Page 644: © SW Productions/Photodisc/Getty Images RF; 18.14: © Meckes/Ottawa/Science Source; Page 653 (bottom right): © C. Sherburne/Photo Link/Getty Images RF; Page 655 (top right): © Burke/Triolo Productions/Brand X Pictures/Getty Images RF; Page 655 (top right): © Michael Lamotte/Cole Group/Photodisc/Getty Images RF; Page 655 (center right): © McGraw-Hill Education/Bob Coyle, photographer; Page 656: © Laszlo Selly/Getty Images; Page 657: © Vibe Images/Getty Images.

Chapter 19

Opener: Courtesy Dr. Robert E. Shoemaker; Page 667: © Comstock Images/Alamy RF;

Page 672: Courtesy of Adam Melonas/www .madridlab.net/melonas/adam-melonas; Page 678: © Laguna Design/Getty Images.

Chapter 20

Opener: © Barbara Penoyar/Photodisc/Getty Images RF; Page 706: © Randy Allbritton/Photodisc/Getty Images RF; Page 726: © Photodisc Collection/Getty Images RF; Page 731 (bottom left): Dr. Charles S. Helling/USDA; Page 731 (top right): © Courtesy of Orchid Cellmark, Germantown, Maryland.

Chapter 21

Opener: © Chris Falkenstein/Photodisc/Getty Images RF; Page 744: © Barry Gregg/Spirit/Corbis RF; Page 755: © Karl Weatherly/Photodisc/Getty Images RF; Page 757: © McGraw-Hill Education/Louis Rosenstock, photographer; Page 761: Tech. Sgt. Tracy L. DeMarc/U.S. Air Force.

Chapter 22

Opener: © Chris Falkenstein/Photodisc/Getty Images RF; 22.1a: © CNRI/Phototake; Page 778 (top left): © Digital Vision/Getty Images RF; Page 778 (top right): © Photodisc/Photo Link/Getty Images RF.

Chapter 23

Opener: © Emilio Simion/Getty Images RF; Page 819: © Ryan McVay/Photodisc/Getty Images RF.

Text Credits

Chapter 1

Page 8, Figure 1.5: Janice Smith, *General, Organic, & Biological Chemistry*, 3e, 2015, pg. 4, figure 1.3. Copyright © 2015 by McGraw-Hill Education. All rights reserved. Used with permission.

Chapter 2

Page 53, Figure 2.7: Raymond Chang & Kenneth Goldsby, *General Chemistry The Essential Concepts*, 7e, 2013, p. 216, figure 7.3. Copyright © 2013 by McGraw-Hill Education. All rights reserved. Used with permission.

Chapter 5

Page 189, Figure 5.8: Rich Bauer, James Birk & Pamela Marks, *Introduction to Chemistry*, 2e, p. 388, figure 10.15. Copyright © 2009 by McGraw-Hill Education. All rights reserved. Used with permission.

Chapter 10

Page 354: Francis Carey & Robert Giuliano, *Organic Chemistry* 9e, 2013, p. 97. Copyright © 2013 by McGraw-Hill Education. All rights reserved. Used with permission.

Chapter 12

Page 417: Francis Carey & Robert Giuliano, Organic Chemistry 9e, 2013, p. 138. Copyright © 2013 by McGraw-Hill Education. All rights reserved. Used with permission.

Index

PRINCIPAL FUNCTIONAL GROUPS IN ORGANIC COMPOUNDS

Type of Compound	Structural Formula	Condensed Formula	Chapter Reference	Example		
				Structural Formula	IUPAC Name	Common Name
Alcohol	R—O—H	ROH	12	CH_3CH_2—O—H	Ethanol	Ethyl alcohol
Aldehyde	R—C(=O)—H	RCHO	13	CH_3C(=O)—H	Ethanal	Acetaldehyde
Amide	R—C(=O)—N(H)—H	$RCONH_2$	15	CH_3C(=O)—N(H)—H	Ethanamide	Acetamide
Amine	R—N(H)—H	RNH_2	15	CH_3CH_2N(H)—H	Ethanamine	Ethylamine
Carboxylic acid	R—C(=O)—O—H	RCOOH	14	CH_3C(=O)—O—H	Ethanoic acid	Acetic acid
Ester	R—C(=O)—O—R'	RCOOR'	14	CH_3C(=O)—OCH_3	Methyl ethanoate	Methyl acetate
Ether	R—O—R'	ROR'	12	CH_3OCH_3	Methoxymethane	Dimethyl ether
Halide	—Cl (or —Br, —F, —I)	RCl	10	CH_3CH_2Cl	Chloroethane	Ethyl chloride
Ketone	R—C(=O)—R'	RCOR'	13	CH_3CCH_3 (C=O)	Propanone	Acetone

METRIC PREFIXES

Multiple	Prefix	Symbol	Submultiple	Prefix	Symbol
10^{12}	tera	T	10^{-1}	deci	d
10^{9}	giga	G	10^{-2}	centi	c
10^{6}	mega	M	10^{-3}	milli	m
10^{3}	kilo	k	10^{-6}	micro	μ
10^{2}	hecto	h	10^{-9}	nano	n
10^{1}	deka	da	10^{-12}	pico	p

CONVERSION FACTORS

Length:
1 meter (m) = 39.4 inches (in)
1 inch (in) = 2.54 centimeters (cm)
1 Ångstrom (Å) = 10^{-10} meter (m)

Mass:
1 kilogram (kg) = 2.20 pounds (lb)
1 pound (lb) = 454 grams (g)
1 atomic mass unit (amu)
= 1.6605×10^{-24} grams (g)

Volume:
1 liter (L) = 1000 milliliters (mL)
= 1000 cm^3
1 liter (L) = 1.06 quarts (qt)

Energy:
1 calorie (cal) = 4.18 joules (J)

Temperature:
$T_{°F} = (1.8 \times T_{°C}) + 32$
$T_{°C} = \dfrac{(T_{°F} - 32)}{1.8}$
$T_K = T_{°C} + 273.15$

Pressure:
1 atmosphere (atm) = 14.7 lbs/in^2 (psi)
1 atm = 760 millimeters of mercury
(760 mm Hg = 760 torr)

PHYSICAL CONSTANTS

Avogadro's number: 6.022×10^{23} units/mol

Speed of light: 3.0×10^{8} m/sec

Gas constant (R): $0.0821 \text{ L} \cdot \text{atm} \cdot \text{K}^{-1} \cdot \text{mol}^{-1}$

Mass of an electron: 9.1094×10^{-28} g
or
5.486×10^{-4} amu

Mass of a proton: 1.6726×10^{-24} g
or
1.007 amu

Mass of a neutron: 1.6750×10^{-24} g
or
1.009 amu

Volume of one mole of ideal gas: 22.4 L (@ 273 K and 1 atm)